CONSTANTS

Universal Gas Constant	$\bar{R} = \begin{cases} 8.314 \text{ kJ/kmol} \cdot \text{K} \\ 1545 \text{ ft} \cdot \text{lbf/lbmol} \cdot {}^\circ\text{R} \\ 1.986 \text{ Btu/lbmol} \cdot {}^\circ\text{R} \end{cases}$
Standard Acceleration of Gravity	$g = \begin{cases} 9.80665 \text{ m/s}^2 \\ 32.174 \text{ ft/s}^2 \end{cases}$
Standard atmospheric pressure	$1 \text{ atm} = \begin{cases} 1.01325 \text{ bar} \\ 14.696 \text{ lbf/in}^2 \end{cases}$
Temperature Relations	$T({}^\circ\text{R}) = 1.8\, T(\text{K})$ $T({}^\circ\text{C}) = T(\text{K}) - 273.15$ $T({}^\circ\text{F}) = T({}^\circ\text{R}) - 459.67$

SECOND EDITION

FUNDAMENTALS OF ENGINEERING THERMODYNAMICS

MICHAEL J. MORAN
The Ohio State University

HOWARD N. SHAPIRO
Iowa State University of Science and Technology

JOHN WILEY & SONS, INC.
New York • Chichester • Brisbane • Toronto • Singapore

COVER	Illustration by Roy Wieman
ACQUISITIONS EDITOR	Cliff Robichaud
PRODUCTION MANAGER	Katharine Rubin
DESIGNER	Ann Marie Renzi
PRODUCTION SUPERVISOR	Sandra Russell
MANUFACTURING MANAGER	Lorraine Fumoso
COPY EDITING MANAGER	Deborah Herbert
ILLUSTRATION	Ishaya Monokoff/Sigmund Malinowski

Recognizing the importance of preserving what has been written, it is a policy of John Wiley & Sons, Inc. to have books of enduring value published in the United States printed on acid-free paper, and we exert our best efforts to that end.

Library of Congress Cataloging in Publication Data:

Moran, Michael J.

Fundamentals of engineering thermodynamics / Michael J. Moran, Howard N. Shapiro.—2nd ed.

Includes index.

ISBN 0-471-53984-8

ISBN 0-471-57117-2 (disk version)

1. Thermodynamics. I. Shapiro, Howard N. II. Title.

TJ265.M66 1992 91-23752

621.402′1—dc20 CIP

Printed in the United States of America

10 9 8 7 6 5 4 3 2

PREFACE

In this second edition we have retained the basic objectives of the first edition: to present a comprehensive and rigorous treatment of engineering thermodynamics from the classical viewpoint, to provide a sound basis for subsequent courses in fluid mechanics and heat transfer, and to prepare students to use thermodynamics in professional practice. Enough material is provided for an introductory course in engineering thermodynamics and for a follow-up course dealing mainly with applications. A background in elementary physics and calculus is presumed. The organization of the text and level of presentation of this edition remain the same as that of the first edition.

New Features

- Over 230 new end-of-chapter problems, bringing the total of these problems to more than 1300.
- Over 100 new *design and open-ended* problems provided under a separate heading at the close of each chapter.
- New material on the International Temperature Scale and refrigerants.
- A second color, used functionally throughout to highlight special aspects of the presentation.
- Interactive computer software designed to reinforce important concepts and to sharpen students' problem-solving skills.
- Appendix tables, figures, and charts available in a separate booklet for use during closed-book examinations.

A significant change in this second edition is the increase in the number of problems. The more than 1300 end-of-chapter problems of the second edition are sequenced to correlate with the subject matter and are listed in order of increasing difficulty. They range from confidence-building exercises, which illustrate basic concepts, to more challenging problems that may involve systems with several components. A number of problems are included for which the use of a computer is either required or advisable, and many problems provide the student with opportunities to apply engineering judgment.

In addition, responding to the renewed emphasis on the design component of engineering curricula being called for by many in academia and industry, we provide a special set of problems at the end of *each* chapter. These *design and open-ended* problems are intended as brief *design experiences* that afford students opportunities to develop their creativity and engineering judgment, formulate design task statements, apply realistic constraints, and consider alternatives. In the earlier chapters of the book, these problems are more open-ended than design oriented, owing to the limited backgrounds of the students. The primary emphasis of the design and open-ended problems is on the subject matter of the text, but students may have to do some collateral reading before a solution can be developed. Instructors may elect to narrow the focus of the problems to allow individual students to achieve a result with a relatively modest expenditure of effort or may decide to use the problems as points

of departure for more extensive group-type projects. An important feature of many of the design and open-ended problems is that students are explicitly required to develop their communications skills by presenting results in the form of written reports, memoranda, schematics, and graphs.

The main text of the second edition is predominantly the same as that of the first edition. New material has been provided, however, in two areas. The new temperature scale, ITS-90, is included in Sec. 5.5, and in Sec. 10.3 the discussion of refrigerants has been expanded to consider the ozone depleting potential of halogenated hydrocarbons like Refrigerant 12. In this second edition, thermodynamic property data are provided for *both* Refrigerant 12 and Refrigerant 134a, which contains no halogens. The new data are used in many end-of-chapter problems. Finally, in this second edition, color has been used functionally to highlight special features of the presentation and to give a more pleasing appearance.

Teaching Aids and Supplements

Accompanying this textbook is an improved and expanded set of supplements consisting of instructional aids for students and instructors. For the student, computer software is now available. The software can be used to evaluate properties in either SI or English units of water, Refrigerant 12, and air. The software also provides the student with several scenarios that relate to specific concepts developed in the text. In a typical scenario the student is introduced to a problem and then is guided through a series of structured, interactive steps that are tied to a specific aspect of engineering thermodynamics. The scenario concludes by giving the student an opportunity to apply newly learned concepts. The software is tailored to the text and features relevance to real-world problems, a high degree of interaction, and extensive use of graphics and animation. It is easy to use, understand, and run.

The following supplements are available to adopters: (1) a comprehensive *Instructor's Manual;* (2) overhead projector *Transparencies* of selected figures and tables; and (3) *booklets* containing the appendix tables, figures, and charts for student use during closed-book examinations. The Instructor's Manual accompanying the textbook provides solutions to all end-of-chapter problems in the same format used in the solved examples. Solutions may be photocopied for posting or preparing transparencies for use in the classroom, eliminating the drudgery of problem solving for the instructor. For the design and open-ended problems, only brief discussions and/or literature references are provided to assist instructors in guiding students to achieve *individual* outcomes. The Instructor's Manual also provides the following instructional aids: sample syllabi for two-course sequences and one-term survey courses in engineering thermodynamics on both quarter and semester bases, and a survey of software currently available for bringing computers into the teaching of engineering thermodynamics.

Features Retained from the First Edition

- A clear and concise presentation.
- A problem-solving methodology that encourages systematic thinking.
- A thorough development of the second law of thermodynamics, featuring the entropy production concept.
- An up-to-date presentation of availability analysis, including an introduction to chemical availability.

- Sound developments of the application areas.
- Flexibility in units, allowing either an SI or a mixed SI/English presentation.

This book has evolved over many years of teaching the subject matter at both the undergraduate and graduate levels. Complete explanations without wordiness together with numerous well-explained examples make the text user-friendly and nearly self-instructive. This frees the instructor from lecturing exclusively and allows class time to be used in other beneficial ways. Our aim has been a clear and concise presentation without sacrificing completeness. We have attempted to make the material interesting and easy to read. Favorable evaluations from instructors who have used the first edition in a wide range of engineering programs indicates that these objectives have been met.

One of our primary goals in this textbook is to encourage students to develop an orderly approach to problem solving. To this end, a formal problem analysis and solution format that helps students think systematically about engineering systems is used throughout the text. Solutions begin by listing assumptions, proceed step-by-step using fundamentals, and conclude with comments that identify key aspects of the solution. Unit conversions are explicitly included when numerical evaluations are made. The solution methodology is illustrated by 138 solved examples, which are set apart from the main text so they can be easily identified. The methodology we use resembles those of two other well-established Wiley titles: *Introduction to Fluid Mechanics* by R. W. Fox and A. T. McDonald and *Fundamentals of Heat Transfer* by F. P. Incropera and D. P. De Witt. With this choice of solution format a series of three books similar in presentation, level, and rigor are available that cover the core thermodynamics, fluid mechanics, and heat transfer sequence common to many curriculums.

As there is greater interest today in entropy and availability principles than ever before, a thorough development of the second law of thermodynamics is provided in Chapters 5 to 7. The importance of the second law is conveyed throughout by stressing its relevance to the goal of proper energy resource utilization. A special feature is the use of the entropy production concept that allows the second law to be applied effectively in ways readily mastered by students (Chapter 6). Another special feature is an up-to-date introduction to availability analysis, including second law efficiencies (Chapter 7). Entropy and availability balances are introduced and applied in ways that parallel the use of the energy balances developed for closed systems and control volumes, unifying the application of the first and second laws. Once introduced, second law concepts are integrated throughout the text in solved examples and end-of-chapter problems. The presentation is structured to allow instructors who wish to omit availability analysis to do so.

Emphasis has been placed on sound developments and careful sequencing of the application areas. Chapters 8 to 14, dealing with applications, allow some flexibility in the order and the amount of material covered. For example, vapor and gas power systems are discussed in Chapters 8 and 9, and refrigeration and heat pump systems are the subject of Chapter 10. But instructors who prefer to treat all vapor cycle material at one time can include vapor-compression and absorption refrigeration with Chapter 8. Advanced and innovative energy systems, such as cogeneration systems, combined power cycles, and cascade refrigeration cycles, are incorporated in Chapters 8 to 10 where they fall appropriately and are not relegated to a catchall chapter. As the study of gas flows falls naturally with the subject of gas turbines and turbojet engines, an introduction to one-dimensional compressible flow has been included in Chapter 9. The chapters dealing with applications provide illustrations of the use of availability and irreversibility principles.

The text has other special features. The development of the first law of thermodynamics in Chapter 2 begins with energy and work concepts familiar to students from earlier physics and engineering mechanics courses and proceeds operationally to the closed system energy balance. Thermodynamic cycles are introduced in Chapter 2, together with the definition of the thermal efficiency of power cycles and coefficients of performance of refrigerators and heat pumps. This permits elementary problem solving with cycles using the first law before cycles are considered in depth in later chapters.

Property relations and data for pure, simple compressible substances are introduced in Chapter 3 *after* the energy concept has been developed in Chapter 2. This arrangement has the following advantages: it reinforces the fact that the energy concept applies to systems generally and it is not limited to instances involving simple compressible substances; it provides the instructor with opportunities for sparking students' interest while they study Chapter 2 by assigning elementary energy analysis problems in which essential property data are provided; and it allows students further practice in applying the energy concept while learning about property relations and data in Chapter 3. Using the compressibility factor as a point of departure, we introduce ideal gas property relations and data in Chapter 3 *following* the discussion of the steam tables. This organization of topics emphasizes to students, usually for the first time, the limitations of the ideal gas model.

The principles of conservation of mass and energy are extended to control volumes in Chapter 4. The primary emphasis is on cases in which one-dimensional flow is assumed, but mass and energy balances are also presented in integral forms that provide a link to material covered in subsequent fluid mechanics and heat transfer courses. Control volumes at steady state are featured, but in-depth discussions of transient cases are also provided. Whether problems are steady state or transient in character, the appropriate thermodynamic models are obtained by deduction from general expressions of the conservation of mass and conservation of energy principles.

The text has been written to allow flexibility in the use of units. It can be studied using SI units only, or using a mix of SI and English units. Well-organized tables and charts are provided in *both* sets of units. Proper use of unit conversion factors is emphasized throughout the text. The force-mass conversion constant g_c is treated as implicit, and equations involving kinetic or potential energy are handled consistently irrespective of the unit system used. The SI property tables of this edition give specific volume in m^3/kg, which is a change from the cm^3/g of the first edition.

Acknowledgments

We thank the many users of our first edition, located at more than 100 universities and colleges in the United States and Canada, who contributed to this revision through their comments and constructive criticism. Special thanks are owed to the individual faculty whose written responses to a questionnaire asking them to evaluate the first edition provided valuable guidance in developing this second edition.

We are extremely gratified by how well the first edition of this book was received and hope that the improvements provided in this second edition will make for an even more effective presentation. Your comments, criticism, and suggestions will be greatly appreciated.

Michael J. Moran
Howard N. Shapiro

CONTENTS

CHAPTER 1

INTRODUCTORY CONCEPTS AND DEFINITIONS

The word thermodynamics stems from the Greek words *therme* (heat) and *dynamis* (force). Although various aspects of what is now known as thermodynamics have been of interest since antiquity, the formal study of thermodynamics began in the early nineteenth century through consideration of the motive power of *heat:* the capacity of hot bodies to produce *work*. Today the scope is larger, dealing generally with *energy* and with relationships among the *properties* of matter.

Thermodynamics is both a branch of physics and an engineering science. The scientist is normally interested in gaining a fundamental understanding of the physical and chemical behavior of fixed, quiescent quantities of matter and uses the principles of thermodynamics to relate the properties of matter. Engineers are generally interested in studying *systems* and how they interact with their *surroundings*. To facilitate this, engineers extend the subject of thermodynamics to the study of systems through which matter flows.

Engineers use principles drawn from thermodynamics and other engineering sciences, such as fluid mechanics and heat and mass transfer, to analyze and design things intended to meet human needs. The wide realm of application of these principles is suggested by Table 1.1, which lists a few of the areas where engineering thermodynamics is important. Engineers seek to achieve improved designs and better performance, as measured by factors such as an increase in the output of some desired product, a reduced input of a scarce resource, a reduction in total costs, or a lesser environmental impact. The principles of engineering thermodynamics play an important part in achieving these goals.

The objective of this chapter is to introduce you to some of the fundamental concepts and definitions that are used in our study of engineering thermodynamics. In most instances the introduction is brief, and further elaboration is provided in subsequent chapters.

TABLE 1.1 Selected Areas of Application of Engineering Thermodynamics

Automotive engines
Turbines
Compressors, pumps
Fossil- and nuclear-fueled power stations
Propulsion systems for aircraft and rockets
Combustion systems
Cryogenic systems, gas separation and liquefaction
Heating, ventilating, and air-conditioning systems
Vapor compression and absorption refrigeration
Heat pumps
Alternative energy systems
Fuel cells
Thermoelectric and thermionic devices
Magnetohydrodynamic (MHD) converters
Solar-activated heating, cooling, and power generation
Geothermal systems
Ocean thermal, wave, and tidal power generation
Wind power
Biomedical applications
Life-support systems
Artificial organs

1.1 THERMODYNAMIC SYSTEMS

An important step in any engineering analysis is to describe precisely what is being studied. In mechanics, if the motion of a body is to be determined, normally the first step is to define a *free body* and identify all the forces exerted on it by other bodies. Newton's second law of motion is then applied. In thermodynamics the term *system* is used to identify the subject of the analysis. Once the system is defined and the relevant interactions with other systems are identified, one or more physical laws or relations are applied.

The system is whatever we want to study. It may be as simple as a free body or as complex as an entire chemical refinery. We may want to study a quantity of matter contained within a closed, rigid-walled tank, or we may want to consider something such as a gas pipeline through which matter flows. Even a vacuum, which is devoid of matter, may be the focus of interest. The composition of the matter inside the system may be fixed or may be changing through chemical or nuclear reactions. The shape or volume of the system being analyzed is not necessarily constant, as when a gas in a cylinder is compressed by a piston or a balloon is inflated.

Everything external to the system is considered to be part of the system's *surroundings*. The system is distinguished from its surroundings by a specified *boundary,* which may be at rest or in motion. You will see that the interactions between a system and its surroundings, which take place across the boundary, play an important part in engineering thermodynamics. It is essential for the boundary to be delineated carefully before proceeding with any thermodynamic analysis. However, since the same physical phenomena often can be analyzed in terms of alternative choices of the system, boundary, and surroundings, the choice of a particular boundary defining a particular system is governed by the convenience it allows in the subsequent analysis.

1.1.1 Types of Systems

Two basic kinds of systems are distinguished in this book. These are referred to, respectively, as *closed systems* and *control volumes*. A closed system refers to a fixed quantity of matter, whereas a control volume is a region of space through which mass may flow.

A closed system is defined when a particular quantity of matter is under study. A closed system always contains the same matter. There can be no transfer of mass across its boundary. A special type of closed system that does not interact in any way with its surroundings is called an *isolated* system.

Figure 1.1 shows a gas in a piston–cylinder assembly. Let us consider the gas to be a closed system. The boundary lies just inside the piston and cylinder walls, as shown by the dashed lines on the figure. If the cylinder were placed over a flame, the gas would expand, raising the piston. The portion of the boundary between the gas and the piston moves with the piston. No mass would cross this or any other part of the boundary.

In subsequent sections of this book, thermodynamic analyses are made of devices such as turbines and pumps through which mass flows. These analyses can be conducted in principle by studying a particular quantity of matter, a closed system, as it passes through the device. In most cases it is simpler to think instead in terms of a given region of space through which mass flows. With this approach, a *region* within a prescribed boundary is studied. The region is called a control volume. Mass may cross the boundary of a control volume.

A schematic diagram of an engine is shown in Fig. 1.2. The dashed line defines a control volume that surrounds the engine. Observe that air, fuel, and exhaust gases cross the boundary.

The term *control mass* is sometimes used in place of closed system, and the term *open system* is used interchangeably with control volume. When the terms control mass and control volume are used, the system boundary is often referred to as a *control surface*.

1.1.2 Macroscopic and Microscopic Views of Thermodynamics

Systems can be studied from a macroscopic or a microscopic point of view. The macroscopic approach to thermodynamics is concerned with the gross or overall behavior. This is sometimes called *classical* thermodynamics. No model of the structure of matter at the molecular, atomic, and subatomic levels is directly used in classical thermodynamics. Although the behavior of systems is affected by molecu-

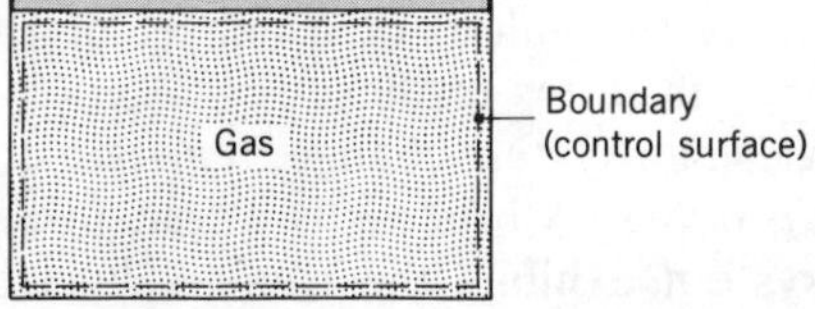

Figure 1.1 Example of a closed system (control mass). A gas in a piston–cylinder assembly.

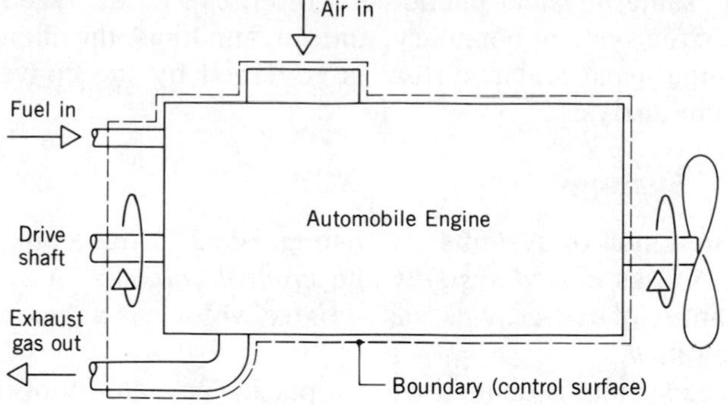

Figure 1.2 Example of a control volume (open system). An automobile engine.

lar structure, classical thermodynamics allows important aspects of system behavior to be evaluated from observations of the overall system.

The microscopic approach to thermodynamics, known as *statistical* thermodynamics, is concerned directly with the structure of matter. The objective of statistical thermodynamics is to characterize by statistical means the average behavior of the particles making up a system of interest and relate this information to the observed macroscopic behavior of the system.

For applications involving lasers, plasmas, high-speed gas flows, chemical kinetics, very low temperatures (cryogenics), and others, the methods of statistical thermodynamics are essential. Moreover, the microscopic approach is instrumental in developing certain property data, for example, ideal gas specific heats (Sec. 3.5.1). For the great majority of engineering applications, however, classical thermodynamics not only provides a considerably more direct approach for analysis and design but also requires far fewer mathematical complications. For these reasons the macroscopic viewpoint is the one adopted in this book. When it serves to promote understanding, however, concepts are interpreted from the microscopic point of view. Finally, we note that relativity effects are not significant for the systems under consideration in this book.

1.2 PROPERTY, STATE, PROCESS, AND EQUILIBRIUM

To describe a system and predict its behavior requires knowledge of its properties and how those properties are related. *Properties* are macroscopic characteristics of a system such as mass, volume, energy (Sec. 2.3), pressure (Sec. 1.4), and temperature (Sec. 1.5) to which numerical values can be assigned at a given time without knowledge of the *history* of the system. Many other properties are considered during the course of our study of engineering thermodynamics. Thermodynamics also deals with quantities that are not properties, such as mass flow rates and energy transfers by work and heat. Additional examples of quantities that are not properties are provided in subsequent chapters. A way to distinguish *non*properties from properties is given shortly.

The word *state* refers to the condition of a system as described by its properties. Since there are normally relations among the properties of a system, the state often can be specified by providing the values of a subset of the properties. All other properties can be determined in terms of these few.

When any of the properties of a system change, the state changes and the system is said to have undergone a *process*. A process is a transformation from one state to another. However, if a system exhibits the same values of its properties at two different times, it is in the same state at these times. A system is said to be at *steady state* if none of its properties changes with time.

A *thermodynamic cycle* is a sequence of processes that begins and ends at the same state. At the conclusion of a cycle all properties have the same values they had at the beginning. Consequently, over the cycle the system experiences no *net* change of state. Cycles that are repeated periodically play prominent roles in many areas of application. For example, steam circulating through an electrical power plant executes a cycle.

At a given state each property has a definite value that can be assigned without knowledge of how the system arrived at that state. Therefore, the change in value of a property as the system is altered from one state to another is determined solely by the two end states and is independent of the particular way the change of state occurred. That is, the change is independent of the details, or *history,* of the process. Conversely, if the value of a quantity is independent of the process between two states, then that quantity is the change in a property. This provides a test that is both necessary and sufficient for determining whether a quantity is a property: *A quantity is a property if, and only if, its change in value between two states is independent of the process.* It follows that if the value of a particular quantity depends on the details of the process, and not solely on the end states, that quantity cannot be a property.

1.2.1 Extensive and Intensive Properties

Thermodynamic properties can be placed in two general classes: extensive and intensive. A property is called *extensive* if its value for an overall system is the sum of its values for the parts into which the system is divided. Mass, volume, energy, and several other properties introduced later are extensive. Extensive properties depend on the size or extent of a system. The extensive properties of a system can change with time, and many thermodynamic analyses consist mainly of carefully accounting for changes in extensive properties such as mass and energy as a system interacts with its surroundings.

Intensive properties are not additive in the sense previously considered. Their values are independent of the size or extent of a system and may vary from place to place within the system at any moment. Thus, intensive properties may be functions of both position and time, whereas extensive properties vary at most with time. Specific volume (Sec. 1.4), pressure, and temperature are important intensive properties; several other intensive properties are introduced in subsequent chapters.

To illustrate the difference between extensive and intensive properties, consider an amount of matter that is uniform in temperature, and imagine that it is divided into two parts. The mass of the whole is the sum of the masses of the parts, and the overall volume is the sum of the volumes of the parts. However, the temperature of the whole is not the sum of the temperatures of the parts; it is the same for each part. Mass and volume are extensive, but temperature is intensive.

1.2.2 Phase and Pure Substance

The term *phase* refers to a quantity of matter that is homogeneous throughout in both chemical composition and physical structure. Homogeneity in physical structure means that the matter is all *solid,* or all *liquid,* or all *vapor* (or equivalently all *gas*). A system can contain one or more phases. For example, a system of liquid water and water vapor (steam) contains *two* phases. When more than one phase is present, the phases are separated by *phase boundaries.* Note that gases, say oxygen and nitrogen, can be mixed in any proportion to form a *single* gas phase. Certain liquids, such as alcohol and water, can be mixed to form a *single* liquid phase. But liquids such as oil and water, which are not miscible, form *two* liquid phases.

A *pure substance* is one that is uniform and invariable in chemical composition. A pure substance can exist in more than one phase, but its chemical composition must be the same in each phase. For example, if liquid water and water vapor form a system with two phases, the system can be regarded as a pure substance because each phase has the same composition. A uniform mixture of gases can be regarded as a pure substance provided it remains a gas and does not react chemically. Changes in composition due to chemical reaction are considered in Chap. 13. A system consisting of air can be regarded as a pure substance as long as it is a mixture of gases; but if a liquid phase should form on cooling, the liquid would have a different composition from the gas phase, and the system would no longer be considered a pure substance.

1.2.3 Equilibrium

Classical thermodynamics places primary emphasis on equilibrium states and changes from one equilibrium state to another. Thus, the concept of *equilibrium* is fundamental. In mechanics, equilibrium means a condition of balance maintained by an equality of opposing forces. In thermodynamics, the concept is more far-reaching, including not only a balance of forces but also a balance of other influences. Each kind of influence refers to a particular aspect of thermodynamic, or complete, equilibrium. Accordingly, several types of equilibrium must exist individually to fulfill the condition of complete equilibrium; among these are mechanical, thermal, phase, and chemical equilibrium. Criteria for these four types of equilibrium are considered in subsequent discussions. For the present we may think of testing to see if a system is in thermodynamic equilibrium by the following procedure: Isolate the system from its surroundings and watch for changes in its observable properties. If there are no changes, it may be concluded that the system was in equilibrium at the moment it was isolated. The system can be said to be at an *equilibrium state*.

When a system is isolated, it cannot interact with its surroundings; however, its state can change as a consequence of spontaneous events occurring internally as its intensive properties, such as temperature and pressure, tend toward uniform values. When all such changes cease, the system is in equilibrium. Hence, for a system to be in equilibrium it must be a single phase or consist of a number of phases that have no tendency to change their conditions when the overall system is isolated from its surroundings. At equilibrium, temperature is uniform throughout the system. Also, pressure can be regarded as uniform throughout as long as the effect of gravity is not significant; otherwise a pressure variation can exist, as in a vertical column of liquid.

In subsequent sections of this book an idealized type of process called a *quasiequilibrium* (or quasistatic) process is considered. A quasiequilibrium process is one in which the departure from thermodynamic equilibrium is at most infinitesimal. All states through which the system passes in a quasiequilibrium process may be considered equilibrium states. Because nonequilibrium effects are inevitably present during actual processes, systems of engineering interest can at best approach, but never realize, this idealized type of process. Our interest in the quasiequilibrium process concept stems mainly from two considerations. First, simple thermodynamic models giving at least *qualitative* information about the behavior of actual systems of interest can often be developed using the quasiequilibrium process concept. This is akin to the use of idealizations such as the point mass or the frictionless pulley in mechanics for the purpose of simplifying an analysis. Second, the quasiequilibrium process concept is instrumental in deducing relationships that exist among the properties of systems at equilibrium (Chaps. 3, 6, and 11).

There is no requirement that a system undergoing an actual process be in equilibrium *during* the process. Some or all of the intervening states may be nonequilibrium states. For many such processes we are limited to knowing the state before the process occurs and the state after the process is completed. However, even if the intervening states of the system are not known, it is often possible to evaluate certain *overall* effects that occur during the process. Examples are provided in the next chapter in the discussions of *work* and *heat*. Typically, nonequilibrium states exhibit spatial variations in intensive properties at a given time. Also, at a specified position intensive properties may vary with time, sometimes chaotically. Spatial and temporal variations in properties such as temperature, pressure, and velocity can be measured accurately in certain cases. It may also be possible to obtain this information by solving appropriate governing equations, expressed in the form of differential equations, either analytically or by computer.

1.3 UNITS FOR MASS, LENGTH, TIME, AND FORCE

When engineering calculations are performed, it is necessary to be concerned with the *units* of the physical quantities involved. A unit is any specified amount of a quantity by comparison with which any other quantity of the same kind is measured. For example, meters, centimeters, kilometers, feet, inches, and miles are all *units of length*. Seconds, minutes, and hours are alternative *time units*.

Because physical quantities are related by definitions and laws, a relatively small number of them suffice to conceive of and measure all others. These may be called *primary dimensions*. The others may be measured in terms of the primary dimensions and are called *secondary*. For example, if length and time were regarded as primary, velocity and area would be secondary. Two commonly used sets of primary dimensions that suffice for applications in *mechanics* are (1) mass, length, and time and (2) force, mass, length, and time. Additional primary dimensions are required when additional physical phenomena come under consideration. Temperature is included for thermodynamics, and electric current is introduced for applications involving electricity.

Once a set of primary dimensions is adopted, a *base unit* for each primary dimension is specified. Units for all other quantities are then derived in terms of the base units. Let us illustrate these ideas by considering briefly two systems of units: SI units and English Engineering units.

1.3.1 SI Units

In the present discussion we consider the system of units called SI, which takes mass, length, and time as primary dimensions and regards force as secondary. SI is the abbreviation for Système International d'Unités (International System of Units). It is the legally accepted system in many countries and is gradually coming into use in the United States. The conventions of the SI are published and controlled by an international treaty organization. The SI *base units* for mass, length, and time are listed in Table 1.2 and discussed in the following paragraphs.

TABLE 1.2 SI Units for Mass, Length, Time, and Force

Quantity	Unit	Symbol
mass	kilogram	kg
length	meter	m
time	second	s
force	newton (= 1 kg·m/s^2)	N

The SI base unit of length is the meter (metre), m, defined as 1,650,763.73 wavelengths in vacuum of the orange-red line of the spectrum of krypton-86. The base unit of time is the second, s. The second is defined as the duration of 9,192,631,770 cycles of the radiation associated with a specified transition of the cesium atom.

The SI base unit of mass is the kilogram, kg. It is equal to the mass of a particular cylinder of platinum–iridium alloy kept by the International Bureau of Weights and Measures near Paris. A duplicate maintained by the U.S. National Bureau of Standards serves as the mass standard for the United States. The kilogram is the only base unit still defined relative to a fabricated object.

The amount of a substance can also be given in terms of the mole (mol), an SI base unit. One mole of any substance is the amount of that substance that contains as many elementary entities (atoms, molecules, ions, electrons, other particles, or specified groups of such particles) as there are atoms in 12 grams (0.012 kilogram) of carbon-12. The gram molecular weight of a substance is the number of grams per mole of that substance. Thus, the gram molecular weight of carbon-12 is 12 g/mol. In this text it is more convenient to work in terms of the kilomole (kmol). One kilomole of a substance is the amount of that substance in kilograms numerically equal to its molecular weight. The number of kilomoles of a substance, n, is obtained by dividing the mass in kilograms by the molecular weight, in kg/kmol. That is

$$n = \frac{m}{M} \tag{1.1}$$

where m and M denote the mass and molecular weight, respectively.

Since it is frequently necessary to work with extremely large or small values when using the SI unit system, a set of standard prefixes is provided in Table 1.3 to simplify matters. For example, km denotes kilometer, that is, 10^3 m.

The SI unit of force, called the newton, is a secondary unit, defined in terms of the base units for mass, length, and time. Newton's second law of motion states that the net force acting on a body is proportional to the product of the mass and the acceleration, written $F \propto ma$. The newton is defined so that the proportionality constant in the expression is equal to unity. That is, Newton's second law is expressed as the equality

$$F = ma \tag{1.2}$$

The newton, N, is the force required to accelerate a mass of 1 kilogram at the rate of 1 meter per second per second. With Eq. 1.2

$$1\ \text{N} = (1\ \text{kg})(1\ \text{m/s}^2) = 1\ \text{kg}\cdot\text{m/s}^2 \tag{1.3}$$

To illustrate the use of the SI units introduced thus far, let us determine the weight in newtons of an object whose mass is 1000 kg, at a place on the earth's surface where the acceleration due to gravity equals a *standard* value defined as 9.80665 m/s^2. Inserting values into Eq. 1.2

TABLE 1.3 SI Unit Prefixes

Factor	Prefix	Symbol	Factor	Prefix	Symbol
10^{12}	tera	T	10^{-2}	centi	c
10^{9}	giga	G	10^{-3}	milli	m
10^{6}	mega	M	10^{-6}	micro	μ
10^{3}	kilo	k	10^{-9}	nano	n
10^{2}	hecto	h	10^{-12}	pico	p

$$F = ma$$
$$= (1000\ \text{kg})(9.80665\ \text{m/s}^2) = 9806.65\ \text{kg}\cdot\text{m/s}^2$$

This force can be expressed in terms of the newton by using Eq. 1.3 as a unit conversion factor. Thus

$$F = \left(9806.65\ \frac{\text{kg}\cdot\text{m}}{\text{s}^2}\right)\left(\frac{1\ \text{N}}{1\ \text{kg}\cdot\text{m/s}^2}\right) = 9806.65\ \text{N}$$

Recall that the *weight* of a body always refers to the force of gravity. When we say a body weighs a certain amount, we mean that this is the force with which the body is attracted to the earth or some other body. The weight is calculated in terms of the mass and the local acceleration due to gravity. Thus, the weight of an object can vary because of the variation of the acceleration of gravity with location, but its mass remains constant. Accordingly, if the object considered previously was on the surface of a planet at a point where the acceleration of gravity is, say, one-tenth of the value used in the above calculation, the mass would remain the same but the weight would be one-tenth of the calculated value.

SI units for other physical quantities are also derived in terms of the SI base units. Some of the derived units occur so frequently that they are given special names and symbols, such as the newton. SI units for quantities pertinent to thermodynamics are given as they are introduced in the text. For further details regarding the SI, consult one of several publications devoted to this topic.

1.3.2 English Engineering Units

Although SI units are intended as a worldwide standard, at the present time many segments of the engineering community in the United States regularly use some other units. A large portion of America's stock of tools and industrial machines and much valuable engineering data utilize units other than SI units. For many years to come, engineers in the United States will have to be conversant with a variety of units.

In this section we consider a system of units that is commonly used in the United States, called the English Engineering system. This system takes mass, length, time, *and* force as primary dimensions. The base units employed for these are listed in Table 1.4 and discussed in the following paragraphs. English units for other quantities pertinent to thermodynamics are given as they are introduced in the text.

The base unit for length is the foot, ft, defined in terms of the meter as

$$1\ \text{ft} = 0.3048\ \text{m} \tag{1.4}$$

The inch, in., is defined in terms of the foot

$$12\ \text{in.} = 1\ \text{ft}$$

One inch equals 2.54 cm. Although units such as the minute and the hour are often

TABLE 1.4 English Units for Mass, Length, Time, and Force

Quantity	Unit	Symbol
mass	pound mass	lb
length	foot	ft
time	second	s
force	pound force ($= 32.1740\ \text{lb}\cdot\text{ft/s}^2$)	lbf

used in engineering, it is convenient to select the second as the English Engineering base unit for time.

The English Engineering base unit of mass is the pound mass, lb, defined in terms of the kilogram as

$$1 \text{ lb} = 0.45359237 \text{ kg} \tag{1.5}$$

The symbol lbm also may be used to denote the pound mass. The amount of a substance can be given in terms of the pound mole (lbmol). One pound mole of a substance is the amount of that substance in pounds mass numerically equal to its molecular weight. Accordingly, when m is in pounds and M is the molecular weight, lb/lbmol, Eq. 1.1 gives the number of pound moles of the substance, n.

In the English Engineering system of units force is regarded as a primary dimension, so a base unit for force is adopted independently of those specified for the other primary dimensions. The base unit for force in the English Engineering system of units is the pound force, lbf, defined as the force with which 1 pound mass would be attracted to the earth at a location where the acceleration of gravity is the standard value, 32.1740 ft/s^2 (9.80665 m/s^2). In this system of units the base unit of force is not obtained with reference to Newton's second law, as in the SI system of units, but is instead established independently.

The concepts of mass, length, time, and force are interrelated by Newton's second law of motion. When all of these are regarded as primary dimensions, it is necessary to introduce explicitly the proportionality constant in Newton's second law, as follows:

$$F = \frac{1}{g_c} ma \tag{1.6}$$

where g_c is a fundamental physical constant that expresses the proportionality between force and the product of mass and acceleration. In the English Engineering system of units

$$1 \text{ lbf} = \frac{(1 \text{ lb})(32.1740 \text{ ft/s}^2)}{g_c}$$

so

$$g_c = 32.1740 \frac{\text{lb} \cdot \text{ft}}{\text{lbf} \cdot \text{s}^2} \tag{1.7}$$

In this system of units the proportionality constant in Newton's second law has both a numerical value different from unity and dimensions.

The pound force, lbf, is not equal to the pound mass, lb, introduced previously. Force and mass are fundamentally different, as are these units. However, the double use of the word "pound" can be confusing, and care must be taken to avoid error. To show the use of these units in a single calculation, let us determine the weight of an object whose mass is 1000 lb at a location where the local acceleration of gravity is 32.0 ft/s^2. By inserting values into Eq. 1.6

$$F = \frac{1}{g_c} ma = \frac{(1000 \text{ lb})(32.0 \text{ ft/s}^2)}{(32.1740 \text{ lb} \cdot \text{ft/lbf} \cdot \text{s}^2)} = 994.59 \text{ lbf}$$

Observe that although 1 pound mass weighs approximately 1 pound force at locations where the acceleration of gravity is not too different from the standard value (32.1740 ft/s^2), the pound force is a unit of force distinct from the pound mass, a unit of mass.

Once base units have been specified for mass, length, and time in the English Engineering system of units, a unit of force could be introduced in a way that *parallels* the introduction of the newton. That is, a force unit could be defined with the proportionality constant in Newton's second law set to unity. With this approach force would be regarded as secondary. From this perspective, we could have *defined* the pound force as

$$1 \text{ lbf} = 32.1740 \frac{\text{lb} \cdot \text{ft}}{\text{s}^2} \tag{1.8}$$

Whether force is regarded as primary or secondary is strictly a matter of viewpoint. Those who elect to regard force, mass, length, and time as primary would show g_c explicitly in Newton's second law, and all expressions derived from it, and use the value of g_c given by Eq. 1.7. However, if force is regarded as secondary, Newton's second law would be written as Eq. 1.2. Equation 1.8 would then be employed as a unit conversion factor relating the pound force to the pound mass, foot, and second in exactly the same way as Eq. 1.3 is used as a unit conversion factor relating the newton to the kilogram, meter, and second. This is the approach we will follow throughout the remainder of the text. Equation 1.8 is employed as a unit conversion factor. The constant g_c will not be included explicitly in our equations.

1.4 SPECIFIC VOLUME AND PRESSURE

Three intensive properties that are particularly important in engineering thermodynamics are specific volume, pressure, and temperature. In this section specific volume and pressure are considered. Temperature is the subject of Sec. 1.5.

From the macroscopic perspective, the description of matter is simplified by considering it to be distributed continuously throughout a region. The correctness of this idealization, known as the *continuum* hypothesis, is inferred from the fact that for an extremely large class of phenomena of engineering interest the resulting description of the behavior of matter is in agreement with measured data.

1.4.1 Density and Specific Volume

When substances can be treated as continua, it is possible to speak of their intensive thermodynamic properties "at a point." Thus, at any instant the density ρ at a point is defined as

$$\rho = \lim_{V \to V'} \left(\frac{m}{V} \right) \tag{1.9}$$

where V' is the smallest volume for which a definite value of the ratio exists. The volume V' contains enough particles for statistical averages to be significant. It is the smallest volume for which the matter can be considered a continuum and is normally small enough that it can be considered a "point." With density defined by Eq. 1.9, density can be described mathematically as a continuous function of position and time.

The density, or local mass per unit volume, is an intensive property that may vary from point to point within a system. Thus, the mass associated with a particular volume V is determined in principle by integration

$$m = \int_V \rho \, dV \tag{1.10}$$

and *not* simply as the product of density and volume.

The *specific volume* v is defined as the reciprocal of the density, $v = 1/\rho$. It is the volume per unit mass. Like density, specific volume is an intensive property and may vary from point to point. SI units for density and specific volume are kg/m^3 and m^3/kg, respectively. However, they are also often expressed, respectively, as g/cm^3 and cm^3/g. English units used for density and specific volume in this text are lb/ft^3 and ft^3/lb, respectively.

In certain circumstances it is convenient to express properties on a per mole basis rather than on a unit mass basis. To signal this, a bar is used over the symbol. Thus, $\bar{v}$ signifies the volume per kmol or lbmol. In this text the units used for $\bar{v}$ are $m^3/kmol$ and $ft^3/lbmol$. Since $n = m/M$, the relationship between $\bar{v}$ and v is

$$\bar{v} = Mv \tag{1.11}$$

where M is the molecular weight in kg/kmol or lb/lbmol, as appropriate.

1.4.2 Pressure

In this section the concept of pressure is introduced from the continuum viewpoint. Let us begin by considering a small area A passing through a point in a fluid at rest. The fluid on one side of the area exerts a compressive force on it that is normal to the area, F_{normal}. An equal but oppositely directed force is exerted on the area by the fluid on the other side. For a fluid at rest, no other forces than these act on the area. The *fluid pressure p* at the specified point is defined as the limit

$$p = \lim_{A \to A'} \left(\frac{F_{normal}}{A} \right) \tag{1.12}$$

where A′ is the area at the "point" in the same limiting sense as used in the definition of density.

If the area A′ was given new orientations by rotating it around the given point, and the pressure determined for each new orientation, it would be found that the pressure at the point is the same in all directions *as long as the fluid is at rest*. This is a consequence of the equilibrium of forces acting on an element of volume surrounding the point. However, the pressure can vary from point to point within a quiescent fluid; examples are the variation of atmospheric pressure with elevation and the pressure variation with depth in oceans, lakes, and other bodies of water.

Consider next a fluid in motion. In this case the force exerted on an area passing through a point in the fluid may be resolved into three mutually perpendicular components: one normal to the area and two in the plane of the area. When expressed on a unit area basis, the component normal to the area is called the *normal stress*, and the two components in the plane of the area are termed *shear stresses*. The magnitudes of the stresses generally vary with the orientation of the area. The state of stress in a fluid in motion is a topic that is normally treated thoroughly in *fluid mechanics*. The deviation of a normal stress from the pressure, the normal stress that would exist were the fluid at rest, is typically very small. In this book we assume that the normal stress at a point is equal to the pressure at that point. This assumption yields results of acceptable accuracy for the applications considered.

The SI unit of pressure and stress is the pascal.

$$1 \text{ pascal} = 1 \text{ N/m}^2$$

However, in this text it is convenient to work with multiples of the pascal: the kilopascal, the bar, and the megapascal.

$$1 \text{ kPa} = 10^3 \text{ N/m}^2$$
$$1 \text{ bar} = 10^5 \text{ N/m}^2$$
$$1 \text{ MPa} = 10^6 \text{ N/m}^2$$

Commonly used English units for pressure and stress are pounds force per square foot, lbf/ft^2, and pounds force per square inch, $lbf/in.^2$ Although atmospheric pressure varies with location on the earth, a standard reference value can be defined and used to express other pressures.

$$1 \text{ standard atmosphere (atm)} = \begin{cases} 1.01325 \times 10^5 \text{ N/m}^2 \\ 14.696 \text{ lbf/in.}^2 \end{cases}$$

Pressure as discussed above is called *absolute* pressure. Throughout this book the term pressure refers to absolute pressure unless explicitly stated otherwise. Although absolute pressures must be used in thermodynamic relations, pressure-measuring devices often indicate the *difference* between the absolute pressure in a system and the absolute pressure of the atmosphere existing outside the measuring device. The magnitude of the difference is called a *gage* pressure or a *vacuum* pressure. The term gage pressure is applied when the pressure in the system is greater than the local atmospheric pressure, p_{atm}.

$$p(\text{gage}) = p(\text{absolute}) - p_{atm}(\text{absolute}) \tag{1.13}$$

When the local atmospheric pressure is greater than the pressure in the system, the term vacuum pressure is used.

$$p(\text{vacuum}) = p_{atm}(\text{absolute}) - p(\text{absolute}) \tag{1.14}$$

The relationship among the various ways of expressing pressure measurements is shown in Fig. 1.3. Engineers frequently use the letters a and g to distinguish between absolute and gage pressures. For example, the absolute and gage pressures in pounds force per square inch are written as psia and psig, respectively. In this book, gage and vacuum pressures are denoted as $lbf/in.^2$ (gage) or $lbf/in.^2$ (vacuum), respectively.

Two commonly used devices for measuring pressure are the manometer and the Bourdon tube. Manometers measure pressure differences in terms of the length of a column of liquid such as water, mercury, or oil. The manometer shown in Fig. 1.4 has one end open to the atmosphere and the other attached to a closed vessel

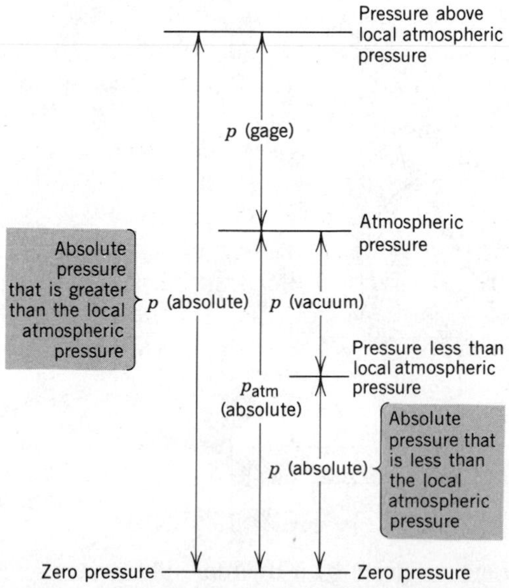

Figure 1.3 Relationships among the absolute, atmospheric, gage, and vacuum pressures.

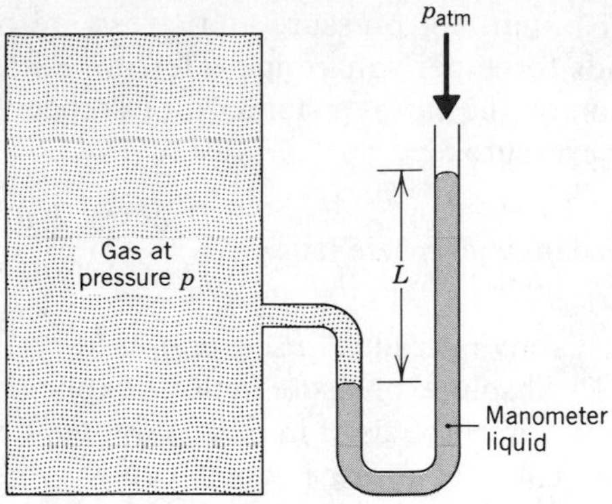

Figure 1.4 Pressure measurement by a manometer.

containing a gas at uniform pressure. The difference between the gas pressure and that of the atmosphere is

$$p - p_{atm} = \rho g L \tag{1.15}$$

where ρ is the density of the manometer liquid, g the acceleration of gravity, and L the difference in the liquid levels. For short columns of liquid ρ and g may be taken as constant. Because of this proportionality between pressure difference and manometer fluid length, pressures are often expressed in terms of millimeters of mercury, inches of water, and so on. It is left as an exercise to develop Eq. 1.15 using an elementary force balance.

A Bourdon tube gage is shown in Fig. 1.5. The figure shows a curved tube having an elliptical cross section with one end attached to the pressure to be measured and the other end connected to a pointer by a mechanism. When fluid under pressure fills the tube, the elliptical section tends to become circular, and the tube straightens. This motion is transmitted by the mechanism to the pointer. By calibrating the deflection of the pointer for known pressures, a graduated scale can be determined from which any applied pressure can be read in suitable units. Because of its construction, the Bourdon tube measures the pressure relative to the pressure of the surroundings existing at the instrument. Accordingly, the dial reads zero when the inside and outside of the tube are at the same pressure.

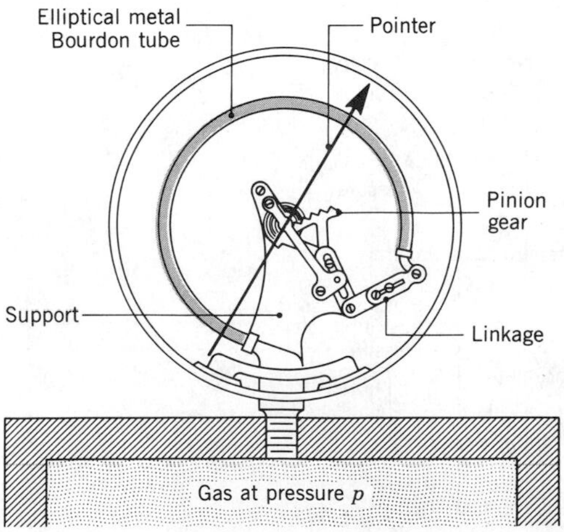

Figure 1.5 Pressure measurement by a Bourdon tube gage.

Pressure can be measured by other means as well. An important class of sensors utilize the *piezoelectric* effect: A charge is generated within certain solid materials when they are deformed. This mechanical input/electrical output provides the basis for pressure measurement as well as displacement and force measurements. Another important type of sensor employs a diaphragm that deflects when a force is applied, altering an inductance, resistance, or capacitance.

1.5 TEMPERATURE

In this section the intensive property temperature is considered along with means of measuring it. Like force, a concept of temperature originates with our sense perceptions. It is rooted in the notion of the "hotness" or "coldness" of a body. We use our sense of touch to distinguish hot bodies from cold bodies and to arrange bodies in their order of "hotness," deciding that 1 is hotter than 2, 2 hotter than 3, and so on. But however sensitive the human body may be, we are unable to gauge this quality precisely. Accordingly, thermometers and temperature scales have been devised to measure it.

1.5.1 Thermal Equilibrium

As for mass, length, and time, it is difficult to give a definition of temperature in terms of concepts that are independently defined or accepted as primitive. However, it is possible to arrive at an objective understanding of *equality* of temperature by using the fact that when the temperature of a body changes, other properties also change.

To illustrate this, consider two copper blocks, and suppose that our senses tell us that one is warmer than the other. If the blocks were brought into contact and isolated from their surroundings, they would interact in a way that can be described as a *thermal* (or *heat*) interaction. During this interaction, it would be observed that the volume of the warmer block decreases somewhat with time, while the volume of the colder block increases with time. Eventually, no further changes in volume would be observed, and the blocks would feel equally warm. Similarly, we would be able to observe that the electrical resistance of the warmer block decreases with time, and that of the colder block increases with time; eventually the electrical resistances would become constant also. When all changes in such observable properties cease, the interaction is at an end. The two blocks are then in *thermal* equilibrium. Considerations such as these lead us to infer that the blocks have a physical property that determines whether they will be in thermal equilibrium. This property is called *temperature,* and we may postulate that when the two blocks are in thermal equilibrium, their temperatures are equal.

The *rate* at which the blocks approach thermal equilibrium with one another can be slowed by separating them with a thick layer of polystyrene foam, rock wool, cork, or other insulating material. Although the rate at which equilibrium is approached can be reduced, no actual material can prevent the blocks from interacting until they attain the same temperature. However, by extrapolating from experience, an *ideal* insulator can be imagined that would preclude them from interacting thermally. Any such ideal insulator is called an *adiabatic wall.* When a system undergoes a process while enclosed by an adiabatic wall, that process is an *adiabatic process.* A process that occurs at constant temperature is an *isothermal process.* An adiabatic process is not necessarily an isothermal process, nor is an isothermal process necessarily adiabatic.

It is a matter of experience that when two bodies are in thermal equilibrium with a third body, they are in thermal equilibrium with one another. This statement, which is sometimes called the *zeroth law of thermodynamics,* is tacitly assumed in every measurement of temperature. Thus, if we want to know if two bodies are at the same temperature, it is not necessary to bring them into contact and see whether their observable properties change with time, as described previously. It is necessary only to see if they are individually in thermal equilibrium with a third body. The third body is usually a *thermometer*.

1.5.2 Thermometers

Any body with at least one measurable property that changes as its temperature changes can be used as a thermometer. Such a property is called a *thermometric* property. The particular substance that exhibits changes in the thermometric property is known as a *thermometric* substance.

A familiar device for temperature measurement is the liquid-in-glass thermometer pictured in Fig. 1.6, which consists of a glass capillary tube connected to a bulb filled with a liquid such as mercury or alcohol and sealed at the other end. The space above the liquid is occupied by the vapor of the liquid or an inert gas. As temperature increases, the liquid expands in volume and rises in the capillary. The length L of the liquid in the capillary depends on the temperature. Accordingly, the liquid is the thermometric substance and L is the thermometric property. Although this type of thermometer is commonly used for ordinary temperature measurements, it is not well suited for applications where extreme accuracy is required.

The gas thermometer shown in Fig. 1.7 is so exceptional in terms of precision and accuracy that it has been adopted internationally as the standard instrument for calibrating other thermometers. The thermometric substance is the gas (normally hydrogen or helium), and the thermometric property is the pressure exerted by the gas. As shown in the figure, the gas is contained in a bulb, and the pressure exerted by it is measured by an open-tube mercury manometer. As temperature increases, the gas expands, forcing mercury up in the open tube. The gas is kept at constant volume by raising or lowering the reservoir. The gas thermometer is used as a standard worldwide by bureaus of standards and research laboratories. However, because gas thermometers require elaborate apparatus and are large, slowly re-

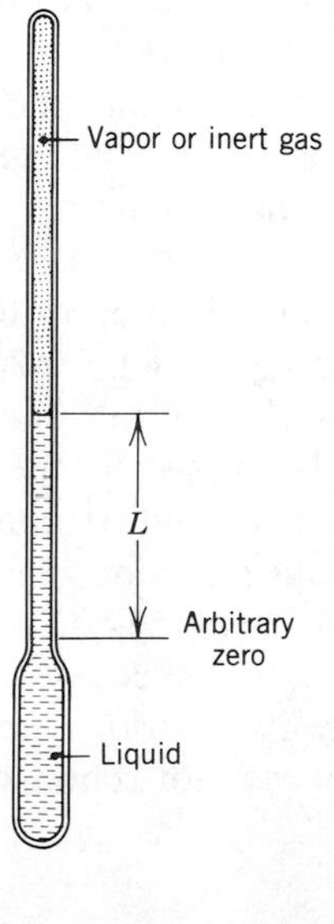

Figure 1.6 Liquid-in-glass thermometer.

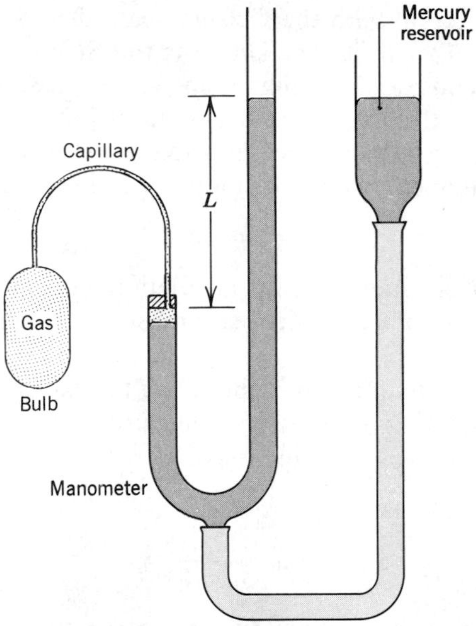

Figure 1.7 Constant-volume gas thermometer.

sponding devices that demand painstaking experimental procedures, smaller, more rapidly responding thermometers are used for most temperature measurements and they are calibrated (directly or indirectly) against gas thermometers.

Sensors known as *thermocouples* are based on the principle that when two dissimilar metals are joined, an electromotive force (emf) that is primarily a function of temperature will exist in a circuit. In certain thermocouples, one thermocouple wire is platinum of a specified purity and the other is an alloy of platinum and rhodium. Thermocouples also utilize copper and constantan (an alloy of copper and nickel), iron and constantan, as well as several other pairs of materials. Electrical-resistance sensors are another important class of temperature measurement devices. These sensors are based on the fact that the electrical resistance of various materials changes in a predictable manner with temperature. The materials used for this purpose are normally conductors (such as platinum, nickel, or copper) or semiconductors. Devices using conductors are known as *resistance temperature detectors* and semiconductor types are called *thermistors*. A variety of instruments measure temperature by sensing radiation. They are known by terms such as *radiation pyrometers* and *optical pyrometers*. This type of thermometer differs from those previously considered in that it does not actually come in contact with the body whose temperature is to be determined, an advantage when dealing with moving objects or bodies at extremely high temperatures.

1.5.3 Gas Temperature Scale and Kelvin Scale

Temperature scales are defined by the numerical value assigned to a *standard fixed point*. By international agreement the standard fixed point is the easily reproducible *triple point of water:* the state of equilibrium between steam, ice, and liquid water (Sec. 3.2). As a matter of convenience, the temperature at this standard fixed point is defined as 273.16 kelvins, abbreviated as 273.16 K. This makes the temperature interval from the *ice point*[1] (273.15 K) to the *steam point*[2] equal to 100 K and thus in

[1] The state of equilibrium between ice and air-saturated water at a pressure of 1 atm.

[2] The state of equilibrium between steam and liquid water at a pressure of 1 atm.

agreement over the interval with the Celsius scale discussed in Sec. 1.5.4, which assigns 100 Celsius degrees to it. The kelvin is the SI base unit for temperature.

It is instructive to consider how numerical values are associated with levels of temperature by the gas thermometer. Let p stand for the pressure in a constant-volume gas thermometer in thermal equilibrium with, say, a bath. A value can be assigned to the bath temperature very simply by a linear relation.

$$T = \alpha p \tag{1.16}$$

where α is an arbitrary constant. The linear relationship is an arbitrary choice; other selections for the correspondence between pressure and temperature could also be made.

The value of α may be determined by inserting the thermometer into another bath maintained at the triple point of water and measuring the pressure, call it p_{tp}, of the confined gas at the triple point temperature. Substituting values into Eq. 1.16 and solving for α

$$\alpha = \frac{273.16}{p_{tp}}$$

The temperature of the original bath, at which the pressure of the confined gas is p, is then

$$T = 273.16\left(\frac{p}{p_{tp}}\right) \tag{1.17}$$

However, since the values of both pressures, p and p_{tp}, depend *in part* on the amount of gas in the bulb, the value assigned by Eq. 1.17 to the bath temperature varies with the amount of gas in the thermometer. This difficulty is overcome in precision thermometry by repeating the measurements (in the original bath and the reference bath) several times with less gas in the bulb in each successive attempt. For each trial the ratio p/p_{tp} is calculated from Eq. 1.17 and plotted versus the corresponding reference pressure p_{tp} of the gas at the triple point temperature. When several such points have been plotted, the resulting curve is extrapolated to the ordinate where $p_{tp} = 0$. This is illustrated in Fig. 1.8 for constant-volume thermometers with a number of different gases. Inspection of the figure shows an important result. At each nonzero value of the reference pressure, the p/p_{tp} values differ with

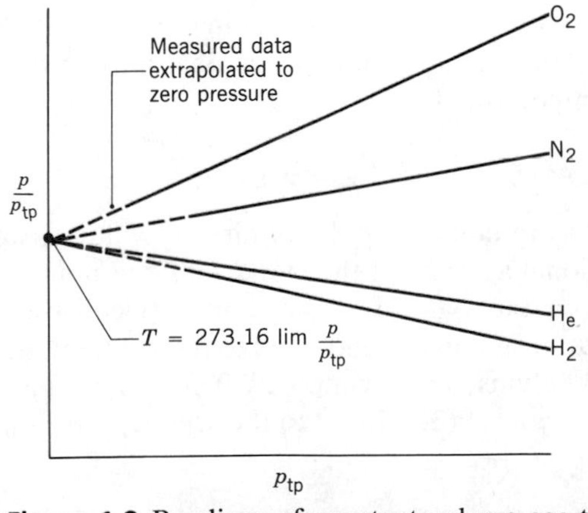

Figure 1.8 Readings of constant-volume gas thermometers for a fixed level of temperature versus p_{tp}, when several gases are used.

the gas employed in the thermometer. However, as pressure decreases, the p/p_{tp} values from thermometers with different gases approach one another, and in the limit as pressure tends to zero, *the same value for p/p_{tp} is obtained for each gas*. Based on these general results, the gas temperature scale is defined by the relationship

$$T = 273.16 \lim \frac{p}{p_{tp}} \tag{1.18}$$

where "lim" means that both p and p_{tp} tend to zero. It should be evident that the determination of temperatures by this means requires extraordinarily careful and elaborate experimental procedures.

Although the temperature scale of Eq. 1.18 is independent of the properties of any one gas, it still depends on the properties of gases in general. Accordingly, the measurement of low temperatures requires a gas that does not condense at these temperatures, and this imposes a limit on the range of temperatures that can be measured by a gas thermometer. The lowest temperature that can be measured with such an instrument is about 1 K, obtained with helium. At high temperatures gases dissociate, and therefore these temperatures also cannot be determined by a gas thermometer. Other empirical means, utilizing the properties of other substances, must be employed to measure temperature in ranges where the gas thermometer is inadequate.

In view of these limitations, it is desirable to have a means for assigning temperature values that depends in no way on the properties of any particular substance or class of substances. Such a scale is called a *thermodynamic* temperature scale. The *Kelvin scale* is an absolute thermodynamic temperature scale that provides a continuous definition of temperature, valid over all ranges of temperature. Empirical measures of temperature, with different thermometers, can be related to the Kelvin scale. To develop the Kelvin scale, it is necessary to use the conservation of energy principle and the second law of thermodynamics; therefore, a detailed account of this is deferred to Sec. 5.5 after these principles have been introduced. However, we note here that the Kelvin scale has a zero of 0 K, and lower temperatures than this are not defined.

The gas scale and Kelvin scale can be shown to be *identical* in the temperature range in which a gas thermometer can be used. For this reason we may write K after a temperature determined by means of Eq. 1.18. Moreover, until the concept of temperature is reconsidered in more detail in Chap. 5, we assume that all temperatures referred to in the interim are in accord with values given by a constant-volume gas thermometer.

1.5.4 Celsius, Rankine, and Fahrenheit Scales

The *Celsius temperature scale* (also called the centigrade scale) uses the unit degree Celsius (°C), which has the same magnitude as the kelvin. Thus, temperature *differences* are identical on both scales. However, the zero point on the Celsius scale is shifted to 273.15 K, as shown by the following relationship between the Celsius temperature and the Kelvin temperature:

$$T(°\text{C}) = T(\text{K}) - 273.15 \tag{1.19}$$

From this it can be seen that on the Celsius scale the triple point of water is 0.01°C and that 0 K corresponds to −273.15°C.

The Celsius scale is defined so that the temperature at the *ice point,* 273.15 K, is 0.00°C, and the temperature at the *steam point,* 373.15 K, is 100.00°C. Accordingly, there are 100 Celsius degrees in this interval of 100 kelvins, a correspondence that is contrived through the selection of 273.16 K as the triple point temperature. Observe that since the temperatures at the ice and steam points are experimental values subject to revision in light of more precise determinations, the only Celsius temperature that is fixed *by definition* is that at the triple point of water.

Two other temperature scales are in common use in engineering in the United States. By definition, the *Rankine scale,* the unit of which is the degree rankine (°R), is proportional to the Kelvin temperature according to

$$T(°\mathrm{R}) = 1.8T(\mathrm{K}) \tag{1.20}$$

As evidenced by Eq. 1.20, the Rankine scale is also an absolute thermodynamic scale with an absolute zero that coincides with the absolute zero of the Kelvin scale. In thermodynamic relationships, temperature is always in terms of the Kelvin or Rankine scale unless specifically stated otherwise.

A degree of the same size as that on the Rankine scale is used in the *Fahrenheit scale,* but the zero point is shifted according to the relation

$$T(°\mathrm{F}) = T(°\mathrm{R}) - 459.67 \tag{1.21}$$

Substituting Eqs. 1.19 and 1.20 into Eq. 1.21, it follows that

$$T(°\mathrm{F}) = 1.8T(°\mathrm{C}) + 32 \tag{1.22}$$

This equation shows that the Fahrenheit temperature of the ice point (0°C) is 32°F and of the steam point (100°C) is 212°F. The 100 Celsius or Kelvin degrees between the ice point and steam point correspond to 180 Fahrenheit or Rankine degrees, as shown in Fig. 1.9, where the Kelvin, Celsius, Rankine, and Fahrenheit scales are compared.

Finally, when making engineering calculations, it is common to round off the last numbers in Eqs. 1.19 and 1.21 to 273 and 460, respectively. This is frequently done in subsequent sections of the text.

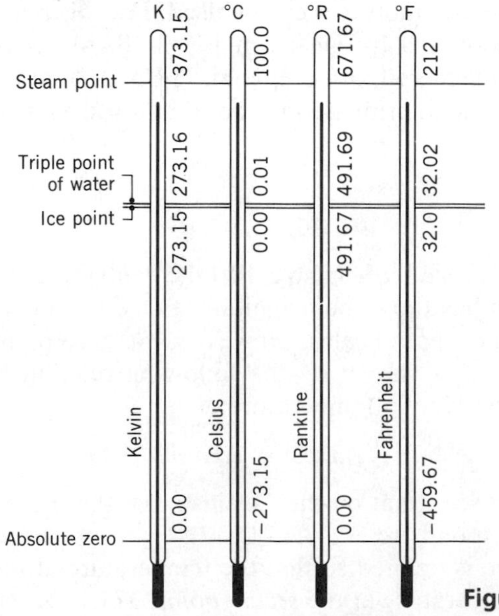

Figure 1.9 Comparison of temperature scales.

1.6 METHODOLOGY FOR SOLVING THERMODYNAMICS PROBLEMS

The first steps in a thermodynamic analysis are definition of the system and identification of the relevant interactions with the surroundings. Attention then turns to the pertinent physical laws and relationships that allow the behavior of the system to be described. Most analyses use, directly or indirectly, one or more of three basic laws. These laws, which are independent of the particular substance or substances under consideration, are

- the conservation of mass principle
- the conservation of energy principle
- the second law of thermodynamics

In addition, relationships among the properties of the particular substance or substances considered are usually necessary (Chap. 3). Newton's second law of motion (Chaps. 2 and 9) and relations such as Fourier's conduction model (Chap. 2) may also play a part.

A major goal of this textbook is to help you learn how to solve engineering problems that involve thermodynamic principles. To this end numerous solved examples and end-of-chapter problems are provided. It is extremely important for you to study the examples *and* solve problems, for mastery of the fundamentals comes only through practice.

To maximize the results and rewards of your efforts, it is necessary to develop a systematic approach. You must think carefully about your solutions and avoid the temptation of starting problems *in the middle* by selecting some seemingly appropriate equation, substituting in numbers, and quickly "punching up" a result on your calculator. Such a haphazard problem-solving approach can lead to difficulties as problems become more complicated. Accordingly, we strongly recommend that problem solutions be organized using the six steps in the box below, which are employed in the solved examples of this text.

The problem solution format used in this text is intended to *guide* your thinking, not substitute for it. Accordingly, you are cautioned to avoid the rote application of these six steps, for this alone would provide few benefits. In some of the earlier examples and end-of-chapter problems, the solution format may seem unnecessary or unwieldy. However, as the problems become more complicated you will see that it reduces errors, saves time, and provides a deeper understanding of the problem at hand.

Known: State briefly in your own words what is known. This requires that you read the problem carefully *and* think about it.

Find: State concisely in your own words what is to be determined.

Schematic and Given Data: Draw a sketch of the system to be considered. Decide whether a closed system or control volume is appropriate for the analysis, and then carefully identify the boundary. Label the diagram with relevant information from the problem statement.

Record all property values you are given or anticipate may be required for subsequent calculations. Sketch appropriate property diagrams (see Sec. 3.2), locating key state points and indicating, if possible, the processes executed by the system.

The importance of good sketches of the system and property diagrams cannot be overemphasized. They are often instrumental in enabling you to think clearly about the problem.

Assumptions: To form a record of how you *model* the problem, list all simplifying assumptions and idealizations made to reduce it to one that is manageable. Sometimes this information can also be noted on the sketches of the previous step.

Analysis: Using your assumptions and idealizations, reduce the appropriate governing equations and relationships to forms that will produce the desired results.

It is advisable to work with equations as long as possible before substituting numerical data. When the equations are reduced to final forms, consider them to determine what additional data may be required. Identify the tables, charts, or property equations that provide the required values. Additional property diagram sketches may be helpful at this point to clarify states and processes.

When all equations and data are in hand, substitute numerical values into the equations. Carefully check that a consistent and appropriate set of units is being employed. Then perform the needed calculations.

Finally, consider whether the magnitudes of the numerical values seem reasonable and the algebraic signs associated with the numerical values are correct.

Comments: When appropriate, discuss your results briefly. Comment on what was learned, identify key aspects of the solution, discuss how better results might be obtained by relaxing certain assumptions, and so on.

As a particular solution evolves you may have to return to an earlier step and revise it in light of a better understanding of the problem. For example, it might be necessary to add or delete an assumption, revise a sketch, determine additional property data, and so on.

PROBLEMS

1.1 Write out the meanings of the following terms: system, control volume, property, intensive property, equilibrium state, process, thermodynamic cycle, phase, adiabatic process.

1.2 Write out the meanings of the following terms: closed system, isolated system, steady state, extensive property, macroscopic viewpoint, pure substance, quasiequilibrium process, primary dimensions, zeroth law of thermodynamics.

1.3 Sketch possible system boundaries for studying each of the following:

(a) a bicycle tire inflating.
(b) a kettle of water boiling.
(c) a household refrigerator in operation.
(d) a jet engine in flight.
(e) a hot steel billet quenching in an oil bath.
(f) a residential gas furnace in operation.
(g) a rocket launching.

Indicate whether the systems should be classed as closed systems or control volumes, and identify locations where the systems interact with their surroundings.

1.4 As illustrated in Fig. P1.4, electric current from a storage battery runs an electric motor. The shaft of the motor is connected to a pulley–mass assembly that raises a mass. Considering the motor as a system, identify locations on the system boundary where the system interacts with its surroundings and describe changes that occur within the system with time. Repeat for an enlarged system that also includes the battery and pulley–mass assembly.

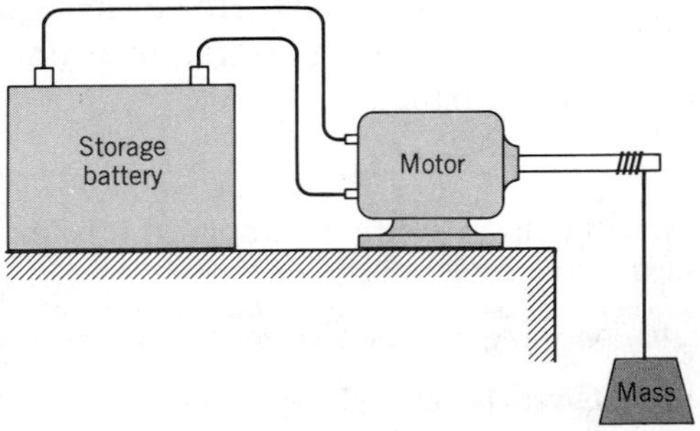

Figure P1.4

1.5 Referring to Fig. 1.2, identify locations on the control volume boundary where the system interacts with its surroundings.

1.6 A system consists of liquid water in equilibrium with a gaseous mixture of air and water vapor. How many phases are present? Does the system consist of a pure substance? Explain.

1.7 A system consists of liquid oxygen in equilibrium with oxygen vapor. How many phases are present? The system undergoes a process during which some of the liquid is vaporized. Can the system be viewed as being a pure substance during the process? Explain

1.8 Consider a system consisting of liquid water and ice undergoing a process. At the end of the process, the ice has melted and only liquid water remains. Can the system be viewed as being a pure substance during the process? Explain.

1.9 An open beaker of liquid ethanol (ethyl alcohol) is placed on a table in a room. After a while, all of the ethanol evaporates. Taking the ethanol and the air in the room as the system, explain whether or not the system can be regarded as a pure substance during the process.

1.10 Determine the weight, in newtons, of an object whose mass is 10 kg at a location where the acceleration of gravity is 9.81 m/s^2.

1.11 Determine the mass, in pounds mass, of an object whose weight is 20 lbf at a location where the acceleration of gravity is 31.1 ft/s^2.

1.12 An object whose mass is 10 kg weighs 90 N. Determine

(a) the local acceleration of gravity, in m/s^2.
(b) the mass, in kg, and the weight, in N, of the object at a location where g = 9.81 m/s.

1.13 An object occupies a volume of 25 ft^3 and weighs 20 lbf at a location where the acceleration of gravity is 31.0 ft/s^2. Determine its weight, in lbf, and its average density, in lb/ft^3, on the moon, where g = 5.57 ft/s^2.

1.14 Atomic and molecular weights of some common substances are listed in Appendix Tables A-1. Using data from the appropriate table, determine

(a) the number of kmol in 50 kg of each of the following: H_2, N_2, NH_3, C_3H_8.
(b) the mass, in lb, of 20 lbmol of each of the following: air, C, H_2O, and CO_2.

1.15 A spring stretches 0.2 cm per newton of applied force. An object is suspended from the spring and a deflection of 3 cm is observed. If g = 9.81 m/s^2, what is the mass of the object, in kg?

1.16 A spring compresses in length by 0.04 in. for every 1 lbf of applied force. Determine the deflection, in inches, of the spring caused by the weight of an object whose mass is 50 lb. The local acceleration of gravity is g = 31.4 ft/s^2.

1.17 You may have used the mass unit *slug* in previous physics or engineering classes. By definition, a mass of 1 slug is accelerated at a rate of 1 ft/s^2 by a force of 1 lbf. Explain why this is a convenient mass unit.

1.18 Acceleration is sometimes measured in g's, or multiples of the standard acceleration of gravity. Determine the net upward force, in newtons, that an astronaut whose mass is 68 kg experiences if the acceleration on lift-off is 10 g's.

1.19 An object whose mass is 3 kg is subjected to an upward force of 50 N. The only other force acting on the object is the force of gravity. Determine the acceleration of the object, in m/s^2, assuming the acceleration of gravity is constant, g = 9.81 m/s^2.

1.20 An object whose mass is 7 lb is subjected to an upward force. The only other force acting on the object is the force of gravity. The acceleration of the object is upward with a magnitude of 60 ft/s^2. The acceleration of gravity is g = 32.2 ft/s^2. Determine the magnitude of the upward force, in lbf.

1.21 An object weighs 15 lbf at a location where the acceleration of gravity is g = 30.5 ft/s^2. Determine the magnitude of the net force, in lbf, required to accelerate the object at 20 ft/s^2.

1.22 An object has a mass of 7 kg. Determine

(a) its weight at a location where the acceleration of gravity is g = 9.7 m/s^2.
(b) the magnitude of the net force, in N, required to accelerate the object at 6 m/s^2.

1.23 An automobile weighs 2200 lbf. What is the magnitude of the net force, in lbf, required to accelerate it at a constant rate of 5 ft/s^2? The acceleration of gravity is g = 32.2 ft/s^2.

1.24 An object weighs 1000 N at a location where the acceleration of gravity is g = 9.4 m/s^2. Determine the net force, in N, required to impart a net acceleration of 5.0 m/s^2.

1.25 A tank contains 0.3 kmol of carbon dioxide (CO_2) gas. The volume occupied by the gas is 2.5 m^3. Determine the mass of CO_2, in kg, and the specific volume on a molar basis, in m^3/kmol.

1.26 The following table lists temperatures and specific volumes of water vapor at two pressures:

p = 0.1 MPa		p = 0.12 MPa	
T (°C)	v (m^3/kg)	T (°C)	v (m^3/kg)
200	2.172	200	1.808
240	2.359	240	1.965
280	2.546	280	2.120

Data encountered in solving problems often do not fall exactly on the grid of values provided by property tables, and *linear interpolation* between adjacent table entries becomes necessary. Using the data provided here, estimate

(a) the specific volume at T = 200°C, p = 0.113 MPa, in m^3/kg.

(b) the temperature at $p = 0.12$ MPa, $v = 1.85$ m^3/kg, in °C.
(c) the temperature at $p = 0.11$ MPa, $v = 2.20$ m^3/kg, in K.

1.27 The following table lists temperatures and specific volumes of water vapor at two pressures:

p = 100 lbf/in.^{2}		p = 120 lbf/in.^{2}	
T (°F)	v (ft^3/lb)	T (°F)	v (ft^3/lb)
350	4.592	350	3.783
400	4.934	400	4.079
450	5.265	450	4.360

Data encountered in solving problems often do not fall exactly on the grid of values provided by property tables, and *linear interpolation* between adjacent table entries becomes necessary. Using the data provided here, estimate

(a) the specific volume at $T = 400$°F, $p = 107$ lbf/in.^{2}, in ft^3/lb.
(b) the temperature at $p = 100$ lbf/in.^{2}, $v = 5.11$ ft^3/lb, in °F.
(c) the specific volume at $T = 380$°F, $p = 111.5$ lbf/in.^{2}, in ft^3/lb.

1.28 A system consisting of 1 kg of a gas undergoes a process during which the relationship between pressure and volume is $pV^{1.3}$ = constant. The process begins with $p_1 = 1$ bar, $V_1 = 1$ m^3 and ends with $V_2 = 3$ m^3. Determine the final pressure, p_2, in bars, and plot the process on a graph of pressure versus volume.

1.29 A system consisting of 1 lb of a gas undergoes a process during which the relation between pressure and volume is pV = constant. The process begins with $p_1 =$ 100 lbf/in.^{2}, $V_1 = 1.5$ ft^3 and ends with $p_2 = 40$ lbf/in.^{2} Determine the final volume, V_2, in ft^3, and plot the process on a graph of pressure versus volume.

1.30 A system consists of air in a piston–cylinder assembly, initially at $p_1 = 20$ lbf/in.^{2}, and occupying a volume of 1.5 ft^3. The air is compressed to a final volume of 0.5 ft^3. During the process, the relation between pressure and volume is $pV^{1.4}$ = constant. Determine the final pressure, in lbf/in.^{2}

1.31 One kilogram of a refrigerant is contained in a piston–cylinder assembly. The refrigerant is compressed from state 1, where $p_1 = 2$ bars, $v_1 = 83.54$ cm^3/g, to state 2, where $p_2 = 10$ bars, $v_2 = 21.34$ cm^3/g. During the process, the relationship between pressure and specific volume is of the form pv^n = constant. Determine the value of the constant n.

1.32 A manometer is attached to a tank of gas in which the pressure is greater than that of the surroundings. The manometer liquid is mercury, with a density of 13.59 g/cm^3. The difference in mercury levels in the manometer is 2 cm. The acceleration of gravity is $g = 9.81$ m/s^2. The atmospheric pressure is 93.0 kPa. Calculate in kPa

(a) the gage pressure of the gas.
(b) the absolute pressure of the gas.

1.33 A vacuum gage attached to a power plant condenser gives a reading of 27.86 in. of mercury. The surrounding atmospheric pressure is 14.66 lbf/in.^{2} Determine the absolute pressure inside the condenser, in lbf/in.^{2} Take the density of mercury as 848 lb/ft^3 and the acceleration of gravity as $g = 32.0$ ft/s^2.

1.34 The absolute pressure inside a tank of gas is 0.05 MPa, and the surrounding atmospheric pressure is 101 kPa. What reading would a Bourdon gage mounted in the tank wall give, in kPa? Is this a *gage* or *vacuum* reading?

1.35 Determine the gage pressure, in lbf/in.^{2}, equivalent to a manometer reading of 1 in. of

(a) water (density = 62.4 lb/ft^3).
(b) mercury (the density of mercury is 13.59 times that of water).

1.36 Determine the gage pressure, in bars, equivalent to a manometer reading of 1 cm of

(a) water (density = 1000 kg/m^3).
(b) mercury (the density of mercury is 13.59 times that of water).

1.37 A glass tube is attached to a pipe carrying water as shown in Fig. P1.37. The pressure of water in the pipe supports a column of water 2 ft high. If the atmospheric pressure is 14.42 lbf/in.^{2}, determine the absolute pressure of the water at the bottom of the tube, in lbf/in.^{2} The specific volume of water is 0.0164 ft^3/lb, and the acceleration of gravity is $g = 32.2$ ft/s^2.

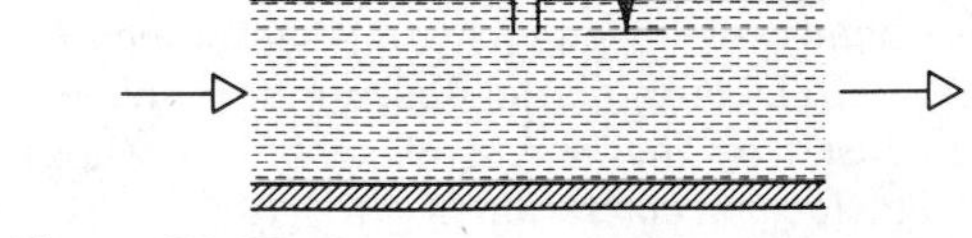

Figure P1.37

1.38 As shown in Fig. P1.38, pressure-measuring devices are placed upstream and downstream of a pump. A U-tube manometer is placed at the upstream location, and a pressure gage is mounted downstream. Using data on the diagram, determine the absolute pressures at the two locations and the difference in gage pressures between them, each in lbf/in.^{2}

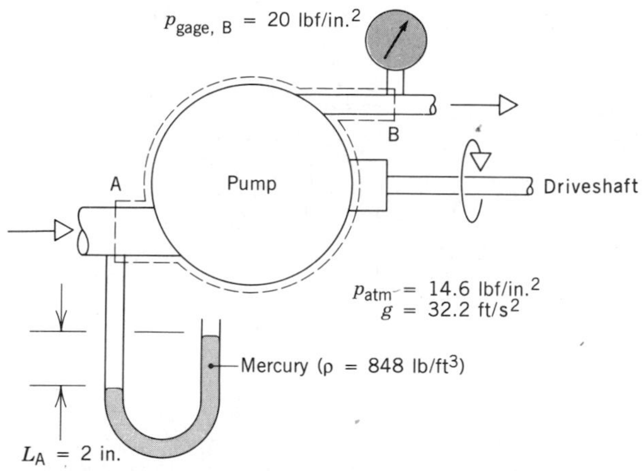

Figure P1.38

1.39 The air in a closed chamber has an absolute pressure of 80 kPa. The pressure of the surrounding atmosphere is equivalent to a 750-mm column of mercury. The density of mercury is 13.59 g/cm^3 and the acceleration of gravity is g = 9.81 m/s^2. Determine the gage (or vacuum) pressure within the chamber, in bars.

1.40 The absolute pressure of a gas entering a compressor is 0.5 bar. At the compressor exit, the gage pressure is 0.8 MPa. The atmospheric pressure is 99 kPa. Determine the change in absolute pressure from inlet to exit, in kPa.

1.41 A vacuum gage at the exit of a steam turbine reads 12.7 lbf/in.^{2} (vacuum), and a pressure gage at the turbine inlet reads 600 lbf/in.^{2} (gage). Determine the change in absolute pressure from inlet to exit, in lbf/in.^{2}

1.42 Determine the absolute pressure exerted on an object submerged 1000 m below the surface of the sea, in bars. The density of seawater is 1020 kg/m^3 and the acceleration of gravity is g = 9.7 m/s^2. The local atmospheric pressure is 0.98 bar.

1.43 An inclined manometer illustrated in Fig. P1.43 is used to measure the pressure of a gas. The liquid inside the manometer has a density of 0.8 g/cm^3, and the manometer reading is shown on the figure. If the atmospheric pressure is 101 kPa and the acceleration of gravity is g = 9.81 m/s^2, what is the absolute pressure of the gas, in kPa?

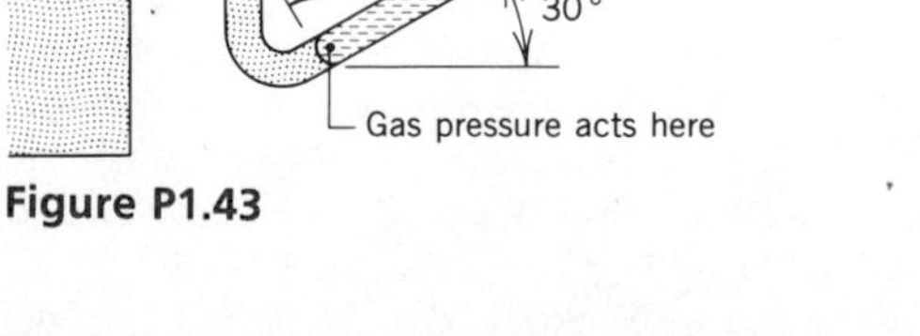

Figure P1.43

1.44 Write a computer program that converts input values from English to SI units or from SI to English units, at the discretion of the user. Include all of the conversions in the table of conversion factors located inside the front cover of this book.

1.45 Write a computer program that converts temperatures from °F, °R, °C, or K to any of the other temperature units, as specified by the user.

1.46 Convert the following temperatures from °C to °F: **(a)** 21°C, **(b)** 0°C, **(c)** −50°C, **(d)** 300°C, **(e)** 100°C, **(f)** −273.15°C. Convert each temperature to the Kelvin scale.

1.47 Convert the following temperatures from °F to °C: **(a)** 70°F, **(b)** 0°F, **(c)** −30°F, **(d)** 500°F, **(e)** 212°F, **(f)** −459.67°F. Convert each temperature to the Rankine scale.

1.48 Two temperature measurements are taken with a thermometer marked with the Celsius scale. Show that the *difference* between the two readings would be the same if the temperatures were converted to the Kelvin scale. Show that the difference would be the same if the Fahrenheit and Rankine scales were used, respectively.

1.49 The relation between resistance R and temperature T for a thermistor closely follows

$$R = R_0 \exp\left[\beta\left(\frac{1}{T} - \frac{1}{T_0}\right)\right]$$

where R_0 is the resistance, in ohms (Ω), measured at temperature T_0 (K) and β is a material constant with units of K. For a particular thermistor R_0 = 2.2 Ω at T_0 = 310 K. From a calibration test, it is found that R = 0.31 Ω at T = 422 K. Determine the value of β for the thermistor and make a plot of resistance versus temperature.

1.50 Over a limited temperature range, the relation between electrical resistance R and temperature T for a resistance temperature detector is

$$R = R_0[1 + \alpha(T - T_0)]$$

where R_0 is the resistance, in ohms (Ω), measured at reference temperature T_0 (in °C) and α is a material constant with units of (°C)$^{-1}$. The following data are obtained for a particular resistance thermometer:

	T (°C)	R (Ω)
Test 1	0	51.39
Test 2	91	51.72

What resistance reading would correspond to a temperature of 50°C on this thermometer?

1.51 Two liquid-in-glass thermometers have Celsius and Fahrenheit scales, respectively. Both thermometers are used to measure the temperature of an object. If each thermometer indicates the same numerical reading, what is the temperature, in kelvins?

1.52 A new absolute temperature scale is proposed. On this scale the ice point of water is 150°S and the steam point is 300°S. Determine the temperatures in °C that correspond to 100° and 400°S, respectively. What is the ratio of the size of the °S to the kelvin?

1.53 As shown in Fig. P1.53, water vapor flows through a valve and enters a turbine. The turbine drives an electrical generator. Mass exits the turbine at a rate of 10,000 kg/h. Using data from the figure as needed,

(a) convert the rate of mass flow to kg/s.
(b) express p_2 in MPa.
(c) express T_1 in K.
(d) express p_1 in bars.

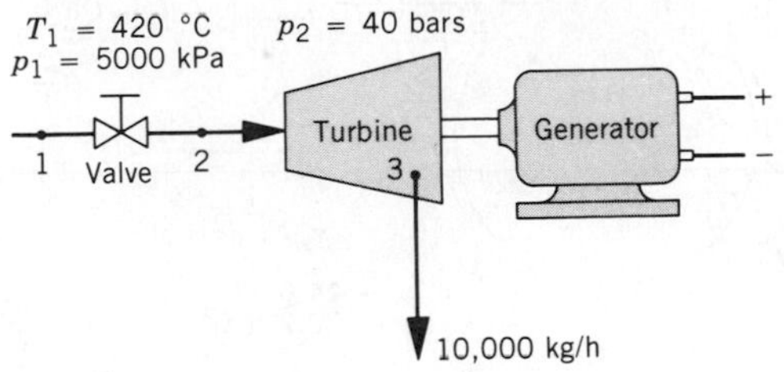

Figure P1.53

DESIGN AND OPEN ENDED PROBLEMS

1.1D Write a memorandum comparing the coverage of the topics listed in the table of contents of this book with the same topics in the college chemistry and physics texts you used. Which topics are emphasized more in your engineering thermodynamics course than in your introductory science courses?

1.2D Energy resource and environmental issues are often in the news these days. During the current term, make a file of newspaper and news magazine clippings on these topics, and prepare a report on a specific aspect of an issue that interests you.

1.3D Write a report explaining the principles and objectives of *statistical thermodynamics*. How does the macroscopic approach to thermodynamics of the present text differ from this? Explain.

1.4D Taking an ordinary drinking glass filled with water, place a card over the top and invert the glass. Upon releasing the card, the water remains in the glass, and the card says in place. Perform appropriate calculations to explain this observation. Repeat with a square of aluminum foil replacing the card and comment. Can you think of any practical applications of this?

1.5D Obtain manufacturers' data on three different types of pressure-measuring sensors for pressures in the range of 0 to 100 lbf/in.^{2} (~70 kPa). Explain the basic operating principles of each sensor and compare the advantages and disadvantages of each device. Consider sensitivity, accuracy, calibration, and cost.

1.6D Obtain manufacturers' data on thermocouple and thermistor temperature sensors for measuring temperatures of hot combustion gases from a furnace. Explain the basic operating principles of each sensor and compare the advantages and disadvantages of each device. Consider sensitivity, accuracy, calibration, and cost.

CHAPTER 2

ENERGY AND THE FIRST LAW OF THERMODYNAMICS

Energy is a fundamental concept of thermodynamics and one of the most significant aspects of engineering analysis. In this chapter we discuss energy and develop equations for applying the principle of conservation of energy. The current presentation is limited to closed systems. In Chap. 4 the development is extended to control volumes.

Energy is a familiar notion, and you already know a great deal about it. In the present chapter several important aspects of the energy concept are developed. Some of these you have encountered before. A basic idea is that energy can be *stored* within systems in various macroscopic forms. Energy can also be *transformed* from one form to another and *transferred* between systems. For closed systems, energy can be transferred by *work* and *heat transfer*. The total amount of energy is *conserved* in all transformations and transfers.

The purpose of this chapter is to organize these ideas into forms suitable for engineering analysis. The presentation begins with a review of energy concepts from mechanics. The thermodynamic concept of energy is then introduced as an extension of the concept of energy in mechanics.

2.1 MECHANICAL CONCEPTS OF ENERGY

Building on the contributions of Galileo and others, Newton formulated a general description of the motions of objects under the influence of applied forces. Newton's laws of motion, which provide the basis for classical mechanics, led to the concepts of *work, kinetic energy,* and *potential energy,* and these led eventually to a broadened concept of energy. The present discussion begins with an application of Newton's second law of motion.

2.1.1 Work and Kinetic Energy

The curved line in Fig. 2.1 represents the path of a body of mass m (a closed system) moving relative to the x–y coordinate frame shown. The velocity of the center of

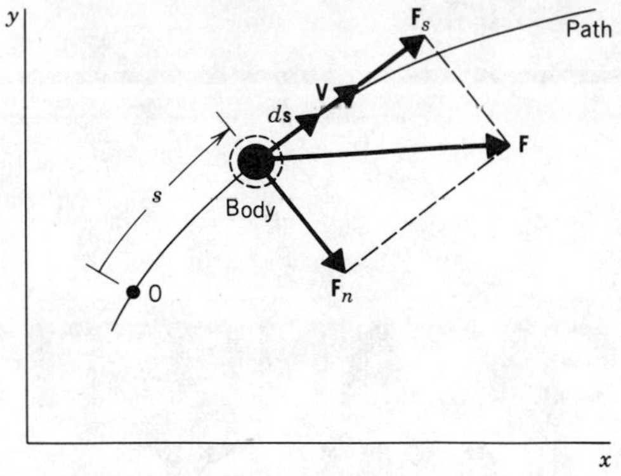

Figure 2.1 Forces acting on a moving system.

mass of the body is denoted by $\mathbf{V}$.[1] The body is acted on by a resultant force $\mathbf{F}$, which may vary in magnitude from location to location along the path. The resultant force is resolved into a component $\mathbf{F}_s$ along the path and a component $\mathbf{F}_n$ normal to the path. The effect of the component $\mathbf{F}_s$ is to change the magnitude of the velocity, whereas the effect of the component $\mathbf{F}_n$ is to change the direction of the velocity. As shown in Fig. 2.1, s is the instantaneous position of the body measured along the path from some fixed point denoted by 0. Since the magnitude of $\mathbf{F}$ can vary from location to location along the path, the magnitudes of $\mathbf{F}_s$ and $\mathbf{F}_n$ are, in general, functions of s.

Let us consider the body as it moves from $s = s_1$, where the magnitude of its velocity is V_1, to $s = s_2$, where its velocity is V_2. Assume for the present discussion that the only interaction between the body and its surroundings involves the force $\mathbf{F}$. By Newton's second law of motion, the magnitude of the component $\mathbf{F}_s$ is related to the change in the magnitude of $\mathbf{V}$ by

$$F_s = m\frac{dV}{dt} \tag{2.1}$$

Using the chain rule, this can be written as

$$F_s = m\frac{dV}{ds}\frac{ds}{dt} = mV\frac{dV}{ds} \tag{2.2}$$

where $V = ds/dt$. Rearranging Eq. 2.2 and integrating from s_1 to s_2 gives

$$\int_{V_1}^{V_2} mV\,dV = \int_{s_1}^{s_2} F_s\,ds \tag{2.3}$$

The integral on the left of Eq. 2.3 is evaluated as follows:

$$\int_{V_1}^{V_2} mV\,dV = \frac{1}{2}mV^2\Big]_{V_1}^{V_2} = \frac{1}{2}m(V_2^2 - V_1^2) \tag{2.4}$$

The quantity $\frac{1}{2}mV^2$ is the *kinetic energy*, KE, of the body. Kinetic energy is a scalar quantity. The *change* in kinetic energy, ΔKE, of the body is[2]

$$\Delta KE = KE_2 - KE_1 = \frac{1}{2}m(V_2^2 - V_1^2) \tag{2.5}$$

The integral on the right of Eq. 2.3 is the *work* of the force F_s as the body moves from s_1 to s_2 along the path. Work is also a scalar quantity.

[1] Boldface symbols denote vectors. Vector magnitudes are shown in lightface type.

[2] The symbol Δ always means "final value minus initial value."

With Eq. 2.4, Eq. 2.3 becomes

$$\frac{1}{2}m(V_2^2 - V_1^2) = \int_{s_1}^{s_2} \mathbf{F} \cdot d\mathbf{s} \tag{2.6}$$

where the expression for work has been written in terms of the scalar product (dot product) of the force vector **F** and the displacement vector $d\mathbf{s}$. Equation 2.6 states that the work of the resultant force on the body equals the change in its kinetic energy. When the body is accelerated by the resultant force, the work done on the body can be considered a *transfer* of energy *to* the body, where it is *stored* as kinetic energy.

Kinetic energy can be assigned a value knowing only the mass of the body and the magnitude of its instantaneous velocity relative to a specified coordinate frame, without regard for how this velocity was attained. Hence, kinetic energy is a *property* of the body. Since kinetic energy is associated with the body as a whole, it is an *extensive* property.

Work has units of force times distance. The units of kinetic energy are the same as for work. In SI, the work unit is the newton-meter, N·m, called the joule, J. In this book it is often convenient to use the kilojoule, kJ. Commonly used English units for work and kinetic energy are the foot-pound force, ft·lbf, and the British thermal unit, Btu.

2.1.2 Potential Energy

Equation 2.6 is the principal result of the previous section. Derived from Newton's second law, the equation gives a relationship between two *defined* concepts: kinetic energy and work. In this section it is used as a point of departure to extend the concept of energy. To begin, refer to Fig. 2.2, which shows a body of mass m that moves vertically from an elevation z_1 to an elevation z_2 relative to the surface of the earth. Two forces are shown acting on the system: a downward force due to gravity with magnitude mg and a vertical force with magnitude R representing the resultant of all *other* forces acting on the system.

The work of each force acting on the body shown in Fig. 2.2 can be determined by using the definition previously given. The total work is the algebraic sum of these individual values. In accordance with Eq. 2.6, the total work equals the change in kinetic energy. That is

$$\frac{1}{2}m(V_2^2 - V_1^2) = \int_{z_1}^{z_2} R\, dz - \int_{z_1}^{z_2} mg\, dz \tag{2.7}$$

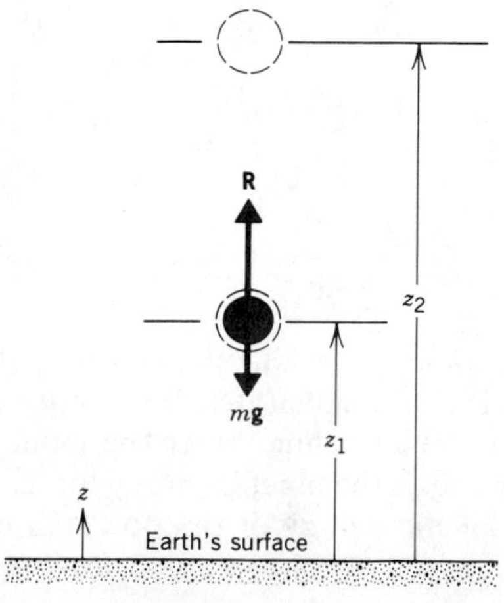

Figure 2.2 Illustration used to introduce the potential energy concept.

A minus sign is introduced before the second term on the right because the gravitational force is directed downward and the assumed displacement is upward.

The first integral on the right of Eq. 2.7 represents the work done by the force **R** on the body as it moves vertically from z_1 to z_2. The second integral can be evaluated as follows:

$$\int_{z_1}^{z_2} mg\,dz = mg(z_2 - z_1) \tag{2.8}$$

where the acceleration of gravity has been assumed to be constant with elevation. By incorporating Eq. 2.8 into Eq. 2.7 and rearranging

$$\frac{1}{2}m(V_2^2 - V_1^2) + mg(z_2 - z_1) = \int_{z_1}^{z_2} R\,dz \tag{2.9}$$

The quantity mgz is the *gravitational potential energy,* PE. The *change* in gravitational potential energy, ΔPE, is

$$\Delta PE = PE_2 - PE_1 = mg(z_2 - z_1) \tag{2.10}$$

The units for potential energy in any system of units are the same as those for kinetic energy and work.

Potential energy is associated with the force of gravity and is therefore an attribute of a system consisting of the body and the earth together. However, evaluating the force of gravity as mg enables the gravitational potential energy to be determined for a specified value of g knowing only the mass of the body and its elevation. With this view, potential energy is regarded, like kinetic energy, as an *extensive property* of the body. Throughout this book it is assumed that elevation differences are small enough that the gravitational force can be considered constant. The concept of gravitational potential energy can be formulated to account for the variation of this force with elevation, however.

Equation 2.9 states that the total work of all forces acting on the body from the surroundings, with the exception of the gravitational force, equals the sum of the changes in the kinetic and potential energies of the body. When the resultant force causes the elevation to be increased, the body to be accelerated, or both, the work done by the force can be considered a *transfer* of energy *to* the body, where it is stored as gravitational potential energy and/or kinetic energy. The notion of energy as a conserved extensive property underlies this interpretation.

The interpretation of Eq. 2.9 as an expression of the conservation of energy principle can be reinforced by considering the special case of a body on which the only force acting is that due to gravity, for then the right side of the equation vanishes and the equation reduces to

$$\frac{1}{2}m(V_2^2 - V_1^2) + mg(z_2 - z_1) = 0$$

or

$$\frac{1}{2}mV_2^2 + mgz_2 = \frac{1}{2}mV_1^2 + mgz_1 \tag{2.11}$$

Under these conditions, the *sum* of the kinetic and gravitational potential energies *remains constant*. Equation 2.11 also illustrates that energy can be *transformed* from one form to another: For an object falling under the influence of gravity *only,* the potential energy would decrease as the kinetic energy increases by an equal amount.

To assign a value to the kinetic energy or the potential energy of a system, it is necessary to assume a datum and specify a value for the quantity at the datum.

Values of kinetic and potential energy are then determined relative to this arbitrary choice of datum and reference value. However, since only *changes* in kinetic and potential energy between two states are required, these arbitrary reference specifications cancel. In this book, the kinetic energy of a system with zero velocity relative to the earth is assumed to be zero. The value of potential energy is taken to be zero when the system is at some specified reference elevation.

2.1.3 Closure

The presentation thus far has centered on systems for which applied forces affect only their overall velocity and position. However, systems of engineering interest normally interact with their surroundings in more complicated ways, with changes in other properties as well. To analyze such systems, the concepts of kinetic and potential energy alone do not suffice, nor does the rudimentary conservation of energy principle introduced in this section. In thermodynamics the concept of energy is broadened to account for other observed changes, and the principle of conservation of energy is extended to include a wide variety of ways in which systems interact with their surroundings. The basis for such generalizations is experimental evidence. These extensions of the concept of energy are developed in the remainder of the chapter, beginning in the next section with a fuller discussion of work.

2.2 ENERGY TRANSFER BY WORK

The work W done by, or on, a system evaluated in terms of macroscopically observable forces and displacements is

$$W = \int_{s_1}^{s_2} \mathbf{F} \cdot d\mathbf{s} \tag{2.12}$$

This relationship is important in thermodynamics. In the present section it is used to evaluate the work done in the compression or expansion of a gas (or liquid), the extension of a solid bar, and the stretching of a liquid film. However, thermodynamics also deals with phenomena not included within the scope of mechanics, so it is necessary to adopt a broader interpretation of work, as follows.

A particular interaction is categorized as a work interaction if it satisfies the following criterion, which can be considered the thermodynamic definition of work: *Work is done by a system on its surroundings if the sole effect on everything external to the system could have been the raising of a weight.* Notice that the raising of a weight is, in effect, a force acting through a distance, so the concept of work in thermodynamics is a natural extension of the concept of work in mechanics. However, the test of whether a work interaction has taken place is not that the elevation of a weight has actually taken place, or that a force has actually acted through a distance, but that the sole effect *could have been* an increase in the elevation of a weight.

Work is a means for transferring energy. Accordingly, the term work does not refer to what is being transferred between systems or to what is stored within systems. Energy is transferred and stored when work is done.

2.2.1 Sign Convention and Notation

Engineering thermodynamics is frequently concerned with devices such as internal combustion engines and turbines whose purpose is to do work. Hence, in contrast to the approach generally taken in mechanics, it is often convenient to consider such work as positive. That is

$W > 0$: work done *by* the system

$W < 0$: work done *on* the system

This sign convention is used throughout the book. In certain instances, however, it is convenient to regard the work done *on* the system to be positive, as has been done in the discussion of Sec. 2.1. To reduce the possibility of misunderstanding in any such case, the direction of energy transfer is shown by an arrow on a sketch of the system, and work is regarded as positive in the direction of the arrow.

To evaluate the integral in Eq. 2.12, it is necessary to know how the force varies with the displacement. This brings out an important idea about work: The value of W depends on the details of the interactions taking place between the system and surroundings during a process and not just the initial and final states of the system. It follows that work is *not a property* of the system or the surroundings. In addition, the limits on the integral of Eq. 2.12 mean "from state 1 to state 2" and cannot be interpreted as the *values* of work at these states. The notion of work at a state *has no meaning,* so the value of this integral should never be indicated as $W_2 - W_1$.

The differential of work, δW, is said to be *inexact* because, in general, the following integral cannot be evaluated without specifying the details of the process:

$$\int_1^2 \delta W = W$$

On the other hand, the differential of a property is said to be *exact* because the change in a property between two particular states depends in no way on the details of the process linking the two states. For example, the change in volume between two states can be determined by integrating the differential dV, without regard for the details of the process, as follows:

$$\int_{V_1}^{V_2} dV = V_2 - V_1$$

where V_1 is the volume *at* state 1 and V_2 is the volume *at* state 2. The differential of every property is exact. Exact differentials are written, as above, using the symbol d. To stress the difference between exact and inexact differentials, the differential of work is written as δW. The symbol δ is also used to identify other inexact differentials encountered later.

Many thermodynamic analyses are concerned with the time rate at which energy transfer occurs. The rate of energy transfer by work is called *power* and is denoted by $\dot{W}$. When a work interaction involves a macroscopically observable force, as in Eq. 2.12, the rate of energy transfer by work is equal to the product of the force and the velocity at the point of application of the force

$$W = \mathbf{F} \cdot \mathbf{V} \tag{2.13}$$

A dot appearing over a symbol, as in $\dot{W}$, is used throughout this book to indicate a time rate. In principle, Eq. 2.13 can be integrated from time t_1 to time t_2 to get the total work done during the time interval

$$W = \int_{t_1}^{t_2} \dot{W}\, dt = \int_{t_1}^{t_2} \mathbf{F} \cdot \mathbf{V}\, dt \tag{2.14}$$

The same sign convention applies for $\dot{W}$ as for W. Since power is a time rate of doing work, it can be expressed in terms of any units for energy and time. In SI, the unit for power is J/s, called the watt. In this book the kilowatt, kW, is also used. Commonly used English units for power are ft·lbf/s, Btu/h, and horsepower, hp.

There are many ways in which work can be done by or on a system. The remainder of this section is devoted to considering several examples, beginning with the important case of the work done when the volume of a quantity of a gas (or liquid) changes by expansion or compression.

2.2.2 Expansion or Compression Work

Let us evaluate the work done by the closed system shown in Fig. 2.3 consisting of a gas (or liquid) contained in a piston–cylinder assembly as the gas expands. During the process the gas pressure exerts a normal force on the piston. Let p denote the pressure acting at the interface between the gas and the piston. The force exerted by the gas on the piston is simply the product $p\mathrm{A}$, where A is the area of the piston face. The work done by the system as the piston is displaced a distance dx is

$$\delta W = p\mathrm{A}\, dx \tag{2.15}$$

The product A dx in Eq. 2.15 equals the change in volume of the system, dV. Thus, the work expression can be written as

$$\delta W = p\, dV \tag{2.16}$$

Since dV is positive when volume increases, the work at the moving boundary is positive when the gas expands. For a compression, dV is negative, and so is work found from Eq. 2.16. These signs are in agreement with the previously stated sign convention for work.

For a change in volume from V_1 to V_2, the work is obtained by integrating Eq. 2.16

$$W = \int_{V_1}^{V_2} p\, dV \tag{2.17}$$

Although Eq. 2.17 is derived for the case of a gas (or liquid) in a piston–cylinder assembly, it is applicable to systems of *any* shape provided the pressure is uniform with position over the moving boundary.

To perform the integral of Eq. 2.17 requires a relationship between the gas pressure *at the moving boundary* and the system volume. In an actual expansion or compression of a gas this information might be difficult or even impossible to obtain. For example, in a cylinder of an automobile engine, combustion and other nonequilibrium effects give rise to nonuniformities in pressure throughout the cylinder volume that may preclude determining the pressure at the moving boundary. Still, if the pressure measured elsewhere in the cylinder is essentially equal to that exerted at the

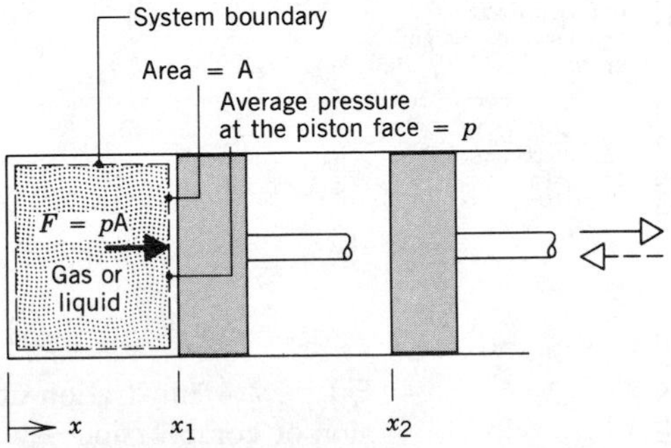

Figure 2.3 Expansion or compression of a gas or liquid.

face of the piston, Eq. 2.17 can be employed along with experimental data to obtain at least a plausible estimate of the work. A pressure transducer whose output is displayed by an oscilloscope trace is one way in which the required data might be obtained. We will see later that in some cases where lack of the required pressure–volume relationship keeps us from evaluating the work from Eq. 2.17, the work can be determined alternatively from an *energy balance* (Sec. 2.5).

2.2.3 Work in Quasiequilibrium Expansion or Compression Processes

An idealized type of process called a *quasiequilibrium* process is introduced in Sec. 1.2. A quasiequilibrium process is one in which all states through which the system passes may be considered equilibrium states. A particularly important aspect of the quasiequilibrium process concept is that the values of the intensive properties are uniform throughout the system, or every phase present in the system, at each state visited.

To consider how a gas (or liquid) might be expanded or compressed in a quasiequilibrium fashion, refer to Fig. 2.4 that shows a system consisting of a gas initially at an equilibrium state. As shown in the figure, the gas pressure is maintained uniform throughout by a number of small masses resting on the freely moving piston. Imagine that one of the masses is removed, allowing the piston to move upward as the gas expands slightly. During such an expansion the state of the gas would depart only slightly from equilibrium. The system would eventually come to a new equilibrium state, where the pressure and all other intensive properties would again be uniform in value. Moreover, were the mass replaced, the gas would be restored to its initial state, while again the departure from equilibrium would be slight. If several of the masses were removed one after another, the gas would pass through a sequence of equilibrium states without ever being far from equilibrium. In the limit as the increments of mass are made vanishingly small, the gas would undergo a quasiequilibrium expansion process. A quasiequilibrium compression can be visualized with similar considerations.

Equation 2.17 can be applied to evaluate the work in quasiequilibrium expansion or compression processes. For such processes the pressure p in the equation is the pressure of the entire quantity of gas (or liquid) undergoing the process, and not just the pressure at the moving boundary. The relationship between the pressure and volume may be graphical or analytical.

A graphical relationship is shown in the pressure–volume diagram (p–V diagram) of Fig. 2.5. Initially, the piston face is at position x_1, and the gas pressure is p_1; at the conclusion of a quasiequilibrium expansion process the piston face is at position x_2, and the pressure is reduced to p_2. At *each* intervening piston position, the uniform

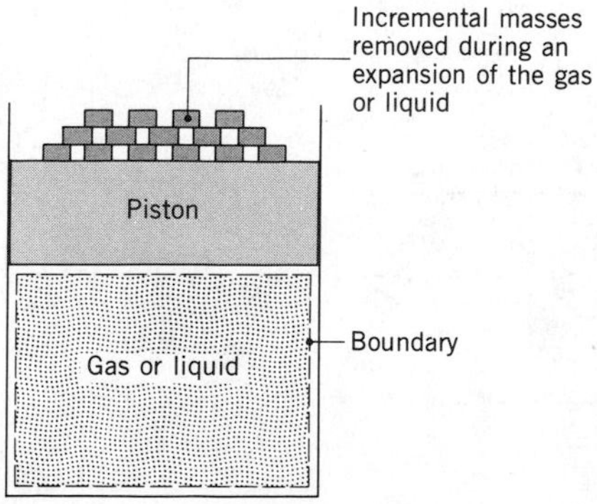

Figure 2.4 Illustration of a quasiequilibrium expansion or compression.

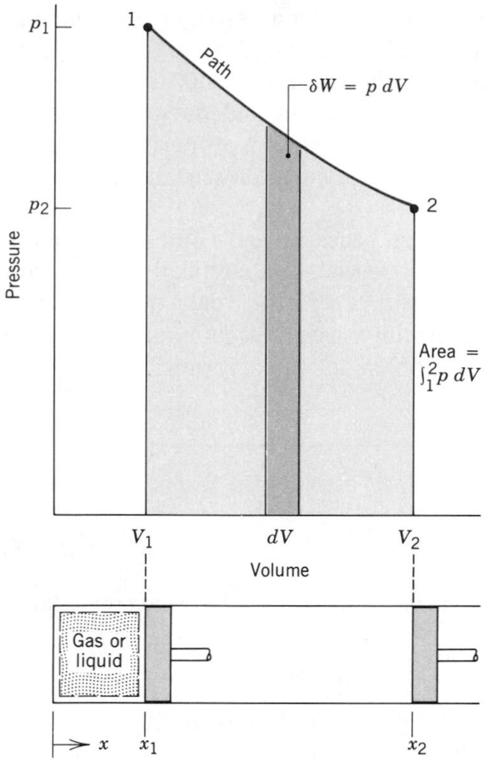

Figure 2.5 Work of a quasiequilibrium expansion or compression process.

pressure throughout the gas is shown as a point on the diagram. The curve, or *path*, connecting states 1 and 2 on the diagram represents the equilibrium states through which the system has passed during the process. The work done by the gas on the piston during the expansion is given by $\int p\,dV$. From calculus it is known that the integral can be interpreted as the area under the curve of pressure versus volume. Thus, the shaded area on Fig. 2.5 is equal to the work for the process. Had the gas been *compressed* from 2 to 1 along the same path on the p–V diagram, the *magnitude* of the work would be the same, but the sign would be negative, indicating that for the compression the energy transfer was from the piston to the fluid.

The area interpretation of work in a quasiequilibrium expansion or compression process allows a simple demonstration of the idea that work depends on the process and therefore is not a property. This can be brought out by referring to Fig. 2.6.

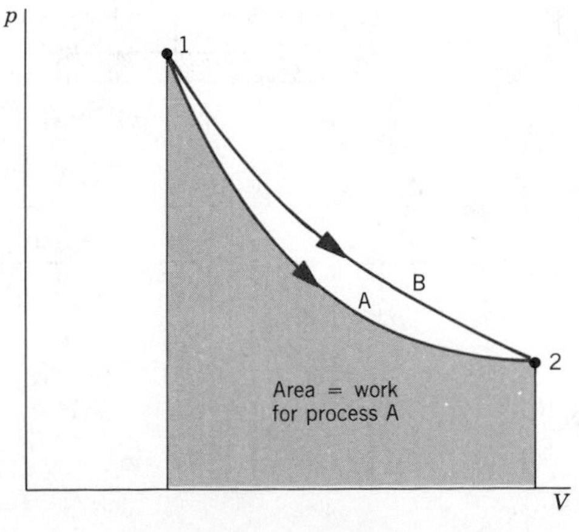

Figure 2.6 Illustration that work depends on the process.

Suppose the gas in a piston–cylinder assembly goes from an initial equilibrium state 1 to a final equilibrium state 2 along two different paths, labeled A and B on Fig. 2.6. Since the area beneath each path represents the work for that process, the work depends on the details of the process as defined by the particular curve and not just on the end states. Recalling the test for a property given in Sec. 1.2, we can conclude that *work is not a property*. The value of work depends on the nature of the process between the end states.

The relationship between pressure and volume during an expansion or compression process also can be described analytically. An example is provided by the expression pV^n = constant, where the value of n is a constant for the particular process. A quasiequilibrium process described by such an expression is called a *polytropic* process. Additional analytical forms for the pressure–volume relationship may also be considered.

EXAMPLE 2.1

PROBLEM

A gas in a piston–cylinder assembly undergoes an expansion process for which the relationship between pressure and volume is given by

$$pV^n = \text{constant}$$

The initial pressure is 3 bars, the initial volume is 0.1 m^3, and the final volume is 0.2 m^3. Determine the work for the process, in kJ, if (a) $n = 1.5$, (b) $n = 1.0$, and (c) $n = 0$.

SOLUTION

Known: A gas in a piston–cylinder assembly undergoes an expansion for which pV^n = constant.

Find: Evaluate the work if (a) $n = 1.5$, (b) $n = 1.0$, (c) $n = 0$.

Schematic and Given Data: The given p–V relationship and the given data for pressure and volume can be used to construct the accompanying pressure–volume diagram of the process.

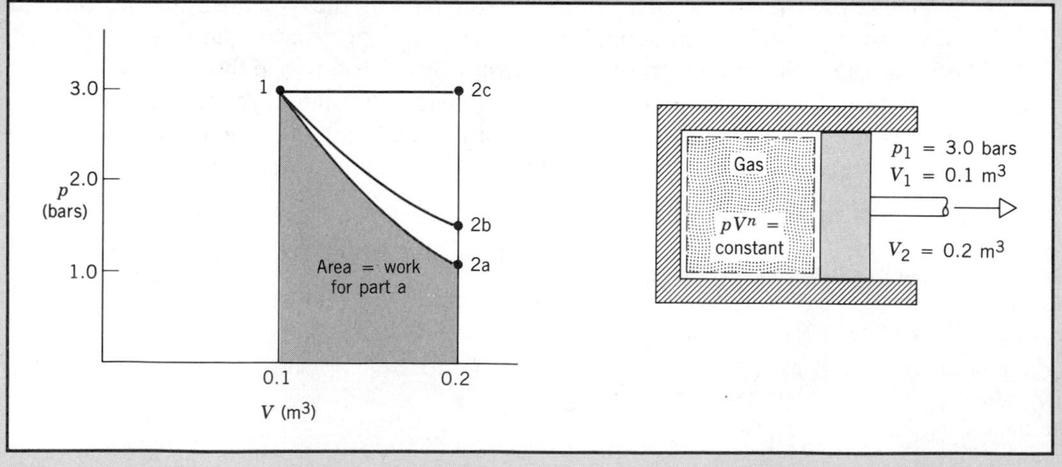

Figure E2.1

Assumptions:

1. The gas is a closed system.
2. The moving boundary is the only work mode.
3. The expansion is a polytropic process.

Analysis: The required values for the work are obtained by integration of Eq. 2.17 using the given pressure–volume relation.

(a) Introducing the relationship $p = \text{constant}/V^n$ into Eq. 2.17 and performing the integration

$$W = \int_{V_1}^{V_2} p\, dV = \int_{V_1}^{V_2} \frac{\text{constant}}{V^n}\, dV$$
$$= \text{constant}\left(\frac{V_2^{1-n} - V_1^{1-n}}{1-n}\right)$$

The constant in this expression can be evaluated at either end state: constant $= p_1V_1^n = p_2V_2^n$. The work expression then becomes

$$W = \frac{(p_2V_2^n)V_2^{1-n} - (p_1V_1^n)V_1^{1-n}}{1-n} = \frac{p_2V_2 - p_1V_1}{1-n}$$

This expression is valid for all values of n except $n = 1.0$. The case $n = 1.0$ is taken up in part (b).

To evaluate W, the pressure at state 2 is required. This can be found by using $p_1V_1^n = p_2V_2^n$, which on rearrangement yields

$$p_2 = p_1\left(\frac{V_1}{V_2}\right)^n = (3\text{ bars})\left(\frac{0.1}{0.2}\right)^{1.5} = 1.06\text{ bars}$$

Accordingly

$$W = \left(\frac{(1.06\text{ bars})(0.2\text{ m}^3) - (3)(0.1)}{1 - 1.5}\right)\left(\frac{10^5\text{ N/m}^2}{1\text{ bar}}\right)\left(\frac{1\text{ kJ}}{10^3\text{ N}\cdot\text{m}}\right)$$
$$= +17.6\text{ kJ}$$

(b) For $n = 1.0$, the pressure–volume relationship is $pV = \text{constant}$ or $p = \text{constant}/V$. The work is

$$W = \text{constant}\int_{V_1}^{V_2}\frac{dV}{V} = (\text{constant})\ln\frac{V_2}{V_1} = (p_1V_1)\ln\frac{V_2}{V_1}$$

Substituting values

$$W = (3\text{ bars})(0.1\text{ m}^3)\left(\frac{10^5\text{ N/m}^2}{1\text{ bar}}\right)\left(\frac{1\text{ kJ}}{10^3\text{ N}\cdot\text{m}}\right)\ln\left(\frac{0.2}{0.1}\right) = +20.79\text{ kJ}$$

(c) For $n = 0$, the pressure–volume relation reduces to $p = \text{constant}$, and the integral becomes $W = p(V_2 - V_1)$, which is a special case of the expression found in part (a). Substituting values and converting units as above, $W = +30$ kJ.

Comments:

1. In each case, the work for the process can be interpreted as the area under the curve representing the process on the accompanying p–V diagram. Note that the relative areas are in agreement with the numerical results.
2. It is not necessary to identify the gas (or liquid) contained within the piston–cylinder assembly. The calculated values for W are determined by the process path and the end states. However, if it is desired to evaluate other properties such as temperature, both the nature and amount of the substance must be provided because appropriate relations among the properties of the particular substance would then be required.
3. The assumption of a polytropic process is significant. If the given pressure–volume relationship were obtained as a fit to experimental pressure–volume data, the value of $\int p\, dV$ would provide a plausible estimate of the work only when the measured pressure is essentially equal to that exerted at the piston face.

2.2.4 Further Examples of Work

The only example of work considered thus far has been the work done by or on a gas (or liquid) when it expands or is compressed. Let us now briefly consider several other examples.

Extension of a Solid Bar. Consider a system consisting of a solid bar under tension, as shown in Fig. 2.7. The bar is fixed at $x = 0$, and a force F is applied at the other end. Let the force be represented as $F = \sigma\mathrm{A}$, where A is the cross-sectional area of the bar and σ the *normal stress acting at the end* of the bar. The work done as the end of the bar moves a distance dx is given by $\delta W = -\sigma\mathrm{A}\, dx$. The minus sign is required because work is done *on* the bar when dx is positive. The work for a change in length from x_1 to x_2 is found by integration

$$W = -\int_{x_1}^{x_2} \sigma\mathrm{A}\, dx \tag{2.18}$$

Equation 2.18 for a solid is the counterpart of Eq. 2.17 for a gas undergoing an expansion or compression.

Stretching of a Liquid Film. Figure 2.8 shows a system consisting of a liquid film suspended on a wire frame. The two surfaces of the film support the thin liquid layer inside by the effect of *surface tension,* owing to microscopic forces between molecules near the liquid–air interfaces. These forces give rise to a macroscopically measurable force perpendicular to any line in the surface. The force per unit length across such a line is the surface tension. Denoting the surface tension *acting at the movable wire* by τ, the force F indicated on the figure can be expressed as $F = 2l\tau$, where the factor 2 is introduced because two film surfaces act at the wire. If the movable wire is displaced by dx, the work is given by $\delta W = -2l\tau\, dx$. The minus sign is required because work is done *on* the system when dx is positive. Corresponding to a displacement dx is a change in the total area of the surfaces in contact with the wire of $d\mathrm{A} = 2l\, dx$, so the expression for work can be written alternatively as $\delta W = -\tau\, d\mathrm{A}$. The work for an increase in surface area from A_1 to A_2 is found by integrating this expression

$$W = -\int_{\mathrm{A}_1}^{\mathrm{A}_2} \tau\, d\mathrm{A} \tag{2.19}$$

Power Transmitted by a Shaft. A rotating shaft is a commonly encountered machine element. Consider a shaft rotating with angular velocity ω and exerting a torque $\mathcal{T}$ on its surroundings. Let the torque be expressed in terms of a tangential force F_t and radius R: $\mathcal{T} = F_t R$. The velocity at the point of application of the force is $\mathrm{V} = R\omega$, where ω is in radians per unit time. Using these relations with Eq. 2.13, we obtain an expression for the *power* transmitted from the shaft to the surroundings

$$\dot{W} = F_t\mathrm{V} = (\mathcal{T}/R)(R\omega) = \mathcal{T}\omega \tag{2.20}$$

A related case involving a gas stirred by a paddle wheel is shown in Fig. 2.9. In principle, the power transmitted by the paddle wheel to the gas could be evaluated in terms of forces and motions at the interface between the paddle wheel and the gas. However, because of complexities associated with gas motions and stresses, this

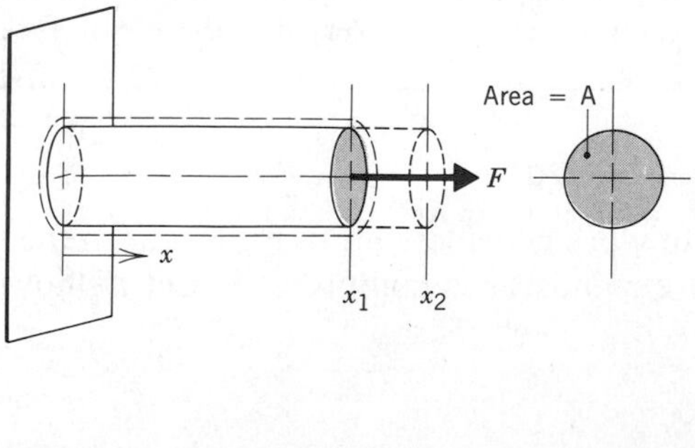

Figure 2.7 Elongation of a solid bar.

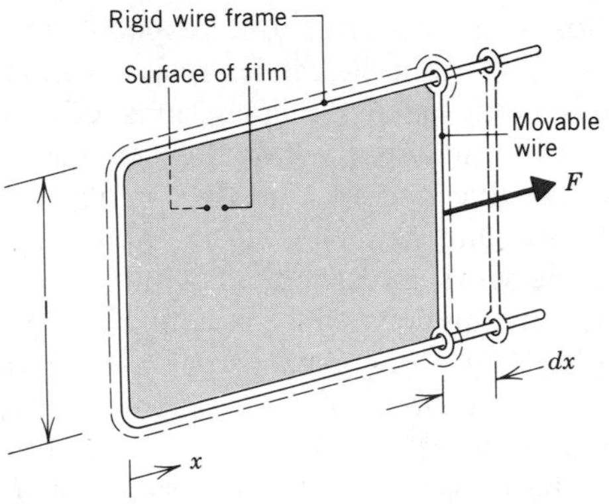

Figure 2.8 Stretching of a liquid film.

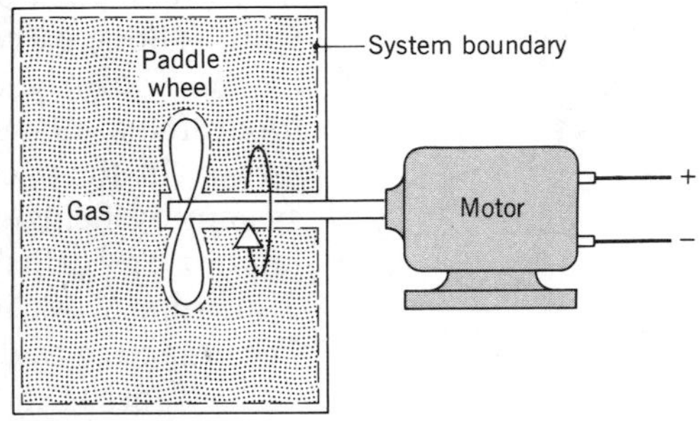

Figure 2.9 Power transmitted to a gas by a paddle wheel.

type of evaluation is deferred to a course in engineering fluid mechanics, where stresses within moving fluids are normally studied in detail.

Electrical Work. Shown in Fig. 2.10 is a system consisting of an electrolytic cell. The cell is connected to an external circuit through which an electric current is flowing. Work is done on the system whenever electrons cross the boundary of the system in response to the force associated with an electric field. Macroscopically, the flow of electrons is manifested as the electric current, i, driven by the electrical potential difference $\mathscr{E}$ existing across the terminals labeled a and b. That this type of interaction can be classed as work follows from the thermodynamic criterion given previously: Whatever the nature of the actual system under consideration, it can be envisioned that the current is supplied to an electric motor that lifts a mass in the surroundings.

The rate at which work is done, or the power, is

$$\dot{W} = -\mathscr{E}i \tag{2.21}$$

The minus sign is required to be in accord with our previously stated sign convention for work. Since the current i equals dZ/dt, the work can be expressed in differential form as

$$\delta W = -\mathscr{E}\, dZ \tag{2.22}$$

where dZ is the amount of electrical charge that flows into the system. When the power is evaluated in terms of the watt, and the unit of current is the ampere (an SI base unit), the unit of electric potential is the volt, defined as 1 watt per ampere.

Work due to Polarization or Magnetization. Let us next refer briefly to the types of work that can be done on systems residing in electric or magnetic fields, known as

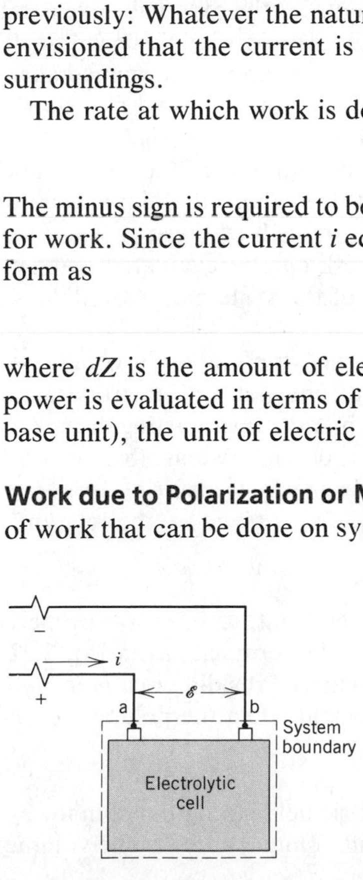

Figure 2.10 Electrolytic cell used to discuss electrical work.

the work of polarization and magnetization, respectively. From the microscopic viewpoint, electrical dipoles within dielectrics resist turning, so work is done when they are aligned by an electric field. Similarly, magnetic dipoles resist turning, so work is done on certain other materials when their magnetization is changed. Polarization and magnetization give rise to *macroscopically* detectable changes in the total dipole moment as the particles making up the material are given new alignments. In these cases the work is associated with forces imposed on the overall system by fields in the surroundings. Forces acting on the material in the system interior are called *body forces*. For such forces the appropriate displacement in evaluating work is the displacement of the matter on which the body force acts. Forces acting at the boundary are called *surface forces*. Examples of work done by surface forces include the expansion or compression of a gas (or liquid) and the extension of a solid.

2.2.5 Further Examples of Work in Quasiequilibrium Processes

Systems other than a gas or liquid in a piston–cylinder assembly can also be envisioned as undergoing processes in a quasiequilibrium fashion. To apply the quasiequilibrium process concept in any such case, it is necessary to conceive of an *ideal situation* in which the external forces acting on the system can be varied so slightly that the resulting imbalance is infinitesimal. As a consequence, the system undergoes a process without ever departing significantly from thermodynamic equilibrium.

The extension of a solid bar and the stretching of a liquid surface can readily be envisioned to occur in a quasiequilibrium manner by direct analogy to the piston–cylinder case. For the bar in Fig. 2.7 the external force can be applied in such a way that it differs only slightly from the opposing force within. The normal stress is then essentially uniform throughout and can be determined as a function of the instantaneous length: $\sigma = \sigma(x)$. Similarly, for the liquid film shown in Fig. 2.8 the external force can be applied to the movable wire in such a way that the force differs only slightly from the opposing force within the film. During such a process, the surface tension is essentially uniform throughout the film and is functionally related to the instantaneous area: $\tau = \tau(A)$. In each of these cases, once the required functional relationship is known, the work can be evaluated using Eq. 2.18 or 2.19, respectively, in terms of properties of the system as a whole as it passes through equilibrium states.

Other systems can also be imagined as undergoing quasiequilibrium processes. For example, it is possible to envision an electrolytic cell being charged or discharged in a quasiequilibrium manner by adjusting the potential difference across the terminals to be slightly greater, or slightly less, than an ideal potential called the cell *electromotive force* (emf). The energy transfer by work for passage of a differential quantity of charge *to* the cell, dZ, is given by the relation

$$\delta W = -\mathscr{E}\, dZ \tag{2.23}$$

In this equation $\mathscr{E}$ denotes the cell emf, an intensive property of the cell, and not just the potential difference across the terminals as in Eq. 2.22.

Consider next a dielectric material residing in a *uniform electric field*. The energy transferred by work from the field when the polarization is increased slightly is

$$\delta W = -\mathbf{E} \cdot d(V\mathbf{P}) \tag{2.24}$$

where the vector $\mathbf{E}$ is the electric field strength within the system, the vector $\mathbf{P}$ is the electric dipole moment per unit volume, and V is the volume of the system. A similar equation for energy transfer by work from a *uniform magnetic field* when the magnetization is increased slightly is

$$\delta W = -\mu_0 \mathbf{H} \cdot d(V\mathbf{M}) \tag{2.25}$$

where the vector **H** is the magnetic field strength within the system, the vector **M** is the magnetic dipole moment per unit volume, and μ_0 is a constant, the permeability of free space. The minus signs appearing in the last three equations are in accord with our previously stated sign convention for work: W takes on a negative value when the energy transfer is *into* the system.

The similarity between the expressions for work in the quasiequilibrium processes considered thus far should be noted. In each case, the work expression is written in the form of an intensive property and the differential of an extensive property. This is brought out by the following expression, which allows for one or more of these work modes to be involved in a process:

$$\delta W = p\,dV - \sigma\,d(\mathrm{A}x) - \tau\,d\mathrm{A} - \mathscr{E}\,dZ - \mathbf{E} \cdot d(V\mathbf{P}) - \mu_0 \mathbf{H} \cdot d(V\mathbf{M}) + \cdots \tag{2.26}$$

where the last three dots represent other products of an intensive property and the differential of a related extensive property that account for work. Because of the notion of work being a product of force and displacement, the intensive property in these relations is sometimes referred to as a ''generalized'' force and the extensive property as a ''generalized'' displacement, even though the quantities making up the work expressions may not bring to mind actual forces and displacements.

Owing to the underlying quasiequilibrium restriction, Eq. 2.26 does not represent every type of work of practical interest. An example is provided by a paddle wheel that stirs a gas or liquid taken as the system. Whenever any shearing action takes place, the system necessarily passes through nonequilibrium states. To appreciate more fully the implications of the quasiequilibrium process concept requires consideration of the second law of thermodynamics, so this concept is discussed again in Chap. 5 after the second law has been introduced.

2.3 ENERGY OF A SYSTEM

We saw in Sec. 2.1 that the kinetic and gravitational potential energies of a system can be changed as a result of work due to external forces. The definition of work was extended in Sec. 2.2 to include a variety of interactions between a system and its surroundings. In the present section, the extended work concept is used to obtain a broader understanding of the energy of a system. A central role is played in these considerations by the *first law of thermodynamics,* which is a generalization of experimental findings.

2.3.1 First Law of Thermodynamics

To introduce the first law of thermodynamics, let us select from among all processes by which a closed system can be taken from one equilibrium state to another those that involve only work interactions between the system and its surroundings. Any such process will be called an *adiabatic* process. Subsequent developments confirm that the present usage of the term adiabatic process is fully in accord with the concept introduced in Sec. 1.5.1.

Many different adiabatic processes are possible between a given pair of end states. However, it is found by experiment that for all adiabatic processes between two states the value of the net work done by or on the system is the same. That is, the value of the net work done by or on a closed system undergoing an adiabatic process between two given states *depends solely on the end states* and not on the details of the adiabatic process.

The foregoing statement is made on the basis of experimental evidence, beginning with the experiments of Joule in the early part of the nineteenth century. Because of inevitable experimental uncertainties, it is not possible to prove by measurements that the net work is exactly the same for all adiabatic processes between the same end states. However, the preponderance of experimental findings supports this conclusion, so it is adopted as a fundamental principle that the work actually is the same. This principle, called the *first law of thermodynamics,* applies regardless of the type of work interaction or the nature of the closed system.

2.3.2 Definition of Energy Change

A general definition of the *change* in energy of a closed system between two equilibrium states is introduced in this section by using the first law of thermodynamics, together with the test for a property given in Sec. 1.2.

Since the net work is the same for all adiabatic processes of a closed system between a given pair of end states, it can be concluded from the property test that the net work for such a process defines the change in some property of the system. This property is called *energy*. Selecting the symbol E to denote the energy of a system, the change in energy between two states is *defined* by

$$E_2 - E_1 = -W_{\text{ad}} \tag{2.27}$$

where W_{ad} denotes the net work for *any* adiabatic process between the two states. The minus sign before the work term in Eq. 2.27 is in accordance with the previously stated sign convention for work.

Since any arbitrary value E_1 can be assigned to the energy of a system at a given state 1, no particular significance can be attached to the value of the energy at state 1 or at *any* other state. Only *changes* in the energy of a system have significance.

Equation 2.27 giving the change in the energy of a system as a result of work done by it, or on it, during an adiabatic process is an expression of the principle of conservation of energy for this type of process. Equations 2.6 and 2.9, introducing changes in kinetic energy and gravitational potential energy, respectively, are special cases of Eq. 2.27.

2.3.3 Internal Energy

The symbol E introduced above denotes the *total* energy of a system. The total energy includes kinetic energy, gravitational potential energy, and other forms of energy. The examples to follow illustrate some of these forms of energy. Many other examples could be provided that enlarge on the same idea.

When work is done to compress a spring, energy is stored within the spring. When a battery is charged, the energy stored within it is increased. And when a gas (or liquid) initially at an equilibrium state in a closed, insulated vessel is stirred vigorously and allowed to come to a final equilibrium state, the energy of the gas is increased in the process. In each of these examples the change in system energy cannot be attributed to changes in the system's kinetic or gravitational potential energy. The change in energy can be accounted for in terms of *internal energy,* however.

In engineering thermodynamics the change in the total energy of a system is considered to be made up of three macroscopic contributions. One is the change in kinetic energy, associated with the motion of the system *as a whole* relative to an external coordinate frame. Another is the change in gravitational potential energy, associated with the position of the system *as a whole* in the earth's gravitational field. All other energy changes are lumped together in the *internal energy* of the system. Like kinetic energy and gravitational potential energy, internal energy is an extensive property of the system, as is the total energy.

Internal energy is represented by the symbol U, and the change in internal energy in a process is $U_2 - U_1$. The specific internal energy is symbolized by u or $\bar{u}$, respectively, depending on whether it is expressed on a unit mass or per mole basis.

The change in the total energy of a system is

$$E_2 - E_1 = (\mathrm{KE}_2 - \mathrm{KE}_1) + (\mathrm{PE}_2 - \mathrm{PE}_1) + (U_2 - U_1)$$

or (2.28)

$$\Delta E = \Delta\mathrm{KE} + \Delta\mathrm{PE} + \Delta U$$

All quantities in Eq. 2.28 are expressed in terms of the energy units previously introduced.

The identification of internal energy as a macroscopic form of energy is a significant step in the present development, for it sets the concept of energy in thermodynamics apart from that of mechanics. In Chap. 3 we will learn how to evaluate changes in internal energy for practically important cases involving gases, liquids, and solids by using empirical data.

To further our understanding of internal energy, consider a system we will often encounter in subsequent sections of the book, a system consisting of a gas contained in a tank. Let us think of the energy attributed to the motions and configurations of the individual molecules, atoms, and subatomic particles making up the matter in the system. Gas molecules move about, encountering other molecules or the walls of the container. Part of the internal energy of the gas is the *translational* kinetic energy of the molecules. Other contributions to the internal energy include the kinetic energy due to *rotation* of the molecules relative to their centers of mass and the kinetic energy associated with *vibrational* motions within the molecules. In addition, energy is stored in the chemical bonds between the atoms that make up the molecules. Energy storage on the atomic level includes energy associated with electron orbital states, nuclear spin, and binding forces in the nucleus. In dense gases, liquids, and solids, intermolecular forces play an important role in affecting the internal energy.

2.3.4 Conservation of Energy Principle for Closed Systems

Thus far, we have considered quantitatively only those interactions between a system and its surroundings that can be classed as work. However, closed systems can also interact with their surroundings in a way that cannot be categorized as work. An example is provided by a gas (or liquid) contained in a closed vessel undergoing a process while in contact with a flame at a temperature greater than that of the gas. This type of interaction is called a *heat* interaction, and the process can be referred to as a *nonadiabatic* process. The change in system energy in nonadiabatic processes cannot be accounted for in terms of work alone. The object of the present section is to introduce means for evaluating nonadiabatic processes quantitatively in terms of energy. This results in an expression of the conservation of energy principle that is particularly convenient for applications of engineering interest.

Refer to Fig. 2.11, which shows an adiabatic process and two *different* nonadiabatic processes, labeled A and B, between the *same two end* states, which are denoted by 1 and 2. Experiments would show that the work for each of the two

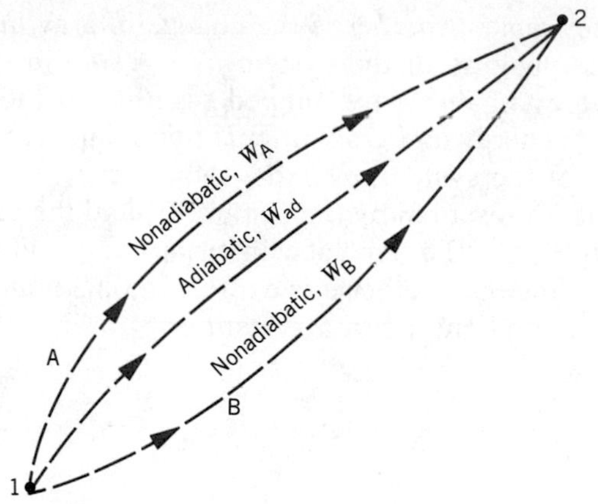

Figure 2.11 Adiabatic and nonadiabatic processes between the same pair of end states.

nonadiabatic processes would be different from the work for the adiabatic process: $W_A \neq W_{ad}$ and $W_B \neq W_{ad}$. The work for each nonadiabatic process might be different as well: $W_A \neq W_B$. However, because the end states are the same, the system would experience precisely the same energy change in each of the processes. Hence, we would have

$$E_2 - E_1 = -W_{ad}$$

but

$$E_2 - E_1 \neq -W_A, \qquad E_2 - E_1 \neq -W_B$$

For the nonadiabatic processes, the energy change of the system is not accounted for by the energy transferred by work.

A fundamental aspect of the energy concept is that energy is conserved. Thus, for the system to experience precisely the same energy change during the nonadiabatic processes as during the adiabatic process, the *net* energy transfer to the system in each of these processes must be the same. It follows that the heat interactions involve energy transfer. Further, the amount of energy Q transferred *to* the closed system by means other than work must equal the sum of the energy change of the system and the amount of energy transferred *from* the system by work. That is

$$Q = (E_2 - E_1) + W$$

This expression can be rewritten as

$$E_2 - E_1 = Q - W \tag{2.29}$$

which states that the change in system energy equals the net energy transfer to the system, as concluded above.

Equation 2.29 summarizes the conservation of energy principle for closed systems of all kinds. Abundant experimental evidence supports this assertion. Application of this equation is discussed in Sec. 2.5 after the energy transfer denoted by Q is considered in more detail in the next section.

2.4 ENERGY TRANSFER BY HEAT

The quantity denoted by Q accounts for the amount of energy transferred to a closed system during a process by means other than work. On the basis of experiment it is known that such an energy transfer is induced only as a result of a temperature difference between the system and its surroundings and occurs only in the direction of decreasing temperature. This means of energy transfer is called *energy transfer by heat*. Because the underlying concept is so important in engineering thermodynamics, this section is devoted to a further consideration of energy transfer by heat.

2.4.1 Sign Convention and Notation

The symbol Q denotes an amount of energy transferred across the boundary of a system in a heat interaction with the system's surroundings. Heat transfer *into* a system is taken to be *positive*, and heat transfer *from* a system is taken as *negative*.

$Q > 0$: heat transfer *to* the system

$Q < 0$: heat transfer *from* the system

This sign convention for heat transfer is used throughout the book. However, as was indicated for work, it is sometimes convenient to show the direction of energy transfer by an arrow on a sketch of the system. Then the heat transfer is regarded as positive in the direction of the arrow. In an adiabatic process there is no energy transfer by heat.

The sign convention for heat transfer is just the *reverse* of the one adopted for work, where a positive value for W signifies an energy transfer *from* the system to the surroundings. These signs for heat and work are a legacy from engineers and scientists who were concerned mainly with steam engines and other devices that develop a work output from an energy input by heat transfer. For such applications, it was convenient to regard both the work developed and the energy input by heat transfer as positive quantities.

The value of a heat transfer depends on the details of a process and not just the end states. Thus, like work, heat is not a property, and its differential is written as δQ. The amount of energy transfer by heat for a process is given by the integral

$$Q = \int_1^2 \delta Q$$

where the limits mean "from state 1 to state 2" and do not refer to the values of heat at those states. As for work, the notion of "heat" at a state has no meaning, and the integral should *never* be evaluated as $Q_2 - Q_1$.

The net *rate* of heat transfer is denoted by $\dot{Q}$. In principle, the amount of energy transfer by heat during a period of time can be found by integrating from time t_1 to time t_2

$$Q = \int_{t_1}^{t_2} \dot{Q}\, dt \tag{2.30}$$

To perform the integration, it would be necessary to know how the rate of heat transfer varies with time.

In some cases it is convenient to use the heat *flux*, $\dot{q}$, which is the heat transfer rate per unit of system surface area. The net rate of heat transfer, $\dot{Q}$, is related to the heat flux $\dot{q}$ by the integral

$$\dot{Q} = \int_A \dot{q}\, dA \tag{2.31}$$

where A represents the area on the boundary of the system where heat transfer occurs. The units for Q and $\dot{Q}$ are the same as those introduced previously for W and $\dot{W}$, respectively. The units for the heat flux are those of the heat transfer rate per unit area: W/m^2 or $Btu/h \cdot ft^2$.

2.4.2 Heat Transfer Modes

Methods based on experiment are available for evaluating energy transfer by heat. These methods recognize two basic transfer mechanisms: *conduction* and *thermal radiation*. In addition, empirical relationships are available for evaluating energy transfer involving certain *combined* modes. A brief description of each of these is given next. A more detailed consideration is left to a course in engineering heat transfer, where these topics are studied in depth.

Energy transfer by *conduction* can take place in solids, liquids, and gases. Conduction can be thought of as the transfer of energy from the more energetic particles of a substance to adjacent particles that are less energetic due to interactions between particles. The time rate of energy transfer by conduction is quantified macroscopically by *Fourier's law*. As an illustration, consider the system shown in Fig. 2.12, in which the temperature distribution is $T(x)$. The time rate at which energy enters the system by conduction through a plane area A perpendicular to the coordinate x is by Fourier's law

$$\dot{Q}]_x = -\kappa A \left.\frac{dT}{dx}\right]_x \tag{2.32}$$

which emphasizes that the driving influence for conduction is a temperature *gradient*. The proportionality factor κ, which may vary with position, is a property of the material called the *thermal conductivity*. Substances with large values of thermal conductivity such as copper are good conductors, and those with small conductivities (cork and polystyrene foam) are good insulators. The minus sign in Eq. 2.32 is a consequence of the requirement that the energy flow is in the direction of decreasing temperature.

Thermal radiation is emitted by matter as a result of changes in the electronic configurations of the atoms or molecules within it. The energy is transported by electromagnetic waves (or photons). Unlike conduction, thermal radiation requires no intervening medium to propagate and can even take place in a vacuum. Solid surfaces, gases, and liquids all emit, absorb, and transmit thermal radiation to varying degrees. The rate at which energy is emitted, $\dot{Q}_e$, *from* a system with a surface

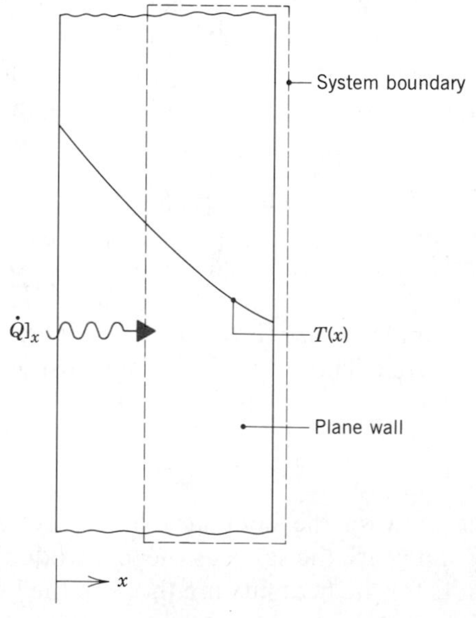

Figure 2.12 Illustration of Fourier's conduction law.

area A is quantified macroscopically by a modified form of the *Stefan–Boltzmann law*

$$\dot{Q}_e = \varepsilon \sigma A T_b^4 \tag{2.33}$$

which shows that thermal radiation is associated with the fourth power of the absolute temperature of the surface, T_b. The emissivity, ε, is a property of the surface that indicates how effectively the surface radiates ($0 \leq \varepsilon \leq 1.0$), and σ is the Stefan–Boltzmann constant. In general, the *net* rate of energy transfer by thermal radiation between two surfaces involves complicated relationships among the properties of the surfaces, their orientations with respect to each other, the extent to which the intervening medium scatters, emits, and absorbs thermal radiation, and other factors.

Energy transfer between a solid surface at one temperature and an adjacent moving gas or liquid at another temperature plays a prominent role in the performance of many devices of practical interest. This is commonly referred to as *convection*. As an illustration, consider the system shown in Fig. 2.13. In this case energy conducted from the system to the adjacent moving fluid is carried away by the *combined* effects of conduction within the fluid and the bulk motion of the fluid. The rate of energy transfer from the system to the fluid can be quantified by the following *empirical* expression:

$$\dot{Q} = hA(T_b - T_f) \tag{2.34}$$

known as *Newton's law of cooling*. In Eq. 2.34, A is the surface area, T_b is the temperature on the surface, and T_f is the fluid temperature away from the surface. For $T_b > T_f$, energy is transferred in the direction indicated by the arrow on Fig. 2.13. The proportionality factor h in Eq. 2.34 is called the *heat transfer coefficient*. The heat transfer coefficient is *not* a thermodynamic property. It is an empirical parameter that incorporates into the heat transfer relationship the nature of the fluid flow pattern near the surface, the fluid properties, and the geometry of the system. When fans or pumps cause the fluid to move, the value of the heat transfer coefficient is generally greater than when relatively slow buoyancy-induced motions occur. These two general categories are called *forced* and *free* (or natural) convection, respectively. It is necessary to study the subject of fluid mechanics to understand more fully the fluid flow aspects of convection heat transfer.

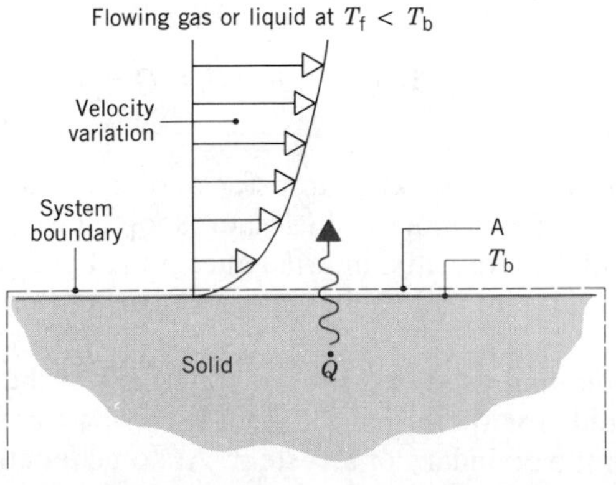

Figure 2.13 Illustration of Newton's law of cooling

2.4.3 Closure

The first step in a thermodynamic analysis is to define the system. It is only after the system boundary has been specified that possible heat interactions with the surroundings are considered, for these are *always* evaluated at the system boundary. In ordinary conversation, the term *heat* is often used when the word *energy* would be more correct thermodynamically. For example, one might hear, "Please close the door or 'heat' will be lost." In *thermodynamics,* heat refers only to a particular means whereby energy is transferred. It does not refer to what is being transferred between systems or to what is stored within systems. Energy is transferred and stored, not heat.

Sometimes the heat transfer of energy to, or from, a system can be neglected. This might occur for several reasons related to the mechanisms for heat transfer discussed above. One might be that the materials surrounding the system are good insulators, or heat transfer might not be significant because there is a small temperature difference between the system and its surroundings. A third reason is that there might not be enough surface area to allow significant heat transfer to occur. When heat transfer is neglected, it is because one or more of these considerations apply.

In the discussions to follow the value of Q is provided, or it is an unknown in the analysis. When Q is provided, it can be assumed that the value has been determined by the methods introduced above. When Q is the unknown, its value is found by using the *energy balance,* discussed next.

2.5 ENERGY BALANCE FOR CLOSED SYSTEMS

The object of this section is to discuss the application of Eq. 2.29, which is simply an expression of the conservation of energy principle for closed systems. The equation can be stated in words as follows:

$$\begin{bmatrix} \textit{change} \text{ in the amount} \\ \text{of energy contained} \\ \text{within the system} \\ \text{during some time} \\ \text{interval} \end{bmatrix} = \begin{bmatrix} \textit{net} \text{ amount of energy} \\ \text{transferred } \textit{in} \text{ across} \\ \text{the system boundary by} \\ \textit{heat} \text{ transfer during} \\ \text{the time interval} \end{bmatrix} - \begin{bmatrix} \textit{net} \text{ amount of energy} \\ \text{transferred } \textit{out} \text{ across} \\ \text{the system boundary} \\ \text{by } \textit{work} \text{ during the} \\ \text{time interval} \end{bmatrix}$$

This word statement emphasizes that Eq. 2.29 is just an accounting balance for energy, an energy balance. It requires that in any process of a closed system the energy of the system increases or decreases by an amount equal to the net amount of energy transferred across its boundary.

Introducing Eq. 2.28 into Eq. 2.29, an alternative form of the energy balance is obtained

$$\Delta \text{KE} + \Delta \text{PE} + \Delta U = Q - W \tag{2.35}$$

This equation shows that an energy transfer across the system boundary is manifested in a change in one or more of the macroscopic energy forms: kinetic energy, gravitational potential energy, and internal energy. All previous references to energy as a conserved quantity are included as special cases of this equation, as can readily be verified.

The phrase *net amount* used in the word statement of the energy balance must be carefully interpreted, for there may be heat or work transfers of energy at many different places on the boundary of a system. At some locations the energy transfers may be into the system, whereas at others they are out of the system. The two terms on the right side account for the *net* results of all the energy transfers by heat and work, respectively, taking place during the time interval under consideration. Note that the algebraic signs before the heat and work terms are different. This follows from the sign conventions previously adopted. A minus sign appears before W in Eq. 2.35 because energy transfer by work *from* the system *to* the surroundings is taken to be positive. A plus sign appears before Q because it is regarded to be positive when the heat transfer of energy is *into* the system *from* the surroundings.

2.5.1 Forms of the Energy Balance

Various special forms of the energy balance can be written. For example, the energy balance in differential form is

$$dE = \delta Q - \delta W \tag{2.36}$$

where dE is the differential of energy, a property. Since Q and W are not properties, their differentials are written as δQ and δW, respectively.

The energy balance can also be written on a time rate basis as follows. By dividing by the time interval Δt, an expression is obtained for the *average* rate of change of energy in terms of *average* rates of energy transfer by heat and work during time interval Δt

$$\frac{\Delta E}{\Delta t} = \frac{Q}{\Delta t} - \frac{W}{\Delta t}$$

Then, in the limit as Δt tends to zero

$$\lim_{\Delta t \to 0} \left(\frac{\Delta E}{\Delta t}\right) = \lim_{\Delta t \to 0} \left(\frac{Q}{\Delta t}\right) - \lim_{\Delta t \to 0} \left(\frac{W}{\Delta t}\right)$$

the instantaneous time rate form of the energy balance is obtained as

$$\frac{dE}{dt} = \dot{Q} - \dot{W} \tag{2.37}$$

The rate form of the energy balance expressed in words is

$$\begin{bmatrix} \text{time } \textit{rate of change} \\ \text{of the energy} \\ \text{contained within} \\ \text{the system } \textit{at} \\ \textit{time t} \end{bmatrix} = \begin{bmatrix} \text{net } \textit{rate} \text{ at which} \\ \text{energy is being} \\ \text{transferred in} \\ \text{by heat transfer} \\ \textit{at time t} \end{bmatrix} - \begin{bmatrix} \text{net } \textit{rate} \text{ at which} \\ \text{energy is being} \\ \text{transferred out} \\ \text{by work } \textit{at} \\ \textit{time t} \end{bmatrix}$$

Since the time rate of change of energy is given by

$$\frac{dE}{dt} = \frac{d\text{KE}}{dt} + \frac{d\text{PE}}{dt} + \frac{dU}{dt}$$

Eq. 2.37 can be expressed alternatively as

$$\frac{d\text{KE}}{dt} + \frac{d\text{PE}}{dt} + \frac{dU}{dt} = \dot{Q} - \dot{W} \tag{2.38}$$

Equations 2.35 through 2.38 provide alternative forms of the energy balance that may be convenient starting points when applying the principle of conservation of energy to closed systems. In Chap. 4 the conservation of energy principle is expressed in forms suitable for the analysis of control volumes. When applying the energy balance in *any* of its forms, it is important to be careful about signs and units and to distinguish carefully between rates and amounts. In addition, it is important to recognize that the location of the system boundary can be relevant in determining whether a particular energy transfer is regarded as heat or work. To illustrate this, consider Fig. 2.14, in which three alternative systems are shown that include a quantity of a gas (or liquid) in a rigid, well-insulated container. In Fig. 2.14*a*, the gas itself is the system. As current flows through the copper plate, there is an energy transfer from the copper plate to the gas. Since this energy transfer occurs as a result of the temperature difference between the plate and the gas, it is classified as a heat transfer. Next, refer to Fig. 2.14*b*, where the boundary is drawn to include the copper plate. It follows from the thermodynamic definition of work that the energy transfer that occurs as current crosses the boundary of this system must be regarded as work. Finally, in Fig. 2.14*c*, the boundary is located so that no energy is transferred across it by heat or work.

Thus far, we have been careful to emphasize that the quantities symbolized by W and Q in the foregoing equations account for transfers of *energy* and not transfers of work and heat, respectively. The terms work and heat denote different *means* whereby energy is transferred and not *what* is transferred. However, to achieve

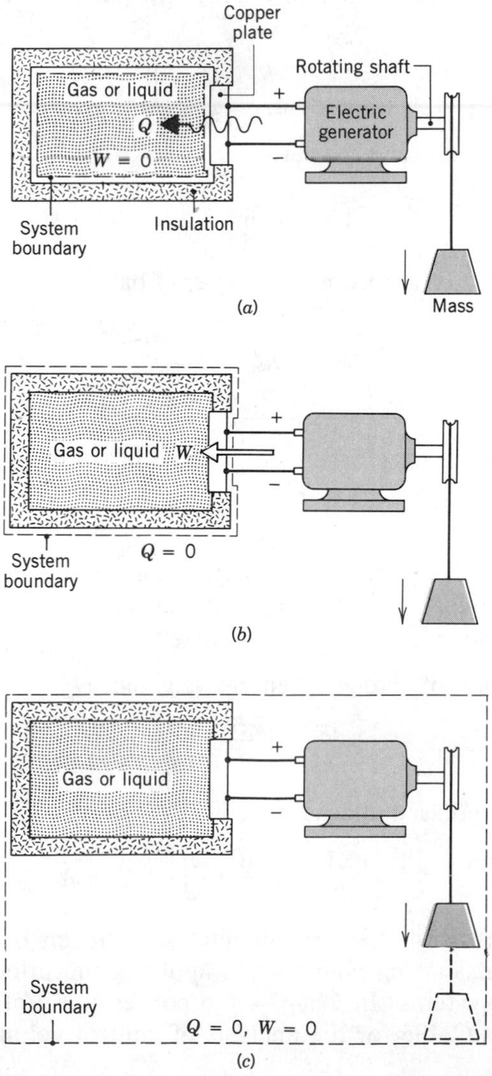

Figure 2.14 Alternative choices for system boundaries.

economy of expression in subsequent discussions, W and Q are often referred to simply as work and heat transfer, respectively. This less formal manner of speaking is commonly used in engineering practice.

2.5.2 Illustrations

The examples that follow bring out many important ideas about energy and the energy balance. They should be studied carefully, and similar approaches should be used when solving the end-of-chapter problems.

In the first example, a system undergoes a process in which changes in kinetic and potential energy are significant. In this text most applications of the energy balance will not involve significant kinetic or potential energy changes, however. Thus, to expedite the solutions of many subsequent examples and end-of-chapter problems, we indicate in the problem statement that such changes can be neglected. If this is not made explicit in a problem statement, you should decide on the basis of the problem at hand how best to handle the kinetic and potential energy terms of the energy balance.

EXAMPLE 2.2

PROBLEM

A closed system initially at rest on the surface of the earth undergoes a process for which there is a net energy transfer *to* the system by work of magnitude 200 Btu. During the process there is a net heat transfer of energy *from* the system to its surroundings of 30 Btu. At the end of the process, the system has a velocity of 200 ft/s at an elevation of 200 ft. The mass of the system is 50 lb, and the local acceleration of gravity is $g = 32.0 \text{ ft/s}^2$. Determine the change of internal energy of the system for the process, in Btu.

SOLUTION

Known: A system of known mass undergoes a process with the amounts of work and heat transfer given. The initial and final velocities and elevations are also specified.

Find: Determine the change in internal energy of the system.

Schematic and Given Data:

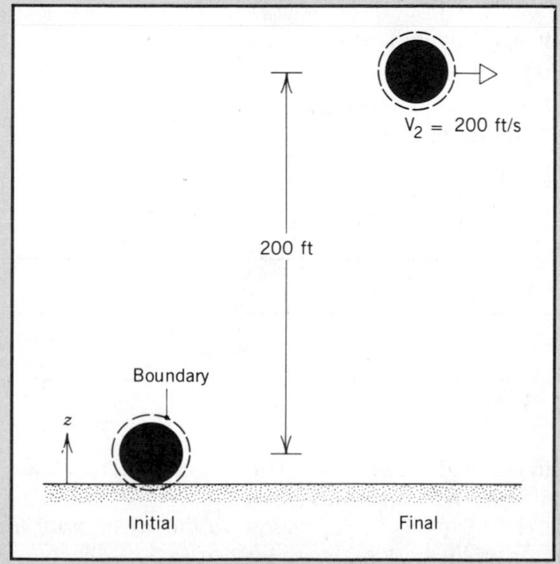

Figure E2.2

Assumptions:

1. A closed system is under consideration.
2. At the end of the process, the system is moving with a uniform velocity.
3. The local acceleration of gravity is constant at $g = 32.0\ \text{ft/s}^2$.

Analysis: An energy balance for the closed system takes the form

$$\Delta\text{KE} + \Delta\text{PE} + \Delta U = Q - W$$

where the kinetic and potential energy changes can be expressed, respectively, as

$$\Delta\text{KE} = \frac{1}{2} m(\text{V}_2^2 - \text{V}_1^2)$$

$$\Delta\text{PE} = mg(z_2 - z_1)$$

Solving for ΔU

$$\Delta U = Q - W - \frac{1}{2} m(\text{V}_2^2 - \text{V}_1^2) - mg(z_2 - z_1)$$

Inserting the values $Q = -30$ Btu, $W = -200$ Btu, $\text{V}_1 = 0$, $\text{V}_2 = 200$ ft/s, $z_2 - z_1 = 200$ ft

$$\begin{aligned}\Delta U &= (-30\ \text{Btu}) - (-200\ \text{Btu}) \\ &\quad - \frac{1}{2}(50\ \text{lb}) \frac{(200^2)\ \text{ft}^2/\text{s}^2}{\left(\dfrac{32.2\ \text{lb}\cdot\text{ft/s}^2}{1\ \text{lbf}}\right)} \left(\frac{1\ \text{Btu}}{778\ \text{ft}\cdot\text{lbf}}\right) \\ &\quad - \frac{(50\ \text{lb})(32.0\ \text{ft/s}^2)(200\ \text{ft})}{\left(\dfrac{32.2\ \text{lb}\cdot\text{ft/s}^2}{1\ \text{lbf}}\right)} \left(\frac{1\ \text{Btu}}{778\ \text{ft}\cdot\text{lbf}}\right) \\ &= (-30\ \text{Btu}) - (-200\ \text{Btu}) - (39.9\ \text{Btu}) - (12.8\ \text{Btu}) \\ &= +117.3\ \text{Btu}\end{aligned}$$

Comments:

1. The positive sign for ΔU indicates that the internal energy of the system increases during the process.
2. Carefully note the unit conversions required when using the English system. The conversion factors 32.174 and 778.17 have been rounded, respectively, to 32.2 and 778. This is done throughout the text.
3. The analysis can be summarized by the following energy "balance sheet" in terms of energy *magnitudes:*

Input	**Change in System Energy**
200 Btu (work)	+39.9 Btu (kinetic energy)
Output	+12.8 Btu (potential energy)
30 Btu (heat transfer)	+117.3 Btu (internal energy)
	+170.0 Btu

The net input exceeds the net output by 170 Btu, and the system energy increases by this amount.

EXAMPLE 2.3

PROBLEM

Consider 5 kg of steam (water vapor) contained within a piston–cylinder assembly. The steam undergoes an expansion from state 1, where the specific internal energy (the internal energy per unit mass) is $u_1 = 2709.9$ kJ/kg, to state 2, where $u_2 = 2659.6$ kJ/kg. During the process, there is a heat transfer of energy *to* the steam with a magnitude of 80 kJ. Also, a paddle wheel transfers energy *to* the steam by work in the amount of 18.5 kJ. There is no significant change in the kinetic or potential energy of the steam. Determine the amount of energy transfer by work from the steam to the piston during the process, in kJ.

SOLUTION

Known: A fixed quantity of steam undergoes an expansion in a piston–cylinder assembly from state 1 to state 2. During the process there is a known heat transfer and a known work transfer of energy by a paddle wheel. The change in specific internal energy of the steam is also known.

Find: Determine the amount of energy transfer by work to the piston during the expansion.

Schematic and Given Data:

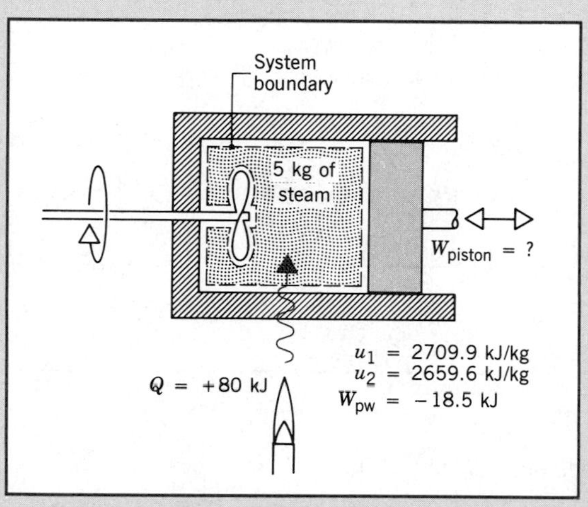

Figure E2.3

Assumptions:

1. The steam is a closed system.
2. There is no change in the kinetic or potential energy of the steam.

Analysis: An energy balance for the closed system takes the form

$$\cancelto{0}{\Delta KE} + \cancelto{0}{\Delta PE} + \Delta U = Q - W$$

where the kinetic and potential energy terms drop out by assumption 2. Then, writing ΔU in terms of specific internal energies, the energy balance becomes

$$m(u_2 - u_1) = Q - W$$

where m is the system mass.

The symbol W represents the *net* work from the system. The net work is the sum of the work associated with the paddle wheel W_{pw} and the work associated with the piston W_{piston}

$$W = W_{pw} + W_{piston}$$

From the given information, $W_{pw} = -18.5$ kJ, where the minus sign is required because the energy flow is *into* the system. The sign for W_{piston} will be interpreted once the final value is calculated. Incorporating the above expression for the work into the energy balance results in

$$m(u_2 - u_1) = Q - (W_{pw} + W_{piston})$$

Solving for W_{piston}

$$W_{piston} = Q - W_{pw} - m(u_2 - u_1)$$

Inserting values

$$W_{piston} = (+80 \text{ kJ}) - (-18.5 \text{ kJ}) - (5 \text{ kg})(2659.6 - 2709.9) \text{ kJ/kg}$$

$$= (+80 \text{ kJ}) - (-18.5 \text{ kJ}) - (-251.5 \text{ kJ})$$
$$= +350 \text{ kJ}$$

Comments:

1. The positive sign for W_{piston} indicates that the energy transfer is from the steam to the piston as the system *expands* during this process.
2. In principle, the energy transfer by work to the piston could be found from $\int p\, dV$. This approach cannot be implemented here, however, because data relating the steam pressure at the moving piston to the volume occupied by the steam are not known.
3. The analysis can be summarized by the following energy "balance sheet" in terms of energy *magnitudes:*

	Input	Output
	18.5 (work, paddle wheel)	350 (work, piston)
	80.0 (heat transfer)	—
Total:	98.5	350

The total output exceeds the total input, and the system energy decreases by the difference: $\Delta U = 98.5 - 350 = -251.5$ kJ.

EXAMPLE 2.4

PROBLEM

Four kilograms of a certain gas is contained within a piston–cylinder assembly. The gas undergoes a process for which the pressure–volume relationship is

$$pV^{1.5} = \text{constant}$$

The initial pressure is 3 bars, the initial volume is 0.1 m^3, and the final volume is 0.2 m^3. The change in specific internal energy of the gas in the process is $u_2 - u_1 = -4.6$ kJ/kg. There are no significant changes in kinetic or potential energy. Determine the net heat transfer for the process, in kJ.

SOLUTION

Known: A gas within a piston–cylinder assembly undergoes an expansion process for which the pressure–volume relation and the change in specific internal energy are specified.

Find: Determine the net heat transfer for the process.

Schematic and Given Data:

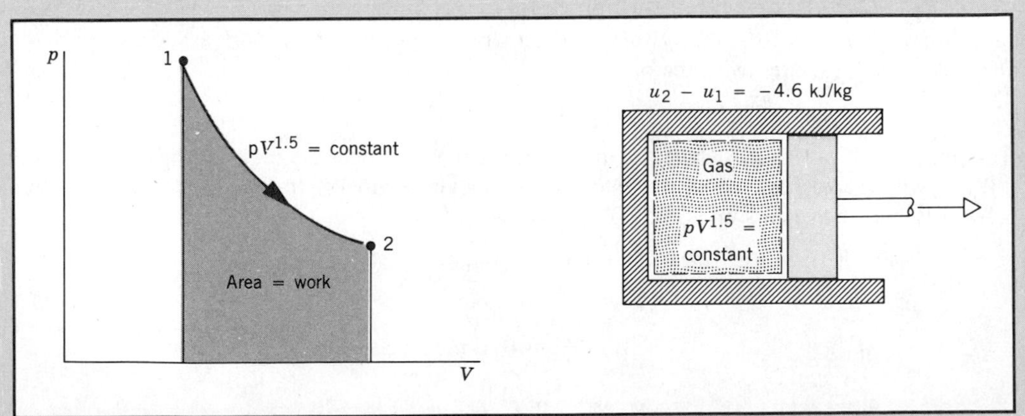

Figure E2.4

Assumptions:

1. The gas is a closed system.
2. The expansion is a polytropic process.
3. There is no change in the kinetic or potential energy of the system.

Analysis: Since there is no change in kinetic or potential energy, the energy balance for the closed system reduces to

$$\Delta U = Q - W$$

Solving for Q

$$Q = \Delta U + W$$

The value of the work for this process is determined in the solution to part (a) of Example 2.1: $W = +17.6$ kJ. The change in internal energy is obtained using given data as

$$\Delta U = m(u_2 - u_1) = 4 \text{ kg} \left(-4.6 \frac{\text{kJ}}{\text{kg}}\right) = -18.4 \text{ kJ}$$

Substituting values

$$Q = \Delta U + W = -18.4 + 17.6 = -0.8 \text{ kJ}$$

Comments:

1. The minus sign for the value of Q means that a net amount of energy has been transferred from the system to its surroundings by heat transfer.
2. The given relationship between pressure and volume allows the process to be represented by the path shown on the accompanying diagram. The area under the curve represents the work. Since they are not properties, the values of the work and heat transfer depend on the details of the process and cannot be determined from the end states only.
3. See Comment 3 of Example 2.1.

EXAMPLE 2.5

PROBLEM

The rate of heat transfer between a certain electric motor and its surroundings varies with time as

$$\dot{Q} = -0.2[1 - e^{(-0.05t)}]$$

where t is in seconds and $\dot{Q}$ is in kilowatts. The shaft of the motor rotates at a constant speed of $\omega = 100$ rad/s (about 955 revolutions per minute or RPM) and applies a constant torque of $\mathcal{T} = 18$ N·m to an external load. The motor draws a constant electric power input equal to 2.0 kW. Obtain an expression for the time rate of change of energy of the motor.

SOLUTION

Known: A motor operates with constant electric power input, shaft speed, and applied torque. The time-varying rate of heat transfer between the motor and its surroundings is given.

Find: Obtain an expression for the time rate of change of energy.

Schematic and Given Data:

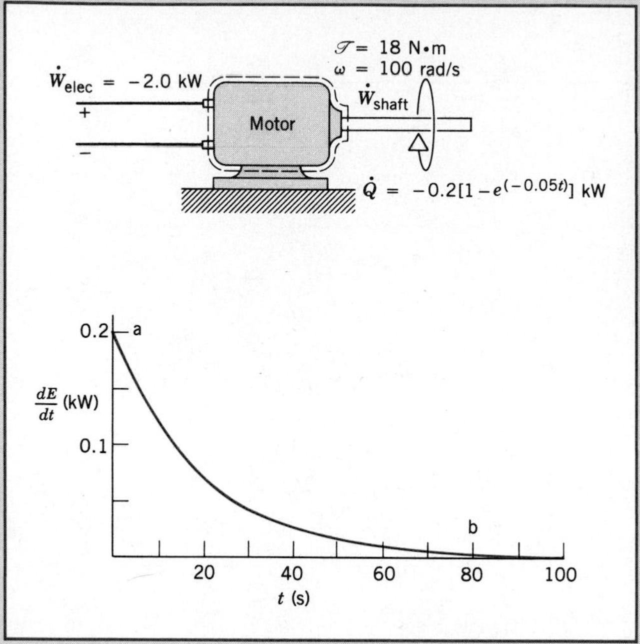

Figure E2.5

Assumption: The system shown in the accompanying sketch is a closed system.

Analysis: The time rate of change of system energy is

$$\frac{dE}{dt} = \dot{Q} - \dot{W}$$

$\dot{W}$ represents the *net* power *from* the system: the sum of the power associated with the rotating shaft, $\dot{W}_{\text{shaft}}$, and the power associated with the electricity flow, $\dot{W}_{\text{elec}}$

$$\dot{W} = \dot{W}_{\text{shaft}} + \dot{W}_{\text{elec}}$$

The rate $\dot{W}_{\text{elec}}$ is known from the problem statement: $\dot{W}_{\text{elec}} = -2.0$ kW, where the negative sign is required because energy is carried into the system by electrical work. The term $\dot{W}_{\text{shaft}}$ can be evaluated with Eq. 2.20 as

$$\dot{W}_{\text{shaft}} = \mathcal{T}\omega = (18\ \text{N}\cdot\text{m})(100\ \text{rad/s}) = 1800\ \text{W} = +1.8\ \text{kW}$$

Because energy exits the system along the rotating shaft, this energy transfer rate is positive.
In summary

$$\dot{W} = \dot{W}_{\text{elec}} + \dot{W}_{\text{shaft}} = (-2.0\ \text{kW}) + (+1.8\ \text{kW}) = -0.2\ \text{kW}$$

where the minus sign means that the electrical power input is greater than the power transferred out along the shaft. With the foregoing result for $\dot{W}$ and the given expression for $\dot{Q}$, the energy rate balance becomes

$$\frac{dE}{dt} = -0.2[1 - e^{(-0.05t)}] - (-0.2) = 0.2e^{(-0.05t)}$$

Comments:

1. The expression obtained for the time rate of change of system energy is graphed in the accompanying figure. The figure shows that initially there is a *transient* period of operation, segment a–b of the curve (approximately), where the energy of the system varies with time. Thereafter, the system energy is constant: The rate of energy input electrically equals the total rate at which energy exits along the rotating shaft and by heat transfer.
2. After the transient period, the system energy would be constant with time and the energy rate balance would reduce as follows:

$$\overset{0}{\cancel{\frac{dE}{dt}}} = \dot{Q} - \dot{W}$$

Thus, the heat transfer would equal the net work

$$\dot{Q} = \dot{W}_{elec} + \dot{W}_{shaft}$$

Inserting previously determined values

$$\dot{Q} = -0.2 \text{ kW}$$

(The same result is obtained by taking the limit as $t \rightarrow \infty$ in the given expression for the variation of $\dot{Q}$ with time.) This constant value for the heat transfer rate can be thought of as the portion of the electrical power input that is not obtained as a mechanical power output because of effects within the motor such as electrical resistance and friction.

2.6 ENERGY ANALYSIS OF CYCLES

In this section the energy concepts developed thus far are illustrated further by application to systems undergoing thermodynamic cycles. Recall from Sec. 1.2 that when a system at a given initial state goes through a sequence of processes and finally returns to that state, the system has executed a thermodynamic cycle. The study of systems undergoing cycles has played an important role in the development of the subject of engineering thermodynamics. Both the first and second laws of thermodynamics have roots in the study of cycles. In addition, there are many important practical applications involving power generation, vehicle propulsion, and refrigeration for which an understanding of thermodynamic cycles is necessary. In this section, cycles are considered from the perspective of the conservation of energy principle. Cycles are studied in greater detail in subsequent chapters, using both the conservation of energy principle and the second law of thermodynamics.

2.6.1 Preliminaries

The energy balance for any system undergoing a thermodynamic cycle takes the form

$$\Delta E_{cycle} = Q_{cycle} - W_{cycle} \tag{2.39}$$

where Q_{cycle} and W_{cycle} represent *net* amounts of energy transfer by heat and work, respectively, for the cycle. Since the system is returned to its initial state after the cycle, there is no *net* change in its energy. Therefore, the left side of Eq. 2.39 equals zero, and the equation reduces to

$$W_{cycle} = Q_{cycle} \tag{2.40}$$

Equation 2.40 is an expression of the conservation of energy principle that must be satisfied by *every* thermodynamic cycle, regardless of the sequence of processes followed by the system undergoing the cycle or the nature of the substances making up the system.

Figure 2.15 provides simplified schematics of two general classes of cycles considered in this book: power cycles and refrigeration and heat pump cycles. In each case pictured, a system undergoes a cycle while communicating thermally with two bodies, one hot and the other cold. These bodies are systems located in the surroundings of the system undergoing the cycle. During each cycle there is also a net amount of energy exchanged with the surroundings by work. Carefully observe that in using the symbols Q_{in} and Q_{out} on Fig. 2.15 we have departed from the previously stated

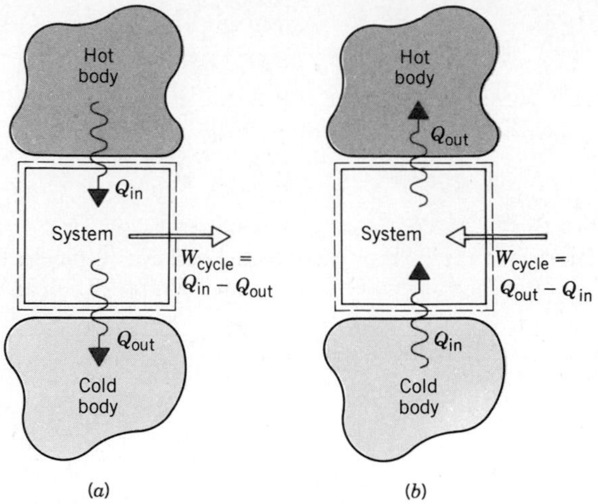

Figure 2.15 Schematic diagrams of two important classes of cycles. (*a*) Power cycles. (*b*) Refrigeration and heat pump cycles.

sign convention for heat transfer. In this section it is advantageous to regard Q_{in} and Q_{out} as transfers of energy in the *directions indicated by the arrows*. The direction of the net work of the cycle, W_{cycle}, is *also indicated by an arrow*. Finally, note that the directions of the energy transfers shown in Fig. 2.15*b* are opposite to those of Fig. 2.15*a*. Let us now briefly consider each type of cycle pictured in Fig. 2.15.

2.6.2 Power Cycles

Systems undergoing cycles of the type shown in Fig. 2.15*a* deliver a net work transfer of energy to their surroundings during each cycle. Any such cycle is called a *power cycle*. From Eq. 2.40, the net work output equals the net heat transfer to the cycle, or

$$W_{cycle} = Q_{in} - Q_{out} \qquad \text{(power cycle)} \tag{2.41}$$

where Q_{in} represents the heat transfer of energy *into* the system from the hot body, and Q_{out} represents heat transfer *out* of the system to the cold body. From Eq. 2.41 it is clear that Q_{in} must be greater than Q_{out} for a *power* cycle. The energy supplied by heat transfer to a system undergoing a power cycle is normally derived from the combustion of fuel or a moderated nuclear reaction; it can also be obtained from solar radiation. The energy Q_{out} is generally discharged to the surrounding atmosphere or a nearby body of water.

The performance of a system undergoing a *power cycle* can be described in terms of the extent to which the energy added by heat, Q_{in}, is *converted* to a net work output, W_{cycle}. The extent of the energy conversion from heat to work is expressed by the following ratio, commonly called the *thermal efficiency:*

$$\eta = \frac{W_{cycle}}{Q_{in}} \qquad \text{(power cycle)} \tag{2.42}$$

Introducing Eq. 2.41, an alternative form is obtained as

$$\eta = \frac{Q_{in} - Q_{out}}{Q_{in}} = 1 - \frac{Q_{out}}{Q_{in}} \qquad \text{(power cycle)} \tag{2.43}$$

Since energy is conserved, it follows that the thermal efficiency can never be greater than unity (100%). However, experience with *actual* power cycles shows that the value of thermal efficiency is invariably *less* than unity. That is, not all the energy added to the system by heat transfer is converted to work; a portion is discharged to the cold body by heat transfer. Using the second law of thermodynamics, we will show in Chap. 5 that the conversion from heat to work cannot be fully accomplished by any power cycle. The thermal efficiency of *every* power cycle must be less than unity.

2.6.3 Refrigeration and Heat Pump Cycles

Next, consider the refrigeration and heat pump cycles shown in Fig. 2.15*b*. For cycles of this type, Q_{in} is the energy transferred by heat *into* the system undergoing the cycle *from* the cold body, and Q_{out} is the energy discharged by heat transfer *from* the system *to* the hot body. To accomplish these energy transfers requires a net work *input*, W_{cycle}. The quantities Q_{in}, Q_{out}, and W_{cycle} are related by the energy balance, which for refrigeration and heat pump cycles takes the form

$$W_{cycle} = Q_{out} - Q_{in} \quad \text{(refrigeration and heat pump cycles)} \tag{2.44}$$

Since W_{cycle} is positive in this equation, it follows that Q_{out} is greater than Q_{in}. Note that the system undergoing the cycle must be at a temperature *below* that of the cold body during the portion of the cycle where the heat transfer Q_{in} occurs. The temperature of the system must be *above* that of the hot body during the portion of the cycle where Q_{out} takes place.

Although we have treated them as the same to this point, refrigeration and heat pump cycles actually have different objectives. The objective of a refrigeration cycle is to cool a refrigerated space or to maintain the temperature within a dwelling or other building *below* that of the surroundings. The objective of a heat pump is to maintain the temperature within a dwelling or other building *above* that of the surroundings or to provide heating for certain industrial processes that occur at elevated temperatures. Since refrigeration and heat pump cycles have different objectives, their performance parameters, called *coefficients of performance,* are defined differently.

The performance of *refrigeration cycles* can be described as the ratio of the amount of energy received by the system undergoing the cycle from the cold body, Q_{in}, to the net work transfer of energy into the system to accomplish this effect, W_{cycle}. Thus, the coefficient of performance, β, is

$$\beta = \frac{Q_{in}}{W_{cycle}} \quad \text{(refrigeration cycle)} \tag{2.45}$$

Introducing Eq. 2.44, an alternative expression for β is obtained as

$$\beta = \frac{Q_{in}}{Q_{out} - Q_{in}} \quad \text{(refrigeration cycle)} \tag{2.46}$$

For a household refrigerator, Q_{out} is discharged to the space in which the refrigerator is located. W_{cycle} is usually provided in the form of electricity to run the motor that drives the refrigerator.

The performance of *heat pumps* can be described as the ratio of the amount of energy discharged from the system undergoing the cycle to the hot body, Q_{out}, to the net work transfer of energy into the system to accomplish this effect, W_{cycle}. Thus, the coefficient of performance, γ, is

$$\gamma = \frac{Q_{out}}{W_{cycle}} \quad \text{(heat pump cycle)} \tag{2.47}$$

Introducing Eq. 2.44, an alternative expression for this coefficient of performance is obtained as

$$\gamma = \frac{Q_{out}}{Q_{out} - Q_{in}} \quad \text{(heat pump cycle)} \tag{2.48}$$

From this equation it can be seen that the value of γ is never less than unity. For residential heat pumps, the energy quantity Q_{in} is normally drawn from the surrounding atmosphere, the ground, or a nearby body of water. W_{cycle} is usually provided by electricity.

The coefficients of performance β and γ are defined as ratios of the desired heat transfer effect to the cost in terms of work to accomplish that effect. Based on the definitions, it is desirable thermodynamically that these coefficients have values that are as large as possible. However, as discussed in Chap. 5, coefficients of performance must satisfy restrictions imposed by the second law of thermodynamics.

PROBLEMS

2.1 An automobile has a mass of 1200 kg. What is its kinetic energy, in kJ, relative to the road when traveling at a velocity of 100 km/h? If the vehicle decelerates to 50 km/h, what is the change in kinetic energy, kJ?

2.2 Determine the gravitational potential energy, in kJ, of 2 m^3 of liquid water at an elevation of 30 m above the surface of the earth. The acceleration of gravity is constant at g = 9.7 m/s^2 and the density of the water is uniform throughout at 1000 kg/m^3. Determine the change in gravitational potential energy, in kJ, if the elevation were decreased to 15 m.

2.3 An object whose mass is 100 lb undergoes a decrease in its kinetic energy of 500 ft · lbf owing to the action of a resultant force. Initially, the velocity is 50 ft/s. There is no change in elevation, and there are no other interactions between the object and its surroundings. Determine the final velocity, in ft/s.

2.4 A body whose mass is 5 lb experiences a decrease in its gravitational potential energy of 250 ft · lbf. If the local acceleration of gravity is constant at g = 31.0 ft/s^2, determine the change in elevation, in ft.

2.5 What is the change in potential energy, in ft · lbf, of an automobile weighing 2600 lbf at sea level when it travels from sea level to an elevation of 1000 ft? Assume the acceleration of gravity is constant at g = 32 ft/s^2.

2.6. An object of mass 10 kg, initially at rest, accelerates to a final velocity of 100 m/s. What is the change in kinetic energy of the object, in kJ?

2.7 An object whose mass is 2000 kg moves at a velocity of 50 m/s and an altitude of 400 m, both measured relative to the surface of the earth. The acceleration of gravity can be taken as constant at g = 9.7 m/s^2.

(a) If the kinetic energy increased by 2400 kJ with no change in elevation, what would be the final velocity, in m/s?
(b) If the potential energy increased by 2400 kJ, with no change in velocity, what would be the final elevation, in m?

2.8 Consider an object whose mass is 1 lb moving with a velocity of 10 ft/s. Determine

(a) the change in velocity, in ft/s, associated with a change in kinetic energy of 1 ft · lbf.
(b) the change in potential energy, in ft · lbf, associated with a 1-ft change in elevation. Use g = 32.2 ft/s^2.

2.9 An object whose mass is 50 kg is accelerated from rest to a velocity of 15 m/s by the action of a resultant force. Determine the work done by the resultant force, in kJ, if there are no other interactions between the object and its surroundings.

2.10 An object whose mass is 150 kg undergoes a change in kinetic energy owing to the action of a resultant force. There is no change in elevation. The work done by the force is 100 kJ, and there are no other interactions between the object and its surroundings. If the initial velocity of the object is 50 m/s, what is the final velocity, in m/s?

2.11 A disk-shaped flywheel, of uniform density ρ, outer radius R, and thickness w, rotates with an angular velocity ω. Obtain an expression for the moment of inertia, I =

$\int_{vol} \rho r^2\, dV$, in terms of ρ, R, and ω. Show that the kinetic energy of the flywheel is given by the relationship KE = $\frac{1}{2} I\omega^2$.

2.12 An object of mass 10 lb moves along a straight line with a velocity of 100 ft/s. Determine the rotational speed, in RPM, of a flywheel whose rotational kinetic energy is equal in magnitude to the linear kinetic energy of the object. The moment of inertia of the flywheel is 150 lb · ft².

2.13 An object whose mass is 40 kg falls freely under the influence of gravity from an elevation of 100 m above the earth's surface. The initial velocity is directed downward with a magnitude of 100 m/s. Ignoring the effect of air resistance, what is the magnitude of the velocity, in m/s, of the object just before it strikes the earth? The acceleration of gravity is $g = 9.81$ m/s².

2.14 An object whose mass is 50 lb is projected upward with an initial velocity of 200 ft/s. Ignoring the effect of air resistance, what is the velocity of the object, in ft/s, when its elevation has increased by 20 ft? The acceleration of gravity is $g = 31.5$ ft/s².

2.15 A block of mass 10.0 kg is pushed along an incline a distance of 5.0 m as its center of gravity is elevated by 3.0 m. The block is acted on by a constant force **R** parallel to the incline having a magnitude of 70 N and by the force of gravity. Assume frictionless surfaces and $g = 9.8$ m/s². Determine, in J,

(a) the work done on the block by the constant force *R*.
(b) the change in potential energy of the block.
(c) the change in kinetic energy of the block.

2.16 An object of mass 20 kg slides down a ramp, beginning from rest. The ramp is inclined at an angle of 30° from the horizontal. When the object reaches the bottom of the ramp, its speed is 7 m/s. If air resistance and friction between the object and the ramp are negligible, determine the distance the object traveled *along the ramp*, in m. The acceleration of gravity is $g = 9.81$ m/s².

2.17 Figure P2.17 shows a ramp used to transfer a box from one conveyer belt to another that is 5 ft lower. The box and its contents have a mass of 25 lb and approach the ramp with a horizontal velocity of 2 ft/s. At the base of the ramp, the magnitude of the velocity has increased to 3.6 ft/s and is directed along the ramp. The acceleration of gravity is g = 32.0 ft/s². Determine for the box

(a) the change in potential energy, in ft · lbf.
(b) the change in kinetic energy, in ft · lbf.

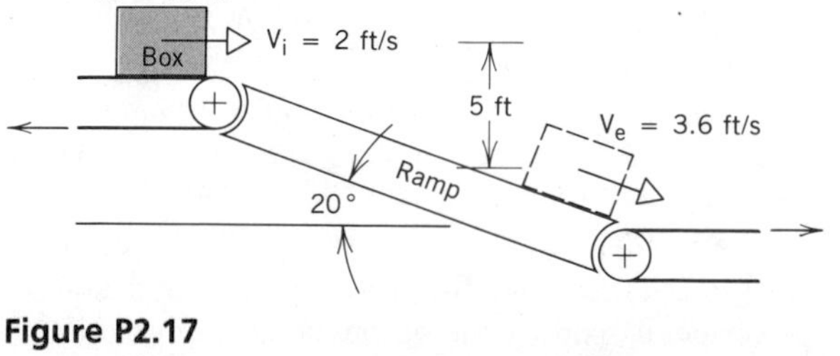

Figure P2.17

2.18 A system with a mass of 10 kg, initially at rest, experiences a constant horizontal acceleration of 4 m/s² due to the action of a resultant force applied for 20 s. Determine the total amount of energy transfer by work, in kJ.

2.19 An object of mass 20 lb, initially moving horizontally with a velocity of 50 ft/s, experiences a constant horizontal acceleration of 5 ft/s² due to the action of a resultant force. The total amount of energy transfer to the object by work is 1100 ft · lbf. Determine the length of time, in s, that the resultant force was applied.

2.20 A gas contained within a piston–cylinder assembly expands in a constant-pressure process at 4 bars from $V_1 = 0.15$ m³ to a final volume of $V_2 = 0.36$ m³. Calculate the work, in kJ.

2.21 Air is compressed in a piston–cylinder assembly from an initial state where $p_1 = 30$ lbf/in.² and $V_1 = 25$ ft³. The relationship between pressure and volume during the process is $pV^{1.4}$ = constant. For the air as the system, the work is −62 Btu. Determine the final volume, in ft³, and the final pressure, in lbf/in.²

2.22 A gas is compressed from $V_1 = 0.09$ m³ to $V_2 = 0.03$ m³. The relation between pressure and volume during the process is $p = -14V + 2.44$, where the units of p and V are bars and m³, respectively. For the gas, find the work, in kJ.

2.23 Carbon dioxide gas in a piston–cylinder assembly expands from an initial state where $p_1 = 60$ lbf/in.², $V_1 = 1.78$ ft³ to a final pressure of $p_2 = 20$ lbf/in.² The relationship between pressure and volume during the process is $pV^{1.2}$ = constant. For the gas, calculate the work done, in ft · lbf. Convert your answer to Btu.

2.24 A gas expands from an initial state where $p_1 = 500$ kPa and $V = 0.1$ m³ to a final state where $p_2 = 100$ kPa. The relationship between pressure and volume during the process is pV = constant. Sketch the process on a p–V diagram and determine the work, in kJ.

2.25 Air undergoes a process from an initial state where $p_1 = 14.0$ lbf/in.², $V_1 = 500$ in.³ to a final state where $p_2 = 60$ lbf/in.², $V_2 = 160$ in.³ The relationship between pressure and volume during the process is pV^n = constant. Determine the value of the constant n and calculate the work, in Btu.

2.26 Air is trapped in a piston–cylinder assembly oriented horizontally as shown in Fig. P2.26. Initially, $p_1 = 100$ kPa, $V_1 = 2 \times 10^{-3}$ m³, and the face of the piston is at $x = 0$. The spring exerts no force on the piston in the initial position. The atmospheric pressure is 100 kPa, and the area of the piston face is 0.018 m². The air expands slowly until its volume is $V_2 = 3 \times 10^{-3}$ m³. During the process, the spring exerts a force on the piston that varies with x according to $F = kx$, where $k = 16.2 \times 10^3$ N/m. There is no friction between the piston and the cylinder wall. Determine the final pressure of the air, in kPa, and the work done by the air on the piston, in kJ.

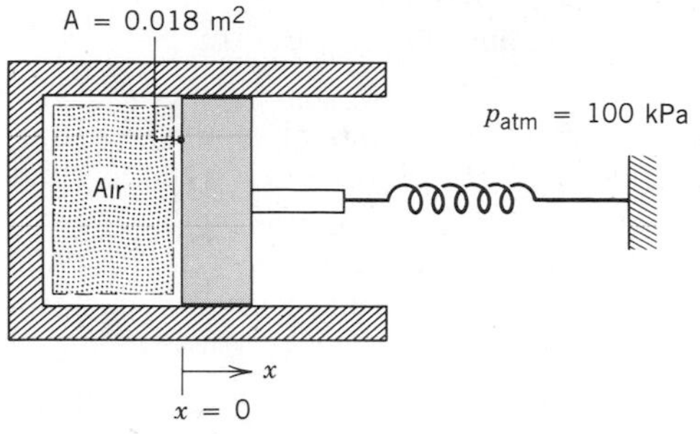

Figure P2.26

2.27 Air undergoes two processes in series.

Process 1–2: expansion from $p_1 = 300$ kPa, $v_1 = 0.019$ m³/kg to $p_2 = 150$ kPa during which the pressure–volume relationship is $pv =$ constant

Process 2–3: constant-pressure compression to $v_3 = v_1$

Sketch the processes on a p–v diagram and determine the work per unit mass of air, in kJ/kg.

2.28 A gas undergoes two processes in series.

Process 1–2: constant volume from $p_1 = 50$ lbf/in.² to $p_2 = 10$ lbf/in.²

Process 2–3: compression with $pV^{1.3}$ = constant from state 2 to $p_3 = 50$ lbf/in.² and $V_3 = 1$ ft³

Sketch the processes on a p–V diagram and determine the work, in Btu.

2.29 Measured data for pressure versus volume during the expansion of gases within the cylinder of an internal combustion engine are given in the table below. Using data from the table, complete the following:

(a) Estimate the work done by the gases on the piston during the expansion, in kJ. Why is this an estimate?
(b) Prepare a plot of pressure versus volume on log–log coordinates, and fit a straight line to the data. Determine the slope of the line.

Data point	p (bars)	V (cm³)
1	20.0	454
2	16.1	540
3	12.2	668
4	9.9	780
5	6.0	1175
6	3.1	1980

2.30 Measured data for pressure versus volume during the compression of a refrigerant within the cylinder of a refrigeration compressor are given in the table below. Using data from the table, complete the following:

(a) Estimate the work on the refrigerant done during the compression, in ft·lbf. Why is this an estimate?
(b) Prepare a plot of pressure versus volume on log–log coordinates, and fit a straight line to the data. Determine the slope of the line.

Data point	p (lbf/in.²)	V (in.³)
1	91	13.0
2	129	11.0
3	193	9.0
4	209	7.0
5	282	5.0
6	376	3.0

2.31 Figure P2.31 shows an object whose mass is 5 lb attached to a rope wound around a pulley. The radius of the pulley is 3 in. If the mass falls at a constant velocity of 5 ft/s, determine the power transmitted to the pulley, in horsepower, and the rotational speed of the pulley, in RPM. The acceleration of gravity is $g = 32.2$ ft/s².

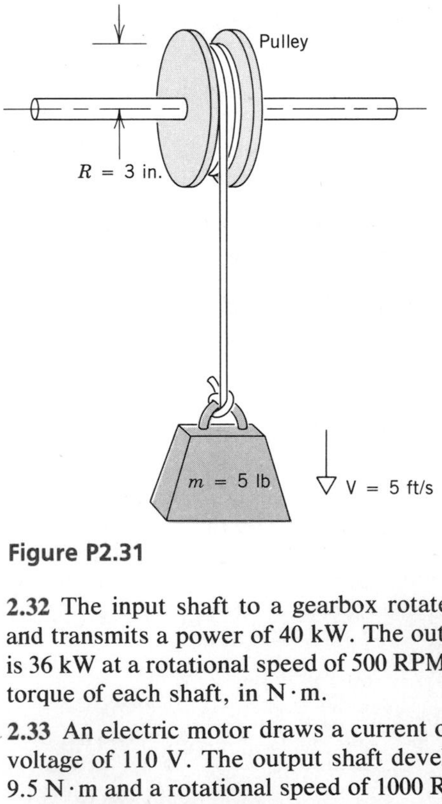

Figure P2.31

2.32 The input shaft to a gearbox rotates at 2000 RPM and transmits a power of 40 kW. The output shaft power is 36 kW at a rotational speed of 500 RPM. Determine the torque of each shaft, in N·m.

2.33 An electric motor draws a current of 10 amp with a voltage of 110 V. The output shaft develops a torque of 9.5 N·m and a rotational speed of 1000 RPM. All operating data are constant with time. Determine

(a) the electric power required by the motor and the power developed by the output shaft, each in kW.
(b) the net power input to the motor, in kW.

(c) the amount of energy transferred to the motor by electrical work and the amount of energy transferred out of the motor by the shaft, in kW·h and Btu, during 2 h of operation.

2.34 A 12-V automotive storage battery is charged with a constant current of 10 amp for 3 h. Determine the amount of energy supplied to the battery, in kW·h. Convert your answer to Btu.

2.35 An electric heater draws a constant current of 6 amp, with an applied voltage of 220 V, for 10 h. Determine the total amount of energy supplied to the heater by electrical work, in kW·h.

2.36 The drag force, F_d, imposed by the surrounding air on an automobile moving with velocity V is given by

$$F_d = C_d A \tfrac{1}{2} \rho V^2$$

where C_d is a constant called the drag coefficient, A is the projected frontal area of the vehicle, and ρ is the air density. For $C_d = 0.42$, $A = 2\ \text{m}^2$, and $\rho = 1.23\ \text{kg/m}^3$, calculate the power required, in kW, to overcome drag at a constant velocity of 100 km/h.

2.37 A solid cylindrical bar (see Fig. 2.7) of diameter 5 mm is slowly stretched from an initial length of 10 cm to a final length of 10.1 cm. The normal stress acting at the end of the bar varies according to $\sigma = C(x - x_1)/x_1$, where x is the position of the end of the bar, x_1 is the initial length, and C is a material constant (Young's modulus). For $C = 2 \times 10^7$ kPa, determine the work done on the bar, in J.

2.38 A steel wire suspended vertically has a cross-sectional area of 0.1 in.^{2} and an initial length of 10 ft. A downward force applied to the end of the wire causes the wire to stretch. The force varies linearly with the length of the wire from zero initially to 2500 lbf when the length has increased by 0.01 ft. Determine

(a) the normal stress, in lbf/in.^{2}, at the end of the wire as a function of the length of the wire.
(b) the work done in stretching the wire, in ft·lbf.

2.39 A soap film is suspended on a 5 cm × 5 cm wire frame, as shown in Fig. 2.8. The movable wire is displaced 1 cm by an applied force, while the surface tension of the soap film remains constant at 25×10^{-5} N/cm. Determine the work done in stretching the film, in J.

2.40 A liquid film is suspended on a rectangular wire frame, as shown in Fig. 2.8. The length of the movable wire is 2 in., and the other dimension initially is 6 in. The movable wire is displaced 1 in. by an applied force, while the surface tension of the liquid film remains constant at 2.5×10^{-4} lbf/in. Determine the work done in stretching the film, in ft·lbf.

2.41 A brick wall of an industrial furnace is 0.1 m thick and has a thermal conductivity of 1.5 W/m·K. At steady state, the temperatures of the inner and outer wall are 1500 and 1200 K, respectively. The temperature distribution is linear through the wall. For an area of 10 m^2, calculate the rate of energy transfer through the wall by conduction, in kW.

2.42 An insulated frame wall of a house has an average thermal conductivity of 0.0318 Btu/h·ft·°R. The thickness of the wall is 6 in. At steady state, the rate of energy transfer by conduction through an area of 160 ft^2 is 400 Btu/h, and the temperature decreases linearly from the inner surface to the outer surface. If the outside surface temperature of the wall is 30°F, what is the inner surface temperature, in °F?

2.43 A surface of area 2 m^2 emits thermal radiation. The emissivity of the surface is $\varepsilon = 0.8$. Determine the rate at which energy is emitted, in kW, for surface temperatures of 200, 300, and 500 K. The value of the Stefan–Boltzmann constant, σ, is 5.67×10^{-8} W/m^2·K^4.

2.44 A sphere of surface area 0.08 ft^2 emits thermal radiation at a rate of 100 Btu/h with a surface temperature of 1000°R. Determine the emissivity, ε, of the surface. The value of the Stefan–Boltzmann constant, σ, is 0.1714×10^{-8} Btu/h·ft^2·°R^4.

2.45 Air at 500 K flows over a flat surface whose temperature at steady state is 400 K. The heat flux to the plate is 2.5 kW/m^2. What is the value of the heat transfer coefficient, h, in W/m^2·K?

2.46 An insulated frame wall of a house has an average thermal conductivity of 0.0318 Btu/h·ft·°R. The thickness of the wall is 6 in. and the area is 160 ft^2. The inside air temperature is 70°F, and the heat transfer coefficient for convection between the inside air and the wall is 1.5 Btu/h·ft^2·°R. On the outside, the heat transfer coefficient is 6 Btu/h·ft^2·°R and the air temperature is −10°F. Ignoring radiation, determine at steady state the rate of heat transfer through the wall, in Btu/h.

2.47 A flat surface is covered with insulation with a thermal conductivity of 0.08 W/m·K. The temperature at the interface between the surface and the insulation is 300°C. The outside of the insulation is exposed to air at 30°C, and the heat transfer coefficient for convection between the insulation and the air is 10 W/m^2·K. Ignoring radiation, determine the minimum thickness of insulation, in m, such that the outside of the insulation is no hotter than 60°C at steady state.

2.48 Each line in the table below gives information about a process of a closed system. Every entry has the same energy units. Fill in the blank spaces in the table.

Process	Q	W	E_1	E_2	ΔE
a	+50	−20	+20		
b		+20		+50	+30
c	−25	−80		+160	
d		−90	+50	−50	0
e		+150	+20	50	−100

2.49 A closed system of mass 2 kg undergoes a process in which there is heat transfer of magnitude 25 kJ from the system to the surroundings. The magnitude of the work done on the system during the process is 100 kJ. The specific internal energy of the system *decreases* by 15 kJ/kg and the elevation of the system increases by 1000 m. The acceleration of gravity is constant at $g = 9.6$ m/s^2. Determine the change in kinetic energy of the system, in kJ.

2.50 A closed system of mass 5 kg undergoes a process in which there is a heat transfer of 150 kJ from the system to the surroundings. The work done by the system is 75 kJ. If the initial specific internal energy of the system is 450 kJ/kg, what is the final specific internal energy, in kJ/kg? Neglect changes in kinetic and potential energy.

2.51 A closed system of mass 5 lb undergoes a process in which there is a heat transfer of 200 ft·lbf from the system to the surroundings. There is no work during the process. The velocity of the system changes from 10 ft/s to 50 ft/s, and the elevation decreases by 150 ft. The acceleration of gravity is g = 32.0 ft/s^2. Determine the change in internal energy of the system, in ft·lbf and Btu.

2.52 A gas in a piston–cylinder assembly undergoes two processes in series. From state 1 to state 2 there is energy transfer by heat to the gas with a magnitude of 500 kJ, and the gas does work on the piston with a magnitude of 800 kJ. The second process, from state 2 to state 3, is a constant-pressure compression at 400 kPa during which there is a heat transfer from the gas with a magnitude of 450 kJ. The following data are also known: $U_1 = 2000$ kJ and $U_3 = 3500$ kJ. Neglecting changes in kinetic and potential energy, calculate the change in volume of the gas during process 2–3, in m^3.

2.53 An electric generator coupled to a windmill produces an average electric power output of 5 kW. The power is used to charge a storage battery. Heat transfer from the battery to the surroundings occurs at a constant rate of 0.6 kW. Determine the total amount of energy stored in the battery, in kJ, in 8 h of operation.

2.54 An electric motor operating at steady state requires an electric power input of 1 Btu/s. Heat transfer occurs from the motor to the surroundings at temperature T_o at a rate of $hA(T_b - T_o)$ where T_b is the average surface temperature of the motor, hA = 10 Btu/h·°R, and $T_o = 80$°F. The torque developed by the shaft of the motor is 12.8 ft·lbf at a rotational speed of 500 RPM. Determine T_b, in °F.

2.55 A closed system undergoes a process during which there is heat transfer to the system at a constant rate of 3 kW, and the power developed by the system varies with time according to

$$\dot{W} = \begin{cases} +2.4t & 0 < t \leq 1 \text{ h} \\ +2.4 & t > 1 \text{ h} \end{cases}$$

where t is in hours and $\dot{W}$ is in kW.

(a) What is the time rate of change of system energy at $t = 0.6$ h, in kW?

(b) Determine the change in system energy after 2 h, in kJ.

2.56 A gas in a piston–cylinder assembly is compressed from $p_1 = 3.4$ bars, $V_1 = 0.0283$ m^3 to $p_2 = 8.2$ bars in a process for which the relation between pressure and volume is $pV^{1.2}$ = constant. The mass of the gas is 0.183 kg. During the process, the heat transfer from the gas is 2.2 kJ. Determine the change in specific internal energy of the gas, in kJ/kg. Kinetic and potential energy effects are negligible.

2.57 Air is contained in a rigid well-insulated tank with a volume of 0.2 m^3. The tank is fitted with a paddle wheel that transfers energy to the air at a constant rate of 4W for 20 min. The initial density of the air is 1.2 kg/m^3. If no changes in kinetic or potential energy occur, determine

(a) the specific volume at the final state, in m^3/kg.

(b) the change in specific internal energy of the air, in kJ/kg.

2.58 A gas is contained in a closed rigid tank fitted with a paddle wheel. The paddle wheel stirs the gas for 20 min, with the power varying with time according to $\dot{W} = -10t$, where $\dot{W}$ is in watts, and t is in min. Heat transfer from the gas to the surroundings takes place at a *constant* rate of 50 W. Determine

(a) the rate of change of energy of the gas at time t = 10 min, in watts.

(b) the net change in energy of the gas after 20 min, in kJ.

2.59 Steam in a piston–cylinder assembly expands from $p_1 = 35$ bars to $p_2 = 7$ bars. The pressure–volume relation during the process is pV^2 = constant. The mass of steam is 2.3 kg. Other steam properties at the initial state are $u_1 = 3282.1$ kJ/kg and $v_1 = 113.24$ cm^3/g. At the final state $u_2 = 2124.6$ kJ/kg. Neglecting changes in kinetic and potential energy, calculate the heat transfer, in kJ, for the steam as the system.

2.60 A gas undergoes a process from state 1, where p_1 = 20 lbf/in.^{2}, to state 2, where $p_2 = 60$ lbf/in.^{2} and $V_2 = 1.5$ ft^3. The pressure–volume relation during the process is pV = constant. The internal energy of the gas increases by 10,000 ft·lbf, and there are no changes in kinetic or potential energy during the process. Determine the heat transfer, in Btu.

2.61 A gas contained within a piston–cylinder assembly is shown in Fig. P2.61. Initially, the piston face is at x = 0, and the spring exerts no force on the piston. As a result of heat transfer, the gas expands, raising the piston until it hits the stops. At this point the piston face is located at $x = 0.05$ m, and the heat transfer ceases. The force exerted by the spring on the piston as the gas expands varies linearly with x according to

$$F_{spring} = kx$$

where $k = 10{,}000$ N/m. Friction between the piston and the cylinder wall can be neglected. The acceleration of gravity is $g = 9.81$ m/s^2. Additional information is given on Fig. P2.61.

(a) What is the initial pressure of the gas, in kPa?
(b) Determine the work done by the gas on the piston, in J.
(c) If the specific internal energies of the gas at the initial and final states are 214 and 337 kJ/kg, respectively, calculate the heat transfer, in J.

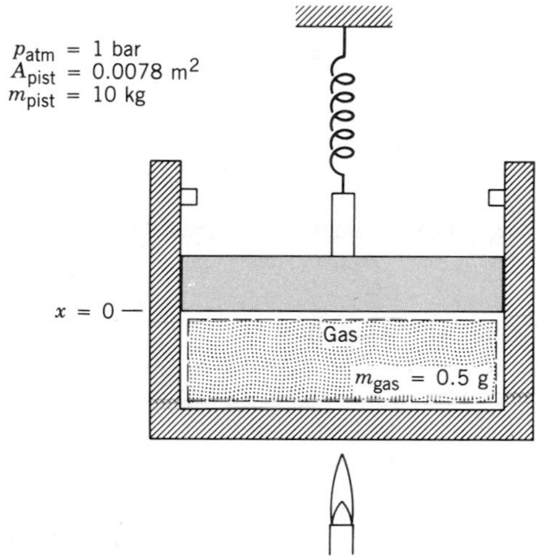

Figure P2.61

2.62 A gas is contained in a vertical piston–cylinder assembly by a piston weighing 675 lbf and having an area of 8 in.^{2} The atmosphere exerts a pressure of 14.7 lbf/in.^{2} on the top of the piston. An electrical resistor transfers energy to the gas in the amount of 3 Btu, and the internal energy of the gas increases by 1 Btu. Neglecting friction between the piston and the cylinder wall and assuming no heat transfer between the gas and the piston–cylinder assembly, determine the change in elevation of the piston, in ft.

2.63 A gas undergoes a thermodynamic cycle consisting of three processes beginning at an initial state where $p_1 =$ 1 bar, $V_1 = 1.5$ m^3, and $U_1 = 512$ kJ. The processes are as follows:

Process 1–2: compression with pV = constant to p_2 = 2 bars, $U_2 = 690$ kJ
Process 2–3: $W_{23} = 0$, $Q_{23} = -150$ kJ
Process 3–1: $W_{31} = +50$ kJ

There are no significant changes in kinetic or potential energy. Determine the heat transfers Q_{12} and Q_{31}, each in kJ.

2.64 A gas undergoes a thermodynamic cycle consisting of the following processes:

Process 1–2: constant pressure, $p = 1.4$ bars, $V_1 = 0.028$ m^3, $W_{12} = 10.5$ kJ
Process 2–3: compression with pV = constant, $U_3 = U_2$
Process 3–1: constant volume, $U_1 - U_3 = -26.4$ kJ

There are no significant changes in kinetic or potential energy.

(a) Sketch the cycle on a p–V diagram.
(b) Calculate the net work for the cycle, in kJ.
(c) Calculate the heat transfer for process 1–2, in kJ.

2.65 A closed system undergoes a thermodynamic cycle consisting of the following processes:

Process 1–2: adiabatic compression with $pV^{1.4}$ = constant from $p_1 = 50$ lbf/in.^{2}, $V_1 = 3$ ft^3 to V_2 = 1 ft^3
Process 2–3: constant volume
Process 3–1: constant pressure, $U_1 - U_3 = 46.7$ Btu

There are no significant changes in kinetic or potential energy.

(a) Sketch the cycle on a p–V diagram.
(b) Calculate the net work for the cycle, in Btu.
(c) Calculate the heat transfer for Process 2-3, in Btu.

2.66 The net work of a power cycle operating as in Fig. 2.15*a* is 10,000 kJ, and the thermal efficiency is 0.4. Determine the heat transfers Q_{in} and Q_{out}, each in kJ.

2.67 For a power cycle operating as shown in Fig. 2.15*a*, the total heat transfer to the cycle, Q_{in}, is 500 MJ. What is the net work developed, in MJ, if the thermal efficiency is 30%?

2.68 The net work of a power cycle operating as in Fig. 2.15*a* is 5×10^6 Btu, and the heat transfer Q_{out} is 12×10^6 Btu. What is the thermal efficiency of the power cycle?

2.69 For a coal-burning electric power plant, what plays the roles of the hot body and the cold body shown in Fig. 2.15*a*?

2.70 For a household refrigerator, what plays the roles of the hot body and the cold body shown in Fig. 2.15*b*?

2.71 For a window air conditioner, what plays the roles of the hot body and the cold body shown in Fig. 2.15*b*?

2.72 A power cycle with a thermal efficiency of 28% operates as shown in Fig. 2.15*a*. For $Q_{in} = 600$ Btu, what is Q_{out}, in Btu?

2.73 A power cycle operates continuously and receives energy by heat transfer from the combustion of fuel at a rate of 75 kW. If the thermal efficiency of the cycle is 33%, at what rate is power developed, in kW?

2.74 A refrigeration cycle operating as in Fig. 2.15*b* has heat transfers $Q_{in} = 2000$ Btu and $Q_{out} = 3200$ Btu. Determine the net work required, in Btu, and the coefficient of performance for the cycle.

2.75 The coefficient of performance of a heat pump cycle operating as in Fig. 2.15*b* is 3.5, and the net work *input* is 5000 kJ. Determine the heat transfers Q_{in} and Q_{out}, in kJ.

2.76 A refrigeration cycle operates continuously and removes energy from the refrigerated space at a rate of 12,000 Btu/h. For a coefficient of performance of 2.6, determine the net rate of work input required, in Btu/h. Convert your answer to horsepower.

DESIGN AND OPEN ENDED PROBLEMS

2.1D The effective use of our energy resources is an important societal goal.

(a) Summarize in a *pie chart* the data on the use of fuels in your state in the residential, commercial, industrial, and transportation sectors. What factors may affect the future availability of these fuels? Does your state have a written energy policy? Discuss.
(b) Determine the present uses of solar energy, hydropower, and wind energy in your area. Discuss factors that affect the extent to which these *renewable* resources are utilized.

2.2D The careful study of objects moving under the influence of gravity often has had significant and far-reaching consequences. Write brief discussions of the following:

(a) When a falling object impacts the earth and comes to rest, it no longer has any kinetic or potential energy relative to the earth's surface. Has energy been destroyed? Has a violation of the energy conservation principle occurred? Discuss.
(b) Using the concepts of heat, work, and stored energy, discuss the events that occur when a meteor enters the earth's atmosphere and eventually falls to earth.

2.3D Write an interactive computer program that determines any one of p_1, V_1, p_2, V_2 or n for a polytropic process of a gas in a piston–cylinder assembly, given values of the other four quantities. The program should also calculate the work. Allow the user to specify English or SI units. Using your program, develop data and construct a plot of work, in kJ, versus polytropic exponent ranging from zero to 10 for $p_1 = 100$ kPa, $V_1 = 1$ m^3, and $p_2 = 300$ kPa.

2.4D James Prescott Joule is sometimes credited with providing the experimental basis for the first law of thermodynamics as we know it today. Write a paper summarizing the empirical basis for the first law. Provide a detailed description of Joule's work and give an overview of the principal experimental investigations that preceded his.

2.5D Obtain data from the engineering literature and develop a table of thermal conductivity values, in SI and English units, for common building materials such as wood, concrete, gypsum wallboard, and various types of insulation. Compare these values with those of ordinary metals, such as copper, steel, and aluminum. How sensitive are the thermal conductivities to temperature? Discuss.

2.6D The outside surfaces of small gasoline engines are often covered with *fins* that enhance the heat transfer between the hot surface and the surrounding air. Larger engines, like automobile engines, have a liquid coolant flowing through passages in the engine block. The coolant then passes through a finned-tube heat exchanger where air is blown over the tubes to accomplish the needed cooling. Considering appropriate data for heat transfer coefficients, engine size, and other design issues related to engine cooling, explain why some engines use liquid coolants and others do not.

2.7D Common vacuum-type *thermos bottles* can keep beverages hot or cold for many hours. Describe the construction of such bottles and explain the basic principles that make them effective.

CHAPTER 3

PROPERTIES OF A PURE, SIMPLE COMPRESSIBLE SUBSTANCE

To apply the energy balance to a system of interest requires knowledge of the properties of the system and how the properties are related. The objective of this chapter is to introduce property relations relevant to engineering thermodynamics. The presentation begins with the *state principle,* which is concerned with some general aspects of the relationships among the properties of systems. Consideration is then given over to the properties of *simple compressible systems* consisting of *pure* substances, for this case is especially important in engineering practice. As part of the presentation, several examples are provided that illustrate the use of the closed system energy balance introduced in Chap. 2, together with the property relations introduced in this chapter.

3.1 STATE PRINCIPLE

The state of a closed system at equilibrium is its condition as described by the values of its thermodynamic properties. From observation of many thermodynamic systems, it is known that not all of these properties are independent of one another, and the state can be uniquely determined by giving the values of the independent properties. Values for all other thermodynamic properties are determined from this independent subset. A general rule known as the *state principle* has been developed as a guide in determining the number of independent properties needed to specify the state of a system. In this section the state principle is considered for systems of specified mass and composition when the influences of overall motion and the earth's gravitational field are excluded.

Based on considerable empirical evidence, it can be concluded that there is one independent property for each way a system's energy can be varied independently. We saw in Chap. 2 that the energy of a closed system can be altered independently by heat or by work. Accordingly, an independent property can be associated with heat transfer as one way of varying the energy, and another independent property can be counted for each relevant way the energy can be changed through work. On the basis of experimental evidence, therefore, it can be stated that the number of independent properties for systems subject to the limitations stated above is one plus the number of *relevant* work interactions. *This is the state principle*. Experience also indicates that, when counting the number of relevant work interactions, it suffices to consider only those that can occur as the system undergoes quasiequilibrium processes.

The term *simple system* is applied when there is only *one* way the system energy can be significantly altered by work as the system undergoes a quasiequilibrium process. Therefore, counting one independent property for heat transfer and another for the single work mode gives a total of two independent properties needed to fix the state of a simple system. *This is the state principle for simple systems*. Although no system is ever truly simple, many systems can be modeled as simple systems for the purpose of thermodynamic analysis. The most important of these models for the applications considered in this book is the *simple compressible system*. Other types of simple systems are simple *elastic* systems and simple *magnetic* systems.

As suggested by the name, changes in volume can have a significant influence on the energy of simple compressible systems. The only mode of energy transfer by work that can occur as a simple compressible system undergoes *quasiequilibrium* processes is associated with volume change and is given by $\int p\, dV$. Since the influence of the earth's gravitational field is assumed negligible, pressure is uniform throughout the system. The simple compressible system model may seem highly restricted; however, experience shows that it is useful for a wide range of engineering applications even when electrical, magnetic, surface tension, and other effects are present to some degree.

Although any two independent properties are sufficient to fix the state of a simple compressible system at equilibrium, it is often convenient to think in terms of a unit mass of substance, 1 kg or 1 lb. The state principle can then be interpreted in terms of *intensive* properties. For example, temperature, T, and another intensive property such as specific volume, v, might be selected as the two independent properties. The state principle asserts that pressure, specific internal energy, and all other pertinent *intensive* properties could be determined as functions of T and v: $p = p(T, v)$, $u = u(T, v)$, and so on. The functional relations would be developed using experimental data and would depend explicitly on the particular chemical and physical makeup of the substances from which the system is composed. The development of such functions is discussed in Chap. 11.

This chapter is concerned with the thermodynamic properties of simple compressible systems consisting of *pure* substances. A pure substance is one of uniform and invariable chemical composition. Property relations for systems in which composition changes by chemical reaction are introduced in Chap. 13.

3.2 *p–v–T* RELATION

In this section we begin consideration of the properties of pure, simple compressible substances and the relations among these properties, starting with pressure, specific volume, and temperature. From experiment it is known that temperature and specific volume can be regarded as independent and pressure determined as a function of these two: $p = p(T, v)$. The graph of such a function is a *surface,* the p–v–T surface.

3.2.1 *p–v–T* Surface

Figure 3.1 is the p–v–T surface of a substance such as water that expands on freezing. Figure 3.2 is for a substance that contracts on freezing, and most sub-

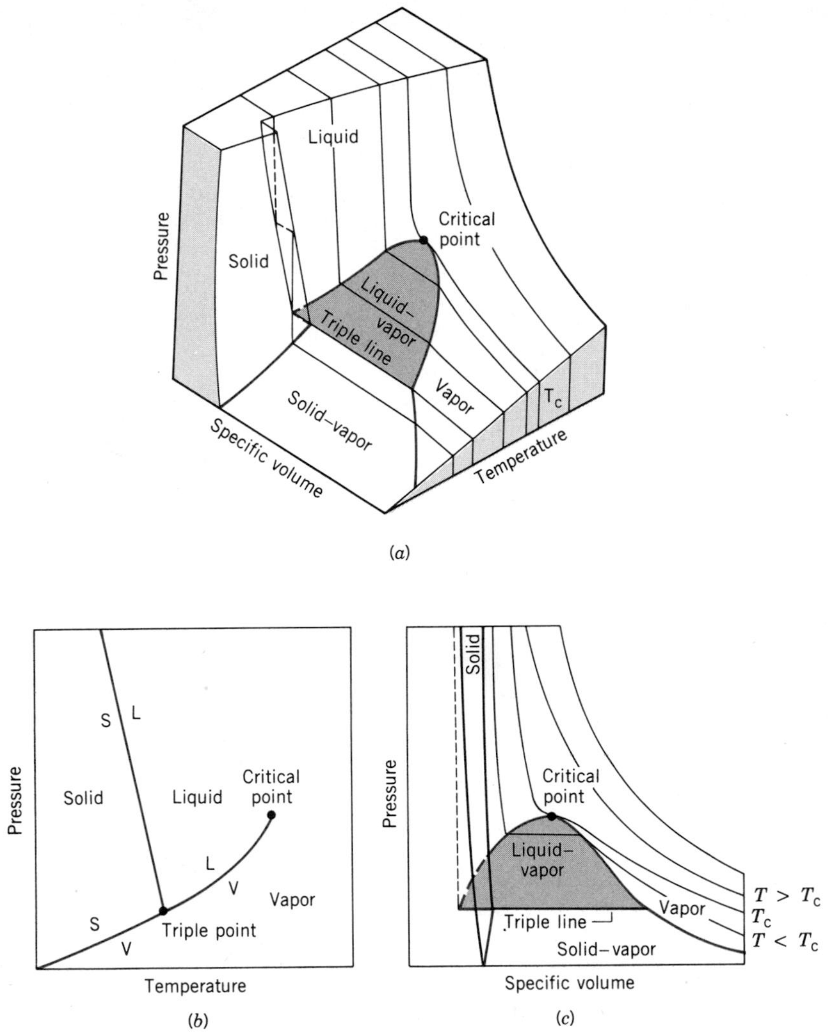

Figure 3.1 p–v–T surface and projections for a substance that expands on freezing. (*a*) Three-dimensional view. (*b*) Phase diagram. (*c*) p–v diagram.

stances exhibit this characteristic. The coordinates of a point on the p–v–T surfaces represent the values that pressure, specific volume, and temperature would assume when the substance is at equilibrium.

There are regions on the p–v–T surfaces of Figs. 3.1 and 3.2 labeled *solid, liquid,* and *vapor*. In these *single-phase* regions, the state is fixed by *any* two of the properties pressure, specific volume, and temperature, since all of these are independent when there is a single phase present. Located between the single-phase regions are *two-phase* regions where two phases exist in equilibrium: liquid–vapor, solid–liquid, and solid–vapor. Two phases can coexist during changes in phase such as vaporization, melting, and sublimation. Within the two-phase regions pressure and temperature are not independent; one cannot be changed without changing the other. In these regions the state cannot be fixed by temperature and pressure alone; however, the state can be fixed by specific volume and either pressure or temperature. Three phases can exist in equilibrium along the line labeled *triple line*.

A state at which a phase change begins or ends is called a *saturation* state. The dome-shaped region composed of the two-phase liquid–vapor states is called the *vapor dome*. The lines bordering the vapor dome are called saturated liquid and saturated vapor lines. At the top of the dome, where the saturated liquid and satu-

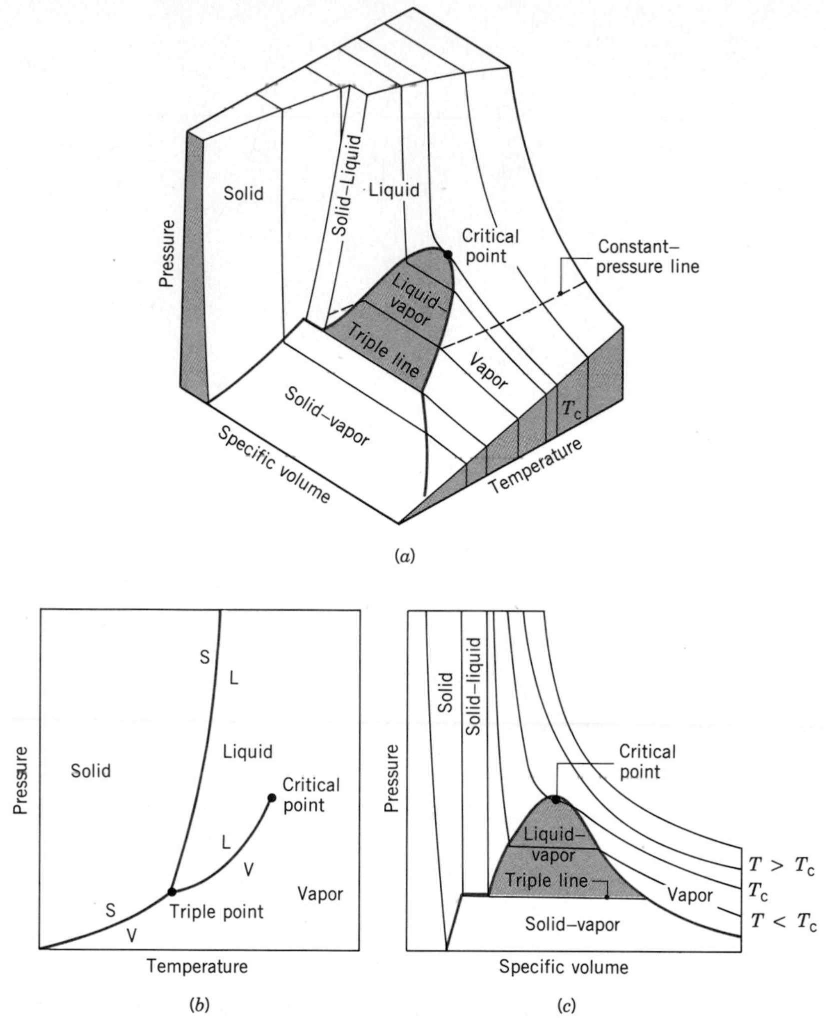

Figure 3.2 p–v–T surface and projections for a substance that contracts on freezing. (*a*) Three-dimensional view. (*b*) Phase diagram. (*c*) p–v diagram.

rated vapor lines meet, is the *critical point*. The *critical temperature* T_c of a pure substance is the maximum temperature at which liquid and vapor phases can coexist in equilibrium. The pressure at the critical point is called the *critical pressure*, p_c. The specific volume at this state is the *critical specific volume*. Values of the critical point properties for a number of substances are given in Tables A-1 located in the Appendix.

The three-dimensional p–v–T surface is useful for bringing out the general relationships among the three phases of matter normally under consideration. However, it is often more convenient to work with two-dimensional projections of the surface. These projections are considered next.

3.2.2 Projections of the p–v–T Surface

The Phase Diagram. If the p–v–T surface is projected onto the pressure–temperature plane, a property diagram known as a *phase* diagram results. As illustrated by Figs. 3.1*b* and 3.2*b*, when the surface is projected in this way, the two-phase *regions* reduce to *lines*. A point on any of these lines represents all two-phase mixtures at that particular temperature and pressure.

The term *saturation temperature* designates the temperature at which a phase change takes place at a given pressure, and this pressure is called the *saturation pressure* for the given temperature. It is apparent from the phase diagrams that for each saturation pressure there is a unique saturation temperature, and conversely.

The triple *line* of the three-dimensional p–v–T surface projects onto a *point* on the phase diagram. This is called the *triple point*. Recall that the triple point of water is used as a reference in defining temperature scales (Sec. 1.5). By agreement, the temperature *assigned* to the triple point of water is 273.16 K (491.69°R). The *measured* pressure at the triple point of water is 0.6113 kPa (0.00602 atm).

The line representing the two-phase solid–liquid region on the phase diagram slopes to the left for substances that expand on freezing and to the right for those that contract. Although a single solid phase region is shown on the phase diagrams of Figs. 3.1 and 3.2, solids can exist in different solid phases. For example, seven different crystalline forms have been identified for water as a solid (ice).

p–v Diagram. Projecting the p–v–T surface onto the pressure-specific volume plane results in a p–v diagram, as shown by Figs. 3.1*c* and 3.2*c*. The figures are labeled with terms that have already been introduced.

When solving problems, a sketch of the p–v diagram is frequently convenient. To facilitate the use of such a sketch, note the appearance of constant-temperature lines (isotherms). By inspection of Figs. 3.1*c* and 3.2*c*, it can be seen that for any specified temperature *less than* the critical temperature, pressure remains constant as the two-phase liquid–vapor region is traversed, but in the single-phase liquid and vapor regions the pressure decreases at fixed temperature as specific volume increases. For temperatures greater than or equal to the critical temperature, pressure decreases continuously at fixed temperature as specific volume increases. There is no passage across the two-phase liquid–vapor region. The critical isotherm passes through a point of inflection at the critical point and the slope is zero there.

T–v Diagram. Projecting the liquid, two-phase liquid–vapor, and vapor regions of the p–v–T surface onto the temperature-specific volume plane results in a T–v diagram as in Fig. 3.3. Since consistent patterns are revealed in the p–v–T behavior of all pure substances, Fig. 3.3 showing a T–v diagram for water can be regarded as representative.

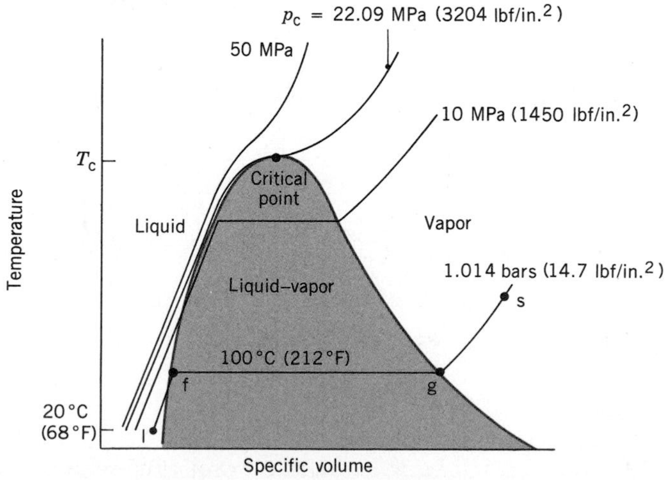

Figure 3.3 Sketch of a temperature-specific volume diagram for water showing the liquid, two-phase liquid–vapor, and vapor regions (not to scale).

As for the p–v diagram, a sketch of the T–v diagram is often convenient for problem solving. To facilitate the use of such a sketch, note the appearance of constant-pressure lines (isobars). For pressures *less than* the critical pressure, such as the 10 MPa (98.7 atm) isobar on Fig. 3.3, the pressure remains constant with temperature as the two-phase region is traversed. In the single-phase liquid and vapor regions the temperature increases at fixed pressure as the specific volume increases. For pressures greater than or equal to the critical pressure, such as the one marked 50 MPa on Fig. 3.3, temperature increases continuously at fixed pressure as the specific volume increases. There is no passage across the two-phase liquid–vapor region.

3.2.3 Phase Change

It is instructive to study some of the events that occur as a pure substance undergoes a phase change. To begin the discussion, consider a closed system consisting of a unit mass (1 kg or 1 lb) of liquid water at 20°C (68°F) contained within a piston–cylinder assembly, as illustrated in Fig. 3.4*a*. This state is represented by point 1 on Fig. 3.3. Suppose the water is slowly heated while its pressure is kept constant and uniform throughout at 1.014 bars (14.7 lbf/in.^{2}).

As the system is heated at constant pressure, the temperature increases considerably while the specific volume increases slightly. Eventually, the system is brought to the state represented by f on Fig. 3.3. This is the saturated liquid state corresponding to the specified pressure. For water at 1.014 bars (14.7 lbf/in.^{2}) the saturation temperature is 100°C (212°F). The liquid states along the line segment 1–f of Fig. 3.3 are sometimes referred to as *subcooled* liquid states because the temperature at these states is less than the saturation temperature at the given pressure. These states are also referred to as *compressed* liquid states because the pressure at each state is higher than the saturation pressure corresponding to the temperature at the state. The names liquid, subcooled liquid, and compressed liquid are used interchangeably.

When the system is at the saturated liquid state (state f of Fig. 3.3), additional heat transfer at fixed pressure results in the formation of vapor without any change in temperature but with a considerable increase in specific volume. As shown in Fig. 3.4*b*, the system would now consist of a two-phase liquid–vapor mixture. When a mixture of liquid and vapor exists in equilibrium, the liquid phase is a saturated liquid and the vapor phase is a saturated vapor. If the system is heated further until the last bit of liquid has vaporized, it is brought to point g on Fig. 3.3, the saturated vapor state. The intervening two-phase liquid–vapor *mixtures* can be distinguished from one another by the *quality,* an intensive property.

For a two-phase liquid–vapor mixture, the ratio of the mass of vapor present to the total mass of the mixture is its quality, x. In symbols,

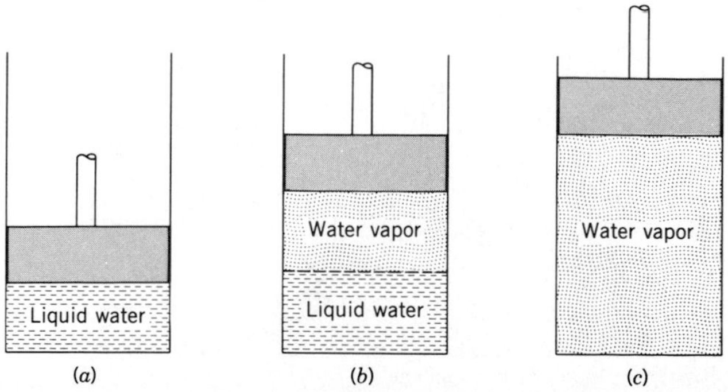

Figure 3.4 Illustration of constant-pressure change from liquid to vapor for water.

$$x = \frac{m_{\text{vapor}}}{m_{\text{liquid}} + m_{\text{vapor}}} \tag{3.1}$$

The value of the quality ranges from zero to unity: at saturated liquid states, $x = 0$, and at saturated vapor states, $x = 1.0$. Although defined as a ratio, the quality is frequently given as a percentage. Examples illustrating the use of quality are provided in Sec. 3.3. Similar parameters can be defined for two-phase solid–vapor and two-phase solid–liquid mixtures.

Let us return to a consideration of Figs. 3.3 and 3.4. When the system is at the saturated vapor state (state g on Fig. 3.3), further heating at fixed pressure results in increases in both temperature and specific volume. The condition of the system would now be as shown in Fig. 3.4*c*. The state labeled s on Fig. 3.3 is representative of the states that would be attained by further heating while keeping the pressure constant. A state such as s is often referred to as a *superheated* vapor state because the system would be at a temperature greater than the saturation temperature corresponding to the given pressure.

Consider next the same thought experiment at two other constant pressures labeled on Fig. 3.3, 10 MPa (1450 lbf/in.^{2}) and 22.09 MPa (3204 lbf/in.^{2}). The first of these pressures is less than the critical pressure of water and the second is the critical pressure. As before, let the system initially contain a liquid at 20°C (68°F). First, let us study the system if it were heated slowly at 10 MPa (1450 lbf/in.^{2}). At this pressure, vapor would form at a higher temperature than in the previous example, because the saturation pressure is higher (refer to Fig. 3.3). In addition, there would be somewhat less of an increase in specific volume from saturated liquid to vapor, as evidenced by the narrowing of the vapor dome. Apart from this, the general behavior would be the same as before. Next, consider the behavior of the system were it heated at the critical pressure. As seen by following the critical isobar on Fig. 3.3, there would be no change in phase from liquid to vapor. At all states there would be only one phase. Vaporization (and the inverse process of condensation) can occur only when the pressure is less than the critical pressure. When the temperature and pressure are above the respective critical values, the terms liquid and vapor tend to lose their significance. At such states a substance is more aptly described simply as a *fluid*.

Although the phase change from liquid to vapor (vaporization) is the one of principal interest in this book, it is also instructive to consider the phase changes from solid to liquid (melting) and from solid to vapor (sublimation). To study these transitions, consider a system consisting of a unit mass of ice at a temperature below the triple point temperature. Let us begin with the case where the system is at state *a* of Fig. 3.5, where the pressure is greater than the triple point pressure. Suppose the system is slowly heated while maintaining the pressure constant and uniform throughout. The temperature increases with heating until point *b* on Fig. 3.5 is attained. At this state the ice is a saturated solid. Additional heat transfer at fixed pressure results in the formation of liquid without any change in temperature. As the system is heated further, the ice continues to melt until eventually the last bit melts, and the system contains only saturated liquid. During the melting process the temperature and pressure remain constant. For most substances, the specific volume increases during melting, but for water the specific volume of the liquid is less than the specific volume of the solid. Further heating at fixed pressure results in an increase in temperature as the system is brought to point *c* on Fig. 3.5. Next, consider the case where the system is initially at state *a*′ of Fig. 3.5, where the

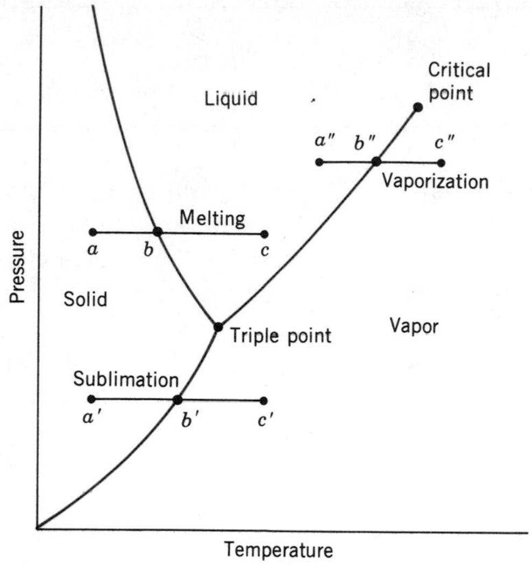

Figure 3.5 Phase diagram for water (not to scale).

pressure is less than the triple point pressure. In this case, if the system is heated at constant pressure it passes through the two-phase solid–vapor region into the vapor region along the line $a'–b'–c'$ shown on Fig. 3.5. The case of vaporization discussed previously is shown on Fig. 3.5 by the line $a''–b''–c''$.

3.3 THERMODYNAMIC PROPERTY DATA

Thermodynamic property data for engineering use appear in various forms, including tables, graphs, and equations. In addition, values for the thermodynamic properties of a growing number of substances are becoming available as microcomputer programs. The emphasis of the present section is on the use of *tables* of thermodynamic properties, since tables are commonly available for pure, simple compressible substances of engineering interest and the use of these tables is an important skill. Since tables for different substances are frequently set up in the same general format, the present discussion centers mainly on Tables A-2 through A-6 giving the properties of water; these are commonly referred to as the *steam tables*. Tables A-7 through A-9 for Refrigerant 12 and Tables A-10 through A-12 for Refrigerant 134a are used similarly, as are tables for other substances found in the engineering literature. Tables are provided in the appendix in SI and English units. Tables in English units are designated with a letter E. For example, the steam tables in English units are Tables A-2E through A-6E.

3.3.1 Pressure, Specific Volume, and Temperature

Vapor and Liquid Tables. The properties of water vapor are listed in Tables A-4 and of liquid water in Tables A-5. These are often referred to as the *superheated* vapor tables and *compressed* liquid tables, respectively. The sketch of the phase diagram shown in Fig. 3.6 brings out the structure of these tables. Since pressure and temperature are independent properties in the single-phase liquid and vapor regions, they can be used to fix the state in these regions. Accordingly, Tables A-4 and A-5 are set up to give values of several properties as functions of pressure and temperature. The first property listed is specific volume. The remaining properties are discussed in subsequent sections. To become familiar with the tables, verify the follow-

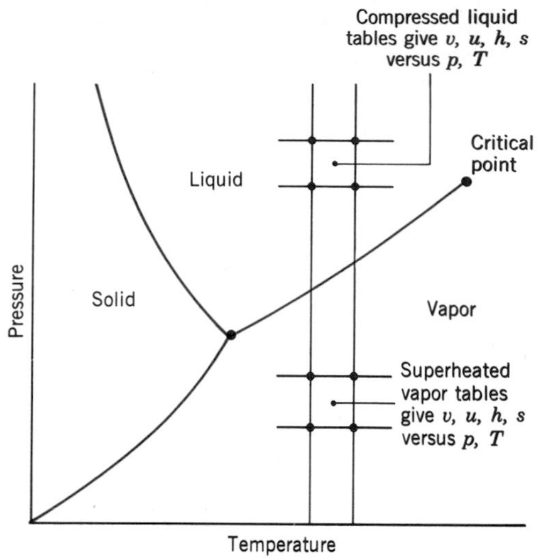

Figure 3.6 Sketch of the phase diagram for water used to discuss the structure of the superheated vapor and compressed liquid tables (not to scale).

ing. Table A-4 gives the specific volume of water vapor at 10.0 MPa and 600°C as 0.03837 m^3/kg. At 10.0 MPa and 100°C, Table A-5 gives the specific volume of liquid water as 1.0385×10^{-3} m^3/kg. Table A-4E gives the specific volume of water vapor at 500 lbf/in.^{2} and 600°F as 1.158 ft^3/lb. At 500 lbf/in.^{2} and 100°F, Table A-5E gives the specific volume of liquid water as 0.016106 ft^3/lb.

For each pressure listed, the values given in the superheated vapor table *begin* with the saturated vapor state and then proceed to higher temperatures. The value shown in parentheses after the pressure in the table heading is the corresponding saturation temperature. For example, in Table A-4 at a pressure of 0.10 MPa, the saturation temperature is listed as 99.63°C. In Table A-4E, the saturation temperature is listed as 212.0°F at a pressure of 14.7 lbf/in.^{2} The data in the compressed liquid table (Tables A-5) *end* with saturated liquid states. That is, for a given pressure the property values are given as the temperature increases to the saturation temperature. As in Tables A-4, the saturation temperature is shown in parentheses following the pressure.

The states encountered when solving problems often do not fall exactly on the grid of values provided by property tables. *Interpolation* between adjacent table entries then becomes necessary. Care should always be exercised when interpolating table values. The tables provided in the appendix are extracted from more extensive tables that are set up so that *linear* interpolation can be used with acceptable accuracy. Linear interpolation is assumed to remain valid when using the abridged tables of the text for the solved examples and end-of-chapter problems.

Saturation Tables. The saturation tables, Tables A-2 and A-3, list property values for the saturated liquid and vapor states. The property values at these states are denoted by the subscripts f and g, respectively. Table A-2 is called the *temperature table,* because temperatures are listed in the first column in convenient increments. The second column gives the corresponding saturation pressures. The next two columns give, respectively, the specific volume of saturated liquid, v_f, and the specific volume of saturated vapor, v_g. Table A-3 is called the *pressure table,* because pressures are listed in the first column in convenient increments. The corresponding saturation temperatures are given in the second column. The next two columns give v_f and v_g, respectively.

The specific volume of a two-phase liquid–vapor mixture can be determined by using the saturation tables and the definition of quality given by Eq. 3.1 as follows. The total volume of the mixture is the sum of the volumes of the liquid and vapor phases

$$V = V_{liq} + V_{vap}$$

Dividing by the total mass of the mixture, m, an *average* specific volume for the mixture is obtained

$$v = \frac{V}{m} = \frac{V_{liq}}{m} + \frac{V_{vap}}{m}$$

Since the liquid phase is a saturated liquid and the vapor phase is a saturated vapor, $V_{liq} = m_{liq}v_f$ and $V_{vap} = m_{vap}v_g$, so

$$v = \left(\frac{m_{liq}}{m}\right) v_f + \left(\frac{m_{vap}}{m}\right) v_g$$

Introducing the definition of quality, $x = m_{vap}/m$, and noting that $m_{liq}/m = 1 - x$, the above expression becomes

$$v = (1 - x)v_f + xv_g = v_f + x(v_g - v_f) \tag{3.2}$$

The increase in specific volume on vaporization $(v_g - v_f)$ is often denoted by v_{fg}.

To illustrate the use of Eq. 3.2, consider a system consisting of a two-phase liquid–vapor mixture of water at 100°C and a quality of 0.9. From Table A-2 at 100°C, $v_f = 1.0435 \times 10^{-3}$ m³/kg and $v_g = 1.673$ m³/kg. The specific volume of the mixture is

$$v = v_f + x(v_g - v_f) = 1.0435 \times 10^{-3} + (0.9)(1.673 - 1.0435 \times 10^{-3})$$
$$= 1.506 \text{ m}^3/\text{kg}$$

Similarly, the specific volume of a two-phase liquid–vapor mixture of water at 212°F and a quality of 0.9 is

$$v = v_f + x(v_g - v_f) = 0.01672 + (0.9)(26.80 - 0.01672)$$
$$= 24.12 \text{ ft}^3/\text{lb}$$

where the v_f and v_g values are obtained from Table A-2E.

To facilitate locating states in the tables, it is often convenient to use values from the saturation tables together with a sketch of a T–v or p–v diagram. For example, if the specific volume v and temperature T are known, refer to the appropriate temperature table, Table A-2 or A-2E, and determine the values of v_f and v_g. A T–v diagram illustrating these data is given in Fig. 3.7. If the given specific volume falls between

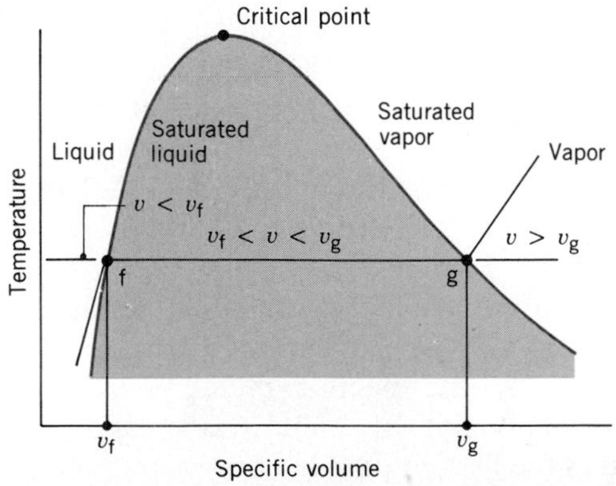

Figure 3.7 Sketch of a temperature-specific volume diagram for water used to discuss the use of the steam tables.

the two specific volume values v_f and v_g, the system consists of a two-phase liquid–vapor mixture, and the pressure is the saturation pressure corresponding to the given temperature. The quality for the system can be found by solving Eq. 3.2. If the given specific volume is greater than v_g, the state is in the superheated vapor region. Then, by interpolating in Table A-4 or A-4E, the pressure and other properties listed can be determined. If the given specific volume is less than v_f, Table A-5 or A-5E would be used to determine the pressure and other properties.

EXAMPLE 3.1

PROBLEM

Water vapor initially at 3.0 MPa and 300°C (state 1) is contained within a piston–cylinder assembly. The water is cooled at constant volume until its temperature reaches 200°C (state 2). The water is then compressed isothermally to a state where the pressure is 2.5 MPa (state 3).

(a) Locate states 1, 2, and 3 on T–v and p–v diagrams.

(b) Determine the specific volume at states 1, 2, and 3 in m³/kg and the quality at state 2.

SOLUTION

Known: Water vapor at 3.0 MPa and 300°C (state 1) is cooled at constant volume to 200°C (state 2) and then is compressed isothermally to a state where the pressure is 2.5 MPa.

Find:

(a) Locate states 1, 2, and 3 on T–v and p–v diagrams.

(b) Determine the specific volume at each of these states and the quality at state 2.

Schematic and Given Data:

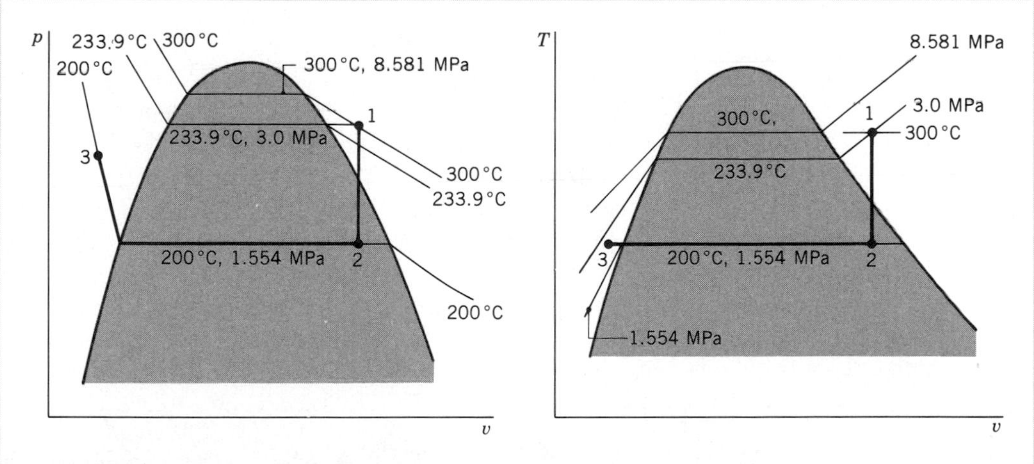

Figure E3.1

Assumptions:

1. The water vapor is a closed system.
2. States 1, 2, and 3 are equilibrium states.

Analysis: To locate state 1, observe that p_1 is *less* than the saturation pressure corresponding to $T_1 = 300°C$ (8.581 MPa). As shown by the accompanying property diagrams, this state is located in the superheated vapor region. Alternatively, note that T_1 is *greater* than the saturation temperature corresponding to $p_1 = 3.0$ MPa (233.9°C). Using this, we reach the same conclusion: State 1 is in the superheated vapor region. Interpolating in Table A-4 at 3.0 MPa between 280°C and 320°C, gives $v_1 = 0.0811$ m³/kg.

To locate state 2, use $v_2 = v_1 = 0.0811$ m³/kg. From Table A-2 at $T_2 = 200°C$, $v_f = 1.1565 \times 10^{-3}$ and $v_g = 0.1274$ m³/kg. Therefore, since $v_f < v_2 < v_g$, state 2 is in the two-phase liquid–vapor region. Using Eq. 3.2, the quality at state 2 is

$$x_2 = \frac{0.0811 - 1.1565 \times 10^{-3}}{0.1274 - 1.1565 \times 10^{-3}} = 0.633$$

To locate state 3, observe that p_3 is *greater* than the saturation pressure corresponding to $T_3 = 200°C$ (1.554 MPa). As shown by the accompanying property diagrams, this state is located in the liquid region. Alternatively, note that the temperature at state 3 is *less* than the saturation temperature corresponding to $p_3 = 2.5$ MPa (224.0°C). Using this, we reach the same conclusion: State 3 is in the liquid region. Turning to Table A-5, we get $v_3 = 1.1555 \times 10^{-3}$ m³/kg.

Comment: Sketches of property diagrams locating states of interest are frequently helpful in solving problems in thermodynamics.

3.3.2 Specific Internal Energy and Enthalpy

In many thermodynamic analyses the sum of the internal energy U and the product of pressure p and volume V appears. Because the sum $U + pV$ occurs so frequently in subsequent discussions, it is convenient to give the combination a name, *enthalpy*, and a distinct symbol, H. By definition

$$H = U + pV \tag{3.3}$$

Since U, p, and V are all properties, this combination is also a property. Enthalpy can be expressed on a unit mass basis

$$h = u + pv \tag{3.4}$$

and per mole

$$\bar{h} = \bar{u} + p\bar{v} \tag{3.5}$$

Units for enthalpy are the same as those for internal energy.

The property tables introduced in Sec. 3.3.1 giving pressure, specific volume, and temperature also provide values of specific internal energy u, enthalpy h, and entropy s. Use of these tables to evaluate u and h is described in the present section; the consideration of entropy is deferred until it is introduced in Chap. 6.

Data for specific internal energy u and enthalpy h are retrieved from the property tables in the same way as for specific volume. For saturation states, the values of u_f and u_g, as well as h_f and h_g, are tabulated versus both saturation pressure and saturation temperature. The specific internal energy for a two-phase liquid–vapor mixture is calculated for a given quality in the same way the specific volume is calculated

$$u = (1 - x)u_f + xu_g = u_f + x(u_g - u_f) \tag{3.6}$$

The increase in specific internal energy on vaporization $(u_g - u_f)$ is often denoted by u_{fg}. Similarly, the specific enthalpy for a two-phase liquid–vapor mixture is given in terms of the quality by

$$h = (1 - x)h_f + xh_g = h_f + x(h_g - h_f) \tag{3.7}$$

The increase in enthalpy during vaporization ($h_g - h_f$) is often tabulated for convenience under the heading h_{fg}.

To illustrate the use of Eqs. 3.6 and 3.7, we will determine the specific enthalpy of Refrigerant 12 when its temperature is 12°C and its specific internal energy is 111.07 kJ/kg. Referring to Table A-7, the given internal energy value falls between u_f and u_g at 12°C, so the state is a two-phase liquid–vapor mixture. The quality of the mixture is found by using Eq. 3.6 and data from Table A-7 as follows:

$$x = \frac{u - u_f}{u_g - u_f} = \frac{111.07 - 46.93}{175.2 - 46.93} = 0.5$$

Then, with the values from Table A-7, Eq. 3.7 gives

$$\begin{aligned} h &= (1 - x)h_f + xh_g \\ &= (1 - 0.5)(47.26) + 0.5(192.56) = 119.91 \text{ kJ/kg} \end{aligned}$$

As a similar example in English units, let us determine the specific enthalpy of Refrigerant 12 when its temperature is 10°F and its specific internal energy is 40.89 Btu/lb. Since this value falls between u_f and u_g at 10°F (Table A-7E), the state is a two-phase liquid–vapor mixture. The quality of the mixture is

$$x = \frac{u - u_f}{u_g - u_f} = \frac{40.89 - 10.62}{71.15 - 10.62} = 0.5$$

Then, with enthalpy values from Table A-7E

$$\begin{aligned} h &= (1 - x)h_f + h_g = (0.5)(10.68) + (0.5)(78.33) \\ &= 44.5 \text{ Btu/lb} \end{aligned}$$

In the superheated vapor tables for water and Refrigerant 12, u and h are tabulated along with v as functions of temperature and pressure. As an illustration, let us evaluate T, v, and h for water at 0.10 MPa and a specific internal energy of 2537.3 kJ/kg. Turning to Table A-3, note that the given value of u is greater than u_g at 0.1 MPa (u_g = 2506.1 kJ/kg). This suggests that the state lies in the superheated vapor region. From Table A-4 it is found that T = 120°C, v = 1.793 m³/kg, and h = 2716.6 kJ/kg. Alternatively, h and u are related by the definition of h

$$\begin{aligned} h &= u + pv \\ &= 2537.3 \frac{\text{kJ}}{\text{kg}} + \left(10^5 \frac{\text{N}}{\text{m}^2}\right)\left(1.793 \frac{\text{m}^3}{\text{kg}}\right)\left(\frac{1 \text{ kJ}}{10^3 \text{ N}\cdot\text{m}}\right) \\ &= 2537.3 + 179.3 = 2716.6 \text{ kJ/kg} \end{aligned}$$

To illustrate the use of the superheat tables in English units, consider water at a state fixed by a pressure equal to 14.7 lbf/in.² and a temperature of 250°F. From Table A-4E, v = 28.42 ft³/lb, u = 1091.5 Btu/lb, and h = 1168.8 Btu/lb. As above, h may be calculated from u, or conversely. Thus

$$\begin{aligned} h &= u + pv \\ &= 1091.5 \frac{\text{Btu}}{\text{lb}} + \left(14.7 \frac{\text{lbf}}{\text{in.}^2}\right)\left(\frac{144 \text{ in.}^2}{1 \text{ ft}^2}\right)\left(28.42 \frac{\text{ft}^3}{\text{lb}}\right)\left(\frac{1 \text{ Btu}}{778 \text{ ft}\cdot\text{lbf}}\right) \\ &= 1091.5 + 77.3 = 1168.8 \text{ Btu/lb} \end{aligned}$$

Specific internal energy and enthalpy data for liquid states of water are presented in Tables A-5. The format of these tables is the same as that of the superheated vapor tables considered previously. Accordingly, property values for liquid states are retrieved in the same manner as those of vapor states.

For water, Tables A-6 give the equilibrium properties of saturated solid and saturated vapor. The first column lists the temperature, and the second column gives the corresponding saturation pressure. These states are at pressures and temperatures *below* those at the triple point. The next two columns give the specific volume of saturated solid, v_i, and saturated vapor, v_g, respectively. The table also provides the specific internal energy, enthalpy, and entropy values for the saturated solid and the saturated vapor at each of the temperatures listed.

Reference States and Reference Values. The values of u, h, and s given in the property tables are not obtained by direct measurement but are calculated from other data that can be more readily determined experimentally. The computational procedures require use of the second law of thermodynamics, so consideration of these procedures is deferred to Chap. 11 after the second law has been introduced. However, because u, h, and s are calculated, the matter of reference states and reference values becomes important and is considered briefly in the following paragraphs.

When applying the energy balance, it is *differences* in internal, kinetic, and potential energy between two states that are important, and *not* the values of these energy quantities at each of the two states. Consider the case of potential energy as an illustration. The numerical value of potential energy measured from the surface of the earth is different from the value relative to the top of a flagpole at the same location. However, the difference in potential energy between any two elevations is precisely the same regardless of the datum selected, because the datum cancels in the calculation. Similarly, values can be assigned to specific internal energy and enthalpy relative to arbitrary reference values at arbitrary reference states. The use of values of a particular property determined relative to an arbitrary reference is unambiguous as long as the calculations being performed involve only differences in that property, for then the reference value cancels. When chemical reactions take place among the substances under consideration, special attention must be given to the matter of reference states and values, however. A discussion of how property values are assigned when analyzing reactive systems is given in Chap. 13.

The tabular values of u and h for water and Refrigerants 12 and 134a provided in the appendix are relative to the following reference states and values. For water, the reference state is saturated liquid at 0.01°C (32.02°F). At this state, the specific internal energy is set to zero. Values of the specific enthalpy are calculated from $h = u + pv$, using the tabulated values for p, v, and u. For the refrigerants, the reference state is saturated liquid at −40°C (−40°F for the tables with English units). At this reference state the specific enthalpy is set to zero. Values of specific internal energy are calculated from $u = h - pv$ by using the tabulated values for p, v, and h. Notice in Table A-7 that this leads to a negative value for internal energy at the reference state, which emphasizes that it is not the numerical values assigned to u and h at a given state that are important but their *differences* between states. The values assigned to particular states change if the reference state or reference values change, but the differences remain the same.

EXAMPLE 3.2

PROBLEM

A well-insulated rigid tank having a volume of 10 ft^3 contains saturated water vapor at 212°F. The water is rapidly stirred until the pressure is 20 lbf/in.^{2} Determine the temperature at the final state, in °F, and the work during the process, in Btu.

SOLUTION

Known: By rapid stirring, water vapor in a well-insulated rigid tank is brought from the saturated vapor state at 212°F to a pressure of 20 lbf/in.^{2}

Find: Determine the temperature at the final state and the work.

Schematic and Given Data:

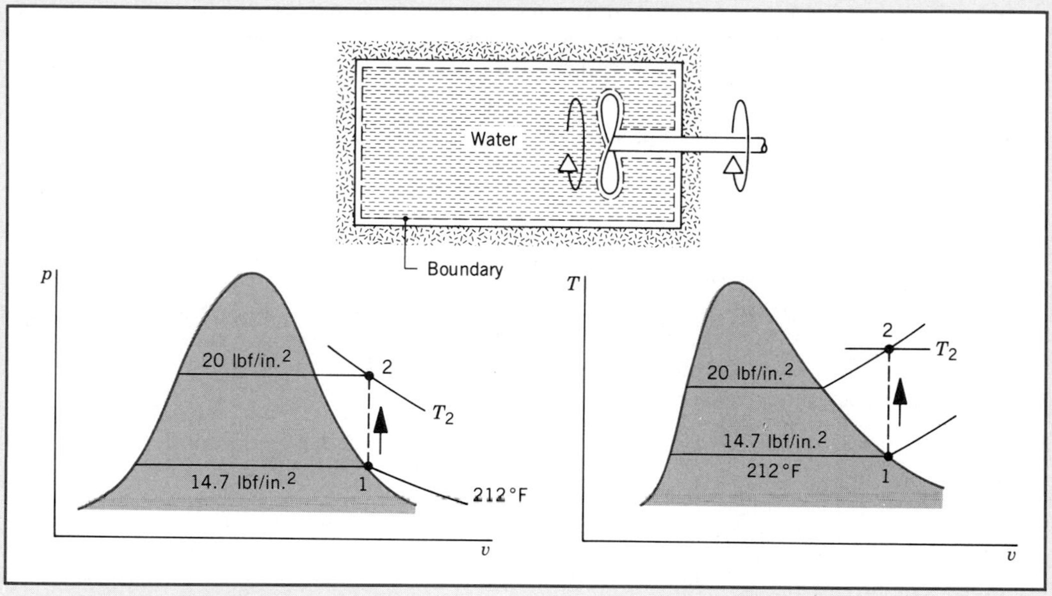

Figure E3.2

Assumptions:

1. The water is a closed system.
2. The initial and final states are at equilibrium. There is no net change in kinetic or potential energy.
3. There is no heat transfer with the surroundings.
4. The tank volume remains constant.

Analysis: To determine the final equilibrium state, the values of two independent intensive properties are required. One of these is pressure, p_2 = 20 lbf/in.^{2}, and the other is the specific volume: $v_2 = v_1$. The initial and final specific volumes are equal because the total mass and total volume are unchanged in the process. The initial and final states are located on the accompanying T–v and p–v diagrams.

From Table A-2E, $v_1 = v_g(212°\text{F}) = 26.80\ \text{ft}^3/\text{lb}$, $u_1 = u_g(212°\text{F}) = 1077.6\ \text{Btu/lb}$. By using $v_2 = v_1$ and interpolating in Table A-4E at p_2 = 20 lbf/in.^{2}

$$T_2 = 445°\text{F}, \qquad u_2 = 1161.6\ \text{Btu/lb}$$

Next, with assumptions 2 and 3 an energy balance for the system reduces to

$$\Delta U + \cancelto{0}{\Delta KE} + \cancelto{0}{\Delta PE} = \cancelto{0}{Q} - W$$

On rearrangement

$$W = -(U_2 - U_1) = -m(u_2 - u_1)$$

To evaluate W requires the system mass. This can be determined from the volume and specific volume

$$m = \frac{V}{v_1} = \left(\frac{10\ \text{ft}^3}{26.8\ \text{ft}^3/\text{lb}}\right) = 0.373\ \text{lb}$$

Finally, by inserting values into the expression for W

$$W = -(0.373 \text{ lb})(1161.6 - 1077.6) \text{ Btu/lb} = -31.3 \text{ Btu}$$

where the minus sign signifies that the energy transfer by work is to the system.

Comments:

1. Although the initial and final states are equilibrium states, the intervening states are not at equilibrium. To emphasize this, the process has been indicated on the T–v and p–v diagrams by a dashed line. Solid lines on property diagrams are reserved for processes that pass through equilibrium states only (quasiequilibrium processes).
2. The analysis illustrates the importance of carefully sketched property diagrams as an adjunct to problem solving.

EXAMPLE 3.3

PROBLEM

Water contained in a piston–cylinder assembly undergoes two processes in series from an initial state where the pressure is 10.0 bars and the temperature is 400°C.

Process 1–2: The water is cooled as it is compressed at constant pressure to the saturated vapor state at 10.0 bars.

Process 2–3: The water is cooled at constant volume to 150°C.

(a) Sketch both processes on T–v and p–v diagrams.

(b) For the overall process determine the work, in kJ/kg.

(c) For the overall process determine the heat transfer, in kJ/kg.

SOLUTION

Known: Water contained in a piston–cylinder assembly undergoes two processes: It is cooled and compressed while keeping the pressure constant, and then cooled at constant volume.

Find: Sketch both processes on T–v and p–v diagrams. Determine the net work and net heat transfer for the overall process per unit of mass contained within the piston–cylinder assembly.

Schematic and Given Data:

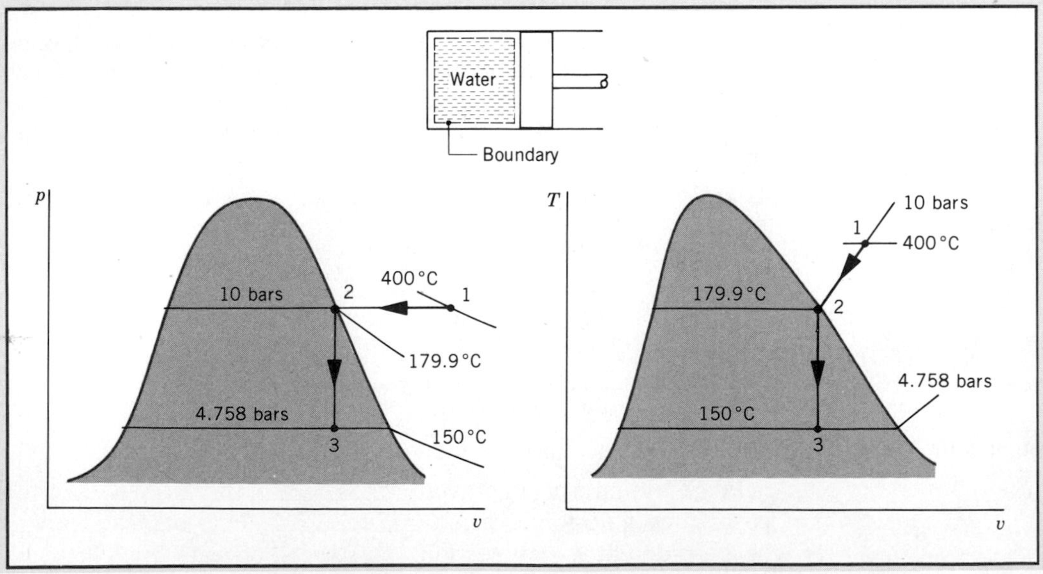

Figure E3.3

Assumptions:

1. The water is a closed system.
2. The piston is the only work mode.
3. There are no changes in kinetic or potential energy.

Analysis:

(a) The accompanying T–v and p–v diagrams show the two processes. Since the temperature at state 1, T_1 = 400°C, is greater than the saturation temperature corresponding to p_1 = 10.0 bars: 179.9°C, state 1 is located in the superheat region.

(b) Since the piston is the only work mechanism

$$W = \int_1^3 p\,dV = \int_1^2 p\,dV + \cancelto{0}{\int_2^3 p\,dV}$$

The second integral vanishes because the volume is constant in Process 2–3. Dividing by the mass and noting that the pressure is constant for Process 1–2

$$\frac{W}{m} = p(v_2 - v_1)$$

The specific volume at state 1 is found from Table A-4 using p_1 = 10.0 bars and T_1 = 400°C: $v_1 = 0.3066\ \text{m}^3/\text{kg}$. Also, u_1 = 2957.3 kJ/kg. The specific volume at state 2 is the saturated vapor value at 10.0 bars: $v_2 = 0.1944\ \text{m}^3/\text{kg}$, from Table A-3. Hence

$$\frac{W}{m} = \left(10 \times 10^5 \frac{\text{N}}{\text{m}^2}\right)(0.1944 - 0.3066)\left(\frac{\text{m}^3}{\text{kg}}\right)\left(\frac{1\ \text{kJ}}{10^3\ \text{N}\cdot\text{m}}\right)$$

$$= -112.2\ \text{kJ/kg}$$

The minus sign indicates that work is done *on* the water vapor by the piston.

(c) An energy balance for the *overall* process reduces to

$$m(u_3 - u_1) = Q - W$$

By rearranging

$$\frac{Q}{m} = (u_3 - u_1) + \frac{W}{m}$$

To evaluate the heat transfer requires u_3, the specific internal energy at state 3. Since T_3 is given and $v_3 = v_2$, two independent intensive properties are known that together fix state 3. To find u_3, first solve for the quality

$$x_3 = \frac{v_3 - v_{f3}}{v_{g3} - v_{f3}} = \frac{0.1944 - 1.0905 \times 10^{-3}}{0.3928 - 1.0905 \times 10^{-3}} = 0.494$$

where v_{f3} and v_{g3} are from Table A-2 at 150°C. Then

$$u_3 = u_{f3} + x_3(u_{g3} - u_{f3}) = 631.68 + 0.494(2559.5 - 631.98)$$
$$= 1583.9\ \text{kJ/kg}$$

where u_{f3} and u_{g3} are from Table A-2 at 150°C.

Substituting values into the energy balance

$$\frac{Q}{m} = 1583.9 - 2957.3 + (-112.2) = -1485.6\ \text{kJ/kg}$$

The minus sign shows that energy is transferred *out* by heat transfer.

3.3.3 Specific Heats c_v and c_p

Several properties related to internal energy are important in thermodynamics. One of these is the property enthalpy introduced in Sec. 3.3.2. Two others, known as *specific heats,* are considered in this section. The specific heats are particularly useful for thermodynamic calculations involving the *ideal gas model* introduced in Sec. 3.5.

The intensive properties c_v and c_p are defined for pure, simple compressible substances as partial derivatives of the functions $u(T, v)$ and $h(T, p)$, respectively

$$c_v = \left.\frac{\partial u}{\partial T}\right)_v \tag{3.8}$$

$$c_p = \left.\frac{\partial h}{\partial T}\right)_p \tag{3.9}$$

where the subscripts v and p denote, respectively, the variables held fixed during differentiation. Values for c_v and c_p can be obtained via statistical mechanics using *spectroscopic* measurements. They can also be determined macroscopically through exacting property measurements. Since u and h can be expressed either on a unit mass basis or per mole, values of the specific heats can be similarly expressed. SI units are kJ/kg·K or kJ/kmol·K. English units are Btu/lb·°R or Btu/lbmol·°R.

The property k, called the specific heat ratio, is simply the ratio

$$k = \frac{c_p}{c_v} \tag{3.10}$$

The properties c_v and c_p are referred to as *specific heats* (or *heat capacities*) because under certain *special conditions* they relate the temperature change of a system to the amount of energy added by heat transfer. However, it is generally preferable to think of c_v and c_p in terms of their definitions, Eqs. 3.8 and 3.9, and not with reference to this limited interpretation involving heat transfer.

In general, c_v is a function of v and T (or p and T), and c_p depends on both p and T (or v and T). Figure 3.8 shows how c_p for water vapor varies as a function of temperature and pressure. The vapor phases of other substances exhibit similar behavior. Note that the figure gives the variation of c_p with temperature in the limit as pressure tends to zero. In this limit, c_p increases with increasing temperature, which is a characteristic exhibited by other gases as well. We will refer again to such *zero-pressure* values for c_v and c_p in Sec. 3.5.1.

Specific heat data are available for common gases, liquids, and solids. Data for gases are introduced in Sec. 3.5 as a part of the discussion of the ideal gas model. Specific heat values for some common liquids and solids are introduced in Sec. 3.3.5 as a part of the discussion of the incompressible substance model.

3.3.4 Approximations for Liquids Using Saturated Liquid Data

Approximate values for v, u, and h at liquid states can be obtained using saturated liquid data. To illustrate, refer to the compressed liquid tables, Tables A-5. These tables show that the specific volume and specific internal energy change very little with pressure *at a fixed temperature*. Because the values of v and u vary only gradually as pressure changes at fixed temperature, the following approximations are reasonable for many engineering calculations:

$$v(T, p) \approx v_f(T) \tag{3.11}$$

$$u(T, p) \approx u_f(T) \tag{3.12}$$

That is, for liquids v and u may be evaluated at the saturated liquid state corresponding to the temperature at the given state.

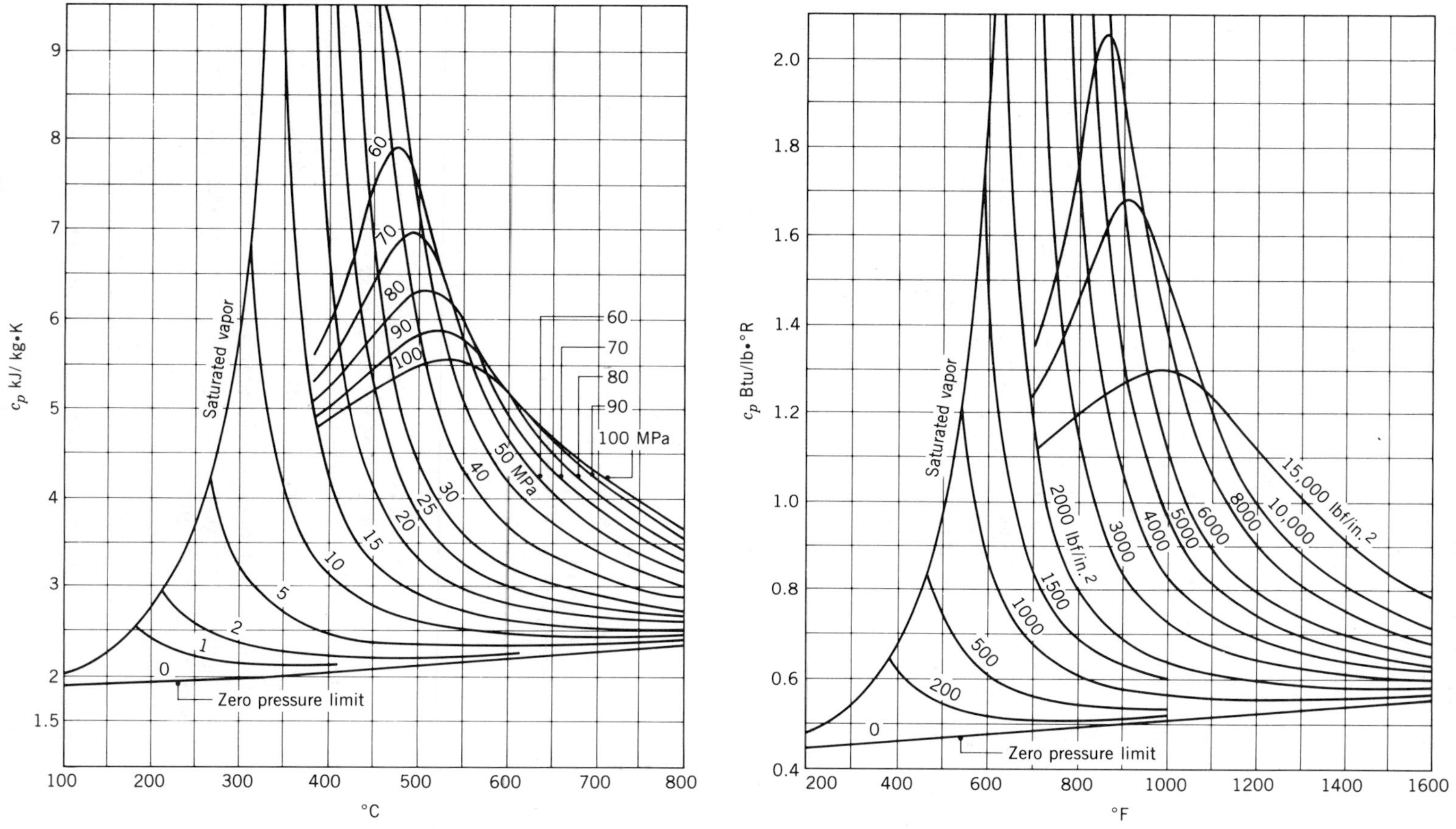

Figure 3.8 c_p of water vapor as a function of temperature and pressure.

An approximate value of h at liquid states can be obtained by using Eqs. 3.11 and 3.12 in the definition $h = u + pv$; thus

$$h(T, p) \approx u_f(T) + pv_f(T) \tag{3.13}$$

Equation 3.13 can be expressed alternatively as

$$h(T, p) \approx h_f(T) + v_f(T)[p - p_{sat}(T)] \tag{3.14}$$

where p_{sat} denotes the saturation pressure at the given temperature. The derivation is left as an exercise. When the contribution of the last term of Eq. 3.14 is small, the specific enthalpy can be approximated by the saturated liquid value, as for v and u. That is, $h(T, p) \approx h_f(T)$.

Although the approximations given here have been presented with reference to liquid water, they also provide plausible approximations for other substances *when the only liquid data available are for saturated liquid states*. In this text, compressed liquid data are presented only for water. When greater accuracy is required than provided by these approximations, other data sources should be consulted for more complete property compilations for the substance under consideration.

3.3.5 Incompressible Substance Model

As noted above, there are regions where the specific volume of liquid water varies little and the specific internal energy varies mainly with temperature. The same general behavior is exhibited by the liquid phases of other substances and by solids. The approximations of Sec. 3.3.4 are based on these observations, as is the *incompressible substance model* under present consideration.

To simplify evaluations involving liquids or solids, the specific volume (density) is often assumed to be constant and the specific internal energy assumed to vary only with temperature. A substance idealized in this way is called *incompressible*.

Since the specific internal energy of a substance modeled as incompressible depends only on temperature, the specific heat c_v is also a function of temperature alone

$$c_v(T) = \frac{du}{dT} \quad \text{(incompressible)} \tag{3.15}$$

This is expressed as an ordinary derivative because u depends only on T.

Although the specific volume is constant and internal energy depends on temperature only, enthalpy varies with both pressure and temperature according to

$$h(T, p) = u(T) + pv \quad \text{(incompressible)} \tag{3.16}$$

For a substance modeled as incompressible, the specific heats c_v and c_p are equal. This is seen by differentiating Eq. 3.16 with respect to temperature while holding pressure fixed to obtain

$$\left.\frac{\partial h}{\partial T}\right)_p = \frac{du}{dT}$$

The left side of this expression is c_p by definition (Eq. 3.9), so using Eq. 3.15 on the right side gives

$$c_p = c_v \quad \text{(incompressible)} \tag{3.17}$$

Thus, for an incompressible substance it is unnecessary to distinguish between c_p and c_v, and both can be represented by the same symbol, c. Specific heats of some common liquids and solids are given versus temperature in Tables A-13. Over limited temperature intervals the variation of c with temperature can be small. In such instances, the specific heat c can be treated as constant without a serious loss of accuracy.

Using Eqs. 3.15 and 3.16, the changes in specific internal energy and specific enthalpy between two states are given, respectively, by

$$u_2 - u_1 = \int_{T_1}^{T_2} c(T)\, dT \qquad \text{(incompressible)} \tag{3.18}$$

$$h_2 - h_1 = u_2 - u_1 + v(p_2 - p_1)$$

$$= \int_{T_1}^{T_2} c(T)\, dT + v(p_2 - p_1) \qquad \text{(incompressible)} \tag{3.19}$$

If the specific heat c is taken as constant, Eqs. 3.18 and 3.19 become, respectively,

$$\left.\begin{aligned} u_2 - u_1 &= c(T_2 - T_1) \\ h_2 - h_1 &= c(T_2 - T_1) + v(p_2 - p_1) \end{aligned}\right\} \qquad \text{(incompressible, constant } c\text{)} \tag{3.20}$$

3.4 p–v–T RELATION FOR GASES

Systems of engineering interest often involve substances in the vapor (gas) phase. Accordingly, evaluation of the properties of gases is important. The object of the present section is to gain a better understanding of the relationship among pressure, specific volume, and temperature of gases. This is important not only as a basis for analyses involving gases but also for the discussions of the next section, where the *ideal gas model* is introduced. The current presentation is conducted in terms of the *compressibility factor* and begins with the introduction of the *universal gas constant.*

3.4.1 Universal Gas Constant

Let a gas be confined in a cylinder by a piston and the entire assembly held at a constant temperature. The piston can be moved to various positions so that a series of equilibrium states at constant temperature can be visited. Suppose the pressure and specific volume are measured at each state and the value of the ratio $p\bar{v}/T$ ($\bar{v}$ is volume per mole) determined. These ratios can then be plotted versus pressure at constant temperature. The results for several temperatures are sketched in Fig. 3.9.

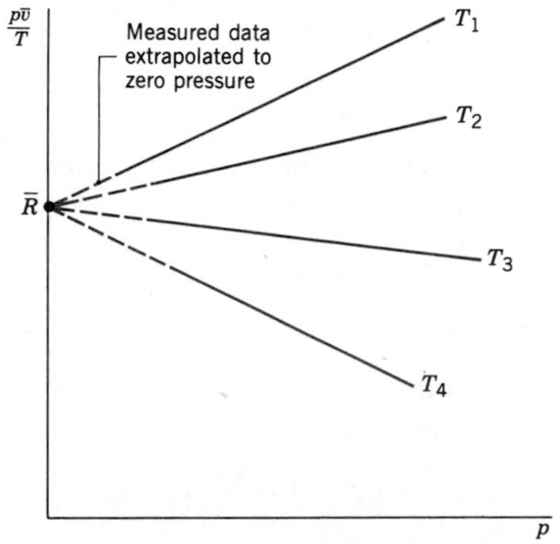

Figure 3.9 Sketch of $p\bar{v}/T$ versus pressure for a gas at several specified values of temperature.

When the ratios are extrapolated to zero pressure, *precisely the same limiting value is obtained* for each curve. That is,

$$\lim_{p \to 0} \frac{p\bar{v}}{T} = \bar{R} \tag{3.21}$$

where $\bar{R}$ denotes the common limit for all temperatures. If this procedure were repeated for other gases, it would be found in every instance that the limit of the ratio $p\bar{v}/T$ as p tends to zero at fixed temperature is the same, namely $\bar{R}$. Since the same limiting value is exhibited by all gases, $\bar{R}$ is called the *universal gas constant*. Its value as determined experimentally is

$$\bar{R} = \begin{cases} 8.314 \text{ kJ/kmol} \cdot \text{K} \\ 1.986 \text{ Btu/lbmol} \cdot {}^\circ\text{R} \\ 1545 \text{ ft} \cdot \text{lbf/lbmol} \cdot {}^\circ\text{R} \end{cases} \tag{3.22}$$

Having introduced the universal gas constant, we turn next to the compressibility factor.

3.4.2 Compressibility Factor

The dimensionless ratio $p\bar{v}/\bar{R}T$ is called the *compressibility factor* and is denoted by Z. That is,

$$Z = \frac{p\bar{v}}{\bar{R}T} \tag{3.23}$$

As illustrated by subsequent calculations, when values for p, $\bar{v}$, $\bar{R}$, and T are used in consistent units, Z is unitless.

Since $\bar{v} = Mv$, where M is the atomic or molecular weight, the compressibility factor can be expressed alternatively as

$$Z = \frac{pv}{RT} \tag{3.24}$$

where

$$R = \frac{\bar{R}}{M} \tag{3.25}$$

R is a constant for the particular gas whose molecular weight is M. Alternative units for R are kJ/kg·K, Btu/lb·°R, and ft·lbf/lb·°R.

Equation 3.21 can be expressed in terms of the compressibility factor as

$$\lim_{p \to 0} Z = 1 \tag{3.26}$$

That is, the compressibility factor Z tends to unity as pressure tends to zero at fixed temperature. This can be illustrated by reference to Fig. 3.10, which shows Z for hydrogen plotted versus pressure at a number of different temperatures.

Considering the curves of Fig. 3.10, it is reasonable to think that the variation with pressure and temperature of the compressibility factor for gases might be expressible as an equation, at least for certain intervals of p and T. Two expressions can be written that enjoy a theoretical basis. One gives the compressibility factor as an infinite series expansion in pressure:

$$Z = 1 + \hat{B}(T)p + \hat{C}(T)p^2 + \hat{D}(T)p^3 + \cdots \tag{3.27}$$

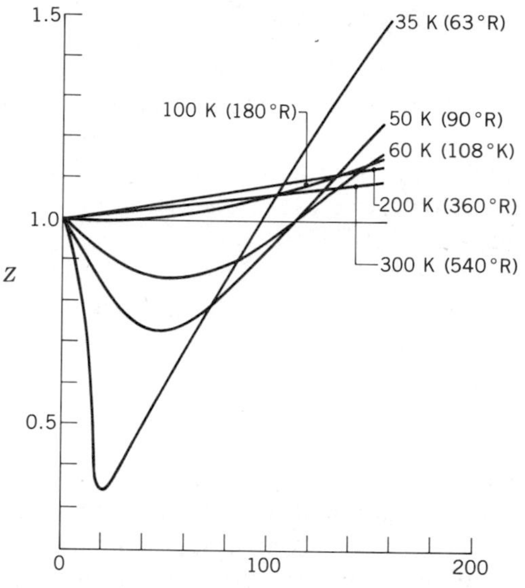

Figure 3.10 Variation of the compressibility factor of hydrogen with pressure at constant temperature.

where the coefficients $\hat{B}$, $\hat{C}$, $\hat{D}$, . . . depend on temperature only. The dots in Eq. 3.27 represent higher-order terms. The other is a series form entirely analogous to Eq. 3.27 but expressed in terms of $1/\bar{v}$ instead of p

$$Z = 1 + \frac{B(T)}{\bar{v}} + \frac{C(T)}{\bar{v}^2} + \frac{D(T)}{\bar{v}^3} + \cdots \tag{3.28}$$

Equations 3.27 and 3.28 are known as *virial expansions,* and the coefficients $\hat{B}$, $\hat{C}$, $\hat{D}$, . . . and B, C, D, . . . are called *virial coefficients*. The word *virial* stems from the Latin word for force. In the present usage it is force interactions among molecules that are intended.

The virial expansions can be derived by the methods of statistical mechanics, and physical significance can be attributed to the coefficients: $B/\bar{v}$ accounts for two-molecule interactions, $C/\bar{v}^2$ accounts for three-molecule interactions, etc. In principle, the virial coefficients can be calculated by using expressions from statistical mechanics derived from consideration of the force fields around the molecules of a gas. The virial coefficients also can be determined from experimental p–v–T data. The virial expansions are used in Sec. 11.1 as a point of departure for the further study of analytical representations of the p–v–T relationship of gases known as *equations of state*.

The virial expansions and the physical significance attributed to the terms making up the expansions can be used to clarify the nature of gas behavior in the limit as pressure tends to zero at fixed temperature. From Eq. 3.27 it is seen that if pressure decreases at fixed temperature, the terms $\hat{B}p$, $\hat{C}p^2$, etc. accounting for various molecular interactions tend to decrease, suggesting that the force interactions become weaker under these circumstances. In the limit as pressure approaches zero, these terms vanish, and the equation reduces to $Z = 1$ in accordance with Eq. 3.26. Similarly, since volume increases when the pressure decreases at fixed temperature, the terms $B/\bar{v}$, $C/\bar{v}^2$, etc. of Eq. 3.28 also vanish in the limit, giving $Z = 1$ when the force interactions between molecules are no longer significant.

3.4.3 Generalized Compressibility Chart

Figure 3.10 is a chart giving the compressibility factor for hydrogen versus pressure at specified values of temperature. Similar charts have been prepared for other gases. When these charts are studied, they are found to be *qualitatively* similar. Further study shows that when the coordinates are suitably modified, the curves for several different gases coincide closely when plotted together on the same coordinate axes, and so *quantitative* similarity also can be achieved. This is referred to as the *principle of corresponding states*. In one such approach, the compressibility factor Z is plotted versus a *reduced* pressure p_R and a *reduced* temperature T_R, defined as

$$p_R = \frac{p}{p_c} \quad \text{and} \quad T_R = \frac{T}{T_c} \tag{3.29}$$

where p_c and T_c denote the critical pressure and temperature, respectively. This results in a *generalized* compressibility chart of the form $Z = f(p_R, T_R)$. Figure 3.11 shows experimental data for 10 different gases on a chart of this type. The solid lines denoting reduced isotherms represent the best curves fitted to the data.

A generalized chart more suitable for problem solving than Fig. 3.11 is given in the appendix as Figs. A-1, A-2, and A-3. In Fig. A-1, p_R ranges from 0 to 1.0; in Fig. A-2, p_R ranges from 0 to 10.0; and in Fig. A-3, p_R ranges from 10.0 to 40.0. At any one temperature, the deviation of observed values from those of the generalized chart increases with pressure. However, for the 30 gases used in developing the chart, the deviation is *at most* on the order of 5% and for most ranges is much less.[1] From Figs. A-1 and A-2 it can be seen that the value of Z tends to unity for all temperatures as pressure tends to zero, in accord with Eq. 3.26. Figure A-3 shows that Z also approaches unity for all pressures at very high temperatures.

Values of specific volume are included on the generalized chart through the variable v'_R, called the *pseudoreduced* specific volume, defined by

$$v'_R = \frac{\bar{v}}{\bar{R}T_c/p_c} \tag{3.30}$$

For correlation purposes, the pseudoreduced specific volume has been found to be preferable to the *reduced* specific volume $v_R = \bar{v}/\bar{v}_c$, where $\bar{v}_c$ is the critical specific volume.

Using the critical pressure and critical temperature of a substance of interest, the generalized chart can be entered with various pairs of the variables T_R, p_R, and v'_R: (T_R, p_R), (p_R, v'_R), or (T_R, v'_R). Tables A-1 list the critical constants for several substances. The merit of the generalized chart for evaluating p, v, and T for gases is simplicity coupled with accuracy. However, the generalized compressibility chart should not be used as a substitute for accurate p–v–T data for a given substance as provided by a table or computer software. The chart is mainly useful for obtaining reasonable estimates in the absence of more accurate data.

The principle of corresponding states underlying the generalized compressibility chart is only approximate in nature. This can be brought out by noting that for all substances the chart gives a common value of about 0.27 for the compressibility factor at the critical point. However, the critical compressibility factor for different substances actually varies from 0.23 to 0.33, so the chart is inaccurate in the vicinity

[1] To determine Z for hydrogen, helium, and neon above a T_R of 5, the reduced temperature and pressure should be calculated using $T_R = T/(T_c + 8)$ and $p_R = p/(p_c + 8)$, where temperatures are in K and pressures are in atm.

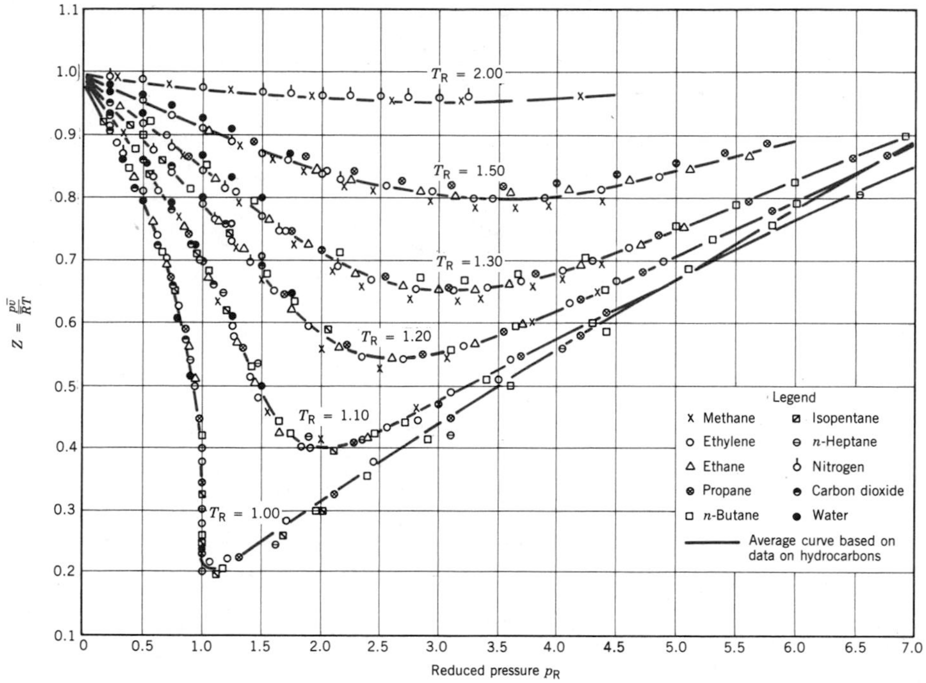

Figure 3.11 Generalized compressibility chart for various gases.

of the critical point. This source of inaccuracy can be removed by restricting the correlation to substances having essentially the same Z_c values, which is equivalent to including the critical compressibility factor in a principle of corresponding states of the form

$$Z = f(T_R, p_R, Z_c) \tag{3.31}$$

Principles of corresponding states also have been proposed using variables other than the ones introduced in the present discussion.

EXAMPLE 3.4

PROBLEM

A fixed amount of water vapor, initially at 20 MPa, 520°C, is cooled at constant volume until its temperature reaches 400°C. Using the compressibility chart, determine

(a) the specific volume of the water vapor in m^3/kg at the initial state.

(b) the pressure in MPa at the final state.

Compare the results of parts (a) and (b) with the values obtained from the superheated vapor table, Table A-4.

SOLUTION

Known: Water vapor is cooled at constant volume from 20 MPa, 520 to 400°C.

Find: Use the compressibility chart and the superheated vapor table to determine the specific volume and final pressure and compare the results.

Schematic and Given Data:

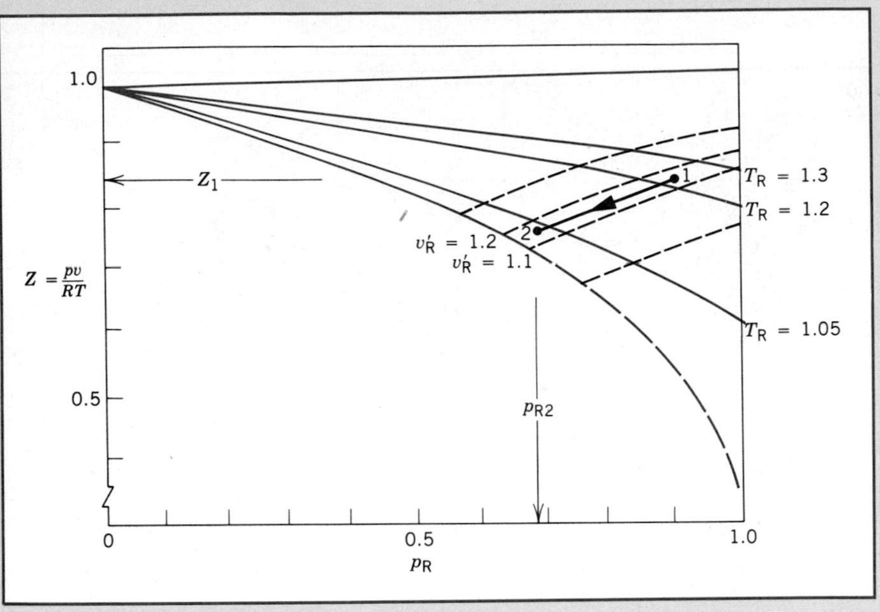

Figure E3.4

Assumptions:

1. The water is a closed system.
2. The initial and final states are at equilibrium.

Analysis:

(a) From Table A-1, $T_c = 647.3$ K and $p_c = 22.09$ MPa for water. Thus

$$T_{R1} = \frac{793}{647.3} = 1.23, \qquad p_{R1} = \frac{20}{22.09} = 0.91$$

With these values for the reduced temperature and reduced pressure, the value of Z obtained from Fig. A-1 is approximately 0.83. Since $Z = pv/RT$, the specific volume at state 1 can be determined as follows:

$$v_1 = Z_1 \frac{RT_1}{p_1} = 0.83 \frac{\bar{R}T_1}{Mp_1}$$

$$= 0.83 \left(\frac{8314 \dfrac{\text{N}\cdot\text{m}}{\text{kmol}\cdot\text{K}}}{18.02 \dfrac{\text{kg}}{\text{kmol}}} \right) \left(\frac{793 \text{ K}}{20 \times 10^6 \dfrac{\text{N}}{\text{m}^2}} \right) = 0.0152 \text{ m}^3/\text{kg}$$

The molecular weight of water is from Table A-1.

Turning to Table A-4, the specific volume at the initial state is 0.01551 m^3/kg, which is in good agreement with the compressibility chart value, as expected.

(b) Since the water vapor is cooled at constant specific volume, the process takes place at constant v_R', as shown on the accompanying figure. Using the value for specific volume determined in part (a)

$$v_R' = \frac{vp_c}{RT_c} = \frac{\left(0.0152 \dfrac{\text{m}^3}{\text{kg}}\right)\left(22.09 \times 10^6 \dfrac{\text{N}}{\text{m}^2}\right)}{\left(\dfrac{8314}{18.02} \dfrac{\text{N}\cdot\text{m}}{\text{kg}\cdot\text{K}}\right)(647.3 \text{ K})} = 1.12$$

At state 2

$$T_{R2} = \frac{673}{647.3} = 1.04$$

Locating the point on the compressibility chart where $v_R' = 1.12$ and $T_R = 1.04$, the corresponding value for p_R is about 0.69. Accordingly

$$p_2 = p_c(p_{R2}) = (22.09 \text{ MPa})(0.69) = 15.24 \text{ MPa}$$

Interpolating in the superheated vapor tables gives $p_2 = 15.16$ MPa. As before, the compressibility chart value is in good agreement with the table value.

Comments:

1. *Absolute* temperature and *absolute* pressure must be used in evaluating the compressibility factor Z, the reduced temperature T_R, and reduced pressure p_R.
2. When values for p, v, R, and T are used in consistent units, the value of Z is unitless.

3.5 IDEAL GAS MODEL

By considering the generalized compressibility data introduced in Sec. 3.4, it can be concluded that at many states the value of the compressibility factor Z is approximately unity. This is the case when the reduced pressure p_R is less than about 0.05. The compressibility factor is also approximately unity when the reduced temperature T_R is greater than 15. For T_R in the interval from about 2 to 3, Z is closely equal to unity over a wide range of reduced pressure. Hence, the pressure, specific volume, and temperature relationship for gases at many states is given approximately by

$$Z = \frac{pv}{RT} = 1$$

or

$$pv = RT \tag{3.32}$$

Equation 3.32 is referred to as the *ideal gas* equation of state. Under appropriate conditions the p–v–T relationship for gases can be accurately represented by this simple expression.

Alternative forms of the same basic relationship among pressure, specific volume, and temperature are obtained as follows. With $v = V/m$, Eq. 3.32 can be expressed as

$$pV = mRT \tag{3.33}$$

In addition, since $v = \bar{v}/M$ and $R = \bar{R}/M$, where M is the atomic or molecular weight, Eq. 3.32 can be expressed as

$$p\bar{v} = \bar{R}T \tag{3.34}$$

or, with $\bar{v} = V/n$, as

$$pV = n\bar{R}T \tag{3.35}$$

For any gas whose equation of state is given *exactly* by $pv = RT$, the specific internal energy depends on temperature *only*. This conclusion is demonstrated formally in Sec. 11.4. It is also supported by experimental observations, beginning with the work of Joule, who showed in 1843 that the internal energy of air at low density depends primarily on temperature. Further motivation from the microscopic viewpoint is provided shortly. The specific enthalpy of a gas described by $pv = RT$ also depends on temperature only, as can be shown by combining the definition of enthalpy, $h = u + pv$, with $u = u(T)$ and the ideal gas equation of state to obtain $h = u(T) + RT$. Taken together, these specifications constitute the *ideal gas model,* summarized as follows:

$$\text{Ideal gas model:} \quad \begin{cases} pv = RT & (3.32) \\ u = u(T) & (3.36) \\ h = h(T) = u(T) + RT & (3.37) \end{cases}$$

The specific internal energy and enthalpy of gases generally depend on two independent properties, not just temperature as presumed by the ideal gas model. Moreover, the ideal gas equation of state does not provide an acceptable approximation at all states. Accordingly, whether the ideal gas model is used depends on the error acceptable in a given calculation. Still, gases often do *approach* ideal gas behavior, and a particularly simplified description is obtained with the ideal gas model.

To verify that a gas can be modeled as an ideal gas, the states of interest can be located on a compressibility chart to determine how well $Z = 1$ is satisfied. As shown in subsequent discussions, other tabular or graphical property data also can be used to determine the suitability of the ideal gas model. To expedite the solutions of many subsequent examples and end-of-chapter problems, we indicate in the problem statements that the ideal gas model should be used when the suitability of this assumption could readily be verified by reference to a compressibility chart or other data. If this simplifying assumption is not made explicit in a problem statement, the ideal gas model should be invoked only after its appropriateness has been checked.

A picture of the dependence of the internal energy of gases on temperature at low density can be obtained with reference to the discussion of the virial expansions in Sec. 3.4. As $p \to 0$ or $v \to \infty$, the force interactions between molecules of a gas become weaker, and the virial expansions approach $Z = 1$ in the limit. The study of gases from the microscopic point of view shows that the dependence of the internal energy of a gas on pressure, or specific volume, at a specified temperature arises primarily because of molecular interactions. Accordingly, as the density of a gas decreases at fixed temperature, there comes a point where the effects of intermolecular forces are minimal. The internal energy is then determined principally by the temperature. From the microscopic point of view, the ideal gas model adheres to several idealizations: The gas consists of molecules that are in random motion and obey the laws of mechanics; the total number of molecules is large, but the volume of the molecules is a negligibly small fraction of the volume occupied by the gas; and no appreciable forces act on the molecules except during collisions.

EXAMPLE 3.5

PROBLEM

One pound of air undergoes a thermodynamic cycle consisting of three processes.

Process 1–2: constant specific volume
Process 2–3: constant-temperature expansion
Process 3–1: constant-pressure compression

At state 1, the temperature is 540°R, and the pressure is 1 atm. At state 2, the pressure is 2 atm. Employing the ideal gas equation of state,

(a) sketch the cycle on p–v coordinates.

(b) determine the temperature at state 2, in °R.

(c) determine the specific volume at state 3, in ft^3/lb.

SOLUTION

Known: Air executes a thermodynamic cycle consisting of three processes: Process 1–2, v = constant; Process 2–3, T = constant; Process 3–1, p = constant. Values are given for T_1, p_1, and p_2.

Find: Sketch the cycle on p–v coordinates and determine T_2 and v_3.

Schematic and Given Data:

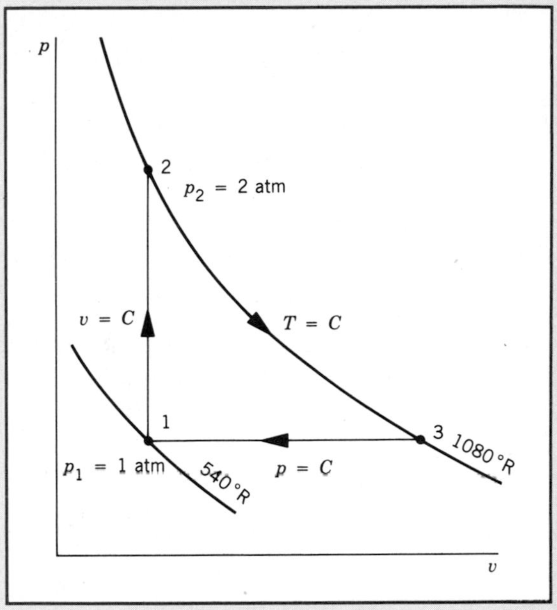

Figure E3.5

Assumptions:

1. The air is a closed system.
2. The air behaves as an ideal gas.

Analysis:

(a) The cycle is shown on p–v coordinates in the accompanying figure. Note that since $p = RT/v$ and temperature is constant, the variation of p with v for the process from 2 to 3 is nonlinear.

(b) Using $pv = RT$, the temperature at state 2 is

$$T_2 = p_2 v_2 / R$$

To obtain the specific volume v_2 required by this relationship, note that $v_2 = v_1$, so

$$v_2 = RT_1 / p_1$$

Combining these two results gives

$$T_2 = \frac{p_2}{p_1} T_1 = \left(\frac{2 \text{ atm}}{1 \text{ atm}}\right)(540\text{°R}) = 1080\text{°R}$$

(c) Since $pv = RT$, the specific volume at state 3 is

$$v_3 = RT_3 / p_3$$

Noting that $T_3 = T_2$, $p_3 = p_1$, and $R = \bar{R}/M$

$$v_3 = \frac{\bar{R} T_2}{M p_1}$$

$$= \left(\frac{1545 \dfrac{\text{ft} \cdot \text{lbf}}{\text{lbmol} \cdot {}^\circ\text{R}}}{28.97 \dfrac{\text{lb}}{\text{lbmol}}} \right) \frac{(1080^\circ\text{R})}{(14.7 \text{ lbf/in.}^2)(144 \text{ in.}^2/\text{ft}^2)}$$

$$= 27.2 \text{ ft}^3/\text{lb}$$

where the molecular weight of air is from Table A-1E.

Comments:

1. Table A-1E gives $p_c = 37.2$ atm, $T_c = 239°\text{R}$ for air. Therefore, $p_{R2} = 0.054$, $T_{R2} = 4.52$. Referring to Fig. A-1, the value of the compressibility factor at this state is $Z \approx 1$. The same conclusion results when states 1 and 3 are checked. Accordingly, $pv = RT$ adequately describes the p–v–T relation for the gas at these states.
2. Carefully note that the equation of state $pv = RT$ requires the use of *absolute* temperature T and *absolute* pressure p.

3.5.1 Internal Energy, Enthalpy, and Specific Heats of Ideal Gases

For a gas obeying the ideal gas model, specific internal energy depends only on temperature. Hence, the specific heat c_v, defined by Eq. 3.8, is also a function of temperature alone. That is,

$$c_v(T) = \frac{du}{dT} \quad \text{(ideal gas)} \tag{3.38}$$

This is expressed as an ordinary derivative because u depends only on T.

By separating variables in Eq. 3.38

$$du = c_v(T)\, dT \quad \text{(ideal gas)} \tag{3.39}$$

On integration,

$$u(T_2) - u(T_1) = \int_{T_1}^{T_2} c_v(T)\, dT \quad \text{(ideal gas)} \tag{3.40}$$

Similarly, for a gas obeying the ideal gas model, the specific enthalpy depends only on temperature, so the specific heat c_p, defined by Eq. 3.9, is also a function of temperature alone. That is

$$c_p(T) = \frac{dh}{dT} \quad \text{(ideal gas)} \tag{3.41}$$

Separating variables in Eq. 3.41

$$dh = c_p(T)\, dT \tag{3.42}$$

On integration

$$h(T_2) - h(T_1) = \int_{T_1}^{T_2} c_p(T)\, dT \quad \text{(ideal gas)} \tag{3.43}$$

An important relationship between the ideal gas specific heats can be developed by differentiating Eq. 3.37 with respect to temperature

$$\frac{dh}{dT} = \frac{du}{dT} + R$$

and introducing Eqs. 3.38 and 3.41 to obtain

$$c_p(T) = c_v(T) + R \qquad \text{(ideal gas)} \tag{3.44}$$

On a molar basis, this is written as

$$\bar{c}_p(T) = \bar{c}_v(T) + \bar{R} \qquad \text{(ideal gas)} \tag{3.45}$$

Although each of the two ideal gas specific heats is a function of temperature, Eqs. 3.44 and 3.45 show that the specific heats differ by just a constant: the gas constant. Knowledge of either specific heat for a particular gas allows the other to be calculated by using only the gas constant. The above equations also show that $c_p > c_v$ and $\bar{c}_p > \bar{c}_v$, respectively.

For an ideal gas, the specific heat ratio, k, is also a function of temperature only

$$k = \frac{c_p(T)}{c_v(T)} \qquad \text{(ideal gas)} \tag{3.46}$$

Since $c_p > c_v$, it follows that $k > 1$. Combining Eqs. 3.44 and 3.46 results in

$$c_p(T) = \frac{kR}{k-1} \tag{3.47a}$$

$$\text{(ideal gas)}$$

$$c_v(T) = \frac{R}{k-1} \tag{3.47b}$$

Similar expressions can be written for the specific heats on a molar basis, with R being replaced by $\bar{R}$.

The foregoing expressions require the ideal gas specific heats as functions of temperature. These relations are available for gases of practical interest in various forms, including graphs, tables, and equations. Figure 3.12 illustrates the variation of

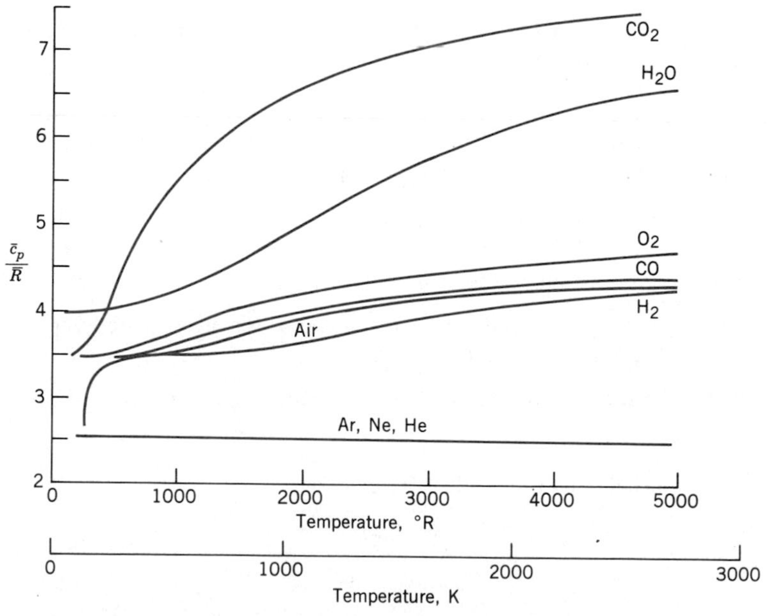

Figure 3.12 Variation of $\bar{c}_p/\bar{R}$ with temperature for a number of gases modeled as ideal gases.

$\bar{c}_p$ (molar basis) with temperature for a number of common gases. In the range of temperature shown, $\bar{c}_p$ increases with temperature for all gases, except for the monatonic gases Ar, Ne, and He. For these, $\bar{c}_p$ is closely constant at the value predicted by kinetic theory: $\bar{c}_p = \frac{5}{2}\bar{R}$. Tabular specific heat data for selected gases are presented versus temperature in Tables A-14. Specific heat data are also available in equation form. Several alternative forms of such equations are found in the engineering literature. An equation that is relatively easy to integrate is the polynomial form

$$\frac{\bar{c}_p}{\bar{R}} = \alpha + \beta T + \gamma T^2 + \delta T^3 + \varepsilon T^4 \tag{3.48}$$

Values of the constants α, β, γ, δ, and ε are listed in Tables A-15 for several gases in the temperature range 300 to 1000 K (540 to 1800°R).

The source of the ideal gas specific heat data introduced above is experiment. Specific heats can be determined macroscopically from painstaking property measurements. In the limit as pressure tends to zero, the properties of a gas tend to merge into those of its ideal gas model, so macroscopically determined specific heats of a gas extrapolated to very low pressures may be called either *zero-pressure* specific heats or *ideal gas* specific heats. Although zero-pressure specific heats can be obtained by extrapolating macroscopically determined experimental data, this is rarely done nowadays because ideal gas specific heats can readily be calculated with expressions from statistical mechanics by using *spectral* data, which can be obtained experimentally with precision. The determination of ideal gas specific heats is one of the important areas where the microscopic approach contributes significantly to the application of thermodynamics.

EXAMPLE 3.6

PROBLEM

Determine the change in specific enthalpy, in kJ/kg, for water vapor from a state where $T_1 = 400$ K and $p_1 = 0.1$ MPa to a state where $T_2 = 900$ K and $p_2 = 0.5$ MPa, by means of

(a) the superheated vapor table for water.

(b) integration using the ideal gas specific heat given by Eq. 3.48 together with coefficient data from Table A-15.

(c) Repeat parts (a) and (b) for a final pressure of 10.0 MPa.

SOLUTION

Known: Water vapor undergoes processes between specified states.

Find: Determine the change in specific enthalpy by means of (a) the superheated vapor tables and (b) integration using Eq. 3.48 together with coefficient data from Table A-15.

Assumptions: In part (b), the ideal gas model applies for steam.

Analysis:

(a) Interpolating in the superheated vapor table, Table A-4

$$h_1 = 2730.5 \text{ kJ/kg}, \qquad h_2 = 3762.2 \text{ kJ/kg}$$

Therefore

$$h_2 - h_1 = 1031.7 \text{ kJ/kg}$$

(b) Inserting the expression for $\bar{c}_p(T)$ given by Eq. 3.48 into Eq. 3.43 and integrating with respect to temperature

$$h_2 - h_1 = \frac{\bar{R}}{M}\int_{T_1}^{T_2} (\alpha + \beta T + \gamma T^2 + \delta T^3 + \varepsilon T^4)\, dT$$

$$= \frac{\bar{R}}{M}\left[\alpha(T_2 - T_1) + \frac{\beta}{2}(T_2^2 - T_1^2) + \frac{\gamma}{3}(T_2^3 - T_1^3) + \frac{\delta}{4}(T_2^4 - T_1^4) + \frac{\varepsilon}{5}(T_2^5 - T_1^5)\right]$$

where the molecular weight M has been introduced to obtain the result on a unit mass basis. With values for the constants from Table A-15

$$h_2 - h_1 = \frac{8.314}{18.02}\left\{4.070(900 - 400) - \frac{1.108}{2(10)^3}[(900)^2 - (400)^2]\right.$$
$$+ \frac{4.152}{3(10)^6}[(900)^3 - (400)^3] - \frac{2.964}{4(10)^9}[(900)^4 - (400)^4]$$
$$\left. + \frac{0.807}{5(10)^{12}}[(900)^5 - (400)^5]\right\}$$
$$= 1025.0 \text{ kJ/kg}$$

This value agrees closely with that determined in part (a).

(c) Consider next the evaluation of the change in specific enthalpy from a state where $T_1 = 400$ K and $p_1 = 0.1$ MPa to a state where $T_2 = 900$ K and $p_2 = 10.0$ MPa. As in part (a), Table A-4 gives $h_1 = 2730.53$ kJ/kg. At $T_2 = 900$ K and $p_2 = 10.0$ MPa, Table A-4 gives $h_2 = 3691.72$ kJ/kg. Thus

$$h_2 - h_1 = 961.19 \text{ kJ/kg}$$

When the ideal gas model is assumed, the change in specific enthalpy is the same as that found in part (b): $h_2 - h_1 = 1025.0$ kJ/kg. This value is unchanged because the ideal gas model specific enthalpy depends on temperature only, and the temperatures at states 1 and 2 of part (c) are the same as those in parts (a) and (b). When $p_2 = 10.0$ MPa, the value for Δh determined with the ideal gas model is 7% greater than the value obtained from the steam tables.

Comment: By reference to the compressibility chart, Fig. A-1, it can be verified that when $p_2 = 0.5$ MPa, both end states are in the region of the vapor phase of water where the ideal gas model provides an accurate description. When $p_2 = 10.0$ MPa, the reduced pressure and temperature at the final state are $p_{R2} = 0.453$ and $T_{R2} = 1.39$, and the compressibility factor is closely $Z_2 = 0.96$.

3.5.2 Ideal Gas Tables

For a number of common gases, evaluations of specific internal energy and enthalpy changes are facilitated by the use of the *ideal gas tables,* Tables A-16 through A-22, which give u and h (or $\bar{u}$ and $\bar{h}$) versus temperature.

To obtain enthalpy versus temperature, write Eq. 3.43 as

$$h(T) = \int_{T_{\text{ref}}}^{T} c_p(T)\, dT + h(T_{\text{ref}})$$

where T_{ref} is an arbitrary reference temperature and $h(T_{\text{ref}})$ is an arbitrary value for enthalpy at the reference temperature. Tables A-16 through A-22 are based on the selection $h = 0$ at $T_{\text{ref}} = 0$ K. Accordingly, a tabulation of enthalpy versus temperature is developed through the integral[2]

$$h(T) = \int_0^T c_p(T)\, dT \tag{3.49}$$

[2] The simple specific heat variation given by Eq. 3.48 is valid only for a limited temperature range, so tabular enthalpy values are calculated from Eq. 3.49 using other expressions that enable the integral to be evaluated accurately over wider ranges of temperature.

Tabulations of internal energy versus temperature are obtained from the tabulated enthalpy values by using $u = h - RT$.

For air as an ideal gas, h and u are given in Table A-16 with units of kJ/kg and in Table A-16E in units of Btu/lb. Values of molar specific enthalpy $\bar{h}$ and internal energy $\bar{u}$ for several other common gases modeled as ideal gases are given in Tables A-17 through A-22 with units of kJ/kmol or Btu/lbmol. Quantities other than specific internal energy and enthalpy appearing in these tables are introduced in Chap. 6 and should be ignored at present.

Tables A-16 through A-22 are convenient for evaluations involving ideal gases, not only because the variation of the specific heats with temperature is accounted for automatically but also because the tables are easy to use, as shown by examples to follow. Values of the thermodynamic properties of a number of gases modeled as ideal gases also can be determined by using computer programs. The principal advantage of such programs is that properties can be determined quickly and accurately without the tedious interpolations often required with tables.

3.5.3 Assumption of Constant Specific Heats

When the specific heats are taken as constants, Eqs. 3.40 and 3.43 reduce, respectively, to

$$u(T_2) - u(T_1) = c_v(T_2 - T_1) \tag{3.50}$$

$$h(T_2) - h(T_1) = c_p(T_2 - T_1) \tag{3.51}$$

Equations 3.50 and 3.51 are often used for thermodynamic analyses involving ideal gases because they enable simple closed-form equations to be developed for many processes.

The constant values of c_v and c_p in Eqs. 3.50 and 3.51 are, strictly speaking, mean values calculated as follows:

$$c_v = \frac{\int_{T_1}^{T_2} c_v(T)\, dT}{T_2 - T_1}, \qquad c_p = \frac{\int_{T_1}^{T_2} c_p(T)\, dT}{T_2 - T_1}$$

However, when the variation of c_v or c_p over a given temperature interval is slight, little error is normally introduced by taking the specific heat required by Eq. 3.50 or 3.51 as the arithmetic average of the specific heat values at the two end temperatures. Alternatively, the specific heat at the average temperature over the interval can be used. These methods are particularly convenient when tabular specific heat data are available, as in Tables A-14, for then the *constant* specific heat values often can be determined by inspection.

EXAMPLE 3.7

PROBLEM

Determine the change in specific enthalpy, in kJ/kg, for water vapor from a state where $T_1 = 400$ K and $p_1 = 0.1$ MPa to a state where $T_2 = 900$ K and $p_2 = 0.5$ MPa, by means of the ideal gas table for water vapor. Compare with the results of parts (a) and (b) of Example 3.6.

SOLUTION

Known: Water vapor undergoes a process between specific states.

Find: Determine the change in specific enthalpy using the ideal gas table for water vapor. Compare with the results of parts (a) and (b) of Example 3.6.

Assumption: The ideal gas model applies for steam at the indicated states.

Analysis: From the ideal gas table for water vapor, Table A-19, $\bar{h}_1 = 13{,}356$ kJ/kmol and $\bar{h}_2 = 31{,}828$ kJ/kmol. Dividing by the molecular weight of water, we obtain the enthalpies on a unit mass basis

$$h_1 = 741.18 \text{ kJ/kg}, \qquad h_2 = 1766.26 \text{ kJ/kg}$$

Thus

$$h_2 - h_1 = 1025.1 \text{ kJ/kg}$$

This value agrees well with those determined in parts (a) and (b) of Example 3.6.

Comments: When expressed on a unit mass basis, the *magnitudes* of the specific enthalpy values obtained from the ideal gas table at states 1 and 2 differ from the values at these states determined from Table A-4. This is a consequence of the different reference states and reference values used in constructing Tables A-4 and A-19. Although the magnitudes differ, the change in specific enthalpy is the same, for the datums cancel when differences are calculated.

EXAMPLE 3.8

PROBLEM

A piston–cylinder assembly contains 2 lb of air at a temperature of 540°R and a pressure of 1.0 atm. The air is compressed to a state where the temperature is 840°R and the pressure is 6.0 atm. During the compression, there is a heat transfer from the air to the surroundings equal to 20 Btu. Using the ideal gas model for air, determine the work during the process, in Btu.

SOLUTION

Known: Two pounds of air are compressed between two specified states while there is heat transfer from the air of a known amount.

Find: Determine the work, in Btu.

Schematic and Given Data:

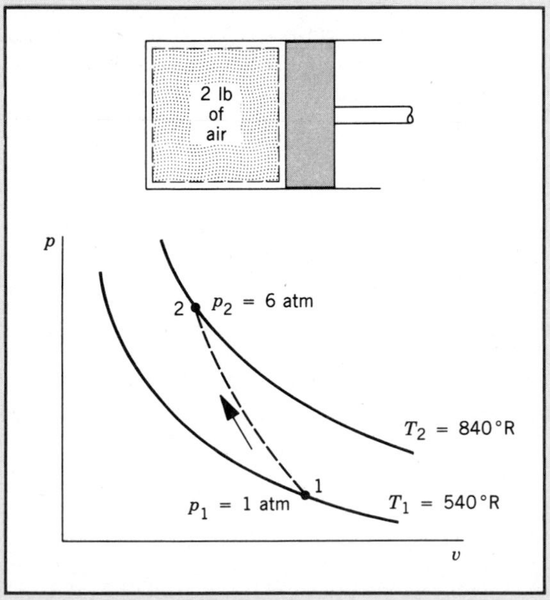

Figure E3.8

Assumptions:

1. The air is a closed system.

2. The initial and final states are equilibrium states. There is no change in kinetic or potential energy.
3. The air is modeled as an ideal gas.

Analysis: An energy balance for the closed system is

$$\Delta \cancelto{0}{KE} + \Delta \cancelto{0}{PE} + \Delta U = Q - W$$

where the kinetic and potential energy terms vanish by assumption 2. Solving for W

$$W = Q - \Delta U = Q - m(u_2 - u_1)$$

From the problem statement, $Q = -20$ Btu. Also, from Table A-16E at $T_1 = 540°\text{R}$, $u_1 = 92.04$ Btu/lb, and at $T_2 = 840°\text{R}$, $u_2 = 143.98$ Btu/lb. Accordingly

$$W = -20 - (2)(143.98 - 92.04) = -123.9 \text{ Btu}$$

The minus sign indicates that work is done on the system in the process.

Comments:

1. Although the initial and final states are assumed to be equilibrium states, the intervening states are not necessarily equilibrium states, so the process has been indicated on the accompanying p–v diagram by a dashed line. This dashed line does not define a "path" for the process.
2. In principle, the work could be evaluated through $\int p\, dV$, but because the variation of pressure at the piston face with volume is not known, the integration cannot be performed without more information.
3. Table A-1E gives $p_c = 37.2$ atm, $T_c = 239°\text{R}$ for air. Therefore at state 1, $p_{R1} = 0.03$, $T_{R1} = 2.26$, and at state 2, $p_{R2} = 0.16$, $T_{R2} = 3.51$. Referring to Fig. A-1, we conclude that at these states $Z \approx 1$, as assumed in the solution.

EXAMPLE 3.9

PROBLEM

Two tanks are connected by a valve. One tank contains 2 kg of carbon monoxide gas at 77°C and 0.7 bar. The other tank holds 8 kg of the same gas at 27°C and 1.2 bars. The valve is opened and the gases are allowed to mix while receiving energy by heat transfer from the surroundings. The final equilibrium temperature is 42°C. Using the ideal gas model, determine

(a) the final equilibrium pressure, in bars. **(b)** the heat transfer for the process, in kJ.

SOLUTION

Known: Two tanks containing different amounts of carbon monoxide gas at initially different states are connected by a valve. The valve is opened and the gas allowed to mix while receiving a certain amount of energy by heat transfer. The final equilibrium temperature is known.

Find: Determine the final pressure and the heat transfer for the process.

Schematic and Given Data:

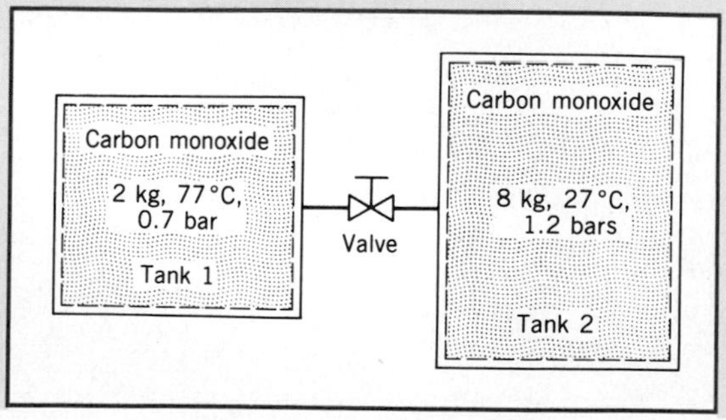

Figure E3.9

Assumptions:

1. The total amount of carbon monoxide gas is a closed system.
2. The gas is modeled as an ideal gas.
3. The gas initially in each tank is in equilibrium. The final state is an equilibrium state.
4. No energy is transferred to, or from, the gas by work.
5. There is no change in kinetic or potential energy.

Analysis:

(a) The final equilibrium pressure p_f can be determined from the ideal gas equation of state

$$p_f = \frac{mRT_f}{V}$$

where m is the sum of the initial amounts of mass present in the two tanks, V is the total volume of the two tanks, and T_f is the final equilibrium temperature. Thus

$$p_f = \frac{(m_1 + m_2)RT_f}{V_1 + V_2}$$

Denoting the initial temperature and pressure in tank 1 as T_1 and p_1, respectively, $V_1 = m_1RT_1/p_1$. Similarly, if the initial temperature and pressure in tank 2 are T_2 and p_2, $V_2 = m_2RT_2/p_2$. Thus, the final pressure is

$$p_f = \frac{(m_1 + m_2)RT_f}{\left(\dfrac{m_1RT_1}{p_1}\right) + \left(\dfrac{m_2RT_2}{p_2}\right)} = \frac{(m_1 + m_2)T_f}{\left(\dfrac{m_1T_1}{p_1}\right) + \left(\dfrac{m_2T_2}{p_2}\right)}$$

Inserting values

$$p_f = \frac{(10 \text{ kg})(315 \text{ K})}{\dfrac{(2 \text{ kg})(350 \text{ K})}{0.7 \text{ bar}} + \dfrac{(8 \text{ kg})(300 \text{ K})}{1.2 \text{ bars}}} = 1.05 \text{ bars}$$

(b) The heat transfer can be found from an energy balance, which reduces with assumptions 4 and 5 to give

$$\Delta U = Q - \cancelto{0}{W}$$

or

$$Q = U_f - U_i$$

U_i is the initial internal energy, given by

$$U_i = m_1u(T_1) + m_2u(T_2)$$

where T_1 and T_2 are the initial temperatures of the CO in tanks 1 and 2, respectively. The final internal energy is U_f

$$U_f = (m_1 + m_2)u(T_f)$$

Introducing these expressions for internal energy, the energy balance becomes

$$Q = m_1[u(T_f) - u(T_1)] + m_2[u(T_f) - u(T_2)]$$

Since the specific heat c_v of CO varies little over the temperature interval from 300 to 350 K (Table A-14), it can be treated as a constant. The expression for Q is then

$$Q = m_1c_v(T_f - T_1) + m_2c_v(T_f - T_2)$$

Evaluating c_v as the mean of the values listed in Table A-14 at 300 K and 350 K, $c_v = 0.745$ kJ/kg·K. Hence

$$Q = (2\text{ kg})\left(0.745\,\frac{\text{kJ}}{\text{kg}\cdot\text{K}}\right)(315\text{ K} - 350\text{ K})$$

$$+ (8\text{ kg})\left(0.745\,\frac{\text{kJ}}{\text{kg}\cdot\text{K}}\right)(315\text{ K} - 300\text{ K})$$

$$= +37.25\text{ kJ}$$

The plus sign indicates that the heat transfer is into the system.

Comments:

1. As an exercise, evaluate Q using specific internal energy values from the ideal gas table for CO, Table A-20. Observe that specific internal energy is given in Table A-20 with units of kJ/kmol.
2. By referring to a generalized compressibility chart, it can be verified that the ideal gas equation of state is appropriate for CO in this range of temperature and pressure.

3.5.4 Polytropic Process of an Ideal Gas

Recall that a *polytropic* process of a closed system is described by a pressure–volume relationship of the form

$$pV^n = \text{constant} \tag{3.52}$$

where n is a constant (Sec. 2.2). For a polytropic process between two states

$$p_1V_1^n = p_2V_2^n$$

or

$$\frac{p_2}{p_1} = \left(\frac{V_1}{V_2}\right)^n \tag{3.53}$$

The exponent n may take on any value from $-\infty$ to $+\infty$, depending on the particular process. When $n = 0$, the process is an isobaric (constant-pressure) process, and when $n = \pm\infty$, the process is an isometric (constant-volume) process.

For a polytropic process

$$\int_1^2 p\,dV = \frac{p_2V_2 - p_1V_1}{1 - n} \qquad (n \neq 1) \tag{3.54}$$

for any exponent n except $n = 1$. When $n = 1$,

$$\int_1^2 p\,dV = p_1V_1 \ln\frac{V_2}{V_1} \qquad (n = 1) \tag{3.55}$$

Example 2.1 provides the details of these integrations.

Equations 3.52 through 3.55 apply to *any* gas (or liquid) undergoing a polytropic process. When the *additional* idealization of ideal gas behavior is appropriate, further relations can be derived. Thus, when the ideal gas equation of state is introduced into Eqs. 3.53, 3.54, and 3.55, the following expressions are obtained, respectively:

$$\frac{T_2}{T_1} = \left(\frac{p_2}{p_1}\right)^{(n-1)/n} = \left(\frac{V_1}{V_2}\right)^{n-1} \qquad \text{(ideal gas)} \tag{3.56}$$

$$\int_1^2 p\,dV = \frac{mR(T_2 - T_1)}{1 - n} \qquad (\text{ideal gas}, n \neq 1) \tag{3.57}$$

$$\int_1^2 p\,dV = mRT \ln\frac{V_2}{V_1} \qquad (\text{ideal gas}, n = 1) \tag{3.58}$$

For an ideal gas, the case $n = 1$ corresponds to an isothermal (constant-temperature) process, as can readily be verified. In addition, when the specific heats are constant, the value of the exponent n corresponding to an adiabatic polytropic process of an ideal gas is the specific heat ratio k. This is demonstrated formally in Sec. 6.7.

EXAMPLE 3.10

PROBLEM

Air undergoes a polytropic compression in a piston–cylinder assembly from $p_1 = 1$ atm, $T_1 = 70°F$ to $p_2 = 5$ atm. Employing the ideal gas model, determine the work and heat transfer per unit mass, in Btu/lb, if $n = 1.3$.

SOLUTION

Known: Air undergoes a polytropic compression process from a given initial state to a specified final pressure.

Find: Determine the work and heat transfer, each in Btu/lb.

Schematic and Given Data:

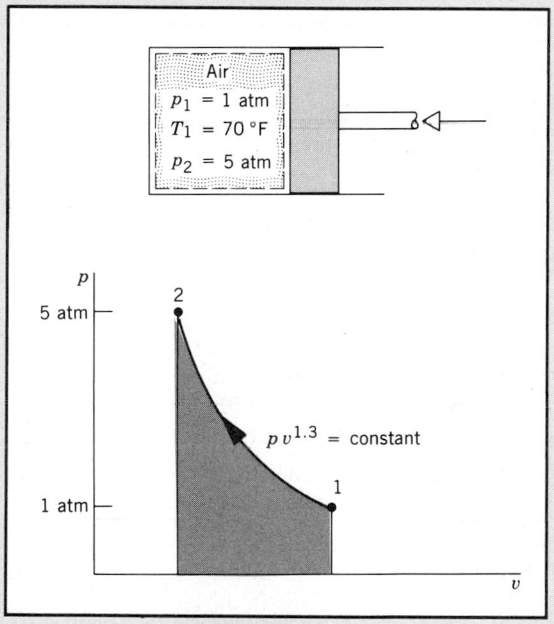

Figure E3.10

Assumptions:

1. The air is a closed system.
2. The air behaves as an ideal gas.
3. The compression is polytropic with $n = 1.3$.
4. There is no change in kinetic or potential energy.

Analysis: The work can be evaluated in this case from the expression

$$W = \int_1^2 p\, dV$$

With Eq. 3.57

$$\frac{W}{m} = \frac{R(T_2 - T_1)}{1 - n}$$

The temperature at the final state, T_2, is required. This can be evaluated from Eq. 3.56

$$T_2 = T_1 \left(\frac{p_2}{p_1}\right)^{(n-1)/n} = 530 \left(\frac{5}{1}\right)^{(1.3-1)/1.3} = 768°\text{R}$$

The work is then

$$\frac{W}{m} = \frac{R(T_2 - T_1)}{1 - n} = \left(\frac{1.986}{28.97} \frac{\text{Btu}}{\text{lb} \cdot °\text{R}}\right)\left(\frac{768 - 530}{1 - 1.3}\right)$$
$$= -54.39 \text{ Btu/lb}$$

The heat transfer can be evaluated from an energy balance. Thus

$$\frac{Q}{m} = \frac{W}{m} + (u_2 - u_1) = -54.39 + (131.38 - 90.33)$$
$$= -13.34 \text{ Btu/lb}$$

where the specific internal energy values are obtained from Table A-16E.

Comment: The states visited in the polytropic compression process are shown by the curve on the accompanying p–v diagram. The magnitude of the work per unit of mass is represented by the shaded area *below* the curve.

PROBLEMS

3.1 Write out the meanings of the following terms: pure substance, simple compressible system, critical point, saturated vapor, saturation pressure, subcooled liquid, superheated vapor, quality, triple point, saturated solid.

3.2 Determine the phase or phases in a system consisting of H_2O at the following conditions and locate the states on carefully labeled p–v and T–v diagrams.

(a) $p = 500$ kPa, $T = 200$°C.
(b) $p = 5$ MPa, $T = 264$°C.
(c) $T = 180$°C, $p = 0.9$ MPa.
(d) $p = 20$ MPa, $T = 100$°C.
(e) $T = -10$°C, $p = 1.0$ kPa.

3.3 Determine the phase or phases in a system consisting of H_2O at the following conditions and locate the states on carefully labeled p–v and T–v diagrams:

(a) $p = 680$ lbf/in.^{2}, $T = 500$°F.
(b) $T = 300$°F, $p = 50$ lbf/in.^{2}
(c) $p = 500$ lbf/in.^{2}, $T = 600$°F.
(d) $T = 5$°F, $p = 20$ lbf/in.^{2}
(e) $p = 1000$ lbf/in.^{2}, $T = 200$°F.

3.4 Plot the pressure–temperature relationship for two-phase liquid–vapor mixtures of water from the triple point temperature to the critical point temperature. Use a logarithmic scale for pressure, in bars, and a linear scale for temperature, in °C.

3.5 Plot the pressure–temperature relationship for two-phase liquid–vapor mixtures of Refrigerant 12 from a temperature of -40°F to the critical temperature, with pressure in lbf/in.^{2} and temperature in °F. Use a logarithmic scale for pressure and a linear scale for temperature.

3.6 Plot the pressure–temperature relationship for two-phase liquid–vapor mixtures of Refrigerant 134a from a temperature of -40 to 100°C, with pressure in kPa and temperature in °C. Use a logarithmic scale for pressure and a linear scale for temperature.

3.7 For H_2O, plot the following on a p–v diagram drawn to scale on log–log coordinates:

(a) the saturated liquid and saturated vapor lines from the triple point to the critical point, with pressure in MPa and specific volume in m^3/kg.
(b) lines of constant temperature at 100 and 300°C.

3.8 For H_2O, plot the following on a p–v diagram drawn to scale on log–log coordinates:

(a) the saturated liquid and saturated vapor lines from the triple point to the critical point, with pressure in lbf/in.^{2} and specific volume in ft^3/lb.
(b) lines of constant temperature at 300 and 1000°F.

3.9 Determine the quality of a two-phase liquid–vapor mixture of

(a) H_2O at 200°C with a specific volume of 0.1 m^3/kg.
(b) Refrigerant 12 at 2.0 bars with a specific volume of 0.07 m^3/kg.
(c) Refrigerant 134a at -40°C with a specific volume of 0.3569 m^3/kg.

3.10 For nitrogen at 100 K the values of specific volume for saturated liquid and saturated vapor are $v_f = 1.452 \times 10^{-3}$ m³/kg and $v_g = 31.31 \times 10^{-3}$ m³/kg, respectively. Determine the quality of 22 kg of a two-phase liquid–vapor mixture at 100 K in a tank with a volume of 0.5 m³.

3.11 Determine the quality of a two-phase liquid–vapor mixture of

(a) H_2O at 100 lbf/in.² with a specific volume of 3.0 ft³/lb.
(b) Refrigerant 12 at 0°F with a specific volume of 1.0 ft³/lb.
(c) Refrigerant 134a at −40°F with a specific volume of 5.7173 ft³/lb.

3.12 A two-phase liquid–vapor mixture of H_2O at 200 lbf/in.² has a specific volume of 1.5 ft³/lb. Determine the specific volume of a two-phase liquid–vapor mixture at 100 lbf/in.² with the same quality.

3.13 Determine the volume, in m³, occupied by 2 kg of H_2O at 4 MPa, 420°C.

3.14 Determine the volume, in ft³, occupied by 4 lb of H_2O at a pressure of 500 lbf/in.² and a temperature of

(a) 800°F.
(b) 200°F.

3.15 A closed vessel with a volume of 0.018 m³ contains 1.2 kg of Refrigerant 12 at 10 bars. Determine the temperature, in °C.

3.16 A closed vessel with a volume of 2 ft³ contains 5 lb of Refrigerant 134a at 70°F. Determine the *gage pressure,* in lbf/in.², within the vessel if the atmospheric pressure is 14.4 lbf/in.²

3.17 Calculate the volume, in m³, occupied by 2 kg of a two-phase liquid–vapor mixture of Refrigerant 12 at −10°C with a quality of 80%.

3.18 Refrigerant 12 is stored in a tank with a volume of 0.21 m³. Determine the mass, in kg, assuming saturated liquid at 20°C. What is the pressure, in kPa?

3.19 A two-phase liquid–vapor mixture of H_2O has a temperature of 300°C and a quality of 75%. The mixture occupies a volume of 0.05 m³. Determine the masses of saturated liquid and saturated vapor present, each in kg.

3.20 A closed system consists of a two-phase liquid–vapor mixture of H_2O in equilibrium at 400°F. The quality of the mixture is 0.2 (20%) and the mass of liquid water present is 0.1 lb. Determine the total volume of the system, in ft³.

3.21 A storage tank in a refrigeration system has a volume of 0.006 m³ and contains a two-phase liquid–vapor mixture of Refrigerant 12 at 180 kPa with quality $x = 0.9$ (90%). Determine the masses of saturated liquid and saturated vapor present, in kg, and the fraction of the total volume occupied by each phase.

3.22 A rigid tank contains 5 lb of water vapor, initially at 60 lbf/in.² and 320°F. The temperature of the water drops to 260°F as a result of heat transfer to the surroundings. Determine the final pressure, in lbf/in.², and the mass of liquid water present at the final state, in lb.

3.23 Five kilograms of saturated water vapor are contained in a closed rigid tank at an initial pressure of 40 bars. The pressure of the water drops to 20 bars as a result of heat transfer to the surroundings. Determine the volume of the tank, in m³, and the quality at the final state.

3.24 Steam is contained in a closed rigid container. Initially, the pressure and temperature of the steam are 15 bars and 240°C, respectively. The temperature drops to 20°C as a result of heat transfer to the surroundings. Determine at the final state

(a) the pressure, in kPa.
(b) the fraction of the total mass that condenses.
(c) the percentages of the volume occupied by saturated liquid and saturated vapor, respectively.

3.25 Saturated liquid water contained in a closed, rigid tank is cooled to a final state where the temperature is 50°C and the masses of saturated vapor and liquid present are 0.03 and 1999.97 kg, respectively. Determine the initial temperature, in °C, and the volume of the tank, in m³.

3.26 Water undergoes an isothermal process from saturated vapor at 400°F to a final specific volume of 4.5 ft³/lb. Determine the final pressure, in lbf/in.², and sketch the process on T–v and p–v diagrams.

3.27 Water vapor is cooled at constant specific volume from an initial state where $p_1 = 400$ lbf/in.² and $T_1 = 1000$°F to a final state where $T_2 = 500$°F. Determine the pressure at the final state, in lbf/in.²

3.28 Water vapor initially at 100 lbf/in.², 400°F, is cooled at constant specific volume. Determine the pressure, in lbf/in.², and the temperature, in °F, at which the water vapor becomes saturated.

3.29 Water vapor initially at 100 lbf/in.², 400°F, is compressed slowly. Heat transfer occurs to maintain the pressure constant. If the mass of water vapor is 3 lb, what is its volume, in ft³, when the vapor becomes saturated? Determine the work for the process, in Btu.

3.30 Refrigerant 134a undergoes a constant-pressure process from saturated vapor at −18.8°C to a final temperature of 20°C. Determine the work for the process, in kJ per kg of refrigerant.

3.31 Two pounds mass of Refrigerant 134a, initially at $p_1 = 180$ lbf/in.² and $T_1 = 120$°F, undergo a constant-pressure process until the final specific volume is $v_2 = 0.2$ ft³/lb. Determine the work for the process, in Btu.

3.32 Water vapor initially at 100 lbf/in.², 400°F, is compressed slowly. Heat transfer occurs to maintain the temperature constant. Determine

(a) the pressure, in lbf/in.^{2}, and the specific volume, in ft^3/lb, when the vapor becomes saturated.
(b) the work, in Btu/lb.

3.33 A two-phase liquid–vapor mixture of H_2O is initially at a pressure of 10 bars. If on heating at fixed volume, the critical point is attained, determine the quality at the initial state.

3.34 A two-phase liquid–vapor mixture of H_2O is initially at a pressure of 150 lbf/in.^{2} If on heating at fixed volume, the critical point is attained, determine the quality at the initial state.

3.35 Refrigerant 12 undergoes a process for which the pressure–volume relation is pv^n = constant. The initial and final states of the refrigerant are fixed by p_1 = 2 bars, T_1 = 10°C and p_2 = 10 bars, T_2 = 60°C, respectively. Evaluate the constant n for this process.

3.36 Refrigerant 134a undergoes a process for which the pressure–volume relation is pv^n = constant. The initial and final states of the refrigerant are fixed by p_1 = 200 kPa, T_1 = −10°C and p_2 = 1000 kPa, T_2 = 50°C, respectively. Calculate the work for the process, in kJ per kg of refrigerant.

3.37 A piston–cylinder assembly contains 0.04 lb of Refrigerant 12. The refrigerant is compressed from an initial state where p_1 = 40 lbf/in.^{2} and T_1 = 48°F to a final state where p_2 = 160 lbf/in.^{2} During the compression process, the pressure and specific volume are related by $pv^{1.05}$ = constant. Determine the work, in Btu, for the refrigerant.

3.38 Determine the specified property data for H_2O at the states indicated and locate each state on carefully labeled p–v and T–v diagrams.

(a) At p = 1.5 bars, T = 280°C, find v in m^3/kg and u in kJ/kg.
(b) At p = 1.5 bars, v = 0.9 m^3/kg, find T in °C and u in kJ/kg.
(c) At T = 400°C, p = 12 bars, find v in m^3/kg and h in kJ/kg.
(d) At T = 320°C, v = 0.30 m^3/kg, find p in MPa and u in kJ/kg.
(e) At p = 32 MPa, T = 400°C, find v in m^3/kg and h in kJ/kg.
(f) At T = 200°C, x = 80%, find p in bars and v in m^3/kg.
(g) At T = −10°C, v = 1.0891 × 10^{-3} m^3/kg, find p in kPa and h in kJ/kg.
(h) At p = 2.5 MPa, T = 140°C, find v in m^3/kg and u in kJ/kg.

3.39 Determine the specified property data for H_2O at the states indicated and show each state on carefully labeled p–v and T–v diagrams.

(a) At p = 20 lbf/in.^{2}, T = 500°F, find v in ft^3/lb and u in Btu/lb.
(b) At p = 20 lbf/in.^{2}, v = 14 ft^3/lb, find T in °F and u in Btu/lb.
(c) At T = 800°F, p = 170 lbf/in.^{2}, find v in ft^3/lb and h in Btu/lb.
(d) At T = 600°F, v = 0.5 ft^3/lb, find p in lbf/in.^{2} and u in Btu/lb.
(e) At p = 700 lbf/in.^{2}, T = 750°F, find v in ft^3/lb and h in Btu/lb.
(f) At T = 400°F, x = 80%, find p in lbf/in.^{2} and v in ft^3/lb
(g) At T = 15°F, v = 0.01745 ft^3/lb, find p in lbf/in.^{2} and h in Btu/lb.
(h) At p = 500 lbf/in.^{2}, T = 320°F, find v in ft^3/lb and u in Btu/lb.

3.40 For Refrigerant 12, determine the values of the specified properties at each of the following conditions:

(a) At p = 200 kPa and T = 88°C, determine v in m^3/kg and u in kJ/kg.
(b) At T = 60°C and v = 0.072 m^3/kg, determine p in kPa and h in kJ/kg.
(c) At p = 800 kPa and v = 0.005 m^3/kg, determine T in °C and u in kJ/kg.

3.41 For Refrigerant 134a, determine the values of the specified properties at each of the following conditions:

(a) At p = 140 lbf/in.^{2} and v = 0.3610 ft^3/lb, determine T in °F and u in Btu/lb.
(b) At p = 140 lbf/in.^{2} and h = 100 Btu/lb, determine T in °F and v in ft^3/lb.
(c) At T = 0°F and v = 3.0 ft^3/lb, determine p in lbf/in.^{2} and h in Btu/lb.

3.42. A quantity of water is at 10 MPa and 180°C. Evaluate the specific volume, in m^3/kg, and the specific enthalpy, in kJ/kg, using

(a) data from Table A-5.
(b) saturated liquid data from Table A-2.

3.43 Evaluate the specific enthalpy, in kJ/kg, of water at 100°C and a pressure of 15 MPa.

3.44 Evaluate the specific volume, in ft^3/lb, and the specific enthalpy, in Btu/lb, of Refrigerant 12 at 95°F and 150 lbf/in.^{2}

3.45 Evaluate the specific volume, in m^3/kg, and the specific enthalpy, in kJ/kg, of Refrigerant 12 at 40°C and 2000 kPa.

3.46 Evaluate the specific volume, in m^3/kg, and the specific enthalpy, in kJ/kg, of Refrigerant 134a at 40°C and 1.4 MPa.

3.47 A rigid, well-insulated tank contains 3 kg of a two-phase liquid–vapor mixture of water, initially at 200 kPa with a quality of 84%. A paddle wheel stirs the mixture until only saturated vapor remains in the tank. Kinetic and potential energy effects are negligible. For the water, determine the amount of energy transferred by work, in kJ.

3.48 A two-phase liquid–vapor mixture of H_2O, initially at 1.0 MPa with a quality of 90%, is contained in a rigid, well-insulated tank. The mass of H_2O is 2 kg. An electric resistance heater in the tank transfers energy to the water at a constant rate of 60 W. Determine the time, in h, when the temperature in the tank reaches 200°C.

3.49 Refrigerant 12 in a piston–cylinder assembly is compressed with no heat transfer from 30 lbf/in.², 20°F to 160 lbf/in.², 150°F. The mass of refrigerant is 0.04 lb. Kinetic and potential energy effects are negligible. For the refrigerant, determine the work, in Btu.

3.50 Refrigerant 12 vapor in a piston–cylinder assembly undergoes a process at constant pressure from an initial state fixed by 8 bars and 50°C to a final state at which the refrigerant is a saturated vapor. For the refrigerant, determine the work and heat transfer, per unit mass, each in kJ/kg. Changes in kinetic and potential energy are negligible.

3.51 Calculate the heat transfer, in kJ, for the process described in Problem 3.25.

3.52 Calculate the heat transfer, in kJ per kg of refrigerant, for the process described in Problem 3.36.

3.53 Saturated water vapor initially at 40 lbf/in.² is contained in a closed, rigid tank with a volume of 1.5 ft³. Heat transfer from the water occurs, and its pressure drops to 20 lbf/in.² Calculate the amount of heat transfer, in Btu, and the masses of saturated liquid and saturated vapor present at the final state, in lb. Kinetic and potential energy effects are negligible.

3.54 A piston–cylinder assembly contains a two-phase liquid–vapor mixture of H_2O initially at 500 kPa with a quality of 98%. Expansion occurs to a state where the pressure is 150 kPa. During the process the pressure and specific volume are related by pv = constant. For the water, determine the work and heat transfer per unit mass, each in kJ/kg.

3.55 Two pounds mass of a two-phase liquid–vapor mixture of H_2O, initially at 100 lbf/in.² and a quality of 50%, are confined to one side of a rigid, well-insulated container by a partition. The other side of the container is initially evacuated. The partition is removed and the water expands to fill the entire container. The pressure at the final equilibrium state is 40 lbf/in.² Determine the change in volume of the H_2O, in ft³.

3.56 Two kilograms of water, initially a saturated liquid at 100 kPa, are heated to saturated vapor while the pressure is maintained constant. Determine the work and heat transfer for the process, each in kJ. Show that the heat transfer equals the change in enthalpy of the water in this case.

3.57 A two-phase liquid–vapor mixture of H_2O with an initial quality of 25% is contained in a piston–cylinder assembly as shown in Fig. P3.57. The mass of the piston is 40 kg, and its diameter is 10 cm. The atmospheric pressure of the surroundings is 1 bar. The initial and final positions of the piston are shown on the diagram. As the water is heated, the pressure inside the cylinder remains constant until the piston hits the stops. Heat transfer to the water continues until its pressure is 3 bars. Friction between the piston and the cylinder wall is negligible. Determine the total amount of heat transfer, in J. Let g = 9.81 m/s².

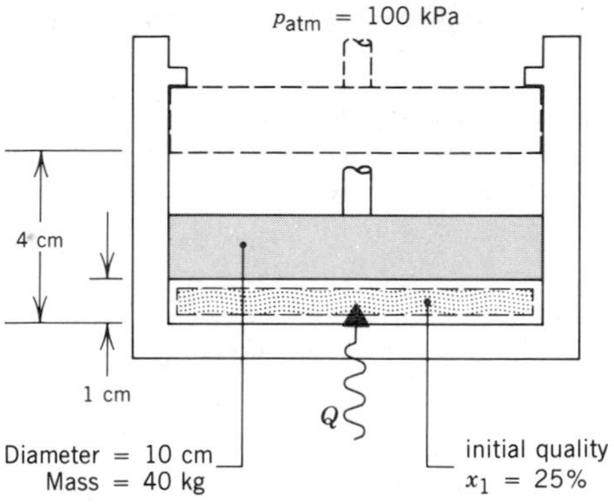

Figure P3.57

3.58 A system consisting of 2 lb of water vapor initially at state 1, where p_1 = 120 lbf/in.², is cooled at a constant volume of 9.05 ft³ to state 2, where T_2 = 300°F. The system then expands isothermally to state 3, where the volume is V_3 = 20 ft³. During the isothermal expansion there is heat transfer of 600 Btu *into* the system. Kinetic and potential energy effects are negligible. Determine

(a) the specific volumes and specific internal energies at each of the three states, in ft³/lb and Btu/lb, respectively.
(b) the heat transfer for Process 1–2 and the work for Process 2–3, each in Btu.

3.59 Two kilograms of Refrigerant 12, initially at 1 MPa and 100°C, undergo a compression at constant pressure until the final volume is half the initial volume. Kinetic and potential energy effects are negligible. Determine the work and heat transfer for the process, each in kJ.

3.60 Refrigerant 12 in a piston–cylinder assembly undergoes two processes in series. At the initial state, p_1 = 90 lbf/in.² and the quality is 100%. Process 1–2 occurs at constant volume until the temperature is T_2 = 200°F. The second process, from state 2 to state 3, occurs at constant temperature, with Q_{23} = −60.9 Btu, until V_3 = 0.6 ft³. Kinetic and potential energy effects are negligible. For 2.2 lb of refrigerant, determine, in Btu

(a) the heat transfer for Process 1–2.
(b) the work for Process 2–3.

3.61 Water vapor in a piston–cylinder assembly undergoes a process from saturated vapor at 150°C to a pressure of 3 bars. During the process, the pressure and specific volume are related by $pv^{1.2}$ = constant. Neglecting kinetic and potential energy effects, determine the heat transfer and the work per unit mass of water vapor, each in kJ/kg.

3.62 A system consisting of 2 kg of H_2O undergoes a cycle composed of the following processes:

Process 1–2: expansion with pv = constant from saturated vapor at 100 bars to 10 bars
Process 2–3: constant-pressure process to $v_3 = v_1$
Process 3–1: constant-volume heating

Sketch the cycle on p–v and T–v diagrams. Neglecting kinetic and potential energy effects, determine the net work for the cycle and the heat transfer for each process, all in kJ.

3.63 A system consisting of 1 kg of H_2O undergoes a cycle composed of the following processes:

Process 1–2: constant-volume heating from p_1 = 5 bars, T_1 = 160°C to p_2 = 10 bars
Process 2–3: constant-pressure process to saturated vapor
Process 3–4: constant-volume cooling
Process 4–1: isothermal expansion with Q_{41} = 815.8 kJ

Sketch the cycle on T–v and p–v diagrams. Neglecting kinetic and potential energy effects, determine the net work for the cycle and the heat transfer for each process, all in kJ.

3.64 A system consisting of 2 lb of H_2O undergoes a cycle composed of the following processes:

Process 1–2: adiabatic compression from p_1 = 200 lbf/in.^{2}, x_1 = 90% to p_2 = 600 lbf/in.^{2}, T_2 = 500°F
Process 2–3: isothermal process with W_{23} = 236.2 Btu to p_3 = 200 lbf/in.^{2}
Process 3–1: constant-pressure process

Neglecting kinetic and potential energy effects, determine the net work for the cycle and the heat transfer for each process, all in Btu.

3.65 An isolated system consists of a copper tank of mass 13 kg containing 4 kg of liquid water. Initially, the temperature of the copper is 27°C and the temperature of the water is 50°C. What is the temperature, in °C, when equilibrium is reached?

3.66 A steel rivet of mass 2 lb, initially at 1000°F, is placed in a large tank containing 5 ft^3 of liquid water initially at 70°F. Eventually, the rivet and water cool back to 70°F as a result of heat transfer to the surroundings. Taking the rivet and water as the system, determine the heat transfer, in Btu. For the steel, c = 0.11 Btu/lb·°R.

3.67 An isolated system consists of a 10-kg copper slab, initially at 30°C, and 0.2 kg of saturated water vapor, initially at 130°C. Assuming no volume change, determine the final equilibrium temperature of the isolated system, in °C.

3.68 A system consists of a liquid, considered incompressible with constant specific heat c, filling a rigid tank whose inner surface area is A. Energy transfer by work from a paddle wheel to the liquid occurs at a constant rate. Energy transfer by heat between the surroundings and the liquid occurs at a rate given by h$A(T_0 - T)$, where T is the instantaneous temperature of the liquid, T_0 is the temperature of the surroundings, and h is an overall heat transfer coefficient. At the initial time, t = 0, the tank and its contents are at the temperature of the surroundings. Obtain a differential equation for temperature T in terms of time t and relevant parameters. Solve the differential equation to obtain $T(t)$.

3.69 Apply an energy rate balance to an element of differential thickness within the plane wall of Fig. 2.12. Introducing the density, specific heat, and Fourier's law, derive a differential equation for temperature as a function of time within the wall.

3.70 Determine the compressibility factor for water vapor at 6 MPa and 600 K, using

(a) data from the compressibility chart.
(b) data from the steam tables.

3.71 Nitrogen (N_2) occupies a volume of 90 L at 27 MPa and 200 K. Determine the mass of nitrogen, in kg.

3.72 Determine the volume, in ft^3, occupied by 5 lbmol of nitrogen (N_2) at 360°R, 3000 lbf/in.^{2}

3.73 Determine the volume, in ft^3, occupied by 30 lb of carbon dioxide (CO_2) at 700 lbf/in.^{2} and 600°R.

3.74 A rigid tank contains 0.5 kg of oxygen (O_2) initially at 40 bars and 180 K. The gas is cooled and the pressure drops to 33 bars. Determine the volume of the tank, in m^3, and the final temperature, in K.

3.75 Butane (C_4H_{10}) undergoes a process in a piston–cylinder assembly from p_1 = 5 MPa, T_1 = 500 K to p_2 = 3 MPa during which the relationship between pressure and volume is $pv^{1.05}$ = constant. The mass of butane is 5 kg. Determine the final temperature, in K, and the work, in kJ.

3.76 Ethylene (C_2H_4) is compressed at constant pressure from an initial state where p_1 = 213 lbf/in.^{2} and T_1 = 612°R to a final temperature of T_2 = 460°R. Determine the work per mole of ethylene, in Btu/lbmol.

3.77 A tank contains 0.042 m^3 of oxygen at 21°C and 15 MPa. Determine the mass of oxygen, in kg, using

(a) the ideal gas model.
(b) data from the compressibility chart.

Comment on the applicability of the ideal gas model for oxygen at this state.

3.78 Show that water vapor can be accurately modeled as an ideal gas at temperatures below about 60°C (140°F).

3.79 For air, determine the pressure range, in kPa, for which the value of the compressibility factor Z remains in the interval 0.95 to 1.05, at temperatures of

(a) $T = 270$ K.
(b) $T = 330$ K.
(c) $T = 400$ K.

3.80 Repeat Problem 3.79 with pressure in lbf/in.^{2} for temperatures of

(a) $T = 480$°R.
(b) $T = 600$°R.
(c) $T = 720$°R.

3.81 Check the applicability of the ideal gas model for

(a) water vapor at 600 lbf/in.^{2}, 700°F.
(b) carbon dioxide (CO_2) at 200 lbf/in.^{2}, 520°R.
(c) air at 14.7 lbf/in.^{2}, 100°F.
(d) air at 1000 lbf/in.^{2}, 100°F.
(e) propane (C_3H_8) at 200 lbf/in.^{2}, 600°R.

3.82 Check the applicability of the ideal gas model for Refrigerant 134a at a temperature of 80°C and a pressure of

(a) 1.6 MPa.
(b) 0.10 MPa.

3.83 Determine the pressure, in kPa, of nitrogen (N_2) at −80°C and a specific volume of 0.0045 m^3/kg using generalized compressibility data and compare with the value obtained using the ideal gas model.

3.84 Determine the temperature, in K, of 5 kg of air at a pressure of 200 kPa and a volume of 1.8 m^3. Verify that ideal gas behavior can be assumed for air under these conditions.

3.85 A tank contains 10 lb of air at 560°R with a pressure of 50 lbf/in.^{2} Determine the volume of the air, in ft^3. Verify that ideal gas behavior can be assumed for air under these conditions.

3.86 A helium-filled balloon has a volume of 0.2 m^3. The temperature of the gas is 288 K and the pressure is 115 kPa. Assuming ideal gas behavior, determine the mass of helium, in kg.

3.87 Assuming the ideal gas model, determine the specific volume, in ft^3/lb, of carbon dioxide (CO_2) gas at 200 lbf/in.^{2} and 600°R.

3.88 One kilomole of air occupies a volume of 25 m^3 at a pressure of 100 kPa. Assuming ideal gas behavior, determine the temperature, in K, and the density, in kg/m^3.

3.89 Assuming the ideal gas model, determine the volume, in ft^3, and the density, in lb/ft^3, of 1 lbmol of air at 540°R and 14.7 lbf/in.^{2}

3.90 Assuming the ideal gas model, calculate the changes in specific internal energy and enthalpy, in kJ/kg, of methane (CH_4) gas from $T_1 = 320$ K to $T_2 = 800$ K.

3.91 Show that the specific heat ratio of a monatomic ideal gas is equal to 5/3.

3.92 A gas having a molecular weight of 16 is contained in a rigid, well-insulated vessel. The initial pressure of the gas is 20 lbf/in.^{2}, the mass is 0.12 lb, and the volume is 2 ft^3. The gas is stirred by a paddle wheel, resulting in a transfer of energy to the gas of magnitude 13,000 ft·lbf. Assuming the ideal gas model with $\bar{c}_p = 4.52 + 0.00737T$, where T is in °R and $\bar{c}_p$ has units of Btu/lbmol·°R, determine the final temperature of the gas, in °R. Neglect kinetic and potential energy effects.

3.93 Consider a gas mixture whose *apparent* molecular weight is 33, initially at 3 bars and 300 K, and occupying a volume of 0.1 m^3. The gas undergoes an expansion to 0.2 m^3 during which the pressure–volume relation is $pV^{1.3}$ = constant. Assume the ideal gas model with $c_v = 0.6 + (2.5 \times 10^{-4})T$, where T is in K and c_v has units of kJ/kg·K. Neglecting kinetic and potential energy effects, determine

(a) the mass of gas, in kg.
(b) the final pressure, in bars.
(c) the final temperature, in K.
(d) the work and heat transfer, each in kJ.

3.94 A rigid, well-insulated tank, with a volume of 2 ft^3, contains air at $p_1 = 20$ lbf/in.^{2}, $T_1 = 500$°R. The air is stirred by a paddle wheel, resulting in an energy transfer to the gas of magnitude 5000 ft·lbf. Assuming ideal gas behavior for the air, determine the final temperature, in °R, and the final pressure, in lbf/in.^{2} Neglect kinetic and potential energy effects.

3.95 One kilogram of air, initially at 5 bars, 350 K, and 3 kg of carbon dioxide (CO_2), initially at 2 bars, 450 K, are confined to opposite sides of a rigid, well-insulated container, as illustrated in Fig. P3.95. The partition is free to move and allows conduction from one gas to the other without energy storage in the partition itself. The air and carbon dioxide each behave as ideal gases. Determine the final equilibrium temperature, in K, and the final pressure, in bars, assuming constant specific heats.

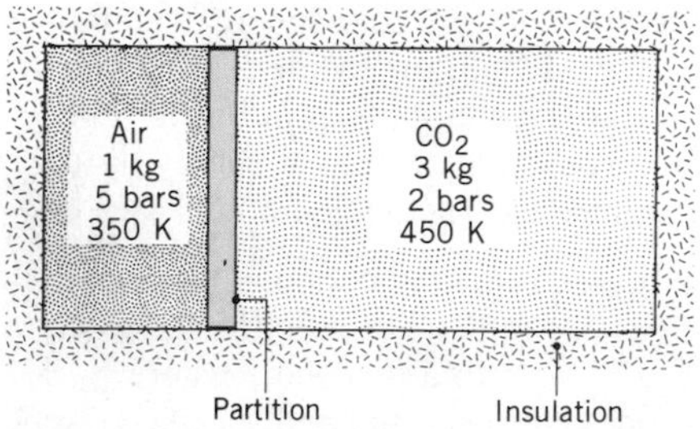

Figure P3.95

3.96 Argon (Ar) gas initially at 500 kPa, 100 K undergoes a polytropic process, with $n = k$, to a final temperature of 300 K. Determine the work and heat transfer for the process, each in kJ per kg of argon. Assume ideal gas behavior.

3.97 Two well-insulated tanks contain air. Initially, tank A holds 1 lb of air at 1440°R, and tank B has 2 lb of air at some unspecified initial temperature. A valve in the line connecting the two tanks is opened and the contents are allowed to mix until equilibrium is attained at a temperature of 1080°R. Assuming the ideal gas model, determine the initial temperature in tank B, in °R.

3.98 Two kilograms of a gas with molecular weight 28 are contained in a closed, rigid tank fitted with an electric resistor. The resistor draws a constant current of 10 amp at a voltage of 12 V for 10 min. Measurements indicate that when equilibrium is reached, the temperature of the gas has increased by 40.3°C. Heat transfer to the surroundings is estimated to occur at a constant rate of 20 W. Assuming ideal gas behavior, determine an average value of the specific heat c_p, in kJ/kg·K, of the gas in this temperature interval based on the measured data.

3.99 Carbon dioxide (CO_2) gas, initially at $T_1 = 530°R$, $p_1 = 15$ lbf/in.^{2}, and $V_1 = 1$ ft^3, is compressed in a piston–cylinder assembly. During the process, the pressure and specific volume are related by $pv^{1.2}$ = constant. The amount of energy transfer *to the gas* by work is 45 Btu per lb of CO_2. Assuming ideal gas behavior, determine the heat transfer, in Btu per lb of gas.

3.100 A gas is confined to one side of a rigid, insulated container divided into equal volumes by a partition. The other side is initially evacuated. The following data are known for the initial state of the gas: $p_1 = 3$ bars, $T_1 = 380$ K, and $V_1 = 0.025$ m^3. The partition is removed, and the gas expands to fill the entire container. Assuming ideal gas behavior, determine the final equilibrium pressure, in bars.

3.101 A rigid tank initially contains 3 kg of air at 500 kPa, 290 K. The tank is connected by a valve to a piston–cylinder assembly oriented vertically and containing 0.05 m^3 of air initially at 200 kPa, 290 K. Although the valve is closed, a slow leak allows air to flow into the cylinder until the tank pressure falls to 200 kPa. The weight of the piston and the pressure of the atmosphere maintain a constant pressure of 200 kPa in the cylider; and owing to heat transfer, the temperature stays constant at 290 K. For the air, determine the total amount of energy transfer by heat, in kJ. Assume ideal gas behavior.

3.102 A piston–cylinder assembly contains 1 kg of nitrogen gas (N_2). The gas expands from an initial state where $T_1 = 700$ K and $p_1 = 5$ bars to a final state where $p_2 = 2$ bars. During the process the pressure and specific volume are related by $pv^{1.3}$ = constant. Assuming ideal gas behavior and neglecting kinetic and potential energy effects, determine the heat transfer during the process, in kJ.

3.103 Air initially at 0.75 bar, 1000 K, and occupying a volume of 0.12 m^3 undergoes two processes. The air is compressed isothermally until the volume is halved. Then it undergoes a constant-pressure process until the volume is halved again. Assuming ideal gas behavior,

(a) sketch the processes on a p–V diagram.

(b) determine the total work for the two processes, in kJ.

(c) determine the total heat transfer for the two processes, in kJ.

3.104 A system consists of 2 kg of carbon dioxide gas initially at state 1, where $p_1 = 1$ bar, $T_1 = 300$ K. The system undergoes a power cycle consisting of the following processes:

Process 1–2: constant volume to $p_2 = 4$ bars
Process 2–3: expansion with $pv^{1.28}$ = constant
Process 3–1: constant-pressure compression

Assuming the ideal gas model and neglecting kinetic and potential energy effects,

(a) sketch the cycle on a p–v diagram.

(b) calculate the thermal efficiency of the cycle.

3.105 One pound of air undergoes a cycle consisting of the following processes:

Process 1–2: constant-pressure expansion with $p = 20$ lbf/in.^{2} from $T_1 = 500°R$ to $v_2 = 1.4v_1$
Process 2–3: adiabatic compression to $v_3 = v_1$ and $T_3 = 820°R$
Process 3–1: constant-volume process

Sketch the cycle on a carefully labeled p–v diagram. Assuming ideal gas behavior, determine the energy transfers by heat and work for each process, in Btu.

3.106 A closed system consists of an ideal gas with mass m and constant specific heat ratio k. If kinetic and potential energy changes are negligible,

(a) show that for *any* adiabatic process the work is

$$W = \frac{mR(T_2 - T_1)}{1 - k}$$

(b) show that an adiabatic *polytropic* process in which work is done only at a moving boundary is described by pV^k = constant.

3.107 Write a computer subroutine that returns an unknown pressure, specific volume, or temperature of an ideal gas given any two as input values. Include a calling parameter that allows any of the gases in Table A-15 to be specified. Allow the user to specify whether English or SI units are to be used, and to obtain specific volume on either a mass or molar basis.

3.108 Write a computer subroutine that returns specific internal energy and enthalpy values for each of the gases in Table A-15 for a given temperature as the input value. Let the reference values for internal energy and enthalpy be zero at a temperature of 300 K. Allow the user to specify whether English or SI units are to be used, and to obtain specific internal energy or enthalpy on either a mass or molar basis.

DESIGN AND OPEN ENDED PROBLEMS

3.1D A vessel is being designed to contain 10 kmol of mercury vapor at 3 MPa, and temperatures from 900 to 1000 K. Specify the vessel volume, wall thickness, and material. What environmental restrictions must be considered when designing a system that utilizes mercury? The p–v–T relation for mercury vapor can be expressed as

$$p = RT/v - (T/v^2)\exp(10.3338 - 0.0312095/T - 2.07950 \ln T)$$

where T is in K, v is in m^3/kg, and p is in Pa.

3.2D Owing to concern about the effect of *chlorofluorocarbon* (*CFC*) compounds, such as Refrigerant 12, on the earth's protective ozone layer, the use of CFCs is being discontinued. Refrigerant 134a is one of the substances that is replacing Refrigerant 12. Compare the pressures, temperatures, and h_{fg} values of saturated Refrigerants 12 and 134a using data from the appendix. Do the refrigerants appear to have comparable saturation properties? Obtaining information from other references, discuss the advantages and disadvantages of other candidate substances to replace Refrigerant 12 in terms of chemical stability, toxicity, flammability, and solubility with mineral oils used to lubricate refrigeration compressors.

3.3D A generalized pressure-temperature relationship for two-phase liquid–vapor mixtures is given in terms of reduced pressure and temperature as

$$p_R = \exp\left(\sum_{i=1}^{6} F_i X^i\right)$$

where $X = 1 - T_R$. The values of the dimensionless constants in this expression are

$$F_1 = -5.7896051 \qquad F_4 = -192.38731$$
$$F_2 = -5.5773833 \qquad F_5 = 550.74771$$
$$F_3 = 14.160472 \qquad F_6 = -761.16542$$

Investigate the accuracy of using this expression for determining saturation pressures and temperature of water, Refrigerant 12, and Refrigerant 134a by comparison with data from the appendix. Obtain data from other references and make similar comparisons for ammonia (NH_3), methane (CH_4), and octane (C_3H_8).

3.4D The following expression represents the pressure-temperature relation for two-phase liquid–vapor mixtures of Refrigerant 12:

$$\ln p_{sat} = F_1 + F_2/T + F_3 \ln T + F_4 T$$

where

$$F_1 = 93.3438056$$
$$F_2 = -4.39618785 \times 10^3$$
$$F_3 = -12.4715334$$
$$F_4 = 1.96060432 \times 10^{-2}$$

for pressure in Pa and temperature in K.

(a) Using data from Table A-7, verify the accuracy of this expression.
(b) Write a computer program that outputs either the saturation pressure or temperature, depending which is input by the user. Allow the user to specify English or SI units.
(c) Modify your program to output values of dp/dT. Use these values together with v_{fg} data from Table A-7 and the *Clapeyron equation* (Eq. 11.40) to evaluate h_{fg}. Compare with h_{fg} data from Table A-7.

3.5D One method of modeling gas behavior from the microscopic viewpoint is known as the *kinetic theory of gases*. Using kinetic theory, derive the ideal gas equation of state and explain the variation of the ideal gas specific heat c_v with temperature. Is the use of kinetic theory limited to ideal gas behavior? Discuss.

3.6D Write an interactive computer program that calculates the work and heat transfer, each per unit mass, for a polytropic process of an ideal gas. Input quantities should include the gas, initial pressure and temperature, polytropic exponent, and final pressure. Interface this routine with those you developed in Problems 2.3D, 3.107, and 3.108. Include an option to base the calculations on a constant specific heat as specified by the user. Using your program, investigate the error introduced by assuming constant specific heats evaluated at 70°F for the polytropic compression of air between 14.7 lbf/in.^{2}, 70°F and 100 lbf/in.^{2} Consider polytropic exponents ranging from $n = 1.0$ to $n = 1.6$.

3.7D A 12 mm-diameter steel ball initially at 1000 K slowly cools to 400 K in still air at 300 K. Estimate the time required to cool the ball.

CHAPTER 4

CONTROL VOLUME ENERGY ANALYSIS

Although devices such as turbines, pumps, and compressors through which mass flows can be analyzed in principle by studying a particular quantity of matter, a closed system, as it passes through the device, it is normally simpler to adopt the *control volume* point of view. A *control volume* is a region of space studied in a particular analysis. The surface that bounds the control volume may be called a boundary or a control surface. The boundary, which is defined relative to a specified coordinate frame, may be fixed or may deform. Mass *and* energy may cross the boundary. As in the case of a closed system, energy transfer can occur by means of work and heat. In addition, another type of energy transfer must be accounted for: the energy accompanying mass as it enters or exits.

The primary objective of this chapter is to develop and illustrate the use of the control volume forms of the conservation of mass and conservation of energy principles. The control volume forms are obtained by transforming the closed system forms.

4.1 CONSERVATION OF MASS FOR A CONTROL VOLUME

In this section an expression of the conservation of mass principle for control volumes is developed and illustrated. As a part of the presentation, the one-dimensional flow model is introduced.

4.1.1 Development of the Mass Rate Balance

The conservation of mass principle for a control volume is introduced using Fig. 4.1, which shows a system consisting of a fixed quantity of matter m that occupies different regions at time t and a later time $t + \Delta t$. At time t, the amount of mass under consideration is the sum

$$m = m_{\mathrm{cv}}(t) + m_i \tag{4.1}$$

where $m_{\mathrm{cv}}(t)$ is the mass contained within the control volume, and m_i is the mass

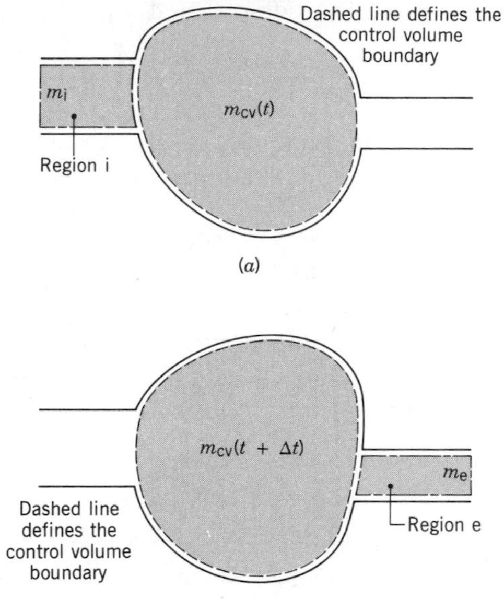

Figure 4.1 Illustration used to develop the conservation of mass principle for a control volume. (*a*) Time t. (*b*) Time $t + \Delta t$.

within the small region labeled i adjacent to the control volume, as shown in Fig. 4.1*a*. Let us study the fixed quantity of matter m as time elapses.

In a time interval Δt all the mass in region i crosses the control volume boundary, while some of the mass, call it m_e, initially contained within the control volume exits to fill the region labeled e adjacent to the control volume, as shown in Fig. 4.1*b*. At time $t + \Delta t$ the amount of mass under consideration can be expressed as

$$m = m_{cv}(t + \Delta t) + m_e \tag{4.2}$$

Observe that the amounts of mass in regions i and e are not necessarily equal and that the amount of mass contained within the control volume may have changed.

Although the system under consideration occupies different regions of space at different times, it consists of the same quantity of matter. Accordingly

$$m_{cv}(t) + m_i = m_{cv}(t + \Delta t) + m_e$$

or on rearrangement

$$m_{cv}(t + \Delta t) - m_{cv}(t) = m_i - m_e \tag{4.3}$$

Equation 4.3 is an *accounting* balance for mass which states that the change in mass of the control volume during time interval Δt equals the amount of mass that enters less the amount of mass that exits.

Equation 4.3 can be expressed on a time rate basis. First, divide by Δt to obtain

$$\frac{m_{cv}(t + \Delta t) - m_{cv}(t)}{\Delta t} = \frac{m_i}{\Delta t} - \frac{m_e}{\Delta t} \tag{4.4}$$

The left side of this equation is the average *rate of change of mass* in the control volume during Δt. The terms of the right side involving mass crossing the boundary of the control volume are average *rates of mass flow* during the time interval. The instantaneous rate form of Eq. 4.4 is obtained by taking the limit as Δt tends to zero. The system and control volume coincide in this limit.

The limit of the term on the left side of Eq. 4.4 is

$$\lim_{\Delta t \to 0} \left[\frac{m_{\text{cv}}(t + \Delta t) - m_{\text{cv}}(t)}{\Delta t}\right] = \frac{dm_{\text{cv}}}{dt}$$

In this expression dm_{cv}/dt is the time rate of change of mass contained within the control volume at time t. In the limit as Δt approaches 0, the terms on the right side become, respectively

$$\lim_{\Delta t \to 0} \frac{m_i}{\Delta t} = \dot{m}_i \quad \text{and} \quad \lim_{\Delta t \to 0} \frac{m_e}{\Delta t} = \dot{m}_e$$

In these expressions $\dot{m}_i$ and $\dot{m}_e$ are the instantaneous *mass flow rates* at the inlet and exit, respectively. As for the symbols $\dot{W}$ and $\dot{Q}$, the dots in the quantities $\dot{m}_i$ and $\dot{m}_e$ denote time rates of transfer. In summary, the limit of Eq. 4.4 as Δt tends to zero is

$$\frac{dm_{\text{cv}}}{dt} = \dot{m}_i - \dot{m}_e \tag{4.5}$$

In general, there may be several locations on the boundary through which mass enters or exits. This may be accounted for by summing, as follows:

$$\frac{dm_{\text{cv}}}{dt} = \sum_i \dot{m}_i - \sum_e \dot{m}_e \tag{4.6}$$

Equation 4.6 is the mass rate balance for control volumes with several inlets and exits. It is the form of the conservation of mass principle employed in subsequent control volume analyses. Expressed in words

$$\begin{bmatrix} \text{time } \textit{rate of change} \text{ of} \\ \text{mass contained within} \\ \text{the control volume } \textit{at time t} \end{bmatrix} = \begin{bmatrix} \text{total } \textit{rate} \text{ of flow} \\ \text{of mass } \textit{in} \text{ across} \\ \text{all inlets } \textit{at time t} \end{bmatrix} - \begin{bmatrix} \text{total } \textit{rate} \text{ of flow} \\ \text{of mass } \textit{out} \text{ across} \\ \text{all exits } \textit{at time t} \end{bmatrix}$$

In SI, all terms in Eq. 4.6 are expressed in kg/s. When English units are employed, all terms are expressed in lb/s.

The following equation accounts for the change of mass of a control volume with several inlets and exits over a time interval:

$$\Delta m_{\text{cv}} = \sum_i m_i - \sum_e m_e \tag{4.7}$$

where m_i and m_e denote, respectively, the amount of mass that enters at i and exits at e during the time interval. Equation 4.7 can be obtained by integrating each term of Eq. 4.6 over the time interval.

4.1.2 Forms of the Mass Rate Balance

The mass rate balance, Eq. 4.6, is a form that is useful for control volume analysis. In many cases, however, it is convenient to express the mass balance in terms of local properties. This is the subject of the present discussion. The total mass contained within the control volume at an instant t can be related to the local density as follows:

$$m_{\text{cv}}(t) = \int_V \rho \, dV \tag{4.8}$$

where the integration is over the volume at time t. Expressions can also be developed relating the mass flow rates to the properties of the matter crossing the boundary and the areas through which the mass enters and exits the control volume. This is considered next.

An expression for the mass flow rate $\dot{m}$ of the matter entering or exiting a control volume can be derived in terms of local properties by considering a small quantity of matter flowing with velocity V across an incremental area dA in a time interval Δt, as shown in Fig. 4.2. Since the portion of the control volume boundary through which mass flows is not necessarily at rest, the velocity shown in the figure is understood to be the velocity *relative* to the area dA. The velocity can be resolved into components normal and tangent to the plane containing dA. In the following development V_n denotes the component of the relative velocity normal to dA in the direction of flow.

The *volume* of the matter crossing dA during the time interval Δt shown in Fig. 4.2 is an oblique cylinder with a volume equal to the product of the area of its base dA and its altitude $V_n \Delta t$. Multiplying by the density ρ gives the amount of mass that crosses dA in time Δt

$$\begin{bmatrix} \text{amount of mass} \\ \text{crossing } dA \text{ during} \\ \text{the time interval } \Delta t \end{bmatrix} = \rho(V_n \Delta t)\, dA$$

Dividing both sides of this equation by Δt and taking the limit as Δt goes to zero, the instantaneous mass flow rate across incremental area dA is

$$\begin{bmatrix} \text{instantaneous rate} \\ \text{of mass flow} \\ \text{across } dA \end{bmatrix} = \rho V_n\, dA$$

When this is integrated over the area A through which mass passes, an expression for the mass flow rate is obtained

$$\dot{m} = \int_A \rho V_n\, dA \tag{4.9}$$

Equation 4.9 can be applied at the inlets and exits to account for the rates of mass flow into and out of the control volume.

With Eqs. 4.8 and 4.9, the mass rate balance Eq. 4.6 can be written as

$$\frac{d}{dt}\int_V \rho\, dV = \sum_i \left(\int_A \rho V_n\, dA\right)_i - \sum_e \left(\int_A \rho V_n\, dA\right)_e \tag{4.10}$$

where the area integrals are over the areas through which mass enters and exits the control volume, respectively. The product ρV_n appearing in this equation, known as the *mass flux*, gives the time rate of mass flow per unit of area. To evaluate the terms of the right side of Eq. 4.10 requires information about the variation of the mass flux over the flow areas. The form of the conservation of mass principle given by Eq. 4.10

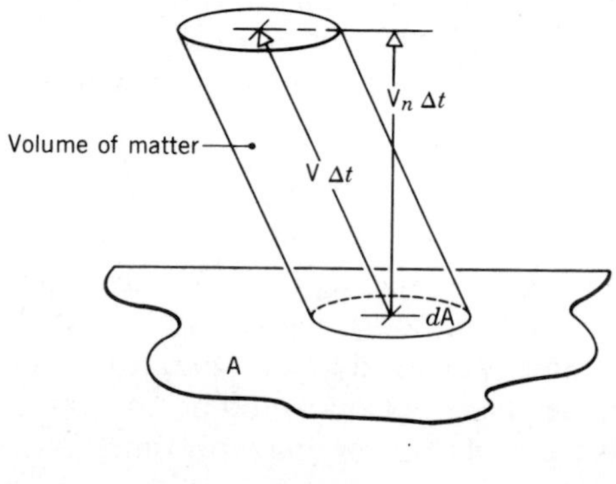

Figure 4.2 Illustration used to develop an expression for mass flow rate in terms of local fluid properties.

is usually considered in detail in fluid mechanics. In thermodynamics, a simplified form is normally employed, as described next.

One-Dimensional Flow. When a flowing stream of matter entering or exiting a control volume adheres to the following idealizations, the flow is said to be *one-dimensional:* (1) The flow is normal to the boundary at locations where mass enters or exits the control volume. (2) *All* intensive properties, including velocity and density, are *uniform with position* (bulk average values) over each inlet or exit area through which matter flows. In subsequent control volume analyses it is tacitly assumed that the boundary of the control volume can be selected so that these idealizations are justified.

When the flow is one-dimensional, Eq. 4.9 for the mass flow rate becomes

$$\dot{m} = \rho \mathrm{AV} \qquad \text{(one-dimensional flow)} \tag{4.11a}$$

or in terms of specific volume

$$\dot{m} = \frac{\mathrm{AV}}{v} \qquad \text{(one-dimensional flow)} \tag{4.11b}$$

where the symbol V now simply denotes the single value that represents the velocity of the flowing gas or liquid at the inlet or exit under consideration. The product AV in this expression is referred to as the *volumetric flow* rate in subsequent discussions. The volumetric flow rate is expressed in units of m^3/s or ft^3/s.

Substituting Eq. 4.11b into Eq. 4.6 results in an expression for the conservation of mass principle for control volumes limited to the case of one-dimensional flow at the inlets and exits

$$\frac{dm_{\text{cv}}}{dt} = \sum_i \frac{\mathrm{A}_i \mathrm{V}_i}{v_i} - \sum_e \frac{\mathrm{A}_e \mathrm{V}_e}{v_e} \qquad \text{(one-dimensional flow)} \tag{4.12}$$

Note that Eq. 4.12 involves summations over the inlets and exits of the control volume. Each individual term in either of these sums applies to a particular inlet or exit. Accordingly, the area, velocity, and specific volume appearing in a term refer only to the corresponding inlet or exit.

Steady State. Many engineering systems can be idealized as being at *steady state,* meaning that *all* properties are unchanging in time. For a control volume at steady state, the identity of the matter within the control volume changes continuously, but the total amount present at any instant remains constant, so $dm_{\text{cv}}/dt = 0$ and Eq. 4.6 reduces to

$$\sum_i \dot{m}_i = \sum_e \dot{m}_e$$

That is, the total incoming and outgoing rates of mass flow are equal.

Equality of total incoming and outgoing rates of mass flow does not necessarily mean that a control volume is at steady state, for although the total amount of mass within the control volume at any instant would be constant, other properties such as temperature and pressure might be varying with time. When a control volume is at steady state, *every* property is independent of time. Note that the steady-state assumption and the one-dimensional flow assumption are independent idealizations. One does not imply the other.

EXAMPLE 4.1

PROBLEM

A feedwater heater operating at steady state has two inlets and one exit. At inlet 1, water vapor enters at $p_1 = 7$ bars, $T_1 = 200°C$ with a mass flow rate of 40 kg/s. At inlet 2, liquid water at $p_2 = 7$ bars, $T_2 = 40°C$ enters through an area $A_2 = 25\ cm^2$. Saturated liquid at 7 bars exits at 3 with a volumetric flow rate of $0.06\ m^3/s$. Determine the mass flow rates at inlet 2 and at the exit, in kg/s, and the velocity at inlet 2, in m/s.

SOLUTION

Known: A stream of water vapor mixes with a liquid water stream to produce a saturated liquid stream at the exit. The states at the inlets and exit are specified. Mass flow rate and volumetric flow rate data are given at one inlet and at the exit, respectively.

Find: Determine the mass flow rates at inlet 2 and at the exit, and the velocity V_2.

Schematic and Given Data:

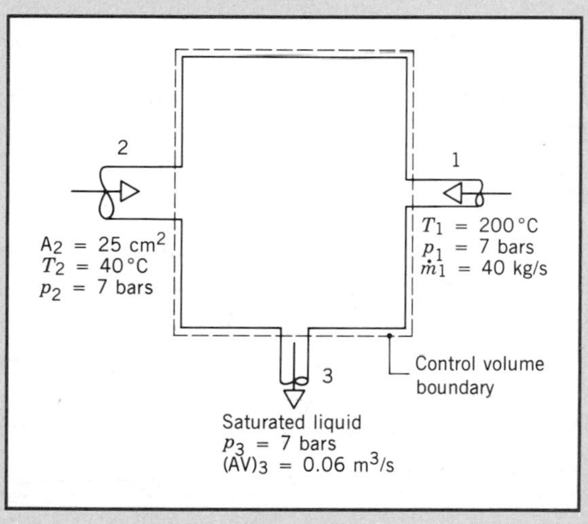

Figure E4.1

Assumptions:

1. The control volume shown on the accompanying figure is at steady state.

2. There is one-dimensional flow at both inlets and the exit.

Analysis: The principal relations to be employed are the mass rate balance (Eq. 4.6) and the expression $\dot{m} = AV/v$, (Eq. 4.11b). At steady state the mass rate balance becomes

$$\cancelto{0}{\frac{dm_{cv}}{dt}} = \dot{m}_1 + \dot{m}_2 - \dot{m}_3$$

Solving for $\dot{m}_2$

$$\dot{m}_2 = \dot{m}_3 - \dot{m}_1$$

The mass flow rate $\dot{m}_1$ is given. The mass flow rate at the exit can be evaluated from the given volumetric flow rate

$$\dot{m}_3 = \frac{(AV)_3}{v_3}$$

where v_3 is the specific volume at the exit. In writing this expression, one-dimensional flow is assumed. From Table A-3, $v_3 = 1.108 \times 10^{-3}$ m³/kg. Hence

$$\dot{m}_3 = \frac{0.06 \text{ m}^3/\text{s}}{(1.108 \times 10^{-3} \text{ m}^3/\text{kg})} = 54.15 \text{ kg/s}$$

The mass flow rate at inlet 2 is then

$$\dot{m}_2 = \dot{m}_3 - \dot{m}_1 = 54.15 - 40 = 14.15 \text{ kg/s}$$

For one-dimensional flow at 2, $\dot{m}_2 = A_2V_2/v_2$, so

$$V_2 = \dot{m}_2 v_2/A_2$$

State 2 is a compressed liquid. The specific volume at this state can be approximated by $v_2 \approx v_f(T_2)$ (Eq. 3.11). From Table A-2 at 40°C, $v_2 = 1.0078 \times 10^{-3}$ m³/kg. So

$$V_2 = \frac{(14.15 \text{ kg/s})(1.0078 \times 10^{-3} \text{ m}^3/\text{kg})}{25 \text{ cm}^2}\left(\frac{10^4 \text{ cm}^2}{1 \text{ m}^2}\right) = 5.7 \text{ m/s}$$

Comment: At steady state the mass flow rate at the exit equals the sum of the mass flow rates at the inlets. It is left as an exercise to show that the volumetric flow rate at the exit does not equal the sum of the volumetric flow rates at the inlets.

4.2 CONSERVATION OF ENERGY FOR A CONTROL VOLUME

In this section the rate form of the energy balance for control volumes is obtained using a development paralleling that given in Sec. 4.1 for the mass rate balance.

4.2.1 Development of the Control Volume Energy Rate Balance

The conservation of energy principle for a control volume can be introduced using Fig. 4.3, which shows a system consisting of a fixed quantity of matter m that occupies different regions at time t and a later time $t + \Delta t$. At time t, the energy of the system under consideration is

$$E(t) = E_{cv}(t) + m_i\left(u_i + \frac{V_i^2}{2} + gz_i\right) \tag{4.13}$$

where $E_{cv}(t)$ is the sum of the internal, kinetic, and gravitational potential energies of

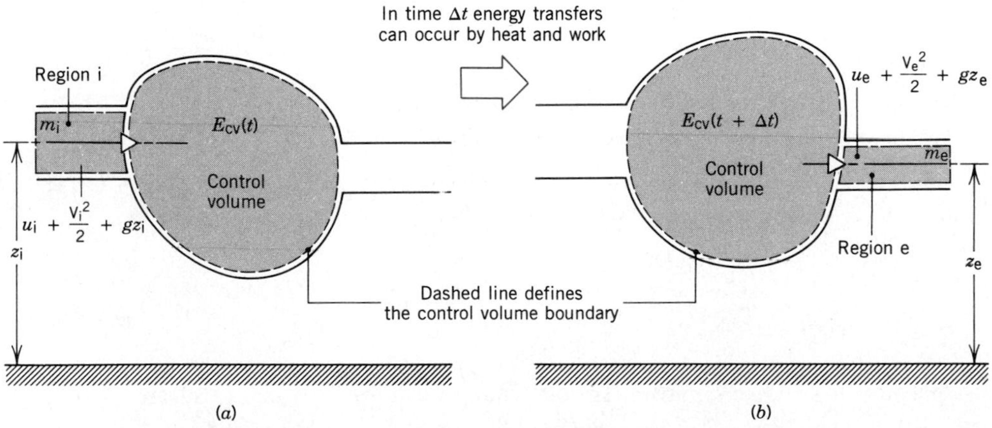

Figure 4.3 Illustration used to develop the conservation of energy principle for a control volume. (*a*) Time t. (*b*) Time $t + \Delta t$.

the mass contained within the control volume at time t. The second term on the right of Eq. 4.13 accounts for the energy associated with the mass m_i contained in the region i adjacent to the control volume. The specific energy of the mass m_i is $(u_i + \mathrm{V}_i^2/2 + gz_i)$. Let us study the fixed quantity of matter m as time elapses.

In a time interval Δt all the mass in region i crosses the control volume boundary while an amount of mass m_e initially in the control volume exits to fill region e, as shown in Fig. 4.3b. During this interval there may be energy transfers to or from the system under consideration by heat and work. At time $t + \Delta t$, the energy of the system is

$$E(t + \Delta t) = E_{\mathrm{cv}}(t + \Delta t) + m_e\left(u_e + \frac{\mathrm{V}_e^2}{2} + gz_e\right) \tag{4.14}$$

Observe that the mass and energy within the control volume may have changed over the time interval and that the masses m_i and m_e are not necessarily equal, nor are their energies necessarily the same. In writing Eqs. 4.13 and 4.14, the intensive properties of the masses m_i and m_e have been assumed to be uniform throughout.

Although the total mass m under consideration occupies different regions of space at different times, it is the same quantity of matter. Consequently, the *closed system* energy balance can be applied

$$E(t + \Delta t) - E(t) = Q - W \tag{4.15}$$

Introducing Eqs. 4.13 and 4.14

$$\left[E_{\mathrm{cv}}(t + \Delta t) + m_e\left(u_e + \frac{\mathrm{V}_e^2}{2} + gz_e\right)\right] - \left[E_{\mathrm{cv}}(t) + m_i\left(u_i + \frac{\mathrm{V}_i^2}{2} + gz_i\right)\right] = Q - W$$

Rearranging this equation gives

$$E_{\mathrm{cv}}(t + \Delta t) - E_{\mathrm{cv}}(t) = Q - W + m_i\left(u_i + \frac{\mathrm{V}_i^2}{2} + gz_i\right) - m_e\left(u_e + \frac{\mathrm{V}_e^2}{2} + gz_e\right) \tag{4.16}$$

The control volume form of the energy rate equation can be obtained by applying a limiting procedure analogous to that employed in Sec. 4.1 to obtain the control volume mass rate balance. First, divide each term of Eq. 4.16 by the time interval Δt:

$$\frac{E_{\mathrm{cv}}(t + \Delta t) - E_{\mathrm{cv}}(t)}{\Delta t} = \frac{Q}{\Delta t} - \frac{W}{\Delta t} + \frac{m_i(u_i + \mathrm{V}_i^2/2 + gz_i)}{\Delta t} - \frac{m_e(u_e + \mathrm{V}_e^2/2 + gz_e)}{\Delta t} \tag{4.17}$$

Then consider the limit of each term as Δt approaches zero.

In the limit as Δt approaches zero, the left side of Eq. 4.17 becomes

$$\lim_{\Delta t \to 0} \frac{E_{\mathrm{cv}}(t + \Delta t) - E_{\mathrm{cv}}(t)}{\Delta t} = \frac{dE_{\mathrm{cv}}}{dt}$$

where dE_{cv}/dt is the time rate of change of energy contained within the control volume.

Consider next the heat transfer term. As Δt approaches zero, the boundaries of the system and control volume coincide, so the heat transfer to the system is also the heat transfer to the control volume. In the limit

$$\lim_{\Delta t \to 0} \frac{Q}{\Delta t} = \dot{Q}$$

Similarly, for the work term

$$\lim_{\Delta t \to 0} \frac{W}{\Delta t} = \dot{W}$$

In these expressions $\dot{Q}$ and $\dot{W}$ are the net rates of energy transfer by heat and by work, respectively, across the boundary of the control volume at time t.

Finally, in the limit as Δt approaches zero, the remaining terms of Eq. 4.17 become, respectively

$$\lim_{\Delta t \to 0} \left[\frac{m_i(u_i + \mathrm{V}_i^2/2 + gz_i)}{\Delta t}\right] = \dot{m}_i \left(u_i + \frac{\mathrm{V}_i^2}{2} + gz_i\right)$$

$$\lim_{\Delta t \to 0} \left[\frac{m_e(u_e + \mathrm{V}_e^2/2 + gz_e)}{\Delta t}\right] = \dot{m}_e \left(u_e + \frac{\mathrm{V}_e^2}{2} + gz_e\right)$$

where $\dot{m}_i$ and $\dot{m}_e$ are the mass flow rates. The terms $(u_i + \mathrm{V}_i^2/2 + gz_i)$ and $(u_e + \mathrm{V}_e^2/2 + gz_e)$ are the specific energies of the flowing matter evaluated at the inlet and exit of the control volume. Recall that we originally assumed uniform properties for each of the masses m_i and m_e crossing the boundary of the control volume. Consequently, in taking the limits above, this corresponds to the assumption of one-dimensional flow through the inlet and exit flow areas.

In summary, the energy rate equation for the control volume of Fig. 4.3 is

$$\frac{dE_{\mathrm{cv}}}{dt} = \dot{Q} - \dot{W} + \dot{m}_i \left(u_i + \frac{\mathrm{V}_i^2}{2} + gz_i\right) - \dot{m}_e \left(u_e + \frac{\mathrm{V}_e^2}{2} + gz_e\right) \tag{4.18}$$

Equation 4.18 shows that, in addition to heat and work, there is another way energy can be transferred across the boundary of a control volume: by energy accompanying mass as it enters or exits the control volume. These energy transfer terms have the form $\dot{m}(u + \mathrm{V}^2/2 + gz)$ when one-dimensional flow is assumed. If there is no mass flow into or out of a control volume, the last two terms of Eq. 4.18 drop out. The equation then reduces to the rate form of the energy balance for a closed system.

4.2.2 Work for a Control Volume

Next, we will place Eq. 4.18 in an alternative form that is more convenient for subsequent applications. This will be accomplished primarily by recasting the work term $\dot{W}$, which represents the net rate of energy transfer by work across *all* portions of the boundary of the control volume.

Because work is always done on or by a control volume where matter flows across the boundary, it is convenient to separate the work term $\dot{W}$ into *two contributions*. One is the work associated with the fluid pressure as mass is introduced at inlets and removed at exits. The other contribution, denoted by $\dot{W}_{\mathrm{cv}}$, includes *all other* work effects, such as those associated with rotating shafts, displacement of the boundary, electrical, magnetic, and surface tension effects.

Consider the work at an exit e associated with the pressure of the flowing matter. Recall from Sec. 2.2 that the rate of energy transfer by work can be expressed as the product of a force and the velocity at the point of application of the force. Accordingly, the rate at which work is done at the exit by the normal force (normal to the exit area in the direction of flow) due to pressure is the product of the normal force, $p_e\mathrm{A}_e$, and the fluid velocity, V_e. That is

$$\left[\begin{array}{c}\text{time rate of energy transfer}\\ \text{by work } \textit{from} \text{ the control}\\ \text{volume at exit } e\end{array}\right] = (p_e\mathrm{A}_e)\mathrm{V}_e \tag{4.19}$$

where p_e is the pressure, A_e is the area, and V_e is the velocity at exit e, respectively. In writing this, the pressure and velocity are assumed to be uniform with position over the flow area. A similar expression can be written for the rate of energy transfer by work into the control volume at inlet i.

With these considerations, the work term W of the energy rate equation, Eq. 4.18, can be written as

$$\dot{W} = \dot{W}_{cv} + (p_e A_e)V_e - (p_i A_i)V_i \tag{4.20}$$

where, in accordance with the sign convention for work, the term at the inlet has a negative sign because energy is transferred into the control volume there. A positive sign precedes the work term at the exit because energy is transferred out of the control volume there. With $AV = \dot{m}v$ from Eq. 4.11b, the above expression for work can be written as

$$\dot{W} = \dot{W}_{cv} + \dot{m}_e(p_e v_e) - \dot{m}_i(p_i v_i) \tag{4.21}$$

where $\dot{m}_i$ and $\dot{m}_e$ are the mass flow rates and v_i and v_e are the specific volumes evaluated at the inlet and exit, respectively. In Eq. 4.21, the terms $\dot{m}_i(p_i v_i)$ and $\dot{m}_e(p_e v_e)$ account for the work associated with the pressure at the inlet and exit, respectively. They are commonly referred to as *flow work*. The term $\dot{W}_{cv}$ accounts for *all other* energy transfers by work across the boundary of the control volume.

4.2.3 Forms of the Control Volume Energy Rate Balance

Substituting Eq. 4.21 in Eq. 4.18 and collecting all terms referring to the inlet and the exit into separate expressions, the following form of the control volume energy rate balance results:

$$\frac{dE_{cv}}{dt} = \dot{Q}_{cv} - \dot{W}_{cv} + \dot{m}_i\left(u_i + p_i v_i + \frac{V_i^2}{2} + gz_i\right) - \dot{m}_e\left(u_e + p_e v_e + \frac{V_e^2}{2} + gz_e\right) \tag{4.22}$$

The subscript "cv" has been added to $\dot{Q}$ to emphasize that this is the heat transfer rate over the boundary (control surface) of the *control volume*.

The last two terms of Eq. 4.22 can be rewritten using the specific enthalpy h introduced in Sec. 3.3.2. With $h = u + pv$, the energy rate balance becomes

$$\frac{dE_{cv}}{dt} = \dot{Q}_{cv} - \dot{W}_{cv} + \dot{m}_i\left(h_i + \frac{V_i^2}{2} + gz_i\right) - \dot{m}_e\left(h_e + \frac{V_e^2}{2} + gz_e\right) \tag{4.23}$$

The appearance of the sum $u + pv$ in the control volume energy equation is the principal reason for introducing enthalpy previously. It is brought in solely as a *convenience:* The algebraic form of the energy rate balance is simplified by the use of enthalpy and, as we have seen, enthalpy is normally tabulated along with other properties.

In practice there may be several locations on the boundary through which mass enters or exits. This can be accounted for by introducing summations as in the mass balance. Accordingly

$$\frac{dE_{cv}}{dt} = \dot{Q}_{cv} - \dot{W}_{cv} + \sum_i \dot{m}_i\left(h_i + \frac{V_i^2}{2} + gz_i\right) - \sum_e \dot{m}_e\left(h_e + \frac{V_e^2}{2} + gz_e\right) \tag{4.24}$$

Equation 4.24 is an *accounting* balance for the energy of the control volume. It states that the rate of energy increase or decrease within the control volume equals the difference between the rates of energy transfer in and out across the boundary. The mechanisms of energy transfer are heat and work, as for closed systems, and the energy that accompanies the mass entering and exiting.

Equation 4.24 is the most general form of the conservation of energy principle for control volumes used for problem solving in this book. It serves as the starting point for applying the conservation of energy principle to control volumes in the analyses to follow. However, as for the case of the mass rate balance, the energy rate balance can be expressed in terms of local properties to obtain forms that are more generally applicable. Thus, the term $E_{cv}(t)$, representing the total energy associated with the control volume at time t, can be written as a volume integral

$$E_{cv}(t) = \int_V \rho e \, dV = \int_V \rho \left(u + \frac{V^2}{2} + gz\right) dV \tag{4.25}$$

Similarly, the terms accounting for the energy transfers accompanying mass flow and flow work at inlets and exits can be expressed as shown in the following form of the energy rate equation:

$$\begin{aligned} \frac{d}{dt} \int_V \rho e \, dV = \dot{Q}_{cv} - \dot{W}_{cv} &+ \sum_i \left[\int_A \left(h + \frac{V^2}{2} + gz\right) \rho V_n \, dA \right]_i \\ &- \sum_e \left[\int_A \left(h + \frac{V^2}{2} + gz\right) \rho V_n \, dA \right]_e \end{aligned} \tag{4.26}$$

Additional forms of the energy rate balance can be obtained by expressing the heat transfer rate $\dot{Q}_{cv}$ as the integral of the *heat flux* over the boundary of the control volume, and the work $\dot{W}_{cv}$ in terms of normal and shear stresses at the moving portions of the boundary.

In principle, the change in the energy of a control volume over a time period can be obtained by integration of the energy rate equation with respect to time. Such integrations would require information about the time dependences of the work and heat transfer rates, the various mass flow rates, and the states at which mass enters and leaves the control volume. Examples of this type of analysis are presented in Sec. 4.4. In Sec. 4.3 to follow, we consider forms that the energy rate balance takes for control volumes at steady state, for these are frequently used in practice.

4.3 ANALYSIS OF CONTROL VOLUMES AT STEADY STATE

In this section steady-state forms of the mass and energy rate balances are developed and applied to a variety of cases of engineering interest, including nozzles, turbines, compressors, and heat exchangers. The steady-state forms obtained do not apply to the transient startup or shutdown periods of operation of such devices, but only to periods of steady operation. This situation is commonly encountered in engineering.

4.3.1 Steady-State Forms of the Mass and Energy Rate Balances

For a control volume at steady state, the condition of the mass within the control volume and at the boundary does not vary with time. The mass flow rates and the rates of energy transfer by heat and work are also constant with time. There can be no accumulation of mass within the control volume, so $dm_{cv}/dt = 0$ and the conservation of mass principle takes the form

$$\underset{\text{(mass rate in)}}{\sum_i \dot{m}_i} = \underset{\text{(mass rate out)}}{\sum_e \dot{m}_e} \tag{4.27}$$

Furthermore, at steady state $dE_{cv}/dt = 0$, so Eq. 4.24 can be written as

$$0 = \dot{Q}_{cv} - \dot{W}_{cv} + \sum_i \dot{m}_i \left(h_i + \frac{V_i^2}{2} + gz_i\right) - \sum_e \dot{m}_e \left(h_e + \frac{V_e^2}{2} + gz_e\right) \tag{4.28a}$$

Alternatively

$$\dot{Q}_{cv} + \sum_i \dot{m}_i \left(h_i + \frac{V_i^2}{2} + gz_i\right) = \dot{W}_{cv} + \sum_e \dot{m}_e \left(h_e + \frac{V_e^2}{2} + gz_e\right) \tag{4.28b}$$

(energy rate in) (energy rate out)

Equation 4.27 states that at steady state the total rate at which mass enters the control volume equals the total rate at which mass exits. Similarly, Eqs. 4.28 state that the total rate at which energy is transferred into the control volume equals the total rate at which energy is transferred out.

Many important applications involve one-inlet, one-exit control volumes at steady state. It is instructive to apply the mass and energy rate balances to this special case. The mass rate balance reduces simply to $\dot{m}_1 = \dot{m}_2$. That is, the mass flow must be the same at the exit, 2, as it is at the inlet, 1. The common mass flow rate is designated simply by $\dot{m}$. Next, applying the energy rate balance and factoring the mass flow rate gives

$$0 = \dot{Q}_{cv} - \dot{W}_{cv} + \dot{m}\left[(h_1 - h_2) + \frac{(V_1^2 - V_2^2)}{2} + g(z_1 - z_2)\right] \tag{4.29a}$$

Or, dividing by the mass flow rate

$$0 = \frac{\dot{Q}_{cv}}{\dot{m}} - \frac{\dot{W}_{cv}}{\dot{m}} + (h_1 - h_2) + \frac{(V_1^2 - V_2^2)}{2} + g(z_1 - z_2) \tag{4.29b}$$

The enthalpy, kinetic energy, and potential energy terms all appear in Eqs. 4.29 as *differences* between their values at the inlet and exit. This illustrates that the datums used to assign values to specific enthalpy, velocity, and elevation cancel, provided the same ones are used at the inlet and exit. In Eq. 4.29b, the ratios $\dot{Q}_{cv}/\dot{m}$ and $\dot{W}_{cv}/\dot{m}$ are rates of energy transfer *per unit mass flowing through the control volume.*

The foregoing steady-state forms of the energy rate balance relate only energy transfer quantities evaluated at the *boundary* of the control volume. No details concerning properties *within* the control volume are required by, or can be determined with, these equations. When applying the energy rate balance in any of its forms, it is necessary to use the same units for all terms in the equation. For instance, *every* term in Eq. 4.29b must have a unit such as kJ/kg or Btu/lb. Appropriate unit conversions are emphasized in examples to follow.

4.3.2 Illustrations

Several examples are given in this section showing the use of the principles of conservation of mass and energy, together with relationships among properties for the analysis of control volumes *at steady state*. The examples are drawn from applications of general interest to engineers and are chosen to illustrate points common to all such analyses. Before studying them, it is recommended that you review the methodology for problem solving outlined in Sec. 1.6. As problems become more complicated, the use of a systematic problem-solving approach becomes increasingly important.

When the mass and energy rate balances are applied to a control volume, simplifications are normally needed to make the analysis tractable. That is, the control volume of interest is *modeled* by making assumptions. The *careful* and *conscious* step of listing assumptions is necessary in every engineering analysis. Therefore, an important part of this section is devoted to considering various assumptions that are commonly made when applying the conservation principles to different types of devices. When you study the examples presented, it is important to be as conscious of the role played by careful assumption making in arriving at solutions as it is to study other aspects of the solutions. In each case considered, steady-state operation is assumed. The flow is regarded as one-dimensional at places where mass enters and exits the control volume. Also, at each of these locations equilibrium property relations are assumed to apply.

In several of the examples to follow, the heat transfer term $\dot{Q}_{cv}$ is set to zero in the energy rate balance because it is small relative to other energy transfers across the boundary. This may be the result of one or more of the following factors: (1) The outer surface of the control volume is well insulated. (2) The outer surface area is too small for there to be effective heat transfer. (3) The temperature difference between the control volume and its surroundings is so small that the heat transfer can be ignored. (4) The gas or liquid passes through the control volume so quickly that there is not enough time for significant heat transfer to occur. The work term $\dot{W}_{cv}$ drops out of the energy rate balance when there are no rotating shafts, displacements of the boundary, electrical effects, or other work mechanisms associated with the control volume being considered. The kinetic and potential energies of the matter entering and exiting the control volume are neglected when they are small relative to other energy transfers.

Nozzles and Diffusers. A *nozzle* is a flow passage of varying cross-sectional area in which the velocity of a gas or liquid increases in the direction of flow. In a *diffuser,* the gas or liquid decelerates in the direction of flow. Figure 4.4 shows a nozzle in which the cross-sectional area decreases in the direction of flow and a diffuser in which the walls of the flow passage diverge. Nozzles and diffusers for high-speed gas flows also can be formed from a converging section followed by diverging section (Sec. 9.13).

For nozzles and diffusers, the only work is *flow work* at locations where mass enters and exits the control volume, so the term $\dot{W}_{cv}$ drops out of the energy rate equation for these devices. The change in potential energy from inlet to exit is negligible under most conditions. At steady state the mass and energy rate balances reduce, respectively, to

$$\cancelto{0}{\frac{dm_{cv}}{dt}} = \dot{m}_1 - \dot{m}_2$$

$$\cancelto{0}{\frac{dE_{cv}}{dt}} = \dot{Q}_{cv} - \cancelto{0}{\dot{W}_{cv}} + \dot{m}_1\left(h_1 + \frac{V_1^2}{2} + gz_1\right) - \dot{m}_2\left(h_2 + \frac{V_2^2}{2} + gz_2\right)$$

where 1 denotes the inlet and 2 the exit. By combining these into a single expression and dropping the potential energy change from inlet to exit

$$0 = \frac{\dot{Q}_{cv}}{\dot{m}} + (h_1 - h_2) + \left(\frac{V_1^2 - V_2^2}{2}\right)$$

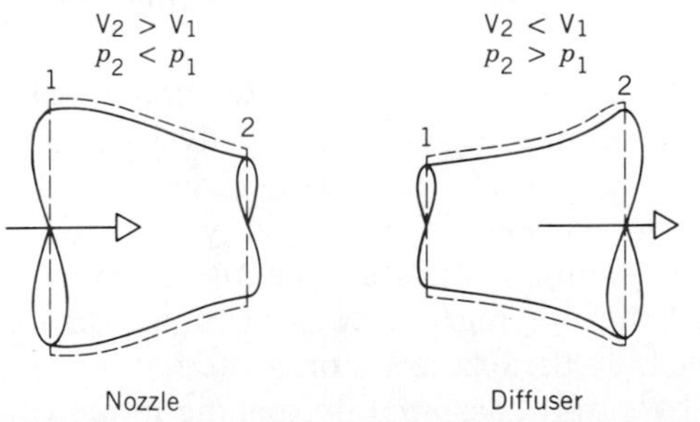

Figure 4.4 Illustration of a nozzle and a diffuser.

where $\dot{m}$ is the mass flow rate. The term $\dot{Q}_{cv}/\dot{m}$ representing heat transfer with the surroundings per unit of mass flowing through the nozzle or diffuser is often small enough relative to the enthalpy and kinetic energy changes that it can be dropped.

EXAMPLE 4.2

PROBLEM

Steam enters a nozzle operating at steady state with $p_1 = 40$ bars, $T_1 = 400°C$, and a velocity of 10 m/s. The steam flows through the nozzle with negligible heat transfer and no significant change in potential energy. At the exit, $p_2 = 15$ bars, and the velocity is 665 m/s. The mass flow rate is 2 kg/s. Determine the exit area of the nozzle, in m^2.

SOLUTION

Known: Steam flows at steady state through a nozzle with known properties at the inlet and exit, a known mass flow rate, and negligible effects of heat transfer and potential energy.

Find: Determine the exit area.

Schematic and Given Data:

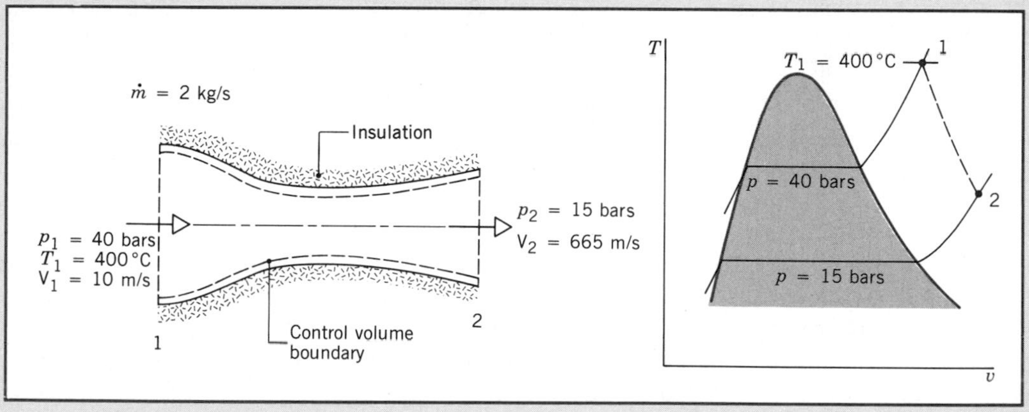

Figure E4.2

Assumptions:

1. The control volume shown on the accompanying figure is at steady state.
2. Heat transfer is negligible and $\dot{W}_{cv} = 0$.
3. The change in potential energy from inlet to exit can be neglected.

Analysis: The exit area can be determined from the mass flow rate $\dot{m}$ and Eq. 4.11b, which can be arranged to read

$$A_2 = \frac{\dot{m}v_2}{V_2}$$

To evaluate A_2 from this equation requires the specific volume v_2 at the exit, and this requires that the exit state be fixed.

The state at the exit is fixed by the values of two independent intensive properties. One is the pressure p_2, which is known. The other is the specific enthalpy h_2, determined from the steady-state energy rate balance

$$0 = \overset{0}{\cancel{\dot{Q}_{cv}}} - \overset{0}{\cancel{\dot{W}_{cv}}} + \dot{m}\left(h_1 + \frac{V_1^2}{2} + gz_1\right) - \dot{m}\left(h_2 + \frac{V_2^2}{2} + gz_2\right)$$

where $\dot{Q}_{cv}$ and $\dot{W}_{cv}$ are deleted by assumption 2. The change in specific potential energy drops out in accordance with assumption 3 and $\dot{m}$ cancels, leaving

$$0 = (h_1 - h_2) + \left(\frac{V_1^2 - V_2^2}{2}\right)$$

Solving for h_2

$$h_2 = h_1 + \left(\frac{V_1^2 - V_2^2}{2}\right)$$

From Table A-4, $h_1 = 3213.6$ kJ/kg. The velocities V_1 and V_2 are given. Inserting values and converting the units of the kinetic energy terms to kJ/kg results in

$$h_2 = 3213.6 \text{ kJ/kg} + \left[\frac{(10)^2 - (665)^2}{2}\right]\left(\frac{\text{m}^2}{\text{s}^2}\right)\left(\frac{1 \text{ N}}{1 \text{ kg}\cdot\text{m/s}^2}\right)\left(\frac{1 \text{ kJ}}{10^3 \text{ N}\cdot\text{m}}\right)$$

$$= 3213.6 - 221.1 = 2992.5 \text{ kJ/kg}$$

Finally, referring to Table A-4 at $p_2 = 15$ bars with $h_2 = 2992.5$ kJ/kg, the specific volume at the exit is $v_2 = 0.1627$ m³/kg. The exit area is then

$$A_2 = \frac{(2 \text{ kg/s})(0.1627 \text{ m}^3/\text{kg})}{665 \text{ m/s}} = 4.89 \times 10^{-4} \text{ m}^2$$

Comments:

1. Care must be taken in converting the units for specific kinetic energy to kJ/kg.
2. Although equilibrium property relations apply at the inlet and exit of the control volume, the intervening states of the fluid are not necessarily equilibrium states. Accordingly, the expansion through the nozzle is represented on the T–v diagram as a dashed line.
3. The area at the nozzle inlet can be found similarly, using $A_1 = \dot{m}v_1/V_1$.

Turbines. A *turbine* is a device in which work is developed as a result of a gas or liquid passing through a set of blades attached to a shaft free to rotate. A schematic of an axial-flow steam or gas turbine is shown in Fig. 4.5. Turbines are widely used in vapor power plants, gas turbine power plants, and aircraft engines (Chaps. 8 and 9). In these applications, superheated steam or a gas enters the turbine and expands to a lower exit pressure as work is developed.

For a turbine at steady state the mass and energy rate balances reduce to give Eq. 4.29b. In most cases, and particularly when gases are under consideration, the potential energy change is negligible. With a proper selection of the boundary of the control volume enclosing the turbine, the kinetic energy change is usually small enough to be neglected. The only heat transfer between the turbine and surroundings would be unavoidable heat transfer, and this is often small relative to the work and enthalpy terms.

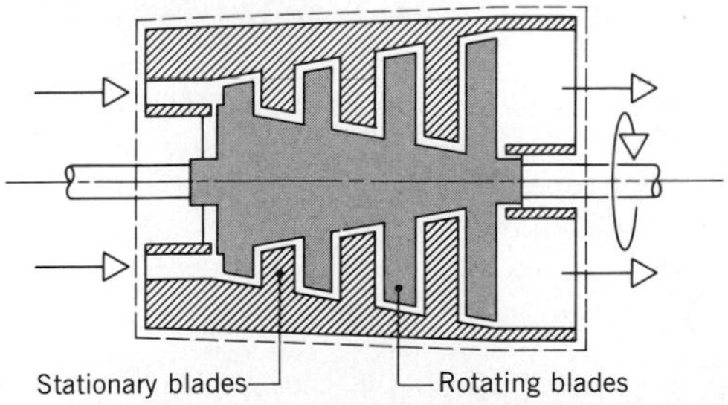

Figure 4.5 Schematic of an axial-flow turbine.

EXAMPLE 4.3

PROBLEM

Steam enters a turbine operating at steady state with a mass flow rate of 4600 kg/h. The turbine develops a power output of 1000 kW. At the inlet, the pressure is 60 bars, the temperature is 400°C, and the velocity is 10 m/s. At the exit, the pressure is 0.1 bar, the quality is 0.9 (90%), and the velocity is 50 m/s. Calculate the rate of heat transfer between the turbine and surroundings, in kW.

SOLUTION

Known: A steam turbine operates at steady state. The mass flow rate, power output, and states of the steam at the inlet and exit are known.

Find: Calculate the rate of heat transfer.

Schematic and Given Data:

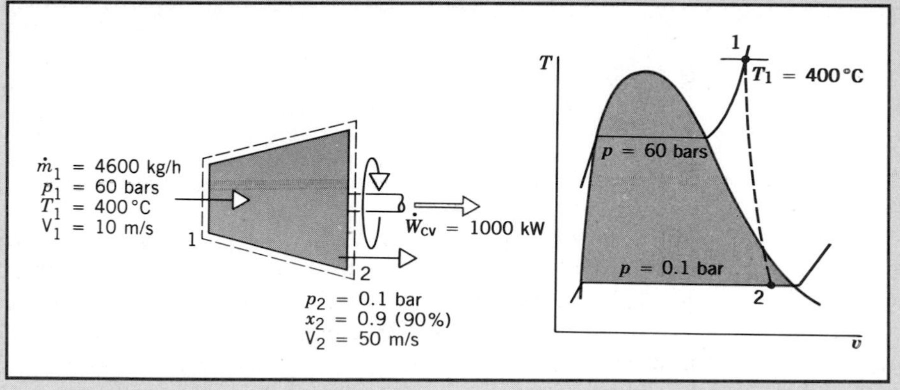

Figure E4.3

Assumptions:

1. The control volume shown on the accompanying figure is at steady state.
2. The change in potential energy from inlet to exit can be neglected.

Analysis: To calculate the heat transfer rate, begin with the one-inlet, one-exit form of the energy rate balance for a control volume at steady state

$$0 = \dot{Q}_{cv} - \dot{W}_{cv} + \dot{m}\left(h_1 + \frac{V_1^2}{2} + gz_1\right) - \dot{m}\left(h_2 + \frac{V_2^2}{2} + gz_2\right)$$

where $\dot{m}$ is the mass flow rate. By solving for $\dot{Q}_{cv}$ and dropping the potential energy change from inlet to exit

$$\dot{Q}_{cv} = \dot{W}_{cv} + \dot{m}\left[(h_2 - h_1) + \left(\frac{V_2^2 - V_1^2}{2}\right)\right]$$

To compare the magnitudes of the enthalpy and kinetic energy terms, and stress the unit conversions needed, each of these terms is evaluated separately.

First, the specific *enthalpy difference* $h_2 - h_1$ is found. Using Table A-4, $h_1 = 3177.2$ kJ/kg. State 2 is a two-phase liquid–vapor mixture, so with data from Table A-3 and the given quality

$$h_2 = h_{f2} + x_2(h_{g2} - h_{f2})$$
$$= 191.83 + (0.9)(2392.8) = 2345.4 \text{ kJ/kg}$$

Hence

$$h_2 - h_1 = 2345.4 - 3177.2 = -831.8 \text{ kJ/kg}$$

Consider next the specific *kinetic energy difference*. Using the given values for the velocities

$$\left(\frac{V_2^2 - V_1^2}{2}\right) = \left[\frac{(50)^2 - (10)^2}{2}\right]\left(\frac{m^2}{s^2}\right)\left(\frac{1\ N}{1\ kg \cdot m/s^2}\right)\left(\frac{1\ kJ}{10^3\ N \cdot m}\right)$$
$$= 1.2\ kJ/kg$$

Calculating $\dot{Q}_{cv}$ from the above expression

$$\dot{Q}_{cv} = (1000\ kW) + \left(4600\ \frac{kg}{h}\right)\left(\frac{1\ h}{3600\ s}\right)(-831.8 + 1.2)\left(\frac{kJ}{kg}\right)\left(\frac{1\ kW}{1\ kJ/s}\right)$$
$$= -61.3\ kW$$

Comments:

1. The negative value of $\dot{Q}_{cv}$ means that there is heat transfer from the turbine to its surroundings, as would be expected.
2. The magnitude of $\dot{Q}_{cv}$ is small relative to the power developed. Also, the magnitude of the change in specific kinetic energy from inlet to exit is very much smaller than the specific enthalpy change.

Compressors and Pumps. *Compressors* are devices in which work is done on a *gas* passing through them in order to raise the pressure. In *pumps,* the work input is used to change the state of a *liquid* passing through. A reciprocating compressor is shown in Fig. 4.6. Figure 4.7 gives schematic diagrams of three different rotating compressors: an axial-flow compressor, a centrifugal compressor, and a Roots-type.

The mass and energy rate balances reduce for compressors and pumps at steady state, as for the case of turbines considered previously. For compressors, the changes in specific kinetic and potential energies from inlet to exit are often small relative to the work done per unit of mass passing through the device. Heat transfer with the surroundings is frequently a secondary effect in both compressors and pumps.

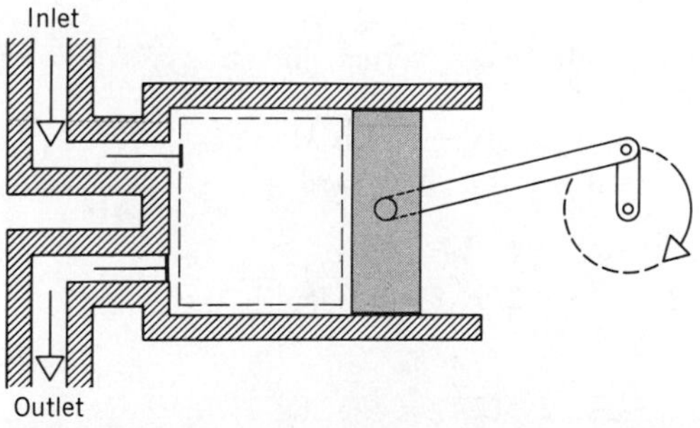

Figure 4.6 Reciprocating compressor.

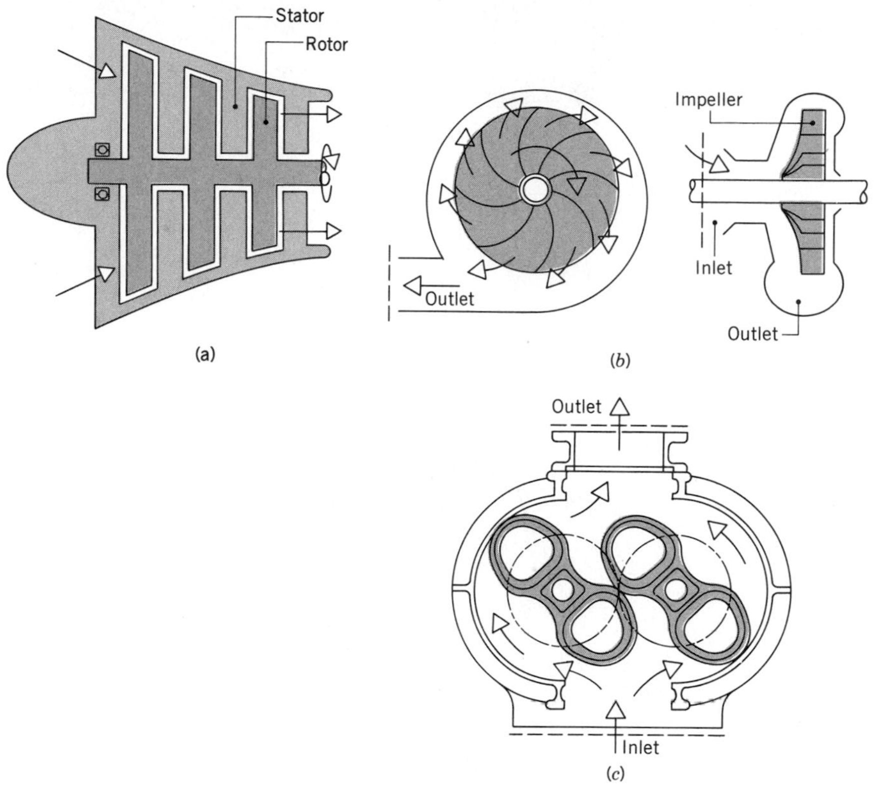

Figure 4.7 Rotating compressors. (*a*) Axial-flow. (*b*) Centrifugal. (*c*) Roots-type.

EXAMPLE 4.4

PROBLEM

Air enters a compressor operating at steady state at a pressure of 1 bar, a temperature of 290 K, and a velocity of 6 m/s through an inlet with an area of 0.1 m^2. At the exit, the pressure is 7 bars, the temperature is 450 K, and the velocity is 2 m/s. Heat transfer from the compressor to its surroundings occurs at a rate of 180 kJ/min. Employing the ideal gas model, calculate the power input to the compressor, in kW.

SOLUTION

Known: An air compressor operates at steady state with known inlet and exit states and a known heat transfer rate.

Find: Calculate the power required by the compressor.

Schematic and Given Data:

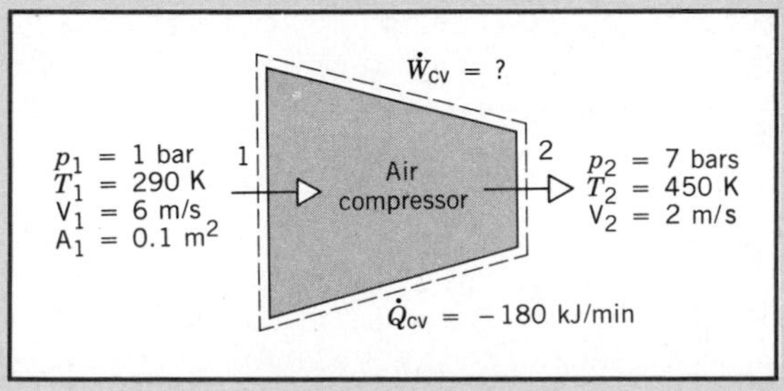

Figure E4.4

Assumptions:

1. The control volume shown on the accompanying figure is at steady state.
2. The change in potential energy from inlet to exit can be neglected.
3. The ideal gas model applies for the air.

Analysis: To calculate the power input to the compressor, begin with the energy rate balance for the one-inlet, one-exit control volume at steady state:

$$0 = \dot{Q}_{cv} - \dot{W}_{cv} + \dot{m}\left(h_1 + \frac{V_1^2}{2} + gz_1\right) - \dot{m}\left(h_2 + \frac{V_2^2}{2} + gz_2\right)$$

Solving

$$\dot{W}_{cv} = \dot{Q}_{cv} + \dot{m}\left[(h_1 - h_2) + \left(\frac{V_1^2 - V_2^2}{2}\right)\right]$$

The change in potential energy from inlet to exit drops out by assumption 2.

The mass flow rate $\dot{m}$ can be evaluated with given data at the inlet and the ideal gas equation of state.

$$\dot{m} = \frac{A_1V_1}{v_1} = \frac{A_1V_1p_1}{(\bar{R}/M)T_1} = \frac{(0.1\ \text{m}^2)(6\ \text{m/s})(10^5\ \text{N/m}^2)}{\left(\dfrac{8314}{28.97}\ \dfrac{\text{N}\cdot\text{m}}{\text{kg}\cdot\text{K}}\right)(290\ \text{K})} = 0.72\ \text{kg/s}$$

The specific enthalpies h_1 and h_2 can be found from Table A-16. At 290 K, $h_1 = 290.16$ kJ/kg. At 450 K, $h_2 = 451.8$ kJ/kg. Substituting values into the expression for $\dot{W}_{cv}$

$$\dot{W}_{cv} = \left(-180\ \frac{\text{kJ}}{\text{min}}\right)\left(\frac{\text{min}}{60\ \text{s}}\right) + 0.72\ \frac{\text{kg}}{\text{s}}\left[(290.16 - 451.8)\ \frac{\text{kJ}}{\text{kg}}\right.$$
$$\left. + \left(\frac{(6)^2 - (2)^2}{2}\right)\left(\frac{\text{m}^2}{\text{s}^2}\right)\left(\frac{1\ \text{N}}{1\ \text{kg}\cdot\text{m/s}^2}\right)\left(\frac{1\ \text{kJ}}{10^3\ \text{N}\cdot\text{m}}\right)\right]$$
$$= -\frac{3\ \text{kJ}}{\text{s}} + 0.72\ \frac{\text{kg}}{\text{s}}(-161.64 + 0.02)\ \frac{\text{kJ}}{\text{kg}}$$
$$= -119.4\ \frac{\text{kJ}}{\text{s}}\left(\frac{1\ \text{kW}}{1\ \text{kJ/s}}\right) = -119.4\ \text{kW}$$

Comments:

1. In this example $\dot{Q}_{cv}$ and $\dot{W}_{cv}$ have negative values, indicating that the direction of the heat transfer is *from* the compressor and work is done *on* the air passing through the compressor. The magnitude of the power *input* to the compressor is 119.4 kW.
2. The contribution of the kinetic energy is negligible in this case. Also, the heat transfer rate is seen to be small relative to the power input.
3. The applicability of the ideal gas model can be checked by reference to the generalized compressibility chart.

EXAMPLE 4.5

PROBLEM

A pump steadily draws water at a mass flow rate of 20 lb/s through a piping arrangement. At the pipe inlet, the pressure is 14.7 lbf/in.^{2}, the temperature is 68°F, and the velocity is 10 ft/s. At the pipe exit, the pressure is 20 lbf/in.^{2}, the temperature is 68°F, and the velocity is 40 ft/s. The exit is located 50 ft above the inlet. Determine the power required by the pump, in Btu/s and horsepower. The local acceleration of gravity is 32.0 ft/s^2.

SOLUTION

Known: Water is steadily pumped through a piping arrangement to a higher elevation where it is discharged with a known mass flow rate. Conditions are specified at the pipe inlet and exit.

Find: Determine the power required by the pump.

Schematic and Given Data:

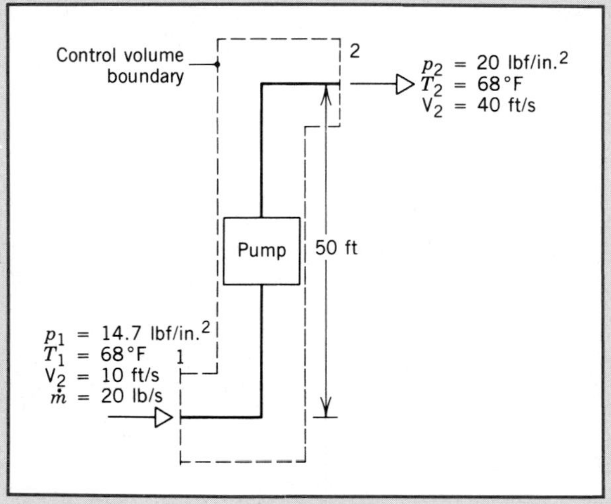

Figure E4.5

Assumptions:

1. The control volume shown on the accompanying figure is at steady state.
2. Heat transfer between the control volume and the surroundings can be ignored.
3. The acceleration of gravity is $g = 32.0$ ft/s^2.

Analysis: At steady state the energy rate balance reduces with assumption 2 to

$$0 = \overset{0}{\cancel{\dot{Q}_{cv}}} - \dot{W}_{cv} + \dot{m}\left(h_1 + \frac{V_1^2}{2} + gz_1\right) - \dot{m}\left(h_2 + \frac{V_2^2}{2} + gz_2\right)$$

where $\dot{m}$ is the mass flow rate. Solving

$$-\dot{W}_{cv} = \dot{m}\left[(h_2 - h_1) + \frac{V_2^2 - V_1^2}{2} + g(z_2 - z_1)\right]$$

To compare the magnitudes of the terms in brackets in this equation and stress the unit conversions required, each term is evaluated separately.

With Eq. 3.14 and $T_2 = T_1$, the change in specific enthalpy from inlet to exit is

$$\begin{aligned} h_2 - h_1 &\approx [h_f + v_f(p - p_{sat})]_2 - [h_f + v_f(p - p_{sat})]_1 \\ &= v_f(p_2 - p_1) \end{aligned}$$

where v_f, h_f, and p_{sat} are the specific volume, enthalpy, and saturation pressure at 68°F. With v_f from Table A-2E

$$\begin{aligned} h_2 - h_1 &= \left(0.01605\ \frac{\text{ft}^3}{\text{lb}}\right)(20 - 14.7)\left(\frac{\text{lbf}}{\text{in.}^2}\right)\left(\frac{144\ \text{in.}^2}{1\ \text{ft}^2}\right)\left(\frac{1\ \text{Btu}}{778\ \text{ft}\cdot\text{lbf}}\right) \\ &= 0.0157\ \text{Btu/lb} \end{aligned}$$

The change in specific kinetic energy from inlet to exit is

$$\frac{V_2^2 - V_1^2}{2} = \frac{(40\ \text{ft/s})^2 - (10\ \text{ft/s})^2}{(2)\left(\dfrac{32.2\ \text{lb}\cdot\text{ft/s}^2}{1\ \text{lbf}}\right)\left(\dfrac{778\ \text{ft}\cdot\text{lbf}}{1\ \text{Btu}}\right)} = 0.0299\ \text{Btu/lb}$$

The change in specific potential energy from inlet to exit is

$$g(z_2 - z_1) = \frac{(32.0\ \text{ft/s}^2)(50\ \text{ft})}{\left(\dfrac{32.2\ \text{lb}\cdot\text{ft/s}^2}{1\ \text{lbf}}\right)\left(\dfrac{778\ \text{ft}\cdot\text{lbf}}{1\ \text{Btu}}\right)} = 0.0639\ \text{Btu/lb}$$

Collecting terms

$$-\dot{W}_{cv} = \left(20\ \frac{\text{lb}}{\text{s}}\right)(0.0157 + 0.0299 + 0.0639)\left(\frac{\text{Btu}}{\text{lb}}\right) = 2.19\ \text{Btu/s}$$

Converting to horsepower

$$-\dot{W}_{cv} = \left(2.19\ \frac{\text{Btu}}{\text{s}}\right)\left(\frac{3600\ \text{s}}{1\ \text{h}}\right)\left(\frac{1\ \text{hp}}{2545\ \text{Btu/h}}\right) = 3.10\ \text{horsepower}$$

Comments:

1. The magnitude of the work *input* to the pump is 3.10 horsepower.
2. The largest portion of the work input goes into increasing the potential energy of the water flowing from the inlet to the exit.
3. Note the use of conversion factors in this example.
4. The specific enthalpy change, $h_2 - h_1$, also could be evaluated on the basis of an incompressible liquid, using Eq. 3.20. This is left as an exercise.

Heat Exchangers. Devices that transfer energy between fluids at different temperatures by heat transfer modes such as discussed in Sec. 2.4.2 are called *heat exchangers*. One common type of heat exchanger is a vessel in which hot and cold streams are mixed directly as shown in Fig. 4.8*a*. An open feedwater heater is an example of this type of device. Another common type of heat exchanger is one in which a gas or liquid is *separated* from another gas or liquid by a wall through which energy passes by conduction. These heat exchangers, known as recuperators, take many different forms. Counterflow and parallel tube-within-a-tube configurations are shown in Figs. 4.8*b* and 4.8*c*, respectively. Other configurations include cross-flow, as in automobile radiators, and multiple-pass shell-and-tube condensers and evaporators. Figure 4.8*d* illustrates a cross-flow heat exchanger.

The only work interaction at the boundary of a control volume enclosing a heat exchanger is flow work at the places where matter enters and exits, so the term $\dot{W}_{cv}$ of the energy rate balance can be set to zero. Although high rates of energy transfer may be achieved from stream to stream, the heat transfer from the outer surface of the heat exchanger to the surroundings is often small enough to be neglected. In addition, the kinetic and potential energies of the flowing streams can often be ignored at the inlets and exits. The next two examples illustrate how the mass and energy rate balances are applied to heat exchangers at steady state.

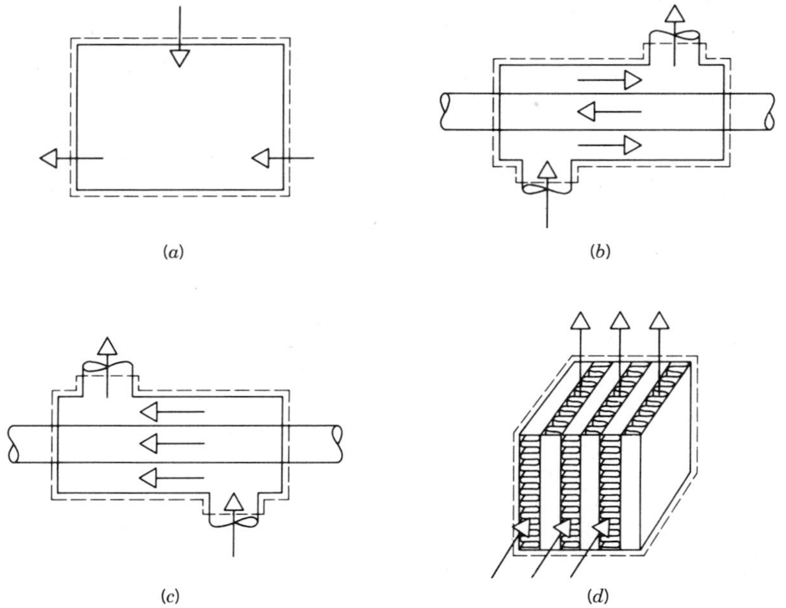

Figure 4.8 Common heat exchanger types. (*a*) Direct contact heat exchanger. (*b*) Tube-within-a-tube counterflow heat exchanger. (*c*) Tube-within-a-tube parallel flow heat exchanger. (*d*) Cross-flow heat exchanger.

EXAMPLE 4.6

PROBLEM

An open feedwater heater in a vapor power plant operates at steady state with liquid entering at inlet 1 with $T_1 = 40°C$ and $p_1 = 7.0$ bars. Water vapor at $T_2 = 200°C$ and $p_2 = 7.0$ bars enters at inlet 2. Saturated liquid water exits with a pressure of $p_3 = 7.0$ bars. Ignoring heat transfer with the surroundings and all kinetic and potential energy effects, determine the ratio of mass flow rates, $\dot{m}_1/\dot{m}_2$.

SOLUTION

Known: An open feedwater heater having two inlets and one exit operates at steady state without significant heat transfer to its surroundings. The states of the fluid entering and leaving are specified.

Find: Determine the ratio of the mass flow rates at the two inlets, $\dot{m}_1/\dot{m}_2$.

Schematic and Given Data:

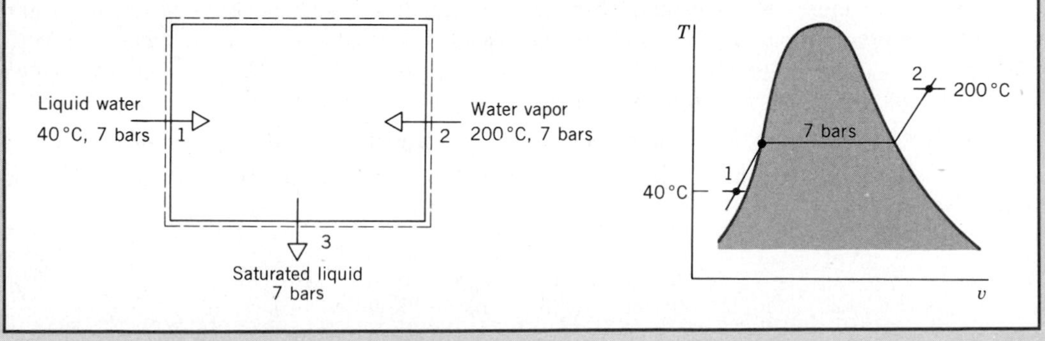

Figure E4.6

Assumptions:

1. The control volume shown on the accompanying figure is at steady state.
2. There is no heat transfer with the surroundings and $\dot{W}_{cv} = 0$.
3. Kinetic and potential energy effects can be neglected.

Analysis: By combining the mass and energy rate balances an expression for the mass flow rate ratio $\dot{m}_1/\dot{m}_2$ can be obtained as follows: The mass rate balance reduces at steady state to give

$$\dot{m}_3 = \dot{m}_1 + \dot{m}_2$$

The steady-state form of the energy rate balance is

$$0 = \underline{\dot{Q}_{cv}} - \underline{\dot{W}_{cv}} + \dot{m}_1\left(h_1 + \underline{\frac{V_1^2}{2} + gz_1}\right) + \dot{m}_2\left(h_2 + \underline{\frac{V_2^2}{2} + gz_2}\right) - \dot{m}_3\left(h_3 + \underline{\frac{V_3^2}{2} + gz_3}\right)$$

The underlined terms drop out by assumptions 2 and 3, leaving

$$0 = \dot{m}_1 h_1 + \dot{m}_2 h_2 - \dot{m}_3 h_3$$

Introducing $\dot{m}_3 = \dot{m}_1 + \dot{m}_2$

$$0 = \dot{m}_1 h_1 + \dot{m}_2 h_2 - (\dot{m}_1 + \dot{m}_2) h_3$$

Dividing by $\dot{m}_2$ and solving for $\dot{m}_1/\dot{m}_2$

$$\frac{\dot{m}_1}{\dot{m}_2} = \frac{h_2 - h_3}{h_3 - h_1}$$

From Table A-4, $h_2 = 2844.8$ kJ/kg, and from Table A-3, $h_3 = 697.22$ kJ/kg. At inlet 1 the water is a liquid. Using Eq. 3.14, together with data from Table A-2, the specific enthalpy at 1 is

$$\begin{aligned} h_1 &\approx h_f(T_1) + v_f(T_1)[p_1 - p_{sat}(T_1)] \\ &= 167.57 \text{ kJ/kg} + \left(1.0078 \times 10^{-3} \frac{\text{m}^3}{\text{kg}}\right)(7 - 0.07384)\left(10^5 \frac{\text{N}}{\text{m}^2}\right)\left(\frac{1 \text{ kJ}}{10^3 \text{ N}\cdot\text{m}}\right) \\ &= 167.57 + 0.7 = 168.27 \text{ kJ/kg} \end{aligned}$$

Finally, calculating the mass flow rate ratio

$$\frac{\dot{m}_1}{\dot{m}_2} = \frac{2844.8 - 697.22}{697.22 - 168.27} = 4.06$$

Comments:

1. This example illustrates the application of the mass and energy rate balances to a control volume with multiple inlets and exits.
2. In this case the second term on the right of Eq. 3.14 does not contribute significantly to the value of h_1, so we could have written $h_1 \approx h_f(T_1)$.

EXAMPLE 4.7

PROBLEM

Steam enters the condenser of a vapor power plant at 0.1 bar with a quality of 0.95 and condensate exits at 0.1 bar and 45°C. Cooling water enters the condenser in a separate stream as a liquid at 20°C and exits as a liquid at 35°C with no change in pressure. Heat transfer from the outside of the condenser and changes in the kinetic and potential energies of the flowing streams can be ignored. For steady-state operation, determine

(a) the ratio of the mass flow rate of the cooling water to the mass flow rate of the condensing steam.

(b) the rate of energy transfer from the condensing steam to the cooling water, in kJ per kg of steam passing through the condenser.

SOLUTION

Known: Steam is condensed at steady state by interacting with a separate liquid water stream.

Find: Determine the ratio of the mass flow rate of the cooling water to the mass flow rate of the steam and the rate of energy transfer from the steam to the cooling water.

Schematic and Given Data:

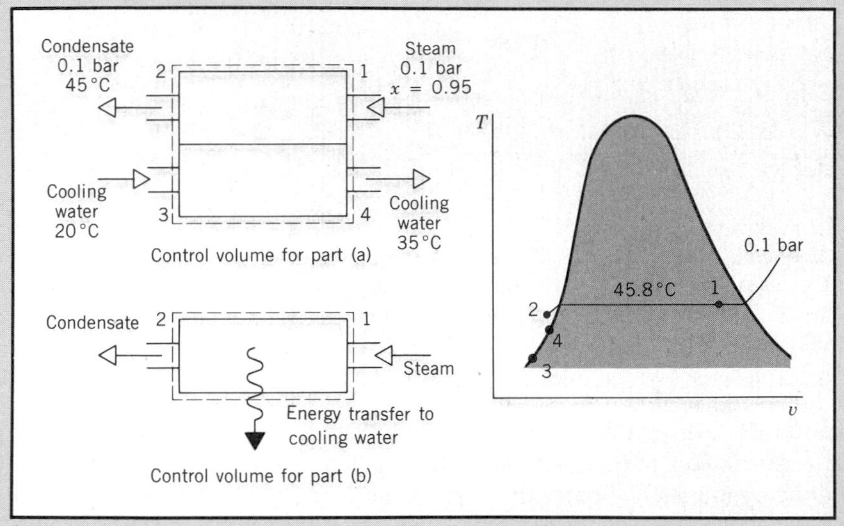

Figure E4.7

Assumptions:

1. Each of the two control volumes shown on the accompanying sketch is at steady state.
2. There is no significant heat transfer between the overall condenser and its surroundings, and $\dot{W}_{cv} = 0$.
3. Changes in the kinetic and potential energies of the flowing streams from inlet to exit can be ignored.
4. The incompressible liquid model applies for the cooling water, for which pressure remains constant.

Analysis: The steam and the cooling water streams do not mix. Thus, the mass rate balances for each of the two streams reduce at steady state to give

$$\dot{m}_1 = \dot{m}_2 \qquad \text{and} \qquad \dot{m}_3 = \dot{m}_4$$

(a) The ratio of the mass flow rate of the cooling water to the mass flow rate of the condensing steam, $\dot{m}_3/\dot{m}_1$, can be found from the steady-state form of the energy rate balance applied to the overall condenser as follows:

$$0 = \underline{\dot{Q}_{cv}} - \underline{\dot{W}_{cv}} + \dot{m}_1\left(h_1 + \underline{\frac{V_1^2}{2} + gz_1}\right) + \dot{m}_3\left(h_3 + \underline{\frac{V_3^2}{2} + gz_3}\right) - \dot{m}_2\left(h_2 + \underline{\frac{V_2^2}{2} + gz_2}\right) - \dot{m}_4\left(h_4 + \underline{\frac{V_4^2}{2} + gz_4}\right)$$

The underlined terms drop out by assumptions 2 and 3. With these simplifications, together with the above mass flow rate relations, the energy rate balance becomes simply

$$0 = \dot{m}_1(h_1 - h_2) + \dot{m}_3(h_3 - h_4)$$

Solving, we get

$$\frac{\dot{m}_3}{\dot{m}_1} = \frac{h_1 - h_2}{h_4 - h_3}$$

The specific enthalpy h_1 can be determined using the given quality and data from Table A-3. From Table A-3 at 0.1 bar, $h_f = 191.83$ kJ/kg and $h_g = 2584.7$ kJ/kg, so

$$h_1 = 191.83 + 0.95(2584.7 - 191.83) = 2465.1 \text{ kJ/kg}$$

The specific enthalpy at 2 is given by $h_2 \approx h_f(T_2) = 188.45$ kJ/kg. With $c = 4.18$ kJ/kg·K from Table A-13, Eq. 3.20 gives $h_4 - h_3 = 62.7$ kJ/kg. Thus

$$\frac{\dot{m}_3}{\dot{m}_1} = \frac{2465.1 - 188.45}{62.7} = 36.3$$

(b) For a control volume enclosing the steam side of the condenser only, the steady-state form of energy rate balance is

$$0 = \dot{Q}_{cv} - \underline{\dot{W}_{cv}} + \dot{m}_1\left(h_1 + \underline{\frac{V_1^2}{2} + gz_1}\right) - \dot{m}_2\left(h_2 + \underline{\frac{V_2^2}{2} + gz_2}\right)$$

The underlined terms drop out by assumptions 2 and 3. Combining this equation with $\dot{m}_1 = \dot{m}_2$, the following expression for the rate of energy transfer between the condensing steam and the cooling water results:

$$\dot{Q}_{cv} = \dot{m}_1(h_2 - h_1)$$

Dividing by the mass flow rate of the steam, $\dot{m}_1$, and inserting values

$$\frac{\dot{Q}_{cv}}{\dot{m}_1} = h_2 - h_1 = 188.45 - 2465.1 = -2276.7 \text{ kJ/kg}$$

where the minus sign signifies that energy is transferred *from* the condensing steam *to* the cooling water.

Comment: Depending on where the boundary of the control volume is located, two different formulations of the energy rate balance are obtained. In part (a), both streams are included in the control volume. Energy transfer between them occurs internally and not across the boundary of the control volume, so the term $\dot{Q}_{cv}$ drops out of the energy rate balance. With the control volume of part (b), however, the term $\dot{Q}_{cv}$ must be included.

Throttling Devices. A significant reduction in pressure can be achieved simply by introducing a restriction into a line through which a gas or liquid flows. This is commonly done by means of a partially opened valve or a porous plug, as illustrated in Fig. 4.9.

For a control volume enclosing such a device, the mass and energy rate balances reduce at steady state to

$$0 = \dot{m}_1 - \dot{m}_2$$

$$0 = \dot{Q}_{cv} - \cancelto{0}{\dot{W}_{cv}} + \dot{m}_1\left(h_1 + \frac{V_1^2}{2} + gz_1\right) - \dot{m}_2\left(h_2 + \frac{V_2^2}{2} + gz_2\right)$$

There is usually no significant heat transfer with the surroundings and the change in potential energy from inlet to exit is negligible. With these idealizations, the mass and energy rate balances combine to give

$$h_1 + \frac{V_1^2}{2} = h_2 + \frac{V_2^2}{2}$$

Although velocities may be relatively high in the vicinity of the restriction, measurements made upstream and downstream of the reduced flow area show in most cases

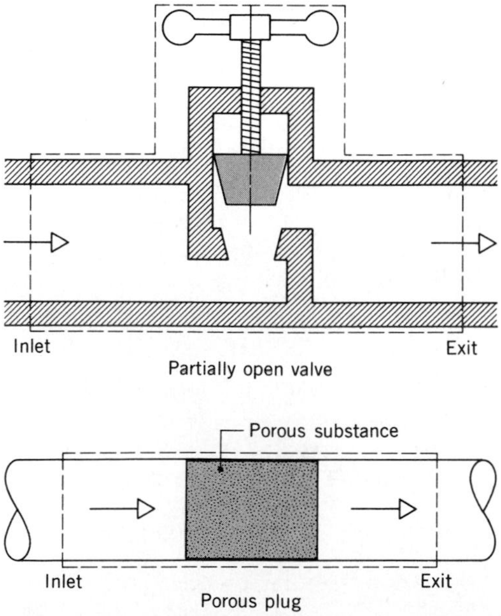

Figure 4.9 Examples of throttling devices.

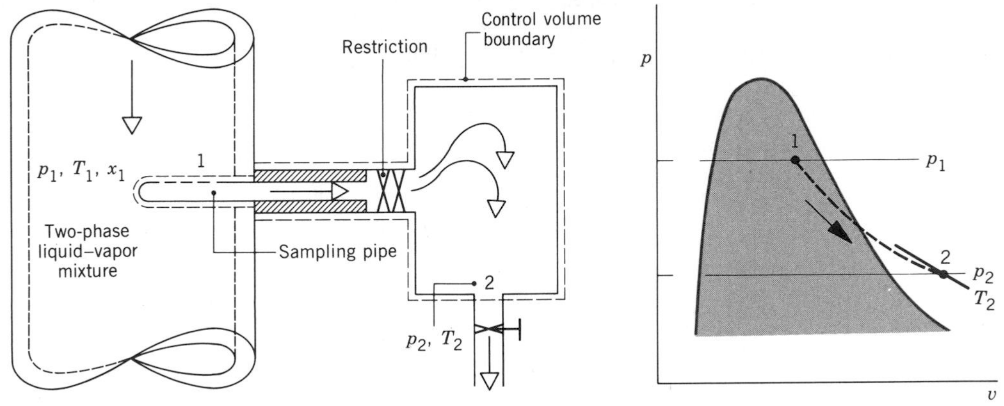

Figure 4.10 Throttling calorimeter and accompanying property diagram.

that the change in the specific kinetic energy of the gas or liquid between these locations can be neglected. With this further simplification, the last equation reduces to

$$h_1 = h_2 \tag{4.30}$$

When the flow through a valve or other restriction is idealized in this way, the process is called a *throttling process*.

An application of the throttling process concept involves the *throttling calorimeter,* which is a device for determining the quality of a two-phase liquid–vapor mixture. Figure 4.10 shows a throttling calorimeter connected to a line carrying a two-phase liquid–vapor mixture at pressure p_1, temperature T_1, and quality x_1. To determine the quality, suppose a small quantity of the two-phase liquid–vapor mixture in the line is steadily diverted through the throttling calorimeter, as indicated in Fig. 4.10. If the pressure of the flowing stream is reduced sufficiently by the restriction, the stream would pass into the superheated vapor region. Measurement of the pressure p_2 and the temperature T_2 at the exit of the calorimeter would determine the state there. With p_2 and T_2 known, the specific enthalpy h_2 could be determined from the superheated-vapor table for the substance, and this would allow the quality to be determined as illustrated in the next example.

Another application of the throttling process occurs in vapor-compression refrigeration systems. A valve is used in such a system to reduce the pressure of the refrigerant from the pressure at the exit of the *condenser* to the lower pressure existing in the *evaporator*. We consider this further in Chap. 10. The throttling process also plays a role in the *Joule–Thomson* expansion considered in Chap. 11.

EXAMPLE 4.8

PROBLEM

A supply line carries a two-phase liquid–vapor mixture of steam at 300 lbf/in.[2] A small fraction of the flow in the line is diverted through a throttling calorimeter and exhausted to the atmosphere at 14.7 lbf/in.[2] The temperature of the exhaust steam is measured as 250°F. Determine the quality of the steam in the supply line.

SOLUTION

Known: Steam diverted from a supply line expands in a throttling process through a throttling calorimeter and is exhausted to the atmosphere.

Find: Determine the quality of the steam in the supply line.

Schematic and Given Data:

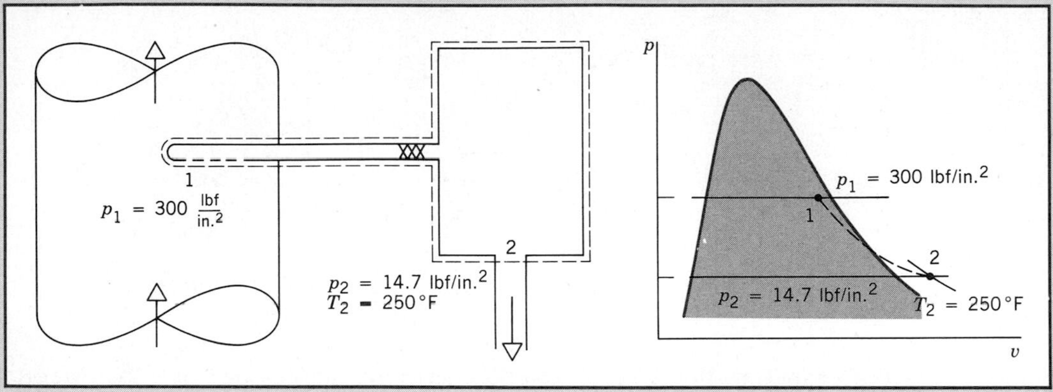

Figure E4.8

Assumptions:

1. The control volume shown on the accompanying figure is at steady state.
2. There is no significant heat transfer with the surroundings, and $\dot{W}_{cv} = 0$.
3. The changes in kinetic and potential energy across the throttling calorimeter are negligible.

Analysis: In accordance with the assumptions listed, $h_1 = h_2$. In addition, for the two-phase liquid–vapor mixture

$$h_1 = h_{f1} + x_1(h_{g1} - h_{f1})$$

Combining these relations and solving for x_1

$$x_1 = \frac{h_2 - h_{f1}}{h_{g1} - h_{f1}}$$

From Table A-3E at 300 lbf/in.², h_{g1} = 1203.9 Btu/lb and h_{f1} = 394.1 Btu/lb. State 2 is in the superheated vapor region, so from Table A-4E at 14.7 lbf/in.² and 250°F, h_2 = 1168.8 Btu/lb. Inserting values in the above expression, the quality in the line is $x_1 = 0.957$ (95.7%).

Comment: The use of the throttling calorimeters is limited to qualities greater than about 94% to ensure that the steam leaving the calorimeter is superheated. Small corrections can be made to account for the effects of unavoidable stray heat transfer with the surroundings.

4.4 TRANSIENT ANALYSIS

Many devices undergo periods of transient operation in which the state changes with time as mass enters or exits. Examples include the startup or shutdown of turbines, compressors, and boilers. Additional examples are provided by vessels that are filled with, or emptied of, gases or liquids. Because the work and heat transfer rates, mass flow rates, and properties may vary with time during transient operation, the steady-state assumption is not appropriate when analyzing cases such as these.

The following examples provide illustrations of the transient analysis of control volumes using the conservation of mass and energy principles. In each case considered, one-dimensional flow is assumed at places where mass crosses the boundary of the control volume, and the mass is assumed to be described by equilibrium property relations at each of these locations. In addition, idealizations are made about the state of the mass within the control volume, the state of the mass entering or exiting the control volume, or both. These idealizations make the transient analyses tractable.

EXAMPLE 4.9

PROBLEM

Two-tenths of a pound of air at 2.0 atm and 540°R is trapped within a syringe by a plunger at one end and a stopcock at the other. The stopcock is opened and the plunger moved to inject the trapped gas into a container that initially holds 1 lb of air at 1.0 atm and 540°R. The plunger maintains the state of the injected gas constant until all has passed the stopcock. Ignoring heat transfer with the surroundings, and kinetic and potential energy effects, determine the final equilibrium temperature, in °R, and pressure, in lbf/in.^{2}, of the gas in the container following injection of the trapped air. Employ the ideal gas model for the air.

SOLUTION

Known: A small amount of air at a given state is injected into a container initially holding air at a different state.

Find: Determine the final equilibrium temperature and pressure of the air in the container after all the injected air has entered.

Schematic and Given Data:

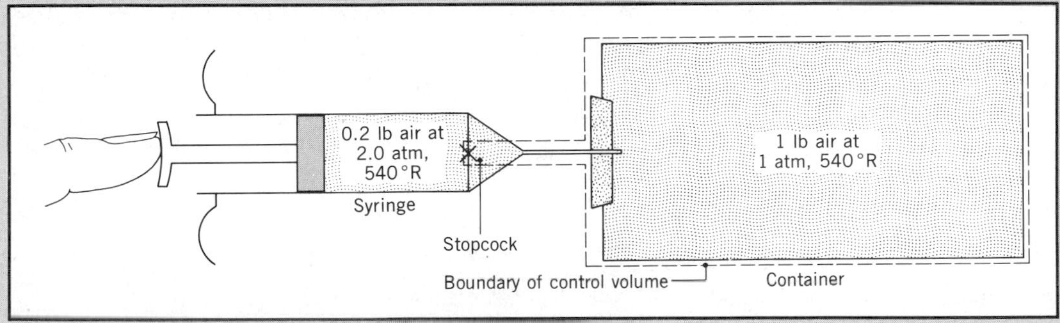

Figure E4.9

Assumptions:

1. The control volume is defined by the dashed line on the accompanying diagram.
2. The state of the injected air remains constant until it passes the stopcock.
3. There is no heat transfer with the surroundings, $\dot{W}_{cv} = 0$, and all kinetic and potential energy effects can be ignored.
4. The ideal gas model applies for the air.
5. The initial and final states of the air within the container are equilibrium states.

Analysis: Since the control volume has a single inlet and no exits, the mass rate balance takes the form

$$\frac{dm_{cv}}{dt} = \dot{m}_i$$

Similarly, the energy rate balance is

$$\frac{dE_{cv}}{dt} = \dot{Q}_{cv} - \dot{W}_{cv} + \dot{m}_i\left(h_i + \frac{V_i^2}{2} + gz_i\right)$$

With assumption 3, $\dot{Q}_{cv} = \dot{W}_{cv} = 0$, and all kinetic and potential energy effects can be neglected. Thus

$$\frac{dU_{cv}}{dt} = \dot{m}_i h_i$$

Combining the mass and energy rate balances gives

$$\frac{dU_{cv}}{dt} = h_i \frac{dm_{cv}}{dt}$$

In accordance with assumption 2, the specific enthalpy of the air entering the control volume is constant: $h_i = h(T_S)$, where T_S is the temperature of the air within the syringe. Thus, on integration, the above equation becomes

$$\Delta U_{cv} = h(T_S)\, \Delta m_{cv}$$

where ΔU_{cv} and Δm_{cv} denote, respectively, the changes in the internal energy and mass of the control volume. The change in the mass contained within the control volume equals the mass initially in the syringe

$$\Delta m_{cv} = m_S$$

The change in internal energy is

$$\Delta U_{cv} = (m_C + m_S)u(T) - m_C u(T_C)$$

where m_C and T_C denote, respectively, the mass and temperature of the air initially in the container, and T is the final temperature of the air within the container.

Collecting the last three equations and solving

$$u(T) = \frac{m_C u(T_C) + m_S h(T_S)}{m_C + m_S}$$

where $m_C = 1$ lb and $m_S = 0.2$ lb. With $T_C = T_S = 540°R$, Table A-16E gives $u = 92.04$ Btu/lb and $h = 129.06$ Btu/lb. Thus

$$u(T) = \frac{(1)(92.04) + (0.2)(129.06)}{1.2} = 98.21 \text{ Btu/lb}$$

Interpolating in Table A-16E with this value of specific internal energy, the final temperature of the air in the container is $T = 576°R$ (116°F).

The final pressure p is found by using the ideal gas equation of state

$$p = (m_C + m_S)RT/V$$

The volume of the container, V, can be calculated from the ideal gas equation of state

$$V = m_C R T_C / p_C$$

where p_C is the pressure of the air initially in the container. Combining the last two equations

$$p = \frac{(m_C + m_S)RT}{(m_C R T_C / p_C)} = p_C \left(\frac{T}{T_C}\right)\left(\frac{m_C + m_S}{m_C}\right)$$

Substituting values

$$p_f = (14.7 \text{ lbf/in.}^2)(576/540)(1.2/1) = 18.8 \text{ lbf/in.}^2$$

Comments:

1. The analysis is simplified by assumption 2, which fixes the state of the air entering the control volume.
2. The plunger does not directly enter the analysis, for in this control volume formulation it is located in the surroundings. However, there is *flow work* at the place where air enters the control volume. This is incorporated in the pv term of the specific enthalpy at the inlet. The final temperature of the air within the container is higher than the initial temperature because energy enters the control volume by flow work.
3. This problem can be solved on a closed system basis by taking the system to consist of *all* the air initially present in the syringe *and* container. The work of this system as the plunger moves is $W = -p_S V_S$, where p_S denotes the constant pressure of the air in the syringe and V_S is the volume occupied by the air initially in the syringe. A minus sign appears because work is done *on* the system by the plunger. The remainder of the closed system analysis is left as an exercise.

EXAMPLE 4.10

PROBLEM

Steam at a pressure of 15 bars and a temperature of 320°C is contained in a large tank. Connected to the tank through a valve is a turbine followed by a small initially evacuated vessel with a volume of 0.6 m^3. The valve is opened and the vessel fills with steam until the pressure is 15 bars and the temperature is 400°C. The valve is then closed. The filling process takes place adiabatically and kinetic and potential energy effects are negligible. Determine the amount of work developed by the turbine, in kJ.

SOLUTION

Known: Steam contained in a large tank at a known state passes from the tank through a turbine into a small vessel of known volume until a specified final condition is attained in the vessel.

Find: Determine the work developed by the turbine.

Schematic and Given Data:

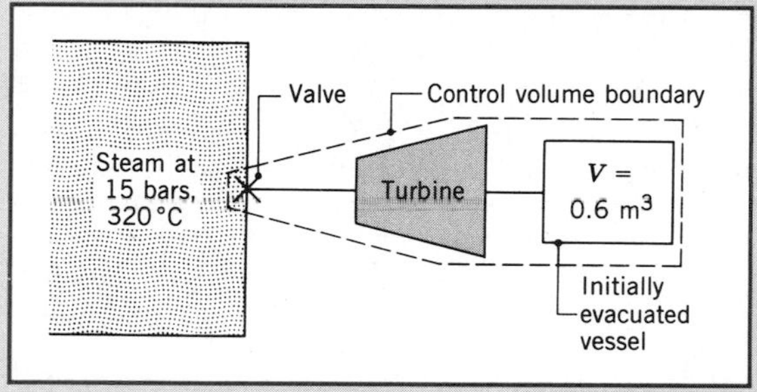

Figure E4.10

Assumptions:

1. The control volume is defined by the dashed line on the accompanying diagram.
2. For the control volume, $\dot{Q}_{cv} = 0$ and kinetic and potential energy effects are negligible.
3. The state of the steam within the large tank remains constant. The final state of the steam in the smaller vessel is an equilibrium state.
4. The amount of mass stored within the turbine and the interconnecting piping at the end of the filling process is negligible.

Analysis: Since the control volume has a single inlet and no exits, the mass rate balance reduces to

$$\frac{dm_{cv}}{dt} = \dot{m}_i$$

The energy rate balance reduces with assumption 2 to

$$\frac{dU_{cv}}{dt} = -\dot{W}_{cv} + \dot{m}_i h_i$$

Combining the mass and energy rate balances gives

$$\frac{dU_{cv}}{dt} = -\dot{W}_{cv} + h_i \frac{dm_{cv}}{dt}$$

Integrating

$$\Delta U_{cv} = -W_{cv} + h_i \, \Delta m_{cv}$$

In accordance with assumption 3, the specific enthalpy of the steam entering the control volume is constant at the value corresponding to the state in the large tank.

Solving for W_{cv}

$$W_{cv} = h_i \, \Delta m_{cv} - \Delta U_{cv}$$

ΔU_{cv} and Δm_{cv} denote, respectively, the changes in internal energy and mass of the control volume. With assumption 4, these terms can be identified with the small vessel only.

Since the small vessel is initially evacuated, the terms ΔU_{cv} and Δm_{cv} reduce to the internal energy and mass within the vessel at the end of the process. That is

$$\Delta U_{cv} = (m_2 u_2) - \cancelto{0}{(m_1 u_1)}, \qquad \Delta m_{cv} = m_2 - \cancelto{0}{m_1}$$

where 1 and 2 denote the initial and final states within the vessel, respectively.

Collecting results yields

$$W_{cv} = m_2(h_i - u_2)$$

The mass within the vessel at the end of the process can be evaluated from the known volume and the specific volume of steam at 15 bars and 400°C from Table A-4

$$m_2 = \frac{V}{v_2} = \frac{0.6 \text{ m}^3}{(0.203 \text{ m}^3/\text{kg})} = 2.96 \text{ kg}$$

The specific internal energy of steam at 15 bars and 400°C from Table A-4 is 2951.3 kJ/kg. Also, at 15 bars and 320°C, $h_i = 3081.9$ kJ/kg.

Substituting values into the expression for work

$$W_{cv} = 2.96 \text{ kg}(3081.9 - 2951.3)\text{kJ/kg} = 386.6 \text{ kJ}$$

Comments:

1. As for Example 4.9, a significant aspect of this example is the energy transfer into the control volume by flow work, incorporated in the pv term of the specific enthalpy at the inlet.
2. If the turbine were removed and steam allowed to flow adiabatically into the smaller vessel without doing work, the final steam temperature in the vessel would be 477°C, as may be verified.

EXAMPLE 4.11

PROBLEM

A vessel having a volume of 0.85 m³ initially contains water as a two-phase liquid–vapor mixture at 260°C and a quality of 0.7. Saturated water vapor at 260°C is slowly withdrawn through a pressure-regulating valve at the top of the tank as energy is transferred by heat to maintain the pressure constant in the tank. This continues until the tank is filled with saturated vapor at 260°C. Determine the amount of heat transfer, in kJ. Neglect all kinetic and potential energy effects.

SOLUTION

Known: A vessel initially holding a two-phase liquid–vapor mixture is heated while saturated water vapor is slowly removed. This continues at constant pressure until the tank is filled only with saturated vapor.

Find: Determine the amount of heat transfer.

Schematic and Given Data:

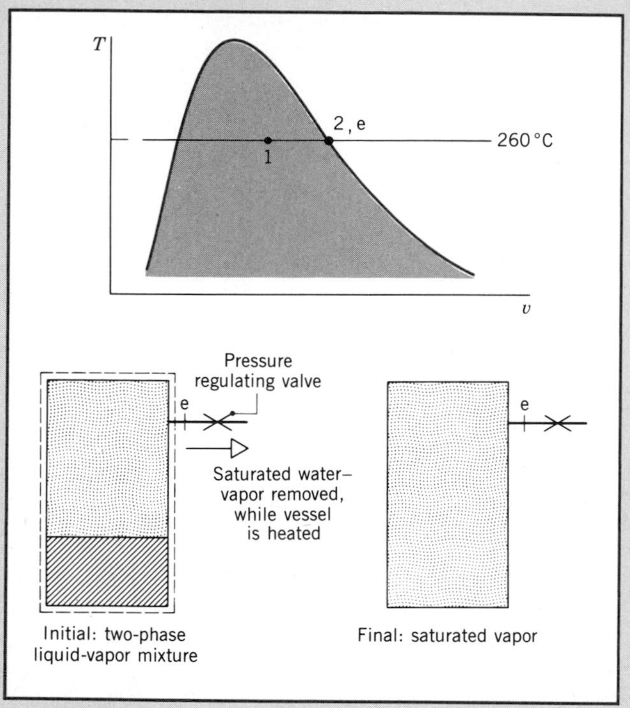

Figure E4.11

Assumptions:

1. The control volume is defined by the dashed line on the accompanying diagram.
2. For the control volume, $\dot{W}_{cv} = 0$ and kinetic and potential energy effects can be neglected.
3. At the exit the state remains constant.
4. The initial and final states of the mass within the vessel are equilibrium states.

Analysis: Since there is a single exit and no inlet, the mass rate balance takes the form

$$\frac{dm_{cv}}{dt} = -\dot{m}_e$$

With assumption 2, the energy rate balance reduces to

$$\frac{dU_{cv}}{dt} = \dot{Q}_{cv} - \dot{m}_e h_e$$

Combining the mass and energy rate balances results in

$$\frac{dU_{cv}}{dt} = \dot{Q}_{cv} + h_e \frac{dm_{cv}}{dt}$$

By assumption 3, the specific enthalpy at the exit is constant. Accordingly, integration of the last equation gives

$$\Delta U_{cv} = Q_{cv} + h_e \,\Delta m_{cv}$$

Solving for the heat transfer Q_{cv}

$$Q_{cv} = \Delta U_{cv} - h_e \,\Delta m_{cv}$$

or

$$Q_{cv} = (m_2 u_2 - m_1 u_1) - h_e(m_2 - m_1)$$

where m_1 and m_2 denote, respectively, the initial and final amounts of mass within the vessel.

The terms u_1 and m_1 of the foregoing equation can be evaluated with property values from Table A-2 at 260°C and the given value for quality. Thus

$$\begin{aligned} u_1 &= u_f + x_1(u_g - u_f) \\ &= 1128.4 + (0.7)(2599.0 - 1128.4) = 2157.8 \text{ kJ/kg} \end{aligned}$$

Also,

$$\begin{aligned} v_1 &= v_f + x_1(v_g - v_f) \\ &= 1.2755 \times 10^{-3} + (0.7)(0.04221 - 1.2755 \times 10^{-3}) = 29.93 \times 10^{-3} \text{ m}^3/\text{kg} \end{aligned}$$

Using the specific volume v_1, the mass initially contained in the tank is

$$m_1 = \frac{V}{v_1} = \frac{0.85 \text{ m}^3}{(29.93 \times 10^{-3} \text{ m}^3/\text{kg})} = 28.4 \text{ kg}$$

The final state of the mass in the vessel is saturated vapor at 260°C, so Table A-2 gives

$$u_2 = u_g(260°\text{C}) = 2599.0 \text{ kJ/kg}, \qquad v_2 = v_g(260°\text{C}) = 42.21 \times 10^{-3} \text{ m}^3/\text{kg}$$

The mass contained within the tank at the end of the process is

$$m_2 = \frac{V}{v_2} = \frac{0.85 \text{ m}^3}{42.21 \times 10^{-3} \text{ m}^3/\text{kg}} = 20.14 \text{ kg}$$

Table A-2 also gives $h_e = h_g(260°\text{C}) = 2796.6$ kJ/kg.

Substituting values into the expression for the heat transfer yields

$$\begin{aligned} Q_{cv} &= (20.14)(2599.0) - (28.4)(2157.8) - 2796.6(20.14 - 28.4) \\ &= 14{,}162 \text{ kJ} \end{aligned}$$

EXAMPLE 4.12

PROBLEM

Water flows at a constant mass flow rate of 7 lb/s into a vertical cylindrical tank. Water exits the tank near its base with a mass flow rate proportional to the height of the liquid in the tank: $\dot{m}_e = 1.4L$ lb/s, where L is the instantaneous liquid height, in ft. The area of the circular base is 0.2 ft². The water density is constant at 62.4 lb/ft³. If the tank is empty initially, determine the variation of the liquid height with time.

SOLUTION

Known: Water enters and exits an initially empty cylindrical tank. The mass flow rate at the inlet is constant. At the exit, the mass flow rate is proportional to the height of liquid in the tank.

Find: Determine the variation of the liquid height L with time.

Schematic and Given Data:

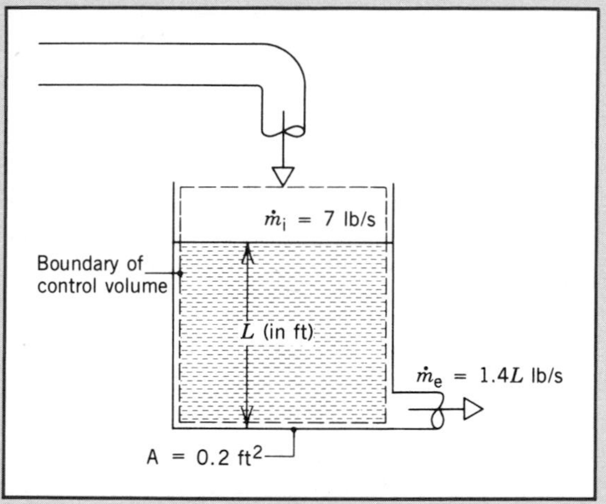

Figure E4.12

Assumptions:

1. The control volume is defined by the dashed line on the accompanying diagram.
2. The water density is constant.

Analysis: As there is a single inlet and a single exit, the mass rate balance reduces to

$$\frac{dm_{cv}}{dt} = \dot{m}_i - \dot{m}_e$$

The mass contained within the tank at time t is given by

$$m_{cv}(t) = \rho A L(t)$$

where ρ is density, A is the area of the circular base, and $L(t)$ is the instantaneous liquid height. Substituting this into the mass rate equation together with the given mass flow rates

$$\frac{d(\rho A L)}{dt} = 7 - 1.4L$$

As density and area are constant, this equation can be written as

$$\frac{dL}{dt} + \left(\frac{1.4}{\rho A}\right) L = \frac{7}{\rho A}$$

which is a first-order, ordinary differential equation with constant coefficients. The solution is

$$L = 5 + C \exp\left(-\frac{1.4t}{\rho A}\right)$$

where C is a constant of integration. The solution can be verified by substitution into the differential equation.

To evaluate the constant C, use the initial condition: at $t = 0$, $L = 0$. Thus, $C = -5$ and the solution becomes

$$L = 5[1 - \exp(-1.4t/\rho A)]$$

Substituting $\rho = 62.4$ lb/ft^3 and A = 0.2 ft^2 results in

$$L = 5[1 - \exp(-0.112t)]$$

where t is in seconds.

Comment: As $t \to \infty$, $L \to 5$. That is, the liquid level remains constant. In this limit, the exiting mass flow rate approaches the inlet value.

EXAMPLE 4.13

PROBLEM

A tank containing 45 kg of liquid water initially at 45°C has one inlet and one exit with equal mass flow rates. Liquid water enters at 45°C and a mass flow rate of 270 kg/h. A cooling coil immersed in the water removes energy at a rate of 7.6 kW. The water is well mixed by a paddle wheel so that the water temperature is uniform throughout. The power input to the water from the paddle wheel is 0.6 kW. The pressures at the inlet and exit are equal and all kinetic and potential energy effects can be ignored. Determine the variation of water temperature with time.

SOLUTION

Known: Liquid water flows into and out of a well-stirred tank with equal mass flow rates as the water in the tank is cooled by a cooling coil.

Find: Determine the variation with time of the temperature of the water within the tank.

Schematic and Given Data:

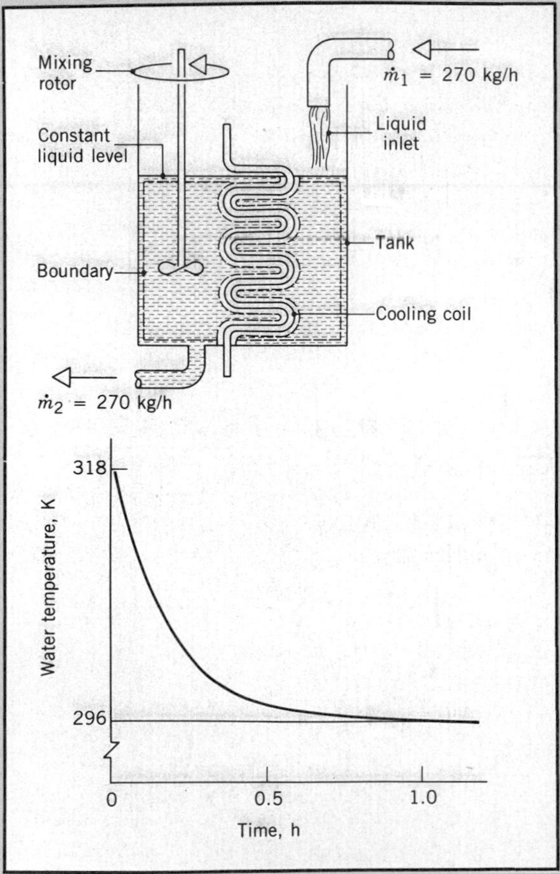

Figure E4.13

Assumptions:

1. The control volume is defined by the dashed line on the accompanying diagram.
2. For the control volume, the only significant heat transfer is with the cooling coil. Kinetic and potential energy effects can be neglected.
3. The water temperature is uniform with position throughout: $T = T(t)$.
4. The water in the tank is incompressible, and the change in pressure between inlet and exit can be ignored.

Analysis: The energy rate balance reduces with assumption 2 to

$$\frac{dU_{cv}}{dt} = \dot{Q}_{cv} - \dot{W}_{cv} + \dot{m}(h_1 - h_2)$$

where $\dot{m}$ denotes the mass flow rate.

The mass contained within the control volume remains constant with time, so the term on the left side of the energy rate balance can be expressed as

$$\frac{dU_{cv}}{dt} = \frac{d(m_{cv}u)}{dt} = m_{cv}\frac{du}{dt}$$

Since the water is assumed incompressible, the specific internal energy depends on temperature only. Hence, the chain rule can be used to write

$$\frac{du}{dt} = \frac{du}{dT}\frac{dT}{dt} = c\frac{dT}{dt}$$

where c is the specific heat. Collecting results

$$\frac{dU_{cv}}{dt} = m_{cv}c\frac{dT}{dt}$$

With Eq. 3.20 the enthalpy term of the energy rate balance can be expressed as

$$h_1 - h_2 = c(T_1 - T_2) + v\cancelto{0}{(p_1 - p_2)}$$

where the pressure term is dropped by assumption 4. Since the water is well mixed, the temperature at the exit equals the temperature of the overall quantity of liquid in the tank, so

$$h_1 - h_2 = c(T_1 - T)$$

where T represents the uniform water temperature at time t.

With the foregoing considerations the energy rate balance becomes

$$m_{cv}c\frac{dT}{dt} = \dot{Q}_{cv} - \dot{W}_{cv} + \dot{m}c(T_1 - T)$$

As can be verified by direct substitution, the solution of this first-order, ordinary differential equation is

$$T = C_1 \exp\left(-\frac{\dot{m}}{m_{cv}}t\right) + \left(\frac{\dot{Q}_{cv} - \dot{W}_{cv}}{\dot{m}c}\right) + T_1$$

The constant C_1 is evaluated using the initial condition: at $t = 0$, $T = T_1$. Finally

$$T = T_1 + \left(\frac{\dot{Q}_{cv} - \dot{W}_{cv}}{\dot{m}c}\right)\left[1 - \exp\left(-\frac{\dot{m}}{m_{cv}}t\right)\right]$$

Substituting given numerical values together with the specific heat c for liquid water from Table A-13

$$T = 318 \text{ K} + \left[\frac{[-7.6 - (-0.6)] \text{ kJ/s}}{\left(\frac{270}{3600}\frac{\text{kg}}{\text{s}}\right)\left(4.2\frac{\text{kJ}}{\text{kg}\cdot\text{K}}\right)}\right]\left[1 - \exp\left(-\frac{270}{45}t\right)\right]$$

$$= 318 - 22[1 - \exp(-6t)]$$

where t is in hours. Using this expression, we can construct the accompanying plot showing the variation of temperature with time.

Comments:

1. As $t \to \infty$, $T \to 296$ K. That is, the water temperature approaches a constant value after sufficient time has elapsed. From the accompanying plot it can be seen that the temperature reaches its constant limiting value in about 1 h.
2. In this case idealizations are made about the state of the mass contained within the system *and* the states of the liquid entering and exiting. These idealizations make the transient analysis tractable.

PROBLEMS

4.1 Mass enters a one-inlet, one-exit control volume at a constant rate of 100 kg/h. At the exit, the mass flow rate varies with time according to $\dot{m}_e = 100(1 - e^{-2t})$, where $\dot{m}_e$ has units of kg/h and t is in h. If the initial mass in the control volume is 50 kg, determine the amount of mass in the control volume at $t = 2$ h.

4.2 A control volume has one inlet and one exit. The mass flow rates in and out are, respectively, $\dot{m}_i = 1.5$ and $\dot{m}_e = 1.5(1 - e^{-0.002t})$, where t is in seconds and $\dot{m}$ is in kg/s. What is the time rate of change of mass in the control volume at $t = 120$ s? Plot the net change in the amount of mass in the control volume from $t = 0$ to any time t.

4.3 A tank contains 2 ft³ of air initially at 50 lbf/in.² and 70°F. A leak develops, and air flows out with a constant mass flow rate of 0.01 lb/s. Using the ideal gas model, determine

(a) the rate of change of mass within the tank at the moment the leak occurs, in lb/s.
(b) the mass of air in the tank after 20 s, in lb.

4.4 For the tank described in Problem 4.3, determine the pressure, in lbf/in.², of the air remaining in the tank after 20 s if the temperature is 70°F.

4.5 A tank providing water to a pump is shown in Fig. P4.5. Water enters the tank through a 1-in.-diameter supply pipe at a constant mass flow rate of 6.8 lb/s and exits to the pump through a 1-in.-diameter pipe. The diameter of the tank is 18 in., and the top of the 2-in.-diameter overflow pipe is 2 ft from the base of the tank. The velocity V, in ft/s, of the water exiting to the pump varies with the height z of the water surface, in ft, according to

$$V = 8.16z^{1/2}$$

Determine the time, in seconds, for a control volume enclosing the initially empty tank to reach steady state. At steady state, what is the mass flow rate, in lb/s, that water spills out through the overflow pipe?

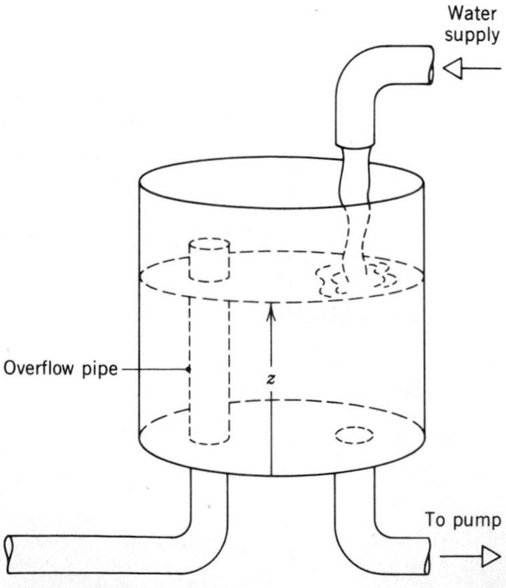

Figure P4.5

4.6 Write out the meanings of the terms one-dimensional flow, volumetric flow rate, and control volume at steady state.

4.7 A pipe carrying an incompressible liquid contains an expansion chamber as illustrated in Fig. P4.7. Develop an expression for the time rate of change of liquid level in the chamber, dL/dt, in terms of the diameters D_1, D_2, and D, and the velocities V_1 and V_2.

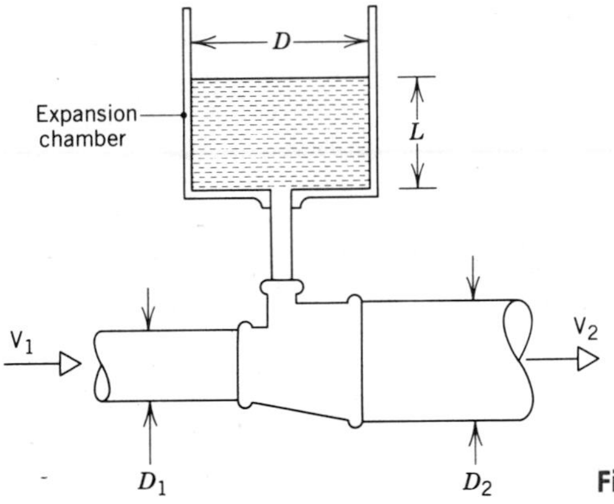

Figure P4.7

4.8 The velocity distribution for flow in a circular pipe of radius R carrying an incompressible liquid is given by the expression $V = V_0[1 - (r/R)^2]$, where r is the radial distance from the pipe centerline and V_0 is the centerline velocity. Plot V/V_0 versus r/R. Derive expressions for the mass flow rate and the average velocity of the flow, in terms of V_0, R, and the liquid density, ρ.

4.9 An incompressible liquid enters a pipe with a uniform velocity of V_1 and exits with the velocity distribution $V_2 = V_0[1 - (r/R)]^{1/7}$, where r is the radial distance from the pipe centerline, R is the radius of the pipe, and V_0 is the centerline velocity. For steady-state operation, develop an expression for V_0 in terms of V_1. Plot V_2/V_0 versus r/R.

4.10 An ideal gas mixture with an *apparent* molecular weight of 48 enters a device at 3 bars, 157°C, and 180 m/s through a flow area of 20 cm². At the exit, the pressure is 1.5 bars, the velocity is 120 m/s, and the flow area is 40 cm². For steady-state operation, determine at the exit

(a) the mass flow rate, in kg/s.
(b) the temperature, in °C.

4.11 *Infiltration* of outside air into a building through miscellaneous cracks can represent a significant load on the heating or cooling equipment. A particular office building has a total crack length of 440 ft around its doors and windows. On windy days, about 0.4 ft^3/min of air enters per foot of crack. In addition, door openings account for about 100 ft^3/min of outside air infiltration on the average. The internal volume of the building is 20,000 ft^3. Assuming ideal gas behavior, estimate the number of times per hour, at steady state, that the air within the building is changed due to infiltration.

4.12 Liquid water enters a boiler at 1000 lbf/in.^{2} and 120°F with a mass flow rate of 3×10^6 lb/h and flows through a number of 1-in.-diameter tubes. At steady state, how many tubes are needed if the velocity in each tube must be less than 20 ft/s?

4.13 Refrigerant 12 enters a device at 7 bars, 70°C with a velocity of 152 m/s. At the exit, the pressure and velocity of the refrigerant are 3 bars and 122 m/s, respectively, the mass flow rate is 1 kg/s, and the area is 1.10 cm^2. For steady-state operation, determine

(a) the inlet area, in cm^2.
(b) the exit temperature, in °C.

4.14 Steam enters a turbine through a duct with a diameter of 0.2 m. The steam velocity is 100 m/s, the pressure is 14 MPa, and the temperature is 600°C. Steam exits the turbine through a duct of diameter 0.8 m with a pressure of 500 kPa and a temperature of 180°C. Determine for operation at steady state

(a) the velocity of the steam at the exit, in m/s.
(b) the mass flow rate of the steam, in kg/h.

4.15 A substance flows through a $\frac{3}{4}$-in.-diameter pipe with a velocity of 15 ft/s at a particular location. Determine the mass flow rate, in lb/s, if the substance is

(a) water at 30 lbf/in.^{2}, 60°F.
(b) air as an ideal gas at 100 lbf/in.^{2}, 100°F.
(c) Refrigerant 134a at 100 lbf/in.^{2}, 100°F.

4.16 Air enters a compressor operating at steady state with a pressure of 14.7 lbf/in.^{2}, a temperature of 80°F, and a volumetric flow rate of 1000 ft^3/min. At the exit, the pressure is 100 lbf/in.^{2} and the temperature is 400°F. Using the ideal gas model, determine the diameter of the exit duct so the exit air velocity does not exceed 700 ft/s.

4.17 Air enters a 0.6-m-diameter fan at 16°C, 101 kPa, and is discharged at 18°C, 105 kPa, with a volumetric flow rate of 0.5 m^3/s. Assuming ideal gas behavior, determine for steady-state operation

(a) the mass flow rate of air, in kg/s.
(b) the volumetric flow rate of air at the inlet, in m^3/s.
(c) the inlet and exit velocities, in m/s.

4.18 Liquid water at 70°F enters a pump with a volumetric flow rate of 7 ft^3/min through an inlet pipe having a diameter of 6 in. The pump operates at steady state and supplies water to two exit pipes having diameters of 2 and 4 in., respectively. The mass flow rate of water in the smaller of the two exit pipes is 4 lb/s, and the temperature of the water exiting each pipe is 72°F. Determine the water velocity in each of the exit pipes, in ft/s.

4.19 A control volume operating at steady state has one inlet and two exits, as shown in Fig. P4.19. Using data on the figure, determine **(a)** the mass flow rate at the inlet, in kg/s, **(b)** the mass flow rate out at 2, in kg/s, and **(c)** the velocity at 3, in m/s.

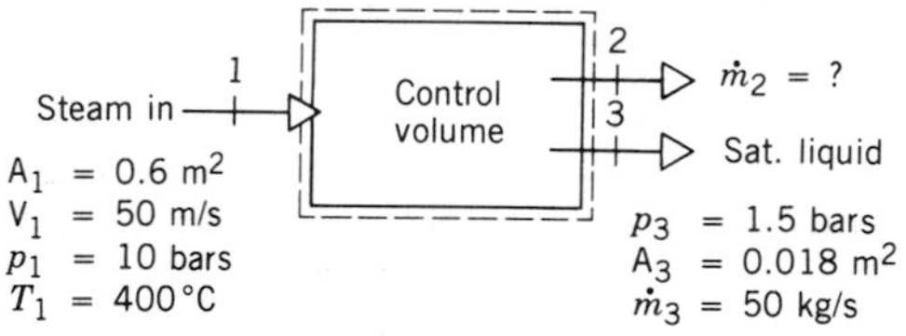

Figure P4.19

4.20 Air enters a device operating at steady state at 10 bars, 580 K, and leaves at 1 bar, 500 K, with a volumetric flow rate of 1.8 m^3/s. The rate of heat transfer to the air is 347 kJ per kg of air flowing. Neglecting kinetic and potential energy effects, determine the power developed, in kW, and the volumetric flow rate at the inlet, in m^3/s.

4.21 Air as an ideal gas at 70°F, 14.7 lbf/in.^{2}, enters a control volume operating at steady state with a volumetric flow rate of 2000 ft^3/min and exits at 40°F. A separate stream of liquid water enters at 90°F and exits at 105°F. The power input is 10 horsepower, and there is no heat transfer from the control volume to its surroundings. Neglecting kinetic and potential energy effects, determine the mass flow rate of the water stream, in lb/min.

4.22 Steam enters a nozzle operating at steady state at 3 MPa and 320°C with negligible velocity and exits at 1.5 MPa and a velocity of 500 m/s. The mass flow rate is 2.22 kg/s. Neglecting heat transfer and potential energy, determine

(a) the exit temperature, in °C.
(b) the exit area, in cm^2.

4.23 Steam enters a well-insulated nozzle with negligible velocity at 200 lbf/in.^{2} and 500°F and exits at 100 lbf/in.^{2} and 400°F. For steady-state operation and neglecting potential energy effects, determine the velocity of the steam exiting the nozzle, in ft/s.

4.24 Methane (CH_4) gas enters a nozzle at 0.5 MPa, 80°C, with a velocity of 10 m/s and exits at 0.3 MPa, with a velocity of 550 m/s. Potential energy and heat transfer effects are negligible, and ideal gas behavior can be assumed for the methane. For steady-state operation, determine the exit temperature, in °C.

4.25 Argon gas flows at steady state through a well-insulated nozzle. The temperature at the inlet is 600°R, and the temperature and velocity at the exit are 460°R and 950 ft/s, respectively. Using the ideal gas model and neglecting potential energy effects, determine the inlet velocity, in ft/s.

4.26 A well-insulated nozzle operates at steady state with air entering at a pressure of 300 lbf/in.^{2}, a temperature of 800°R, and a velocity of 300 ft/s. Air exits the nozzle with a pressure of 100 lbf/in.^{2} and a temperature of 600°R. Neglecting potential energy effects, determine the ratio of the exit area to the inlet area.

4.27 Steam enters a diffuser operating at steady state with a pressure of 14.7 lbf/in.^{2}, a temperature of 300°F, and a velocity of 500 ft/s. Steam exits the diffuser as a saturated vapor, with negligible kinetic energy. Heat transfer occurs from the steam to its surroundings at a rate of 19.59 Btu per lb of steam flowing. Neglecting potential energy effects, determine the exit pressure, in lbf/in.^{2}

4.28 Air enters an insulated diffuser operating at steady state with a pressure of 0.7 bar, a temperature of 57°C, and a velocity of 200 m/s. At the exit, the pressure is 1.0 bar. The exit flow area is 20% greater than the inlet flow area. Potential energy effects can be neglected. Using the ideal gas model with a constant specific heat c_p evaluated at the inlet temperature, determine at the exit

(a) the temperature, in K.
(b) the velocity, in m/s.

4.29 The inlet ducting to a jet engine forms a diffuser that decelerates the entering air to zero velocity relative to the engine before the air enters the compressor. Consider a jet airplane flying at 1000 km/h where the local atmospheric pressure is 0.6 bar and the air temperature is 8°C. Assuming ideal gas behavior and neglecting potential energy effects, determine the temperature, in °C, of the air entering the compressor.

4.30 Refrigerant 12 enters an insulated diffuser as a saturated vapor at 80°F with a velocity of 420 ft/s. At the exit, the pressure and temperature are 100 lbf/in.^{2} and 100°F, respectively. The diffuser operates at steady state and the potential energy change from inlet to exit is negligible. For an exit area of 10 in.^{2}, determine

(a) the exit velocity, in ft/s.
(b) the mass flow rate, in lb/s.

4.31 Carbon dioxide gas is decelerated to negligible velocity in a well-insulated diffuser. The pressure and temperature at the inlet are 20 lbf/in.^{2} and 500°R, and the pressure and temperature at the exit are 25 lbf/in.^{2} and 540°R, respectively. The potential energy change from inlet to exit is negligible. Determine for steady-state operation

(a) the inlet velocity, in ft/s.
(b) the inlet area, in in.^{2}, for a mass flow rate of 1 lb/s.

4.32 Air expands with a mass flow rate of 10 kg/s through a turbine from 5 bars, 900 K to 1 bar, 600 K. The inlet velocity is small compared to the exit velocity of 100 m/s. The turbine operates at steady state. Heat transfer from the turbine to the surroundings and potential energy effects are negligible. Calculate the power developed by the turbine, in kW, and the exit area, in m^2.

4.33 Air expands through a turbine operating at steady state. At the inlet, p_1 = 150 lbf/in.^{2}, T_1 = 1500°R, and at the exit, p_2 = 14.5 lbf/in.^{2} The volumetric flow rate of air entering the turbine is 3000 ft^3/min, and the power developed is 3400 horsepower. Neglecting heat transfer and kinetic and potential energy effects, determine the exit temperature, T_2, in °R.

4.34 At steady state, a well-insulated steam turbine develops 10,000 kW of power for a steam flow rate of 12 kg/s. The steam enters at 4 MPa and leaves at 4 kPa with a quality of 90%. Kinetic and potential energy changes from inlet to exit can be neglected. What is the inlet temperature of the steam, in °C?

4.35 Steam enters a turbine operating at steady state at 600°F and 200 lbf/in.^{2} with a velocity of 80 ft/s and leaves as saturated vapor at 5 lbf/in.^{2} with a velocity of 300 ft/s. The power developed by the turbine is 200 horsepower. Heat transfer from the turbine to the surroundings occurs at a rate of 50,000 Btu/h. Neglecting potential energy effects, determine the mass flow rate of steam, in lb/s.

4.36 Nitrogen gas flows at steady state through a turbine that develops a power output of 50 horsepower. The gas enters with a velocity of 200 ft/s, a pressure of 50 lbf/in.^{2}, and a temperature of 1000°R. At the exit, the velocity is negligible, the pressure is 20 lbf/in.^{2}, and the temperature is 700°R. Heat transfer from the surface of the turbine to the surroundings occurs at a rate of 16 Btu per lb of nitrogen flowing. Neglecting potential energy effects, determine the diameter of the inlet, in ft.

4.37 Steam enters a well-insulated turbine operating at steady state with negligible velocity at 4 MPa, 320°C. The steam expands to an exit pressure of 0.07 MPa, a specific volume of 2.19 m^3/kg, and a velocity of 90 m/s. The diameter of the exit is 0.6 m. Neglecting potential energy effects, determine the power developed by the turbine, in kW.

4.38 The inlet to a turbine in a hydroelectric plant is located at an elevation of 100 m above the exit. The inlet and exit ducts each have diameters of 1 m. Water enters with a velocity of 3 m/s at a temperature of 20°C and passes through the turbine with negligible changes in temperature and pressure. Heat transfer between the turbine and its surroundings can also be neglected. The acceleration of gravity is constant at g = 9.81 m/s^2. What power, in kW, is developed by the turbine at steady state?

4.39 A well-insulated turbine operating at steady state is sketched in Fig. P4.39. Steam enters at 3 MPa, 400°C,

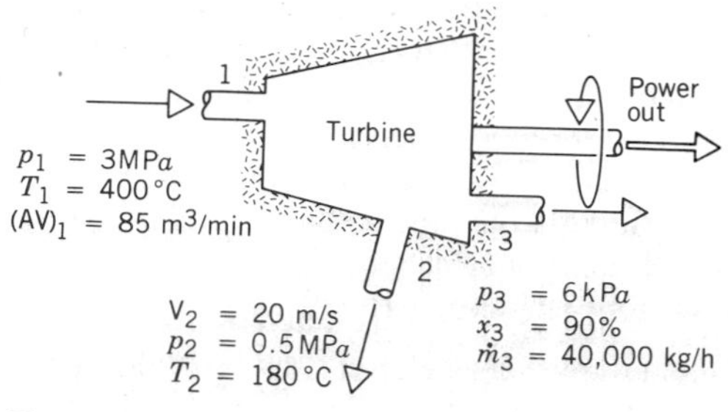

Figure P4.39

with a volumetric flow rate of 85 m^3/min. Some steam is extracted from the turbine at a pressure of 0.5 MPa and a temperature of 180°C. The rest expands to a pressure of 6 kPa and exits with a mass flow rate of 40,000 kg/h and a quality of 90%. Kinetic and potential energy effects can be neglected. Determine

(a) the diameter, in m, of the duct through which steam is extracted, if the velocity there is 20 m/s.
(b) the power developed by the turbine, in kW and Btu/h.

4.40 Steam at 600 lbf/in.^{2}, 1000°F, enters a turbine operating at steady state with a mass flow rate of 2000 lb/min. Some steam is extracted from the turbine through a duct with a cross-sectional area of 1.6 ft^2. The pressure of the extracted steam is 100 lbf/in.^{2} and the velocity is 25.9 ft/s. The rest of the steam exits as a two-phase liquid–vapor mixture with a temperature of 50°F, a quality of 80%, and a volumetric flow rate of 36,350 ft^3/s. Kinetic and potential energy effects and heat transfer between the turbine and its surroundings are negligible. Determine

(a) the volumetric flow rate of the incoming steam, in ft^3/min.
(b) the temperature of the extracted steam, in °F.
(c) the power produced by the turbine, in Btu/h.

4.41 Air is compressed at steady state from 1 bar, 300 K, to 6 bars, 440 K with a mass flow rate of 4 kg/s. Heat transfer occurs at a rate of 42.3 kJ per kg of air flowing to cooling water circulating in a water jacket enclosing the compressor. If kinetic and potential energy changes of the air from inlet to exit are negligible, calculate the compressor power, in kW.

4.42 A well-insulated compressor operating at steady state takes in air at 80°F, 14.5 lbf/in.^{2}, with a volumetric flow rate of 600 ft^3/min, and compresses it to 60 lbf/in.^{2} The power input to the compressor is 70 horsepower. Kinetic and potential energy changes from inlet to exit can be neglected. Determine the exit temperature, in °F.

4.43 Air enters a compressor with a pressure of 14.7 lbf/in.^{2} and a temperature of 70°F. The volumetric flow rate at the inlet is 166 ft^3/s, and the flow area is 0.26 ft^2. At the exit, the pressure is 35 lbf/in.^{2}, the temperature is 280°F, and the velocity is 500 ft/s. Heat transfer from the compressor occurs at a rate of 750 Btu/min. Potential energy effects are negligible. Determine the compressor power, in Btu/s, for steady-state operation.

4.44 A compressor operates at steady state with Refrigerant 134a as the working fluid. The refrigerant enters at 0.2 MPa, 0°C, with a volumetric flow rate of 0.6 m^3/min. The diameters of the inlet and exit pipes are 3 and 1.5 cm, respectively. At the exit, the pressure is 1.0 MPa and the temperature is 50°C. If the compressor power input is 3 kW, what is the rate of heat transfer, in kW?

4.45 A compressor operates at steady state with Refrigerant 12 as the working fluid. The pressures and temperatures at the inlet and exit are 2 bars, 0°C, and 12 bars, 50°C, respectively. The volumetric flow rate of refrigerant entering is 0.5 m^3/min. Heat transfer from the compressor occurs at the rate of 5 kJ per kg of refrigerant flowing. Neglecting kinetic and potential energy effects, calculate the power, in kW.

4.46 Carbon dioxide gas is compressed at steady state from a pressure of 20 lbf/in.^{2} and a specific volume of 6 ft^3/lb to a temperature of 580°R. The gas enters the compressor through a 6-in.-diameter duct with a velocity of 30 ft/s and leaves with a velocity of 80 ft/s. The work input per unit mass of carbon dioxide flowing is 34 Btu/lb. Using the ideal gas model and neglecting potential energy effects, determine the heat transfer rate from the compressor to its surroundings, in Btu/min.

4.47 A compressor operating at steady state takes in 45 kg/min of methane gas (CH_4) at 1 bar, 25°C, 15 m/s, and compresses it with negligible heat transfer to 2 bars, 360 K, 90 m/s. Potential energy effects are negligible. Using the ideal gas model, calculate the power required, in kW.

4.48 Ammonia vapor (NH_3) is compressed at steady state from a pressure of 10 lbf/in.^{2} and a density of 0.03 lb/ft^3 to a pressure of 100 lbf/in.^{2} The mass flow rate is 2 lb/min, and the power input is 380 Btu/min. The compressor is well insulated and kinetic and potential energy effects are negligible. Using the ideal gas model, determine the temperature, in °R, and the density, in lb/ft^3, of the ammonia exiting the compressor.

4.49 Refrigerant 12 is compressed at steady state from 2.4 bars, 0°C, to 12 bars, 50°C. Refrigerant enters the compressor with a volumetric flow rate of 0.38 m^3/min, and the power input to the compressor is 2.6 kW. Cooling water circulates through a water jacket enclosing the compressor with a mass flow rate of 0.04 kg/s and negligible change in pressure. Heat transfer from the outside of the water jacket and all kinetic and potential energy effects can be neglected. Determine the temperature rise of the cooling water, in °C.

4.50 Air enters a compressor operating at steady state with a volumetric flow rate of 37 m^3/min at 136 kPa, 305 K and exits with a pressure of 680 kPa and a specific volume of 0.1686 m^3/kg. The work input to the compressor is 161.5 kJ per kg of air flowing. Energy transfer from the air to cooling water circulating in a water jacket enclosing the compressor results in an increase in the temperature of the cooling water of 11°C with no change in pressure. Neglecting heat transfer from the outside of the jacket as well as all kinetic and potential energy effects, what is the mass flow rate of the cooling water, in kg/min?

4.51 A pump steadily delivers water at a volumetric flow rate of 0.05 m^3/s through a pipe of diameter 18 cm located 100 m above the inlet pipe, which has a diameter of 15 cm. The pressure is nearly equal to 1 bar at both the inlet and the exit, and the temperature is nearly constant at 20°C throughout. Determine the power required by the pump, in kW. The acceleration of gravity is $g = 9.81$ m/s^2.

4.52 A water pump operating at steady state has 3-in.-diameter inlet and exit pipes, each at the same elevation. For a power input of 2 horsepower there is a pressure rise of 12 $lbf/in.^2$ from inlet to exit. If the water temperature remains nearly constant temperature at 70°F, determine the mass flow rate, in lb/s, and the volumetric flow rate in gal/min.

4.53 An oil pump operating at steady state requires a power input of 1 horsepower and delivers oil at a rate of 11 lb/s through a 1-in.-diameter pipe. There is no significant elevation difference between inlet and exit, and the inlet kinetic energy is negligible. Heat transfer between the pump and its surroundings is negligible, and there is no significant change in temperature as the oil passes through the pump. If the oil has a density of 100 lb/ft^3, determine the pressure change from inlet to exit, in $lbf/in.^2$

4.54 Refrigerant 134a enters a heat exchanger operating at steady state as a superheated vapor at 10 bars, 60°C, where it is cooled and condensed to saturated liquid at 10 bars. The mass flow rate of the refrigerant is 10 kg/min. A separate stream of air enters the heat exchanger at 37°C with a mass flow rate of 80 kg/min. Ignoring heat transfer from the outside of the heat exchanger and neglecting kinetic and potential energy effects, determine the exit air temperature, in °C.

4.55 Steam at 1 $lbf/in.^2$ and 120°F enters a condenser operating at steady state and is condensed to saturated liquid at 1 $lbf/in.^2$ on the outside of tubes through which cooling water flows. The mass flow rate of steam is 200,000 lb/h. In passing through the tubes, the cooling water increases in temperature by 18°F and experiences no pressure drop. Neglecting kinetic and potential energy effects and ignoring heat transfer from the outside of the condenser, determine

(a) the mass flow rate of cooling water, in lb/h.
(b) the rate of energy transfer, in Btu/h, from the condensing steam to the cooling water.

4.56 A steam boiler tube is designed to produce a stream of saturated vapor at 300 kPa from saturated liquid entering at the same pressure. At steady state, the flow rate is 0.004 kg/s. The boiler is constructed from a well-insulated stainless steel pipe through which the steam flows. Electrodes clamped to the pipe at each end cause a 10-V direct current to pass through the pipe material. Determine the required size of the power supply, in kW, and the expected current draw, in amperes.

4.57 Air is heated as it flows at steady state through a constant area duct. At the inlet, the pressure is 2 bars, the temperature is 300 K, and the velocity is 87 m/s. At the exit, the pressure and velocity are 1.45 bars and 360 m/s, respectively. Neglecting potential energy effects, determine the heat transfer per unit mass of air flowing through the duct, in kJ/kg.

4.58 Refrigerant 12 enters a heat exchanger in a refrigeration system operating at steady state as saturated vapor at 30°F and exits at 40 $lbf/in.^2$ and 40°F. A separate liquid stream of refrigerant passes in counterflow to the vapor stream, entering as saturated liquid at 105°F and exiting at a lower temperature while experiencing no pressure drop. The outside of the heat exchanger is well insulated and the streams have equal mass flow rates. Neglecting kinetic and potential energy effects, determine the exit temperature of the liquid stream, in °F.

4.59 The cooling coil of an air-conditioning system is a heat exchanger in which air passes over tubes through which a refrigerant flows. Air enters with a volumetric flow rate of 40 m^3/min at 40°C, 1.1 bars, and exits at 20°C, 1 bar. Refrigerant 12 enters the tubes with a quality of 40% at 10°C and exits as saturated vapor at 10°C. Ignoring heat transfer from the outside of the heat exchanger and neglecting kinetic and potential energy effects, determine at steady state

(a) the mass flow rate of refrigerant, in kg/min.
(b) the rate of energy transfer, in kJ/min, from the air to the refrigerant.

4.60 Refrigerant 12 flows at steady state through a long horizontal pipe having an inside diameter of 4 cm, entering as saturated vapor at −10°C with a mass flow rate of 17 kg/min. At the exit, the temperature is 5°C and the pressure is 2 bars. Determine the change in kinetic energy from inlet to exit, per unit mass of refrigerant flow, in kJ/kg, and the rate of heat transfer between the refrigerant and its surroundings, in kW.

4.61 Figure P4.61 shows a solar collector panel with a surface area of 32 ft^2. The panel receives energy from the sun at a rate of 150 Btu/h per ft^2 of collector surface. Forty percent of the incoming energy is lost to the surroundings. The remainder is used to warm liquid water from 130 to 160°F. The water passes through the solar collector with a negligible pressure drop. Neglecting kinetic and potential energy effects, determine at steady state the mass flow rate of the water, in lb/min. How

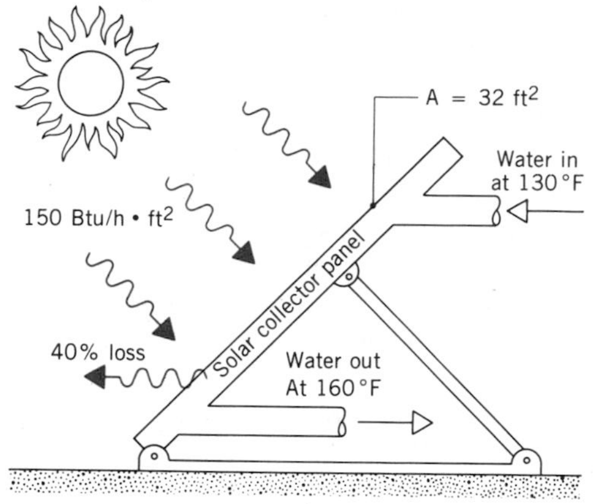

Figure P4.61

many solar collectors would be needed to provide a total of 40 gal of 160°F water in 30 min?

4.62 As shown in Fig. P4.62, steam enters a de-superheater operating at steady state at 30 bars, 320°C, where it is mixed with liquid water at 25 bars, 200°C to produce saturated vapor at 20 bars. Heat transfer between the device and its surroundings and kinetic and potential energy effects can be neglected. Determine the mass flow rate of liquid, in kg/h, for a steam mass flow rate of 50,000 kg/h.

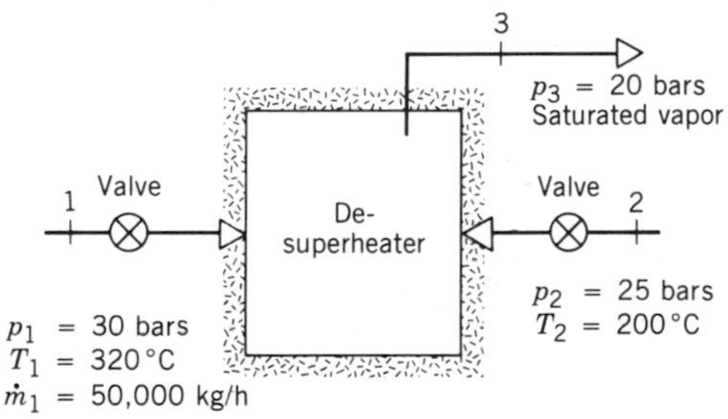

Figure P4.62

4.63 An open feedwater heater operates at steady state with liquid water entering at inlet 1 at 50°C and 1.0 MPa. A separate stream of water enters at inlet 2 with $\dot{m}_2/\dot{m}_1 = 0.22$ and a pressure of 1.0 MPa. Saturated liquid water exits with a pressure of 1.0 MPa. Ignore heat transfer with the surroundings and neglect kinetic and potential energy effects. For the stream entering at inlet 2, determine the temperature, in °C, if superheated, or the quality if saturated.

4.64 The device shown in Fig. P4.64 operates at steady state. Refrigerant 12 enters as a two-phase liquid–vapor mixture at location 1 and is mixed with a stream of vapor entering at location 2. Saturated vapor and saturated liquid leave at locations 3 and 4, respectively. The mass flow rates at locations 1 and 3 are the same. The flow rates at locations 2 and 4 each equal 16 lb/min. The pressure is

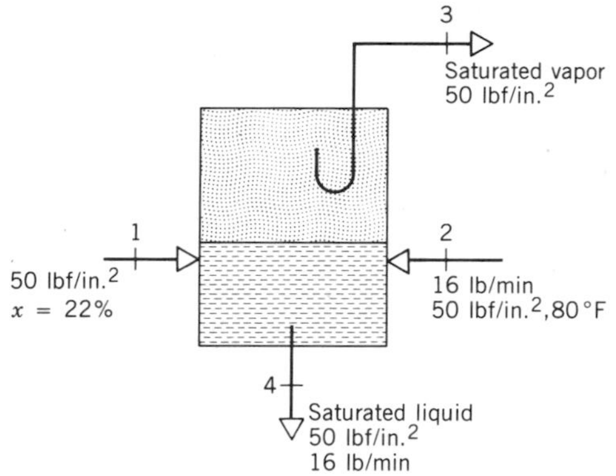

Figure P4.64

maintained constant at 50 lbf/in.² Ignoring heat transfer with the surroundings and neglecting kinetic and potential energy effects, what is the mass flow rate at location 1, in lb/min?

4.65 Refrigerant 134a enters the expansion valve of a refrigeration system with a pressure of 1.2 MPa and exits at 0.24 MPa with a quality of $x = 0.3$ (30%). If the refrigerant undergoes a throttling process, what is the temperature, in °C, of the refrigerant entering the expansion valve?

4.66 In a steam power plant, a valve controls the flow into the main turbine. Steam enters the valve at 8 MPa, 600°C, and leaves at 6 MPa. If the steam undergoes a throttling process, what is the temperature of the steam leaving the valve, in °C?

4.67 Steam at 500 lbf/in.², 500°F, enters a well-insulated valve operating at steady state and expands to 200 lbf/in.² with no significant changes in kinetic or potential energy. Determine the temperature at the exit of the valve, in °F, and the ratio of inlet to exit volumetric flow areas.

4.68 A large pipe carries steam as a two-phase liquid–vapor mixture at 1.0 MPa. A small quantity is withdrawn through a throttling calorimeter, where it undergoes a throttling process to an exit pressure of 0.1 MPa. The temperature at the exit of the calorimeter is observed to be 112°C. What is the quality of the steam in the pipeline?

4.69 A two-phase liquid–vapor mixture of H_2O enters a valve with a quality of 98% and exits as saturated vapor at 100 lbf/in.² If the steam undergoes a throttling process, determine the pressure at the inlet, in lbf/in.² Also determine the change in specific internal energy from inlet to exit, in Btu/lb, and comment on the result.

4.70 Steam flows at steady state through a well-insulated pressure-reducing valve. At the inlet, the pressure is 4 MPa, the temperature is 320°C, and the velocity is 60 m/s. The pressure of the steam at the exit is 0.7 MPa. The inlet and exit have equal diameters, and potential energy effects are negligible. Determine the specific enthalpy, in kJ/kg, and the velocity, in m/s, at the exit of the valve.

4.71 Figure P4.71 shows a part of a refrigeration system consisting of a suction line heat exchanger and an evapo-

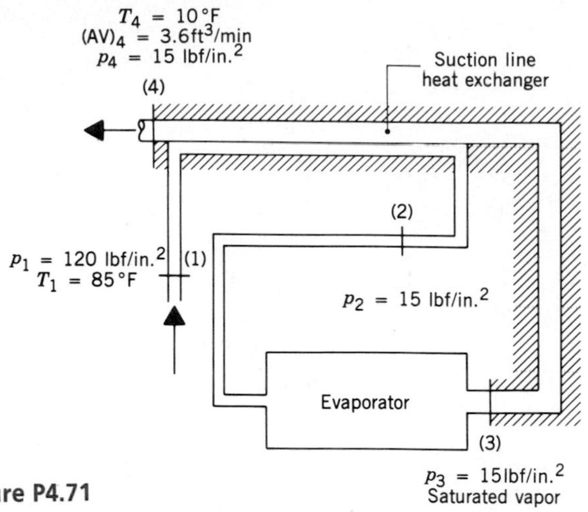

Figure P4.71

rator. Data for steady-state operation with Refrigerant 134a are given on the figure. There is no significant heat transfer between the heat exchanger and its surroundings. For a control volume enclosing the evaporator, determine the heat transfer rate, in Btu/h. Kinetic and potential energy effects are negligible.

4.72 Refrigerant 12 enters the flash chamber shown in Fig. P4.72 at 10 bars, 36°C, with a mass flow rate of 482 kg/h. Saturated liquid and saturated vapor exit as separate streams at 4 bars. Heat transfer to the surroundings and kinetic and potential energy effects can be ignored. For steady-state operation, determine the mass flow rates of the liquid and vapor streams, in kg/h.

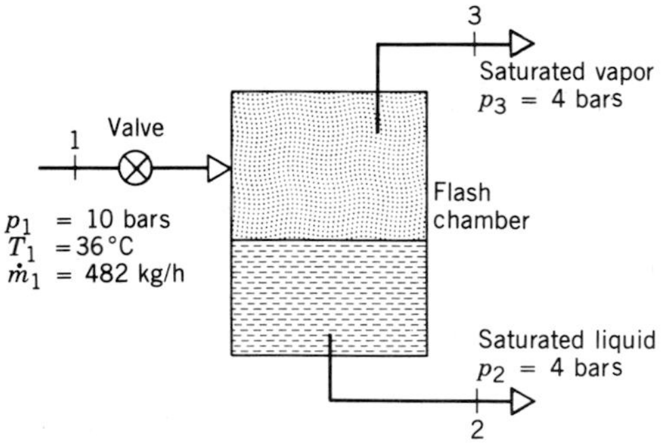

Figure P4.72

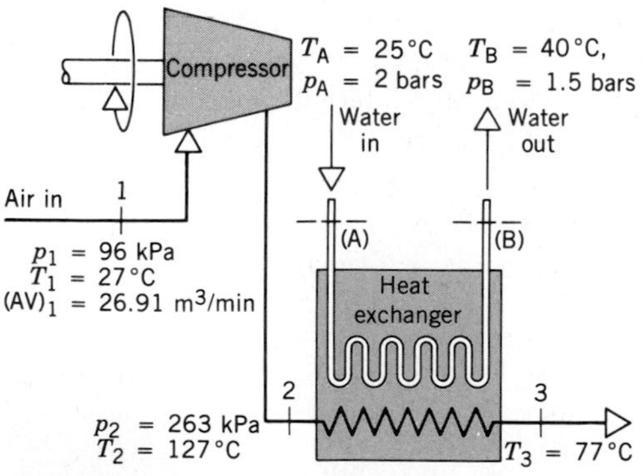

Figure P4.73

4.73 Air as an ideal gas flows at steady state through the compressor and heat exchanger shown in Fig. P4.73. Data for the various flow streams are shown on the figure. Heat transfer to the surroundings can be neglected, as can all kinetic and potential energy effects. Calculate the compressor power, in kW, and the mass flow rate of the cooling water, in kg/s.

4.74 Water enters the steam generator of the power plant shown in Fig. P4.74 at 1000 lbf/in.^{2}, 120°F, and exits at 1000 lbf/in.^{2}, 1100°F. The steam is then fed into a well-insulated turbine, developing a power output of 30,000 Btu/s before exiting the turbine as a two-phase liquid–vapor mixture at 1 lbf/in.^{2} with a quality of 90%. Combustion gases enter the steam generator at 1800°R and are cooled to 800°R. For steady-state operation, determine the mass flow rate of the combustion gases, in lb/s, and the rate of energy transfer between the gas and vapor streams in the steam generator, in Btu/s. The properties of the combustion gases can be taken as those of air as an ideal gas. Ignore all stray heat transfers with the surroundings and all kinetic and potential energy effects.

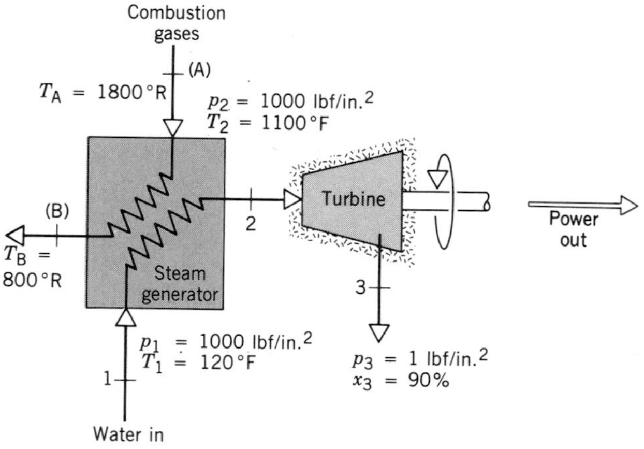

Figure P4.74

4.75 A residential heat pump system operating at steady state is shown schematically in Fig. P4.75. Refrigerant 12 circulates through the components of the system, and property data at the numbered locations are given on the figure. The mass flow rate of the refrigerant is 4.6 kg/min. Kinetic and potential energy effects are negligible. Determine the rates of heat transfer, in kJ/min,

(a) from the refrigerant to the return air passing through the condenser.
(b) between the compressor and the surroundings.
(c) from the outside air to the refrigerant passing through the evaporator.

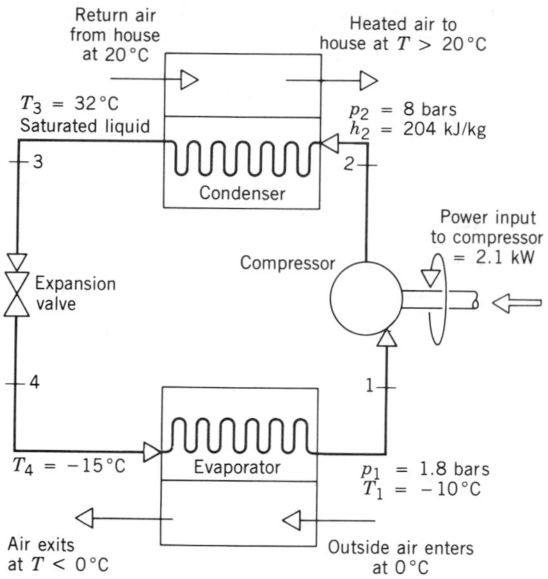

Figure P4.75

4.76 Figure P4.76 shows a simple vapor power plant operating at steady state with water circulating through the components. Relevant data at key locations are given on the figure. The mass flow rate of the water is 130 kg/s. Kinetic and potential energy effects are negligible. Determine

(a) the heat transfer rate to the steam passing through the steam generator, in kW.

(b) the *net* power developed for adiabatic operation of the turbine and pump, in kW.

(c) the mass flow rate of the cooling water passing through the condenser, in kg/s, if the cooling water experiences a temperature rise of 15°C and a negligible pressure drop.

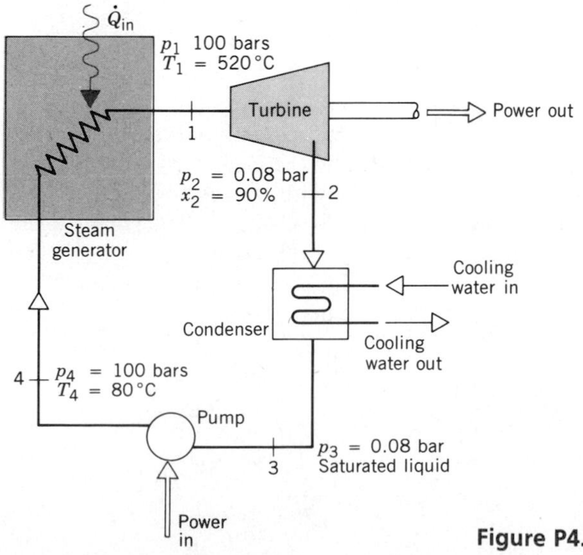

Figure P4.76

4.77 A simple gas turbine power plant operating at steady state is illustrated schematically in Fig. P4.77. The power plant consists of an air compressor mounted on the same shaft as the turbine. Relevant data are given on the figure. Kinetic and potential energy effects are negligible. Using the ideal gas model, determine for adiabatic operation of the compressor and turbine

(a) the work input to the compressor, in Btu per lb of air flowing.

(b) the temperature at the turbine inlet, in °R.

(c) the heat transfer rate, $\dot{Q}_{in}$, in Btu/s.

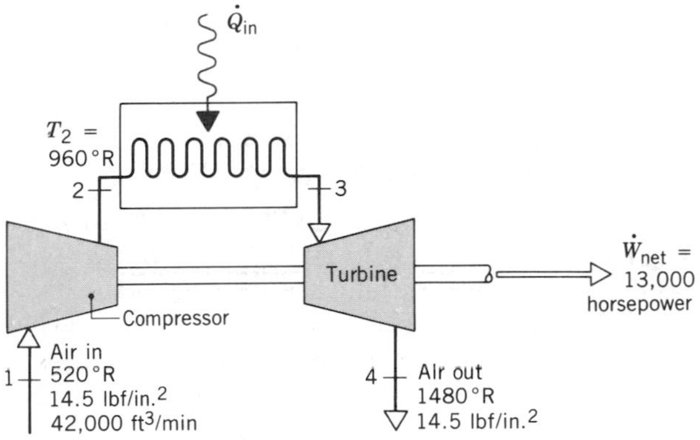

Figure P4.77

4.78 The rigid tank illustrated in Fig. P4.78 has a volume of 0.06 m^3 and initially contains a two-phase liquid–vapor mixture of H_2O at a pressure of 15 bars and a quality of 20%. As the tank contents are heated, a pressure-regulating valve keeps the pressure constant in the tank by allowing saturated vapor to escape. Heating continues until no liquid remains in the tank. Neglecting kinetic and potential energy effects, determine

(a) the amount of heat transfer, in kJ.

(b) the mass of vapor that escapes, in kg.

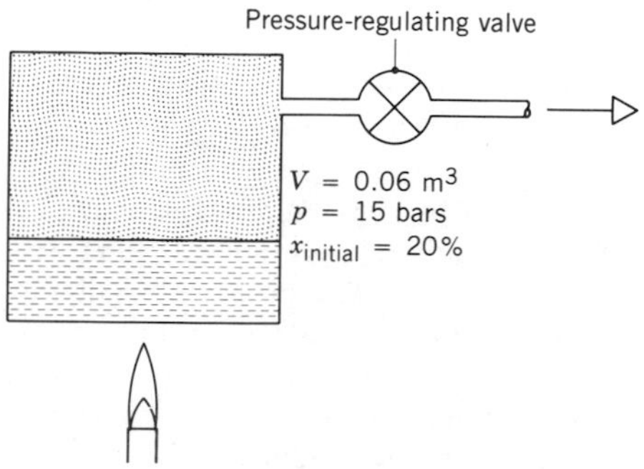

Figure P4.78

4.79 A rigid tank of volume 0.5 m^3 is initially evacuated. A tiny hole develops in the wall, and air from the surroundings at 1 bar, 21°C leaks in. Eventually, the pressure in the tank reaches 1 bar. The process occurs slowly enough that heat transfer between the tank and the surroundings keeps the temperature of the air inside the tank constant at 21°C. Determine the amount of heat transfer, in kJ.

4.80 A well-insulated rigid tank of volume 10 m^3 is connected to a large steam line through which steam flows at 1.5 MPa and 280°C. The tank is initially evacuated. Steam is allowed to flow into the tank until the pressure inside is 1.5 MPa. Calculate the amount of mass that flows into the tank, in kg.

4.81 A well-insulated rigid tank of volume 7 ft^3 initially contains helium at 160°F and 30 lbf/in.^{2} A valve connected to the tank is opened, and helium is withdrawn slowly until the pressure within the tank drops to 18 lbf/in.^{2} An electrical resistor inside the tank maintains the temperature at 160°F. Determine

(a) the mass of helium withdrawn, in lb.
(b) the energy input to the resistor, in Btu.

4.82 A well-insulated rigid tank of volume 10 ft^3 contains carbon dioxide, initially at 30 lbf/in.^{2} and 60°F. The tank is connected to a large supply line carrying carbon dioxide at 120 lbf/in.^{2}, 100°F. A valve between the line and the tank is opened and gas flows into the tank until the pressure reaches 120 lbf/in.^{2} Determine

(a) the final temperature in the tank, in °F.
(b) the mass of carbon dioxide that enters, in lb.

4.83 A tank of volume 1 m^3 initially contains steam at 6 MPa and 320°C. Steam is withdrawn slowly from the tank until the pressure drops to 1.5 MPa. Heat transfer to the tank contents maintains the temperature constant at 320°C. Neglecting all kinetic and potential energy effects, determine the heat transfer, in kJ.

4.84 A 1 m^3 tank initially contains air at 300 kPa, 300 K. Air slowly escapes from the tank until the pressure drops to 100 kPa. The air that remains in the tank undergoes a process described by $pv^{1.2}$ = constant. For a control volume enclosing the tank, determine the heat transfer, in kJ. Assume ideal gas behavior with constant specific heats.

4.85 A well-insulated tank contains 25 kg of Refrigerant 12, initially at 300 kPa with a quality of 0.8 (80%). The pressure is maintained by nitrogen gas acting against a flexible bladder, as shown in Fig. P4.85. The valve is opened between the tank and a supply line carrying Refrigerant 12 at 1.0 MPa, 120°C. The pressure regulator allows the pressure in the tank to remain at 300 kPa as the bladder expands. The valve between the line and the tank is closed at the instant when all the liquid has vaporized. Determine the amount of refrigerant admitted to the tank, in kg.

4.86 A well-insulated piston–cylinder assembly is connected by a valve to an air supply at 100 lbf/in.^{2}, 80°F, as shown in Fig. P4.86. The air inside the cylinder is initially at 14.7 lbf/in.^{2}, 80°F, and occupies a volume of 0.1 ft^3. Initially, the piston face is located at $x = 0$ and the spring

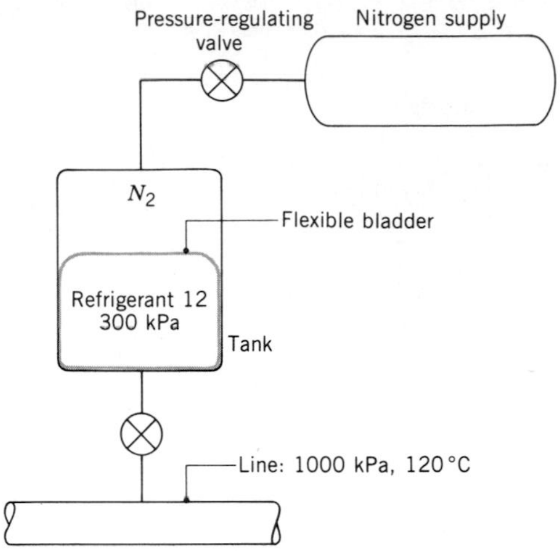

Figure P4.85

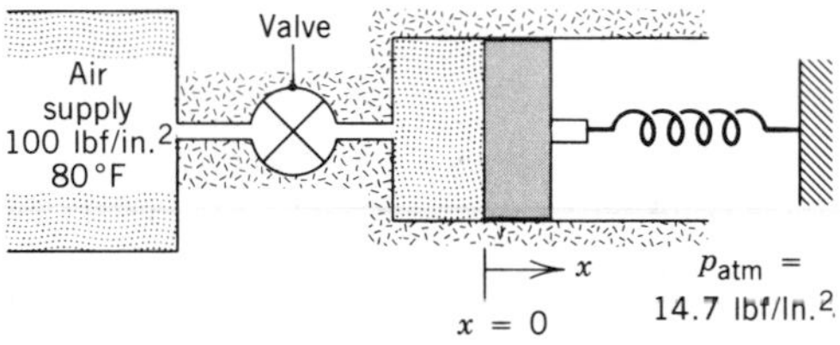

Figure P4.86

exerts no force on the piston. The atmospheric pressure is 14.7 lbf/in.^{2}, and the area of the piston face is 0.22 ft^2. The valve is opened, and air is admitted slowly until the volume of the air inside the cylinder is 0.4 ft^3. During the process, the spring exerts a force on the piston that varies according to $F = kx$, where $k = 700$ lbf/ft. There is no friction between the piston and the cylinder wall. Using the ideal gas model, determine for the air inside the cylinder

(a) the final pressure, in lbf/in.^{2}
(b) the final temperature, in °F.

4.87 A well-insulated chamber of volume 1 ft^3 is shown in Fig. P4.87. Initially, the chamber contains air at 14.7 lbf/in.^{2} and 100°F. Connected to the chamber are supply and discharge pipes equipped with valves that control the flow rates into and out of the chamber. The supply air is at 30 lbf/in.^{2}, 200°F. Both valves are opened simulta-

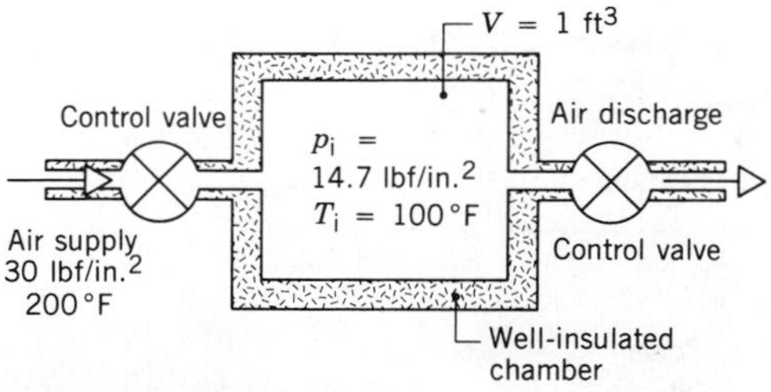

Figure P4.87

neously, allowing air to flow with a constant mass flow rate of 1 lb/min through each valve. The air within the chamber is well mixed, so the temperature and pressure at any time can be taken as uniform throughout. Determine the temperature, in °F, and the pressure, in lbf/in.^{2}, of the air in the chamber as functions of time. Neglect kinetic and potential energy effects and use the ideal gas model with constant specific heats.

DESIGN AND OPEN ENDED PROBLEMS

4.1D Household water-heating bills can be reduced by using flow restrictors on shower heads, lowering the temperature setting of the water heater, and enclosing the water heater in an insulating blanket. Estimate the annual energy savings for a *typical* U.S. household by implementing each of these changes. Consider both gas-fired and electrically heated water heaters.

4.2D A number of devices of engineering significance have been considered in the present chapter. For each of the devices listed below, give examples of important applications and discuss why each device is particularly suited to that use. Also, with the aid of diagrams, explain the operating principles of each device.

(a) *reciprocating, axial-flow,* and *centrifugal* compressors.

(b) *direct-contact, counterflow* and *parallel flow tube-within-a-tube,* and *cross-flow* heat exchangers.

4.3D Velocity distributions for *laminar* and *turbulent* flow in circular pipes are given, respectively, in Problems 4.8 and 4.9. For each velocity distribution, derive an expression for the *specific* kinetic energy carried by an incompressible fluid through an area A normal to the flow. What is the percent error in each case if the specific kinetic energy is evaluated in terms of the *average* velocity as $(V_{ave})^2/2$? Which distribution adheres most closely to the idealizations of one-dimensional flow? Under what circumstances would turbulent flow exist in a pipe? What significance does the velocity distribution have for volumetric and mass flow rate measurements in pipes.

4.4D A manufacturing facility has a 500-ft run of *nominal* $1\frac{1}{2}$ in., *Schedule* 40, uninsulated steel pipe. Steam enters the pipe as saturated vapor at 12 lbf/in.^{2} (gage) with a mass flow rate of 80,000 lb/h. How would you estimate the energy savings associated with insulating the bare steam pipe? Does insulating the pipe affect the exit state of the steam? Explain fully. Comment on the economic trade-offs involved in deciding whether to implement the insulation project.

4.5D A factory utilizes compressed air at a pressure of 85 lbf/in.^{2}(gage) to actuate valves and control processes. A plant *energy audit* identifies ten $\frac{1}{4}$-in.-diameter and twenty $\frac{1}{32}$-in.-diameter holes in various air lines throughout the plant. The volumetric flow rate of air exiting each hole is estimated to be

$$(AV) = (56.4 \times 10^3)\ D^2$$

where D is the diameter in in., and the volumetric flow rate is in ft^3/h. The factory air compressors operate 8760 h annually. Would a significant reduction in compressor power be achieved if all the leaks were repaired? Explain.

4.6D *Low-head* dams (3 to 10 m), commonly used for flood control on many rivers, provide an opportunity for electric power generation using hydraulic turbine-generators. Estimates of this hydroelectric potential must take into account the available head and the river flow, each of which vary considerably throughout the year. Using U.S. Geological Survey data, determine the typical variations in head and flow for a river in your locale. Based on this information, estimate the total annual electric generation of a hydraulic turbine placed on the river. Does the peak generating capacity occur at the same time of year as peak electrical demand in your area? Would you recommend that your local utility take advantage of this opportunity for electric power generation? Discuss.

4.7D Figure 4.7D illustrates an experimental apparatus to test *shell-and-tube* evaporators in which a refrigerant evaporates on the outside of tubes through which a mixture of water and ethylene glycol flows. As shown by dashed lines in the figure, two subsystems provide refrigerant and water-glycol such that the test evaporator operates at steady state. The refrigerant subsystem must deliver saturated liquid Refrigerant 12 or 134a to the evaporator at temperatures as low as −20°C. The maximum rate of heat transfer between the two streams is 100 kW. Draw schematics showing appropriate layouts within each subsystem of heat exchangers, pumps, and connecting piping. For each component, write preliminary specifications for the mass flow and energy transfer rates. For a control volume enclosing the evaporator, specify the types and locations of sensors needed to obtain data to perform energy and mass balances.

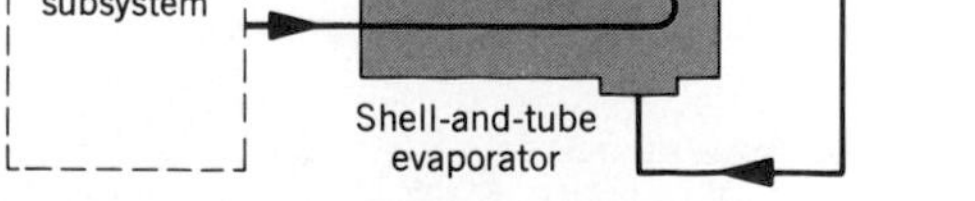

Figure 4.7D

CHAPTER 5

THE SECOND LAW OF THERMODYNAMICS

The presentation to this point has emphasized the use of the conservation of mass and conservation of energy principles, together with property relations for thermodynamic analysis. In Chaps. 2 through 4 these fundamentals are applied to increasingly complex situations. The conservation principles do not always suffice, however, and often the second law of thermodynamics is also required for thermodynamic analysis. In this chapter the second law of thermodynamics is introduced. A number of deductions that may be called corollaries of the second law are also considered. The current presentation provides the basis for subsequent developments involving the second law in Chaps. 6 and 7.

5.1 INTRODUCTION

The objective of the present section is to motivate the need for and the usefulness of the second law. The discussion shows why not one but a number of alternative formulations of the second law have been advanced.

5.1.1 Direction of Processes

It is a matter of everyday experience that there is a definite direction for *spontaneous* processes. This can be brought out by considering the three systems pictured in Fig. 5.1.

An object at an elevated temperature T_i placed in contact with atmospheric air at temperature T_0 would eventually cool to the temperature of its much larger surroundings, as illustrated in Fig. 5.1*a*. In conformity with the conservation of energy principle, the decrease in internal energy of the body would appear as an increase in the internal energy of the surroundings. The reverse process would not take place *spontaneously,* even though energy could be conserved: The internal energy of the surroundings would not decrease spontaneously while the body warmed from T_0 to its initial temperature.

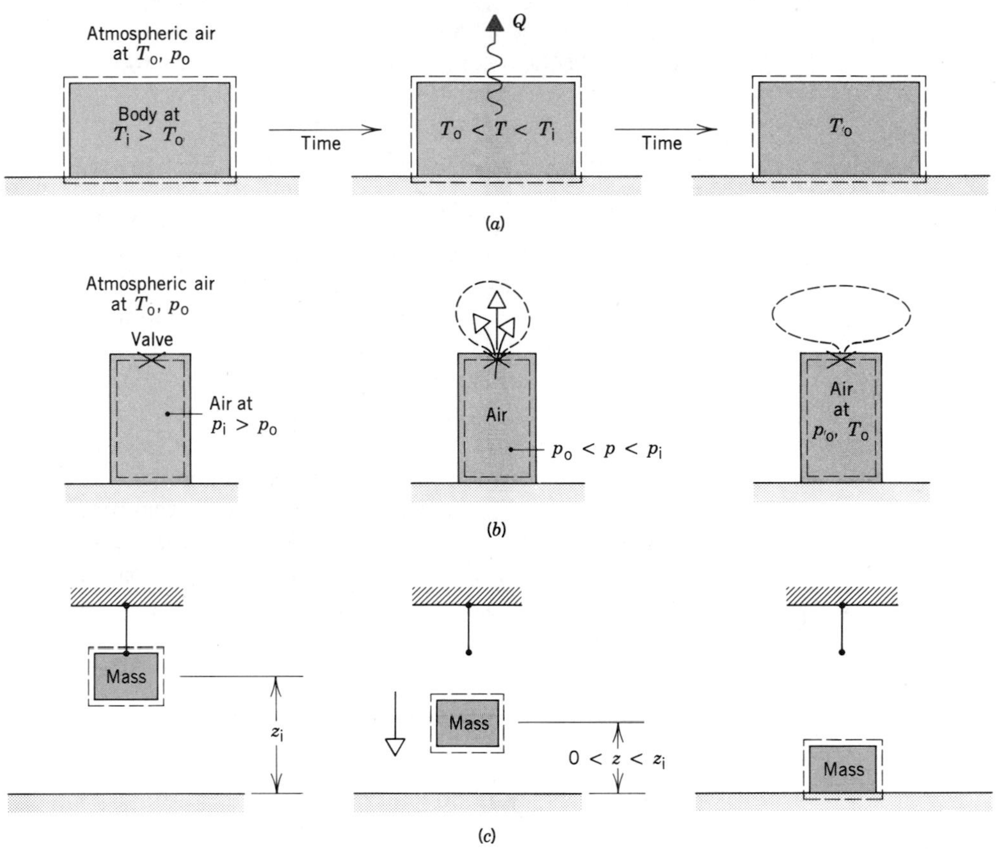

Figure 5.1 Illustrations of spontaneous processes and the eventual attainment of equilibrium with the surroundings. (*a*) Spontaneous heat transfer. (*b*) Spontaneous expansion. (*c*) Falling mass.

Air held at a high pressure p_i in a closed tank would flow spontaneously to the lower pressure surroundings at p_0 if the interconnecting valve were opened, as illustrated in Fig. 5.1*b*. Eventually fluid motions would cease and all of the air would be at the same pressure as the surroundings. Drawing on experience, it should be clear that the reverse process would not take place *spontaneously,* even though energy could be conserved: Air would not flow spontaneously from the surroundings at p_0 into the tank, returning the pressure to its initial value.

A mass suspended by a cable at elevation z_i would fall when released, as illustrated in Fig. 5.1*c*. When it comes to rest, the potential energy of the mass in its initial condition would appear as an increase in the internal energy of the mass and its surroundings, in accordance with the conservation of energy principle. Eventually, the mass also would come to the temperature of its much larger surroundings. The reverse process would not take place *spontaneously,* even though energy could be conserved: The mass would not return unaided to its initial elevation while its internal energy or that of its surroundings decreased.

In each of the cases considered, the initial condition of the system can be restored, but not in a spontaneous process. Some auxiliary devices would be required. By such auxiliary means the object could be reheated to its initial temperature, the air could be returned to the tank and restored to its initial pressure, and the mass could be lifted to its initial height. In each case, a fuel or electrical input normally would be required for the auxiliary devices to function, so a permanent change in the condition of the surroundings would result.

The foregoing examples indicate that not every process consistent with the principle of energy conservation can occur. Generally, an energy balance alone neither enables the preferred direction to be predicted nor permits the processes that can occur to be distinguished from those that cannot. In elementary cases such as the ones considered, experience can be drawn upon to deduce whether particular spontaneous processes occur and to deduce their directions. For more complex cases, where experience is lacking or uncertain, a guiding principle would be helpful. This is provided by the *second law*.

When left to themselves, systems tend to undergo spontaneous changes until a condition of equilibrium is achieved, both internally and with their surroundings. In some cases equilibrium is reached quickly, in others it is achieved slowly. For example, some chemical reactions reach equilibrium in fractions of seconds; an ice cube requires a few minutes to melt; and it may take years for an iron bar to rust away. Whether the process is rapid or slow, it must of course satisfy the conservation of energy principle. However, this principle alone would be insufficient for determining the final equilibrium state. Another general principle is required. This is also provided by the *second law*.

5.1.2 Opportunities for Developing Work

By exploiting the spontaneous processes shown in Fig. 5.1, it is possible, in principle, for work to be developed as equilibrium is attained. For example, instead of permitting the body of Fig. 5.1*a* to cool spontaneously with no other result, energy could be delivered by heat transfer to a system undergoing a power cycle that would develop a net amount of work (Sec. 2.6). Once the object attained equilibrium with the surroundings, the process would cease. Although there is an *opportunity* for developing work in this case, the opportunity would be wasted if the body were permitted to cool without developing any work. Referring again to the compressed air case of Fig. 5.1*b*, instead of permitting the air to expand aimlessly into the lower-pressure surroundings, the stream could be passed through a turbine-wheel and work developed. Accordingly, in this case there is also a possibility for developing work that would not be exploited in an uncontrolled process. In the case of Fig. 5.1*c*, instead of permitting the mass to fall in an uncontrolled way, it could be lowered gradually while turning a wheel, lifting another mass, and so on.

Underlying the possibility for developing work is an intensive property difference. In the first case shown in Fig. 5.1, a difference in *temperature* provides the impetus for a heat transfer from one region to another and thereby the possibility for work to be developed. It is not essential for T_i to be greater than T_0 as shown in Fig. 5.1*a*. If the body were initially at a temperature less than that of the surroundings, $T_i < T_0$, energy could flow by heat transfer *from* the surroundings *to* it, still giving the possibility of developing work by means of a system undergoing a cycle. In the case shown in Fig. 5.1*b*, a difference in *pressure* provides the impetus for air to flow and thereby the opportunity for work to be developed. This possibility exists whether $p_i > p_0$, as shown in Fig. 5.1*b*, or $p_i < p_0$. If $p_i < p_0$, air would flow *into* the tank from the surroundings. For the third case, shown in Fig. 5.1*c*, the *elevation* difference provides the possibility for developing work.

The foregoing discussion can be summarized by noting that when an imbalance exists between two systems, there is an opportunity for developing work that would be irrevocably lost if the systems were allowed to come into equilibrium in an uncontrolled way. Recognizing this possibility for work, we can pose two questions: (1) What is the theoretical maximum value for the work that could be obtained? (2) What are the factors that would preclude the realization of the maximum value? That there should be a maximum value is fully in accord with experience, for if it were possible to develop unlimited work, few concerns would be voiced over our dwindling fuel supplies. Also in accord with experience is the idea that even the best devices would be subject to factors such as friction, which would preclude the attainment of the theoretical maximum work. The second law of thermodynamics provides the means for determining the theoretical maximum and evaluating quantitatively the factors that preclude attaining the maximum.

5.1.3 Aspects of the Second Law

The preceding discussion can be summarized by noting that the second law and deductions from it are useful because they provide means for (1) predicting the direction of processes, (2) establishing conditions for equilibrium, (3) determining the best *theoretical* performance of cycles, engines, and other devices, and (4) evaluating quantitatively the factors that preclude the attainment of the best theoretical performance level. Additional uses of the second law include its roles in (5) defining a temperature scale independent of the properties of any thermometric substance and (6) developing means for evaluating properties such as u and h in terms of properties that are more readily obtained experimentally. Scientists and engineers have found many additional applications of the second law and deductions from it. It also has been used in economics, philosophy, and other areas far removed from engineering thermodynamics.

The six points listed can be thought of as aspects of the second law of thermodynamics and not as independent and unrelated ideas. Nonetheless, given the variety of these topic areas, it should be easy to understand why there is no one simple statement of the second law that brings out each one clearly. There are several alternative, yet equivalent, formulations of the second law. In the next section, two equivalent statements of the second law are introduced as a *point of departure* for our study of the second law and its consequences. Although the exact relationship of these particular formulations to each of the second law aspects listed above may not be immediately apparent, all aspects listed can be obtained by deduction from these formulations or their corollaries. It is important to add that in every instance where a consequence of the second law has been tested directly or indirectly by experiment, it has been unfailingly verified. Accordingly, the basis of the second law of thermodynamics, like every other physical law, is experimental evidence.

5.2 STATEMENTS OF THE SECOND LAW

Among many alternative statements of the second law, two are frequently used in engineering thermodynamics. They are the *Clausius* and *Kelvin–Planck* statements. The objective of this section is to introduce these two second law statements and demonstrate that they are equivalent.

The Clausius statement has been selected as a point of departure for the study of the second law and its consequences because it is in accord with experience and therefore easy to accept. Though somewhat more abstract, the Kelvin–Planck statement has the advantage that it provides an efficient means for bringing out important second law deductions related to systems undergoing thermodynamic cycles. One of these deductions, the Clausius inequality (Sec. 6.1), leads directly to the property entropy and to formulations of the second law convenient for the analysis of closed systems and control volumes as they undergo processes that are not necessarily cycles.

5.2.1 Clausius and Kelvin–Planck Statements of the Second Law

The *Clausius statement* of the second law may be given as follows: *It is impossible for any system to operate in such a way that the sole result would be an energy transfer by heat from a cooler to a hotter body.*

The Clausius statement does not rule out the possibility of transferring energy by heat from a cooler body to a hotter body, for this is exactly what refrigerators and heat pumps accomplish. However, as the words "sole result" in the statement suggest, when a heat transfer from a cooler body to a hotter body occurs, there must be some *other effect* within the system accomplishing the heat transfer, its surroundings, or both. If the system operates in a thermodynamic cycle, its initial state is restored after each cycle, so the only place that must be examined for such *other* effects is its surroundings. For example, cooling in the home is accomplished by refrigerators, which are driven by electric motors that require work from their surroundings to operate. The Clausius statement implies that it is impossible to construct a refrigeration cycle that operates without an input of work.

Before giving the Kelvin–Planck statement of the second law, the concept of a *thermal reservoir* must be introduced. A thermal reservoir, or simply a reservoir, is a special kind of closed system that always remains at constant temperature even though energy is added or removed by heat transfer. A reservoir is an idealization, of course, but such a system can be approximated in a number of ways—by the earth's atmosphere, large bodies of water (lakes, oceans), a large block of copper, and so on. Another example is provided by a system consisting of two phases: Although the ratio of the masses of the two phases changes as the system is heated or cooled at constant pressure, the temperature remains constant as long as both phases coexist. Extensive properties of a thermal reservoir such as internal energy can change in interactions with other systems even though the reservoir temperature remains constant.

Having introduced the thermal reservoir concept, we may give the *Kelvin–Planck* statement of the second law as follows: *It is impossible for any system to operate in a thermodynamic cycle and deliver a net amount of work to its surroundings while receiving energy by heat transfer from a single thermal reservoir.*

The Kelvin–Planck statement does not rule out the possibility of a system developing a net amount of work from a heat transfer drawn from a single reservoir. It only denies this possibility if the system undergoes a thermodynamic cycle. An analytical formulation of the Kelvin–Planck statement is provided in Sec. 5.3.4.

5.2.2 Equivalence of the Clausius and Kelvin–Planck Statements

The equivalence of the Clausius and Kelvin–Planck statements can be demonstrated by showing that the violation of each statement implies the violation of the other. That a violation of the Clausius statement implies a violation of the Kelvin–Planck statement is readily shown using Fig. 5.2, which pictures a hot reservoir, a cold reservoir, and two systems. The system on the left transfers energy Q_C from the cold reservoir to the hot reservoir by heat transfer without other effects occurring and thus *violates the Clausius statement*. The system on the right operates in a cycle while receiving Q_H (greater than Q_C) from the hot reservoir, rejecting Q_C to the cold reservoir, and delivering work W_{cycle} to the surroundings. The energy flows labeled on Fig. 5.2 are in the directions indicated by the arrows.

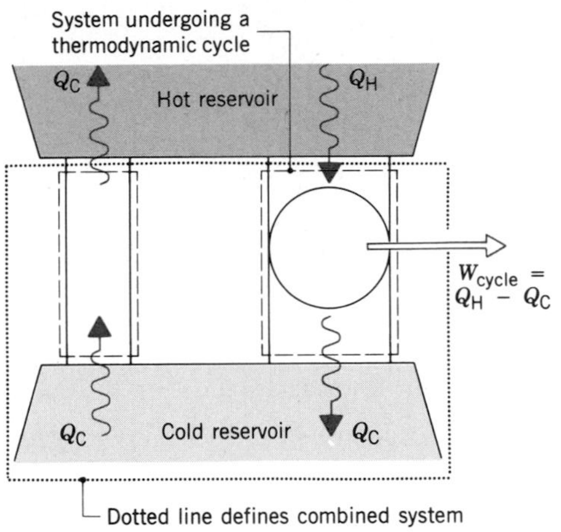

Figure 5.2 Illustration used to demonstrate the equivalence of the Clausius and Kelvin–Planck statements of the second law.

Consider the *combined* system shown by a dotted line on Fig. 5.2, which consists of the cold reservoir and the two devices. The combined system can be regarded as executing a cycle because one part undergoes a cycle and the other two parts experience no net change in their conditions. Moreover, the combined system receives energy ($Q_H - Q_C$) by heat transfer from a single reservoir, the hot reservoir, and produces an equivalent amount of work. Accordingly, the combined system violates the Kelvin–Planck statement. Thus, a violation of the Clausius statement implies a violation of the Kelvin–Planck statement. The equivalence of the two second law statements is demonstrated completely when it is also shown that a violation of the Kelvin–Planck statement implies a violation of the Clausius statement. This is left as an exercise.

5.3 IRREVERSIBLE AND REVERSIBLE PROCESSES

One of the important uses of the second law of thermodynamics in engineering is to determine the best theoretical performance of systems. By comparing actual performance with the theoretical optimum, insights often can be gained into the potential for doing better. As might be surmised, the best performance is evaluated in terms of idealized processes. In this section such idealized processes are introduced and distinguished from actual processes.

5.3.1 Irreversible Processes

A process is called *irreversible* if the system and all parts of its surroundings cannot be exactly restored to their respective initial states after the process has occurred. A process is *reversible* if both the system and surroundings can be returned to their initial states. Irreversible processes are the subject of the present discussion. Reversible processes are discussed in Sec. 5.3.2.

A system that has undergone an irreversible process is not necessarily precluded from being restored to its initial state. However, were the system restored to its initial state, it would not also be possible to return the surroundings to the state they were in initially. As illustrated below, the second law can be used to determine whether both the system and surroundings can be returned to their initial states after a process has occurred. That is, the second law can be used to determine whether a given process is reversible or irreversible.

It should be apparent from the discussion of the Clausius statement of the second law that any process involving a spontaneous heat transfer from a hotter body to a cooler body is irreversible. Otherwise, it would be possible to return this energy from the cooler body to the hotter body with no other effects within the two bodies or their surroundings. However, this possibility is denied by the Clausius statement. In addition to spontaneous heat transfer, processes involving other kinds of spontaneous events are irreversible, such as an unrestrained expansion of a gas or liquid. There are also many other effects whose presence during a process renders it irreversible. Friction, electrical resistance, hysteresis, and inelastic deformation are important examples. In summary, irreversible processes normally include one or more of the following:

- Heat transfer through a finite temperature difference
- Unrestrained expansion of a gas or liquid to a lower pressure
- Spontaneous chemical reaction
- Spontaneous mixing of matter at different compositions or states
- Friction—sliding friction as well as friction in the flow of fluids
- Electric current flow through a resistance
- Magnetization or polarization with hysteresis
- Inelastic deformation

The second law can be used to demonstrate that a process involving any of these is, in fact, irreversible. This reasoning will soon be introduced and illustrated for a case involving friction.

Although the foregoing list is not exhaustive, it does suggest that *all actual processes are irreversible*. That is, every process involves effects such as those listed, whether it is a naturally occurring process or one involving a device of our construction, from the simplest mechanism to the largest industrial plant. The term "irreversibility" will be used to identify any of these effects. The list given previously comprises a few of the irreversibilities that are commonly encountered.

As a system undergoes a process, irreversibilities may be found within the system as well as within its surroundings, although in certain instances they may be located predominately in one place or the other. For many analyses it is convenient to divide the irreversibilities present into two classes. *Internal irreversibilities* are those that occur within the system. *External irreversibilities* are those that occur within the surroundings, often the immediate surroundings. As this distinction depends solely on the location of the boundary, there is some arbitrariness in the classification, for by extending the boundary to take in a portion of the surroundings, all irreversibilities become "internal." Nonetheless, as shown by subsequent developments, this distinction between irreversibilities is often useful.

Engineers should be able to recognize irreversibilities, evaluate their influence, and develop practical means for reducing them. However, certain systems, such as brakes, rely on the effect of friction or other irreversibilities in their operation. The need to achieve profitable rates of production, high heat transfer rates, rapid accelerations, and so on invariably dictates the presence of significant irreversibilities. Furthermore, irreversibilities are tolerated to some degree in every type of system because the changes in design and operation required to reduce them would be too costly. Accordingly, although improved thermodynamic performance can accompany the reduction of irreversibilities, steps taken in this direction are constrained by a number of practical factors often related to costs.

Whenever any irreversibility is present during a process, the process must necessarily be irreversible. However, the irreversibility of the process can be *demonstrated* using the Kelvin–Planck statement of the second law and the following procedure: (1) Assume there is a way to return the system and surroundings to their respective initial states. (2) Show that as a consequence of this assumption, it would be possible to devise a cycle that produces work while no effect occurs other than a heat transfer from a single reservoir. Since the existence of such a cycle is denied by the Kelvin–Planck statement, the initial assumption must be in error and it follows that the process is irreversible. Example 5.1 illustrates this reasoning for a simplified case involving friction.

The approach illustrated by Example 5.1 also can be employed to demonstrate that processes involving the unrestrained expansion of a gas or liquid to a lower pressure and other effects from the list given previously are irreversible. However, in most instances the use of the Kelvin–Planck statement to demonstrate the irreversibility of processes is cumbersome. It is normally easier to use the entropy production concept (Sec. 6.5).

5.3.2 Reversible Processes

A process of a system is *reversible* if the system and all parts of its surroundings can be exactly restored to their respective initial states after the process has taken place. It should be evident from the discussion of irreversible processes that reversible processes are purely hypothetical. Clearly, no process can be reversible that involves spontaneous heat transfer through a finite temperature difference, an unrestrained expansion of a gas or liquid, friction, or any of the other effects listed previously. In a strict sense of the word, a reversible process is one that is *perfectly* executed.

All actual processes are irreversible. Reversible processes do not occur. Even so, certain processes that do occur are approximately reversible. The passage of a gas through a properly designed nozzle or diffuser is an example (Sec. 6.8). Many other devices also can be made to approach reversible operation by taking measures to reduce the significance of irreversibilities, such as lubricating surfaces to reduce friction. A reversible process is the *limiting case* as irreversibilities, both internal and external, are reduced further and further.

Although reversible processes do not actually occur, numerous reversible processes can be imagined. A particularly elementary example is a pendulum oscillating in an evacuated space. The pendulum motion approaches reversibility more closely as friction at the pivot point is reduced. In the limit as friction is eliminated, the states of both the pendulum and its surroundings would be completely restored at the end of each period of motion. By definition, such a process is reversible.

A system consisting of a gas being alternately compressed and expanded in a piston–cylinder assembly provides another example. For simplicity, we assume no friction between the piston and the cylinder walls and no heat transfer with the surroundings. With a very small increase in the external pressure, the piston would compress the gas slightly. At each intermediate volume during the compression, the intensive properties T, p, v, etc. would be uniform throughout: The gas would pass through a series of equilibrium states. With a small decrease in the external pressure, the piston would slowly move out as the gas expands. At each intermediate volume of the expansion, the intensive properties of the gas would be at the same uniform values they had at the corresponding step during the compression. When the gas volume returned to its initial value, all properties would be restored to their initial values as well. The work done *on* the gas during the compression would equal the work done *by* the gas during the expansion. If the work between the system and its surroundings were delivered to, and received from, a frictionless pulley–mass assembly, or the equivalent, there would also be no net change in the surroundings. This process would be reversible. In contrast, if the gas were rapidly compressed, the pressure near the piston face would be higher than that elsewhere in the gas.

Spatial variations in other intensive properties might also occur. The intermediate states of the compression process would therefore not be equilibrium states. Moreover, even if the gas were restored to its initial state without additional irreversibilities, more work would have been required to compress the gas than would be returned to the surroundings in the subsequent expansion of the gas. Since a permanent change in the surroundings would occur, this process would be irreversible.

As a final example of a reversible process, consider two bodies at different temperatures that are able to communicate thermally. With a *finite* temperature difference between them, a spontaneous heat transfer would take place and, as discussed previously, this would be irreversible. It might be expected that the importance of this irreversibility would diminish as the temperature difference narrows, and this is the case. As the difference in temperature between the bodies approaches zero, the heat transfer would approach reversibility. From the discussion of heat transfer modes in Sec. 2.4.2, we know that the transfer of a finite amount of energy by heat between bodies whose temperatures differ only slightly would require a considerable amount of time, a large heat transfer surface area, or both. To approach reversibility, therefore, a heat transfer would require an infinite amount of time and/or an infinite surface area.

EXAMPLE 5.1

PROBLEM

A system consists of a block of mass m and an inclined plane. Initially the block is at rest on the inclined plane. A process occurs in which the block slides down the plane, eventually coming to rest again at a lower elevation. There is no significant heat transfer between the system and its surroundings during the process. Use the Kelvin–Planck statement of the second law to demonstrate that this process involving friction is irreversible.

SOLUTION

Known: A block slides down an inclined plane under the influence of gravity to a lower elevation.

Find: Using the Kelvin–Planck statement, show that the process is irreversible.

Schematic and Given Data:

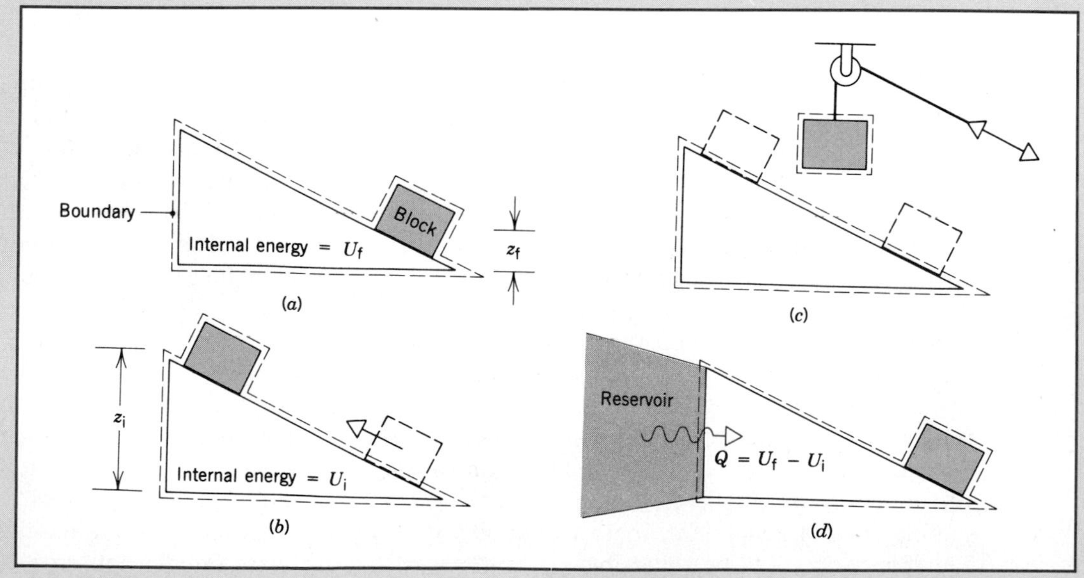

Figure E5.1

Assumptions:

1. The block and inclined plane form a closed system.
2. The block is at rest initially and finally.
3. During the process there are no significant work or heat interactions between the system and its surroundings.
4. As illustrated in the accompanying figure, a pulley–cable arrangement and a reservoir are present as aids in demonstrating that the process is irreversible.

Analysis: Taking the block and inclined plane as the system, the closed system energy balance reduces with assumptions 2 and 3 to

$$(U_f - U_i) + mg(z_f - z_i) + \cancelto{0}{(KE_f - KE_i)} = \cancelto{0}{Q} - \overset{0}{W}$$

or

$$U_f - U_i = mg(z_i - z_f)$$

Thus, friction between the block and plane during the process acts to convert the potential energy decrease of the block to internal energy of the block–plane system. Since no work or heat interactions occur between the system and its surroundings, the condition of the surroundings remains unchanged during the process. This allows attention to be centered on the system only in demonstrating that the process is irreversible.

When the block is at rest after sliding down the plane, its elevation is z_f and the internal energy of the block–plane system is U_f. To demonstrate that the process is irreversible using the Kelvin–Planck statement, let us take this condition of the system, shown in Fig. E5.1*a*, as the initial state of a cycle consisting of three processes.

Process 1: Assume the reverse process can occur with no change in the surroundings. That is, as shown in Fig. E5.1*b*, assume the block returns spontaneously to its initial elevation and the internal energy of the system decreases to its initial value, U_i. (This is the process we want to demonstrate is impossible.)

Process 2: As shown in Fig. E5.1*c*, use the pulley–cable arrangement provided to lower the block from z_i to z_f, allowing the decrease in potential energy to do work by lifting another mass located in the surroundings. The work done by the system equals the decrease in the potential energy of the block: $mg(z_i - z_f)$.

Process 3: The internal energy of the system can be increased from U_i to U_f by bringing it into communication with the reservoir, as shown in Fig. E5.1*d*. The heat transfer required is $Q = U_f - U_i$. Or with the result of the energy balance on the system given above, $Q = mg(z_i - z_f)$. At the conclusion of this process the block is again at elevation z_f and the internal energy of the block–plane system is restored to U_f.

The net result of this cycle is to draw energy from a single reservoir by heat transfer and produce an equivalent amount of work. There are no other effects. However, such a cycle is denied by the Kelvin–Planck statement. Since both the heating of the system by the reservoir (Process 3) and the lowering of the mass by the pulley–cable while work is done (Process 2) are possible, it can be concluded that it is Process 1 that is impossible. Since Process 1 is the reverse of the original process in which the block slides down the plane, it follows that the original process is irreversible.

5.3.3 Internally Reversible Processes

In an irreversible process, irreversibilities are present within the system, its surroundings, or both. A reversible process is one in which there are no internal *or* external irreversibilities. A related concept useful in many following discussions is that of the *internally reversible* process. A system is said to have undergone an internally reversible process if nothing takes place within the system to make the process irreversible: *There are no internal irreversibilities*. Irreversibilities may be located within the surroundings, however, as when there is heat transfer between a portion of the boundary that is at one temperature and the surroundings at another.

At every intermediate state of an internally reversible process of a closed system, all intensive properties are uniform throughout each phase present. That is, the temperature, pressure, specific volume, and other intensive properties do not vary with position. If there were a spatial variation in temperature, say, there would be a tendency for a spontaneous energy transfer by conduction to occur *within* the system in the direction of decreasing temperature. For reversibility, however, no spontaneous processes can be present. From these considerations it can be concluded that the internally reversible process consists of a series of equilibrium states: It is a quasiequilibrium process. To avoid having two terms that refer to the same thing, in subsequent discussions we will refer to *any* such process as an internally reversible process.

The use of the internally reversible process concept in thermodynamics is comparable to the idealizations so often made in mechanics: point masses, frictionless pulleys, rigid beams, and so on. In much the same way as these are used in mechanics to simplify an analysis and arrive at a tractable model, simple thermodynamic models of complex situations can be obtained through the use of internally reversible processes. Initial calculations are often based on internally reversible processes. The calculated results may then be adjusted with efficiencies or correction factors to obtain reasonable estimates of actual performance under various operating conditions. Internally reversible processes are also useful in determining the best thermodynamic performance of systems.

The internally reversible process concept can be employed to refine the definition of the thermal reservoir introduced in Sec. 5.2.1. In subsequent discussions we assume that no internal irreversibilities are present within a thermal reservoir. Accordingly, every process of a thermal reservoir is an internally reversible process.

5.3.4 Analytical Form of the Kelvin–Planck Statement

An analytical expression of the Kelvin–Planck statement of the second law is developed in the present section. This expression is employed in subsequent sections to obtain several important deductions.

Let us begin by recalling that the conservation of energy principle imposes a constraint on the *magnitudes* of the net work and heat transfer of *every* cycle, namely

$$W_{\text{cycle}} = Q_{\text{cycle}}$$

In words, the net work done by (or on) the system undergoing a cycle equals the net heat transfer to (or from) the system. For the discussion to follow, it is particularly important to note that if W_{cycle} is negative, then Q_{cycle} is also negative. That is, if a net amount of energy is transferred by work *to* the system during the cycle, then an equal amount of energy is transferred by heat *from* the system during the cycle.

Now consider the *special case* of a system that undergoes a cycle while exchanging energy by heat transfer with a single reservoir only, as shown in Fig. 5.3. Let us suppose that work is delivered to, or received from, the pulley–mass assembly located in the surroundings. A flywheel, spring, or some other device also can perform the same function. Any cycle operating as shown in Fig. 5.3 must satisfy the above constraint on the *magnitudes* of the net work and heat transfer between the system and its surroundings during the cycle. The cycle also must satisfy a constraint imposed by the second law on the *direction* of these energy transfers: According to the Kelvin–Planck statement, a system undergoing a cycle while communicating thermally with a single reservoir *cannot* deliver a net amount of work to its surroundings. That is, the net work of the cycle *cannot be positive*. However, the Kelvin–

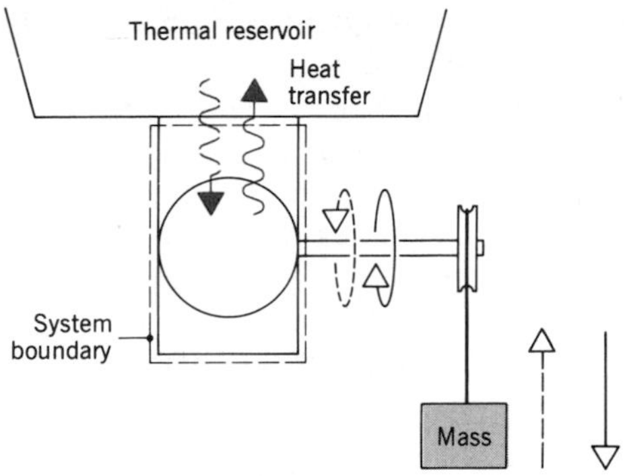

Figure 5.3 System executing a cycle while exchanging energy by heat transfer with a single thermal reservoir.

Planck statement does not rule out the possibility that there might be a net work transfer of energy *to* the system during the cycle or that the net work is zero. These deductions can be summarized as follows:

$$W_{\text{cycle}} \leq 0 \qquad \text{(single reservoir)} \tag{5.1}$$

where the words *single reservoir* are added to emphasize that the system communicates thermally only with a single reservoir as it executes the cycle. Combining Eq. 5.1 with $W_{\text{cycle}} = Q_{\text{cycle}}$ gives $Q_{\text{cycle}} \leq 0$. Each of these inequalities can be regarded as an analytical expression of the Kelvin–Planck statement of the second law of thermodynamics.

The "less than" sign of Eq. 5.1 applies when irreversibilities are present as a system of interest undergoes a cycle while communicating thermally with a single reservoir. The "equal to" sign applies only when no irreversibilities are present. In subsequent applications of Eq. 5.1, the irreversibilities of primary interest are those that occur *within* the system (internal irreversibilities). To eliminate extraneous factors in such applications, therefore, it is assumed that these are the only irreversibilities that might be present. Hence, the pulley–mass assembly, flywheel, or other device to which work is delivered, or from which it is received, is idealized as free of irreversibilities. Thermal reservoirs previously have been specified as free of irreversibilities.

To demonstrate the correspondence of the "less than" and "equal to" signs of Eq. 5.1 with the presence and absence of irreversibilities, respectively, consider a cycle operating as shown in Fig. 5.3 for which the equality applies. At the conclusion of one cycle:

- The system would necessarily be returned to its initial state.
- Since $W_{\text{cycle}} = 0$, there would be no *net* change in the elevation of the mass used to store energy in the surroundings.
- Since $W_{\text{cycle}} = Q_{\text{cycle}}$, it follows that $Q_{\text{cycle}} = 0$, so there also would be no *net* change in the condition of the reservoir.

Thus, the system and all elements of its surroundings would be exactly restored to their respective initial conditions. By definition, such a cycle is reversible. Accordingly, there can be no irreversibilities present within the system or its surroundings. It is left as an exercise to show the converse: If the cycle occurs reversibly, the equality applies. Since a cycle is either reversible or irreversible, it follows that the sign of inequality implies the presence of irreversibilities. Conversely, the inequality must be used whenever irreversibilities are present.

Equation 5.1 is employed in subsequent sections to obtain a number of significant deductions. In each of these applications, the idealizations used in the present discussion are assumed: The thermal reservoir and the portion of the surroundings with which work occurs are free of irreversibilities. This allows the "less than" sign to be associated with irreversibilities *within* the system of interest. The "equal to" sign is employed only when no irreversibilities of any kind are present.

5.4 SECOND LAW COROLLARIES FOR THERMODYNAMIC CYCLES

Several important corollaries of the second law related to power cycles and refrigeration and heat pump cycles are presented in this section. These results further our understanding of the implications of the second law and provide the basis for additional important deductions from the second law introduced in subsequent sections. To maintain the presentation at an introductory level, only systems undergoing thermodynamic cycles while communicating thermally with *two* thermal reservoirs are considered. Because familiarity with thermodynamic cycles is required, it is recommended that you review Sec. 2.6, where cycles are considered from the perspective of the conservation of energy principle and the thermal efficiency of power cycles and coefficients of performance for refrigeration and heat pump cycles are introduced.

5.4.1 Second Law Limitations on Power Cycles. The Carnot Corollaries

A significant limitation on the performance of systems undergoing power cycles can be brought out using the Kelvin–Planck statement of the second law. Consider Fig. 5.4, which shows a system that executes a cycle while communicating thermally with two thermal reservoirs, a hot reservoir and a cold reservoir, and developing net work W_{cycle}. The thermal efficiency of the cycle is

$$\eta = \frac{W_{\text{cycle}}}{Q_{\text{H}}} = 1 - \frac{Q_{\text{C}}}{Q_{\text{H}}} \tag{5.2}$$

where Q_{H} is the amount of energy received by the system from the hot reservoir by heat transfer and Q_{C} is the amount of energy discharged from the system to the cold reservoir by heat transfer. The energy transfers labeled on Fig. 5.4 are in the directions indicated by the arrows.

If the value of Q_{C} were zero, the system of Fig. 5.4 would withdraw energy Q_{H} from the hot reservoir and produce an equal amount of work, while undergoing a cycle. The thermal efficiency of such a cycle would have a value of unity (100%). However, this method of operation would violate the Kelvin–Planck statement and

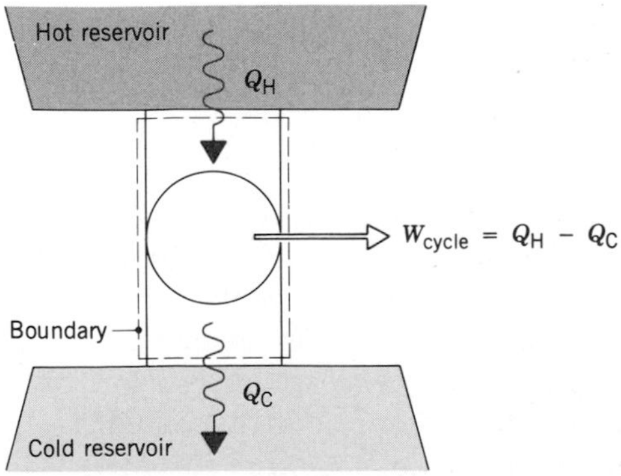

Figure 5.4 Representation of a system undergoing a power cycle while exchanging energy by heat transfer with two reservoirs.

thus is not allowed. It follows that for any system executing a power cycle while operating between two reservoirs, only a portion of the heat transfer Q_H can be obtained as work, and the remainder, Q_C, must be discharged by heat transfer to the cold reservoir. That is, the thermal efficiency must be less than 100%. In arriving at this conclusion it was *not* necessary to (1) identify the nature of the substance contained within the system, (2) specify the exact series of processes making up the cycle, or (3) indicate whether the processes are actual processes or somehow idealized. The conclusion that the thermal efficiency must be less than 100% applies to *all* power cycles whatever their details of operation. This may be regarded as a corollary of the second law.

Since no power cycle can have a thermal efficiency of 100%, it is of interest to determine whether there is a maximum theoretical efficiency. The maximum theoretical efficiency for systems undergoing power cycles while communicating thermally with two thermal reservoirs at different temperatures is evaluated in Sec. 5.6 with reference to the following two corollaries of the second law, called the *Carnot* corollaries.

Corollary 1: The thermal efficiency of an irreversible power cycle is always less than the thermal efficiency of a reversible power cycle when each operates between the same two thermal reservoirs.

Corollary 2: All reversible power cycles operating between the same two thermal reservoirs have the same thermal efficiency.

A cycle is considered *reversible* when there are no irreversibilities within the system as it undergoes the cycle and heat transfers between the system and reservoirs occur reversibly.

The idea underlying Corollary 1 is in agreement with expectations stemming from the discussion of the second law up to this point in the book. Namely, the presence of irreversibilities during the execution of a cycle is expected to exact a penalty. If two systems operating between the same reservoirs each receive the same amount of energy Q_H and one executes a reversible cycle while the other executes an irreversible cycle, it is in accord with intuition that the net work developed by the irreversible cycle will be less, and it will therefore have the smaller thermal efficiency.

Corollary 1 can be demonstrated using the arrangement of Fig. 5.5. A reversible power cycle R and an irreversible power cycle I operate between the same two reservoirs and each receives the same amount of energy Q_H from the hot reservoir. The reversible cycle produces work W_R while the irreversible cycle produces work W_I. In accord with the conservation of energy principle, each cycle discharges energy to the cold reservoir equal to the difference between Q_H and the work pro-

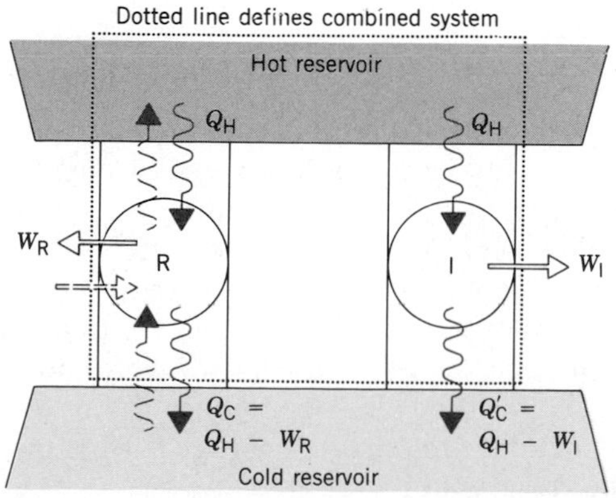

Figure 5.5 Sketch for demonstrating that a reversible cycle R is more efficient than an irreversible cycle I when they operate between the same two reservoirs.

duced. Let R now operate in the opposite direction as a refrigeration (or heat pump) cycle. Since R is reversible, the magnitudes of the energy transfers W_R, Q_H, and Q_C remain the same, but the energy transfers are oppositely directed, as shown by the dashed lines on Fig. 5.5. Moreover, with R operating in the opposite direction, the hot reservoir would experience *no net change* in its condition since it would receive Q_H *from* R while passing Q_H *to* I.

The demonstration of Corollary 1 is completed by considering the *combined system* shown by the dotted line on Fig. 5.5, which consists of the two cycles and the hot reservoir. Since its parts execute cycles or experience no net change, the combined system operates in a cycle. Moreover, it exchanges energy by heat transfer with a single reservoir: the cold reservoir. Accordingly, the combined system must satisfy Eq. 5.1 expressed as

$$W_{\text{cycle}} < 0 \qquad \text{(single reservoir)}$$

where the inequality is used because the combined system is irreversible in its operation since irreversible cycle I is one of its parts. Evaluating W_{cycle} for the combined system in terms of the work amounts W_I and W_R, the above inequality becomes

$$W_I - W_R < 0$$

which shows that W_I must be less than W_R. Since each cycle receives the same energy input, Q_H, it follows that $\eta_I < \eta_R$ and this completes the demonstration of Corollary 1.

Corollary 2 can be demonstrated in a parallel way by considering any two reversible cycles R_1 and R_2 operating between the same two reservoirs. Then, letting R_1 play the role of R and R_2 the role of I in the previous development, a combined system consisting of the two cycles and the hot reservoir may be formed that must obey Eq. 5.1. However, in applying Eq. 5.1 to this combined system, the equality is used because the system is reversible in operation. Thus, it can be concluded that $W_{R1} = W_{R2}$, and therefore, $\eta_{R1} = \eta_{R2}$. The details are left as an exercise.

Corollary 2 refers only to reversible cycles. All processes of a reversible cycle are in a strict sense of the word perfect. Accordingly, if two reversible cycles operating between the same reservoirs each receive the same amount of energy Q_H but one could produce more work than the other, it could only be as a result of more advantageous selections for the substance making up the system (it is conceivable that, say, air might be better than water vapor) *or* the series of processes making up the cycle (nonflow processes might be preferable to flow processes). Corollary 2 denies both possibilities and indicates that the cycles must have the same efficiency whatever the choices for the working substance or the series of processes.

5.4.2 Second Law Limitations on Refrigeration and Heat Pump Cycles

The second law of thermodynamics places limits on the performance of refrigeration and heat pump cycles as it does for power cycles. Consider Fig. 5.6, which shows a system undergoing a cycle while communicating thermally with two thermal reservoirs, a hot and a cold reservoir. The energy transfers labeled on the figure are in the directions indicated by the arrows. In accord with the conservation of energy principle, the cycle discharges energy Q_H by heat transfer to the hot reservoir equal to the sum of the energy Q_C received by heat transfer from the cold reservoir and the net work input. This cycle might be a refrigeration cycle or a heat pump cycle, depending on whether its function is to remove energy Q_C from the cold reservoir or deliver energy Q_H to the hot reservoir.

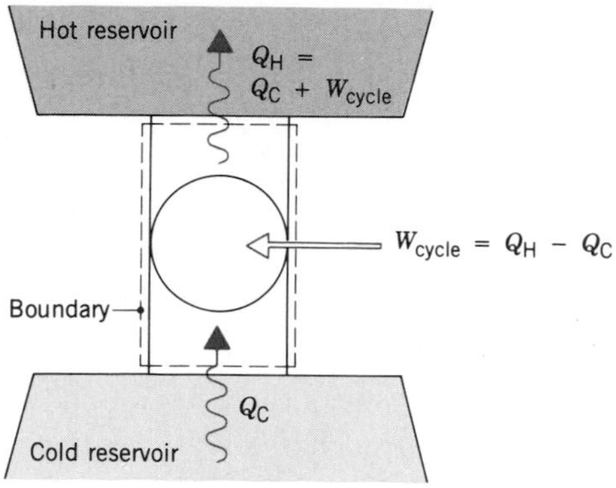

Figure 5.6 Representation of a system undergoing a refrigeration or heat pump cycle while exchanging energy by heat transfer with two reservoirs.

For a refrigeration cycle the coefficient of performance is

$$\beta = \frac{Q_C}{W_{cycle}} = \frac{Q_C}{Q_H - Q_C} \tag{5.3}$$

The coefficient of performance for a heat pump cycle is

$$\gamma = \frac{Q_H}{W_{cycle}} = \frac{Q_H}{Q_H - Q_C} \tag{5.4}$$

As the net work input to the cycle W_{cycle} tends to zero, the coefficients of performance given by Eqs. 5.3 and 5.4 approach a value of infinity. If W_{cycle} were identically zero, the system of Fig. 5.6 would withdraw energy Q_C from the cold reservoir and deliver energy Q_C to the hot reservoir, while undergoing a cycle. However, this method of operation would violate the Clausius statement of the second law and thus is not allowed. It follows that these coefficients of performance must invariably be finite in value.

Since the coefficients of performance of refrigeration and heat pump cycles must be finite, it is of interest to determine whether there are maximum theoretical values. The maximum theoretical coefficients of performance for systems undergoing refrigeration and heat pump cycles while communicating thermally with two reservoirs at different temperatures are evaluated in Sec. 5.6, with reference to the corollaries of the second law introduced below.

Figure 5.7 shows a reversible refrigeration cycle R and an irreversible refrigeration cycle I operating between the same two reservoirs. Each cycle removes the same

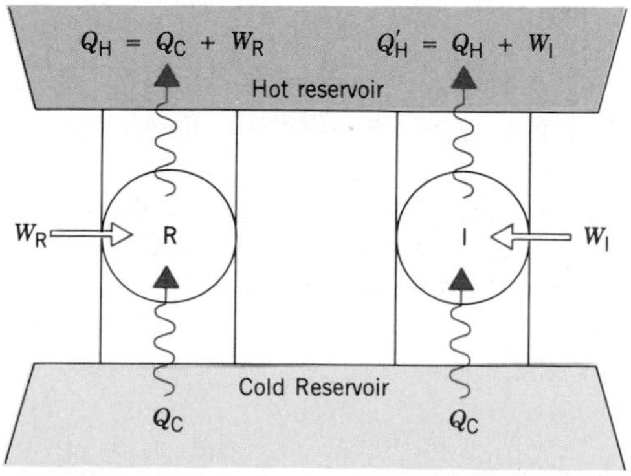

Figure 5.7 Sketch for demonstrating that a reversible refrigeration cycle R has a greater coefficient of performance than an irreversible cycle I when they operate between the same two reservoirs.

energy Q_C from the cold reservoir. The net work input required to operate R is W_R, while the net work input for I is W_I. Each cycle discharges energy by heat transfer to the hot reservoir equal to the sum of Q_C and the net work input. The directions of energy transfer are shown by arrows on Fig. 5.7. The presence of irreversibilities during the operation of a refrigeration cycle is expected to exact a penalty. If two refrigerators working between the same reservoirs each receive an identical energy transfer from the cold reservoir, Q_C, and one executes a reversible cycle while the other executes an irreversible cycle, it is expected that the irreversible cycle requires a greater net work input and thus has the smaller coefficient of performance. From these considerations the following corollary of the second law can be proposed: *The coefficient of performance of an irreversible refrigeration cycle is always less than the coefficient of performance of a reversible refrigeration cycle when both exchange energy by heat transfer with the same two thermal reservoirs.* This corollary can be proved using Eq. 5.1 and a procedure closely patterned after that employed for the Carnot corollaries. The details are left as an exercise. By a simple extension it may also be shown that all reversible refrigeration cycles operating between the same two reservoirs have the same coefficient of performance.

Next, consider a reversible heat pump cycle R and an irreversible heat pump cycle I operating between the same two reservoirs. Each cycle delivers the same amount of energy Q_H by heat transfer to the hot reservoir. The net work input required to operate R is W_R while the net work input for I is W_I. As for the case of refrigeration cycles, a penalty is exacted by irreversibilities: The irreversible cycle requires more work to deliver the same energy Q_H to the hot reservoir. Hence, *the coefficient of performance of the irreversible heat pump is less than that of the reversible heat pump*. It can also be concluded that all reversible heat pump cycles operating between the same two reservoirs have the same coefficient of performance. These conclusions can be demonstrated formally using Eq. 5.1 and a procedure like that used for the corollaries considered previously.

The discussion of second law limitations on power, refrigeration, and heat pump cycles initiated in the present section continues in Sec. 5.6 after the Kelvin temperature scale is formally introduced.

5.5 KELVIN TEMPERATURE SCALE

Using the second Carnot corollary, we are able to reconsider a point only touched on in Sec. 1.5, namely, the development of means for assigning temperature values independent of the properties of any thermometric substance. These considerations lead to the Kelvin temperature scale. The Kelvin scale provides a continuous definition of temperature connecting the several empirical methods for assigning temperatures. The Kelvin scale also plays an important role in developing further aspects of the second law, including the evaluation of maximum performance measures for cycles in the next section.

From the second Carnot corollary we can conclude that all reversible power cycles operating between the same two reservoirs have the same thermal efficiency, regardless of the nature of the substance making up the system executing the cycle or the series of processes. Since the efficiency is independent of these factors, its value can be related only to the nature of the reservoirs themselves. Noting that it is the difference in *temperature* between the two reservoirs that provides the impetus for heat transfer between them, and thereby for the production of work during the cycle, we reason that the efficiency depends *only* on the temperatures of the two reservoirs.

To pursue this line of thinking, consider a system undergoing a reversible power cycle while operating between reservoirs at temperatures θ_H and θ_C *on a scale to be defined*. Based on the foregoing reasoning, the cycle thermal efficiency depends only on the two temperatures

$$\eta = \eta(\theta_C, \theta_H)$$

Combining this with Eq. 5.2 gives

$$\eta(\theta_C, \theta_H) = 1 - \frac{Q_C}{Q_H}$$

or on rearrangement

$$\frac{Q_C}{Q_H} = 1 - \eta(\theta_C, \theta_H)$$

This can be expressed more concisely as

$$\left(\frac{Q_C}{Q_H}\right)_{\substack{\text{rev}\\\text{cycle}}} = \psi(\theta_C, \theta_H) \tag{5.5}$$

where the function ψ is for the present unspecified. Note that the words "rev cycle" are added to this expression to emphasize that it applies only to systems undergoing reversible cycles while operating between two reservoirs. Equation 5.5 shows that for such cycles the ratio of the heat transfers Q_C/Q_H is related only to the reservoir temperatures.

Equation 5.5 provides a basis for defining a *thermodynamic* temperature scale: a scale independent of the properties of any substance. There are alternative choices for the function ψ that lead to this end. The *Kelvin scale* is obtained by making a particularly simple choice, namely, $\psi = T_C/T_H$, where T is the symbol used to denote temperatures on the Kelvin scale. With this, Eq. 5.5 becomes

$$\left(\frac{Q_C}{Q_H}\right)_{\substack{\text{rev}\\\text{cycle}}} = \frac{T_C}{T_H} \tag{5.6}$$

Thus, two temperatures on the Kelvin scale are in the same ratio as the values of the heat transfers absorbed and rejected, respectively, by a system undergoing a reversible cycle while communicating thermally with reservoirs at these temperatures.

If a reversible power cycle were operated in the opposite direction as a refrigeration or heat pump cycle, the magnitudes of the energy transfers Q_C and Q_H would remain the same, but the energy transfers would be oppositely directed. Accordingly, Eq. 5.6 applies to each type of cycle considered thus far, provided the system undergoing the cycle operates between two thermal reservoirs and the cycle is reversible.

Equation 5.6 gives only a ratio of temperatures. To complete the definition of the Kelvin scale, it is necessary to proceed as in Sec. 1.5 by assigning the value 273.16 K to the temperature at the triple point of water. Then, if a reversible cycle is operated between a reservoir at 273.16 K and another reservoir at temperature T, the two temperatures are related according to

$$T = 273.16\left(\frac{Q}{Q_{tp}}\right)_{\substack{\text{rev}\\\text{cycle}}} \tag{5.7}$$

where Q_{tp} and Q are the heat transfers between the cycle and reservoirs at 273.16 K and temperature T, respectively. In the present case the heat transfer Q plays the role of the thermometric property. However, since the performance of a reversible cycle is independent of the make-up of the system executing the cycle, the definition of temperature given by Eq. 5.7 depends in no way on the properties of any substance or class of substances.

In Sec. 1.5 we noted that the Kelvin scale has a zero of 0 K, and lower temperatures than this are not defined. Let us take up these points by considering a reversible power cycle operating between reservoirs at 273.16 K and a lower temperature T. Referring to Eq. 5.7, we know that the energy rejected from the cycle by heat transfer Q would not be negative, so T must be nonnegative. Equation 5.7 also shows that the smaller the value of Q, the lower the value of T, and conversely. Accordingly, as Q approaches zero the temperature T approaches zero. It can be concluded that a temperature of zero on the Kelvin scale is the lowest conceivable temperature. This temperature is called the *absolute* zero, and the Kelvin scale is called an absolute temperature scale.

When numerical values of the thermodynamic temperature are to be determined, it is not possible to use reversible cycles, for these exist only in our imaginations. However, temperatures evaluated using the constant-volume gas thermometer introduced in Sec. 1.5 are identical to those of the Kelvin scale in the range of temperatures where the gas thermometer can be used. Other empirical approaches can be employed for temperatures above and below the range accessible to gas thermometry. The Kelvin scale provides a continuous definition of temperature valid over all ranges and provides an essential connection between the several empirical measures of temperature.

To provide a standard for temperature measurement taking into account both theoretical and practical considerations, the International Temperature Scale (ITS) was adopted in 1927. This scale has been refined and extended in several revisions,

TABLE 5.1 Defining Fixed Points of the International Temperature Scale of 1990

T(K)	Substance[a]	State[b]
3 to 5	He	Vapor pressure point
13.8033	e-H_2	Triple point
≈17	e-H_2	Vapor pressure point
≈20.3	e-H_2	Vapor pressure point
24.5561	Ne	Triple point
54.3584	O_2	Triple point
83.8058	Ar	Triple point
234.3156	Hg	Triple point
273.16	H_2O	Triple point
302.9146	Ga	Melting point
429.7485	In	Freezing point
505.078	Sn	Freezing point
692.677	Zn	Freezing point
933.473	Al	Freezing point
1234.93	Ag	Freezing point
1337.33	Au	Freezing point
1357.77	Cu	Freezing point

[a] He denotes ^3He or ^4He; e-H_2 is hydrogen at the equilibrium concentration of the ortho- and para-molecular forms.

[b] Triple point: temperature at which the solid, liquid, and vapor phases are in equilibrium. Melting point, freezing point: temperature, at a pressure of 101.325 kPa, at which the solid and liquid phases are in equilibrium.

Source: H. Preston-Thomas, "The International Temperature Scale of 1990 (ITS-90)," *Metrologia* 27, 3–10 (1990).

most recently in 1990. The International Temperature Scale of 1990 (ITS-90) is defined in such a way that the temperature measured on it conforms with the thermodynamic temperature, the unit of which is the kelvin, to within the limits of accuracy of measurement obtainable in 1990. The ITS-90 is based on the assigned values of temperature of a number of reproducible *fixed points* (Table 5.1). Interpolation between the fixed-point temperatures is accomplished by formulas that give the relation between readings of standard instruments and values of the ITS. In the range from 0.65 to 5.0 K, ITS-90 is defined by equations giving the temperature as functions of the vapor pressures of particular helium isotopes. The range from 3.0 to 24.5561 K is based on measurements using a helium constant-volume gas thermometer. In the range from 13.8033 to 1234.93 K, ITS-90 is defined by means of certain platinum resistance thermometers. Above 1234.9 K the temperature is defined using *Planck's equation for blackbody radiation* and measurements of the intensity of visible-spectrum radiation.

5.6 MAXIMUM PERFORMANCE MEASURES FOR POWER, REFRIGERATION, AND HEAT PUMP CYCLES OPERATING BETWEEN TWO RESERVOIRS

The discussion of Sec. 5.4 continues in this section with the development of expressions for the maximum thermal efficiency of power cycles and the maximum coefficients of performance of refrigeration and heat pump cycles in terms of reservoir temperatures evaluated on the Kelvin scale. These expressions can be used as standards of comparison for actual power, refrigeration, and heat pump cycles.

5.6.1 Power Cycles

The use of Eq. 5.6 in Eq. 5.2 results in an expression for the thermal efficiency of a system undergoing a reversible *power cycle* while operating between thermal reservoirs at temperatures T_H and T_C. That is

$$\eta_{max} = 1 - \frac{T_C}{T_H} \tag{5.8}$$

As temperatures on the Rankine scale differ from Kelvin temperatures only by the factor 1.8, the T's in Eq. 5.8 may be on either scale of temperature.

Recalling the two Carnot corollaries, it should be evident that the efficiency given by Eq. 5.8 is the thermal efficiency of *all* reversible power cycles operating between two reservoirs at temperatures T_H and T_C, and the *maximum* efficiency *any* power cycle can have while operating between the two reservoirs. It is common to refer to the maximum thermal efficiency given by Eq. 5.8 as the *Carnot* efficiency. By inspection, the value of the Carnot efficiency increases as T_H increases and/or T_C decreases.

Equation 5.8 is presented graphically in Fig. 5.8. The temperature T_C used in constructing the figure is 298 K in recognition that actual power cycles discharge energy by heat transfer at about the temperature of the local atmosphere or cooling water drawn from a nearby river or lake. Note that the possibility of increasing the thermal efficiency by reducing T_C below that of the environment is not practical, for maintaining T_C lower than the ambient temperature would require a refrigerator that would have to be supplied work to operate.

Figure 5.8 shows that the thermal efficiency increases with T_H. Referring to segment a–b of the curve, where T_H and η are relatively low, we can see that η increases

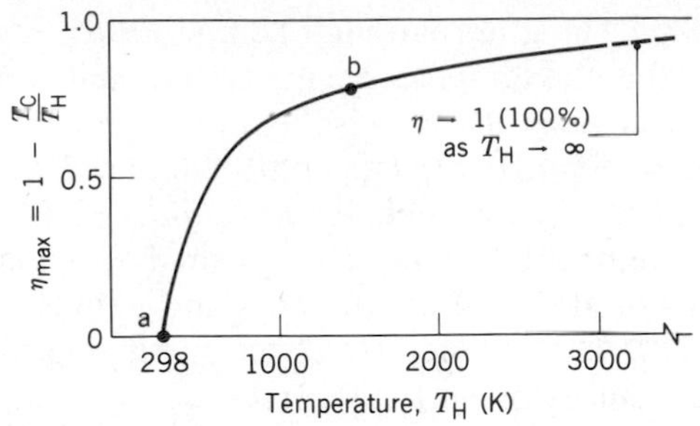

Figure 5.8 Carnot efficiency versus T_H, for $T_C = 298$K.

rapidly as T_H increases, showing that in this range even a small increase in T_H can have a large effect on efficiency. Though these conclusions, drawn as they are from Fig. 5.8, apply strictly only to systems undergoing reversible cycles, they are qualitatively correct for actual power cycles. The thermal efficiencies of actual cycles are observed to increase as the *average* temperature at which energy is added by heat transfer increases and/or the *average* temperature at which energy is discharged by heat transfer is reduced.

Conventional power-producing cycles have thermal efficiencies ranging up to about 40%. This value may seem low, but the comparison should be made with an appropriate limiting value and not 100%. For example, consider a system executing a power cycle for which the average temperature of heat addition is 745 K and the average temperature at which heat is discharged is 298 K. For a reversible cycle receiving and discharging energy by heat transfer at these temperatures, the thermal efficiency given by Eq. 5.8 is 60%. When compared to this value, an actual thermal efficiency of 40% does not appear to be so low. The cycle would be operating at two-thirds of the theoretical maximum. A more complete discussion of power cycles is provided in Chaps. 8 and 9.

5.6.2 Refrigeration and Heat Pump Cycles

Equation 5.6 is also applicable to reversible refrigeration and heat pump cycles operating between two thermal reservoirs, but for these Q_C represents the heat added to the cycle from the cold reservoir at temperature T_C on the Kelvin scale and Q_H is the heat discharged to the hot reservoir at temperature T_H. Introducing Eq. 5.6 in Eq. 5.3 results in the following expression for the coefficient of performance of any system undergoing a reversible refrigeration cycle while operating between the two reservoirs:

$$\beta_{max} = \frac{T_C}{T_H - T_C} \tag{5.9}$$

Similarly, substitution of Eq. 5.6 in Eq. 5.4 gives the following expression for the coefficient of performance of any system undergoing a reversible heat pump cycle while operating between the two reservoirs:

$$\gamma_{max} = \frac{T_H}{T_H - T_C} \tag{5.10}$$

The development of Eqs. 5.9 and 5.10 is left as an exercise. Note that the temperatures used to evaluate β_{max} and γ_{max} must be absolute temperatures, degrees Kelvin or degrees Rankine.

From the discussion of Sec. 5.4.2, it follows that Eqs. 5.9 and 5.10 are the maximum coefficients of performance that any refrigeration and heat pump cycles can have while operating between reservoirs at temperatures T_H and T_C. As for the case of the Carnot efficiency, these expressions can be used as standards of comparison for actual refrigerators and heat pumps. A more complete discussion of refrigeration and heat pump cycles is provided in Chap. 10.

EXAMPLE 5.2

PROBLEM

An inventor claims to have developed a power cycle capable of delivering a net work output of 410 kJ for an energy input by heat transfer of 1000 kJ. The system undergoing the cycle receives the heat transfer from hot gases at a temperature of 500 K and discharges energy by heat transfer to the atmosphere at 300 K. Evaluate this claim.

SOLUTION

Known: A system operates in a cycle and produces a net amount of work while receiving and discharging energy by heat transfer at fixed temperatures.

Find: Evaluate the claim that the cycle can develop 410 kJ of work for an energy input by heat of 1000 kJ.

Schematic and Given Data:

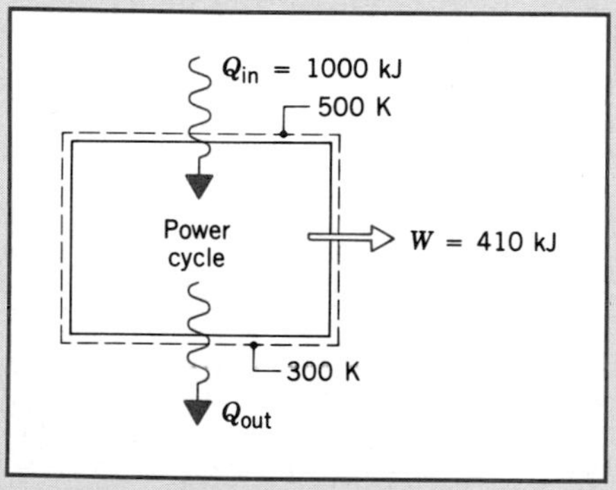

Figure E5.2

Assumptions:

1. The system is shown on the accompanying figure.
2. The hot gases and the atmosphere play the roles of hot and cold reservoirs, respectively.

Analysis: Using the values supplied by the inventor, the cycle thermal efficiency is

$$\eta = \frac{410}{1000} = 0.41\ (41\%)$$

The maximum thermal efficiency *any* power cycle can have while operating between reservoirs at $T_H = 500$ K and $T_C = 300$ K is given by

$$\eta_{max} = 1 - \frac{T_C}{T_H} = 1 - \frac{300}{500} = 0.40\ (40\%)$$

Since the thermal efficiency of the actual cycle exceeds the maximum theoretical value, the claim cannot be valid.

Comment: The temperatures used in evaluating η_{max} *must* be absolute temperatures, degrees Kelvin or degrees Rankine.

EXAMPLE 5.3

PROBLEM

By steadily circulating a refrigerant at low temperature through passages in the walls of the freezer compartment, a refrigerator maintains the freezer compartment at −5°C when the air surrounding the refrigerator is at 22°C. The rate of heat transfer from the freezer compartment to the refrigerant is 8000 kJ/h and the power input required to operate the refrigerator is 3200 kJ/h. Determine the coefficient of performance of the refrigerator and compare with the coefficient of performance of a reversible refrigeration cycle operating between reservoirs at the same two temperatures.

SOLUTION

Known: A refrigerator maintains a freezer compartment at a specified temperature. The rate of heat transfer from the refrigerated space, the power input to operate the refrigerator, and the ambient temperature are known.

Find: Determine the coefficient of performance and compare with that of a reversible refrigerator operating between reservoirs at the same two temperatures.

Schematic and Given Data:

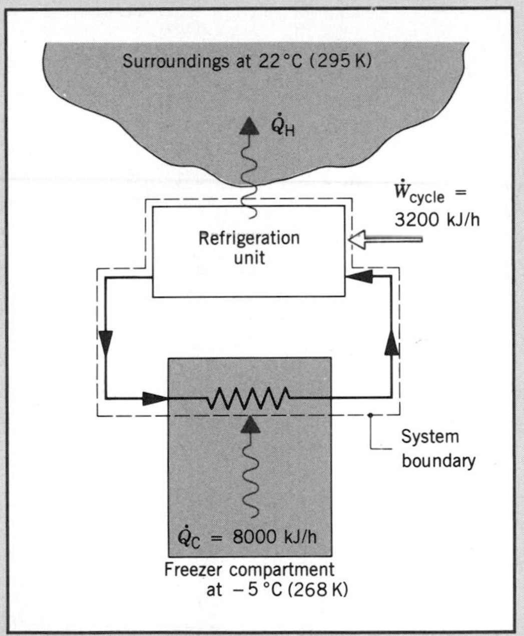

Figure E5.3

Assumptions:

1. The system shown on the accompanying figure is at steady state.
2. The freezer compartment and the surrounding air play the roles of cold and hot reservoirs, respectively.

Analysis: Using the given operating data, the coefficient of performance of the refrigerator is

$$\beta = \frac{\dot{Q}_C}{\dot{W}_{cycle}} = \frac{8000 \text{ kJ/h}}{3200 \text{ kJ/h}} = 2.5$$

Substituting values into Eq. 5.9 gives the coefficient of performance of a reversible refrigeration cycle operating between reservoirs at $T_C = 268$ K and $T_H = 295$ K

$$\beta_{max} = \frac{T_C}{T_H - T_C} = \frac{268 \text{ K}}{295 \text{ K} - 268 \text{ K}} = 9.9$$

Comment: The difference between the actual and maximum coefficients of performance suggests that there may be some potential for improving the thermodynamic performance. This objective should be approached cautiously, however, for improved performance may require increases in size, complexity, and cost.

EXAMPLE 5.4

PROBLEM

A dwelling requires 5×10^5 Btu per day to maintain its temperature at 70°F when the outside temperature is 50°F. If a heat pump cycle is used to supply this energy, determine the minimum theoretical work input for one day of operation, in Btu.

SOLUTION

Known: A heat pump cycle maintains a dwelling at a specified temperature. The energy supplied to the dwelling and the ambient temperature are known.

Find: Determine the *minimum* theoretical work required by the heat pump.

Schematic and Given Data:

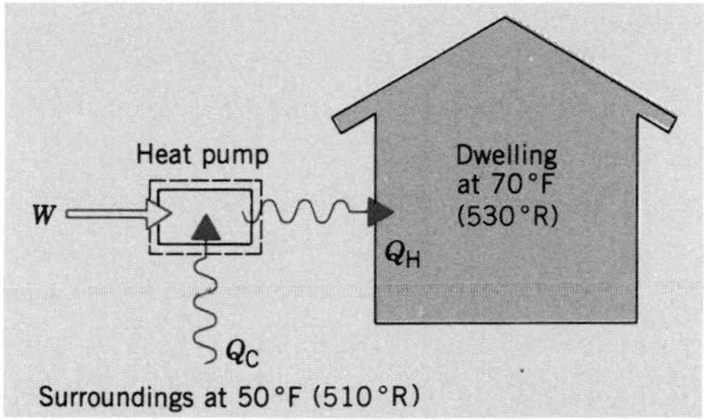

Figure E5.4

Assumptions:

1. The system is shown on the accompanying figure.
2. The dwelling and the outside air play the roles of hot and cold reservoirs, respectively.

Analysis: Noting that the directions of energy transfer are shown by the arrows on the accompanying figure, the energy balance for the heat pump cycle is

$$W = Q_H - Q_C$$

The condition of minimum work corresponds to reversible operation. For the reversible operation of a heat pump between reservoirs at T_C and T_H, the heat transfers are related by Eq. 5.6

$$\left(\frac{Q_C}{Q_H}\right)_{\substack{\text{rev}\\ \text{cycles}}} = \frac{T_C}{T_H}$$

Eliminating Q_C between these two equations, the following expression for the minimum work is obtained:

$$W_{min} = \left(1 - \frac{T_C}{T_H}\right) Q_H$$

Inserting values

$$W_{min} = \left(1 - \frac{510}{530}\right) (5 \times 10^5 \text{ Btu}) = 1.89 \times 10^4 \text{ Btu}$$

Comments:

1. Because of irreversibilities, an actual heat pump must be supplied more work than W_{min} to accomplish the same heating effect.
2. The energy equation used in the solution applies to *any* heat pump cycle, reversible or irreversible, but Eq. 5.6 applies only to reversible cycles operating between reservoirs at two fixed temperatures.

5.7 CARNOT CYCLE

The Carnot cycle introduced in this section provides a specific example of a reversible power cycle operating between two thermal reservoirs. Two other examples are provided in Chap. 9: the Ericsson and Stirling cycles. Each of these cycles exhibits the Carnot efficiency given by Eq. 5.8.

In a Carnot cycle, the system executing the cycle undergoes a series of four internally reversible processes: two adiabatic processes alternated with two isothermal processes. Figure 5.9 shows the p–v diagram of a Carnot power cycle in which the system is a gas in a piston–cylinder assembly. Figure 5.10 provides details of how the cycle is executed. The piston and cylinder walls are nonconducting. The heat transfers are in the directions of the arrows. Also note that there are two reservoirs at temperatures T_H and T_C, respectively, and an insulating stand. Initially, the piston–cylinder assembly is on the insulating stand and the system is at state 1. The four processes of the cycle are

Process 1–2: The gas is compressed *adiabatically* to state 2, where the temperature is T_H.

Process 2–3: The assembly is placed in contact with the reservoir at T_H. The gas expands *isothermally* while receiving energy Q_H from the hot reservoir by heat transfer.

Process 3–4: The assembly is again placed on the insulating stand and the gas is allowed to continue to expand *adiabatically* until the temperature drops to T_C.

Process 4–1: The assembly is placed in contact with the reservoir at T_C. The gas is compressed *isothermally* to its initial state while it discharges energy Q_C to the cold reservoir by heat transfer.

For the heat transfer during Process 2–3 to be reversible, the difference between the gas temperature and the temperature of the hot reservoir must be vanishingly small. Since the reservoir temperature remains constant, this implies that the temperature of the gas also remains constant during Process 2–3. The same can be concluded for Process 4–1.

For each of the four internally reversible processes of the Carnot cycle, the work can be represented as an area on Fig. 5.9. The area under the adiabatic process line 1–2 represents the work done per unit of mass to compress the gas in this process. The areas under process lines 2–3 and 3–4 represent the work done per unit of mass by the gas as it expands in these processes. The area under process line 4–1 is the work done per unit of mass to compress the gas in this process. The enclosed area on

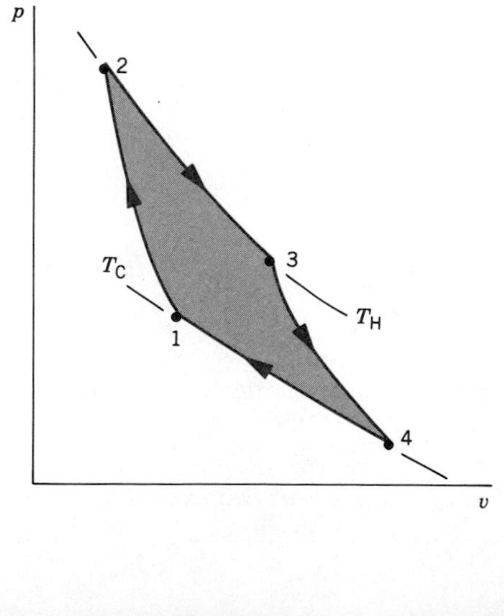

Figure 5.9 p–v diagram for a Carnot power cycle executed by a gas.

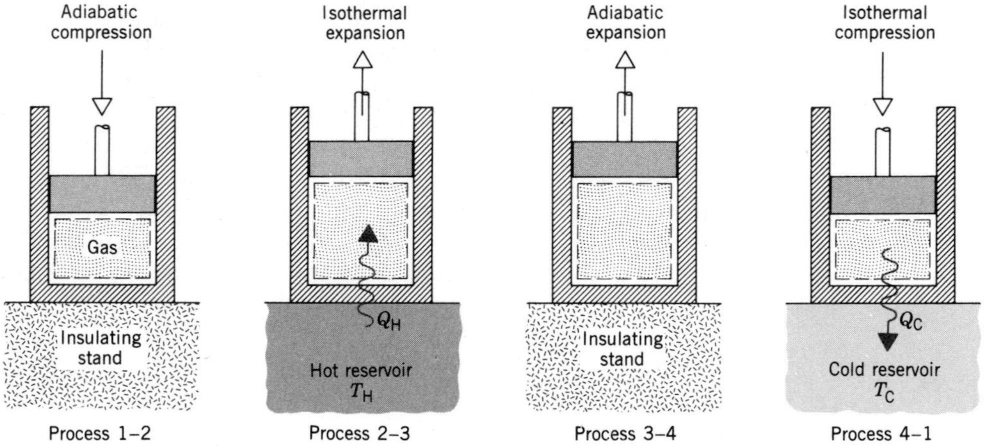

Figure 5.10 Carnot power cycle executed by a gas in a piston–cylinder assembly.

the p–v diagram, shown shaded, is the net work developed by the cycle per unit of mass.

The Carnot cycle is not limited to processes of a closed system taking place in a piston–cylinder assembly. Figure 5.11 shows the schematic and accompanying p–v diagram of a Carnot cycle executed by water steadily circulating through a series of

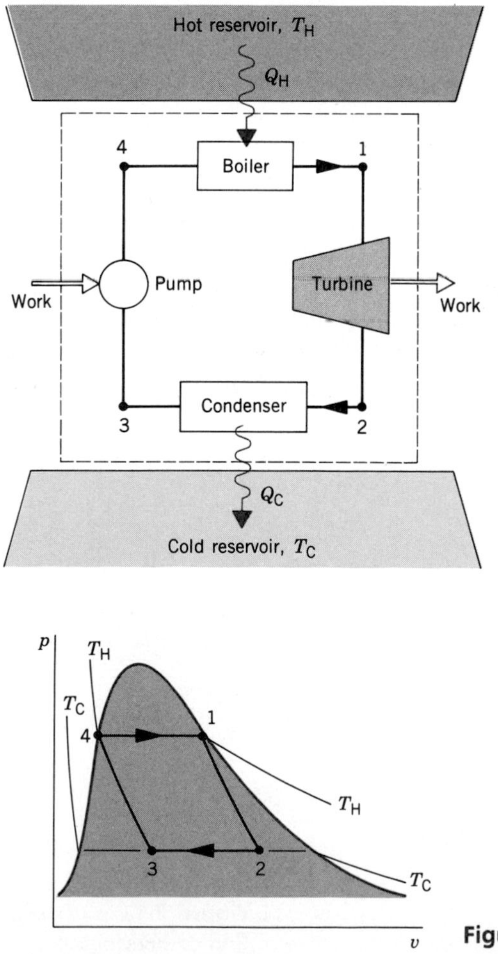

Figure 5.11 Carnot vapor power cycle.

four interconnected components that has features in common with a simple vapor power plant. As the water flows through the boiler, a *change of phase* from liquid to vapor at constant temperature T_H occurs as a result of heat transfer from the hot reservoir. Since temperature remains constant, pressure also remains constant during the phase change. The steam exiting the boiler expands adiabatically through the turbine and work is developed. In this process the temperature decreases to the temperature of the cold reservoir, T_C, and there is an accompanying decrease in pressure. As the steam passes through the condenser, a heat transfer to the cold reservoir occurs and some of the vapor condenses at constant temperature T_C. Since temperature remains constant, pressure also remains constant as the water passes through the condenser. The fourth component is a pump, or compressor, that receives a two-phase liquid–vapor mixture from the condenser and returns it adiabatically to the state at the boiler entrance. During this process, which requires a work input to increase the pressure, the temperature increases from T_C to T_H. Carnot cycles also can be devised that are composed of processes in which a capacitor is charged and discharged, a paramagnetic substance is magnetized and demagnetized, and so on. However, regardless of the type of process selected or the working substance used, the Carnot cycle always has the same four internally reversible processes: two adiabatic processes alternated with two isothermal processes. Moreover, the thermal efficiency is always given by Eq. 5.8 in terms of the temperatures of the two reservoirs evaluated on the Kelvin or Rankine scale.

If a Carnot power cycle is operated in the reverse direction, the magnitudes of all energy transfers remain the same but the energy transfers are oppositely directed. In the reverse direction, the cycle may be regarded as a reversible refrigeration or heat pump cycle, for which the coefficients of performance are given by Eqs. 5.9 and 5.10, respectively. A Carnot refrigeration or heat pump cycle executed by a gas in a piston–cylinder assembly is shown in Fig. 5.12. The cycle consists of the following four processes in series:

Process 1–2: The gas expands *isothermally* at T_C while *receiving* energy Q_C from the cold reservoir by heat transfer.

Process 2–3: The gas is compressed *adiabatically* until its temperature is T_H.

Process 3–4: The gas is compressed *isothermally* at T_H while it *discharges* energy Q_H to the hot reservoir by heat transfer.

Process 4–1: The gas expands *adiabatically* until its temperature decreases to T_C.

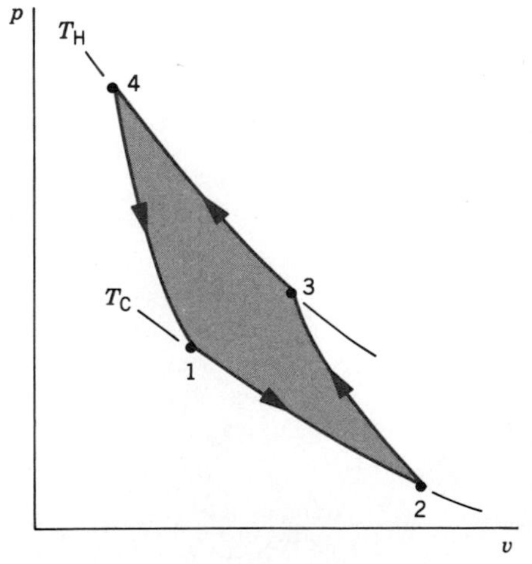

Figure 5.12 p–v diagram for a Carnot refrigeration or heat pump cycle executed by a gas.

It will be recalled that a refrigeration or heat pump effect can be accomplished in a cycle only if a net work input is supplied to the system executing the cycle. In the case of the cycle shown in Fig. 5.12, the shaded area represents the net work input per unit of mass.

PROBLEMS

5.1 For each of the following, explain how work might be developed as the system comes to equilibrium with its surroundings. Propose a means whereby the initial condition of the system might be restored.

(a) The system consists of an ice cube placed in a cup of tap water. The ice cube melts and eventually equilibrium is attained.
(b) The system consists of an initially evacuated tank surrounded by atmospheric air. An interconnecting valve is opened and air fills the tank. Eventually equilibrium is attained.
(c) The system consists of a marble. When released at a height above the bottom of a bowl and allowed to roll down the bowl's side, the marble eventually comes to rest.

5.2 An ordinary household refrigerator receives electrical work from its surroundings while discharging energy by heat transfer to its surroundings (the kitchen, say). Is this in violation of the Kelvin–Planck statement of the second law of thermodynamics? Explain.

5.3 One kilogram of air as an ideal gas expands isothermally at 20°C from a volume of 1 m^3 to a volume of 2 m^3. During this process there is heat transfer to the air from a single reservoir (the surrounding atmosphere) and the air does work. Evaluate the work and heat transfer for the process, in kJ. Is this process in violation of the Kelvin–Planck statement of the second law of thermodynamics? Explain.

5.4 Show that a violation of the Kelvin–Planck statement of the second law of thermodynamics implies a violation of the Clausius statement of the second law.

5.5 A closed system undergoes a thermodynamic cycle while communicating thermally with two reservoirs. The system receives energy Q_C from the cold reservoir and discharges energy Q_H to the hot reservoir while delivering a net amount of work to its surroundings. Prove that such a cycle is impossible using

(a) the Clausius statement of the second law of thermodynamics.
(b) the Kelvin–Planck statement of the second law of thermodynamics.

5.6 Consider as the system an automobile engine. List the irreversibilities present during operation.

5.7 Consider as the system an ordinary household refrigerator. List the irreversibilities present during operation.

5.8 Consider as the system an ordinary forced air, natural gas-fired household furnace. List the irreversibilities present during operation.

5.9 A hot thermal reservoir is separated from a cold thermal reservoir by a cylindrical rod insulated on its lateral surface. Energy transfer by conduction between the two reservoirs takes place through the rod, which remains at steady state. Using the Kelvin–Planck statement of the second law, demonstrate that such a process is irreversible.

5.10 A rigid insulated tank is divided into halves by a partition. On one side of the partition is a gas. The other side is initially evacuated. A valve in the partition is opened and the gas expands to fill the entire volume. Using the Kelvin–Planck statement of the second law, demonstrate that this process is irreversible.

5.11 A gas contained within a piston–cylinder assembly expands adiabatically. The gas is subsequently compressed back to its initial state. Does the gas undergo a reversible process? Discuss.

5.12 A gas within a piston–cylinder assembly is allowed to expand against the piston in such a way that all intensive properties of the gas are uniform with position throughout the gas during the expansion. Is this process internally reversible? Is this process reversible? Discuss.

5.13 Show that if a system undergoes a thermodynamic cycle reversibly while communicating thermally with a single reservoir, the equality in Eq. 5.1 applies.

5.14 A power cycle I and a reversible power cycle R operate between the same two reservoirs, as shown in Fig. 5.5. If cycle I has a smaller thermal efficiency than cycle R, prove that cycle I must be irreversible.

5.15 A reversible power cycle R and an irreversible power cycle I operate between the same two reservoirs. If each cycle receives the same amount of energy Q_H from the hot reservoir, show that cycle I necessarily discharges more energy Q_C to the cold reservoir than cycle R. Discuss the implications of this for actual power cycles.

5.16 Using the Kelvin–Planck statement of the second law of thermodynamics, prove that all reversible power cycles operating between the same two reservoirs have the same thermal efficiency.

5.17 Using the Kelvin–Planck statement of the second law of thermodynamics, prove the following:

(a) The coefficient of performance of an irreversible refrigeration cycle is always less than the coefficient of performance of a reversible refrigeration cycle when both exchange energy by heat transfer with the same two reservoirs.
(b) All reversible refrigeration cycles operating between the same two reservoirs have the same coefficient of performance.

5.18 Using the Kelvin–Planck statement of the second law of thermodynamics, prove the following:

(a) The coefficient of performance of an irreversible heat pump cycle is always less than the coefficient of performance of a reversible heat pump cycle when both exchange energy by heat transfer with the same two reservoirs.
(b) All reversible heat pump cycles operating between the same two reservoirs have the same coefficient of performance.

5.19 Before introducing the temperature scale now known as the Kelvin scale, Kelvin suggested a *logarithmic* scale in which the function ψ of Eq. 5.5 takes the form

$$\psi = \exp \theta_C / \exp \theta_H$$

where θ_H and θ_C denote, respectively, the temperatures of the hot and cold reservoirs on this scale.

(a) Show that the relation between the Kelvin temperature T and the temperature θ on the logarithmic scale is

$$\theta = \ln T + C$$

where C is a constant.
(b) On the Kelvin scale, temperatures vary from 0 to $+\infty$. Determine the range of temperature values on the logarithmic scale.
(c) Obtain an expression for the thermal efficiency of any system undergoing a reversible power cycle while operating between reservoirs at temperatures θ_H and θ_C on the logarithmic scale.

5.20 Demonstrate that the *gas temperature scale* (Sec. 1.5.3) is identical to the *Kelvin temperature scale*.

5.21 A reversible power cycle receives 1000 Btu of energy by heat transfer from a reservoir at 1000°F and discharges energy by heat transfer to a reservoir at 300°F. Determine the thermal efficiency and the net work developed, in Btu.

5.22 A reversible power cycle operates between a reservoir at temperature T and a lower-temperature reservoir at 280 K. At steady state, the cycle develops 40 kW of power while rejecting 1000 kJ/min of energy by heat transfer to the cold reservoir. Determine T, in K.

5.23 A certain reversible power cycle has the same thermal efficiency for hot and cold reservoirs at 1000 and 500°R, respectively, as for hot and cold reservoirs at T(°R) and 1000°R. Determine T, in °R.

5.24 A reversible power cycle whose thermal efficiency is 50% operates between a reservoir at 540°F and a reservoir at a lower temperature T. Determine T, in °R.

5.25 To increase the thermal efficiency of a reversible power cycle operating between reservoirs at T_H and T_C, would you increase T_H while keeping T_C constant, or decrease T_C while keeping T_H constant? Discuss.

5.26 Two reversible power cycles are arranged in series. The first cycle receives energy by heat transfer from a reservoir at temperature T_H and rejects energy to a reservoir at an intermediate temperature T. The second cycle receives the energy rejected by the first cycle from the reservoir at temperature T and rejects energy to a reservoir at temperature T_C lower than T. Derive an expression for the intermediate temperature T in terms of T_H and T_C when

(a) the net work of the two power cycles is equal.
(b) the thermal efficiencies of the two power cycles are equal.

5.27 If the thermal efficiency of a reversible power cycle operating between two reservoirs is denoted by η_{max}, develop an expression in terms of η_{max} for the coefficient of performance of

(a) a reversible refrigeration cycle operating between the same two reservoirs.
(b) a reversible heat pump operating between the same two reservoirs.

5.28 A power cycle operating between two reservoirs receives energy Q_H by heat transfer from a hot reservoir at $T_H = 2000$ K and rejects energy Q_C by heat transfer to a cold reservoir at $T_C = 400$ K. For each of the following cases determine whether the cycle operates reversibly, irreversibly, or is impossible:

(a) $Q_H = 1000$ kJ, $W_{cycle} = 850$ kJ.
(b) $Q_H = 2000$ kJ, $Q_C = 400$ kJ.
(c) $W_{cycle} = 1600$ kJ, $Q_C = 500$ kJ.
(d) $Q_H = 1000$ kJ, $\eta = 30\%$.

5.29 The data listed below are claimed for a power cycle operating between reservoirs at 727 and 127°C. For each case, determine if any principles of thermodynamics would be violated.

(a) $Q_H = 300$ kJ, $W_{cycle} = 160$ kJ, $Q_C = 140$ kJ.
(b) $Q_H = 300$ kJ, $W_{cycle} = 180$ kJ, $Q_C = 120$ kJ.
(c) $Q_H = 300$ kJ, $W_{cycle} = 170$ kJ, $Q_C = 140$ kJ.

5.30 An inventor claims to have developed a device that executes a power cycle while operating between reservoirs at 500 and 300 K that has a thermal efficiency of **(a)** 45%, **(b)** 40%. Evaluate the claim for each case.

5.31 At steady state, a new power cycle is claimed by its inventor to develop 4 horsepower for a heat addition rate of 300 Btu/min. If the cycle operates between reservoirs at 3000 and 1000°R, evaluate this claim.

5.32 At steady state, a cycle develops a power output of 10 kW for heat addition at a rate of 10 kJ *per cycle of operation* from a source at 3000 K. Energy is rejected by heat transfer to cooling water at 300 K. Determine the *minimum* theoretical number of cycles required per minute.

5.33 A proposed power cycle is to have a thermal efficiency of 30% while receiving energy by heat transfer from condensing steam and discharging energy by heat transfer to a nearby lake at 70°F. Determine the *lowest* possible condensing steam pressure, in lbf/in.^2

5.34 At steady state, a power cycle generates 100 MW of electricity while discharging energy by heat transfer at a rate of 55×10^7 Btu/h to cooling water at an average temperature of 70°F. The average temperature of the steam passing through the boiler is 900°F. The plant requires a power input of 1500 horsepower to drive pumps and other components.

(a) Determine the *net* power output and the thermal efficiency.
(b) For the stated conditions, determine the *minimum* theoretical rate at which energy could be discharged to the cooling water, in Btu/h. Compare with the actual rate and discuss.

5.35 Electrical power can be developed in principle by utilizing the naturally occurring decrease with depth of the temperature of ocean water. At one location the ocean surface temperature is 60°F, while at a depth of 1/3 mile the temperature is 35°F. Determine the maximum possible thermal efficiency for any power cycle operating between these temperatures.

5.36 Geothermal power plants harness underground sources of hot water or steam for the production of electricity. One such plant receives a supply of hot water at 340°F and rejects energy by heat transfer to the atmosphere, which is at 40°F. Determine the maximum possible thermal efficiency for any power cycle operating between these temperatures.

5.37 Figure P5.37 shows a system for collecting solar radiation and utilizing it for the production of electricity by a power cycle. The solar collector receives solar radiation at the rate of 0.315 kW per m^2 of area and provides energy to a storage unit whose temperature remains constant at 500 K. The power cycle receives energy by heat transfer from the storage unit, generates electricity at the rate 1000 kW, and discharges energy by heat transfer to the surroundings at 20°C. For operation at steady state, determine the minimum theoretical collector area required, in m^2.

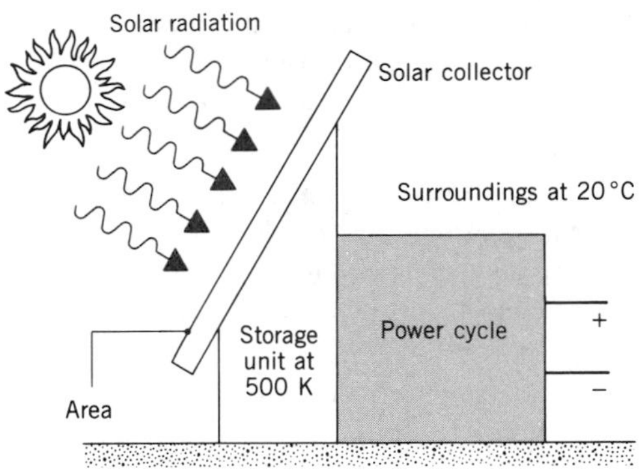

Figure P5.37

5.38 Referring to Problem 5.37, determine the collector area required if 15% of the energy incident on the collector is lost in unavoidable heat transfer with the surroundings and the thermal efficiency of the power cycle is 27%.

5.39 A certain system collects solar radiation to provide energy by heat transfer at 100°C to a power cycle that discharges energy by heat transfer to the surroundings at 20°C. Another system focuses solar radiation to provide energy at 200°C. On the basis of thermodynamic reasoning, which of the systems would be preferred? Discuss.

5.40 At steady state, a reversible refrigeration cycle removes 12,000 Btu/h of energy by heat transfer from a reservoir at 0°F and discharges energy by heat transfer to a reservoir at 70°F. Determine

(a) the coefficient of performance.
(b) the power input required, in horsepower.

5.41 A refrigeration cycle operating between two reservoirs receives energy Q_C from a cold reservoir at T_C = 250 K and rejects energy Q_H to a hot reservoir at T_H = 300 K. For each of the following cases determine whether the cycle operates reversibly, irreversibly, or is impossible:

(a) $Q_C = 1000$ kJ, $W_{\text{cycle}} = 400$ kJ.
(b) $Q_C = 2000$ kJ, $Q_H = 2200$ kJ.
(c) $Q_H = 3000$ kJ, $W_{\text{cycle}} = 500$ kJ.
(d) $W_{\text{cycle}} = 400$ kJ, $\beta = 6$.

5.42 At steady state, a refrigeration cycle maintains a freezer compartment at −3°C by removing energy by heat transfer from the food inside at a rate of 1000 kJ/h. If the cycle discharges energy by heat transfer to the surroundings at 23°C, determine the minimum theoretical power input required by the refrigeration cycle, in kW.

5.43 Determine if a tray of ice cubes could remain frozen when placed in a food freezer having a coefficient of performance of 9 operating in a room where the temperature is 32°C (90°F).

5.44 The surroundings of a freezer are at a temperature of 27°C. To maintain the freezer compartment at −40°C, it is necessary to remove energy by heat transfer at the rate of 1.25 kW. Determine

(a) the maximum possible coefficient of performance.
(b) the minimum theoretical power input required, in kW, for any refrigeration cycle operating under these conditions.

5.45 At steady state, a refrigeration cycle driven by a 1-horsepower motor removes 200 Btu/min of energy by heat transfer from a space maintained at 20°F and discharges energy by heat transfer to surroundings at 75°F.

(a) Determine the coefficient of performance of the refrigerator.
(b) Determine the rate at which energy is discharged to the surroundings, in Btu/min.
(c) Evaluate the maximum coefficient of performance any refrigeration cycle can have while operating between these two temperatures.

5.46 At steady state, a refrigerator whose coefficient of performance is 2.5 removes energy by heat transfer from a freezer compartment at 0°C at the rate of 8000 kJ/h and discharges energy by heat transfer to the surroundings, which are at 20°C. Determine the power input to the refrigerator and compare with the power input required by a reversible refrigeration cycle operating between reservoirs at these two temperatures.

5.47 At steady state, a refrigeration cycle removes 600 Btu/min of energy by heat transfer from a space maintained at −60°F and discharges energy by heat transfer to surroundings at 60°F. If the coefficient of performance of the cycle is two-thirds of that of a reversible refrigeration cycle operating between these two temperatures, determine the power input to the cycle, in horsepower.

5.48 A refrigeration cycle having a coefficient of performance of 3 maintains a computer laboratory at 18°C on a day when the outside temperature is 30°C. The *thermal load* at steady state consists of energy entering through the walls and windows at a rate of 30,000 kJ/h and from the occupants, computers, and lighting at a rate of 6000 kJ/h. Determine the power required by this air conditioner and compare this with the *minimum* theoretical power required for any refrigeration cycle operating under these conditions, each in kW.

5.49 A heat pump operating at steady state in its summer cooling mode requires a power input of 1.25 horsepower to maintain a dwelling at 70°F. If heat transfer through the walls and roof is 1500 Btu/h per degree temperature difference between inside and outside, determine the *highest* theoretical outside temperature for which the cooling requirements can be met.

5.50 A reversible power cycle receives Q_H from a hot reservoir at temperature T_H and rejects energy by heat transfer to the surroundings at temperature T_0. The work developed by the power cycle is used to drive a refrigeration cycle that removes Q_C from a cold reservoir at temperature T_C and discharges energy by heat transfer to the same surroundings at T_0. Develop an expression for the ratio Q_C/Q_H in terms of the temperatures T_H, T_0, and T_C.

5.51 Plot the coefficient of performance β_{max} given by Eq. 5.9 for T_H = 298 K versus T_C ranging between 235 and 298 K. Discuss the practical implications of the decrease in the coefficient of performance with decreasing temperature T_C.

5.52 If heat transfer through the walls and roof of a dwelling is 2×10^6 kJ per day, determine the *minimum* theoretical power, in kW, to drive a heat pump operating at steady state between the dwelling at 20°C and

(a) the outdoor air at 0°C.
(b) a pond at 5°C.
(c) the ground at 13°C.

5.53 A heat pump provides 3×10^4 kJ/h to maintain a dwelling at 23°C on a day when the outside temperature is 0°C. The power input to the heat pump is 4 kW. Determine the coefficient of performance of the heat pump and compare it with the coefficient of performance of a reversible heat pump cycle operating between reservoirs at the same two temperatures.

5.54 When the outside temperature is −10°C, a residential heat pump must provide 3.5×10^6 kJ per day to a dwelling to maintain its temperature at 20°C. If electricity costs 8 cents per kilowatt-hour, determine the minimum theoretical operating cost for each day of operation.

5.55 By supplying 90000 Btu/h, a heat pump maintains the temperature of a dwelling at 70°F when the outside air is at 32°F. If electricity costs 8 cents per kilowatt-hour, determine the minimum theoretical operating cost for each day of operation.

5.56 A heat pump driven by a 0.4-kW electric motor provides heating for a building on a day when the outside air is at −10°C and energy is lost through the walls and roof at a rate of 16000 kJ/h. What is the *maximum* theoretical temperature that can be maintained within the building, in °C, and would this temperature be suitable for office workers?

5.57 A heat pump requires a power input of 1 kW to maintain a dwelling at 20°C. If the heat transfer rate through the walls and roof is 3000 kJ/h per degree temperature difference between the inside and outside, determine the *minimum* theoretical outside temperature for which the heating requirement can be met.

5.58 A heat pump maintains a dwelling at 70°F when the outside temperature is 40°F. The heat transfer rate through the walls and roof is 1500 Btu/h per degree temperature difference between the inside and outside. Determine the *minimum* theoretical power required to drive the heat pump, in horsepower.

5.59 A dwelling for which the heat transfer rate through the walls and roof is 1400 Btu/h per degree temperature difference between the inside and outside is to be maintained at 68°F. For a day when the outside temperature is 28°F, determine the power required, in kW, to heat the house using electrical resistance elements and compare with the *minimum* theoretical power that would be required by a heat pump. Repeat the comparison using typical manufacturer's data for the heat pump coefficient of performance.

5.60 A reversible power cycle receives energy Q_H from a reservoir at temperature T_H and rejects Q_C to a reservoir at temperature T_C. The work developed by the power cycle is used to drive a reversible heat pump that removes energy Q'_C from a reservoir at temperature T'_C and rejects energy Q'_H to a reservoir at temperature T'_H.

(a) Develop an expression for the ratio Q'_H/Q_H in terms of the temperatures of the four reservoirs.
(b) What must be the relationship of the temperatures T_H, T_C, T'_C, and T'_H for Q'_H/Q_H to exceed a value of unity?

5.61 Plot the coefficient of performance γ_{max} given by Eq. 5.10 for $T_H = 535°R$ versus T_C ranging between 425 and 535°R. Discuss the practical implications of the decrease in the coefficient of performance with decreasing temperature T_C.

5.62 One kilogram of water executes a Carnot power cycle. During the isothermal expansion, the water is heated until it is a saturated vapor from an initial state where the pressure is 15 bars and the quality is 25%. The vapor then expands adiabatically to a pressure of 1 bar and a quality of 84.9%.

(a) Sketch the cycle on p–v coordinates.
(b) Evaluate the heat and work for each process, in kJ.
(c) Evaluate the thermal efficiency.

5.63 One-tenth pound of water executes a Carnot power cycle. During the isothermal expansion, the water is heated at 640°F from a saturated liquid to a saturated vapor. The vapor then expands adiabatically to a temperature of 40°F and a quality of 59%.

(a) Sketch the cycle on p–v coordinates.
(b) Evaluate the heat and work for each process, in Btu.
(c) Evaluate the thermal efficiency.

5.64 One-half kilogram of air as an ideal gas executes a Carnot power cycle having a thermal efficiency of 50%. The heat transfer to the air during the isothermal expansion is 40 kJ. At the beginning of the isothermal expansion, the pressure is 7 bars and the volume is 0.12 m³. Determine

(a) the maximum and minimum temperatures for the cycle, in K.
(b) the volume at the end of the isothermal expansion, in m³.
(c) the work and heat transfer for each of the four processes, in kJ.
(d) Sketch the cycle on p–v coordinates.

5.65 One pound of air as an ideal gas executes a Carnot power cycle. The isothermal expansion occurs at 2000°R from a pressure of 2710 lbf/in.² to a pressure of 1882 lbf/in.² The isothermal compression occurs at 537°R from a pressure of 14.73 lbf/in.² to a pressure of 21.17 lbf/in.² Determine

(a) the heat transfer to the air during the isothermal expansion, in Btu.
(b) the heat transfer from the air during the isothermal compression, in Btu.
(c) the work done by, or on, the air in each of the four processes, in Btu.
(d) the cycle thermal efficiency.
(e) Sketch the cycle on p–v coordinates.

5.66 The pressure–volume diagram of a Carnot power cycle executed by an ideal gas with constant specific heat ratio k is shown in Fig. P5.66. Demonstrate that

(a) $V_4V_2 = V_1V_3$.
(b) $T_2/T_3 = (p_2/p_3)^{(k-1)/k}$.

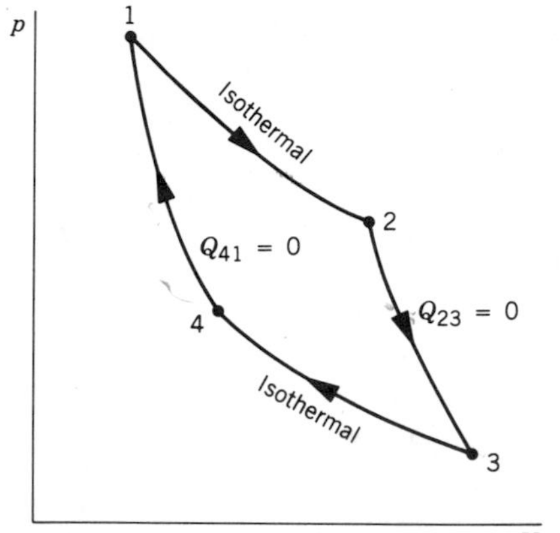

Figure P5.66

5.67 Two pounds of air as an ideal gas execute a Carnot cycle while operating between thermal reservoirs at 500 and 70°F. The pressures at the initial and final states of the isothermal expansion are 400 and 200 lbf/in.², respectively. The specific heat ratio is $k = 1.4$. Using the results of Problem 5.66 as needed, determine

(a) the pressures at the initial and final states of the isothermal compression, in lbf/in.²
(b) the work and heat transfer for each of the four processes, in Btu.
(c) the thermal efficiency.

5.68 Evaluate the coefficient of performance for the refrigeration cycle obtained on reversing the power cycle described in

(a) Problem 5.62. **(c)** Problem 5.64.
(b) Problem 5.63. **(d)** Problem 5.65.

5.69 Evaluate the coefficient of performance for the heat pump cycle obtained on reversing the power cycle described in

(a) Problem 5.62. **(c)** Problem 5.64.
(b) Problem 5.63. **(d)** Problem 5.65.

DESIGN AND OPEN ENDED PROBLEMS

5.1D The heat transfer rate through the walls and roof of a building is 3570 kJ/h per degree temperature difference between the inside and outside. For outdoor temperatures ranging from 15 to −30°C, make a comparison of the energy cost, in \$/h, of maintaining the building at 20°C by means of an electric heat pump, direct electric resistance heating, and a conventional oil-fired furnace. Report your results graphically.

5.2D Rhyolitic magma with a temperature of 900°C lies 7 km below a site in California. It has been proposed to drill a well into the magma chamber and insert two coaxial pipes. Cold water forced down the annular region between the two pipes would push through the hot magma and evaporate. The steam generated would flow up the center pipe and into a steam power plant. The cost of *magma power* is expected to fall between 9 and 22 cents/kW · h. How does this compare with the costs of electricity generated in plants using fossil or nuclear fuel? What are some of the principal factors that would result in the actual thermal efficiency of the magma power plant being considerably less than the maximum theoretical value?

5.3D To maintain the passenger compartment of an automobile traveling at 30 mi/h at 70°F when the surrounding air temperature is 90°F, the vehicle's air conditioner removes 18,000 Btu/h by heat transfer. Estimate the amount of engine horsepower required to drive the air conditioner. Referring to typical manufacturer's data, compare your estimate with the actual horsepower requirement. Discuss the tradeoffs between the initial *unit cost* of an automobile air-conditioning system and its *operating cost.*

5.4D An 800-MW nuclear power plant is to be constructed within 125 miles of your campus. To achieve the required cooling, one option under consideration is to locate the plant on a river. Identify rivers located within the indicated geographical region that may be suitable as sites for the plant. Taking the average temperature of the steam passing through the boiler as (a) 600°F, (b) 700°F, (c) 800°F, estimate the temperature rise of each candidate river. Discuss the environmental impact of such a temperature rise. Make a recommendation as to the viability of this option for your region. What alternative means might also be considered for achieving the required cooling?

5.5D Arguing that the kitchen is often one of the warmest locations in any dwelling, an inventor has proposed a new form of home appliance that combines within a *single unit* a room air conditioner and a refrigerator-freezer. Develop sketches showing how this form of appliance might be configured. The appliance cannot have a total volume that exceeds the volume of commonly available upright refrigerators by more than 10%.

5.6D Figure P5.6D shows that the typical thermal efficiency of U.S. power plants increased rapidly from 1925 to 1960, but has increased only gradually since then. Discuss the most important factors contributing to this plateauing of efficiency and the most promising near-term and long-term technologies that might lead to appreciable efficiency gains.

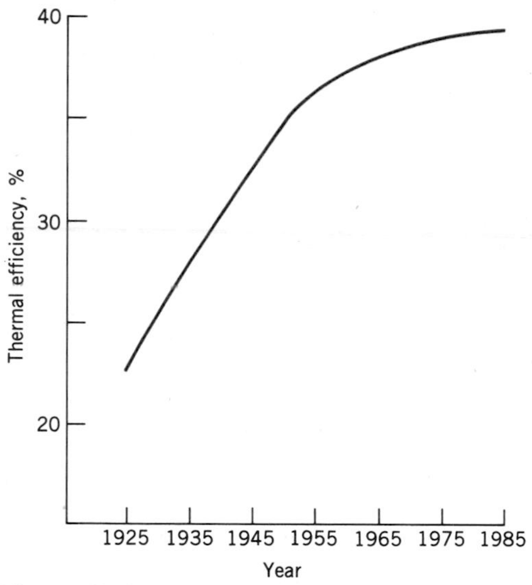

Figure P5.6D

5.7D Abandoned coal mines filled with water lie below a town in Nova Scotia. The mines, which were once the world's largest and deepest coal mines, are a nearly inexhaustible source of water at temperatures in the range 17 to 20°C. Develop specific *cost-effective* proposals for employing this resource for the heating and cooling of the town's industrial and commercial buildings.

5.8D On observing one of those bobbing toy birds that seemingly take an endless series of sips from a bowl filled with water, an inventor suggests that this behavior can be modeled as a power cycle and that a commercial scale version be developed. Critically evaluate these suggestions.

CHAPTER 6

ENTROPY

Up to this point, our study of the second law has been concerned primarily with what it says about systems undergoing thermodynamic cycles. In this chapter means are introduced for analyzing systems from the second law perspective as they undergo processes that are not necessarily cycles. The property *entropy* plays a prominent part in these considerations.

The word *energy* is so much a part of the language that you were undoubtedly familiar with the term before encountering it in early science courses. This familiarity probably facilitated the study of energy in these courses and in the current course in engineering thermodynamics. In the present chapter you will see that the study of systems from a second law perspective is conveniently accomplished in terms of the property *entropy*. Energy and entropy are both abstract concepts. However, unlike energy, the word entropy is seldom heard in everyday conversation, and you may never have dealt with it quantitatively before. The principal object of the present chapter, therefore, is to introduce entropy and show its use for thermodynamic analysis. Energy and entropy play important roles in the remaining chapters of this book.

6.1 CLAUSIUS INEQUALITY

Corollaries of the second law are developed in Chap. 5 for systems undergoing cycles while communicating thermally with *two* reservoirs, a hot reservoir and a cold reservoir. In the present section a corollary of the second law known as the Clausius inequality is introduced that is applicable to *any* cycle without regard for the body, or bodies, from which the cycle receives energy by heat transfer or to which the cycle rejects energy by heat transfer. The Clausius inequality provides the basis for introducing two ideas instrumental for quantitative evaluations of both closed systems and control volumes from a second law perspective: the property *entropy* (Sec. 6.2) and the *entropy production* concept (Sec. 6.5).

The Clausius inequality states that

$$\oint \left(\frac{\delta Q}{T}\right)_{\text{b}} \le 0 \tag{6.1}$$

where δQ represents the heat transfer at a part of the system boundary during a portion of the cycle, and T is the absolute temperature at that part of the boundary. The subscript "b" serves as a reminder that the integrand is evaluated at the boundary of the system executing the cycle. The symbol $\oint$ indicates that the integral is to be performed over all parts of the boundary and over the entire cycle. In Eq. 6.1, the equality applies when there are no internal irreversibilities as the system executes the cycle, and the inequality applies when internal irreversibilities are present.

The Clausius inequality can be demonstrated using the arrangement of Fig. 6.1. A system receives energy δQ at a location on its boundary where the absolute temperature is T while the system develops work δW. In keeping with our sign convention for heat transfer, the phrase *receives energy* δQ includes the possibility of heat transfer *from* the system. The energy δQ is received from (or absorbed by) a thermal reservoir at T_{res}. To ensure that no irreversibility is introduced as as result of heat transfer between the reservoir and the system, let it be accomplished through an intermediary system that undergoes a cycle without irreversibilities of any kind. The cycle receives energy $\delta Q'$ from the reservoir and supplies δQ to the system while producing work $\delta W'$. From the definition of the Kelvin scale (Eq. 5.6), we have the following relationship between the heat transfers and temperatures:

$$\frac{\delta Q'}{T_{\text{res}}} = \left(\frac{\delta Q}{T}\right)_{\text{b}} \tag{6.2}$$

As temperature may vary, a multiplicity of such reversible cycles may be required.

Consider next the combined system shown by the dotted line on Fig. 6.1. An energy balance for the combined system is

$$dE_{\text{C}} = \delta Q' - \delta W_{\text{C}}$$

where δW_{C} is the total work of the combined system, the sum of δW and $\delta W'$, and dE_{C} denotes the change in energy of the combined system. Solving the energy equation for δW_{C} and using Eq. 6.2 to eliminate $\delta Q'$ from the resulting expression yields

$$\delta W_{\text{C}} = T_{\text{res}} \left(\frac{\delta Q}{T}\right)_{\text{b}} - dE_{\text{C}}$$

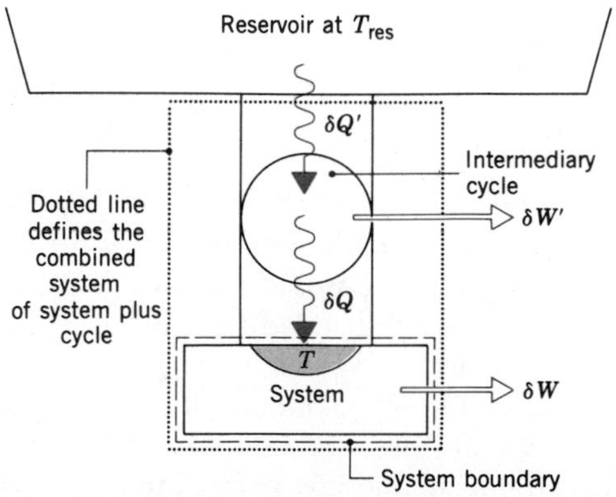

Figure 6.1 Illustration used to develop the Clausius inequality.

Let the system undergo a single cycle, while the intermediary system undergoes one or more cycles. The total work of the combined system is

$$W_C = \oint T_{res} \left(\frac{\delta Q}{T}\right)_b - \cancelto{0}{\oint dE_C} = T_{res} \oint \left(\frac{\delta Q}{T}\right)_b \tag{6.3}$$

Since the reservoir temperature is constant, T_{res} can be brought outside the integral. The term involving the energy of the combined system vanishes because the energy change for any cycle is zero. The combined system operates in a cycle because its parts execute cycles. Since the combined system undergoes a cycle and exchanges energy by heat transfer with a single reservoir, Eq. 5.1 expressing the Kelvin–Planck statement of the second law must be satisfied. Using this, Eq. 6.3 reduces to give Eq. 6.1, where the equality applies when there are *no irreversibilities within the system* as it executes the cycle and the inequality applies when *internal irreversibilities are present*. This interpretation actually refers to the combination of system plus intermediary cycle. However, the intermediary cycle is regarded as free of irreversibilities, so the only possible site of irreversibilities is the system alone.

The inequality in Eq. 6.1 can be eliminated by writing the expression as

$$\oint \left(\frac{\delta Q}{T}\right)_b = -\sigma_{cycle} \tag{6.4}$$

where σ_{cycle} can be viewed as representing the "strength" of the inequality. The value of σ_{cycle} is positive when internal irreversibilities are present, zero when no internal irreversibilities are present, and can never be negative. In summary, the nature of a cycle executed by a system is indicated by the value for σ_{cycle} as follows:

$\sigma_{cycle} = 0$	no irreversibilities present within the system
$\sigma_{cycle} > 0$	irreversibilities present within the system
$\sigma_{cycle} < 0$	impossible

Accordingly, σ_{cycle} is a measure of the effect of the irreversibilities present within the system executing the cycle. This point is developed further in Sec. 6.5, where σ_{cycle} is identified as the *entropy produced* by internal irreversibilities during the cycle.

6.2 DEFINITION OF ENTROPY CHANGE

A quantity is a property if, and only if, its change in value between two states is independent of the process (Sec. 1.2). This aspect of the property concept is used in the present section together with Eq. 6.4 to introduce entropy.

Two cycles executed by a closed system are represented in Fig. 6.2. One cycle consists of an internally reversible process A from state 1 to state 2, followed by internally reversible process C from state 2 to state 1. The other cycle consists of an internally reversible process B from state 1 to state 2, followed by the same process C from state 2 to state 1 as in the first cycle. For the first cycle, Eq. 6.4 takes the form

$$\left(\int_1^2 \frac{\delta Q}{T}\right)_A + \left(\int_2^1 \frac{\delta Q}{T}\right)_C = -\cancelto{0}{\sigma_{cycle}} \tag{6.5}$$

and for the second cycle

$$\left(\int_1^2 \frac{\delta Q}{T}\right)_B + \left(\int_2^1 \frac{\delta Q}{T}\right)_C = -\cancelto{0}{\sigma_{cycle}} \tag{6.6}$$

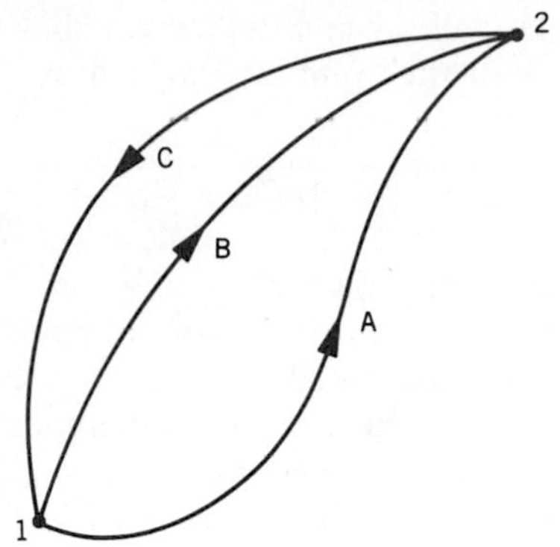

Figure 6.2 Two internally reversible cycles used to demonstrate that entropy is a property.

In writing Eqs. 6.5 and 6.6, the term σ_{cycle} has been set to zero since the cycles are composed of internally reversible processes.

When Eq. 6.6 is subtracted from Eq. 6.5

$$\left(\int_1^2 \frac{\delta Q}{T}\right)_A = \left(\int_1^2 \frac{\delta Q}{T}\right)_B$$

This shows that the integral of $\delta Q/T$ is the same for both processes. Since A and B are arbitrary, it follows that the integral of $\delta Q/T$ has the same value for *any* internally reversible process between the two states. In other words, the value of the integral depends on the end states only. It can be concluded, therefore, that the integral defines the change in some property of the system. Selecting the symbol S to denote this property, its change is given by

$$S_2 - S_1 = \left(\int_1^2 \frac{\delta Q}{T}\right)_{\substack{\text{int}\\\text{rev}}} \tag{6.7}$$

where the subscript "int rev" is added as a reminder that the integration is carried out for any internally reversible process linking the two states. This property is called *entropy*. Entropy is an *extensive* property. The SI unit for entropy is J/K. However, in this book it is convenient to work in terms of kJ/K. A commonly employed English unit for entropy is Btu/°R. Units in SI for *specific* entropy are kJ/kg·K for s and kJ/kmol·K for $\bar{s}$. Commonly used English units for *specific* entropy are Btu/lb·°R and Btu/lbmol·°R.

Since entropy is a property, the change in entropy of a system in going from one state to another is the same for *all* processes, both internally reversible and irreversible, between these two states. Thus, Eq. 6.7 allows the determination of the change in entropy, and once it has been evaluated, this is the magnitude of the entropy change for all processes of the system between the two states. The evaluation of entropy changes is discussed further in the next section.

It should be clear that entropy is defined and evaluated in terms of a particular integral for which *no accompanying physical picture is given*. We encountered this previously with the property enthalpy. Enthalpy is introduced without physical motivation in Sec. 3.3.2. Then, in Chap. 4, enthalpy is shown to be useful for thermodynamic analysis. As for the case of enthalpy, to gain an appreciation for entropy you need to understand *how* it is used and *what* it is used for.

6.3 ENTROPY OF A PURE, SIMPLE COMPRESSIBLE SUBSTANCE

The defining equation for entropy change, Eq. 6.7, serves as the basis for constructing tables and diagrams for the evaluation of entropy. This is done by assigning a value to entropy at a reference state. Both the entropy value and the reference state can be selected arbitrarily. The value of entropy at any state y relative to the value at the reference state x is obtained in principle by integration

$$S_y = S_x + \left(\int_x^y \frac{\delta Q}{T} \right)_{\substack{\text{int} \\ \text{rev}}} \tag{6.8}$$

where S_x is the value of entropy specified at the reference state.

The use of entropy values determined relative to an arbitrary reference state is satisfactory as long as they are used in calculations involving entropy differences, for then the reference value cancels. This approach suffices for applications involving pure, simple compressible systems. When chemical reactions occur, it is necessary to work in terms of *absolute* values of entropy determined using the *third law of thermodynamics* (Chap. 13).

6.3.1 Tabular Entropy Data

Tables of thermodynamic data are introduced in Sec. 3.3. The emphasis there is on evaluating the properties p, v, T, u, and h, which are required for application of the conservation of mass and energy principles. For application of the second law, entropy values are often required. Specific entropy is tabulated in the same way as the properties v, u, and h.

In the superheat regions of the tables for water and Refrigerants 12 and 134a, specific entropy is tabulated along with v, u, and h versus temperature and pressure. To illustrate the use of the superheated vapor tables for evaluating entropy, consider two states of water. At state 1 the pressure is 3 MPa and the temperature is 500°C. At state 2 the pressure is $p_2 = 0.3$ MPa and the specific entropy is the same as at state 1, $s_2 = s_1$. The object is to determine the temperature at state 2. Using T_1 and p_1, we find the specific entropy at state 1 from Table A-4 as $s_1 = 7.2338$ kJ/kg·K. State 2 is fixed by the pressure, $p_2 = 0.3$ MPa, and the specific entropy, $s_2 = 7.2338$ kJ/kg·K. Returning to Table A-4 at 0.3 MPa and interpolating with s_2 between 160 and 200°C results in $T_2 = 183$°C.

As a similar example in English units, consider two states of water. At state 1 the pressure is 450 lbf/in.^{2} and the temperature is 900°F. At state 2 the pressure is $p_2 = 40$ lbf/in.^{2} and the specific entropy is the same as at state 1, $s_2 = s_1$. As before, the object is to determine T_2. Using T_1 and p_1, we find the specific entropy at state 1 from Table A-4E as $s_1 = 1.7113$ Btu/lb·°R. State 2 is fixed by the pressure, $p_2 = 40$ lbf/in.^{2}, and the specific entropy, $s_2 = 1.7113$ Btu/lb·°R. Returning to Table A-4E at 40 lbf/in.^{2} and interpolating with s_2 between 300 and 350°F results in $T_2 = 319$°F.

For saturation states, the values of s_f and s_g are tabulated as a function of either saturation pressure or saturation temperature. The specific entropy of a two-phase liquid–vapor mixture is calculated using the quality

$$s = (1 - x)s_f + xs_g = s_f + x(s_g - s_f) \tag{6.9}$$

These relations are identical in form to those for v, u, and h (Sec. 3.3). As an illustration, we will determine the specific entropy of Refrigerant 12 at a state where the temperature is 0°C and the specific internal energy is 103.14 kJ/kg. Referring to

Table A-7, we see that the given value for u is located between u_f and u_g at 0°C, so the system is a two-phase liquid–vapor mixture. The quality of the mixture can be determined from the known specific internal energy

$$x = \frac{u - u_f}{u_g - u_f} = \frac{103.14 - 35.83}{170.44 - 35.83} = 0.5$$

Then with values from Table A-7

$$\begin{aligned} s &= (1 - x)s_f + xs_g \\ &= (0.5)(0.1420) + (0.5)(0.6965) = 0.4193 \text{ kJ/kg} \cdot \text{K} \end{aligned}$$

To illustrate the use of Eq. 6.9 with data in English units, let us determine the specific entropy of Refrigerant 12 at a state where the temperature is 0°F and the specific internal energy is 45.49 Btu/lb. Referring to Table A-7E, the given value for u is located between u_f and u_g at 0°F, so the system is a two-phase liquid–vapor mixture. The quality of the mixture can be determined from the known specific internal energy

$$x = \frac{u - u_f}{u_g - u_f} = \frac{45.49 - 8.47}{70.17 - 8.47} = 0.6$$

Then with values from Table A-7E

$$\begin{aligned} s &= (1 - x)s_f + xs_g \\ &= (0.4)(0.0193) + (0.6)(0.1689) = 0.1091 \text{ Btu/lb} \cdot {}^\circ\text{R} \end{aligned}$$

Compressed liquid data are presented for water in Tables A-5. In these tables s, v, u, and h are tabulated versus temperature and pressure as in the superheat tables, and the tables are used similarly. In the absence of compressed liquid data, the value of the specific entropy can be estimated in the same way as estimates for v and u are obtained for liquid states (Sec. 3.3.4), by using the saturated liquid value at the given temperature

$$s(T, p) \approx s_f(T) \tag{6.10}$$

To illustrate the use of Eq. 6.10, suppose the value of specific entropy is required for water at 25 bars, 200°C. From Table A-5 the specific entropy is obtained as $s = 2.3294$ kJ/kg · K. Using the saturated liquid value for specific entropy at 200°C from Table A-2, the specific entropy is approximated with Eq. 6.10 as $s = 2.3309$ kJ/kg · K, which agrees with the previous value. As a similar illustration in English units, let us determine the value of specific entropy for water at 500 lbf/in.^{2} and 300°F. From Table A-5E the specific entropy is obtained as $s = 0.43641$ Btu/lb · °R. Using the saturated liquid value for specific entropy at 300°F from Table A-2E, we can approximate the specific entropy with Eq. 6.10 as $s = 0.4372$ Btu/lb · °R, which is in agreement with the value from the compressed liquid table.

The specific entropy values for water and the refrigerants given in Tables A-2 through A-12 are relative to the following reference states and values. For water, the entropy of saturated liquid at 0.01°C (32.02°F) is set to zero. For the refrigerants, the entropy of the saturated liquid at −40°C, or −40°F, is assigned a value of zero.

6.3.2 Graphical Entropy Data

The use of property diagrams as an adjunct to problem solving is emphasized throughout this book. When applying the second law, it is frequently helpful to locate states and plot processes on diagrams having entropy as a coordinate. Two commonly used figures having entropy as one of the coordinates are the temperature–entropy diagram and the enthalpy–entropy diagram.

Temperature–Entropy Diagram. The main features of a temperature–entropy diagram are shown in Fig. 6.3. For detailed figures for water in SI and English units, see Figs. A-7. Observe that lines of constant enthalpy are shown on these figures. Also note that in the superheated vapor region constant specific volume lines have a steeper slope than constant pressure lines. Lines of constant quality are shown in the two-phase liquid–vapor region. On some figures, lines of constant quality are marked as *percent-moisture* lines. The percent-moisture is defined as the ratio of the mass of liquid to the total mass. To become familiar with the temperature–entropy diagrams for water provided in the appendix, use them to repeat the exercises considered earlier in this section using the superheated vapor tables.

In the superheated vapor region of the T–s diagram, constant specific enthalpy lines become nearly horizontal as pressure is reduced. These states are shown as the shaded area on Fig. 6.3. For states in this region of the diagram, the enthalpy is determined primarily by the temperature. The variation in pressure between states has little effect: $h(T, p) \approx h(T)$. This is the region of the chart where the ideal gas model provides a reasonable approximation. For superheated vapor states outside the shaded area, both temperature and pressure are required to evaluate enthalpy, and the ideal gas model is not suitable.

Enthalpy–Entropy Diagram. The essential features of an enthalpy–entropy diagram, commonly known as a Mollier diagram, are shown in Fig. 6.4. For detailed figures for water in SI and English units, see Figs. A-8. Note the location of the critical point and the appearance of lines of constant temperature and constant pressure. Lines of constant quality are shown in the two-phase liquid–vapor region (some figures give lines of constant percent-moisture). The figure is intended for

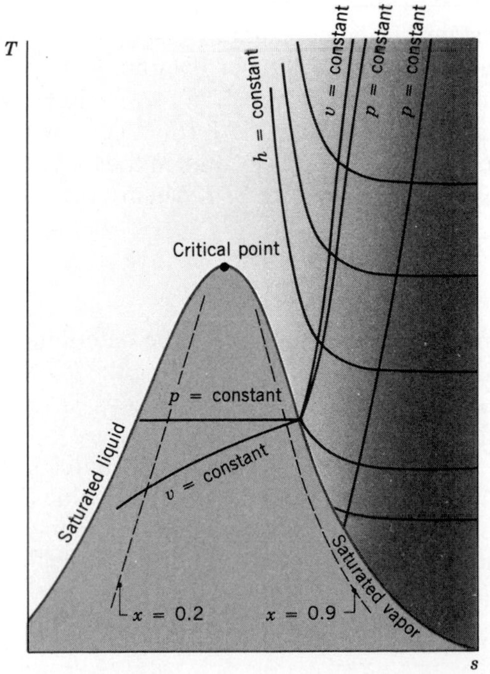

Figure 6.3 Temperature–entropy diagram.

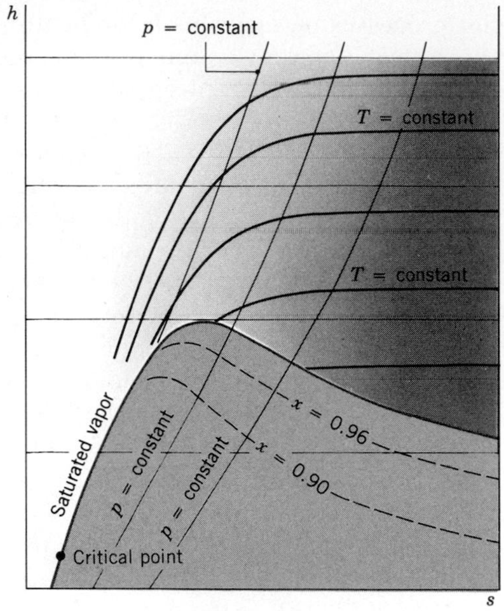

Figure 6.4 Enthalpy–entropy diagram.

evaluating properties at superheated vapor states and for two-phase liquid–vapor mixtures. Liquid data are seldom shown. In the superheated vapor region, constant-temperature lines become nearly horizontal as pressure is reduced. These states are shown, approximately, as the shaded area on Fig. 6.4. This area corresponds to the shaded area on the temperature–entropy diagram of Fig. 6.3, where the ideal gas model provides a reasonable approximation.

To illustrate the use of the Mollier diagram in SI units, consider two states of water. At state 1, $T_1 = 240°C$, $p_1 = 0.10$ MPa. The specific enthalpy and quality are required at state 2, where $p_2 = 0.01$ MPa and $s_2 = s_1$. Turning to Fig. A-8, state 1 is located in the superheated vapor region. Dropping a vertical line into the two-phase liquid–vapor region, state 2 is located, and by inspection $x_2 = 0.98$ and $h_2 = 2537$ kJ/kg. As a similar example using the Mollier diagram in English units, consider two states of water. At state 1, $T_1 = 450°F$ and $p_1 = 14.7$ lbf/in.^{2} The specific enthalpy and quality are required at state 2, where $p_2 = 2$ lbf/in.^{2} Turning to Fig. A-8E, state 1 is in the superheated vapor region. State 2 is located in the two-phase liquid–vapor region on a vertical line passing through state 1. By inspection $x_2 = 0.99$ and $h_2 = 1106$ Btu/lb.

6.3.3 *T dS* Equations

The change in entropy between two states can be determined in principle using Eq. 6.7. Such evaluations can also be conducted using the $T\,dS$ equations developed in this section. The importance of the $T\,dS$ equations is greater than their role in assigning entropy values, however. In Chap. 11 they are used as a point of departure for deriving many important property relations for pure, simple compressible systems, including means for constructing the property tables giving u, h, and s.

Consider a pure, simple compressible system undergoing an internally reversible process. In the absence of overall system motion and the effect of gravity, an energy balance in differential form is

$$(\delta Q)_{\substack{\text{int}\\\text{rev}}} = dU + (\delta W)_{\substack{\text{int}\\\text{rev}}} \tag{6.11}$$

By definition of simple compressible system (Sec. 3.1), the work is given by

$$(\delta W)_{\substack{\text{int}\\\text{rev}}} = p\, dV \tag{6.12}$$

On a differential basis, the defining equation for entropy change, Eq. 6.7, takes the form

$$dS = \left(\frac{\delta Q}{T}\right)_{\substack{\text{int}\\\text{rev}}} \tag{6.13}$$

or, on rearrangement

$$(\delta Q)_{\substack{\text{int}\\\text{rev}}} = T\, dS \tag{6.14}$$

Substituting Eqs. 6.12 and 6.14 into Eq. 6.11, the *first T dS* equation results:

$$T\, dS = dU + p\, dV \tag{6.15}$$

The *second T dS* equation is obtained from Eq. 6.15 using $H = U + pV$. Forming the differential

$$dH = dU + d(pV) = dU + p\, dV + V\, dp$$

On rearrangement

$$dU + p\, dV = dH - V\, dp$$

Substituting this into Eq. 6.15 gives the *second T dS* equation:

$$T\, dS = dH - V\, dp \tag{6.16}$$

The $T\, dS$ equations can be written on a unit mass basis as

$$T\, ds = du + p\, dv \tag{6.17a}$$

$$T\, ds = dh - v\, dp \tag{6.17b}$$

or on a per mole basis as

$$T\, d\bar{s} = d\bar{u} + p\, d\bar{v} \tag{6.18a}$$

$$T\, d\bar{s} = d\bar{h} - \bar{v}\, dp. \tag{6.18b}$$

Although the $T\, dS$ equations were derived by considering an internally reversible process, an entropy change for a system obtained by integrating these equations is the change for *any* process of the system, reversible or irreversible, between two equilibrium states. Because entropy is a property, the change in entropy between two states is independent of the details of the process linking the states.

The $T\, dS$ equations allow entropy changes to be evaluated from other more readily determined property data. Their use to evaluate entropy changes for ideal gases is illustrated in Sec. 6.3.4 and for incompressible substances in Sec. 6.3.5. In Chap. 11 means based on the $T\, dS$ equations are introduced for evaluating changes in entropy, enthalpy, and internal energy for the general case of pure, simple compressible systems.

As an exercise illustrating the use of the $T\,dS$ equations, consider a change in phase from saturated liquid to saturated vapor at constant temperature and pressure. Since pressure is constant, Eq. 6.17b reduces to give

$$ds = \frac{dh}{T}$$

Then, because temperature is also constant during the phase change,

$$s_g - s_f = \frac{h_g - h_f}{T}$$

This relationship shows how $s_g - s_f$ is calculated for tabulation in property tables. As an application, consider Refrigerant 12 at 0°C. From Table A-7, $h_g - h_f = 151.48$ kJ/kg. Therefore

$$s_g - s_f = \frac{151.48}{273.15} = 0.5546\ \frac{\text{kJ}}{\text{kg}\cdot\text{K}}$$

which, allowing for round-off, is the value calculated using s_f and s_g from the table. To give a similar example in English units, consider Refrigerant 12 at 0°F. From Table A-7E, $h_g - h_f = 68.75$ Btu/lb, so

$$s_g - s_f = \frac{68.75}{459.67} = 0.1496\ \frac{\text{Btu}}{\text{lb}\cdot{}^\circ\text{R}}$$

which agrees with the table data.

6.3.4 Entropy Change of an Ideal Gas

In this section the $T\,dS$ equations are used to evaluate the entropy change between two states of an ideal gas. It is convenient to begin with Eqs. 6.17 expressed as

$$ds = \frac{du}{T} + \frac{p}{T}\,dv \tag{6.19}$$

$$ds = \frac{dh}{T} - \frac{v}{T}\,dp \tag{6.20}$$

For an ideal gas, $du = c_v(T)\,dT$, $dh = c_p(T)\,dT$, and $pv = RT$. With these relations, Eqs. 6.19 and 6.20 become, respectively

$$ds = c_v(T)\,\frac{dT}{T} + R\,\frac{dv}{v} \quad \text{and} \quad ds = c_p(T)\,\frac{dT}{T} - R\,\frac{dp}{p} \tag{6.21}$$

Since R is a constant, the last terms of Eqs. 6.21 can be integrated directly. However, because c_v and c_p are functions of temperature for ideal gases, it is necessary to have information about the functional relationships before the integration of the first term in these equations can be performed. The two specific heats are related by

$$c_p(T) = c_v(T) + R \tag{3.44}$$

where R is the gas constant, so knowledge of either specific heat function suffices.

On integration, Eqs. 6.21 give, respectively

$$s(T_2, v_2) - s(T_1, v_1) = \int_{T_1}^{T_2} c_v(T)\,\frac{dT}{T} + R \ln \frac{v_2}{v_1} \tag{6.22}$$

$$s(T_2, p_2) - s(T_1, p_1) = \int_{T_1}^{T_2} c_p(T)\,\frac{dT}{T} - R \ln \frac{p_2}{p_1} \tag{6.23}$$

As for internal energy and enthalpy changes, the evaluation of entropy changes for ideal gases can be reduced to a convenient tabular approach. To introduce this, we begin by selecting a reference state and reference value: The value of the specific entropy is set to zero at the state where the temperature is 0 K and the pressure is 1 atmosphere. Then, using Eq. 6.23, the specific entropy at a state where the temperature is T and the pressure is 1 atm is determined relative to this reference state and reference value as

$$s^\circ(T) = \int_0^T \frac{c_p(T)}{T}\, dT \tag{6.24}$$

The symbol $s^\circ(T)$ denotes the specific entropy at temperature T and *a pressure of 1 atm*. Because s° depends only on temperature, it can be tabulated versus temperature, like h and u. For air as an ideal gas, s° with units of kJ/kg·K or Btu/lb·°R is given in Tables A-16. Values of $\bar{s}^\circ$ for several other common gases are given in Tables A-17 through A-22 with units of kJ/kmol·K or Btu/lbmol·°R. Since the integral of Eq. 6.23 can be expressed in terms of s°

$$\begin{aligned}\int_{T_1}^{T_2} c_p \frac{dT}{T} &= \int_0^{T_2} c_p \frac{dT}{T} - \int_0^{T_1} c_p \frac{dT}{T} \\ &= s^\circ(T_2) - s^\circ(T_1)\end{aligned}$$

it follows that Eq. 6.23 can be written as

$$s(T_2, p_2) - s(T_1, p_1) = s^\circ(T_2) - s^\circ(T_1) - R \ln \frac{p_2}{p_1} \tag{6.25a}$$

or on a per mole basis as

$$\bar{s}(T_2, p_2) - \bar{s}(T_1, p_1) = \bar{s}^\circ(T_2) - \bar{s}^\circ(T_1) - \bar{R} \ln \frac{p_2}{p_1} \tag{6.25b}$$

By the use of Eqs. 6.25 and the tabulated values for s° or $\bar{s}^\circ$, entropy changes can be determined that account explicitly for the variation of specific heat with temperature. If a table is not available for a particular gas of interest, the integrals of Eqs. 6.22 and 6.23 can be evaluated analytically or numerically using specific heat data such as provided in Tables A-14 and A-15. This approach is illustrated in Example 6.1.

When the specific heat is taken as constant, Eqs. 6.22 and 6.23 become, respectively

$$s(T_2, v_2) - s(T_1, v_1) = c_v \ln \frac{T_2}{T_1} + R \ln \frac{v_2}{v_1} \tag{6.26}$$

$$s(T_2, p_2) - s(T_1, p_1) = c_p \ln \frac{T_2}{T_1} - R \ln \frac{p_2}{p_1} \tag{6.27}$$

Although the simple forms of Eqs. 6.26 and 6.27 may be convenient for certain evaluations, the use of Eqs. 6.25 is generally more accurate because the variation of specific heats with temperature is included.

6.3.5 Entropy Change of an Incompressible Substance

The incompressible substance model introduced in Sec. 3.3.5 assumes that the specific volume (density) is constant and the specific heat depends solely on temperature, $c_v = c(T)$. Accordingly, the differential change in specific internal energy is $du = c(T)\, dT$ and Eq. 6.19 reduces to

EXAMPLE 6.1

PROBLEM

Water vapor at 400 K and 0.1 MPa undergoes a process to 900 K and 0.5 MPa. Determine the change in specific entropy, in kJ/kmol · K, using (a) the superheated vapor table, (b) the ideal gas table for water vapor, (c) integration with $\bar{c}_p(T)$ from Table A-15.

SOLUTION

Known: Water vapor undergoes a process from 400 K, 0.1 MPa to 900 K, 0.5 MPa.

Find: Determine the change in specific entropy using the superheated vapor table, the ideal gas table for water vapor, and integration with $\bar{c}_p$ from Table A-15.

Assumptions: In parts (b) and (c) the ideal gas model for water vapor is to be used.

Analysis:

(a) Interpolating in Table A-4, $s_1 = 7.501$ kJ/kg · K and $s_2 = 8.418$ kJ/kg · K. Thus,

$$s_2 - s_1 = 0.917 \text{ kJ/kg} \cdot \text{K}$$

Using the molecular weight, we can express this on a molar basis as

$$\bar{s}_2 - \bar{s}_1 = (0.917 \text{ kJ/kg} \cdot \text{K})(18.02 \text{ kg/kmol}) = 16.524 \text{ kJ/kmol} \cdot \text{K}$$

(b) As can readily be verified by reference to a generalized compressibility chart, the two states under consideration are in the region of the vapor phase of water where the ideal gas model provides a reasonable approximation. Referring to the ideal gas table for water vapor, Table A-19, at $T_1 = 400$ K, $\bar{s}_1^\circ = 198.673$ kJ/kmol · K. At $T_2 = 900$ K, $\bar{s}_2^\circ = 228.321$ kJ/kmol · K. Inserting these into Eq. 6.25b with $\bar{R} = 8.314$ kJ/kmol · K

$$\bar{s}_2 - \bar{s}_1 = 228.321 - 198.673 - 8.314 \ln 5 = 16.267 \text{ kJ/kmol} \cdot \text{K}$$

This agrees with the value found in part (a).

(c) Table A-15 provides the following expression for $\bar{c}_p(T)$ of water vapor modeled as an ideal gas:

$$\frac{\bar{c}_p}{\bar{R}} = \alpha + \beta T + \gamma T^2 + \delta T^3 + \varepsilon T^4$$

Inserting this into Eq. 6.23, written on a molar basis, and integrating

$$\frac{\bar{s}_2 - \bar{s}_1}{\bar{R}} = \int_{T_1}^{T_2} \left(\frac{\alpha + \beta T + \gamma T^2 + \delta T^3 + \varepsilon T^4}{T} \right) dT - \ln \frac{p_2}{p_1}$$

$$= \alpha \ln \frac{T_2}{T_1} + \beta(T_2 - T_1) + \frac{\gamma}{2}(T_2^2 - T_1^2) + \frac{\delta}{3}(T_2^3 - T_1^3) + \frac{\varepsilon}{4}(T_2^4 - T_1^4) - \ln \frac{p_2}{p_1}$$

The values for the constants α, β, γ, δ, and ε for water are obtained from Table A-15: $\alpha = 4.07$, $\beta = -1.108 \times 10^{-3}$, $\gamma = 4.152 \times 10^{-6}$, $\delta = -2.964 \times 10^{-9}$, $\varepsilon = 0.807 \times 10^{-12}$. Hence

$$\frac{\bar{s}_2 - \bar{s}_1}{\bar{R}} = 4.07 \ln \frac{900}{400} - 1.108 \left[\frac{900 - 400}{10^3} \right] + \frac{4.152}{2} \left[\frac{(900)^2 - (400)^2}{10^6} \right]$$

$$- \frac{2.964}{3} \left[\frac{(900)^3 - (400)^3}{10^9} \right] + \frac{0.807}{4} \left[\frac{(900)^4 - (400)^4}{10^{12}} \right] - \ln 5$$

Calculating terms

$$\frac{\bar{s}_2 - \bar{s}_1}{\bar{R}} = 3.3005 - 0.554 + 1.3494 - 0.657 + 0.1272 - 1.6094$$
$$= 1.9567$$

With $\bar{R} = 8.314$ kJ/kmol·K, the foregoing equation gives

$$\bar{s}_2 - \bar{s}_1 = 16.27 \text{ kJ/kmol·K}$$

This agrees with the values found in parts (a) and (b).

Comment: For the states of this example, the ideal gas model is accurate for water vapor. This is not generally the case, however.

$$ds = \frac{c(T)\,dT}{T} + \frac{p\,\overset{0}{\cancel{dv}}}{T} = \frac{c(T)\,dT}{T}$$

On integration, the change in specific entropy is

$$s_2 - s_1 = \int_{T_1}^{T_2} \frac{c(T)}{T}\,dT \qquad \text{(incompressible)}$$

When the specific heat is assumed constant, this becomes

$$s_2 - s_1 = c \ln \frac{T_2}{T_1} \qquad \text{(incompressible, constant } c\text{)} \tag{6.28}$$

Specific heats of some common liquids and solids are given in Table A-13.

6.4 ENTROPY CHANGE IN INTERNALLY REVERSIBLE PROCESSES

In this section the relationship between entropy change and heat transfer for internally reversible processes is considered. The concepts introduced have important applications in subsequent sections of the book. The present discussion is limited to the case of closed systems. Similar considerations for control volumes are presented in Sec. 6.9.

As a closed system undergoes an internally reversible process, its entropy can increase, decrease, or remain constant. This can be brought out using

$$dS = \left(\frac{\delta Q}{T}\right)_{\substack{\text{int}\\ \text{rev}}} \tag{6.13}$$

Equation 6.13 indicates that when a closed system undergoing an internally reversible process receives energy by heat transfer, it experiences an increase in entropy. Conversely, when energy is removed from the system by heat transfer, the entropy of the system decreases. This can be interpreted to mean that an entropy transfer *accompanies* heat transfer. The direction of the entropy transfer is the same as that of the heat transfer. In an *adiabatic* internally reversible process, the entropy would remain constant. A constant-entropy process is called an *isentropic* process.

On rearrangement, Eq. 6.13 becomes

$$(\delta Q)_{\substack{\text{int}\\ \text{rev}}} = T\,dS \tag{6.14}$$

Integrating from an initial state 1 to a final state 2

$$Q_{\substack{\text{int}\\\text{rev}}} = \int_1^2 T\,dS \tag{6.29}$$

From Eq. 6.29 it can be concluded that an energy transfer by heat to a closed system during an internally reversible process can be represented as an area on a temperature–entropy diagram. Figure 6.5 illustrates the area interpretation of heat transfer for an arbitrary internally reversible process in which temperature varies. Carefully note that temperature must be in degrees Kelvin or degrees Rankine, and the area is the entire area under the curve (shown shaded). Also note that the area interpretation of heat transfer is not valid for irreversible processes, as will be demonstrated later.

To provide an example illustrating both the entropy change that accompanies heat transfer and the area interpretation of heat transfer, consider Fig. 6.6*a*, which shows a *Carnot power cycle* (Sec. 5.7). The cycle consists of four internally reversible processes in series: two isothermal processes alternated with two adiabatic processes. In Process 2–3, heat transfer to the system occurs while the temperature of the system remains constant at T_H. The system entropy increases due to the accompanying entropy transfer. For this process, Eq. 6.29 gives $Q_{23} = T_H(S_3 - S_2)$, so area 2–3–a–b–2 on Fig. 6.6*a* represents the heat transfer during the process. Process 3–4 is an adiabatic and internally reversible process and thus is an isentropic (constant-entropy) process. Process 4–1 is an isothermal process at T_C during which heat is transferred *from* the system. Since entropy transfer accompanies the heat transfer, system entropy decreases. For this process, Eq. 6.29 gives $Q_{41} = T_C(S_1 - S_4)$, which is negative in value. Area 4–1–b–a–4 on Fig. 6.6*a* represents the *magnitude* of the heat transfer Q_{41}. Process 1–2, which completes the cycle, is adiabatic and internally reversible (isentropic). The net work of any cycle is equal to the net heat transfer, so *enclosed* area 1–2–3–4–1 represents the net work of the cycle. The thermal efficiency of the cycle may also be expressed in terms of areas:

$$\eta = \frac{W_{\text{cycle}}}{Q_{23}} = \frac{\text{area } 1\text{–}2\text{–}3\text{–}4\text{–}1}{\text{area } 2\text{–}3\text{–a–b–}2}$$

The numerator of this expression is $(T_H - T_C)(S_3 - S_2)$ and the denominator is $T_H(S_3 - S_2)$, so the thermal efficiency can be given in terms of temperatures only as $\eta = 1 - T_C/T_H$. If the cycle were reversed as shown in Fig. 6.6*b*, the result would be a Carnot refrigeration or heat pump cycle. In the reversed direction, heat is transferred to the system while its temperature remains at T_C, so entropy increases during process 1–2. In Process 3–4 heat is transferred from the system while the temperature remains constant at T_H and entropy decreases.

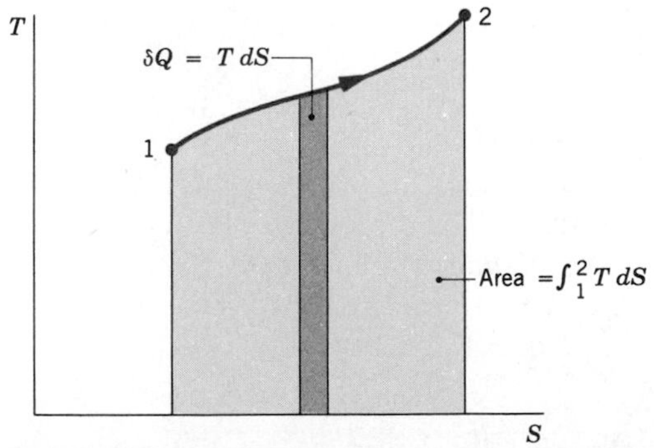

Figure 6.5 Area representation of heat transfer for an internally reversible process.

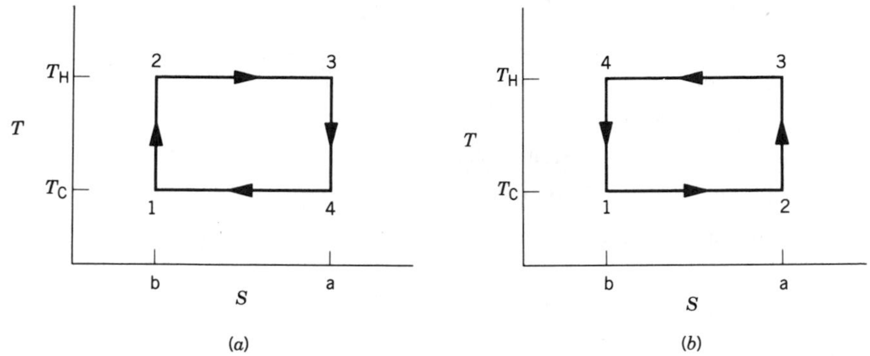

Figure 6.6 Carnot cycles on the temperature–entropy diagram. (*a*) Power cycle. (*b*) Refrigeration or heat pump cycle.

EXAMPLE 6.2

PROBLEM

Water initially a saturated liquid at 100°C is contained in a piston–cylinder assembly. The water undergoes a process to the corresponding saturated vapor state, during which the piston moves freely in the cylinder. If the change of state is brought about by heating the water as it undergoes an internally reversible process at constant pressure and temperature, determine the work and heat transfer per unit of mass, each in kJ/kg.

SOLUTION

Known: Water contained in a piston–cylinder assembly undergoes an internally reversible process at 100°C from saturated liquid to saturated vapor.

Find: Determine the work and heat transfer per unit mass.

Schematic and Given Data:

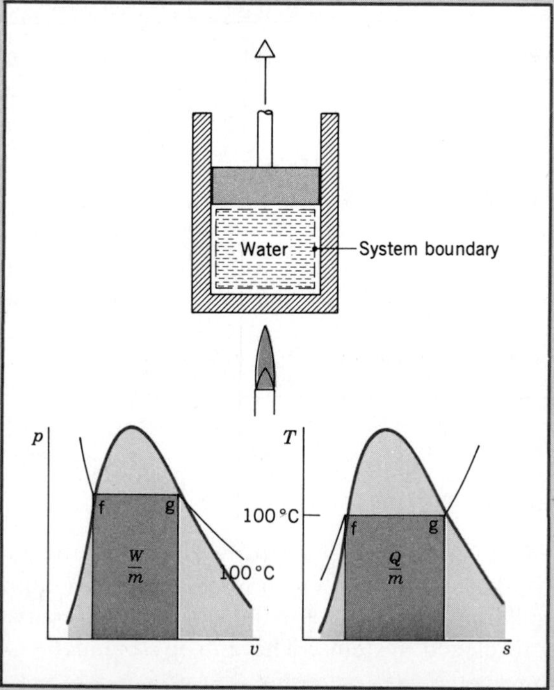

Figure E6.2

Assumptions:

1. The water in the piston–cylinder assembly is a closed system.

2. The process is internally reversible.

3. Temperature and pressure are constant during the process.

4. There is no change in kinetic or potential energy between the two end states.

Analysis: At constant pressure the work is

$$\frac{W}{m} = \int_{\mathrm{f}}^{\mathrm{g}} p\, dv = p(v_{\mathrm{g}} - v_{\mathrm{f}})$$

With values from Table A-2

$$\frac{W}{m} = \left(1.014 \times 10^5 \frac{\mathrm{N}}{\mathrm{m}^2}\right)(1.673 - 1.0435 \times 10^{-3})\left(\frac{\mathrm{m}^3}{\mathrm{kg}}\right)\left(\frac{1\ \mathrm{kJ}}{10^3\ \mathrm{N \cdot m}}\right)$$

$$= 170\ \mathrm{kJ/kg}$$

Since the process is internally reversible and at constant temperature, Eq. 6.29 gives

$$Q = \int_{\mathrm{f}}^{\mathrm{g}} T\, dS = m \int_{\mathrm{f}}^{\mathrm{g}} T\, ds$$

or

$$\frac{Q}{m} = T(s_{\mathrm{g}} - s_{\mathrm{f}})$$

With values from Table A-2

$$\frac{Q}{m} = (373.15\ \mathrm{K})(7.3549 - 1.3069)\ \mathrm{kJ/kg \cdot K} = 2257\ \mathrm{kJ/kg}$$

As shown in the accompanying figure, the work and heat transfer can be represented as areas on p–v and T–s diagrams, respectively.

Comment: The heat transfer can be evaluated alternatively from an energy balance written on a unit mass basis as

$$u_{\mathrm{g}} - u_{\mathrm{f}} = \frac{Q}{m} - \frac{W}{m}$$

Introducing $W/m = p(v_{\mathrm{g}} - v_{\mathrm{f}})$ and solving

$$\frac{Q}{m} = (u_{\mathrm{g}} - u_{\mathrm{f}}) + p(v_{\mathrm{g}} - v_{\mathrm{f}})$$

$$= (u_{\mathrm{g}} + pv_{\mathrm{g}}) - (u_{\mathrm{f}} + pv_{\mathrm{f}})$$

$$= h_{\mathrm{g}} - h_{\mathrm{f}}$$

From Table A-2 at 100°C, $h_{\mathrm{g}} - h_{\mathrm{f}} = 2257$ kJ/kg, which is the same value for Q/m as obtained in the solution.

6.5 ENTROPY BALANCE FOR CLOSED SYSTEMS

In this section, Eq. 6.4 and the defining equation for entropy change are used to develop the *entropy balance* for closed systems. The entropy balance is an expression of the second law that is particularly convenient for thermodynamic analysis. The current presentation is limited to closed systems. The entropy balance is extended to control volumes in Sec. 6.6.

6.5.1 Development of the Entropy Balance

Shown in Fig. 6.7 is a cycle executed by a closed system. The cycle consists of process I, during which internal irreversibilities may be present, followed by internally reversible process R. For this cycle, Eq. 6.4 takes the form

$$\int_1^2 \left(\frac{\delta Q}{T}\right)_b + \int_2^1 \left(\frac{\delta Q}{T}\right)_{\substack{\text{int}\\ \text{rev}}} = -\sigma \tag{6.30}$$

where the first integral is for process I and the second is for process R. The subscript b in the first integral serves as a reminder that the integrand is evaluated at the system boundary. The subscript is not required in the second integral because temperature is uniform throughout the system at each intermediate state of an internally reversible process. Since no irreversibilities are associated with process R, the term σ_{cycle} of Eq. 6.4, which accounts for the effect of irreversibilities during the cycle, refers only to process I and is shown in Eq. 6.30 simply as σ.

Applying the definition of entropy change, we can express the second integral of Eq. 6.30 as

$$S_1 - S_2 = \int_2^1 \left(\frac{\delta Q}{T}\right)_{\substack{\text{int}\\ \text{rev}}}$$

With this, Eq. 6.30 becomes

$$\int_1^2 \left(\frac{\delta Q}{T}\right)_b + (S_1 - S_2) = -\sigma$$

Finally, on rearranging the last equation, the closed system entropy balance results:

$$\underbrace{S_2 - S_1}_{\substack{\text{entropy}\\ \text{change}}} = \underbrace{\int_1^2 \left(\frac{\delta Q}{T}\right)_b}_{\substack{\text{entropy}\\ \text{transfer}}} + \underbrace{\sigma}_{\substack{\text{entropy}\\ \text{production}}} \tag{6.31}$$

If the end states are fixed, the entropy change on the left side of Eq. 6.31 can be evaluated independently of the details of the process. However, the two terms on the right side depend explicitly on the nature of the process and cannot be determined solely from knowledge of the end states. The first term on the right side of Eq. 6.31 is associated with heat transfer to or from the system during the process. This term can be interpreted as the *entropy transfer accompanying heat transfer*. The direction of

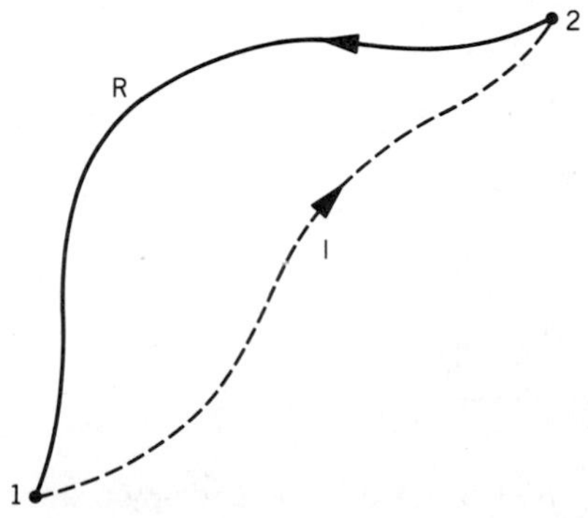

Figure 6.7 Cycle used to develop the entropy balance for a closed system.

entropy transfer is the same as the direction of the heat transfer, and the same sign convention applies as for heat transfer: A positive value means that entropy is transferred into the system, and a negative value means that entropy is transferred out. When there is no heat transfer, there is no entropy transfer.

The entropy change of a system is not accounted for solely by the entropy transfer, but is due in part to the second term on the right side of Eq. 6.31 denoted by σ. The term σ is positive when internal irreversibilities are present during the process and vanishes when no internal irreversibilities are present. This can be described by saying that entropy is *produced* within the system by the action of irreversibilities. The second law of thermodynamics can be interpreted as requiring that entropy is produced by irreversibilities and conserved only in the limit as irreversibilities are reduced to zero. Since σ measures the effect of irreversibilities present within the system during a process, its value depends on the nature of the process and not solely on the end states. It is *not* a property.

For a simple illustration of the entropy transfer and entropy production concepts, as well as the accounting nature of the entropy balance, consider Fig. 6.8. The figure shows a system consisting of a gas or liquid in a rigid container stirred by a paddle wheel while receiving a heat transfer Q from a reservoir. The temperature at the portion of the boundary where heat transfer occurs is the same as the constant temperature of the reservoir, T_b. By definition, the reservoir is free of irreversibilities; however, the system is not without irreversibilities, for fluid friction is evidently present, and there may be other irreversibilities within the system.

Let us now apply the entropy balance to the system and to the reservoir. Since T_b is constant, the integral in Eq. 6.31 is readily evaluated, and the entropy balance for the *system* reduces to

$$S_2 - S_1 = \frac{Q}{T_b} + \sigma \tag{6.32}$$

where Q/T_b accounts for entropy transfer into the system accompanying heat transfer Q. The entropy balance for the *reservoir* takes the form

$$\Delta S]_{res} = \frac{Q_{res}}{T_b} + \overset{0}{\not{\sigma}_{res}}$$

where the entropy production term is set equal to zero because the reservoir is without irreversibilities. Since $Q_{res} = -Q$, the last equation becomes

$$\Delta S]_{res} = -\frac{Q}{T_b} \tag{6.33}$$

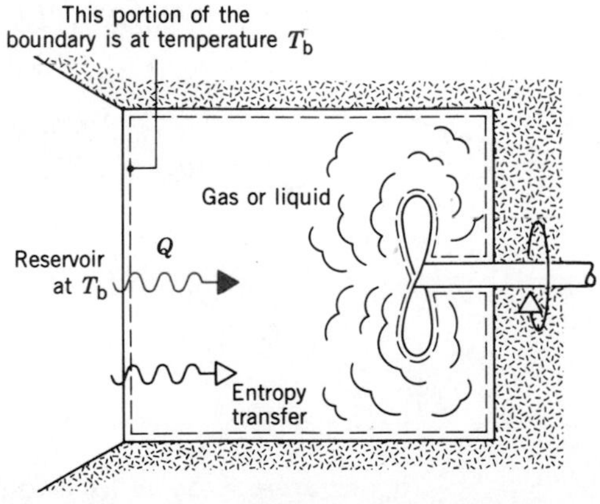

Figure 6.8 Illustration of the entropy transfer and entropy production concepts.

The minus sign signals that entropy is carried out of the reservoir accompanying heat transfer. Hence, the entropy of the reservoir *decreases* by an amount equal to the entropy transferred from it to the system. However, as shown by Eq. 6.32, the entropy change of the system *exceeds* the amount of entropy transferred to it because of entropy production within the system.

If the heat transfer were oppositely directed, passing instead from the system to the reservoir, the magnitude of the entropy transfer would remain the same, but its direction would be reversed. In this case, the entropy of the system might *decrease*. This would occur if the amount of entropy transferred *from* the system *to* the reservoir *exceeded* the amount of entropy produced within the system due to irreversibilities. Finally, observe that there is no entropy transfer associated with work.

When applying the entropy balance to a closed system, it is essential to remember the requirements imposed by the second law on entropy production: The second law requires that entropy *production* be positive, or zero, in value

$$\sigma: \begin{cases} > 0 & \text{irreversibilities present within the system} \\ = 0 & \text{no irreversibilities present within the system} \end{cases} \tag{6.34}$$

The value of the entropy production cannot be negative. By contrast, the *change* in entropy of the system may be positive, negative, or zero:

$$S_2 - S_1: \begin{cases} > 0 \\ = 0 \\ < 0 \end{cases} \tag{6.35}$$

Like other properties, entropy change can be determined without knowledge of the details of the process.

6.5.2 Forms of the Entropy Balance for a Closed System

The entropy balance can be expressed in various forms that may be convenient for particular analyses. For example, if heat transfer takes place at several locations on the boundary of a system where the temperatures do not vary with position or time, the entropy transfer term can be expressed as a sum, so Eq. 6.31 takes the form

$$S_2 - S_1 = \sum_j \frac{Q_j}{T_j} + \sigma \tag{6.36}$$

where Q_j/T_j is the amount of entropy transferred through the portion of the boundary at temperature T_j.

The entropy balance can also be expressed on a time rate basis as

$$\frac{dS}{dt} = \sum_j \frac{\dot{Q}_j}{T_j} + \dot{\sigma} \tag{6.37}$$

where dS/dt is the time rate of change of entropy of the system. The term $\dot{Q}_j/T_j$ represents the time rate of entropy transfer through the portion of the boundary whose instantaneous temperature is T_j. The term $\dot{\sigma}$ accounts for the time rate of entropy production due to irreversibilities within the system.

It is sometimes convenient to use the entropy balance expressed in differential form:

$$dS = \left(\frac{\delta Q}{T}\right)_\text{b} + \delta\sigma \tag{6.38}$$

Note that the differentials of the nonproperties Q and σ are shown, respectively, as

δQ and $\delta\sigma$. When there are no internal irreversibilities, $\delta\sigma$ vanishes and Eq. 6.38 reduces to Eq. 6.13.

Regardless of the form taken by the entropy balance, the objective in many applications is to evaluate the entropy production term. However, the value of the entropy production for a given process of a system often does not have much significance by itself. The significance is normally determined through comparison. For example, the entropy production within a given component might be compared to the entropy production values of the other components included in an overall system formed by these components. By comparing entropy production values, the components where appreciable irreversibilities occur can be identified and rank ordered. This allows attention to be focused on the components that contribute most heavily to inefficient operation of the overall system.

To evaluate the entropy transfer term of the entropy balance requires information regarding both the heat transfer taking place across each portion of the boundary of a system and the temperature on the boundary where the heat transfer occurs. In many cases, the entropy transfer term is not subject to direct evaluation because the required information is either unknown or not defined, such as when the system passes through states sufficiently far from equilibrium. By enlarging the system to take in some of the immediate surroundings, a boundary often can be defined on which the entropy transfer can be evaluated. However, the irreversibilities present would not be just those for the system of interest but those for the enlarged system. Accordingly, the entropy production term would account for the effects of internal irreversibilities within the system *and* external irreversibilities present within that portion of the surroundings included within the enlarged system.

6.5.3 Illustrations

The following examples illustrate the use of the energy and entropy balances for the analysis of closed systems. Property relations and property diagrams also contribute significantly in developing solutions.

EXAMPLE 6.3

PROBLEM

Water initially a saturated liquid at 100°C is contained within a piston–cylinder assembly. The water undergoes a process to the corresponding saturated vapor state, during which the piston moves freely in the cylinder. There is no heat transfer with the surroundings. If the change of state is brought about by the action of a paddle wheel, determine the work per unit mass, in kJ/kg, and the amount of entropy produced per unit mass, in kJ/kg·K.

SOLUTION

Known: Water contained in a piston–cylinder assembly undergoes an adiabatic process from saturated liquid to saturated vapor at 100°C. During the process, the piston moves freely, and the water is rapidly stirred by a paddle wheel.

Find: Determine the work per unit mass and the entropy produced per unit mass.

Schematic and Given Data:

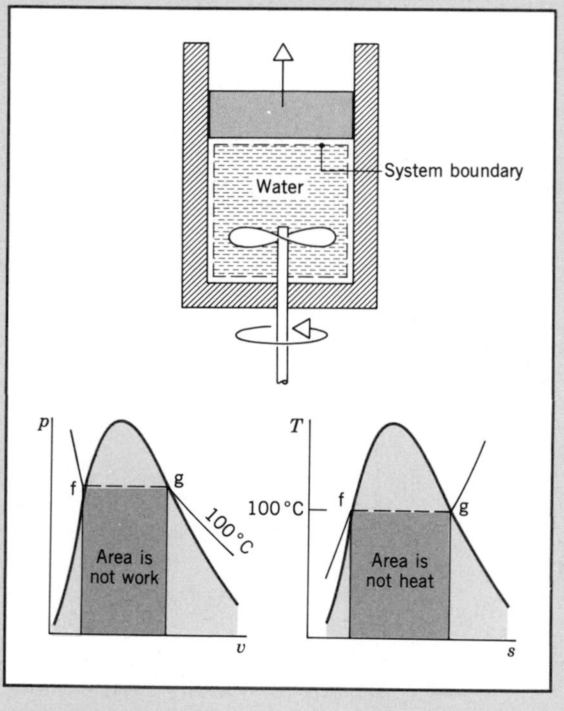

Figure E6.3

Assumptions:

1. The water in the piston–cylinder assembly is a closed system.
2. There is no heat transfer with the surroundings.
3. The system is at an equilibrium state initially and finally. There is no change in kinetic or potential energy between these two states.

Analysis: As the volume of the system increases during the process, there is an energy transfer by work from the system during the expansion, as well as an energy transfer by work to the system via the paddle wheel. The *net* work can be evaluated from an energy balance, which reduces with assumptions 2 and 3 to

$$\Delta U + \cancelto{0}{\Delta KE} + \cancelto{0}{\Delta PE} = \cancelto{0}{Q} - W$$

On a unit mass basis, the energy balance reduces to

$$\frac{W}{m} = -(u_g - u_f)$$

With specific internal energy values from Table A-2 at 100°C

$$\frac{W}{m} = -2087.56 \, \frac{\text{kJ}}{\text{kg}}$$

The minus sign indicates that the work input by stirring is greater in magnitude than the work done by the water as it expands.

The amount of entropy produced is evaluated by applying an entropy balance. Since there is no heat transfer, the term accounting for entropy transfer vanishes

$$\Delta S = \cancelto{0}{\int_1^2 \left(\frac{\delta Q}{T}\right)_b} + \sigma$$

On a unit mass basis, this becomes on rearrangement

$$\frac{\sigma}{m} = s_g - s_f$$

With specific entropy values from Table A-2 at 100°C

$$\frac{\sigma}{m} = 6.048 \frac{\text{kJ}}{\text{kg} \cdot \text{K}}$$

Comments:

1. Although each end state is an equilibrium state at the same pressure and temperature, the pressure and temperature are not necessarily uniform throughout the system at *intervening* states, nor are they necessarily constant in value during the process. Accordingly, there is no well-defined "path" for the process. This is emphasized by the use of dashed lines to represent the process on the accompanying p–v and T–s diagrams. The dashed lines indicate only that a process has taken place, and no "area" should be associated with them. In particular, note that the process is adiabatic, so the "area" below the dashed line on the T–s diagram can have no significance as heat transfer. Similarly, the work cannot be associated with an area on the p–v diagram.
2. The change of state is the same in the present example as in Example 6.2. However, in Example 6.2 the change of state is brought about by heat transfer while the system undergoes an internally reversible process. Accordingly, the value of entropy production for the process of Example 6.2 is zero. Here, irreversibilities are present during the process and the entropy production is positive in value. Accordingly, different values of entropy production are obtained for two processes between the *same* end states. This reinforces the idea that entropy production is not a property.

EXAMPLE 6.4

PROBLEM

Refrigerant 12 is compressed adiabatically in a piston–cylinder assembly from saturated vapor at 10°F to a final pressure of 120 lbf/in.^{2} Determine the minimum theoretical work input required per unit mass of refrigerant, in Btu/lb.

SOLUTION

Known: Refrigerant 12 is compressed without heat transfer from a specified initial state to a specified final pressure.

Find: Determine the minimum theoretical work input required per unit of mass.

Schematic and Given Data:

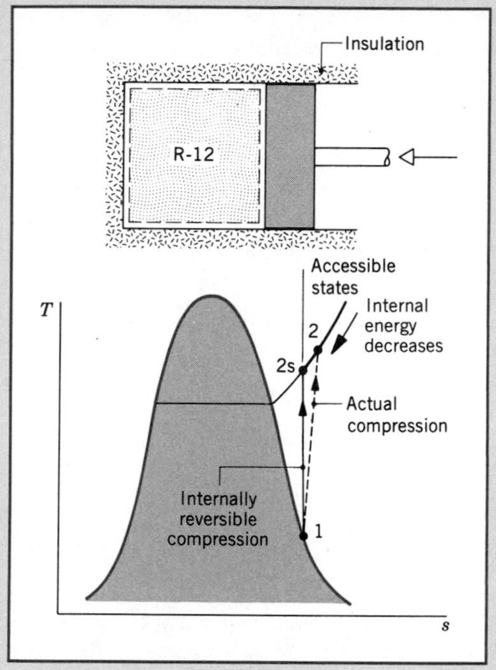

Figure E6.4

Assumptions:

1. The Refrigerant 12 is a closed system.
2. There is no heat transfer with the surroundings.
3. The initial and final states are equilibrium states. There is no change in kinetic or potential energy between these states.

Analysis: An expression for the work can be obtained from an energy balance. By applying assumptions 2 and 3

$$\Delta U + \cancelto{0}{\Delta KE} + \cancelto{0}{\Delta PE} = \cancelto{0}{Q} - W$$

When written on a unit mass basis, this becomes

$$\left(-\frac{W}{m}\right) = u_2 - u_1$$

The specific internal energy u_1 can be obtained from Table A-7E as $u_1 = 71.15$ Btu/lb. Since u_1 is known, the value for the work input depends on the specific internal energy u_2. The minimum work input corresponds to the smallest allowed value for u_2, determined using the second law as follows.

Applying an entropy balance

$$\Delta S = \cancelto{0}{\int_1^2 \left(\frac{\delta Q}{T}\right)_b} + \sigma$$

where the entropy transfer term is set equal to zero because the process is adiabatic. Thus, the *allowed* final states must satisfy

$$s_2 - s_1 = \frac{\sigma}{m} \geq 0$$

The restriction indicated by the foregoing equation can be interpreted using the accompanying T–s diagram. Since σ cannot be negative, states with $s_2 < s_1$ are not accessible adiabatically. When irreversibilities are present during the compression, entropy is produced, so $s_2 > s_1$. The state labeled as 2s on the diagram would be attained in the limit as irreversibilities are reduced to zero. This state corresponds to an *isentropic* compression.

By inspection of Table A-9E, we see that when pressure is fixed, the specific internal energy decreases as temperature decreases. Thus, the smallest allowed value for u_2 corresponds to state 2s. Interpolating in Table A-9E at 120 lbf/in.^{2}, with $s_{2s} = s_1 = 0.1680$ Btu/lb·°R, we find that $u_{2s} = 81.17$ Btu/lb. Finally

$$\left(-\frac{W}{m}\right)_{\min} = u_{2s} - u_1 = 81.17 - 71.15 = 10.02 \text{ Btu/lb}$$

Comment: The effect of irreversibilities exacts a penalty on the work input required: A greater work input is needed for the actual adiabatic compression process than for an internally reversible adiabatic process between the same initial state and the same final pressure.

EXAMPLE 6.5

PROBLEM

During steady-state operation, a parallel-shaft gearbox receives 600 kW through the high-speed shaft, but owing to friction and other irreversibilities, delivers 588 kW through the low-speed shaft. The gearbox is cooled on its outer surface according to

$$Q = -\text{hA}(T_b - T_0)$$

where h is the *heat transfer coefficient*, A is the outer surface area, T_b is the uniform temperature of the outer surface, and T_0 is the uniform temperature of the surroundings away from the immediate vicinity of the gearbox. Evaluate the rate of entropy production $\dot{\sigma}$, in kW/K, for (a) the gearbox as the system and (b) an enlarged system consisting of the gearbox and enough of its surroundings that heat transfer occurs at temperature T_0. Let h = 0.17 kW/m^2·K, A = 1.8 m^2, and $T_0 = 293$ K.

SOLUTION

Known: A gearbox operates at steady state with a known power input through the high-speed shaft and a known power output through the low-speed shaft. An expression for the heat transfer rate from the outer surface is also known.

Find: Evaluate the entropy production rate $\dot{\sigma}$ for each of two specified systems.

Schematic and Given Data:

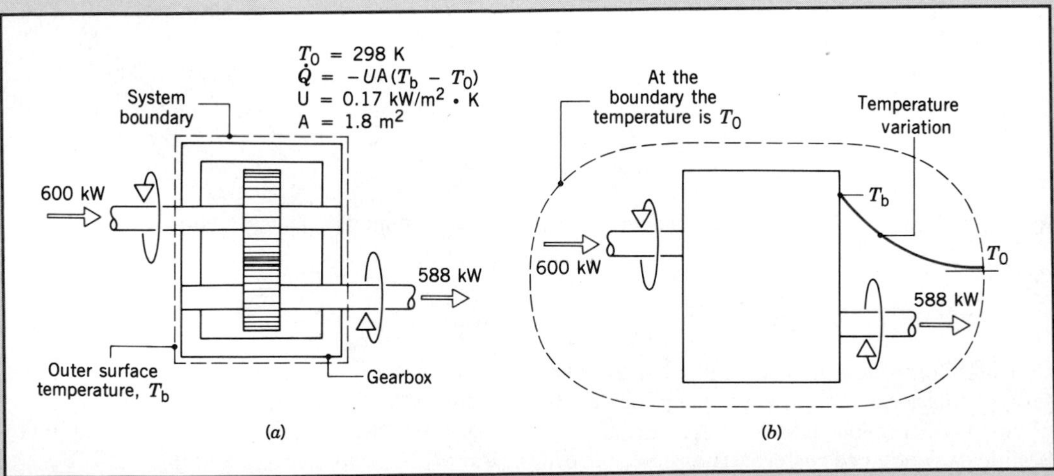

Figure E6.5

Assumptions:

1. In part (a) the gearbox is taken as a closed system operating at steady state, as shown on the accompanying sketch.
2. In part (b) the gearbox and a portion of its surroundings are taken as a closed system, as shown on the accompanying sketch.
3. The temperature of the outer surface of the gearbox and the temperature of the surroundings are each uniform.

Analysis:

(a) To obtain an expression for the entropy production rate, begin with the entropy balance for a closed system on a time rate basis, Eq. 6.37. Since heat transfer takes place only at temperature T_b, the entropy rate balance reduces at steady state to

$$\cancelto{0}{\frac{dS}{dt}} = \frac{\dot{Q}}{T_b} + \dot{\sigma}$$

Solving

$$\dot{\sigma} = -\frac{\dot{Q}}{T_b}$$

To evaluate $\dot{\sigma}$ requires the heat transfer rate $\dot{Q}$ and the surface temperature T_b.

The heat transfer rate can be obtained by applying the rate form of the energy balance, which at steady state reduces to

$$\cancelto{0}{\frac{dE}{dt}} = \dot{Q} - \dot{W} \quad \text{or} \quad \dot{Q} = \dot{W}$$

Using given data

$$\dot{Q} = (588 \text{ kW}) + (-600 \text{ kW}) = -12 \text{ kW}$$

The surface temperature can be evaluated by rearranging the given expression for the heat transfer rate at the outer surface to read

$$T_b = \frac{-\dot{Q}}{\text{hA}} + T_0$$

Substituting values

$$T_b = \frac{-(-12 \text{ kW})}{(0.17 \text{ kW/m}^2\cdot\text{K})(1.8 \text{ m}^2)} + 293 \text{ K} = 332 \text{ K}$$

Finally, inserting values into the expression for $\dot{\sigma}$ gives

$$\dot{\sigma} = -\frac{(-12 \text{ kW})}{(332 \text{ K})} = 0.0361 \text{ kW/K}$$

(b) For the enlarged system, the entropy rate balance reduces at steady state to

$$\cancelto{0}{\frac{dS}{dt}} = \frac{\dot{Q}}{T_0} + \dot{\sigma}$$

Solving

$$\dot{\sigma} = -\frac{\dot{Q}}{T_0}$$

From the solution of part (a), $\dot{Q} = -12$ kW. Then, with $T_0 = 293$ K

$$\dot{\sigma} = -\frac{(-12 \text{ kW})}{293 \text{ K}} = 0.0410 \text{ kW/K}$$

Comment: The value of the entropy production rate calculated in part (a) gauges the significance of irreversibilities *within* the gearbox. In part (b) an additional source of irreversibility is included in the enlarged system, namely the irreversibility associated with the heat transfer from the outer surface of the gearbox at T_b to the surroundings at T_0. Accordingly, the numerical value for $\dot{\sigma}$ in part (b) is greater than the value in part (a).

6.5.4 Increase of Entropy Principle

Our study of the second law began in Sec. 5.1 with a discussion of the directionality of processes. In the present development, it is shown that the energy and entropy balances can be used together to determine direction.

The present discussion centers on an enlarged system comprising a system and that portion of the surroundings affected by the system as it undergoes a process. Since all energy and mass transfers taking place are included within the boundary of the enlarged system, the enlarged system can be regarded as an *isolated* system.

An energy balance for the isolated system reduces to

$$\Delta E]_{\text{isol}} = 0 \tag{6.39a}$$

because no energy transfers take place across its boundary. Thus, the energy of the isolated system remains constant. Since energy is an extensive property, its value for the isolated system is the sum of its values for the system and surroundings, respectively, so Eq. 6.39a can be written as

$$\Delta E]_{\text{system}} + \Delta E]_{\text{surr}} = 0 \tag{6.39b}$$

In either of these forms, the conservation of energy principle places a constraint on the processes that can occur. For a process to take place, it is necessary for the energy of the system plus the surroundings to remain constant. However, not all processes for which this constraint is satisfied can actually occur. Processes also must satisfy the second law, as discussed next.

An entropy balance for the isolated system reduces to

$$\Delta S]_{\text{isol}} = \cancelto{0}{\int_1^2 \left(\frac{\delta Q}{T}\right)_b} + \sigma_{\text{isol}}$$

or

$$\Delta S]_{\text{isol}} = \sigma_{\text{isol}} \tag{6.40a}$$

where σ_{isol} is the total amount of entropy produced within the system and its surroundings. Since entropy is produced in all actual processes, the only processes that can occur are those for which the entropy of the isolated system increases. This is known as the *increase of entropy principle*. The increase of entropy principle is sometimes adopted as a statement of the second law.

Since entropy is an extensive property, its value for the isolated system is the sum of its values for the system and surroundings, respectively, so Eq. 6.40a can be written as

$$\Delta S]_{\text{system}} + \Delta S]_{\text{surr}} = \sigma_{\text{isol}} \tag{6.40b}$$

Notice that this equation does not require the entropy change to be positive for both the system and surroundings but only that the *sum* of the changes is positive. In either of these forms, the increase of entropy principle dictates the direction in which any process can proceed: the direction that causes the total entropy of the system plus surroundings to increase.

We noted previously the tendency of systems left to themselves to undergo processes until a condition of equilibrium is attained (Sec. 5.1). The increase of entropy principle suggests that the entropy of an isolated system increases as the state of equilibrium is approached, with the equilibrium state being attained when the entropy reaches a maximum. This interpretation is considered again in Sec. 14.1.1, which deals with equilibrium criteria.

In *statistical thermodynamics,* entropy is associated with the notion of *disorder* and the second law statement that the entropy of an isolated system undergoing a spontaneous process tends to increase is equivalent to saying that the disorder of the isolated system tends to increase. Let us conclude the present discussion with a brief summary of concepts from the microscopic viewpoint related to these ideas.

Viewed macroscopically, an equilibrium state of a system appears to be unchanging, but on the microscopic level the particles making up the matter are continually in motion. Accordingly, a vast number of possible microscopic states correspond to any given macroscopic equilibrium state. The total number of possible microscopic states available to a system is called the thermodynamic probability, w. Entropy is related to w by the Boltzmann relation

$$S = \mathrm{k} \ln w \tag{6.41}$$

where k is Boltzmann's constant. From this equation we see that any process that increases the number of possible microscopic states of a system increases its entropy, and conversely. Hence, for an isolated system, processes only occur in such a way that the number of microscopic states available to the system increases. The number w is referred to as the *disorder* of the system. We can say, then, that the only processes an isolated system can undergo are those that increase the disorder of the system.

EXAMPLE 6.6

PROBLEM

A 0.8-lb bar of metal initially at 1900°R is removed from an oven and quenched by immersing it in a closed tank containing 20 lb of water initially at 530°R. Each substance can be modeled as incompressible. An appropriate constant specific heat value for the water is $c_w = 1.0$ Btu/lb·°R, and an appropriate value for the metal is $c_m = 0.1$ Btu/lb·°R. Heat transfer from the tank contents can be neglected. Determine (a) the final equilibrium temperature of the metal bar and the water, in °R, and (b) the amount of entropy produced, in Btu/°R.

SOLUTION

Known: A hot metal bar is quenched by immersing it in a tank containing water.

Find: Determine the final equilibrium temperature of the metal bar and the water, and the amount of entropy produced.

Schematic and Given Data:

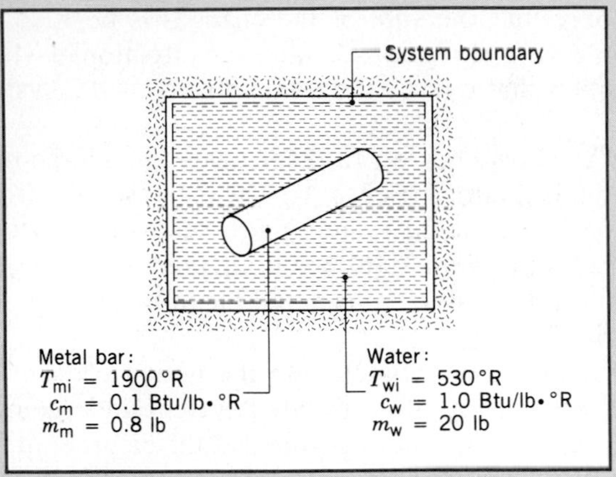

Figure E6.6

Assumptions:

1. The metal bar and the water within the tank form a system, as shown on the accompanying sketch.
2. The system is isolated.
3. There is no change in kinetic or potential energy.
4. The water and metal bar are each modeled as incompressible with known specific heats.

Analysis:

(a) The final equilibrium temperature can be evaluated from an energy balance for the closed system

$$\Delta \cancelto{0}{KE} + \Delta \cancelto{0}{PE} + \Delta U = \cancelto{0}{Q} - \cancelto{0}{W}$$

where the indicated terms vanish by assumptions 2 and 3. Since internal energy is an extensive property, its value for the overall system is the sum of the values for the water and metal, respectively. Thus, the energy balance becomes

$$\Delta U]_{\text{water}} + \Delta U]_{\text{metal}} = 0$$

Evaluating the internal energy changes of the water and metal in terms of the constant specific heats

$$m_w c_w (T_f - T_{wi}) + m_m c_m (T_f - T_{mi}) = 0$$

where T_f is the final equilibrium temperature, and T_{wi} and T_{mi} are the initial temperatures of the water and metal, respectively. By solving for T_f and inserting values

$$T_f = \frac{m_w(c_w/c_m)T_{wi} + m_m T_{mi}}{m_w(c_w/c_m) + m_m}$$

$$= \frac{(20 \text{ lb})(10)(530°\text{R}) + (0.8 \text{ lb})(1900°\text{R})}{(20 \text{ lb})(10) + (0.8 \text{ lb})}$$

$$= 535°\text{R}$$

(b) The amount of entropy production can be evaluated from an entropy balance. Since no heat transfer occurs between the system and its surroundings, there is no accompanying entropy transfer, and an entropy balance for the overall system reduces to

$$\Delta S = \cancelto{0}{\int_1^2 \left(\frac{\delta Q}{T}\right)_b} + \sigma$$

Entropy is an extensive property, so its value for the system is the sum of its values for the water and the metal, respectively, and the entropy balance becomes

$$\Delta S]_{\text{water}} + \Delta S]_{\text{metal}} = \sigma$$

Evaluating the entropy changes using Eq. 6.28 for incompressible substances, the foregoing equation can be written as

$$\sigma = m_w c_w \ln \frac{T_f}{T_{wi}} + m_m c_m \ln \frac{T_f}{T_{mi}}$$

Inserting values

$$\sigma = (20 \text{ lb}) \left(1.0 \frac{\text{Btu}}{\text{lb} \cdot {}^\circ\text{R}}\right) \ln \frac{535}{530} + (0.8 \text{ lb}) \left(0.1 \frac{\text{Btu}}{\text{lb} \cdot {}^\circ\text{R}}\right) \ln \frac{535}{1900}$$

$$= \left(0.1878 \frac{\text{Btu}}{{}^\circ\text{R}}\right) + \left(-0.1014 \frac{\text{Btu}}{{}^\circ\text{R}}\right) = 0.0864 \text{ Btu/}{}^\circ\text{R}$$

Comment: The metal bar experiences a *decrease* in entropy. The entropy of the water *increases*. In accord with the increase of entropy principle, the entropy of the isolated system *increases*.

6.6 ENTROPY RATE BALANCE FOR CONTROL VOLUMES

Thus far the discussion of the entropy balance concept has been restricted to the case of closed systems. In the present section the entropy balance is extended to control volumes.

We begin by noting that the control volume entropy rate balance can be derived by an approach closely paralleling that employed in Secs. 4.1 and 4.2, where the control volume forms of the mass and energy balances were obtained by transforming the closed system forms. The present development proceeds less formally by arguing that, like mass and energy, entropy is an extensive property, so it too can be transferred into or out of a control volume by streams of matter. Since this is the principal difference between the closed system and control volume forms, the control volume entropy rate balance can be obtained by modifying Eq. 6.37 to account for these entropy transfers. The result is

$$\underbrace{\frac{dS_{cv}}{dt}}_{\substack{\text{rate of}\\ \text{entropy}\\ \text{change}}} = \underbrace{\sum_j \frac{\dot{Q}_j}{T_j} + \sum_i \dot{m}_i s_i - \sum_e \dot{m}_e s_e}_{\substack{\text{rates of}\\ \text{entropy}\\ \text{transfer}}} + \underbrace{\dot{\sigma}_{cv}}_{\substack{\text{rate of}\\ \text{entropy}\\ \text{production}}} \tag{6.42}$$

where dS_{cv}/dt represents the time rate of change of entropy within the control volume. The terms $\dot{m}_i s_i$ and $\dot{m}_e s_e$ account, respectively, for rates of entropy *transfer* into and out of the control volume accompanying mass flow. In writing Eq. 6.42, one-dimensional flow is assumed at locations where mass enters and exits. The term $\dot{Q}_j$ represents the time rate of heat transfer at the location on the boundary where the instantaneous temperature is T_j. The ratio $\dot{Q}_j/T_j$ accounts for the accompanying rate of entropy *transfer*. The term $\dot{\sigma}_{cv}$ denotes the time rate of entropy *production* due to irreversibilities *within* the control volume.

Equation 6.42 is the most general form of the entropy rate balance used for problem solving in this book. However, as for the cases of the control volume mass and energy rate balances, the entropy rate balance can be expressed in terms of local properties to obtain forms that are more generally applicable. Thus, the term $S_{cv}(t)$, representing the total entropy associated with the control volume at time t, can be written as a volume integral

$$S_{cv}(t) = \int_V \rho s \, dV$$

where ρ and s denote, respectively, the local density and specific entropy. The rate of entropy transfer accompanying heat transfer can be expressed more generally as an integral over the surface of the control volume

$$\begin{bmatrix} \text{time rate of entropy} \\ \text{transfer accompanying} \\ \text{heat transfer} \end{bmatrix} = \int_A \left(\frac{\dot{q}}{T}\right)_b dA$$

where $\dot{q}$ is the *heat flux,* the time rate of heat transfer per unit of surface area, through the location on the boundary where the instantaneous temperature is T. The subscript "b" is added as a reminder that the integrand is evaluated on the boundary of the control volume. In addition, the terms accounting for entropy transfer accompanying mass flow can be expressed as integrals over the inlet and exit flow areas, resulting in the following form of the entropy rate equation:

$$\frac{d}{dt}\int_V \rho s \, dV = \int_A \left(\frac{\dot{q}}{T}\right)_b dA + \sum_i \left(\int_A s\rho V_n \, dA\right)_i - \sum_e \left(\int_A s\rho V_n \, dA\right)_e + \dot{\sigma}_{cv} \quad (6.43)$$

where V_n denotes the normal component in the direction of flow of the velocity relative to the flow area. In some cases, it is also convenient to express the entropy production rate as a volume integral of the local volumetric rate of entropy production within the control volume. The study of Eq. 6.43 brings out the assumptions underlying Eq. 6.42. Finally, note that for a closed system the sums accounting for entropy transfer at inlets and exits drop out, and Eq. 6.43 reduces to give a more general form of Eq. 6.37.

6.6.1 Analysis of Control Volumes at Steady State

Since a great many engineering analyses involve control volumes at steady state, it is instructive to list steady-state forms of the balances developed for mass, energy, and entropy. At steady state, the conservation of mass principle takes the form

$$\sum_i \dot{m}_i = \sum_e \dot{m}_e \quad (4.27)$$

The energy rate balance at steady state is

$$0 = \dot{Q}_{cv} - \dot{W}_{cv} + \sum_i \dot{m}_i \left(h_i + \frac{V_i^2}{2} + gz_i\right) - \sum_e \dot{m}_e \left(h_e + \frac{V_e^2}{2} + gz_e\right) \quad (4.28a)$$

Finally, at steady state, the entropy rate balance, Eq. 6.42, reduces to

$$0 = \sum_j \frac{\dot{Q}_j}{T_j} + \sum_i \dot{m}_i s_i - \sum_e \dot{m}_e s_e + \dot{\sigma}_{cv} \quad (6.44)$$

These equations often must be solved simultaneously, together with appropriate property relations.

Mass and energy are conserved quantities, but entropy is not generally conserved. Equation 4.27 indicates that at steady state the total rate of mass flow into the control volume equals the total rate of mass flow out of the control volume. Similarly, Eq. 4.28a states that the total rate of energy transfer into the control volume equals the total rate of energy transfer out of the control volume. However, Eq. 6.44 shows that the rate at which entropy is transferred out must *exceed* the rate at which entropy enters, the difference being the rate of entropy production within the control volume owing to irreversibilities.

Since many important applications involve one-inlet, one-exit control volumes at steady state, let us also list the form of the entropy rate balance for this case:

$$0 = \sum_j \frac{\dot{Q}_j}{T_j} + \dot{m}(s_1 - s_2) + \dot{\sigma}_{cv}$$

Or, on dividing by the mass flow rate $\dot{m}$ and rearranging

$$s_2 - s_1 = \frac{1}{\dot{m}}\left(\sum_j \frac{\dot{Q}_j}{T_j}\right) + \frac{\dot{\sigma}_{cv}}{\dot{m}} \tag{6.45}$$

The two terms on the right side of Eq. 6.45 denote, respectively, the rate of entropy transfer accompanying heat transfer and the rate of entropy production within the control volume, each *per unit of mass flowing through the control volume*. From Eq. 6.45 it can be concluded that the entropy of a unit of mass passing from inlet to exit can increase, decrease, or remain the same. However, because the value of the second term on the right can never be negative, a decrease in the specific entropy from inlet to exit can be realized only when more entropy is transferred out of the control volume accompanying heat transfer than is produced by irreversibilities within the control volume. When the value of this entropy transfer term is positive, the specific entropy at the exit is greater than the specific entropy at the inlet whether internal irreversibilities are present or not. In the special case where there is no entropy transfer accompanying heat transfer, Eq. 6.45 reduces to

$$s_2 - s_1 = \frac{\dot{\sigma}_{cv}}{\dot{m}} \tag{6.46}$$

Accordingly, when irreversibilities are present within the control volume, the entropy of a unit of mass increases as it passes from inlet to exit. In the limiting case in which no irreversibilities are present, the unit mass passes through the control volume with no change in its entropy—that is, isentropically.

6.6.2 Illustrations

The following examples illustrate the use of the mass, energy, and entropy balances for the analysis of control volumes at steady state. Carefully note that property relations and property diagrams also play important roles in arriving at solutions.

EXAMPLE 6.7

PROBLEM

Steam enters a turbine with a pressure of 30 bars, a temperature of 400°C, and a velocity of 160 m/s. Saturated vapor at 100°C exits with a velocity of 100 m/s. At steady state, the turbine develops work equal to 540 kJ per kg of steam flowing through the turbine. Heat transfer between the turbine and its surroundings occurs at an average outer surface temperature of 500 K. Determine the rate at which entropy is produced within the turbine per kilogram of steam flowing, in kJ/kg·K. Neglect the change in potential energy between inlet and exit.

SOLUTION

Known: Steam expands through a turbine at steady state for which data are provided.

Find: Determine the rate of entropy production per kg of steam flowing.

Schematic and Given Data:

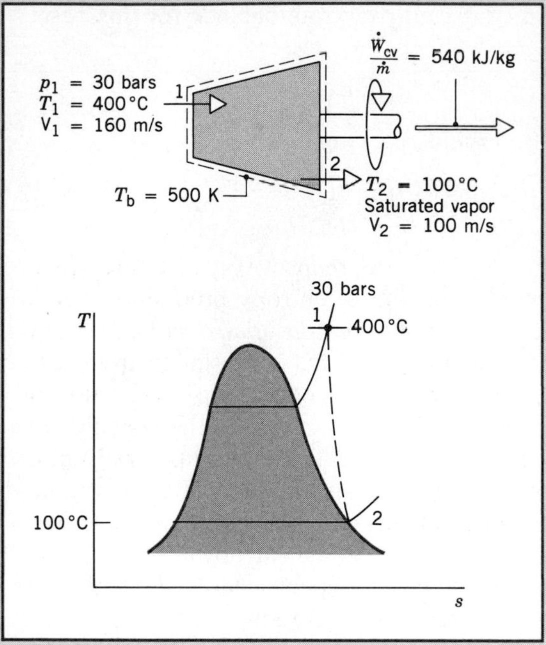

Figure E6.7

Assumptions:

1. The control volume shown on the accompanying sketch is at steady state.
2. Heat transfer between the turbine and the surroundings occurs at a known temperature.
3. The change in potential energy between inlet and exit can be neglected.

Analysis: To determine the entropy production per unit mass flowing through the turbine, begin with mass and entropy rate balances for the one-inlet, one-exit control volume at steady state:

$$0 = \dot{m}_1 - \dot{m}_2$$

$$0 = \sum_j \frac{\dot{Q}_j}{T_j} + \dot{m}_1 s_1 - \dot{m}_2 s_2 + \dot{\sigma}_{cv}$$

Since heat transfer occurs only at $T_b = 500$ K, the first term on the right side of the entropy rate balance reduces to $\dot{Q}_{cv}/T_b$. Combining the mass and entropy rate balances

$$0 = \frac{\dot{Q}_{cv}}{T_b} + \dot{m}(s_1 - s_2) + \dot{\sigma}_{cv}$$

where $\dot{m}$ is the mass flow rate. Solving for $\dot{\sigma}_{cv}/\dot{m}$

$$\frac{\dot{\sigma}_{cv}}{\dot{m}} = -\frac{\dot{Q}_{cv}/\dot{m}}{T_b} + (s_2 - s_1)$$

The heat transfer rate required by this expression is evaluated next.
Reduction of the mass and energy rate balances results in

$$\frac{\dot{Q}_{cv}}{\dot{m}} = \frac{\dot{W}_{cv}}{\dot{m}} + (h_2 - h_1) + \left(\frac{V_2^2 - V_1^2}{2}\right)$$

where the potential energy change from inlet to exit is dropped by assumption 3. From Table A-4 at 30 bars, 400°C, $h_1 = 3230.9$ kJ/kg, and from Table A-2, $h_2 = h_g(100°C) = 2676.1$ kJ/kg. Thus

$$\frac{\dot{Q}_{cv}}{\dot{m}} = 540\ \frac{\text{kJ}}{\text{kg}} + (2676.1 - 3230.9)\left(\frac{\text{kJ}}{\text{kg}}\right) + \left[\frac{(100)^2 - (160)^2}{2}\right]\left(\frac{\text{m}^2}{\text{s}^2}\right)\left(\frac{1\ \text{N}}{1\ \text{kg}\cdot\text{m/s}^2}\right)\left(\frac{1\ \text{kJ}}{10^3\ \text{N}\cdot\text{m}}\right)$$

$$= 540 - 554.8 - 7.8 = -22.6\ \text{kJ/kg}$$

From Table A-2, $s_2 = 7.3549$ kJ/kg·K, and from Table A-4, $s_1 = 6.9212$ kJ/kg·K. By inserting values into the expression for entropy production

$$\frac{\dot{\sigma}_{cv}}{\dot{m}} = -\frac{(-22.6 \text{ kJ/kg})}{500 \text{ K}} + (7.3549 - 6.9212)\left(\frac{\text{kJ}}{\text{kg} \cdot \text{K}}\right)$$

$$= 0.0452 + 0.4337 = 0.4789 \text{ kJ/kg} \cdot \text{K}$$

Comment: If the boundary were located to include a portion of the immediate surroundings so heat transfer would take place at the temperature of the surroundings, say, $T_0 = 293$ K, the entropy production for the enlarged control volume would be 0.511 kJ/kg·K. It is left as an exercise to verify this value and to explain why the entropy production for the enlarged control volume would be greater than for a control volume consisting of the turbine only.

EXAMPLE 6.8

PROBLEM

Two alternative systems are under consideration for bringing a stream of air from 17 to 52°C at an essentially constant pressure of 1 bar.

System 1: The air temperature is increased as a consequence of the stirring of a liquid surrounding the line carrying the air.

System 2: The air temperature is increased by passing it through one side of a counterflow heat exchanger. On the other side, steam condenses at a pressure of 1 bar from saturated vapor to saturated liquid.

Both systems operate at steady state. All kinetic and potential energy effects can be ignored and no significant heat transfer with the surroundings occurs. For each of the two systems, calculate the rate of entropy production, in kJ/K per kg of air passing through the system.

SOLUTION

Known: Two alternative systems operating at steady state increase the temperature of an air stream with no change in its pressure.

Find: For each system, calculate the rate of entropy production per kg of air passing through the system.

Schematic and Given Data:

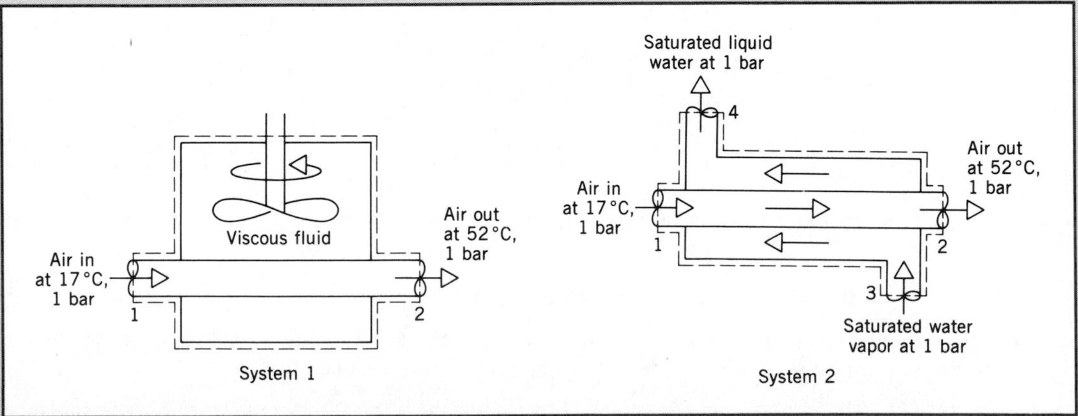

Figure E6.8

Assumptions:

1. Each of the two control volumes shown on the accompanying figure is at steady state.
2. There are no heat transfers with the surroundings. For system 2, $\dot{W}_{cv} = 0$.
3. All kinetic and potential energy effects can be ignored.
4. The air is modeled as an ideal gas.
5. The pressure of the condensing steam remains constant. The pressure of the air also remains constant.

Analysis:

System 1

At steady state, the mass rate balance reduces to $\dot{m}_1 = \dot{m}_2 = \dot{m}$. The entropy rate balance reduces to

$$0 = \sum_j \frac{\dot{Q}_j}{T_j}^{0} + \dot{m}_1 s_1 - \dot{m}_2 s_2 + \dot{\sigma}_{cv}$$

Combining these results and solving

$$\frac{\dot{\sigma}_{cv}}{\dot{m}} = s_2 - s_1$$

The change in specific entropy of the air between inlet and exit can be determined using

$$s_2 - s_1 = s^\circ(T_2) - s^\circ(T_1) - R \ln(p_2/p_1)^{0}$$

where the last term drops out by assumption 5. Selecting values for s° from Table A-16 at $T_1 = 290$ K and $T_2 = 325$ K

$$s_2 - s_1 = (1.78249 - 1.66802) \text{ kJ/kg} \cdot \text{K}$$

Thus

$$\frac{\dot{\sigma}_{cv}}{\dot{m}} = 0.1145 \text{ kJ/kg} \cdot \text{K}$$

System 2

Since the air and water streams are separate, conservation of mass at steady state gives $\dot{m}_1 = \dot{m}_2$ and $\dot{m}_3 = \dot{m}_4$. The entropy rate balance reduces at steady state with assumption 2 to

$$0 = \sum_j \frac{\dot{Q}_j}{T_j}^{0} + \dot{m}_1 s_1 + \dot{m}_3 s_3 - \dot{m}_2 s_2 - \dot{m}_4 s_4 + \dot{\sigma}_{cv}$$

Combining mass and entropy rate balances

$$0 = \dot{m}_1(s_1 - s_2) + \dot{m}_3(s_3 - s_4) + \dot{\sigma}_{cv}$$

and solving

$$\frac{\dot{\sigma}_{cv}}{\dot{m}_1} = (s_2 - s_1) + \left(\frac{\dot{m}_3}{\dot{m}_1}\right)(s_4 - s_3)$$

The ratio of mass flow rates, $\dot{m}_3/\dot{m}_1$, can be determined by means of an energy rate balance. At steady state

$$0 = \dot{Q}_{cv}^{0} - \dot{W}_{cv}^{0} + \dot{m}_1 h_1 + \dot{m}_3 h_3 - \dot{m}_2 h_2 - \dot{m}_4 h_4$$

where the heat transfer and work terms drop out by assumption 2, and the kinetic and potential energy terms are omitted in accordance with assumption 3. With $\dot{m}_1 = \dot{m}_2$ and $\dot{m}_3 = \dot{m}_4$, this reduces to give

$$\frac{\dot{m}_3}{\dot{m}_1} = \frac{h_2 - h_1}{h_3 - h_4}$$

From Table A-16, $h_1 = 290.16$ kJ/kg and $h_2 = 325.31$ kJ/kg. From Table A-3, $h_3 = 2675.5$ kJ/kg and $h_4 = 417.46$ kJ/kg. Thus

$$\frac{\dot{m}_3}{\dot{m}_1} = \frac{325.31 - 290.16}{2675.5 - 417.46} = 0.0156$$

From Table A-3 at 1 bar, $s_4 = 1.3026$ kJ/kg·K and $s_3 = 7.3594$ kJ/kg·K. Thus, $s_4 - s_3 = -6.0568$ kJ/kg·K. The value of $s_2 - s_1$ is the same as that determined for system 1. By inserting values into the expression for entropy production

$$\frac{\dot{\sigma}_{cv}}{\dot{m}_1} = 0.1145 + (0.0156)(-6.0568) = 0.02 \text{ kJ/kg·K}$$

Comments:

1. The rate of entropy production in system 1 is considerably greater than in system 2. Of the two alternatives, system 2 is better *thermodynamically*, as might have been anticipated.
2. Economic considerations normally play a decisive role in practice when deciding among alternatives.

6.7 ISENTROPIC PROCESSES

The term *isentropic* means constant entropy. Isentropic processes are encountered in many subsequent parts of this book. The object of the present section is to explain how properties are evaluated at any two states of a process in which there is no change in specific entropy.

6.7.1 Use of Tabular and Graphical Property Data

As illustrated by Fig. 6.9, temperature–entropy and enthalpy–entropy diagrams are particularly convenient for determining properties at states having the same value of specific entropy. All states on a vertical line passing through a given state have the same entropy. If state 1 on Fig. 6.9 is fixed by pressure p_1 and temperature T_1, states 2 and 3 are readily located once one additional property, such as pressure or temperature, is specified. The values of several other properties at states 2 and 3 can then be read directly from the figures. Illustrations of this procedure are provided in Sec. 6.3.2.

Tabular data also can be used to relate two states having the same specific entropy. For the case shown in Fig. 6.9, the specific entropy at state 1 could be determined from the superheated vapor table. Then, with $s_2 = s_1$ and one other property value, such as p_2 or T_2, state 2 could be located in the superheated vapor table. The values

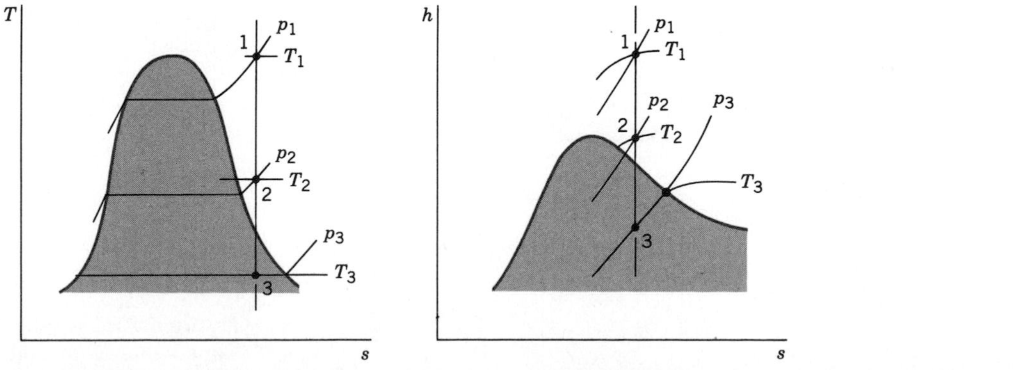

Figure 6.9 T–s and h–s diagrams showing states having the same value of specific entropy.

of the properties v, u, and h at state 2 can then be read from the table. An illustration of this procedure is given in Sec. 6.3.1. Note that state 3 falls in the two-phase liquid–vapor regions of Fig. 6.9. Since $s_3 = s_1$, the quality at state 3 could be determined using Eq. 6.9. With the quality known, other properties such as v, u, and h could then be evaluated.

6.7.2 Use of the Ideal Gas Model

Figure 6.10 shows two states of an ideal gas having the same value of specific entropy. Let us consider the relations among pressure, specific volume, and temperature at these states.

For two states having the same specific entropy, Eq. 6.25a reduces to

$$0 = s^\circ(T_2) - s^\circ(T_1) - R \ln \frac{p_2}{p_1} \tag{6.47}$$

Equation 6.47 involves four property values: p_1, T_1, p_2, and T_2. If any three are known, the fourth can be determined. If, for example, the temperature at state 1 and the pressure ratio p_2/p_1 are known, the temperature at state 2 can be determined from

$$s^\circ(T_2) = s^\circ(T_1) + R \ln \frac{p_2}{p_1}$$

Since T_1 is known, $s^\circ(T_1)$ would be obtained from the appropriate table, the value of $s^\circ(T_2)$ would be calculated, and then temperature T_2 would be determined by interpolation. If p_1, T_1, and T_2 are specified and the pressure at state 2 is the unknown, Eq. 6.47 would be solved to obtain

$$p_2 = p_1 \exp\left[\frac{s^\circ(T_2) - s^\circ(T_1)}{R}\right]$$

The above equation serves as the basis of an alternative tabular approach for relating the temperatures and pressures at two states of an ideal gas having the same specific entropy. To introduce this, rewrite the equation as

$$\frac{p_2}{p_1} = \frac{\exp[s^\circ(T_2)/R]}{\exp[s^\circ(T_1)/R]}$$

The quantity $\exp[s^\circ(T)/R]$ appearing in this expression is solely a function of temperature. It is commonly called the *relative pressure* and is given the symbol $p_r(T)$. A

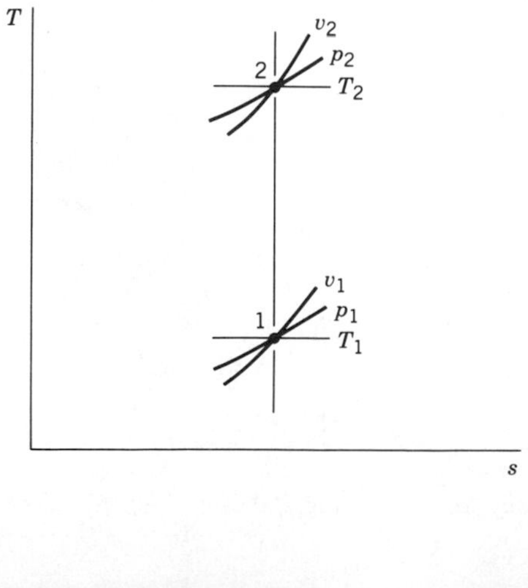

Figure 6.10 Two states of an ideal gas having the same value of specific entropy.

tabulation of p_r versus temperature for *air* is provided in Tables A-16.[1] In terms of relative pressures, the last equation becomes

$$\frac{p_2}{p_1} = \frac{p_{r2}}{p_{r1}} \qquad (s_1 = s_2) \tag{6.48}$$

where $p_{r1} = p_r(T_1)$ and $p_{r2} = p_r(T_2)$. Observe that p_r is not truly a pressure, so the name relative pressure is misleading. Also, be careful not to confuse p_r with the reduced pressure of the compressibility diagram.

A relation between specific volumes and temperatures for two states of an ideal gas having the same specific entropy can also be developed. With the ideal gas equation of state, $v = RT/p$, the ratio of the specific volumes is

$$\frac{v_2}{v_1} = \left(\frac{RT_2}{p_2}\right)\left(\frac{p_1}{RT_1}\right)$$

Then, since the two states have the same specific entropy, Eq. 6.48 can be introduced to give

$$\frac{v_2}{v_1} = \left[\frac{RT_2}{p_r(T_2)}\right]\left[\frac{p_r(T_1)}{RT_1}\right]$$

The ratio $RT/p_r(T)$ appearing on the right side of the last equation is solely a function of temperature. It is commonly called the *relative volume* and is given the symbol $v_r(T)$. Values of v_r are tabulated versus temperature in this text for *air* in Tables A-16. In terms of relative volumes, the last equation becomes

$$\frac{v_2}{v_1} = \frac{v_{r2}}{v_{r1}} \qquad (s_1 = s_2) \tag{6.49}$$

where $v_{r1} = v_r(T_1)$ and $v_{r2} = v_r(T_2)$. Despite the name given to it, $v_r(T)$ is not truly a volume. Also, be careful not to confuse it with the pseudoreduced specific volume of the compressibility diagram.

Let us consider next how properties are related for isentropic processes of an ideal gas when the specific heats are constants. For any such case, Eqs. 6.26 and 6.27 reduce to the equations

$$0 = c_p \ln\frac{T_2}{T_1} - R\ln\frac{p_2}{p_1}$$

$$0 = c_v \ln\frac{T_2}{T_1} + R\ln\frac{v_2}{v_1}$$

Introducing the ideal gas relations

$$c_p = \frac{kR}{k-1}, \qquad c_v = \frac{R}{k-1} \tag{3.47}$$

these equations can be solved, respectively, to give

$$\frac{T_2}{T_1} = \left(\frac{p_2}{p_1}\right)^{(k-1)/k} \qquad (s_1 = s_2,\ \text{constant } k) \tag{6.50}$$

$$\frac{T_2}{T_1} = \left(\frac{v_1}{v_2}\right)^{k-1} \qquad (s_1 = s_2,\ \text{constant } k) \tag{6.51}$$

[1] The values of the relative pressure determined with this definition are inconveniently large, so they are divided by a scale factor before tabulating to give a convenient range of numbers.

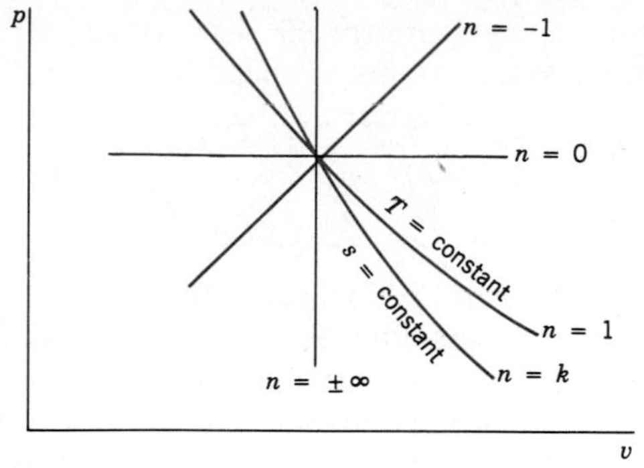

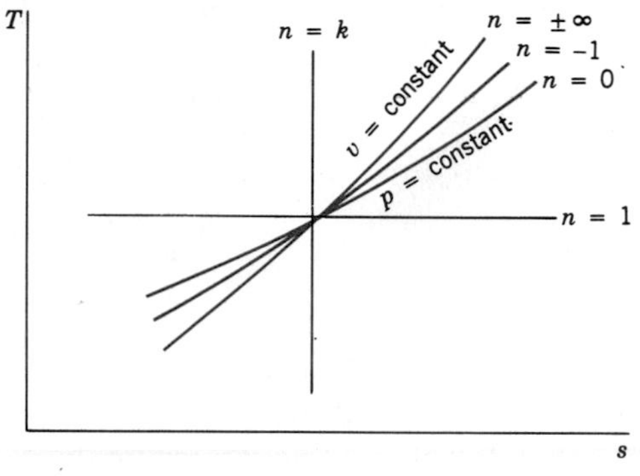

Figure 6.11 Polytropic processes on p–v and T–s diagrams.

When the specific heat ratio k is constant, Eqs. 6.50 and 6.51 are the counterparts, respectively, of Eqs. 6.48 and 6.49 for the case of variable specific heats.

The following relation can be obtained by eliminating the temperature ratio from Eqs. 6.50 and 6.51:

$$\frac{p_2}{p_1} = \left(\frac{v_1}{v_2}\right)^k \qquad (s_1 = s_2,\ \text{constant } k) \tag{6.52}$$

From the form of Eq. 6.52, it can be concluded that a polytropic process pv^k = constant of an ideal gas with constant k is an isentropic process. We noted in Sec. 3.5.4 that a polytropic process of an ideal gas for which $n = 1$ is an isothermal (constant-temperature) process. For *any* fluid, $n = 0$ corresponds to an isobaric (constant-pressure) process and $n = \pm\infty$ corresponds to an isometric (constant-volume) process. Polytropic processes corresponding to these values of n are shown in Fig. 6.11 on p–v and T–s diagrams.

EXAMPLE 6.9

PROBLEM

Air undergoes an isentropic process from p_1 = 1 atm, T_1 = 540°R to a final state where the temperature is T_2 = 1160°R. Employing the ideal gas model, determine the final pressure p_2, in atm. Solve using (a) p_r data from Table A-16E and (b) a constant specific heat ratio k evaluated at the mean temperature, 850°R, from Table A-14E.

SOLUTION

Known: Air undergoes an isentropic process from a state where pressure and temperature are known to a state where the temperature is specified.

Find: Determine the final pressure using (a) p_r data and (b) a constant value for the specific heat ratio k.

Schematic and Given Data:

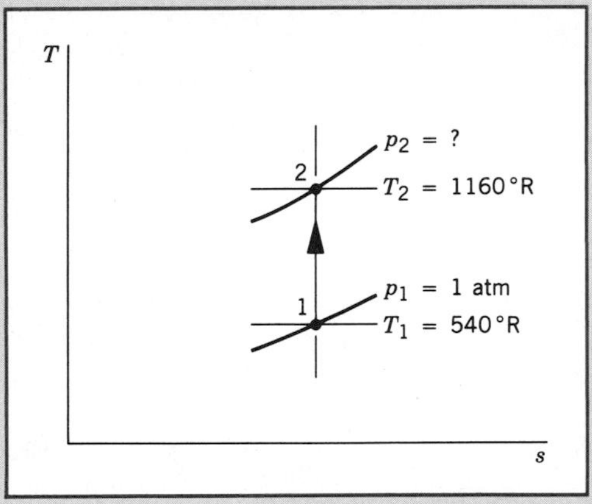

Figure E6.9

Assumptions:

1. A quantity of air as the system undergoes an isentropic process.

2. The air can be modeled as an ideal gas.

3. In part (b) the specific heat ratio is constant.

Analysis:

(a) The pressures and temperatures at two states of an ideal gas having the same specific entropy are related by Eq. 6.48

$$\frac{p_2}{p_1} = \frac{p_{r2}}{p_{r1}}$$

Solving

$$p_2 = p_1 \frac{p_{r2}}{p_{r1}}$$

With p_r values from Table A-16E

$$p = (1 \text{ atm}) \frac{21.18}{1.3860} = 15.28 \text{ atm}$$

(b) When the specific heat ratio k is assumed constant, the temperatures and pressures at two states of an ideal gas having the same specific entropy are related by Eq. 6.50. Thus

$$p_2 = p_1 \left(\frac{T_2}{T_1}\right)^{k/(k-1)}$$

From Table A-14E at 390°F (850°R), $k = 1.39$. Inserting values into the above expression

$$p_2 = (1 \text{ atm}) \left(\frac{1160}{540}\right)^{1.39/0.39} = 15.26 \text{ atm}$$

Comment: The close agreement between the answers obtained in parts (a) and (b) is attributable to the use of an appropriate value for the specific heat ratio k.

EXAMPLE 6.10

PROBLEM

A rigid, well-insulated tank is filled initially with 5.0 kg of air at a pressure of 5 bars and a temperature of 500 K. A leak develops, and air slowly escapes until the pressure of the air remaining in the tank is 1 bar. Employing the ideal gas model, determine the amount of mass remaining in the tank and its temperature.

SOLUTION

Known: A leak develops in a rigid, insulated tank initially containing air at a known state. Air slowly escapes until the pressure in the tank is reduced to a specified value.

Find: Determine the amount of mass remaining in the tank and its temperature.

Schematic and Given Data:

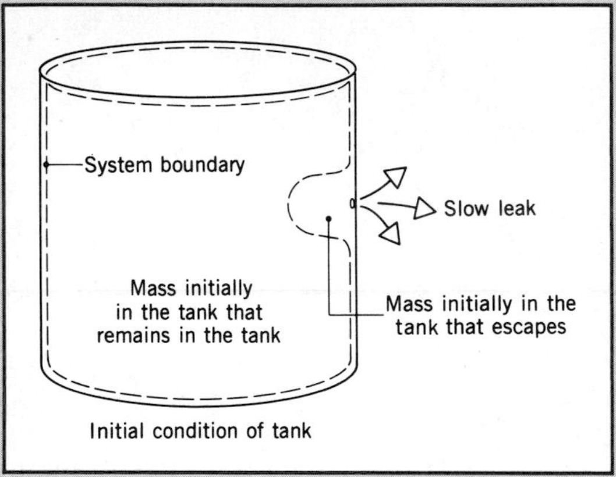

Figure E6.10

Assumptions:

1. As shown on the accompanying sketch, the closed system is the mass initially in the tank that remains in the tank.
2. There is no significant heat transfer between the system and its surroundings.
3. Irreversibilities within the tank can be ignored as the air slowly escapes.
4. The air is modeled as an ideal gas.

Analysis: With the ideal gas equation of state, the mass remaining in the tank at the end of the process is

$$m_2 = \frac{p_2 V}{(\bar{R}/M)T_2}$$

where p_2 and T_2 are the final pressure and temperature, respectively. The initial amount of mass within the tank, m_1, is

$$m_1 = \frac{p_1 V}{(\bar{R}/M)T_1}$$

where p_1 and T_1 are the initial pressure and temperature, respectively. Eliminating volume between these two expressions gives

$$m_2 = \left(\frac{p_2}{p_1}\right)\left(\frac{T_1}{T_2}\right) m_1$$

Except for the final temperature of the air remaining in the tank, T_2, all required values are known. The remainder of the solution mainly concerns the evaluation of T_2.

For the closed system under consideration, there are no significant irreversibilities (assumption 3), and no heat transfer occurs (assumption 2). For this system, the entropy balance reduces to

$$\Delta S = \int_1^2 \left(\frac{\delta Q}{T}\right)_b^{\,0} + \sigma^{0} = 0$$

Since the system mass remains constant, $\Delta S = m\,\Delta s$, so

$$\Delta s = 0$$

That is, the initial and final states of the system have the same value of *specific* entropy.

Using Eq. 6.48

$$p_{r2} = \left(\frac{p_2}{p_1}\right) p_{r1}$$

where $p_1 = 5$ bars and $p_2 = 1$ bar. With $p_{r1} = 8.411$ from Table A-16 at 500 K, the previous equation gives $p_{r2} = 1.6822$. Using this to interpolate in Table A-16, $T_2 = 317$ K.

Finally, inserting values into the expression for system mass

$$m_2 = \left(\frac{1 \text{ bar}}{5 \text{ bars}}\right)\left(\frac{500 \text{ K}}{317 \text{ K}}\right)(5.0 \text{ kg}) = 1.58 \text{ kg}$$

Comment: This problem also could be solved by considering a control volume enclosing the tank. The state of the control volume would change with time as air escapes. The details of the analysis are left as an exercise.

6.8 ISENTROPIC EFFICIENCIES OF TURBINES, NOZZLES, COMPRESSORS, AND PUMPS

Engineers make frequent use of efficiencies and many different efficiency definitions are employed. In the present section *isentropic* efficiencies for turbines, nozzles, compressors, and pumps are introduced. These efficiencies are frequently used in subsequent sections of the book.

Isentropic efficiencies involve a comparison between the actual performance of a device and the performance that would be achieved under idealized circumstances for the same inlet state and the same exit pressure. To consider this for a turbine, refer to Fig. 6.12, which shows a turbine expansion on a Mollier diagram. The state

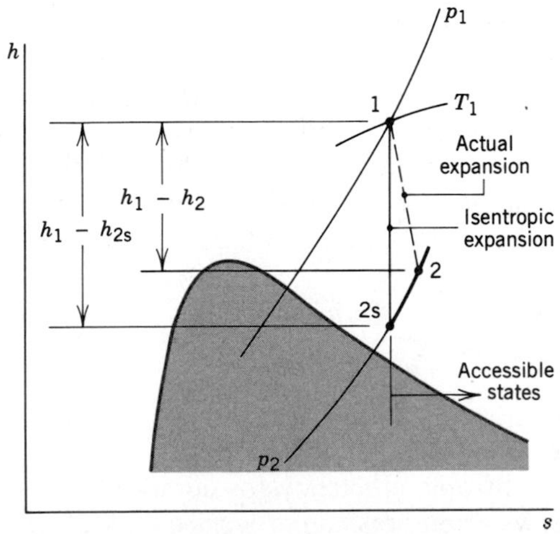

Figure 6.12 Comparison of actual and isentropic expansions through a turbine.

of the matter entering the turbine and the exit pressure are fixed. Heat transfer between the turbine and its surroundings is ignored, as are kinetic and potential energy effects. With these assumptions, the mass and energy rate balances reduce, at steady state, to give the work developed per unit of mass flowing through the turbine

$$\frac{\dot{W}_{cv}}{\dot{m}} = h_1 - h_2$$

Since state 1 is fixed, the specific enthalpy h_1 is known. Accordingly, the value of the work depends on the specific enthalpy h_2 only, and increases as h_2 is reduced. The *maximum* value for the turbine work corresponds to the smallest *allowed* value for the specific enthalpy at the turbine exit. This can be determined using the second law. The allowed exit states are constrained by

$$\frac{\dot{\sigma}_{cv}}{\dot{m}} = s_2 - s_1 \geq 0$$

which is obtained by reduction of the entropy rate balance. Because the entropy production $\dot{\sigma}_{cv}/\dot{m}$ cannot be negative, states with $s_2 < s_1$ are not accessible in an adiabatic expansion. The only states that can be attained in an actual expansion are those with $s_2 > s_1$. The state labeled "2s" on Fig. 6.12 would be attained only in the limit of no internal irreversibilities. This corresponds to an isentropic expansion through the turbine. For fixed exit pressure, the specific enthalpy h_2 decreases as the specific entropy s_2 decreases. Therefore, the *smallest allowed* value for h_2 corresponds to state 2s, and the *maximum* value for the turbine work is

$$\left(\frac{\dot{W}_{cv}}{\dot{m}}\right)_s = h_1 - h_{2s}$$

In an actual expansion through the turbine $h_2 > h_{2s}$, and thus less work than the maximum would be developed. This difference can be gauged by the *isentropic turbine efficiency* defined by

$$\eta_t = \frac{\dot{W}_{cv}/\dot{m}}{(\dot{W}_{cv}/\dot{m})_s} \tag{6.53}$$

Both the numerator and denominator of this expression are evaluated for the same inlet state and the same exit pressure. The value of η_t is typically 0.7 to 0.9 (70–90%).

A similar approach to that for turbines can be used to introduce the isentropic efficiency of nozzles operating at steady state. The *isentropic nozzle efficiency* is defined as the ratio of the actual specific kinetic energy of the gas leaving the nozzle, $V_2^2/2$, to the kinetic energy at the exit that would be achieved in an isentropic expansion between the same inlet state and the same exhaust pressure, $(V_2^2/2)_s$:

$$\eta_{nozzle} = \frac{V_2^2/2}{(V_2^2/2)_s} \tag{6.54}$$

Nozzle efficiencies of 95% or more are common, indicating that well-designed nozzles are nearly free of internal irreversibilities.

The form of the isentropic efficiency for compressors is taken up next. Refer to Fig. 6.13, which shows a compression process on a Mollier diagram. The state of the matter entering the compressor and the exit pressure are fixed. For negligible heat

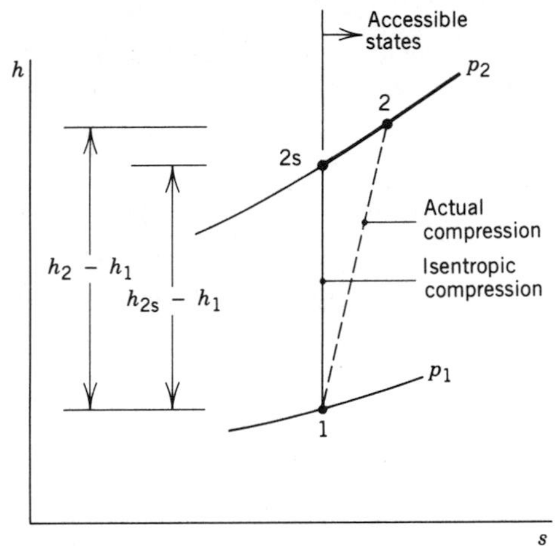

Figure 6.13 Comparison of actual and isentropic compressions.

transfer with the surroundings and no appreciable kinetic and potential energy effects, the work *input* per unit of mass flowing through the compressor is

$$\frac{\dot{W}_{cv}}{\dot{m}} = h_2 - h_1$$

Since state 1 is fixed, the specific enthalpy h_1 is known. Accordingly, the value of the work input depends on the specific enthalpy at the exit, h_2. The above expression shows that the magnitude of the work input decreases as h_2 decreases. The *minimum* work input corresponds to the smallest *allowed* value for the specific enthalpy at the compressor exit. With similar reasoning as for the turbine, this is the enthalpy at the exit state that would be achieved in an isentropic compression from the specified inlet state to the specified exit pressure. The minimum work input is given, therefore, by

$$\left(-\frac{\dot{W}_{cv}}{\dot{m}}\right)_s = h_{2s} - h_1$$

In an actual compression, $h_2 > h_{2s}$, and thus more work than the minimum would be required. This difference can be gauged by the *isentropic compressor efficiency* defined by

$$\eta_c = \frac{(\dot{W}_{cv}/\dot{m})_s}{\dot{W}_{cv}/\dot{m}} \tag{6.55}$$

Both the numerator and denominator of this expression are evaluated for the same inlet state and the same exit pressure. The value of η_c is typically 75 to 85% for compressors. An *isentropic pump efficiency*, η_p, is defined similarly.

EXAMPLE 6.11

PROBLEM

A steam turbine operates at steady state with inlet conditions of $p_1 = 5$ bars, $T_1 = 320°C$. Steam leaves the turbine at a pressure of 1 bar. There is no significant heat transfer between the turbine and its surroundings, and kinetic and potential energy changes between inlet and exit are negligible. If the isentropic turbine efficiency is 75%, determine the work developed per unit mass of steam flowing through the turbine, in kJ/kg.

SOLUTION

Known: Steam expands through a turbine operating at steady state from a specified inlet state to a specified exit pressure. The turbine efficiency is known.

Find: Determine the work developed per unit mass of steam flowing through the turbine.

Schematic and Given Data:

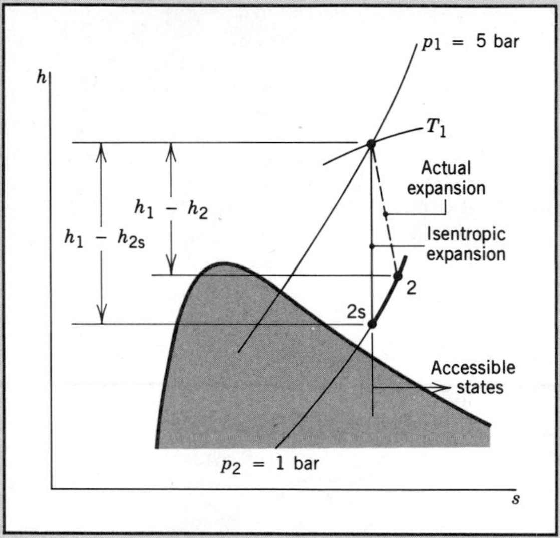

Figure E6.11

Assumptions:

1. A control volume enclosing the turbine is at steady state.
2. The expansion is adiabatic and changes in kinetic and potential energy between the inlet and exit can be neglected.

Analysis: The work developed can be determined using the isentropic turbine efficiency, Eq. 6.53, which on rearrangement gives

$$\frac{\dot{W}_{cv}}{\dot{m}} = \eta_t \left(\frac{\dot{W}_{cv}}{\dot{m}}\right)_s = \eta_t(h_1 - h_{2s})$$

From Table A-4, $h_1 = 3105.6$ kJ/kg and $s_1 = 7.5308$ kJ/kg K. The exit state for an isentropic expansion is fixed by $p_2 = 1$ bar and $s_{2s} = s_1$. Interpolating with specific entropy in Table A-4 at 1 bar gives $h_{2s} = 2743.0$ kJ/kg. Substituting values

$$\frac{\dot{W}_{cv}}{\dot{m}} = 0.75(3105.6 - 2743.0) = 271.95 \text{ kJ/kg}$$

Comment: The effect of irreversibilities is to exact a penalty on the work output of the turbine: The work is only 75% of what it would be for an isentropic expansion between the given inlet state and the turbine exhaust pressure. This is clearly illustrated in terms of enthalpy differences on the accompanying h–s diagram.

EXAMPLE 6.12

PROBLEM

A turbine operating at steady state receives air at a pressure of $p_1 = 3.0$ bars and a temperature of $T_1 = 390$ K. Air exits the turbine at a pressure of $p_2 = 1.0$ bar. The work developed is measured as 74 kJ per kg of air flowing through the turbine. The turbine operates adiabatically, and changes in kinetic and potential energy between inlet and exit can be neglected. Using the ideal gas model for air, determine the turbine efficiency.

SOLUTION

Known: Air expands through a turbine at steady state from a specified inlet state to a specified exit pressure. The work developed per kg of air flowing through the turbine is known.

Find: Determine the turbine efficiency.

Schematic and Given Data:

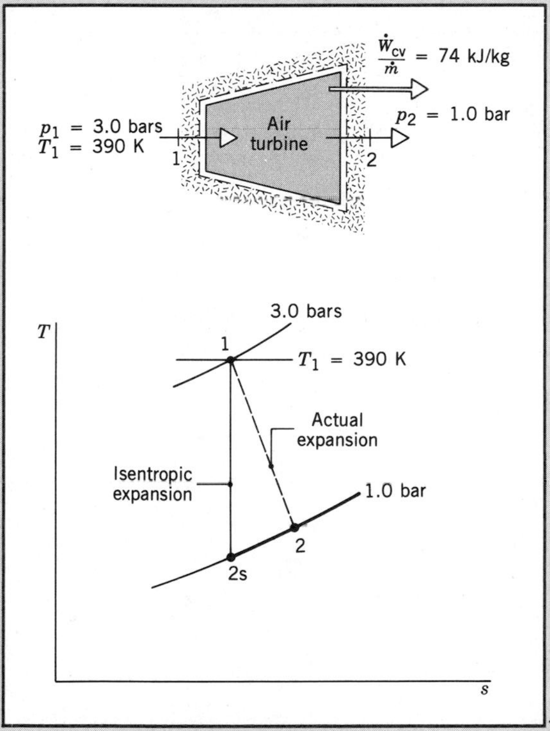

Figure E6.12

Assumptions:

1. The control volume shown on the accompanying sketch is at steady state.
2. The expansion is adiabatic and changes in kinetic and potential energy between inlet and exit can be neglected.
3. The air is modeled as an ideal gas.

Analysis: The numerator of the isentropic turbine efficiency, Eq. 6.53, is known. The denominator is evaluated as follows.

The work developed in an isentropic expansion from the given inlet state to the specified exit pressure is

$$\left(\frac{\dot{W}_{cv}}{\dot{m}}\right)_s = h_1 - h_{2s}$$

From Table A-16 at 390 K, $h_1 = 390.88$ kJ/kg. To determine h_{2s}, use Eq. 6.48

$$p_r(T_{2s}) = \left(\frac{p_2}{p_1}\right) p_r(T_1)$$

With $p_2 = 1.0$ bar, $p_1 = 3.0$ bars, and $p_{r1} = 3.481$ from Table A-16 at 390 K

$$p_r(T_{2s}) = \left(\frac{1.0}{3.0}\right)(3.481) = 1.1603$$

Interpolation in Table A-16 gives $h_{2s} = 285.27$ kJ/kg. Thus

$$\left(\frac{\dot{W}_{cv}}{\dot{m}}\right)_s = 390.88 - 285.27 = 105.6 \text{ kJ/kg}$$

Substituting values into Eq. 6.53

$$\eta_t = \frac{\dot{W}_{cv}/\dot{m}}{(\dot{W}_{cv}/\dot{m})_s} = \frac{74}{105.6} = 0.70 \; (70\%)$$

EXAMPLE 6.13

PROBLEM

Steam enters a nozzle operating at steady state at $p_1 = 140$ lbf/in.^{2} and $T_1 = 600$°F with a velocity of 100 ft/s. The pressure and temperature at the exit are $p_2 = 40$ lbf/in.^{2} and $T_2 = 350$°F. There is no significant heat transfer between the nozzle and its surroundings, and changes in potential energy between inlet and exit can be neglected. Determine the nozzle efficiency.

SOLUTION

Known: Steam expands through a nozzle at steady state from a specified inlet state to a specified exit state. The velocity at the inlet is known.

Find: Determine the nozzle efficiency.

Schematic and Given Data:

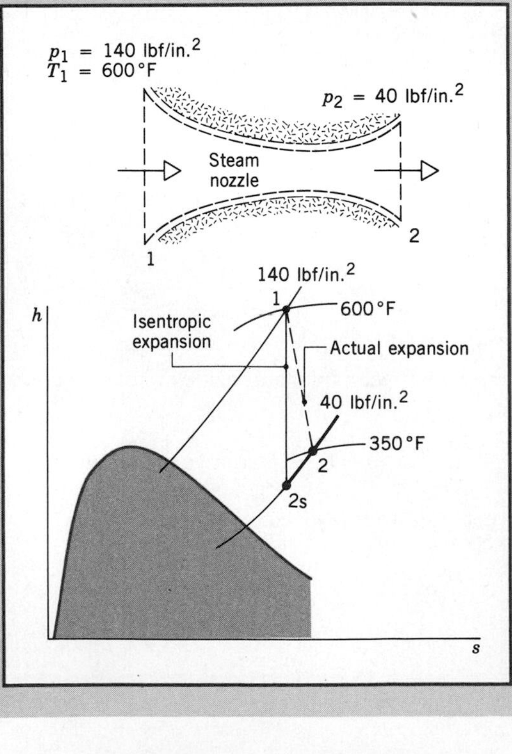

Figure E6.13

Assumptions:

1. The control volume shown on the accompanying sketch operates adiabatically at steady state.
2. For the control volume, $\dot{W}_{cv} = 0$ and the change in potential energy between inlet and exit can be neglected.

Analysis: The nozzle efficiency given by Eq. 6.54 requires the actual specific kinetic energy at the nozzle exit and the specific kinetic energy that would be achieved at the exit in an isentropic expansion from the given inlet state to the given exit pressure. The mass and energy rate balances for the one-inlet, one-exit control volume at steady state reduce to give

$$\frac{V_2^2}{2} = h_1 - h_2 + \frac{V_1^2}{2}$$

This equation applies for both the actual expansion and the isentropic expansion.

From Table A-4E at $T_1 = 600°F$ and $p_1 = 140$ lbf/in.^{2}, $h_1 = 1326.4$ Btu/lb, $s_1 = 1.7191$ Btu/lb·°R. Also, with $T_2 = 350°F$ and $p_2 = 40$ lbf/in.^{2}, $h_2 = 1211.8$ Btu/lb. Thus, the actual specific kinetic energy at the exit in Btu/lb is

$$\frac{V_2^2}{2} = 1326.4\,\frac{\text{Btu}}{\text{lb}} - 1211.8\,\frac{\text{Btu}}{\text{lb}} + \frac{(100\text{ ft/s})^2}{(2)\left(\dfrac{32.2\text{ lb}\cdot\text{ft/s}^2}{1\text{ lbf}}\right)\left(\dfrac{778\text{ ft}\cdot\text{lbf}}{1\text{ Btu}}\right)}$$

$$= 114.8\,\frac{\text{Btu}}{\text{lb}}$$

Interpolating in Table A-4E at 40 lbf/in.^{2}, with $s_{2s} = s_1 = 1.7191$ Btu/lb·°R, results in $h_{2s} = 1202.3$ Btu/lb. Accordingly, the specific kinetic energy at the exit for an isentropic expansion is

$$\left(\frac{V_2^2}{2}\right)_s = 1326.4 - 1202.3 + \frac{(100)^2}{(2)(32.2)(778)} = 124.3\text{ Btu/lb}$$

Substituting values into Eq. 6.54

$$\eta_{\text{nozzle}} = \frac{(V_2^2/2)}{(V_2^2/2)_s} = \frac{114.8}{124.3} = 0.924\ (92.4\%)$$

Comment: An important irreversibility in nozzles is friction between the flowing gas or liquid and the nozzle wall. The effect of friction is that a smaller exit kinetic energy, and thus a smaller exit velocity, is realized than would have been obtained in an isentropic expansion.

EXAMPLE 6.14

PROBLEM

An air compressor operating at steady state receives air at $p_1 = 0.95$ bar and $T_1 = 22°C$. The ratio of the pressure at the exit to the pressure at the inlet is 6. There is no significant heat transfer between the compressor and its surroundings, and kinetic and potential energy changes between the inlet and the exit are negligible. If the isentropic compressor efficiency is 82%, determine the temperature of the air at the compressor exit. Use the ideal gas model for air.

SOLUTION

Known: Air is compressed from a specified inlet state through a specified pressure ratio. The isentropic compressor efficiency is known.

Find: Determine the temperature of the air at the exit.

Schematic and Given Data:

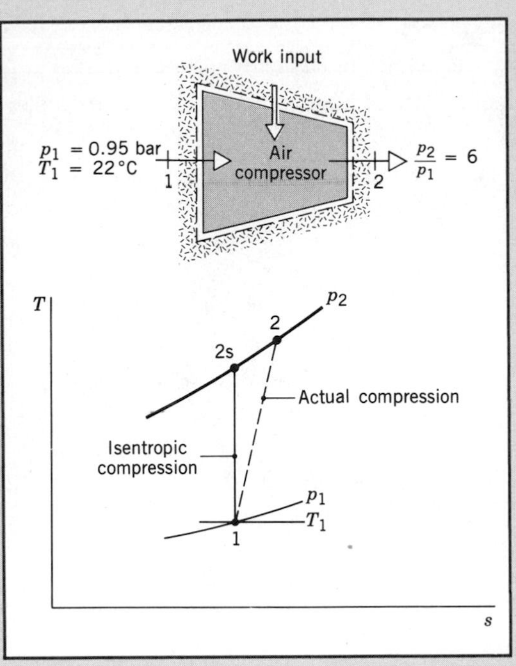

Figure E6.14

Assumptions:

1. The control volume shown on the accompanying sketch operates adiabatically at steady state.
2. The changes in kinetic and potential energy between inlet and exit can be neglected.
3. The air is modeled as an ideal gas.

Analysis: The temperature at the exit can be determined from the specific enthalpy at the exit. The exit specific enthalpy h_2 can be found using the isentropic compressor efficiency

$$\eta_c = \frac{h_{2s} - h_1}{h_2 - h_1}$$

Solving for h_2

$$h_2 = h_1 + \frac{(h_{2s} - h_1)}{\eta_c}$$

Table A-16 gives $h_1 = 295.17$ kJ/kg at $T_1 = 295$ K. To find h_{2s}, use Eq. 6.48.

$$p_r(T_{2s}) = \left(\frac{p_2}{p_1}\right) p_r(T_1) = (6)(1.3068) = 7.841$$

where p_{r1} is obtained from Table A-16 at 295 K. The specific enthalpy at state 2s is then found by interpolation in Table A-16: $h_{2s} = 493.0$ kJ/kg. Accordingly, the specific enthalpy at the exit is

$$h_2 = 295.17 + \frac{(493.0 - 295.17)}{0.82} = 536.4 \text{ kJ/kg}$$

Interpolating in Table A-16, we get $T_2 = 532$ K.

Comment: Because of irreversibilities, the compressor requires a greater work *input* than would be required in an isentropic compression to the specified exit pressure. As shown by the accompanying property diagram, the air exits at a temperature higher than would be achieved in an isentropic compression.

6.9 HEAT TRANSFER AND WORK IN INTERNALLY REVERSIBLE, STEADY-STATE FLOW PROCESSES

This section concerns one-inlet, one-exit control volumes at steady state. The objective is to derive expressions for the heat transfer and the work in the absence of internal irreversibilities. The resulting expressions have several important applications.

For a control volume at steady state in which the flow is both *isothermal* and *internally reversible,* the appropriate form of the entropy rate balance is

$$0 = \frac{\dot{Q}_{cv}}{T} + \dot{m}(s_1 - s_2) + \overset{0}{\cancel{\dot{\sigma}_{cv}}}$$

where 1 and 2 denote the inlet and exit, respectively, and $\dot{m}$ is the mass flow rate. Solving this equation, the heat transfer per unit of mass passing through the control volume is

$$\frac{\dot{Q}_{cv}}{\dot{m}} = T(s_2 - s_1)$$

More generally, the temperature would vary as the gas or liquid flows through the control volume. However, we can consider the temperature variation to consist of a series of infinitesimal steps. Then, the heat transfer per unit of mass would be given as

$$\left(\frac{\dot{Q}_{cv}}{\dot{m}}\right)_{\substack{\text{int}\\\text{rev}}} = \int_1^2 T\,ds \tag{6.56}$$

The subscript "int rev" serves to remind us that the expression applies only to control volumes in which there are no internal irreversibilities. The integral of Eq. 6.56 is performed from inlet to exit. When the states visited by a unit mass as it passes reversibly from inlet to exit are described by a curve on a T–s diagram, the magnitude of the heat transfer per unit of mass flowing can be represented as the area *under* the curve.

The work per unit of mass passing through the control volume can be found from an energy rate balance, which reduces at steady state to give

$$\frac{\dot{W}_{cv}}{\dot{m}} = \frac{\dot{Q}_{cv}}{\dot{m}} + (h_1 - h_2) + \left(\frac{V_1^2 - V_2^2}{2}\right) + g(z_1 - z_2) \tag{6.57}$$

This equation is a statement of the conservation of energy principle that applies when irreversibilities are present within the control volume as well as when they are absent. However, if consideration is restricted to the internally reversible case, Eq. 6.56 can be introduced to obtain

$$\left(\frac{\dot{W}_{cv}}{\dot{m}}\right)_{\substack{\text{int}\\\text{rev}}} = \int_1^2 T\,ds + (h_1 - h_2) + \left(\frac{V_1^2 - V_2^2}{2}\right) + g(z_1 - z_2) \tag{6.58}$$

where the subscript "int rev" has the same significance as before. Since internal irreversibilities are absent, a unit of mass traverses a sequence of equilibrium states as it passes from inlet to exit. Entropy, enthalpy, and pressure changes are therefore related by Eq. 6.17b

$$T\,ds = dh - v\,dp$$

which on integration gives

$$\int_1^2 T\,ds = (h_2 - h_1) - \int_1^2 v\,dp$$

Introducing this relation, Eq. 6.58 becomes

$$\left(\frac{\dot{W}_{cv}}{\dot{m}}\right)_{\substack{\text{int}\\\text{rev}}} = -\int_1^2 v\,dp + \left(\frac{V_1^2 - V_2^2}{2}\right) + g(z_1 - z_2) \tag{6.59}$$

When the states visited by a unit of mass as it passes reversibly from inlet to exit are described by a curve on a p–v diagram as shown in Fig. 6.14, the magnitude of the integral $\int v\,dp$ is represented by the shaded area *behind* the curve.

Equation 6.59 can also be applied to study the performance of control volumes at steady state in which $\dot{W}_{cv}$ is zero, as in the case of nozzles and diffusers. For any such case, the equation becomes

$$\int_1^2 v\,dp + \left(\frac{V_2^2 - V_1^2}{2}\right) + g(z_2 - z_1) = 0 \tag{6.60}$$

This is a form of the *Bernoulli* equation frequently used in fluid mechanics.

Equation 6.59 is often applied to devices such as turbines, compressors, and pumps. In many of these cases, there is no significant change in kinetic or potential energy from inlet to exit, so

$$\left(\frac{\dot{W}_{cv}}{\dot{m}}\right)_{\substack{\text{int}\\\text{rev}}} = -\int_1^2 v\,dp \qquad (\Delta\text{ke} = \Delta\text{pe} = 0) \tag{6.61}$$

This expression shows that the work is related to the magnitude of the specific volume of the gas or liquid as it flows from inlet to exit. To develop this idea further, consider two devices: a pump through which liquid water passes and a compressor through which water vapor passes. For the *same pressure rise,* the pump would require a much smaller work *input* per unit of mass flowing than would the compressor because the liquid specific volume is much smaller than that of vapor. This conclusion is qualitatively correct for actual pumps and compressors, where irreversibilities are present during operation.

Equation 6.61 is frequently used in one of several special forms. For example, if the specific volume remains approximately constant, as in many applications with liquids,

Figure 6.14 Area interpretation of $\int v\,dp$.

$$\left(\frac{\dot{W}_{cv}}{\dot{m}}\right)_{\substack{int\\rev}} = -v(p_2 - p_1) \qquad (v = \text{constant}, \Delta ke = \Delta pe = 0) \tag{6.62}$$

Additional special forms of Eq. 6.61 result when each unit of mass undergoes a *polytropic* process as it passes through the control volume. For such a process, the relationship between pressure and specific volume is

$$pv^n = \text{constant} \tag{6.63}$$

By introducing this into Eq. 6.61 and performing the integration

$$\left(\frac{\dot{W}_{cv}}{\dot{m}}\right)_{\substack{int\\rev}} = -\int_1^2 v\,dp = -(\text{constant})^{1/n}\int_1^2 \frac{dp}{p^{1/n}}$$

$$= -\frac{n}{n-1}(p_2v_2 - p_1v_1) \qquad (\text{polytropic}, n \neq 1) \tag{6.64}$$

for any value of n except $n = 1$. When $n = 1$, $pv =$ constant, and the work is

$$\left(\frac{\dot{W}_{cv}}{\dot{m}}\right)_{\substack{int\\rev}} = -\int_1^2 v\,dp = -\text{constant}\int_1^2 \frac{dp}{p}$$

$$= -(p_1v_1)\ln(p_2/p_1) \qquad (\text{polytropic}, n = 1) \tag{6.65}$$

Equations 6.64 and 6.65 apply generally to polytropic processes of *any* gas (or liquid). For the special case of an ideal gas, Eq. 6.64 becomes

$$\left(\frac{\dot{W}_{cv}}{\dot{m}}\right)_{\substack{int\\rev}} = -\frac{nR}{n-1}(T_2 - T_1) \qquad (\text{ideal gas}, n \neq 1) \tag{6.66a}$$

For a polytropic process of an ideal gas (Sec. 3.5.4)

$$\frac{T_2}{T_1} = \left(\frac{p_2}{p_1}\right)^{(n-1)/n}$$

Thus, Eq. 6.66a can be expressed alternatively as

$$\left(\frac{\dot{W}_{cv}}{\dot{m}}\right)_{\substack{int\\rev}} = -\frac{nRT_1}{n-1}\left[\left(\frac{p_2}{p_1}\right)^{(n-1)/n} - 1\right] \qquad (\text{ideal gas}, n \neq 1) \tag{6.66b}$$

For the case of an ideal gas, Eq. 6.65 becomes

$$\left(\frac{\dot{W}_{cv}}{\dot{m}}\right)_{\substack{int\\rev}} = -RT\ln(p_2/p_1) \qquad (\text{ideal gas}, n = 1) \tag{6.67}$$

EXAMPLE 6.15

PROBLEM

An air compressor operates at steady state with air entering at $p_1 = 1$ bar, $T_1 = 20°C$, and exiting at $p_2 = 5$ bars. Determine the work and heat transfer per unit of mass passing through the device, in kJ/kg, if the air undergoes a polytropic process with $n = 1.3$. Neglect changes in kinetic and potential energy between the inlet and the exit. Use the ideal gas model for air.

SOLUTION

Known: Air is compressed in a polytropic process from a specified inlet state to a specified exit pressure.

Find: Determine the work and heat transfer per unit of mass passing through the device.

Schematic and Given Data:

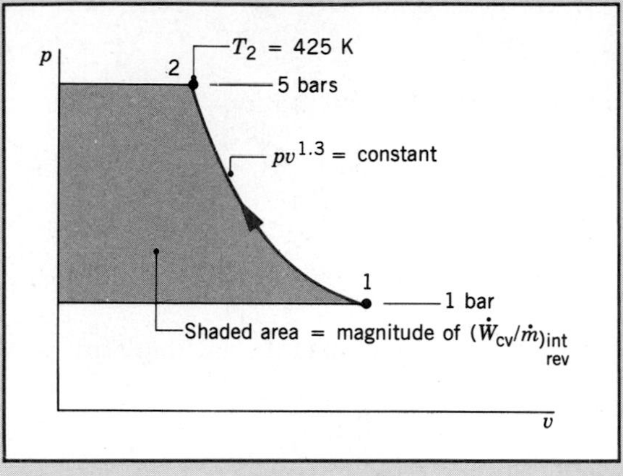

Figure E6.15

Assumptions:

1. A control volume enclosing the compressor is at steady state.
2. The air undergoes a polytropic process with $n = 1.3$.
3. The air behaves as an ideal gas.
4. Changes in kinetic and potential energy from inlet to exit can be neglected.

Analysis: The temperature at the exit can be found using

$$T_2 = T_1 \left(\frac{p_2}{p_1}\right)^{(n-1)/n} = 293 \left(\frac{5}{1}\right)^{(1.3-1)/1.3} = 425 \text{ K}$$

Substituting known values into Eq. 6.66a then gives

$$\frac{\dot{W}_{cv}}{\dot{m}} = -\frac{nR}{n-1}(T_2 - T_1) = -\frac{1.3}{1.3-1}\left(\frac{8.314}{28.97}\frac{\text{kJ}}{\text{kg}\cdot\text{K}}\right)(425 - 293)\text{ K}$$
$$= -164.2 \text{ kJ/kg}$$

The heat transfer can be evaluated by reducing the mass and energy rate balances with the appropriate assumptions to obtain

$$\frac{\dot{Q}_{cv}}{\dot{m}} = \frac{\dot{W}_{cv}}{\dot{m}} + h_2 - h_1$$

Using the temperatures T_1 and T_2, the required specific enthalpy values are obtained from Table A-16 as $h_1 = 293.17$ kJ/kg and $h_2 = 426.35$ kJ/kg. Thus

$$\frac{\dot{Q}_{cv}}{\dot{m}} = -164.15 + (426.35 - 293.17) = -31 \text{ kJ/kg}$$

Comment: The states visited in the polytropic compression process are shown by the curve on the accompanying p–v diagram. The magnitude of the work per unit of mass passing through the compressor is represented by the shaded area *behind* the curve.

PROBLEMS

6.1 A system executes a power cycle while receiving 2000 Btu at a temperature of 1000°R from a reservoir at 1000°R and discharging energy at a temperature of 500°R to a reservoir at 500°R. There are no other heat transfers. Determine σ_{cycle} if

(a) the cycle is reversible.
(b) the thermal efficiency is 25%.

6.2 A reversible power cycle R and an irreversible power cycle I operate between the same two reservoirs. Each receives Q_H from the hot reservoir. The reversible cycle develops work W_R, while the irreversible cycle develops work W_I.

(a) Evaluate σ_{cycle} for cycle I in terms of W_I, W_R, and temperature T_C of the cold reservoir only.
(b) Demonstrate that σ_{cycle} must be positive.

6.3 A reversible refrigeration cycle R and an irreversible refrigeration cycle I operate between the same two reservoirs and each removes Q_C from the cold reservoir. The net work input required by R is W_R, while the net work input for I is W_I. Show that $W_I > W_R$.

6.4 A reversible power cycle receives energy Q_1 and Q_2 from hot reservoirs at temperatures T_1 and T_2, respectively, and discharges energy Q_3 to a cold reservoir at temperature T_3.

(a) Obtain an expression for the thermal efficiency in terms of the ratios T_1/T_3, T_2/T_3, $q = Q_2/Q_1$.
(b) Discuss the result of part (a) in each of these limits: $\lim q \to 0$, $\lim q \to \infty$, $\lim T_1 \to \infty$.

6.5 Complete the following involving reversible and irreversible cycles:

(a) Reversible and irreversible *power* cycles each discharge energy Q_C to a cold reservoir at temperature T_C and receive energy Q_H from hot reservoirs at temperatures T_H and T'_H, respectively. There are no other heat transfers. Show that $T'_H > T_H$.
(b) Reversible and irreversible *refrigeration* cycles each discharge energy Q_H to a hot reservoir at temperature T_H and receive energy Q_C from cold reservoirs at temperatures T_C and T'_C, respectively. There are no other heat transfers. Show that $T'_C > T_C$.
(c) Reversible and irreversible *heat pump* cycles each receive energy Q_C from a cold reservoir at temperature T_C and discharge energy Q_H to hot reservoirs at temperatures T_H and T'_H, respectively. There are no other heat transfers. Show that $T'_H < T_H$.

6.6 The system shown schematically in Fig. P6.6 undergoes a cycle while receiving energy at the rate $\dot{Q}_0$ from the surroundings at temperature T_0, $\dot{Q}_s$ from a source at temperature T_s, and delivering energy at the rate $\dot{Q}_u$ at a *use* temperature T_u. There are no other energy transfers. For $T_s > T_u > T_0$, obtain an expression for the *maximum* theoretical value of $\dot{Q}_u$ in terms of $\dot{Q}_s$ and the temperatures T_s, T_u, and T_0.

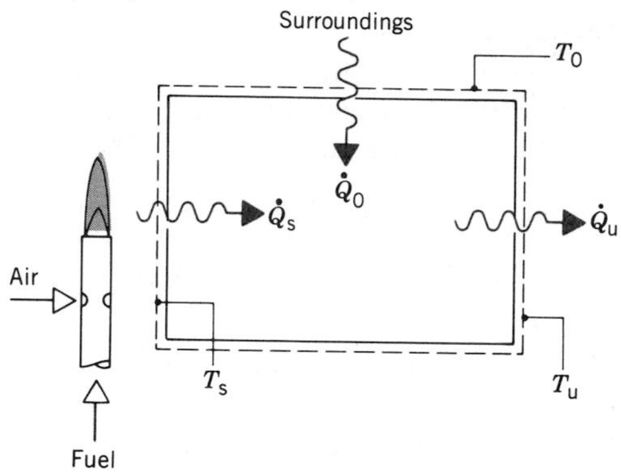

Figure P6.6

6.7 Determine the specific entropy, in kJ/kg·K, at the following states from the steam tables:

(a) water at p = 2.5 MPa, T = 400°C.
(b) water at p = 2.5 MPa, T = 200°C.
(c) water at p = 2.5 MPa, u = 1500 kJ/kg.

6.8 Determine the specific entropy, in Btu/lb·°R, at the following states from the steam tables:

(a) p = 500 lbf/in.², T = 575°F.
(b) p = 500 lbf/in.², T = 150°F.
(c) p = 500 lbf/in.², h = 827 Btu/lb.

6.9 Determine the change in specific entropy between the specified states, in kJ/kg·K.

(a) water, p_1 = 10 MPa, T_1 = 400°C, p_2 = 10 MPa, T_2 = 100°C.
(b) Refrigerant 12, h_1 = 84.8 kJ/kg, T_1 = −40°C, saturated vapor at p_2 = 5 bars.
(c) air as an ideal gas, T_1 = 7°C, p_1 = 2 bars, T_2 = 327°C, p_2 = 1 bar.
(d) hydrogen (H_2) as an ideal gas, T_1 = 727°C, p_1 = 1 bar, T_2 = 25°C, p_2 = 3 bars.

6.10 Determine the change in specific entropy between the specified states, in Btu/lb·°R.

(a) water, p_1 = 1000 lbf/in.², T_1 = 800°F, p_2 = 1000 lbf/in.², T_2 = 100°F.
(b) Refrigerant 12, h_1 = 36.46 Btu/lb, T_1 = −40°F, saturated vapor at p_2 = 40 lbf/in.²
(c) air as an ideal gas, T_1 = 40°F, p_1 = 2 atm, T_2 = 420°F, p_2 = 1 atm.
(d) carbon dioxide as an ideal gas, T_1 = 820°F, p_1 = 1 atm, T_2 = 77°F, p_2 = 3 atm.

6.11 One kilogram of Refrigerant 12 undergoes a process from 4 bars, 100°C to a state where the pressure is 1 bar. During the process there is a change in specific entropy, $s_2 - s_1 = -0.05$ kJ/kg·K. Determine the temperature at the final state, in °C.

6.12 One pound of water undergoes a process with no change in specific entropy from an initial state where $p_1 =$ 100 lbf/in.^{2}, $T_1 = 650$°F to a state where $p_2 = 5$ lbf/in.^{2}

(a) Using data from the steam tables, determine the quality at state 2, x_2.
(b) Determine x_2 using the Mollier diagram, Fig. A-8E.

6.13 Employing the ideal gas model, determine the change in specific entropy between the indicated states, in kJ/kmol·K. Solve two ways: Use the appropriate ideal gas table and use a constant specific heat value from Table A-14.

(a) air, $p_1 = 100$ kPa, $T_1 = 20$°C, $p_2 = 100$ kPa, $T_2 =$ 100°C.
(b) air, $p_1 = 1$ bar, $T_1 = 27$°C, $p_2 = 3$ bars, $T_2 = 377$°C.
(c) carbon dioxide, $p_1 = 150$ kPa, $T_1 = 30$°C, $p_2 = 300$ kPa, $T_2 = 300$°C.
(d) carbon monoxide, $T_1 = 300$ K, $v_1 = 1.1$ m^3/kg, $T_2 = 500$ K, $v_2 = 0.75$ m^3/kg.
(e) nitrogen, $p_1 = 2$ MPa, $T_1 = 800$ K, $p_2 = 1$ MPa, $T_2 = 300$ K.

6.14 Employing the ideal gas model, determine the change in specific entropy between the indicated states, in Btu/lbmol·°R. Solve two ways: Use the appropriate ideal gas table and use a constant specific heat value from Table A-14E.

(a) air, $p_1 = 1$ atm, $T_1 = 40$°F, $p_2 = 1$ atm, $T_2 = 400$°F.
(b) air, $p_1 = 20$ lbf/in.^{2}, $T_1 = 100$°F, $p_2 = 60$ lbf/in.^{2}, $T_2 = 300$°F.
(c) carbon dioxide, $p_1 = 1$ atm, $T_1 = 40$°F, $p_2 = 3$ atm, $T_2 = 500$°F.
(d) carbon dioxide, $T_1 = 200$°F, $v_1 = 20$ ft^3/lb, $T_2 =$ 400°F, $v_2 = 15$ ft^3/lb.
(e) nitrogen, $p_1 = 2$ atm, $T_1 = 800$°F, $p_2 = 1$ atm, $T_2 =$ 200°F.

6.15 In each of the following cases, determine the indicated property for a process in which there is no change in specific entropy between state 1 and state 2:

(a) water, $p_1 = 14.7$ lbf/in.^{2}, $T_1 = 500$°F, $p_2 = 100$ lbf/in.^{2} Find T_2 in °F.
(b) water, $T_1 = 10$°C, $x_1 = 0.75$, saturated vapor at state 2. Find p_2 in bars.
(c) air as an ideal gas, $T_1 = 27$°C, $p_1 = 1.5$ bars, $T_2 =$ 127°C. Find p_2 in bars.
(d) air as an ideal gas, $T_1 = 100$°F, $p_1 = 3$ atm, $p_2 = 2$ atm. Find T_2 in °F.
(e) Refrigerant 12, $T_1 = 100$°C, $p_1 = 4$ bars, $p_2 = 1$ bar. Find v_2 in m^3/kg.

6.16 One kilogram of oxygen (O_2) undergoes a process from 300 K, 2 bars to 1500 K, 1.5 bars. Determine the change in specific entropy, in kJ/kg·K, using

(a) Equation 6.23 with $\bar{c}_p(T)$ from Table A-15.
(b) Equation 6.25b with $\bar{s}^\circ$ from Table A-18.
(c) Equation 6.27 with c_p at 900 K from Table A-14.

6.17 Two kilograms of water undergo a process from an initial state where the pressure is 2.5 MPa and the temperature is 400°C to a final state of 2.5 MPa, 100°C. Determine the entropy change of the water, in kJ/K, assuming the process is

(a) irreversible.
(b) internally reversible.

6.18 Answer the following true or false. If false, explain why.

(a) The change of entropy of a closed system is the same for every process between two specified states.
(b) The entropy of a closed system cannot decrease.
(c) The entropy of a fixed amount of an ideal gas decreases in every isothermal compression.
(d) The specific internal energy, enthalpy, and entropy of an ideal gas are each functions of temperature alone.
(e) One of the $T\,ds$ equations has the form $T\,ds = dh - v\,dp$.
(f) The entropy of a fixed amount of an incompressible substance increases in every process in which temperature increases.

6.19 One-half kilogram of carbon dioxide gas (CO_2) undergoes a process from $p_1 = 1.5$ bars, $T_1 = 300$ K to $p_2 = 5$ bars, $T_2 = 370$ K. For the process $W = -33.07$ kJ. Employing the ideal gas model, determine

(a) the heat transfer, in kJ.
(b) the change in entropy, in kJ/K.
(c) Show the initial and final states on a T–s diagram.

6.20 Methane gas (CH_4) enters a compressor at 298 K, 1 bar and exits at 360 K, 2 bars. Employing the ideal gas model, determine the change in specific entropy from inlet to exit of the compressor, in kJ/kg·K.

6.21 A quantity of air undergoes a thermodynamic cycle consisting of three processes in series.

Process 1–2: constant-volume heating from $p_1 = 0.1$ MPa, $T_1 = 15$°C, $V_1 =$ 0.02 m^3 to $p_2 = 0.42$ MPa
Process 2–3: constant-pressure cooling
Process 3–1: isothermal heating to the initial state

Employing the ideal gas model with $c_p = 1$ kJ/kg·K, evaluate the change in entropy for each process. Sketch the cycle on p–v coordinates.

6.22 A closed system consists of an ideal gas with constant specific heat ratio k.

(a) The gas undergoes a process in which temperature increases from T_1 to T_2. Show that the entropy change for the process is greater if the change in state occurs at constant pressure than if it occurs at constant volume. Sketch the processes on p–v and T–s coordinates.

(b) The gas undergoes a process in which pressure increases from p_1 to p_2. Show that the ratio of the entropy change for an isothermal process to the entropy change for a constant-volume process is $(1 - k)$. Sketch the processes on p–v and T–s coordinates.

6.23 A quantity of liquid water undergoes a process from 80°C, 7.5 MPa to saturated liquid at 40°C. Determine the change in specific entropy, in kJ/kg·K, using

(a) Tables A-2 and A-5.
(b) saturated liquid data only from Table A-2.
(c) the incompressible liquid model with a constant specific heat from Table A-13.

6.24 One kilogram of liquid water initially saturated at 160°C expands to a final pressure of 1.5 bars in an isothermal internally reversible process. Determine the work and heat transfer, each in kJ. Sketch the process on p–v and T–s coordinates. Associate the work and heat transfer with areas on these diagrams.

6.25 Two kilograms of water initially at 160°C and 1.0 MPa undergo an isothermal internally reversible expansion during which 2700 kJ is received by heat transfer. Determine

(a) the final pressure, in MPa.
(b) the work, in kJ.

6.26 One pound of air initially at 100 lbf/in.², 500°F undergoes an internally reversible process to 45 lbf/in.² For the following cases, determine the heat transfer and the work, each in Btu, and show the process on p–v and T–s coordinates. The process is

(a) isothermal.
(b) adiabatic.
(c) constant volume.

6.27 A gas initially at 2.8 bars and 60°C is compressed to a final pressure of 14 bars in an isothermal internally reversible process. Determine the heat transfer and the work, each in kJ per kg of gas, if the gas is **(a)** Refrigerant 12, **(b)** air as an ideal gas. Sketch the process on p–v and T–s coordinates.

6.28 Reconsider the data of Problem 6.27, but now suppose the gas is compressed to 14 bars adiabatically without internal irreversibilities. Determine the work, in kJ per kg of gas, for **(a)** Refrigerant 12, **(b)** air as an ideal gas. Sketch the process on p–v and T–s coordinates.

6.29 One pound of argon initially at 15 lbf/in.², 80°F undergoes an internally reversible compression to 90 lbf/in.² Employing the ideal gas model, determine the work, in Btu, and the change in entropy, in Btu/°R, if the process is **(a)** isothermal, **(b)** polytropic with $n = 1.3$, **(c)** adiabatic. Sketch the processes on p–v and T–s coordinates.

6.30 Air initially occupying 1 m³ at 1.5 bars, 20°C undergoes an internally reversible compression during which pV^n = constant to a final state where the pressure is 6 bars and the temperature is 120°C. Determine

(a) the value of n.
(b) the work and heat transfer, each in kJ.
(c) the entropy change, in kJ/K.

6.31 Air initially occupying a volume of 1 m³ at 1 bar, 20°C undergoes two internally reversible processes in series

Process 1–2: compression to 5 bars during which $pV^{1.2}$ = constant
Process 2–3: adiabatic expansion to 1 bar.

(a) Sketch the two processes on p–v and T–s coordinates.
(b) Determine the temperatures at states 2 and 3, in °C.
(c) Determine the net work, in kJ.

6.32 Air initially at 14.7 lbf/in.², 70°F undergoes two internally reversible processes in series.

Process 1–2: isothermal compression to 90 lbf/in.²
Process 2–3: adiabatic expansion to 14.7 lbf/in.²

(a) Sketch the processes on p–v and T–s coordinates.
(b) Determine the temperature at state 3, in °F.
(c) Determine the net work, in Btu per lb.

6.33 The two cycles shown in Fig. P6.33 are composed of internally reversible processes. Determine an expression for the thermal efficiency of each cycle in terms of the temperatures T_1 and T_3.

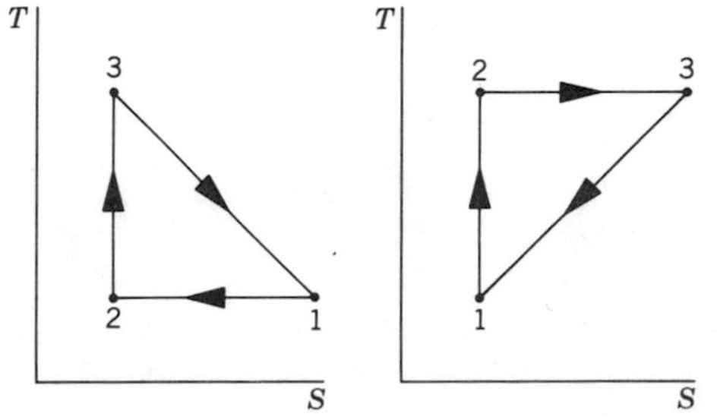

Figure P6.33

6.34 One kilogram of water executes a Carnot power cycle. At the beginning of the isothermal expansion, the temperature is 160°C and the volume is 0.1 m³. The isothermal expansion continues until the volume becomes

0.3 m^3. The temperature at the conclusion of the adiabatic expansion is 50°C.

(a) Sketch the cycle on *T–s* and *p–v* coordinates.
(b) Determine the heat added and net work, each in kJ.
(c) Evaluate the thermal efficiency.

6.35 A quantity of air undergoes a Carnot power cycle. At the beginning of the isothermal expansion, the temperature is 300°C and the pressure is 10 bars. At the beginning of the isothermal compression, the temperature is 100°C and the pressure is 0.1 bar. Employing the ideal gas model, determine

(a) the pressures at the end of the isothermal expansion and compression processes, each in bars.
(b) the heat added and the net work developed per cycle, each in kJ per kg of air.
(c) the thermal efficiency.

6.36 A quantity of air undergoes a thermodynamic cycle consisting of three internally reversible processes in series.

Process 1–2: isothermal compression from $p_1 = 1$ bar to $p_2 = 4.75$ bars
Process 2–3: constant pressure expansion to $T_3 = 390$ K
Process 3–1: adiabatic expansion to the initial state

Employing the ideal gas model,

(a) sketch the cycle on *p–v* and *T–s* coordinates.
(b) If the cycle is a power cycle, determine its thermal efficiency. If the cycle is a refrigeration cycle, determine its coefficient of performance.

6.37 A quantity of air undergoes a thermodynamic cycle consisting of three internally reversible processes in series.

Process 1–2: constant-pressure compression from $p_1 = 12$ lbf/in.^{2}, $T_1 = 80$°F
Process 2–3: constant-volume heat addition to 540°F
Process 3–1: adiabatic expansion to the initial state

Employing the ideal gas model,

(a) sketch the cycle on *p–v* and *T–s* coordinates.
(b) If the cycle is a power cycle, determine its thermal efficiency. If the cycle is a refrigeration cycle, determine its coefficient of performance.

6.38 Answer the following true or false. If false, explain why.

(a) One corollary of the second law of thermodynamics states that the change in entropy of a closed system must be greater than zero or equal to zero.
(b) When a net amount of energy is added by heat transfer to a closed system undergoing an internally reversible process, a net amount of work is also done.
(c) A process that violates the second law of thermodynamics violates the first law of thermodynamics.
(d) A closed system can experience a decrease in entropy only when there is energy transfer by heat from the system during the process.
(e) Entropy is produced in every actual process of a closed system.
(f) If there is no change in entropy between two states of a closed system, the process is necessarily adiabatic and internally reversible.
(g) The energy of an isolated system must remain constant, but the entropy can only decrease.

6.39 A closed system undergoes a process in which the work done by the system is 5 kJ and the heat transfer Q occurs only at temperature T_b. For each case, determine whether the entropy change of the system is positive, negative, zero, or indeterminate.

(a) internally reversible process, $Q = +10$ kJ.
(b) internally reversible process, $Q = 0$.
(c) internally reversible process, $Q = -10$ kJ.
(d) internal irreversibilities present, $Q = +10$ kJ.
(e) internal irreversibilities present, $Q = 0$.
(f) internal irreversibilities present, $Q = -10$ kJ.

6.40 For each of the following processes, specify whether the entropy change of the closed system is positive, negative, zero, or indeterminate:

(a) Two kilograms of water vapor undergoing an adiabatic process.
(b) One pound of nitrogen undergoing an internally reversible process.
(c) Three kilograms of Refrigerant 134a undergoing an adiabatic process during which it is stirred by a paddle wheel.
(d) One pound of carbon dioxide undergoing an isothermal process.
(e) Two pounds of oxygen modeled as an ideal gas undergoing a constant pressure process to a lower temperature.
(f) One kilogram of argon modeled as an ideal gas undergoing an isothermal process to a higher pressure.

6.41 A fixed mass of water m, initially a saturated liquid, is brought to a saturated vapor condition while its pressure and temperature remain constant.

(a) Derive expressions for the work and heat transfer in terms of the mass m and properties that can be obtained directly from the *steam tables*.
(b) Demonstrate that this process is internally reversible.

6.42 A quantity of air initially at 1 atm is shown in Fig. 6.8. Consider a process in which the temperature of the air increases from 40 to 600°F by some combination of stirring by the paddle wheel and heating from the reservoir that is at 540°F. Suggest how this might be done with

(a) minimum entropy production.
(b) maximum entropy production.

6.43 Taken together, a certain closed system and its surroundings make up an isolated system. Answer the following true or false. If false, explain why.

(a) No process is allowed in which the entropies of both the system and the surroundings increase.
(b) During a process, the entropy of the system might decrease, while the entropy of the surroundings increases, and conversely.
(c) No process is allowed in which the entropies of both the system and the surroundings remain unchanged.
(d) A process can occur in which the entropies of both the system and the surroundings decrease.

6.44 One pound of oxygen at 40°F and 100 lbf/in.^{2} expands isothermally to a final pressure of 14 lbf/in.^{2} while receiving energy by heat transfer from a reservoir at 80°F.

(a) For the oxygen as the system, evaluate the work and heat transfer, each in Btu, and the amount of entropy produced, in Btu/°R.
(b) Evaluate the entropy production for an enlarged system that includes the oxygen and a *thin* wall separating the oxygen from the reservoir. Compare with the entropy production of part (a) and comment on the difference.

6.45 A piston–cylinder assembly initially contains 0.04 m^3 of water at 1.0 MPa, 320°C. The water expands adiabatically to a final pressure of 0.15 MPa. Determine the maximum theoretical work that can be developed by the water in such an expansion, in kJ.

6.46 Two pounds of steam expand adiabatically from state 1, where p_1 = 100 lbf/in.^{2}, T_1 = 500°F, to state 2, where p_2 = 10 lbf/in.^{2}, s_2 = 1.7998 Btu/lb·°R.

(a) Calculate the work, in Btu.
(b) What is the maximum theoretical work for an expansion from the same initial state to the same final pressure, in Btu?

6.47 An insulated piston–cylinder assembly contains Refrigerant 12, initially occupying 3 ft^3 at 90 lbf/in.^{2}, 100°F. The refrigerant expands to a final state where the pressure is 30 lbf/in.^{2} The work developed by the refrigerant is measured as 60 Btu. Can this value be correct?

6.48 One pound of air is initially at 20 lbf/in.^{2} and 140°F. Can a final state at 15 lbf/in.^{2} and 60°F be attained in an adiabatic process?

6.49 One kilogram of Refrigerant 12 contained within a piston–cylinder assembly undergoes a process from a state where the pressure is 6 bars and the quality is 50% to a state where the temperature is 16°C and the refrigerant is saturated liquid. Determine the change in specific entropy of the refrigerant, in kJ/kg·K. Can this process be accomplished adiabatically?

6.50 Air is compressed from a state where the pressure is 1 bar and the temperature is 27°C to a state where the pressure is 5 bars and the temperature is 177°C. Can this process occur adiabatically? If yes, determine the work per unit mass of air, in kJ/kg. If no, determine the direction of the heat transfer.

6.51 Air as an ideal gas with c_p = 0.241 Btu/lb·°R is compressed from a state where the pressure is 15 lbf/in.^{2} and the temperature is 80°F to a state where the pressure is 50 lbf/in.^{2} and the temperature is 240°F. Can this process occur adiabatically? If yes, determine the work per unit mass of air, in Btu/lb. If no, determine the direction of the heat transfer.

6.52 A piston–cylinder assembly contains 0.1 lb of Refrigerant 12 initially as saturated vapor at −10°F. The refrigerant is compressed adiabatically to a final volume 0.021 ft^3. Determine if it is possible for the pressure of the refrigerant at the final state to be

(a) 200 lbf/in.^{2}
(b) 300 lbf/in.^{2}

6.53 Two kilograms of Refrigerant 12 initially at 1.4 bars, 60°C are compressed to saturated vapor at 60°C. During this process, the temperature of the refrigerant departs by no more than 0.1 from 60°C. The heat transfer from the refrigerant during the process is reported to be 100 kJ. Can this value be correct?

6.54 A patent application describes a device which at steady state receives a heat transfer at the rate 500 Btu/s at a temperature of 540°F and develops power equal to 500 Btu/s. There are no other energy transfers. Does the claimed performance violate any principles of thermodynamics? Explain.

6.55 One-tenth kilogram of water initially at 3 bars, 200°C undergoes a process to 15 bar, 210°C while being stirred by a paddle wheel rotating at 100 RPM. Heat transfer with the surroundings at 22°C occurs through a *thin* wall. The net work is measured as −17.5 kJ. Kinetic and potential energy effects can be ignored. Determine whether it is possible for the work measurement to be correct.

6.56 A gearbox operating at steady state receives 0.1 kW along the input shaft and delivers 0.095 kW along the output shaft. The outer surface of the gearbox is at 50°C. For the gearbox, determine

(a) the rate of heat transfer, in kW.
(b) the rate at which entropy is produced, in kW/K.

6.57 One pound of air is compressed adiabatically from 40°F, 1 atm to 5 atm. The compression requires 20% more work than if it were accomplished from the same initial state to the same final pressure without irreversibilities. For the air, determine

(a) the work required, in Btu.
(b) the amount of entropy produced, in Btu/°R.

6.58 A thermally insulated 30-ohm resistor receives a current of 6 amps for 3 s. The mass of the resistor is 0.1 lb, its specific heat is 0.2 Btu/lb·°R, and its initial temperature is 70°F. For the resistor, determine

(a) the final temperature, in °F.
(b) the amount of entropy produced, in Btu/°R.

6.59 At steady state, an electric motor develops power along its output shaft at the rate of 2 kW while drawing 20 amps at 120 V. The outer surface of the motor is at 50°C. For the motor, determine the rate of heat transfer, in kW, and the rate of entropy production, in kW/K.

6.60 An isolated system consists of a closed copper vessel of mass 0.1 kg containing 1 kg of liquid water, each initially at 5°C, immersed in 20 kg of liquid water, initially at 20°C. The system is allowed to come to equilibrium. Determine

(a) the final temperature, in °C.
(b) the entropy change, in kJ/K, for the copper vessel and each of the liquid water masses.
(c) the amount of entropy produced, in kJ/K.

6.61 At steady state, work at a rate of 0.3 kW is done by a paddle wheel on a liquid contained within a closed, rigid tank. Heat transfer occurs at a temperature of 60°C to surroundings that, away from the immediate vicinity of the tank, are at 20°C. Determine the rate of entropy production, in kW/K,

(a) for the tank and its contents as the system.
(b) for an enlarged system including the tank and enough of the nearby surroundings for the heat transfer to occur at 20°C.

6.62 A system initially contains 5 lb of liquid water at 80°F and 1 lb of ice at 32°F. The system attains an equilibrium state, while pressure remains constant at 1 atm. If heat transfer with the surroundings is negligible, determine

(a) the final temperature, in °F.
(b) the amount of entropy produced in the process, in Btu/°R.

For water, the specific enthalpy change for a phase change from solid to liquid at 1 atm in 144 Btu/lb.

6.63 A system initially consists of a rivet at 1800°F whose mass is 1 lb and a two-phase solid–liquid mixture of water at 32°F, 1 atm in which the mass of ice is 5 lb and the mass of liquid is 10 lb. The specific heat of the rivet is 0.12 Btu/lb·°R. The system attains an equilibrium state while pressure remains constant. If heat transfer with the surroundings is negligible, determine

(a) the final temperature, in °F.
(b) the amount of entropy produced, in Btu/°R.

For water, the specific enthalpy change for a phase change from solid to liquid at 1 atm is 144 Btu/lb.

6.64 A system consists of 2 m^3 of hydrogen gas, initially at 35°C, 215 kPa, contained within a closed rigid tank. Energy is transferred to the system from a reservoir at 300°C until the temperature of the hydrogen is 160°C. The temperature of the system boundary where heat transfer occurs is 300°C. Modeling the hydrogen as an ideal gas, determine the heat transfer, in kJ, the change in entropy, in kJ/K, and the amount of entropy produced, in kJ/K. For the reservoir, determine the change in entropy, in kJ/K. Why do these two entropy changes differ?

6.65 An isolated system of total mass m is formed by mixing two equal masses of the same liquid initially at the temperatures T_1 and T_2. Eventually, the system attains an equilibrium state. Each mass is incompressible with constant specific heat c.

(a) Show that the amount of entropy produced is

$$\sigma = mc \ln \left[\frac{T_1 + T_2}{2(T_1 T_2)^{1/2}} \right]$$

(b) Demonstrate that σ must be positive.

6.66 A system consisting of air initially at 300 K and 1 bar experiences the two different types of interactions described below. In each case, the system is brought from the initial state to a state where the temperature is 500 K, while volume remains constant. For each case, determine the amount of entropy produced, in kJ/kg·K, and comment on the difference between the values.

(a) The temperature rise is brought about adiabatically by stirring the air with a paddle wheel.
(b) The temperature rise is brought about by heat transfer from a reservoir at 600 K. The temperature of the system where heat transfer occurs is 600 K.

6.67 An insulated vessel is divided into two compartments connected by a valve. Initially, one compartment contains steam at 1.0 MPa, 500°C, and the other is evacuated. The valve is opened and the steam is allowed to fill the entire volume, achieving a final pressure of 0.1 MPa. Determine

(a) the final temperature, in °C.
(b) the percentage of the vessel volume initially occupied by steam.
(c) the amount of entropy produced, in kJ/kg·K.

6.68 Two insulated tanks are connected by a valve. One tank initially contains 0.5 kg of air at 80°C, 1 bar, and the other contains 1.0 kg of air at 50°C, 2 bars. The valve is opened and the two quantities of air are allowed to mix until equilibrium is attained. Employing the ideal gas model, determine

(a) the final temperature, in °C.
(b) the final pressure, in bars.
(c) the amount of entropy produced, in kJ/K.

6.69 An electric motor operating at steady state draws a current of 10 amp with a voltage of 220 V. The output shaft rotates at 1000 RPM with a torque of 16 N·m applied to an external load. The rate of heat transfer from the motor to its surroundings is related to the surface temperature T_b and the ambient temperature T_0 by $\dot{Q} = hA(T_b - T_0)$, where h = 100 W/m²·K, A = 0.195 m², and T_0 = 293 K. Energy transfers are considered positive in the directions indicated by the arrows on Fig. P6.69.

(a) Determine the temperature T_b, in K.
(b) For the motor as the system, determine the rate of entropy production, in kW/K.
(c) If the system boundary is located to take in enough of the nearby surroundings for heat transfer to take place at temperature T_0, determine the rate of entropy production, in kW/K, for the enlarged system.

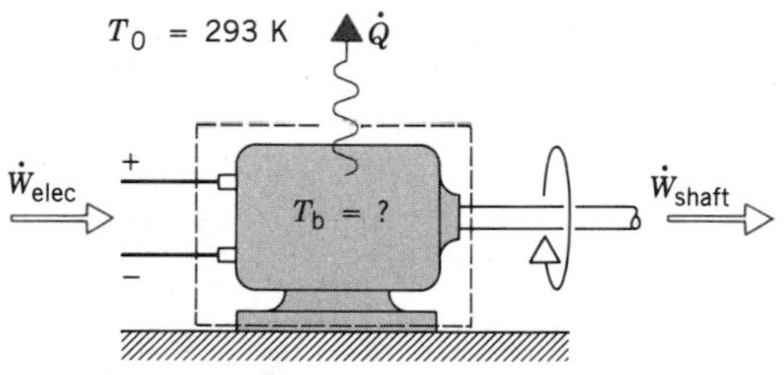

Figure P6.69

6.70 An insulated cylinder is initially divided into halves by a frictionless, thermally conducting piston. On one side of the piston is 10 ft³ of air at 600°R and 2 atm. On the other side is 10 ft³ of air at 600°R and 1 atm. The piston is released and equilibrium is attained, with the piston experiencing no change of state. Employing the ideal gas model, determine for the air

(a) the final pressure, in atm.
(b) the final temperature, in °R.
(c) the amount of entropy produced, in Btu/°R.

6.71 An insulated, rigid tank is divided into two compartments by a frictionless, thermally conducting piston. One compartment initially contains 1 m³ of saturated water vapor at 4 MPa and the other compartment contains 1 m³ of water vapor at 20 MPa, 800°C. The piston is released and equilibrium is attained, with the piston experiencing no change of state. For the water as the system, determine the amount of entropy produced, in kJ/K.

6.72 A cylindrical copper rod of base area A and length L is insulated on its lateral surface. One end of the rod is in contact with a wall at temperature T_H. The other end is in contact with a wall at a lower temperature T_C. At steady state, the rate at which energy is conducted into the rod from the hot wall is

$$\dot{Q}_H = \frac{\kappa A(T_H - T_C)}{L}$$

where κ is the thermal conductivity of the copper rod.

(a) For the rod as the system, obtain an expression for the time rate of entropy production in terms of A, L, T_H, T_C, and κ.
(b) If T_H = 227°C, T_C = 77°C, κ = 0.4 kW/m·K, A = 0.1 m², L = 1 m, calculate the heat transfer rate $\dot{Q}_H$, in kW, and the time rate of entropy production, in kW/K.

6.73 A cylindrical rod of length L insulated on its lateral surface is initially in contact at one end with a wall at temperature T_H and at the other end with a wall at a lower temperature T_C. The temperature within the rod initially varies linearly with position z according to

$$T(z) = T_H - \left(\frac{T_H - T_C}{L}\right) z$$

The rod is insulated on its ends and eventually comes to a final equilibrium state where the temperature is T_f. Evaluate T_f in terms of T_H and T_C and show that the amount of entropy produced is

$$\sigma = mc\left(1 + \ln T_f + \frac{T_C}{T_H - T_C} \ln T_C - \frac{T_H}{T_H - T_C} \ln T_H\right)$$

where c is the specific heat of the rod.

6.74 Figure P6.74 gives the temperature–entropy diagram for a closed system undergoing a power cycle. Internal irreversibilities are present during the adiabatic process from state 1 to state 2. All other processes are internally reversible. Determine the thermal efficiency and compare with the thermal efficiency of a Carnot power cycle operating between reservoirs at the same maximum and minimum temperatures.

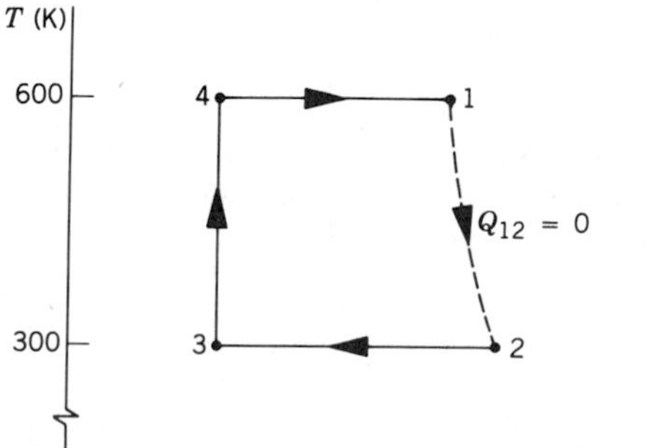

Figure P6.74

6.75 A system undergoing a thermodynamic cycle receives Q_H at temperature T'_H and discharges Q_C at temperature T'_C. There are no other heat transfers.

(a) Show that the net work developed per cycle is given by

$$W_{cycle} = Q_H \left(1 - \frac{T'_C}{T'_H}\right) - T'_C \sigma$$

where σ is the amount of entropy produced per cycle owing to irreversibilities within the system.

(b) If the heat transfers Q_H and Q_C are with hot and cold reservoirs, respectively, what is the relationship of T'_H to the temperature of the hot reservoir T_H and the relationship of T'_C to the temperature of the cold reservoir T_C?

(c) Obtain an expression for W_{cycle} if there are (i) no internal irreversibilities, (ii) no internal *or* external irreversibilities.

6.76 A system undergoes a reversible power cycle while communicating thermally with three thermal reservoirs at 1500, 1000, and 500°R, respectively. The system receives 3000 Btu from the reservoir at 1500°R and discharges 500 Btu to the reservoir at 500°R. Energy is also discharged to the reservoir at 1000°R. Determine the thermal efficiency.

6.77 A system undergoes a thermodynamic cycle while receiving energy by heat transfer from a tank of liquid water initially at 200°F and rejecting energy by heat transfer to the surroundings at 60°F. If the final water temperature is 60°F, determine the *minimum* theoretical volume of water in the tank, in gal, for the cycle to produce net work equal to 1.5×10^5 Btu.

6.78 A system undergoes a thermodynamic cycle while receiving energy by heat transfer from an incompressible body of mass m and specific heat c initially at temperature T_H. The system undergoing the cycle discharges energy by heat transfer to another incompressible body of mass m and specific heat c initially at a lower temperature T_C. Work is developed by the cycle until the temperature of each of the two bodies is the same. Show that the maximum theoretical amount of work that can be developed is

$$W_{max} = mc[T_H + T_C - 2(T_H T_C)^{1/2}]$$

6.79 From the result of Problem 6.78, it can be inferred that the minimum theoretical work *input* required by a refrigeration cycle to bring the two bodies from the same initial temperature to the respective final temperatures of T_H and T_C ($T_C < T_H$) is given by

$$W_{min} = mc[2(T_H T_C)^{1/2} - T_H - T_C]$$

Derive this expression.

6.80 The temperature of a 12-oz (0.354-L) can of soft drink is reduced from 20 to 5°C by a device undergoing a refrigeration cycle. The device receives energy by heat transfer from the soft drink and discharges energy by heat transfer to the surroundings, which are at 20°C. There are no other heat transfers. Determine the minimum theoretical work required, in kJ, assuming the soft drink is an incompressible liquid with the properties of liquid water. Ignore the aluminum can.

6.81 A gas flows through a one-inlet, one-exit control volume operating at steady state. Heat transfer at the rate $\dot{Q}_{cv}$ takes place only at a location on the boundary where the temperature is T_b. For each of the following cases, determine whether the specific entropy of the gas at the exit is greater than, equal to, or less than the specific entropy of the gas at the inlet:

(a) no internal irreversibilities, $\dot{Q}_{cv} > 0$.
(b) no internal irreversibilities, $\dot{Q}_{cv} < 0$.
(c) no internal irreversibilities, insulated.
(d) internal irreversibilities, insulated.
(e) internal irreversibilities, $\dot{Q}_{cv} > 0$.

What can be said when internal irreversibilities are present and $\dot{Q}_{cv} < 0$?

6.82 Steam at 3.0 MPa, 500°C, 70 m/s enters an insulated turbine operating at steady state and exits at 0.3 MPa, 140 m/s. The work developed per kg of steam flowing is claimed to be 667 kJ/kg. Can this claim be correct? Discuss.

6.83 Air enters a turbine operating at steady state at 1.2 bars, 580°C and exits at 0.24 bar, 190°C. Can this expansion be adiabatic? Explain.

6.84 Air enters an insulated turbine operating at steady state at 6 bars, 597°C and exits at 1 bar, 297°C. Neglecting kinetic and potential energy changes, determine

(a) the work developed, in kJ per kg of air flowing through the turbine.
(b) whether the expansion is internally reversible, irreversible, or impossible.

6.85 The following steady-state data are claimed for an insulated steam turbine. At the inlet, the steam is at 100 lbf/in.^{2} and 500°F. At the exit, the pressure is 2 lbf/in.^{2} The mass flow rate is 30,000 lb/h and the power developed is 3800 horsepower. Evaluate this claim. Kinetic and potential energy changes can be neglected.

6.86 A refrigerator compressor operating at steady state receives saturated Refrigerant 12 vapor at 5°F and delivers vapor at 140 lbf/in.^{2}, 120°F. What conclusion can be reached regarding the direction of heat transfer between the compressor and its surroundings?

6.87 Methane gas (CH_4) enters a compressor operating at steady state at 280 K, 0.7 bar and exits at 380 K, 3.4 bars. Employing the ideal gas model, determine the change in specific entropy of each unit of mass passing through the compressor, in kJ/kg·K. Can the compressor be operating adiabatically?

6.88 Air flowing through a horizontal, insulated duct was studied by students in a laboratory. One student group measured the pressure, temperature, and velocity at a location in the duct as 0.95 bar, 67°C, 75 m/s. At another location, the respective values were found to be 0.8 bar, 22°C, 310 m/s. The group neglected to note the direction of the flow, however. Using the known data, determine the direction.

6.89 Nitrogen gas (N_2) at 6 bars, 21°C enters an insulated control volume operating at steady state for which $\dot{W}_{cv} = 0$. Half of the nitrogen exits the device at 1 bar and 82°C and the other half exits at 1 bar and −40°C. The effects of kinetic and potential energy are negligible. Employing the ideal gas model with constant c_p, decide whether the device can operate as described.

6.90 An inventor claims to have developed a device requiring no work input or heat transfer, yet able to produce at steady state hot and cold air streams as shown in Fig. P6.90. Evaluate this claim, employing the ideal gas model for air and ignoring kinetic and potential energy effects.

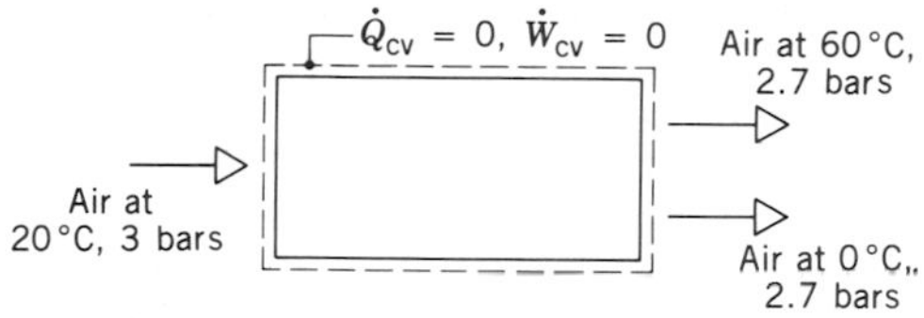

Figure P6.90

6.91 Steam at 14.7 lbf/in.², 250°F enters a compressor operating at steady state and exits at 160 lbf/in.², 400°F. The ambient temperature is 70°F, and changes in kinetic and potential energy can be ignored. The work input is claimed to be 120 Btu per lb of steam flowing through the compressor. Determine whether this claim can be correct.

6.92 Refrigerant 12 enters a valve as a saturated liquid at 7 bars and is steadily throttled to a pressure of 1 bar. Determine the rate of entropy production per unit mass of refrigerant flowing, in kJ/kg·K. If the valve is replaced by a turbine operating at steady state, determine the maximum theoretical work that can be developed, in kJ/kg. In each case, ignore heat transfer with the surroundings and changes in kinetic and potential energy.

6.93 Air at 100 lbf/in.², 100°F enters a turbine operating at steady state and expands adiabatically to 25 lbf/in.² Employing the ideal gas model and neglecting kinetic and potential energy changes, determine

(a) the lowest temperature that could exist at the exit.
(b) the highest temperature that could exist at the exit.

For each case, determine the work developed, in Btu per lb of air flowing, and the rate of entropy production, in Btu/°R per lb of air flowing.

6.94 Air enters an insulated diffuser operating at steady state at 0.6 bar, −3°C, and 260 m/s and exits with a velocity of 130 m/s. Employing the ideal gas model and ignoring potential energy, determine

(a) the temperature of the air at the exit, in °C.
(b) the maximum attainable exit pressure, in bars.

6.95 At steady state, an insulated mixing chamber receives two liquid streams of the same substance at temperatures T_1 and T_2 and mass flow rates $\dot{m}_1$ and $\dot{m}_2$, respectively, A single stream exits at T_3 and $\dot{m}_3$. Using the incompressible substance model with constant specific heat c, obtain an expression for

(a) T_3 in terms of T_1, T_2, and the ratio of mass flow rates $\dot{m}_1/\dot{m}_3$.
(b) the rate of entropy production per unit of mass exiting the chamber in terms of c, T_1/T_2, and $\dot{m}_1/\dot{m}_3$.
(c) For fixed values of c and T_1/T_2, determine the value of $\dot{m}_1/\dot{m}_3$ for which the rate of entropy production is a maximum.

6.96 According to test data, a new type of engine takes in streams of water at 400°F, 40 lbf/in.² and 200°F, 40 lbf/in.² The mass flow rate of the higher temperature stream is twice that of the other. A single stream exits at 40 lbf/in.² with a mass flow rate of 3 lb/s. There is no significant heat transfer between the engine and its surroundings, and kinetic and potential energy effects are negligible. For operation at steady state, determine the *maximum* theoretical rate that power can be developed, in horsepower.

6.97 An inventor seeks to patent a device that at steady state would receive a stream of air at 600°R, 5 atm with a mass flow rate of 100 lb/h and a stream of air at 2000°R, 5 atm with a mass flow rate of 120 lb/h. A single stream exits. The device is well insulated, there is no energy transfer by work, and kinetic and potential energy effects can be ignored. Determine the *maximum* theoretical pressure that the exiting stream could have, in atm. How might this be accomplished?

6.98 An inventor is seeking to patent a device that at steady state would receive a stream of saturated water vapor at 210°C and discharge a condensate stream at 20°C, 0.1 MPa while delivering energy by heat transfer at 300°C at a rate of 2000 kJ per kg of steam entering. The only other energy transfer would involve heat rejection at 20°C. Kinetic and potential energy effects would be negligible. Does the claimed performance violate any principles of thermodynamics? Explain. What is the *maximum* theoretical amount of energy, in kJ per kg of steam entering, that could be delivered at 300°C?

6.99 A patent application describes a device for chilling water which at steady state receives energy by heat transfer at a location on its surface where the temperature is 540°F and discharges energy by heat transfer to the surroundings at another location on its surface where the temperature is 100°F. A warm liquid water stream enters at 100°F, 1 atm and a cool stream exits at 70°F, 1 atm. The device requires no power input to operate, and there are no significant effects of kinetic and potential energy. Determine the minimum theoretical heat addition required, in Btu per lb of cool water exiting the device.

6.100 Figure P6.100 shows a proposed device to develop power using energy discharged from a high-temperature industrial process. The figure provides property data for steady-state operation. All surfaces are well insulated except for one at 227°C, through which heat transfer to the steam occurs at a rate of 100 kJ per kg of steam flowing through the device. Ignoring changes in kinetic and potential energy, evaluate the maximum theoretical work that can be developed, in kJ per kg of steam.

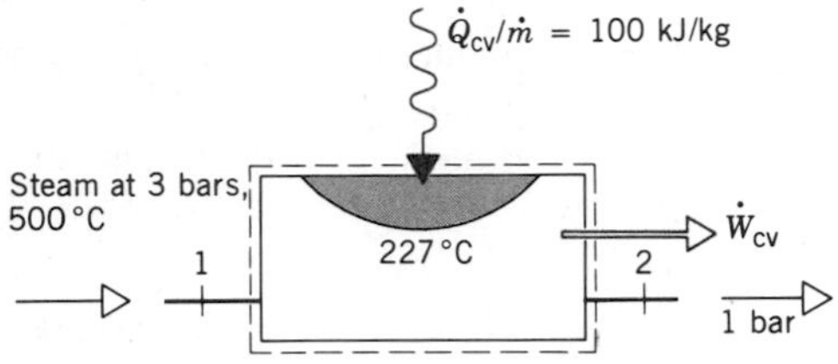

Figure P6.100

6.101 Figure P6.101 shows a gas turbine power plant operating at steady state consisting of a compressor, a heat exchanger, and a turbine. Air enters the compressor at 0.95 bar, 22°C and exits the turbine at 0.95 bar, 421°C. Heat transfer to the air as it flows through the heat exchanger occurs at an average temperature of 488°C. The compressor and turbine operate adiabatically. Determine the maximum theoretical value for the *net* work that can be developed by the power plant, in kJ per kg of air flowing.

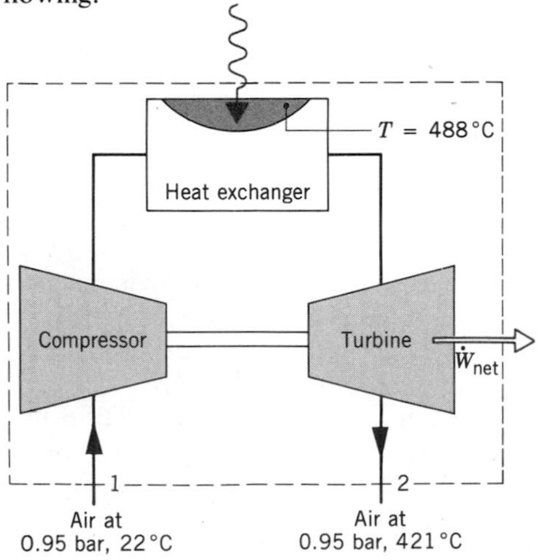

Figure P6.101

6.102 Figure P6.102 shows a 30-ohm electrical resistor located in an insulated duct carrying a stream of air. At steady state, an electric current of 15 amp passes through the resistor, whose temperature remains constant at 28°C. The air enters the duct at 15°C, 1 atm and exits at 25°C with a negligible change in pressure. Kinetic and potential energy changes can be ignored.

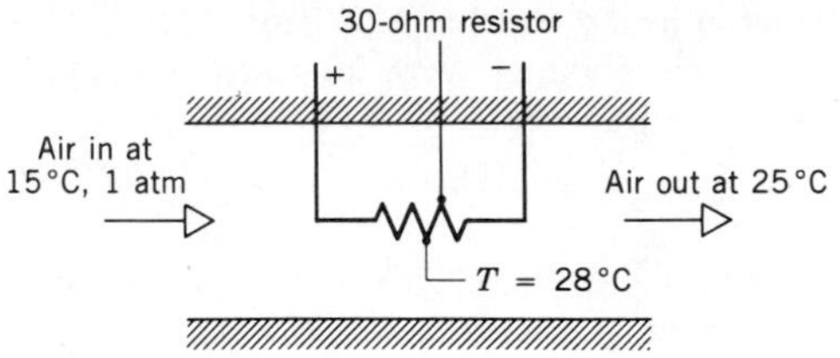

Figure P6.102

(a) For the resistor as the system, determine the rate of entropy production, in kW/K.
(b) For a control volume enclosing the air in the duct and the resistor, determine the mass flow rate of the air, in kg/s, and rate of entropy production, in kW/K.

6.103 Steam at 1000 lbf/in.^{2}, 1100°F enters an insulated turbine with a mass flow rate of 50,000 lb/h. At steady state, 20% of the total flow is extracted at 300 lbf/in.^{2}, 800°F and diverted for use in another process. The remainder of the total flow continues to expand through the turbine, exiting at 1 lbf/in.^{2} and 99% quality. Changes in kinetic and potential energy can be ignored. For the turbine, determine

(a) the power developed, in Btu/h.
(b) the rate of entropy production, in Btu/h·°R.

6.104 Steam enters a turbine operating at steady state at a pressure of 3 MPa, a temperature of 400°C, and a velocity of 160 m/s. Saturated vapor exits at 100°C, with a velocity of 100 m/s. Heat transfer from the turbine to its surroundings takes place at the rate of 30 kJ per kg of steam at a location where the average surface temperature is 400K. Determine the work developed, in kJ, and the rate at which entropy is produced within the turbine, in kJ/K, each per kg of steam flowing.

6.105 The steam turbine of Problem 6.104 is located in a factory space where the ambient temperature is 27°C. Determine the rate of entropy production, in kJ/K per kg of steam flowing, for an enlarged control volume that includes the turbine and enough of its immediate surroundings so that heat transfer takes place from the control volume at the ambient temperature. Explain why the entropy production value differs from that calculated in Problem 6.104.

6.106 Air enters a turbine operating at steady state with a pressure of 75 lbf/in.^{2}, a temperature of 800°R, and a velocity of 400 ft/s. At the turbine exit, the conditions are 15 lbf/in.^{2}, 600°R, and 100 ft/s. Heat transfer from the turbine to its surroundings takes place at a location where the average surface temperature is 620°R. The rate of heat transfer is 10 Btu per lb of air passing through the turbine. Determine the work developed, in Btu, and the rate at which entropy is produced within the turbine, in Btu/°R, each per lb of air flowing.

6.107 Referring to Problem 6.106, locate the boundary of the control volume to include the turbine and a portion of its immediate surroundings so that the heat transfer occurs at the ambient temperature, 40°F. For this enlarged control volume, determine the rate of entropy production in Btu/°R per lb of air passing through the turbine. Explain why the entropy production value differs from that calculated in Problem 6.106.

6.108 Oxygen enters a nozzle operating at steady state at 3.8 MPa, 387°C, and 10 m/s. At the nozzle exit, the conditions are 150 kPa, 37°C, and 750 m/s, respectively.

(a) For a control volume enclosing the nozzle only, determine the heat transfer, in kJ, and the change in specific entropy, in kJ/K, each per kg of oxygen flowing through the nozzle. What additional information would be required to evaluate the rate of entropy production?

(b) Evaluate the rate of entropy production, in kJ/K per kg of oxygen flowing, for an enlarged control volume enclosing the nozzle and a portion of its immediate surroundings so that the heat transfer occurs at the ambient temperature, 20°C.

6.109 Air enters a compressor operating at steady state at 35°C and 4 bars with a volumetric flow rate of 2 m^3/min and is compressed isothermally to 18 bars. There are no irreversibilities within the compressor. Employing the ideal gas model and ignoring kinetic and potential energy effects, obtain the following results:

(a) For a control volume enclosing the compressor only, determine the heat transfer rate and the power input, each in kW.

(b) Calculate the rate of entropy production, in kW/K, for an enlarged control volume enclosing the compressor and a portion of its immediate surroundings so that heat transfer occurs at the ambient temperature, 20°C. Explain why the entropy production value differs from that of part (a).

6.110 Air is compressed in an axial-flow compressor operating at steady state from 27°C, 1.3 bars to a pressure of 2.7 bars. The work input required is 94.6 kJ per kg of air passing through the compressor. Heat transfer from the compressor occurs at the rate of 14 kJ per kg of air at a location on the compressor's surface where the temperature is 40°C. Kinetic and potential energy changes can be ignored. Determine

(a) the temperature of the air at the exit, in °C.

(b) the rate at which entropy is produced within the compressor, in kJ/K per kg of air flowing.

6.111 Air enters a compressor operating at steady state at 1 bar, 20°C with a volumetric flow rate of 9 m^3/min and exits at 5 bars, 160°C. Cooling water is circulated through a water jacket enclosing the compressor at a rate of 10 kg/min, entering at 15°C, and exiting at 23°C with a negligible change in pressure. There is no significant heat transfer from the outer surface of the water jacket, and all kinetic and potential effects are negligible. For the water-jacketed compressor as the control volume, determine the power required, in kW, and the rate of entropy production, in kW/K.

6.112 Refrigerant 12 enters a counterflow heat exchanger at −20°C, with a quality of 35%, and leaves as saturated vapor at −20°C. Air at 1 atm enters the heat exchanger in a separate stream with a flow rate of 4 kg/s and is cooled from 300 to 285 K with no significant change in pressure. The heat exchanger is at steady state, and there is no appreciable heat transfer from its outer surface. Neglecting kinetic and potential energy effects, calculate the rate of entropy production within the heat exchanger, in kW/K.

6.113 A counterflow heat exchanger operates at steady state with negligible kinetic and potential energy effects. In one stream, liquid water enters at 15°C and exits at 25°C with a negligible change in pressure. In the other stream, Refrigerant 12 enters at 14 bars, 80°C with a mass flow rate of 5 kg/min and exits as saturated liquid at 52°C. Heat transfer from the outer surface of the heat exchanger can be ignored. Determine

(a) the mass flow rate of the liquid water stream, in kg/min.

(b) the rate of entropy production within the heat exchanger, in kW/K.

6.114 Steam at 0.7 MPa, 355°C enters an open feedwater heater operating at steady state. A separate stream of liquid water enters at 0.7 MPa, 35°C. A single mixed stream exits as saturated liquid at 0.7 MPa. Heat transfer with the surroundings and kinetic and potential energy effects can be ignored. Determine

(a) the ratio of the mass flow rates of the incoming streams.

(b) the rate at which entropy is produced within the feedwater heater, in kJ/K per kg of liquid exiting.

6.115 At steady state, steam with a mass flow rate of 10 lb/s enters a turbine at 800°F and 600 lbf/in.^{2} and expands to 60 lbf/in.^{2} The power developed by the turbine is 2852 horsepower. The steam then passes through a counterflow heat exchanger with a negligible change in pressure, exiting at 800°F. Air enters the heat exchanger in a separate stream at 20 lbf/in.^{2}, 1020°F and exits at 18 lbf/in.^{2}, 620°F. Kinetic and potential energy changes can be ignored and there is no significant heat transfer between either component and its surroundings. Determine

(a) the mass flow rate of air, in lb/s.

(b) the rate of entropy production in the turbine, in Btu/s · °R.

(c) the rate of entropy production in the heat exchanger, in Btu/s · °R.

6.116 For the control volume of Example 4.10, determine the amount of entropy produced during filling, in kJ/K.

Repeat for the case where no work is developed by the turbine.

6.117 Steam is contained in a large vessel at 100 lbf/in.^{2}, 450°F. Connected to the vessel by a valve is an initially evacuated tank having a volume of 1 ft^3. The valve is opened until the tank is filled with steam at a pressure of 100 lbf/in.^{2} If the filling is adiabatic, kinetic and potential energy effects are negligible, and the state of the large vessel remains constant, determine

(a) the final temperature of the steam within the tank, in °F.
(b) the amount of entropy produced within the tank, in Btu/°R.

6.118 Neglecting internal irreversibilities, solve Example 4.11 using mass and entropy rate balances.

6.119 A 370-ft^3 tank initially filled with air at 1 atm and 80°F is evacuated by pumping out the air, while the tank contents are maintained at 80°F by heat transfer through the walls. The compressor discharges air to the surroundings that are at 1 atm and 80°F. Determine the *minimum* theoretical work required, in Btu.

6.120 A piston–cylinder assembly initially contains 0.1 m^3 of carbon dioxide gas at 0.3 bar and 400 K. The gas is compressed isentropically to a state where the temperature is 560 K. Employing the ideal gas model and neglecting kinetic and potential energy effects, determine the final pressure, in bars, and the work in kJ, using

(a) data from Table A-21.
(b) a constant specific heat ratio from Table A-14 at the mean temperature, 480 K.
(c) a constant specific heat ratio from Table A-14 at 300 K.

6.121 Air enters a turbine operating at steady state at 6 bars and 1100 K and expands isentropically to a state where the temperature is 700 K. Employing the ideal gas model and ignoring kinetic and potential energy changes, determine the pressure at the exit, in bars, and the work, in kJ per kg of air flowing, using

(a) data from Table A-16.
(b) a constant specific heat ratio from Table A-14 at the mean temperature, 900 K.
(c) a constant specific heat ratio from Table A-14 at 300 K.

6.122 A gas undergoes an isentropic expansion from a state where the temperature is T_1 and the pressure is p_1. Write a computer program that allows

(a) p_2 to be found when T_2 is specified.
(b) T_2 to be found when p_2 is specified.

Employ the ideal gas model with $\bar{c}_p$ from Table A-15 for any of the gases listed, as specified by the user.

6.123 An ideal gas with constant specific heat ratio k enters a nozzle operating at steady state at pressure p_1, temperature T_1, and velocity V_1. The air expands isentropically to a pressure of p_2. Develop an expression for the velocity at the exit, V_2, in terms of k, R, V_1, T_1, p_1, and p_2, only.

6.124 An ideal gas undergoes a polytropic process from T_1, p_1 to a state where the temperature is T_2.

(a) Derive an expression for the change in specific entropy in terms of n, R, T_1, T_2, $s°(T_1)$, and $s°(T_2)$.
(b) Using the result of part (a), develop an expression for n if $s_1 = s_2$.

6.125 A rigid, well-insulated tank with a volume of 10 ft^3 is initially filled with water vapor at 700°F, 60 lbf/in.^{2} A leak develops and vapor slowly escapes until the pressure of the vapor remaining in the tank is 1 atm. Determine

(a) the final temperature of the water vapor within the tank, in °F.
(b) the initial and final amounts of water vapor present in the tank, in lb.

6.126 A rigid well-insulated tank having a volume of 10 m^3 is filled initially with water vapor at a pressure of 0.7 MPa and a temperature of 240°C. A leak develops and steam slowly escapes until the pressure within the tank becomes 0.15 MPa. Determine

(a) the final temperature of the water within the tank, in °C.
(b) the amount of mass that exits the tank, in kg.

6.127 A rigid tank is filled initially with 5.0 kg of air at a pressure of 0.5 MPa and a temperature of 500 K. The air is allowed to discharge through a turbine into the atmosphere, developing work until the pressure in the tank has fallen to the atmospheric level of 0.1 MPa. Determine the *maximum* theoretical amount of work that could be developed, in kJ, ignoring heat transfer with the atmosphere.

6.128 A compressed air bottle initially containing air at 30 atm and 540°F is to be used to drive a small turbine as an emergency starter for an engine. The turbine is required to produce work in the amount of 100 Btu. The pressure in the bottle is allowed to fall to 3 atm during the process and the turbine exhausts to the atmosphere at 1 atm. Determine the *minimum* theoretical volume of the bottle, in ft^3, that would suffice for this purpose. Ignore heat transfer with the atmosphere.

6.129 Air enters a 3600-kW turbine operating at steady state with a mass flow rate of 18 kg/s at a temperature of 800°C and a velocity of 100 m/s. The air expands adiabatically through the turbine and exits at a velocity of 150 m/s. The air then enters a diffuser where it is decelerated isentropically to a velocity of 10 m/s and a pressure of 1 atm. Employing the ideal gas model,

(a) determine the pressure of the air between the turbine and diffuser, in atm.
(b) show the processes on a T–s diagram.

6.130 Steam at 3.0 MPa, 500°C enters an insulated turbine operating at steady state with a mass flow rate of 7 kg/s and exits at 0.3 MPa. Kinetic and potential energy effects are negligible.

(a) Determine the maximum theoretical power that can be developed by the turbine, in kW, and the corresponding exit temperature, in °C.
(b) If the steam exits the turbine at 240°C, determine the isentropic turbine efficiency.

6.131 Steam at 1.0 MPa and 300°C enters an insulated turbine operating at steady state and exits at 15 kPa. Kinetic and potential energy effects are negligible. If the work developed by the turbine is measured as 630 kJ per kg of steam flowing through the turbine, evaluate the isentropic turbine efficiency.

6.132 Hydrogen at 2.28 atm and 400°F enters an insulated turbine operating at steady state and expands to 1 atm, 260°F. Kinetic and potential energy effects can be ignored. Determine the isentropic turbine efficiency.

6.133 Water vapor at 2 MPa and 440°C enters an insulated turbine operating at steady state and expands to 70 kPa. The isentropic turbine efficiency is 80%. Kinetic and potential energy effects can be ignored. Determine the exit temperature if superheated vapor exits or the quality if a two-phase liquid–vapor mixture exits.

6.134 Air at 4.5 bars, 550 K enters an insulated turbine operating at steady state and exits at 1.5 bars. Kinetic and potential energy effects are negligible. If the isentropic turbine efficiency is 85%, determine

(a) the temperature at the turbine exit, in K.
(b) the work developed, in kJ per kg of air flowing.

6.135 Steam at 400°F enters an insulated turbine operating at steady state and exits at 3 $lbf/in.^2$ If the isentropic turbine efficiency is 72%, determine the range of turbine inlet pressures that ensures the quality of the steam at the exit will be at least 90%.

6.136 Figure P6.136 shows a throttling valve in parallel with a steam turbine having an isentropic efficiency of 90%. For operation at steady state, determine the mass flow rate through the turbine, in lb/s, and the power developed by the turbine, in horsepower. Heat transfer with the surroundings and kinetic and potential energy effects can be neglected. Locate the numbered states on a Mollier diagram.

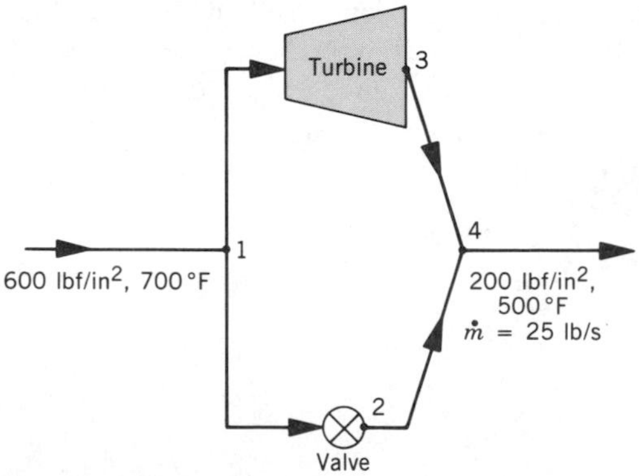

Figure P6.136

6.137 Water vapor enters an insulated nozzle operating at steady state at 60 $lbf/in.^2$, 350°F, 10 ft/s and exits at 35 $lbf/in.^2$, 1440 ft/s. Determine the isentropic nozzle efficiency.

6.138 Refrigerant 12 enters an insulated nozzle operating at steady state at 120 $lbf/in.^2$, 220°F, 10 ft/s and expands to 30 $lbf/in.^2$ If the isentropic nozzle efficiency is 95%, determine the velocity at the exit, in ft/s.

6.139 Air enters an insulated nozzle operating at steady state at 80 $lbf/in.^2$, 120°F, 10 ft/s with a mass flow rate of 0.4 lb/s. The exit pressure is 50 $lbf/in.^2$ If the isentropic nozzle efficiency is 95%, determine

(a) the exit velocity, in ft/s.
(b) the exit area, in ft^2.

6.140 Argon enters an insulated nozzle at 2.77 bars, 1300 K, 10 m/s and exits at 1 bar, 900 K. For steady-state operation, determine

(a) the exit velocity, in m/s.
(b) the isentropic nozzle efficiency.
(c) the rate of entropy production, in kJ/K per kg of argon flowing.

6.141 Refrigerant 12 enters a compressor operating at steady state as saturated vapor at −5°C and exits at a pressure of 8 bars. There is no significant heat transfer with the surroundings, and kinetic and potential energy effects can be ignored.

(a) Determine the minimum theoretical work input required, in kJ per kg of refrigerant flowing through the compressor, and the corresponding exit temperature, in °C.
(b) If the refrigerant exits at a temperature of 50°C, determine the isentropic compressor efficiency.

6.142 Air enters an insulated compressor operating at steady state at 1.05 bars, 23°C with a mass flow rate of 1.8 kg/s and exits at 2.9 bars. Kinetic and potential energy effects are negligible.

(a) Determine the minimum theoretical power input required, in kW, and the corresponding exit temperature, in °C.
(b) If the isentropic compressor efficiency is 80%, determine the power input, in kW, and the temperature at the compressor exit, in °C.

6.143 Refrigerant 12 enters a compressor operating at steady state as saturated vapor at −5°C and exits at a pressure of 14 bars. The isentropic compressor efficiency is 75%. Heat transfer between the compressor and its surroundings can be ignored. Kinetic and potential energy effects are also negligible. Determine

(a) the exit temperature, in °C.
(b) the work input, in kJ per kg of refrigerant flowing.

6.144 Air at 40°F, 1 atm enters a compressor operating at steady state and exits at 620°F, 8.6 atm. Heat transfer with the surroundings is negligible, and kinetic and potential energy effects can be ignored. Determine

(a) the isentropic compressor efficiency.
(b) the rate of entropy production, in Btu/°R per lb of air flowing.

6.145 Refrigerant 12 enters an insulated compressor operating at steady state as saturated vapor at −10°C at a rate of 18 kg/min. At the compressor exit, the pressure of the refrigerant is 5 bars. Kinetic and potential energy effects can be neglected. If the rate of entropy production must be kept less than 0.389 kJ/min·K, determine the allowed range of values for the isentropic compressor efficiency.

6.146 A compressor operating at steady state takes in atmospheric air at a rate of 1 kg/s and discharges air at 5 bars. Estimate the power required, in kW, and the exit temperature, in °C. Take particular care in listing your assumptions.

6.147 Steam at 1000 lbf/in.^{2}, 900°F enters a valve with a mass flow rate of 100,000 lb/h and undergoes a throttling process to 900 lbf/in.^{2} The steam then enters a two-stage turbine, through which it expands adiabatically. At the exit of the first turbine stage, the steam pressure is 200 lbf/in.^{2} and 20% of the flow is withdrawn for some other use. The rest of the steam expands through the second turbine stage to an exhaust pressure of 1 lbf/in.^{2} The isentropic turbine efficiency of each turbine stage is 85%. For steady-state operation and negligible kinetic and potential energy effects, determine the power output of the turbine, in Btu/h. Show the principal states on T–s and h–s diagrams.

6.148 In a gas turbine operating at steady state, air enters the compressor at 0.95 bar and 22°C and exits at 5.7 bars. The air then passes through a heat exchanger before entering the turbine at 1100 K, 5.7 bars. Air exits the turbine at 0.95 bar. The compressor and turbine operate adiabatically and kinetic and potential energy effects can be ignored. Determine the *net* work developed by the plant, in kJ per kg of air flow, if

(a) the compressor and turbine operate without internal irreversibilities.
(b) the compressor and turbine isentropic efficiencies are 82 and 85%, respectively.

6.149 Figure P6.149 shows liquid water at 80 lbf/in.^{2}, 300°F entering a flash chamber through a valve at the rate of 8000 lb/h. At the valve exit, the pressure is 42 lbf/in.^{2} Saturated liquid at 40 lbf/in.^{2} exits from the bottom of the flash chamber and saturated vapor at 40 lbf/in.^{2} exits from near the top. The vapor stream is fed to a steam turbine having an isentropic efficiency of 80% and an exit pressure of 2 lbf/in.^{2} For steady-state operation, negligible heat transfer with the surroundings, and no significant kinetic and potential energy effects, determine the rates of entropy production for the valve, the flash chamber, and the turbine, each in Btu/h·°R. Discuss.

6.150 Air enters a compressor operating at steady state at 17°C, 1 bar and exits at a pressure of 5 bars. Kinetic and potential energy changes can be ignored. If there are no internal irreversibilities, evaluate the work and heat transfer, each in kJ per kg of air flowing, for the following cases:

(a) isothermal compression.
(b) polytropic compression with $n = 1.3$.
(c) adiabatic compression.

Sketch the processes on p–v and T–s coordinates and associate areas on the diagrams with the work and heat transfer in each case.

6.151 Air enters a compressor operating at steady state at 15 lbf/in.^{2}, 60°F and exits at 75 lbf/in.^{2} Kinetic and potential energy changes can be ignored. If there are no internal irreversibilities, evaluate the work and heat transfer, each in Btu per lb of air flowing, for the following cases: **(a)** isothermal compression, **(b)** polytropic compression with $n = 1.3$, **(c)** adiabatic compression. Sketch the processes on p–v and T–s coordinates and associate areas on the diagrams with the work and heat transfer of each case.

6.152 Air enters a compressor operating at steady state with a volumetric flow rate of 30 m^3/min at 35°C and 4 bars. The air is compressed isothermally without internal irreversibilities, exiting at 18 bars. Kinetic and potential

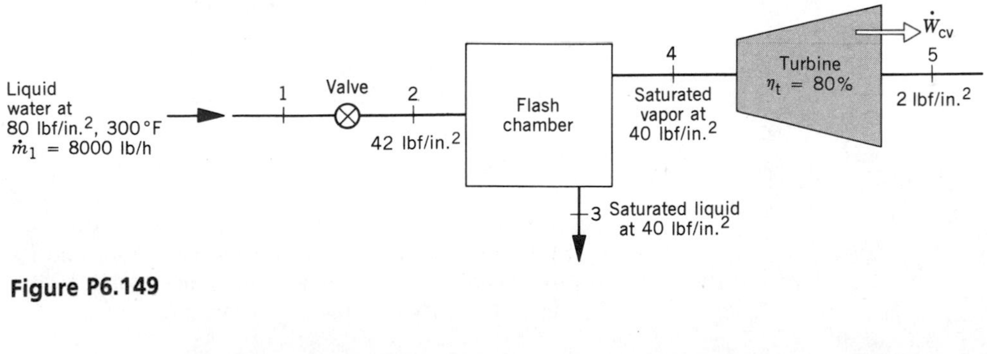

Figure P6.149

energy effects can be ignored. Evaluate the work required and the heat transfer, each in kW.

6.153 Refrigerant 12 enters a compressor operating at steady state as saturated vapor at 2 bars with a volumetric flow rate of 1.9×10^{-3} m^3/s. The refrigerant is compressed to a pressure of 8 bars in an internally reversible process according to $pv^{1.05}$ = constant. Neglecting kinetic and potential energy effects, determine

(a) the power required, in kW.
(b) the rate of heat transfer, in kW.

6.154 Water behind a dam enters an intake pipe at a pressure of 1.05 bars and velocity of 1 m/s, flows through a hydraulic turbine, and exits at a point 120 m below the intake at 1 bar, 15°C, 12 m/s. The diameter of the exit pipe is 2 m and the local acceleration of gravity is 9.8 m/s^2. Estimate the maximum theoretical power that can be developed by the turbine at steady state, in kW.

6.155 Compare the work required at steady state to compress *water vapor* isentropically to 3 MPa from the saturated vapor state at 0.1 MPa to the work required to pump *liquid water* isentropically to 3 MPa from the saturated liquid state at 0.1 MPa, each in kJ per kg of water flowing through the device. Kinetic and potential energy effects can be ignored.

6.156 Saturated liquid water at 2.0 lbf/in.^{2} enters a pump operating at steady state with a mass flow rate of 270,000 lb/h and exits as a liquid at 14.7 lbf/in.^{2} The isentropic pump efficiency is 60%. Ignoring kinetic and potential energy effects, estimate the power required by the pump, in horsepower.

6.157 Figure P6.157 shows three devices operating at steady state: a pump, a boiler, and a turbine. The turbine provides the power required to drive the pump and also supplies power to other devices. For adiabatic operation of the pump and turbine, and ignoring kinetic and potential energy effects, determine, in kJ per kg of steam flowing

(a) the work required by the pump.
(b) the *net* work developed by the turbine.
(c) the heat transfer to the boiler.

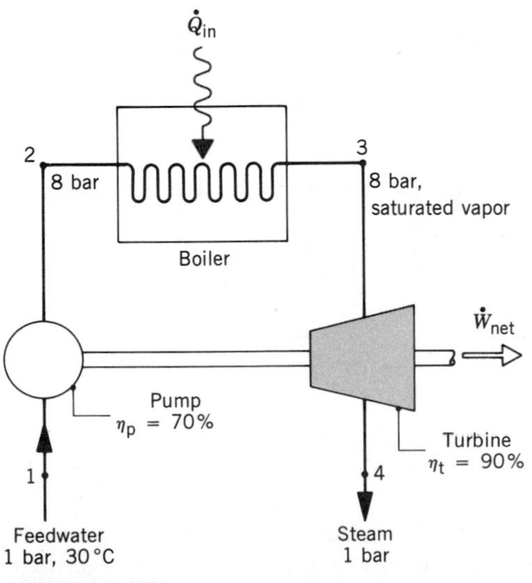

Figure P6.157

6.158 Liquid water enters a pipe at 2 bars, 15°C, 2 m/s and is pumped to a location where liquid is discharged at 6 bars, 8 m/s. The exit is located 50 m vertically above the inlet. The diameter of the exit pipe is 2 cm. Estimate the minimum power input required to drive the pump at steady state. The local acceleration of gravity is 9.8 m/s^2.

6.159 Liquid water at 70°F, 10 ft/s enters a pipe and flows to a location where the pressure is 14.7 lbf/in.^{2}, the velocity is 20 ft/s, and the elevation is 40 ft above the inlet. The local acceleration of gravity is 32 ft/s^2. Estimate the minimum pressure required at the pipe inlet.

6.160 A pump operating at steady state draws water from 20 ft underground where the pressure is 15 lbf/in.^{2} and delivers it 2 ft above ground at a pressure of 45 lbf/in.^{2} and a mass flow rate of 30 lb/s. Determine the *minimum* theoretical power required by the pump, in horsepower, ignoring kinetic energy effects. The local acceleration of gravity is 32.2 ft/s^2.

6.161 A 4-horsepower pump operating at steady state draws in liquid water at 1 atm, 60°F and delivers it at 5 atm at an elevation 20 ft above the inlet. There is no significant change in velocity between the inlet and exit, and the local acceleration of gravity is 32.2 ft/s^2. Determine the *minimum* theoretical time, in min, required to pump 1000 gal.

6.162 A 4-kW pump operating at steady state draws in liquid water at 1 bar, 16°C with a mass flow rate of 4.5 kg/s. If there are no significant kinetic and potential energy effects, determine the *maximum* theoretical pressure the liquid water can have at the exit, in bar. The local acceleration of gravity is 9.81 m/s^2.

6.163 A 1.15-horsepower pump operating at steady state delivers 10,000 lb/h from a location 30 ft underground to an elevation above the pump where the flow area is 1.0 in.^{2} and the water pressure is 50 lbf/in.^{2} At the inlet underground, the water is at 55°F, 15 lbf/in.^{2}, and has a velocity of 2 ft/s. Determine the *maximum* theoretical elevation above the pump at which the water can be delivered. The local acceleration of gravity is 32.2 ft/s^2.

6.164 Air enters a nozzle operating at steady state at 75 lbf/in.^{2}, 400°F, 10 ft/s and undergoes a polytropic expansion to 15 lbf/in.^{2} with $n = 1.4$. Using the ideal gas model and ignoring potential energy effects, determine

(a) the exit velocity, in ft/s.
(b) the rate of heat transfer between the air and its surroundings, in Btu per lb of air flowing.

DESIGN AND OPEN ENDED PROBLEMS

6.1D Flywheel storage systems can be *charged* using relatively inexpensive *off-peak* electricity generated during the hours between midnight and dawn, say, and *discharged* to generate electricity during *peak* demand periods when electricity is more costly. Specify the preliminary design (configuration, size, materials) of a flywheel system that would store 125 kW·h of energy over night. What are some potential advantages and disadvantages of flywheel storage systems in comparison to battery storage systems? Check to see if your local utility offers reduced off-peak rates for some consumers.

6.2D The preliminary design of a solar-activated system that produces hot water and generates electricity is shown in Fig. P6.2D. At the lower elevation, a shallow pond receives solar energy through a plastic sheet. Ambient air entering as shown is first heated and humidified, and then rises in the chimney owing to buoyancy to a higher elevation where about 50% of the airborne moisture condenses, providing the energy input to the power plant. The electricity generated by the power plant is stored in a battery and the condensate is stored in a tank. Cooling water at 60°F is available from the bottom of the reservoir. The water evaporation rate at the lower elevation is 1 lb per day for each ft^2 of collector area. For a hot water storage capacity of 1000 gal, estimate the collector area required, in ft^2, and the amount of electricity stored, in kW·h. Would you recommend the development of such a system? Explain.

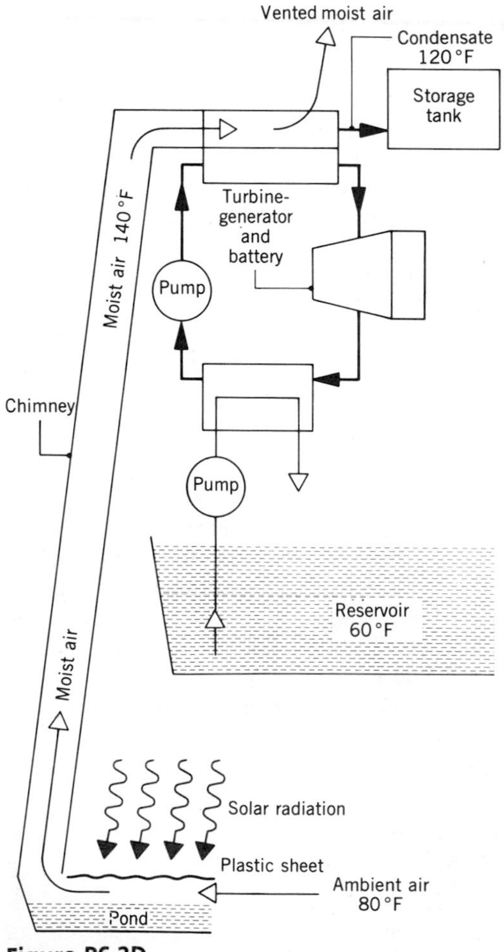

Figure P6.2D

6.3D A proposed solar-activated power plant is shown in Fig. P6.3D. Photovoltaic solar cells would produce DC electric current to drive an electrolysis generator that converts water into oxygen and hydrogen. The gases would be delivered to a water-cooled burner chamber where they react to form water vapor at a relatively high temperature and pressure. The water vapor would expand through a turbine driving a generator, producing electricity. Critically evaluate this proposal.

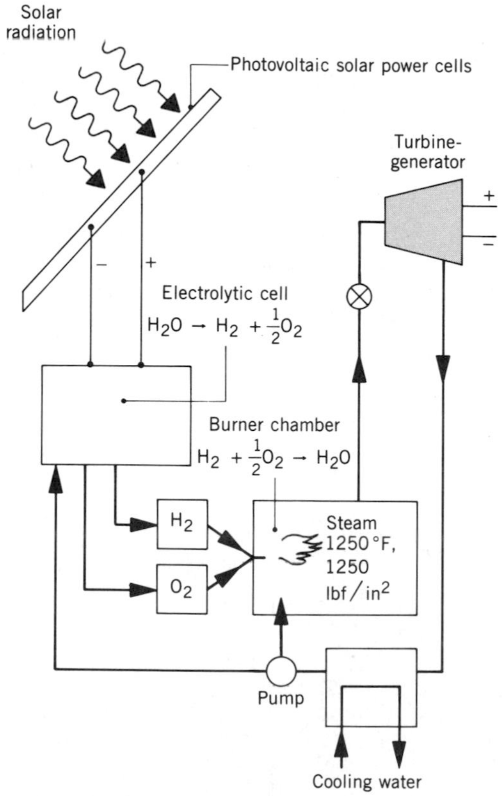

Figure P6.3D

6.4D Figure P6.4D illustrates an *ocean thermal energy* power plant that generates power by exploiting the naturally occurring decrease of the temperature of ocean water with depth. Warm surface water enters the evaporator with a mass flow rate of $\dot{m}_w$ at temperature $T_w = 28°C$ and exits at $T_1 < T_w$. Cool water brought from a depth of 600 m enters the condenser with a mass flow rate of $\dot{m}_c$ at temperature $T_c = 5°C$ and exits at $T_2 > T_c$. The pumps for the ocean water flows and other auxiliary equipment typically require 15% of the gross power generated. *Estimate* the mass flow rates $\dot{m}_w$ and $\dot{m}_c$, in kg/s, for a desired net power output of 125 MW.

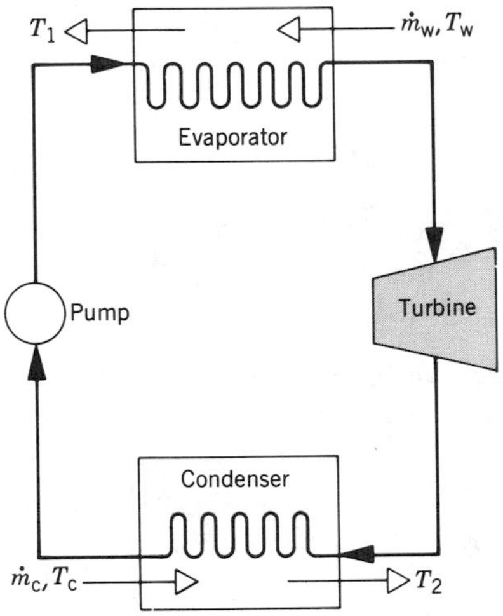

Figure P6.4D

6.5D Electricity is generated *hydroelectrically* by allowing dammed water to flow down a sluice through a hydraulic turbine. Determine for 700 ft of fall the minimum amount of water that would be required, in gal, to provide the annual electricity requirement of an all-electric dwelling having a living area of about 2000 ft^2. Estimate the *actual* amount of water required. What would your local utility charge for the annual electricity requirement? How does this compare with the annual cost of electricity generated hydroelectrically? Explain the difference, if any.

6.6D Owing to leaks from a petroleum storage terminal over the last 50 years, an estimated 14.5 million gal of oil, gasoline, and naptha sit atop an aquifier located below the streets of Brooklyn, New York. Estimate the cost of the power required to pump this material to the surface. Also comment on other pumping costs that might be incurred. What considerations would be entailed in disposing of the petroleum after it has been brought to the surface?

6.7D By combining the features of a *venturi* flowmeter and a *gamma ray* densitometer, devise an experimental means for determining the mass flow rate of a flowing *two-phase* liquid–vapor mixture of water.

6.8D Figure P6.8D illustrates a typical source of irreversibility in power plants owing to the need for *practical rates* of heat transfer and *practical-sized* heat exchangers: The temperature T'_H at which the system undergoing the cycle receives energy is less than the reservoir temperature T_H, and the temperature T'_C at which energy is discharged from the system is greater than the reservoir temperature T_C. The heat transfer rates indicated on the figure can be described simply by

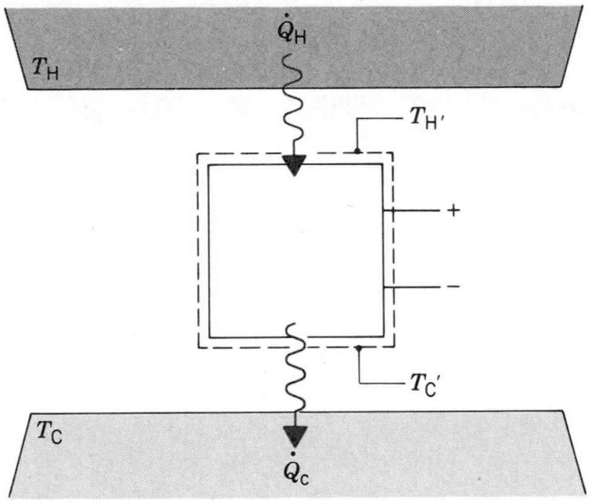

Figure P6.8D

$$\dot{Q}_H = (hA)_H(T_H - T'_H)$$
$$\dot{Q}_C = (hA)_C(T'_C - T_C)$$

where h and A denote, respectively, heat transfer coefficient and area. If there are no *internal* irreversibilities as the system undergoes the cycle, the thermal efficiency would be $\eta = 1 - T'_C/T'_H$. This efficiency is bounded from above by $\eta_{max} = 1 - T_C/T_H$, applying when *external* irreversibilities are also absent: when $T_H = T'_H$ and $T_C = T'_C$.

(a) Denoting the value of electricity per kW·h by c, show that the rate at which the plant would produce income is

$$\dot{\$} = \frac{\gamma(\eta_{max} - \eta)\eta}{1 - \eta}$$

where

$$\gamma = \frac{c\,T_H}{\left(\frac{1}{hA}\right)_H + \left(\frac{1}{hA}\right)_C}$$

(b) Holding all parameters fixed except T'_C and T'_H, plot the ratio $\dot{\$}/\gamma$ versus η for $0 \le \eta \le \eta_{max}$. Determine analytically the optimum value for $\dot{\$}$. Does this correspond to *fully reversible* operation? Discuss.

(c) To achieve the rate of income $\dot{\$}$, various costs would necessarily be incurred. Discuss the principal costs for a power plant from a *life-cycle cost* viewpoint.

CHAPTER 7

AVAILABILITY (EXERGY) ANALYSIS

This chapter introduces a method that uses the conservation of mass and conservation of energy principles together with the second law of thermodynamics for the design and analysis of thermal systems. The method is identified here as *availability* analysis. Another term frequently used to identify it is *exergy* analysis.

The importance of developing thermal systems that make effective use of nonrenewable energy resources such as oil, natural gas, and coal is apparent. The method of availability analysis is particularly suited for furthering the goal of more efficient energy resource use, for it enables the locations, types, and true magnitudes of waste and loss to be determined. This information can be used in the design of thermal systems and to guide efforts to reduce sources of inefficiency in existing systems.

7.1 INTRODUCTION

Energy is conserved in every device or process. It cannot be destroyed. Energy entering with fuel, electricity, flowing streams of matter, and so on can be accounted for in the products and by-products. However, as illustrated by the discussion to follow, the energy conservation idea alone is inadequate for depicting some important aspects of energy resource utilization.

Figure 7.1*a* shows an isolated system consisting initially of a small container of fuel surrounded by air in abundance. Suppose the fuel burns (Fig. 7.1*b*) so that finally there is a slightly warm mixture of combustion products and air as shown in Fig. 7.1*c*. Although the total *quantity* of energy associated with the system would be unchanged, the initial fuel–air combination would be intrinsically more useful than the final warm mixture. For instance, the fuel might be used in some device to generate electricity or produce superheated steam, whereas the uses to which the slightly warm combustion products can be put would be far more limited in scope. We might say that the system has a greater *potential for use* initially than it has finally. Since nothing but a final warm mixture would be achieved in the process, this potential would be largely wasted. More precisely, the initial potential would be

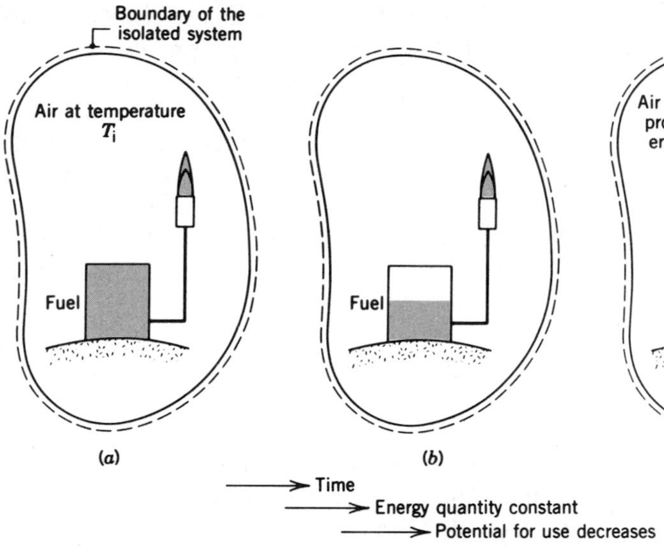

Figure 7.1 Illustration used to introduce availability.

largely *destroyed* because of the irreversible nature of the process. Anticipating the main results of this chapter, we can read the property *availability* as *potential for use* wherever it appears in the discussion. This example illustrates that, unlike energy, availability is not conserved.

Subsequent discussion shows that availability not only can be destroyed by irreversibilities but also can be transferred to or from a system, as in losses accompanying heat transfers to the surroundings. Improved energy resource utilization can be realized by reducing availability destruction within a system and/or losses. The objective in an availability analysis is to identify sites where availability destructions and losses occur and rank order them for significance. This allows attention to be centered on the aspects of system operation that offer the greatest opportunities for improvement.

7.2 AVAILABILITY

The basis for the availability concept is present in the introduction to the second law provided in Chap. 5. A principal conclusion of Sec. 5.1 is that an opportunity exists for doing work whenever two systems at different states are brought into communication, for in principle work can be developed as the systems are allowed to come into equilibrium. When one of the two systems is a suitably idealized system called an *environment* and the other is some closed system of interest, *availability,* by definition, is the *maximum theoretical work* obtainable as they interact to equilibrium.

The definition of availability will not be complete, however, until we define the environment and show how numerical values for availability can be determined. These tasks are closely related because the numerical value of availability depends on the state of a closed system of interest, as well as the condition of the environment.

7.2.1 Environment

Any system, whether a component in a larger system such as a steam turbine in a power plant or the larger system (power plant) itself, operates within surroundings of some kind. It is important to distinguish between the environment and the system's surroundings. As used in this book, the term *surroundings* refers to *everything* not included in the system. The term *environment* applies to some *portion* of the surroundings, the intensive properties of each phase of which are uniform and do not change significantly as a result of any process under consideration. The environment is regarded as free of irreversibilities. All significant irreversibilities are located within the system and its immediate surroundings. Internal irreversibilities are those located within the system. External irreversibilities reside in the immediate surroundings. The environment and immediate surroundings for the case of a simple power plant are illustrated in Fig. 7.2.

The physical world is complicated, and to include every detail in an analysis is not practical. Accordingly, in describing the environment, simplifications are made, and a model results. The validity and utility of an analysis using any model are, of course, restricted by the idealizations made in formulating the model. In this book the environment is regarded to be a simple compressible system that is *large* in extent and *uniform* in temperature, T_0, and pressure, p_0. In keeping with the idea that the environment has to do with the actual physical world, the values for p_0 and T_0 used throughout a particular analysis are normally taken as typical environmental conditions, such as 1 atm and 25°C (77°F). Although its intensive properties do not change, the environment can experience changes in its extensive properties as a result of interactions with other systems. Changes in the extensive properties internal energy U_e, entropy S_e, and volume V_e of the environment are related through the *first T dS* equation, Eq. 6.15. Since T_0 and p_0 are constant, this equation takes the form

$$\Delta U_e = T_0\,\Delta S_e - p_0\,\Delta V_e \tag{7.1}$$

In this chapter kinetic and potential energies are evaluated relative to the environment, all parts of which are considered to be at rest with respect to one another. Accordingly, as indicated by the foregoing equation, a change in the energy of the environment can be a change in its internal energy only. Equation 7.1 is used below to develop an expression for evaluating availability. In Chap. 13 the environment concept is extended to allow for the possibility of chemical reactions, which are excluded from the present considerations.

We consider next the concept of the *dead state*, which is also important in completing our understanding of the property availability.

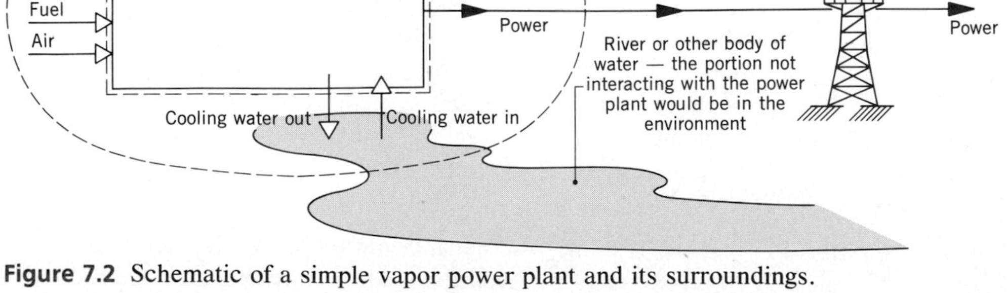

Figure 7.2 Schematic of a simple vapor power plant and its surroundings.

7.2.2 Dead State

If the state of a fixed quantity of matter, a closed system, departs from that of the environment, an opportunity exists for developing work. However, as the system changes state toward that of the environment, the opportunity diminishes, ceasing to exist when the two are in equilibrium with one another. This state of the system is called the *dead state*. At the dead state, the fixed quantity of matter under consideration is imagined to be sealed in an envelope impervious to mass flow, at rest relative to the environment, and internally in equilibrium at the temperature T_0 and pressure p_0 of the environment. At the dead state, both the closed system and environment possess energy, but the value of availability is zero because there is no possibility of a spontaneous change within the closed system or the environment, nor can there be an interaction between them.

With the introduction of the concepts of environment and dead state, we are in a position to show how a numerical value can be determined for availability. This is considered next.

7.2.3 Evaluation of Availability

The object of this section is to show that the availability of a closed system at a specified state is given by the expression

$$A = (E - U_0) + p_0(V - V_0) - T_0(S - S_0) \tag{7.2}$$

where $E(= U + \text{KE} + \text{PE})$, V, and S denote, respectively, the energy, volume, and entropy of the closed system, and U_0, V_0, and S_0 are the values of the same properties of the closed system when at the dead state. By inspection of Eq. 7.2, the units of availability are seen to be the same as those of energy.

Equation 7.2 can be derived by applying energy and entropy balances to the *combined system* shown in Fig. 7.3, which consists of a closed system and an environment. Availability is the maximum theoretical work that could be done by the combined system if the closed system were to come into equilibrium with the environment—that is, if the closed system passed to the dead state. Since the objective is to evaluate the maximum work that could be developed by the combined system, the boundary of the combined system is located so that the only energy transfers across it are work transfers of energy. This ensures that the work developed by the com-

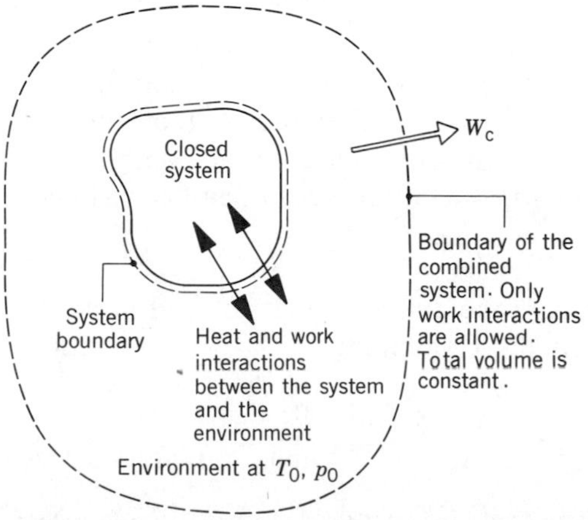

Figure 7.3 Combined system of closed system and environment.

bined system is not affected by heat transfer to or from it. Moreover, although the volumes of the closed system and the environment can vary, the boundary of the combined system is located so that the total volume of the combined system remains constant. This ensures that the work developed by the combined system is fully available for lifting a weight in its surroundings, say, and is not expended in merely displacing the surroundings of the combined system.

An energy balance for the combined system reduces to

$$\Delta E_c = \overset{0}{\cancel{Q_c}} - W_c \tag{7.3}$$

where W_c is the work developed by the combined system, and ΔE_c is the energy change of the combined system, equal to the sum of the energy changes of the closed system and the environment. The energy of the closed system initially is denoted by E, which includes the kinetic energy, potential energy, and internal energy of the system. As noted above, kinetic energy and potential energy are evaluated rėlative to the environment, so the energy of the closed system when at the dead state would be just its internal energy, U_0. Accordingly, ΔE_c can be expressed as

$$\Delta E_c = (U_0 - E) + \Delta U_e$$

Using Eq. 7.1 to replace ΔU_e, the expression becomes

$$\Delta E_c = (U_0 - E) + (T_0 \, \Delta S_e - p_0 \, \Delta V_e) \tag{7.4}$$

Substituting Eq. 7.4 into Eq. 7.3 and solving for W_c gives

$$W_c = (E - U_0) - (T_0 \, \Delta S_e - p_0 \, \Delta V_e)$$

As noted previously, the total volume of the combined system is constant. Hence, the change in volume of the environment is equal in magnitude but opposite in sign to the volume change of the closed system: $\Delta V_e = -(V_0 - V)$. With this substitution, the above expression for work becomes

$$W_c = (E - U_0) + p_0(V - V_0) - T_0 \, \Delta S_e \tag{7.5}$$

This equation gives the work developed by the combined system as the closed system passes to the dead state while interacting only with the environment. The maximum theoretical value for the work is determined using the entropy balance as follows.

The entropy balance for the combined system reduces to give

$$\Delta S_c = \sigma_c$$

where the entropy transfer term is omitted because no heat transfer takes place across the boundary of the combined system, and σ_c accounts for entropy production due to irreversibilities as the closed system comes into equilibrium with the environment. ΔS_c is the entropy change of the combined system, equal to the sum of the entropy changes for the closed system and environment, respectively

$$\Delta S_c = (S_0 - S) + \Delta S_e$$

where S and S_0 denote the entropy of the closed system at the given state and the dead state, respectively. Combining the last two equations

$$(S_0 - S) + \Delta S_e = \sigma_c \tag{7.6}$$

Eliminating ΔS_e between Eqs. 7.5 and 7.6 results in

$$W_c = \underline{(E - U_0) + p_0(V - V_0) - T_0(S - S_0)} - T_0\sigma_c \tag{7.7}$$

The value of the underlined term in Eq. 7.7 is determined by the two end states of the closed system—the given state and the dead state—and is independent of the details of the process linking these states. However, the value of the term $T_0\sigma_c$ depends on the nature of the process as the closed system passes to the dead state. In accordance with the second law, $T_0\sigma_c$ is positive when irreversibilities are present and vanishes in the limiting case where there are no irreversibilities. The value of $T_0\sigma_c$ cannot be negative. Hence, the *maximum* theoretical value for the work of the combined system is obtained by setting $T_0\sigma_c$ to zero in Eq. 7.7:

$$W_{c,max} = (E - U_0) + P_0(V - V_0) - T_0(S - S_0)$$

By definition, the extensive property availability, A, is this maximum value. Accordingly, Eq. 7.2 is seen to be the appropriate expression for evaluating availability.

The value of availability cannot be negative. If the closed system were at any state other than the dead state, the system would be able to change its condition *spontaneously* toward the dead state; this tendency would cease when the dead state was reached. No work must be done to effect such a spontaneous change. Accordingly, since any change in state of the closed system to the dead state can be accomplished with *at least zero* work being developed, the *maximum* work (availability) cannot be negative.

Availability is not generally conserved but is destroyed by irreversibilities. A limiting case is when availability is completely destroyed, as would occur if the closed system were permitted to undergo a spontaneous change to the dead state with no provision to obtain work. The potential to develop work that existed originally would be completely wasted in such a spontaneous process.

The property availability has been viewed thus far as the *maximum* theoretical work obtainable from the combined system of system plus environment as a closed system passes *from* a given state *to* the dead state while interacting with the environment only. Alternatively, availability can be regarded as the magnitude of the *minimum* theoretical work *input* required to bring the closed system *from* the dead state *to* the given state. Using energy and entropy balances as above, we can readily develop Eq. 7.2 from this viewpoint.

7.2.4 Further Aspects of Availability

Availability is a measure of the departure of the state of a closed system from that of the environment. It is therefore an attribute of the closed system and environment together. However, once the environment is specified, a value can be assigned to availability in terms of property values for the closed system only, so availability can be regarded as a property of the closed system.

Although availability is an extensive property, it is often convenient to work with it on a unit mass or molar basis. The specific availability on a mass basis, a, is given by

$$a = (e - u_0) + p_0(v - v_0) - T_0(s - s_0) \tag{7.8}$$

where e, v, and s are the specific energy, volume, and entropy, respectively, at a given state; u_0, v_0, and s_0 are the same specific properties evaluated at the dead state. With $e = u + \mathrm{V}^2/2 + gz$,

$$a = [(u + \mathrm{V}^2/2 + gz) - u_0] + p_0(v - v_0) - T_0(s - s_0)$$

or

$$a = (u - u_0) + p_0(v - v_0) - T_0(s - s_0) + \mathrm{V}^2/2 + gz \tag{7.9}$$

The units of specific availability are the same as those of specific energy.

Using Eq. 7.2, we can determine the *change* in the availability between two states of a closed system as the difference

$$A_2 - A_1 = (E_2 - E_1) + p_0(V_2 - V_1) - T_0(S_2 - S_1) \tag{7.10}$$

where the values of p_0 and T_0 are determined by the state of the environment.

When a system is at the dead state, it is in *thermal* and *mechanical* equilibrium with the environment, and the value of availability is zero. We might say more precisely that the *thermomechanical* contribution to availability is zero. This modifying term distinguishes the availability concept of the present chapter from another aspect introduced in Sec. 13.6, where the contents of a system at the dead state are permitted to enter into chemical reaction with environmental components and in so doing develop additional work. As illustrated by subsequent discussions, the thermomechanical availability concept suffices for a wide range of thermodynamic evaluations, however.

The following examples illustrate the calculation of availability using relations introduced in this section.

EXAMPLE 7.1

PROBLEM

Two kilograms of water undergo a process from an initial state where the water is saturated vapor at 120°C, the velocity is 30 m/s, and the elevation is 6 m to a final state where the water is saturated liquid at 10°C, the velocity is 25 m/s, and the elevation is 3 m. Determine in kJ (a) the availability at the initial state, (b) the availability at the final state, and (c) the change in availability. Take $T_0 = 25°C$, $p_0 = 1$ atm. Let $g = 9.8$ m/s^2.

SOLUTION

Known: Two kilograms of water undergo a process between two specified states.

Find: Determine the availability at the initial and final states and the change in availability.

Schematic and Given Data:

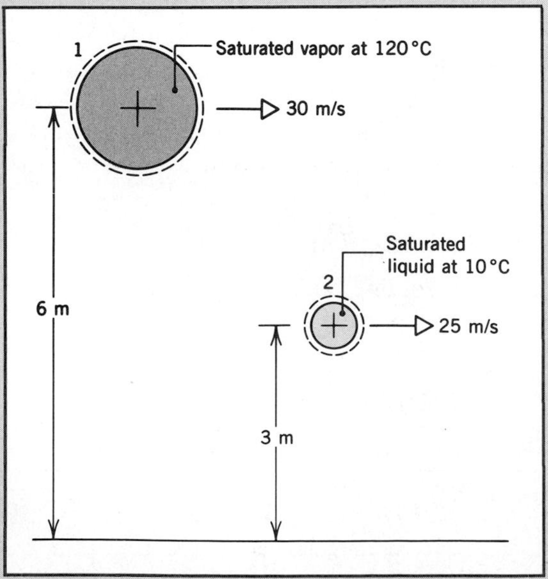

Figure E7.1

Assumptions:

1. The water is a closed system at equilibrium states initially and finally.
2. The velocities and elevations are measured relative to the environment.
3. $T_0 = 25°C$, $p_0 = 1$ atm, and $g = 9.8$ m/s^2.

Solution: The availability at the initial and final states can be determined using

$$A = m\left[(u - u_0) + p_0(v - v_0) - T_0(s - s_0) + \frac{V^2}{2} + gz\right]$$

To evaluate availability requires values of the specific volume, specific internal energy, and specific entropy of water at the dead state. At the dead state, where $T_0 = 25°C$ and $p_0 = 1$ atm, water is a liquid. Thus, with Eqs. 3.11, 3.12, and 6.10 and values from Table A-2, $v_0 = 1.0029 \times 10^{-3}$ m^3/kg, $u_0 = 104.88$ kJ/kg, $s_0 = 0.3674$ kJ/kg·K.

(a) For water as saturated vapor at 120°C, Table A-2 gives $p = 1.985$ bars, $v = 0.8919$ m^3/kg, $u = 2529.3$ kJ/kg, $s = 7.1296$ kJ/kg·K. Substituting into the above expression for availability, the availability at the initial state is

$$\begin{aligned} A_1 = (2\text{ kg})\Bigg\{&(2529.3 - 104.88)\frac{\text{kJ}}{\text{kg}} \\ &+ \left[\left(1.013 \times 10^5 \frac{\text{N}}{\text{m}^2}\right)(0.8919 - 1.0029 \times 10^{-3})\text{m}^3/\text{kg}\right]\left(\frac{1\text{ kJ}}{10^3\text{ N}\cdot\text{m}}\right) \\ &- (298\text{ K})(7.1296 - 0.3674)\frac{\text{kJ}}{\text{kg}\cdot\text{K}} \\ &+ \left[\frac{(30\text{ m/s})^2}{2} + (9.8\text{ m/s}^2)(6\text{ m})\right]\left(\frac{1\text{ N}}{1\text{ kg}\cdot\text{m/s}^2}\right)\left(\frac{1\text{ kJ}}{10^3\text{ N}\cdot\text{m}}\right)\Bigg\} \\ = {}& 2(2424.42 + 90.25 - 2015.14 + 0.45 + 0.06) \\ = {}& 1000.1\text{ kJ} \end{aligned}$$

(b) For water as saturated liquid at 10°C, $p = 0.01228$ bar, $v = 1.0004 \times 10^{-3}$ m^3/kg, $u = 42$ kJ/kg, $s = 0.151$ kJ/kg·K. The availability at the final state is $A_2 = 3.9$ kJ.

(c) The change in the availability of the system is

$$A_2 - A_1 = 3.9 - 1000.1 = -996.2\text{ kJ}$$

The minus sign indicates that the availability *decreases* in this process.

Comments:

1. Availability is a measure of the departure of the state of the system from that of the environment. At all states, $A \geq 0$. This applies when $T > T_0$ and $p > p_0$ as in part (a) and when $T < T_0$ and $p < p_0$ as in part (b).
2. The kinetic and potential energies of the system measured relative to the environment contribute their full values to the availability magnitude, for in principle each could be completely converted to work were the system brought to rest at zero elevation relative to the environment.
3. The change in availability can be determined more simply with Eq. 7.10. Equation 7.10 requires dead state property values only for T_0 and p_0. With the approach used in the solution, values for u_0, v_0, and s_0 are also required.

EXAMPLE 7.2

PROBLEM

A cylinder of an internal combustion engine contains 2450 cm^3 of gaseous combustion products at a pressure of 7 bars and a temperature of 867°C just before the exhaust valve opens. Determine the specific availability of the gas, in kJ/kg. Ignore the effects of motion and gravity, and model the combustion products as air as an ideal gas. Take $T_0 = 27°C$ and $p_0 = 1$ atm.

SOLUTION

Known: Gaseous combustion products at a specified state are contained in the cylinder of an internal combustion engine.

Find: Determine the specific availability.

Schematic and Given Data:

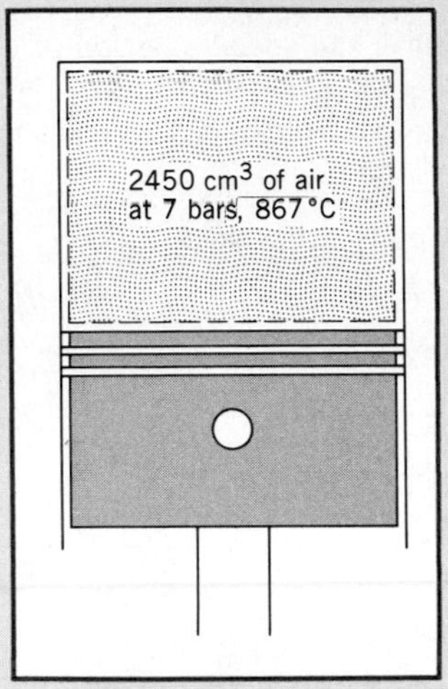

Figure E7.2

Assumptions:

1. The gaseous combustion products are a closed system.
2. The combustion products are modeled as air as an ideal gas.
3. The effects of motion and gravity can be ignored.
4. $T_0 = 27°C$ and $p_0 = 1$ atm.

Solution: With assumption 3, the specific availability is

$$a = u - u_0 + p_0(v - v_0) - T_0(s - s_0)$$

The internal energy and entropy terms are evaluated using Table A-16, as follows:

$$\begin{aligned} u - u_0 &= 880.35 - 214.07 \\ &= 666.28 \text{ kJ/kg} \\ s - s_0 &= s°(T) - s°(T_0) - \frac{\bar{R}}{M} \ln \frac{p}{p_0} \\ &= 3.11883 - 1.70203 - \left(\frac{8.314}{28.97}\right) \ln \left(\frac{7}{1.013}\right) \\ &= 0.8621 \text{ kJ/kg·K} \\ T_0(s - s_0) &= (300 \text{ K})(0.8621 \text{ kJ/kg·K}) \\ &= 258.62 \text{ kJ/kg} \end{aligned}$$

The $p_0(v - v_0)$ term is evaluated using the ideal gas equation of state: $v = (\bar{R}/M)T/p$ and $v_0 = (\bar{R}/M)T_0/p_0$, so

$$\begin{aligned} p_0(v - v_0) &= \frac{\bar{R}}{M}\left(\frac{p_0 T}{p} - T_0\right) \\ &= \frac{8.314}{28.97}\left[\frac{(1.013)(1140)}{7} - 300\right] \\ &= -38.75 \text{ kJ/kg} \end{aligned}$$

Substituting values into the above expression for the specific availability

$$a = 666.28 + (-38.75) - 258.62$$
$$= 368.91 \text{ kJ/kg}$$

Comment: If the gases are discharged directly to the surroundings, the potential for developing work quantified by the availability value determined in the solution is wasted. However, by venting the gases through a turbine some work could be developed. This principle is utilized by the *turbochargers* added to some internal combustion engines.

7.3 AVAILABILITY BALANCE FOR CLOSED SYSTEMS

A closed system at a given state can attain new states through work and heat interactions with other systems. Since the availability value associated with a new state would generally differ from the value at the initial state, transfers of availability across the system boundary can be inferred to *accompany* heat and work interactions. The change in availability of a system during a process would not equal the net availability transferred, however, for availability would be destroyed by irreversibilities present within the system during the process. The concepts of availability change, availability transfer, and availability destruction are related by the closed system availability balance introduced in this section. The availability balance concept is extended to control volumes in Sec. 7.5. These balances are expressions of the second law of thermodynamics and provide the basis for availability analysis.

7.3.1 Development of the Availability Balance

The availability balance for a closed system is developed by combining the closed system energy and entropy balances. The forms of the energy and entropy balances used in the development are, respectively

$$E_2 - E_1 = \int_1^2 \delta Q - W$$

$$S_2 - S_1 = \int_1^2 \left(\frac{\delta Q}{T}\right)_b + \sigma$$

where W and Q represent, respectively, work and heat transfers between the system under study and its surroundings. These interactions do not necessarily involve the environment. In the entropy balance, T_b denotes the temperature on the boundary where δQ is received and the term σ accounts for internal irreversibilities.

As the first step in deriving the availability balance, multiply the entropy balance by the temperature T_0 and subtract the resulting expression from the energy balance to obtain

$$(E_2 - E_1) - T_0(S_2 - S_1) = \int_1^2 \delta Q - T_0 \int_1^2 \left(\frac{\delta Q}{T}\right)_b - W - T_0\sigma$$

Collecting the terms involving δQ and introducing Eq. 7.10 on the left side, we can rewrite this expression as

$$(A_2 - A_1) - p_0(V_2 - V_1) = \int_1^2 \left(1 - \frac{T_0}{T_b}\right) \delta Q - W - T_0\sigma$$

Rearranging, the *closed system availability balance* results

$$\underbrace{A_2 - A_1}_{\text{availability change}} = \underbrace{\int_1^2 \left(1 - \frac{T_0}{T_b}\right) \delta Q - [W - p_0(V_2 - V_1)]}_{\text{availability transfers}} - \underbrace{T_0 \sigma}_{\text{availability destruction}} \tag{7.11}$$

Since Eq. 7.11 is obtained by deduction from the energy and entropy balances, it is not an independent result, but can be used in place of the entropy balance as an expression of the second law.

For specified end states and given values of p_0 and T_0, the availability change $A_2 - A_1$ on the left side of Eq. 7.11 can be evaluated from Eq. 7.10. The underlined terms on the right depend explicitly on the nature of the process, however, and cannot be determined by knowing only the end states and the values of p_0 and T_0. The first underlined term on the right side of Eq. 7.11 is associated with heat transfer to or from the system during the process. It can be interpreted as the *availability transfer accompanying heat*. That is

$$\begin{bmatrix}\text{availability transfer}\\ \text{accompanying heat}\end{bmatrix} = \int_1^2 \left(1 - \frac{T_0}{T_b}\right) \delta Q \tag{7.12}$$

The second underlined term on the right side of Eq. 7.11 is associated with work. It can be interpreted as the *availability transfer accompanying work*. That is

$$\begin{bmatrix}\text{availability transfer}\\ \text{accompanying work}\end{bmatrix} = [W - p_0(V_2 - V_1)] \tag{7.13}$$

The availability transfer expressions are discussed further in Sec. 7.3.2. The third underlined term on the right side of Eq. 7.11 accounts for the destruction of availability due to irreversibilities within the system. It is called the *irreversibility* and is symbolized by I.

$$I = T_0 \sigma \tag{7.14}$$

To summarize, Eq. 7.11 states that the change in availability of a closed system can be accounted for in terms of availability transfers and the destruction of availability due to irreversibilities within the system.

When applying the availability balance, it is essential to remember the requirements imposed by the second law on the irreversibility I: In accordance with the second law, the irreversibility is positive when irreversibilities are present within the system during the process and vanishes in the limiting case where there are no irreversibilities. That is

$$I: \begin{cases} > 0 & \text{irreversibilities present with the system} \\ = 0 & \text{no irreversibilities present within the system} \end{cases} \tag{7.15}$$

The value of the irreversibility cannot be negative. It is *not* a property. By contrast, availability is a property, and like other properties, the *change* in availability of a system can be positive, negative, or zero

$$A_2 - A_1: \begin{cases} > 0 \\ = 0 \\ < 0 \end{cases} \tag{7.16}$$

As in the case of the mass, energy, and entropy balances, the availability balance can be expressed in various forms that may be more appropriate for particular analyses. A form of the availability balance for closed systems that is sometimes convenient is the rate equation

$$\frac{dA}{dt} = \sum_j \left(1 - \frac{T_0}{T_j}\right) \dot{Q}_j - \left(\dot{W} - p_0 \frac{dV}{dt}\right) - \dot{I} \tag{7.17}$$

where dA/dt is the time rate of change of availability. The term $(1 - T_0/T_j)\dot{Q}_j$ represents the time rate of availability transfer accompanying heat transfer at the rate $\dot{Q}_j$ occurring at the location on the boundary where the instantaneous temperature is T_j. The term $\dot{W}$ represents the time rate of energy transfer by work. The accompanying availability transfer is given by $(\dot{W} - p_0\, dV/dt)$, where dV/dt is the time rate of change of system volume. The term $\dot{I}$ accounts for the time rate of availability destruction due to irreversibilities within the system and is related to the rate of entropy production within the system by $\dot{I} = T_0\dot{\sigma}$.

For an *isolated* system, no heat or work interactions with the surroundings occur, and thus there are no transfers of availability between the system and its surroundings. Accordingly, the availability balance reduces to give

$$\Delta A]_{\text{isol}} = -I_{\text{isol}} \tag{7.18}$$

Since I_{isol} must be positive in any actual process, the only processes of an isolated system allowed by the second law are those for which the availability of the isolated system *decreases*. For availability, this conclusion is the counterpart of the increase of entropy principle (Sec. 6.5.4) and, like the increase of entropy principle, can be regarded as an alternative statement of the second law.

To close our introduction to the availability balance concept, we note that most thermal systems are supplied with availability inputs derived directly or indirectly from the consumption of fossil fuels or other energy resources. Accordingly, destructions and avoidable losses of availability represent waste of some of the potential for use of these energy resources. By devising ways to reduce sources of inefficiency, better use can be made of fuels. The availability balance can be used to determine the locations, types, and magnitudes of waste of the potential of energy resources, and thus can play an important part in developing strategies for more effective fuel use.

7.3.2 Availability Transfer

Before taking up examples illustrating the use of the closed system availability balance, we consider why the availability transfer expressions take the forms they do. This is accomplished through simple thought experiments. Consider first a large metal part initially at the dead state. If the part was hoisted from a factory floor into a heat-treating furnace, the availability of the metal part would increase because its elevation would be increased. As the metal part was heated in the furnace, the availability of the part would increase further as its temperature increased because of heat transfer with the hot furnace gases. In a subsequent quenching process, the metal part would experience a decrease in availability as its temperature decreased due to heat transfer with the quenching medium. In each of these processes, the metal part would not actually interact with the environment used to assign availability values. However, like the availability values at the states visited by the metal part, the availability transfers taking place between the part and its surroundings would be evaluated *relative to the environment* used to define availability. The following paragraphs provide means for conceptualizing the availability transfers that accompany heat transfer and work.

Consider next a closed system undergoing a process in which a heat transfer Q takes place across a portion of the system boundary where the temperature T_b is constant at $T_b > T_0$. In accordance with Eq. 7.12, the accompanying availability transfer is given by

$$\begin{bmatrix}\text{availability transfer}\\ \text{accompanying heat}\end{bmatrix} = \left(1 - \frac{T_0}{T_b}\right) Q \tag{7.19}$$

The right side of this equation is recognized from Sec. 5.6 as the work that *could* be developed by a reversible power cycle receiving Q at temperature T_b and discharging energy by heat transfer to the environment at T_0. Accordingly, without regard for the nature of the surroundings with which the system is *actually* interacting, we may interpret the *magnitude* of an availability transfer accompanying heat transfer as the work that *could* be developed by supplying the heat transfer to a reversible power cycle operating between T_b and T_0. This interpretation also applies for heat transfer below T_0, but then we think of the *magnitude* of an availability transfer accompanying heat as the work that *could* be developed by a reversible power cycle receiving a heat transfer from the environment at T_0 and discharging Q at temperature $T_b < T_0$.

Thus far, we have considered only the *magnitude* of an availability transfer accompanying heat. It is necessary to account also for the *direction*. The form of Eq. 7.19 shows that when T_b is greater than T_0, the heat transfer and accompanying availability transfer would be in the same direction: Both quantities would be positive, or negative. However, when T_b is less than T_0, the sign of the availability transfer would be opposite to the sign of the heat transfer, so the heat transfer and accompanying availability transfer would be oppositely directed. To illustrate this

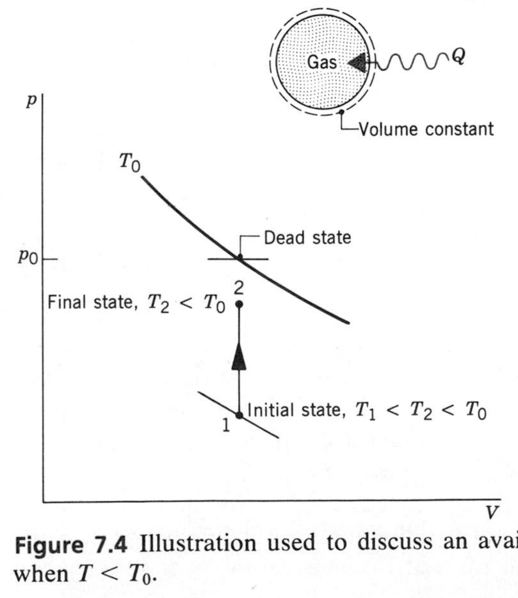

Figure 7.4 Illustration used to discuss an availability transfer accompanying heat transfer when $T < T_0$.

simply, refer to Fig. 7.4, which shows a system consisting of a gas heated at constant volume. As indicated by the p–V diagram, the initial and final temperatures of the gas are each less than T_0. Since the state of the system is brought closer to the dead state in this process, the availability of the system must decrease as it is heated. Conversely, were the gas cooled from state 2 to state 1, the availability of the system would increase because the state of the system would be moved farther from the dead state. Accordingly, when the temperature at the location where heat transfer occurs is *less* than the temperature of the environment, the heat transfer and accompanying availability transfer are *oppositely directed.* This becomes significant when studying the performance of refrigerators and heat pumps, where heat transfers can occur at temperatures below that of the environment.

We conclude the present discussion by taking up a simple example that motivates the form taken by the expression accounting for an availability transfer accompanying work, Eq. 7.13. Consider a closed system that does work W while undergoing an adiabatic process in which the system volume increases: $V_2 > V_1$. Although the system would not necessarily interact with the environment, the *magnitude* of the availability transfer is evaluated as the *maximum* work that *could* be obtained *were* the system and environment interacting. As illustrated by Fig. 7.5, all the work W of the system in the process would not be available for delivery from a combined system of system plus environment because a portion would be spent in pushing aside the environment, whose pressure is uniform at p_0. Since the system would do work on the surroundings equal to $p_0(V_2 - V_1)$, the maximum amount of work that could be derived from the combined system would thus be

$$W_c = W - p_0(V_2 - V_1)$$

which is in accordance with the form of Eq. 7.13. If there were no change in the system volume during the process, the transfer of availability accompanying work would equal the work W of the system.

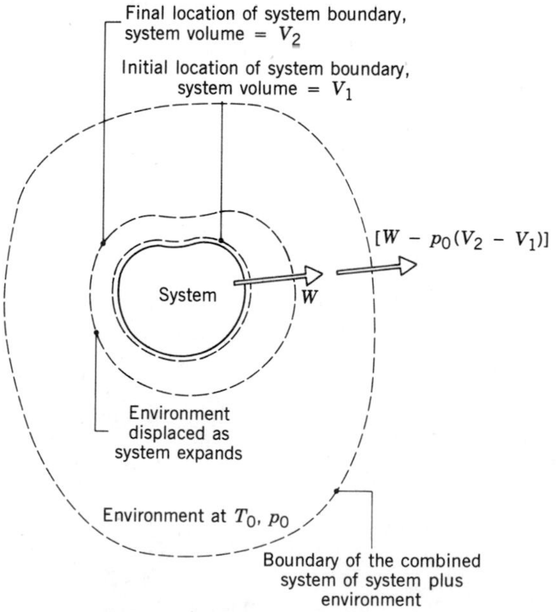

Figure 7.5 Illustration used to discuss the expression for an availability transfer accompanying work.

7.3.3 Illustrations

Further consideration of the availability balance and the availability transfer and destruction concepts is provided by the examples that follow.

EXAMPLE 7.3

PROBLEM

A 0.8-lb metal bar initially at 1900°R is removed from an oven and quenched by immersing it in a rigid tank containing 20 lb of water initially at 530°R. Each substance can be modeled as incompressible. An appropriate constant specific heat for the water is $c_w = 1.0$ Btu/lb·°R, and an appropriate value for the metal is $c_m = 0.1$ Btu/lb·°R. Heat transfer from the tank contents can be neglected. Determine the amount of availability destroyed, in Btu. Let $T_0 = 77°F$.

SOLUTION

Known: A hot metal bar is quenched by immersing it in a tank containing water.

Find: Determine the amount of availability destroyed.

Schematic and Given Data: See Fig. E6.6.

Assumptions:

1. The metal bar and the water within the tank form a system, as shown in the figure accompanying Example 6.6.
2. The system is isolated.
3. There are no significant kinetic or potential energy effects.
4. The water and metal bar are each modeled as incompressible with known specific heats.
5. $T_0 = 77°F$.

Analysis: Taking the metal bar and the water as the closed system, an availability balance reduces to

$$\Delta A = \overbrace{\int_1^2 \left(1 - \frac{T_0}{T_b}\right) \delta Q}^{0} - (W - \overbrace{p_0 \, \Delta V}^{0}) - I$$

The indicated terms vanish because the system is isolated (assumption 2). The availability balance becomes simply

$$I = -\Delta A$$

Since availability is an extensive property

$$\Delta A = \Delta A]_{\text{metal}} + \Delta A]_{\text{water}}$$

With Eq. 7.10

$$\Delta A = [\Delta U + \overbrace{p_0 \Delta V}^{0} - T_0 \, \Delta S]_{\text{metal}} + [\Delta U + \overbrace{p_0 \Delta V}^{0} - T_0 \, \Delta S]_{\text{water}}$$

where the kinetic and potential energy terms are dropped by assumption 3, and $\Delta V = 0$ because the metal and water are incompressible (assumption 4). From an energy balance, $\Delta U]_{\text{metal}} + \Delta U]_{\text{water}} = 0$, so the above expression for ΔA reduces to

$$\Delta A = -T_0(\Delta S]_{\text{metal}} + \Delta S]_{\text{water}})$$

Combining this with the availability balance gives

$$I = T_0(\Delta S]_{\text{metal}} + \Delta S]_{\text{water}})$$

The term within parentheses is the amount of entropy produced in the process, which is evaluated in the solution to Example 6.6 as

$$\Delta S]_{\text{metal}} + \Delta S]_{\text{water}} = 0.0864 \text{ Btu/°R}$$

Hence, the irreversibility is

$$I = (537°\text{R})(0.0864 \text{ Btu/°R}) = 46.4 \text{ Btu}$$

Comments:

1. The energy of the isolated system remains constant in this process. However, the availability of the isolated system decreases. The decrease in the availability of the system is due to availability destruction.
2. The irreversibility I can be found alternatively using $I = T_0\sigma$, where the value of the entropy production is obtained from an entropy balance.

EXAMPLE 7.4

PROBLEM

Water initially a saturated liquid at 100°C is contained in a piston–cylinder assembly. The water undergoes a process to the corresponding saturated vapor state, during which the piston moves freely in the cylinder. For each of the two processes described below, determine on a unit of mass basis the change in availability, the availability transfer accompanying work, the availability transfer accompanying heat, and the irreversibility, each in kJ/kg. Let $T_0 = 20°\text{C}$, $p_0 = 1.014$ bars.

(a) The change in state is brought about by heating the water as it undergoes an internally reversible process at constant temperature and pressure.

(b) The change in state is brought about adiabatically by the stirring action of a paddle wheel.

SOLUTION

Known: Saturated liquid at 100°C undergoes a process to the corresponding saturated vapor state.

Find: Determine the change in availability, the availability transfers accompanying work and heat, and the irreversibility for each of two specified processes.

Schematic and Given Data: See Figs. E6.2 and E6.3.

Assumptions:

1. For part (a), see the assumptions listed for Example 6.2. For part (b), see the assumptions listed for Example 6.3.
2. $T_0 = 20°\text{C}$, $p_0 = 1.014$ bars.

Analysis:

(a) The change in specific availability is

$$\Delta a = u_g - u_f + p_0(v_g - v_f) - T_0(s_g - s_f)$$

Using data from Table A-2

$$\Delta a = 2087.56 \frac{\text{kJ}}{\text{kg}} + \frac{(1.014 \times 10^5 \text{ N/m}^2)(1.672 \text{ m}^3/\text{kg})}{(10^3 \text{ N}\cdot\text{m/kJ})} - (293.15 \text{ K})(6.048 \text{ kJ/kg}\cdot\text{K})$$

$$= 484 \text{ kJ/kg}$$

Using the expression for work obtained in the solution to Example 6.2, $W/m = pv_{fg}$, the transfer of availability accompanying work is

$$\begin{bmatrix}\text{transfer of availability}\\ \text{accompanying work}\end{bmatrix} = \frac{W}{m} - p_0(v_g - v_f)$$

$$= (p - p_0)v_{fg} = 0$$

Although the work has a nonzero value, there is no accompanying availability transfer because $p = p_0$.

Using the heat transfer value calculated in Example 6.2, the transfer of availability accompanying heat transfer in the constant-temperature process is

$$\begin{bmatrix}\text{transfer of availability}\\ \text{accompanying heat transfer}\end{bmatrix} = \left(1 - \frac{T_0}{T}\right)\frac{Q}{m}$$

$$= \left(1 - \frac{293.15\ \text{K}}{373.15\ \text{K}}\right)\left(2257\ \frac{\text{kJ}}{\text{kg}}\right)$$

$$= 484\ \text{kJ/kg}$$

The positive value indicates that availability transfer occurs in the same direction as the heat transfer.

Since the process is accomplished without irreversibilities, the irreversibility I is necessarily zero in value. This can be verified by inserting the three availability quantities evaluated above into an availability balance.

(b) Since the end states are the same as in part (a), the change in availability is the same. Moreover, because there is no heat transfer, there is no availability transfer accompanying heat. The availability transfer accompanying work is

$$\begin{bmatrix}\text{transfer of availability}\\ \text{accompanying work}\end{bmatrix} = \frac{W}{m} - p_0(v_g - v_f)$$

With the net work value determined in Example 6.3 and evaluating the change in specific volume as in part (a)

$$\begin{bmatrix}\text{transfer of availability}\\ \text{accompanying work}\end{bmatrix} = -2087.56 - \frac{(1.014 \times 10^5)(1.672)}{10^3}$$

$$= -2257\ \text{kJ/kg}$$

The minus sign indicates that the net transfer of availability accompanying work is into the system.

Finally, the irreversibility is determined from an availability balance. Solving for the irreversibility per unit of mass

$$\frac{I}{m} = -\Delta a - \left[\frac{W}{m} - p_0(v_g - v_f)\right] = -484 - (-2257) = 1773\ \text{kJ/kg}$$

The numerical values obtained can be interpreted as follows: 2257 kJ/kg of availability is transferred into the system accompanying work; of this, 1773 kJ/kg is destroyed by irreversibilities, leaving a net increase of only 484 kJ/kg.

Comments:

1. This example illustrates that the availability change during a process is determined by the end states, but the values of the availability transfers and the irreversibility depend on the nature of the process.
2. The irreversibility value of part (b) can also be determined using $I/m = T_0(\sigma/m)$, where the entropy produced per unit of mass, σ/m, is obtained from the solution to Example 6.3.

EXAMPLE 7.5

PROBLEM

A heat transfer equal to 5 Btu takes place through the inner surface of a wall. Measurements made during steady-state operation reveal temperatures of $T_1 = 2500°\text{R}$ and $T_2 = 1000°\text{R}$ at the inner and outer surfaces, respectively. Determine in Btu (a) the amount of availability transfer accompanying heat at the inner and outer surfaces of the wall and (b) the irreversibility. Let $T_0 = 500°\text{R}$.

SOLUTION

Known: There is heat transfer into and out of a wall for which the temperatures at the inner and outer surfaces are known.

Find: Determine the amounts of availability transfer accompanying heat at the inner and outer surfaces of the wall and the irreversibility.

Schematic and Given Data:

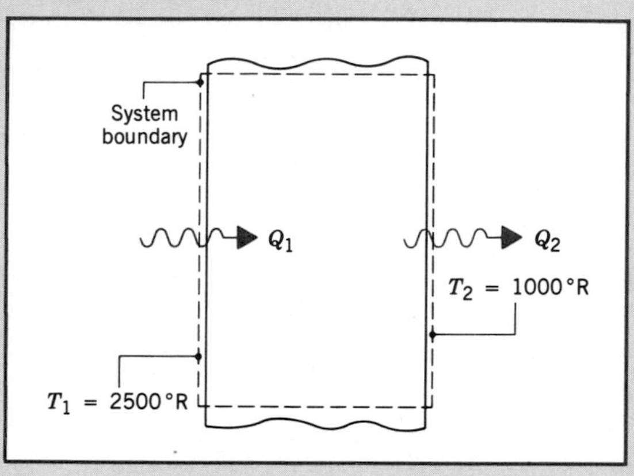

Figure E7.5

Assumptions:

1. The system shown in the accompanying figure is at steady state.
2. $T_0 = 500°R$.

Analysis:

(a) At steady state, no property changes with time, so $\Delta E = 0$. As there is also no work, an energy balance reduces to give $Q_1 = Q_2 = 5$ Btu. The directions of these heat transfers are shown by arrows on the accompanying figure.

Since the temperature T_1 is uniform over the inner surface, the accompanying availability transfer is

$$\begin{bmatrix}\text{availability transfer} \\ \text{by heat at the} \\ \text{inner surface}\end{bmatrix} = \left(1 - \frac{T_0}{T_1}\right) Q_1$$

$$= \left(1 - \frac{500}{2500}\right) (5 \text{ Btu}) = 4 \text{ Btu}$$

This value is positive, indicating that the availability transfer is in the same direction as the heat transfer.

The availability transfer at the outer surface is

$$\begin{bmatrix}\text{availability transfer} \\ \text{by heat at the} \\ \text{outer surface}\end{bmatrix} = \left(1 - \frac{T_0}{T_2}\right) Q_2$$

$$= \left(1 - \frac{500}{1000}\right) (5 \text{ Btu}) = 2.5 \text{ Btu}$$

(b) The irreversibility can be determined from an availability balance. At steady state, $\Delta A = \Delta V = 0$. There is no work. Hence,

$$I = \left(1 - \frac{T_0}{T_1}\right) Q_1 - \left(1 - \frac{T_0}{T_2}\right) Q_2$$

Inserting values from part (a)

$$I = 4 \text{ Btu} - 2.5 \text{ Btu} = 1.5 \text{ Btu}$$

Comments:

1. The term $[1 - (T_0/T_1)]Q_1$ is the work that could be obtained by supplying Q_1 to a reversible power cycle operating between T_1 and T_0. Likewise, $[1 - (T_0/T_2)]Q_2$ is the work that could be obtained by supplying Q_2 to a reversible power cycle operating between T_2 and T_0. The difference between these two values, the irreversibility I, represents an opportunity to develop work that is irrevocably lost in the spontaneous conduction process.
2. Although the same amount of energy is transferred through the inner and outer surfaces, the availability transfer at the outer surface is less because the difference between the environment temperature T_0 and T_2 is smaller than between T_0 and T_1.
3. If the availability entering the wall is derived from the combustion of a fuel, the value of the irreversibility represents a waste of the fuel's potential for use.

EXAMPLE 7.6

PROBLEM

During steady-state operation, a parallel-shaft gearbox receives 600 kW through the high-speed shaft but, owing to friction and other irreversibilities, delivers 588 kW through the low-speed shaft. The gearbox is cooled on its outer surface according to $\dot{Q} = -hA(T_b - T_0)$, where h is the *heat transfer coefficient,* A the outer surface area, T_b the uniform temperature of the outer surface, and T_0 the uniform temperature of the surroundings far from the gearbox (the temperature of the environment). For the gearbox, determine in kW (a) the time rate of availability transfer accompanying heat, (b) the time rate of availability transfer accompanying work, and (c) the irreversibility rate. Let $h = 0.17\ \text{kW/m}^2\cdot\text{K}$, $A = 1.8\ \text{m}^2$, and $T_0 = 293$ K.

SOLUTION

Known: A gearbox operates at steady state with a known power input through the high-speed shaft and a known power output through the low-speed shaft. An expression for the heat transfer rate from the outer surface is also known.

Find: Determine the time rates of availability transfer accompanying heat and work and the irreversibility rate.

Schematic and Given Data: See Fig. E6.5.

Assumptions:

1. See the solution to Example 6.5.
2. $T_0 = 293$ K.

Analysis:

(a) Since the temperature T_b at the outer surface of the gearbox is uniform with position and constant with time,

$$\begin{bmatrix}\text{time rate of availability}\\ \text{transfer accompanying heat}\end{bmatrix} = \left(1 - \frac{T_0}{T_b}\right)\dot{Q}$$

From the solution to Example 6.5, $\dot{Q} = -12$ kW and $T_b = 332$ K. Hence

$$\begin{bmatrix}\text{time rate of availability}\\ \text{transfer accompanying heat}\end{bmatrix} = \left(1 - \frac{293}{332}\right)(-12\ \text{kW}]$$

$$= -1.41\ \text{kW}$$

where the minus sign denotes a transfer from the system.

(b) At steady state there is no volume change, so

$$\begin{bmatrix}\text{time rate of availability}\\ \text{transfer accompanying work}\end{bmatrix} = \dot{W} - p_0 \overset{0}{\cancel{\frac{dV}{dt}}}$$
$$= \dot{W} = -12 \text{ kW}$$

The minus sign indicates that a net amount of availability is transferred into the system accompanying work: 600 kW is carried *in* through the high-speed shaft and 588 kW is carried *out* through the low-speed shaft.

(c) For the closed system at steady state, the availability rate balance reduces to

$$\overset{0}{\cancel{\frac{dA}{dt}}} = \left(1 - \frac{T_0}{T_b}\right)\dot{Q} - \left(\dot{W} - p_0 \overset{0}{\cancel{\frac{dV}{dt}}}\right) - \dot{I}$$

Solving for $\dot{I}$

$$\dot{I} = \left(1 - \frac{T_0}{T_b}\right)\dot{Q} - \dot{W} = -1.41 - (-12) = 10.59 \text{ kW}$$

Comments:

1. The analysis can be summarized by the following availability *balance sheet* in terms of availability *magnitudes* on a rate basis:

Rate of Availability In		Rate of Availability Out	
High-speed shaft:	600 kW	Low-speed shaft:	588.00 kW
Heat transfer:	0	Heat transfer:	1.41 kW
Total:	600 kW	Total:	589.41 kW

The difference between these totals is the rate at which availability is destroyed: 10.59 kW.

2. The availability transfer to the surroundings accompanying heat has a potential for use. For example, the gearbox casing could be water-cooled and the heated water used in some way. When it is impractical to utilize such an availability transfer, it would then be accounted for as a *loss* of availability charged against system performance.

7.4 FLOW AVAILABILITY

The objective of the present section is to develop the *flow availability* concept. This concept is important for the control volume form of the availability rate balance introduced in Sec. 7.5 to follow. As a prerequisite to consideration of the flow availability, the concept of an availability transfer accompanying *flow work* is introduced.

7.4.1 Availability Transfer Accompanying Flow Work

When one-dimensional flow is assumed, the work at the inlet or exit of a control volume, the *flow work*, is given on a time rate basis by $\dot{m}(pv)$, where $\dot{m}$ is the mass flow rate, p is the pressure, and v is the specific volume at the inlet or exit (Sec. 4.2.2). The objective of the present discussion is to introduce the following expression, which accounts for the availability transfer accompanying flow work:

$$\begin{bmatrix}\text{time rate of availability transfer}\\ \text{accompanying flow work}\end{bmatrix} = \dot{m}(pv - p_0 v) \tag{7.20}$$

Let us develop Eq. 7.20 for the case pictured in Fig. 7.6. The figure shows a closed system that occupies different regions at time t and a later time $t + \Delta t$. During the

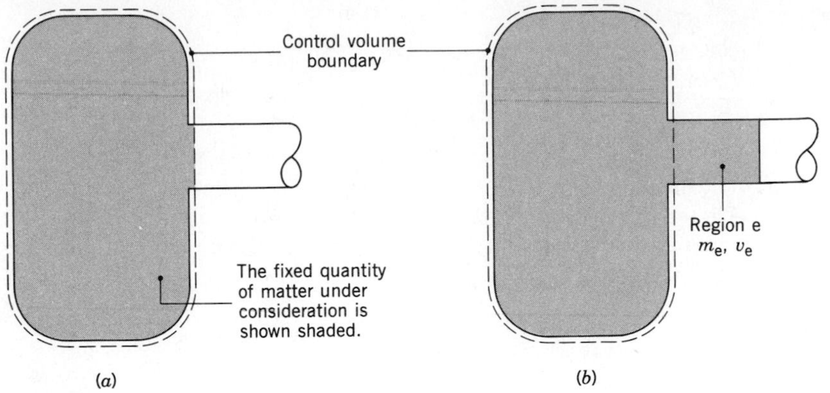

Figure 7.6 Illustration used to introduce the flow availability concept. (*a*) Time t. (*b*) Time $t + \Delta t$.

time interval Δt, some of the mass initially within the region labeled *control volume* exits to fill the small region e adjacent to the control volume, as shown in Fig. 7.6*b*. We assume that the increase in the volume of the closed system in the time interval Δt is equal to the volume of region e and, for further simplicity, that the only work is associated with this volume change. With Eq. 7.13, the availability transfer accompanying work is

$$\begin{bmatrix}\text{availability transfer}\\ \text{accompanying work}\end{bmatrix} = W - p_0\,\Delta V \tag{7.21}$$

where ΔV is the volume change of the system. The volume change of the system equals the volume of region e. Thus, we may write $\Delta V = m_e v_e$, where m_e is the mass within region e and v_e is the specific volume, which is regarded as uniform throughout region e. With this expression for ΔV, Eq. 7.21 becomes

$$\begin{bmatrix}\text{availability transfer}\\ \text{accompanying work}\end{bmatrix} = W - m_e(p_0 v_e) \tag{7.22}$$

Equation 7.22 can be placed on a time rate basis by dividing each term by the time interval Δt and taking the limit as Δt approaches zero. That is

$$\begin{bmatrix}\text{time rate of availability}\\ \text{transfer accompanying work}\end{bmatrix} = \lim_{\Delta t\to 0}\left(\frac{W}{\Delta t}\right) - \lim_{\Delta t\to 0}\left[\frac{m_e}{\Delta t}(p_0 v_e)\right] \tag{7.23}$$

In the limit as Δt approaches zero, the boundaries of the closed system and control volume coincide. Accordingly, in this limit the rate of energy transfer by work from the closed system is also the rate of energy transfer by work from the control volume. For the present case, this is just the flow work. Thus, the first term on the right side of Eq. 7.23 becomes

$$\lim_{\Delta t\to 0}\left(\frac{W}{\Delta t}\right) = \dot{m}_e(p_e v_e) \tag{7.24}$$

where $\dot{m}_e$ is the mass flow rate at the exit of the control volume. In the limit as Δt approaches zero, the second term on the right side of Eq. 7.23 becomes

$$\lim_{\Delta t\to 0}\left[\frac{m_e}{\Delta t}(p_0 v_e)\right] = \dot{m}_e(p_0 v_e) \tag{7.25}$$

In this limit, the assumption of uniform specific volume throughout region e corresponds to the assumption of uniform specific volume across the exit (one-dimensional flow).

Substituting Eqs. 7.24 and 7.25 into Eq. 7.23 gives

$$\begin{bmatrix}\text{time rate of availability transfer}\\ \text{accompanying flow work}\end{bmatrix} = \dot{m}_e(p_e v_e) - \dot{m}_e(p_0 v_e) \tag{7.26}$$
$$= \dot{m}_e(p_e v_e - p_0 v_e)$$

Extending the reasoning given here, it can be shown that an expression having the same form as Eq. 7.26 accounts for the transfer of availability accompanying flow work at inlets to control volumes as well. The general result applying at both inlets and exits is given by Eq. 7.20.

7.4.2 Development of the Flow Availability Concept

With the introduction of the expression for the availability transfer accompanying flow work, attention now turns to the flow availability. When mass flows across the boundary of a control volume, there is an accompanying energy transfer given by

$$\begin{bmatrix}\text{time rate of energy transfer}\\ \text{accompanying mass flow}\end{bmatrix} = \dot{m}e \tag{7.27}$$
$$= \dot{m}\left(u + \frac{V^2}{2} + gz\right)$$

where e is the specific energy evaluated at the inlet or exit under consideration. Likewise, when mass enters or exits a control volume, there is an accompanying availability transfer given by

$$\begin{bmatrix}\text{time rate of availability transfer}\\ \text{accompanying mass flow}\end{bmatrix} = \dot{m}a \tag{7.28}$$
$$= \dot{m}[(e - u_0) + p_0(v - v_0) - T_0(s - s_0)]$$

where a is the specific availability at the inlet or exit under consideration. In writing Eqs. 7.27 and 7.28, one-dimensional flow is assumed. In addition to an energy transfer accompanying mass flow, energy transfer by flow work also takes place at locations where mass enters or exits a control volume. Transfers of availability accompanying flow work are accounted for by Eq. 7.20.

Since transfers of availability accompanying mass flow and flow work occur at locations where mass enters or exits a control volume, a single expression giving the sum of these effects is convenient. Thus, with Eqs. 7.20 and 7.28,

$$\begin{bmatrix}\text{time rate of availability transfer}\\ \text{accompanying mass flow } \textit{and} \text{ flow work}\end{bmatrix} = \dot{m}[a + (pv - p_0 v_0)] \tag{7.29}$$
$$= \dot{m}[\underline{(e - u_0) + p_0(v - v_0) - T_0(s - s_0)}$$
$$+ \underline{(pv - p_0 v)}]$$

The underlined terms in Eq. 7.29 represent, per unit of mass, the availability transfer accompanying mass flow and flow work, respectively. The sum identified by underlining is called the *specific flow availability* and given the symbol a_f. Thus, the specific flow availability is

$$a_f = (e - u_0) + p_0(v - v_0) - T_0(s - s_0) + (pv - p_0 v) \tag{7.30}$$

The specific flow availability can be placed in a more convenient form for calculation by introducing $e = u + V^2/2 + gz$ in Eq. 7.30 and simplifying to obtain

$$a_f = \left(u + \frac{V^2}{2} + gz - u_0\right) + (pv - p_0 v_0) - T_0(s - s_0)$$

With $h = u + pv$ and $h_0 = u_0 + p_0v_0$, this becomes

$$a_f = h - h_0 - T_0(s - s_0) + \frac{V^2}{2} + gz \tag{7.31}$$

In Eq. 7.31, h and s represent the specific enthalpy and entropy, respectively, at the inlet or exit under consideration; h_0 and s_0 represent the values of these properties when evaluated at the dead state. Equation 7.31 giving the flow availability is the principal result of this section and is used in the next section to formulate the availability rate balance for control volumes. The flow availability is also commonly referred to as *flow exergy*, or simply as *exergy*.

A comparison of the current development with that of Sec. 4.2 shows that the flow availability evolves here in the same way as does enthalpy in the development of the control volume energy rate balance, and they have similar interpretations: Each quantity is a sum consisting of a term associated with the flowing mass (specific internal energy for enthalpy, specific availability for flow availability) and a contribution associated with flow work at the inlet or exit under consideration.

7.5 AVAILABILITY RATE BALANCE FOR CONTROL VOLUMES

In this section, the availability balance is extended to a form applicable to control volumes. The control volume form is generally the most useful for engineering analysis.

The availability rate balance for a control volume can be derived using an approach like that employed in Secs. 4.1 and 4.2, where the control volume forms of the mass and energy rate balances are obtained by transforming the closed system forms. However, as in the development of the entropy rate balance for control volumes (Sec. 6.6), the present derivation is conducted less formally by modifying the closed system rate form, Eq. 7.17, to account for the availability transfers accompanying mass flow and flow work at the inlets and exits. The result is

$$\underbrace{\frac{dA_{cv}}{dt}}_{\substack{\text{rate of}\\\text{availability}\\\text{change}}} = \underbrace{\sum_j \left(1 - \frac{T_0}{T_j}\right)\dot{Q}_j - \left(\dot{W}_{cv} - p_0\frac{dV_{cv}}{dt}\right) + \sum_i \dot{m}_i a_{fi} - \sum_e \dot{m}_e a_{fe}}_{\substack{\text{rates of}\\\text{availability}\\\text{transfer}}} - \underbrace{\dot{I}_{cv}}_{\substack{\text{rate of}\\\text{availability}\\\text{destruction}}} \tag{7.32}$$

As for control volume rate balances considered previously, i denotes inlets and e denotes exits.

In Eq. 7.32 the term dA_{cv}/dt represents the time rate of change of the availability of the control volume. The term $\dot{Q}_j$ represents the time rate of heat transfer at the location on the boundary where the instantaneous temperature is T_j. The accompanying availability transfer is given by $(1 - T_0/T_j)\dot{Q}_j$. The term $\dot{W}_{cv}$ represents the time rate of energy transfer by work *other than flow work*. The accompanying availability transfer is given by $(\dot{W}_{cv} - p_0\, dV_{cv}/dt)$, where dV_{cv}/dt is the time rate of change of volume. The term $\dot{m}_i a_{fi}$ accounts for the time rate of availability transfer accompanying mass flow *and* flow work at inlet i. Similarly, $\dot{m}_e a_{fe}$ accounts for the time rate of availability transfer accompanying mass flow *and* flow work at exit e. The flow availabilities a_{fi} and a_{fe} appearing in these expressions are evaluated from Eq. 7.31. Finally, the term $\dot{I}_{cv}$ accounts for the time rate of availability destruction due to irreversibilities *within* the control volume.

Since a great many engineering analyses involve control volumes at steady state, it is convenient to list steady-state forms of the availability rate balance. At steady state, $dA_{cv}/dt = dV_{cv}/dt = 0$, so Eq. 7.32 reduces to

$$0 = \sum_j \left(1 - \frac{T_0}{T_j}\right) \dot{Q}_j - \dot{W}_{cv} + \sum_i \dot{m}_i a_{fi} - \sum_e \dot{m}_e a_{fe} - \dot{I}_{cv} \tag{7.33}$$

This equation states that the rate at which availability is transferred into the control volume must *exceed* the rate at which availability is transferred out, the difference being the rate at which availability is destroyed within the control volume due to irreversibilities.

If there is a single inlet and a single exit, denoted by 1 and 2, respectively, Eq. 7.33 becomes

$$0 = \sum_j \left(1 - \frac{T_0}{T_j}\right) \dot{Q}_j - \dot{W}_{cv} + \dot{m}(a_{f1} - a_{f2}) - \dot{I}_{cv} \tag{7.34}$$

where $\dot{m}$ is the mass flow rate. The term $(a_{f1} - a_{f2})$ is evaluated using Eq. 7.31 as

$$a_{f1} - a_{f2} = (h_1 - h_2) - T_0(s_1 - s_2) + \frac{V_1^2 - V_2^2}{2} + g(z_1 - z_2) \tag{7.35}$$

The following examples illustrate the use of the mass, energy, and availability rate balances for the analysis of control volumes at steady state. Property data also play an important role in arriving at solutions. When the irreversibility $\dot{I}_{cv}$ is the objective, it can be determined either from an availability balance or from $\dot{I}_{cv} = T_0\dot{\sigma}_{cv}$, where $\dot{\sigma}_{cv}$ is the rate of entropy production evaluated from an entropy balance. The second of these procedures normally requires less property evaluation and computation.

EXAMPLE 7.7

PROBLEM

Superheated water vapor enters a valve at 500 lbf/in.², 500°F and exits at a pressure of 80 lbf/in.² The expansion is a throttling process. Determine the specific flow availability at the inlet and exit and the irreversibility per unit of mass flowing, each in Btu/lb. Let $T_0 = 77°F$, $p_0 = 1$ atm.

SOLUTION

Known: Water vapor expands in a throttling process through a valve from a specified inlet state to a specified exit pressure.

Find: Determine the specific flow availability at the inlet and exit of the valve and the irreversibility per unit of mass flowing.

Schematic and Given Data:

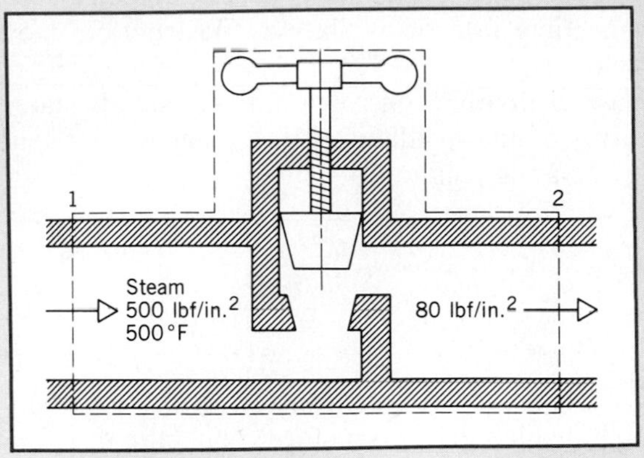

Figure E7.7

Assumptions:

1. The control volume shown in the accompanying figure is at steady state.
2. For the throttling process, $\dot{Q}_{cv} = \dot{W}_{cv} = 0$, and kinetic and potential energy effects can be ignored.
3. $T_0 = 77°F$, $p_0 = 1$ atm.

Analysis: The state at the inlet is specified. The state at the exit can be fixed by reducing the steady-state mass and energy rate balances to obtain

$$h_2 = h_1$$

Thus, the exit state is fixed by p_2 and h_2. From Table A-4E, $h_1 = 1231.5$ Btu/lb, $s_1 = 1.4923$ Btu/lb·°R. Interpolating at a pressure of 80 lbf/in.^{2} with $h_2 = h_1$, the specific entropy at the exit is $s_2 = 1.680$ Btu/lb·°R. Evaluating h_0 and s_0 at the saturated liquid state corresponding to T_0, Table A-2E gives $h_0 = 45.09$ Btu/lb, $s_0 = 0.0878$ Btu/lb·°R.

Dropping $V^2/2$ and gz, we obtain the specific flow availability from Eq. 7.31 as

$$a_f = h - h_0 - T_0(s - s_0)$$

Substituting values into the expression for a_f, the flow availability at the inlet is

$$a_{f1} = (1231.5 - 45.09) - 537(1.4923 - 0.0878) = 432.2 \text{ Btu/lb}$$

At the exit

$$a_{f2} = (1231.5 - 45.09) - 537(1.680 - 0.0878) = 331.4 \text{ Btu/lb}$$

With assumptions listed, the steady-state form of the availability rate balance, Eq. 7.34, reduces to

$$0 = \sum_j \cancelto{0}{\left(1 - \frac{T_0}{T_j}\right) \dot{Q}_j} - \cancelto{0}{\dot{W}_{cv}} + \dot{m}(a_{f1} - a_{f2}) - \dot{I}_{cv}$$

Dividing by the mass flow rate $\dot{m}$ and solving for the irreversibility

$$\frac{\dot{I}_{cv}}{\dot{m}} = (a_{f1} - a_{f2})$$

Inserting values

$$\frac{\dot{I}_{cv}}{\dot{m}} = 432.2 - 331.4 = 100.8 \text{ Btu/lb}$$

Comments:

1. Energy is conserved in the throttling process, but availability is destroyed. The source of the availability destruction is the uncontrolled expansion that occurs.

2. Since $h_1 = h_2$, the expression for the irreversibility developed in the solution reduces to

$$\frac{\dot{I}_{cv}}{\dot{m}} = T_0(s_2 - s_1)$$

With this expression, the irreversibility can be determined knowing only T_0 and the specific entropies s_1 and s_2. The above equation can be obtained alternatively beginning with the relationship $\dot{I}_{cv} = T_0\dot{\sigma}_{cv}$ and then evaluating the rate of entropy production $\dot{\sigma}_{cv}$ from an entropy balance.

EXAMPLE 7.8

PROBLEM

Steam enters a turbine with a pressure of 30 bars, a temperature of 400°C, and a velocity of 160 m/s. Steam exits as saturated vapor at 100°C with a velocity of 100 m/s. At steady state, the turbine develops work at a rate of 540 kJ per kg of steam flowing through the turbine. Heat transfer between the turbine and its surroundings occurs at an average outer surface temperature of 500 K. Determine the irreversibility per unit mass of steam flowing through the turbine, in kJ/kg. Neglect the change in potential energy between inlet and exit. Let $T_0 = 25°C$, $p_0 = 1$ atm.

SOLUTION

Known: Steam expands through a turbine for which steady-state data are provided.

Find: Determine the irreversibility per unit mass of steam flowing through the turbine.

Schematic and Given Data: See Fig. E6.7.

Assumptions:

1. The turbine is at steady state.
2. Heat transfer between the turbine and the surroundings occurs at a known temperature.
3. The change in potential energy between inlet and exit can be neglected.
4. $T_0 = 25°C$, $p_0 = 1$ atm.

Analysis: The irreversibility can be determined by rearranging the steady-state form of the availability rate balance, Eq. 7.34, to give

$$\frac{\dot{I}_{cv}}{\dot{m}} = \left(1 - \frac{T_0}{T_b}\right)\left(\frac{\dot{Q}_{cv}}{\dot{m}}\right) - \frac{\dot{W}_{cv}}{\dot{m}} + (a_{f1} - a_{f2})$$

where T_b denotes the temperature on the boundary where heat transfer occurs.

The quantity $\dot{Q}_{cv}/\dot{m}$ is evaluated in the solution to Example 6.7 using the steady-state forms of the mass and energy rate balances: $\dot{Q}_{cv}/\dot{m} = -22.6$ kJ/kg. The term $\dot{W}_{cv}/\dot{m}$ is given as 540 kJ/kg. The difference in flow availability is evaluated with Eq. 7.35 as

$$a_{f1} - a_{f2} = (h_1 - h_2) - T_0(s_1 - s_2) + \left(\frac{V_1^2 - V_2^2}{2}\right)$$

where the potential energy term is dropped by assumption 3. From Table A-4, $h_1 = 3230.9$ kJ/kg, $s_1 = 6.9212$ kJ/kg·K. From Table A-2, $h_2 = 2676.1$ kJ/kg, $s_2 = 7.3549$ kJ/kg·K. Hence

$$a_{f1} - a_{f2} = \left[(3230.9 - 2676.1) - 298(6.9212 - 7.3549) + \frac{(160)^2 - (100)^2}{2(10^3)}\right]$$

$$= 691.84 \text{ kJ/kg}$$

Substituting values into the expression for irreversibility

$$\frac{\dot{I}_{cv}}{\dot{m}} = \left(1 - \frac{298}{500}\right)(-22.6) - 540 + 691.84$$

$$= -9.13 - 540 + 691.84 = 142.71 \text{ kJ/kg}$$

Comments:

1. The availability values determined in the solution can be summarized as follows.

   Net availability supplied to the turbine:

$$a_{f1} - a_{f2} = 691.84 \text{ kJ/kg}$$

   Disposition of the availability supplied:

work =	540.00 kJ/kg	(77.3%)
availability destroyed =	142.71 kJ/kg	(20.4%)
transfer out accompanying heat transfer =	9.13 kJ/kg	(1.3%)
	691.84 kJ/kg	(100.0%)

2. The irreversibility can be determined alternatively using $\dot{I}_{cv} = T_0\dot{\sigma}_{cv}$, where $\dot{\sigma}_{cv}$ is the rate of entropy production from an entropy balance. The solution to Example 6.7 provides $\dot{\sigma}_{cv}/\dot{m} = 0.4789$ kJ/kg·K.

EXAMPLE 7.9

PROBLEM

Determine the irreversibility for the open feedwater heater of Example 4.6, in kJ per kg of liquid exiting. Let $T_0 =$ 25°C, $p_0 = 1$ atm.

SOLUTION

Known: An open feedwater heater has water entering as a liquid at $T_1 = 40$°C, $p_1 = 7.0$ bars and as a vapor at $T_2 =$ 200°C, $p_2 = 7.0$ bars. A single saturated liquid water stream exits with a pressure $p_3 = 7.0$ bars.

Find: Determine the irreversibility of the feedwater heater per kg of liquid exiting.

Schematic and Given Data: See Fig. E4.6.

Assumptions:

1. The feedwater heater operates at steady state.
2. There is no heat transfer with the surroundings and $\dot{W}_{cv} = 0$.
3. Kinetic and potential energy effects are negligible.
4. $T_0 = 25$°C, $p_0 = 1$ atm.

Analysis: The irreversibility can be determined from an availability rate balance. Alternatively, the rate of entropy production can be found from an entropy rate balance and the irreversibility calculated from $\dot{I}_{cv} = T_0\dot{\sigma}_{cv}$. The second of these methods is used in the solution to follow.

Since there is no heat transfer with the surroundings, the entropy rate balance reduces at steady state to

$$0 = \dot{m}_1 s_1 + \dot{m}_2 s_2 - \dot{m}_3 s_3 + \dot{\sigma}_{cv}$$

Solving for $\dot{\sigma}_{cv}/\dot{m}_3$

$$\frac{\dot{\sigma}_{cv}}{\dot{m}_3} = s_3 - \left(\frac{\dot{m}_1}{\dot{m}_3} s_1 + \frac{\dot{m}_2}{\dot{m}_3} s_2\right)$$

The mass flow rate ratios required by the above equation can be found from a mass rate balance

$$\dot{m}_3 = \dot{m}_1 + \dot{m}_2$$

together with the result of Example 4.6: $\dot{m}_1/\dot{m}_2 = 4.06$. Thus, $\dot{m}_1/\dot{m}_3 = 0.8024$ and $\dot{m}_2/\dot{m}_3 = 0.1976$.

The specific entropy values required by the expression for the rate of entropy production can be obtained from the steam tables. With $s_1 = s_f(T_1)$, Table A-2 gives $s_1 = 0.5725$ kJ/kg·K. From Table A-3, $s_3 = 1.9922$ kJ/kg·K. Table A-4 gives $s_2 = 6.8865$ kJ/kg·K.

Inserting values

$$\frac{\dot{\sigma}_{cv}}{\dot{m}_3} = 1.9922 - [0.8024(0.5725) + 0.1976(6.8865)]$$

$$= 0.1721 \text{ kJ/kg·K}$$

Finally, the irreversibility is

$$\frac{\dot{I}_{cv}}{\dot{m}_3} = T_0\left(\frac{\dot{\sigma}_{cv}}{\dot{m}_3}\right) = (298 \text{ K})(0.1721 \text{ kJ/kg·K})$$

$$= 51.29 \text{ kJ/kg}$$

Comment: Evaluating the irreversibility rate $\dot{I}_{cv}$ from the rate of entropy production $\dot{\sigma}_{cv}$ normally involves less property evaluation and computation than required by an availability balance. It is left as an exercise to solve this problem by using an availability rate balance.

7.6 SECOND LAW EFFICIENCY

The objective of this section is to show the use of the availability concept in assessing the effectiveness of energy resource utilization. As part of the presentation, the second law efficiency concept is introduced and illustrated.

7.6.1 Matching End Use to Source

Tasks such as space heating, heating in industrial furnaces, and process steam generation commonly involve the combustion of coal, oil, or natural gas. When the products of combustion are at a temperature significantly greater than required by a given task, the end use is not well matched to the source and the result is inefficient use of the fuel burned. To illustrate this simply, refer to Fig. 7.7, which shows a closed system receiving a heat transfer at the rate $\dot{Q}_s$ at a *source* temperature T_s and delivering $\dot{Q}_u$ at a *use* temperature T_u. Energy is lost to the surroundings by heat transfer at a rate $\dot{Q}_l$ across a portion of the surface at T_l. All energy transfers shown on the figure are in the directions indicated by the arrows.

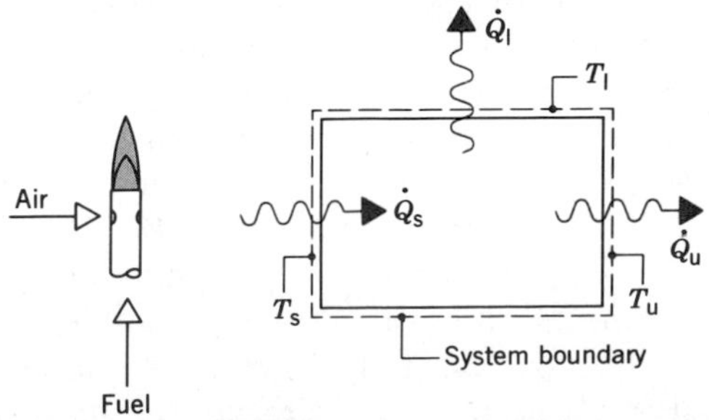

Figure 7.7 Schematic used to discuss the efficient use of fuel.

Assuming that the system of Fig. 7.7 operates at steady state and there is no work, the closed system energy and availability rate balances reduce, respectively, to

$$\overset{0}{\cancel{\frac{dE}{dt}}} = (\dot{Q}_s - \dot{Q}_u - \dot{Q}_l) - \overset{0}{\cancel{\dot{W}}}$$

$$\overset{0}{\cancel{\frac{dA}{dt}}} = \left[\left(1 - \frac{T_0}{T_s}\right)\dot{Q}_s - \left(1 - \frac{T_0}{T_u}\right)\dot{Q}_u - \left(1 - \frac{T_0}{T_l}\right)\dot{Q}_l\right] - \left[\overset{0}{\cancel{\dot{W}}} - p_0 \overset{0}{\cancel{\frac{dV}{dt}}}\right] - \dot{I}$$

These equations can be rewritten as follows:

$$\dot{Q}_s = \dot{Q}_u + \dot{Q}_l \tag{7.36}$$

$$\left(1 - \frac{T_0}{T_s}\right)\dot{Q}_s = \left(1 - \frac{T_0}{T_u}\right)\dot{Q}_u + \left(1 - \frac{T_0}{T_l}\right)\dot{Q}_l + \dot{I} \tag{7.37}$$

Equation 7.36 indicates that the energy carried in by heat transfer, $\dot{Q}_s$, is either used, $\dot{Q}_u$, or lost to the surroundings, $\dot{Q}_l$. This can be described by an efficiency in terms of energy rates in the form product/input as

$$\eta = \frac{\dot{Q}_u}{\dot{Q}_s} \tag{7.38}$$

In principle, the value of η can be increased by applying insulation to reduce the loss. The limiting value, when $\dot{Q}_l = 0$, is $\eta = 1$ (100%).

Equation 7.37 shows that the availability carried into the system accompanying the heat transfer $\dot{Q}_s$ is either transferred from the system accompanying the heat transfers $\dot{Q}_u$ and $\dot{Q}_l$ or destroyed by irreversibilities within the system. This can be described by an efficiency in the form product/input as

$$\varepsilon = \frac{(1 - T_0/T_u)\dot{Q}_u}{(1 - T_0/T_s)\dot{Q}_s} \tag{7.39}$$

Introducing Eq. 7.38 into Eq. 7.39 results in

$$\varepsilon = \eta\left(\frac{1 - T_0/T_u}{1 - T_0/T_s}\right) \tag{7.40}$$

Note that η and ε each gauge how effectively the input is converted to the product. The parameter η does this on an energy basis, whereas ε does it on an availability basis. The parameter ε, defined with reference to the availability concept, may be called a *second law* efficiency.

Equation 7.40 indicates that a value for η as close to unity as practical is important for proper utilization of the availability transferred from the hot combustion gas to the system. However, this alone would not ensure effective utilization. The temperatures T_s and T_u are also important, with availability utilization improving as the use temperature T_u approaches the source temperature T_s. For proper utilization of availability, therefore, it is desirable to have a value for η as close to unity as practical and also a good *match* between the source and use temperatures.

To emphasize the central role of temperature in second law efficiency considerations, a graph of Eq. 7.40 is provided in Fig. 7.8. The figure gives the second law efficiency ε versus the use temperature T_u for an assumed source temperature T_s = 2200 K (3960°R). Figure 7.8 shows that ε tends to unity (100%) as the use temperature approaches T_s. In most cases, however, the use temperature is substantially below T_s. Indicated on the graph are efficiencies for three applications: space heating at T_u = 320 K (576°R), process steam generation at T_u = 480 K (864°R), and heating in

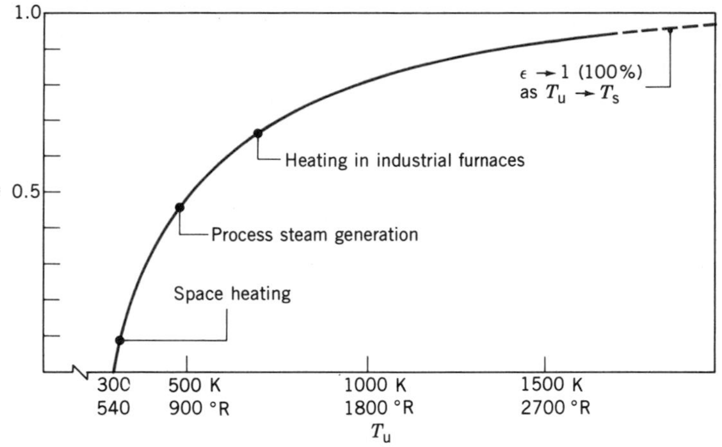

Figure 7.8 Effect of use temperature T_u on the second law efficiency ε (T_s = 2200 K, η = 100%).

industrial furnaces at T_u = 700 K (1260°R). These efficiency values suggest that fuel is used far more effectively in the higher use-temperature industrial applications than in the lower use-temperature space heating. The especially low second law efficiency for space heating reflects the fact that fuel is consumed to produce only slightly warm air, which from an availability perspective has considerably less utility. The efficiencies given on Fig. 7.8 are actually on the *high* side, for in constructing the figure we have assumed η to be unity (100%). Moreover, as additional destruction and loss of availability would be associated with combustion, the overall efficiency from fuel input to end use would be less than indicated by the values shown on the figure.

Let us conclude the present discussion by noting that improved fuel utilization might be achieved by interposing a power cycle between the source and use temperatures. The overall system would then receive a heat transfer at T_s and discharge a heat transfer at T_u while producing work. This method of sequentially producing power and a heat transfer for some specified use is called *cogeneration*. The advantage of cogeneration is that both power and a heat transfer for some intended use can be developed with a total fuel expenditure that is less than would be required to develop them individually. Further discussion of cogeneration is provided in Sec. 8.5. Cogeneration is an example of a *multiple-use method*. Two other multiple-use methods that can be employed to improve energy resource utilization are *power recovery* and *waste heat recovery*. Power recovery can be accomplished by inserting a turbine into a pressurized gas or liquid stream to capture some of the availability that would otherwise be destroyed in a spontaneous expansion. Waste heat recovery contributes to overall efficiency by using some of the availability that would otherwise be discarded to the surroundings, as in the exhaust gases of large internal combustion engines.

EXAMPLE 7.10

PROBLEM

Measurements during steady-state operation indicate that warm air exits a hand-held hair dryer at a temperature of 83°C with a velocity of 9.1 m/s through an area of 18.7 cm^2. Air enters the dryer at a temperature of 22°C and a pressure of 1 bar with a velocity of 3.7 m/s. No significant change in pressure between inlet and exit is observed. Also, no significant heat transfer between the dryer and its surroundings occurs, and potential energy effects can be ignored. Let $T_0 = 22°C$. For the hair dryer (a) evaluate the power $\dot{W}_{cv}$, in kW, and (b) devise and evaluate a second law efficiency.

SOLUTION

Known: A hand-held hair dryer has air entering and exiting at known states.

Find: For the hair dryer, evaluate the power $\dot{W}_{cv}$ and devise and evaluate a second law efficiency.

Schematic and Given Data:

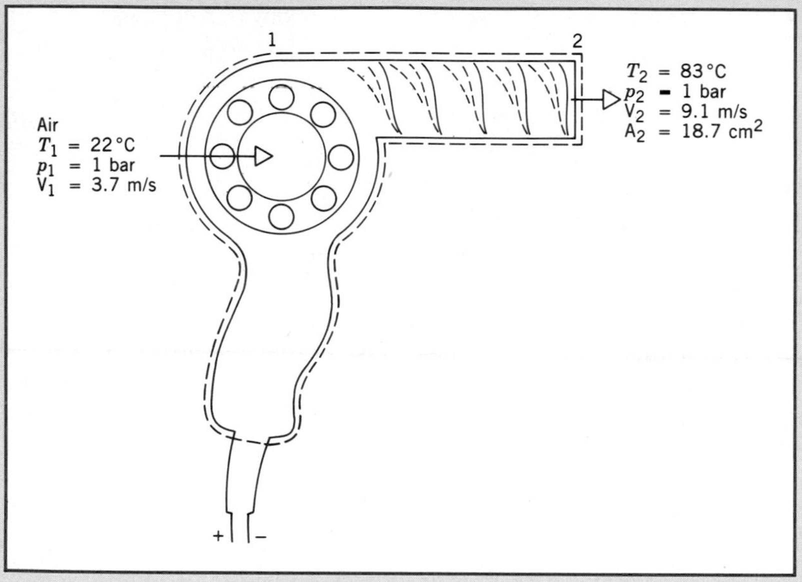

Figure E7.10

Assumptions:

1. The control volume shown in the accompanying figure is at steady state with $\dot{Q}_{cv} = 0$ and negligible potential energy effects.
2. There is no change in pressure between inlet and exit.
3. The air is modeled as an ideal gas.
4. $T_0 = 22°C$.

Analysis:

(a) The power $\dot{W}_{cv}$ for the control volume in the accompanying figure can be evaluated using mass and energy rate balances. At steady state, the mass rate balance reduces to give $\dot{m}_1 = \dot{m}_2 = \dot{m}$. Using this relationship between the mass flow rates together with assumption 1, the energy rate balance reduces at steady state to give

$$\dot{W}_{cv} = \dot{m}\left(h_1 - h_2 + \frac{V_1^2 - V_2^2}{2}\right)$$

The mass flow rate $\dot{m}$ can be evaluated from the expression

$$\dot{m} = \frac{A_2V_2}{v_2}$$

Using the ideal gas model for air, we get $v_2 = RT_2/p_2$. Hence

$$\dot{m} = \frac{A_2V_2p_2}{RT_2}$$

Inserting values

$$\dot{m} = \frac{\left[\dfrac{18.7\ \text{cm}^2}{(100\ \text{cm/m})^2}\right]\left(9.1\ \dfrac{\text{m}}{\text{s}}\right)\left(10^5\ \dfrac{\text{N}}{\text{m}^2}\right)}{\left(\dfrac{8314}{28.97}\ \dfrac{\text{N}\cdot\text{m}}{\text{kg}\cdot\text{K}}\right)(356\ \text{K})} = 0.0167\ \text{kg/s}$$

Then, with specific enthalpy values from Table A-16, $h_1 = 295.2$ kJ/kg and $h_2 = 356.5$ kJ/kg, the value of $\dot{W}_{cv}$ is obtained as follows:

$$\dot{W}_{cv} = 0.0167\ \frac{\text{kg}}{\text{s}}\left[295.2\ \frac{\text{kJ}}{\text{kg}} - 356.5\ \frac{\text{kJ}}{\text{kg}} + \frac{(3.7\ \text{m/s})^2 - (9.1\ \text{m/s})^2}{2}\left(\frac{1\ \text{N}}{1\ \text{kg}\cdot\text{m/s}^2}\right)\left(\frac{1\ \text{kJ}}{10^3\ \text{N}\cdot\text{m}}\right)\right]$$

$$= 0.0167\ \frac{\text{kg}}{\text{s}}\left(-61.3\ \frac{\text{kJ}}{\text{kg}} - 0.03\ \frac{\text{kJ}}{\text{kg}}\right)\left(\frac{1\ \text{kW}}{1\ \text{kJ/s}}\right) = -1.02\ \text{kW}$$

The minus sign reminds us that an electrical power *input* is required to operate the hair dryer. The change in kinetic energy between inlet and exit is seen to be negligible and is ignored in the remainder of the solution.

(b) Since $\dot{Q}_{cv} = 0$ and $\dot{m}_1 = \dot{m}_2$, the steady-state availability rate balance reduces as follows:

$$0 = \overbrace{\sum_j \left(1 - \frac{T_0}{T_j}\right)\dot{Q}_j}^{0} - \dot{W}_{cv} + \dot{m}(a_{f1} - a_{f2}) - \dot{I}_{cv}$$

On rearrangement

$$(-\dot{W}_{cv}) = \dot{m}(a_{f2} - a_{f1}) + \dot{I}_{cv}$$

The left side of this equation is the rate at which availability is supplied to the hair dryer. The equation shows that the availability supplied either goes into increasing the flow availability or is destroyed by irreversibilities within the dryer. Accordingly, a second law efficiency can be defined as the ratio of the flow availability increase to the availability supplied. That is

$$\varepsilon = \frac{\dot{m}(a_{f2} - a_{f1})}{(-\dot{W}_{cv})}$$

By introducing the expression for $\dot{W}_{cv}$ from the energy rate balance and using Eq. 7.35 to evaluate the flow availability increase

$$\varepsilon = \frac{\dot{m}[h_2 - h_1 - T_0(s_2 - s_1)]}{\dot{m}(h_2 - h_1)} = 1 - \frac{T_0(s_2 - s_1)}{(h_2 - h_1)}$$

Note that the kinetic energy change between inlet and exit is omitted from this expression.

Since there is no change in pressure from inlet to exit (assumption 2), $s_2 - s_1 = s_2^\circ - s_1^\circ$, where s_1° and s_2° are obtained from Table A-16 using T_1 and T_2, respectively, as $s_1^\circ = 1.685$ kJ/kg·K, $s_2^\circ = 1.874$ kJ/kg·K. Hence

$$\varepsilon = 1 - \frac{(295\ \text{K})(1.874 - 1.685)\ \text{kJ/kg}\cdot\text{K}}{61.3\ \text{kJ/kg}}$$

$$= 1 - \frac{55.76\ \text{kJ/kg}}{61.3\ \text{kJ/kg}} = 0.09\ (9\%)$$

Comment: In this device, a relatively high potential for use electrical input is used to obtain a stream of slightly warm air. The low value of the efficiency ε emphasizes that source and end use are not well matched.

7.6.2 Further Illustrations of Second Law Efficiencies

Second law efficiency expressions can take many different forms. Several examples are given in the current section for thermal system components of practical interest. In every instance, the efficiency is derived by the use of the availability rate balance. The approach used here serves as a model for the development of second law efficiency expressions for other components. Each of the cases considered involves a control volume at steady state, and we assume that there is no heat transfer between the control volume and its surroundings. The current presentation is not exhaustive. Many other second law efficiency expressions can be written.

Turbines, Compressors, and Pumps. For a turbine operating at steady state with no heat transfer with its surroundings, the steady-state form of the availability rate balance, Eq. 7.34, reduces as follows:

$$0 = \cancelto{0}{\sum_j \left(1 - \frac{T_0}{T_j}\right) \dot{Q}_j} - \dot{W}_{cv} + \dot{m}(a_{f1} - a_{f2}) - \dot{I}_{cv}$$

This equation can be rearranged to read

$$a_{f1} - a_{f2} = \frac{\dot{W}_{cv}}{\dot{m}} + \frac{\dot{I}_{cv}}{\dot{m}} \tag{7.41}$$

The term on the left of Eq. 7.41 is the decrease in flow availability from turbine inlet to exit. The equation shows that the flow availability decreases because the turbine develops work, $\dot{W}_{cv}/\dot{m}$, and availability is destroyed, $\dot{I}_{cv}/\dot{m}$. A parameter that gauges how effectively the flow availability decrease is converted to the desired product is

$$\varepsilon = \frac{\dot{W}_{cv}/\dot{m}}{a_{f1} - a_{f2}} \tag{7.42}$$

This particular second law efficiency is sometimes referred to as the *turbine effectiveness*. It is left as an exercise to verify that the effectiveness of the turbine considered in Example 6.11 is 81.2% when $T_0 = 25°C$. Carefully note that the turbine effectiveness is defined differently from the isentropic turbine efficiency introduced in Sec. 6.8.

For a compressor or pump operating at steady state with no heat transfer with its surroundings, the availability rate balance can be placed in the form

$$\left(-\frac{\dot{W}_{cv}}{\dot{m}}\right) = a_{f2} - a_{f1} + \frac{\dot{I}_{cv}}{\dot{m}}$$

Thus, the availability input to the device, $-\dot{W}_{cv}/\dot{m}$, is accounted for either as an increase in the flow availability between inlet and exit or as availability destroyed. The effectiveness of the conversion from work input to flow availability increase is gauged by

$$\varepsilon = \frac{a_{f2} - a_{f1}}{(-\dot{W}_{cv}/\dot{m})} \tag{7.43}$$

Heat Exchanger without Mixing. The heat exchanger shown in Fig. 7.9 operates at steady state with no heat transfer with its surroundings. The availability rate balance, Eq. 7.33, reduces to

$$0 = \cancelto{0}{\sum_j \left(1 - \frac{T_0}{T_j}\right) \dot{Q}_j} - \cancelto{0}{\dot{W}_{cv}} + (\dot{m}_h a_{f1} + \dot{m}_c a_{f3}) - (\dot{m}_h a_{f2} + \dot{m}_c a_{f4}) - \dot{I}_{cv}$$

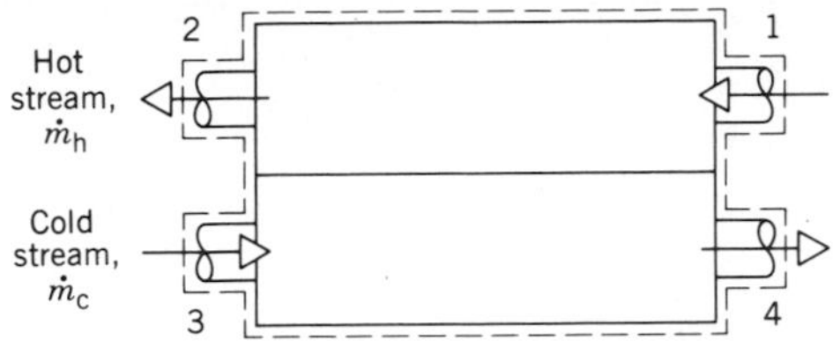

Figure 7.9 Counterflow heat exchanger.

where $\dot{m}_h$ is the mass flow rate of the hot stream and $\dot{m}_c$ is the mass flow rate of the cold stream. This can be rearranged to read

$$\dot{m}_h(a_{f1} - a_{f2}) = \dot{m}_c(a_{f4} - a_{f3}) + \dot{I}_{cv} \tag{7.44}$$

The term on the left of Eq. 7.44 accounts for the decrease in the availability of the hot stream. The first term on the right accounts for the increase in availability of the cold stream. Regarding the hot stream as supplying the availability increase of the cold stream as well as the availability destroyed, we can write an efficiency expression as

$$\varepsilon = \frac{\dot{m}_c(a_{f4} - a_{f3})}{\dot{m}_h(a_{f1} - a_{f2})} \tag{7.45}$$

Heat Exchanger with Mixing. The direct contact heat exchanger shown in Fig. 7.10 operates at steady state with no heat transfer with its surroundings. The availability rate balance, Eq. 7.33, reduces to

$$0 = \sum_j \left(1 - \frac{T_0}{T_j}\right)\dot{Q}_j^{\,0} - \dot{W}_{cv}^{\,0} + \dot{m}_1 a_{f1} + \dot{m}_2 a_{f2} - \dot{m}_3 a_{f3} - \dot{I}_{cv}$$

With $\dot{m}_3 = \dot{m}_1 + \dot{m}_2$ from a mass rate balance, this can be written as

$$\dot{m}_1(a_{f1} - a_{f3}) = \dot{m}_2(a_{f3} - a_{f2}) + \dot{I}_{cv} \tag{7.46}$$

The term on the left of Eq. 7.46 accounts for the decrease in the availability of the hot stream between inlet and exit. The first term on the right accounts for the increase in the availability of the cold stream between inlet and exit. Regarding the hot stream as supplying the availability increase of the cold stream as well as the availability destroyed by irreversibilities, we can write an efficiency expression as

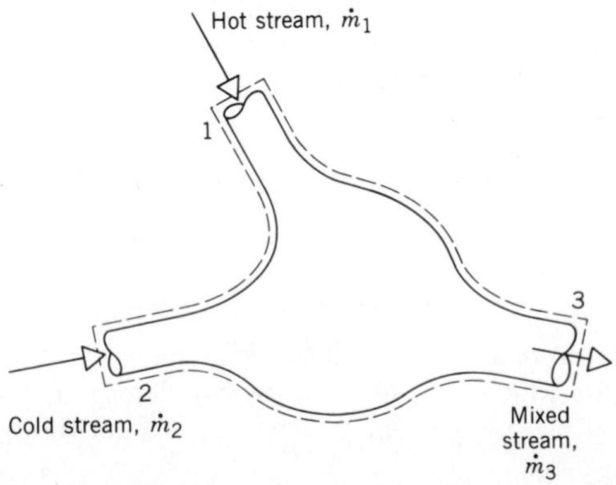

Figure 7.10 Direct contact heat exchanger.

$$\varepsilon = \frac{\dot{m}_2(a_{f3} - a_{f2})}{\dot{m}_1(a_{f1} - a_{f3})} \tag{7.47}$$

7.6.3 Closure

The concept of a second law efficiency is useful for distinguishing means for utilizing energy resources that are thermodynamically effective from those that are less so. Second law efficiencies also can be used to evaluate the effectiveness of engineering measures taken to improve the performance of a thermal system. This is done by comparing the efficiency values determined before and after modifications have been made to show how much improvement has been achieved. Moreover, second law efficiencies can be used to gauge the potential for improvement in the performance of a given thermal system by comparing the efficiency of the system to the efficiency of like systems. A significant difference between these values would suggest that improved performance might be achievable.

It is important to recognize that the limit of 100% second law efficiency should not be regarded as a practical objective. This theoretical limit could be attained only if there were no availability destructions or losses. To achieve such idealized processes would require extremely long times to execute processes and/or complex devices, both of which are at odds with the objective of profitable operation. In practice, decisions are usually made on the basis of *total* costs. An increase in efficiency to reduce fuel consumption, or otherwise utilize energy resources better, normally requires additional expenditures for facilities and operations. Accordingly, an improvement generally would not be implemented if an increase in total cost would result. The trade-off between fuel savings and additional investment invariably dictates a lower efficiency than might be achieved *theoretically* and may even result in a lower efficiency than could be achieved using the *best available* technology.

PROBLEMS

7.1 Write a definition of availability. Is it possible for availability to be negative? Discuss.

7.2 A system consists of 10 kg of water at 5°C and 1 bar. Determine the availability, in kJ, if the system is at rest and zero elevation relative to the availability reference environment for which T_0 = 20°C, p_0 = 1 bar.

7.3 Determine the availability, in kJ, for each of the following:

(a) Two kilograms of water vapor at 6 MPa and 450°C.
(b) Three kilograms of Refrigerant 12 at 1.4 MPa and 90°C.
(c) One kilogram of air as an ideal gas at 7.5 MPa and 475°C.

In each case, the mass is at rest and zero elevation relative to the availability reference environment for which T_0 = 20°C, p_0 = 0.1 MPa.

7.4 Determine the specific availability, in kJ/kg, of

(a) saturated water vapor at 100°C.
(b) saturated liquid water at 0.1°C.
(c) Refrigerant 12 at −10°C, 1 bar.

In each case, consider a fixed mass at rest and zero elevation relative to the availability reference environment for which T_0 = 20°C, p_0 = 1 bar.

7.5 Determine the specific availability, in Btu/lb, of

(a) saturated liquid water at 32°F.
(b) saturated water vapor at 212°F.

In each case, consider a fixed mass at rest and zero elevation relative to the availability reference environment for which T_0 = 60°F, p_0 = 1 atm.

7.6 A vessel contains air at 90 lbf/in.^{2} and 80°F. Using the ideal gas model, determine the specific availability of the air, in ft·lbf/lb. The air is at rest and zero elevation relative to the availability reference environment for which $T_0 = 80°F$, $p_0 = 15$ lbf/in.^{2}

7.7 A balloon filled with helium at 20°C and 1 bar is moving with a velocity of 15 m/s at an elevation of 0.5 km relative to the availability reference environment for which $T_0 = 20°C$, $p_0 = 1$ bar. Using the ideal gas model, determine the specific availability of the helium, in kJ/kg.

7.8 A vessel contains 1 lb of air at 58.8 lbf/in.^{2} and 200°F. Using the ideal gas model, determine the specific availability of the air, in Btu/lb. The air is at rest and negligible elevation relative to the availability reference environment for which $T_0 = 60°F$, $p_0 = 14.7$ lbf/in.^{2}

7.9 Oxygen at 600°R and 1 atm fills a balloon at rest on the surface of the earth at a location where the ambient temperature is 40°F and the ambient pressure is 1 atm. Using the ideal gas model, determine the specific availability of the oxygen, in Btu/lb, relative to the earth and its atmosphere at this location.

7.10 Consider water as a saturated solid at 0°C. Determine the specific availability, in kJ/kg, if $T_0 = 20°C$, $p_0 =$ 1 atm, and there are no significant effects of motion or gravity.

7.11 Consider water as a saturated solid at 32°F. Determine the specific availability, in Btu/lb, if $T_0 = 70°F$, $p_0 =$ 1 atm, and there are no significant effects of motion or gravity.

7.12 The contents of a storage tank have an availability of 7500 kJ. Determine the volume of the tank in m^3 if the tank is filled with

- **(a)** air as an ideal gas at 500°C and 3 bars.
- **(b)** water vapor at 500°C and 3 bars.

Ignore the effects of motion and gravity and let $T_0 =$ 22°C, $p_0 = 1$ atm.

7.13 Air as an ideal gas is stored in a closed vessel of volume V at temperature T_0 and pressure p.

(a) Ignoring motion and gravity, obtain the following expression for the availability of the air:

$$A = p_0 V\left(1 - \frac{p}{p_0} + \frac{p}{p_0}\ln\frac{p}{p_0}\right)$$

(b) Using the result of part (a), evaluate V in ft^3 if $A =$ 3412 Btu, $p = 1000$ lbf/in.^{2}, $p_0 = 1$ atm, $T_0 = 60°F$.

7.14 An ideal gas is stored in a closed vessel at pressure p and temperature T_0. Derive an expression for the specific availability in terms of p, p_0, T_0, and the gas constant R. Ignore the effects of motion and gravity.

7.15 An ideal gas is stored in a closed vessel at temperature T and pressure p_0. Derive an expression for the specific availability in terms of T, T_0, and the specific heat c_p, which can be taken as constant. Ignore the effects of motion and gravity.

7.16 Show that the availability associated with an evacuated space of volume V is given by $p_0 V$, where p_0 is the pressure of the availability reference environment.

7.17 Water vapor initially at 8 MPa and 400°C fills a rigid vessel. The water vapor is cooled until the pressure becomes 4 MPa. There is no work during the process. For the water, determine the heat transfer per unit mass of water vapor and the change in specific availability, each in kJ/kg. Let $T_0 = 25°C$, $p_0 = 0.1$ MPa.

7.18 Two pounds of air initially at 200°F and 50 lbf/in.^{2} undergo a process to a final state at −10°F and 10 lbf/in.^{2} Employing the ideal gas model, determine, in Btu,

- **(a)** the availability at the initial state.
- **(b)** the availability at the final state.
- **(c)** the change in availability.

Let $T_0 = 77°F$, $p_0 = 14.7$ lbf/in.^{2}, and ignore the effects of motion and gravity.

7.19 Two pounds of oxygen initially at 60 lbf/in.^{2}, 10°F undergo a process to a final state at 15 lbf/in.^{2}, 150°F. Employing the ideal gas model, determine the change in availability, in Btu. Let $T_0 = 77°F$, $p_0 = 15$ lbf/in.^{2} and ignore the effects of motion and gravity.

7.20 Consider 1 lb of steam initially at 200 lbf/in.^{2} and 500°F as the system. Determine the change in availability, in Btu, for each of the following processes:

- **(a)** The system is heated at constant pressure until its volume is increased by 50%.
- **(b)** The system expands isothermally until its volume is increased by 50%.

Let $T_0 = 60°F$, $p_0 = 1$ atm.

7.21 A flywheel with a moment of inertia of 160 lb·ft^2 rotates at 3000 RPM. As the flywheel is braked to rest, its rotational kinetic energy is converted entirely to internal energy of the brake lining. The brake lining has a mass of 5 lb and can be regarded as an incompressible solid with a specific heat $c = 1$ Btu/lb·°R. There is no significant heat transfer with the surroundings. **(a)** Determine the final temperature of the brake lining, in °F, if its initial temperature is 60°F. **(b)** Determine the maximum possible rotational speed, in RPM, that could be attained by the flywheel using energy stored in the brake lining after the flywheel has been braked to rest. Let $T_0 = 60°F$.

7.22 Water vapor initially at 3 bars and 240°C cools at constant pressure with no internal irreversibilities until its temperature is 20°C. For 1 kg of water, determine the work, the heat transfer, and the amounts of availability transfer accompanying work and heat transfer, each in kJ. Let $T_0 = 20°C$, $p_0 = 1$ bar.

7.23 Air initially at 2 bars and 27°C is heated at constant pressure with no internal irreversibilities to a final temperature of 227°C. For 1 kg of air, determine the work, the heat transfer, and the amounts of availability transfer accompanying work and heat transfer, each in kJ. Let $T_0 = 280$ K, $p_0 = 1$ bar.

7.24 One pound of air is contained in a closed, rigid, insulated tank. Initially the temperature is 500°R and the pressure is 1 atm. Eighty-five Btu of work are done on the air by a paddle wheel. For the air as the system, determine the change in availability, the transfer of availability accompanying work, and the irreversibility, all in Btu. Neglect kinetic and potential energy and let $T_0 = 500°R$, $p_0 = 1$ atm.

7.25 Air initially at 27°C and 1 bar is contained within a rigid, insulated tank. The air is stirred by a paddle wheel until its pressure is 1.2 bars. For 1 kg of air, determine the work and the irreversibility, each in kJ. Neglect kinetic and potential energy and let $T_0 = 27°C$, $p_0 = 1$ bar.

7.26 A rigid, insulated vessel contains 10 ft^3 of air initially at 100°F and 1 atm. The air is stirred by a paddle wheel until its temperature is 400°F. Employing the ideal gas model and neglecting kinetic and potential energy effects, determine the work, the change in availability, and the irreversibility, each in Btu. Let $T_0 = 70°F$, $p_0 = 1$ atm.

7.27 A rigid, well-insulated tank consists of two compartments, each having a volume of 1 m^3, separated by a valve. Initially, one of the compartments is evacuated and the other contains nitrogen gas at 600 kPa and 80°C. The valve is opened and the nitrogen expands to fill the total volume, eventually achieving an equilibrium state.

(a) Determine the final temperature, in °C, and final pressure, in kPa.
(b) Evaluate the irreversibility, in kJ.

Employ the ideal gas model and let $T_0 = 20°C$, $p_0 = 0.1$ MPa.

7.28 Two solid blocks, each having mass m and specific heat c, and initially at temperatures T_1 and T_2, respectively, are brought into contact, insulated on their outer surfaces, and allowed to come into thermal equilibrium. Derive an expression for the availability destroyed in terms of m, c, T_1, T_2, and the temperature of the environment, T_0. Demonstrate that the availability destruction cannot be negative.

7.29 One kilogram of Refrigerant 134a is compressed adiabatically from the saturated vapor state at −10°C to a final state where the pressure is 8 bars and the temperature is 50°C. Determine the work and the irreversibility, each in kJ/kg. Let $T_0 = 20°C$, $p_0 = 1$ bar.

7.30 A gearbox operating at steady state receives 20 horsepower along its input shaft, delivers power along its output shaft, and is cooled on its outer surface according to $\dot{Q} = -hA(T_b - T_0)$, where $T_b = 110°F$ is the temperature of the outer surface and $T_0 = 40°F$ is the temperature of the surroundings far from the gearbox. The product of the heat transfer coefficient h and outer surface area A is 35Btu/h·°R. For the gearbox, determine the power delivered along the output shaft, the rate of availability transfer accompanying heat, and the irreversibility rate, each in Btu/h. Express each quantity as a percentage of the input power.

7.31 Determine whether the following devices can operate at steady state as indicated. Let $T_0 = 27°C$.

(a) Electricity is supplied to a device at the rate of 10 kW. Heat transfer from the device occurs at a place on its surface where the temperature is 327°C. There are no other energy transfers.
(b) Heat transfer to a device occurs at a place on its surface where the temperature is 327°C. The device delivers electricity to its surroundings at the rate of 10 kW. There are no other energy transfers.

7.32 Ten pounds of air initially at 1 atm and 600°R are contained within a closed, rigid tank. Energy is transferred to the air by heat transfer from a thermal reservoir at 900°R until the air temperature is 800°R. During heat transfer, the temperature of the air where heat transfer occurs is 900°R. There is no work. Employing the ideal gas model, determine the heat transfer, the change in availability, the availability transfer accompanying heat, and the irreversibility, each in Btu. Let $T_0 = 40°F$, $p_0 = 1$ atm.

7.33 For the electric motor of Problem 6.59, evaluate the irreversibility rate and the rate of availability transfer accompanying heat, each in kW. Express each quantity as a percentage of the electrical power supplied to the motor. Let $T_0 = 20°C$.

7.34 A quantity of air initially at 300 K and 1 bar experiences the two different types of interactions described below. In each case, the air is brought from the initial state to a state where the temperature is 500 K, while volume remains constant. Kinetic and potential energy effects are negligible. For each case, determine the irreversibility per unit mass of air present, in kJ/kg. Comment on the difference between the irreversibility values. Let $T_0 = 300$ K, $p_0 = 1$ atm.

(a) The temperature rise is brought about adiabatically by stirring the air with a paddle wheel.
(b) The temperature rise is brought about by heat transfer from a thermal reservoir at 600 K. The temperature of the air at the location where heat transfer occurs is 600 K.

7.35 A thermal reservoir at 1000 K is separated from another thermal reservoir at 500 K by a cylindrical rod insulated on its lateral surfaces. At steady state, energy transfer by conduction takes place through the rod at a rate of 10 kW. Evaluate the irreversibility rate for the rod, in kW. Let $T_0 = 300$ K.

7.36 For the electric motor of Problem 6.69, evaluate the irreversibility rate and the rate of availability transfer accompanying heat, each in kW. Express each quantity as a percentage of the electrical power supplied to the motor. Let $T_0 = 293$ K.

7.37 An electrical resistor is used in a furnace, as illustrated in Fig. P7.37. At steady state, electrical power must be supplied to the resistor at a rate of 8.5 kW per meter of length to maintain it at 1500 K when the furnace walls are at 500 K. Let $T_0 = 300$ K.

(a) For the resistor as the system, determine the rate of availability transfer accompanying heat and the irreversibility rate, each in kW per meter of length.
(b) For the space between the resistor and the walls as the system, evaluate the irreversibility rate, in kW per meter of length.

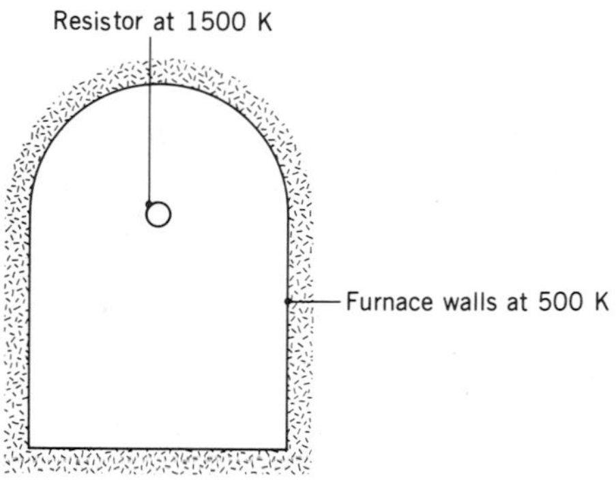

Figure P7.37

7.38 0.2 kg of water initially at 0.9 MPa and a quality of 40% undergoes an internally reversible expansion at constant pressure and temperature to saturated vapor while receiving energy through a thin wall from a thermal reservoir at 600 K. Let $T_0 = 25°C$, $p_0 = 0.1$ MPa.

(a) For the water as the system, evaluate the work, the heat transfer, the availability transfer accompanying work, and the availability transfer accompanying heat, each in kJ.
(b) Evaluate the irreversibility, in kJ, for an enlarged system that includes the water and the wall. Compare with the irreversibility of part (a) and comment on the difference.

7.39 Five kilograms of water initially at 160°C and 10 bars undergo an internally reversible expansion at constant temperature to 1.5 bars while receiving energy through a thin wall from a thermal reservoir at 900 K. Let $T_0 = 25°C$, $p_0 = 1$ bar.

(a) For the water as the system, evaluate the work, the heat transfer, the availability transfer accompanying work, and the availability transfer accompanying heat, each in kJ.
(b) Evaluate the irreversibility, in kJ, for an enlarged system that includes the water and the wall. Compare with the irreversibility of part (a) and comment on the difference.

7.40 A system undergoes a power cycle while receiving energy Q_H by heat transfer at temperature T_H and discharging energy Q_C by heat transfer at a lower temperature T_C. There are no other heat transfers.

(a) Show that the thermal efficiency of the cycle can be expressed as

$$\eta = 1 - \frac{T_C}{T_H} - \frac{T_C I}{T_0 Q_H}$$

where T_0 is the temperature of the availability reference environment and I is the irreversibility for the cycle.
(b) Obtain an expression for the maximum theoretical value for the thermal efficiency.
(c) Derive an expression for the irreversibility for the case in which no net work is developed by the cycle. Discuss.

7.41 A rigid, well-insulated tank is filled initially with 5.0 kg of air at a pressure of 0.5 MPa and a temperature of 500 K. A leak develops and air slowly escapes to the surroundings, which are at 22°C, 0.1 MPa, until the pressure of the air remaining in the tank is 0.1 MPa. Determine the change in the amount of availability stored in the tank, in kJ. Does availability destruction occur? Where? Explain. Let $T_0 = 22°C$, $p_0 = 0.1$ MPa and ignore the effects of motion and gravity.

7.42 Is it possible for the specific flow availability to be negative? Discuss.

7.43 Determine the specific availability and the specific flow availability, each in Btu/lb, for steam at 1200 lbf/in.^{2}, 800°F, V = 800 ft/s, and z = 1000 ft. The velocity and elevation are relative to the availability reference environment for which $T_0 = 60°F$, $p_0 = 1$ atm. Let g = 32.2 ft/s^2.

7.44 For each of the following cases, evaluate the specific availability and specific flow availability, each in kJ/kg. The velocity is relative to the availability reference environment for which $T_0 = 25°C$, $p_0 = 1$ bar. The effect of gravity can be neglected.

(a) water vapor at 30 MPa, 800°C, 135 m/s.
(b) Refrigerant 12 at 1 MPa, 80°C, 5 m/s.
(c) nitrogen as an ideal gas at 6.2 MPa, 527°C, 290 m/s.

7.45 A stream of hot water at 300°F, 500 lbf/in.^{2}, and a velocity of 20 ft/s is obtained from a geothermal supply. Determine the specific flow availability, in Btu/lb. The velocity is relative to the availability reference environment for which $T_0 = 77°F$, $p_0 = 1$ atm. Neglect the effect of gravity.

7.46 For an ideal gas with constant specific heat ratio k, show that in the absence of significant effects of motion and gravity the specific flow availability can be expressed as

$$\frac{a_f}{c_p T_0} = \frac{T}{T_0} - 1 - \ln\frac{T}{T_0} + \ln\left(\frac{p}{p_0}\right)^{(k-1)/k}$$

Develop plots for $k = 1.2$, 1.3, and 1.4 of a_f/c_pT_0 versus T/T_0 for $p/p_0 = 0.25$, 0.5, 1, 2, 4. The specific flow availability can take on negative values when $p/p_0 < 1$. What does a negative value mean physically?

7.47 The gas turbine power plant illustrated in Fig. P6.101 develops *net* work equal to 230 kJ/kg. Determine the specific flow availability exiting at state 2 with the exhaust, in kJ/kg. Is this amount significant? Explain. Let $T_0 = 22°C$, $p_0 = 0.095$ MPa and ignore the effects of motion and gravity.

7.48 The state of a flowing gas is defined by h, s, V, and z, where velocity and elevation are relative to the availability reference environment for which the temperature is T_0 and the pressure is p_0. Determine the *maximum* theoretical work, per unit mass of gas flowing, that could be developed by any one-inlet, one-exit control volume at steady state that would reduce the stream to the dead state at the exit while allowing heat transfer only at T_0. Discuss.

7.49 Steam exits a turbine with a mass flow rate of 40,000 kg/h at a pressure of 0.0040 MPa, a quality of 94%, and a velocity of 140 m/s. Determine the *maximum* theoretical power that could be developed, in kW, by any one-inlet, one-exit control volume at steady state that would reduce the steam to the dead state at the exit while allowing heat transfer only at temperature T_0. The velocity is relative to the availability reference environment for which $T_0 = 15°C$, $p_0 = 0.1$ MPa. Neglect the effect of gravity.

7.50 Water vapor enters a valve at a temperature of 500°F and a pressure of 500 lbf/in.^{2} and undergoes a throttling process to 80 lbf/in.^{2}

(a) Determine the flow availability at the valve inlet and exit and the irreversibility rate, all in Btu per lb of water vapor passing through the valve.
(b) Evaluating availability at 8 cents per kW·h, estimate the annual cost of this source of irreversibility for a mass flow rate of 1 lb/s.

Let $T_0 = 77°F$, $p_0 = 1$ atm.

7.51 Determine the irreversibility rate, in Btu per lb flowing, in a throttling process to 20 lbf/in.^{2} for **(a)** steam entering a throttling valve as saturated vapor at 100 lbf/in.^{2}, **(b)** air as an ideal gas entering a throttling valve at 100 lbf/in.^{2} Let $T_0 = 60°F$, $p_0 = 1$ atm.

7.52 At steady state, an insulated steam turbine develops work at a rate of 298 Btu per lb of steam flowing through the turbine. Steam enters at 500 lbf/in.^{2} and 900°F and exits at 14.7 lbf/in.^{2} Evaluating availability at 8 cents per kW·h, determine for a mass flow rate of 1 lb/s the value of the power produced and the cost of the availability destroyed, each in \$/h. Neglect kinetic and potential energy effects. Let $T_0 = 60°F$, $p_0 = 14.7$ lbf/in.^{2}

7.53 Steam enters a turbine operating at steady state at a pressure of 3 MPa, a temperature of 400°C, and a velocity of 160 m/s. Steam exits as saturated vapor at 100°C with a velocity of 100 m/s. There is no significant change in elevation. Heat transfer from the turbine at the rate of 30 kJ per kg of steam flowing takes place at an average surface temperature of 400 K. For the turbine, determine, in kJ per kg of steam

(a) the work developed.
(b) the magnitude and direction of the availability transfer accompanying heat transfer.
(c) the irreversibility.

Let $T_0 = 22°C$, $p_0 = 0.1$ MPa.

7.54 The steam turbine of Problem 7.53 is located in an industrial plant where the ambient temperature is 27°C. Determine the irreversibility rate, in kJ per kg of steam flowing, for a control volume that includes the turbine and enough of its immediate surroundings so that heat transfer takes place at the ambient temperature. Explain why the irreversibility value is different from that calculated in Problem 7.53.

7.55 Air enters a turbine operating at steady state at a pressure of 75 lbf/in.^{2}, a temperature of 800°R, and a velocity of 400 ft/s. At the exit, the conditions are 15 lbf/in.^{2}, 600°R, and 100 ft/s. There is no significant change in elevation. Heat transfer from the turbine to its surroundings at a rate of 10 Btu per lb of air flowing takes place at an average surface temperature of 620°R. Determine, in Btu per lb of air passing through the turbine, the work developed and the irreversibility rate. Let $T_0 = 40°F$, $p_0 = 15$ lbf/in.^{2}

7.56 Referring to Problem 7.55, locate the boundary of the control volume to include the turbine and a portion of its immediate surroundings so that heat transfer occurs at the temperature T_0. For this enlarged control volume, determine the irreversibility rate in Btu per lb of air passing through the turbine. Explain why the irreversibility value is different from that calculated in Problem 7.55.

7.57 An insulated turbine operating at steady state receives steam at 400 lbf/in.^{2}, 600°F and exhausts at 50 lbf/in.^{2} The irreversibility rate is 34 Btu per lb of steam expanding through the turbine. Kinetic and potential energy effects can be ignored and $T_0 = 60°F$, $p_0 = 1$ atm. Determine the work developed, in Btu per lb of steam expanding through the turbine.

7.58 Air at 1.4 bars, 17°C, 70 m/s enters an insulated compressor operating at steady state and exits at 4.2 bars, 147°C, 110 m/s. Determine, in kJ per kg of air flowing through the compressor, the work and the irreversibility. Express the irreversibility rate as a percentage of the power required by the compressor. Potential energy effects are negligible. Let $T_0 = 280°C$, $p_0 = 1$ bar.

7.59 Refrigerant 12 at 0°C, 200 kPa, and a mass flow rate of 90 kg/h enters an insulated compressor operating at steady state and exits at 1.2 MPa. The isentropic compressor efficiency is 83.3%. Determine

(a) the temperature of the refrigerant exiting the compressor, in °C.
(b) the power input to the compressor, in kW.
(c) the irreversibility rate expressed as a percentage of the power required by the compressor.

Neglect kinetic and potential energy effects and let $T_0 =$ 20°C, $p_0 =$ 1 bar.

7.60 Water vapor at 0.4 MPa and 400°C enters an insulated turbine operating at steady state and expands to 0.1 MPa and 100°C. Kinetic and potential energy effects can be neglected.

(a) Determine the work developed, in kJ per kg of water vapor passing through the turbine.
(b) Determine the *maximum* theoretical work per unit of mass flowing, in kJ/kg, that could be developed by any one-inlet, one-exit control volume at steady state that has water vapor entering and exiting at the specified states, while allowing heat transfer only at temperature T_0.

Compare the results of parts (a) and (b) and comment on the difference. Let T_0 = 27°C, p_0 = 0.1 MPa.

7.61 Air enters an insulated turbine operating at steady state at 5 bars, 400 K, and 150 m/s. At the turbine exit the conditions are 1 bar, 300 K, and 70 m/s. There is no significant change in elevation.

(a) Determine the work developed, in kJ per kg of air passing through the turbine.
(b) Determine the *maximum* theoretical work per unit of mass flowing, in kJ/kg, that could be developed by any one-inlet, one-exit control volume at steady state that has air entering and exiting at the specified states, while allowing heat transfer only at temperature T_0.

Compare the results of parts (a) and (b) and comment on the difference. Let T_0 = 27°C, p_0 = 1 atm.

7.62 For the compressor of Problem 6.110, determine the irreversibility rate and the rate of availability transfer accompanying heat, each in kJ per kg of air flow. Express the irreversibility rate as a percentage of the work input to the compressor. Let T_0 = 20°C, p_0 = 1 atm.

7.63 A compressor fitted with a water jacket and operating at steady state takes in air with a volumetric flow rate of 9 m^3/min at 20°C, 100 kPa and discharges air at 160°C, 500 kPa. Cooling water enters the water jacket at 15°C, 100 kPa with a mass flow rate of 10 kg/min and exits at 25°C and essentially the same pressure. There is no significant heat transfer from the outer surface of the water jacket to its surroundings, and kinetic and potential energy effects can be ignored. For the water-jacketed compressor, determine the power input required and the irreversibility rate, each in kW. Let T_0 = 20°C, p_0 = 1 atm.

7.64 Steam at 1.4 MPa and 350°C with a mass flow rate of 0.125 kg/s enters an insulated turbine operating at steady state and exhausts at 100 kPa. For values of the isentropic turbine efficiency of 100%, 80%, 40%, and 0%, determine the temperature of the exhaust steam, in °C, the power developed by the turbine, in kW, and the irreversibility rate, in kW. Neglect kinetic and potential energy effects. Let T_0 = 20°C, p_0 = 0.1 MPa.

7.65 Steam enters an insulated turbine operating at steady state at 100 lbf/in.^{2}, 500°F, with a mass flow rate of 3 × 10^5 lb/h and expands to a pressure of 1 atm. The isentropic turbine efficiency is 80%. If availability is valued at 8 cents per kW·h, determine

(a) the value of the power produced, in \$/h.
(b) the cost of the availability destroyed, in \$/h.
(c) Repeat for isentropic turbine efficiencies of 90 and 100%. Would you recommend that a more efficient turbine be used? What are some of the factors that might enter into your decision?

Let T_0 = 70°F, p_0 = 1 atm.

7.66 For the compressor of Problem 6.111, determine the irreversibility rate, in kW. Express the irreversibility rate as a percentage of the power input to the compressor. Let T_0 = 20°C, p_0 = 1 bar.

7.67 Evaluating availability at 8 cents per kW·h, determine the hourly costs of the power input and irreversibility for

(a) Problem 7.63.
(b) Problem 7.66.

7.68 An insulated steam turbine at steady state can be operated at part-load conditions by throttling the steam to a lower pressure before it enters the turbine. Before throttling, the steam is at 200 lbf/in.^{2}, 600°F. After throttling, the pressure is 150 lbf/in.^{2} At the turbine exit, the steam is at 1 lbf/in.^{2} and a quality of 95%. Determine the irreversibility rates, each in Btu per lb of steam flowing, for the throttling valve and the turbine. Neglect kinetic and potential energy effects and let T_0 = 60°F, p_0 = 1 atm.

7.69 A counterflow heat exchanger operating at steady state has water entering as saturated vapor at 1 bar and exiting as saturated liquid at 1 bar. In a separate stream, air enters at 300 K, 1 bar and exits at 335 K with a negligible change of pressure. Determine

(a) the mass flow rate of the air required, in kg per kg of steam condensed.
(b) the change in the flow availability of the steam, in kJ per kg of steam condensed.
(c) the change in the flow availability of the air, in kJ per kg of steam condensed.
(d) the irreversibility of the heat exchanger, in kJ per kg of steam condensed.

Heat transfer between the heat exchanger and its surroundings is negligible, as are changes in kinetic and potential energy. Let T_0 = 300 K, p_0 = 1 bar.

7.70 Saturated water vapor at 0.008 MPa and a mass flow

rate of 2.6×10^5 kg/h enters the condenser of a 100-MW power plant and exits as a saturated liquid at 0.008 MPa. The cooling water stream enters at 15°C and exits at 35°C with a negligible change in pressure. At steady state, determine

(a) the *net* rate energy exits the plant with the cooling water stream, in MW.
(b) the *net* rate availability exits the plant with the cooling water stream, in MW.

Compare and discuss these values. Is the *loss* with the cooling water significant? Explain. What are some possible uses for the exiting cooling water? Let T_0 = 20°C, p_0 = 0.1 MPa.

7.71 Air enters a counterflow heat exchanger operating at steady state at 22°C, 0.1 MPa and exits at 7°C. Refrigerant 134a enters as a saturated liquid at 0°C with a mass flow rate of 30 kg/h and exits as a saturated vapor. There is no significant change in pressure for either stream.

(a) For the Refrigerant 134a stream, determine the rate of heat transfer and evaluate the accompanying rate of availability transfer, each in kJ/h.
(b) For each of the two streams, evaluate the flow availability change $\dot{m}(a_{fe}-a_{fi})$, in kJ/h, and comment.

Let T_0 = 22°C, p_0 = 0.1 MPa and ignore the effects of motion and gravity.

7.72 Argon gas enters an insulated nozzle operating at steady state at 1027°C, 4 bars, and 10 m/s. At the exit, the temperature and pressure of the argon are 627°C and 1.45 bars, respectively. Determine

(a) the exit velocity, in m/s.
(b) the isentropic nozzle efficiency.
(c) the irreversibility, in kJ per kg of argon flowing through the nozzle.

Assume the ideal gas model for argon and ignore potential energy. Let T_0 = 20°C, p_0 = 1 atm.

7.73 Oxygen enters a nozzle operating at steady state at 3.8 MPa, 387°C, and 10 m/s. At the nozzle exit, the conditions are 150 kPa, 37°C, and 750 m/s.

(a) For a control volume enclosing the nozzle, determine the heat transfer, in kJ per kg of oxygen flowing. What additional information would be required to evaluate the irreversibility?
(b) For a control volume including the nozzle and a portion of its immediate surroundings so that the heat transfer occurs at temperature T_0, evaluate the irreversibility, in kJ per kg of oxygen flowing.

Neglect potential energy. Let T_0 = 20°C, p_0 = 1 atm.

7.74 Air enters a compressor operating at steady state at 35°C and 400 kPa with a volumetric flow rate of 0.3 m³/min and is compressed isothermally with no internal irreversibilities to 1.8 MPa. Kinetic and potential energy effects are negligible. Let T_0 = 20°C, p_0 = 1 atm.

(a) For a control volume enclosing the compressor only, determine the availability transfers accompanying heat and work and the irreversibility, all in kW.
(b) Repeat part (a) for a control volume enclosing the compressor and a portion of its immediate surroundings so that heat transfer occurs at T_0.

7.75 An industrial process requires a liquid water stream at 180°F, 1 atm. It is proposed to obtain this condition at steady state by mixing a stream of liquid water at 50°F, 1 atm with steam supplied from a boiler as saturated vapor at 1 atm. If the irreversibility of the mixing chamber cannot be permitted to exceed 145,000 Btu/h, determine the range of values allowed for the mass flow rate of the liquid water stream. Neglect kinetic and potential energy and heat transfer with the surroundings. Let T_0 = 77°F, p_0 = 1 atm.

7.76 For the open feedwater heater of Problem 6.114, determine the irreversibility rate, in kJ per kg of liquid exiting. Let T_0 = 20°C, p_0 = 1 atm.

7.77 Steam at 3 MPa and 700°C is available at one location in an industrial plant. At another location, steam at 2 MPa and 400°C is required for use in a certain process. An engineer suggests that steam at this condition can be provided by allowing the higher-pressure steam to expand through a valve to 2 MPa and then cool to 400°C through heat transfer to the surroundings, which are at 20°C.

(a) Evaluate this suggestion using availability principles.
(b) Evaluating availability at 8 cents per kW·h, estimate the annual cost of this alternative for a mass flow rate of 1 kg/s.
(c) Suggest one or more alternative methods for obtaining steam at the desired condition that would be preferable thermodynamically. What economic considerations might be involved in making the final decision regarding which alternative to choose?

Let T_0 = 20°C, p_0 = 0.1 MPa.

7.78 For the turbine and heat exchanger of Problem 6.115, determine the irreversibility rates, each in Btu/s. Evaluating availability at 8 cents per kW·h, determine the hourly cost of each of these quantities. Let T_0 = 40°F, p_0 = 1 atm.

7.79 Two alternative systems operating at steady state are under consideration for bringing air from 17 to 52°C at an essentially constant pressure of 1 bar. The systems operate as follows:

System 1: The air temperature is increased as a consequence of the stirring of a liquid surrounding a line containing the air.

System 2: The air temperature is increased by passing it through one side of a counterflow heat exchanger. On the other side, steam condenses at a pressure of 1 bar from saturated vapor to saturated liquid.

On the basis of thermodynamic principles, which alternative would be preferred? Ignore all kinetic and potential energy effects and assume that no significant heat transfer with the surroundings occurs. Use the ideal gas model for the air. Let $T_0 = 17°C$, $p_0 = 1$ bar.

7.80 Two alternative systems operating at steady state are under consideration for bringing air from 50 to 120°F at an essentially constant pressure of 1 atm. The systems operate as follows:

System 1: The air temperature is increased by means of an electric resistor.

System 2: The air temperature is increased by passing it through one side of a counterflow heat exchanger. On the other side, steam condenses at a pressure of 1 atm from saturated vapor to saturated liquid.

On the basis of thermodynamic principles, which alternative would be preferred? Ignore all kinetic and potential energy effects. Assume that no significant heat transfer with the surroundings occurs, the ideal gas model applies for the air. Let $T_0 = 70°F$, $p_0 = 1$ atm.

7.81 For the compressor and turbine of Problem 6.148(b), determine the irreversibility rates, each in kJ per kg of air flowing. Express each as a percentage of the net work developed by the power plant. Let $T_0 = 22°C$, $p_0 = 0.95$ bar.

7.82 For the valve, flash chamber, and turbine of Problem 6.149, determine the irreversibility rates, each in Btu/h. Express each as a percentage of the power developed by the turbine. Let $T_0 = 40°F$, $p_0 = 1$ atm.

7.83 Three devices, each operating at steady state, are under consideration for delivering a heat transfer at a use temperature $T_u = 140°F$. In each case, the energy loss by heat transfer from the device to its surroundings is 10% of the supply. Energy is supplied to the devices as follows:

Device 1: A solar collector system that supplies heat at $T_s = 240°F$.

Device 2: A solar collector system that supplies heat at $T_s = 340°F$.

Device 3: An electrical resistance element that supplies heat at $T_s = 540°F$.

For each case, evaluate the second law efficiency as given by Eq. 7.40. Which alternative is preferred thermodynamically? Let $T_0 = 40°F$.

7.84 The temperature of water contained in a closed, well-insulated tank is increased from 60 to 120°F by passing an electric current through a resistor within the tank. Devise and evaluate a second law efficiency for this water heater. Assume that the water is incompressible and the states of the resistor and the enclosing tank do not change. Let $T_0 = 60°F$.

7.85 From an input of electricity, an electric resistance furnace operating at steady state delivers energy by heat transfer to a dwelling at the rate $\dot{Q}_u$ at a *use* temperature T_u. There are no other significant energy transfers.

(a) Devise a second law efficiency for the furnace.
(b) Evaluate the efficiency obtained in part (a) for a use temperature of 35°C and $T_0 = 20°C$.

7.86 Air enters the insulated duct shown in Fig. P7.86 at a temperature of 60°F and a pressure of 1 atm and exits at a temperature of 140°F and a pressure only slightly less than 1 atm. Electricity is provided to the resistor a rate of 0.1 kW. Kinetic and potential energy effects can be ignored. For steady state operation,

(a) determine the irreversibility rate, in kW.
(b) devise and evaluate a second law efficiency for the heater.

Let $T_0 = 60°F$, $p_0 = 1$ atm.

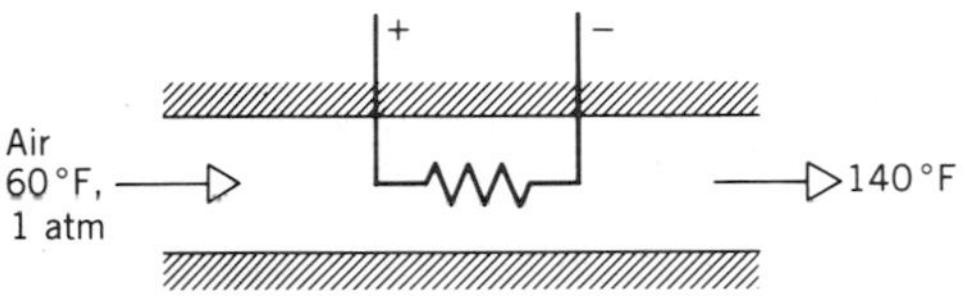

Figure P7.86

7.87 For each case considered in Problem 7.34, devise and evaluate a second law efficiency.

7.88 For the turbine of Problem 7.52, determine the turbine effectiveness and the isentropic turbine efficiency.

7.89 For the turbine of Problem 7.68, determine the turbine effectiveness.

7.90 For the heat exchanger of Problem 7.69, determine the second law efficiency as given by Eq. 7.45.

7.91 Superheated water vapor at 400 lbf/in.^{2}, 600°F enters a turbine and exits at a pressure of 50 lbf/in.^{2} and a quality of 99.5%. The turbine operates at steady state. Heat transfer between the turbine and its surroundings can be neglected, as can kinetic and potential energy effects. Let $T_0 = 60°F$, $p_0 = 1$ atm. Determine

(a) the irreversibility rate, in Btu per lb of steam expanding through the turbine.
(b) the isentropic turbine efficiency.
(c) the turbine effectiveness.

7.92 Saturated water vapor at 400 lbf/in.^{2} enters an insulated turbine operating at steady state. At the turbine exit, the pressure is 0.6 lbf/in.^{2} The work developed equals 306 Btu per lb of steam passing through the turbine. Kinetic and potential energy effects can be neglected. Let $T_0 = 60°F$, $p_0 = 1$ atm. Determine

(a) the irreversibility rate, in Btu per lb of steam expanding through the turbine.

(b) the isentropic turbine efficiency.
(c) the turbine effectiveness.

7.93 Air enters an insulated turbine operating at steady state with a pressure of 5 bars, a temperature of 450 K, and a velocity of 130 m/s. At the turbine exit, the conditions are 1 bar, 330 K, and 30 m/s. There is no significant elevation change between inlet and exit. Determine

(a) the work developed and the irreversibility, each in kJ per kg of air flowing through the turbine.
(b) the turbine effectiveness.

Let $T_0 = 300$ K, $p_0 = 1$ atm.

7.94 Argon enters an insulated turbine operating at steady state at 1000°C and 2 MPa and exhausts at 350 kPa. The mass flow rate is 0.5 kg/s and the turbine develops power at the rate of 120 kW. Determine

(a) the temperature of the argon at the turbine exit, in °C.
(b) the irreversibility rate of the turbine, in kW.
(c) the turbine effectiveness.

Neglect kinetic and potential energy effects. Let $T_0 =$ 20°C, $p_0 = 1$ bar.

7.95 A steam turbine operating at steady state develops 9750 horsepower. The turbine receives 100,000 lb/h of steam at 400 lbf/in.^{2} and 600°F. At a point in the turbine where the pressure is 60 lbf/in.^{2} and the temperature is 300°F, steam is bled off at the rate of 25,000 lb/h. The remaining steam continues to expand through the turbine, exiting at 2 lbf/in.^{2} and 90% quality. **(a)** Determine the rate of heat transfer between the turbine and its surroundings, in Btu/h. **(b)** Devise and evaluate a second law efficiency for the turbine. Kinetic and potential energy effects can be ignored. Let $T_0 = 77$°F, $p_0 = 1$ atm.

7.96 Air enters an insulated compressor operating at steady state at 20 lbf/in.^{2}, 50°F, 200 ft/s, and exits at 50 lbf/in.^{2}, 260°F, 350 ft/s. There is no significant elevation change. Taking $T_0 = 40$°F, $p_0 = 14.7$ lbf/in.^{2}, evaluate

(a) the work and irreversibility, each in Btu per lb of air passing through the compressor.
(b) a second law efficiency for the compressor.

7.97 A compressor operating at steady state takes in 1 kg/s of air at 1 bar and 25°C and compresses it to 8 bars and 160°C. Heat transfer from the compressor to its surroundings occurs at a rate of 100 kW.

(a) Determine the power input, in kW.
(b) Evaluate a second law efficiency for the compressor.

Neglect kinetic and potential energy changes. Let $T_0 =$ 25°C, $p_0 = 1$ bar.

7.98 A counterflow heat exchanger operates at steady state. Air flows on both sides with equal mass flow rates. On one side, air enters at 850°R, 60 lbf/in.^{2} and exits at 1000°R, 50 lbf/in.^{2} On the other side, air enters at 1300°R, 16 lbf/in.^{2} and exits at temperature T and 14.7 lbf/in.^{2} Heat transfer between the heat exchanger and its surroundings can be ignored, as can all effects of kinetic and potential energy. Evaluate

(a) the temperature, T, in °R.
(b) the irreversibility rate, in Btu per lb of air flowing.
(c) the second law efficiency given by Eq. 7.45.

Let $T_0 = 60$°F, $p_0 = 1$ atm.

7.99 A counterflow heat exchanger operating at steady state has oil and liquid water flowing in separate streams. The oil and water can each be regarded as incompressible with constant specific heats 2.1 and 4.2 kJ/kg·K, respectively. Neither experiences a significant pressure change in passing through the heat exchanger. The oil is cooled from 440 to 320 K, while the water temperature increases from 290 K to temperature T. The respective mass flow rates of the oil and water are 800 and 3200 kg/h. Heat transfer between the heat exchanger and its surroundings can be ignored, as can all effects of kinetic and potential energy. Determine

(a) the temperature T, in K.
(b) the irreversibility rate, in kJ/h.
(c) the second law efficiency given by Eq. 7.45.

Let $T_0 = 17$°C, $p_0 = 1$ atm.

7.100 In the boiler of a power plant are tubes through which water flows as it is brought from 0.8 MPa, 150°C to 240°C at essentially constant pressure. Combustion gases passing over the tubes cool from 1067 to 547°C at essentially constant pressure. The combustion gases can be modeled as air as an ideal gas. There is no significant heat transfer from the boiler to its surroundings. Assuming steady state and neglecting kinetic and potential energy effects, determine

(a) the mass flow rate of the combustion gases, in kg per kg of steam flowing.
(b) the irreversibility rate, in kJ per kg of steam flowing.
(c) the second law efficiency given by Eq. 7.45.

Let $T_0 = 25$°C, $p_0 = 1$ atm.

7.101 In the boiler of a power plant are tubes through which water flows as it is brought from a saturated liquid condition at 1200 lbf/in.^{2} to 1400°F at essentially constant pressure. Combustion gases passing over the tubes cool from 2300 to 1800°F at essentially constant pressure. The combustion gases can be modeled as air as an ideal gas. There is no significant heat transfer from the boiler to its surroundings. Assuming steady state and neglecting kinetic and potential energy effects, determine

(a) the mass flow rate of the combustion gases, in lb per lb of steam flowing.
(b) the irreversibility, in Btu per lb of steam flowing.
(c) the second law efficiency given by Eq. 7.45.

Let $T_0 = 70$°F, $p_0 = 1$ atm.

DESIGN AND OPEN ENDED PROBLEMS

7.1D Four options are under consideration to provide *spot* cooling for machining operations using compressed air. These include expansion of compressed air at 70°F to atmospheric pressure through **(a)** a throttling valve, **(b)** a nozzle, **(c)** an air turbine, **(d)** a vortex tube. A vortex tube is a device having no moving mechanical parts that converts an inlet stream of compressed air into two air streams, one cold and one hot. The accompanying table provides manufacturer's data for a vortex tube adequate for the present application. Critically evaluate these options and prepare a memorandum summarizing your recommendations.

Compressed Air Inlet Temperature = 70°F								
		Exit Temperature, °F						
Compressed Air Inlet Pressure lbf/in.^{2} (gage)		**Fraction of the Inlet Air Volume Released through the Cold Outlet, %**						
		20	**30**	**40**	**50**	**60**	**70**	**80**
20	Hot	84.5	94.5	106.0	119.5	134.0	152.5	177.0
	Cold	8.5	10.5	14.5	19.5	26.5	34.0	42.5
60	Hot	93.5	110.0	128.5	150.0	174.0	202.0	238.0
	Cold	−34.0	−30.0	−23.0	−14.0	−3.0	10.5	25.5
100	Hot	96.0	115.0	136.5	161.0	189.0	221.0	262.0
	Cold	−53.0	−48.0	−40.0	−29.0	−16.0	−0.5	17.0
140	Hot	95.5	116.0	140.5	166.0	194.0	226.0	263.0
	Cold	−65.0	−59.0	−51.0	−39.0	−24.0	−6.0	13.5

7.2D A supermarket has an annual electrical requirement of 2×10^6 kW·h for lighting, space heating, refrigeration, and air conditioning. What would your local utility charge for this annual electricity requirement? If the requirement were to be met by wind turbines coupled to electrical generators, determine the *minimum* theoretical number of turbines that would be needed for conditions typical in your locale. Estimate the *actual* number needed. Would you recommend the use of wind turbines for this application? Discuss.

7.3D A combination *venturi scrubber* and *cyclone* for cleaning gases bearing minute foreign particles is to be inserted in a line carrying a gas consisting mainly of nitrogen (N_2) from one processing site to another. Figure P7.3D provides a schematic and operating data. For a *first cost* of \$10,000 and 60 h of operation per week, *annualize* the cost of owning and operating the device over a *life* of five years, assuming no *salvage* value after five years.

7.4D In a *cogeneration* system that produces both electric power at a rate of 10,000 kW and steam at 65 lbf/in.^{2} for some industrial process, the turbine receives steam from a boiler at 750°F, 650 lbf/in.^{2} with a mass flow rate of 1.8×10^5 lb/h. The unit cost of the steam exiting the boiler is 0.3 cents per lb and the *annualized* cost of the turbine is

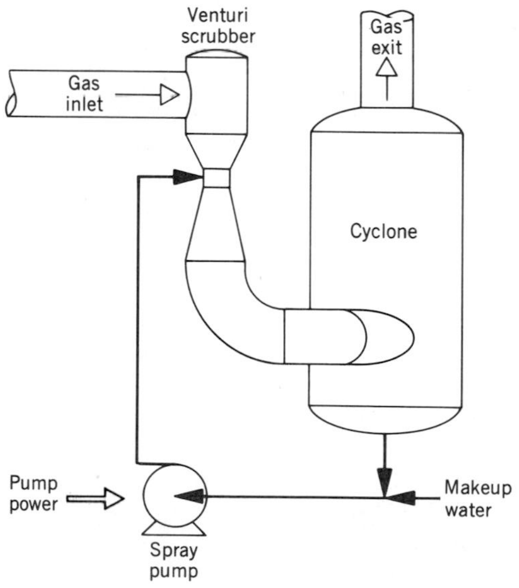

Scrubber pressure drop:	55.1 kPa
Spray pump power:	0.15 kW
Gas flow rate:	24,755 kg/h
Gas temperature:	
Inlet	37.8 °C
Exit	36.1 °C
Inlet gas pressure	474.3 kPa

Figure P7.3D

\$60,000/yr. Evaluate the cost of the power developed, in cents per kW·h, and the cost of the process steam, in cents per lb. Condenser cooling water is provided at 50°F, 1 atm.

7.5D A proposed automobile fuel-saving measure would employ an *availability reclaiming* system that uses the vehicle's engine to compress air as the vehicle decelerates to rest. When the vehicle is underway again, the stored compressed air would drive an air motor that assists the engine in accelerating the vehicle. Specify the design of air storage vessels: size, shape, and materials that would allow the air motor *alone* to bring the vehicle back up to at least 75% of its speed before the vehicle comes to a stop. Incorporate safety features in your design. Prepare a sketch that shows the placement of the principal components onboard the vehicle.

7.6D Power plants usually operate large coal or nuclear power units to meet their *baseloads* because these units are more reliable and economical when run continuously. In periods of *peak demand*, combustion turbines burning costly oil or natural gas are normally employed to supplement the power generated. The use of such premium fuels makes *peak power* relatively expensive. One concept for more economic operation overall is to run the baseload units at full capacity, even when demand is low, and to store the excess availability for use during peak-demand times. Write a paper in which the underlying principles are discussed and costs are compared for the following *availability storage* means: pumped-hydroelectric storage, compressed-air storage, battery storage, superconducting magnetic storage. Draw schematic diagrams showing the principal components of each of these storage types.

7.7D Compressed-air storage power plants use *off-peak baseload* electricity to compress air into underground reservoirs that measure 10 to 20 million ft^3. When electricity is needed to meet *peak demand*, the compressed air is withdrawn, heated by burning natural gas or oil, and run through a turbine to produce power. Compressed air can be stored in several types of natural underground enclosures. Determine if suitable natural sites exist in your locale for compressed-air storage power plants. Suggest means for reducing the loss and destruction of availability attending the charging and discharging of the underground enclosure.

7.8D Using manufacturer's data, carefully define and evaluate a second law efficiency for a common domestic appliance or system of your choice. If the manufacturer's data are insufficient to perform the evaluation, conduct a test of your own using appropriate instrumentation. Estimate the overall experimental uncertainty of your test.

CHAPTER 8

VAPOR POWER SYSTEMS

An important engineering goal is to devise systems that accomplish desired types of energy conversion. The present chapter and the next are concerned with several types of power-generating systems, each of which produces a net power output from an energy source of chemical, nuclear, or solar origin. The objective is to describe some of the practical arrangements employed for power production and illustrate how such power plants can be modeled thermodynamically. The discussion is organized into three main areas of application: vapor power plants, gas turbine power plants, and internal combustion engines. These power systems, together with hydroelectric power plants, produce virtually all of the electrical and mechanical power used worldwide. The present chapter deals with *vapor* power plants in which the *working fluid* is alternately vaporized and condensed. Chapter 9 is concerned with gas turbines and internal combustion engines in which the working fluid remains a gas.

The processes taking place in power-generating systems are sufficiently complicated that idealizations are required to develop tractable thermodynamic models. Such modeling is an important initial step in engineering design. Although the study of simplified models generally leads only to qualitative conclusions about the performance of the corresponding actual devices, models often allow deductions about how changes in major operating parameters affect actual performance. They also provide relatively simple settings in which to discuss the functions and benefits of features intended to improve overall performance.

Several concepts related to thermodynamic *power cycles* introduced in previous chapters are important for the present discussions. You will recall that the conservation of energy principle requires that the net work developed by a power cycle equals the net heat added. An important deduction from the second law is that the thermal efficiency, which indicates the extent to which the heat added is converted to a net work output, must be less than 100%. Previous discussions also have indicated that improved thermodynamic performance accompanies the reduction of irreversibilities. The extent to which irreversibilities can be reduced in power-generating systems depends on thermodynamic, economic, and other factors, however.

8.1 PRELIMINARIES

The vast majority of electrical generating plants are variations of vapor power plants in which water is the working fluid. The basic components of a simplified fossil-fuel vapor power plant are shown schematically in Fig. 8.1. To facilitate thermodynamic analysis, the overall plant can be broken down into the four major subsystems identified by the letters A through D on the diagram. The focus of our considerations in this chapter is subsystem A, where the important energy conversion from *heat to work* occurs. But first, let us briefly consider the other subsystems.

The function of subsystem B is to supply the energy required to vaporize the water passing through the boiler. In fossil-fuel plants, this is accomplished by heat transfer *to* the working fluid passing through tubes and drums in the boiler *from* the hot gases produced by the combustion of a fossil fuel. In nuclear plants, the origin of the energy is a controlled nuclear reaction taking place in an isolated reactor building. Pressurized water or a liquid metal reactor coolant is used to transfer energy released in the nuclear reaction to the working fluid in specially designed heat exchangers. Solar power plants have receivers for concentrating and collecting solar radiation to vaporize the working fluid. Regardless of the energy source, the vapor produced in the boiler passes through a turbine, where it expands to a lower pressure. The shaft of the turbine is connected to an electric generator (subsystem D). The vapor leaving the turbine passes through the condenser, where it condenses on the outside of tubes carrying cooling water. The cooling water circuit comprises subsystem C. For the plant shown, the cooling water is sent to a cooling tower, where energy taken up in the condenser is rejected to the atmosphere. The cooling water is then recirculated through the condenser.

Concern for the environment and safety considerations govern what is allowable in the interactions between subsystems B and C and their surroundings. One of the major difficulties in finding a site for a vapor power plant is access to sufficient

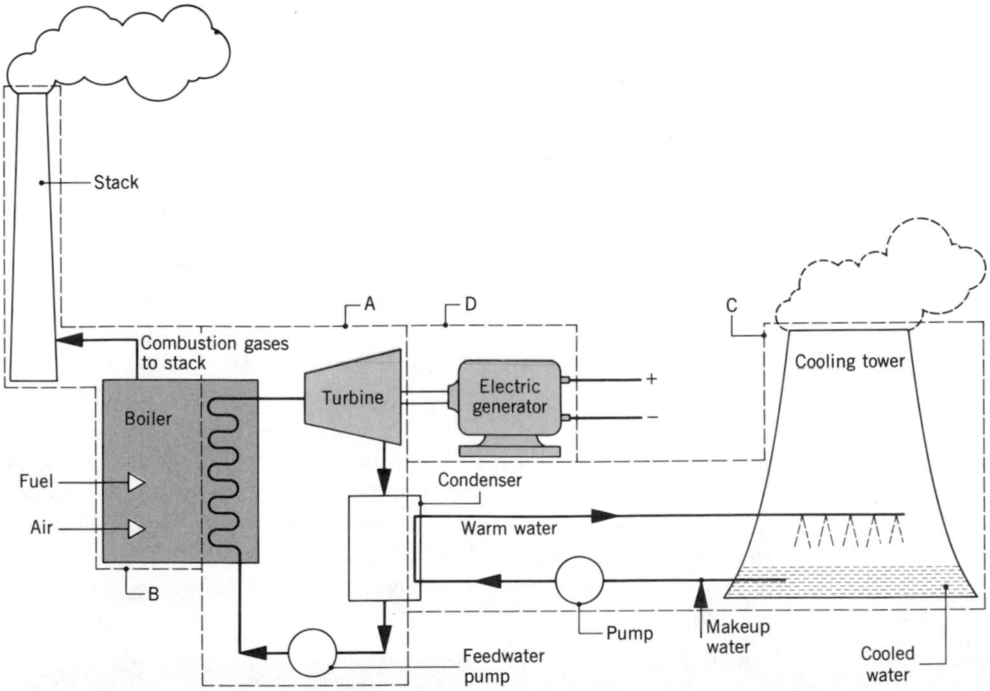

Figure 8.1 Components of a simple vapor power plant.

quantities of cooling water. For this reason and to minimize *thermal pollution* effects, most power plants now employ cooling towers. In addition to the question of cooling water, the safe processing and delivery of fuel, the control of pollutant discharges, and the disposal of wastes are issues that must be dealt with in both fossil-fueled and nuclear-fueled plants to ensure safety and operation with an acceptable level of environmental impact. Solar power plants are generally regarded as nonpolluting and safe but are presently too costly to be widely used.

8.2 RANKINE CYCLE

All of the fundamentals required for the thermodynamic analysis of power-generating systems already have been introduced. They include the conservation of mass and conservation of energy principles, the second law of thermodynamics, and thermodynamic data. These principles apply to individual plant components such as turbines, pumps, and heat exchangers as well as to the most complicated overall power plants. The object of this section is to introduce the Rankine cycle, which is a thermodynamic cycle that models the subsystem labeled A on Fig. 8.1. The presentation begins by considering the thermodynamic analysis of this subsystem.

8.2.1 Principal Work and Heat Transfers

The principal work and heat transfers of subsystem A are illustrated in Fig. 8.2. In subsequent discussions, these energy transfers are taken to be positive in the directions of the arrows. The unavoidable stray heat transfer that takes place between the plant components and their surroundings is neglected here for simplicity. Kinetic and potential energy changes are also ignored. Each component is regarded as operating at steady state. Using the conservation of mass and conservation of energy principles together with these idealizations, let us develop expressions for the energy transfers shown on Fig. 8.2 beginning at state 1 and proceeding through each component in turn. Note that a unit of mass following this path periodically undergoes a thermodynamic cycle.

Vapor from the boiler at state 1, having an elevated temperature and pressure, expands through the turbine to produce work and then is discharged to the condenser at state 2 with relatively low pressure. Neglecting heat transfer with the

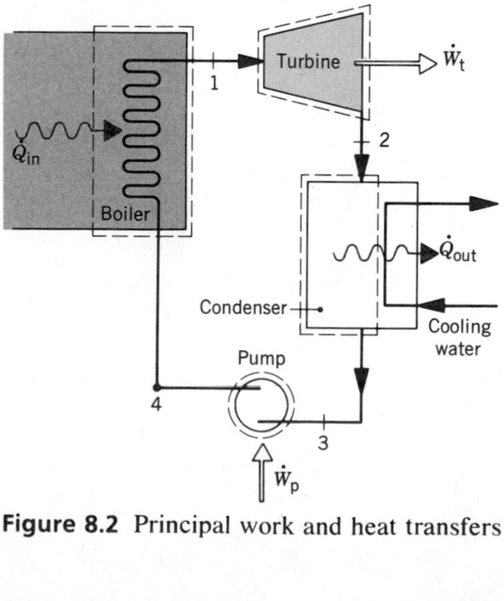

Figure 8.2 Principal work and heat transfers of subsystem A.

surroundings, the mass and energy rate balances for a control volume around the turbine reduce at steady state to give

$$0 = \cancelto{0}{\dot{Q}_{cv}} - \dot{W}_t + \dot{m}\left[h_1 - h_2 + \cancelto{0}{\frac{V_1^2 - V_2^2}{2}} + \cancelto{0}{g(z_1 - z_2)}\right]$$

or

$$\frac{\dot{W}_t}{\dot{m}} = h_1 - h_2 \tag{8.1}$$

where $\dot{m}$ denotes the mass flow rate of the working fluid, and $\dot{W}_t/\dot{m}$ is the rate at which work is developed per unit of mass of steam passing through the turbine. As noted above, kinetic and potential energy changes are ignored.

In the condenser there is heat transfer from the vapor to cooling water flowing in a separate stream. The vapor condenses and the temperature of the cooling water increases. At steady state, mass and energy rate balances for a control volume enclosing the condensing side of the heat exchanger give

$$\frac{\dot{Q}_{out}}{\dot{m}} = h_2 - h_3 \tag{8.2}$$

where $\dot{Q}_{out}/\dot{m}$ is the rate at which energy is transferred by heat *from* the working fluid to the cooling water per unit mass of working fluid passing through the condenser. This energy transfer is positive in the direction of the arrow on Fig. 8.2.

The liquid condensate leaving the condenser at 3 is pumped from the condenser pressure to the boiler pressure. Taking a control volume around the pump and assuming no heat transfer with the surroundings, mass and energy rate balances give

$$\frac{\dot{W}_p}{\dot{m}} = h_4 - h_3 \tag{8.3}$$

where $\dot{W}_p/\dot{m}$ is the rate of power *input* per unit of mass passing through the pump. This energy transfer is positive in the direction of the arrow on Fig. 8.2.

The working fluid completes a cycle as the liquid leaving the pump at 4, called the boiler *feedwater*, is heated to saturation and evaporated in the boiler. Taking a control volume enclosing the boiler tubes and drums carrying the feedwater from state 4 to state 1, mass and energy rate balances give

$$\frac{\dot{Q}_{in}}{\dot{m}} = h_1 - h_4 \tag{8.4}$$

where $\dot{Q}_{in}/\dot{m}$ is the rate of heat transfer from the energy source into the working fluid per unit mass passing through the boiler.

The thermal efficiency gauges the extent to which the energy input to the working fluid passing through the boiler is converted to the *net* work output. Using the quantities and expressions just introduced, the thermal efficiency of the power cycle of Fig. 8.2 is

$$\eta = \frac{\dot{W}_t/\dot{m} - \dot{W}_p/\dot{m}}{\dot{Q}_{in}/\dot{m}} = \frac{(h_1 - h_2) - (h_4 - h_3)}{h_1 - h_4} \tag{8.5a}$$

The net work output equals the net heat input. Thus, the thermal efficiency can be expressed alternatively as

$$\eta = \frac{\dot{Q}_{in}/\dot{m} - \dot{Q}_{out}/\dot{m}}{\dot{Q}_{in}/\dot{m}} = 1 - \frac{\dot{Q}_{out}/\dot{m}}{\dot{Q}_{in}/\dot{m}}$$

$$= 1 - \frac{(h_2 - h_3)}{(h_1 - h_4)} \tag{8.5b}$$

Another parameter used to describe power plant performance is the *back work ratio,* or bwr, defined as the ratio of the pump work input to the work developed by the turbine. With Eqs. 8.1 and 8.3, the back work ratio for the power cycle of Fig. 8.2 is

$$\text{bwr} = \frac{\dot{W}_p/\dot{m}}{\dot{W}_t/\dot{m}} = \frac{(h_4 - h_3)}{(h_1 - h_2)} \tag{8.6}$$

Examples to follow illustrate that the change in specific enthalpy for the expansion of vapor through the turbine is normally many times greater than the increase in enthalpy for the liquid passing through the pump. Hence, the back work ratio is characteristically quite low for vapor power plants.

Provided states 1 through 4 are fixed, Eqs. 8.1 through 8.6 can be applied to determine the thermodynamic performance of a simple vapor power plant. Since these equations have been developed from mass and energy rate balances, they apply equally for actual performance when irreversibilities are present and for idealized performance in the absence of such effects. It might be surmised that the irreversibilities of the various power plant components can affect overall performance, and this is the case. Even so, it is instructive to consider an idealized cycle in which they are assumed absent, for such a cycle establishes an *upper limit* on the performance of the Rankine cycle. The ideal cycle also provides a simple setting in which to study various aspects of vapor power plant performance.

8.2.2 Ideal Rankine Cycle

If the working fluid passes through the various components of the simple vapor power cycle without irreversibilities, frictional pressure drops would be absent from the boiler and condenser, and the working fluid would flow through these components at constant pressure. Also, in the absence of irreversibilities and heat transfer with the surroundings, the processes through the turbine and pump would be isentropic. A cycle adhering to these idealizations is the *ideal Rankine cycle* shown in Fig. 8.3.

Referring to Fig. 8.3, we see that the working fluid undergoes the following series of internally reversible processes:

Process 1–2: Isentropic expansion of the working fluid through the turbine from saturated vapor at state 1 to the condenser pressure.

Process 2–3: Heat transfer *from* the working fluid as it flows at constant pressure through the condenser with saturated liquid at state 3.

Process 3–4: Isentropic compression in the pump to state 4 in the compressed liquid region.

Process 4–1: Heat transfer *to* the working fluid as it flows at constant pressure through the boiler to complete the cycle.

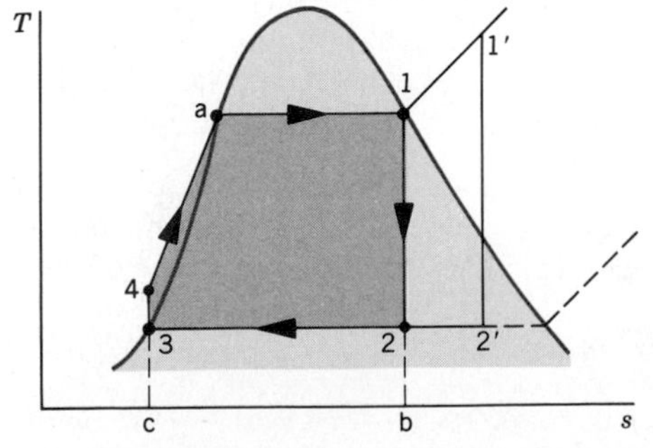

Figure 8.3 Temperature–entropy diagram of the ideal Rankine cycle.

The ideal Rankine cycle also includes the possibility of superheating the vapor, as in cycle 1′–2′–3–4–1′. The importance of superheating is discussed in Sec. 8.3.

Because the pump is idealized as operating without irreversibilities, Eq. 6.61 can be invoked as an alternative to Eq. 8.3 for evaluating the pump work. That is,

$$\left(\frac{\dot{W}_p}{\dot{m}}\right)_{\substack{\text{int}\\ \text{rev}}} = \int_3^4 v\, dp \tag{8.7a}$$

where the minus sign has been dropped so that the integral takes on positive values in accordance with the sign for the pump work adopted previously. The subscript "int rev" has been retained as a reminder that this expression is restricted to an internally reversible process through the pump. No such designation is required by Eq. 8.3, however, because it expresses the conservation of mass and energy principles and thus is not restricted to processes that are internally reversible.

Evaluation of the integral of Eq. 8.7a requires a relationship between the specific volume and pressure for the process. Because the specific volume of the liquid normally varies only slightly as the liquid flows from the inlet to the exit of the pump, an approximation to the value of the integral can be had by taking the specific volume at the pump inlet, v_3, as typical for the process. Then

$$\left(\frac{\dot{W}_p}{\dot{m}}\right)_{\substack{\text{int}\\ \text{rev}}} \approx v_3(p_4 - p_3) \tag{8.7b}$$

The next example is an illustration of the analysis of an ideal Rankine cycle. Note that a minor departure from our usual problem-solving methodology is used in this example and examples to follow. In the ***Analysis*** portion of the solution, attention is focused initially on the systematic evaluation of specific enthalpies and other required property values at each numbered state in the cycle. This eliminates the need to interrupt the solution repeatedly with property determinations and reinforces what is known about the processes in each component, since given information and assumptions are normally required to fix each of the numbered states.

EXAMPLE 8.1

PROBLEM

Steam is the working fluid in an ideal Rankine cycle. Saturated vapor enters the turbine at 8.0 MPa and saturated liquid exits the condenser at a pressure of 0.008 MPa. The *net* power output of the cycle is 100 MW. Determine for the cycle (a) the thermal efficiency, (b) the back work ratio, (c) the mass flow rate of the steam, in kg/h, (d) the rate of heat transfer, $\dot{Q}_{in}$, into the working fluid as it passes through the boiler, in MW, (e) the rate of heat transfer, $\dot{Q}_{out}$, from the condensing steam as it passes through the condenser, in MW, (f) the mass flow rate of the condenser cooling water, in kg/h, if cooling water enters the condenser at 15°C and exits at 35°C.

SOLUTION

Known: An ideal Rankine cycle operates with steam as the working fluid. The boiler and condenser pressures are specified, and the net power output is given.

Find: Determine the thermal efficiency, the back work ratio, the mass flow rate of the steam, in kg/h, the rate of heat transfer to the working fluid as it passes through the boiler, in MW, the rate of heat transfer from the condensing steam as it passes through the condenser, in MW, the mass flow rate of the condenser cooling water, which enters at 15°C and exits at 35°C.

Schematic and Given Data:

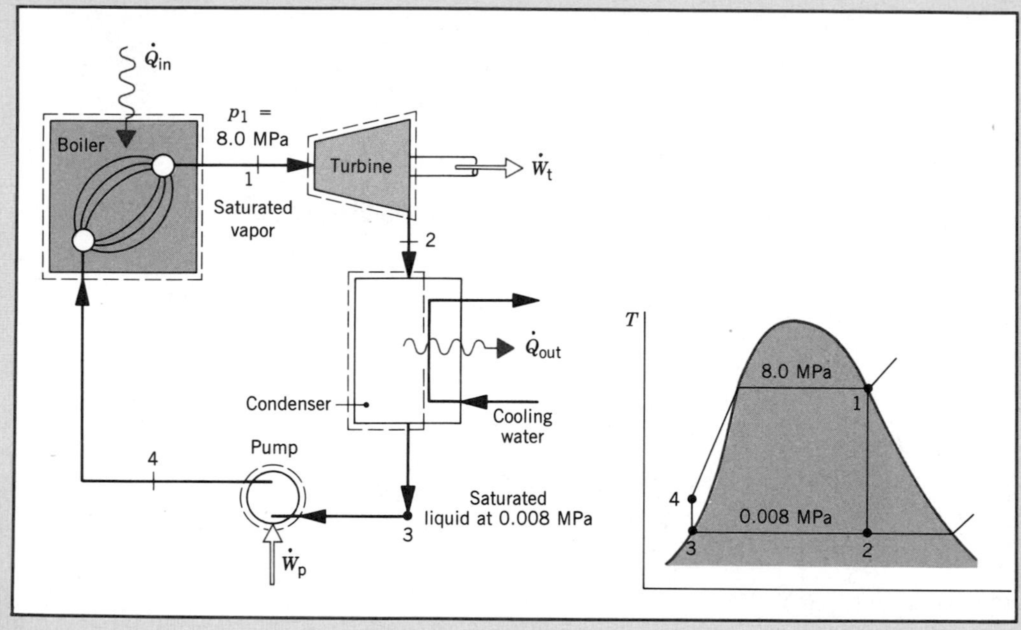

Figure E8.1

Assumptions:

1. Each component of the cycle is analyzed as a control volume at steady state. The control volumes are shown on the accompanying sketch by dashed lines.
2. All processes of the working fluid are internally reversible.
3. The turbine and pump operate adiabatically.
4. Kinetic and potential energy effects are negligible.
5. Saturated vapor enters the turbine. Condensate exits the condenser as saturated liquid.

Analysis: To begin the analysis, let us fix each of the principal states located on the accompanying schematic and T–s diagrams. Starting at the inlet to the turbine, the pressure is 8.0 MPa and the steam is a saturated vapor, so from Table A-3, $h_1 = 2758.0$ kJ/kg and $s_1 = 5.7432$ kJ/kg·K.

State 2 is fixed by $p_2 = 0.008$ MPa and the fact that the specific entropy is constant for the adiabatic, internally reversible expansion through the turbine. Using saturated liquid and saturated vapor data from Table A-3, we find that the quality at state 2 is

$$x_2 = \frac{s_2 - s_f}{s_g - s_f} = \frac{5.7432 - 0.5926}{7.6361} = 0.6745$$

The enthalpy is then

$$h_2 = h_f + x_2 h_{fg} = 173.88 + (0.6745)2403.1$$
$$= 1794.8 \text{ kJ/kg}$$

State 3 is saturated liquid at 0.008 MPa, so $h_3 = 173.88$ kJ/kg.

State 4 is fixed by the boiler pressure p_4 and the specific entropy $s_4 = s_3$. In principle, the specific enthalpy h_4 can be found by interpolation in the compressed liquid tables. However, because compressed liquid data are relatively sparse, it is more convenient to solve Eq. 8.3 for h_4, using Eq. 8.7b to approximate the pump work. With this approach

$$h_4 = h_3 + \dot{W}_p/\dot{m} = h_3 + v_3(p_4 - p_3)$$

By inserting property values from Table A-3

$$h_4 = 173.88 \text{ kJ/kg} + (1.0084 \times 10^{-3} \text{ m}^3/\text{kg})(8.0 - 0.008) \text{ MPa} \left(\frac{10^6 \text{ N/m}^2}{1 \text{ MPa}}\right)\left(\frac{\text{kJ}}{10^3 \text{ N}\cdot\text{m}}\right)$$

$$= 173.88 + 8.06 = 181.94 \text{ kJ/kg}$$

(a) The *net* power developed by the cycle is

$$\dot{W}_{cycle} = \dot{W}_t - \dot{W}_p$$

Mass and energy rate balances for control volumes around the turbine and pump give, respectively

$$\frac{\dot{W}_t}{\dot{m}} = h_1 - h_2 \quad \text{and} \quad \frac{\dot{W}_p}{\dot{m}} = h_4 - h_3$$

where $\dot{m}$ is the mass flow rate of the steam. The rate of heat transfer to the working fluid as it passes through the boiler is determined using mass and energy rate balances as

$$\frac{\dot{Q}_{in}}{\dot{m}} = h_1 - h_4$$

The thermal efficiency is then

$$\eta = \frac{\dot{W}_t - \dot{W}_p}{\dot{Q}_{in}} = \frac{(h_1 - h_2) - (h_4 - h_3)}{h_1 - h_4}$$

$$= \frac{(2758.0 - 1794.8) - (181.94 - 173.88) \text{ kJ/kg}}{(2758.0 - 181.94) \text{ kJ/kg}}$$

$$= 0.371 \ (37.1\%)$$

(b) The back work ratio is

$$\text{bwr} = \frac{\dot{W}_p}{\dot{W}_t} = \frac{h_4 - h_3}{h_1 - h_2} = \frac{(181.94 - 173.88) \text{ kJ/kg}}{(2758.0 - 1794.8) \text{ kJ/kg}}$$

$$= \frac{8.06}{963.2} = 8.37 \times 10^{-3} \ (0.84\%)$$

(c) The mass flow rate of the steam can be obtained from the expression for the net power given in part (a). Thus

$$\dot{m} = \frac{\dot{W}_{cycle}}{(h_1 - h_2) - (h_4 - h_3)}$$

$$= \frac{(100 \text{ MW})(10^3 \text{ kW/MW})(3600 \text{ s/h})}{(963.2 - 8.06) \text{ kJ/kg}}$$

$$= 3.77 \times 10^5 \text{ kg/h}$$

(d) With the expression for $\dot{Q}_{in}$ from part (a) and previously determined specific enthalpy values

$$\dot{Q}_{in} = \dot{m}(h_1 - h_4)$$

$$= \frac{(3.77 \times 10^5 \text{ kg/h})(2758.0 - 181.94) \text{ kJ/kg}}{(3600 \text{ s/h})(1000 \text{ kW/MW})}$$

$$= 269.77 \text{ MW}$$

(e) Mass and energy rate balances applied to a control volume enclosing the steam side of the condenser give

$$\dot{Q}_{out} = \dot{m}(h_2 - h_3)$$

$$= \frac{(3.77 \times 10^5 \text{ kg/h})(1794.8 - 173.88) \text{ kJ/kg}}{(3600 \text{ s/h})(1000 \text{ kW/MW})}$$

$$= 169.75 \text{ MW}$$

Note that the ratio of $\dot{Q}_{out}$ to $\dot{Q}_{in}$ is 0.629 (62.9%).

Alternatively, $\dot{Q}_{out}$ can be determined from an energy rate balance on the *overall* vapor power plant. At steady state, the net power developed equals the net rate of heat transfer to the plant

$$\dot{W}_{cycle} = \dot{Q}_{in} - \dot{Q}_{out}$$

Rearranging this expression and inserting values

$$\dot{Q}_{out} = \dot{Q}_{in} - \dot{W}_{cycle} = 269.77 \text{ MW} - 100 \text{ MW} = 169.77 \text{ MW}$$

The slight difference from the above value is due to round-off.

(f) Taking a control volume around the condenser, the mass and energy rate balances give at steady state

$$0 = \overset{0}{\cancel{\dot{Q}_{cv}}} - \overset{0}{\cancel{\dot{W}_{cv}}} + \dot{m}_{cw}(h_{cw,in} - h_{cw,out}) + \dot{m}(h_2 - h_3)$$

where $\dot{m}_{cw}$ is the mass flow rate of the cooling water. Solving for $\dot{m}_{cw}$

$$\dot{m}_{cw} = \frac{\dot{m}(h_2 - h_3)}{(h_{cw,out} - h_{cw,in})}$$

The numerator in this expression is evaluated in part (e). For the cooling water, $h \approx h_f(T)$, so with saturated liquid enthalpy values from Table A-2 at the entering and exiting temperatures of the cooling water

$$\dot{m}_{cw} = \frac{(169.75\ \text{MW})(10^3\ \text{kW/MW})(3600\ \text{s/h})}{(146.68 - 62.99)\ \text{kJ/kg}} = 7.3 \times 10^6\ \text{kg/h}$$

Comments:

1. A slightly revised problem-solving methodology is introduced in this example problem.
2. Note that the back work ratio is relatively low for the Rankine cycle. In the present case, the work required to operate the pump is less than 1% of the turbine output.
3. In this example, 62.9% of the energy added to the working fluid by heat transfer is subsequently discharged to the cooling water. Although considerable energy is carried away by the cooling water, its *potential for use* is limited because the water exits at a temperature only a few degrees greater than that of the surroundings. See Sec. 8.6 for further discussion.

8.2.3 Effects of Boiler and Condenser Pressures on the Rankine Cycle

Since the ideal Rankine cycle consists entirely of internally reversible processes, an expression for thermal efficiency can be obtained in terms of *average* temperatures during the heat interaction processes. Let us begin the development of this expression by noting that areas under the process lines of Fig. 8.3 can be interpreted as the heat transfer per unit of mass flowing through the respective components. For example, the total area 1–b–c–4–a–1 represents the heat transfer into the working fluid per unit of mass passing through the boiler. In symbols,

$$\left(\frac{\dot{Q}_{in}}{\dot{m}}\right)_{\substack{\text{int}\\\text{rev}}} = \int_4^1 T\,ds = \text{area } 1\text{–b–c–4–a–1}$$

The integral can be written in terms of an average temperature of heat addition, $\bar{T}_{in}$, as follows:

$$\left(\frac{\dot{Q}_{in}}{\dot{m}}\right)_{\substack{\text{int}\\\text{rev}}} = \bar{T}_{in}(s_1 - s_4)$$

where the overbar denotes *average*. Similarly, area 2–b–c–3–2 represents the heat transfer from the condensing steam per unit of mass passing through the condenser

$$\left(\frac{\dot{Q}_{out}}{\dot{m}}\right)_{\substack{\text{int}\\\text{rev}}} = T_{out}(s_2 - s_3) = \text{area } 2\text{–b–c–3–2}$$

$$= T_{out}(s_1 - s_4)$$

where T_{out} denotes the temperature on the steam side of the condenser of the ideal Rankine cycle pictured in Fig. 8.3. The thermal efficiency of the ideal Rankine cycle can be expressed in terms of these heat transfers as

$$\eta_{\text{ideal}} = 1 - \frac{(\dot{Q}_{\text{out}}/\dot{m})_{\substack{\text{int}\\ \text{rev}}}}{(\dot{Q}_{\text{in}}/\dot{m})_{\substack{\text{int}\\ \text{rev}}}} = 1 - \frac{T_{\text{out}}}{\bar{T}_{\text{in}}} \tag{8.8}$$

By the study of Eq. 8.8, we conclude that the thermal efficiency of the ideal cycle tends to increase as the average temperature at which energy is added by heat transfer increases and/or the temperature at which energy is rejected decreases. With similar reasoning, these conclusions can be shown to apply to the other ideal cycles considered in this chapter and the next.

Equation 8.8 can be employed to study the effects on performance of changes in the boiler and condenser pressures. Although these findings are obtained with reference to the ideal Rankine cycle, they also hold qualitatively for actual vapor power plants. Figure 8.4*a* shows two ideal cycles having the same condenser pressure but different boiler pressures. By inspection, the average temperature of heat addition is seen to be greater for the higher-pressure cycle 1′–2′–3–4′–1′ than for cycle 1–2–3–4–1. It follows that increasing the boiler pressure of the ideal Rankine cycle tends to increase the thermal efficiency.

Figure 8.4*b* shows two cycles with the same boiler pressure but two different condenser pressures. One condenser operates at atmospheric pressure and the other at *less than* atmospheric pressure. The temperature of heat discharge for cycle 1–2–3–4–1 condensing at atmospheric pressure is 100°C (212°F). The temperature of heat discharge for the lower-pressure cycle 1–2″–3″–4″–1 is correspondingly lower, so this cycle has the greater thermal efficiency. It follows that decreasing the condenser pressure tends to increase the thermal efficiency.

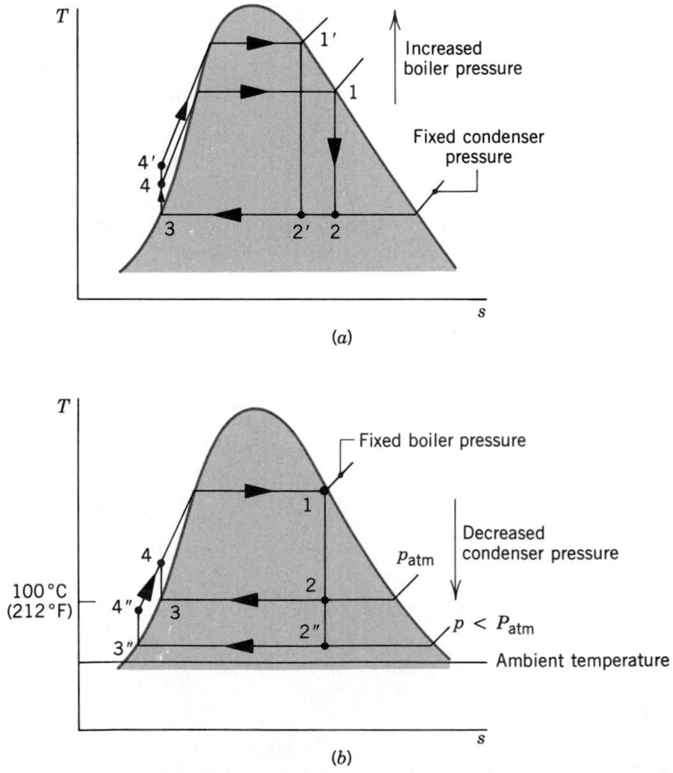

Figure 8.4 Effects of varying operating pressures on the ideal Rankine cycle. (*a*) Effect of boiler pressure. (*b*) Effect of condenser pressure.

The lowest feasible condenser pressure is the saturation pressure corresponding to the ambient temperature, for this is the lowest possible temperature for heat rejection to the surroundings. The goal of maintaining the lowest practical turbine exhaust (condenser) pressure is a primary reason for including the condenser in a power plant. Liquid water at atmospheric pressure could be drawn into the boiler by a pump, and steam could be discharged directly to the atmosphere at the turbine exit. However, by including a condenser in which the steam side is operated at a pressure *below atmospheric*, the turbine has a lower-pressure region in which to discharge, resulting in a significant increase in net work and thermal efficiency. The addition of a condenser also allows the working fluid to flow in a closed loop. This arrangement permits continual circulation of the working fluid, so purified water that is less corrosive than tap water can be used.

The form of Eq. 8.8 is similar to the expression used to determine the thermal efficiency of the Carnot vapor power cycle introduced in Sec. 5.7. However, by referring to Fig. 8.5, it is evident that the ideal Rankine cycle 1–2–3–4–4′–1 has a lower thermal efficiency than the Carnot cycle 1–2–3′–4′–1 having the same maximum temperature T_H and minimum temperature T_C because the average temperature between 4 and 4′ is less than T_H. Despite the greater thermal efficiency of the Carnot cycle, it has two shortcomings as a model for the simple vapor power cycle. First, the heat passing to the working fluid of a vapor power plant is usually obtained from hot products of combustion cooling at approximately constant pressure. To exploit fully the energy released on combustion, the hot products should be cooled as much as possible. The first portion of the heating process of the Rankine cycle shown in Fig. 8.5, Process 4–4′, is achieved by cooling the combustion products *below* the maximum temperature T_H. With the Carnot cycle, however, the combustion products would be cooled *at the most* to T_H. Thus, a smaller portion of the energy released on combustion would be used. The second shortcoming of the Carnot vapor power cycle involves the pumping process. Note that the state 3′ of Fig. 8.5 is a two-phase liquid–vapor mixture. Significant practical problems are encountered in developing pumps that handle two-phase mixtures, as would be required by Carnot cycle 1–2–3′–4′–1. It is far easier to condense the vapor completely and handle only liquid

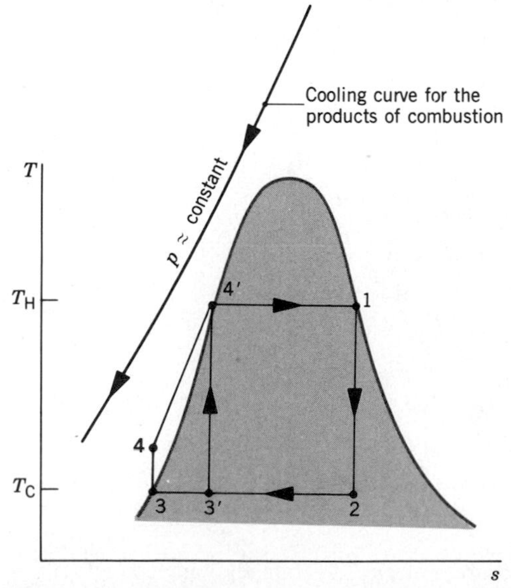

Figure 8.5 Illustration used to compare the ideal Rankine cycle with the Carnot cycle.

in the pump, as is done in the Rankine cycle. Pumping from 3 to 4 and constant-pressure heating without work from 4 to 4′ are processes that can be closely achieved in practice.

8.2.4 Principal Irreversibilities and Losses

Irreversibilities and losses are associated with each of the four subsystems shown in Fig. 8.1. Some of these effects have a more pronounced influence on performance than others. Let us consider the irreversibilities and losses experienced by the working fluid as it circulates through the closed circuit of the Rankine cycle.

The principal irreversibility experienced by the working fluid is associated with the expansion through the turbine. Heat transfer from the turbine to the surroundings represents a loss, but since it is usually of secondary importance, this loss is ignored in subsequent discussions. As illustrated by Process 1–2 of Fig. 8.6, an actual adiabatic expansion through the turbine is accompanied by an increase in entropy. The work developed per unit of mass in this process is less than for the corresponding isentropic expansion 1–2s. The turbine efficiency η_t introduced in Sec. 6.8 allows the effect of irreversibilities within the turbine to be accounted for in terms of the actual and isentropic work amounts. Designating the states as in Fig. 8.6, the isentropic turbine efficiency is

$$\eta_t = \frac{(\dot{W}_t/\dot{m})}{(\dot{W}_t/\dot{m})_s} = \frac{h_1 - h_2}{h_1 - h_{2s}} \tag{8.9}$$

where the numerator is the actual work developed per unit of mass passing through the turbine and the denominator is the work for an isentropic expansion from the turbine inlet state to the turbine exhaust pressure. Irreversibilities within the turbine significantly reduce the net power output of the plant.

The work input to the pump required to overcome frictional effects also reduces the net power output of the plant. In the absence of heat transfer to the surroundings, there would be an increase in entropy across the pump. Process 3–4 of Fig. 8.6 illustrates the actual pumping process. The work input for this process is *greater* than for the corresponding isentropic process 3–4s. The pump efficiency η_p introduced in Sec. 6.8 allows the effect of irreversibilities within the pump to be accounted for in terms of the actual and isentropic work amounts. Designating the states as in Fig. 8.6, the isentropic pump efficiency is

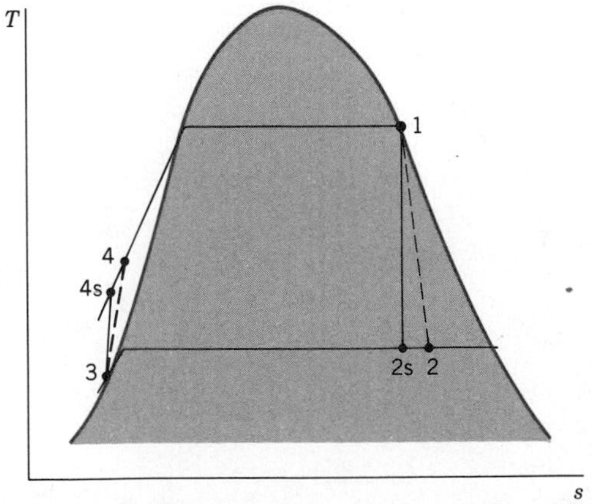

Figure 8.6 Temperature–entropy diagram showing the effects of turbine and pump irreversibilities.

$$\eta_p = \frac{(\dot{W}_p/\dot{m})_s}{(\dot{W}_p/\dot{m})} = \frac{h_{4s} - h_3}{h_4 - h_3} \tag{8.10}$$

In this expression, the pump work for the isentropic process appears in the numerator. The actual pump work, being the larger quantity, is the denominator. Because the pump work is so much less than the turbine work, irreversibilities in the pump have a much smaller impact on the net work of the cycle than do irreversibilities in the turbine.

In addition to the foregoing, there are several other relatively less important sources of nonideality. Stray heat transfers from the outside surfaces of the plant components have detrimental effects on performance, since such losses reduce the extent of conversion from heat input to work output. Frictional effects resulting in pressure drops are sources of irreversibility as the working fluid flows through the boiler, condenser, and piping connecting the various components. Detailed thermodynamic analyses would account for these effects. For simplicity, however, they are ignored in the subsequent discussions. Thus, Fig. 8.6 shows no pressure drops for flow through the boiler and condenser or between plant components. Another effect on performance is suggested by the placement of state 3 on Fig. 8.6. At this state, the temperature of the working fluid exiting the condenser would be lower than the saturation temperature corresponding to the condenser pressure. This is disadvantageous because a greater heat transfer would be required in the boiler to bring the water to saturation.

The irreversibilities mentioned above are *internal* irreversibilities experienced by the working fluid as it flows around the closed loop of the Rankine cycle. The most significant sources of irreversibility for an overall fossil-fueled vapor power plant, however, are associated with the combustion of the fuel and the subsequent heat transfer from the hot combustion products to the cycle working fluid. These effects occur in the surroundings of the subsystem labeled A on Fig. 8.1 and thus are *external* irreversibilities for the Rankine cycle. These irreversibilities are considered further in Secs. 8.6 and 13.6.

Another effect that occurs in the surroundings is the energy discharge to the cooling water as the working fluid condenses. Although considerable energy is carried away by the cooling water, its *utility* is limited. For condensers in which steam condenses near the ambient temperature, the cooling water experiences a temperature rise of only a few degrees over the temperature of the surroundings in passing through the condenser and thus has limited usefulness. Accordingly, the significance of this loss is far less than suggested by the magnitude of the energy transferred to the cooling water. The utility of condenser cooling water is evaluated in Sec. 8.6 using the availability concept.

EXAMPLE 8.2

PROBLEM

Reconsider the vapor power cycle of Example 8.1, but include in the analysis that the turbine and the pump each have an isentropic efficiency of 85%. Determine for the modified cycle (a) the thermal efficiency, (b) the mass flow rate of steam, in kg/h, for a net power output of 100 MW, (c) the rate of heat transfer $\dot{Q}_{in}$ into the working fluid as it passes through the boiler, in MW, (d) the rate of heat transfer $\dot{Q}_{out}$ from the condensing steam as it passes through the condenser, in MW, (e) the mass flow rate of the condenser cooling water, in kg/h, if cooling water enters the condenser at 15°C and exits at 35°C.

SOLUTION

Known: A vapor power cycle operates with steam as the working fluid. The turbine and pump both have efficiencies of 85%.

Find: Determine the thermal efficiency, the mass flow rate, in kg/h, the rate of heat transfer to the working fluid as it passes through the boiler, in MW, the heat transfer rate from the condensing steam as it passes through the condenser, in MW, and the mass flow rate of the condenser cooling water, in kg/h.

Schematic and Given Data:

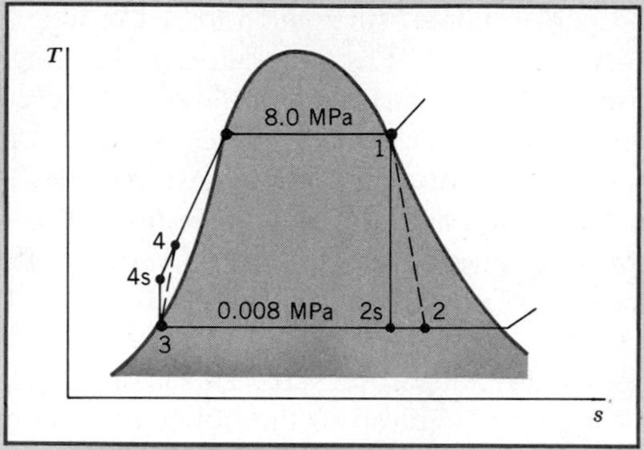

Figure E8.2

Assumptions:

1. Each component of the cycle is analyzed as a control volume at steady state.
2. The working fluid passes through the boiler and condenser at constant pressure. Saturated vapor enters the turbine. The condensate is saturated at the condenser exit.
3. The turbine and pump each operate adiabatically with an efficiency of 85%.
4. Kinetic and potential energy effects are negligible.

Analysis: Owing to the presence of irreversibilities during the expansion of the steam through the turbine, there is an increase in specific entropy from turbine inlet to exit, as shown on the accompanying *T–s* diagram. Similarly, there is an increase in specific entropy from pump inlet to exit. Let us begin the analysis by fixing each of the principal states. State 1 is the same as in Example 8.1, so $h_1 = 2758.0$ kJ/kg and $s_1 = 5.7432$ kJ/kg·K.

The specific enthalpy at the turbine exit, state 2, can be determined using the turbine efficiency

$$\eta_t = \frac{\dot{W}_t/\dot{m}}{(\dot{W}_t/\dot{m})_s} = \frac{h_1 - h_2}{h_1 - h_{2s}}$$

where h_{2s} is the specific enthalpy at state 2s on the accompanying *T–s* diagram. From the solution to Example 8.1, $h_{2s} = 1794.8$ kJ/kg. Solving for h_2 and inserting known values

$$\begin{aligned} h_2 &= h_1 - \eta_t(h_1 - h_{2s}) \\ &= 2758 - 0.85(2758 - 1794.8) = 1939.3 \text{ kJ/kg} \end{aligned}$$

State 3 is the same as in Example 8.1, so $h_3 = 173.88$ kJ/kg.

To determine the specific enthalpy at the pump exit, state 4, reduce mass and energy rate balances for a control volume around the pump to obtain $\dot{W}_p/\dot{m} = h_4 - h_3$. On rearrangement, the specific enthalpy at state 4 is

$$h_4 = h_3 + \dot{W}_p/\dot{m}$$

To determine h_4 from this expression requires the pump work, which can be evaluated using the pump efficiency η_p, as follows. By definition

$$\eta_p = \frac{(\dot{W}_p/\dot{m})_s}{(\dot{W}_p/\dot{m})}$$

The term $(\dot{W}_p/\dot{m})_s$ can be evaluated using Eq. 8.7b. Then solving for $\dot{W}_p/\dot{m}$ results in

$$\frac{\dot{W}_p}{\dot{m}} = \frac{v_3(p_4 - p_3)}{\eta_p}$$

The numerator of this expression was determined in the solution to Example 8.1. Accordingly,

$$\frac{\dot{W}_p}{\dot{m}} = \frac{8.06 \text{ kJ/kg}}{0.85} = 9.48 \text{ kJ/kg}$$

The specific enthalpy at the pump exit is then

$$h_4 = h_3 + \dot{W}_p/\dot{m} = 173.88 + 9.48 = 183.36 \text{ kJ/kg}$$

(a) The net power developed by the cycle is

$$\dot{W}_{cycle} = \dot{W}_t - \dot{W}_p = \dot{m}[(h_1 - h_2) - (h_4 - h_3)]$$

The rate of heat transfer to the working fluid as it passes through the boiler is

$$\dot{Q}_{in} = \dot{m}(h_1 - h_4)$$

Thus, the thermal efficiency is

$$\eta = \frac{(h_1 - h_2) - (h_4 - h_3)}{h_1 - h_4}$$

Inserting values

$$\eta = \frac{(2758 - 1939.3) - 9.48}{2758 - 183.36} = 0.314 \text{ (31.4\%)}$$

(b) With the net power expression of part (a), the mass flow rate of the steam is

$$\dot{m} = \frac{\dot{W}_{cycle}}{(h_1 - h_2) - (h_4 - h_3)}$$

$$= \frac{(100 \text{ MW})(3600 \text{ s/h})(10^3 \text{ kW/MW})}{(818.7 - 9.48) \text{ kJ/kg}} = 4.449 \times 10^5 \text{ kg/h}$$

(c) With the expression for $\dot{Q}_{in}$ from part (a) and previously determined specific enthalpy values

$$\dot{Q}_{in} = \dot{m}(h_1 - h_4)$$

$$= \frac{(4.449 \times 10^5 \text{ kg/h})(2758 - 183.36) \text{ kJ/kg}}{(3600 \text{ s/h})(10^3 \text{ kW/MW})} = 318.2 \text{ MW}$$

(d) The rate of heat transfer from the condensing steam to the cooling water is

$$\dot{Q}_{out} = \dot{m}(h_2 - h_3)$$

$$= \frac{(4.449 \times 10^5 \text{ kg/h})(1939.3 - 173.88) \text{ kJ/kg}}{(3600 \text{ s/h})(10^3 \text{ kW/MW})} = 218.2 \text{ MW}$$

(e) The mass flow rate of the cooling water can be determined from

$$\dot{m}_{cw} = \frac{\dot{m}(h_2 - h_3)}{(h_{cw,out} - h_{cw,in})}$$

$$= \frac{(218.2 \text{ MW})(10^3 \text{ kW/MW})(3600 \text{ s/h})}{(146.68 - 62.99) \text{ kJ/kg}} = 9.39 \times 10^6 \text{ kg/h}$$

Comment: The effect of irreversibilities within the turbine and pump can be gauged by comparing the present values with their counterparts in Example 8.1. In this example, the turbine work per unit of mass is less and the pump work per unit of mass is greater than in Example 8.1. The thermal efficiency in the present case is less than in the ideal case of the previous example. For a fixed net power output (100 MW), the smaller net work output per unit mass in the present case dictates a greater mass flow rate of steam. The magnitude of the heat transfer to the cooling water is greater in this example than in Example 8.1; consequently, a greater mass flow rate of cooling water would have to be accommodated.

8.3 SUPERHEAT AND REHEAT

The representations of the vapor power cycle considered thus far do not depict actual vapor power plants faithfully, for various modifications are usually incorporated to improve overall performance. In this section we consider two cycle modifications known as *superheat* and *reheat*. Both features are normally incorporated into vapor power plants.

Let us begin the discussion by noting that an increase in the boiler pressure or a decrease in the condenser pressure may result in a reduction of the steam quality at the exit of the turbine. This can be seen by comparing states 2′ and 2″ of Figs. 8.4*a* and 8.4*b* to the corresponding state 2 of each diagram. If the quality of the mixture passing through the turbine becomes too low, the impact of liquid droplets on the turbine blades can erode them, causing a decrease in the turbine efficiency and an increased need for maintenance. Accordingly, common practice is to maintain at least 90% quality ($x_2 = 0.9$) at the turbine exit. The cycle modifications known as *superheat* and *reheat* permit advantageous operating pressures in the boiler and condenser and yet offset the problem of low quality of the turbine exhaust.

First, let us consider superheat. As we are not limited to having saturated vapor at the turbine inlet, further energy can be added by heat transfer to the steam, bringing it to a superheated vapor condition at the turbine inlet. This is accomplished in a separate heat exchanger called a superheater. The combination of boiler and superheater is referred to as a *steam generator*. Figure 8.3 shows an ideal Rankine cycle with superheated vapor at the turbine inlet: cycle 1′–2′–3–4–1′. The cycle with superheat has a higher average temperature of heat addition than the cycle without superheating (cycle 1–2–3–4–1), so the thermal efficiency is higher. Moreover, the quality at turbine exhaust state 2′ is greater than at state 2, which would be the turbine exhaust state without superheating. Accordingly, superheating also tends to alleviate the problem of low steam quality at the turbine exhaust. With sufficient superheating, the turbine exhaust state may even fall in the superheated vapor region.

A further modification normally employed in vapor power plants is *reheat*. With reheat, a power plant can take advantage of the increased efficiency that results with higher boiler pressures and yet avoid low-quality steam at the turbine exhaust. In the

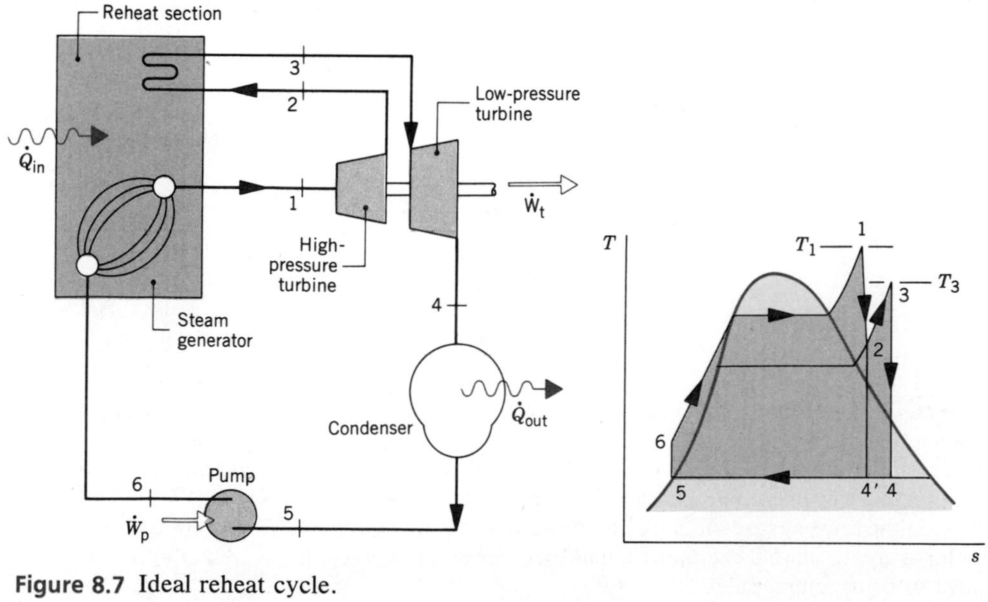

Figure 8.7 Ideal reheat cycle.

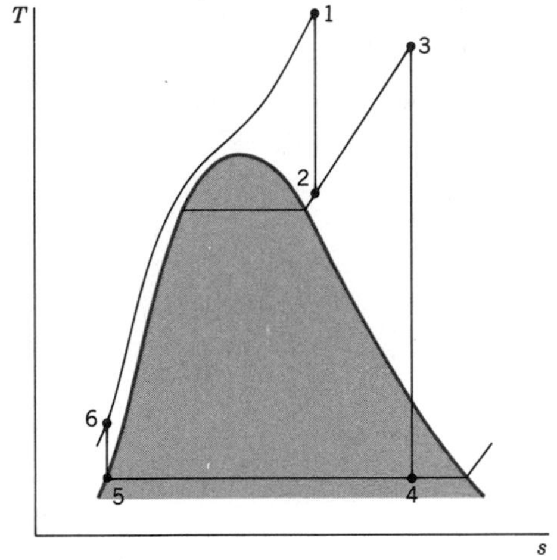

Figure 8.8 Supercritical ideal reheat cycle.

ideal reheat cycle shown in Fig. 8.7, the steam does not expand to the condenser pressure in a single stage. The steam expands through a first-stage turbine (Process 1–2) to some pressure between the steam generator and condenser pressures. The steam is then reheated in the steam generator (Process 2–3). Ideally, there would be no pressure drop as the steam is reheated. After reheating, the steam expands in a second-stage turbine to the condenser pressure (Process 3–4). The principal advantage of reheat is to increase the quality of the steam at the turbine exhaust. This can be seen from the *T–s* diagram of Fig. 8.7 by comparing state 4 with state 4′, the turbine exhaust state without reheating. When computing the thermal efficiency of a reheat cycle, it is necessary to account for the work output of both turbine stages as well as the total heat addition occurring in the vaporization/superheating and reheating processes. This calculation is illustrated in Example 8.3.

The temperature of the steam entering the turbine is restricted by metallurgical limitations imposed by the materials used to fabricate the superheater, reheater, and turbine. High pressure in the steam generator also requires piping that can withstand great stresses at elevated temperatures. Although these factors limit the gains that can be realized through superheating and reheating, improved materials and methods of fabrication have permitted significant increases over the years in the maximum allowed cycle temperatures and steam generator pressures, with corresponding increases in thermal efficiency. This has progressed to the extent that vapor power plants can be designed to operate with steam generator pressures exceeding the critical pressure of water (22.1 MPa, 3203.6 lbf/in.^{2}) and turbine inlet temperatures of about 600°C (1100°F). Figure 8.8 shows an ideal cycle with a supercritical steam generator pressure. Observe that no phase change occurs during the heat addition process from 6 to 1.

EXAMPLE 8.3

PROBLEM

Steam is the working fluid in an ideal Rankine cycle with superheat and reheat. Steam enters the first-stage turbine at 8.0 MPa, 480°C, and expands to 0.7 MPa. It is then reheated to 440°C before entering the second-stage turbine, where it expands to the condenser pressure of 0.008 MPa. The *net* power output is 100 MW. Determine (a) the thermal efficiency of the cycle, (b) the mass flow rate of steam, in kg/h, (c) the rate of heat transfer $\dot{Q}_{out}$ from the condensing steam as it passes through the condenser, in MW.

SOLUTION

Known: An ideal reheat cycle operates with steam as the working fluid. Operating pressures and temperatures are specified, and the net power output is given.

Find: Determine the thermal efficiency, the mass flow rate of the steam, in kg/h, and the heat transfer rate from the condensing steam as it passes through the condenser, in MW.

Schematic and Given Data:

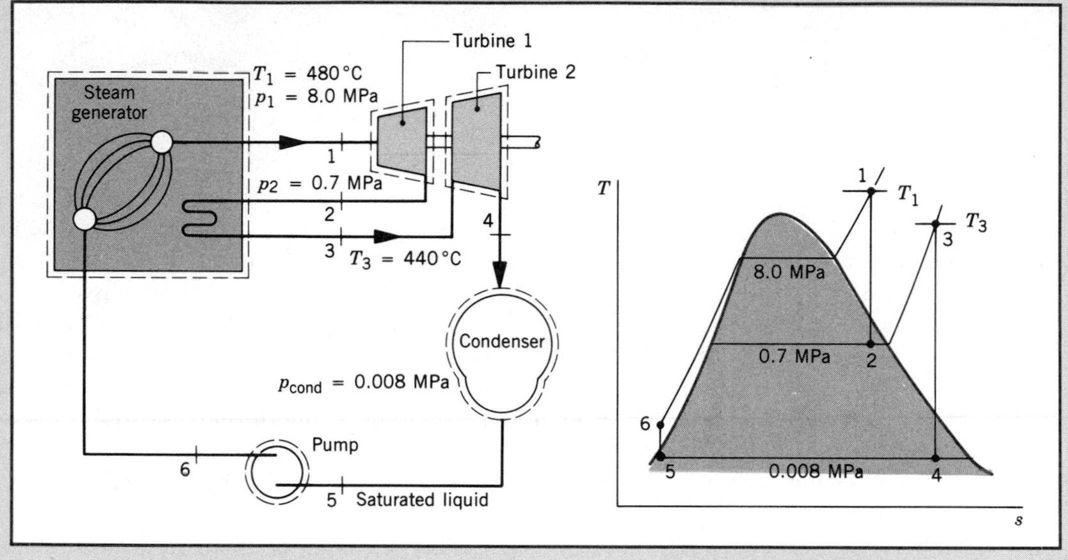

Figure E8.3

Assumptions:

1. Each component in the cycle is analyzed as a control volume at steady state. The control volumes are shown on the accompanying sketch by dashed lines.
2. All processes of the working fluid are internally reversible.
3. The turbine and pump operate adiabatically.
4. Condensate exits the condenser as saturated liquid.
5. Kinetic and potential energy effects are negligible.

Analysis: To begin, let us fix each of the principal states. Starting at the inlet to the first turbine stage, the pressure is 8.0 MPa and the temperature is 480°C, so the steam is a superheated vapor. From Table A-4, $h_1 = 3348.4$ kJ/kg and $s_1 = 6.6586$ kJ/kg·K.

State 2 is fixed by $p_2 = 0.7$ MPa and $s_2 = s_1$ for the isentropic expansion through the first-stage turbine. Using saturated liquid and saturated vapor data from Table A-3, the quality at state 2 is

$$x_2 = \frac{s_2 - s_f}{s_g - s_f} = \frac{6.6586 - 1.9922}{6.708 - 1.9922} = 0.9895$$

The specific enthalpy is then

$$\begin{aligned} h_2 &= h_f + x_2 h_{fg} \\ &= 697.22 + (0.9895)2066.3 = 2741.8 \text{ kJ/kg} \end{aligned}$$

State 3 is superheated vapor with $p_3 = 0.7$ MPa and $T_3 = 440$°C, so from Table A-4, $h_3 = 3353.3$ kJ/kg and $s_3 =$ 7.7571 kJ/kg·K.

To fix state 4, use $p_4 = 0.008$ MPa and $s_4 = s_3$ for the isentropic expansion through the second-stage turbine. With data from Table A-3, the quality at state 4 is

$$x_4 = \frac{s_4 - s_f}{s_g - s_f} = \frac{7.7571 - 0.5926}{8.2287 - 0.5926} = 0.9382$$

The specific enthalpy is

$$h_4 = 173.88 + (0.9382)2403.1 = 2428.5 \text{ kJ/kg}$$

State 5 is saturated liquid at 0.008 MPa, so $h_5 = 173.88$ kJ/kg. Finally, the state at the pump exit is the same as in Example 8.1, so $h_6 = 181.94$ kJ/kg.

(a) The *net* power developed by the cycle is

$$\dot{W}_{cycle} = \dot{W}_{t1} + \dot{W}_{t2} - \dot{W}_p$$

Mass and energy rate balances for the two turbine stages and the pump reduce to give, respectively

$$\text{Turbine 1:} \quad \dot{W}_{t1}/\dot{m} = h_1 - h_2$$

$$\text{Turbine 2:} \quad \dot{W}_{t2}/\dot{m} = h_3 - h_4$$

$$\text{Pump:} \quad \dot{W}_p/\dot{m} = h_6 - h_5$$

where $\dot{m}$ is the mass flow rate of the steam.

The total rate of heat transfer to the working fluid as it passes through the boiler–superheater and reheater is

$$\frac{\dot{Q}_{in}}{\dot{m}} = (h_1 - h_6) + (h_3 - h_2)$$

Using these expressions, the thermal efficiency is

$$\begin{aligned}\eta &= \frac{(h_1 - h_2) + (h_3 - h_4) - (h_6 - h_5)}{(h_1 - h_6) + (h_3 - h_2)} \\ &= \frac{(3348.4 - 2741.8) + (3353.3 - 2428.5) - (181.94 - 173.88)}{(3348.4 - 181.94) + (3353.3 - 2741.8)} \\ &= \frac{606.6 + 924.8 - 8.06}{3166.5 + 611.5} = \frac{1523.3 \text{ kJ/kg}}{3778 \text{ kJ/kg}} = 0.403 \ (40.3\%)\end{aligned}$$

(b) The mass flow rate of the steam can be obtained with the expression for net power given in part (a).

$$\begin{aligned}\dot{m} &= \frac{\dot{W}_{cycle}}{(h_1 - h_2) + (h_3 - h_4) - (h_6 - h_5)} \\ &= \frac{(100 \text{ MW})(3600 \text{ s/h})(10^3 \text{ kW/MW})}{(606.6 + 924.8 - 8.06) \text{ kJ/kg}} = 2.363 \times 10^5 \text{ kg/h}\end{aligned}$$

(c) The rate of heat transfer from the condensing steam to the cooling water is

$$\begin{aligned}\dot{Q}_{out} &= \dot{m}(h_4 - h_5) \\ &= \frac{2.363 \times 10^5 \text{ kg/h } (2428.5 - 173.88) \text{ kJ/kg}}{(3600 \text{ s/h})(10^3 \text{ kW/MW})} = 148 \text{ MW}\end{aligned}$$

Comment: It is instructive to compare the present values with their counterparts in Example 8.1. With superheat and reheat, the thermal efficiency is increased over that of the cycle of Example 8.1. For a specified net power output (100 MW), a larger thermal efficiency means that a smaller mass flow rate of steam is required. Moreover, with a greater thermal efficiency the rate of heat transfer to the cooling water is also less, resulting in a reduced demand for cooling water. With reheating, the steam quality at the turbine exhaust is substantially increased over the value for the cycle of Example 8.1.

EXAMPLE 8.4

PROBLEM

Reconsider the reheat cycle of Example 8.3, but include in the analysis that each turbine stage has an isentropic efficiency of 85%. Determine the thermal efficiency.

SOLUTION

Known: A reheat cycle operates with steam as the working fluid. Operating pressures and temperatures are specified. Each turbine stage has an efficiency of 85%.

Find: Determine the thermal efficiency.

Schematic and Given Data:

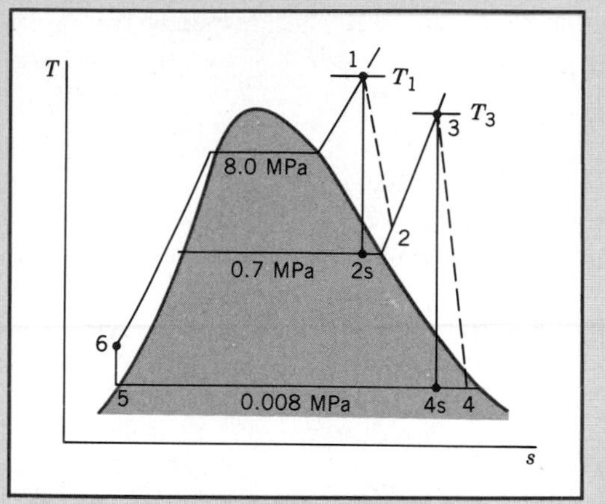

Figure E8.4

Assumptions:

1. As in Example 8.3, each component is analyzed as a control volume at steady state.
2. Except for the two turbine stages, all processes are internally reversible.
3. The turbine and pump operate adiabatically.
4. The condensate exits the condenser as saturated liquid.
5. Kinetic and potential energy effects are negligible.

Analysis: From the solution to Example 8.3, the following specific enthalpy values are known, in kJ/kg: $h_1 = 3348.4$, $h_{2s} = 2741.8$, $h_3 = 3353.3$, $h_{4s} = 2428.5$, $h_5 = 173.88$, $h_6 = 181.94$.

The specific enthalpy at the exit of the first-stage turbine, h_2, can be determined by solving the expression for the turbine efficiency to obtain

$$h_2 = h_1 - \eta_t(h_1 - h_{2s})$$
$$= 3348.4 - 0.85(3348.4 - 2741.8) = 2832.8 \text{ kJ/kg}$$

The specific enthalpy at the exit of the second-stage turbine can be found similarly:

$$h_4 = h_3 - \eta_t(h_3 - h_{4s})$$
$$= 3353.3 - 0.85(3353.3 - 2428.5) = 2567.2 \text{ kJ/kg}$$

The thermal efficiency is then

$$\eta = \frac{(h_1 - h_2) + (h_3 - h_4) - (h_6 - h_5)}{(h_1 - h_6) + (h_3 - h_2)}$$

$$= \frac{(3348.4 - 2832.8) + (3353.3 - 2567.2) - (181.94 - 173.88)}{(3348.4 - 181.94) + (3353.3 - 2832.8)}$$

$$= \frac{1293.6}{3687.0} = 0.351\ (35.1\%)$$

Comment: Owing to the irreversibilities present in the turbine stages, the net work per unit of mass developed in the present case is significantly less than in the case of Example 8.3. The thermal efficiency is also considerably less.

8.4 REGENERATIVE VAPOR POWER CYCLE

Another commonly used method for increasing the thermal efficiency of vapor power plants is *regenerative feedwater heating,* or simply *regeneration.* This is the subject of the present section.

To introduce the principle underlying regenerative feedwater heating, consider Fig. 8.3 once again. In cycle 1–2–3–4–a–1, the working fluid would enter the boiler as a compressed liquid at state 4 and be heated while in the liquid phase to state a. With regenerative feedwater heating, the working fluid would enter the boiler at a state *between* 4 and a. As a result, the average temperature of heat addition would be increased.

8.4.1 Open Feedwater Heaters

Let us consider how regeneration can be accomplished using an *open feedwater* heater, a direct contact-type heat exchanger in which streams at different temperatures mix to form a stream at an intermediate temperature. Shown in Fig. 8.9 are the schematic diagram and the associated T–s diagram for a regenerative vapor power cycle having one open feedwater heater. For this cycle, the working fluid passes

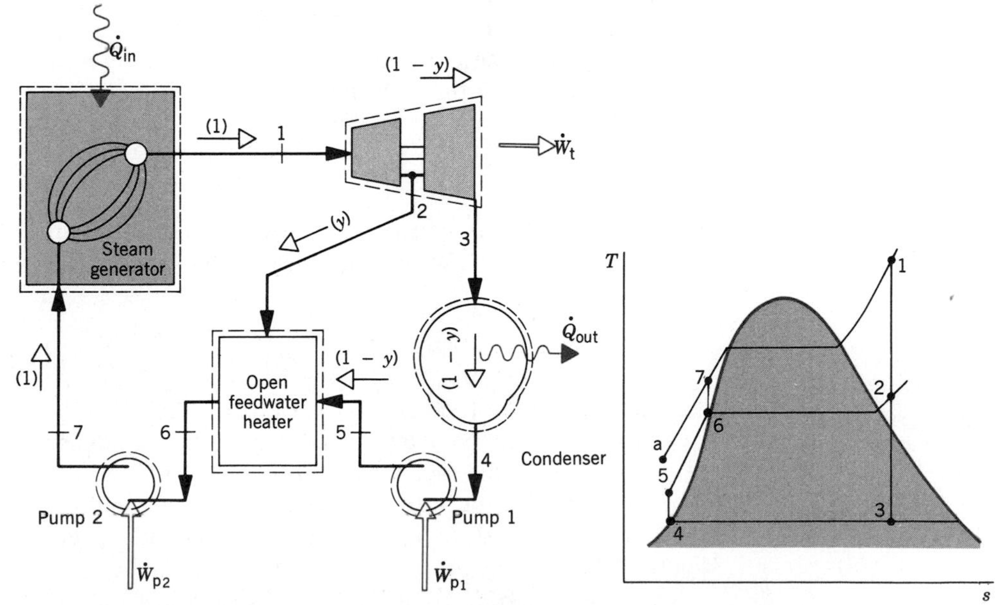

Figure 8.9 Regenerative vapor power cycle with one open feedwater heater.

isentropically through the turbine stages and pumps, and the flow through the steam generator, condenser, and feedwater heater takes place with no pressure drop in any of these components. Steam enters the first-stage turbine at state 1 and expands to state 2, where a fraction of the total flow is *extracted,* or *bled,* into an open feedwater heater operating at the extraction pressure, p_2. The rest of the steam expands through the second-stage turbine to state 3. This portion of the total flow is condensed to saturated liquid, state 4, and then pumped to the extraction pressure and introduced into the feedwater heater at state 5. A single mixed stream exits the feedwater heater at state 6. For the case shown in Fig. 8.9, the mass flow rates of the streams entering the feedwater heater are chosen so that the stream exiting the feedwater heater is a saturated liquid at the extraction pressure. The liquid at state 6 is then pumped to the steam generator pressure and enters the steam generator at state 7. Finally, the working fluid is heated from state 7 to state 1 in the steam generator.

Referring to the T–s diagram of the cycle, note that the heat addition would take place from state 7 to state 1, rather than from state a to state 1, as would be the case without regeneration. Accordingly, the amount of energy that must be supplied from the combustion of a fossil fuel, or another source, to vaporize and superheat the steam would be reduced. This is the desired outcome. Only a portion of the total flow expands through the second-stage turbine (Process 2–3), however, so less work would be developed as well. In practice, operating conditions are chosen so that the reduction in heat added more than offsets the decrease in net work developed, resulting in an increased thermal efficiency in regenerative power plants.

Consider next the thermodynamic analysis of the regenerative cycle illustrated in Fig. 8.9. An important initial step in analyzing any regenerative vapor cycle is the evaluation of the mass flow rates through each of the components. Taking a single control volume enclosing both turbine stages, the mass rate balance reduces at steady state to

$$\dot{m}_2 + \dot{m}_3 = \dot{m}_1$$

where $\dot{m}_1$ is the rate at which mass enters the first-stage turbine at state 1, $\dot{m}_2$ is the rate at which mass is extracted and exits at state 2, and $\dot{m}_3$ is the rate at which mass exits the second-stage turbine at state 3. Dividing by $\dot{m}_1$ places this on the basis of a *unit of mass* passing through the first-stage turbine

$$\frac{\dot{m}_2}{\dot{m}_1} + \frac{\dot{m}_3}{\dot{m}_1} = 1$$

Denoting the fraction of the total flow extracted at state 2 by y ($y = \dot{m}_2/\dot{m}_1$), the fraction of the total flow passing through the second-stage turbine is

$$\frac{\dot{m}_3}{\dot{m}_1} = 1 - y \tag{8.11}$$

The fractions of the total flow at various locations are indicated on Fig. 8.9.

The fraction y can be determined by applying the conservation of mass and conservation of energy principles to a control volume around the feedwater heater. Assuming no heat transfer between the feedwater heater and its surroundings and ignoring kinetic and potential energy effects, the mass and energy rate balances reduce at steady state to give

$$0 = yh_2 + (1 - y)h_5 - h_6$$

Solving for y

$$y = \frac{h_6 - h_5}{h_2 - h_5} \tag{8.12}$$

Equation 8.12 allows the fraction y to be determined when states 2, 5, and 6 are fixed.

Expressions for the principal work and heat transfers of the regenerative cycle can be determined by applying mass and energy rate balances to control volumes around the individual components. Beginning with the turbine, the total work is the sum of the work developed by each turbine stage. Neglecting kinetic and potential energy effects and assuming no heat transfer with the surroundings, we can express the total turbine work on the basis of a unit of mass passing through the first-stage turbine as

$$\frac{\dot{W}_t}{\dot{m}_1} = (h_1 - h_2) + (1 - y)(h_2 - h_3) \tag{8.13}$$

The total pump work is the sum of the work required to operate each pump individually. On the basis of a unit of mass passing through the first-stage turbine, the total pump work is

$$\frac{\dot{W}_p}{\dot{m}_1} = (h_7 - h_6) + (1 - y)(h_5 - h_4) \tag{8.14}$$

The energy added by heat transfer to the working fluid passing through the steam generator, per unit of mass expanding through the first-stage turbine, is

$$\frac{\dot{Q}_{in}}{\dot{m}_1} = h_1 - h_7 \tag{8.15}$$

and the energy rejected by heat transfer to the cooling water is

$$\frac{\dot{Q}_{out}}{\dot{m}_1} = (1 - y)(h_3 - h_4) \tag{8.16}$$

The following example illustrates the analysis of a regenerative cycle with one open feedwater heater and the determination of properties at various state points around the cycle.

EXAMPLE 8.5

PROBLEM

Consider a regenerative vapor power cycle with one open feedwater heater. Steam enters the turbine at 8.0 MPa, 480°C and expands to 0.7 MPa, where some of the steam is extracted and diverted to the open feedwater heater operating at 0.7 MPa. The remaining steam expands through the second-stage turbine to the condenser pressure of 0.008 MPa. Saturated liquid exits the open feedwater heater at 0.7 MPa. The isentropic efficiency of each turbine stage is 85%. If the net power output of the cycle is 100 MW, determine (a) the thermal efficiency and (b) the mass flow rate of steam entering the first turbine stage, in kg/h.

SOLUTION

Known: A regenerative vapor power cycle operates with steam as the working fluid. Operating pressures and temperatures are specified; the efficiency of each turbine stage and the net power output are also given.

Find: Determine the thermal efficiency and the mass flow rate into the turbine, in kg/h.

Schematic and Given Data:

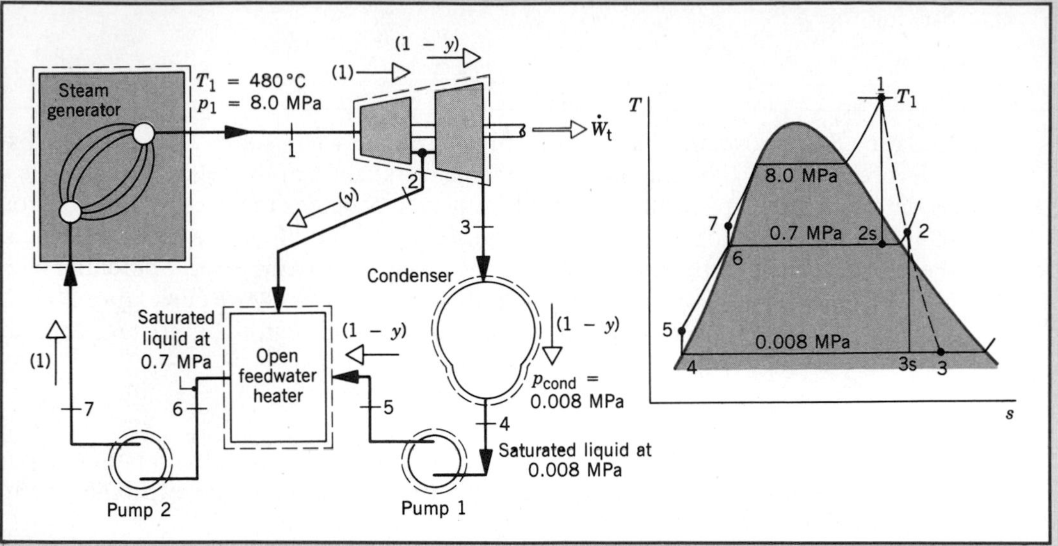

Figure E8.5

Assumptions:

1. Each component in the cycle is analyzed as a steady-state control volume. The control volumes are shown in the accompanying sketch by dashed lines.
2. All processes of the working fluid are internally reversible, except for the expansions through the two turbine stages and mixing in the open feedwater heater.
3. The turbines, pumps, and feedwater heater operate adiabatically.
4. Kinetic and potential energy effects are negligible.
5. Saturated liquid exits the open feedwater heater, and saturated liquid exits the condenser.

Analysis: The specific enthalpy at states 1 and 4 can be read from the steam tables. The specific enthalpy at state 2 is evaluated in the solution to Example 8.4. The specific entropy at state 2 can be obtained from the steam tables using the known values of enthalpy and pressure at this state. In summary, $h_1 = 3348.4$ kJ/kg, $h_2 = 2832.8$ kJ/kg, $s_2 = 6.8606$ kJ/kg·K, $h_4 = 173.88$ kJ/kg.

The specific enthalpy at state 3 can be determined using the efficiency of the second-stage turbine

$$h_3 = h_2 - \eta_t(h_2 - h_{3s})$$

With $s_{3s} = s_2$, the quality at state 3s is $x_{3s} = 0.8208$; using this, we get $h_{3s} = 2146.3$ kJ/kg. Hence

$$h_3 = 2832.8 - 0.85(2832.8 - 2146.3) = 2249.3 \text{ kJ/kg}$$

State 6 is saturated liquid at 0.7 MPa. Thus, $h_6 = 697.22$ kJ/kg.

Since the pumps are assumed to operate with no irreversibilities, the specific enthalpy values at states 5 and 7 can be determined as

$$\begin{aligned} h_5 &= h_4 + v_4(p_5 - p_4) \\ &= 173.88 + (1.0084 \times 10^{-3})(\text{m}^3/\text{kg})(0.7 - 0.008) \text{ MPa} \left(\frac{10^6 \text{ N/m}^2}{1 \text{ MPa}}\right)\left(\frac{\text{kJ}}{10^3 \text{ N}\cdot\text{m}}\right) \\ &= 174.6 \text{ kJ/kg} \\ h_7 &= h_6 + v_6(p_7 - p_6) \\ &= 697.22 + (1.1080 \times 10^{-3})(8.0 - 0.7)(10^3) \\ &= 705.3 \text{ kJ/kg} \end{aligned}$$

Applying mass and energy rate balances to a control volume enclosing the open heater, we find the fraction y of the flow extracted at state 2 from

$$y = \frac{h_6 - h_5}{h_2 - h_5} = \frac{697.22 - 174.6}{2832.8 - 174.6} = 0.1966$$

(a) On the basis of a unit of mass passing through the first-stage turbine, the total turbine work output is

$$\frac{\dot{W}_t}{\dot{m}_1} = (h_1 - h_2) + (1 - y)(h_2 - h_3)$$
$$= (3348.4 - 2832.8) + (0.8034)(2832.8 - 2249.3)$$
$$= 984.4 \text{ kJ/kg}$$

The total pump work per unit of mass passing through the first-stage turbine is

$$\frac{\dot{W}_p}{\dot{m}_1} = (h_7 - h_6) + (1 - y)(h_5 - h_4)$$
$$= (705.3 - 697.22) + (0.8034)(174.6 - 173.88)$$
$$= 8.7 \text{ kJ/kg}$$

The heat added in the steam generator per unit of mass passing through the first-stage turbine is

$$\frac{\dot{Q}_{in}}{\dot{m}_1} = h_1 - h_7 = 3348.4 - 705.3 = 2643.1 \text{ kJ/kg}$$

The thermal efficiency is then

$$\eta = \frac{\dot{W}_t/\dot{m}_1 - \dot{W}_p/\dot{m}_1}{\dot{Q}_{in}/\dot{m}_1} = \frac{984.4 - 8.7}{2643.1} = 0.369 \ (36.9\%)$$

(b) The mass flow rate of the steam entering the turbine, $\dot{m}_1$, can be determined using the given value for the net power output, 100 MW. Since

$$\dot{W}_{cycle} = \dot{W}_t - \dot{W}_p$$

and

$$\frac{\dot{W}_t}{\dot{m}_1} = 984.4 \text{ kJ/kg} \quad \text{and} \quad \frac{\dot{W}_p}{\dot{m}_1} = 8.7 \text{ kJ/kg}$$

it follows that

$$\dot{m}_1 = \frac{(100 \text{ MW})(3600 \text{ s/h})}{(984.4 - 8.7) \text{ kJ/kg}} \left(\frac{10^3 \text{ kJ/s}}{1 \text{ MW}}\right) = 3.69 \times 10^5 \text{ kg/h}$$

Comment: Note that the fractions of the total flow at various locations are labeled on the accompanying figure.

8.4.2 Closed Feedwater Heaters

Regenerative feedwater heating can also be accomplished with *closed feedwater heaters*. Closed heaters are shell-and-tube-type recuperators in which the feedwater temperature increases as the extracted steam condenses on the outside of the tubes carrying the feedwater. Since the two streams do not mix, they can be at different pressures. The diagrams of Fig. 8.10 show two different schemes for removing the condensate from closed feedwater heaters. In Fig. 8.10*a*, this is accomplished by means of a pump whose function is to pump the condensate forward to a higher-pressure point in the cycle. In Fig. 8.10*b*, the condensate is allowed to pass through a *trap* into a feedwater heater operating at a lower pressure or into the condenser. A trap is a type of valve that permits only liquid to pass through to a region of lower pressure.

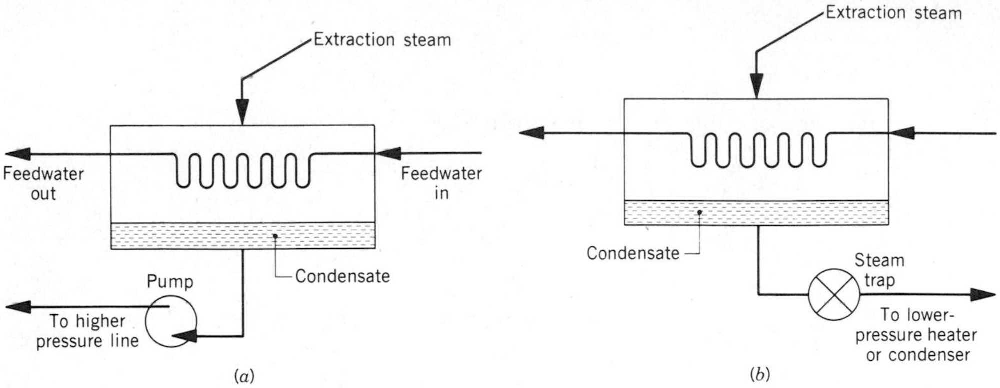

Figure 8.10 Examples of closed feedwater heaters.

A regenerative vapor power cycle having one closed feedwater heater with the condensate trapped into the condenser is shown schematically in Fig. 8.11. For this cycle, the working fluid passes isentropically through the turbine stages and pumps, and there are no pressure drops accompanying the flow through the other components. The *T–s* diagram shows the principal states of the cycle. The total steam flow expands through the first-stage turbine from state 1 to state 2. At this location, a fraction of the flow is bled into the closed feedwater heater, where it condenses. Saturated liquid at the extraction pressure exits the feedwater heater at state 7. The condensate is then trapped into the condenser, where it is reunited with the portion of the total flow passing through the second-stage turbine. The expansion from state 7 to state 8 through the trap is irreversible, so it is shown by a dashed line on the *T–s* diagram. The total flow exiting the condenser as saturated liquid at state 4 is pumped to the steam generator pressure and enters the feedwater heater at state 5. The temperature of the feedwater is increased in passing through the feedwater heater. The feedwater then exits at state 6. The cycle is completed as the working fluid is heated in the steam generator at constant pressure from state 6 to state 1. Although the closed heater shown on the figure operates with no pressure drop in either

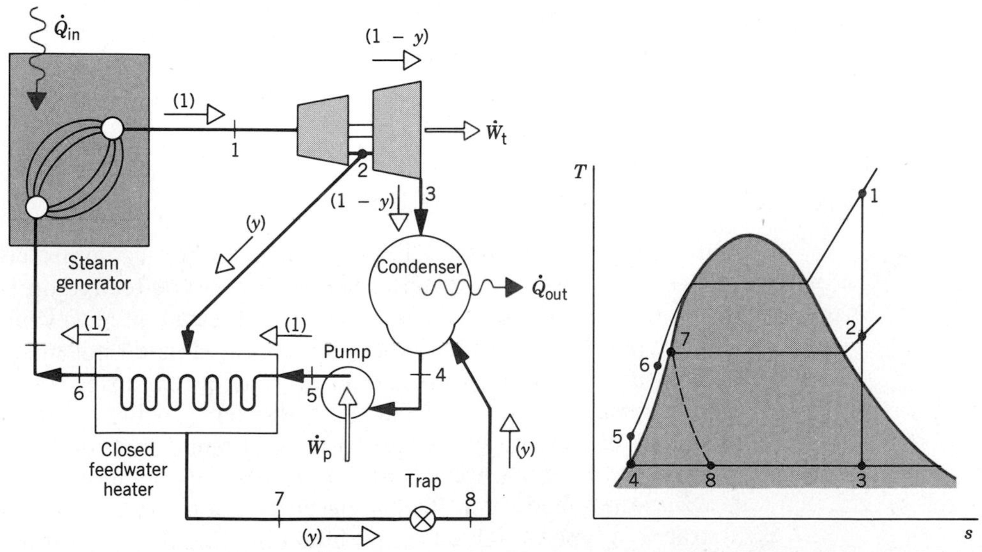

Figure 8.11 Regenerative vapor power cycle with one closed feedwater heater.

stream, there is a source of irreversibility due to the stream-to-stream temperature differences.

The schematic diagram of the cycle shown in Fig. 8.11 is labeled with the fractions of the total flow at various locations. This is usually helpful in analyzing such cycles. The fraction of the total flow extracted, y, can be determined by applying the conservation of mass and conservation of energy principles to a control volume around the closed heater. Assuming no heat transfer between the feedwater heater and its surroundings and neglecting kinetic and potential energy effects, the mass and energy rate balances reduce at steady state to give

$$0 = y(h_2 - h_7) + (h_5 - h_6)$$

Solving for y

$$y = \frac{h_6 - h_5}{h_2 - h_7} \tag{8.17}$$

The principal work and heat transfers are evaluated as discussed previously.

8.4.3 Multiple Feedwater Heaters

The thermal efficiency of the regenerative cycle can be increased by incorporating several feedwater heaters at suitably chosen pressures. The number of feedwater heaters used is based on economic considerations, since incremental increases in thermal efficiency achieved with each additional heater must justify the added capital costs (heater, piping, pumps, etc.). Power plant designers use computer programs to simulate the thermodynamic and economic performance of different designs to help them decide on the number of heaters to use, the types of heaters, and the pressures at which they should operate.

Figure 8.12 shows the layout of a power plant with three closed feedwater heaters and one open heater. Power plants with multiple feedwater heaters ordinarily have at least one open feedwater heater operating at a pressure greater than atmospheric pressure so that oxygen and other dissolved gases can be vented from the cycle. This

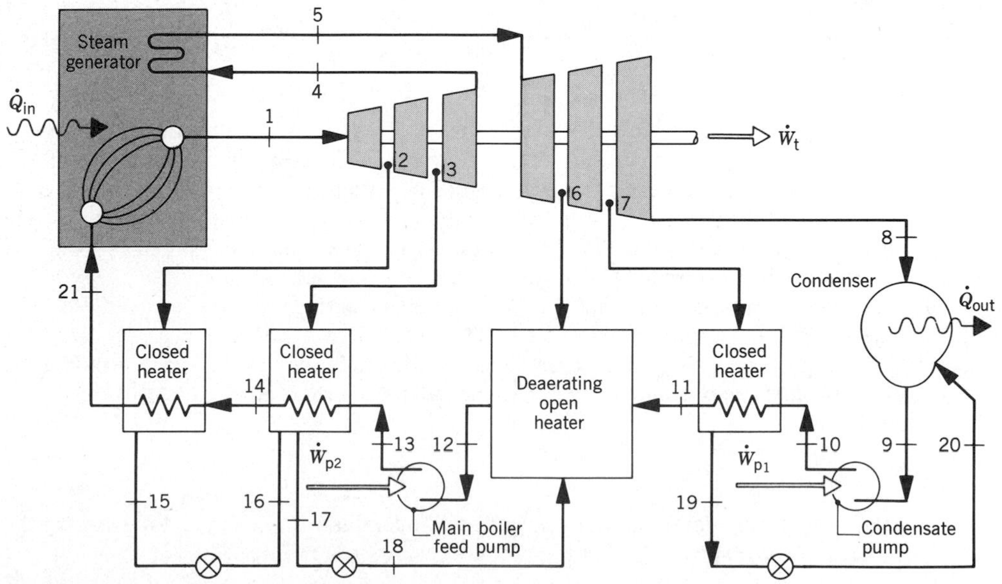

Figure 8.12 Example of a power plant layout.

procedure, known as *deaeration,* is needed to maintain the purity of the working fluid in order to minimize corrosion. Actual power plants have many of the same basic features as the one shown in the figure.

In analyzing regenerative vapor power cycles with multiple feedwater heaters, it is good practice to base the analysis on a unit of mass entering the first-stage turbine. To clarify the quantities of matter flowing through the various plant components, the fractions of the total flow removed at each extraction point and the fraction of the total flow remaining at each state point in the cycle should be labeled on a schematic diagram of the cycle. The fractions extracted are determined from mass and energy rate balances for control volumes around each of the feedwater heaters, starting with the highest-pressure heater and proceeding to each lower-pressure heater in turn. This procedure is used in the next example that involves a reheat–regenerative vapor power cycle with two feedwater heaters.

8.5 WORKING FLUID CHARACTERISTICS, BINARY VAPOR CYCLES, AND COGENERATION

Water is used as the working fluid in the vast majority of vapor power systems because it is plentiful and low in cost, nontoxic, chemically stable, and relatively noncorrosive. In addition, water has a relatively large change in specific enthalpy when it vaporizes at ordinary steam generator pressures, which tends to limit the mass flow rate for a desired power plant output. The properties of liquid water and water vapor are also such that the back work ratios achieved are characteristically quite low, and the techniques of reheat and regeneration are effective for increasing power plant efficiencies.

Water is less satisfactory insofar as some other desirable working fluid characteristics are concerned. For example, the critical temperature of water is only 374.14°C (705.4°F), which is about 225°C (440°F) below the maximum allowable turbine inlet temperatures. Accordingly, to achieve a high average temperature of heat addition

EXAMPLE 8.6

PROBLEM

Consider a reheat–regenerative vapor power cycle with two feedwater heaters, a closed feedwater heater and an open feedwater heater. Steam enters the first turbine at 8.0 MPa, 480°C and expands to 0.7 MPa. The steam is reheated to 440°C before entering the second turbine, where it expands to the condenser pressure of 0.008 MPa. Steam is extracted from the first turbine at 2 MPa and fed to the closed feedwater heater. Feedwater leaves the closed heater at 205°C and 8.0 MPa, and condensate exits as saturated liquid at 2 MPa. The condensate is trapped into the open feedwater heater. Steam extracted from the second turbine at 0.3 MPa is also fed into the open feedwater heater, which operates at 0.3 MPa. The stream exiting the open feedwater heater is saturated liquid at 0.3 MPa. The *net* power output of the cycle is 100 MW. There is no stray heat transfer from any component to its surroundings. If the working fluid experiences no irreversibilities as it passes through the turbines, pumps, steam generator, reheater, and condenser, determine (a) the thermal efficiency, (b) the mass flow rate of the steam entering the first turbine, in kg/h.

SOLUTION

Known: A reheat–regenerative vapor power cycle operates with steam as the working fluid. Operating pressures and temperatures are specified, and the net power output is given.

Find: Determine the thermal efficiency and the mass flow rate entering the first turbine, in kg/h.

Schematic and Given Data:

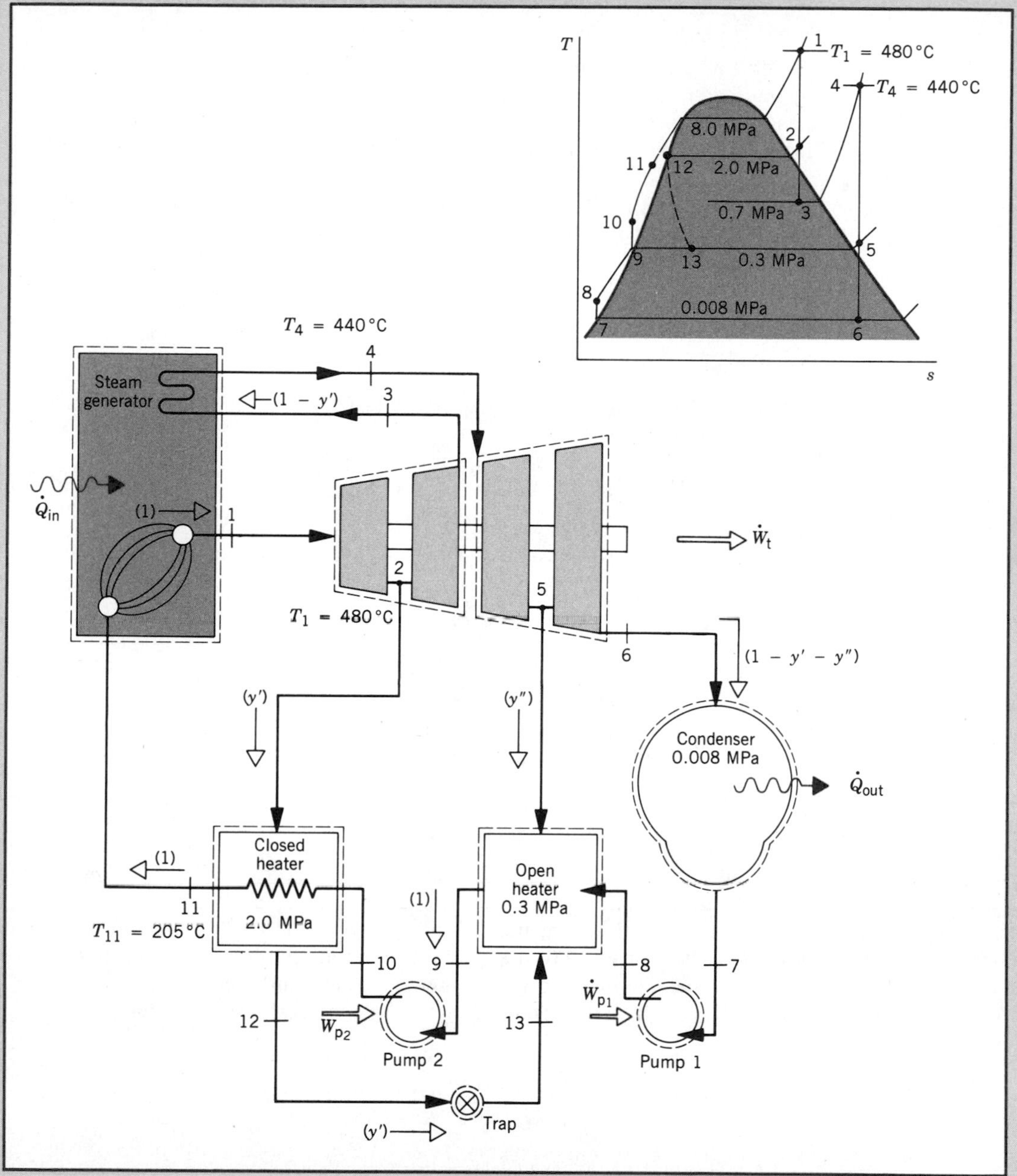

Figure E8.6

Assumptions:

1. Each component in the cycle is analyzed as a control volume at steady state. The control volumes are shown on the accompanying sketch by dashed lines.
2. There is no stray heat transfer from any component to its surroundings.
3. The working fluid undergoes internally reversible processes as it passes through the turbines, pumps, steam generator, reheater, and condenser.
4. The expansion through the trap is a *throttling* process.
5. Kinetic and potential energy effects are negligible.

6. Condensate exits the closed heater as a saturated liquid at 2 MPa. Feedwater exits the open heater as a saturated liquid at 0.3 MPa. Condensate exits the condenser as a saturated liquid.

Analysis: Let us determine the specific enthalpies at the principal states of the cycle. State 1 is the same as in Example 8.3, so $h_1 = 3348.4$ kJ/kg and $s_1 = 6.6586$ kJ/kg·K.

State 2 is fixed by $p_2 = 2.0$ MPa and the specific entropy s_2, which is the same as that of state 1. Interpolating in Table A-4, we get $h_2 = 2963.5$ kJ/kg. The state at the exit of the first turbine is the same as at the exit of the first turbine of Example 8.3, so $h_3 = 2741.8$ kJ/kg.

State 4 is superheated vapor at 0.7 MPa, 440°C. From Table A-4, $h_4 = 3353.3$ kJ/kg and $s_4 = 7.7571$ kJ/kg·K. Interpolating in Table A-4 at $p_5 = 0.3$ MPa and $s_5 = s_4 = 7.7571$ kJ/kg·K, the enthalpy at state 5 is $h_5 = 3101.5$ kJ/kg.

Using $s_6 = s_4$, the quality at state 6 is found to be $x_6 = 0.9382$. So

$$\begin{aligned} h_6 &= h_f + x_6 h_{fg} \\ &= 173.88 + (0.9382)2403.1 = 2428.5 \text{ kJ/kg} \end{aligned}$$

At the condenser exit, $h_7 = 173.88$ kJ/kg. The specific enthalpy at the exit of the first pump is

$$\begin{aligned} h_8 &= h_7 + v_7(p_8 - p_7) \\ &= 173.88 + (1.0084)(0.3 - 0.008) = 174.17 \text{ kJ/kg} \end{aligned}$$

The required unit conversions were considered in previous examples.

The liquid leaving the open feedwater heater at state 9 is saturated liquid at 0.3 MPa. The specific enthalpy is $h_9 = 561.47$ kJ/kg. The specific enthalpy at the exit of the second pump is

$$\begin{aligned} h_{10} &= h_9 + v_9(p_{10} - p_9) \\ &= 561.47 + (1.0732)(8.0 - 0.3) = 569.73 \text{ kJ/kg} \end{aligned}$$

The condensate leaving the closed heater is saturated at 2 MPa. From Table A-3, $h_{12} = 908.79$ kJ/kg. The fluid passing through the trap undergoes a throttling process, so $h_{13} = 908.79$ kJ/kg.

The specific enthalpy of the feedwater exiting the closed heater at 8.0 MPa and 205°C is found using Eq. 3.14 as

$$\begin{aligned} h_{11} &= h_f + v_f(p_{11} - p_{sat}) \\ &= 875.1 + (1.1646)(8.0 - 1.73) = 882.4 \text{ kJ/kg} \end{aligned}$$

where h_f and v_f are the saturated liquid specific enthalpy and specific volume at 205°C, respectively, and p_{sat} is the saturation pressure in MPa at this temperature.

The schematic diagram of the cycle is labeled with the fractions of the total flow into the turbine that remain at various locations. The fractions of the total flow diverted to the closed heater and open heater, respectively, are $y' = \dot{m}_2/\dot{m}_1$ and $y'' = \dot{m}_5/\dot{m}_1$, where $\dot{m}_1$ denotes the mass flow rate entering the first turbine.

The fraction y′ can be determined by application of mass and energy rate balances to a control volume enclosing the closed heater. The result is

$$y' = \frac{h_{11} - h_{10}}{h_2 - h_{12}} = \frac{882.4 - 569.73}{2963.5 - 908.79} = 0.1522$$

The fraction y″ can be determined by application of mass and energy rate balances to a control volume enclosing the open heater, resulting in

$$0 = y''h_5 + (1 - y' - y'')h_8 + y'h_{13} - h_9$$

Solving for y''

$$\begin{aligned} y'' &= \frac{(1 - y')h_8 + y'h_{13} - h_9}{h_8 - h_5} \\ &= \frac{(0.8478)174.17 + (0.1522)908.79 - 561.47}{174.17 - 3101.5} \\ &= 0.0941 \end{aligned}$$

(a) The following work and heat transfer values are expressed on the basis of a unit mass entering the first turbine. The work developed by the first turbine per unit of mass entering is the sum

$$\frac{\dot{W}_{t1}}{\dot{m}_1} = (h_1 - h_2) + (1 - y')(h_2 - h_3)$$

$$= (3348.4 - 2963.5) + (0.8478)(2963.5 - 2741.8)$$

$$= 572.9 \text{ kJ/kg}$$

Similarly, for the second turbine

$$\frac{\dot{W}_{t2}}{\dot{m}_1} = (1 - y')(h_4 - h_5) + (1 - y' - y'')(h_5 - h_6)$$

$$= (0.8478)(3353.3 - 3101.5) + (0.7537)(3101.5 - 2428.5)$$

$$= 720.7 \text{ kJ/kg}$$

For the first pump

$$\frac{\dot{W}_{p1}}{\dot{m}_1} = (1 - y' - y'')(h_8 - h_7)$$

$$= (0.7537)(174.17 - 173.88) = 0.22 \text{ kJ/kg}$$

and for the second pump

$$\frac{\dot{W}_{p2}}{\dot{m}_1} = (h_{10} - h_9)$$

$$= 569.73 - 561.47 = 8.26 \text{ kJ/kg}$$

The total heat added is the sum of the energy added by heat transfer during boiling/superheating and reheating. When expressed on the basis of a unit of mass entering the first turbine, this is

$$\frac{\dot{Q}_{in}}{\dot{m}_1} = (h_1 - h_{11}) + (1 - y')(h_4 - h_3)$$

$$= (3348.4 - 882.4) + (0.8478)(3353.3 - 2741.8)$$

$$= 2984.4 \text{ kJ/kg}$$

With the foregoing values, the thermal efficiency is

$$\eta = \frac{\dot{W}_{t1}/\dot{m}_1 + \dot{W}_{t2}/\dot{m}_1 - \dot{W}_{p1}/\dot{m}_1 - \dot{W}_{p2}/\dot{m}_1}{\dot{Q}_{in}/\dot{m}_1}$$

$$= \frac{572.9 + 720.7 - 0.22 - 8.26}{2984.4} = 0.431 \ (43.1\%)$$

(b) The mass flow rate entering the first turbine can be determined using the given value of the net power output. Thus

$$\dot{m}_1 = \frac{\dot{W}_{cycle}}{\dot{W}_{t1}/\dot{m}_1 + \dot{W}_{t2}/\dot{m}_1 - \dot{W}_{p1}/\dot{m}_1 - \dot{W}_{p2}/\dot{m}_1}$$

$$= \frac{(100 \text{ MW})(3600 \text{ s/h})(1000 \text{ kW/MW})}{1285.1 \text{ kJ/kg}} = 2.8 \times 10^5 \text{ kg/h}$$

Comment: Compared to the corresponding values determined for the simple Rankine cycle of Example 8.1, the thermal efficiency of the present regenerative cycle is substantially greater and the mass flow rate is considerably less.

and realize the attendant higher thermal efficiency, it may be necessary for the steam generator to operate at supercritical pressures. This requires costly piping and heat exchanger tubes capable of withstanding great stresses. Another undesirable characteristic of water is that its saturation pressure at normal condenser temperatures is well below atmospheric pressure. As a result, air can leak into the system, necessitating the use of special ejector pumps attached to the condenser or deaerating feedwater heaters to remove the air.

Although water has some shortcomings as a working fluid, no other single working fluid has been found that is more satisfactory overall for large electrical generating plants. Still, vapor power cycles intended for special uses may employ working fluids that are better matched to the application at hand than water. Cycles that operate at relatively low temperatures may perform best with a refrigerant such as ammonia as the working fluid. Power systems for spaceflight applications may employ a substance such as mercury as the working fluid because it has desirable performance characteristics at the relatively high temperatures at which these power systems operate. Moreover, water may be used together with some other substance in a *binary* vapor cycle to achieve better overall performance than could be realized with water alone.

In a binary vapor power cycle two working fluids are used, one with good high-temperature characteristics and another with good characteristics at the lower-temperature end of the operating range. A schematic diagram and an accompanying T–s diagram of a binary vapor cycle using mercury and water are shown in Fig. 8.13. In this arrangement, two ideal Rankine cycles are combined, with the heat rejection from the high-temperature cycle (the topping cycle) being used as the energy input for the low-temperature cycle. This energy transfer is accomplished in an interconnecting heat exchanger, which serves as the condenser for the mercury cycle and the boiler for the water cycle. Since the increase in the specific enthalpy of the water as it passes through the heat exchanger is several times the magnitude of the specific enthalpy decrease of the mercury, several units of mass of mercury must circulate in the topping cycle for each unit of mass of water in the other cycle. Binary vapor power cycles can operate with higher average temperatures of heat addition than conventional cycles using water only and can attain higher thermal efficiencies. However, the savings in fuel costs that might be achieved in this manner are for the present offset by increased costs related to the construction and operation of the more complex cycle arrangement.

The binary cycle considered above is just one example of how systems can be linked to obtain overall systems that utilize fuel more effectively. Other examples involve a multiple-use strategy known as *cogeneration*. Cogeneration is the subject

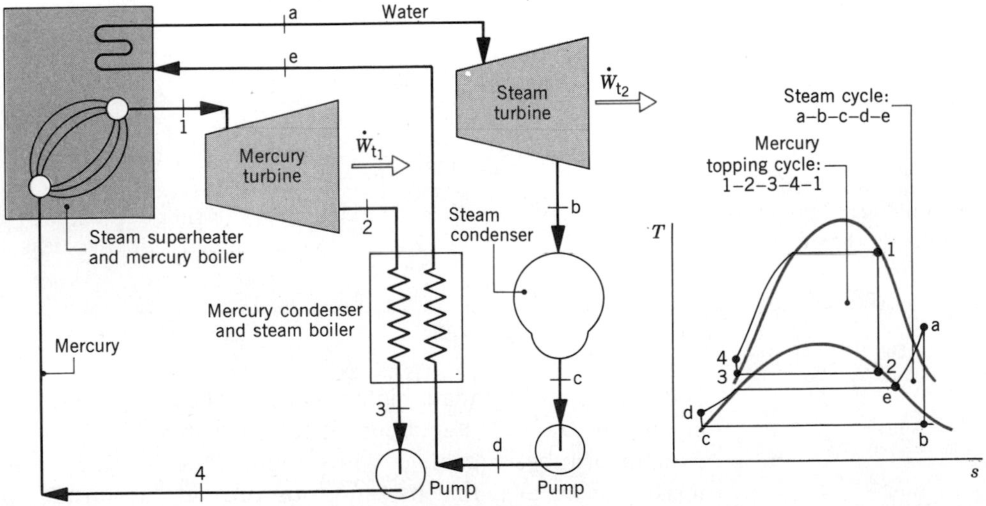

Figure 8.13 Mercury–water binary vapor cycle.

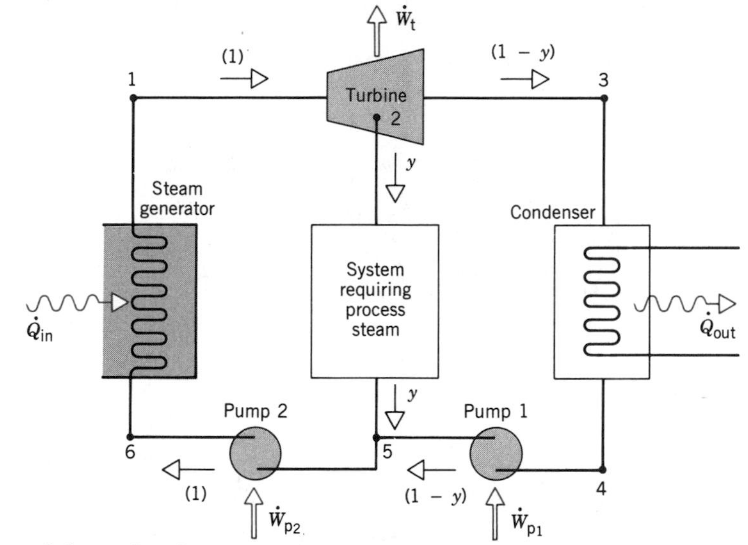

Figure 8.14 Schematic of a cogeneration system in which process steam is bled from the turbine.

of the remainder of the present section. The first application of cogeneration considered involves combustion heating and the development of process steam.

Direct combustion heating and the development of process steam together account for a substantial portion of industrial energy resource use. But because the heat and steam are often required at a relatively low temperature, good use is not made of the relatively high-temperature products of combustion obtained by burning fuel. This source of inefficiency can be reduced with a cogeneration arrangement in which fuel is consumed to produce both electricity and steam (or heat), but with a fuel expenditure less than would be required to produce them individually. A schematic of such a cogeneration system is shown in Fig. 8.14. The figure is labeled with the fractions of the total flow entering the steam turbine that remain at various locations. On the basis of a unit of mass entering the turbine at state 1, a fraction y is extracted at an intermediate point 2 and diverted to some process that requires steam at this condition. The remaining fraction, $1 - y$, expands to the turbine exit at state 3, producing power in addition to that developed by the first turbine stage. Eventually this fraction rejoins the amount extracted and the combined stream is returned to the steam generator. When no process steam is required, all of the steam generated by the steam generator is allowed to expand through the turbine.

Industries such as pulp and paper production and food processing, which require steam for various processes in addition to electricity for operating machines, lighting, etc., are particularly well suited for the use of cogeneration. Another cogeneration arrangement being used increasingly involves *district heating*. In this application a power plant is integrated into a community so that it provides electricity for industrial, commercial, and domestic use together with steam for process needs, space heating, and domestic water heating. District heating is commonly used in northern Europe and is being increasingly used in the United States.

8.6 CASE STUDY: AVAILABILITY ANALYSIS OF A VAPOR POWER PLANT[1]

The discussions to this point show that a useful picture of power plant performance can be obtained with the conservation of mass and conservation of energy principles.

[1] Chapter 7 is a prerequisite for the study of this section.

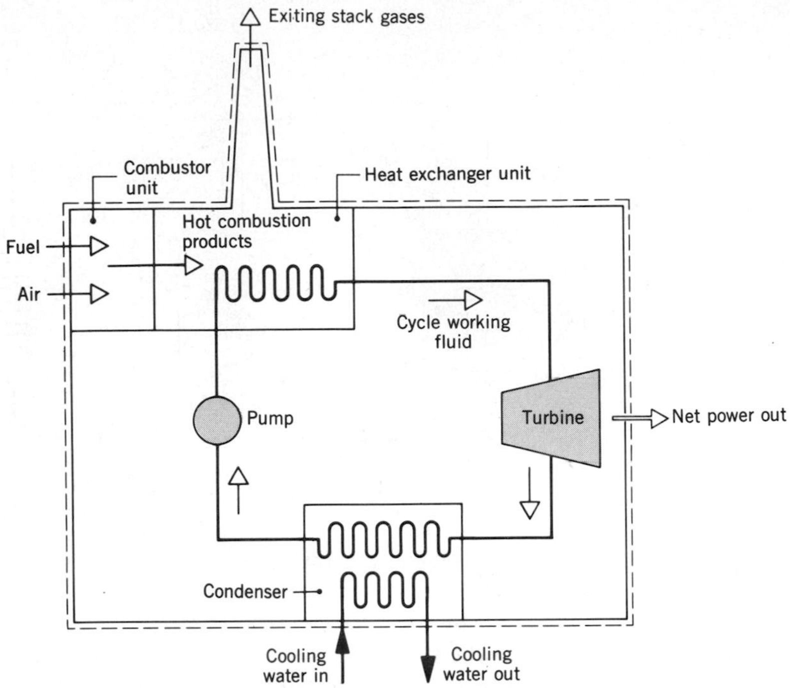

Figure 8.15 Power plant schematic for the availability analysis case study.

However, these principles provide only the *quantities* of energy transferred to and from the plant and do not allow the *utility* of the different types of energy transfer to be assessed. A unit of energy exiting as generated electricity would be considered equivalent to a unit of energy exiting in relatively low-temperature cooling water. Also, nothing can be learned with the conservation principles alone about the relative significance of the irreversibilities present in the various plant components and the losses associated with those components. The method of availability analysis introduced in Chap. 7 allows issues such as these to be dealt with quantitatively.

In this section we consider the disposition of the availability entering a power plant with the fuel. (Means for evaluating the fuel availability are introduced in Sec. 13.6.) A portion of the fuel availability is ultimately returned to the plant surroundings as the net work developed. However, the largest part is either destroyed by irreversibilities within the various plant components or carried from the plant by the cooling water, stack gases, and unavoidable heat transfers with the surroundings. These considerations are illustrated in the present section by three solved examples, treating respectively the boiler, turbine and pump, and condenser of a simple vapor power plant.

The irreversibilities present in each power plant component exact a tariff on the availability supplied to the plant, as measured by the availability destroyed in that component. The component levying the greatest tariff is the boiler, for a significant portion of the fuel availability entering the plant is destroyed by irreversibilities within it. There are two main sources of irreversibility in the boiler. One is the irreversible energy transfer occurring between the hot combustion gases and the working fluid of the vapor power cycle flowing through the boiler tubes. The other is the combustion process itself. To simplify the present discussion, the boiler is considered to consist of a combustor unit in which fuel and air are burned to produce hot combustion gases, followed by a heat exchanger unit where the cycle working fluid is vaporized as the hot gases cool. This idealization is illustrated in Fig. 8.15.

For the purposes of illustration, let us assume that 30% of the availability entering the combustion unit with the fuel is destroyed by the combustion irreversibility and 1% of the fuel availability exits the heat exchanger unit with the stack gases. The corresponding values for an actual power plant might differ substantially from these nominal values, however. (Means for evaluating the combustion irreversibility and the availability accompanying the exiting stack gases are introduced in Sec. 13.6.) Using the foregoing values for the combustion irreversibility and stack gas loss, it follows that a *maximum* of 69% of the fuel availability remains for transfer to the cycle working fluid. It is from this portion of the fuel availability that the net work developed by the plant is obtained. In Examples 8.7 through 8.9, we trace the disposition of the availability supplied by the hot combustion gases passing through the heat exchanger unit. The principal results of this series of examples are reported in Table 8.1. Carefully note that the values of Table 8.1 are keyed to the vapor power plant of Examples 8.7 through 8.9 and thus have only qualitative significance for vapor power plants in general.

The entries of Table 8.1 suggest some general observations about vapor power plant performance. First, the table shows that the irreversibilities are more significant than the plant losses. The largest portion of the availability entering the plant with the fuel is destroyed, with the boiler irreversibility overshadowing all others. By contrast, the loss associated with heat transfer to the cooling water is unimportant. The cycle thermal efficiency (calculated in the solution to Example 8.2) is 31.4%, so over two-thirds (68.6%) of the *energy* supplied to the cycle working fluid is subsequently carried out by the condenser cooling water. The amount of *availability* carried out is relatively minor, however, because the temperature of the cooling water is raised only a few degrees over that of the surroundings and thus has limited utility. The loss amounts to only 1% of the availability entering the plant with the fuel. Similarly, losses accompanying unavoidable heat transfer with the surroundings and the exiting stack gases typically amount only to a few percent of the availability entering the plant with the fuel and are generally overstated when considered from the perspective of the conservation of mass and energy principles alone.

TABLE 8.1 Vapor Power Plant Availability Analysis Summary[a]

Outputs	
Net power out[b]	30%
Losses	
Condenser cooling water[c]	1%
Stack gases (assumed)	1%
Irreversibility	
Boiler	
Combustion unit (assumed)	30%
Heat exchanger unit[d]	30%
Turbine[e]	5%
Pump[f]	—
Condenser[g]	3%
Total	100%

[a] All values are expressed as a percentage of the availability carried into the plant with the fuel. Values are rounded to the nearest full percent. Availability losses associated with stray heat transfer from plant components are ignored.

[b] Example 8.8.

[c] Example 8.9.

[d] Example 8.7.

[e] Example 8.8.

[f] Example 8.8.

[g] Example 8.9.

An availability analysis allows the sites where availability destructions or losses occur to be identified and rank ordered for significance. This knowledge is useful in directing attention to aspects of plant performance that offer the greatest opportunities for improvement through the application of practical engineering measures. However, the decision to adopt any particular modification is governed by economic considerations that take into account both economies in fuel use and the costs incurred to achieve those economies.

The calculations presented in the following examples illustrate the application of availability principles through the analysis of a simple vapor power plant. There are no fundamental difficulties, however, in applying the methodology to actual power plants, including consideration of the combustion process. The same procedures also can be used to analyze the gas turbine power plants considered in Chap. 9 and other types of thermal systems.

EXAMPLE 8.7

PROBLEM

The heat exchanger unit of the boiler of Example 8.2 has a stream of water entering as a liquid at 8.0 MPa and exiting as a saturated vapor at 8.0 MPa. In a separate stream, gaseous products of combustion cool at a constant pressure of 1 atm from 1107 to 547°C. The gaseous stream can be modeled as air as an ideal gas. Let $T_0 = 22°C$, $p_0 =$ 1 atm. Determine (a) the net rate at which availability is carried into the heat exchanger unit by the gas stream, in MW, (b) the net rate at which availability is carried from the heat exchanger by the water stream, in MW, (c) the irreversibility rate, in MW, (d) the second law efficiency given by Eq. 7.45.

SOLUTION

Known: A heat exchanger at steady state has a water stream entering and exiting at known states and a separate gas stream entering and exiting at known states.

Find: Determine the net rate at which availability is carried into the heat exchanger by the gas stream, in MW, the net rate at which availability is carried from the heat exchanger by the water stream, in MW, the irreversibility rate, in MW, and the second law efficiency.

Schematic and Given Data:

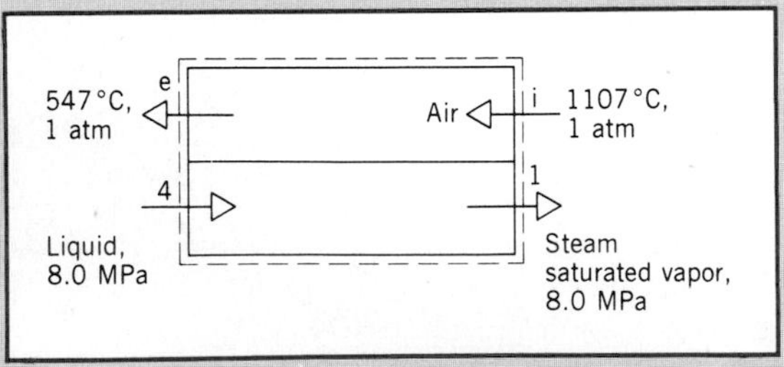

Figure E8.7

Assumptions:

1. The control volume shown in the accompanying figure operates at steady state with $\dot{Q}_{cv} = \dot{W}_{cv} = 0$.
2. Kinetic and potential energy effects can be ignored.
3. The gaseous combustion products are modeled as air as an ideal gas.
4. The air and the water each pass through the steam generator at constant pressure.
5. Only 69% of the fuel availability entering the plant remains after accounting for the stack loss and combustion irreversibility.
6. $T_0 = 22°C$, $p_0 = 1$ atm.

Analysis: Since all results are to be expressed on a time rate basis, the analysis begins by evaluating the mass flow rate of the air in terms of the mass flow rate of the water. The air and water pass through the boiler in separate streams. Hence, at steady state the conservation of mass principle requires

$$\dot{m}_i = \dot{m}_e \qquad \text{(air)}$$

$$\dot{m}_4 = \dot{m}_1 \qquad \text{(water)}$$

Using these relations, an energy rate balance for the overall control volume reduces at steady state to

$$0 = \cancelto{0}{\dot{Q}_{cv}} - \cancelto{0}{\dot{W}_{cv}} + \dot{m}_a(h_i - h_e) + \dot{m}(h_4 - h_1)$$

where $\dot{Q}_{cv} = \dot{W}_{cv} = 0$ by assumption 1, and the kinetic and potential energy terms are dropped by assumption 2. In this equation $\dot{m}_a$ and $\dot{m}$ denote, respectively, the mass flow rates of the air and water. On solving

$$\frac{\dot{m}_a}{\dot{m}} = \frac{h_1 - h_4}{h_i - h_e}$$

The solution to Example 8.2 gives $h_1 = 2758$ kJ/kg and $h_4 = 183.36$ kJ/kg. From Table A-16, $h_i = 1491.44$ kJ/kg and $h_e = 843.98$ kJ/kg. Hence

$$\frac{\dot{m}_a}{\dot{m}} = \frac{2758 - 183.36}{1491.44 - 843.98} = 3.977 \frac{\text{kg (air)}}{\text{kg (steam)}}$$

From Example 8.2, $\dot{m} = 4.449 \times 10^5$ kg/h. Thus, $\dot{m}_a = 17.694 \times 10^5$ kg/h.

(a) The net rate at which availability is carried into the heat exchanger unit by the gaseous stream can be evaluated using Eq. 7.35

$$\begin{bmatrix} \text{net rate at which availability} \\ \text{is carried in by the} \\ \text{gaseous stream} \end{bmatrix} = \dot{m}_a(a_{fi} - a_{fe})$$

$$= \dot{m}_a[\text{h}_i - h_e - T_0(s_i - s_e)]$$

Since the gas pressure remains constant, Eq. 6.25a giving the change in specific entropy of an ideal gas reduces to $s_i - s_e = s_i^\circ - s_e^\circ$. Thus, with h and s° values from Table A-16

$$\begin{aligned} \dot{m}_a(a_{fi} - a_{fe}) &= (17.694 \times 10^5 \text{ kg/h})[(1491.44 - 843.98) \text{ kJ/kg} \\ &\quad - (295 \text{ K})(3.34474 - 2.74504) \text{ kJ/kg}\cdot\text{K}] \\ &= \frac{8.326 \times 10^8 \text{ kJ/h}}{(3600 \text{ s/h})}\left(\frac{1 \text{ MW}}{10^3 \text{ kJ/s}}\right) = 231.28 \text{ MW} \end{aligned}$$

(b) The net rate at which availability is carried out of the boiler by the water stream is determined similarly:

$$\begin{bmatrix} \text{net rate at which availability} \\ \text{is carried out by the} \\ \text{water stream} \end{bmatrix} = \dot{m}(a_{f1} - a_{f4})$$

$$= \dot{m}[h_1 - h_4 - T_0(s_1 - s_4)]$$

From Table A-3, $s_1 = 5.7432$ kJ/kg·K. Double interpolation in Table A-5 at 8.0 MPa and $h_4 = 183.36$ kJ/kg gives $s_4 = 0.5957$ kJ/kg·K. Substituting known values

$$\begin{aligned} \dot{m}(a_{f1} - a_{f4}) &= (4.449 \times 10^5)[(2758 - 183.36) \\ &\quad - 295(5.7432 - 0.5957)] \\ &= \frac{4.699 \times 10^8 \text{ kJ/h}}{(3600 \text{ s/h})}\left(\frac{1 \text{ MW}}{10^3 \text{ kJ/s}}\right) = 130.53 \text{ MW} \end{aligned}$$

(c) The irreversibility can be evaluated by reducing the availability rate balance to obtain

$$\dot{I}_{cv} = \dot{m}_a(a_{fi} - a_{fe}) + \dot{m}(a_{f4} - a_{f1})$$

With the results of parts (a) and (b)

$$\dot{I}_{\text{cv}} = 231.28 \text{ MW} - 130.53 \text{ MW} = 100.75 \text{ MW}$$

(d) The second law efficiency given by Eq. 7.45 is

$$\varepsilon = \frac{\dot{m}(a_{f1} - a_{f4})}{\dot{m}_a(a_{fi} - a_{fe})} = \frac{130.53 \text{ MW}}{231.28 \text{ MW}} = 0.564\ (56.4\%)$$

This calculation indicates that 43.6% of the availability supplied to the heat exchanger unit by the cooling combustion products is destroyed. However, since only 69% of the fuel availability entering the plant is assumed to remain after the stack loss and combustion irreversibility are accounted for (assumption 5), it can be concluded that $0.69 \times 43.6\% = 30\%$ of the availability entering the plant with the fuel is destroyed within the heat exchanger. This is the value listed in Table 8.1.

Comments:

1. Since energy is conserved, the rate at which energy is transferred to the water as it flows through the heat exchanger *exactly equals* the rate at which energy is transferred from the cooling gas passing through the heat exchanger. The decrease in availability of the gas as it passes through the heat exchanger *exceeds* the increase in the availability of the water by the amount of availability destroyed, however.
2. Underlying the assumption that each stream passes through the heat exchanger at constant pressure is the neglect of friction as an irreversibility. Thus, the only contributor to availability destruction in this case is heat transfer from the higher-temperature combustion products to the vaporizing water.
3. The irreversibility can be determined alternatively by evaluating the entropy produced per unit mass of steam, $\dot{\sigma}_{\text{cv}}/\dot{m}$, from an entropy rate balance and multiplying by T_0 to obtain $\dot{I}_{\text{cv}}/\dot{m} = T_0(\dot{\sigma}_{\text{cv}}/\dot{m})$.

EXAMPLE 8.8

PROBLEM

Reconsider the turbine and pump of Example 8.2. Determine for each of these components the rate at which availability is destroyed, in MW. Express each result as a percentage of the availability entering the plant with the fuel. Let $T_0 = 22°\text{C}$, $p_0 = 1$ atm.

SOLUTION

Known: A vapor power cycle operates with steam as the working fluid. The turbine and pump each has an isentropic efficiency of 85%.

Find: For the turbine and the pump individually, determine the rate at which availability is destroyed, in MW. Express each result as a percentage of the availability entering the plant with the fuel.

Schematic and Given Data:

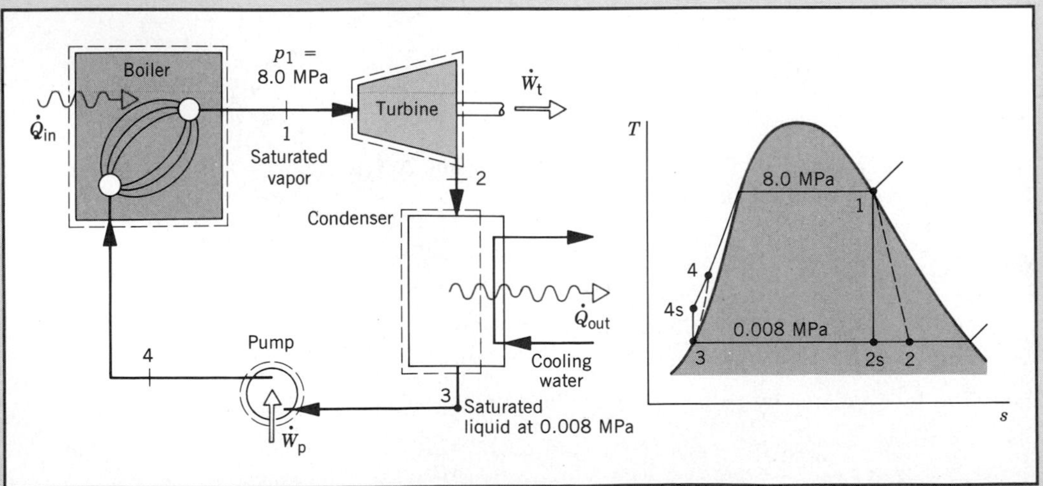

Figure E8.8

Assumptions:

1. The turbine and the pump can each be analyzed as a control volume at steady state.
2. The turbine and pump operate adiabatically and each has an efficiency of 85%.
3. Kinetic and potential energy effects are negligible.
4. Only 69% of the fuel availability entering the plant remains after accounting for the stack loss and combustion irreversibility.
5. $T_0 = 22°C$, $p_0 = 1$ atm.

Analysis: The irreversibility can be found by reducing the availability rate balance or by use of the relationship $\dot{I}_{cv} = T_0\dot{\sigma}_{cv}$, where $\dot{\sigma}_{cv}$ is the rate of entropy production from an entropy rate balance. With either approach, the irreversibility rate for the turbine can be expressed as

$$\dot{I}_{cv} = \dot{m}T_0(s_2 - s_1)$$

From Table A-3, $s_1 = 5.7432$ kJ/kg·K. Using $h_2 = 1939.3$ kJ/kg from the solution to Example 8.2, the value of s_2 can be determined from Table A-3, giving $s_2 = 6.2021$ kJ/kg·K. Substituting values

$$\dot{I}_{cv} = (4.449 \times 10^5 \text{ kg/h})(295 \text{ K})(6.2021 - 5.7432)$$

$$= \left(0.602 \times 10^8 \frac{\text{kJ}}{\text{h}}\right)\left(\frac{\text{h}}{3600 \text{ s}}\right)\left(\frac{1 \text{ MW}}{10^3 \text{ kJ/s}}\right) = 16.72 \text{ MW}$$

From the solution to Example 8.7, the net rate at which availability is supplied by the cooling combustion gases is 231.28 MW. The turbine irreversibility expressed as a percentage of this is (16.72/231.28)(100%) = 7.23%. However, since only 69% of the entering fuel availability remains after the stack loss and combustion irreversibility are accounted for, it can be concluded that 0.69 × 7.23% = 5% of the availability entering the plant with the fuel is destroyed within the turbine. This is the value listed in Table 8.1.

Similarly, the irreversibility rate for the pump is

$$\dot{I}_{cv} = \dot{m}T_0(s_4 - s_3)$$

With s_3 from Table A-3 and s_4 from the solution to Example 8.7

$$\dot{I}_{cv} = (4.449 \times 10^5 \text{ kg/h})(295 \text{ K})(0.5957 - 0.5926)(\text{kJ/kg}\cdot\text{K})$$

$$= \left(4.07 \times 10^5 \frac{\text{kJ}}{\text{h}}\right)\left(\frac{\text{h}}{3600 \text{ s}}\right)\left(\frac{1 \text{ MW}}{10^3 \text{ kJ/s}}\right) = 0.11 \text{ MW}$$

Expressing this as a percentage of the availability entering the plant as calculated above, we have (0.11/231.28)(69%) = 0.03%. This value is rounded to zero in Table 8.1.

Comment: The net power output of the vapor power plant of Example 8.2 is 100 MW. Expressing this as a percentage of the rate at which availability is carried into the plant with the fuel, (100/231.28)(69%) = 30%.

EXAMPLE 8.9

PROBLEM

The condenser of Example 8.2 involves two separate water streams. In one stream a two-phase liquid–vapor mixture enters at 0.008 MPa and exits as a saturated liquid at 0.008 MPa. In the other stream, cooling water enters at 15°C and exits at 35°C. (a) Determine the net rate at which availability is carried from the condenser by the cooling water, in MW. Express this result as a percentage of the availability entering the plant with the fuel. (b) Determine for the condenser the irreversibility rate, in MW. Express this result as a percentage of the availability entering the plant with the fuel. Let $T_0 = 22°C$ and $p_0 = 1$ atm.

SOLUTION

Known: A condenser at steady state has two streams: (1) a two-phase liquid–vapor mixture entering and condensate exiting at known states and (2) a separate cooling water stream entering and exiting at known temperatures.

Find: Determine the net rate at which availability is carried from the condenser by the cooling water stream and the irreversibility rate for the condenser. Express both quantities in MW and as percentages of the availability entering the plant with the fuel.

Schematic and Given Data:

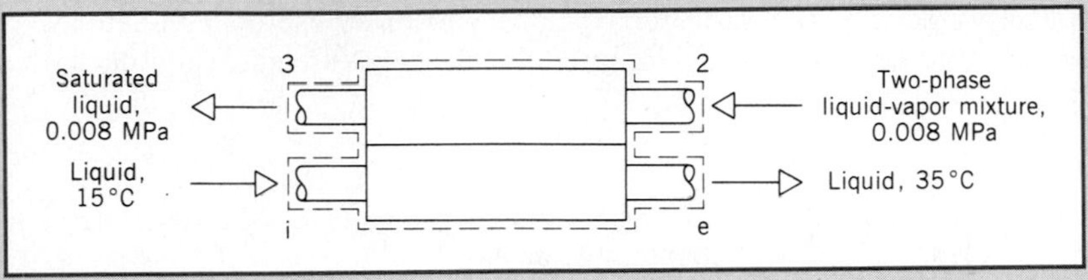

Figure E8.9

Assumptions:

1. The control volume shown on the accompanying figure operates at steady state with $\dot{Q}_{cv} = \dot{W}_{cv} = 0$.
2. Kinetic and potential energy effects can be ignored.
3. Only 69% of the fuel availability remains after accounting for the stack loss and combustion irreversibility.
4. $T_0 = 22°C$, $p_0 = 1$ atm.

Analysis:

(a) The net rate at which availability is carried out of the condenser can be evaluated by using Eq. 7.35:

$$\begin{bmatrix}\text{net rate at which availability} \\ \text{is carried out by the} \\ \text{cooling water}\end{bmatrix} = \dot{m}_{cw}(a_{fe} - a_{fi})$$

$$= \dot{m}_{cw}[h_e - h_i - T_0(s_e - s_i)]$$

where $\dot{m}_{cw}$ is the mass flow rate of the cooling water from the solution to Example 8.2. With saturated liquid values for specific enthalpy and entropy from Table A-2 at the specified inlet and exit temperatures of the cooling water

$$\dot{m}_{cw}(a_{fe} - a_{fi}) = (9.39 \times 10^6 \text{ kg/h})[(146.68 - 62.99) \text{ kJ/kg} - (295 \text{ K})(0.5053 - 0.2245) \text{ kJ/kg·K}]$$

$$= \frac{8.019 \times 10^6 \text{ kJ/h}}{(3600 \text{ s/h})}\left(\frac{1 \text{ MW}}{10^3 \text{ kJ/s}}\right) = 2.23 \text{ MW}$$

Expressing this as a percentage of the availability entering the plant with the fuel, we get (2.23/231.28)(69%) = 1%. This is the value listed in Table 8.1.

(b) The irreversibility rate for the condenser can be evaluated by reducing the availability rate balance. Alternatively, the relationship $\dot{I}_{cv} = T_0\dot{\sigma}_{cv}$ can be employed, where $\dot{\sigma}_{cv}$ is the time rate of entropy production for the condenser determined from an entropy rate balance. With either approach, the irreversibility rate can be expressed as

$$\dot{I}_{cv} = T_0[\dot{m}(s_3 - s_2) + \dot{m}_{cw}(s_e - s_i)]$$

Substituting values

$$\dot{I}_{cv} = 295[(4.449 \times 10^5)(0.5926 - 6.2021) + (9.39 \times 10^6)(0.5053 - 0.2245)]$$

$$= \frac{416.1 \times 10^5 \text{ kJ/h}}{(3600 \text{ s/h})}\left(\frac{1 \text{ MW}}{10^3 \text{ kJ/s}}\right) = 11.56 \text{ MW}$$

Expressing this as a percentage of the availability entering the plant with the fuel, we get (11.56/231.28)(69%) = 3%. This is the value listed in Table 8.1.

Comment: The cycle thermal efficiency evaluated in the solution to Example 8.2 is 31.4%. Thus, it can be concluded that over two-thirds (68.6%) of the *energy* supplied by the cooling combustion gas to the cycle working fluid is subsequently carried out by the condenser cooling water. However, the amount of *availability* carried out with the cooling water is relatively minor because the temperature of the water is raised only a few degrees over that of the surroundings.

PROBLEMS

8.1 Water is the working fluid in an ideal Rankine cycle. The condenser pressure is 8 kPa, and saturated vapor enters the turbine at **(a)** 18 MPa and **(b)** 4 MPa. The net power output of the cycle is 100 MW. Determine for each case the mass flow rate of steam, in kg/h, the heat transfer rates for the working fluid passing through the boiler and condenser, each in kW, and the thermal efficiency.

8.2 Water is the working fluid in an ideal Rankine cycle. Superheated vapor enters the turbine at 8 MPa, 480°C. The condenser pressure is 8 kPa. The net power output of the cycle is 100 MW. Determine for the cycle

(a) the rate of heat transfer to the working fluid passing through the steam generator, in kW.
(b) the thermal efficiency.
(c) the mass flow rate of condenser cooling water, in kg/h, if the cooling water enters the condenser at 15°C and exits at 35°C with negligible pressure change.

8.3 Water is the working fluid in a Carnot vapor power cycle. Saturated liquid enters the boiler at a pressure of 8 MPa, and saturated vapor enters the turbine. The condenser pressure is 8 kPa. Determine

(a) the thermal efficiency.
(b) the back work ratio.
(c) the heat transfer to the working fluid per unit mass passing through the boiler, in kJ/kg.
(d) the heat transfer from the working fluid per unit mass passing through the condenser, in kJ/kg.

8.4 Repeat the analysis of Problem 8.2 for condenser pressures of **(a)** 0.1 MPa and **(b)** 6 kPa. Compare the results with those obtained in Problem 8.2.

8.5 Repeat the analysis of Problem 8.2 for steam generator pressures of **(a)** 4 MPa and **(b)** 24 MPa, while maintaining a turbine inlet temperature of 480°C. Compare the results with those obtained in Problem 8.2.

8.6 Water is the working fluid in an ideal Rankine cycle. Saturated vapor enters the turbine at 1000 lbf/in.^{2} The condenser pressure is 1 lbf/in.^{2} Cooling water passing through the condenser experiences a temperature increase from 60 to 80°F with negligible pressure change. Determine

(a) the back work ratio.
(b) the net work per unit mass of steam flow, in Btu/lb.
(c) the heat transfer to the steam passing through the boiler, in Btu per lb of steam flowing.
(d) the thermal efficiency.
(e) the ratio of the cooling water flow rate to the steam flow rate.

8.7 Water is the working fluid in a Carnot vapor power cycle. Saturated liquid enters the boiler at a pressure of 1000 lbf/in.^{2}, and saturated vapor enters the turbine. The condenser pressure is 1 lbf/in.^{2} Determine

(a) the thermal efficiency.
(b) the back work ratio.
(c) the net work of the cycle per unit mass of water flowing, in Btu/lb.
(d) the heat transfer from the working fluid passing through the condenser, in Btu per lb of steam flowing.

8.8 Repeat the analysis of Problem 8.6 with superheated vapor entering the turbine at 900°F, 1000 lbf/in.^{2} Compare the results with those obtained in Problem 8.6.

8.9 Water is the working fluid in an ideal Rankine cycle. The pressure and temperature at the turbine inlet are 1200 lbf/in.^{2} and 1000°F, respectively, and the condenser pressure is 1 lbf/in.^{2} The mass flow rate of steam entering the turbine is 1.4×10^6 lb/h. The cooling water experiences a temperature increase from 60 to 80°F, with negligible pressure drop, as it passes through the condenser. Determine for the cycle

(a) the net power developed, in Btu/h.
(b) the thermal efficiency.
(c) the mass flow rate of cooling water, in lb/h.

8.10 Repeat the analysis of Problem 8.9 for condenser pressures of 0.4 and 14.7 lbf/in.^{2} The mass flow rate is unchanged. Compare the results with those obtained in Problem 8.9.

8.11 Repeat the analysis of Problem 8.9 for steam generator pressures of **(a)** 600 lbf/in.^{2} and **(b)** 3500 lbf/in.^{2}, while maintaining a turbine inlet temperature of 1000°F. The mass flow rate is unchanged. Compare the results with those obtained in Problem 8.9.

8.12 A power plant based on the Rankine cycle is under development to provide a net power output of 10 MW. Concentrating solar collectors are to be used to generate steam at 300°C and 2 MPa for expansion through the turbine. Cooling water is available at 20°C. Specify the preliminary design of the cycle and estimate the thermal efficiency and the steam and cooling water flow rates, in kg/h.

8.13 Refrigerant 12 is the working fluid in a solar power plant operating on a Rankine cycle. Saturated vapor at 60°C enters the turbine, and the condenser operates at a pressure of 6 bar. The rate of energy input to the collectors from solar radiation is 0.4 kW per m^2 of collector surface area. Determine the minimum possible solar collector surface area, in m^2, per kW of power developed by the plant.

8.14 At a particular location in the ocean, the temperature near the surface is 80°F, and the temperature at a depth of 1500 ft is 46°F. A power plant based on the Rankine cycle, with ammonia as the working fluid, has been proposed to utilize this naturally occurring temperature gradient to produce electrical power. The power to be developed by the turbine is 8.2×10^8 Btu/h. A schematic diagram and a table of property values for ammonia are given in Fig. P8.14. For simplicity, the properties of seawater can be taken as those of pure water.

(a) Calculate the thermal efficiency of the proposed cycle and compare it to the thermal efficiency of a reversible power cycle operating between thermal reservoirs at 80 and 46°F.
(b) Calculate the net power output of the plant, in Btu/h, if the pumps used to circulate seawater through the evaporator and condenser heat exchangers require a total power input of 2.53×10^8 Btu/h.
(c) Determine the seawater flow rates through the boiler and condenser, in lb/h.

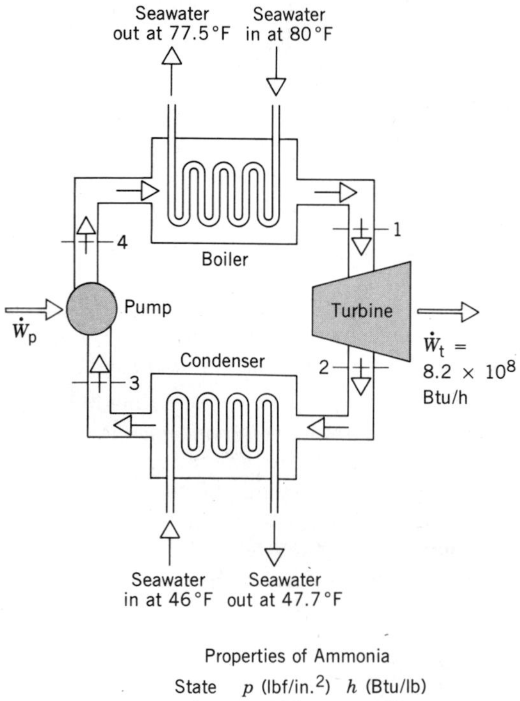

Properties of Ammonia

State	p (lbf/in.^{2})	h (Btu/lb)
1	136	628
2	92	610
3	92	119
4	136	121

Figure P8.14

8.15 The cycle of Problem 8.3 is modified to include the effects of irreversibilities in the adiabatic expansion and compression processes. If the states at the turbine and pump inlets remain unchanged, repeat the analysis of Problem 8.3 for the modified Carnot cycle with $\eta_t = 0.80$ and $\eta_p = 0.70$.

8.16 Steam enters the turbine of a vapor power plant with a pressure of 10 MPa and expands with no heat transfer to 8 kPa. The isentropic turbine efficiency is 85%. Determine the minimum turbine inlet temperature, in °C, to ensure a quality of at least 90% at the exit of the turbine.

8.17 Reconsider the analysis of Problem 8.2, but include in the analysis that the turbine and pump have isentropic efficiencies of 85 and 70%, respectively. Determine for the modified cycle

(a) the thermal efficiency.
(b) the mass flow rate of steam, in kg/h, for a net power output of 100 MW.
(c) the mass flow rate of condenser cooling water, in kg/h, if the cooling water enters the condenser at 15°C and exits at 35°C with negligible pressure change.

8.18 Reconsider the cycle of Problem 8.8, but include in the analysis that the turbine and pump have isentropic efficiencies of 85 and 78%, respectively. Answer the same questions for the modified cycle as in Problem 8.8.

8.19 Reconsider the cycle of Problem 8.9, but include in the analysis that the turbine and pump have isentropic efficiencies of 88%. The mass flow rate is unchanged. Determine for the modified cycle

(a) the net power developed, in Btu/h.
(b) the rate of heat transfer to the working fluid passing through the steam generator, in Btu/h.
(c) the thermal efficiency.
(d) the volumetric flow rate of cooling water entering the condenser, in ft^3/min.

8.20 Reconsider the cycle of Problem 8.19, but insert a throttling valve between the steam generator and the turbine that reduces the turbine inlet pressure to 1000 lbf/in.^{2}

but does not change the mass flow rate. Answer the same questions about the modified cycle as in Problem 8.19.

8.21 Superheated steam at 8 MPa and 480°C leaves the steam generator of a vapor power plant. Heat transfer and frictional effects in the line connecting the steam generator and the turbine reduce the pressure and temperature at the turbine inlet to 7.6 MPa and 440°C, respectively. The pressure at the exit of the turbine is 10 kPa, and liquid leaves the condenser at 8 kPa, 36°C. The pressure is increased to 8.6 MPa across the pump. The turbine and pump have isentropic efficiencies of 88%. Determine, for a net power output of 100 MW,

(a) the mass flow rate of steam, in kg/h.
(b) the thermal efficiency.
(c) the rate of heat transfer from the line connecting the steam generator and turbine, in MW.
(d) the turbine effectiveness.

8.22 Water is the working fluid in a vapor power plant. Superheated steam leaves the steam generator at 1200 lbf/in.^{2}, 1000°F and enters the turbine at 1100 lbf/in.^{2}, 960°F. The steam expands through the turbine, exiting at 1.2 lbf/in.^{2} with a quality of 94%. Condensate leaves the condenser at 0.8 lbf/in.^{2}, 90°F and is pumped to 1250 lbf/in.^{2} before entering the steam generator. The pump efficiency is 80%. Determine

(a) the thermal efficiency of the cycle.
(b) the isentropic turbine efficiency.
(c) the turbine effectiveness.

8.23 In a power plant operating on a Rankine cycle, steam enters the turbine at 500 lbf/in.^{2} and 800°F and expands to 2 lbf/in.^{2} and a quality of 92%. Heat transfer from the turbine to the surroundings occurs at a rate of 50 Btu per lb of steam flowing. All other aspects of the operation follow the ideal cycle. Determine

(a) the net work developed per unit mass of steam flow, in Btu/lb.
(b) the thermal efficiency.
(c) the irreversibility rate per unit mass of steam flow, in Btu/lb, for a control volume enclosing the turbine and a portion of its immediate surroundings so heat transfer occurs at $T_0 = 60°F$.

8.24 A vapor power plant operates with steam entering the turbine at 2500 lbf/in.^{2}, 1000°F and a condenser pressure of 1 lbf/in.^{2} The turbine and pump operate with isentropic efficiencies of 82 and 77%, respectively. For the cycle, determine

(a) the turbine work, in Btu per lb of steam flowing.
(b) the thermal efficiency.
(c) the heat transfer to cooling water passing through the condenser, in Btu per lb of steam condensed.

8.25 In the preliminary design of a power plant, water is chosen as the working fluid and it is determined that the turbine inlet temperature may not exceed 520°C. Based on expected cooling water temperatures, the condenser is to operate at a pressure of 0.06 bar. Determine the steam generator pressure required if the isentropic turbine efficiency is 80% and the quality of steam at the turbine exit must be at least 90%.

8.26 Steam at 10 MPa, 600°C enters the first-stage turbine of an ideal Rankine cycle with reheat. The steam leaving the reheat section of the steam generator is at 500°C, and the condenser pressure is 6 kPa. If the quality at the exit of the second-stage turbine is 90%, determine the cycle thermal efficiency.

8.27 The ideal Rankine cycle of Problem 8.9 is modified to include reheat. In the modified cycle, steam expands through the first-stage turbine to saturated vapor and then is reheated to 900°F. If the mass flow rate of steam in the modified cycle is the same as in Problem 8.9, determine for the modified cycle

(a) the net power developed, in Btu/h.
(b) the rate of heat transfer to the working fluid in the reheat process, in Btu/h.
(c) the thermal efficiency.

8.28 Water is the working fluid in a vapor power cycle with superheat and reheat. Superheated vapor enters the first turbine stage at 8 MPa, 480°C and expands to 0.7 MPa. Then, it is reheated to 480°C before entering the second turbine stage, where it expands to the condenser pressure of 8 kPa. The mass flow rate of steam entering the first turbine stage is 2.63×10^5 kg/h. Each turbine stage operates with an isentropic efficiency of 88%. The pump operates with an efficiency of 80%. Determine for the cycle

(a) the net power developed, in kW.
(b) the thermal efficiency.
(c) the rate of heat transfer to cooling water passing through the condenser, in kW.

8.29 Steam at 10 MPa, 600°C enters the first-stage turbine of a Rankine cycle with reheat. Steam enters the second-stage turbine after being reheated to 500°C. Saturated vapor exits the second-stage turbine. The condenser pressure is 6 kPa. Each turbine stage operates with an isentropic efficiency of 85%. All other aspects of the operation follow the ideal cycle. Determine the cycle thermal efficiency.

8.30 Steam at 32 MPa, 520°C leaves the steam generator of an ideal Rankine cycle modified to include three turbine stages with reheat between the stages. The reheat pressures are 4 and 0.5 MPa, respectively, and steam enters the second-stage turbine at 440°C and the third-stage turbine at 360°C. The condenser pressure is 0.08 bar. Determine for the cycle

(a) the net work per unit mass of steam flowing, in kJ/kg.
(b) the thermal efficiency.

8.31 An ideal Rankine cycle with reheat uses water as the working fluid. The conditions at the inlet to the first-stage turbine are 14 MPa, 600°C and the steam is reheated between the turbine stages to 600°C. For a condenser pressure of 6 kPa, plot the cycle thermal efficiency versus reheat pressure for pressures from 2 to 12 MPa.

8.32 An ideal Rankine cycle with reheat uses water as the working fluid. The conditions at the inlet to the first turbine stage are 1600 lbf/in.^{2}, 1200°F and the steam is reheated between the turbine stages to 1200°F. For a condenser pressure of 1 lbf/in.^{2}, plot the cycle thermal efficiency versus reheat pressure for pressures from 60 to 1200 lbf/in.^{2}

8.33 Modify the ideal Rankine cycle of Problem 8.2 to include one open feedwater heater operating at 0.7 MPa. Saturated liquid exits the feedwater heater at 0.7 MPa. Answer the same questions about the modified cycle as in Problem 8.2 and discuss the results.

8.34 Repeat the analysis of Problem 8.33 for feedwater heater pressures of **(a)** 4 MPa, **(b)** 2 MPa, and **(c)** 0.15 MPa.

8.35 A power plant operates on a regenerative vapor power cycle with one open feedwater heater. Steam enters the first turbine stage at 12 MPa, 520°C and expands to 1 MPa, where some of the steam is extracted and diverted to the open feedwater heater operating at 1 MPa. The remaining steam expands through the second turbine stage to the condenser pressure of 6 kPa. Saturated liquid exits the open feedwater heater at 1 MPa. For isentropic processes in the turbines and pumps, determine for the cycle **(a)** the thermal efficiency and **(b)** the net power developed, in MW, for a mass flow rate of 10^6 kg/h into the first-stage turbine.

8.36 Compare the results of Problem 8.35 with those for an ideal Rankine cycle having the same turbine inlet conditions and condenser pressure, but no regenerator.

8.37 Reconsider the cycle of Problem 8.35, but include in the analysis that each turbine stage has an isentropic efficiency of 82%. The pump efficiencies remain 100%.

8.38 Modify the ideal Rankine cycle of Problem 8.9 to include one open feedwater heater operating at 100 lbf/in.^{2} Saturated liquid exits the open feedwater heater at 100 lbf/in.^{2} The mass flow rate of steam into the first turbine stage is the same as the mass flow rate of steam in Problem 8.9. Answer the same questions about the modified cycle as in Problem 8.9 and discuss the results.

8.39 Reconsider the cycle of Problem 8.38, but include in the analysis that the isentropic efficiency of each turbine stage is 88% and of each pump is 80%.

8.40 A regenerative vapor power cycle has two turbine stages with steam entering the first stage at 2500 lbf/in.^{2}, 1000°F and expanding to 140 lbf/in.^{2} Some of the flow is extracted at 140 lbf/in.^{2} and directed into an open feedwater heater operating at the same pressure. The remainder expands through the second turbine stage to the condenser pressure of 1 lbf/in.^{2} Saturated liquid leaves the feedwater heater at 140 lbf/in.^{2} For isentropic operation of the turbine stages and pumps, determine the thermal efficiency of the cycle.

8.41 Reconsider the cycle of Problem 8.40, but include in the analysis that each turbine stage and pump has an isentropic efficiency of 85%.

8.42 Modify the ideal Rankine cycle of Problem 8.2 to include one closed feedwater heater using extracted steam at 0.7 MPa. Condensate drains from the feedwater heater as saturated liquid at 0.7 MPa and is trapped into the condenser. The feedwater leaves the heater at 8 MPa and a temperature equal to the saturation temperature at 0.7 MPa. Answer the same questions about the modified cycle as in Problem 8.2 and discuss the results.

8.43 Repeat the analysis of Problem 8.42 for feedwater heater pressures of **(a)** 4 MPa, **(b)** 2 MPa, and **(c)** 0.15 MPa.

8.44 A power plant operates on a regenerative vapor power cycle with one closed feedwater heater. Steam enters the first turbine stage at 120 bar, 520°C and expands to 10 bar, where some of the steam is extracted and diverted to a closed feedwater heater. Condensate exiting the feedwater heater as saturated liquid at 10 bar passes through a trap into the condenser. The feedwater exits the heater at 120 bar with a temperature of 170°C. The condenser pressure is 0.06 bar. For isentropic processes in each turbine stage and the pump, determine for the cycle **(a)** the thermal efficiency and **(b)** the net power developed, in kW, for a mass flow rate of 10^6 kg/h into the first-stage turbine.

8.45 Reconsider the cycle of Problem 8.44, but include in the analysis that each turbine stage has an isentropic efficiency of 82%. The pump efficiency remains 100%.

8.46 Figure P8.46 shows a closed feedwater heater, with the condensate pumped into the feedwater line. Using data on the figure, determine the temperature, in °C, at location 6 and show each numbered state on a carefully labeled T–s diagram.

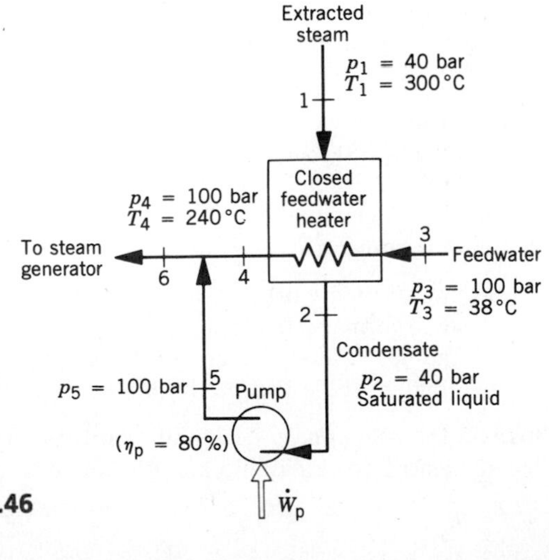

Figure P8.46

8.47 Modify the ideal Rankine cycle of Problem 8.9 to include one closed feedwater heater using extracted steam at 100 lbf/in.^{2} Condensate exiting the heater as saturated liquid at 100 lbf/in.^{2} passes through a trap into the condenser. The feedwater leaves the heater at 1200 lbf/in.^{2} and a temperature equal to the saturation temperature at 100 lbf/in.^{2} The mass flow rate of steam entering the first-stage turbine is the same as the steam flow rate in Problem 8.9. Answer the same questions about the modified cycle as in Problem 8.9 and discuss the results.

8.48 Reconsider the cycle of Problem 8.47, but include in the analysis that each turbine stage has an isentropic efficiency of 88% and the pump efficiency is 80%.

8.49 A regenerative vapor power cycle has two turbine stages with steam entering the first stage at 2500 lbf/in.^{2}, 1000°F and expanding to 140 lbf/in.^{2} Some of the flow is extracted at 140 lbf/in.^{2} and directed to a closed feedwater heater, where it is condensed to saturated liquid at 140 lbf/in.^{2} before being trapped into the condenser. The feedwater leaves the closed heater at 2500 lbf/in.^{2}, 350°F. The condenser pressure is 1 lbf/in.^{2} If the turbine stages and pumps operate isentropically, determine the thermal efficiency of the cycle.

8.50 Reconsider the cycle of Problem 8.49, but include in the analysis that the isentropic efficiencies of the turbine stages and the pump are 85%.

8.51 Consider a regenerative vapor power cycle with two feedwater heaters, a closed one and an open one. Steam enters the first turbine stage at 80 bar, 480°C, and expands to 20 bar. Some steam is extracted at 20 bar and fed to the closed feedwater heater. The remainder expands through the second-stage turbine to 3 bar, where an additional amount is extracted and fed into the open feedwater heater, which operates at 3 bar. The steam expanding through the third-stage turbine exits at the condenser pressure of 0.08 bar. Feedwater leaves the closed heater at 205°C, 80 bar, and condensate exiting as saturated liquid at 20 bar is trapped into the open heater. Saturated liquid at 3 bar leaves the open feedwater heater. The net power output of the cycle is 100 MW. If the turbine stages and pumps are isentropic, determine

(a) the thermal efficiency.

(b) the mass flow rate of steam entering the first turbine, in kg/h.

8.52 Reconsider the cycle of Problem 8.51, but include in the analysis that each turbine stage has an isentropic efficiency of 85% and each pump an efficiency of 82%.

8.53 A power plant operates on a regenerative vapor power cycle with two feedwater heaters. Steam enters the first turbine stage at 12 MPa, 520°C and expands in three stages to the condenser pressure of 6 kPa. Between the first and second stages, some steam is diverted to a closed feedwater heater at 1 MPa, with saturated liquid condensate being pumped ahead into the boiler feedwater line. The feedwater leaves the closed heater at 12 MPa, 170°C. Steam is extracted between the second and third turbine stages at 0.15 MPa and fed into an open feedwater heater operating at that pressure. Saturated liquid at 0.15 MPa leaves the open feedwater heater. For isentropic processes in the pumps and turbines, determine for the cycle **(a)** the thermal efficiency and **(b)** the net power developed, in kW, for a mass flow rate of 10^6 kg/h into the first-stage turbine.

8.54 Reconsider the cycle of Problem 8.53, but include in the analysis that each turbine stage has an isentropic efficiency of 82% and each pump an efficiency of 100%.

8.55 Modify the ideal Rankine cycle of Problem 8.9 to include one closed feedwater heater using extracted steam at 250 lbf/in.^{2} and one open feedwater heater operating at 40 lbf/in.^{2} Saturated liquid condensate drains from the closed heater at 250 lbf/in.^{2} and is trapped into the open heater. The feedwater leaves the closed heater at 1200 lbf/in.^{2}, 392°F, and saturated liquid leaves the open heater at 40 lbf/in.^{2} The mass flow rate of steam entering the first-stage turbine is the same as the steam flow rate in Problem 8.9. Answer the same questions about the modified cycle as in Problem 8.9 and discuss the results.

8.56 Reconsider the cycle of Problem 8.55, but include in the analysis that each turbine stage has an isentropic efficiency of 88% and each pump an efficiency of 80%.

8.57 A regenerative vapor power cycle has three turbine stages with steam entering the first stage at 2500 lbf/in.^{2}, 1000°F. The cycle has two feedwater heaters, a closed feedwater heater using extracted steam at 600 lbf/in.^{2} and an open feedwater heater operating at 60 lbf/in.^{2} Saturated liquid condensate drains from the closed heater at 600 lbf/in.^{2} and passes through a trap into the open heater. The feedwater leaves the closed heater at 2500 lbf/in.^{2}, 478°F. Saturated liquid leaves the open heater at 60 lbf/in.^{2}, and the condenser pressure is 1 lbf/in.^{2} For isentropic operation of the turbine stages and pumps, determine the thermal efficiency of the cycle.

8.58 Modify the regenerative vapor power cycle in Problem 8.51 to include reheat at 2 MPa. The portion of the flow that passes through the second turbine stage is reheated to 440°C before it enters the second turbine stage. Determine the thermal efficiency of the modified cycle.

8.59 Reconsider the cycle of Problem 8.58, but include in the analysis that each turbine stage has an isentropic efficiency of 85% and each pump an efficiency of 82%.

8.60 Modify the regenerative vapor power cycle in Problem 8.55 to include reheat at 250 lbf/in.^{2} The portion of the flow that passes through the second turbine stage is reheated to 900°F before it enters the turbine. Determine the thermal efficiency of the modified cycle.

8.61 Reconsider the cycle of Problem 8.60, but include in the analysis that each turbine stage has an isentropic efficiency of 88% and each pump an efficiency of 80%.

8.62 Steam enters the first turbine stage of a vapor power cycle with reheat and regeneration at 32 MPa, 600°C, and expands to 8 MPa. A portion of the flow is diverted to a closed feedwater heater at 8 MPa, and the remainder is reheated to 560°C before entering the second turbine stage. Expansion through the second turbine stage occurs to 1 MPa, where another portion of the flow is diverted to a second closed feedwater heater at 1 MPa. The remainder of the flow expands through the third turbine stage to 0.15 MPa, where a portion of the flow is diverted to an open feedwater heater operating at 0.15 MPa, and the rest expands through the fourth turbine stage to the condenser pressure of 6 kPa. Condensate leaves each closed feedwater heater as saturated liquid at the respective extraction pressure. The condensate streams from the closed heaters each pass through traps into the next lower-pressure feedwater heater. Saturated liquid exiting the open heater is pumped to the steam generator pressure. For isentropic operation of the turbine stages and pumps,

(a) sketch the layout of the cycle and number the principal state points.
(b) determine the thermal efficiency of the cycle.
(c) calculate the mass flow rate into the first turbine stage, in kg/h, for a net power output of 500 MW.

8.63 Steam enters the first turbine stage of a vapor power plant with reheat and regeneration at 1800 lbf/in.^{2}, 1100°F and expands in five stages to a condenser pressure of 1 lbf/in.^{2} Reheat is at 100 lbf/in.^{2} to 1000°F. The cycle includes three feedwater heaters. Closed heaters operate at 600 and 160 lbf/in.^{2}, with the drains from each trapped into the next lower-pressure feedwater heater. The feedwater leaving each closed heater is at the saturation temperature corresponding to the extraction pressure. An open feedwater heater operates at 20 lbf/in.^{2} The pumps operate isentropically, and each turbine stage has an isentropic efficiency of 88%.

(a) Sketch the layout of the cycle and number the principal state points.
(b) Determine the thermal efficiency of the cycle.
(c) Calculate the mass flow rate into the first turbine stage, in lb/h, for a net power output of 3×10^9 Btu/h.

8.64 A binary vapor power cycle consists of two ideal Rankine cycles with cesium and steam as the working fluids. In the cesium cycle, saturated vapor cesium enters the turbine at 100 lbf/in.^{2}, and saturated liquid exits the condenser at 740°F. Relevant property data for saturated cesium are given in the table below. The heat rejected from the cesium cycle is provided to the steam cycle, producing saturated water vapor at 2250 lbf/in.^{2} The steam is then superheated to 1100°F through further heat addition from another source. Saturated liquid water leaves the condenser at 1 lbf/in.^{2} For a net power output of 3×10^9 Btu/h from the binary cycle, determine

(a) the mass flow rates of cesium and water, in lb/h.
(b) the net rate of heat addition to the overall binary cycle, in Btu/h.
(c) the thermal efficiency.

p (lbf/in.^{2})	T (°F)	v_f (ft^3/lb)	v_g	h_f (Btu/lb)	h_g	s_f (Btu/lb·°R)	s_g
0.306	740	0.009147	310.9	70.0	300.0	0.2105	0.4022
100	2200	0.009911	1.597	127	325.1	0.2451	0.3351

8.65 Figure P8.65 shows a vapor power cycle with reheat and regeneration. The steam generator produces vapor at 1000 lbf/in.^{2}, 800°F. Some of this steam expands through the first turbine stage to 100 lbf/in.^{2} and the remainder is directed to the heat exchanger. The steam exiting the first turbine stage enters the flash chamber. Saturated vapor and saturated liquid at 100 lbf/in.^{2} exit the flash chamber as separate streams. The vapor is reheated in the heat exchanger to 530°F before entering the second turbine stage. The open feedwater heater operates at 100 lbf/in.^{2}, and the condenser pressure is 1 lbf/in.^{2} Each turbine stage

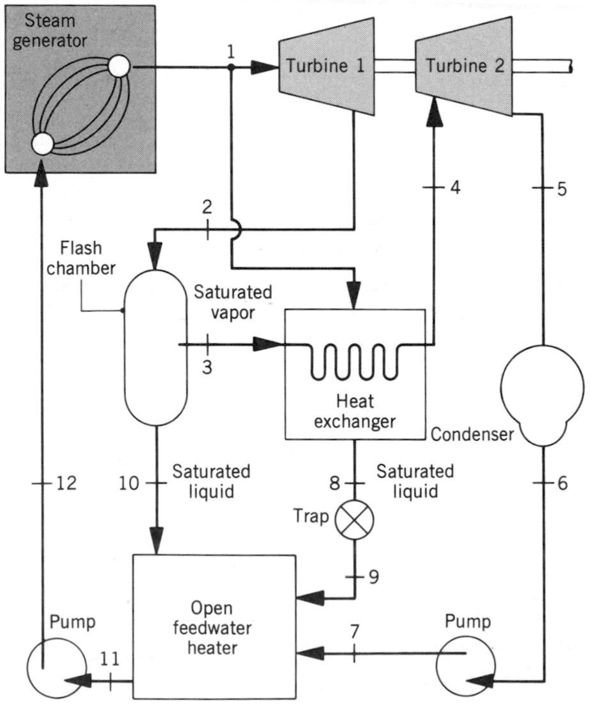

Figure P8.65

has a isentropic efficiency of 88%. For a net power output of 5×10^9 Btu/h, determine

(a) the mass flow rate through the steam generator, in lb/h.
(b) the thermal efficiency of the cycle.
(c) the rate of heat transfer to the cooling water passing through the condenser, in Btu/h.

8.66 Water is the working fluid in a cogeneration cycle that generates electricity and provides heat for campus buildings. Steam at 2 MPa, 320°C, expands through a two-stage turbine. Some steam is extracted between the two stages at 0.15 MPa to provide 10^6 kJ/h for heating, and the remainder expands through the second stage to the condenser pressure of 0.06 bar. The net power developed by the cycle is 600 kW. Condensate returns from the campus buildings at 0.1 MPa, 60°C and passes through a trap into the condenser, where it is reunited with the main feedwater flow. Saturated liquid at 0.06 bar leaves the condenser. Each turbine stage has an isentropic efficiency of 80%, and the pumping process can be considered isentropic. Determine

(a) the mass flow rate of steam into the first turbine stage, in kg/h.
(b) the heat transfer rate to the working fluid passing through the steam generator, in kJ/h.
(c) the rate of heat transfer to the cooling water passing through the condenser, in kJ/h.

8.67 Consider a cogeneration system operating as illustrated in Fig. 8.14. The steam generator provides 10^6 kg/h of steam at 8 MPa, 480°C, of which 4×10^5 kg/h is extracted between the first and second turbine stages at 1 MPa and diverted to a process heating load. Condensate returns from the process heating load at 0.95 MPa, 120°C and is mixed with liquid exiting the lower-pressure pump at 0.95 MPa. The entire flow is then pumped to the steam generator pressure. Saturated liquid at 8 kPa leaves the condenser. The turbine stages and the pumps operate with isentropic efficiencies of 86 and 80%, respectively. Determine

(a) the heating load, in kJ/h.
(b) the power developed by the turbine, in kW.
(c) the rate of heat transfer to the working fluid passing through the steam generator, in kJ/h.

8.68 The steam generator in a cogeneration plant produces 800,000 lb/h of steam at 400 lbf/in.^{2}, 800°F for power production and process heating. A three-stage turbine develops power at a rate of 2×10^8 Btu/h to run electrical generators and also provides process steam at 100 lbf/in.^{2} between the first and second stages. Between the second and third stages, some steam is diverted to an open feedwater heater, which operates at 20 lbf/in.^{2} The condenser pressure is 1 lbf/in.^{2} The process steam is condensed to liquid at 100 lbf/in.^{2}, 150°F and returns to the open feedwater heater after passing through a trap. The turbine stages operate with isentropic turbine efficiencies of 85%. The pumping processes can be assumed isentropic. For the cogeneration plant, determine

(a) the rate of heat transfer to the working fluid passing through the steam generator, in Btu/h.
(b) the mass flow rate of process steam, in lb/h.
(c) the rate of heat transfer from the process steam, in Btu/h.

8.69 Figure P8.69 shows the schematic diagram of a cogeneration cycle. In the steam cycle, superheated vapor enters the turbine with a mass flow rate of 5 kg/s at 40 bar, 440°C and expands isentropically to 1.5 bar. Half of the flow is extracted at 1.5 bar and used for industrial process heating. The rest of the steam passes through a heat exchanger, which serves as the boiler of the Refrigerant 12 cycle and the condenser of the steam cycle. The condensate leaves the heat exchanger as saturated liquid at 1 bar, where it is combined with the return flow from the process, at 60°C and 1 bar, before being pumped isentropically to the steam generator pressure. The Refrigerant 12 cycle is an ideal Rankine cycle with refrigerant entering the turbine at 16 bar, 100°C and saturated liquid leaving the condenser at 9 bar. Determine, in kW,

(a) the rate of heat transfer to the working fluid passing through the steam generator of the steam cycle.

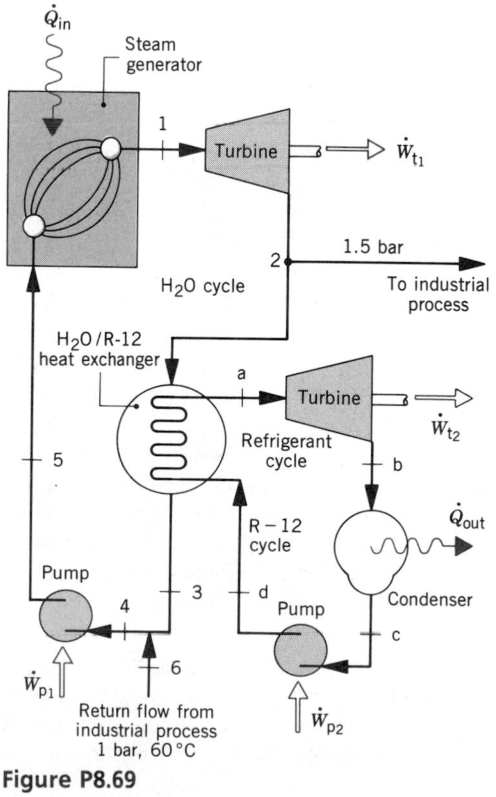

Figure P8.69

(b) the net power output of the binary cycle.
(c) the rate of heat transfer to the industrial process.

8.70 In a *cogeneration* system, a Rankine cycle operates with steam entering the turbine at 800 lbf/in.^{2}, 700°F, and a condenser pressure of 180 lbf/in.^{2} The isentropic turbine efficiency is 80%. Energy rejected by the condensing steam is transferred to a separate process stream of water entering at 250°F, 140 lbf/in.^{2} and exiting as saturated vapor at 140 lbf/in.^{2} Determine the mass flow rate, in lb/h, for the working fluid of the Rankine cycle if the mass flow rate of the process stream is 50,000 lb/h. Devise and evaluate a second law efficiency for the overall cogeneration system. Let T_0 = 70°F, p_0 = 14.7 lbf/in.^{2}

8.71 The steam generator of a vapor power plant can be considered for simplicity to consist of a combustor unit in which fuel and air are burned to produce hot combustion gases, followed by a heat exchanger unit where the cycle working fluid is vaporized and superheated as the hot gases cool. Consider water as the working fluid undergoing the cycle of Problem 8.17. Hot combustion gases, which are assumed to have the properties of air, enter the heat exchanger portion of the steam generator at 1200 K and exit at 500 K with a negligible change in pressure. Determine for the heat exchanger unit

(a) the net rate at which availability is carried in by the gas stream, in kW.
(b) the net rate at which availability is carried out by the water stream, in kW.
(c) the irreversibility rate, in kW.
(d) the second law efficiency given by Eq. 7.45.

Let T_0 = 15°C, p_0 = 0.1 MPa.

8.72 Perform an availability accounting of the cycle of Problem 8.17. Determine, per unit mass of water flowing through each component, the irreversibility rates of **(a)** the turbine, **(b)** the pump, and **(c)** the condenser, each in kJ/kg. Determine the ratio of the net work developed by the cycle to the increase in flow availability of the working fluid passing through the steam generator. Let T_0 = 15°C, p_0 = 0.1 MPa.

8.73 Reconsider the Carnot vapor power cycle of Problem 8.3. Suppose the heat transfer into the working fluid passing through the boiler is from a thermal reservoir at T_H = 927°C and the heat rejection from the working fluid passing through the condenser is to a reservoir at T_C = 15°C. Determine the irreversibility rates of the boiler and condenser, per unit mass of water flow, in kJ/kg. What is the maximum theoretical thermal efficiency of any power cycle operating between thermal reservoirs at these values of T_H and T_C?

8.74 In the steam generator of the cycle of Problem 8.19, the energy input to the working fluid is provided by heat transfer from hot gaseous products of combustion, which cool as a separate stream from 1340 to 360°F with a negligible pressure drop. The gas stream can be modeled as air as an ideal gas. Determine, in Btu/h, the irreversibility rates

(a) of the heat exchanger unit of the steam generator.
(b) of the turbine and pump.
(c) of the condenser.

Also, calculate the net rate at which availability is carried away by the cooling water passing through the condenser, in Btu/h. Let T_0 = 60°F, p_0 = 14.7 lbf/in.^{2}

8.75 For the regenerative vapor power cycle of Problem 8.57, calculate the irreversibility rates of the feedwater heaters per unit mass of steam entering the first turbine stage. Express each as a fraction of the flow availability increase of the working fluid passing through the steam generator. Let T_0 = 60°F, p_0 = 14.7 lbf/in.^{2}

8.76 Determine the rate of availability input, in MW, to the working fluid passing through the steam generator in Problem 8.62. Perform calculations to account for all outputs, losses, and destructions of this availability. Let T_0 = 15°C, p_0 = 1 bar.

8.77 Determine the rate of availability input, in Btu/h, to the working fluid passing through the steam generator in Problem 8.63. Perform calculations to account for all outputs, losses, and destructions of this availability. Let T_0 = 60°F, p_0 = 14.7 lbf/in.^{2}

8.78 Determine the rate of availability transfer, in Btu/h, to the working fluid passing through the steam generator in Problem 8.65. Perform calculations to account for all outputs, losses, and destructions of this availability. Let T_0 = 60°F, p_0 = 14.7 lbf/in.^{2}

8.79 Reconsider the closed feedwater heater, the pump, and the associated piping in Problem 8.46. If the mass flow rate of feedwater to the steam generator at location 6 is 600 kg/s, determine the irreversibility rates for control volumes enclosing

(a) the closed feedwater heater.
(b) the pump.
(c) the location where streams 4 and 5 mix to form stream 6.

Let T_0 = 20°C, p_0 = 1 bar.

8.80 Steam enters the turbine of a vapor power plant at 900°F, 500 lbf/in.^{2} and expands adiabatically, exiting at 1 lbf/in.^{2} with a quality of 97%. Condensate leaves the condenser as saturated liquid at 1 lbf/in.^{2} The isentropic pump efficiency is 80%. The specific flow availability of the fuel entering the combustor unit of the steam generator is estimated to be 22,322 Btu/lb, and no availability is carried in by the combustion air. The flow availability of the stack gases exiting the steam generator is estimated to be 1418 Btu per lb of fuel entering. The mass flow rate of the steam is 12.67 lb per lb of fuel entering the steam generator. Cooling water enters the condenser at T_0 = 77°F, p_0 = 14.7 lbf/in.^{2} and exits at 90°F, 14.7 lbf/in.^{2} Determine, as percentages of the availability entering with the fuel,

(a) the availability exiting with the stack gases.
(b) the availability destroyed in the steam generator.
(c) the net power developed by the cycle.
(d) the availability destroyed in the turbine and the pump.
(e) the availability exiting with the cooling water.
(f) the availability destroyed in the condenser.

DESIGN AND OPEN ENDED PROBLEMS

8.1D A vapor power cycle with a steam generator pressure of 3500 lbf/in.^{2} is under consideration for a 500-MW power plant. The maximum cycle temperature may not exceed 1100°F, and the average cooling water temperature is 60°F. Develop a preliminary design of the cycle, specifying whether reheat and regeneration should be included and, if so, at what pressures. Show a schematic of your proposed cycle layout and provide calculations to justify the configuration that you recommend.

8.2D Figure P8.2D shows a possible arrangement to utilize solar energy for small-scale electric power production using *flat-plate* solar collectors coupled to a *phase-change thermal storage* unit. Thermal storage allows power to be developed by the cycle independently of the solar energy collection. For Refrigerant 12 as the working fluid, specify the key operating temperatures and pressures for a vapor power cycle with a net power output of 0.5 MW. Estimate the solar collector area, the capacity of the thermal storage unit, and the thermal efficiency of your recommended cycle. Justify your design with supporting calculations.

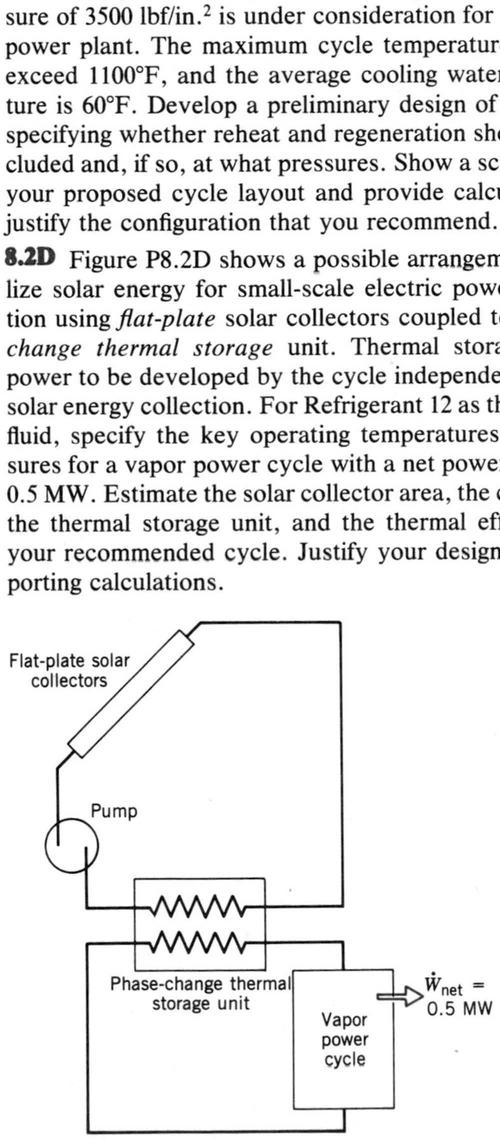

Figure P8.2D

8.3D Vast quantities of water circulate through the condensers of large power plants, exiting at temperatures 10 to 15°C above the ambient temperature. What possible uses could be made of the condenser cooling water? Does this warm water represent a significant resource? What environmental concerns are associated with cooling water? Discuss.

8.4D *Ocean thermal energy conversion* (OTEC) systems have been studied as a way to produce power by utilizing the temperature decrease that occurs naturally with depth in the ocean. What are some of the important design considerations in developing OTEC systems? Consider materials, working fluids, cycle concepts, and economics. What practical considerations arise in generating power far from shore? Write a report of your findings.

8.5D *Nitinol,* any one of several paramagnetic alloys of nickel and titanium, has the characteristic that it deforms when cooled and resumes its previous shape when reheated. Utilize the *shape-memory* feature of nitinol to devise a thermodynamic power cycle that can produce a net power output using a waste gas stream at 50°C as the energy source and rejecting energy by heat transfer to the surroundings at 20°C. Provide schematics to illustrate the operation of your cycle. Estimate the thermal efficiency of such a nitinol engine and comment.

8.6D Many hospitals, industrial facilities, college campuses, and towns employ *cogeneration* plants that provide district heating/cooling and generate electricity. Sketch the main components and layout of such an existing cogeneration system in your locale. Determine the heating and cooling loads served by the system and its electrical generating capacity. Estimate the energy savings obtained through the use of cogeneration compared to meeting these loads individually. Discuss.

8.7D In one *cogeneration* approach, power generation and *flash desalination* of seawater are coupled in a single plant. Specify a possible configuration for such a facility that would generate approximately 80 MW of electricity and estimate the rate of fresh water production that could be achieved. Discuss factors affecting the feasibility of such a scheme, including economics, siting, and accessibility to cooling water.

8.8D A *binary vapor power cycle* is under consideration with a mercury *topping* cycle coupled to a steam cycle. The mercury boiler operates at 120 lbf/in.^{2}, and the steam condenser operates at 1 lbf/in.^{2} The maximum temperature in the steam cycle is 840°F. Estimate the operating pressures for the mercury and water streams in the interconnecting heat exchanger to maximize the thermal efficiency for a nominal power output of 40 MW. Discuss the advantages and disadvantages of using mercury as a working fluid. What other substances might be more suitable as working fluids in the topping cycle?

CHAPTER 9

GAS POWER SYSTEMS

The vapor power systems studied in Chap. 8 use a working fluid that is alternately vaporized and condensed. The present chapter is concerned with power plants utilizing a working fluid that is always a gas. Included in this group of power plants are gas turbines and internal combustion engines of the spark-ignition and compression-ignition types. In the first part of the chapter, internal combustion engines are considered. Gas turbine power plants are discussed in the second part of the chapter. The chapter concludes with a brief study of compressible flow in nozzles and diffusers, which are components in gas turbines for aircraft propulsion and other devices of practical importance.

■ INTERNAL COMBUSTION ENGINES

This part of the chapter deals with *internal* combustion engines. Although most gas turbines are also internal combustion engines, the name is usually applied to *reciprocating* internal combustion engines of the type commonly used in automobiles, trucks, and buses. These engines differ from the power plants considered thus far because the processes occur within reciprocating piston–cylinder arrangements and not in interconnected series of different components.

Two principal types of reciprocating internal combustion engines are the *spark-ignition* engine and the *compression-ignition* engine. In a spark-ignition engine a mixture of fuel and air is ignited by a spark plug. In a compression-ignition engine air is compressed to a high enough pressure and temperature that combustion occurs spontaneously when fuel is injected. Spark-ignition engines have advantages for applications requiring power up to about 225 kW (300 horsepower). Because they are relatively light and lower in cost, spark-ignition engines are particularly suited for use in automobiles. Compression-ignition engines are normally preferred for applications where fuel economy and relatively large amounts of power are required (heavy trucks and buses, locomotives and ships, auxiliary power units). In the middle range, spark-ignition and compression-ignition engines are used.

9.1 PRELIMINARIES

Figure 9.1 is a sketch of a reciprocating internal combustion engine consisting of a piston that moves within a cylinder fitted with two valves. The sketch is labeled with some special terms. The *bore* of the cylinder is its diameter. The *stroke* is the distance the piston moves in one direction. The piston is said to be at *top dead center* when it has moved to a position where the cylinder volume is a minimum. This minimum volume is known as the *clearance* volume. When the piston has moved to the position of maximum cylinder volume, the piston is at *bottom dead center*. The volume swept out by the piston as it moves from the top dead center to the bottom dead center position is called the *displacement volume*. The *compression ratio r* is defined as the volume at bottom dead center divided by the volume at top dead center. The reciprocating motion of the piston is converted to rotary motion by a crank mechanism.

In a *four-stroke* internal combustion engine, the piston executes four distinct strokes within the cylinder for every two revolutions of the crankshaft. Figure 9.2 gives a pressure–displacement diagram such as might be displayed on an oscilloscope. With the intake valve open, the piston makes an *intake stroke* to draw a fresh charge into the cylinder. For spark-ignition engines, the charge is a combustible mixture of fuel and air. Air alone is the charge in compression-ignition engines. Next, with both valves closed, the piston undergoes a *compression stroke,* raising the temperature and pressure of the charge. This requires work input from the piston to the cylinder contents. A combustion process is then initiated (both valves closed), resulting in a high-pressure, high-temperature gas mixture. Combustion is induced near the end of the compression stroke in spark-ignition engines by the spark plug. In compression-ignition engines, combustion is initiated by injecting fuel into the hot compressed air, beginning near the end of the compression stroke and continuing through the first part of the expansion. A *power* stroke follows the compression

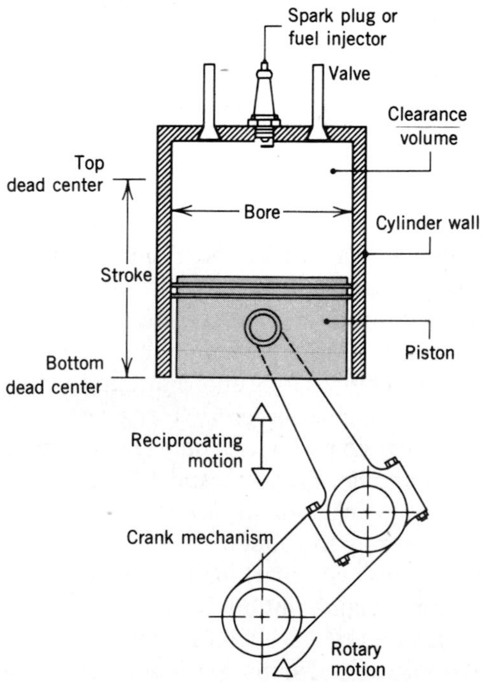

Figure 9.1 Nomenclature for reciprocating piston–cylinder engines.

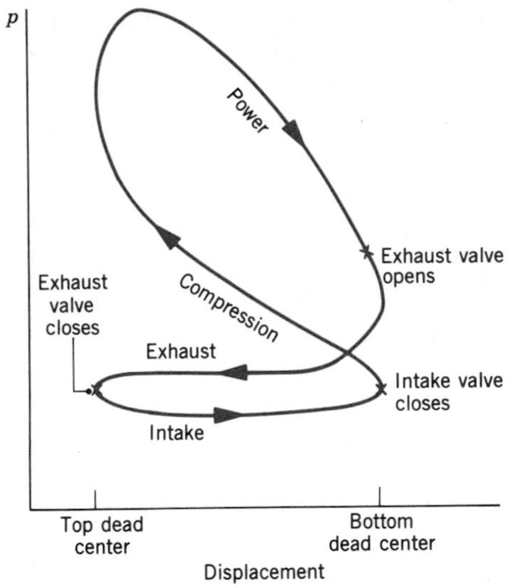

Figure 9.2 Pressure–displacement diagram for a reciprocating internal combustion engine.

stroke, during which the gas mixture expands and work is done on the piston as it returns to bottom dead center. The piston then executes an *exhaust stroke* in which the burned gases are purged from the cylinder through the open exhaust valve. Smaller engines operate on two-stroke cycles. In two-stroke engines, the intake, compression, expansion, and exhaust operations are accomplished in one revolution of the crankshaft. Although internal combustion engines undergo *mechanical* cycles, the cylinder contents do not execute a *thermodynamic* cycle, for matter is introduced with one composition and is later discharged at a different composition.

A parameter used to describe the performance of reciprocating piston engines is the *mean effective pressure,* or mep. The mean effective pressure is the theoretical constant pressure that, if it acted on the piston during the power stroke, would produce the same *net* work as actually developed in one cycle. That is

$$\text{mep} = \frac{\text{net work for one cycle}}{\text{displacement volume}} \tag{9.1}$$

For two engines of equal displacement volume, the one with a higher mean effective pressure would produce the greater net work and, if the engines run at the same speed, greater power.

A detailed study of the performance of a reciprocating internal combustion engine would take into account many features. These would include the combustion process occurring within the cylinder, and the effects of irreversibilities associated with friction and with pressure and temperature gradients. Heat transfer between the gases in the cylinder and the cylinder walls and the work required to charge the cylinder and exhaust the products of combustion also would be considered. Owing to these complexities, accurate modeling of reciprocating internal combustion engines normally involves computer simulation. To conduct *elementary* thermodynamic analyses of internal combustion engines, considerable simplification is required. One procedure is to employ an *air-standard analysis* having the following elements: (1) A fixed amount of air modeled as an ideal gas is the working fluid. (2) The combustion process is replaced by a heat transfer from an external source. (3) There are no exhaust and intake processes as in an actual engine. The cycle is completed by a constant-volume heat transfer process taking place while the piston is at the bottom dead center position. (4) All processes are internally reversible. In addition, in a *cold air-standard analysis*, the specific heats are assumed constant at their ambient temperature values. With an air-standard analysis, it is unnecessary to deal directly with the complexities of the combustion process or the change of composition on combustion.

Although an air-standard analysis simplifies the study of internal combustion engines considerably, values for the mean effective pressure and operating temperatures and pressures calculated on this basis may depart significantly from those of actual engines. Accordingly, air-standard analysis allows internal combustion engines to be examined only qualitatively. Still, insights concerning actual performance can result with such an approach.

In the remainder of this part of the chapter, we consider three cycles that adhere to air-standard cycle idealizations: the Otto, Diesel, and dual cycles. These cycles differ only in the way the heat addition process that replaces combustion in the actual cycle is modeled.

9.2 AIR-STANDARD OTTO CYCLE

The air-standard Otto cycle is an ideal cycle that assumes the heat addition occurs instantaneously while the piston is at top dead center. The Otto cycle is shown on the p–v and T–s diagrams of Fig. 9.3. The cycle consists of four internally reversible processes in series. Process 1–2 is an isentropic compression of the air as the piston moves from bottom dead center to top dead center. Process 2–3 is a constant-volume heat transfer to the air from an external source while the piston is at top dead center. This process is intended to represent the ignition of the fuel–air mixture and the subsequent rapid burning. Process 3–4 is an isentropic expansion (power stroke). The cycle is completed by the constant-volume Process 4–1 in which heat is rejected from the air while the piston is at bottom dead center.

Since the air-standard Otto cycle is composed of internally reversible processes, areas on the T–s and p–v diagrams of Fig. 9.3 can be interpreted as heat and work, respectively. On the T–s diagram, area 2–3–a–b–2 represents the heat added per unit of mass and area 1–4–a–b–1 the heat rejected per unit of mass. On the p–v diagram, area 1–2–a–b–1 represents the work input per unit of mass during the compression process and area 3–4–b–a–3 is the work done per unit of mass in the

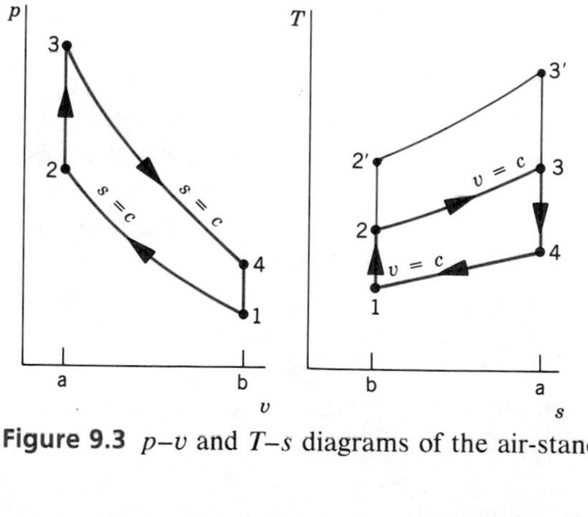

Figure 9.3 p–v and T–s diagrams of the air-standard Otto cycle.

expansion process. The enclosed area of each figure can be interpreted as the net work output or, equivalently, the net heat added.

The air-standard Otto cycle consists of two processes in which there is work but no heat transfer, Processes 1–2 and 3–4, and two processes in which there is heat transfer but no work, Processes 2–3 and 4–1. Expressions for these energy transfers are obtained by reducing the closed system energy balance assuming that changes in kinetic and potential energy can be ignored. The results are

$$\begin{aligned} \frac{W_{12}}{m} &= u_2 - u_1, & \frac{W_{34}}{m} &= u_3 - u_4 \\ \frac{Q_{23}}{m} &= u_3 - u_2, & \frac{Q_{41}}{m} &= u_4 - u_1 \end{aligned} \tag{9.2}$$

Carefully note that in writing Eqs. 9.2, we have departed from our usual sign convention for heat and work. When analyzing cycles, it is frequently convenient to regard all work and heat transfers as positive quantities. Thus, W_{12}/m is a positive number representing the work *input* during compression and Q_{41}/m is a positive number representing the heat *rejected* in Process 4–1. The net work of the cycle is expressed as

$$\frac{W_{\text{cycle}}}{m} = \frac{W_{34}}{m} - \frac{W_{12}}{m} = (u_3 - u_4) - (u_2 - u_1)$$

Alternatively, the net work can be evaluated as the net heat added

$$\frac{W_{\text{cycle}}}{m} = \frac{Q_{23}}{m} - \frac{Q_{41}}{m} = (u_3 - u_2) - (u_4 - u_1)$$

which, on rearrangement, can be placed in the same form as the previous expression for net work.

The thermal efficiency is the ratio of the net work of the cycle to the heat added.

$$\eta = \frac{(u_3 - u_2) - (u_4 - u_1)}{u_3 - u_2} = 1 - \frac{u_4 - u_1}{u_3 - u_2} \tag{9.3}$$

When air table data are used to conduct an analysis involving an air-standard Otto cycle, the specific internal energy values required by Eq. 9.3 can be obtained from Table A-16 or A-16E as appropriate. The following relationships apply for the isentropic processes 1–2 and 3–4:

$$v_{r2} = v_{r1}\left(\frac{V_2}{V_1}\right) = \frac{v_{r1}}{r} \tag{9.4}$$

$$v_{r4} = v_{r3}\left(\frac{V_4}{V_3}\right) = r v_{r3} \tag{9.5}$$

where r denotes the compression ratio. Note that since $V_3 = V_2$ and $V_4 = V_1$, $r = V_1/V_2 = V_4/V_3$. The parameter v_r is tabulated versus temperature for air in Tables A-16.

When the Otto cycle is analyzed on a cold air-standard basis, the following expressions introduced in Sec. 6.7 would be used for the isentropic processes in place of Eqs. 9.4 and 9.5, respectively:

$$\frac{T_2}{T_1} = \left(\frac{V_1}{V_2}\right)^{k-1} = r^{k-1} \quad \text{(constant } k\text{)} \tag{9.6}$$

$$\frac{T_4}{T_3} = \left(\frac{V_3}{V_4}\right)^{k-1} = \frac{1}{r^{k-1}} \quad \text{(constant } k\text{)} \tag{9.7}$$

where k is the specific heat ratio, $k = c_p/c_v$.

Effect of Compression Ratio on Performance. By referring to the T–s diagram of Fig. 9.3, we can conclude that the Otto cycle thermal efficiency increases as the compression ratio increases. An increase in the compression ratio changes the cycle from 1–2–3–4–1 to 1–2′–3′–4–1. Since the average temperature of heat addition is greater in the latter cycle and both cycles have the same heat rejection process, cycle 1–2′–3′–4–1 would have the greater thermal efficiency. The increase in thermal efficiency with compression ratio is also brought out simply by the following development on a cold air-standard basis. For constant c_v, Eq. 9.3 becomes

$$\eta = 1 - \frac{c_v(T_4 - T_1)}{c_v(T_3 - T_2)}$$

On rearrangement

$$\eta = 1 - \frac{T_1}{T_2}\left(\frac{T_4/T_1 - 1}{T_3/T_2 - 1}\right)$$

From Eqs. 9.6 and 9.7 above, $T_4/T_1 = T_3/T_2$, so

$$\eta = 1 - \frac{T_1}{T_2}$$

Finally, introducing Eq. 9.6

$$\eta = 1 - \frac{1}{r^{k-1}} \quad \text{(constant } k\text{)} \tag{9.8}$$

Equation 9.8 indicates that the cold air-standard Otto cycle thermal efficiency is a function of compression ratio only. This relationship is shown in Fig. 9.4 for $k = 1.4$.

The foregoing discussion suggests that it is advantageous for internal combustion engines to have high compression ratios, and this is the case. The possibility of autoignition, or "knock," places an upper limit on the compression ratio of spark-ignition engines, however. After the spark has ignited a portion of the fuel–air mixture, the rise in pressure accompanying combustion compresses the remaining charge. Autoignition can occur if the temperature of the unburned mixture becomes too high before the mixture is consumed by the flame front. Since the temperature attained by the air–fuel mixture during the compression stroke increases as the compression ratio increases, the likelihood of autoignition occurring increases with the compression ratio. Autoignition may result in high-pressure waves in the cylinder (manifested by a knocking or pinging sound) that can lead to loss of power as well as engine damage. Fuels formulated with tetraethyl lead are resistant to autoignition and thus allow relatively high compression ratios. The *unleaded* gasoline in common use today because of environmental concerns over air pollution limits the compression ratios of spark-ignition engines to approximately 9. Higher compression ratios

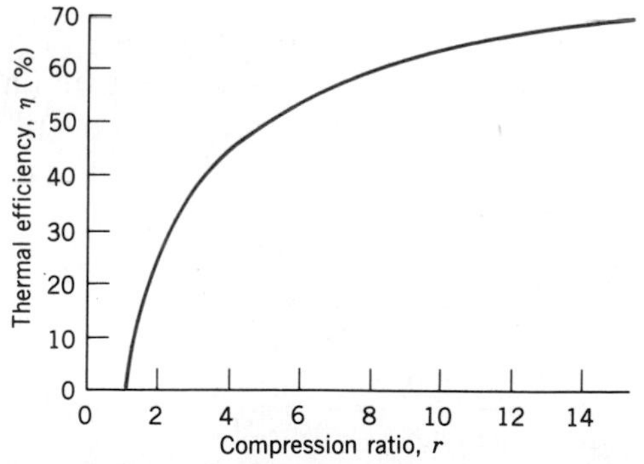

Figure 9.4 Thermal efficiency of the cold air-standard Otto cycle, $k = 1.4$.

can be achieved in compression-ignition engines because air alone is compressed. Compression ratios in the range of 12 to 20 are typical. Compression-ignition engines also can use less refined fuels having higher ignition temperatures than the volatile fuels required by spark-ignition engines.

EXAMPLE 9.1

PROBLEM

The temperature at the beginning of the compression process of an air-standard Otto cycle with a compression ratio of 8 is 540°R, the pressure is 1 atm, and the cylinder volume is 0.02 ft^3. The maximum temperature during the cycle is 3600°R. Determine (a) the temperature and pressure at the end of each process of the cycle, (b) the thermal efficiency, and (c) the mean effective pressure, in atm.

SOLUTION

Known: An air-standard Otto cycle with a given value of compression ratio is executed with specified conditions at the beginning of the compression stroke and a specified maximum temperature during the cycle.

Find: Determine the temperature and pressure at the end of each process, the thermal efficiency, and mean effective pressure, in atm.

Schematic and Given Data:

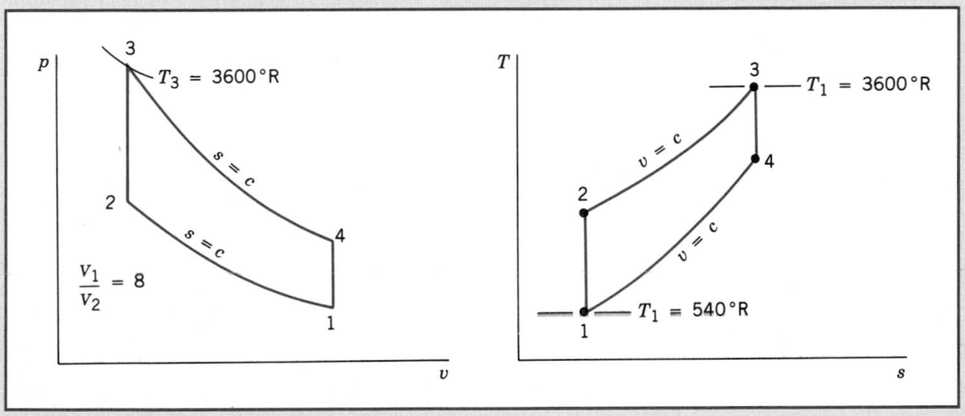

Figure E9.1

Assumptions:

1. The air in the piston–cylinder assembly is the closed system.
2. The compression and expansion processes are adiabatic.
3. All processes are internally reversible.
4. The air is modeled as an ideal gas.
5. Kinetic and potential energy effects are negligible.

Analysis:

(a) The analysis begins by determining the temperature, pressure, and specific internal energy at each principal state of the cycle. At T_1 = 540°R, Table A-16E gives u_1 = 92.04 Btu/lb and v_{r1} = 144.32.

For the isentropic compression Process 1–2

$$v_{r2} = \frac{V_2}{V_1} v_{r1} = \frac{v_{r1}}{r} = \frac{144.32}{8} = 18.04$$

Interpolating with v_{r2} in Table A-16E, we get $T_2 = 1212°R$ and $u_2 = 211.3$ Btu/lb. With the ideal gas equation of state

$$p_2 = p_1 \frac{T_2}{T_1}\frac{V_1}{V_2} = (1 \text{ atm}) \left(\frac{1212}{540}\right) 8 = 17.96 \text{ atm}$$

The pressure at state 2 can be evaluated alternatively by using the isentropic relationship, $p_2 = p_1(p_{r2}/p_{r1})$.
Since Process 2–3 occurs at constant volume, the ideal gas equation of state gives

$$p_3 = p_2 \frac{T_3}{T_2} = (17.96 \text{ atm}) \left(\frac{3600°R}{1212°R}\right) = 53.3 \text{ atm}$$

At $T_3 = 3600°R$, Table A-16E gives $u_3 = 721.44$ Btu/lb and $v_{r3} = 0.6449$.
For the isentropic expansion process 3–4

$$v_{r4} = v_{r3} \frac{V_4}{V_3} = v_{r3} \frac{V_1}{V_2} = 0.6449(8) = 5.16$$

Interpolating in Table A-16E with v_{r4} gives, $T_4 = 1878°R$, $u_4 = 342.2$ Btu/lb. The pressure at state 4 can be found using the isentropic relationship $p_4 = p_3(p_{r4}/p_{r3})$ or the ideal gas equation of state applied at states 1 and 4. With $V_4 = V_1$, the ideal gas equation of state gives

$$p_4 = p_1 \frac{T_4}{T_1} = (1 \text{ atm}) \left(\frac{1878°R}{540°R}\right) = 3.48 \text{ atm}$$

(b) The thermal efficiency is

$$\eta = 1 - \frac{Q_{41}/m}{Q_{23}/m} = 1 - \frac{u_4 - u_1}{u_3 - u_2}$$
$$= 1 - \frac{342.2 - 92.04}{721.44 - 211.3} = 0.51\ (51\%)$$

(c) To evaluate the mean effective pressure requires the net work per cycle. That is

$$W_{cycle} = m[(u_3 - u_4) - (u_2 - u_1)]$$

where m is the mass of the air, evaluated from the ideal gas equation of state as follows:

$$m = \frac{p_1 V_1}{(\bar{R}/M)T_1}$$
$$= \frac{(14.696 \text{ lbf/in.}^2)(144 \text{ in.}^2/\text{ft}^2)(0.02 \text{ ft}^3)}{\left(\frac{1545}{28.97}\frac{\text{ft}\cdot\text{lbf}}{\text{lb}\cdot°R}\right)(540°R)}$$
$$= 1.47 \times 10^{-3} \text{ lb}$$

By inserting values into the expression for W_{cycle}

$$W_{cycle} = (1.47 \times 10^{-3} \text{ lb})[(721.44 - 342.2) - (211.3 - 92.04)] \text{ Btu/lb}$$
$$= 0.382 \text{ Btu}$$

The displacement volume is $V_1 - V_2$, so the mean effective pressure is given by

$$\text{mep} = \frac{W_{cycle}}{V_1 - V_2} = \frac{W_{cycle}}{V_1(1 - V_2/V_1)}$$
$$= \frac{0.382 \text{ Btu}}{(0.02 \text{ ft}^3)(1 - 1/8)} \left(\frac{778 \text{ ft}\cdot\text{lbf}}{\text{Btu}}\right)\left(\frac{\text{ft}^2}{144 \text{ in.}^2}\right)$$
$$= 118 \text{ lbf/in.}^2 = 8.03 \text{ atm}$$

Comment: This solution utilizes Table A-16E for air, which accounts explicitly for the variation of the specific heats with temperature. A solution also can be developed on a cold air-standard basis in which constant specific heats are assumed. This solution is left as an exercise, but the results are presented in the following table for comparison:

Parameter	**Air-Standard Analysis**	**Cold Air-Standard Analysis, $k = 1.4$**
T_2	1212°R	1241°R
T_3	3600°R	3600°R
T_4	1878°R	1567°R
η	0.51 (51%)	0.565 (56.5%)
mep	8.03 atm	7.05 atm

9.3 AIR-STANDARD DIESEL CYCLE

The air-standard Diesel cycle is an ideal cycle that assumes the heat addition occurs during a constant-pressure process that starts with the piston at top dead center. The Diesel cycle is shown on p–v and T–s diagrams in Fig. 9.5. The cycle consists of four internally reversible processes in series. The first process from state 1 to state 2 is the same as in the Otto cycle: an isentropic compression. Heat is not transferred to the working fluid at constant volume as in the Otto cycle, however. In the Diesel cycle, heat is transferred to the working fluid at *constant pressure*. Process 2–3 also makes up the first part of the power stroke. The isentropic expansion from state 3 to state 4 is the remainder of the power stroke. As in the Otto cycle, the cycle is completed by constant-volume Process 4–1 in which heat is rejected from the air while the piston is at bottom dead center. This process replaces the exhaust and intake processes of the actual engine.

Since the air-standard Diesel cycle is composed of internally reversible processes, areas on the T–s and p–v diagrams of Fig. 9.5 can be interpreted as heat and work, respectively. On the T–s diagram, area 2–3–a–b–2 represents the heat added per unit of mass and area 1–4–a–b–1 is the heat rejected per unit of mass. On the p–v diagram, area 1–2–a–b–1 is the work input per unit of mass during the compression process. Area 2–3–4–b–a–2 is the work done per unit of mass as the piston moves

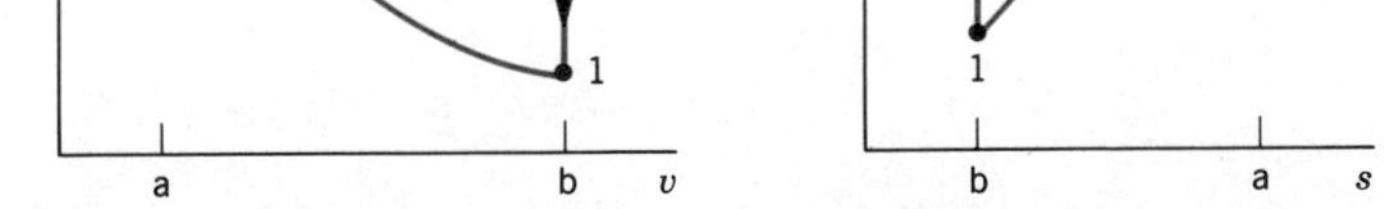

Figure 9.5 p–v and T–s diagrams of the air-standard Diesel cycle.

from top dead center to bottom dead center. The enclosed area of each figure is the net work output, which equals the net heat added.

In the Diesel cycle the heat addition takes place at constant pressure. Accordingly, Process 2–3 involves both work and heat. The work is given by

$$\frac{W_{23}}{m} = \int_2^3 p\,dv = p_2(v_3 - v_2) \tag{9.9}$$

The heat added in Process 2–3 can be found by applying the closed system energy balance

$$m(u_3 - u_2) = Q_{23} - W_{23}$$

Introducing Eq. 9.9 and solving for the heat transfer

$$\frac{Q_{23}}{m} = (u_3 - u_2) + p(v_3 - v_2) = (u_3 + pv_3) - (u_2 + pv_2)$$

$$= h_3 - h_2 \tag{9.10}$$

where the specific enthalpy is introduced to simplify the expression. As in the Otto cycle, the heat rejected in Process 4–1 is given by

$$\frac{Q_{41}}{m} = u_4 - u_1$$

The thermal efficiency is the ratio of the net work of the cycle to the heat added

$$\eta = \frac{W_{\text{cycle}}/m}{Q_{23}/m} = 1 - \frac{Q_{41}/m}{Q_{23}/m} = 1 - \frac{u_4 - u_1}{h_3 - h_2} \tag{9.11}$$

As for the Otto cycle, the thermal efficiency of the Diesel cycle increases with the compression ratio.

To evaluate the thermal efficiency from Eq. 9.11 requires values for u_1, u_4, h_2, and h_3 or equivalently the temperatures at the principal states of the cycle. Let us consider next how these temperatures are evaluated. For a given initial temperature T_1 and compression ratio r, the temperature at state 2 can be found using the following isentropic relationship and v_r data:

$$v_{r2} = \frac{V_2}{V_1} v_{r1} = \frac{1}{r} v_{r1}$$

To find T_3, note that the ideal gas equation of state reduces with $p_3 = p_2$ to give

$$T_3 = \frac{V_3}{V_2} T_2 = r_c T_2$$

where $r_c = V_3/V_2$, called the *cutoff ratio*, has been introduced.

Since $V_4 = V_1$, the volume ratio for the isentropic process 3–4 can be expressed as

$$\frac{V_4}{V_3} = \frac{V_4}{V_2}\frac{V_2}{V_3} = \frac{V_1}{V_2}\frac{V_2}{V_3} = \frac{r}{r_c} \tag{9.12}$$

where the compression ratio r and cutoff ratio r_c have been introduced for conciseness.

Using Eq. 9.12 together with v_{r3} at T_3, the temperature T_4 can be determined by interpolation once v_{r4} is found from the isentropic relationship

$$v_{r4} = \frac{V_4}{V_3} v_{r3} = \frac{r}{r_c} v_{r3}$$

In a *cold air-standard analysis,* the appropriate expression for evaluating T_2 is provided by

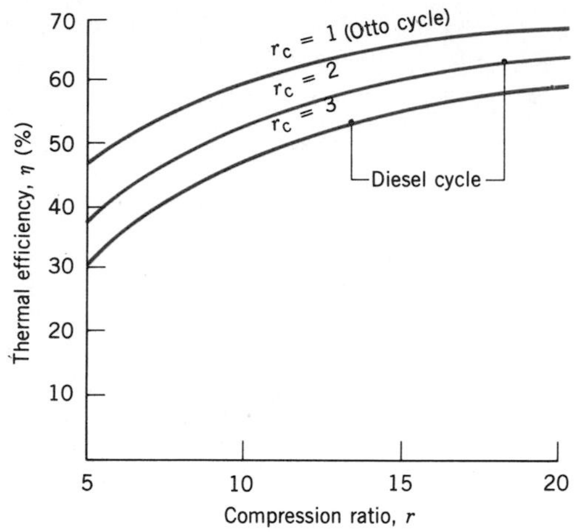

Figure 9.6 Thermal efficiency of the cold air-standard Diesel cycle, $k = 1.4$.

$$\frac{T_2}{T_1} = \left(\frac{V_1}{V_2}\right)^{k-1} = r^{k-1} \qquad \text{(constant } k\text{)}$$

The temperature T_4 is found similarly from

$$\frac{T_4}{T_3} = \left(\frac{V_3}{V_4}\right)^{k-1} = \left(\frac{r_c}{r}\right)^{k-1} \qquad \text{(constant } k\text{)}$$

where Eq. 9.12 has been used to replace the volume ratio.

On a *cold air-standard* basis, the thermal efficiency of the Diesel cycle can be expressed as

$$\eta = 1 - \frac{1}{r^{k-1}}\left[\frac{r_c^k - 1}{k(r_c - 1)}\right] \qquad \text{(constant } k\text{)} \qquad (9.13)$$

where r is the compression ratio and r_c the *cutoff* ratio. The derivation is left as an exercise. This relationship is shown in Fig. 9.6 for $k = 1.4$. Equation 9.13 for the Diesel cycle differs from Eq. 9.8 for the Otto cycle only by the term in brackets, which for $r_c > 1$ is greater than unity. Thus, when the compression ratio is the same, the thermal efficiency of the cold air-standard Diesel cycle would be less than that of the cold air-standard Otto cycle.

EXAMPLE 9.2

PROBLEM

At the beginning of the compression process of an air-standard Diesel cycle operating with a compression ratio of 18, the temperature is 300 K and the pressure is 0.1 MPa. The cutoff ratio for the cycle is 2. Determine (a) the temperature and pressure at the end of each process of the cycle, (b) the thermal efficiency, (c) the mean effective pressure, in MPa.

SOLUTION

Known: An air-standard Diesel cycle is executed with specified conditions at the beginning of the compression stroke. The compression and cutoff ratios are given.

Find: Determine the temperature and pressure at the end of each process, the thermal efficiency, and mean effective pressure.

Schematic and Given Data:

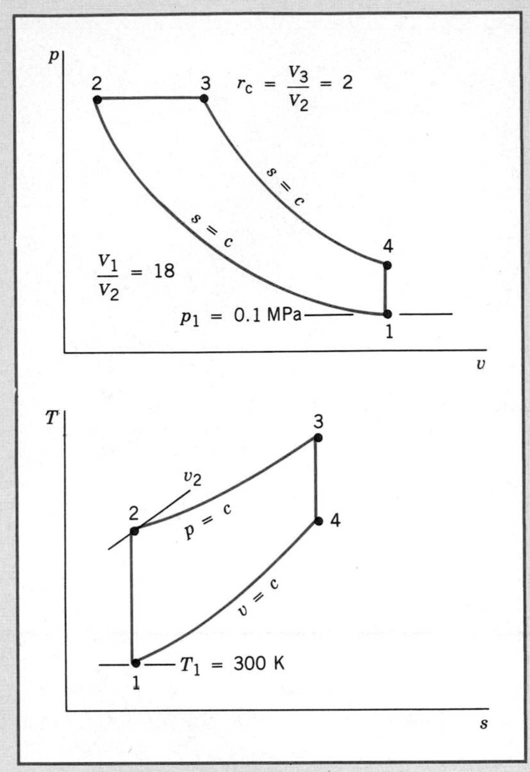

Figure E9.2

Assumptions:

1. The air in the piston–cylinder assembly is the closed system.
2. The compression and expansion processes are adiabatic.
3. All processes are internally reversible.
4. The air is modeled as an ideal gas.
5. Kinetic and potential energy effects are negligible.

Analysis:

(a) The analysis begins by determining properties at each principal state of the cycle. With $T_1 = 300$ K, Table A-16 gives $u_1 = 214.07$ kJ/kg and $v_{r1} = 621.2$. For the isentropic compression process 1–2

$$v_{r2} = \frac{V_2}{V_1} v_{r1} = \frac{v_{r1}}{r} = \frac{621.2}{18} = 34.51$$

Interpolating in Table A-16, we get $T_2 = 898.3$ K and $h_2 = 930.98$ kJ/kg. With the ideal gas equation of state

$$p_2 = p_1 \frac{T_2}{T_1}\frac{V_1}{V_2} = (0.1)\left(\frac{898.3}{300}\right)(18) = 5.39 \text{ MPa}$$

The pressure at state 2 can be evaluated alternatively using the isentropic relationship, $p_2 = p_1(p_{r2}/p_{r1})$.

Since Process 2–3 occurs at constant pressure, the ideal gas equation of state gives

$$T_3 = \frac{V_3}{V_2} T_2$$

Introducing the cutoff ratio, $r_c = V_3/V_2$

$$T_3 = r_c T_2 = 2(898.3) = 1796.6 \text{ K}$$

From Table A-16, $h_3 = 1999.1$ kJ/kg and $v_{r3} = 3.97$.
For the isentropic expansion process 3–4

$$v_{r4} = \frac{V_4}{V_3} v_{r3} = \frac{V_4}{V_2}\frac{V_2}{V_3} v_{r3}$$

Introducing $V_4 = V_1$, the compression ratio r, and the cutoff ratio r_c, we have

$$v_{r4} = \frac{r}{r_c} v_{r3} = \frac{18}{2}(3.97) = 35.73$$

By interpolating in Table A-16 with v_{r4}, $u_4 = 664.3$ kJ/kg and $T_4 = 887.7$ K. The pressure at state 4 can be found using the isentropic relationship $p_4 = p_3(p_{r4}/p_{r3})$ or the ideal gas equation of state applied at states 1 and 4. With $V_4 = V_1$, the ideal gas equation of state gives

$$p_4 = p_1 \frac{T_4}{T_1} = (0.1 \text{ MPa}) \left(\frac{887.7 \text{ K}}{300 \text{ K}}\right) = 0.3 \text{ MPa}$$

(b) The thermal efficiency is found using

$$\eta = 1 - \frac{Q_{41}/m}{Q_{23}/m} = 1 - \frac{u_4 - u_1}{h_3 - h_2}$$

$$= 1 - \frac{664.3 - 214.07}{1999.1 - 930.98} = 0.578 \; (57.8\%)$$

(c) The mean effective pressure written in terms of specific volumes is

$$\text{mep} = \frac{W_{\text{cycle}}/m}{v_1 - v_2} = \frac{W_{\text{cycle}}/m}{v_1(1 - 1/r)}$$

The net work of the cycle equals the net heat added

$$\frac{W_{\text{cycle}}}{m} = \frac{Q_{23}}{m} - \frac{Q_{41}}{m} = (h_3 - h_2) - (u_4 - u_1)$$

$$= (1999.1 - 930.98) - (664.3 - 214.07)$$

$$= 617.9 \text{ kJ/kg}$$

The specific volume at state 1 is

$$v_1 = \frac{(\bar{R}/M)T_1}{p_1} = \frac{\left(\frac{8314}{28.97}\frac{\text{N}\cdot\text{m}}{\text{kg}\cdot\text{K}}\right)(300 \text{ K})}{10^5 \text{ N/m}^2} = 0.861 \text{ m}^3/\text{kg}$$

Inserting values

$$\text{mep} = \frac{617.9 \text{ kJ/kg}}{0.861(1 - 1/18) \text{ m}^3/\text{kg}} \left(\frac{10^3 \text{ N}\cdot\text{m}}{\text{kJ}}\right)\left(\frac{\text{MPa}}{10^6 \text{ N/m}^2}\right)$$

$$= 0.76 \text{ MPa}$$

Comment: This solution uses the air tables, which account explicitly for the variation of the specific heats with temperature. Note that Eq. 9.13 based on the assumption of *constant* specific heats has not been used to determine the thermal efficiency. The cold air-standard solution of this example is left as an exercise.

9.4 AIR-STANDARD DUAL CYCLE

The pressure–volume diagrams of actual internal combustion engines are not described well by the Otto and Diesel cycles. An air-standard cycle that can be made to approximate the pressure variations more closely is the *air-standard dual cycle*. The dual cycle is shown in Fig. 9.7. As in the Otto and Diesel cycles, Process 1–2 is an

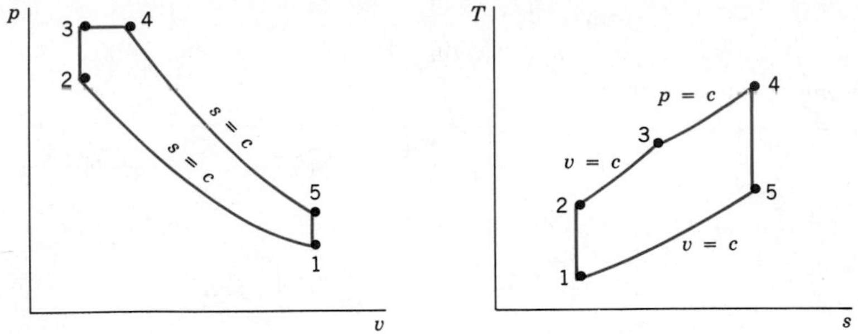

Figure 9.7 p–v and T–s diagrams of the air-standard dual cycle.

isentropic compression. The heat addition occurs in two steps, however: Process 2–3 is a constant-volume heat addition; Process 3–4 is a constant-pressure heat addition. Process 3–4 also makes up the first part of the power stroke. The isentropic expansion from state 4 to state 5 is the remainder of the power stroke. As in the Otto and Diesel cycles, the cycle is completed by a constant-volume heat rejection process, Process 5–1. Areas on the T–s and p–v diagrams can be interpreted as heat and work, respectively, as in the cases of the Otto and Diesel cycles.

Since the dual cycle is composed of the same types of processes as the Otto and Diesel cycles, we can simply write down the appropriate work and heat transfer expressions by reference to the corresponding earlier developments. Thus, during the isentropic compression process 1–2 there is no heat transfer, and the work is

$$\frac{W_{12}}{m} = u_2 - u_1$$

As for the corresponding process of the Otto cycle, in the constant-volume portion of the heat addition process, Process 2–3, there is no work, and the heat transfer is

$$\frac{Q_{23}}{m} = u_3 - u_2$$

In the constant-pressure portion of the heat addition process, Process 3–4, there is both work and heat transfer, as for the corresponding process of the Diesel cycle

$$\frac{W_{34}}{m} = p(v_4 - v_3) \quad \text{and} \quad \frac{Q_{34}}{m} = h_4 - h_3$$

During the isentropic expansion process 4–5 there is no heat transfer, and the work is

$$\frac{W_{45}}{m} = u_4 - u_5$$

Finally, the constant-volume heat rejection process 5–1 that completes the cycle involves heat transfer but no work

$$\frac{Q_{51}}{m} = u_5 - u_1$$

The thermal efficiency is the ratio of the net work of the cycle to the *total* heat added

$$\eta = \frac{W_{\text{cycle}}/m}{(Q_{23}/m + Q_{34}/m)} = 1 - \frac{Q_{51}/m}{(Q_{23}/m + Q_{34}/m)}$$

$$= 1 - \frac{(u_5 - u_1)}{(u_3 - u_2) + (h_4 - h_3)} \tag{9.14}$$

The example to follow provides an illustration of the analysis of an air-standard dual cycle. The analysis exhibits many of the features found in the Otto and Diesel cycle examples considered previously.

EXAMPLE 9.3

PROBLEM

At the beginning of the compression process of an air-standard dual cycle with a compression ratio of 18, the temperature is 300 K and the pressure is 0.1 MPa. The pressure ratio for the constant-volume part of the heating process is 1.5 : 1. The volume ratio for the constant pressure part of the heating process is 1.2 : 1. Determine (a) the thermal efficiency and (b) the mean effective pressure, in MPa.

SOLUTION

Known: An air-standard dual cycle is executed in a piston–cylinder assembly. Conditions are known at the beginning of the compression process, and necessary volume and pressure ratios are specified.

Find: Determine the thermal efficiency and the mep, in MPa.

Schematic and Given Data:

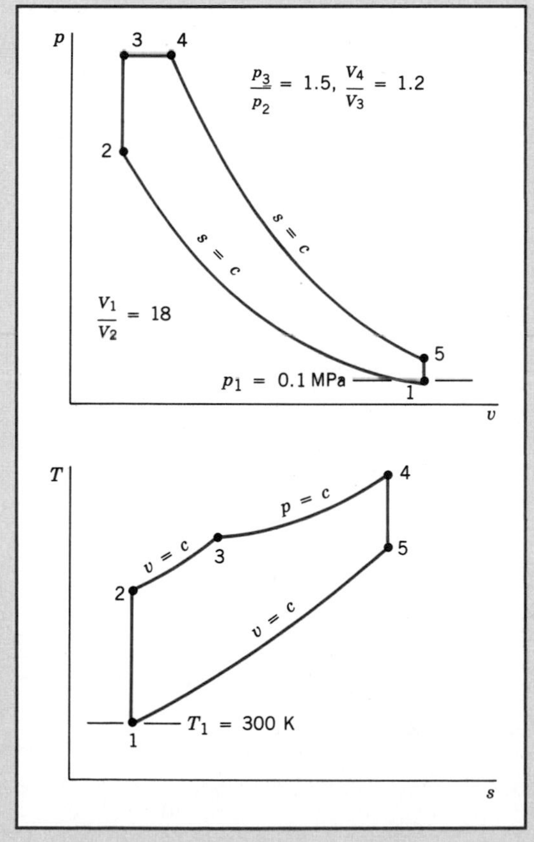

Figure E9.3

Assumptions:

1. The air in the piston–cylinder assembly is the closed system.
2. The compression and expansion processes are adiabatic.
3. All processes are internally reversible.

4. The air is modeled as an ideal gas.
5. Kinetic and potential energy effects are negligible.

Analysis: The analysis begins by determining properties at each principal state of the cycle. States 1 and 2 are the same as in Example 9.2, so $u_1 = 214.07$ kJ/kg, $T_2 = 898.3$ K, $u_2 = 673.2$ kJ/kg. Since Process 2–3 occurs at constant volume, the ideal gas equation of state reduces to give

$$T_3 = \frac{p_3}{p_2} T_2 = (1.5)(898.3) = 1347.5 \text{ K}$$

Interpolating in Table A-16, we get $h_3 = 1452.6$ kJ/kg and $u_3 = 1065.8$ kJ/kg.

Since Process 3–4 occurs at constant pressure, the ideal gas equation of state reduces to give

$$T_4 = \frac{V_4}{V_3} T_3 = (1.2)(1347.5) = 1617 \text{ K}$$

From Table A-16, $h_4 = 1778.3$ kJ/kg and $v_{r4} = 5.609$.

Process 4–5 is an isentropic expansion, so

$$v_{r5} = v_{r4} \frac{V_5}{V_4}$$

The volume ratio V_5/V_4 required by this equation can be expressed as

$$\frac{V_5}{V_4} = \frac{V_5}{V_3}\frac{V_3}{V_4}$$

With $V_5 = V_1$, $V_2 = V_3$, and given volume ratios

$$\frac{V_5}{V_4} = \frac{V_1}{V_2}\frac{V_3}{V_4} = 18\left(\frac{1}{1.2}\right) = 15$$

Inserting this in the above expression for v_{r5}

$$v_{r5} = (5.609)(15) = 84.135$$

Interpolating in Table A-16, we get $u_5 = 475.96$ kJ/kg.

(a) The thermal efficiency is

$$\eta = 1 - \frac{Q_{51}/m}{(Q_{23}/m + Q_{34}/m)} = 1 - \frac{(u_5 - u_1)}{(u_3 - u_2) + (h_4 - h_3)}$$

$$= 1 - \frac{(475.96 - 214.07)}{(1065.8 - 673.2) + (1778.3 - 1452.6)}$$

$$= 0.635 \ (63.5\%)$$

(b) The mean effective pressure is

$$\text{mep} = \frac{W_{\text{cycle}}/m}{v_1 - v_2} = \frac{W_{\text{cycle}}/m}{v_1(1 - 1/r)}$$

The net work of the cycle equals the net heat added, so

$$\text{mep} = \frac{(u_3 - u_2) + (h_4 - h_3) - (u_5 - u_1)}{v_1(1 - 1/r)}$$

The specific volume at state 1 is evaluated in Example 9.2 as $v_1 = 0.861$ m³/kg. Inserting values into the above expression for mep

$$\text{mep} = \frac{[(1065.8 - 673.2) + (1778.3 - 1452.6) - (475.96 - 214.07)]\left(\frac{\text{kJ}}{\text{kg}}\right)\left(\frac{10^3 \text{ N}\cdot\text{m}}{\text{kJ}}\right)\left(\frac{\text{MPa}}{10^6 \text{ N/m}^2}\right)}{0.861(1 - 1/18) \text{ m}^3/\text{kg}} = 0.56 \text{ MPa}$$

GAS TURBINE POWER PLANTS

This part of the chapter deals with gas turbine power plants. Gas turbines tend to be lighter and more compact than the vapor power plants studied in Chap. 8. Gas turbines are used for stationary power generation. In addition, the favorable power-output-to-weight ratio of gas turbines makes them well suited for transportation applications (aircraft propulsion, marine power plants, and so on).

9.5 PRELIMINARIES

Gas turbine power plants may operate on either an open or closed basis. The open mode pictured in Fig. 9.8*a* is more common. This is an engine in which atmospheric air is continuously drawn into the compressor, where it is compressed to a high pressure. The air then enters a combustion chamber, or combustor, where it is mixed with fuel and combustion occurs, resulting in combustion products at an elevated temperature. The combustion products expand through the turbine and are subsequently discharged to the surroundings. Part of the turbine work developed is used to drive the compressor; the remainder is available to generate electricity, to propel a vehicle, or for other purposes. In the system pictured in Fig. 9.8*b*, the working fluid receives an energy input by heat transfer from an external source, for example, a gas-cooled nuclear reactor. The gas exiting the turbine is passed through a heat exchanger, where it is cooled prior to reentering the compressor.

An idealization often used in the study of open gas turbine power plants is that of an *air-standard* analysis. In an air-standard analysis two assumptions are always made: (1) The working fluid is air, which behaves as an ideal gas, and (2) the temperature rise that would be brought about by combustion is accomplished by a heat transfer from an external source. With air-standard analysis, it is unnecessary to deal directly with the complexities of the combustion process or be concerned with the change of composition that occurs during combustion. Additional assumptions introduced later are also used in air-standard analyses of gas turbines. Although an air-standard analysis simplifies the study of gas turbine power plants considerably, numerical values calculated on this basis may provide only qualitative indications of power plant performance. Sufficient information about combustion and the proper-

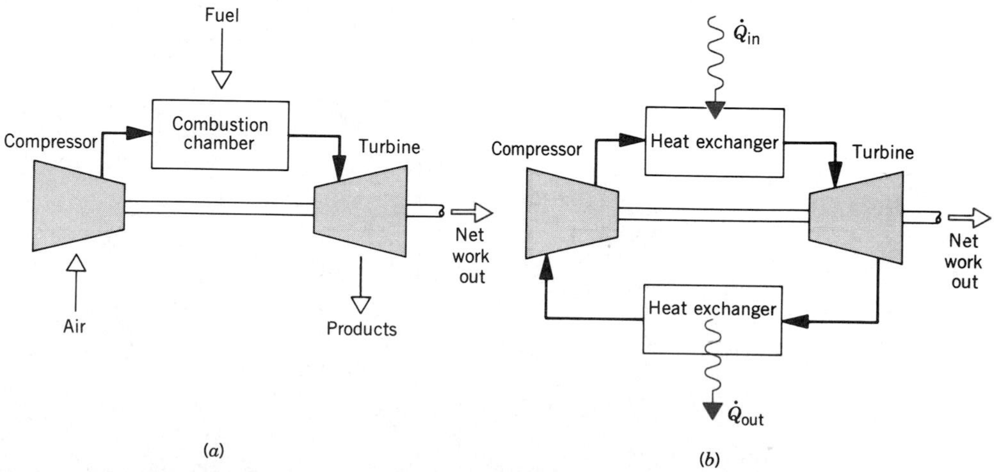

Figure 9.8 Simple gas turbine. (*a*) Open to the atmosphere. (*b*) Closed.

ties of products of combustion is known that the study of gas turbines can be conducted without the foregoing assumptions. Nevertheless, in the interest of simplicity the current presentation proceeds on the basis of an air-standard analysis.

9.6 AIR-STANDARD BRAYTON CYCLE

A schematic diagram of an air-standard gas turbine is shown in Fig. 9.9. The directions of the principal energy transfers are indicated on this figure by arrows. In accordance with the assumptions of an air-standard analysis, the temperature rise that would be achieved in the combustion process is brought about by a heat transfer to the working fluid from an external source and the working fluid is considered to be air as an ideal gas. With the air-standard idealizations, air would be drawn into the compressor at state 1 from the surroundings and later returned to the surroundings at state 4 with a temperature greater than the ambient temperature. After interacting with the surroundings, each unit mass of discharged air would eventually return to the same state as the air entering the compressor, so we may think of the air passing through the components of the gas turbine as undergoing a thermodynamic cycle. A simplified representation of the states visited by the air in such a cycle can be devised by regarding the turbine exhaust air as restored to the compressor inlet state by passing through a heat exchanger where heat rejection to the surroundings occurs. The cycle that results with this further idealization is called an air-standard *Brayton* cycle.

9.6.1 Principal Work and Heat Transfers

The following expressions for the work and heat transfers of energy that occur at steady state are readily derived by reduction of the control volume mass and energy rate balances. These energy transfers are positive in the directions of the arrows in Fig. 9.9. Assuming the turbine operates adiabatically and with negligible effects of kinetic and potential energy, the work developed per unit of mass is

$$\frac{\dot{W}_t}{\dot{m}} = h_3 - h_4 \tag{9.15}$$

where $\dot{m}$ denotes the mass flow rate. With the same assumptions, the compressor work per unit of mass is

$$\frac{\dot{W}_c}{\dot{m}} = h_2 - h_1 \tag{9.16}$$

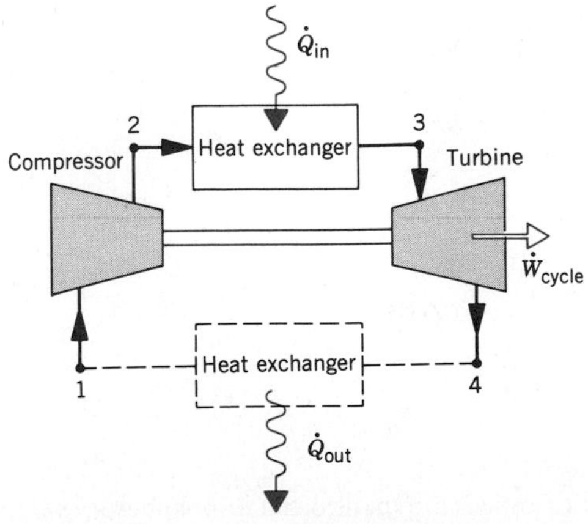

Figure 9.9 Air-standard gas turbine cycle.

The symbol $\dot{W}_c$ denotes work *input* and takes on a positive value. The heat added to the cycle per unit of mass is

$$\frac{\dot{Q}_{in}}{\dot{m}} = h_3 - h_2 \tag{9.17}$$

The heat rejected per unit of mass is

$$\frac{\dot{Q}_{out}}{\dot{m}} = h_4 - h_1 \tag{9.18}$$

where $\dot{Q}_{out}$ is positive in value.

The thermal efficiency of the cycle in Fig. 9.9 is

$$\eta = \frac{\dot{W}_t/\dot{m} - \dot{W}_c/\dot{m}}{\dot{Q}_{in}/\dot{m}} = \frac{(h_3 - h_4) - (h_2 - h_1)}{h_3 - h_2} \tag{9.19}$$

The back work ratio for the cycle is

$$\text{bwr} = \frac{\dot{W}_c/\dot{m}}{\dot{W}_t/\dot{m}} = \frac{h_2 - h_1}{h_3 - h_4} \tag{9.20}$$

For the same pressure rise, a gas turbine compressor would require a much greater work input per unit of mass flow than the pump of a vapor power plant because the average specific volume of the gas flowing through the compressor would be many times greater than that of the liquid passing through the pump (Sec. 6.9). Hence, a relatively large portion of the work developed by the turbine is required to drive the compressor. Typical back work ratios of gas turbines range from 40 to 80%. In comparison, the back work ratios of vapor power plants are normally only 1 or 2%.

If the temperatures at the numbered states of the cycle are known, the specific enthalpies required by the foregoing equations are readily obtained from the ideal gas table for air, Table A-16 or Table A-16E. Alternatively, with the sacrifice of some accuracy, the variation of the specific heats with temperature can be ignored and the specific heats taken as constant. The air-standard analysis is then referred to as a *cold* air-standard analysis. As illustrated by the discussion of internal combustion engines given previously, the chief advantage of the assumption of constant specific heats is that simple expressions for quantities such as thermal efficiency can be derived, and these can be used to deduce qualitative indications of cycle performance without involving tabular data.

Since Eqs. 9.15 through 9.20 have been developed from mass and energy rate balances, they apply equally when irreversibilities are present and in the absence of irreversibilities. Although irreversibilities and losses associated with the various power plant components have a pronounced effect on overall performance, it is instructive to consider an idealized cycle in which they are assumed absent, for such a cycle establishes an upper limit on the performance of the air-standard Brayton cycle. This is considered next.

9.6.2 Ideal Air-Standard Brayton Cycle

Ignoring irreversibilities as the air circulates through the various components of the Brayton cycle, there are no frictional pressure drops, and the air flows at constant pressure through the heat exchangers. If stray heat transfers to the surroundings are also ignored, the processes through the turbine and compressor are isentropic. The ideal cycle shown on the p–v and T–s diagrams in Fig. 9.10 adheres to these idealizations.

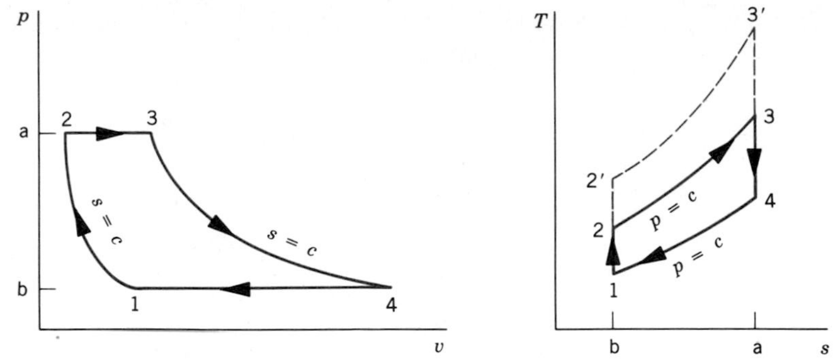

Figure 9.10 Air-standard ideal Brayton cycle.

Areas on the T–s and p–v diagrams of Fig. 9.10 can be interpreted as heat and work, respectively, per unit of mass flowing. On the T–s diagram, area 2–3–a–b–2 represents the heat added per unit of mass and area 1–4–a–b–1 is the heat rejected per unit of mass. On the p–v diagram, area 1–2–a–b–1 represents the compressor work input per unit of mass and area 3–4–b–a–3 is the turbine work output per unit of mass (Sec. 6.9). The enclosed area on each figure can be interpreted as the net work output or, equivalently, the net heat added.

When air table data are used to conduct an analysis involving the ideal Brayton cycle, the following relationships, introduced in Sec. 6.7, apply for the isentropic processes 1–2 and 3–4:

$$p_{r2} = p_{r1}\frac{p_2}{p_1} \tag{9.21}$$

$$p_{r4} = p_{r3}\frac{p_4}{p_3} = p_{r3}\frac{p_1}{p_2} \tag{9.22}$$

Recall that p_r is tabulated versus temperature in Tables A-16. Since the air flows through the heat exchangers of the ideal cycle at constant pressure, it follows that $p_4/p_3 = p_1/p_2$. This relationship has been used in writing Eq. 9.22.

When an ideal Brayton cycle is analyzed on a cold air-standard basis, the specific heats are taken as constant. Equations 9.21 and 9.22 are then replaced, respectively, by the following expressions, introduced in Sec. 6.7:

$$T_2 = T_1\left(\frac{p_2}{p_1}\right)^{(k-1)/k} \tag{9.23}$$

$$T_4 = T_3\left(\frac{p_4}{p_3}\right)^{(k-1)/k} = T_3\left(\frac{p_1}{p_2}\right)^{(k-1)/k} \tag{9.24}$$

where k is the specific heat ratio, $k = c_p/c_v$.

Effect of Pressure Ratio on Performance. Conclusions that are qualitatively correct for actual gas turbines can be drawn from a study of the ideal Brayton cycle. The first of these conclusions is that the thermal efficiency increases with increasing pressure ratio across the compressor. Referring to the T–s diagram of Fig. 9.10, we see that an increase in the pressure ratio changes the cycle from 1–2–3–4–1 to 1–2′–3′–4–1. Since the average temperature of heat addition is greater in the latter cycle and both cycles have the same heat rejection process, cycle 1–2′–3′–4–1 would have the greater thermal efficiency.

The increase in thermal efficiency with the pressure ratio across the compressor is also brought out simply by the following development, in which the specific heat c_p, and thus the specific heat ratio k, is assumed constant. For constant c_p, Eq. 9.19 becomes

$$\eta = \frac{c_p(T_3 - T_4) - c_p(T_2 - T_1)}{c_p(T_3 - T_2)} = 1 - \frac{(T_4 - T_1)}{(T_3 - T_2)}$$

Or, on further rearrangement

$$\eta = 1 - \frac{T_1}{T_2}\left(\frac{T_4/T_1 - 1}{T_3/T_2 - 1}\right)$$

From Eqs. 9.23 and 9.24 above, $T_4/T_1 = T_3/T_2$, so

$$\eta = 1 - \frac{T_1}{T_2}$$

Finally, introducing Eq. 9.23

$$\eta = 1 - \frac{1}{(p_2/p_1)^{(k-1)/k}} \quad (k \text{ constant}) \qquad (9.25)$$

By inspection of Eq. 9.25, it can be seen that the cold air-standard ideal Brayton cycle thermal efficiency is a function of the pressure ratio across the compressor. This relationship is shown in Fig. 9.11 for $k = 1.4$.

There is a limit of about 1700 K (3060°R) imposed by metallurgical considerations on the maximum allowed temperature at the turbine inlet. It is instructive therefore to consider the effect of compressor pressure ratio on thermal efficiency when the turbine inlet temperature is restricted to the maximum allowable temperature. The T–s diagrams of two ideal Brayton cycles having the same turbine inlet temperature but different compressor pressure ratios are shown in Fig. 9.12. Cycle A has a greater pressure ratio than cycle B and thus the greater thermal efficiency. However, cycle B has a larger enclosed area and thus the greater net work developed per unit of mass flow. Accordingly, for cycle A to develop the same net *power* output as cycle B, a larger mass flow rate would be required, and this would dictate a larger system. These considerations are important for gas turbines intended for use in vehicles where engine weight must be kept small. For such applications, it is desirable to operate near the compressor pressure ratio that yields the most work per unit of mass and not the pressure ratio for the greatest thermal efficiency. Example 9.5 provides an illustration of the determination of the pressure ratio for maximum net work per unit of mass flow for the cold air-standard Brayton cycle.

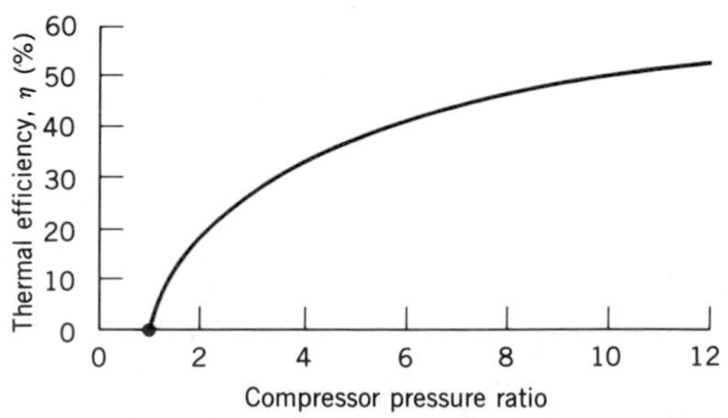

Figure 9.11 Thermal efficiency as a function of compressor pressure ratio for the cold air-standard ideal Brayton cycle, $k = 1.4$.

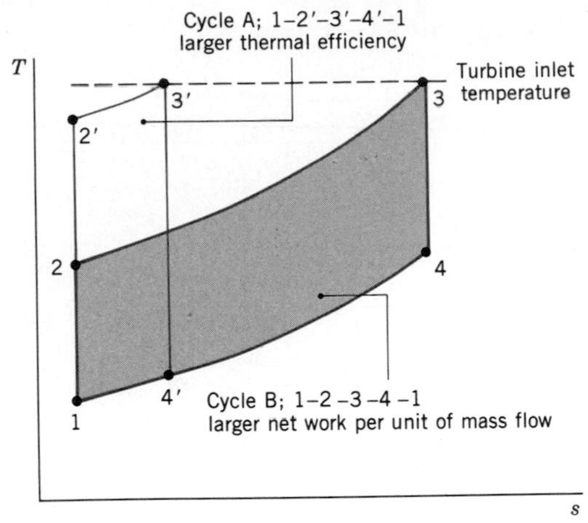

Figure 9.12 Ideal Brayton cycles with different pressure ratios and the same turbine inlet temperatures.

EXAMPLE 9.4

PROBLEM

Air enters the compressor of an ideal air-standard Brayton cycle at 100 kPa, 300 K, with a volumetric flow rate of 5 m^3/s. The compressor pressure ratio is 10. The turbine inlet temperature is 1400 K. Determine (a) the thermal efficiency of the cycle, (b) the back work ratio, (c) the *net* power developed, in kW.

SOLUTION

Known: An ideal air-standard Brayton cycle operates with given compressor inlet conditions, given turbine inlet temperature, and a known compressor pressure ratio.

Find: Determine the thermal efficiency, the back work ratio, and the *net* power developed, in kW.

Schematic and Given Data:

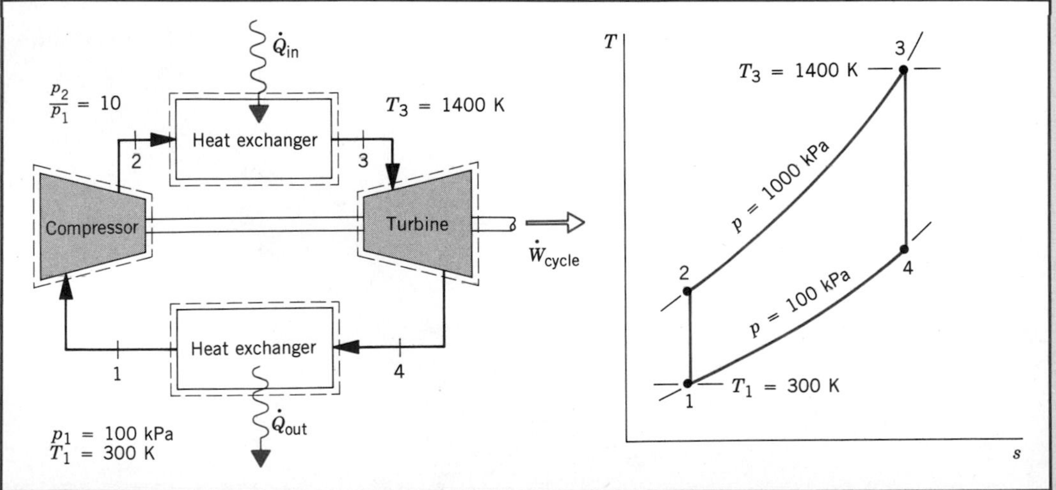

Figure E9.4

Assumptions:

1. Each component is analyzed as a control volume at steady state. The control volumes are shown on the accompanying sketch by dashed lines.
2. The turbine and compressor processes are isentropic.
3. There are no pressure drops for flow through the heat exchangers.
4. Kinetic and potential energy effects are negligible.
5. The working fluid is air modeled as an ideal gas.

Analysis: The analysis begins by determining the specific enthalpy at each numbered state of the cycle. At state 1, the temperature is 300 K. From Table A-16, $h_1 = 300.19$ kJ/kg and $p_{r1} = 1.386$.

Since the compressor process is isentropic, the following relationship can be used to determine h_2:

$$p_{r2} = \frac{p_2}{p_1} p_{r1} = (10)(1.386) = 13.86$$

Then, interpolating in Table A-16, we obtain $h_2 = 579.9$ kJ/kg.

The temperature at state 3 is given as $T_3 = 1400$ K. With this temperature, the specific enthalpy at state 3 from Table A-16 is $h_3 = 1515.4$ kJ/kg. Also, $p_{r3} = 450.5$.

The specific enthalpy at state 4 is found by using the isentropic relation

$$p_{r4} = p_{r3}\frac{p_4}{p_3} = (450.5)(1/10) = 45.05$$

Interpolating in Table A-16, we get $h_4 = 808.5$ kJ/kg.

(a) The thermal efficiency is

$$\eta = \frac{(\dot{W}_t/\dot{m}) - (\dot{W}_c/\dot{m})}{\dot{Q}_{in}/\dot{m}}$$

$$= \frac{(h_3 - h_4) - (h_2 - h_1)}{h_3 - h_2} = \frac{(1515.4 - 808.5) - (579.9 - 300.19)}{1515.4 - 579.9}$$

$$= \frac{706.9 - 279.7}{935.5} = 0.457\ (45.7\%)$$

(b) The back work ratio is

$$\text{bwr} = \frac{\dot{W}_c/\dot{m}}{\dot{W}_t/\dot{m}} = \frac{h_2 - h_1}{h_3 - h_4} = \frac{279.7}{706.9} = 0.396\ (39.6\%)$$

(c) The net power developed is

$$\dot{W}_{cycle} = \dot{m}[(h_3 - h_4) - (h_2 - h_1)]$$

To evaluate the net power requires the mass flow rate $\dot{m}$, which can be determined from the volumetric flow rate and specific volume at the compressor inlet as follows:

$$\dot{m} = \frac{(AV)_1}{v_1}$$

Since $v_1 = (\bar{R}/M)T_1/p_1$, this becomes

$$\dot{m} = \frac{(AV)_1 p_1}{(\bar{R}/M)T_1} = \frac{(5\ \text{m}^3/\text{s})(100 \times 10^3\ \text{N/m}^2)}{\left(\dfrac{8314}{28.97}\dfrac{\text{N}\cdot\text{m}}{\text{kg}\cdot\text{K}}\right)(300\ \text{K})}$$

$$= 5.807\ \text{kg/s}$$

Finally,

$$\dot{W}_{cycle} = (5.807\ \text{kg/s})(706.9 - 279.7)\left(\frac{\text{kJ}}{\text{kg}}\right)\left(\frac{1\ \text{kW}}{1\ \text{kJ/s}}\right) = 2481\ \text{kW}$$

Comments:

1. Compare the value of the back work ratio in the present example with the back work ratio of the simple vapor power cycle of Example 8.1.
2. The use of the ideal gas table for air is featured in this solution. A solution also can be developed on a cold air-standard basis in which constant specific heats are assumed. The details are left as an exercise, but the results are presented in the following table for comparison:

Parameter	Air-Standard Analysis	Cold Air-Standard Analysis, $k = 1.4$
T_2	574.1 K	579.2 K
T_4	787.7 K	725.1 K
η	0.457	0.482
bwr	0.396	0.414
$\dot{W}_{cycle}$	2481 kW	2308 kW

EXAMPLE 9.5

PROBLEM

Determine the pressure ratio across the compressor of an ideal Brayton cycle for the maximum net work output per unit of mass flow if the state at the compressor inlet and the temperature at the turbine inlet are fixed. Use a cold air-standard analysis and ignore kinetic and potential energy effects.

SOLUTION

Known: An ideal Brayton cycle operates with a specified state at the inlet to the compressor and a specified turbine inlet temperature.

Find: Determine the pressure ratio across the compressor for the maximum net work output per unit of mass flow.

Schematic and Given Data:

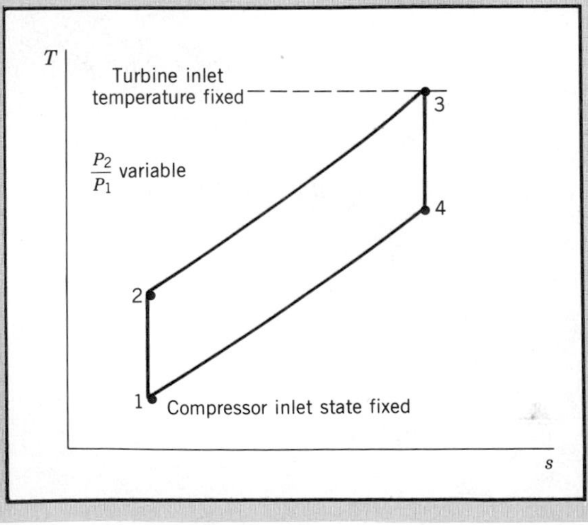

Figure E9.5

Assumptions:

1. Each component is analyzed as a control volume at steady state.
2. The turbine and compressor processes are isentropic.
3. There are no pressure drops for flow through the heat exchangers.
4. Kinetic and potential energy effects are negligible.
5. The working fluid is air modeled as an ideal gas.
6. The specific heat c_p and the specific heat ratio k are constant.

Analysis: The net work of the cycle per unit of mass flow is

$$\frac{\dot{W}_{\text{cycle}}}{\dot{m}} = (h_3 - h_4) - (h_2 - h_1)$$

Since c_p is constant (assumption 6)

$$\frac{\dot{W}_{\text{cycle}}}{\dot{m}} = c_p[(T_3 - T_4) - (T_2 - T_1)]$$

Or on rearrangement

$$\frac{\dot{W}_{\text{cycle}}}{\dot{m}} = c_p T_1 \left(\frac{T_3}{T_1} - \frac{T_4}{T_3}\frac{T_3}{T_1} - \frac{T_2}{T_1} + 1\right)$$

Replacing the temperature ratios T_2/T_1 and T_4/T_3 by using Eqs. 9.23 and 9.24, respectively, gives

$$\frac{\dot{W}_{\text{cycle}}}{\dot{m}} = c_p T_1 \left[\frac{T_3}{T_1} - \frac{T_3}{T_1}\left(\frac{p_1}{p_2}\right)^{(k-1)/k} - \left(\frac{p_2}{p_1}\right)^{(k-1)/k} + 1\right]$$

From this expression it can be concluded that for specified values of T_1, T_3, and c_p, the value of the net work output per unit of mass flow varies with the pressure ratio p_2/p_1 only.

To determine the pressure ratio that maximizes the net work output per unit of mass flow, first form the derivative

$$\frac{\partial(\dot{W}_{\text{cycle}}/\dot{m})}{\partial(p_2/p_1)} = \frac{\partial}{\partial(p_2/p_1)}\left\{c_p T_1\left[\frac{T_3}{T_1} - \frac{T_3}{T_1}\left(\frac{p_1}{p_2}\right)^{(k-1)/k} - \left(\frac{p_2}{p_1}\right)^{(k-1)/k} + 1\right]\right\}$$

$$= c_p T_1 \left(\frac{k-1}{k}\right)\left[\left(\frac{T_3}{T_1}\right)\left(\frac{p_1}{p_2}\right)^{-1/k}\left(\frac{p_1}{p_2}\right)^2 - \left(\frac{p_2}{p_1}\right)^{-1/k}\right]$$

$$= c_p T_1 \left(\frac{k-1}{k}\right)\left[\left(\frac{T_3}{T_1}\right)\left(\frac{p_1}{p_2}\right)^{(2k-1)/k} - \left(\frac{p_2}{p_1}\right)^{-1/k}\right]$$

When the partial derivative is set to zero, the following relationship is obtained:

$$\frac{p_2}{p_1} = \left(\frac{T_3}{T_1}\right)^{k/[2(k-1)]}$$

By checking the sign of the second derivative, we can verify that the net work per unit of mass flow is a maximum when this relationship is satisfied.

Comment: For gas turbines intended for transportation, it is desirable to keep engine size small. Thus, the gas turbine should operate near the compressor pressure ratio that yields the most work per unit of mass flow. The present example provides an elementary illustration of how the compressor pressure ratio for maximum net work per unit of mass flow is determined under the constraint of a fixed turbine inlet temperature.

9.6.3 Principal Irreversibilities and Losses

The principal state points of a simple closed-cycle gas turbine power plant might be shown more realistically as in Fig. 9.13*a*. Because of irreversibilities within the compressor and turbine, the working fluid would experience increases in specific entropy across these components. Owing to irreversibilities, there also would be pressure drops as the working fluid passes through the heat exchangers (or the combustion chamber of the open-cycle gas turbine). However, because frictional pressure drops are less significant sources of irreversibility, we ignore them in subsequent discussions and for simplicity show the flow through the heat exchangers as occurring at constant pressure. This is illustrated by Fig. 9.13*b*. Stray heat transfers from the power plant components to the surroundings represent losses, but these effects are usually of secondary importance and are also ignored in subsequent discussions.

As the effect of irreversibilities in the turbine and compressor becomes more pronounced, the work developed by the turbine decreases and the work input to the compressor increases, resulting in a marked decrease in the net work of the power plant. Accordingly, if an appreciable amount of net work is to be developed by the plant, relatively high turbine and compressor efficiencies are required. After decades of developmental effort, efficiencies of 80 to 90% can now be achieved for the turbines and compressors in gas turbine power plants. Designating the states as in Fig. 9.13*b*, the isentropic turbine and compressor efficiencies are given by

$$\eta_t = \frac{(\dot{W}_t/\dot{m})}{(\dot{W}_t/\dot{m})_s} = \frac{h_3 - h_4}{h_3 - h_{4s}}$$

$$\eta_c = \frac{(\dot{W}_c/\dot{m})_s}{(\dot{W}_c/\dot{m})} = \frac{h_{2s} - h_1}{h_2 - h_1}$$

The example to follow brings out the effect of turbine and compressor irreversibilities on plant performance.

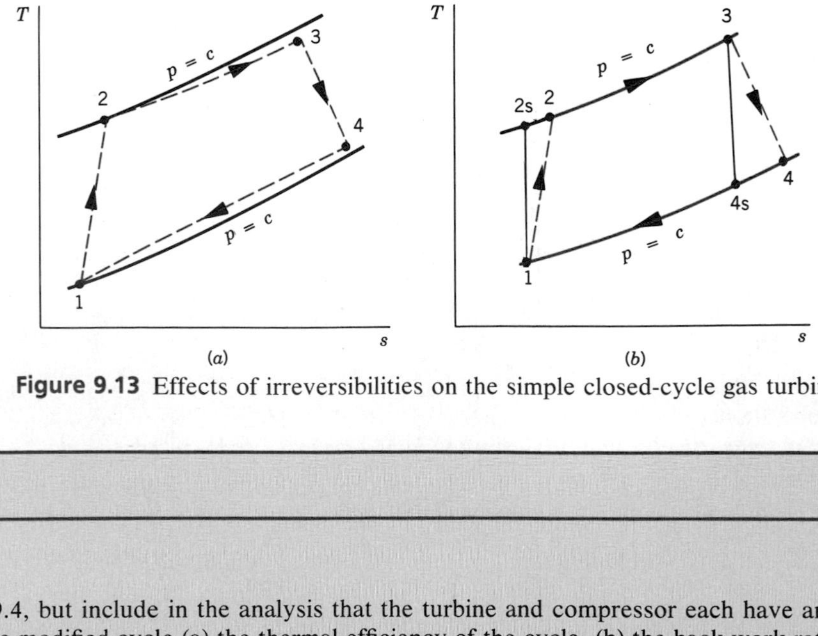

Figure 9.13 Effects of irreversibilities on the simple closed-cycle gas turbine.

EXAMPLE 9.6

PROBLEM

Reconsider Example 9.4, but include in the analysis that the turbine and compressor each have an efficiency of 80%. Determine for the modified cycle (a) the thermal efficiency of the cycle, (b) the back work ratio, (c) the *net* power developed, in kW.

SOLUTION

Known: An air-standard Brayton cycle operates with given compressor inlet conditions, given turbine inlet temperature, and known compressor pressure ratio. The compressor and turbine each have an efficiency of 80%.

Find: Determine the thermal efficiency, the back work ratio, and the net power developed, in kW.

Schematic and Given Data:

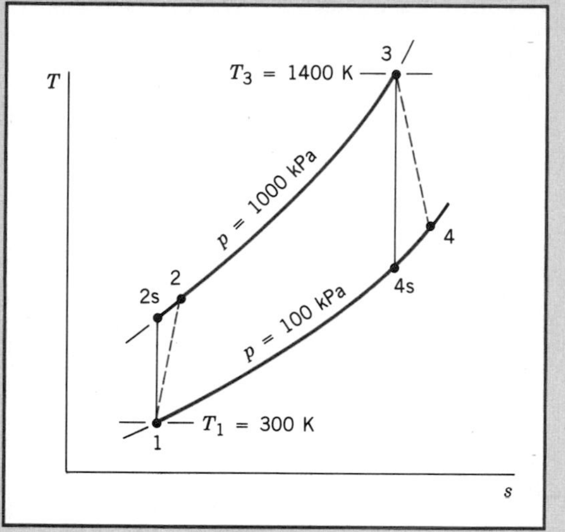

Figure E9.6

Assumptions:

1. Each component is analyzed as a control volume at steady state.
2. The compressor and turbine are adiabatic.
3. There are no pressure drops for flow through the heat exchangers.
4. Kinetic and potential energy effects are negligible.
5. The working fluid is air modeled as an ideal gas.

Analysis:

(a) The thermal efficiency is given by

$$\eta = \frac{(\dot{W}_t/\dot{m}) - (\dot{W}_c/\dot{m})}{\dot{Q}_{in}/\dot{m}}$$

The work terms in the numerator of this expression are evaluated using the given values of the compressor and turbine efficiencies as follows:

The turbine work per unit of mass is

$$\frac{\dot{W}_t}{\dot{m}} = \eta_t \left(\frac{\dot{W}_t}{\dot{m}}\right)_s$$

where η_t is the turbine efficiency. The value of $(\dot{W}_t/\dot{m})_s$ is determined in the solution to Example 9.4 as 706.9 kJ/kg. Thus

$$\frac{\dot{W}_t}{\dot{m}} = 0.8(706.9) = 565.5 \text{ kJ/kg}$$

For the compressor, the work per unit of mass is

$$\frac{\dot{W}_c}{\dot{m}} = \frac{(\dot{W}_c/\dot{m})_s}{\eta_c}$$

where η_c is the compressor efficiency. The value of $(\dot{W}_c/\dot{m})_s$ is determined in the solution to Example 9.4 as 279.7 kJ/kg, so

$$\frac{\dot{W}_c}{\dot{m}} = \frac{279.7}{0.8} = 349.6 \text{ kJ/kg}$$

The specific enthalpy at the compressor exit, h_2, is required to evaluate the denominator of the thermal efficiency expression. This enthalpy can be determined by solving

$$\frac{\dot{W}_c}{\dot{m}} = h_2 - h_1$$

to obtain

$$h_2 = h_1 + \dot{W}_c/\dot{m}$$

Inserting known values

$$h_2 = 300.19 + 349.6 = 649.8 \text{ kJ/kg}$$

The heat transfer to the working fluid per unit of mass flow is then

$$\frac{\dot{Q}_{in}}{\dot{m}} = h_3 - h_2 = 1515.4 - 649.8 = 865.6 \text{ kJ/kg}$$

where h_3 is from the solution to Example 9.4.

Finally, the thermal efficiency is

$$\eta = \frac{565.5 - 349.6}{865.6} = 0.249 \ (24.9\%)$$

(b) The back work ratio is

$$\text{bwr} = \frac{\dot{W}_c/\dot{m}}{\dot{W}_t/\dot{m}} = \frac{349.6}{565.5} = 0.618 \ (61.8\%)$$

(c) The mass flow rate is the same as in Example 9.4. The net power developed by the cycle is then

$$\dot{W}_{\text{cycle}} = \left(5.807 \frac{\text{kg}}{\text{s}}\right)(565.5 - 349.6) \frac{\text{kJ}}{\text{kg}} \left(\frac{1 \text{ kW}}{1 \text{ kJ/s}}\right) = 1254 \text{ kW}$$

Comments:

1. Irreversibilities within the turbine and compressor have a significant impact on the performance of gas turbines. This is brought out by comparing the results of the present example with those of Example 9.4. Irreversibilities result in an increase in the work of compression and a reduction in work output of the turbine. The back work ratio is greatly increased and the thermal efficiency significantly decreased.
2. The solution to this example on a cold air-standard basis is left as an exercise.

9.7 REGENERATIVE GAS TURBINES

The turbine exhaust temperature of a gas turbine is normally well above the ambient temperature. Accordingly, the hot turbine exhaust gas has a potential for use that would be irrevocably lost were the gas discarded directly to the surroundings. One way of utilizing this potential is by means of a heat exchanger called a *regenerator,* which allows the air exiting the compressor to be *preheated* before entering the combustor, thereby reducing the amount of fuel that must be burned in the combustor. The combined cycle arrangement considered in Sec. 9.10 is another way to utilize the hot turbine exhaust gas.

An air-standard Brayton cycle modified to include a regenerator is illustrated in Fig. 9.14. The regenerator shown is a counterflow heat exchanger through which the

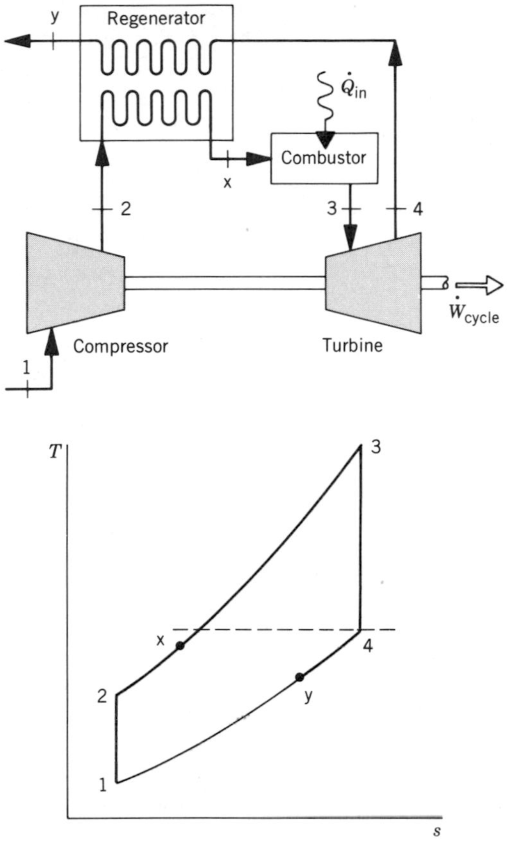

Figure 9.14 Regenerative air-standard gas turbine cycle.

hot turbine exhaust gas and the cooler air leaving the compressor pass in opposite directions. Ideally, no frictional pressure drop occurs in either stream. The turbine exhaust gas is cooled from state 4 to state y, while the air exiting the compressor is heated from state 2 to state x. Hence, a heat transfer from a source external to the cycle is required only to increase the air temperature from state x to state 3, rather than from state 2 to state 3, as would be the case without regeneration. The heat added per unit of mass is then given by

$$\frac{\dot{Q}_{in}}{\dot{m}} = h_3 - h_x \qquad (9.26)$$

The net work developed per unit of mass flow is not altered by the addition of a regenerator. Thus, since the heat added is reduced, the thermal efficiency increases.

From Eq. 9.26 it can be concluded that the external heat transfer required by a gas turbine power plant decreases as the specific enthalpy h_x increases and thus as the temperature T_x increases. Evidently, there is an incentive in terms of fuel saved for selecting a regenerator that provides the greatest practical value for this temperature. To consider the *maximum* theoretical value for T_x, refer to Fig. 9.15*a*, which shows typical temperature variations of the hot and cold streams of a counterflow heat exchanger. Since a finite temperature difference between the streams is required for heat transfer to occur, the temperature of the cold stream at each location, denoted by the coordinate z, is less than that of the hot stream. In particular, the temperature of the colder stream as it exits the heat exchanger is less than the temperature of the incoming hot stream. If the heat transfer area were increased,

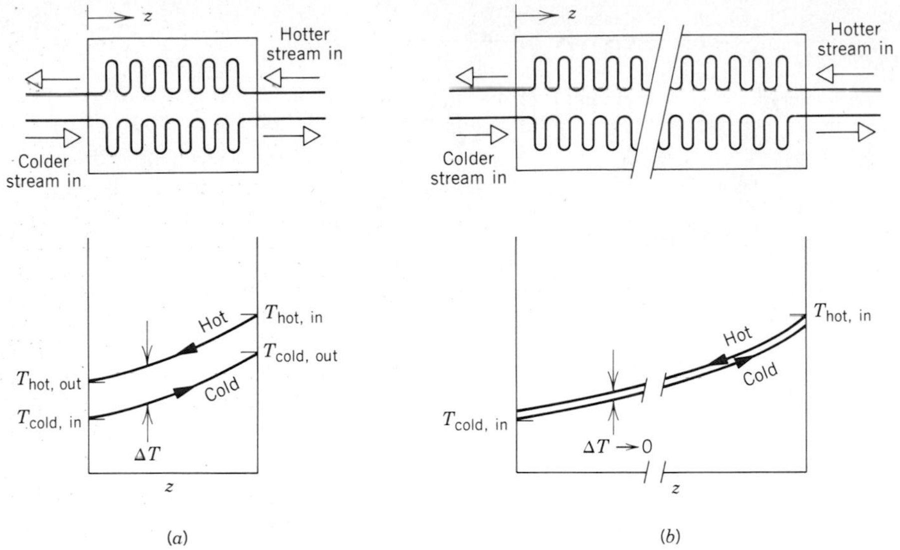

Figure 9.15 Temperature distributions in counterflow heat exchangers. (*a*) Actual. (*b*) Reversible.

providing more opportunity for heat transfer between the two streams, there would be a smaller temperature difference at each location. In the limiting case of infinite heat transfer area, the temperature difference would approach zero at all locations, as illustrated in Fig. 9.15*b*, and the heat transfer would approach reversibility. In this limit, the exit temperature of the colder stream would approach the temperature of the incoming hot stream. Thus, the highest possible temperature that could be achieved by the colder stream is the temperature of the incoming hot gas.

Referring again to the regenerator of Fig. 9.14, we can conclude from the discussion of Fig. 9.15 that the maximum theoretical value for the temperature T_x is the turbine exhaust temperature T_4, obtained if the regenerator were operating reversibly. The *regenerator effectiveness*, η_{reg}, is a parameter that gauges the departure of an actual regenerator from such an ideal regenerator. The regenerator effectiveness is defined as the ratio of the actual enthalpy increase of the air flowing through the compressor side of the regenerator to the maximum theoretical enthalpy increase. That is,

$$\eta_{reg} = \frac{h_x - h_2}{h_4 - h_2} \tag{9.27}$$

As heat transfer approaches reversibility, h_x approaches h_4 and η_{reg} tends to unity (100%).

In practice, regenerator effectiveness values typically range from 60 to 80%, and thus the temperature T_x of the air exiting on the compressor side of the regenerator is normally below the turbine exhaust temperature. To increase the effectiveness above this range can result in equipment costs that cancel any advantage due to fuel savings. Moreover, the greater heat transfer area that would be required for a larger effectiveness can result in a significant frictional pressure drop for flow through the regenerator, thereby affecting overall performance. The decision to add a regenerator is influenced by considerations such as these, and the final decision is primarily an economic one.

EXAMPLE 9.7

PROBLEM

If a regenerator with an effectiveness of 80% is incorporated in the cycle of Example 9.4, determine the thermal efficiency.

SOLUTION

Known: A regenerative gas turbine operates with air as the working fluid. The compressor inlet state, turbine inlet temperature, and compressor pressure ratio are known. Also known is the regenerator effectiveness.

Find: Determine the thermal efficiency.

Schematic and Given Data:

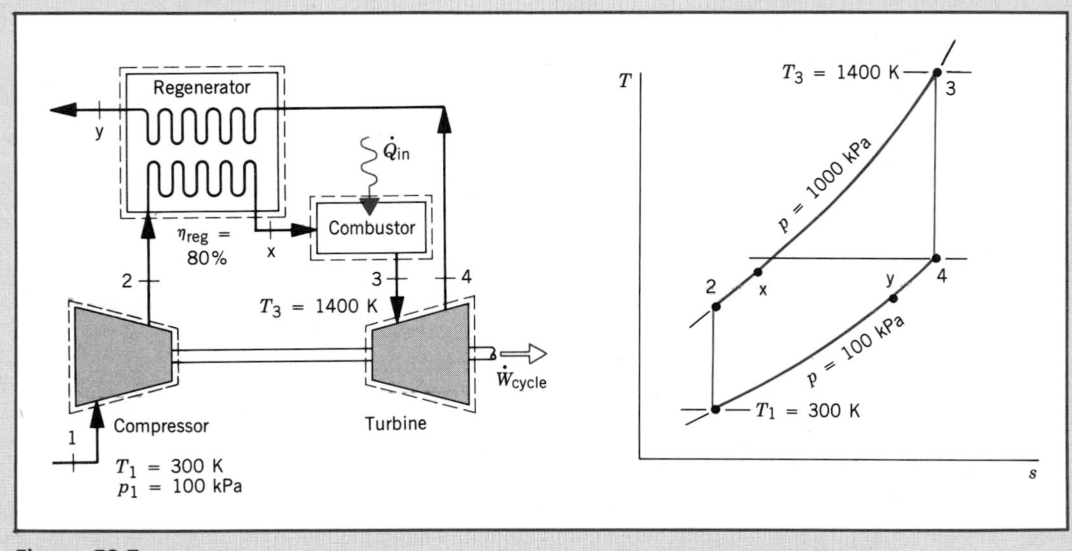

Figure E9.7

Assumptions:

1. Each component is analyzed as a control volume at steady state. The control volumes are shown on the accompanying sketch by dashed lines.
2. The compressor and turbine processes are isentropic.
3. There are no pressure drops for flow through the heat exchangers.
4. The regenerator effectiveness is 80%.
5. Kinetic and potential energy effects are negligible.
6. The working fluid is air modeled as an ideal gas.

Analysis: The specific enthalpy values at the numbered states on the T–s diagram are the same as those in Example 9.4: $h_1 = 300.19$ kJ/kg, $h_2 = 579.9$ kJ/kg, $h_3 = 1515.4$ kJ/kg, $h_4 = 808.5$ kJ/kg.

To find the specific enthalpy h_x, the regenerator effectiveness is used as follows: By definition

$$\eta_{reg} = \frac{h_x - h_2}{h_4 - h_2}$$

Solving for h_x

$$h_x = \eta_{reg}(h_4 - h_2) + h_2$$
$$= (0.8)(808.5 - 579.9) + 579.9 = 762.8 \text{ kJ/kg}$$

With the specific enthalpy values determined above, the thermal efficiency is

$$\eta = \frac{(\dot{W}_t/\dot{m}) - (\dot{W}_c/\dot{m})}{(\dot{Q}_{in}/\dot{m})} = \frac{(h_3 - h_4) - (h_2 - h_1)}{(h_3 - h_x)}$$

$$= \frac{(1515.4 - 808.5) - (579.9 - 300.19)}{(1515.4 - 762.8)}$$

$$= 0.568\ (56.8\%)$$

Comments:

1. Comparing the present thermal efficiency value with the one determined in Example 9.4, it should be evident that the thermal efficiency can be increased significantly by means of regeneration.
2. The values for work per unit of mass flow of the compressor and turbine are unchanged by the addition of the regenerator. Thus, the back work ratio and net work output are not affected by this modification.
3. The regenerator allows improved fuel utilization to be achieved by transferring a portion of the availability in the hot turbine exhaust gas to the cooler air flowing on the other side.

9.8 REGENERATIVE GAS TURBINES WITH REHEAT AND INTERCOOLING

Two modifications of the basic gas turbine that increase the net work developed are multistage expansion with *reheat* and multistage compression with *intercooling*. When used in conjunction with regeneration, these modifications can result in substantial increases in thermal efficiency. The concepts of reheat and intercooling are introduced in this section.

9.8.1 Gas Turbines with Reheat

For metallurgical reasons, the temperature of the gaseous combustion products entering the turbine must be limited. This temperature can be controlled by providing air in excess of the amount required to burn the fuel in the combustor (see Chap. 13). As a consequence, the gases exiting the combustor contain sufficient air to support the combustion of additional fuel. Some gas turbine power plants take advantage of the excess air by means of a multistage turbine with a reheat combustor between the stages. With this arrangement the net work per unit of mass flow can be increased. Let us consider reheat from the vantage point of an air-standard analysis.

The basic features of a two-stage gas turbine with reheat are brought out by considering an ideal air-standard Brayton cycle modified as shown in Fig. 9.16. After expansion from state 3 to state a in the first turbine, the gas is reheated at constant pressure from state a to state b. The expansion is then completed in the second turbine from state b to state 4. The ideal Brayton cycle without reheat, 1–2–3–4′–1, is shown on the same T–s diagram for comparison. Because lines of constant pressure on a T–s diagram diverge slightly with increasing entropy, the total work of the two-stage turbine is greater than that of a single expansion from state 3 to state 4′. Thus, the *net* work for the reheat cycle is greater than that of the cycle without reheat. Despite the increase in net work with reheat, the cycle thermal efficiency would not necessarily increase because a greater total heat addition would be required. However, the temperature at the exit of the turbine is higher with reheat than without reheat, so the potential for regeneration is enhanced. When reheat and regeneration are used together, the thermal efficiency can increase significantly. The following example provides an illustration.

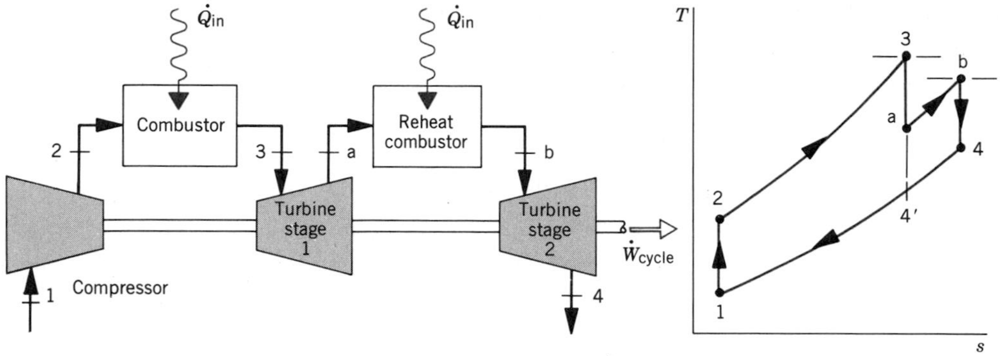

Figure 9.16 Ideal gas turbine with reheat.

EXAMPLE 9.8

PROBLEM

Consider a modification of the cycle of Example 9.4 involving reheat and regeneration. Air enters the compressor at 100 kPa, 300 K and is compressed to 1000 kPa. The temperature at the inlet to the first turbine stage is 1400 K. The expansion takes place isentropically in two stages, with reheat to 1400 K between the stages at a constant pressure of 300 kPa. A regenerator having an effectiveness of 100% is also incorporated in the cycle. Determine the thermal efficiency.

SOLUTION

Known: An ideal air-standard gas turbine cycle operates with reheat and regeneration. Temperatures and pressures at principal states are specified.

Find: Determine the thermal efficiency.

Schematic and Given Data:

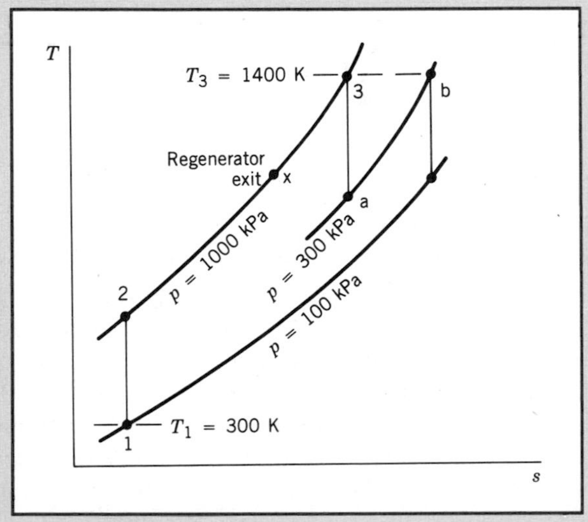

Figure E9.8

Assumptions:

1. Each component of the power plant is analyzed as a control volume at steady state.
2. The compressor and turbine processes are isentropic.

3. There are no pressure drops for flow through the heat exchangers.
4. The regenerator effectiveness is 100%.
5. Kinetic and potential energy effects are negligible.
6. The working fluid is air modeled as an ideal gas.

Analysis: Let us begin by determining the specific enthalpies at each principal state of the cycle. States 1, 2, and 3 are the same as in Example 9.4: $h_1 = 300.19$ kJ/kg, $h_2 = 579.9$ kJ/kg, $h_3 = 1515.4$ kJ/kg. The temperature at state b is the same as at state 3, so $h_b = h_3$.

Since the first turbine process is isentropic, the enthalpy at state a can be determined using p_r data from Table A-16 and the relationship

$$p_{ra} = p_{r3}\frac{p_a}{p_3} = (450.5)\frac{300}{1000} = 135.15$$

Interpolating in Table A-16, we get $h_a = 1095.9$ kJ/kg.

The second turbine process is also isentropic, so the enthalpy at state 4 can be determined similarly. Thus

$$p_{r4} = p_{rb}\frac{p_4}{p_b} = (450.5)\frac{100}{300} = 150.17$$

Interpolating in Table A-16, we obtain $h_4 = 1127.6$ kJ/kg. Since the regenerator effectiveness is 100%, $h_x = h_4 = 1127.6$ kJ/kg.

The thermal efficiency calculation must take into account the compressor work, the work of *each* turbine, and the *total* heat added. Thus, on a unit mass basis

$$\eta = \frac{(h_3 - h_a) + (h_b - h_4) - (h_2 - h_1)}{(h_3 - h_x) + (h_b - h_a)}$$

$$= \frac{(1515.4 - 1095.9) + (1515.4 - 1127.6) - (579.9 - 300.19)}{(1515.4 - 1127.6) + (1515.4 - 1095.9)}$$

$$= 0.654\ (65.4\%)$$

Comment: Comparing the present value with the thermal efficiency determined in Example 9.4, we can conclude that the use of reheat coupled with regeneration can result in a substantial increase in thermal efficiency.

9.8.2 Compression with Intercooling

The net work output of a gas turbine also can be increased by reducing the compressor work input. This can be accomplished by means of multistage compression with intercooling. The present discussion provides an introduction to this subject.

Let us first consider the work input to compressors at steady state, assuming that irreversibilities are absent and changes in kinetic and potential energy from inlet to exit are negligible. The p–v diagram of Fig. 9.17 shows two possible compression paths from a specified inlet state 1 to a specified final pressure p_2. Path 1–2′ is for an adiabatic compression. Path 1–2 corresponds to a compression with heat transfer *from* the working fluid to the surroundings. The area to the left of each curve equals the magnitude of the work per unit mass of the respective process (see Sec. 6.9). The smaller area to the left of Process 1–2 indicates that the work of this process is less than for the adiabatic compression from 1 to 2′. This suggests that cooling a gas *during* compression is advantageous in terms of the work-input requirement.

Although cooling a gas *as it is compressed* would reduce the work, a heat transfer rate high enough to effect a significant reduction in work is difficult to achieve in practice. A practical alternative is to separate the work and heat interactions into

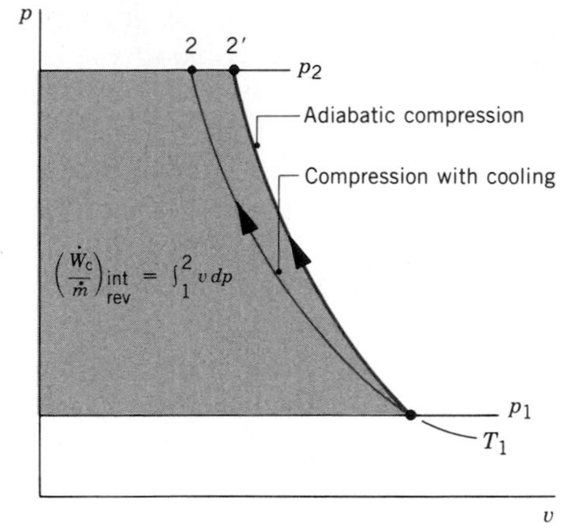

Figure 9.17 Internally reversible compression processes between two fixed pressures.

separate processes by letting compression take place in stages with heat exchangers, called *intercoolers,* cooling the gas between stages. Figure 9.18 illustrates a two-stage compressor with an intercooler. The accompanying p–v and T–s diagrams show the states for internally reversible processes. Process 1–c denotes an isentropic compression from state 1 to state c where the pressure is p_i. In Process c–d the gas is cooled at constant pressure from temperature T_c to T_d. Process d–2 is an isentropic compression to state 2. The work input per unit of mass flow is represented on the p–v diagram by shaded area 1–c–d–2–a–b–1. Without intercooling the gas would be compressed isentropically in a single stage from state 1 to state 2′ and the work would be represented by enclosed area 1–2′–a–b–1. The crosshatched area on the p–v diagram represents the reduction in work that would be achieved with intercooling.

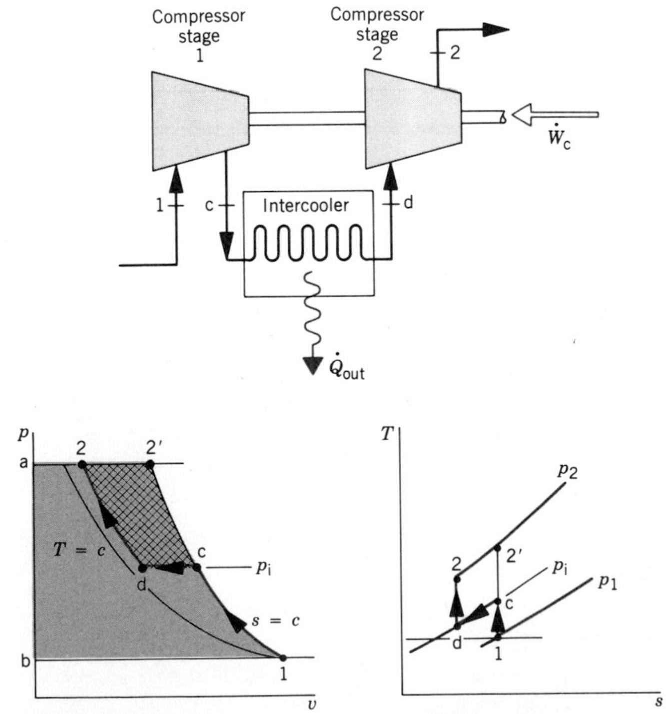

Figure 9.18 Two-stage compression with intercooling.

The size of the crosshatched area on the p–v diagram of Fig. 9.18 representing the reduction in work with intercooling depends on both the temperature T_d at the exit of the intercooler and the intercooler pressure p_i. By properly selecting T_d and p_i, the total work input to the compressor can be minimized. For example, if the pressure p_i is specified, the work input would decrease (crosshatched area would increase) as the temperature T_d approaches T_1, the temperature at the inlet to the compressor. For air entering the compressor from the surroundings, T_1 would be the limiting temperature that could be achieved at state d through heat transfer with the surroundings only. Also, for a specified value of the temperature T_d, the pressure p_i can be selected so that the total work input is a minimum (crosshatched area is a maximum). Example 9.10 provides an illustration of the determination of the intercooler pressure for minimum total work using a cold air-standard analysis.

Some large compressors have several stages of compression with intercooling between stages. The determination of the number of stages and the conditions at which to operate the various intercoolers is a problem in optimization. The use of multistage compression with intercooling in a gas turbine power plant increases the net work developed by reducing the compression work. By itself, though, compression with intercooling would not necessarily increase the thermal efficiency of a gas turbine because the temperature of the air entering the combustor would be reduced (compare temperatures at states 2′ and 2 on the T–s diagram of Fig. 9.18). A lower temperature at the combustor inlet would require additional heat transfer to achieve the desired turbine inlet temperature. The lower temperature at the compressor exit enhances the potential for regeneration, however, so when intercooling is used in conjunction with regeneration, an appreciable increase in thermal efficiency can result.

EXAMPLE 9.9

PROBLEM

Air is compressed from 100 kPa, 300 K to 1000 kPa in a two-stage compressor with intercooling between stages. The intercooler pressure is 300 kPa. The air is cooled back to 300 K in the intercooler before entering the second compressor stage. Each compressor stage is isentropic. For steady-state operation and negligible changes in kinetic and potential energy from inlet to exit, determine (a) the temperature at the exit of the second compressor stage and (b) the total compressor work input per unit of mass flow. (c) Repeat for a single stage of compression from the given inlet state to the final pressure.

SOLUTION

Known: Air is compressed at steady state in a two-stage compressor with intercooling between stages. Operating pressures and temperatures are given.

Find: Determine the temperature at the exit of the second compressor stage and the total work input per unit of mass flow. Repeat for a single stage of compression.

Schematic and Given Data:

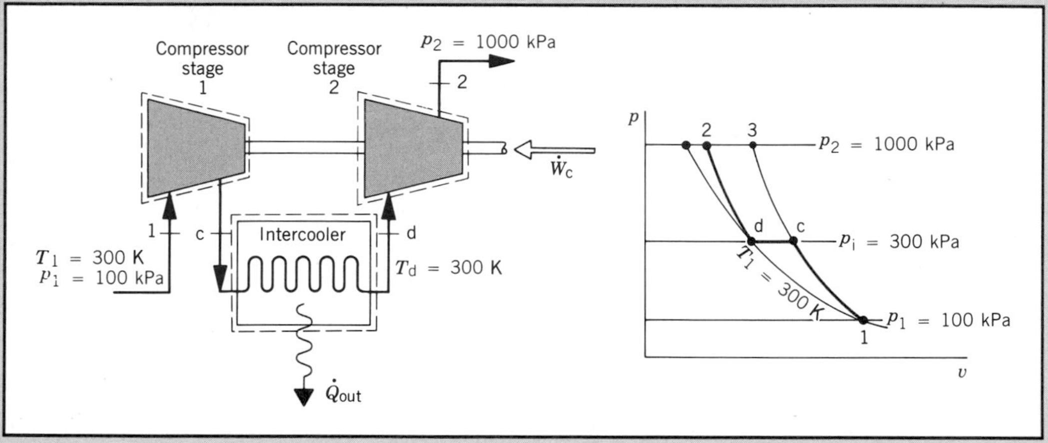

Figure E9.9

Assumptions:

1. The compressor stages and intercooler are analyzed as control volumes at steady state. The control volumes are shown on the accompanying sketch by dashed lines.
2. The compression processes are isentropic.
3. There is no pressure drop for flow through the intercooler.
4. Kinetic and potential energy effects are negligible.
5. The air is modeled as an ideal gas.

Analysis:

(a) The temperature at the exit of the second compressor stage, T_2, can be found using the following relationship for the isentropic process d–2:

$$p_{r2} = p_{rd}\frac{p_2}{p_d}$$

With p_{rd} at $T_d = 300$ K from Table A-16, $p_2 = 1000$ kPa, and $p_d = 300$ kPa,

$$p_{r2} = (1.386)\frac{1000}{300} = 4.62$$

Interpolating in Table A-16, we get $T_2 = 422$ K and $h_2 = 423.8$ kJ/kg

(b) The total compressor work input per unit of mass is the sum of the work inputs for the two stages. That is

$$\frac{\dot{W}_c}{\dot{m}} = (h_c - h_1) + (h_2 - h_d)$$

From Table A-16 at $T_1 = 300$ K, $h_1 = 300.19$ kJ/kg. Since $T_d = T_1$, $h_d = 300.19$ kJ/kg. To find h_c, use p_r data from Table A-16 together with $p_1 = 100$ kPa and $p_c = 300$ kPa to write

$$p_{rc} = p_{r1}\frac{p_c}{p_1} = (1.386)\frac{300}{100} = 4.158$$

Interpolating in Table A-16, we obtain $h_c = 411.3$ kJ/kg. Hence, the total compressor work per unit of mass is

$$\frac{\dot{W}_c}{\dot{m}} = (411.3 - 300.19) + (423.8 - 300.19) = 234.7 \text{ kJ/kg}$$

(c) For a single isentropic stage of compression, the exit state would be state 3 located on the accompanying p–v diagram. The temperature at this state can be determined using

$$p_{r3} = p_{r1}\frac{p_3}{p_1} = (1.386)\frac{1000}{100} = 13.86$$

Interpolating in Table A-16, we get $T_3 = 574$ K and $h_3 = 579.9$ kJ/kg.

The work input for a single stage of compression is then

$$\frac{\dot{W}_c}{\dot{m}} = h_3 - h_1 = 579.9 - 300.19 = 279.7 \text{ kJ/kg}$$

Comment: This example shows that a smaller work input is required with two-stage compression and intercooling than with a single stage of compression. With intercooling, however, a lower gas temperature at the compressor exit is also achieved.

EXAMPLE 9.10

PROBLEM

If the inlet state and the exit pressure are specified for a two-stage compressor operating at steady state, show that the minimum total work input is required when the pressure ratio is the same across each stage. Use a cold air-standard analysis assuming that each compression process is isentropic, there is no pressure drop through the intercooler, and the temperature at the inlet to each compressor stage is the same. Kinetic and potential energy effects can be ignored.

SOLUTION

Known: A two-stage compressor with intercooling operates at steady state under specified conditions.

Find: Show that the minimum total work input is required when the pressure ratio is the same across each stage.

Schematic and Given Data:

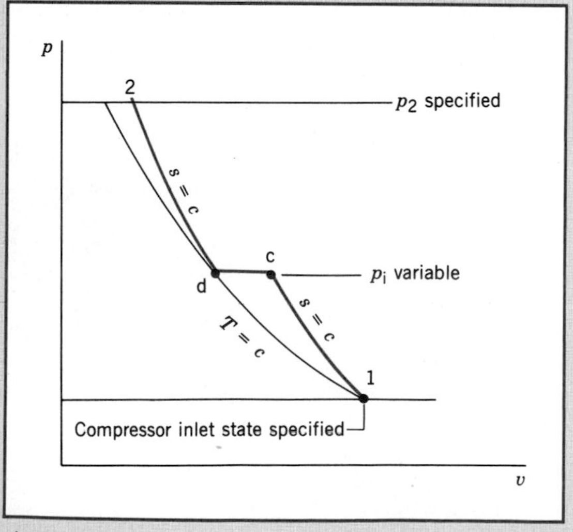

Figure E9.10

Assumptions:

1. The compressor stages and intercooler are analyzed as control volumes at steady state.
2. The compression processes are isentropic.

3. There is no pressure drop for flow through the intercooler.
4. The temperature at the inlet to both compressor stages is the same.
5. Kinetic and potential energy effects are negligible.
6. The working fluid is air modeled as an ideal gas.
7. The specific heat c_p and the specific heat ratio k are constant.

Analysis: The total compressor work input per unit of mass flow is

$$\frac{\dot{W}_c}{\dot{m}} = (h_c - h_1) + (h_2 - h_d)$$

Since c_p is constant

$$\frac{\dot{W}_c}{\dot{m}} = c_p(T_c - T_1) + c_p(T_2 - T_d)$$

With $T_d = T_1$ (assumption 4), this becomes on rearrangement

$$\frac{\dot{W}_c}{\dot{m}} = c_p T_1 \left(\frac{T_c}{T_1} + \frac{T_2}{T_1} - 2\right)$$

Since the compression processes are isentropic and the specific heat ratio k is constant, the pressure and temperature ratios across the compressor stages are related, respectively, by

$$\frac{T_c}{T_1} = \left(\frac{p_i}{p_1}\right)^{(k-1)/k} \quad \text{and} \quad \frac{T_2}{T_d} = \left(\frac{p_2}{p_i}\right)^{(k-1)/k}$$

In the second of these equations, $T_d = T_1$ by assumption 4.

Collecting results

$$\frac{\dot{W}_c}{\dot{m}} = c_p T_1 \left[\left(\frac{p_i}{p_1}\right)^{(k-1)/k} + \left(\frac{p_2}{p_i}\right)^{(k-1)/k} - 2\right]$$

Hence, for specified values of T_1, p_1, p_2, and c_p, the value of the total compressor work input varies with the intercooler pressure only. To determine the pressure p_i that minimizes the total work, form the derivative

$$\frac{\partial(\dot{W}_c/\dot{m})}{\partial p_i} = \frac{\partial}{\partial p_i}\left\{c_p T_1 \left[\left(\frac{p_i}{p_1}\right)^{(k-1)/k} + \left(\frac{p_2}{p_i}\right)^{(k-1)/k} - 2\right]\right\}$$

$$= c_p T_1 \left(\frac{k-1}{k}\right)\left[\left(\frac{p_i}{p_1}\right)^{-1/k}\left(\frac{1}{p_1}\right) + \left(\frac{p_2}{p_i}\right)^{-1/k}\left(-\frac{p_2}{p_i^2}\right)\right]$$

$$= c_p T_1 \left(\frac{k-1}{k}\right)\frac{1}{p_i}\left[\left(\frac{p_i}{p_1}\right)^{(k-1)/k} - \left(\frac{p_2}{p_i}\right)^{(k-1)/k}\right]$$

When the partial derivative is set to zero, the desired relationship is obtained

$$\frac{p_i}{p_1} = \frac{p_2}{p_i}$$

By checking the sign of the second derivative, it can be verified that the total compressor work is a minimum.

Comment: Although the foregoing relationship is developed for a two-stage compressor, similar results can be obtained for multistage compressors.

9.8.3 Reheat and Intercooling

Reheat between turbine stages and intercooling between compressor stages provide two important advantages: The net work output is increased, and the potential for regeneration is enhanced. Accordingly, when reheat and intercooling are used together with regeneration, a substantial improvement in performance can be realized.

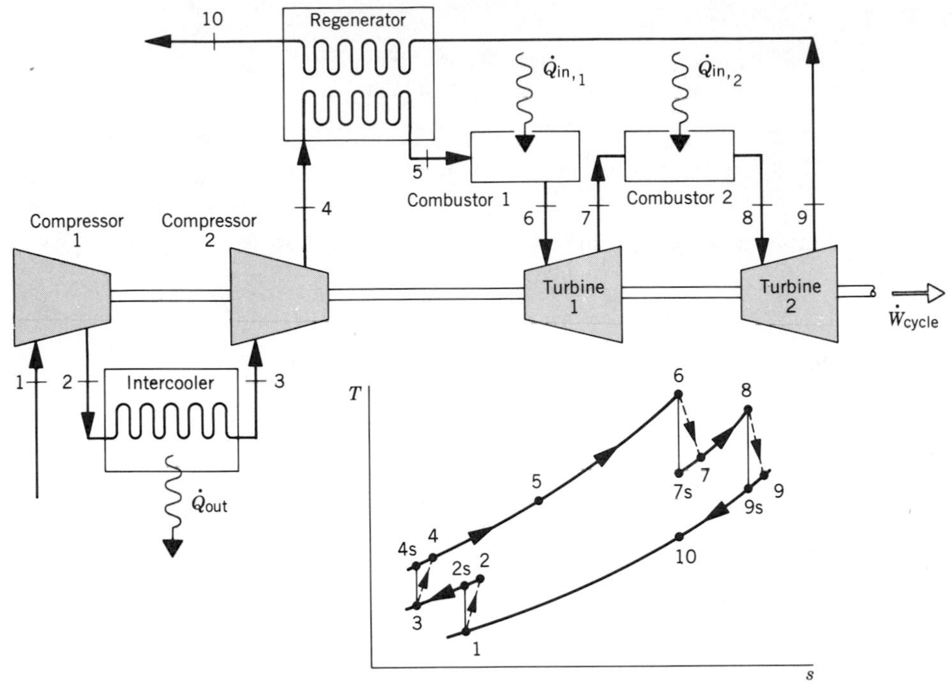

Figure 9.19 Regenerative gas turbine with intercooling and reheat.

One arrangement incorporating reheat, intercooling, and regeneration is shown in Fig. 9.19. This gas turbine has two stages of compression and two turbine stages. The accompanying T–s diagram is drawn to indicate irreversibilities in the compressor and turbine stages. The pressure drops that would occur as the working fluid passes through the intercooler, regenerator, and combustors are not shown.

EXAMPLE 9.11

PROBLEM

A regenerative gas turbine with intercooling and reheat operates at steady state. Air enters the compressor at 100 kPa, 300 K with a mass flow rate of 5.807 kg/s. The pressure ratio across the two-stage compressor is 10. The pressure ratio across the two-stage turbine is also 10. The intercooler and reheater each operate at 300 kPa. At the inlets to the turbine stages, the temperature is 1400 K. The temperature at the inlet to the second compressor stage is 300 K. The efficiency of each compressor and turbine stage is 80%. The regenerator effectiveness is 80%. Determine (a) the thermal efficiency, (b) the back work ratio, (c) the net power developed, in kW.

SOLUTION

Known: An air-standard regenerative gas turbine with intercooling and reheat operates at steady state. Operating pressures and temperatures are specified. Turbine and compressor efficiencies are given and the regenerator effectiveness is known.

Find: Determine the thermal efficiency, back work ratio, and net power developed, in kW.

Schematic and Given Data:

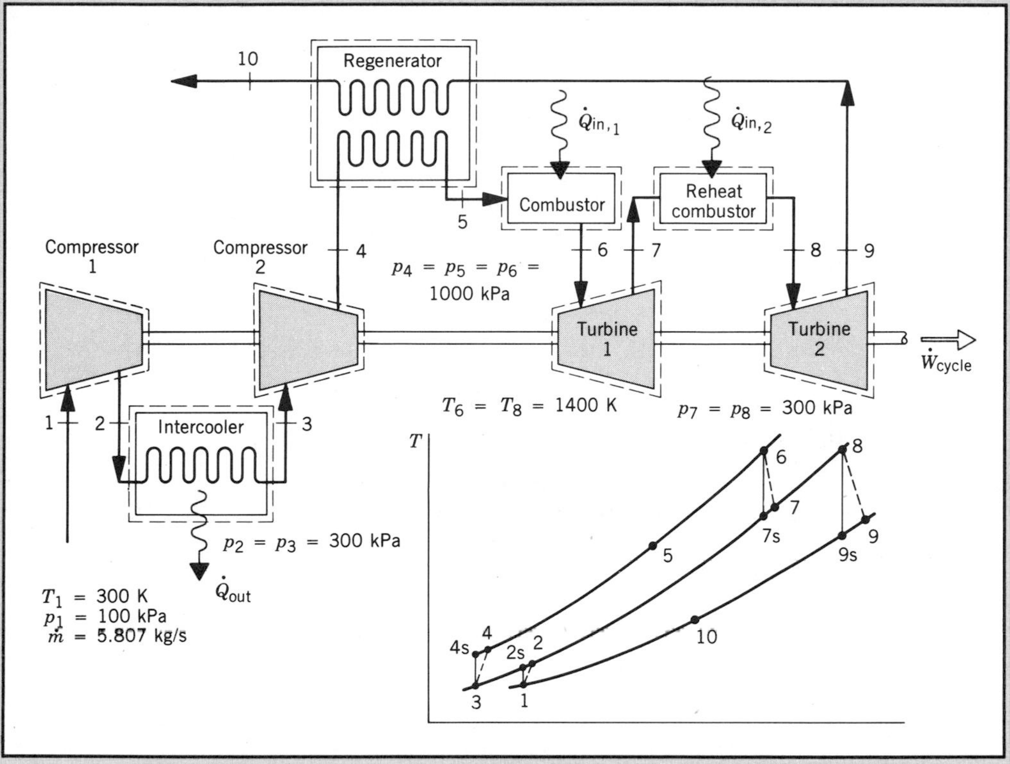

Figure E9.11

Assumptions:

1. Each component is analyzed as a control volume at steady state. The control volumes are shown on the accompanying sketch by dashed lines.
2. There are no pressure drops for flow through the heat exchangers.
3. The compressor and turbine are adiabatic.
4. Kinetic and potential energy effects are negligible.
5. The working fluid is air modeled as an ideal gas.

Analysis: Let us begin by listing the specific enthalpies at the principal states of this cycle. The enthalpies at states 1, 2s, 3, and 4s are obtained from the solution to Example 9.9 where these states are designated as 1, c, d, and 2, respectively. Thus, $h_1 = h_3 = 300.19$ kJ/kg, $h_{2s} = 411.3$ kJ/kg, $h_{4s} = 423.8$ kJ/kg.

The specific enthalpies at states 6, 7s, 8, and 9s are obtained from the solution to Example 9.8, where these states are designated as 3, a, b, and 4, respectively. Thus, $h_6 = h_8 = 1515.4$ kJ/kg, $h_{7s} = 1095.9$ kJ/kg, $h_{9s} = 1127.6$ kJ/kg.

The specific enthalpy at state 4 can be determined using the efficiency of the second compressor stage

$$\eta_c = \frac{h_{4s} - h_3}{h_4 - h_3}$$

Solving for h_4

$$h_4 = h_3 + \frac{h_{4s} - h_3}{\eta_c} = 300.19 + \left(\frac{423.8 - 300.19}{0.8}\right)$$

$$= 454.7 \text{ kJ/kg}$$

Similarly, the specific enthalpy at state 2 is $h_2 = 439.1$ kJ/kg.

The specific enthalpy at state 9 can be determined using the efficiency of the second turbine stage

$$\eta_t = \frac{h_8 - h_9}{h_8 - h_{9s}}$$

Solving for h_9

$$h_9 = h_8 - \eta_t(h_8 - h_{9s}) = 1515.4 - 0.8(1515.4 - 1127.6)$$
$$= 1205.2 \text{ kJ/kg}$$

Similarly, the specific enthalpy at state 7 is $h_7 = 1179.8$ kJ/kg.

The specific enthalpy at state 5 can be determined using the regenerator effectiveness

$$\eta_{reg} = \frac{h_5 - h_4}{h_9 - h_4}$$

Solving for h_5

$$h_5 = h_4 + \eta_{reg}(h_9 - h_4) = 454.7 + 0.8(1205.2 - 454.7)$$
$$= 1055.1 \text{ kJ/kg}$$

(a) The thermal efficiency must take into account the work of both turbine stages, the work of both compressor stages, and the total heat added. The total turbine work per unit of mass flow is

$$\frac{\dot{W}_t}{\dot{m}} = (h_6 - h_7) + (h_8 - h_9)$$
$$= (1515.4 - 1179.8) + (1515.4 - 1205.2) = 645.8 \text{ kJ/kg}$$

The total compressor work input per unit of mass flow is

$$\frac{\dot{W}_c}{\dot{m}} = (h_2 - h_1) + (h_4 - h_3)$$
$$= (439.1 - 300.19) + (454.7 - 300.19) = 293.4 \text{ kJ/kg}$$

The total heat added per unit of mass flow is

$$\frac{\dot{Q}_{in}}{\dot{m}} = (h_6 - h_5) + (h_8 - h_7)$$
$$= (1515.4 - 1055.1) + (1515.4 - 1179.8) = 795.9 \text{ kJ/kg}$$

Calculating the thermal efficiency

$$\eta = \frac{645.8 - 293.4}{795.9} = 0.443 \; (44.3\%)$$

(b) The back work ratio is

$$\text{bwr} = \frac{\dot{W}_c/\dot{m}}{\dot{W}_t/\dot{m}} = \frac{293.4}{645.8} = 0.454 \; (45.4\%)$$

(c) The net power developed is

$$\dot{W}_{cycle} = \dot{m}(\dot{W}_t/\dot{m} - \dot{W}_c/\dot{m})$$
$$= \left(5.807 \frac{\text{kg}}{\text{s}}\right)(645.8 - 293.4)\frac{\text{kJ}}{\text{kg}}\left(\frac{1 \text{ kJ/s}}{1 \text{ kW}}\right) = 2046 \text{ kW}$$

Comment: Comparing the thermal efficiency, back work ratio, and net power values of the current example with the corresponding values of Example 9.6, it should be evident that gas turbine power plant performance can be increased significantly by coupling reheat and intercooling with regeneration.

9.9 GAS TURBINES FOR AIRCRAFT PROPULSION

Gas turbines are particularly suited for aircraft propulsion because of their favorable power-to-weight ratios. The *turbojet engine* is commonly used for this purpose. As illustrated in Fig. 9.20, this type of engine consists of three main sections: the diffuser, the gas generator, and the nozzle. The diffuser placed before the compressor decelerates the incoming air relative to the engine. A pressure rise known as the *ram* effect is associated with this deceleration. The gas generator section consists of a compressor, combustor, and turbine, with the same functions as the corresponding components of a stationary gas turbine power plant. In a turbojet engine, the turbine power output need only be sufficient to drive the compressor and auxiliary equipment, however. The gases leave the turbine at a pressure significantly greater than atmospheric and expand through the nozzle to a high velocity before being discharged to the surroundings. The overall change in the velocity of the gases relative to the engine gives rise to the propulsive force, or thrust. Some turbojets are equipped with an *afterburner,* as shown in Fig. 9.21. This is essentially a reheat device in which additional fuel is injected into the gas exiting the turbine and burned, producing a higher temperature at the nozzle inlet than would be achieved otherwise. As a consequence, a greater nozzle exit velocity is attained, resulting in increased thrust.

The T–s diagram of the processes in an ideal turbojet engine is shown in Fig. 9.20*b*. In accordance with the assumptions of an air-standard analysis, the working fluid is air modeled as an ideal gas. The diffuser, compressor, turbine, and nozzle

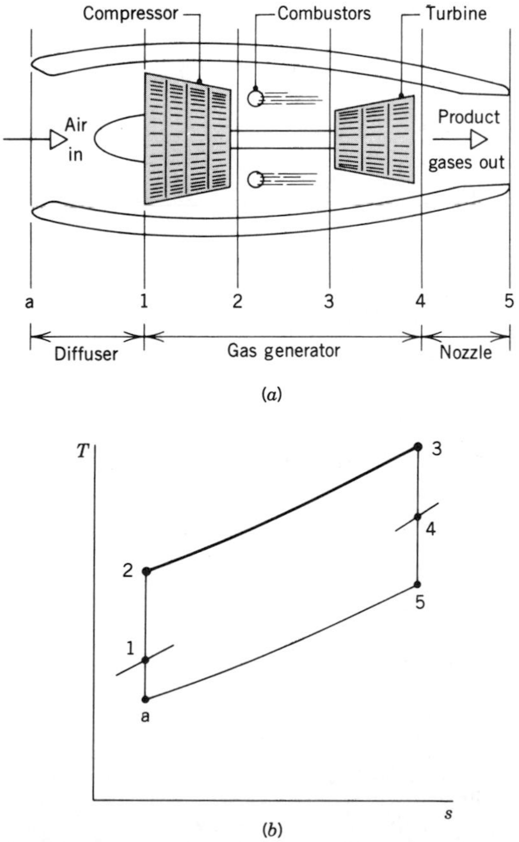

Figure 9.20 Turbojet engine schematic and accompanying ideal T–s diagram.

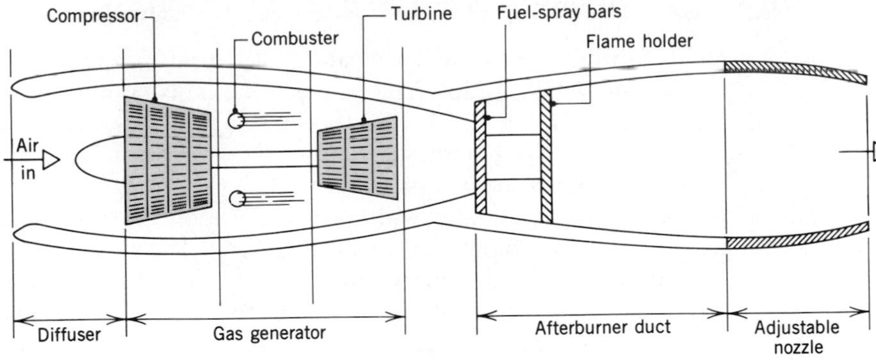

Figure 9.21 Schematic of a turbojet engine with afterburner.

processes are isentropic, and the combustor operates at constant pressure. Isentropic process a–1 shows the pressure rise that occurs in the diffuser as the air decelerates in passing through this component. Process 1–2 is an isentropic compression. Process 2–3 is a constant-pressure heat addition. Process 3–4 is an isentropic expansion through the turbine during which work is developed. Process 4–5 is an isentropic expansion through the nozzle in which the air accelerates and the pressure decreases. Owing to irreversibilities in an actual engine, there would be increases in specific entropy across the diffuser, compressor, turbine, and nozzle. In addition,

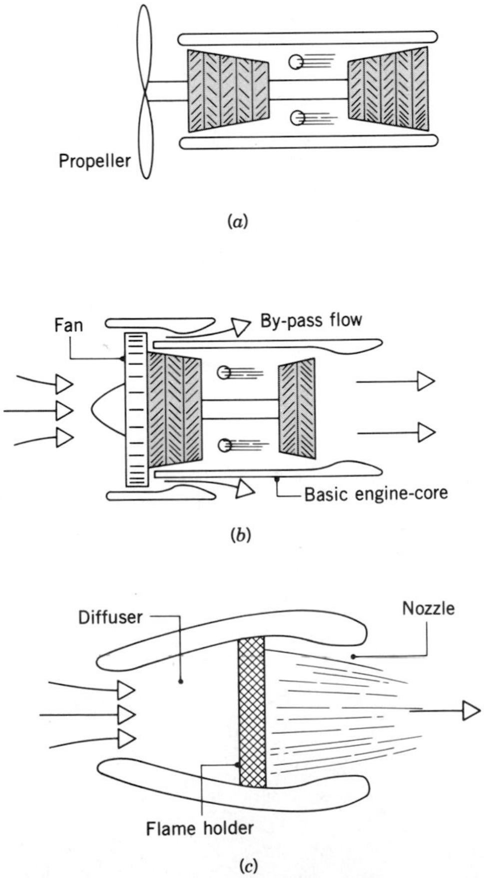

Figure 9.22 Other examples of aircraft engines. (*a*) Turboprop. (*b*) Turbofan. (*c*) Ramjet.

there would be a pressure drop through the combustor of the actual engine. Further details regarding flow through nozzles and diffusers are provided in the third, and final, part of the present chapter. The subject of combustion is discussed in Chap. 13.

In a typical thermodynamic analysis of a turbojet on an air-standard basis the following quantities might be known: the velocity at the diffuser inlet, the compressor pressure ratio, and the turbine inlet temperature. The objective of the analysis would be to determine the velocity at the nozzle exit. Once the nozzle exit velocity is determined, the thrust is determined by applying Newton's second law of motion in a form suitable for a control volume (Sec. 9.12). All principles required for the thermodynamic analysis of turbojet engines on an air-standard basis have been introduced. Example 9.12 provides an illustration.

Other related applications of the gas turbine include *turboprop* and *turbofan* engines. The turboprop engine shown in Fig. 9.22*a* consists of a gas turbine in which the gases are allowed to expand through the turbine to atmospheric pressure. The net power developed is directed to a propeller, which provides thrust to the aircraft. Turboprops are efficient propulsion devices for speeds of up to about 600 km/h (400 miles/h). In the turbofan shown in Fig. 9.22*b*, the core of the engine is much like a turbojet, and some thrust is obtained from expansion through the nozzle. However, a set of large-diameter blades attached to the front of the engine accelerates air around the core. This *bypass flow* provides thrust for takeoff, whereas the core of the engine provides thrust for cruising. Turbofan engines are commonly used for commercial aircraft with flight speeds of up to about 1000 km/h (600 miles/h). A particularly simple type of engine known as a ramjet is shown in Fig. 9.22*c*. This engine requires neither a compressor nor a turbine. A sufficient pressure rise is obtained by decelerating the high-speed incoming air in the diffuser (ram effect). For the ramjet to operate, therefore, the aircraft must already be in flight at a sufficiently great speed. The combustion products exiting the combustor are expanded through the nozzle to produce the thrust.

In each of the engines mentioned thus far, combustion of the fuel is supported by air brought into the engines from the atmosphere. For very high-altitude flight and space travel, where this is no longer possible, *rockets* may be employed. In these applications, both fuel and an oxidizer (such as liquid oxygen) are carried on board the craft. Thrust is developed when the high-pressure gases obtained on combustion are expanded through a nozzle and discharged from the rocket.

EXAMPLE 9.12

PROBLEM

Air enters a turbojet engine at 11.8 lbf/in.², 430°R, and an inlet velocity of 620 miles/h (909.3 ft/s). The pressure ratio across the compressor is 8. The turbine inlet temperature is 2150°R and the pressure at the nozzle exit is 11.8 lbf/in.² The work developed by the turbine just equals the compressor work input. The diffuser, compressor, turbine, and nozzle processes are isentropic, and there is no pressure drop for flow through the combustor. For operation at steady state, determine the velocity at the nozzle exit and the pressure at each principal state. Neglect kinetic energy at the exit of all components except the nozzle and neglect potential energy throughout.

SOLUTION

Known: An ideal turbojet engine operates at steady state. Key operating conditions are specified.

Find: Determine the velocity at the nozzle exit, in ft/s, and the pressure, in lbf/in.², at each principal state.

Schematic and Given Data:

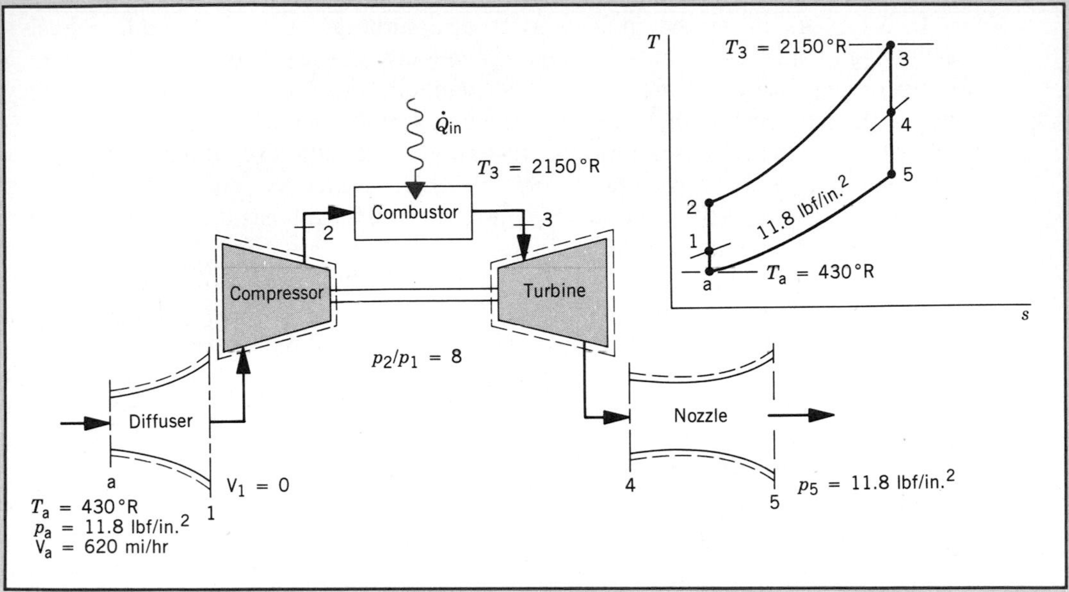

Figure E9.12

Assumptions:

1. Each component is analyzed as a control volume at steady state. The control volumes are shown on the accompanying sketch by dashed lines.
2. The diffuser, compressor, turbine, and nozzle processes are isentropic.
3. There is no pressure drop for flow through the combustor.
4. The turbine work output is just sufficient to drive the compressor.
5. Except at the inlet and exit of the engine, kinetic energy effects can be ignored. Potential energy effects are negligible throughout.
6. The working fluid is air modeled as an ideal gas.

Analysis: To determine the velocity at the exit to the nozzle, the mass and energy rate balances for a control volume enclosing this component reduce at steady state to give

$$0 = \overset{0}{\cancel{\dot{Q}_{cv}}} - \overset{0}{\cancel{\dot{W}_{cv}}} + \dot{m}\left[(h_4 - h_5) + \left(\frac{\overset{0}{\cancel{V_4^2}} - V_5^2}{2}\right) + g(\overset{0}{z_4 \cancel{-} z_5})\right]$$

where $\dot{m}$ is the mass flow rate. The inlet kinetic energy is dropped by assumption 5. Solving for V_5

$$V_5 = \sqrt{2(h_4 - h_5)}$$

This expression requires values for the specific enthalpies h_4 and h_5 at the nozzle inlet and exit, respectively. With the operating parameters specified, the determination of these enthalpy values is accomplished by analyzing each component in turn, beginning with the diffuser. The pressure at each principal state can be evaluated as a part of the analyses required to find the enthalpies h_4 and h_5.

Mass and energy rate balances for a control volume enclosing the diffuser reduce to give

$$h_1 = h_a + \frac{V_a^2}{2}$$

With h_a from Table A-16E and the given value of V_a,

$$h_1 = 102.7 \text{ Btu/lb} + \left[\frac{(909.3)^2}{2}\right]\left(\frac{\text{ft}^2}{\text{s}^2}\right)\left(\frac{1 \text{ lbf}}{32.2 \text{ lb}\cdot\text{ft/s}^2}\right)\left(\frac{1 \text{ Btu}}{778 \text{ ft}\cdot\text{lbf}}\right)$$
$$= 119.2 \text{ Btu/lb}$$

Interpolating in Table A-16E gives $p_{r1} = 1.051$. The flow through the diffuser is isentropic, so pressure p_1 is

$$p_1 = \frac{p_{r1}}{p_{ra}} p_a$$

With p_r data from Table A-16E and the known value of p_a

$$p_1 = \frac{1.051}{0.6268} (11.8 \text{ lbf/in.}^2) = 19.79 \text{ lbf/in.}^2$$

Using the given compressor pressure ratio, the pressure at state 2 is $p_2 = 8(19.79 \text{ lbf/in.}^2) = 158.3 \text{ lbf/in.}^2$

The flow through the compressor is also isentropic. Thus

$$p_{r2} = p_{r1} \frac{p_2}{p_1} = 1.051(8) = 8.408$$

Interpolating in Table A-16E, we get $h_2 = 216.2$ Btu/lb.

At state 3 the temperature is given as $T_3 = 2150°\text{R}$. From Table A-16E, $h_3 = 546.54$ Btu/lb. By assumption 3, $p_3 = p_2$. The work developed by the turbine is just sufficient to drive the compressor (assumption 4). That is

$$\frac{\dot{W}_t}{\dot{m}} = \frac{\dot{W}_c}{\dot{m}}$$

or

$$h_3 - h_4 = h_2 - h_1$$

Solving for h_4

$$h_4 = h_3 + h_1 - h_2 = 546.54 + 119.2 - 216.2$$
$$= 449.5 \text{ Btu/lb}$$

Interpolating in Table A-16E with h_4, gives $p_{r4} = 113.8$

The expansion through the turbine is isentropic, so

$$p_4 = p_3 \frac{p_{r4}}{p_{r3}}$$

With $p_3 = p_2$ and p_r data from Table A-16E

$$p_4 = (158.3 \text{ lbf/in.}^2) \frac{113.8}{233.5} = 77.2 \text{ lbf/in.}^2$$

The expansion through the nozzle is isentropic to $p_5 = 11.8 \text{ lbf/in.}^2$ Thus

$$p_{r5} = p_{r4} \frac{p_5}{p_4} = (113.8) \frac{11.8}{77.2} = 17.39$$

From Table A-16E, $h_5 = 265.8$ Btu/lb, which is the remaining specific enthalpy value required to determine the velocity at the nozzle exit.

Using the values for h_4 and h_5 determined above, the velocity at the nozzle exit is

$$V_5 = \sqrt{2(h_4 - h_5)}$$
$$= \sqrt{2(449.5 - 265.8) \frac{\text{Btu}}{\text{lb}} \left(\frac{32.2 \text{ lb}\cdot\text{ft/s}^2}{1 \text{ lbf}}\right)\left(\frac{778 \text{ ft}\cdot\text{lbf}}{1 \text{ Btu}}\right)}$$
$$= 3034 \text{ ft/s } (2069 \text{ mi/h})$$

Comment: The increase in the velocity of the air as it passes through the engine gives rise to the thrust produced by the engine. A detailed analysis of the forces acting on the engine requires Newton's second law of motion in a form suitable for control volumes.

9.10 COMBINED GAS TURBINE–VAPOR POWER CYCLE

The binary vapor cycle introduced in Sec. 8.5 is an example of a *combined power cycle*. A combined cycle is one based on the coupling of two power cycles such that the heat discharged from one cycle is used partly or wholly as the heat input for the other cycle. In the present section, the combined gas turbine–vapor power cycle is considered.

The exhaust stream at the turbine exit of a gas turbine is at a relatively high temperature. One way that this high-temperature gas stream can be used to improve fuel utilization is by means of a regenerator that allows the turbine exhaust gas to preheat the air between the compressor and combustor (Sec. 9.7). Another method is provided by the combined cycle shown in Fig. 9.23, involving a gas turbine cycle and a vapor power cycle. The two power cycles are coupled so that the heat input to the vapor cycle is derived from the gas turbine cycle, which may be called the *topping* cycle.

The combined cycle would have a thermal efficiency greater than either cycle would have individually. With reference to Fig. 9.23, the thermal efficiency of the combined cycle is

$$\eta = \frac{\dot{W}_{\text{gas}} + \dot{W}_{\text{vap}}}{\dot{Q}_{\text{in}}} \tag{9.28}$$

where $\dot{W}_{\text{gas}}$ is the *net* power developed by the gas turbine, $\dot{W}_{\text{vap}}$ is the *net* power developed by the vapor cycle, and $\dot{Q}_{\text{in}}$ denotes the rate of heat transfer to the gas turbine. The evaluation of the quantities appearing in Eq. 9.28 follows the procedures described in the sections on vapor cycles and gas turbines.

The relation for the energy transferred from the gas cycle to the vapor cycle for the system of Fig. 9.23 can be obtained by applying the mass and energy rate balances to a control volume enclosing the heat exchanger. For steady-state operation, negligi-

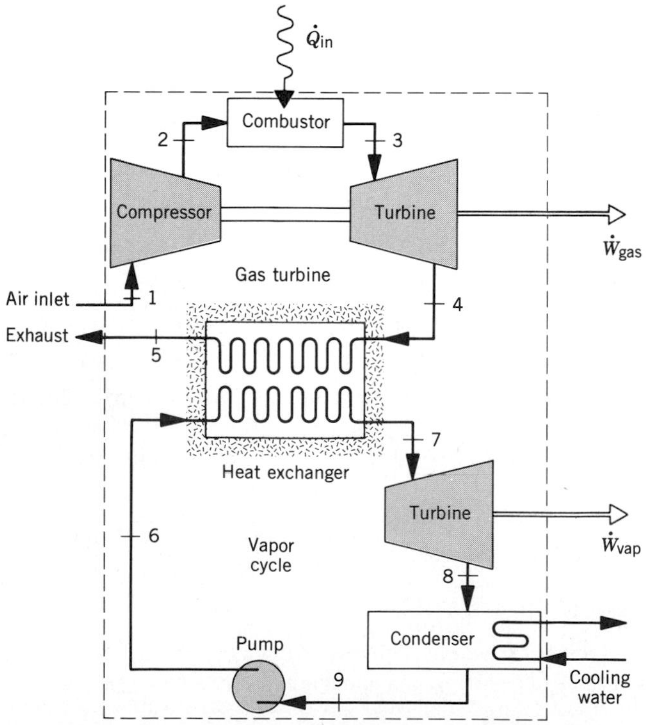

Figure 9.23 Combined gas turbine–vapor power plant.

ble heat transfer with the surroundings, and no significant changes in kinetic and potential energy, the result is

$$\dot{m}_v(h_7 - h_6) = \dot{m}_a(h_4 - h_5) \tag{9.29}$$

where $\dot{m}_a$ and $\dot{m}_v$ are the mass flow rates of the air and vapor, respectively. Because of limitations imposed by the size of the interconnecting heat exchanger, not all of the heat rejected from the gas turbine cycle would appear as the supply to the vapor cycle. But even though the temperature of the gas at state 5 may be higher than the ambient temperature, much of the availability of this stream will have been utilized in vaporizing the working fluid of the vapor cycle.

9.11 ERICSSON AND STIRLING CYCLES

Significant increases in the thermal efficiency of gas turbine power plants can be achieved through intercooling, reheat, and regeneration. There is an economic limit to the number of stages that can be employed, and normally there would be no more than two or three. Nonetheless, it is instructive to consider the situation where the number of stages of both intercooling and reheat becomes indefinitely large. Figure 9.24*a* shows an *ideal* closed regenerative gas turbine cycle with several stages of compression and expansion and a regenerator whose effectiveness is 100%. Each intercooler is assumed to return the working fluid to the temperature T_C at the inlet to the first compression stage and each reheater restores the working fluid to the temperature T_H at the inlet to the first turbine stage. The regenerator allows the heat input for Process 2–3 to be obtained from the heat rejected in Process 4–1. Accordingly, all the heat added *externally* would occur in the reheaters, and all the heat rejected to the surroundings would take place in the intercoolers. In the limit, as an infinite number of reheat and intercooler stages is employed, all the heat added would occur when the working fluid is at its highest temperature, T_H, and all the heat rejected would take place when the working fluid is at its lowest temperature, T_C. The limiting cycle, shown in Fig. 9.24*b*, is called the *Ericsson* cycle. Since irreversibilities are presumed absent and all the heat is supplied and rejected isothermally, the thermal efficiency of the Ericsson cycle equals that of *any* reversible power cycle operating with heat addition at the temperature T_H and heat rejection at the temperature T_C: $\eta_{max} = 1 - T_C/T_H$. This expression was applied previously to evaluate the thermal efficiency of Carnot power cycles. Although the details of the Ericsson cycle differ from those of the Carnot cycle, both cycles have the same value of thermal efficiency when operating between the temperatures T_H and T_C.

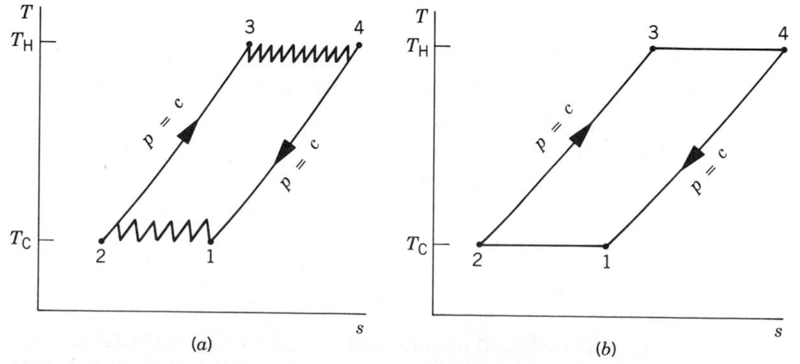

Figure 9.24 Ericsson cycle as a limit of ideal gas turbine operation using multistage compression with intercooling, multistage expansion with reheating, and regeneration.

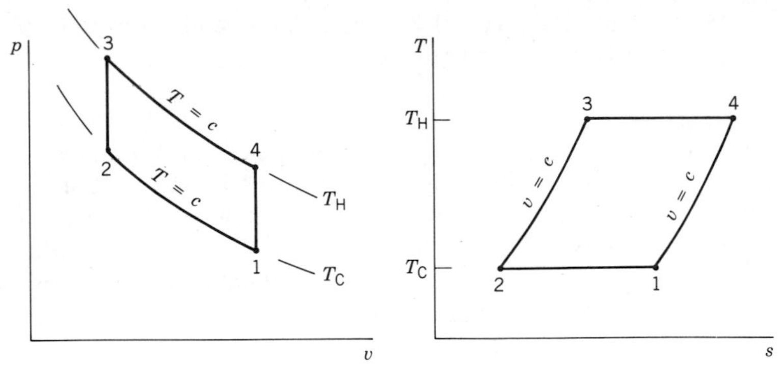

Figure 9.25 p–v and T–s diagrams of the Stirling cycle.

Another cycle that employs a regenerator is the *Stirling* cycle, shown on the p–v and T–s diagrams of Fig. 9.25. The cycle consists of four internally reversible processes in series: isothermal compression from state 1 to state 2 at temperature T_C, constant-volume heating from state 2 to state 3, isothermal expansion from state 3 to state 4 at temperature T_H, and constant-volume cooling from state 4 to state 1 to complete the cycle. A regenerator whose effectiveness is 100% allows the heat rejected during Process 4–1 to be used as the heat input in Process 2–3. Accordingly, all the heat added to the working fluid externally would take place in the isothermal process 3–4 and all the heat rejected to the surroundings would occur in the isothermal process 1–2. It can be concluded, therefore, that the thermal efficiency of the Stirling cycle is given by the same expression as for the Carnot and Ericsson cycles.

The Ericsson and Stirling cycles are principally of theoretical interest as examples of cycles that exhibit the same thermal efficiency as the Carnot cycle. However, a practical engine of the piston–cylinder type that operates on a closed regenerative cycle having features in common with the Stirling cycle has been under study in recent years. This engine is known as the Stirling engine. The Stirling engine offers the opportunity for high efficiency together with reduced emissions from combustion products, because the combustion takes place externally and not within the cylinder as for internal combustion engines. In the Stirling engine, energy is transferred to the working fluid from products of combustion, which are kept separate. It is an *external* combustion engine.

COMPRESSIBLE FLOW THROUGH NOZZLES AND DIFFUSERS

In many applications of engineering interest, gases move at relatively high velocities and exhibit appreciable changes in density. The flows through the nozzles and diffusers of jet engines discussed in Sec. 9.9 are important examples. Other examples are the flows through wind tunnels, shock tubes, and steam ejectors. These flows are known as *compressible* flows. In this part of the chapter, we introduce some of the principles involved in analyzing compressible flows.

9.12 PRELIMINARIES

Concepts introduced in this section play important roles in the study of compressible flows. The momentum equation is introduced in a form applicable to the analysis of control volumes at steady state. The velocity of sound is also defined, and the concepts of Mach number and stagnation state are discussed.

9.12.1 Momentum Equation for Steady One-Dimensional Flow

The analysis of compressible flows requires the principles of conservation of mass and energy, the second law of thermodynamics and relations among the thermodynamic properties of the flowing gas. In addition, Newton's second law of motion is required. Application of Newton's second law of motion to systems of fixed mass (closed systems) involves the familiar form

$$\mathbf{F} = m\mathbf{a}$$

where $\mathbf{F}$ is the resultant force acting *on* a system of mass m and $\mathbf{a}$ is the acceleration. The object of the present discussion is to introduce Newton's second law of motion in a form appropriate for the study of the control volumes considered in subsequent discussions. More general control volume formulations are normally provided in fluid mechanics texts.

Consider the control volume shown in Fig. 9.26, which has a single inlet, designated by 1, and a single exit, designated by 2. The flow is assumed to be one-dimensional at these locations. The energy and entropy rate equations for such a control volume have terms that account for energy and entropy transfers, respectively, at the inlets and exits. Momentum also can be carried into or out of the control volume at the inlets and exits, and such transfers can be accounted for as

$$\begin{bmatrix} \text{time rate of momentum} \\ \text{transfer into or} \\ \text{out of a control volume} \\ \text{accompanying mass flow} \end{bmatrix} = \dot{m}\mathbf{V} \tag{9.30}$$

In this expression, the momentum per unit of mass flowing across the boundary of the control volume is given by the velocity vector $\mathbf{V}$. In accordance with the one-dimensional flow model, the vector is normal to the inlet or exit and oriented in the direction of flow.

In words, Newton's second law of motion for control volumes is

$$\begin{bmatrix} \text{time rate of change} \\ \text{of momentum contained} \\ \text{within the control volume} \end{bmatrix} = \begin{bmatrix} \text{resultant force} \\ \text{acting } on \text{ the} \\ \text{control volume} \end{bmatrix} + \begin{bmatrix} \text{net rate at which momentum is} \\ \text{transferred into the control} \\ \text{volume accompanying mass flow} \end{bmatrix}$$

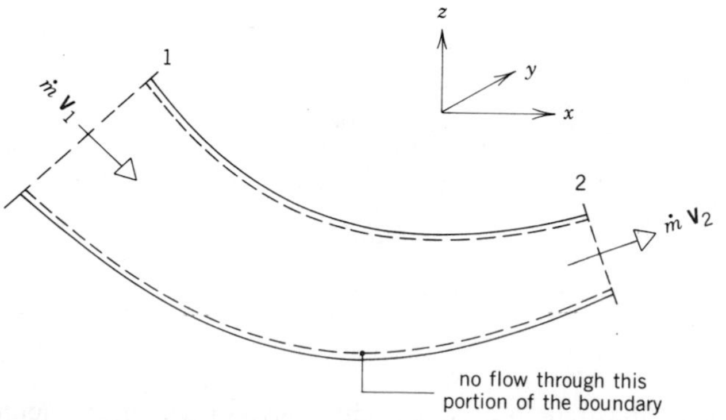

Figure 9.26 One-inlet, one-exit control volume at steady state labeled with momentum transfers accompanying mass flow.

At steady state, the total amount of momentum contained in the control volume is constant with time. Accordingly, when applying Newton's second law of motion to control volumes at steady state, it is necessary to consider only the momentum accompanying the incoming and outgoing streams of matter and the forces acting on the control volume. Newton's law then states that the resultant force **F** acting *on* the control volume equals the difference between the rates of momentum exiting and entering the control volume accompanying mass flow, written

$$\mathbf{F} = \dot{m}_2\mathbf{V}_2 - \dot{m}_1\mathbf{V}_1 = \dot{m}(\mathbf{V}_2 - \mathbf{V}_1) \tag{9.31}$$

Since $\dot{m}_1 = \dot{m}_2$ at steady state, the common mass flow is designated in this expression simply as $\dot{m}$. The resultant force includes the forces due to pressure acting at the inlet and exit, forces acting on the portion of the boundary through which there is no mass flow, and the force of gravity. The expression of Newton's second law of motion given by Eq. 9.31 suffices for subsequent discussions.

9.12.2 Velocity of Sound and Mach Number

A sound wave is a small pressure disturbance that propagates through a gas, liquid, or solid at a velocity c that depends on the properties of the medium. In this section we obtain an expression that relates the *velocity of sound,* or sonic velocity, to other properties. The velocity of sound is an important property in the study of compressible flows.

Let us begin by referring to Fig. 9.27a, which shows a pressure wave moving to the right with a velocity of magnitude c. The wave is generated by a small displacement of the piston. As shown on the figure, the pressure, density, and temperature in the region to the left of the wave depart from the respective values of the undisturbed fluid to the right of the wave, which are designated simply p, ρ, and T. After the wave has passed, the fluid to its left is in steady motion with a velocity of magnitude ΔV.

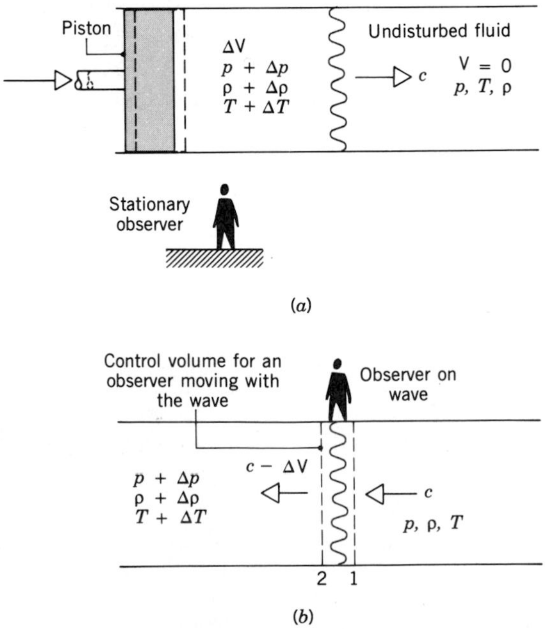

Figure 9.27 Illustrations used to analyze the propagation of a sound wave. (a) Propagation of a pressure wave through a quiescent fluid, relative to a stationary observer. (b) Observer at rest relative to the wave.

Figure 9.27*a* shows the wave from the point of view of a stationary observer. It is easier to analyze this situation from the point of view of an observer at rest relative to the wave, as shown in Fig. 9.27*b*. By adopting this viewpoint, a steady-state analysis can be applied to the control volume identified on the figure. To an observer at rest relative to the wave, it appears as though the fluid is moving toward the stationary wave from the right with velocity c, pressure p, density ρ, and temperature T and moving away on the left with velocity $c - \Delta V$, pressure $p + \Delta p$, density $\rho + \Delta\rho$, and temperature $T + \Delta T$.

At steady state, the conservation of mass principle for the control volume reduces to $\dot{m}_1 = \dot{m}_2$, or

$$\rho A c = (\rho + \Delta\rho) A (c - \Delta V)$$

On rearrangement,

$$0 = c\,\Delta\rho - \rho\,\Delta V - \cancelto{0}{\Delta\rho\,\Delta V} \tag{9.32}$$

The disturbance is weak enough that the third term on the right of Eq. 9.32 can be neglected, leaving

$$\Delta V = (c/\rho)\,\Delta\rho \tag{9.33}$$

Next, the momentum equation, Eq. 9.31, is applied to the control volume under consideration. Since the thickness of the wave is small, shearing forces at the wall are negligible. The effect of gravity is also ignored. Hence, the only significant forces acting on the control volume in the direction of flow are the forces due to pressure at the inlet and exit. With these idealizations, the component of the momentum equation in the direction of flow reduces to

$$\begin{aligned} pA - (p + \Delta p)A &= \dot{m}(c - \Delta V) - \dot{m}c \\ &= \dot{m}(c - \Delta V - c) \\ &= (\rho A c)(-\Delta V) \end{aligned}$$

or

$$\Delta p = \rho c\,\Delta V \tag{9.34}$$

Combining Eqs. 9.33 and 9.34 and solving for c

$$c = \sqrt{\frac{\Delta p}{\Delta\rho}} \tag{9.35}$$

For sound waves, the differences in pressure, density, and temperature across the wave are quite small. Hence, the ratio $\Delta p/\Delta\rho$ in Eq. 9.35 can be interpreted as the derivative of pressure with respect to density across the wave. Furthermore, experiments indicate that the relation between pressure and density across a sound wave is nearly *isentropic*. The expression for the velocity of sound then becomes

$$c = \sqrt{\left(\frac{\partial p}{\partial \rho}\right)_s} \tag{9.36a}$$

or in terms of specific volume

$$c = \sqrt{-v^2\left(\frac{\partial p}{\partial v}\right)_s} \tag{9.36b}$$

The velocity of sound is an intensive property whose value depends on the state of the medium through which sound propagates. Although we have assumed that sound propagates isentropically, the medium itself may be undergoing any process.

Means for evaluating the velocity of sound c for gases, liquids, and solids are introduced in Sec. 11.5. The special case of an ideal gas will be considered here because this case is used extensively later in the chapter. The relationship between pressure and specific volume of an ideal gas at fixed entropy is pv^k = constant, where k is the specific heat ratio. Thus, $(\partial p/\partial v)_s = -kp/v$, and Eq. 9.36b gives $c = \sqrt{kpv}$. With the ideal gas equation of state

$$c = \sqrt{kRT} \qquad \text{(ideal gas)} \tag{9.37}$$

To illustrate the use of Eq. 9.37, let us calculate the velocity of sound in air at 300 K (540°R) and 650 K (1170°R). From Table A-14 at 300 K, $k = 1.4$. Thus

$$c = \sqrt{1.4\left(\frac{8314}{28.97}\,\frac{\text{N}\cdot\text{m}}{\text{kg}\cdot\text{K}}\right)(300\ \text{K})\left(\frac{1\ \text{kg}\cdot\text{m/s}^2}{1\ \text{N}}\right)} = 347\ \frac{\text{m}}{\text{s}}\left(1138\ \frac{\text{ft}}{\text{s}}\right)$$

At 650 K, $k = 1.37$, and $c = 506$ m/s (1660 ft/s), as can be verified. As examples in English units, consider next helium at 495°R (275 K) and 1080°R (600 K). For a monatomic gas, the specific heat ratio is essentially independent of temperature and has the value $k = 1.67$. Thus, at 495°R

$$c = \sqrt{1.67\left(\frac{1545}{4}\,\frac{\text{ft}\cdot\text{lbf}}{\text{lb}\cdot{}^\circ\text{R}}\right)(495^\circ\text{R})\left(\frac{32.2\ \text{lb}\cdot\text{ft/s}^2}{1\ \text{lbf}}\right)} = 3206\ \frac{\text{ft}}{\text{s}}\left(977\ \frac{\text{m}}{\text{s}}\right)$$

At 1080°R, $c = 4736$ ft/s (1444 m/s), as can be verified.

In subsequent discussions, the ratio of the velocity V at a state in a flowing fluid to the value of the sonic velocity c at the same state plays an important role. This ratio is called the *Mach number M*

$$M = \frac{\text{V}}{c} \tag{9.38}$$

When $M > 1$, the flow is said to be *supersonic;* when $M < 1$, the flow is *subsonic;* and when $M = 1$, the flow is *sonic*. The term *hypersonic* is used for flows with Mach numbers much greater than one, and the term *transonic* refers to flows where the Mach number is close to unity.

9.12.3 Stagnation Properties

When dealing with compressible flows, it is often convenient to work with properties evaluated at a reference state known as the *stagnation state*. The stagnation state is the state a flowing fluid would attain if it were decelerated to zero velocity isentropically. We might imagine this as taking place in a diffuser operating at steady state. By reducing an energy balance for such a diffuser, it can be concluded that the enthalpy at the stagnation state associated with an actual state in the flow where the specific enthalpy is h and the velocity is V is given by

$$h_o = h + \frac{\text{V}^2}{2} \tag{9.39}$$

The enthalpy designated here as h_o is called the *stagnation enthalpy*. The pressure p_o and temperature T_o at a stagnation state are called the *stagnation pressure* and *stagnation temperature*, respectively.

9.13 ONE-DIMENSIONAL STEADY FLOW IN NOZZLES AND DIFFUSERS

Although the subject of compressible flow arises in a great many important areas of engineering application, the remainder of this presentation is concerned only with flow through nozzles and diffusers. Texts dealing with compressible flow should be consulted for discussion of other areas of application.

In the present section we determine the shapes required by nozzles and diffusers for subsonic and supersonic flow. This is accomplished using mass, energy, entropy, and momentum principles, together with property relationships. In addition, we study how the flow through nozzles is affected as conditions at the nozzle exit are changed. The presentation concludes with an analysis of normal shocks, which can exist in supersonic flows.

9.13.1 Effects of Area Change in Subsonic and Supersonic Flows

The objective of the present discussion is to establish criteria for determining whether a nozzle or diffuser should have a converging, diverging, or converging–diverging shape. This is approached using differential equations relating the principal variables.

Let us begin by considering a control volume enclosing a nozzle or diffuser. At steady state, the mass flow rate is constant, so

$$\rho \mathrm{AV} = \text{constant}$$

In differential form

$$d(\rho \mathrm{AV}) = 0$$

$$\mathrm{AV}\, d\rho + \rho \mathrm{A}\, d\mathrm{V} + \rho \mathrm{V}\, d\mathrm{A} = 0$$

or on dividing each term by $\rho \mathrm{AV}$

$$\frac{d\rho}{\rho} + \frac{d\mathrm{V}}{\mathrm{V}} + \frac{d\mathrm{A}}{\mathrm{A}} = 0 \tag{9.40}$$

Assuming $\dot{Q}_{cv} = \dot{W}_{cv} = 0$ and negligible potential energy effects, an energy rate balance reduces to give

$$h_2 + \frac{\mathrm{V}_2^2}{2} = h_1 + \frac{\mathrm{V}_1^2}{2}$$

Introducing Eq. 9.39, it follows that the stagnation enthalpies at states 1 and 2 are equal: $h_{o2} = h_{o1}$. Since any state downstream of the inlet can be regarded as state 2, the following relationship between the specific enthalpy and kinetic energy must be satisfied at each state:

$$h + \frac{\mathrm{V}^2}{2} = h_{o1} \qquad \text{(constant)}$$

In differential form this becomes

$$dh = -\mathrm{V}\, d\mathrm{V} \tag{9.41}$$

This equation shows that if the velocity increases (decreases) in the direction of flow, the specific enthalpy must decrease (increase) in the direction of flow, and conversely.

In addition to Eqs. 9.40 and 9.41 expressing conservation of mass and energy, relationships among properties must be taken into consideration. Assuming the flow occurs isentropically, the property relation (Eq. 6.17b)

$$T\,ds = dh - \frac{dp}{\rho}$$

reduces to give

$$dh = \frac{1}{\rho}\,dp \tag{9.42}$$

This equation shows that when pressure increases or decreases in the direction of flow, the specific enthalpy changes in the same way.

Forming the differential of the property relation $p = p(\rho, s)$

$$dp = \left(\frac{\partial p}{\partial \rho}\right)_s d\rho + \left(\frac{\partial p}{\partial s}\right)_\rho ds$$

The second term vanishes in isentropic flow. Introducing Eq. 9.36a, we have

$$dp = c^2\,d\rho \tag{9.43}$$

which shows that when pressure increases or decreases in the direction of flow, density changes in the same way.

Additional conclusions can be drawn by combining the above differential equations. Combining Eqs. 9.41 and 9.42 results in

$$\frac{1}{\rho}\,dp = -\mathrm{V}\,d\mathrm{V} \tag{9.44}$$

which shows that if the velocity increases (decreases) in the direction of flow, the pressure must decrease (increase) in the direction of flow, and conversely.

Eliminating dp between Eqs. 9.43 and 9.44 and combining the result with Eq. 9.40 gives

$$\frac{d\mathrm{A}}{\mathrm{A}} = -\frac{d\mathrm{V}}{\mathrm{V}}\left[1 - \left(\frac{\mathrm{V}}{c}\right)^2\right]$$

or with the *Mach number M*

$$\frac{d\mathrm{A}}{\mathrm{A}} = -\frac{d\mathrm{V}}{\mathrm{V}}(1 - M^2) \tag{9.45}$$

Equation 9.45 shows how area must vary with velocity. The following four cases can be identified:

Case 1: Subsonic nozzle. $d\mathrm{V} > 0$, $M < 1 \Rightarrow d\mathrm{A} < 0$: The duct *converges* in the direction of flow.

Case 2: Supersonic nozzle. $d\mathrm{V} > 0$, $M > 1 \Rightarrow d\mathrm{A} > 0$: The duct *diverges* in the direction of flow.

Case 3: Supersonic diffusers. $d\mathrm{V} < 0$, $M > 1 \Rightarrow d\mathrm{A} < 0$: The duct *converges* in the direction of flow.

Case 4: Subsonic diffuser. $d\mathrm{V} < 0$, $M < 1 \Rightarrow d\mathrm{A} > 0$: The duct *diverges* in the direction of flow.

The conclusions reached above concerning the nature of the flow in subsonic and supersonic nozzles and diffusers are summarized in Fig. 9.28. From Fig. 9.28*a*, we

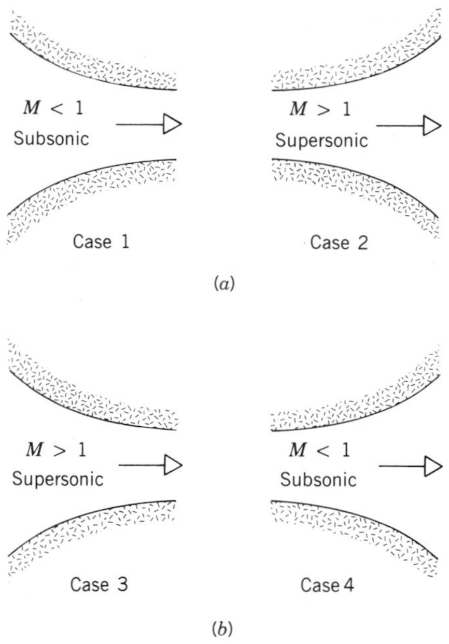

Figure 9.28 Effects of area change in subsonic and supersonic flows. (*a*) Nozzles: V increases; h, p, and ρ decrease. (*b*) Diffusers: V decreases; h, p, and ρ increase.

see that to accelerate a fluid flowing subsonically, a converging nozzle must be used, but once $M = 1$ is achieved, further acceleration can occur only in a diverging nozzle. From Fig. 9.28*b*, we see that a converging diffuser is required to decelerate a fluid flowing supersonically, but once $M = 1$ is achieved, further deceleration can occur only in a diverging diffuser. These findings suggest that a Mach number of unity can occur only at the location in a nozzle or diffuser where the cross-sectional area is a minimum. This location is called the *throat*.

The developments of this section have not required the specification of an equation of state; thus, the conclusions hold for all gases. Moreover, although the conclusions have been drawn under the restriction of isentropic flow through nozzles and diffusers, they are at least qualitatively valid for actual flows because the flow through well-designed nozzles and diffusers is nearly isentropic. Isentropic nozzle efficiencies (Sec. 6.8) in excess of 95% can be attained in practice.

9.13.2 Effects of Back Pressure on Mass Flow Rate

In the present discussion we consider the effect of varying the *back pressure* on the rate of mass flow through nozzles. The *back pressure* is the pressure in the exhaust region outside the nozzle exit. The case of converging nozzles is taken up first and then converging–diverging nozzles are considered.

Converging Nozzles. Figure 9.29 shows a converging duct with stagnation conditions at the inlet, discharging into a region in which the back pressure p_B can be varied. Let us consider how the mass flow rate $\dot{m}$ and nozzle exit pressure p_E vary as the back pressure is decreased while keeping the inlet conditions fixed.

When $p_B = p_E = p_o$, there is no flow, so $\dot{m} = 0$. This corresponds to case a of Fig. 9.29. If the back pressure p_B is decreased, as in cases b and c, there will be flow through the nozzle. As long as the flow is subsonic at the exit, information about changing conditions in the exhaust region can be transmitted upstream. Decreases in back pressure thus result in greater mass flow rates and new pressure variations within the nozzle. In each instance, the velocity is subsonic throughout the nozzle and the exit pressure equals the back pressure. The exit Mach number increases as

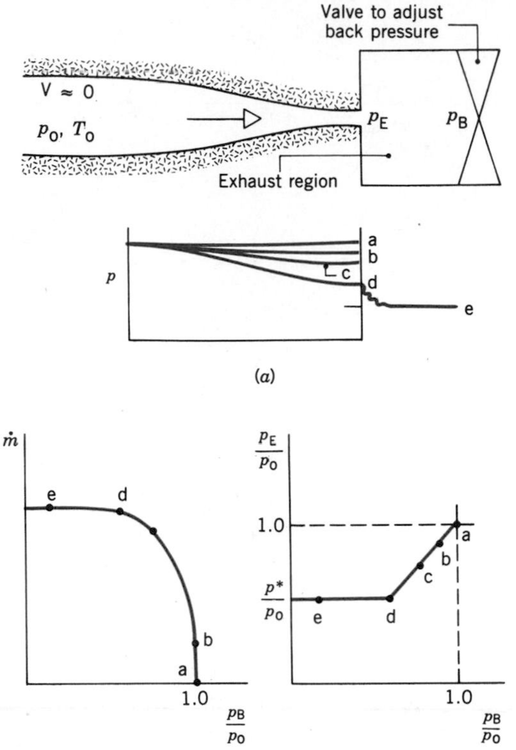

Figure 9.29 Effect of back pressure on the operation of a converging nozzle.

p_B decreases, however, and eventually a Mach number of unity will be attained at the nozzle exit. The corresponding pressure is denoted by p^*, called the *critical pressure*. This case is represented by d on Fig. 9.29.

Recalling that the Mach number cannot increase beyond unity in a converging section, let us consider next what happens when the back pressure is reduced further to a value less than p^*, such as represented by case e. Since the velocity at the exit equals the velocity of sound, information about changing conditions in the exhaust region no longer can be transmitted upstream past the exit plane. Accordingly, reductions in p_B below p^* have no effect on flow conditions in the nozzle. Neither the pressure variation within the nozzle nor the mass flow rate is affected. Under these conditions, the nozzle is said to be *choked*. When a nozzle is choked, the mass flow rate is the maximum possible for the given stagnation conditions. For p_B less than p^*, the flow expands outside the nozzle to match the lower back pressure, as shown by case e of Fig. 9.29. The pressure variation outside the nozzle cannot be predicted using the one-dimensional flow model.

Converging–Diverging Nozzles. Figure 9.30 illustrates the effects of varying back pressure on a *converging–diverging* nozzle. Let us first discuss the cases designated a, b, c, and d. Case a corresponds to $p_B = p_E = p_o$ for which there is no flow. When the back pressure is slightly less than p_o (case b), there is some flow, and the flow is subsonic throughout the nozzle. In accordance with the discussion of Fig. 9.28, the greatest velocity and lowest pressure occur at the throat, and the diverging portion acts as a diffuser in which pressure increases and velocity decreases in the direction of flow. If the back pressure is reduced further, corresponding to case c, the mass flow rate and velocity at the throat are greater than before. Still, the flow remains subsonic throughout and qualitatively the same as case b. As the back pressure is

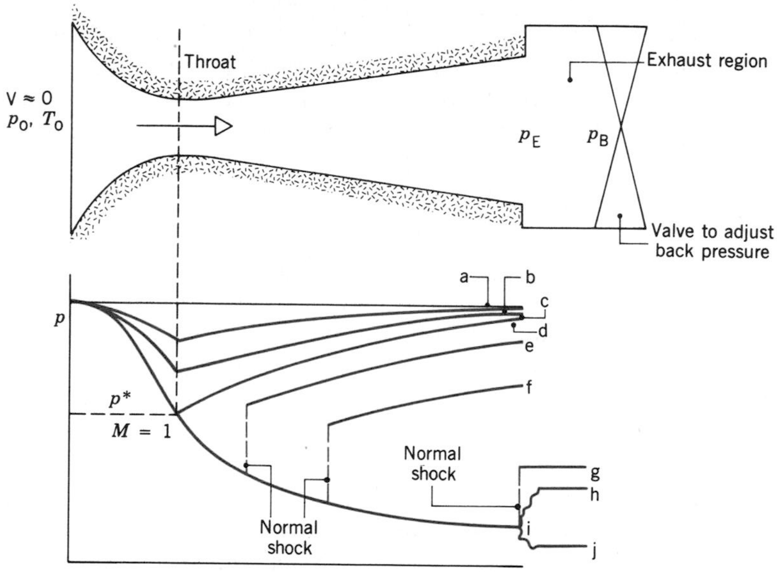

Figure 9.30 Effect of back pressure on the operation of a converging–diverging nozzle.

reduced, the Mach number at the throat increases, and eventually a Mach number of unity is attained there (case d). As before, the greatest velocity and lowest pressure occur at the throat, and the diverging portion remains a subsonic diffuser. However, because the throat velocity is sonic, the nozzle is now *choked:* The maximum mass flow rate has been attained for the given stagnation conditions. Further reductions in back pressure cannot result in an increase in the mass flow rate.

When the back pressure is reduced below that corresponding to case d, the flow through the converging portion and at the throat remains unchanged. Conditions within the diverging portion can be altered, however, as illustrated by cases e, f, and g. In case e, the fluid passing the throat continues to expand and becomes supersonic in the diverging portion just downstream of the throat; but at a certain location an abrupt change in properties occurs. This is called a *normal shock*. Across the shock, there is a rapid and irreversible increase in pressure, accompanied by a rapid decrease from supersonic to subsonic flow. Downstream of the shock, the diverging duct acts as a subsonic diffuser in which the fluid continues to decelerate and the pressure increases to match the back pressure imposed at the exit. If the back pressure is reduced further (case f), the location of the shock moves downstream, but the flow remains qualitatively the same as in case e. With further reductions in back pressure, the shock location moves farther downstream of the throat until it just stands at the exit (case g). In this case, the flow throughout the nozzle is isentropic, with subsonic flow in the converging portion, $M = 1$ at the throat, and supersonic flow in the diverging portion. Since the fluid leaving the nozzle passes through a shock, it is subsonic just downstream of the exit plane.

Finally, let us consider cases h, i, and j where the back pressure is less than that corresponding to case g. In each of these cases, the flow through the nozzle is not affected. The adjustment to changing back pressure occurs outside the nozzle. In case h, the pressure decreases continuously as the fluid expands isentropically through the nozzle and then increases to the back pressure outside the nozzle. The compression that occurs outside the nozzle involves *oblique shock waves*. In case i, the fluid expands isentropically to the back pressure and no shocks occur within or outside the nozzle. In case j, the fluid expands isentropically through the nozzle and then expands outside the nozzle to the back pressure through *oblique expansion* waves. Once $M = 1$ is achieved at the throat, the mass flow rate is fixed at the maximum value for the given stagnation conditions, so the mass flow rate is the same for back pressures corresponding to cases d through j. The pressure variations outside the nozzle involving oblique waves cannot be predicted using the one-dimensional flow model.

9.13.3 Flow Across a Normal Shock

We have seen that under certain conditions a rapid and abrupt change of state called a shock takes place in the diverging portion of a supersonic nozzle. In a *normal* shock, this change of state occurs across a plane normal to the direction of flow. The object of the present discussion is to develop means for determining the change of state across a normal shock.

A control volume enclosing a normal shock is shown in Fig. 9.31. The control volume is assumed to be at steady state with $\dot{W}_{cv} = 0$, $\dot{Q}_{cv} = 0$ and negligible effects of potential energy. The thickness of the shock is very small (on the order of 10^{-5} cm). Thus, there is no significant change in flow area across the shock, even though it may occur in a diverging passage, and the forces acting at the wall can be neglected relative to the pressure forces acting at the upstream and downstream locations denoted by x and y, respectively.

The upstream and downstream states are related by the following equations:

Mass:

$$\rho_x V_x = \rho_y V_y \tag{9.46}$$

Energy:

$$h_x + \frac{V_x^2}{2} = h_y + \frac{V_y^2}{2} \tag{9.47a}$$

or

$$h_{ox} = h_{oy} \tag{9.47b}$$

Momentum:

$$p_x - p_y = \rho_y V_y^2 - \rho_x V_x^2 \tag{9.48}$$

Entropy:

$$s_y - s_x = \dot{\sigma}_{cv}/\dot{m} \tag{9.49}$$

When combined with property relations for the particular fluid under consideration, Eqs. 9.46, 9.47, and 9.48 allow the downstream conditions to be determined for specified upstream conditions. Equation 9.49 leads to the important conclusion that

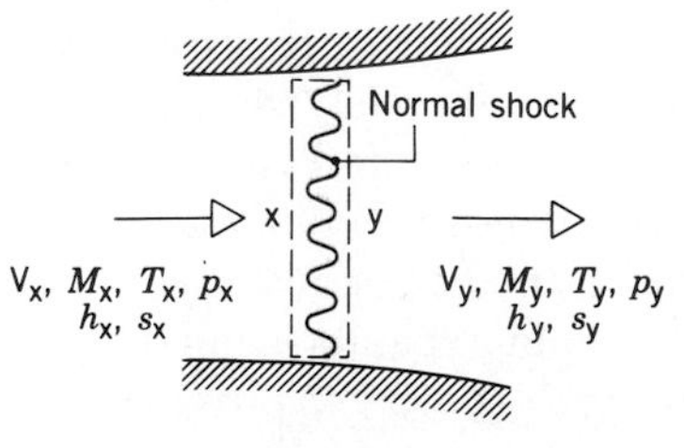

Figure 9.31 Control volume used in the analysis of a normal shock.

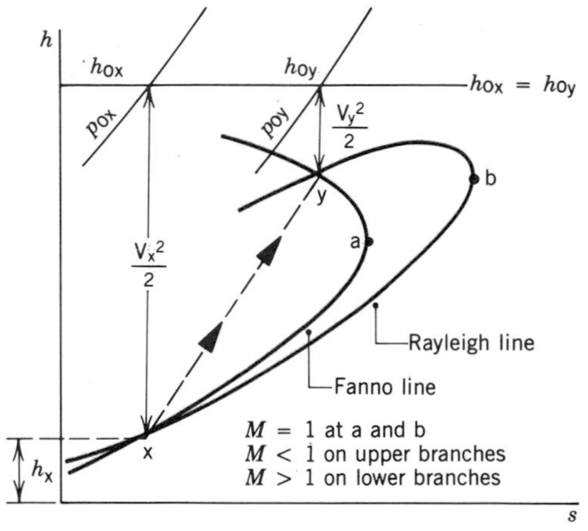

Figure 9.32 Intersection of Fanno and Rayleigh lines as a solution to the normal shock equations.

the downstream state *must* have greater specific entropy than the upstream state, or $s_y > s_x$.

The mass and energy equations, Eqs. 9.46 and 9.47, can be combined with property relations for the particular fluid to give an equation that when plotted on an h–s diagram is called a *Fanno line*. Similarly, the mass and momentum equations, Eqs. 9.46 and 9.48, can be combined to give an equation that when plotted on an h–s diagram is called a *Rayleigh line*. Fanno and Rayleigh lines are sketched on h–s coordinates in Fig. 9.32. It can be shown that the point of maximum entropy on each line, points a and b, corresponds to $M = 1$. It also can be shown that the upper and lower branches of each line correspond, respectively, to subsonic and supersonic velocities.

The downstream state y must satisfy the mass, energy, and momentum equations simultaneously, so state y is fixed by the intersection of the Fanno and Rayleigh lines passing through state x. Since $s_y > s_x$, it can be concluded that the flow across the shock can only pass *from* x *to* y. Accordingly, the velocity changes from supersonic before the shock ($M_x > 1$) to subsonic after the shock ($M_y < 1$). This conclusion is consistent with the discussion of cases e, f, and g in Fig. 9.30. A significant increase in pressure across the shock accompanies the decrease in velocity. Figure 9.32 also locates the stagnation states corresponding to the states upstream and downstream of the shock. The stagnation enthalpy does not change across the shock, but there is a marked decrease in stagnation pressure associated with the irreversible process occurring in the normal shock region.

9.14 FLOW IN NOZZLES AND DIFFUSERS OF IDEAL GASES WITH CONSTANT SPECIFIC HEATS

The discussion of flow in nozzles and diffusers presented in Sec. 9.13 requires no assumption regarding the equation of state, and therefore the results obtained hold generally. Attention is now restricted to ideal gases with constant specific heats. This case is appropriate for many practical problems involving flow through nozzles and diffusers.

The assumption of constant specific heats allows the derivation of relatively simple closed-form equations. Let us begin with Eq. 9.39, which for the case of an ideal gas with constant c_p becomes

$$T_o = T + \frac{V^2}{2c_p}$$

where T_o is the stagnation temperature. Introducing $c_p = kR/(k - 1)$, together with Eqs. 9.37 and 9.38

$$\frac{T_o}{T} = 1 + \frac{k-1}{2} M^2 \tag{9.50}$$

With Eq. 6.50, a relationship between the temperature T and pressure p of the moving stream and the corresponding stagnation temperature T_o and the stagnation pressure p_o is

$$\frac{p_o}{p} = \left(\frac{T_o}{T}\right)^{k/(k-1)}$$

Introducing Eq. 9.50 into this expression gives

$$\frac{p_o}{p} = \left(1 + \frac{k-1}{2} M^2\right)^{k/(k-1)} \tag{9.51}$$

Although sonic conditions may not actually be attained in a particular flow, it is convenient to have an expression relating the area A at a given section to the area A* that *would be* required for sonic flow ($M = 1$) at the same mass flow rate and stagnation state. These areas are related through

$$\rho AV = \rho^* A^* V^*$$

where ρ^* and V^* are the density and velocity, respectively, when $M = 1$. Introducing the ideal gas equation of state, together with Eqs. 9.37 and 9.38, and solving for A/A^*

$$\frac{A}{A^*} = \frac{1}{M}\frac{p^*}{p}\left(\frac{T}{T^*}\right)^{1/2} = \frac{1}{M}\frac{p^*/p_o}{p/p_o}\left(\frac{T/T_o}{T^*/T_o}\right)^{1/2}$$

where T^* and p^* are the temperature and pressure, respectively, when $M = 1$. Then with Eqs. 9.50 and 9.51

$$\frac{A}{A^*} = \frac{1}{M}\left[\left(\frac{2}{k+1}\right)\left(1 + \frac{k-1}{2} M^2\right)\right]^{(k+1)/2(k-1)} \tag{9.52}$$

The variation of A/A^* with M is given in Fig. 9.33 for $k = 1.4$. The figure shows that a unique value of A/A^* corresponds to any choice of M. However, for a given value of A/A^* other than unity, there are two possible values for the Mach number, one subsonic and one supersonic. This is consistent with the discussion of Fig. 9.28, where it was found that a converging–diverging passage with a section of minimum area is required to accelerate a flow from subsonic to supersonic velocity.

Equations 9.50, 9.51, and 9.52 allow the ratios T/T_o, p/p_o, and A/A^* to be computed and tabulated with the Mach number as the single independent variable for a

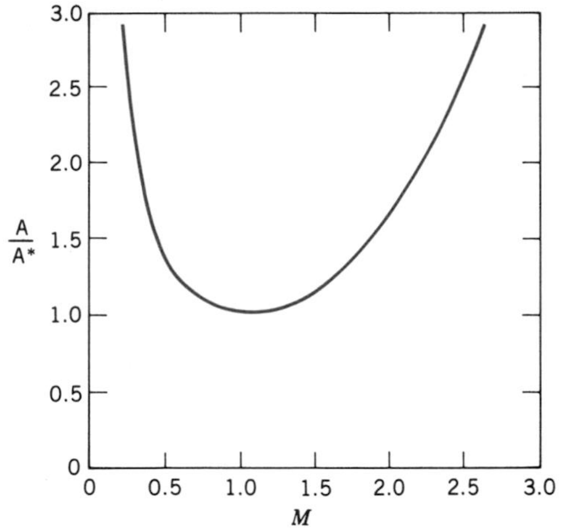

Figure 9.33 Variation of A/A* with Mach number in isentropic flow for $k = 1.4$.

specified value of k. Table 9.1(a) provides a tabulation of this kind for $k = 1.4$. Such tables can facilitate the analysis of isentropic flow through nozzles and diffusers.

Next, let us develop closed-form equations for normal shocks for the case of an ideal gas with constant specific heats. For this case, it follows from the energy equation, Eq. 9.47b, that there is no change in stagnation temperature across the

TABLE 9.1 One-Dimensional Compressible Flow Functions for an Ideal Gas with $k = 1.4$

(a) Isentropic Flow Functions				(b) Normal Shock Functions				
M	T/T_o	p/p_o	A/A^*	M_x	M_y	p_y/p_x	T_y/T_x	p_{oy}/p_{ox}
0	1.000 00	1.000 00	∞	1.00	1.000 00	1.0000	1.0000	1.000 00
0.10	0.998 00	0.993 03	5.8218	1.10	0.911 77	1.2450	1.0649	0.998 92
0.20	0.992 06	0.972 50	2.9635	1.20	0.842 17	1.5133	1.1280	0.992 80
0.30	0.982 32	0.939 47	2.0351	1.30	0.785 96	1.8050	1.1909	0.979 35
0.40	0.968 99	0.895 62	1.5901	1.40	0.739 71	2.1200	1.2547	0.958 19
0.50	0.952 38	0.843 02	1.3398	1.50	0.701 09	2.4583	1.3202	0.929 78
0.60	0.932 84	0.784 00	1.1882	1.60	0.668 44	2.8201	1.3880	0.895 20
0.70	0.910 75	0.720 92	1.094 37	1.70	0.640 55	3.2050	1.4583	0.855 73
0.80	0.886 52	0.656 02	1.038 23	1.80	0.616 50	3.6133	1.5316	0.812 68
0.90	0.860 58	0.591 26	1.008 86	1.90	0.595 62	4.0450	1.6079	0.767 35
1.00	0.833 33	0.528 28	1.000 00	2.00	0.577 35	4.5000	1.6875	0.720 88
1.10	0.805 15	0.468 35	1.007 93	2.10	0.561 28	4.9784	1.7704	0.674 22
1.20	0.776 40	0.412 38	1.030 44	2.20	0.547 06	5.4800	1.8569	0.628 12
1.30	0.747 38	0.360 92	1.066 31	2.30	0.534 41	6.0050	1.9468	0.583 31
1.40	0.718 39	0.314 24	1.1149	2.40	0.523 12	6.5533	2.0403	0.540 15
1.50	0.689 65	0.272 40	1.1762	2.50	0.512 99	7.1250	2.1375	0.499 02
1.60	0.661 38	0.235 27	1.2502	2.60	0.503 87	7.7200	2.2383	0.460 12
1.70	0.633 72	0.202 59	1.3376	2.70	0.495 63	8.3383	2.3429	0.423 59
1.80	0.606 80	0.174 04	1.4390	2.80	0.488 17	8.9800	2.4512	0.389 46
1.90	0.580 72	0.149 24	1.5552	2.90	0.481 38	9.6450	2.5632	0.357 73
2.00	0.555 56	0.127 80	1.6875	3.00	0.475 19	10.333	2.6790	0.328 34
2.10	0.531 35	0.109 35	1.8369	4.00	0.434 96	18.500	4.0469	0.138 76
2.20	0.508 13	0.093 52	2.0050	5.00	0.415 23	29.000	5.8000	0.061 72
2.30	0.485 91	0.079 97	2.1931	10.00	0.387 57	116.50	20.388	0.003 04
2.40	0.464 68	0.068 40	2.4031	∞	0.377 96	∞	∞	0.0

shock, $T_{ox} = T_{oy}$. Then, with Eq. 9.50, the following expression for the ratio of temperatures across the shock is obtained:

$$\frac{T_y}{T_x} = \frac{1 + \frac{k-1}{2} M_x^2}{1 + \frac{k-1}{2} M_y^2} \tag{9.53}$$

Rearranging Eq. 9.48

$$p_x + \rho_x V_x^2 = p_y + \rho_y V_y^2$$

Introducing the ideal gas equation of state, together with Eqs. 9.37 and 9.38, the ratio of the pressure downstream of the shock to the pressure upstream is

$$\frac{p_y}{p_x} = \frac{1 + kM_x^2}{1 + kM_y^2} \tag{9.54}$$

Similarly, Eq. 9.46 becomes

$$\frac{p_y}{p_x} = \sqrt{\frac{T_y}{T_x}} \frac{M_x}{M_y}$$

The following equation relating the Mach numbers M_x and M_y across the shock can be obtained when Eqs. 9.53 and 9.54 are introduced in this expression:

$$M_y^2 = \frac{M_x^2 + \frac{2}{k-1}}{\frac{2k}{k-1} M_x^2 - 1} \tag{9.55}$$

The ratio of stagnation pressures across a shock p_{oy}/p_{ox} is often useful. It is left as an exercise to show that

$$\frac{p_{oy}}{p_{ox}} = \frac{M_x}{M_y} \left(\frac{1 + \frac{k-1}{2} M_y^2}{1 + \frac{k-1}{2} M_x^2} \right)^{(k+1)/2(k-1)} \tag{9.56}$$

Since there is no area change across a shock, Eqs. 9.52 and 9.56 combine to give

$$\frac{A_x^*}{A_y^*} = \frac{p_{oy}}{p_{ox}} \tag{9.57}$$

For specified values of M_x and specific heat ratio k, the Mach number downstream of a shock can be found from Eq. 9.55. Then, with M_x, M_y, and k known, the ratios T_y/T_x, p_y/p_x, and p_{oy}/p_{ox} can be determined from Eqs. 9.53, 9.54, and 9.56. Accordingly, tables can be set up giving M_y, T_y/T_x, p_y/p_x, and p_{oy}/p_{ox} versus the Mach number M_x as the single independent variable for a specified value of k. Table 9.1(b) is a tabulation of this kind for $k = 1.4$.

EXAMPLE 9.13

PROBLEM

A converging nozzle has an exit area of 0.001 m^2. Air enters the nozzle with negligible velocity at a pressure of 1.0 MPa and a temperature of 360 K. For isentropic flow of an ideal gas with $k = 1.4$, determine the mass flow rate, in kg/s, and the exit Mach number for back pressures of (a) 500 kPa and (b) 784 kPa.

SOLUTION

Known: Air flows isentropically from specified stagnation conditions through a converging nozzle with a known exit area.

Find: For back pressures of 500 and 784 kPa, determine the mass flow rate, in kg/s, and the exit Mach number.

Schematic and Given Data:

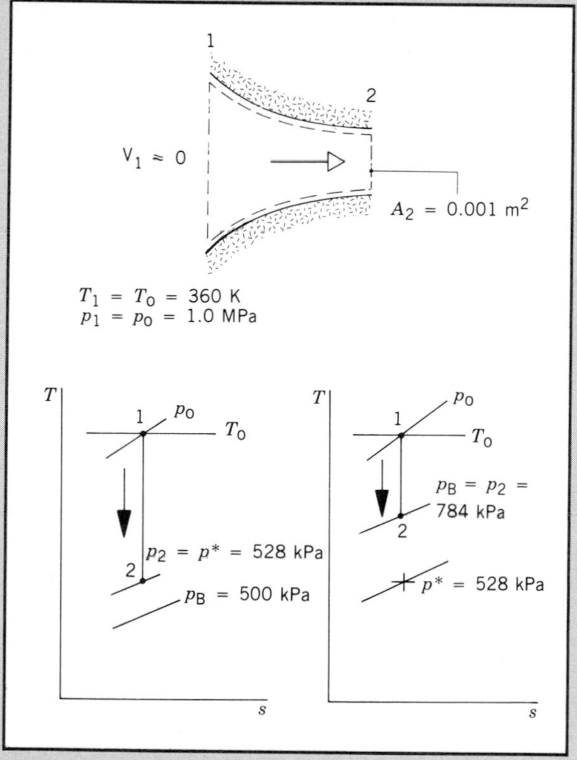

Figure E9.13

Assumptions:

1. The control volume shown in the accompanying sketch operates at steady state.
2. The air is modeled as an ideal gas with $k = 1.4$.
3. Flow through the nozzle is isentropic.

Analysis: The first step is to check whether the flow is choked. With $k = 1.4$ and $M = 1.0$, Eq. 9.51 gives $p^*/p_o = 0.528$. Since $p_o = 1.0$ MPa, the critical pressure is $p^* = 528$ kPa. Thus, for back pressures of 528 kPa or less, the Mach number is unity at the exit and the nozzle is choked.

(a) From the above discussion, it follows that for a back pressure of 500 kPa, the nozzle is choked. At the exit, $M_2 = 1.0$ and the exit pressure equals the critical pressure, $p_2 = 528$ kPa. The mass flow rate is the maximum value that can be attained for the given stagnation properties. With the ideal gas equation of state, the mass flow rate is

$$\dot{m} = \rho_2 A_2 V_2 = \frac{p_2}{RT_2} A_2 V_2$$

The exit area A_2 required by this expression is specified as 10^{-3} m^2. Since $M = 1$ at the exit, the exit temperature T_2 can be found from Eq. 9.50 as 300 K. With Eq. 9.37, the exit velocity V_2 is

$$V_2 = \sqrt{kRT_2}$$

$$= \sqrt{1.4\left(\frac{8314}{28.97}\frac{\text{N}\cdot\text{m}}{\text{kg}\cdot\text{K}}\right)(300\text{ K})\left(\frac{1\text{ kg}\cdot\text{m/s}^2}{1\text{ N}}\right)} = 347.2\text{ m/s}$$

Finally

$$\dot{m} = \frac{(528 \times 10^3\text{ N/m}^2)(10^{-3}\text{ m}^2)(347.2\text{ m/s})}{\left(\frac{8314}{28.97}\frac{\text{N}\cdot\text{m}}{\text{kg}\cdot\text{K}}\right)(300\text{ K})} = 2.13\text{ kg/s}$$

(b) Since the back pressure of 784 kPa is greater than the critical pressure determined above, the flow throughout the nozzle is subsonic and the exit pressure equals the back pressure, $p_2 = 784$ kPa. The exit Mach number can be found by solving Eq. 9.51 to obtain

$$M_2 = \left\{\frac{2}{k-1}\left[\left(\frac{p_o}{p_2}\right)^{(k-1)/k} - 1\right]\right\}^{1/2}$$

Inserting values

$$M_2 = \left\{\frac{2}{1.4-1}\left[\left(\frac{1 \times 10^6}{7.84 \times 10^5}\right)^{0.286} - 1\right]\right\}^{1/2} = 0.6$$

With the exit Mach number known, the exit temperature T_2 can be found from Eq. 9.50 as 336 K. The exit velocity is then

$$V_2 = M_2 c_2 = M_2\sqrt{kRT_2} = 0.6\sqrt{1.4\left(\frac{8314}{28.97}\right)(336)}$$

$$= 220.5\text{ m/s}$$

The mass flow rate is

$$\dot{m} = \rho_2 A_2 V_2 = \frac{p_2}{RT_2} A_2 V_2 = \frac{(784 \times 10^3)(10^{-3})(220.5)}{(8314/28.97)(336)}$$

$$= 1.79\text{ kg/s}$$

Comment: The use of Table 9.1(a) reduces some of the computation required in the solution presented above. It is left as an exercise to develop a solution using this table.

EXAMPLE 9.14

PROBLEM

A converging–diverging nozzle operating at steady state has a throat area of 1.0 in.^{2} and an exit area of 2.4 in.^{2} Air enters the nozzle with a negligible velocity at a pressure of 100 lbf/in.^{2} and a temperature of 500°R. For air as an ideal gas with $k = 1.4$, determine the mass flow rate, in lb/s, the exit pressure, in lbf/in.^{2}, and exit Mach number for each of the five following cases. (a) Isentropic flow with $M = 0.7$ at the throat. (b) Isentropic flow with $M = 1$ at the throat and the diverging portion acting as a diffuser. (c) Isentropic flow with $M = 1$ at the throat and the diverging portion acting as a nozzle. (d) Isentropic flow through the nozzle with a normal shock standing at the exit. (e) A normal shock stands in the diverging section at a location where the area is 2.0 in.^{2} Elsewhere in the nozzle, the flow is isentropic.

SOLUTION

Known: Air flows from specified stagnation conditions through a converging–diverging nozzle having a known throat and exit area.

Find: The mass flow rate, exit pressure, and exit Mach number are to be determined for each of five cases.

Schematic and Given Data:

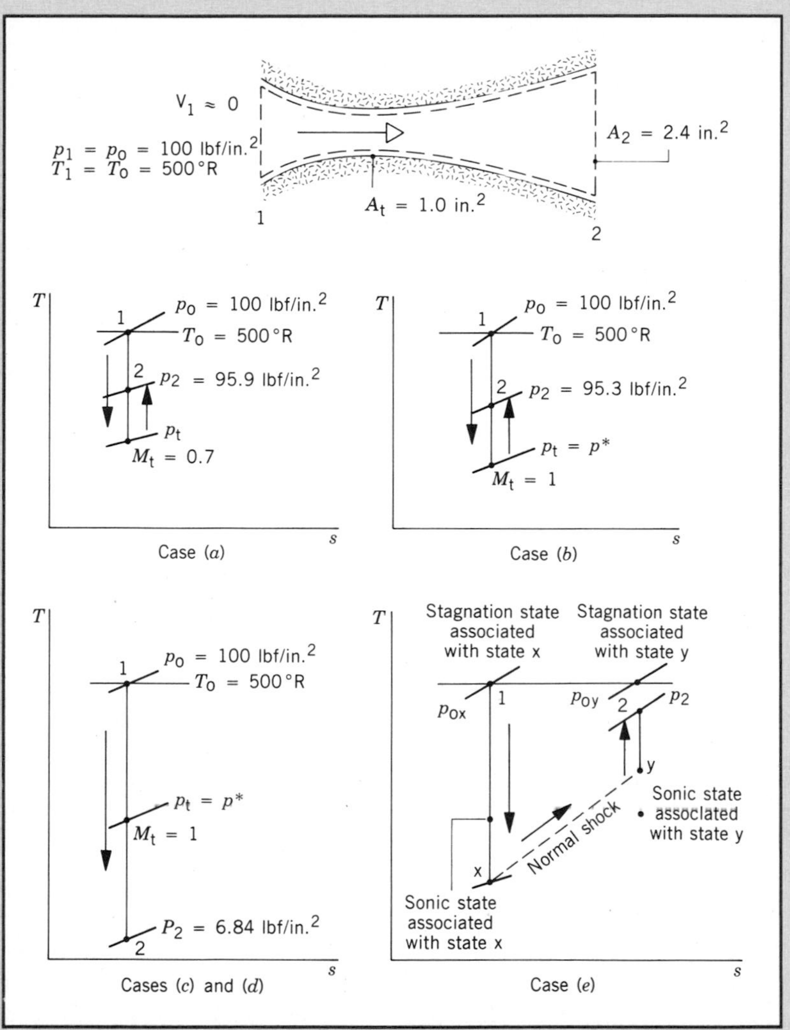

Figure E9.14

Assumptions:

1. The control volume shown in the accompanying sketch operates at steady state. The *T–s* diagrams provided locate states within the nozzle.
2. The air is modeled as an ideal gas with $k = 1.4$.
3. Flow through the nozzle is isentropic throughout, except for case e, where a shock stands in the diverging section.

Analysis:

(a) The accompanying *T–s* diagram shows the states visited by the gas in this case. The following are known: the Mach number at the throat, $M_t = 0.7$, the throat area, $A_t = 1.0$ in.^{2}, and the exit area, $A_2 = 2.4$ in.^{2} The exit

Mach number M_2, exit temperature T_2, and exit pressure p_2 can be determined using the identity

$$\frac{A_2}{A^*} = \frac{A_2}{A_t}\frac{A_t}{A^*}$$

With $M_t = 0.7$, Table 9.1(a) gives $A_t/A^* = 1.09437$. Thus

$$\frac{A_2}{A^*} = \left(\frac{2.4 \text{ in.}^2}{1.0 \text{ in.}^2}\right)(1.09437) = 2.6265$$

The flow throughout the nozzle, including the exit, is subsonic. Accordingly, with this value for A_2/A^*, Table 9.1(a) gives $M_2 \approx 0.24$. For $M_2 = 0.24$, $T_2/T_o = 0.988$, and $p_2/p_o = 0.959$. Since the stagnation temperature and pressure are 500°R and 100 lbf/in.², respectively, it follows that $T_2 = 494$°R and $p_2 = 95.9$ lbf/in.²

The velocity at the exit is

$$\begin{aligned} V_2 &= M_2 c_2 = M_2\sqrt{kRT_2} \\ &= 0.24\sqrt{1.4\left(\frac{1545}{28.97}\frac{\text{ft}\cdot\text{lbf}}{\text{lb}\cdot°\text{R}}\right)(494°\text{R})\left(\frac{32.2 \text{ lb}\cdot\text{ft/s}^2}{1 \text{ lbf}}\right)} \\ &= 262 \text{ ft/s} \end{aligned}$$

The mass flow rate is

$$\begin{aligned} \dot{m} &= \rho_2 A_2 V_2 = \frac{p_2}{RT_2} A_2 V_2 \\ &= \frac{(95.9 \text{ lbf/in.}^2)(2.4 \text{ in.}^2)(262 \text{ ft/s})}{\left(\frac{1545}{28.97}\frac{\text{ft}\cdot\text{lbf}}{\text{lb}\cdot°\text{R}}\right)(494°\text{R})} = 2.29 \text{ lb/s} \end{aligned}$$

(b) The accompanying T–s diagram shows the states visited by the gas in this case. Since $M = 1$ at the throat, we have $A_t = A^*$, and thus $A_2/A^* = 2.4$. Table 9.1(a) gives two Mach numbers for this ratio: $M \approx 0.26$ and $M \approx 2.4$. The diverging portion acts as a diffuser in the present part of the example; accordingly, the subsonic value is appropriate. The supersonic value is appropriate in part (c).

Thus, from Table 9.1(a) we have at $M_2 = 0.26$, $T_2/T_o = 0.986$, and $p_2/p_o = 0.953$. Since $T_o = 500$°R and $p_o = 100$ lbf/in.², it follows that $T_2 = 493$°R and $p_2 = 95.3$ lbf/in.²

The velocity at the exit is

$$\begin{aligned} V_2 &= M_2 c_2 = M_2\sqrt{kRT_2} \\ &= 0.26\sqrt{(1.4)\left(\frac{1545}{28.97}\right)(493)(32.2)} = 283 \text{ ft/s} \end{aligned}$$

The mass flow rate is

$$\dot{m} = \frac{p_2}{RT_2} A_2 V_2 = \frac{(95.3)(2.4)(283)}{\left(\frac{1545}{28.97}\right)(493)} = 2.46 \text{ lb/s}$$

This is the maximum mass flow rate for the specified geometry and stagnation conditions.

(c) The accompanying T–s diagram shows the states visited by the gas in this case. As discussed in part (b), the exit Mach number in the present part of the example is $M_2 = 2.4$. Using this, Table 9.1(a) gives $p_2/p_o = 0.0684$. With $p_o = 100$ lbf/in.², the pressure at the exit is $p_2 = 6.84$ lbf/in.² Since the nozzle is choked, the mass flow rate is the same as found in part (b).

(d) Since a normal shock stands at the exit and the flow upstream of the shock is isentropic, the Mach number M_x and the pressure p_x correspond to the values found in part (c), $M_x = 2.4$, $p_x = 6.84$ lbf/in.² Then, from Table 9.1(b), $M_y \approx 0.52$ and $p_y/p_x = 6.5533$. The pressure downstream of the shock is thus 44.82 lbf/in.² This is the exit pressure. The mass flow is the same as found in part (b).

(e) The accompanying T–s diagram shows the states visited by the gas. It is known that a shock stands in the diverging portion where the area is $A_x = 2.0$ in.² Since a shock occurs, the flow is sonic at the throat, so $A_x^* = A_t$

$= 1.0$ in.^{2} The Mach number M_x can then be found from Table 9.1(a), by using $A_x/A_x^* = 2$, as $M_x = 2.2$.
The Mach number at the exit can be determined using the identity

$$\frac{A_2}{A_y^*} = \left(\frac{A_2}{A_x^*}\right)\left(\frac{A_x^*}{A_y^*}\right)$$

Introducing Eq. 9.57 to replace A_x^*/A_y^*, this becomes

$$\frac{A_2}{A_y^*} = \left(\frac{A_2}{A_x^*}\right)\left(\frac{p_{oy}}{p_{ox}}\right)$$

where p_{ox} and p_{oy} are the stagnation pressures before and after the shock, respectively. With $M_x = 2.2$, the ratio of stagnation pressures is obtained from Table 9.1(b) as $p_{oy}/p_{ox} = 0.62812$. Thus

$$\frac{A_2}{A_y^*} = \left(\frac{2.4 \text{ in.}^2}{1.0 \text{ in.}^2}\right)(0.62812) = 1.51$$

Using this ratio and noting that the flow is subsonic after the shock, Table 9.1(a) gives $M_2 \approx 0.43$, for which $p_2/p_{oy} = 0.88$.

The pressure at the exit can be determined using the identity

$$p_2 = \left(\frac{p_2}{p_{oy}}\right)\left(\frac{p_{oy}}{p_{ox}}\right) p_{ox} = (0.88)(0.628)\left(100 \frac{\text{lbf}}{\text{in.}^2}\right) = 55.3 \text{ lbf/in.}^2$$

Since the flow is choked, the mass flow rate is the same as that found in part (b).

Comment: Part (a) of the present example corresponds to the cases labeled b and c on Fig. 9.30. Part (c) corresponds to case d of Fig. 9.30. Part (d) corresponds to case g of Fig. 9.30 and part (e) corresponds to cases e and f.

PROBLEMS

9.1 An air-standard Otto cycle has a compression ratio of 8.5. At the beginning of compression, $p_1 = 100$ kPa and $T_1 = 300$ K. The heat addition per unit mass of air is 1400 kJ/kg. Determine

(a) the net work, in kJ per kg of air.
(b) the thermal efficiency of the cycle.
(c) the mean effective pressure, in kPa.

9.2 Solve Problem 9.1 on a cold air-standard basis with specific heats evaluated at 300 K.

9.3 At the beginning of the compression process of an air-standard Otto cycle, $p_1 = 1$ bar, $T_1 = 290$ K, $V_1 = 400$ cm^3. The maximum temperature in the cycle is 2200 K and the compression ratio is 8. Determine

(a) the heat addition, in kJ.
(b) the net work, in kJ.
(c) the thermal efficiency.
(d) the mean effective pressure, in bar.
(e) the availability transfer from the air accompanying the heat rejection process, in kJ, for $T_0 = 290$ K, $p_0 = 1$ bar.

9.4 Solve Problem 9.3 on a cold air-standard basis with specific heats evaluated at 300 K.

9.5 Consider the cycle in Problem 9.3 as a model of the processes in each cylinder of a spark-ignition engine. If the engine has four cylinders and the cycle is repeated 1500 times per min in each cylinder, determine the net power output, in kW.

9.6 An air-standard Otto cycle has a compression ratio of 6 and the temperature and pressure at the beginning of the compression process are 520°R and 14.2 lbf/in.^{2}, respectively. The heat addition per unit mass of air is 600 Btu/lb. Determine

(a) the maximum temperature, in °R.
(b) the maximum pressure, in lbf/in.^{2}
(c) the thermal efficiency.

9.7 Solve Problem 9.6 on a cold air-standard basis with specific heats evaluated at 520°R.

9.8 At the beginning of the compression process in an air-standard Otto cycle, $p_1 = 14.7$ lbf/in.^{2} and $T_1 = 530$°R. The compression ratio is 8. Determine the thermal efficiency and mean effective pressure, in lbf/in.^{2}, for maximum cycle temperatures of 2000, 3000, 4000, and 5000°R.

9.9 At the beginning of the compression process of an air-standard Otto cycle, $p_1 = 14.7$ lbf/in.^{2} and $T_1 = 530$°R.

The maximum cycle temperature is 3000°R. Determine the thermal efficiency and mean effective pressure, in lbf/in.^{2}, for compression ratios of 6, 8, and 10.

9.10 Solve Problem 9.9 on a cold air-standard basis using $k = 1.4$.

9.11 An air-standard Otto cycle has a compression ratio of 9. At the beginning of compression, $p_1 = 95$ kPa and $T_1 = 37°C$. The heat addition to the air is 0.75 kJ, and the maximum temperature in the cycle is 1020 K. Determine

(a) the heat rejection, in kJ.
(b) the net work, in kJ.
(c) the thermal efficiency.
(d) the mean effective pressure, in kPa.

9.12 The compression ratio of an air-standard Otto cycle is 8. At the end of the expansion process, the pressure is 90 lbf/in.^{2} and the temperature is 900°R. The heat rejection from the cycle is 70 Btu per lb of air. Determine

(a) the net work, in Btu per lb of air.
(b) the thermal efficiency.
(c) the mean effective pressure, in lbf/in.^{2}

9.13 Consider a modification of the air-standard Otto cycle in which the isentropic compression and expansion processes are each replaced with polytropic processes having $n = 1.3$. The compression ratio is 10 for the modified cycle. At the beginning of compression, $p_1 = 1$ bar and $T_1 = 310$ K. The maximum temperature during the cycle is 2200 K. Determine

(a) the heat transfer and work per unit mass of air, in kJ/kg, for each process in the modified cycle.
(b) the thermal efficiency.

9.14 A four-cylinder, four-stroke internal combustion engine has a bore of 3.7 in. and a stroke of 3.4 in. The clearance volume is 16% of the cylinder volume at bottom dead center and the crankshaft rotates at 2400 RPM. The processes within each cylinder are modeled as an air-standard Otto cycle with a pressure of 14.5 lbf/in.^{2} and a temperature of 60°F at the beginning of compression. The maximum temperature in the cycle is 5200°R. Based on this model, calculate the net work per cycle, in Btu, and the power developed by the engine, in horsepower.

9.15 Write a computer program to calculate the net work per unit mass, the thermal efficiency, and the mean effective pressure of a cold air-standard Otto cycle. Plot each of the above quantities as a function of compression ratio in the range of 5 to 15 for $p_1 = 1$ bar and $T_1 = 300$ K, and a maximum cycle temperature of 2000 K.

9.16 Using the computer program developed in Problem 9.15, investigate the effect of maximum cycle temperature on the net work per unit mass of air for cold air-standard Otto cycles with compression ratios of 6, 8, and 10. Use $p_1 = 14$ lbf/in.^{2} and $T_1 = 530°R$ and let the maximum temperature in each case vary from 2000 to 5000°R.

9.17 The pressure-specific volume diagram of the air-standard *Atkinson* cycle is shown in Fig. P9.17. The cycle consists of isentropic compression, constant volume heat addition, isentropic expansion, and constant-pressure compression. For a particular Atkinson cycle, the compression ratio during isentropic compression is 8.5. At the beginning of this compression process, $p_1 = 100$ kPa and $T_1 = 300$ K. The constant volume heat addition per unit mass of air is 1400 kJ/kg. Determine **(a)** the net work, in kJ per kg of air, **(b)** the thermal efficiency of the cycle, and **(c)** the mean effective pressure, in kPa. Compare your answers with those obtained for the Otto cycle in Problem 9.1 and discuss.

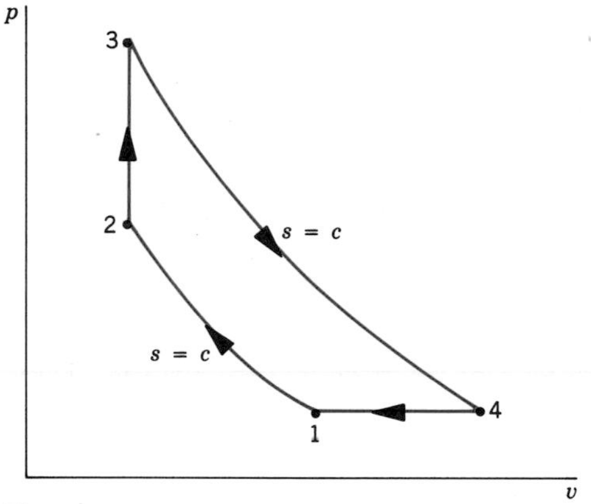

Figure P9.17

9.18 On a cold air-standard basis, derive an expression for the thermal efficiency of the Atkinson cycle (see Fig. P9.17) in terms of the volume ratio during the isentropic compression, the pressure ratio for the constant volume process, and the specific heat ratio. Compare the thermal efficiencies of the cold air-standard Atkinson and Otto cycles, each having the same compression ratio and maximum temperature. Discuss.

9.19 The pressure and temperature at the beginning of compression of an air-standard Diesel cycle are 95 kPa and 290 K, respectively. At the end of the heat addition, the pressure is 6.5 MPa and the temperature is 2000 K. Determine

(a) the compression ratio.
(b) the cutoff ratio.
(c) the thermal efficiency of the cycle.
(d) the mean effective pressure, in kPa.

9.20 Solve Problem 9.19 on a cold air-standard basis with specific heats evaluated at 300 K.

9.21 The compression ratio of an air-standard Diesel cycle is 17 and the conditions at the beginning of compression are $p_1 = 14.0$ lbf/in.^{2}, $V_1 = 2$ ft^3, and $T_1 = 520°R$. The maximum temperature in the cycle is 4000°R. Calculate

(a) the net work for the cycle, in Btu.
(b) the thermal efficiency.
(c) the mean effective pressure, in lbf/in.^{2}

9.22 Solve Problem 9.21 on a cold air-standard basis with specific heats evaluated at 520°R.

9.23 The conditions at the beginning of compression in an air-standard Diesel cycle are fixed by $p_1 = 200$ kPa, $T_1 = 380$ K. The compression ratio is 20 and the heat addition per unit mass is 900 kJ/kg. Determine

(a) the maximum temperature, in K.
(b) the maximum pressure, in kPa.
(c) the cutoff ratio.
(d) the net work per unit mass of air, in kJ/kg.
(e) the thermal efficiency.
(f) the availability transfer from the air accompanying the heat rejection process, in kJ per kg of air. Let $T_0 =$ 300 K, $p_0 = 100$ kPa.

9.24 An air-standard Diesel cycle has a compression ratio of 16 and a cutoff ratio of 2. At the beginning of compression, $p_1 = 14.2$ lbf/in.^{2}, $V_1 = 0.5$ ft^3, and $T_1 = 520$°R. Calculate

(a) the heat added, in Btu.
(b) the heat rejected, in Btu.
(c) the thermal efficiency.
(d) the mean effective pressure, in lbf/in.^{2}
(e) the availability transfer from the air accompanying the heat rejection process, in kJ, for $T_0 = 520$°R, $p_0 =$ 14.2 lbf/in.^{2}

9.25 The displacement volume of an internal combustion engine is 3 L. The processes within each cylinder of the engine are modeled as an air-standard Diesel cycle with a cutoff ratio of 2.5. The state of the air at the beginning of compression is fixed by $p_1 = 95$ kPa, $T_1 = 22$°C, and $V_1 =$ 3.17 L. Determine the net work per cycle, in kJ, and the power developed by the engine, in kW, if the cycle is executed 1000 times per min.

9.26 The state at the beginning of compression of an air-standard Diesel cycle is fixed by $p_1 = 100$ kPa and $T_1 =$ 310 K. The compression ratio is 15. For cutoff ratios of 1.5, 2, and 2.5, determine

(a) the maximum temperature, in K.
(b) the pressure at the end of the expansion, in kPa.
(c) the net work per unit mass of air, in kJ/kg.
(d) the thermal efficiency.

9.27 An air-standard Diesel cycle has a maximum temperature of 1800 K. At the beginning of compression, $p_1 = 95$ kPa and $T_1 = 300$ K. The mass of air is 12 g. For compression ratios of 15, 18, and 21, determine

(a) the net work of the cycle, in kJ.
(b) the thermal efficiency.
(c) the mean effective pressure, in kPa.

9.28 At the beginning of compression in an air-standard Diesel cycle, $p_1 = 96$ kPa, $V_1 = 0.016$ m^3, and $T_1 = 290$ K. The compression ratio is 15 and the heat addition per cycle is 10 kJ. Determine

(a) the mass of air, in kg.
(b) the maximum temperature in the cycle, in K.
(c) the net work, in kJ, and the thermal efficiency.

9.29 Write a computer program to calculate the heat addition per unit mass, the net work per unit mass, the thermal efficiency, and the mean effective pressure of a cold air-standard Diesel cycle. Plot each of the above quantities as functions of the compression ratio for $p_1 = 1$ bar, $T_1 = 300$ K, and for maximum cycle temperatures of 1200, 1500, 1800, and 2100 K.

9.30 An air-standard dual cycle has a compression ratio of 9. At the beginning of compression, $p_1 = 100$ kPa and $T_1 = 300$ K. The heat addition per unit mass of air is 1400 kJ/kg, with two thirds added at constant volume and the rest at constant pressure. Determine

(a) the temperatures at the end of each heat addition process, in K.
(b) the net work of the cycle per unit mass of air, in kJ/kg.
(c) the thermal efficiency.
(d) the mean effective pressure, in kPa.

9.31 Solve Problem 9.30 on a cold air-standard basis with specific heats evaluated at 300 K.

9.32 An air-standard dual cycle has a compression ratio of 17 and a cutoff ratio of 1.2. At the beginning of compression, $p_1 = 95$ kPa and $T_1 = 310$ K. The pressure doubles during the constant volume heat addition process. If the mass of air is 0.5 kg, determine

(a) the heat addition at constant volume and at constant pressure, each in kJ.
(b) the net work of the cycle, in kJ.
(c) the heat rejection, in kJ.
(d) the thermal efficiency.

9.33 The pressure and temperature at the beginning of compression in an air-standard dual cycle are 14.0 lbf/in.^{2} and 520°R, respectively. The compression ratio is 15 and the heat addition per unit mass of air is 800 Btu/lb. At the end of the constant volume heat addition process, the pressure is 1200 lbf/in.^{2} Determine the net work and heat rejection for the cycle, each per unit mass of air, in Btu/lb.

9.34 An air-standard dual cycle has a compression ratio of 16. At the beginning of compression, $p_1 = 14.5$ lbf/in.^{2}, $V_1 = 0.5$ ft^3, and $T_1 = 50$°F. The pressure doubles during the constant volume heat addition process. For maximum cycle temperatures of 3000, 3500, and 4000°R, determine

(a) the heat addition to the cycle, in Btu.
(b) the net work of the cycle, in Btu.

(c) the thermal efficiency.
(d) the mean effective pressure, in $lbf/in.^2$

9.35 Write a computer program to calculate the net work per unit mass, the thermal efficiency, and the mean effective pressure of a cold air-standard dual cycle. Plot each of the above quantities as functions of compression ratio for different fractions of constant volume and constant pressure heat addition. Let $p_1 = 1$ bar, $T_1 = 300$ K, and the total heat addition be 2000 kJ/kg. Consider compression ratios in the range from 10 to 20.

9.36 Air enters the compressor of an ideal air-standard Brayton cycle at 100 kPa, 300 K, with a volumetric flow rate of 5 m^3/s. The compressor pressure ratio is 10. For turbine inlet temperatures of 1000, 1200, and 1600 K, determine

(a) the thermal efficiency of the cycle.
(b) the back work ratio.
(c) the net power developed, in kW.

9.37 Air enters the compressor of an ideal air-standard Brayton cycle at 100 kPa, 300 K, with a volumetric flow rate of 5 m^3/s. The turbine inlet temperature is 1400 K. For compressor pressure ratios of 6, 8, and 12, determine

(a) the thermal efficiency of the cycle.
(b) the back work ratio.
(c) the net power developed, in kW.

9.38 An ideal air-standard Brayton cycle has a compressor pressure ratio of 10. Air enters the compressor at $p_1 =$ 14.7 $lbf/in.^2$, $T_1 = 70$°F, with a mass flow rate of 90,000 lb/h. The turbine inlet temperature is 2200°R. Calculate

(a) the thermal efficiency.
(b) the net power developed, in horsepower.

9.39 The net power developed by an ideal air-standard Brayton cycle is 2.5×10^7 Btu/h. The pressure ratio for the cycle is 12 and the minimum and maximum temperatures are 520°R and 2800°R, respectively. Determine

(a) the thermal efficiency of the cycle.
(b) the mass flow rate of air, in lb/h.

9.40 Solve Problem 9.39 on a cold air-standard basis with specific heats evaluated at 520°R.

9.41 Consider an ideal air-standard Brayton cycle with minimum and maximum temperatures of 300 K and 1500 K, respectively. The pressure ratio is that which maximizes the net work developed by the cycle per unit mass of air flow. On a cold air-standard basis, calculate

(a) the compressor and turbine work per unit mass of air flow, each in kJ/kg.
(b) the thermal efficiency of the cycle.

9.42 On the basis of a cold air-standard analysis, show that the back work ratio of an ideal air-standard Brayton cycle equals the ratio of absolute temperatures at the compressor inlet and the turbine outlet.

9.43 The compressor inlet temperature of an ideal air-standard Brayton cycle is 520°R and the maximum allowable turbine inlet temperature is 2600°R. Plot the net work developed per unit mass of air flow, in Btu/lb, and the thermal efficiency versus compressor pressure ratio for pressure ratios ranging from 12 to 24. Using your plots, estimate the pressure ratio for maximum net work and the corresponding value of thermal efficiency. Compare the results to those obtained in analyzing the cycle on a cold air-standard basis.

9.44 The compressor inlet temperature for an ideal Brayton cycle is T_1 and the turbine inlet temperature is T_3. Using a cold air-standard analysis, show that the temperature T_2 at the compressor exit that maximizes the net work developed per unit mass of air flow is $T_2 = (T_1T_3)^{1/2}$.

9.45 Reconsider Problem 9.36, but include in the analysis that the turbine and compressor each have efficiencies of 80%. Determine for the modified cycle

(a) the thermal efficiency.
(b) the back work ratio.
(c) the net power developed, in kW.
(d) the irreversibility rates in the compressor and turbine, respectively, each in kW, for $T_0 = 300$ K.

9.46 Reconsider Problem 9.38, but include in the analysis that the turbine and compressor have efficiencies of 88 and 84%, respectively. Determine for the modified cycle

(a) the thermal efficiency.
(b) the net power developed, in horsepower.
(c) the irreversibility rates in the compressor and turbine, respectively, each in horsepower, for $T_0 = 70$°F.

9.47 The compressor and turbine of a simple gas turbine each have isentropic efficiencies of 82%. The compressor pressure ratio is 12. The minimum and maximum temperatures are 290 K and 1400 K, respectively. On the basis of an air-standard analysis, compare the values of **(a)** the net work per unit mass of air flowing, in kJ/kg, **(b)** the heat rejected per unit mass of air flowing, in kJ/kg, and **(c)** the thermal efficiency to the same quantities evaluated for an ideal cycle.

9.48 Air enters the compressor of a simple gas turbine at $p_1 = 14$ $lbf/in.^2$, $T_1 = 520$°R, and a volumetric flow rate of 10,000 ft^3/min. The isentropic efficiencies of the compressor and turbine are 83 and 87%, respectively. The compressor pressure ratio is 14 and the temperature at the turbine inlet is 2500°R. On the basis of an air-standard analysis, calculate

(a) the thermal efficiency of the cycle.
(b) the net power developed, in Btu/hr.

9.49 Solve Problem 9.48 on a cold air-standard basis with specific heats evaluated at 520°R.

9.50 Air enters the compressor of a simple gas turbine at 100 kPa, 300 K, with a volumetric flow rate of 5 m^3/s. The compressor pressure ratio is 10 and its isentropic efficiency is 85%. At the inlet to the turbine, the pressure is 950 kPa, and the temperature is 1400 K. The turbine has an isentropic efficiency of 88% and the exit pressure is 100 kPa. On the basis of an air-standard analysis, determine

(a) the thermal efficiency of the cycle.
(b) the net power developed, in kW.

9.51 Air enters the compressor of a simple gas turbine at 14.5 lbf/in.^{2}, 80°F, and exits at 87 lbf/in.^{2}, 514°F. The air enters the turbine at 1540°F, 87 lbf/in.^{2} and expands to 917°F, 14.5 lbf/in.^{2} The compressor and turbine operate adiabatically, and kinetic and potential energy effects are negligible. On the basis of an air-standard analysis, evaluate all availability inputs, destructions, and losses. Let T_0 = 80°F, p_0 = 14.5 lbf/in.^{2}

9.52 Devise and evaluate a second law efficiency for the gas turbine cycle of Problem 9.51. Discuss.

9.53 Write a computer program to calculate the net work per unit mass of air flow and the thermal efficiency of a simple gas turbine on the basis of a cold air-standard analysis. Assume no pressure drops through the heat exchangers and adiabatic processes in the compressor and turbine. Allow for turbine and compressor efficiencies less than 100%. For turbine and compressor efficiencies each equal to **(a)** 100%, **(b)** 90%, and **(c)** 80%, plot thermal efficiency and net work per unit mass of air flow versus compressor pressure ratio for p_1 = 100 kPa and T_1 = 300 K at the compressor inlet and a turbine inlet temperature of 1500 K.

9.54 If a regenerator with an effectiveness of 80% is incorporated into the cycle of Problem 9.45, determine

(a) the thermal efficiency.
(b) the back work ratio.
(c) the net power developed, in kW.
(d) the irreversibility rate of the regenerator, in kW, for T_0 = 300 K.

9.55 If a regenerator with an effectiveness of 80% is incorporated into the cycle of Problem 9.46, determine

(a) the thermal efficiency.
(b) the net power developed, in horsepower.
(c) the irreversibility rate of the regenerator, in horsepower, for T_0 = 70°F.

9.56 Reconsider Problem 9.47, but include a regenerator in the cycle. For regenerator effectiveness values of 100%, 80%, and 60%, calculate

(a) the net work per unit mass of air flowing, in kJ/kg.
(b) the heat rejected per unit mass of air flowing, in kJ/kg.
(c) the thermal efficiency.

9.57 If a regenerator with an effectiveness of 78% is incorporated into the cycle of Problem 9.48, determine

(a) the thermal efficiency.
(b) the percent decrease in heat addition to the air.
(c) the net power developed, in Btu/h.

9.58 On the basis of a cold air-standard analysis, show that the thermal efficiency of an ideal regenerative gas turbine can be expressed as

$$\eta = 1 - \left(\frac{T_1}{T_3}\right)(r)^{(k-1)/k}$$

where r is the compressor pressure ratio, and T_1 and T_3 denote the temperatures at the compressor and turbine inlets, respectively.

9.59 A regenerative gas turbine power plant is shown in Fig. P9.59. Air enters the compressor at 1 bar, 27°C and is compressed to 4 bar. The isentropic efficiency of the compressor is 80%, and the regenerator effectiveness is 90%. All the power developed by the higher-pressure turbine is used to run the compressor and the lower-pressure turbine provides the net power output of 97 kW. Each turbine has an isentropic efficiency of 87% and the temperature at the inlet to the high-pressure turbine is 1200 K. Determine

(a) the mass flow rate of air into the compressor, in kg/s.
(b) the thermal efficiency.
(c) the temperature of the air at the exit of the regenerator, in K.

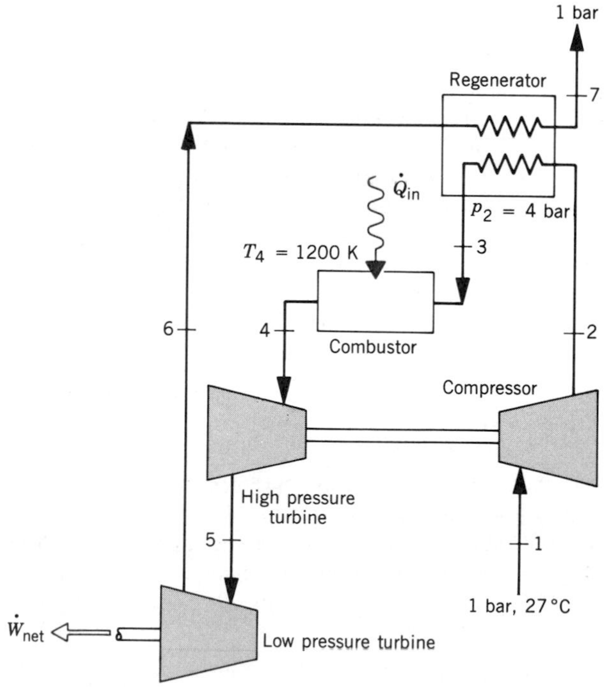

Figure P9.59

9.60 A regenerative gas turbine develops a net power output of 10^7 Btu/h. Air enters the compressor at 14 lbf/

in.^{2}, 540°R, and is compressed to 70 lbf/in.^{2} The air then passes through the regenerator and exits at 1060°R. The temperature at the turbine inlet is 1540°R. The compressor and turbine each have an isentropic efficiency of 80%. Using an air-standard analysis, calculate

(a) the thermal efficiency of the cycle.
(b) the regenerator effectiveness.
(c) the volumetric flow rate of air entering the compressor, in ft^3/min.

9.61 Include a regenerator in the cycle of Problem 9.53 and modify the computer program accordingly. For regenerator effectiveness values of **(a)** 100%, **(b)** 85%, **(c)** 70%, plot thermal efficiency and net work per unit mass of air flow versus pressure ratio. Consider the special case of turbine and compressor efficiencies, each equal to 80%, p_1 = 100 kPa and T_1 = 300 K at the compressor inlet, and a temperature of 1500 K at the inlet to the turbine.

9.62 Air enters the turbine of a gas turbine at 400 kPa, 1200 K, and expands to 100 kPa in two stages. Between the stages, the air is reheated at a constant pressure of 200 kPa to 1200 K. The expansion through each turbine stage is isentropic. Determine, in kJ per kg of air flowing

(a) the work developed by each stage.
(b) the heat transfer for the reheat process.
(c) the increase in net work as compared to a single stage of expansion with no reheat.

9.63 Reconsider Problem 9.62 and include in the analysis that each turbine stage has an isentropic efficiency of 88%.

9.64 Consider a two-stage turbine operating at steady state with reheat at constant pressure between the stages. Show that the maximum work is developed when the pressure ratio is the same across each stage. Use a cold air-standard analysis, assuming the inlet state and the exit pressure are specified, each expansion process is isentropic, and the temperature at the inlet to each turbine stage is the same. Kinetic and potential energy effects can be ignored.

9.65 Repeat the calculations of Example 9.8 for turbine stage efficiencies of 82%, a compressor efficiency of 88%, and a regenerator effectiveness of 80%.

9.66 Consider a modification of the cycle in Problem 9.55 involving a two-stage turbine expansion with reheat between the stages at a constant pressure of 50 lbf/in.^{2} Each turbine stage has an isentropic efficiency of 88% and the temperature of the air entering the second stage is 2000°R. Determine

(a) the thermal efficiency.
(b) the net power developed, in horsepower.

9.67 A two-stage air compressor operates at steady state, compressing 10 m^3/min of air from 1 bar, 300 K, to 10 bar. An intercooler between the two stages cools the air to 300 K at a constant pressure of 3 bar. The compression processes are isentropic. Calculate the power required to run the compressor, in kW, and compare the result to the power required for isentropic compression from the same inlet state to the same final pressure.

9.68 Reconsider Problem 9.67 and include in the analysis that each compressor stage has an isentropic efficiency of 85%.

9.69 Air enters a compressor operating at steady state at 15 lbf/in.^{2}, 60°F, with a volumetric flow rate of 5000 ft^3/min. The compression occurs in two stages, with each stage being a polytropic process with n = 1.3. The air is cooled to 100°F between the stages by an intercooler operating at 40 lbf/in.^{2} Air exits the compressor at 120 lbf/in.^{2} Determine, in Btu per min

(a) the power and heat transfer rate for each compressor stage.
(b) the heat transfer rate for the intercooler.

9.70 The two-stage compressor of Problem 9.68 is part of a regenerative gas turbine. The turbine inlet temperature is 1300 K and the turbine has an isentropic efficiency of 87%. The regenerator effectiveness is 80%. Determine

(a) the thermal efficiency.
(b) the rate of heat addition, in kW.
(c) the net power developed, in kW.

9.71 Modify the cycle of Problem 9.55 by including two-stage compression, with intercooling between the stages at a pressure of 50 lbf/in.^{2} If each compressor stage has an isentropic efficiency of 84% and the temperature of the air entering the second stage is 70°F, determine

(a) the thermal efficiency.
(b) the net power developed, in horsepower.

9.72 Air enters the compressor of a gas turbine at 100 kPa, 300 K, and a volumetric flow rate of 5 m^3/s. The air is compressed in two stages to 1200 kPa, with intercooling to 300 K between the stages at a pressure of 350 kPa. The turbine inlet temperature is 1400 K and the expansion occurs in two stages, with reheat to 1340 K between the stages at a pressure of 350 kPa. The compressor and turbine stage efficiencies are 87 and 85%, respectively. Determine

(a) the thermal efficiency of the cycle.
(b) the back work ratio.
(c) the net power developed, in kW.

9.73 Reconsider Problem 9.72 but include a regenerator with an effectiveness of 80%.

9.74 Modify the cycle of Problem 9.55 by including both compression with intercooling and reheat between turbine stages. Compression occurs in two stages, with intercooling to 70°F between the stages at 50 lbf/in.^{2} The turbine expansion also occurs in two stages, with reheat to 2000°R between the stages at 50 lbf/in.^{2} The isentropic efficiencies of the turbine and compressor stages are 88 and 84%, respectively. Determine for the modified cycle

(a) the thermal efficiency.
(b) the net power developed, in horsepower.

9.75 Air at 22 kPa, 220 K, and 250 m/s enters a turbojet engine in flight at an altitude of 10,000 m (33,000 ft). The pressure ratio across the compressor is 12. The turbine inlet temperature is 1240 K, and the pressure at the nozzle exit is 22 kPa. The diffuser and nozzle processes are isentropic, the compressor and turbine have isentropic efficiencies of 85 and 88%, respectively, and there is no pressure drop for flow through the combustor. On the basis of an air-standard analysis, determine

(a) the pressures and temperatures at each principal state, in kPa and K, respectively.
(b) the velocity at the nozzle exit, in m/s.

Neglect kinetic energy except at the diffuser inlet and the nozzle exit.

9.76 Air enters the diffuser of a turbojet engine with a mass flow rate of 150 lb/s at 12 lbf/in.^{2}, 420°R, and a velocity of 800 ft/s. The pressure ratio for the compressor is 10, and its isentropic efficiency is 87%. Air enters the turbine at 2250°R with the same pressure as at the exit of the compressor. Air exits the nozzle at 12 lbf/in.^{2} The nozzle and diffuser operate isentropically and the turbine has an isentropic efficiency of 90%. On the basis of an air-standard analysis, calculate

(a) the rate of heat addition, in Btu/h.
(b) the pressure at the turbine exit, in lbf/in.^{2}
(c) the compressor power input, in Btu/h.
(d) the velocity at the nozzle exit, in ft/s.

Neglect kinetic energy except at the diffuser inlet and the nozzle exit.

9.77 Consider the addition of an afterburner to the turbojet in Problem 9.75 that raises the temperature at the inlet of the nozzle to 1100 K. Determine the velocity at the nozzle exit, in m/s.

9.78 Air enters the diffuser of a ramjet engine at 6 lbf/in.^{2}, 420°R, with a velocity of 1600 ft/s, and is decelerated to zero velocity. After combustion, the gases reach a temperature of 2200°R before being discharged through the nozzle at 6 lbf/in.^{2} On the basis of an air-standard analysis, determine

(a) the pressure at the diffuser exit, in lbf/in.^{2}
(b) the velocity at the nozzle exit, in ft/s.

Neglect kinetic energy except at the diffuser inlet and the nozzle exit.

9.79 Air enters the diffuser of a ramjet engine at 25 kPa, 220 K, with a velocity of 3080 km/h and is decelerated to negligible velocity. On the basis of an air-standard analysis, the heat addition is 900 kJ per kg of air passing through the engine. Air exits the nozzle at 25 kPa. Determine

(a) the pressure at the diffuser exit, in kPa.
(b) the velocity at the nozzle exit, in m/s.

Neglect kinetic energy except at the diffuser inlet and the nozzle exit.

9.80 In a turboprop engine, the turbine drives a propeller as well as the compressor. Air enters the diffuser with a mass flow rate of 45 kg/s at 40 kPa, 240 K, and a velocity of 180 m/s, and is decelerated to zero velocity. The compressor pressure ratio is 10 and the compressor has an isentropic efficiency of 85%. The turbine inlet temperature is 1140 K, and its isentropic efficiency is 85%. The turbine exit pressure is 50 kPa. Using an air-standard analysis, determine

(a) the power delivered to the propeller, in kW.
(b) the velocity at the nozzle exit, in m/s.

Neglect kinetic energy except at the diffuser inlet and the nozzle exit.

9.81 A combined gas turbine-vapor power plant has a net power output of 10 MW. Air enters the compressor of the gas turbine at 100 kPa, 300 K, and is compressed to 1200 kPa. The isentropic efficiency of the compressor is 84%. The conditions at the inlet to the turbine are 1200 kPa and 1400 K. Air expands through the turbine, that has an isentropic efficiency of 88%, to a pressure of 100 kPa. The air then passes through the interconnecting heat exchanger, and is finally discharged at 480 K. Steam enters the turbine of the vapor power cycle at 8 MPa, 400°C, and expands to the condenser pressure of 8 kPa. Water enters the pump as saturated liquid at 8 kPa. The turbine and pump have isentropic efficiencies of 90 and 80%, respectively. Determine

(a) the mass flow rates of air and water, each in kg/s.
(b) the rate of heat transfer to the combined cycle, in MW.
(c) the thermal efficiency of the combined cycle.

9.82 For the combined cycle of Problem 9.81, determine the availability increase of the air as it passes through the combustor of the gas turbine, in MW. Account for this availability by calculating all availability outputs, destructions, and losses for the combined cycle. Let T_0 = 27°C, p_0 = 100 kPa.

9.83 A simple gas turbine is the topping cycle for a simple vapor power cycle. Air enters the compressor of the gas turbine at 60°F, 14.7 lbf/in.^{2}, with a volumetric flow rate of 40,000 ft^3/min. The compressor pressure ratio is 12 and the turbine inlet temperature is 2600°R. The compressor and turbine each have isentropic efficiencies of 88%. The air leaves the interconnecting heat exchanger at 840°R, 14.7 lbf/in.^{2} Steam enters the turbine of the vapor cycle at 1000 lbf/in.^{2}, 900°F, and expands to the condenser pressure of 1 lbf/in.^{2} Water enters the pump as saturated liquid at 1 lbf/in.^{2} The turbine and pump efficiencies are 90 and 70%, respectively. Cooling water passing through the condenser experiences a temperature rise from 60 to 80°F with a negligible change in pressure. Determine

(a) the power developed by the air and steam turbines, each in Btu/h.
(b) the thermal efficiency of the combined cycle.
(c) the cooling water flow rate, in lb/h.

9.84 For the combined cycle of Problem 9.83, determine the increase in availability of the air as it passes through the combustor of the gas turbine, in Btu/h. Account for this availability by calculating all availability outputs, destructions, and losses for the combined cycle. Let T_0 = 60°F, p_0 = 14.7 lbf/in.^{2}

9.85 Hydrogen enters the turbine of an Ericsson cycle at 920 K, 15 bar, with a mass flow rate of 1 kg/s. The temperature and pressure at the inlet to the compressor are 300 K and 1.5 bar, respectively. Determine

(a) the net power developed, in kW.
(b) the thermal efficiency.
(c) the back work ratio.

9.86 Air is the working fluid in an Ericsson cycle. Expansion through the turbine from 150 lbf/in.^{2} at the inlet to 15 lbf/in.^{2} at the exit takes place at a constant temperature of 2000°R. Heat transfer from the compressor occurs at 520°R. Determine

(a) the net work, in Btu per lb of air flowing.
(b) the thermal efficiency.

9.87 Thirty-six grams of air undergo a Stirling cycle with a compression ratio of six. At the beginning of the isothermal compression, the pressure and volume are 1 bar and 0.03 m^3, respectively. The temperature during the isothermal expansion is 1000 K. Determine

(a) the net work, in kJ.
(b) the thermal efficiency.
(c) the mean effective pressure, in bar.

9.88 Steam is the working fluid in a Stirling cycle. In the isothermal compression, the steam is compressed from 1 lbf/in.^{2}, 400°F, to 100 lbf/in.^{2} The isothermal expansion occurs at 600°F. Determine

(a) the work and heat transfer, in Btu per lb of steam for each process in the cycle.
(b) the thermal efficiency.

9.89 If the mass flow rate of air for the cycle in Problem 9.75 is 50 kg/s, calculate the thrust developed by the engine, in kN.

9.90 Calculate the thrust developed by the turbojet engine in Problem 9.76, in lbf.

9.91 Calculate the ratio of the thrust developed to the mass flow rate of air, in N per kg/s, for the ramjet engine in Problem 9.79.

9.92 Air flows at steady state through a horizontal, well-insulated, constant area duct whose diameter is 0.2 m. At the inlet, p_1 = 380 kPa, T_1 = 400 K, and the velocity is 30 m/s. The temperature of the air leaving the duct is 390 K. Determine

(a) the velocity at the exit, in m/s.
(b) the pressure at the exit, in kPa.
(c) the magnitude of the net horizontal force exerted by the duct wall on the air, in N. In which direction does this force act?

9.93 Air flows at steady state through a horizontal, well-insulated duct of varying cross-sectional area, entering at 10 lbf/in.^{2}, 60°F, with a velocity of 1000 ft/s and a mass flow rate of 75 lb/s. The air undergoes an isentropic process to an exit pressure of 15 lbf/in.^{2} Calculate

(a) the flow area at the exit, in ft^2.
(b) the force exerted by the duct wall on the air in the direction of flow, in lbf.

9.94 Using the ideal gas model, determine the sonic velocity, in m/s, for

(a) air at 1000 K.
(b) carbon dioxide at 500 K.
(c) helium at 300 K.

9.95 Determine the Mach number of air at 1 bar, 400 m/s, and a temperature of

(a) 300 K.
(b) 500 K.
(c) 1000 K.

9.96 Calculate the Mach number of carbon dioxide at 1 atm, 1500 ft/s, and a temperature of

(a) 500°R.
(b) 1000°R.
(c) 1500°R.

9.97 For a stream of air at 100 kPa, 340 K, and 200 m/s, determine

(a) the Mach number.
(b) the stagnation temperature, in K.
(c) the stagnation pressure, in kPa.

9.98 For a stream of carbon dioxide at 30 lbf/in.^{2}, 440°R, and 800 ft/s, determine

(a) the Mach number.
(b) the stagnation temperature, in °R.
(c) the stagnation pressure, in lbf/in.^{2}

9.99 Using the Mollier diagram, Fig. A-8, determine for water vapor at 2 MPa, 520 K, and 300 m/s

(a) the stagnation enthalpy, in kJ/kg.
(b) the stagnation pressure, in MPa.
(c) the stagnation temperature, in K.

9.100 Air enters a nozzle operating at steady state at 3 bar, 440 K, with a velocity of 145 m/s, and expands isentropically to an exit velocity of 460 m/s. Determine

(a) the exit pressure, in bar.
(b) the ratio of the exit area to the inlet area.
(c) whether the nozzle is diverging only, converging only, or converging–diverging in cross section.

9.101 For the same nozzle inlet conditions and exit pressure as in Problem 9.100, calculate the exit velocity, in m/s, for adiabatic operation with an isentropic nozzle efficiency of 90%.

9.102 Steam expands isentropically through a converging nozzle from a large tank at 10 MPa, 360°C. The exit area of the nozzle is 5 cm^2. If the nozzle is choked at an exit pressure of 5.5 MPa, determine the exit velocity, in m/s, the exit plane pressure, in bar, and the mass flow rate of steam, in kg/s, for back pressures of

(a) 8 MPa.
(b) 6 MPa.
(c) 4 MPa.

9.103 Air enters a nozzle operating at steady state, with negligible velocity at 100 lbf/in.^{2}, 860°R, and expands isentropically. For a mass flow rate of 4 lb/s, calculate the velocity, in ft/s, the Mach number, and the cross-sectional area, in ft^2, at locations within the nozzle where the pressure is

(a) 80 lbf/in.^{2}
(b) 60 lbf/in.^{2}
(c) 50 lbf/in.^{2}
(d) 40 lbf/in.^{2}

Plot velocity, Mach number, and area versus pressure, and using the plots, determine the pressure at which the flow is sonic.

9.104 An ideal gas mixture with $k = 1.3$ and a molecular weight of 15 is supplied to a converging nozzle at $p_o = 7$ bar, $T_o = 560$ K, which discharges into a region where the pressure is 1 bar. For steady isentropic flow through the nozzle with a mass flow rate of 0.5 kg/s, determine at the nozzle exit

(a) the temperature of the gas, in K.
(b) the velocity of the gas, in m/s.
(c) the flow area, in m^2.

9.105 If the nozzle of Problem 9.104 has an isentropic nozzle efficiency of 86%, determine the exit temperature, in K, and the exit velocity, in m/s, for adiabatic operation.

9.106 For the isentropic flow of an ideal gas with constant specific heat ratio k, the ratio of the temperature T^* to the stagnation temperature T_o is $T^*/T_o = 2/(k + 1)$. Develop this relationship.

9.107 Air at $p_o = 1.4$ bar, $T_o = 280$ K expands isentropically through a converging nozzle and discharges to the atmosphere at 1 bar. The exit plane area is 0.0013 m^2. Determine the mass flow rate, in kg/s. Is it possible to increase the mass flow rate by increasing the supply region pressure?

9.108 Air as an ideal gas with $k = 1.4$ undergoes a normal shock. The upstream conditions are $p_x = 0.5$ bar, $T_x = 280$ K, and $M_x = 1.8$. Determine

(a) the pressure p_y, in bar.
(b) the stagnation pressure p_{ox}, in bar.
(c) the stagnation temperature T_{ox}, in K.
(d) the change in specific entropy across the shock, in kJ/kg·K.

9.109 A converging–diverging nozzle operating at steady state has an exit area of 2 in.^{2} Air as an ideal gas with $k = 1.4$ flows through the nozzle, discharging to the atmosphere at 14.7 lbf/in.^{2} and 540°R. A normal shock stands at the exit plane with $M_x = 2$. Up to the shock, the flow is isentropic. Determine

(a) the stagnation pressure p_{ox}, in lbf/in.^{2}
(b) the stagnation temperature T_{ox}, in °R.
(c) the mass flow rate of air, in lb/s.

9.110 For the nozzle in Problem 9.109, calculate the throat area, in in.^{2}, and the entropy produced, in Btu/°R per lb of air flowing.

9.111 Air at 3.4 bar, 530 K, and 180 m/s enters a converging–diverging nozzle operating at steady state. A normal shock stands in the diverging section at a location where the Mach number is $M_x = 2$. The flow is isentropic, except where the shock stands. If the air behaves as an ideal gas with $k = 1.4$, determine

(a) the stagnation temperature T_{ox}, in K.
(b) the stagnation pressure p_{ox}, in bar.
(c) the pressure p_x, in bar.
(d) the pressure p_y, in bar.
(e) the stagnation pressure p_{oy}, in bar.
(f) the stagnation temperature T_{oy}, in K.

If the mass flow rate of air is 0.5 kg/s, and the exit plane pressure is 2.4 bar, determine the throat area and the exit plane area, each in m^2.

9.112 Air as an ideal gas with $k = 1.4$ enters a converging–diverging nozzle operating at steady state with negligible velocity at 150 lbf/in.^{2}, 660°R. A normal shock stands at the exit plane, and $p_x = 10.5$ lbf/in.^{2} Up to the shock, the flow is isentropic. For a throat area of 1 in.^{2}, calculate

(a) the mass flow rate of air, in lb/s.
(b) the Mach number M_x, velocity V_x, in ft/s, and the cross-sectional area A_x, in in.^{2}

9.113 For the nozzle in Problem 9.112, determine

(a) the pressure in the back region, in lbf/in.^{2}
(b) the change in stagnation pressure across the shock, in lbf/in.^{2}

(c) the irreversibility, in Btu per lb of air flowing, for $T_0 = 60°F$.

9.114 Derive the following expressions: **(a)** Eq. 9.55, **(b)** Eq. 9.56, **(c)** Eq. 9.57.

9.115 Write a computer program to generate tables of the same isentropic flow functions as in Table 9.1(a) for specific heat ratios of 1.2, 1.3, 1.4, and 1.67 and Mach numbers ranging from 0 to 5.

9.116 Write a computer program to generate tables of the same normal shock functions as in Table 9.1(b) for specific heat ratios of 1.2, 1.3, 1.4, and 1.67 and Mach numbers ranging from 1 to 5.

DESIGN AND OPEN ENDED PROBLEMS

9.1D With the development of high-strength metal–ceramic materials, internal combustion engines can now be built with no cylinder wall cooling. These *adiabatic engines* operate with cylinder wall temperatures as high as 1700°F. What are the important considerations for adiabatic engine design? Are they likely to find widespread application? Discuss.

9.2D An important factor leading to the increases in the performance of gas turbines over the years has been the use of *turbine blade cooling* to allow higher turbine inlet temperatures. With the aid of diagrams, explain the various methods of turbine blade cooling that are commonly used. What are some of the most important research issues in gas turbine blade technology today? Discuss.

9.3D *Steam-injected* gas turbines use hot turbine exhaust gases to produce steam that is injected directly into the gas turbine system. Figure P9.3D illustrates two possible approaches to steam injection. In one approach, steam is injected directly into the combustor. In the other approach, steam is injected into the low-pressure turbine stages.

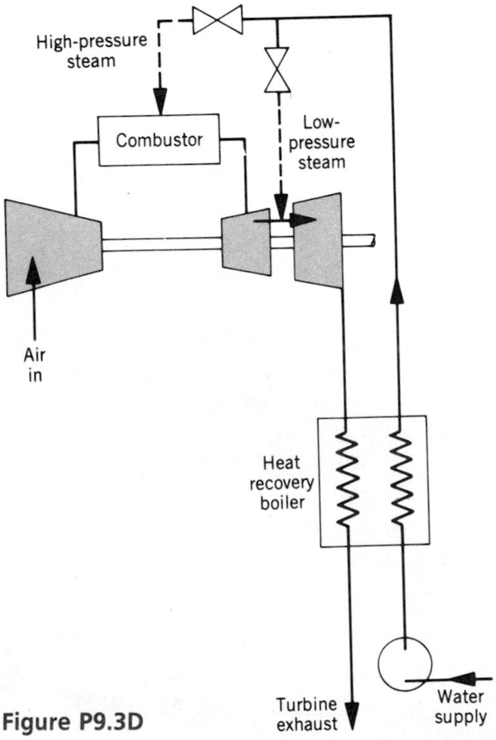

Figure P9.3D

(a) What are the relative advantages and disadvantages of these steam injection approaches? Why does the steam injection concept show promise for better power cycle performance? Discuss.

(b) Each of the steam injection approaches is inherently simpler than combined cycles like the one illustrated in Fig. 9.23. What are the other advantages and disadvantages of direct steam injection compared to combined cycles?

9.4D Using both energy and availability concepts, compare the Rankine and *Kalina* cycles as the bottoming cycle of a combined gas turbine–vapor power plant. Employ an ammonia–water mixture as the working fluid for the Kalina cycle. What are the advantages and disadvantages of each type of bottoming cycle? Does the Kalina cycle hold much promise for future application? Discuss.

9.5D What are the operating principles of *magnetohydrodynamic (MHD)* generators for direct energy conversion? Provide schematics of how MHD could be used in a topping plant for a conventional vapor power plant. What overall thermal efficiencies might be achieved? What barriers must be overcome for MHD to become a practical reality?

9.6D A manufacturing company purchases 80,000 MW·h of electricity annually. A natural gas-fired boiler in the plant provides energy for process loads equal to three times the electrical load. The plant manager has received a proposal for a *cogeneration* system employing a gas turbine fueled by natural gas to run an electric generator instead of purchasing electric power. A heat exchanger would be installed at the turbine exit to recover energy from the exhaust to meet the process loads. Estimate the possible fuel savings that might be achieved with the proposed gas turbine arrangement. Discuss the economic considerations that the manager must consider in making a decision about installing the proposed system.

9.7D Develop sketches of a possible piston–cylinder and regenerator arrangement for a 50-horsepower *Stirling* engine. Include such factors as the number and configuration of cylinders, the method of regeneration, the working fluid, and the operating pressures and temperatures. In what ways would the operation of your engine resemble the Stirling cycle? How would it be different?

9.8D A *wind tunnel* is being designed with a straight test section having a cross-sectional area of 0.1 m^2, as illustrated in Fig. 9.8D. Air is to be brought into the test section at a Mach number of 2 through a converging–diverging nozzle. Downstream of the test section, the air is reduced to a negligible velocity before entering the fan that discharges to the atmosphere. Specify the preliminary design of the tunnel, including the nozzle and diffuser throat areas, the mass flow rate at steady state, and the power requirement. Justify any simplifying assumptions you make.

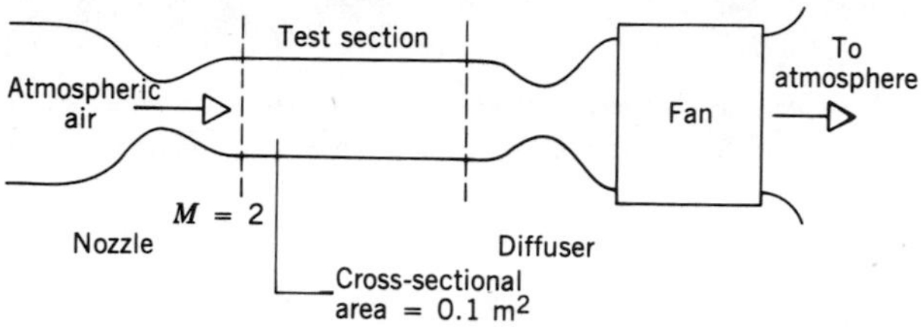

Figure 9.8D

CHAPTER 10

REFRIGERATION AND HEAT PUMP SYSTEMS

Refrigeration systems for food preservation and air conditioning play prominent roles in our everyday lives. Heat pumps are also being utilized increasingly for heating buildings and for the production of industrial process heat. There are many other examples of commercial and industrial uses of refrigeration, including air separation to obtain liquid oxygen and liquid nitrogen, liquefaction of natural gas, and production of ice.

The objective of this chapter is to describe some of the common types of refrigeration and heat pump systems presently in use and to illustrate how such systems can be modeled thermodynamically. The three principal types described are the vapor-compression, absorption, and reversed Brayton cycles. As for the power systems studied in Chaps. 8 and 9, both vapor and gas systems are considered. In vapor systems the refrigerant is alternatively vaporized and condensed. In gas refrigeration systems the refrigerant remains a gas.

10.1 PRELIMINARIES

To introduce some important aspects of refrigeration, let us begin by considering a Carnot vapor refrigeration cycle. This cycle is obtained by reversing the Carnot vapor power cycle introduced in Sec. 5.7. Figure 10.1 shows the schematic and accompanying T–s diagram of a Carnot refrigeration cycle operating between a region at temperature T_C and another region at a higher temperature T_H. The cycle is executed by a refrigerant circulating steadily through a series of components. All processes are internally reversible. Also, since heat transfers between the refrigerant and each region occur with no temperature differences, there are no external irreversibilities. The energy transfers shown on the diagram are positive in the directions indicated by the arrows.

Let us follow the refrigerant as it passes steadily through each of the components in the cycle, beginning at the inlet to the evaporator. The refrigerant enters the evaporator as a two-phase liquid–vapor mixture at state 4. In the evaporator some of the refrigerant changes phase from liquid to vapor as a result of heat transfer from

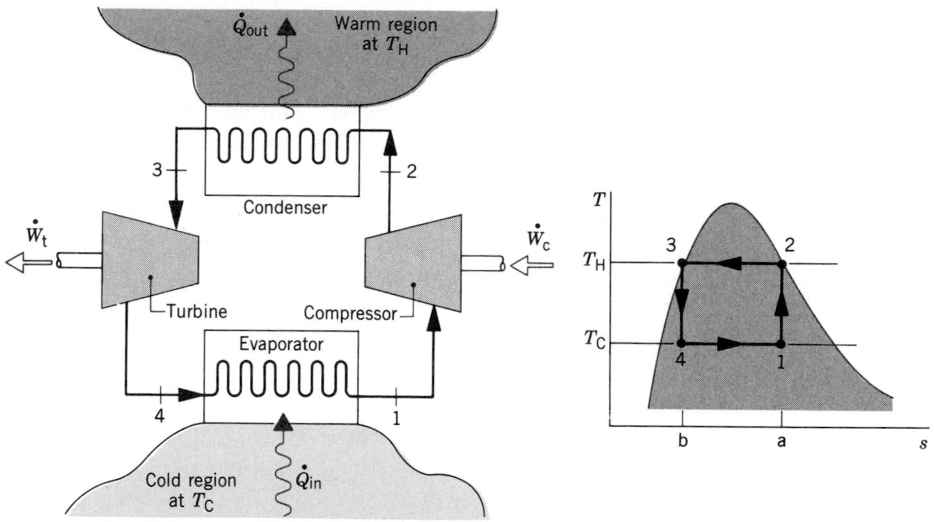

Figure 10.1 Carnot vapor refrigeration cycle.

the region at temperature T_C to the refrigerant. The temperature and pressure of the refrigerant remain constant during the process from state 4 to state 1. The refrigerant is then compressed adiabatically from state 1, where it is a two-phase liquid–vapor mixture, to state 2, where it is a saturated vapor. During this process, the temperature of the refrigerant increases from T_C to T_H, and the pressure also increases. The refrigerant passes from the compressor into the condenser, where it changes phase from saturated vapor to saturated liquid as a result of heat transfer to the region at temperature T_H. The temperature and pressure remain constant in the process from state 2 to state 3. The refrigerant returns to the state at the inlet of the evaporator by expanding adiabatically through a turbine. In this process, from state 3 to state 4 the temperature decreases from T_H to T_C, and there is a decrease in pressure.

Since the Carnot vapor refrigeration cycle is made up of reversible processes, areas on the T–s diagram can be interpreted as heat transfers. Area 1–a–b–4–1 is the heat added to the refrigerant from the cold region per unit mass of refrigerant flowing. Area 2–a–b–3–2 is the heat discharged from the refrigerant to the warm region per unit mass of refrigerant flowing. The enclosed area 1–2–3–4–1 is the *net* heat transfer *from* the refrigerant. The net heat transfer *from* the refrigerant equals the net work *done on* the refrigerant. The net work is the difference between the compressor work input and the turbine work output.

The coefficient of performance β of *any* refrigeration cycle is the ratio of the refrigeration effect to the net work input required to achieve that effect. For the Carnot vapor refrigeration cycle shown in Fig. 10.1, the coefficient of performance is

$$\begin{aligned}\beta_{\max} &= \frac{\dot{Q}_{in}/\dot{m}}{\dot{W}_c/\dot{m} - \dot{W}_t/\dot{m}} \\ &= \frac{\text{area } 1\text{–a–b–4–1}}{\text{area } 1\text{–2–3–4–1}} = \frac{T_C(s_a - s_b)}{(T_H - T_C)(s_a - s_b)} \\ &= \frac{T_C}{T_H - T_C} \qquad (10.1)\end{aligned}$$

This equation, which corresponds to Eq. 5.9, represents the *maximum* theoretical coefficient of performance of *any* refrigeration cycle operating between regions at T_C and T_H.

Actual vapor refrigeration systems depart significantly from the ideal cycle considered above and have coefficients of performance lower than would be calculated from Eq. 10.1. One of the most significant departures from ideality is related to the heat transfers between the refrigerant and the two regions. In actual systems, these heat transfers are not accomplished reversibly as presumed above. In particular, to achieve a rate of heat transfer sufficient to maintain the temperature of the cold region at T_C with a practical-sized evaporator requires the temperature of the refrigerant in the evaporator, T'_C, to be several degrees *below* T_C. This is illustrated by the placement of the temperature T'_C on the T–s diagram of Fig. 10.2. Similarly, to obtain a sufficient heat transfer rate from the refrigerant to the warm region requires that the refrigerant temperature in the condenser, T'_H, be several degrees *above* T_H. This is illustrated by the placement of the temperature T'_H on the T–s diagram of Fig. 10.2.

Maintaining the refrigerant temperatures in the heat exchangers at T'_C and T'_H rather than at T_C and T_H, respectively, has the effect of reducing the coefficient of performance. This can be seen by expressing the coefficient of performance of the refrigeration cycle designated by 1′–2′–3′–4′–1′ on Fig. 10.2 as

$$\beta' = \frac{\text{area } 1'\text{–a–b–}4'\text{–}1}{\text{area } 1'\text{–}2'\text{–}3'\text{–}4'\text{–}1'} = \frac{T'_C}{T'_H - T'_C} \tag{10.2}$$

Comparing the areas underlying the expressions for β_{max} and β' given above, we conclude that the value of β' is less than β_{max}. This conclusion about the effect of refrigerant temperature on the coefficient of performance also applies to other refrigeration cycles considered in the chapter.

Even when the temperature differences between the refrigerant and warm and cold regions are taken into consideration, there are other features that make the Carnot vapor refrigeration cycle impractical as a prototype. Referring again to the Carnot cycle in Fig. 10.1, note that the compression process from state 1 to state 2 occurs with the refrigerant as a two-phase liquid–vapor mixture. This is commonly referred to as *wet compression*. Wet compression is normally avoided because the presence of liquid droplets can damage the compressor. In actual systems, the compressor handles vapor only. This is known as *dry compression*.

Another feature that makes the Carnot cycle impractical is the expansion process from the saturated liquid state 3 on Fig. 10.1 to the low-quality, two-phase liquid–vapor mixture state 4. This expansion produces a relatively small amount of work compared to the work input in the compression process. The work output achieved by an actual turbine would be smaller yet because turbines operating under these conditions typically have low efficiencies. Accordingly, the work output of the turbine is sacrificed by substituting a simple throttling valve for the expansion turbine,

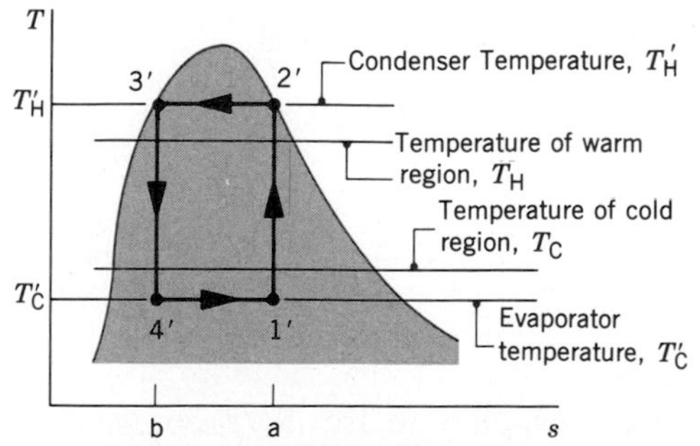

Figure 10.2 Comparison of the condenser and evaporator temperatures with those of the warm and cold regions.

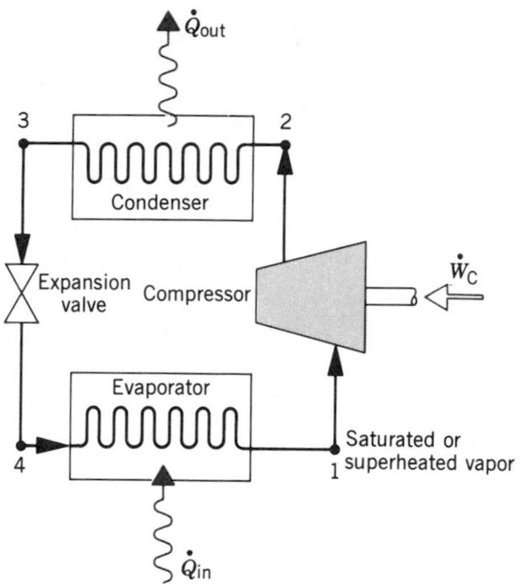

Figure 10.3 Components of a vapor-compression refrigeration system.

with consequent savings in initial and maintenance costs. The components of the resulting cycle are illustrated in Fig. 10.3, where dry compression is presumed. This cycle, known as the *vapor-compression refrigeration cycle,* is the subject of the section to follow.

10.2 VAPOR-COMPRESSION REFRIGERATION

Vapor-compression refrigeration systems are the most common refrigeration systems in use today. The object of this section is to introduce some important features of systems of this type and to illustrate how they are modeled thermodynamically.

10.2.1 Principal Work and Heat Transfers

Let us consider the steady-state operation of the vapor-compression system illustrated in Fig. 10.3. Shown on the figure are the principal work and heat transfers, which are positive in the directions of the arrows. Kinetic and potential energy changes are neglected in the following analyses of the components. We begin with the evaporator, where the desired refrigeration effect is achieved.

As the refrigerant passes through the evaporator, heat transfer from the refrigerated space results in the vaporization of the refrigerant. For a control volume enclosing the refrigerant side of the evaporator, the mass and energy rate balances reduce to give the rate of heat transfer per unit mass of refrigerant flow

$$\frac{\dot{Q}_{in}}{\dot{m}} = h_1 - h_4 \tag{10.3}$$

where $\dot{m}$ is the mass flow rate of the refrigerant. The heat transfer rate $\dot{Q}_{in}$ is referred to as the *refrigeration capacity*. In the SI unit system, the capacity is normally expressed in kW. In the English unit system, the refrigeration capacity may be expressed in Btu/h. Another commonly used unit for the refrigeration capacity is the *ton of refrigeration,* which is equal to 200 Btu/min or about 211 kJ/min.

The refrigerant leaving the evaporator is compressed to a relatively high pressure and temperature by the compressor. Assuming no heat transfer to or from the compressor, the mass and energy rate balances for a control volume enclosing the compressor give

$$\frac{\dot{W}_c}{\dot{m}} = h_2 - h_1 \tag{10.4}$$

where $\dot{W}_c/\dot{m}$ is the rate of power *input* per unit mass of refrigerant.

Next, the refrigerant passes through the condenser, where the refrigerant condenses and there is heat transfer from the refrigerant to the cooler surroundings. For a control volume enclosing the refrigerant side of the condenser, the rate of heat transfer from the refrigerant per unit mass of refrigerant is

$$\frac{\dot{Q}_{out}}{\dot{m}} = h_2 - h_3 \tag{10.5}$$

Finally, the refrigerant at state 3 enters the expansion valve and expands to the evaporator pressure. This process is usually modeled as a *throttling* process, during which

$$h_4 = h_3 \tag{10.6}$$

The refrigerant pressure decreases in the irreversible adiabatic expansion, and there is an accompanying increase in specific entropy. The refrigerant exits the valve at state 4 as a two-phase liquid–vapor mixture.

In the vapor-compression system, the net power input is equal to the compressor power, since the expansion valve involves no power input or output. Using the quantities and expressions introduced above, the coefficient of performance of the vapor-compression refrigeration system of Fig. 10.3 is

$$\beta = \frac{\dot{Q}_{in}/\dot{m}}{\dot{W}_c/\dot{m}} = \frac{h_1 - h_4}{h_2 - h_1} \tag{10.7}$$

Provided states 1 through 4 are fixed, Eqs. 10.3 through 10.7 can be used to evaluate the principal work and heat transfers and the coefficient of performance of the vapor-compression system shown in Fig. 10.3. Since these equations have been developed by reducing mass and energy rate balances, they apply equally for actual performance when irreversibilities are present in the evaporator, compressor, and condenser and for idealized performance in the absence of such effects. Although irreversibilities in the evaporator, compressor, and condenser can have a pronounced effect on overall performance, it is instructive to consider an idealized cycle in which they are assumed absent. Such a cycle establishes an upper limit on the performance of the vapor-compression refrigeration cycle. It is considered next.

10.2.2 Performance of Vapor-Compression Systems

If irreversibilities within the evaporator, compressor, and condenser are ignored, there are no frictional pressure drops, and the refrigerant flows at constant pressure through the two heat exchangers. If stray heat transfers to the surroundings are also ignored, the compression process is isentropic. With these considerations, the *ideal* vapor-compression refrigeration cycle labeled 1–2s–3–4–1 on the *T–s* diagram of Fig. 10.4 results. The cycle consists of the following series of processes:

Process 1–2s: *Isentropic* compression of the refrigerant from state 1 to the condenser pressure at state 2s.

Process 2s–3: Heat transfer *from* the refrigerant as it flows at constant pressure through the condenser. The refrigerant exits as a liquid at state 3.

Process 3–4: *Throttling* process from state 3 to a two-phase liquid–vapor mixture at 4.

Process 4–1: Heat transfer *to* the refrigerant as it flows at constant pressure through the evaporator to complete the cycle.

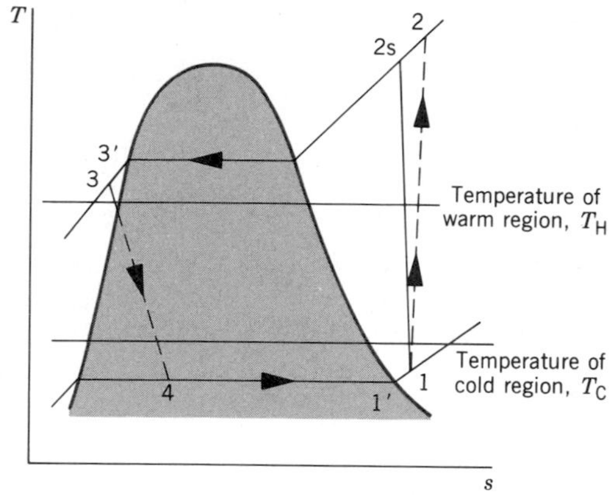

Figure 10.4 *T–s* diagrams of vapor-compression refrigeration cycles.

All of the processes in the above cycle are internally reversible except for the throttling process. Despite the inclusion of this irreversible process, the cycle is commonly referred to as *ideal*. The ideal cycle is often represented with a saturated vapor, state 1′, at the compressor inlet and a saturated liquid, state 3′, at the condenser exit.

The operating temperatures of the vapor-compression refrigeration cycle are established by the temperature T_C to be maintained in the cold region and the temperature T_H of the warm region to which heat is discharged. As shown on Fig. 10.4, the refrigerant temperature in the evaporator must be less than T_C and the refrigerant temperature in the condenser must be greater than T_H. Examples to follow illustrate that the coefficient of performance decreases as the average temperature of the refrigerant in the evaporator decreases and as the average temperature of the refrigerant in the condenser increases.

Also shown in Fig. 10.4 is cycle 1–2–3–4–1, which illustrates a key departure of actual system performance from that of the ideal cycle. This is the departure owing to internal irreversibilities present during compression, suggested by the use of a dashed line for the compression process from state 1 to state 2. The dashed line is drawn to show the increase in specific entropy that would accompany an *adiabatic* irreversible compression. If sufficient heat transfer occurred from the compressor to its surroundings, the specific entropy at state 2 could be less than at state 1, however. Comparing cycle 1–2–3–4–1 with the corresponding ideal cycle 1–2s–3–4–1, we see that the refrigeration capacity is the same for each, but the work input is greater in the case of irreversible compression than in the ideal cycle. Accordingly, the coefficient of performance of cycle 1–2–3–4–1 is less than that of cycle 1–2s–3–4–1. The effect of irreversible compression can be accounted for by using the isentropic compressor efficiency, which for states designated as in Fig. 10.4 is given by

$$\eta_c = \frac{(\dot{W}_c/\dot{m})_s}{(\dot{W}_c/\dot{m})} = \frac{h_{2s} - h_1}{h_2 - h_1}$$

Additional departures from ideality stem from frictional effects that result in pressure drops as the refrigerant flows through the evaporator, condenser, and piping connecting the various components. These pressure drops are not shown on the *T–s* diagram of Fig. 10.4 and are ignored in subsequent discussions for simplicity. Many of the main points brought out in this discussion of refrigeration system performance are illustrated in the following three examples.

EXAMPLE 10.1

PROBLEM

Refrigerant 12 is the working fluid in an ideal vapor-compression refrigeration cycle that communicates thermally with a cold region at 20°C and a warm region at 40°C. Saturated vapor enters the compressor at 20°C and saturated liquid leaves the condenser at 40°C. The mass flow rate of the refrigerant is 0.008 kg/s. Determine (a) the compressor power, in kW, (b) the refrigeration capacity, in tons, (c) the coefficient of performance, and (d) the coefficient of performance of a Carnot refrigeration cycle operating between warm and cold regions at 40 and 20°C, respectively.

SOLUTION

Known: An ideal vapor-compression refrigeration cycle operates with Refrigerant 12. The states of the refrigerant entering the compressor and leaving the condenser are specified, and the mass flow rate is given.

Find: Determine the compressor power, in kW, the refrigeration capacity, in tons, the coefficient of performance, and the coefficient of performance of a Carnot vapor refrigeration cycle operating between warm and cold regions at the specified temperatures.

Schematic and Given Data:

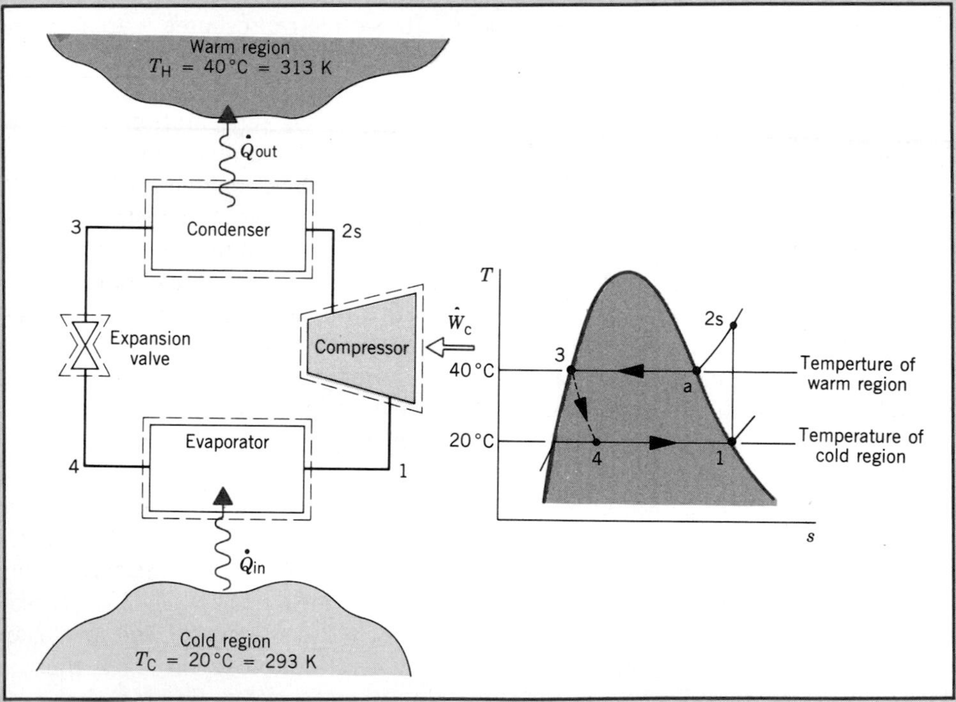

Figure E10.1

Assumptions:

1. Each component of the cycle is analyzed as a control volume at steady state. The control volumes are indicated by dashed lines on the accompanying sketch.
2. Except for the expansion through the valve, which is a throttling process, all processes of the refrigerant are internally reversible.
3. The compressor and expansion valve operate adiabatically.
4. Kinetic and potential energy effects are negligible.
5. Saturated vapor enters the compressor, and saturated liquid leaves the condenser.

Analysis: Let us begin by fixing each of the principal states located on the accompanying schematic and T–s diagrams. At the inlet to the compressor, the refrigerant is a saturated vapor at 20°C, so from Table A-7, $h_1 = 195.78$ kJ/kg and $s_1 = 0.6884$ kJ/kg·K.

The pressure at state 2s is the saturation pressure corresponding to 40°C, or $p_2 = 9.6065$ bar. State 2s is fixed by p_2 and the fact that the specific entropy is constant for the adiabatic, internally reversible compression process. The refrigerant at state 2s is a superheated vapor; interpolating in Table A-9 gives $h_{2s} = 205.1$ kJ/kg.

State 3 is saturated liquid at 40°C, so $h_3 = 74.59$ kJ/kg. The expansion through the valve is a throttling process (assumption 2), so $h_4 = h_3$.

(a) The compressor work input is

$$\dot{W}_c = \dot{m}(h_{2s} - h_1)$$

where $\dot{m}$ is the mass flow rate of refrigerant. Inserting values

$$\dot{W}_c = (0.008 \text{ kg/s})(205.1 - 195.78) \text{ kJ/kg} \left(\frac{1 \text{ kW}}{1 \text{ kJ/s}}\right)$$

$$= 0.075 \text{ kW}$$

(b) The refrigeration capacity is the heat transfer rate to the refrigerant passing through the evaporator. This is given by

$$\dot{Q}_{in} = \dot{m}(h_1 - h_4)$$

$$= (0.008 \text{ kg/s})(60 \text{ s/min})(195.78 - 74.59) \text{ kJ/kg} \left(\frac{1 \text{ ton}}{211 \text{ kJ/min}}\right)$$

$$= 0.276 \text{ ton}$$

(c) The coefficient of performance β is

$$\beta = \frac{\dot{Q}_{in}}{\dot{W}_c} = \frac{h_1 - h_4}{h_{2s} - h_1} = \frac{195.78 - 74.59}{205.1 - 195.78} = 13.0$$

(d) For a Carnot vapor refrigeration cycle operating at $T_H = 313$ K and $T_C = 293$ K, the coefficient of performance determined from Eq. 10.1 is

$$\beta_{max} = \frac{T_C}{T_H - T_C} = 14.65$$

Comments:

1. As expected, the ideal vapor-compression cycle has a lower coefficient of performance than a Carnot cycle operating between the temperatures of the warm and cold regions. The smaller value can be attributed to the effects of the external irreversibility associated with desuperheating the refrigerant in the condenser (Process 2s–a on the T–s diagram) and the internal irreversibility of the throttling process.
2. Double interpolation in Table A-9 is required to determine h_{2s}.

EXAMPLE 10.2

PROBLEM

Modify Example 10.1 to allow for temperature differences between the refrigerant and the warm and cold regions as follows. Saturated vapor enters the compressor at 12°C. Saturated liquid leaves the condenser at a pressure of 1.4 MPa. Determine for the modified vapor-compression refrigeration cycle (a) the compressor power, in kW, (b) the refrigeration capacity, in tons, (c) the coefficient of performance.

SOLUTION

Known: An ideal vapor-compression refrigeration cycle operates with Refrigerant 12 as the working fluid. The evaporator temperature and condenser pressure are specified, and the mass flow rate is given.

Find: Determine the compressor power, in kW, the refrigeration capacity, in tons, and the coefficient of performance.

Schematic and Given Data:

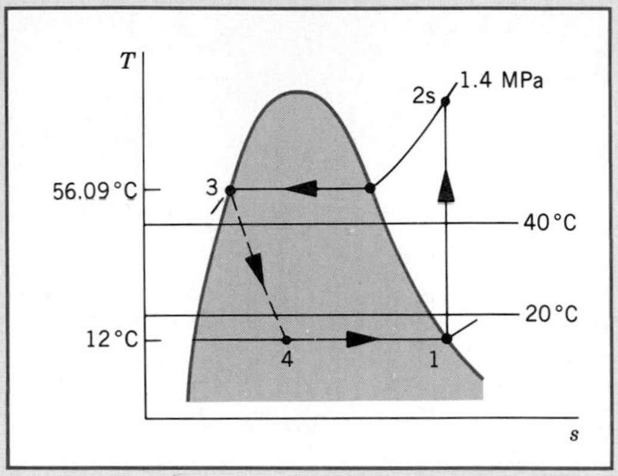

Figure E10.2

Assumptions:

1. Each component of the cycle is analyzed as a control volume at steady state. The control volumes are indicated by dashed lines on the sketch accompanying Example 10.1.
2. Except for the process through the expansion valve, which is a throttling process, all processes of the refrigerant are internally reversible.
3. The compressor and expansion valve operate adiabatically.
4. Kinetic and potential energy effects are negligible.
5. Saturated vapor enters the compressor, and saturated liquid exits the condenser.

Analysis: Let us begin by fixing each of the principal states located on the accompanying *T–s* diagram. Starting at the inlet to the compressor, the refrigerant is a saturated vapor at 12°C, so from Table A-7, $h_1 = 192.56$ kJ/kg and $s_1 = 0.6913$ kJ/kg·K.

The superheated vapor at state 2s is fixed by $p_2 = 1.4$ MPa and the fact that the specific entropy is constant for the adiabatic, internally reversible compression process. Interpolating in Table A-9 gives $h_{2s} = 212.71$ kJ/kg.

State 3 is a saturated liquid at 1.4 MPa, so $h_3 = 91.46$ kJ/kg. The expansion through the valve is a throttling process; thus, $h_4 = h_3$.

(a) The compressor power input is

$$\dot{W}_c = \dot{m}(h_{2s} - h_1)$$

where $\dot{m}$ is the mass flow rate of refrigerant. Inserting values

$$\dot{W}_c = (0.008 \text{ kg/s})(212.71 - 192.56) \text{ kJ/kg} \left(\frac{1 \text{ kW}}{1 \text{ kJ/s}}\right)$$

$$= 0.161 \text{ kW}$$

(b) The refrigeration capacity is

$$\dot{Q}_{in} = \dot{m}(h_1 - h_4)$$

$$= (0.008 \text{ kg/s})(60 \text{ s/min})(192.56 - 91.46) \text{ kJ/kg} \left(\frac{1 \text{ ton}}{211 \text{ kJ/min}}\right)$$

$$= 0.23 \text{ ton}$$

(c) The coefficient of performance β is

$$\beta = \frac{\dot{Q}_{\text{in}}}{\dot{W}_{\text{c}}} = \frac{h_1 - h_4}{h_{2s} - h_1} = \frac{192.56 - 91.46}{212.71 - 192.56} = 5.02$$

Comment: Comparing the results of the present example with those of Example 10.1, we see that the power input required by the compressor is greater in the present case. Furthermore, the refrigeration capacity and coefficient of performance are smaller in this example than in Example 10.1. This illustrates the considerable influence on performance of irreversible heat transfer between the refrigerant and the cold and warm regions.

EXAMPLE 10.3

PROBLEM

Reconsider the vapor-compression refrigeration cycle of Example 10.2, but include in the analysis that the compressor has an efficiency of 80%. Also, let the temperature of the liquid leaving the condenser be 48°C. Determine for the modified cycle (a) the compressor power, in kW, (b) the refrigeration capacity, in tons, (c) the coefficient of performance, and (d) the irreversibility rates of the compressor and expansion valve, in kW, for T_0 = 40°C.

SOLUTION

Known: A vapor-compression refrigeration cycle has a compressor efficiency of 80%.

Find: Determine the compressor power, in kW, the refrigeration capacity, in tons, the coefficient of performance, and the irreversibility rates of the compressor and expansion valve, in kW.

Schematic and Given Data:

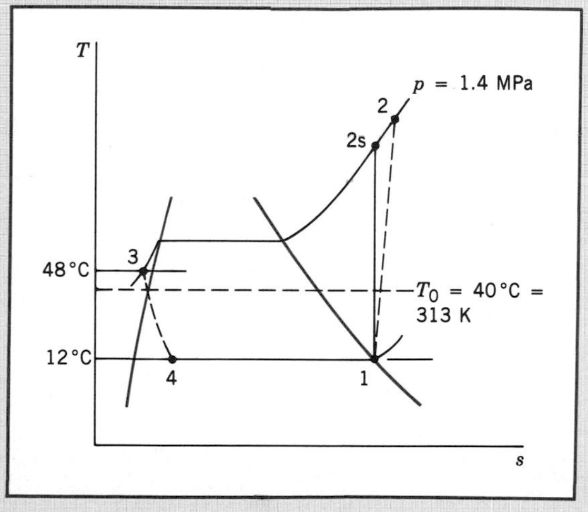

Figure E10.3

Assumptions:

1. Each component of the cycle is analyzed as a control volume at steady state.
2. There are no pressure drops through the evaporator and condenser.
3. The compressor operates adiabatically with an efficiency of 80%. The expansion through the valve is a throttling process.
4. Kinetic and potential energy effects are negligible.
5. Saturated vapor at 12°C enters the compressor, and liquid at 48°C leaves the condenser.
6. The environment temperature for calculating irreversibility rate is T_0 = 40°C.

Analysis: Let us begin by fixing the principal states. State 1 is the same as in Example 10.2, so $h_1 = 192.56$ kJ/kg and $s_1 = 0.6913$ kJ/kg·K.

Owing to the presence of irreversibilities during the adiabatic compression process, there is an increase in specific entropy from compressor inlet to exit. The state at the compressor exit, state 2, can be fixed using the compressor efficiency

$$\eta_c = \frac{(\dot{W}_c/\dot{m})_s}{\dot{W}_c/\dot{m}} = \frac{(h_{2s} - h_1)}{(h_2 - h_1)}$$

where h_{2s} is the specific enthalpy at state 2s, as indicated on the accompanying T–s diagram. From the solution to Example 10.2, $h_{2s} = 212.71$ kJ/kg. By solving for h_2 and inserting known values

$$h_2 = \frac{h_{2s} - h_1}{\eta_c} + h_1 = \frac{(212.71 - 192.56)}{(0.80)} + 192.56 = 217.75 \text{ kJ/kg}$$

State 2 is fixed by the value of specific enthalpy h_2 and the pressure, $p_2 = 1.4$ MPa. Interpolating in Table A-9, the specific entropy is $s_2 = 0.7060$ kJ/kg·K.

The state at the condenser exit, state 3, is in the liquid region. The specific enthalpy is approximated using Eq. 3.14, together with saturated liquid data at 48°C, as follows:

$$\begin{aligned} h_3 &= h_f + v_f(p_3 - p_{sat}) \\ &= 82.83 + (0.8199 \times 10^{-3})\frac{\text{m}^3}{\text{kg}}(1.4 - 1.1639)(10^6)\frac{\text{N}}{\text{m}^2}\left(\frac{\text{kJ}}{10^3\ \text{N}\cdot\text{m}}\right) \\ &= 83.02 \text{ kJ/kg} \end{aligned}$$

Clearly, in this case $h_3 \approx h_f(T_3)$. The specific entropy at state 3 is $s_3 \approx s_f = 0.2973$ kJ/kg·K.

The expansion through the valve is a throttling process; thus, $h_4 = h_3$. The quality and specific entropy at state 4 are, respectively

$$x_4 = \frac{h_4 - h_{f4}}{h_{g4} - h_{f4}} = \frac{83.02 - 47.26}{145.30} = 0.2461$$

and

$$\begin{aligned} s_4 &= s_{f4} + x_4(s_{g4} - s_{f4}) \\ &= 0.1817 + (0.2461)(0.6913 - 0.1817) = 0.3071 \text{ kJ/kg·K} \end{aligned}$$

(a) The compressor power is

$$\begin{aligned} \dot{W}_c &= \dot{m}(h_2 - h_1) \\ &= (0.008 \text{ kg/s})(217.75 - 192.56) \text{ kJ/kg}\left(\frac{1 \text{ kW}}{1 \text{ kJ/s}}\right) = 0.202 \text{ kW} \end{aligned}$$

(b) The refrigeration capacity is

$$\begin{aligned} \dot{Q}_{in} &= \dot{m}(h_1 - h_4) \\ &= (0.008 \text{ kg/s})(60 \text{ s/min})(192.56 - 83.02) \text{ kJ/kg}\left(\frac{1 \text{ ton}}{211 \text{ kJ/min}}\right) \\ &= 0.249 \text{ ton} \end{aligned}$$

(c) The coefficient of performance is

$$\beta = \frac{(h_1 - h_4)}{(h_2 - h_1)} = \frac{(192.56 - 83.02)}{(217.75 - 192.56)} = 4.35$$

(d) The irreversibility rates in the compressor and expansion valve can be found by reducing the availability rate balance or using the relationship $\dot{I}_{cv} = T_0\dot{\sigma}_{cv}$, where $\dot{\sigma}_{cv}$ is the rate of entropy production from an entropy rate balance. With either approach, the irreversibility rates for the compressor and valve are, respectively

$$\dot{I}_c = \dot{m}T_0(s_2 - s_1) \qquad \text{and} \qquad \dot{I}_{valve} = \dot{m}T_0(s_4 - s_3)$$

Substituting values

$$\dot{I}_c = \left(0.008\,\frac{\text{kg}}{\text{s}}\right)(313\text{ K})(0.7060 - 0.6913)\,\frac{\text{kJ}}{\text{kg}\cdot\text{K}}\left(\frac{1\text{ kW}}{1\text{ kJ/s}}\right) = 0.037\text{ kW}$$

and

$$\dot{I}_{\text{valve}} = (0.008)(313)(0.3071 - 0.2973) = 0.025\text{ kW}$$

Comments:

1. Irreversibilities in the compressor result in an increased compressor power requirement compared to the isentropic compression of Example 10.2. As a consequence, in the present case the coefficient of performance is lower.
2. The irreversibility rates calculated in part (d) represent the rates of availability destruction, owing to irreversibilities as the refrigerant flows through the compressor and valve. When expressed as percentages, the portions of the availability input to the compressor as work that are destroyed in these two components are 18.3 and 12.4%, respectively.

10.3 REFRIGERANT PROPERTIES

The most common refrigerants in use today in vapor-compression refrigeration systems are the *halogenated hydrocarbons*. Refrigerant 12, which has been used frequently in previous discussions, is one of these. The chemical name of this refrigerant is dichlorodifluoromethane (CCl_2F_2). It is also known by the trade names Freon-12 and Genatron-12. Two other commonly employed halogenated hydrocarbon refrigerants are Refrigerant 11 and Refrigerant 22. Ammonia is another important refrigerant, particularly in the absorption refrigeration systems discussed in Sec. 10.5.

Owing to concern about the effects of halogenated refrigerants on the earth's protective ozone layer, international agreements have been reached to phase out their use. The chemical tetrafluoroethane (CF_3CFH_2), called Refrigerant 134a, contains no halogens, and as such is considered an environmentally acceptable substitute for Refrigerant 12. Thermodynamic property data for Refrigerant 134a are included in appendix Tables A-10 through A-12. These data allow us to investigate some of the effects on refrigeration cycles of replacing Refrigerant 12 with Refrigerant 134a.

The temperatures of the refrigerant in the evaporator and condenser are fixed by the temperatures of the cold and warm regions, respectively, with which the system interacts thermally. This, in turn, determines the operating pressures in the evaporator and condenser. Consequently, the selection of a refrigerant is based partly on the suitability of its pressure–temperature relationship in the range of the particular application. It is generally desirable to avoid excessively low pressures in the evaporator and excessively high pressures in the condenser. Other considerations in refrigerant selection include chemical stability, toxicity, corrosiveness, and cost. The type of compressor also affects the choice of refrigerant. Centrifugal compressors are best suited for low evaporator pressures and refrigerants with large specific volumes at low pressure. Reciprocating compressors perform better over large pressure ranges and are better able to handle low specific volume refrigerants.

A thermodynamic property diagram widely used in the refrigeration field is the pressure–enthalpy or p–h diagram. Figure 10.5 shows the main features of such a property diagram. The principal states of the vapor-compression cycles of Fig. 10.4 are located on this p–h diagram. It is left as an exercise to sketch the cycles of

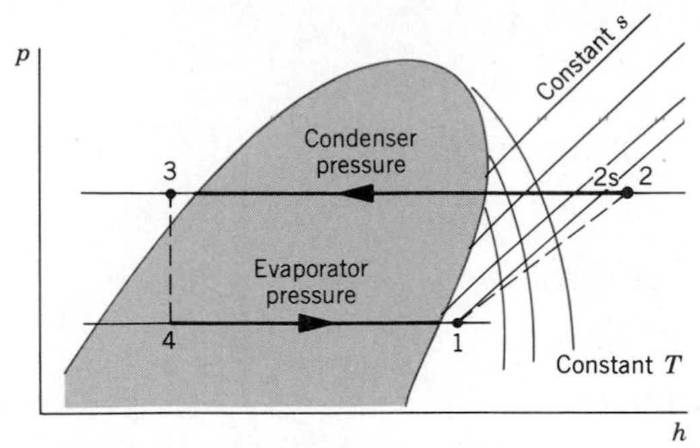

Figure 10.5 Principal features of the pressure–enthalpy diagram for a typical refrigerant, with vapor-compression cycles superimposed.

Examples 10.1, 10.2, and 10.3 on p–h diagrams. Property tables and p–h diagrams for many refrigerants are given in handbooks dealing with refrigeration.

10.4 CASCADE AND MULTISTAGE VAPOR-COMPRESSION SYSTEMS

Variations of the basic vapor-compression refrigeration cycle are used to improve performance or for special applications. Two variations are presented in this section. The first is a *combined cycle* arrangement in which refrigeration at relatively low temperature is achieved through a series of vapor-compression systems, with each normally employing a different refrigerant. In the second variation, the work of compression is reduced through *multistage compression with intercooling* between the stages. These variations are analogous to power cycle modifications considered in Chaps. 8 and 9.

10.4.1 Cascade Cycles

Combined cycle arrangements for refrigeration are called *cascade* cycles. In Fig. 10.6 is shown a *double* cascade cycle in which *two* vapor-compression refrigeration cycles, labeled A and B, are arranged in series with a counterflow heat exchanger linking them. In the intermediate heat exchanger, the energy rejected during condensation of the refrigerant in the lower-temperature cycle A is used to evaporate the refrigerant in the higher-temperature cycle B. The desired refrigeration effect occurs in the low-temperature evaporator, and heat rejection from the overall cycle occurs in the high-temperature condenser. The coefficient of performance is the ratio of the refrigeration effect to the *total* work input:

$$\beta = \frac{\dot{Q}_{\text{in}}}{\dot{W}_{\text{cA}} + \dot{W}_{\text{cB}}}$$

The mass flow rates in cycles A and B normally would be different. However, the mass flow rates are related by mass and energy rate balances on the interconnecting counterflow heat exchanger serving as the condenser for cycle A and the evaporator for cycle B. Although only two cycles are shown in Fig. 10.6, cascade cycles may employ three or more individual cycles.

A major advantage of a cascade system with a closed-type heat exchanger as illustrated in Fig. 10.6 is that the refrigerants in the two or more stages can be selected to have reasonable evaporator and condenser pressures in the two or more temperature ranges. In a double cascade system, a refrigerant would be selected for cycle A that has a saturation pressure–temperature relationship that allows refrigera-

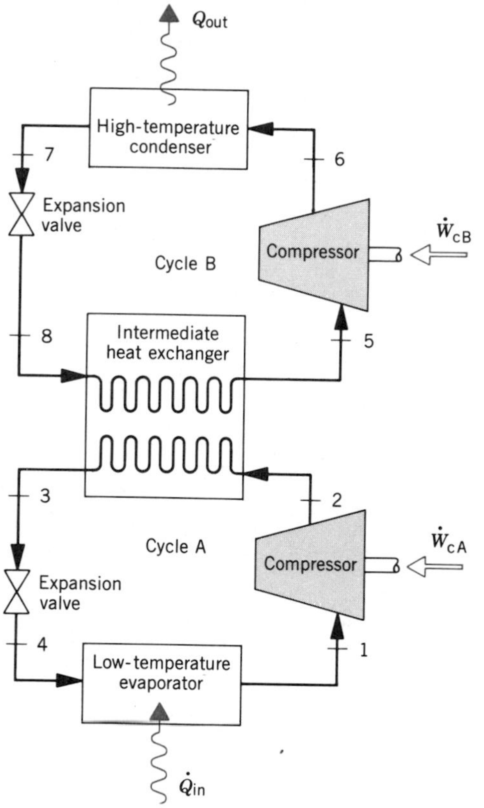

Figure 10.6 Example of a cascade vapor-compression refrigeration cycle.

tion at a relatively low temperature without excessively low evaporator pressures. The refrigerant for cycle B would have saturation characteristics that permit condensation at the required temperature without excessively high condenser pressures.

10.4.2 Multistage Compression with Intercooling

The advantages of multistage compression with intercooling between stages have been cited in Sec. 9.8, dealing with gas power systems. Intercooling is achieved in gas power systems by heat transfer to the lower-temperature surroundings. In refrigeration systems, the refrigerant temperature is below that of the surroundings for much of the cycle, so other means must be employed to accomplish intercooling and achieve the attendant savings in the required compressor work input. One arrangement for two-stage compression using the refrigerant itself for intercooling is shown in Fig. 10.7. The principal states of the refrigerant for an ideal cycle are shown on the accompanying T–s diagram.

Intercooling is accomplished in this cycle by means of a direct contact heat exchanger. Relatively low-temperature saturated vapor enters the heat exchanger at state 9, where it mixes with higher-temperature refrigerant leaving the first compression stage at state 2. A single mixed stream exits the heat exchanger at an intermediate temperature at state 3 and is compressed in the second compressor stage to the condenser pressure at state 4. Less work is required per unit of mass flow for compression from 1 to 2 followed by compression from 3 to 4 than for a single stage of compression 1–2–a. Since the refrigerant temperature entering the condenser at state 4 is lower than for a single stage of compression in which the refrigerant would

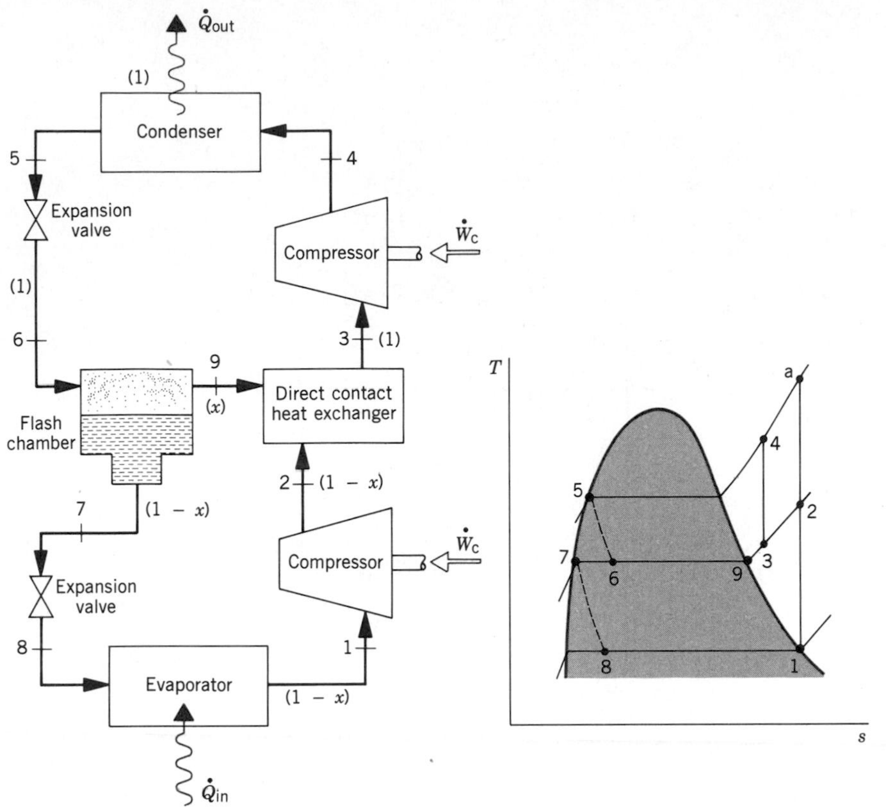

Figure 10.7 Refrigeration cycle with two stages of compression and flash intercooling.

enter the condenser at state a, the external irreversibility associated with heat transfer in the condenser is also reduced.

A central role is played in the cycle of Fig. 10.7 by a liquid–vapor separator, called a *flash* chamber. Refrigerant exiting the condenser at state 5 expands through a valve and enters the flash chamber at state 6 as a two-phase liquid–vapor mixture with quality x. In the flash chamber, the liquid and vapor components separate into two streams. Saturated vapor exiting the flash chamber enters the heat exchanger at state 9, where intercooling is achieved as discussed above. Saturated liquid exiting the flash chamber at state 7 expands through a second valve into the evaporator. On the basis of a unit of mass flowing through the condenser, the fraction of the vapor formed in the flash chamber equals the quality x of the refrigerant at state 6. The fraction of the liquid formed is then $(1 - x)$. The fractions of the total flow at various locations are indicated on Fig. 10.7.

10.5 ABSORPTION REFRIGERATION

Absorption refrigeration cycles are the subject of this section. These cycles have some features in common with the vapor-compression cycles considered previously but differ in two important respects. One is the nature of the compression process. Instead of compressing a vapor between the evaporator and the condenser, the refrigerant of an absorption system is absorbed by a secondary substance, called an absorbent, to form a *liquid solution*. The liquid solution is then *pumped* to the higher pressure. Because the average specific volume of the liquid solution is much less than that of the refrigerant vapor, significantly less work is required (see Sec. 6.9 for discussion). Accordingly, absorption refrigeration systems have the advantage of relatively small work input compared to vapor-compression systems.

The other main difference between absorption and vapor-compression systems is that some means must be introduced in absorption systems to retrieve the refrigerant vapor from the liquid solution before the refrigerant enters the condenser. This involves heat transfer from a relatively high-temperature source. Steam or waste heat that otherwise would be discharged to the surroundings without use is particularly economical for this purpose. Natural gas or some other fuel can be burned to provide the heat source, and there have been practical applications of absorption refrigeration using alternative energy sources such as solar and geothermal energy.

The principal components of an absorption refrigeration system are shown schematically in Fig. 10.8. In this case, ammonia is the refrigerant and water is the absorbent. Ammonia circulates through the condenser, expansion valve, and evaporator in much the same way as in a vapor-compression system. However, the compressor is replaced by the absorber, pump, generator, and valve shown on the right side of the diagram. In the *absorber,* ammonia vapor coming from the evaporator at state 1 is absorbed by liquid water. The formation of this liquid solution is exothermic. Since the amount of ammonia that can be dissolved in water increases as the solution temperature decreases, cooling water is circulated around the absorber to remove the energy released as ammonia goes into solution and maintain the temperature in the absorber as low as possible. The strong ammonia–water solution leaves the absorber at point a and enters the *pump,* where its pressure is increased to that of the generator. In the *generator,* heat transfer from a relatively high-temperature source drives ammonia vapor out of the solution (an endothermic process), leaving a weak ammonia–water solution in the generator. The vapor liberated passes to the

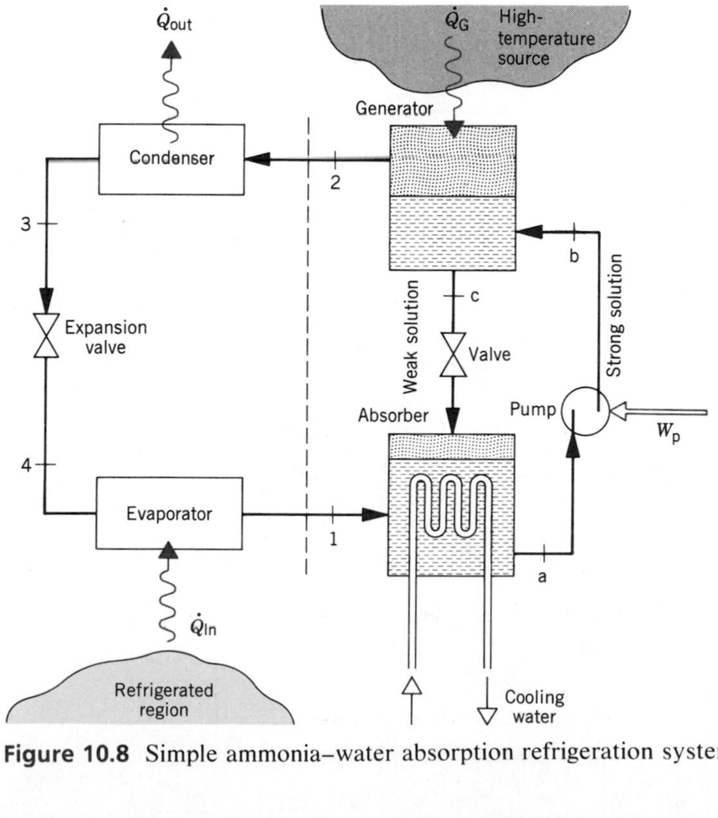

Figure 10.8 Simple ammonia–water absorption refrigeration system.

condenser at state 2, and the remaining weak solution at c flows back to the absorber through a *valve*. The only work input is the power required to operate the pump, and this is small in comparison to the work that would be required to compress refrigerant vapor between the same pressure levels. However, costs associated with the heat source and extra equipment not required by vapor-compressor systems can cancel the advantage of a smaller work input.

Ammonia–water systems normally employ several modifications of the simple absorption cycle considered above. Two common modifications are illustrated in Fig. 10.9. In this cycle, a heat exchanger is included between the generator and the absorber that allows the strong water–ammonia solution entering the generator to be preheated by the weak solution returning from the generator to the absorber, thereby reducing the heat transfer to the generator, $\dot{Q}_G$. The other modification shown on the figure is the *rectifier* placed between the generator and the condenser. The function of the rectifier is to remove any traces of water from the refrigerant before it enters the condenser. This eliminates the possibility of ice formation in the expansion valve and the evaporator.

Another type of absorption system uses *lithium bromide* as the absorbent and *water* as the refrigerant. The basic principle of operation is the same as for ammonia–water systems. To achieve refrigeration at lower temperatures than are possible with water as the refrigerant, a lithium bromide–water absorption system may be com-

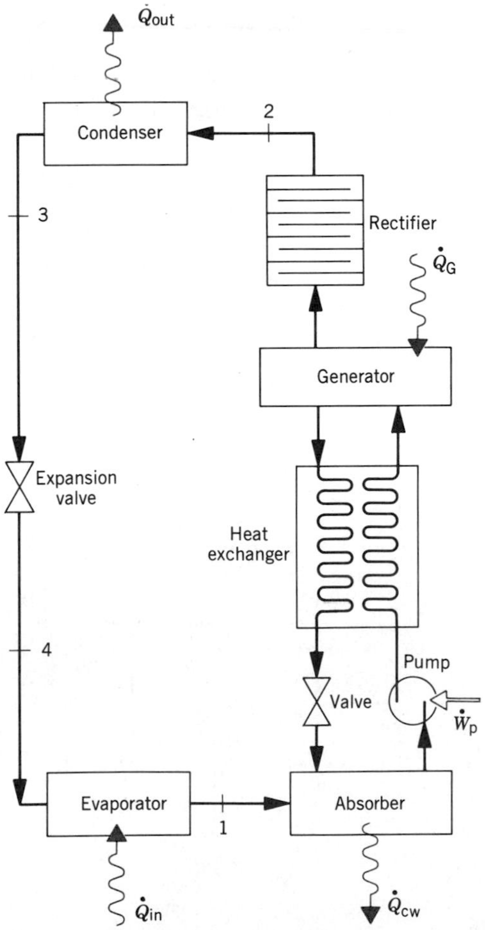

Figure 10.9 Modified ammonia–water absorption system.

bined with another cycle using a refrigerant with good low-temperature characteristics, such as ammonia, to form a cascade refrigeration system.

10.6 HEAT PUMP SYSTEMS

The objective of a heat pump is to maintain the temperature within a dwelling or other building above the temperature of the surroundings or to provide a heat transfer for certain industrial processes that occur at elevated temperatures. Heat pump systems have many features in common with the refrigeration systems considered thus far and may be of the vapor-compression or absorption type. Vapor-compression heat pumps are well suited for space heating applications and are being used increasingly for this purpose. Absorption heat pumps have been developed for industrial applications and are under active study for space heating. To introduce some aspects of heat pump operation, let us begin by considering the Carnot heat pump cycle.

10.6.1 Carnot Heat Pump Cycle

By simply changing our viewpoint, we can regard the cycle shown in Fig. 10.1 as a *heat pump*. The objective of the cycle now, however, is to deliver the heat transfer $\dot{Q}_{out}$ to the warm region, which is the space to be heated. At steady state, the rate at which energy is supplied to the warm region by heat transfer is the sum of the energy supplied to the working fluid from the cold region, $\dot{Q}_{in}$, and the net rate of work input to the cycle, $\dot{W}_{net}$. That is

$$\dot{Q}_{out} = \dot{Q}_{in} + \dot{W}_{net} \tag{10.8}$$

The coefficient of performance of *any* heat pump cycle is defined as the ratio of the heating effect to the net work required to achieve that effect. For the Carnot heat pump cycle of Fig. 10.1

$$\begin{aligned}\gamma_{max} &= \frac{\dot{Q}_{out}/\dot{m}}{\dot{W}_c/\dot{m} - \dot{W}_t/\dot{m}} \\ &= \frac{\text{area 2–a–b–3–2}}{\text{area 1–2–3–4–1}} \\ &= \frac{T_H(s_a - s_b)}{(T_H - T_C)(s_a - s_b)} = \frac{T_H}{T_H - T_C}\end{aligned} \tag{10.9}$$

This equation, which corresponds to Eq. 5.10, represents the *maximum* theoretical coefficient of performance for any heat pump cycle operating between two regions at temperatures T_C and T_H. Actual heat pump systems have coefficients of performance that are lower than would be calculated from Eq. 10.9.

A study of Eq. 10.9 shows that as the temperature T_C of the cold region decreases, the coefficient of performance of the Carnot heat pump decreases. This trait is also exhibited by actual heat pump systems and suggests why heat pumps in which the role of the cold region is played by the local atmosphere (air source heat pumps) normally require backup systems to provide heating on days when the ambient temperature becomes very low. If sources such as well water or the ground itself are used, relatively high coefficients of performance can be achieved despite low ambient air temperatures, and backup systems may not be required.

10.6.2 Vapor-Compression Heat Pumps

Actual heat pump systems depart significantly from the Carnot cycle model. Most systems in common use today are of the vapor-compression type. The method of analysis of vapor-compression heat pumps is the same as that of vapor-compression refrigeration cycles considered previously. Also, the previous discussions concerning the departure of actual systems from ideality apply for vapor-compression heat pump systems as for vapor-compression refrigeration cycles.

As illustrated by Fig. 10.10, a typical vapor-compression heat pump for space heating has the same basic components as the vapor-compression refrigeration system: compressor, condenser, expansion valve, and evaporator. The objective of the system is different, however. In a heat pump system, $\dot{Q}_{\text{in}}$ comes from the surroundings, and $\dot{Q}_{\text{out}}$ is directed to the dwelling as the desired effect. A net work input is required to accomplish this effect.

The coefficient of performance of a simple vapor-compression heat pump with states as designated on Fig. 10.10 is

$$\gamma = \frac{\dot{Q}_{\text{out}}/\dot{m}}{\dot{W}_c/\dot{m}} = \frac{h_2 - h_3}{h_2 - h_1} \tag{10.10}$$

The value of γ can never be less than unity.

Many possible sources are available for heat transfer to the refrigerant passing through the evaporator. These include the outside air, the ground, and water from lakes, rivers, or wells. Liquid circulated through a solar collector and stored in an insulated tank also can be used as a source for a heat pump. Industrial heat pumps employ waste heat or warm liquid or gas streams as the low-temperature source and are capable of achieving relatively high condenser temperatures.

In the most common type of vapor-compression heat pump for space heating, the evaporator communicates thermally with the outside air. Such *air source* heat pumps also can be used to provide cooling in the summer with the use of a reversing valve, as illustrated in Fig. 10.11. The solid lines show the flow path of the refrigerant in the heating mode, as described previously. To use the same components as an air conditioner, the valve is actuated, and the refrigerant follows the path indicated by the dashed line. In the cooling mode, the outside heat exchanger becomes the condenser, and the inside heat exchanger becomes the evaporator. Although heat pumps can be more costly to install and operate than other direct heating systems, they can be competitive when the potential for dual use is considered.

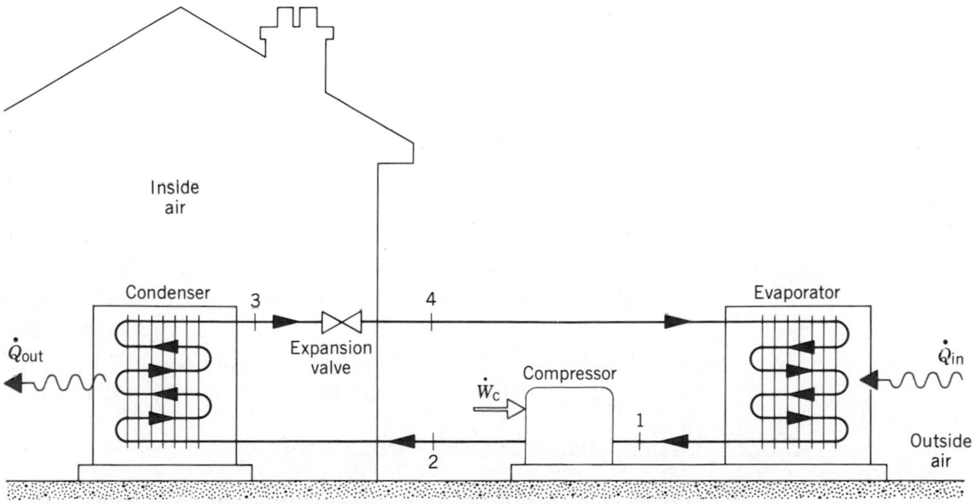

Figure 10.10 Vapor-compression heat pump system for space heating.

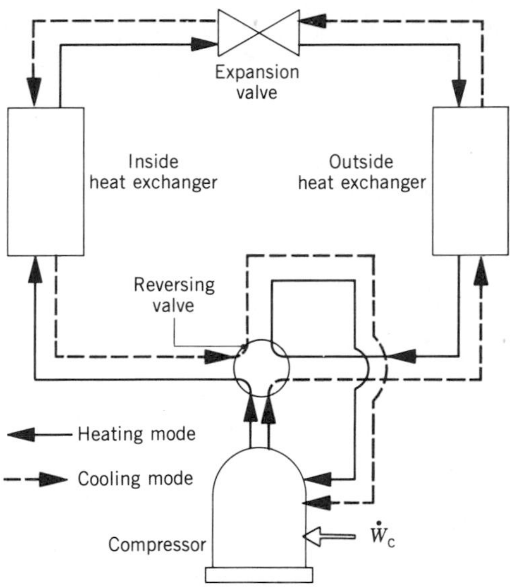

Figure 10.11 Example of an air-to-air reversing heat pump.

10.7 GAS REFRIGERATION SYSTEMS

All refrigeration systems considered thus far involve changes in phase. Let us now turn to gas refrigeration systems in which the working fluid remains a gas throughout. Gas refrigeration systems have a number of important applications. They are used to achieve very low temperatures for the liquefaction of air and other gases and for other specialized applications such as aircraft cabin cooling. The Brayton refrigeration cycle illustrates an important type of gas refrigeration system.

10.7.1 Brayton Refrigeration Cycle

The Brayton refrigeration cycle is the reverse of the closed Brayton power cycle introduced in Sec. 9.6. A schematic of the reversed Brayton cycle is provided in Fig. 10.12*a*. The refrigerant gas, which may be air, enters the compressor at state 1, where its temperature is somewhat below the temperature of the refrigerated region, and is compressed to state 2. The gas is then cooled to state 3 by heat transfer to the surroundings. Next, the gas is expanded to state 4, where its temperature, T_4, is much less than that of the refrigerated region. Refrigeration is achieved through heat transfer from the refrigerated region to the gas as it passes from state 4 to state 1, completing the cycle. The T–s diagram in Fig. 10.12*b* shows an *ideal* Brayton refrigeration cycle, denoted by 1–2s–3–4s–1, in which all processes are assumed to be internally reversible and the processes in the turbine and compressor are adiabatic. Also shown is the cycle 1–2–3–4–1, which suggests the effects of irreversibilities during adiabatic compression and expansion. Frictional pressure drops have been ignored.

The method of analysis of the Brayton refrigeration cycle is similar to that of the Brayton power cycle. Thus, at steady state the work of the compressor and the turbine per unit of mass flow are, respectively,

$$\frac{\dot{W}_c}{\dot{m}} = h_2 - h_1 \qquad \text{and} \qquad \frac{\dot{W}_t}{\dot{m}} = h_3 - h_4$$

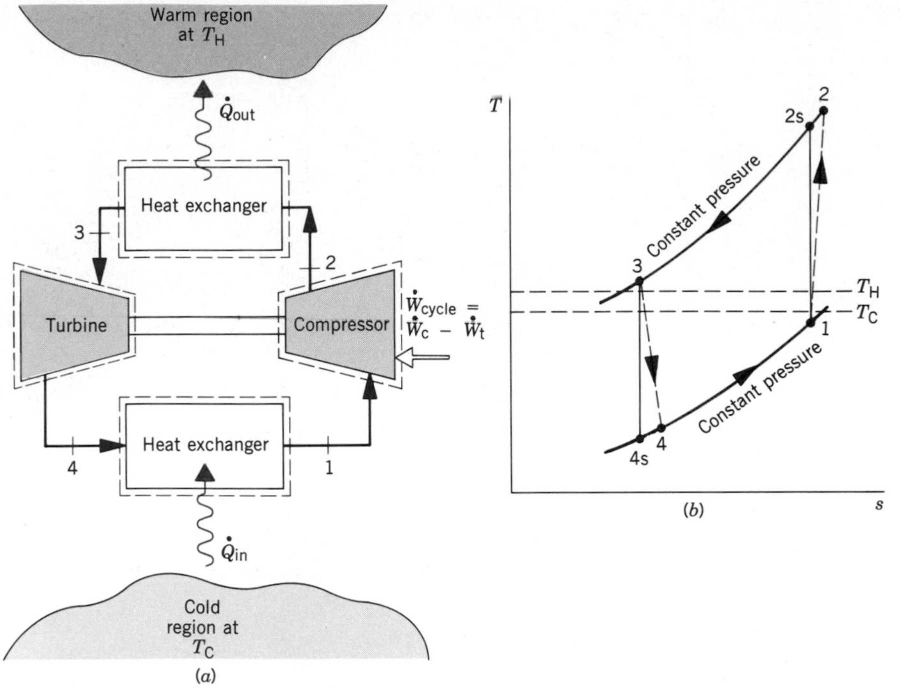

Figure 10.12 Brayton refrigeration cycle.

In obtaining these expressions, heat transfer with the surroundings and changes in kinetic and potential energy have been ignored. In gas refrigeration systems, the work developed by the turbine is considerable and cannot be ignored as in vapor-compression systems.

The heat transfer from the cold region to the refrigerant gas circulating through the low-pressure heat exchanger, the refrigeration effect, is

$$\frac{\dot{Q}_{in}}{\dot{m}} = h_1 - h_4$$

The coefficient of performance is the ratio of the refrigeration effect to the net work input:

$$\beta = \frac{\dot{Q}_{in}/\dot{m}}{\dot{W}_c/\dot{m} - \dot{W}_t/\dot{m}} = \frac{(h_1 - h_4)}{(h_2 - h_1) - (h_3 - h_4)} \tag{10.11}$$

Irreversibilities within the compressor and turbine serve to decrease the coefficient of performance significantly from that of the corresponding ideal cycle because the compressor work requirement is increased and the turbine work output is decreased. This is illustrated in examples to follow.

EXAMPLE 10.4

PROBLEM

Air enters the compressor of an ideal Brayton refrigeration cycle at 1 atm, 480°R, with a volumetric flow rate of 50 ft^3/s. If the compressor pressure ratio is 3 and the turbine inlet temperature is 540°R, determine (a) the *net* power input, in Btu/min, (b) the refrigeration capacity, in Btu/min, (c) the coefficient of performance.

SOLUTION

Known: An ideal Brayton refrigeration cycle operates with air. Compressor inlet conditions, the turbine inlet temperature, and the compressor pressure ratio are given.

Find: Determine the *net* power input, in Btu/min, the refrigeration capacity, in Btu/min, and the coefficient of performance.

Schematic and Given Data:

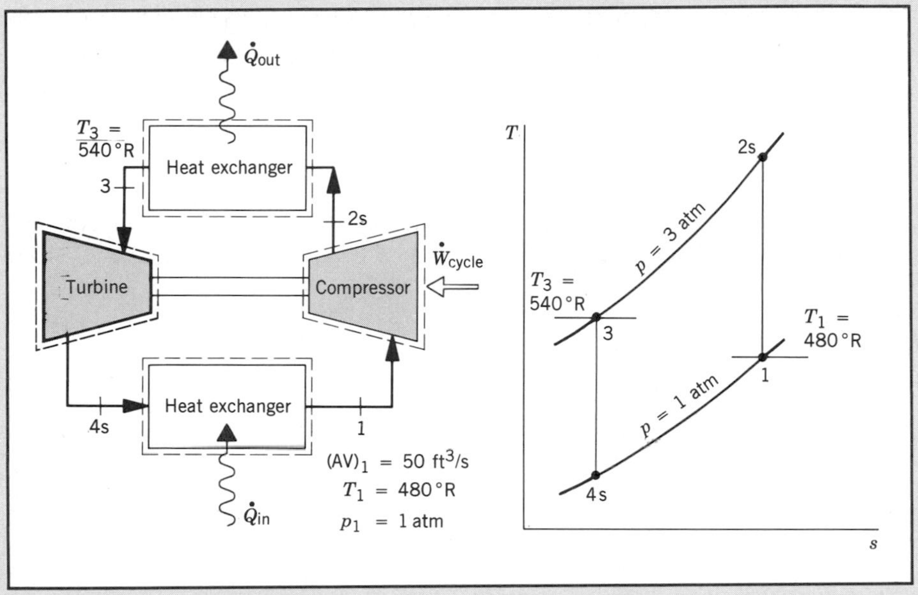

Figure E10.4

Assumptions:

1. Each component of the cycle is analyzed as a control volume at steady state. The control volumes are indicated by dashed lines on the accompanying sketch.
2. The turbine and compressor processes are isentropic.
3. There are no pressure drops through the heat exchangers.
4. Kinetic and potential energy effects are negligible.
5. The working fluid is air modeled as an ideal gas.

Analysis: The analysis begins by determining the specific enthalpy at each numbered state of the cycle. At state 1, the temperature is 480°R. From Table A-16E, $h_1 = 114.69$ Btu/lb, $p_{r1} = 0.9182$. Since the compressor process is isentropic, h_{2s} can be determined by first evaluating p_r at state 2s. That is

$$p_{r2} = \frac{p_2}{p_1} p_{r1} = (3)(0.9182) = 2.755$$

Then, interpolating in Table A-16E, we get $h_{2s} = 157.1$ Btu/lb.

The temperature at state 3 is given as $T_3 = 540°R$. From Table A-16E, $h_3 = 129.06$ Btu/lb, $p_{r3} = 1.3860$. The specific enthalpy at state 4s is found by using the isentropic relation

$$p_{r4} = p_{r3} \frac{p_4}{p_3} = (1.3860)(1/3) = 0.462$$

Interpolating in Table A-16E, we obtain $h_{4s} = 94.1$ Btu/lb.

(a) The net power input is

$$\dot{W}_{cycle} = \dot{m}[(h_{2s} - h_1) - (h_3 - h_{4s})]$$

This requires the mass flow rate $\dot{m}$, which can be determined from the volumetric flow rate and the specific volume at the compressor inlet:

$$\dot{m} = \frac{(\mathrm{AV})_1}{v_1}$$

Since $v_1 = (\bar{R}/M)T_1/p_1$

$$\begin{aligned}\dot{m} &= \frac{(\mathrm{AV})_1 p_1}{(\bar{R}/M)T_1} \\ &= \frac{(50\ \mathrm{ft^3/s})(60\ \mathrm{s/min})(14.7\ \mathrm{lbf/in.^2})(144\ \mathrm{in.^2/ft^2})}{\left(\dfrac{1545\ \mathrm{ft \cdot lbf}}{28.97\ \mathrm{lb \cdot {}^\circ R}}\right)(480^\circ\mathrm{R})} \\ &= 248\ \mathrm{lb/min}\end{aligned}$$

Finally

$$\begin{aligned}\dot{W}_{\mathrm{cycle}} &= (248\ \mathrm{lb/min})[(157.1 - 114.69) - (129.06 - 94.1)]\ \mathrm{Btu/lb} \\ &= (248)(7.45) = 1848\ \mathrm{Btu/min}\end{aligned}$$

(b) The refrigeration capacity is

$$\begin{aligned}\dot{Q}_{\mathrm{in}} &= \dot{m}(h_1 - h_{4s}) \\ &= (248\ \mathrm{lb/min})(114.69 - 94.1)\ \mathrm{Btu/lb} \\ &= (248)(20.59) = 5106\ \mathrm{Btu/min}\end{aligned}$$

(c) The coefficient of performance is

$$\beta = \frac{\dot{Q}_{\mathrm{in}}}{\dot{W}_{\mathrm{cycle}}} = \frac{5106}{1848} = 2.76$$

Comment: Note the relatively low value for the coefficient of performance of this ideal cycle as compared with the ideal vapor-compression cycle of Example 10.1.

EXAMPLE 10.5

PROBLEM

Reconsider Example 10.4, but include in the analysis that the compressor and turbine each have an isentropic efficiency of 80%. Determine for the modified cycle (a) the *net* power input, in Btu/min, (b) the refrigeration capacity, in Btu/min, (c) the coefficient of performance.

SOLUTION

Known: A Brayton refrigeration cycle operates with air. Compressor inlet conditions, the turbine inlet temperature, and the compressor pressure ratio are given. The compressor and turbine each have an efficiency of 80%.

Find: Determine the *net* power input, in Btu/min, the refrigeration capacity, in Btu/min, and the coefficient of performance.

Schematic and Given Data:

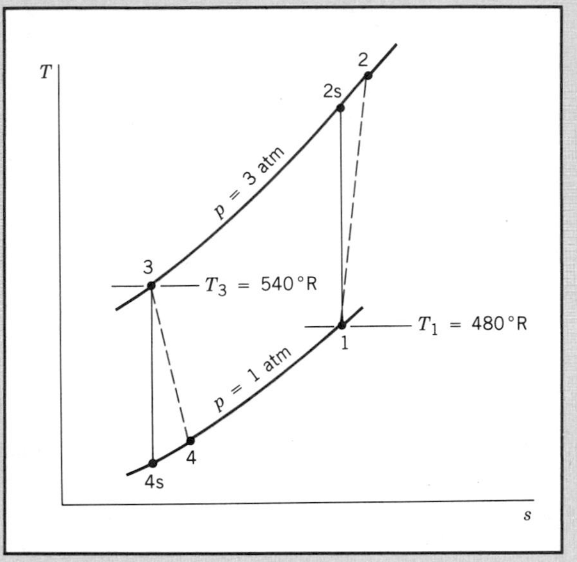

Figure E10.5

Assumptions:

1. Each component of the cycle is analyzed as a control volume at steady state.
2. The compressor and turbine are adiabatic.
3. There are no pressure drops through the heat exchangers.
4. Kinetic and potential energy effects are negligible.
5. The working fluid is air modeled as an ideal gas.

Analysis:

(a) The power input to the compressor is evaluated using the compressor isentropic efficiency, η_c. Thus

$$\frac{\dot{W}_c}{\dot{m}} = \frac{(\dot{W}_c/\dot{m})_s}{\eta_c}$$

The value of the work per unit mass for the isentropic compression, $(\dot{W}_c/\dot{m})_s$, is determined with data from the solution in Example 10.4 as 42.41 Btu/lb. The actual power required is then

$$\dot{W}_c = \frac{\dot{m}(\dot{W}_c/\dot{m})_s}{\eta_c} = \frac{(248 \text{ lb/min})(42.41 \text{ Btu/lb})}{(0.8)}$$
$$= 13{,}147 \text{ Btu/min}$$

The turbine power output is determined in a similar manner, using the turbine isentropic efficiency η_t. Thus, $\dot{W}_t/\dot{m} = \eta_t(\dot{W}_t/\dot{m})_s$. Using data from the solution to Example 10.4 gives $(\dot{W}_t/\dot{m})_s = 34.96$ Btu/lb. The actual turbine work is then

$$\dot{W}_t = \dot{m}\eta_t(\dot{W}_t/\dot{m})_s = (248 \text{ lb/min})(0.8)(34.96 \text{ Btu/lb})$$
$$= 6936 \text{ Btu/min}$$

The *net* power input to the cycle is

$$\dot{W}_{\text{cycle}} = 13{,}147 - 6936 = 6211 \text{ Btu/min}$$

(b) The specific enthalpy at the turbine exit, h_4, is required to evaluate the refrigeration capacity. This enthalpy can be determined by solving $\dot{W}_t = \dot{m}(h_3 - h_4)$ to obtain $h_4 = h_3 - \dot{W}_t/\dot{m}$. Inserting known values

$$h_4 = 129.06 - \left(\frac{6936}{248}\right) = 101.1 \text{ Btu/lb}$$

The refrigeration capacity is then

$$\dot{Q}_{in} = \dot{m}(h_1 - h_4) = (248)(114.69 - 101.1) = 3370 \text{ Btu/min}$$

(c) The coefficient of performance is

$$\beta = \frac{\dot{Q}_{in}}{\dot{W}_{cycle}} = \frac{3370}{6211} = 0.543$$

Comments:

1. Irreversibilities in the compressor and turbine have a significant effect on the performance of gas refrigeration systems. This is brought out by comparing the results of the present example with those of Example 10.4. Irreversibilities result in an increase in the work of compression and a reduction in the work output of the turbine. The refrigeration capacity is reduced. The overall effect is that the coefficient of performance is decreased significantly.
2. The value of the coefficient of performance in this example is less than unity. This means that the refrigeration effect is smaller than the net work required to achieve it.

10.7.2 Additional Gas Refrigeration Applications

To obtain even moderate refrigeration capacities with the Brayton refrigeration cycle, equipment capable of achieving relatively high pressures and volumetric flow rates is needed. For most applications involving air conditioning and for ordinary refrigeration processes, vapor-compression systems can be built more cheaply and can operate with higher coefficients of performance than gas refrigeration systems. With suitable modifications, however, gas refrigeration systems can be used to achieve temperatures of about −150°C (−240°F), which are well below the temperatures normally obtained with vapor systems.

Figure 10.13 shows the schematic and T–s diagram of an ideal Brayton cycle modified by the introduction of a regenerative heat exchanger. The heat exchanger allows the air entering the turbine at state 3 to be cooled *below* the ambient temperature T_H. In the subsequent expansion through the turbine, the air achieves a much lower temperature at state 4 than would have been possible without the regenerative

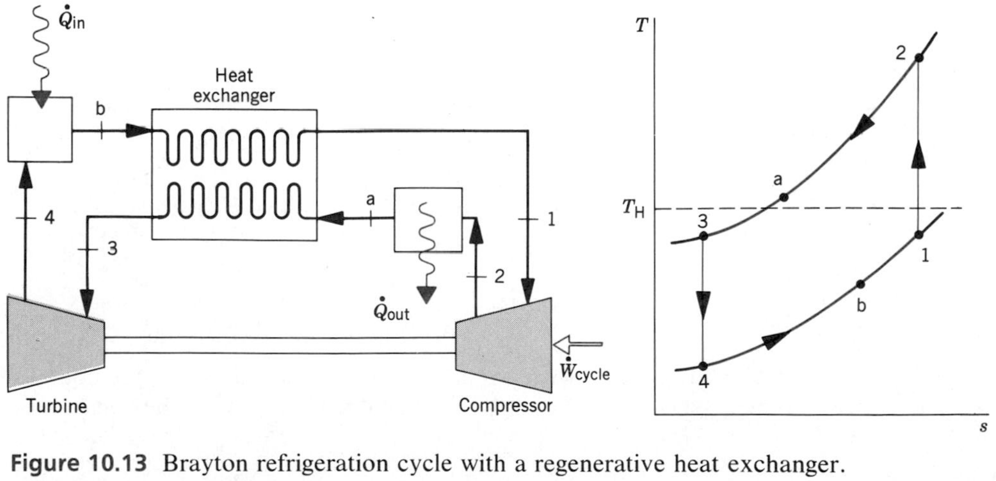

Figure 10.13 Brayton refrigeration cycle with a regenerative heat exchanger.

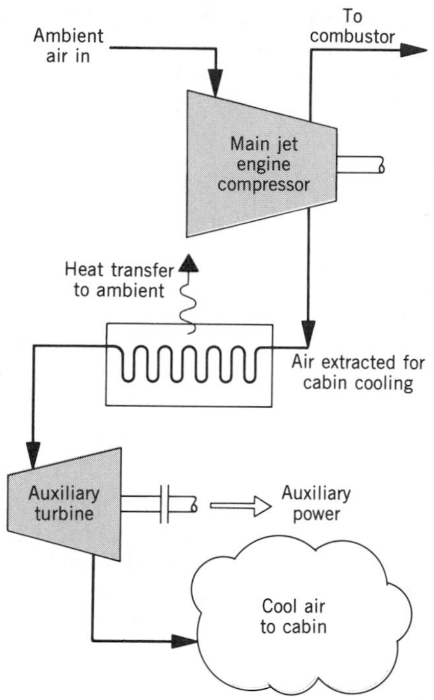

Figure 10.14 An application of gas refrigeration to aircraft cabin cooling.

heat exchanger. Accordingly, the refrigeration effect, achieved from state 4 to state b, occurs at a correspondingly lower average temperature.

An example of the application of gas refrigeration to cabin cooling in an aircraft is illustrated in Fig. 10.14. As shown in the figure, a small amount of high-pressure air is extracted from the main jet engine compressor and cooled by heat transfer to the ambient. The high-pressure air is then expanded through an auxiliary turbine to the pressure maintained in the cabin. The air temperature is reduced in the expansion and thus is able to fulfill its cabin cooling function. As an additional benefit, the turbine expansion can provide some of the auxiliary power needs of the aircraft. Size and weight are important considerations in the selection of equipment for use in aircraft. Open-cycle systems, like the example given here, utilize *compact* high-speed rotary turbines and compressors. Furthermore, since the air for cooling comes directly from the surroundings, there are fewer heat exchangers than would be needed if a separate refrigerant were circulated in a closed vapor-compression cycle.

PROBLEMS

10.1 A Carnot vapor refrigeration cycle uses Refrigerant 12 as the working fluid. The refrigerant enters the condenser as saturated vapor at 28°C and leaves as saturated liquid. The evaporator operates at a temperature of −10°C. Determine, in kJ per kg of refrigerant flow,

(a) the work input to the compressor.
(b) the work developed by the turbine.
(c) the heat transfer to the refrigerant passing through the evaporator.

What is the coefficient of performance of the cycle?

10.2 Refrigerant 12 is the working fluid in a Carnot vapor refrigeration cycle for which the evaporator temperature is 0°C. Saturated vapor enters the condenser at 40°C, and saturated liquid exits at the same temperature. The mass flow rate of refrigerant is 3 kg/min. Determine

(a) the rate of heat transfer to the refrigerant passing through the evaporator, in kW.
(b) the net power input to the cycle, in kW.
(c) the coefficient of performance.

10.3 A Carnot vapor refrigeration cycle operates between thermal reservoirs at 40°F and 90°F. For **(a)** Refrigerant 12, **(b)** Refrigerant 134a, and **(c)** water as the working fluid, determine the operating pressures in the condenser and evaporator, in lbf/in.^{2}, and the coefficient of performance.

10.4 A Carnot vapor refrigeration cycle is used to maintain a cold region at 0°F when the ambient temperature is 70°F. Refrigerant 134a enters the condenser as saturated vapor at 100 lbf/in.^{2} and leaves as saturated liquid at the same pressure. The evaporator pressure is 20 lbf/in.^{2} The mass flow rate of refrigerant is 12 lb/min. Calculate

(a) the compressor and turbine power, each in Btu/min.
(b) the coefficient of performance.

10.5 For the cycle in Problem 10.4, determine

(a) the rates of heat transfer, in Btu/min, for the refrigerant flowing through the evaporator and condenser, respectively.
(b) the rates and directions of availability transfer accompanying each of these heat transfers, in Btu/min. Let T_0 = 70°F.

10.6 An ideal vapor-compression refrigeration cycle operates at steady state with Refrigerant 12 as the working fluid. Saturated vapor enters the compressor at −10°C, and saturated liquid leaves the condenser at 28°C. The mass flow rate of refrigerant is 5 kg/min. Determine

(a) the compressor power, in kW.
(b) the refrigerating capacity, in tons.
(c) the coefficient of performance.

10.7 Modify the cycle in Problem 10.6 to have saturated vapor entering the compressor at 1.6 bar and saturated liquid leaving the condenser at 9 bar. Answer the same questions for the modified cycle as in Problem 10.6.

10.8 An ideal vapor-compression refrigeration system operates at steady state with Refrigerant 134a as the working fluid. Superheated vapor enters the compressor at 30 lbf/in.^{2}, 20°F, and saturated liquid leaves the condenser at 140 lbf/in.^{2} The refrigeration capacity is 5 tons. Determine

(a) the compressor power, in horsepower.
(b) the rate of heat transfer from the working fluid passing through the condenser, in Btu/min.
(c) the coefficient of performance.

10.9 Refrigerant 12 enters the compressor of an ideal vapor-compression refrigeration system as saturated vapor at −15°C with a volumetric flow rate of 1 m^3/min. The refrigerant leaves the condenser at 36°C, 10 bar. Determine

(a) the compressor power, in kW.
(b) the refrigerating capacity, in tons.
(c) the coefficient of performance.

10.10 An ideal vapor-compression refrigeration cycle, with Refrigerant 12 as the working fluid, has an evaporator temperature of −20°C and a condenser pressure of 9 bar. Saturated vapor enters the compressor, and saturated liquid exits the condenser. The mass flow rate of the refrigerant is 3 kg/min. Determine

(a) the coefficient of performance.
(b) the refrigerating capacity, in tons.

10.11 Refrigerant 12 enters the compressor of an ideal vapor-compression refrigeration cycle as saturated vapor at −10°F. The condenser pressure is 160 lbf/in.^{2} The mass flow rate of refrigerant is 6 lb/min. Determine the coefficient of performance and the refrigerating capacity, in tons, for **(a)** saturated liquid, **(b)** compressed liquid at 105°F, and **(c)** compressed liquid at 100°F at the condenser exit.

10.12 To determine the effect of lowering the evaporator temperature on the performance of an ideal vapor-compression refrigeration cycle, recalculate the coefficient of performance and the refrigerating capacity, in tons, for the cycle in Problem 10.10 for saturated vapor entering the compressor at

(a) −10°C.
(b) −30°C.
(c) −40°C.

All other conditions are the same as in Problem 10.10.

10.13 To determine the effect of increasing condenser pressure on the performance of an ideal vapor-compression refrigeration cycle, recalculate the coefficient of performance and the refrigerating capacity, in tons, for the cycle in Problem 10.10 for condenser pressures of

(a) 7 bar.
(b) 10 bar.
(c) 12 bar.

All other conditions are the same as in Problem 10.10.

10.14 Modify the cycle in Problem 10.7 to have an isentropic compressor efficiency of 80% and let the temperature of the liquid leaving the condenser be 34°C. Determine, for the modified cycle,

(a) the compressor power, in kW.
(b) the refrigerating capacity, in tons.
(c) the coefficient of performance.
(d) the irreversibility rates of the compressor and expansion valve, each in kW, for T_0 = 28°C.

10.15 Modify the cycle in Problem 10.8 to have an isentropic compressor efficiency of 85% and let the temperature of the liquid leaving the condenser be 95°F. Determine, for the modified cycle,

(a) the compressor power, in horsepower.
(b) the rate of heat transfer from the working fluid passing through the condenser, in Btu/min.
(c) the coefficient of performance.
(d) the irreversibility rates of the compressor and expansion valve, each in Btu/min, for T_0 = 80°F.

10.16 A vapor-compression refrigeration system circulates Refrigerant 12 at a rate of 6 kg/min. The refrigerant enters the compressor at −20°C, 1.4 bar, and exits at 7 bar. The isentropic compressor efficiency is 67%. There are no appreciable pressure drops as the refrigerant flows through the condenser and evaporator. The refrigerant leaves the condenser as saturated liquid at 7 bar. Ignoring heat transfer between the compressor and its surroundings, determine

(a) the coefficient of performance.
(b) the refrigerating capacity, in tons.
(c) the irreversibility rates of the compressor and expansion valve, each in kW.
(d) the changes in specific flow availability of the refrigerant passing through the evaporator and condenser, respectively, each in kJ/kg.

Let T_0 = 21°C, p_0 = 1 bar.

10.17 A vapor-compression refrigeration system, using Refrigerant 12 as the working fluid, has evaporator and condenser pressures of 1.4 and 12 bar, respectively. The refrigerant passes through each heat exchanger with a negligible pressure drop. At the inlet and exit of the compressor, the temperatures are −10°C and 80°C, respectively. The heat transfer rate from the working fluid passing through the condenser is 15 kW, and saturated liquid exits at 12 bar. If the compressor operates adiabatically, determine

(a) the compressor power input, in kW.
(b) the coefficient of performance.

10.18 In a vapor-compression refrigeration cycle, Refrigerant 12 exits the evaporator as saturated vapor at 20 lbf/in.^{2} The refrigerant enters the condenser at 160 lbf/in.^{2} and 160°F, and saturated liquid exits at 160 lbf/in.^{2} There is no significant heat transfer between the compressor and its surroundings, and the refrigerant passes through the evaporator with a negligible change in pressure. If the refrigerating capacity is 5 tons, determine

(a) the mass flow rate of refrigerant, in lb/min.
(b) the power input to the compressor, in Btu/min.
(c) the coefficient of performance.
(d) the isentropic compressor efficiency.

10.19 A vapor-compression refrigeration system with a capacity of 10 tons has superheated Refrigerant 12 vapor entering the compressor at 15°C, 4 bar, and exiting at 12 bar. The compression process can be taken as polytropic, with n = 1.02. At the condenser exit, the pressure is 11.6 bar, and the temperature is 44°C. The condenser is water-cooled, with water entering at 20°C and leaving at 30°C with a negligible change in pressure. Heat transfer from the outside of the condenser can be neglected. Determine

(a) the power input and the heat transfer rate for the compressor, each in kW.
(b) the coefficient of performance.
(c) the irreversibility rate of the condenser, in kW, for T_0 = 20°C.

10.20 The capacity of a Refrigerant 12 vapor-compression refrigeration system is 10 tons. Saturated vapor at 40 lbf/in.^{2} enters the compressor, and superheated vapor leaves at 110°F, 120 lbf/in.^{2} Heat transfer from the compressor to its surroundings occurs at a rate of 1.2 Btu per lb of refrigerant passing through the compressor. Liquid refrigerant enters the expansion valve at 90°F, 120 lbf/in.^{2} The condenser is water-cooled, with water entering at 60°F and leaving at 75°F with a negligible change in pressure. Determine

(a) the compressor power input, in Btu/min.
(b) the mass flow rate of cooling water through the condenser, in lb/min.
(c) the coefficient of performance.

10.21 A vapor-compression refrigeration system for a household refrigerator has a refrigerating capacity of 1000 Btu/h. Refrigerant enters the evaporator at −10°F and exits at 0°F. The isentropic compressor efficiency is 80%. The refrigerant condenses at 95°F and exits the condenser subcooled at 90°F. There are no significant pressure drops in the flows through the evaporator and condenser. Determine the evaporator and condenser pressures, each in lbf/in.^{2}, the mass flow rate of refrigerant, in lb/min, the compressor power input, in horsepower, and the coefficient of performance for **(a)** Refrigerant 12 and **(b)** Refrigerant 134a as the working fluid.

10.22 Figure P10.22 shows a *steam jet* refrigeration system that produces chilled water in a flash chamber. The

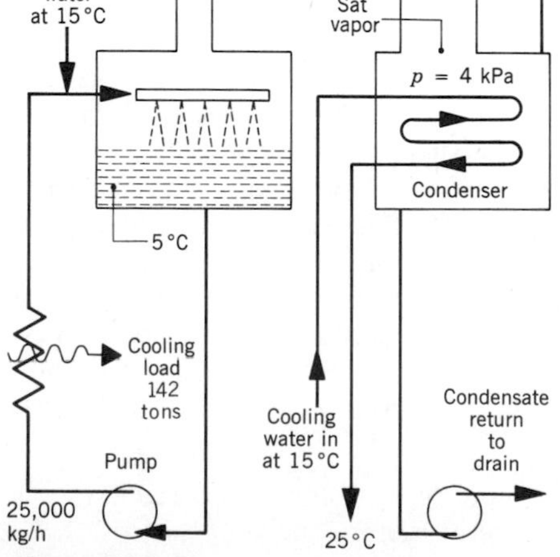

Figure P10.22

chamber is maintained at a vacuum pressure by the steam ejector, which removes the vapor generated by entraining it in the low-pressure jet and discharging into the condenser. The vacuum pump removes air and other noncondensable gases from the condenser shell. For the conditions shown on the figure, determine the make-up water and cooling water flow rates, each in kg/h.

10.23 A vapor-compression refrigeration system with the cascade cycle arrangement of Fig. 10.6 is used to obtain refrigeration at a low temperature. Refrigerant 12 is the working fluid in the high-temperature cycle and Refrigerant 13 (chlorotrifluoromethane) is used in the low-temperature cycle. For the Refrigerant 13 cycle, the working fluid enters the compressor as saturated vapor at −30°F and is compressed isentropically to 250 lbf/in.² Saturated liquid leaves the intermediate heat exchanger at 250 lbf/in.² and enters the expansion valve. For the Refrigerant 12 cycle, the working fluid enters the compressor as saturated vapor at 20°F and is compressed isentropically to 160 lbf/in.² Saturated liquid enters the expansion valve at 160 lbf/in.² The refrigerating capacity of the cascade system is 20 tons. The following table gives data for Refrigerant 13:

Pressure (lbf/in.²)	Temperature (°F)	Specific Enthalpy (Btu/lb)	Specific Entropy (Btu/lb °R)
106.3	$T_{sat} = -30$	$h_f = 2.46$ $h_g = 54.4$	$s_g = 0.1265$
250.0	$T = 40$	$h = 60.0$	$s = 0.1265$
250.0	$T_{sat} = 22.48$	$h_f = 16.2$	$s_f = 0.0352$

Determine

(a) the power input to each compressor, in Btu/min.
(b) the rate of heat transfer from the Refrigerant 12 passing through the high temperature condenser, in Btu/min.
(c) the overall coefficient of performance of the cascade cycle.
(d) the irreversibility rate of the intermediate heat exchanger, in Btu/min, for $T_0 = 90°F$, $p_0 = 14.7$ lbf/in.²

10.24 A vapor-compression refrigeration system uses the arrangement shown in Fig. 10.7 for two-stage compression with intercooling between the stages. Refrigerant 12 is the working fluid. Saturated vapor at −30°C enters the first compressor stage. The flash chamber and direct contact heat exchanger operate at 4 bar, and the condenser pressure is 12 bar. Saturated liquid streams at 12 and 4 bar enter the high- and low-pressure expansion valves, respectively. If each compressor operates isentropically and the refrigerating capacity of the system is 10 tons, determine

(a) the power input to each compressor, in kW.
(b) the coefficient of performance.

10.25 Figure P10.25 shows a two-stage vapor-compression refrigeration system with Refrigerant 12 as the working fluid. The system uses a direct-contact heat exchanger to achieve intercooling. The evaporator has a refrigerating capacity of 30 tons and produces −20°F saturated vapor at its exit. In the first compressor stage, the refrigerant is compressed adiabatically to 50 lbf/in.², which is the pressure in the direct-contact heat exchanger. Saturated vapor at 50 lbf/in.² enters the second compressor stage and is compressed adiabatically to 180 lbf/in.² Each compressor stage has an isentropic efficiency of 85%. There are no significant pressure drops as the refrigerant passes through the heat exchangers. Saturated liquid enters each expansion valve. Determine

(a) the mass flow rate through the evaporator, in lb/min.
(b) the power input to each compressor stage, in horsepower.
(c) the coefficient of performance.

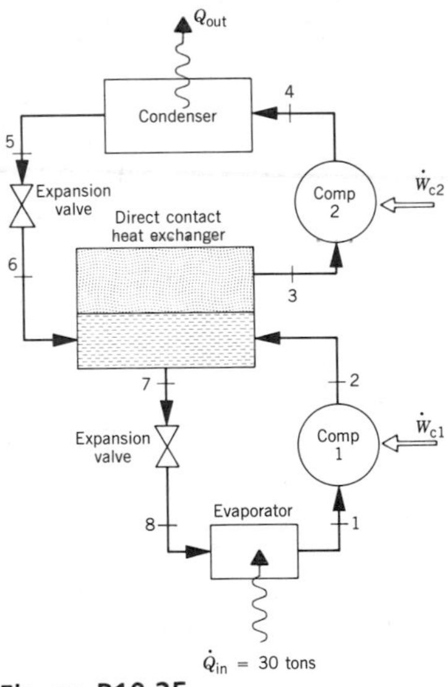

Figure P10.25

10.26 Figure P10.26 shows the schematic diagram of a vapor-compression refrigeration system with two evaporators using Refrigerant 12 as the working fluid. This arrangement is used to achieve refrigeration at two different temperatures with a single compressor and a single condenser. The low-temperature evaporator operates at −18°C with saturated vapor at its exit and has a refrigerating capacity of 3 tons. The higher-temperature evaporator produces saturated vapor at 3.2 bar at its exit and has a refrigerating capacity of 2 tons. Compression is isentropic to the condenser pressure of 10 bar. There are no significant pressure drops in the flows through the condenser and the two evaporators, and the refrigerant leaves the condenser as saturated liquid at 10 bar. Calculate

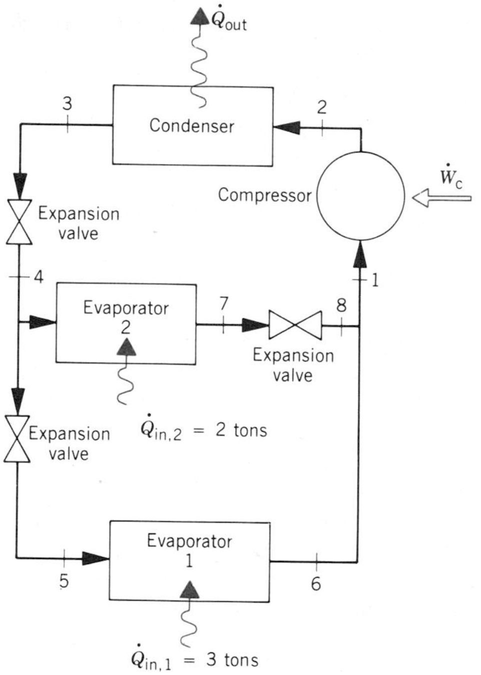

Figure P10.26

(a) the mass flow rates of refrigerant through each evaporator, in kg/min.
(b) the compressor power input, in kW.
(c) the rate of heat transfer from the refrigerant passing through the condenser, in kW.

10.27 An ideal vapor-compression refrigeration cycle is modified to include a counterflow heat exchanger, as shown in Fig. P10.27. Refrigerant 12 leaves the evaporator as saturated vapor at 1.4 bar and is heated at constant pressure to 20°C before entering the compressor. Following isentropic compression to 12 bar, the refrigerant passes through the condenser, exiting at 44°C, 12 bar. The liquid then passes through the heat exchanger, entering the expansion valve at 12 bar. If the mass flow rate of refrigerant is 6 kg/min, determine

(a) the refrigeration capacity, in tons of refrigeration.
(b) the compressor power input, in kW.
(c) the coefficient of performance.

Discuss possible advantages and disadvantages of this arrangement.

10.28 An ideal vapor-compression heat pump cycle with Refrigerant 12 as the working fluid provides 15 kW to maintain a building at 20°C when the outside temperature is 5°C. Saturated vapor at 2.4 bar leaves the evaporator, and saturated liquid at 8 bar leaves the condenser. Calculate

(a) the power input to the compressor, in kW.
(b) the coefficient of performance.
(c) the coefficient of performance of a reversible heat pump cycle operating between thermal reservoirs at 20 and 5°C.

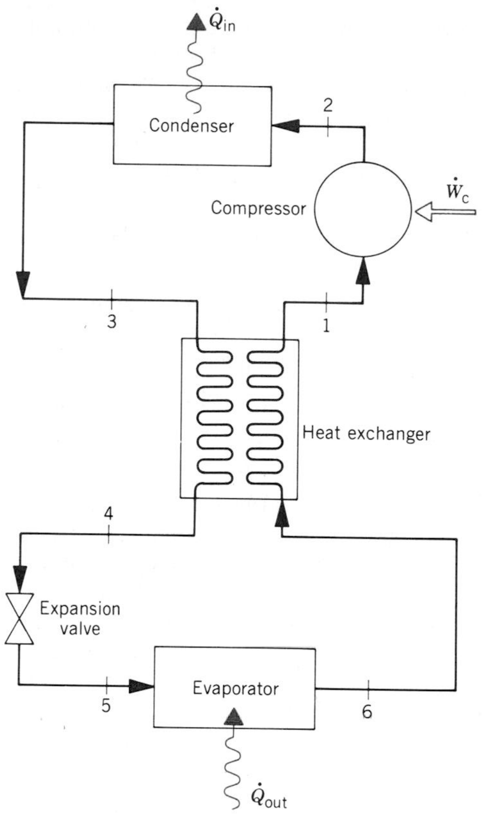

Figure P10.27

10.29 Refrigerant 12 is the working fluid in a vapor-compression heat pump system with a heating capacity of 24,000 Btu/h. The condenser operates at 160 lbf/in.^{2}, and the evaporator temperature is −10°F. The refrigerant is a saturated vapor at the evaporator exit and a liquid at 105°F at the condenser exit. Pressure drops in the flows through the evaporator and condenser are negligible. The compression process is adiabatic, and the temperature at the compressor exit is 160°F. Determine

(a) the mass flow rate of refrigerant, in lb/min.
(b) the compressor power input, in horsepower.
(c) the isentropic compressor efficiency.
(d) the coefficient of performance.

10.30 A vapor-compression heat pump system uses Refrigerant 134a as the working fluid. The refrigerant enters the compressor at 2.4 bar, 0°C, with a volumetric flow rate of 0.6 m^3/min. Compression is adiabatic to 9 bar, 60°C, and saturated liquid exits the condenser at 9 bar. Determine

(a) the power input to the compressor, in kW.
(b) the heating capacity of the system, in kW and tons.
(c) the coefficient of performance.
(d) the isentropic compressor efficiency.

10.31 On a particular day when the outside temperature is 5°C, a house requires a heat transfer rate of 12 kW to maintain the inside temperature at 20°C. A vapor-compression heat pump with Refrigerant 12 as the working fluid is to be used to provide the necessary heating. Specify appropriate evaporator and condenser pressures of a cycle for this purpose. Let the refrigerant be saturated vapor at the evaporator exit and saturated liquid at the condenser exit. Calculate

(a) the mass flow rate of refrigerant, in kg/min.
(b) the compressor power, in kW.
(c) the coefficient of performance.

10.32 Repeat the calculations of Problem 10.31 for Refrigerant 134a as the working fluid. Compare the results with those of Problem 10.31 and discuss.

10.33 A process requires a heat transfer rate of 2×10^6 Btu/h at 150°F. It is proposed that a Refrigerant 12 vapor-compression heat pump be used to develop the process heat using a waste water stream at 100°F as the lower-temperature source. Specify appropriate evaporator and condenser pressures of a cycle for this purpose. Let the refrigerant be saturated vapor at the evaporator exit and saturated liquid at the condenser exit. Calculate

(a) the mass flow rate of refrigerant, in lb/h.
(b) the compressor power, in horsepower.
(c) the coefficient of performance.

10.34 A vapor-compression heat pump with a heating capacity of 500 kJ/min is driven by a power cycle with a thermal efficiency of 25%. For the heat pump, Refrigerant 12 is compressed from saturated vapor at −10°C to the condenser pressure of 10 bar. The isentropic compressor efficiency is 80%. Liquid enters the expansion valve at 9.6 bar, 36°C. For the power cycle, 80% of the heat rejected is transferred to the heated space.

(a) Determine the power input to the heat pump compressor, in kW.
(b) Evaluate the ratio of the total rate that heat is delivered to the heated space to the rate of heat input to the power cycle. Discuss.

10.35 The condenser of a Refrigerant 12 vapor-compression heat pump operating at steady state heats 2000 ft^3/min of inside air from 70°F, 14.5 lbf/in.^{2} to 90°F. Refrigerant enters the compressor at 20 lbf/in.^{2}, 5°F and is compressed to 120 lbf/in.^{2} according to $pv^{1.05}$ = constant. Liquid refrigerant exits the condenser at 85°F. Outside air passing through the evaporator is cooled from 20 to 5°F. Pressure drops for flow through the heat exchangers can be neglected. Calculate

(a) the mass flow rate of refrigerant, in lb/min.
(b) the power input to the compressor, in Btu/min.
(c) the rate of heat transfer from the compressor, in Btu/min.
(d) the heating capacity of the heat pump, in tons.
(e) the mass flow rate of outside air passing through the evaporator, in lb/min.

10.36 Refrigerant 12 enters the compressor of a vapor-compression heat pump at 30°F, 40 lbf/in.^{2} where it is compressed to 200 lbf/in.^{2}, 180°F. The power input to the compressor is 16,000 Btu/h. The refrigerant mass flow rate is 600 lb/h. The state at the exit of the condenser is fixed by 194 lbf/in.^{2}, 124°F. At the exit of the expansion valve, the pressure is 42 lbf/in^2. Let T_0 = 32°F, p_0 = 14.7 lbf/in.^{2}

(a) Determine the changes in specific flow availability between the inlet and exit of each component in the cycle, in Btu/lb. In each case, explain the observed increase or decrease.
(b) Determine the irreversibility rate in the expansion valve, in Btu/h.
(c) Locating the control surface to include a portion of the immediate surroundings so that unavoidable heat transfer takes place at temperature T_0, determine the irreversibility rate for the compressor, in Btu/h.
(d) Verify that the results of parts (a), (b), and (c) account for all the availability entering the cycle accompanying the power input to the compressor.

10.37 Air enters the compressor of an ideal Brayton refrigeration cycle at 100 kPa, 270 K. The compressor pressure ratio is 3, and the temperature at the turbine inlet is 310 K. Determine

(a) the net work input, per unit mass of air flow, in kJ/kg.
(b) the refrigeration capacity, per unit mass of air flow, in kJ/kg.
(c) the coefficient of performance.
(d) the coefficient of performance of a reversible refrigeration cycle operating between thermal reservoirs at T_C = 270 K and T_H = 310 K, respectively.

10.38 Reconsider Problem 10.37, but include in the analysis that the compressor and turbine have isentropic efficiencies of 85 and 88%, respectively. Determine for the modified cycle

(a) the coefficient of performance.
(b) the irreversibility rates, per unit mass of air flow, in the compressor and turbine, each in kJ/kg, for T_0 = 310 K.

10.39 An ideal Brayton refrigeration cycle has a compressor pressure ratio of 4. At the compressor inlet, the pressure and temperature of the entering air are 45 lbf/in.^{2} and 480°R. The temperature at the exit of the turbine is 370°R. For a refrigerating capacity of 5 tons, determine

(a) the net power input, in Btu/min.
(b) the coefficient of performance.
(c) the specific volumes of the air at the compressor and turbine inlets, each in ft^3/lb.

10.40 Reconsider Problem 10.39, but include in the analysis that the compressor and turbine each have isentropic efficiencies of 80%. Determine for the modified cycle

(a) the mass flow rate of air, in lb/min.
(b) the coefficient of performance.

10.41 Air enters the compressor of an ideal Brayton refrigeration cycle at 140 kPa, 270 K, with a volumetric flow rate of 1 m^3/s, and is compressed to 420 kPa. The temperature at the turbine inlet is 320 K. Determine

(a) the net power input, in kW.
(b) the refrigerating capacity, in kW.
(c) the coefficient of performance.
(d) the coefficient of performance of a reversible refrigeration cycle operating between reservoirs at T_C = 270 K and T_H = 320 K.

10.42 Write a computer program to calculate the refrigeration capacity and net work per unit mass of air flow, as well as the coefficient of performance, of a Brayton refrigeration cycle on the basis of constant specific heats. Assume no pressure drops through the heat exchangers and adiabatic processes in the compressor and turbine. Allow for turbine and compressor efficiencies less than 100%. For turbine and compressor efficiencies of **(a)** 100%, **(b)** 90%, and **(c)** 80%, plot refrigeration capacity and net work per unit mass of air flow, as well as the coefficient of performance, versus the compressor pressure ratio for T_1 = 260 K, p_1 = 100 kPa, and a temperature at the turbine inlet of 300 K.

10.43 The Brayton refrigeration cycle of Problem 10.37 is modified by the introduction of a regenerative heat exchanger. In the modified cycle, compressed air enters the regenerative heat exchanger at 310 K and is cooled to 280 K before entering the turbine. Determine, for the modified cycle,

(a) the lowest temperature, in K.
(b) the net work input per unit mass of air flow, in kJ/kg.
(c) the refrigeration capacity, per unit mass of air flow, in kJ/kg.
(d) the coefficient of performance.

10.44 Reconsider Problem 10.43, but include in the analysis that the compressor and turbine have isentropic efficiencies of 85 and 88%, respectively. Answer the same questions as in Problem 10.43.

10.45 Consider a Brayton refrigeration cycle with a regenerative heat exchanger. Air enters the compressor at 540°R, 20 lbf/in.^{2} and is compressed isentropically to 80 lbf/in.^{2} Compressed air enters the regenerative heat exchanger at 600°R and is cooled to 540°R before entering the turbine. The expansion through the turbine is isentropic. If the refrigeration capacity is 15 tons, calculate

(a) the mass flow rate of air, in lb/min.
(b) the coefficient of performance.

10.46 Reconsider Problem 10.45, but include in the analysis that the compressor and turbine each have isentropic efficiencies of 85%. Answer the same questions for the modified cycle as in Problem 10.45.

10.47 When a regenerator is added to the ideal Brayton refrigeration cycle, as in Fig. 10.13, does the coefficient of performance increase or decrease? Explain.

10.48 Air at 2 bar, 380 K is extracted from a main jet engine compressor for cabin cooling. The extracted air enters a heat exchanger where it is cooled at constant pressure to 320 K through heat transfer with the ambient. It then expands adiabatically to 0.95 bar through a turbine and is discharged into the cabin. The turbine has an isentropic efficiency of 75%. If the mass flow rate of the air is 1.0 kg/s, determine

(a) the power developed by the turbine, in kW.
(b) the rate of heat transfer from the air to the ambient, in kW.

10.49 Air at 32 lbf/in.^{2}, 680°R is extracted from a main jet engine compressor for cabin cooling. The extracted air enters a heat exchanger where it is cooled at constant pressure to 600°R through heat transfer with the ambient. It then expands adiabatically to 14 lbf/in.^{2} through a turbine and is discharged into the cabin at 500°R with a mass flow rate of 200 lb/min. Determine

(a) the power developed by the turbine, in horsepower.
(b) the isentropic turbine efficiency.
(c) the rate of heat transfer from the air to the ambient, in Btu/min.

10.50 Air undergoes a *Stirling refrigeration cycle,* which is the reverse of the Stirling power cycle introduced in Sec. 9.11. At the beginning of the isothermal expansion, the pressure and temperature are 100 kPa and 300 K, respectively. The compression ratio is 6, and the temperature during the isothermal expansion is 100 K. Determine

(a) the heat transfer during the isothermal expansion, in kJ per kg of air.
(b) the net work for the cycle, in kJ per kg of air.
(c) the coefficient of performance.

DESIGN AND OPEN ENDED PROBLEMS

10.1D A vapor-compression refrigeration system using Refrigerant 134a is being designed for a household food freezer. The refrigeration system must maintain a temperature of 0°F within the freezer compartment when the temperature of the room is 90°F. Under these conditions, the steady-state heat transfer rate from the room into the freezer compartment is 1500 Btu/h. Specify operating pressures and temperatures at key points within the refrigeration system and estimate the refrigerant mass flow rate and compressor power required.

10.2D Design a bench-top-sized air-to-air vapor-compression refrigeration system to be used in a student laboratory. Include in your design the capability to utilize either of two expansion devices: a *capillary tube* or a *thermostatic expansion valve*. Provide sketches of the layout of your system, including appropriate interconnecting piping. Specify the locations and types of sensors to allow students to measure the electrical power consumption and refrigeration capacity, as well as perform energy balances on the evaporator and condenser.

10.3D An air-conditioning system is under consideration that will use a vapor-compression ice maker during the nighttime hours when electric rates are lowest. The stored ice is then employed to provide for a daytime air-conditioning load of up to 100 tons. Two options have been proposed: (1) use a conventionally sized chiller and shift the entire load to nighttime, or (2) use a smaller chiller that runs 24 h/day. Which option makes more sense thermodynamically? Which is likely to be more economical to operate? Discuss.

10.4D A *ground-coupled heat pump* using Refrigerant 22 ($CHClF_2$) is being designed to provide up to 200,000 Btu/h to maintain a building at 70°F. Owing to a high water table at the proposed location, a *vertical well* design is under consideration for the ground loop, as illustrated in Fig. P10.4D. If the average ground water temperature is 50°F, estimate the number of 200-ft wells needed to provide sufficient heat transfer with the ground. Specify operating pressures and temperatures at key points within the heat pump system and estimate the coefficient of performance.

10.5D A vapor-compression heat pump is being designed for use in a water distiller that will provide 10 gal/day for domestic use. The system is shown schematically in Fig. 10.5D. Water vapor formed on the immersed condenser tubes migrates to the top of the heat exchanger shell, where it condenses at atmospheric pressure on the cold evaporator tubes and is drained away. Determine a suitable refrigerant and specify the operating pressures and temperatures at key points in the system. Estimate the coefficient of performance of the heat pump.

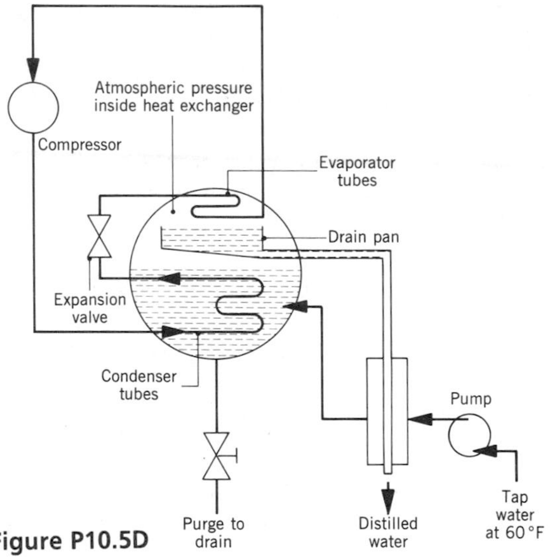

Figure P10.5D

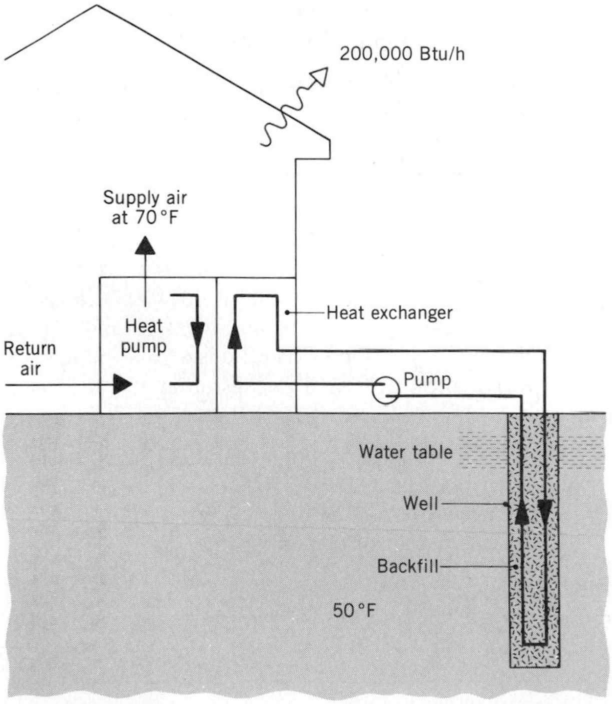

Figure P10.4D

10.6D The *Vuilleumier* cycle combines a refrigeration cycle with a power cycle. Write a description of the operating principles of the cycle. Develop sketches of a possible piston–cylinder and regenerator arrangement for a Vuilleumier refrigeration system. Does the Vuilleumier cycle have promise for practical application? Discuss.

10.7D A manufacturing plant requires 200 tons of refrigeration for air conditioning. Two options are under consideration for using saturated steam at 45 lbf/in.² as an energy source to generate chilled water at 50°F for air conditioning. One is a *steam jet* refrigeration system (see Fig. P10.22), and the other is an *ammonia–water absorption* chiller. Estimate the steam requirements for each option. Which makes more sense thermodynamically? Provide appropriate calculations to justify your answer.

CHAPTER 11

THERMODYNAMIC RELATIONS FOR SIMPLE COMPRESSIBLE SUBSTANCES

Means are available for determining pressure, temperature, volume, and mass experimentally. In addition, the relationships between the specific heats c_v and c_p and temperature at relatively low pressure are accessible experimentally. Values for certain other thermodynamic properties also can be measured without great difficulty. However, specific internal energy, enthalpy, and entropy are among those properties which are not easily obtained experimentally, so we resort to computational procedures to determine values for these. The tabulated values used thus far for *u*, *h*, and *s* are calculated using experimental data on properties that are more readily measured, together with appropriate thermodynamic relations derived from the conservation of energy principle and the second law of thermodynamics. The object of the present chapter is to introduce the thermodynamic relations that allow *u*, *h*, *s*, and other thermodynamic properties of simple compressible systems to be evaluated. Although primary emphasis is on systems involving a single chemical species such as water, an introduction to property relations for mixtures and solutions is also included.

11.1 EQUATIONS OF STATE

An essential ingredient for the calculation of properties such as the specific internal energy, enthalpy, and entropy of a substance is an accurate representation of the relationship among pressure, specific volume, and temperature. The p–v–T relationship can be expressed in three general ways: There are *tabular* representations, as exemplified by the steam tables. The relationship also can be expressed *graphically*, as in the p–v–T surface and compressibility factor charts. *Analytical* formulations, called equations of state, constitute a third general way of expressing the p–v–T relationship. The virial equation and the ideal gas equation are examples of equations of state introduced in previous sections of the book. Analytical formulations of the p–v–T relationship are particularly convenient for performing the mathematical operations required to calculate u, h, s, and other thermodynamic properties. The object of the present section is to expand on the discussion of p–v–T relations for simple compressible substances presented in Chap. 3 by introducing some commonly used equations of state.

11.1.1 Introduction

Recall from Sec. 3.4 that the *virial equation of state* can be derived from the principles of statistical mechanics to relate the p–v–T behavior of a gas to the forces between molecules. In one form, the compressibility factor Z is expanded in inverse powers of specific volume as

$$Z = 1 + \frac{B(T)}{\bar{v}} + \frac{C(T)}{\bar{v}^2} + \frac{D(T)}{\bar{v}^3} + \cdots \tag{11.1}$$

The coefficients B, C, D, etc. are called, respectively, the second, third, fourth, etc. virial coefficients. Each virial coefficient is a function of temperature alone. In principle, the virial coefficients are calculable if a suitable model for describing the forces of interaction between the molecules of the gas under consideration is known. Future advances in refining the theory of molecular interactions may allow the virial coefficients to be predicted with considerable accuracy from the fundamental properties of the molecules involved. However, at present, just the first two or three coefficients can be calculated and only for gases consisting of relatively simple molecules. Equation 11.1 can also be used in an empirical fashion in which the coefficients become parameters whose magnitudes are determined by fitting p–v–T data in particular realms of interest. Only the first few coefficients can be found this way, and the result is a *truncated* equation valid only for certain states.

In the limiting case where the gas molecules are assumed not to interact in any way, the second, third, and higher terms of Eq. 11.1 vanish and the equation reduces to $Z = 1$. Since $Z = p\bar{v}/\bar{R}T$, this gives the ideal gas equation of state $p\bar{v} = \bar{R}T$. The ideal gas equation of state provides an acceptable approximation at many states, including but not limited to states where the pressure is low relative to the critical pressure and/or the temperature is high relative to the critical temperature of the substance under consideration. At many other states, however, the ideal gas equation of state provides a poor approximation.

Over 100 equations of state have been developed in an attempt to improve on the ideal gas equation of state and yet avoid the complexities inherent in a full virial series. In general, these equations exhibit little in the way of fundamental physical significance and are mainly empirical in character. Most are developed for gases, but some describe the p–v–T behavior of the liquid phase, at least qualitatively. Every equation of state is restricted to particular states. This realm of applicability is often indicated by giving an interval of pressure, or density, where the equation can be expected to represent the p–v–T behavior faithfully. When it is not stated, the realm of applicability of a given equation can be approximated by expressing the equation in terms of the compressibility factor Z and the reduced properties p_R, T_R, v'_R and superimposing the result on a generalized compressibility chart or comparing with tabulated compressibility data.[1]

[1] See, for example, R. C. Reid and T. K. Sherwood, *The Properties of Gases and Liquids*, 2nd ed., McGraw-Hill, New York, 1966, pp. 587–595.

Equations of state can be classified by the number of adjustable constants they involve. Let us consider some of the more commonly used equations of state in the order of their increasing complexity, beginning with two-constant equations of state.

11.1.2 Two-Constant Equations of State

An improvement over the ideal gas equation of state based on elementary molecular arguments was suggested in 1873 by van der Waals, who noted that gas molecules actually occupy more than the negligibly small volume presumed by the ideal gas model and also exert long-range attractive forces on one another. Thus, not all of the volume of a container would be available to the gas molecules, and the force they exert on the container wall would be reduced because of the attractive forces that exist between molecules. Employing these elementary molecular arguments, van der Waals proposed the following equation of state:

$$p = \frac{\bar{R}T}{\bar{v} - b} - \frac{a}{\bar{v}^2} \tag{11.2}$$

The constant b is intended to account for the finite volume occupied by the molecules, the term $a/\bar{v}^2$ accounts for the forces of attraction between molecules, and $\bar{R}$ is the universal gas constant. Note that when a and b are set to zero, the ideal gas equation of state results.

The van der Waals equation gives pressure as a function of temperature and specific volume and thus is *explicit* in pressure. Since the equation can be solved for temperature as a function of pressure and specific volume, it is also explicit in temperature. However, the equation is cubic in specific volume, so it cannot generally be solved for specific volume in terms of temperature and pressure. The van der Waals equation is *not* explicit in specific volume.

The van der Waals equation is a *two-constant* equation of state. For a specified substance, values for the constants a and b can be found by fitting the equation to p–v–T data. With this approach several sets of constants might be required to cover all states of interest. Alternatively, a single set of constants for the van der Waals equation can be determined by noting that the critical isotherm passes through a point of inflection at the critical point, and the slope is zero there (Sec. 3.2.2). Expressed mathematically, these conditions are, respectively

$$\left(\frac{\partial^2 p}{\partial \bar{v}^2}\right)_T = 0, \quad \left(\frac{\partial p}{\partial \bar{v}}\right)_T = 0 \qquad \text{(critical point)} \tag{11.3}$$

Although less overall accuracy normally results when the constants a and b are determined using critical point behavior than when they are determined by fitting p–v–T data in a particular region of interest, the advantage of this approach is that the van der Waals constants can be expressed in terms of the critical pressure p_c and critical temperature T_c only, as demonstrated next.

For the van der Waals equation at the critical point

$$p_c = \frac{\bar{R}T_c}{\bar{v}_c - b} - \frac{a}{\bar{v}_c^2}$$

Applying Eqs. 11.3 with the van der Waals equation gives

$$\left(\frac{\partial^2 p}{\partial \bar{v}^2}\right)_T = \frac{2\bar{R}T_c}{(\bar{v}_c - b)^3} - \frac{6a}{\bar{v}_c^4} = 0$$

$$\left(\frac{\partial p}{\partial \bar{v}}\right)_T = -\frac{\bar{R}T_c}{(\bar{v}_c - b)^2} + \frac{2a}{\bar{v}_c^3} = 0$$

Solving the foregoing three equations for a, b, and $\bar{v}_c$ in terms of the critical pressure and critical temperature

$$a = \frac{27}{64}\frac{\bar{R}^2 T_c^2}{p_c} \tag{11.4a}$$

$$b = \frac{\bar{R}T_c}{8p_c} \tag{11.4b}$$

$$\bar{v}_c = \frac{3}{8}\frac{\bar{R}T_c}{p_c} \tag{11.4c}$$

Values of the van der Waals constants a and b determined from Eqs. 11.4a and 11.4b for several common substances are given in Table A-23 for pressure in bars, specific volume in m^3/kmol, and temperature in K. Values of a and b for the same substances are given in Table A-23E for pressure in atm, specific volume in ft^3/lbmol, and temperature in °R.

Introducing the compressibility factor $Z = p\bar{v}/\bar{R}T$, the reduced temperature $T_R = T/T_c$, the pseudoreduced specific volume $v'_R = p_c\bar{v}/\bar{R}T_c$, and the foregoing expressions for a and b, the van der Waals equation can be written in terms of Z, v'_R, and T_R as

$$Z = \frac{v'_R}{v'_R - 1/8} - \frac{27/64}{T_R v'_R} \tag{11.5}$$

or alternatively in terms of Z, T_R, and p_R as

$$Z^3 - \left(\frac{p_R}{8T_R} + 1\right) Z^2 + \left(\frac{27p_R}{64T_R^2}\right) Z - \frac{27p_R^2}{512T_R^3} = 0 \tag{11.6}$$

The details of these developments are left as exercises. Equation 11.5 can be evaluated for specified values of v'_R and T_R and the resultant Z values located on a generalized compressibility chart to show approximately where the equation performs satisfactorily. A similar approach can be taken with Eq. 11.6.

The compressibility factor at the critical point yielded by the van der Waals equation is determined from Eq. 11.4c as

$$Z_c = \frac{p_c\bar{v}_c}{\bar{R}T_c} = 0.375$$

Actually, Z_c varies from about 0.23 to 0.33 for most substances (see Tables A-1). Accordingly, with the set of constants given by Eqs. 11.4, the van der Waals equation is inaccurate in the vicinity of the critical point. Further study would show inaccuracy in other regions as well, so this equation is not suitable for many thermodynamic evaluations. The van der Waals equation is of interest to us primarily because it is the simplest model that accounts for the departure of actual gas behavior from the ideal gas equation of state.

Three other two-constant equations of state that have been widely used are the Berthelot, Dieterici, and Redlich–Kwong equations. The Redlich–Kwong equation, considered by many to be the best of the two-constant equations of state, is

$$p = \frac{\bar{R}T}{\bar{v} - b} - \frac{a}{\bar{v}(\bar{v} + b)T^{1/2}} \tag{11.7}$$

This equation, proposed in 1949, is mainly empirical in nature, with no rigorous justification in terms of molecular arguments. The Redlich–Kwong equation is explicit in pressure but not in specific volume or temperature. Like the van der Waals equation, the Redlich–Kwong equation is cubic in specific volume.

As for the van der Waals equation, the constants a and b in Eq. 11.7 can be determined for a specified substance by fitting the equation to p–v–T data, with several sets of constants required to represent accurately all states of interest. Alternatively, a single set of constants in terms of the critical pressure and critical temperature only can be evaluated using Eqs. 11.3, as for the van der Waals equation. The result is

$$a = a' \frac{\bar{R}^2 T_c^{5/2}}{p_c} \quad \text{and} \quad b = b' \frac{\bar{R} T_c}{p_c} \tag{11.8}$$

where $a' = 0.42748$ and $b' = 0.08664$. Evaluation of these constants is left as an exercise. Values of the Redlich–Kwong constants a and b determined from Eqs. 11.8 for several common substances are given in Table A-23 for pressure in bars, specific volume in m³/kmol, and temperature in K. Values of a and b for the same substances are given in Table A-23E for pressure in atm, specific volume in ft³/lbmol, and temperature in °R.

Introducing the compressibility factor Z, the reduced temperature T_R, the pseudoreduced specific volume v'_R, and the foregoing expressions for a and b, the Redlich–Kwong equation can be written as

$$Z = \frac{v'_R}{v'_R - b'} - \frac{a'}{(v'_R + b')T_R^{3/2}} \tag{11.9}$$

Equation 11.9 can be evaluated at specified values of v'_R and T_R and the resultant Z values located on a generalized compressibility chart to show the regions where the equation performs satisfactorily. With the constants given by Eqs. 11.8, the compressibility factor at the critical point yielded by the Redlich–Kwong equation is $Z_c = 0.333$, which is at the high end of the range of values for most substances, indicating that inaccuracy in the vicinity of the critical point should be expected.

Although the Redlich–Kwong equation is somewhat more difficult to manipulate mathematically than the van der Waals equation, it is more accurate, particularly at higher pressures. In recent years several modified forms of this equation have been proposed to achieve improved accuracy. The two-constant Redlich–Kwong equation performs better than some equations of state having several adjustable constants; still, two-constant equations of state tend to be limited in accuracy as pressure (or density) increases. Increased accuracy normally requires equations with a greater number of adjustable constants.

EXAMPLE 11.1

PROBLEM

A cylindrical tank containing 4.0 kg of carbon monoxide gas at −50°C has an inner diameter of 0.2 m and a length of 1 m. Determine the pressure, in bars, exerted by the gas using (a) the generalized compressibility chart, (b) the ideal gas equation of state, (c) the van der Waals equation of state, (d) the Redlich–Kwong equation of state.

SOLUTION

Known: A cylindrical tank of known dimensions contains 4.0 kg of CO gas at −50°C.

Find: Determine the pressure exerted by the gas using four alternative methods.

Schematic and Given Data:

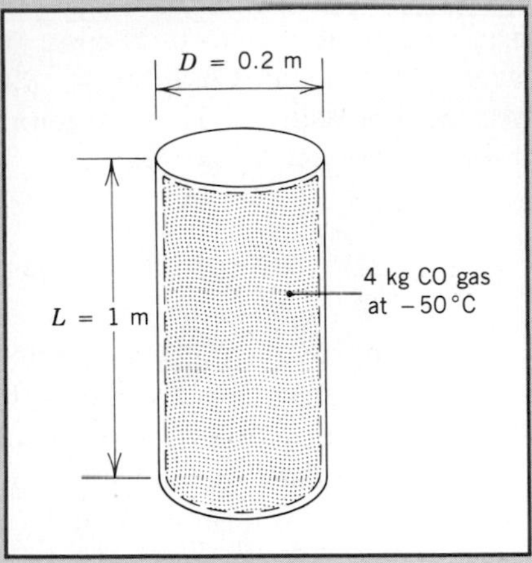

Figure E11.1

Assumptions:

1. As shown in the accompanying figure, the closed system is taken as the gas.
2. The system is at equilibrium.

Analysis: The molar specific volume of the gas is required in each part of the solution. Let us begin by evaluating it. The volume occupied by the gas is

$$V = \left(\frac{\pi D^2}{4}\right) L = \frac{\pi(0.2\ \text{m})^2(1.0\ \text{m})}{4} = 0.0314\ \text{m}^3$$

The molar specific volume is then

$$\bar{v} = Mv = M\left(\frac{V}{m}\right) = \left(28\ \frac{\text{kg}}{\text{kmol}}\right)\left(\frac{0.0314\ \text{m}^3}{4.0\ \text{kg}}\right) = 0.2198\ \frac{\text{m}^3}{\text{kmol}}$$

(a) From Table A-1 for CO, $T_c = 133$ K, $p_c = 35$ bar. Thus, the reduced temperature T_R and pseudoreduced specific volume v'_R are, respectively

$$T_R = \frac{223\ \text{K}}{133\ \text{K}} = 1.68$$

$$v'_R = \frac{\bar{v}p_c}{\bar{R}T_c} = \frac{(0.2198\ \text{m}^3/\text{kmol})(35 \times 10^5\ \text{N/m}^2)}{(8314\ \text{N}\cdot\text{m/kmol}\cdot\text{K})(133\ \text{K})} = 0.696$$

Turning to Fig. A-2, $Z \approx 0.9$. Solving $Z = p\bar{v}/\bar{R}T$ for pressure and inserting known values

$$p = \frac{Z\bar{R}T}{\bar{v}} = \frac{0.9(8314\ \text{N}\cdot\text{m/kmol}\cdot\text{K})(223\ \text{K})}{(0.2198\ \text{m}^3/\text{kmol})}\left(\frac{\text{bar}}{10^5\ \text{N/m}^2}\right) = 75.9\ \text{bar}$$

(b) The ideal gas equation of state gives

$$p = \frac{\bar{R}T}{\bar{v}} = \frac{(8314\ \text{N}\cdot\text{m/kmol}\cdot\text{K})(223\ \text{K})}{(0.2198\ \text{m}^3/\text{kmol})}\left(\frac{\text{bar}}{10^5\ \text{N/m}^2}\right) = 84.4\ \text{bar}$$

(c) For carbon monoxide, the van der Waals constants a and b given by Eqs. 11.4 can be read directly from Table A-23. Thus

$$a = 1.474\ \text{bar}\left(\frac{\text{m}^3}{\text{kmol}}\right)^2 \quad \text{and} \quad b = 0.0395\ \frac{\text{m}^3}{\text{kmol}}$$

Substituting into Eq. 11.2

$$p = \frac{\bar{R}T}{\bar{v} - b} - \frac{a}{\bar{v}^2}$$

$$= \frac{(8314\ \text{N}\cdot\text{m/kmol}\cdot\text{K})(223\ \text{K})}{(0.2198 - 0.0395)(\text{m}^3/\text{kmol})}\left(\frac{\text{bar}}{10^5\ \text{N/m}^2}\right) - \frac{1.474\ \text{bar}(\text{m}^3/\text{kmol})^2}{(0.2198\ \text{m}^3/\text{kmol})^2}$$

$$= 72.3\ \text{bar}$$

Alternatively, the values for v'_R and T_R obtained in the solution of part (a) can be substituted into Eq. 11.5, giving $Z = 0.86$. Then, with $p = Z\bar{R}T/\bar{v}$, $p = 72.5$ bar.

(d) For carbon monoxide, the Redlich–Kwong constants given by Eqs. 11.8 can be read directly from Table A-23. Thus

$$a = \frac{17.22\ \text{bar}(\text{m}^6)(\text{K})^{1/2}}{(\text{kmol})^2} \quad \text{and} \quad b = 0.02737\ \text{m}^3/\text{kmol}$$

Substituting into Eq. 11.7

$$p = \frac{\bar{R}T}{\bar{v} - b} - \frac{a}{\bar{v}(\bar{v} + b)T^{1/2}}$$

$$= \frac{(8314\ \text{N}\cdot\text{m/kmol}\cdot\text{K})(223\ \text{K})}{(0.2198 - 0.02737)\ \text{m}^3/\text{kmol}}\left(\frac{\text{bar}}{10^5\ \text{N/m}^2}\right) - \frac{17.22\ \text{bar}}{(0.2198)(0.24717)(223)^{1/2}}$$

$$= 75.1\ \text{bar}$$

Alternatively, the values for v'_R and T_R obtained in the solution of part (a) can be substituted into Eq. 11.9, giving $Z = 0.89$. Then, with $p = Z\bar{R}T/\bar{v}$, $p = 75.1$ bar.

Comment: In comparison to the value of part (a), the ideal gas equation of state predicts a pressure that is 11% higher and the van der Waals equation gives a value that is 5% lower. The Redlich–Kwong value is about 1% less than the value obtained using the compressibility chart.

11.1.3 Multiconstant Equations of State

To fit the p–v–T data of gases over a wide range of states, Beattie and Bridgeman proposed in 1928 a pressure-explicit equation involving five constants in addition to the gas constant. The Beattie–Bridgeman equation can be expressed in a truncated virial form as

$$p = \frac{\bar{R}T}{\bar{v}} + \frac{\beta}{\bar{v}^2} + \frac{\gamma}{\bar{v}^3} + \frac{\delta}{\bar{v}^4} \tag{11.10}$$

where

$$\begin{aligned} \beta &= B\bar{R}T - A - c\bar{R}/T^2 \\ \gamma &= Bb\bar{R}T + Aa - Bc\bar{R}/T^2 \\ \delta &= Bbc\bar{R}/T^2 \end{aligned} \tag{11.11}$$

The five constants a, b, c, A, and B appearing in these equations are determined by curve fitting to experimental data.

Benedict, Webb, and Rubin extended the Beattie–Bridgeman equation of state to cover a broader range of states. The resulting equation, involving eight constants in addition to the gas constant, has been particularly successful in predicting the p–v–T behavior of *light hydrocarbons*. The Benedict–Webb–Rubin equation is

$$p = \frac{\bar{R}T}{\bar{v}} + \left(B\bar{R}T - A - \frac{C}{T^2}\right)\frac{1}{\bar{v}^2} + \frac{(b\bar{R}T - a)}{\bar{v}^3} + \frac{a\alpha}{\bar{v}^6} + \frac{c}{\bar{v}^3T^2}\left(1 + \frac{\gamma}{\bar{v}^2}\right)\exp\left(-\frac{\gamma}{\bar{v}^2}\right) \tag{11.12}$$

Values of the constants appearing in Eq. 11.12 for five common substances are given in Table A-23 for pressure in bars, specific volume in m^3/kmol, and temperature in K. Values of the constants for the same substances are given in Table A-23E for pressure in atm, specific volume in ft^3/lbmol, and temperature in °R. Because Eq. 11.12 has been so successful, its realm of applicability has been extended by introducing additional constants.

Equations 11.10 and 11.12 are merely representative of multiconstant equations of state. Many other multiconstant equations have been proposed. With the advent of high-speed computers, equations having 50 or more constants have been developed for representing the p–v–T behavior of different substances.

11.2 THERMODYNAMIC FUNCTIONS OF TWO INDEPENDENT VARIABLES

Values of two independent intensive properties are sufficient to fix the state of a simple compressible system of specified mass and composition (Sec. 3.1). All other intensive properties can be determined as functions of the two independent properties: $p = p(T, v)$, $u = u(T, v)$, $h = h(T, v)$, and so on. These are all functions of two independent variables of the form $z = z(x, y)$, with x and y being the independent variables. It might also be recalled that the differential of every property is *exact*. The differentials of nonproperties such as work and heat are inexact. Let us review briefly some concepts from calculus about functions of two independent variables and their differentials.

The *exact differential* of a function z, continuous in the variables x and y, is

$$dz = \left(\frac{\partial z}{\partial x}\right)_y dx + \left(\frac{\partial z}{\partial y}\right)_x dy \tag{11.13a}$$

This can be expressed alternatively as

$$dz = M\,dx + N\,dy \tag{11.13b}$$

where $M = (\partial z/\partial x)_y$ and $N = (\partial z/\partial y)_x$. The coefficient M is the partial derivative of z with respect to x (the variable y being held constant). Similarly, N is the partial derivative of z with respect to y (the variable x being held constant).

If the coefficients M and N have continuous first partial derivatives, the order in which a second partial derivative of the function z is taken is immaterial. That is

$$\frac{\partial}{\partial y}\left[\left(\frac{\partial z}{\partial x}\right)_y\right]_x = \frac{\partial}{\partial x}\left[\left(\frac{\partial z}{\partial y}\right)_x\right]_y \tag{11.14a}$$

or

$$\left(\frac{\partial M}{\partial y}\right)_x = \left(\frac{\partial N}{\partial x}\right)_y \tag{11.14b}$$

In words, Eqs. 11.14 indicate that the mixed second partial derivatives of the function z are equal. The relationship in Eq. 11.14 is both a necessary and sufficient condition for the *exactness* of a differential expression, and it may therefore be used as a *test* for exactness. When an expression such as $M\,dx + N\,dy$ does not meet this test, no function z exists whose differential is equal to this expression. In thermodynamics, Eq. 11.14 is not generally used to test exactness but rather to develop additional property relations. This is illustrated in Sec. 11.3 to follow.

Two other relations among partial derivatives are listed next for which applications are found in subsequent sections of this chapter. These are

$$\left(\frac{\partial x}{\partial y}\right)_z \left(\frac{\partial y}{\partial x}\right)_z = 1 \tag{11.15}$$

and

$$\left(\frac{\partial y}{\partial z}\right)_x \left(\frac{\partial z}{\partial x}\right)_y \left(\frac{\partial x}{\partial y}\right)_z = -1 \tag{11.16}$$

To demonstrate Eqs. 11.15 and 11.16, consider the three quantities x, y, and z, any two of which may be selected as the independent variables. Thus, we can write $x = x(y, z)$ and $y = y(x, z)$. The differentials of these functions are, respectively

$$dx = \left(\frac{\partial x}{\partial y}\right)_z dy + \left(\frac{\partial x}{\partial z}\right)_y dz \qquad \text{and} \qquad dy = \left(\frac{\partial y}{\partial x}\right)_z dx + \left(\frac{\partial y}{\partial z}\right)_x dz$$

Eliminating dy between these two equations results in

$$\left[1 - \left(\frac{\partial x}{\partial y}\right)_z \left(\frac{\partial y}{\partial x}\right)_z\right] dx = \left[\left(\frac{\partial x}{\partial y}\right)_z \left(\frac{\partial y}{\partial z}\right)_x + \left(\frac{\partial x}{\partial z}\right)_y\right] dz \tag{11.17}$$

Since x and z can be varied independently, let us hold z constant and vary x. That is, let $dz = 0$ and $dx \neq 0$. It then follows from Eq. 11.17 that the coefficient of dx must vanish, so Eq. 11.15 must be satisfied. Similarly, when $dx = 0$ and $dz \neq 0$, the coefficient of dz in Eq. 11.17 must vanish. Introducing Eq. 11.15 into the resulting expression and rearranging gives Eq. 11.16. The details are left as an exercise.

An equation of state $p = p(T, v)$ provides a specific example of a function of two independent variables. The partial derivatives $(\partial p/\partial T)_v$ and $(\partial p/\partial v)_T$ of $p(T, v)$ are important for subsequent discussions. The quantity $(\partial p/\partial T)_v$ is the partial derivative of p with respect to T (the variable v being held constant). This partial derivative represents the slope at a point on a line of constant specific volume (isometric) projected onto the p–T plane. Similarly, the partial derivative $(\partial p/\partial v)_T$ is the partial derivative of p with respect to v (the variable T being held constant). This partial derivative represents the slope at a point on a line of constant temperature (isotherm) projected on the p–v plane. The partial derivatives $(\partial p/\partial T)_v$ and $(\partial p/\partial v)_T$ are themselves intensive properties because they have unique values at each state.

The p–v–T surfaces given in Figs. 3.1 and 3.2 are graphical representations of functions of the form $p = p(v, T)$. Figure 11.1 shows the liquid, vapor, and two-phase regions of a p–v–T surface projected onto the p–v and p–T planes. Referring first to Fig. 11.1*a*, note that several isotherms are sketched. In the single-phase regions, the partial derivative $(\partial p/\partial v)_T$ giving the slope is negative at each state along an isotherm except at the critical point, where the partial derivative vanishes. Since the isotherms are horizontal in the two-phase liquid–vapor region, the partial derivative $(\partial p/\partial v)_T$ vanishes there as well. For these states, pressure is independent of specific volume and is a function of temperature only: $p = p_{\text{sat}}(T)$.

Figure 11.1*b* shows the liquid and vapor regions with several isometrics superimposed. In the single-phase regions, the isometrics are nearly straight or are slightly

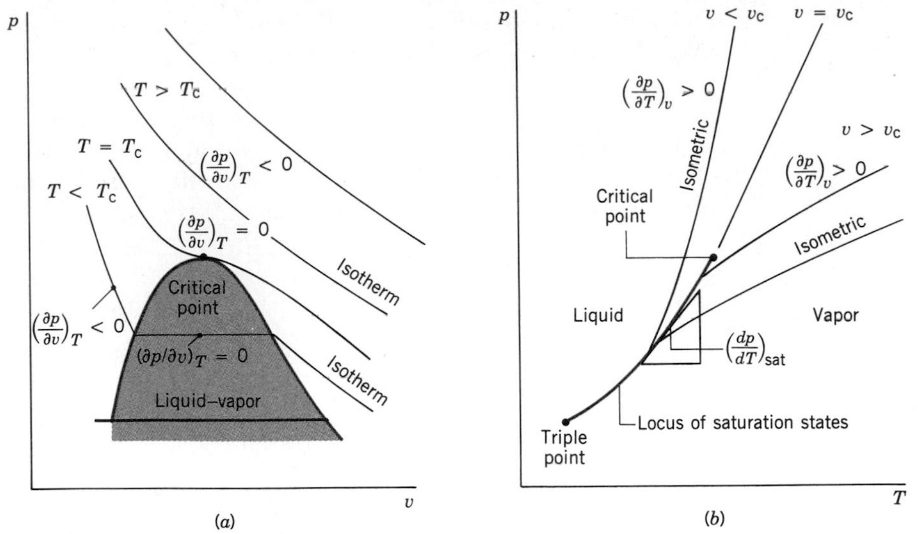

Figure 11.1 Diagrams used to discuss and $(\partial p/\partial v)_T$ $(\partial p/\partial T)_v$. (*a*) p–v diagram. (*b*) Phase diagram.

curved and the partial derivative $(\partial p/\partial T)_v$ is positive at each state along the curves. For the two-phase liquid–vapor states corresponding to a specified value of temperature, the pressure is independent of specific volume and is determined by the temperature only. Hence, the slopes of the isometrics passing through the two-phase states corresponding to a specified temperature are all equal, being given by the slope of the saturation curve at that temperature, denoted simply as $(dp/dT)_{\text{sat}}$. For these two-phase states, $(\partial p/\partial T)_v = (dp/dT)_{\text{sat}}$.

In this section, important aspects of functions of two variables have been introduced. The following example illustrates some of these ideas using the van der Waals equation of state.

EXAMPLE 11.2

PROBLEM

For the van der Waals equation of state, (a) determine an expression for the exact differential dp, (b) show that the mixed second partial derivatives of the result obtained in part (a) are equal, and (c) develop an expression for the partial derivative $(\partial v/\partial T)_p$.

SOLUTION

Known: The equation of state is the van der Waals equation.

Find: Determine the differential dp, show that the mixed second partial derivatives of dp are equal, and develop an expression for $(\partial v/\partial T)_p$.

Analysis:

(a) By definition, the differential of a function $p = p(T, v)$ is

$$dp = \left(\frac{\partial p}{\partial T}\right)_v dT + \left(\frac{\partial p}{\partial v}\right)_T dv$$

The partial derivatives appearing in this expression obtained from the van der Waals equation expressed as $p = RT/(v - b) - a/v^2$ are

$$M = \left(\frac{\partial p}{\partial T}\right)_v = \frac{R}{v - b}, \qquad N = \left(\frac{\partial p}{\partial v}\right)_T = -\frac{RT}{(v - b)^2} + \frac{2a}{v^3}$$

Accordingly, the differential takes the form

$$dp = \left(\frac{R}{v - b}\right) dT + \left[\frac{-RT}{(v - b)^2} + \frac{2a}{v^3}\right] dv$$

(b) Calculating the mixed second partial derivatives

$$\left(\frac{\partial M}{\partial v}\right)_T = \frac{\partial}{\partial v}\left[\left(\frac{\partial p}{\partial T}\right)_v\right]_T = -\frac{R}{(v - b)^2}$$

$$\left(\frac{\partial N}{\partial T}\right)_v = \frac{\partial}{\partial T}\left[\left(\frac{\partial p}{\partial v}\right)_T\right]_v = -\frac{R}{(v - b)^2}$$

Thus, the mixed second partial derivatives are equal.

(c) An expression for $(\partial v/\partial T)_p$ can be derived using Eqs. 11.15 and 11.16. Thus, with $x = p$, $y = v$, and $z = T$, Eq. 11.16 gives

$$\left(\frac{\partial v}{\partial T}\right)_p \left(\frac{\partial p}{\partial v}\right)_T \left(\frac{\partial T}{\partial p}\right)_v = -1$$

or

$$\left(\frac{\partial v}{\partial T}\right)_p = -\frac{1}{(\partial p/\partial v)_T(\partial T/\partial p)_v}$$

Then, with $x = T$, $y = p$, and $z = v$, Eq. 11.15 gives

$$\left(\frac{\partial T}{\partial p}\right)_v = \frac{1}{(\partial p/\partial T)_v}$$

Combining these results

$$\left(\frac{\partial v}{\partial T}\right)_p = -\frac{(\partial p/\partial T)_v}{(\partial p/\partial v)_T}$$

The numerator and denominator of this expression have been evaluated in part (a), so

$$\left(\frac{\partial v}{\partial T}\right)_p = -\frac{R/(v - b)}{[-RT/(v - b)^2 + 2a/v^3]}$$

which is the desired result.

Comment: Since the van der Waals equation is cubic in specific volume, it can be solved for $v(T, p)$ at only certain states. Part (c) shows how the partial derivative $(\partial v/\partial T)_p$ can be evaluated at states where it exists.

11.3 RELATIONS FROM EXACT DIFFERENTIALS

In this section, expressions known as the *Maxwell* relations are derived, and the concept of a *fundamental thermodynamic function* is introduced. These results, which are important for subsequent discussions, are obtained for simple compressible systems of fixed chemical composition using the concept of an exact differential.

11.3.1 Principal Exact Differentials

The principal results of this section are obtained using Eqs. 11.18, 11.19, 11.22, and 11.23. The first two of these equations are derived in Sec. 6.3.3, where they are referred to as the *T ds equations*. For present purposes, it is convenient to express them as

$$du = T\,ds - p\,dv \tag{11.18}$$

$$dh = T\,ds + v\,dp \tag{11.19}$$

The other two equations used to obtain the results of this section involve, respectively, the specific *Helmholtz* function ψ defined by

$$\psi = u - Ts \tag{11.20}$$

and the specific *Gibbs* function g defined by

$$g = h - Ts \tag{11.21}$$

The Helmholtz and Gibbs functions are properties because each is defined in terms of properties. From inspection of Eqs. 11.20 and 11.21, the units of ψ and g are the same as those of u and h. These two new properties are introduced solely because they contribute to the present discussion, and no physical significance need be attached to them at this point.

Forming the differential $d\psi$

$$d\psi = du - d(Ts) = du - T\,ds - s\,dT$$

Substituting Eq. 11.18 into this gives

$$d\psi = -p\,dv - s\,dT \tag{11.22}$$

Similarly, forming the differential dg

$$dg = dh - d(Ts) = dh - T\,ds - s\,dT$$

Substituting Eq. 11.19 into this gives

$$dg = v\,dp - s\,dT \tag{11.23}$$

11.3.2 Maxwell Relations

The four differential equations introduced above, Eqs. 11.18, 11.19, 11.22, and 11.23, provide the basis for a large number of important property relations. Since only properties are involved, each is an exact differential exhibiting the general form $dz = M\,dx + N\,dy$ considered in Sec. 11.2. Underlying these exact differentials are, respectively, functions of the form $u(s, v)$, $h(s, p)$, $\psi(v, T)$, and $g(T, p)$. Let us consider these functions in the order given.

The differential of the function $u = u(s, v)$ is

$$du = \left(\frac{\partial u}{\partial s}\right)_v ds + \left(\frac{\partial u}{\partial v}\right)_s dv$$

By comparison with Eq. 11.18, we conclude that

$$T = \left(\frac{\partial u}{\partial s}\right)_v \tag{11.24}$$

$$-p = \left(\frac{\partial u}{\partial v}\right)_s \tag{11.25}$$

The differential of the function $h = h(s, p)$ is

$$dh = \left(\frac{\partial h}{\partial s}\right)_p ds + \left(\frac{\partial h}{\partial p}\right)_s dp$$

By comparison with Eq. 11.19, we conclude that

$$T = \left(\frac{\partial h}{\partial s}\right)_p \tag{11.26}$$

$$v = \left(\frac{\partial h}{\partial p}\right)_s \tag{11.27}$$

Similarly, the coefficients $-p$ and $-s$ of Eq. 11.22 are partial derivatives of $\psi(v, T)$

$$-p = \left(\frac{\partial \psi}{\partial v}\right)_T \tag{11.28}$$

$$-s = \left(\frac{\partial \psi}{\partial T}\right)_v \tag{11.29}$$

and the coefficients v and $-s$ of Eq. 11.23 are partial derivatives of $g(T, p)$

$$v = \left(\frac{\partial g}{\partial p}\right)_T \tag{11.30}$$

$$-s = \left(\frac{\partial g}{\partial T}\right)_p \tag{11.31}$$

As each of the four differentials introduced above is exact, the second mixed partial derivatives are equal. Thus, in Eq. 11.18, T plays the role of M in Eq. 11.14b and $-p$ plays the role of N in Eq. 11.14b, so

$$\left(\frac{\partial T}{\partial v}\right)_s = -\left(\frac{\partial p}{\partial s}\right)_v \tag{11.32}$$

In Eq. 11.19, T and v play the roles of M and N in Eq. 11.14b, respectively. Thus

$$\left(\frac{\partial T}{\partial p}\right)_s = \left(\frac{\partial v}{\partial s}\right)_p \tag{11.33}$$

Similarly, from Eqs. 11.22 and 11.23 follow

$$\left(\frac{\partial p}{\partial T}\right)_v = \left(\frac{\partial s}{\partial v}\right)_T \tag{11.34}$$

$$\left(\frac{\partial v}{\partial T}\right)_p = -\left(\frac{\partial s}{\partial p}\right)_T \tag{11.35}$$

These four equations are known as the *Maxwell relations*. The utility of the Maxwell relations is demonstrated in subsequent sections of this chapter. However, to give a specific illustration at this point, suppose that the partial derivative $(\partial s/\partial v)_T$ involving entropy is required for a certain purpose. The Maxwell relation Eq. 11.34 would allow the derivative to be determined by evaluating the partial derivative $(\partial p/\partial T)_v$, which can be obtained using p–v–T data only. Further elaboration of this is provided in Example 11.3.

Since each of the properties T, p, v, s appears on the left side of two of the eight equations, Eqs. 11.24 through 11.31, four additional property relations can be obtained by equating such equations. They are

$$\left(\frac{\partial u}{\partial s}\right)_v = \left(\frac{\partial h}{\partial s}\right)_p, \qquad \left(\frac{\partial u}{\partial v}\right)_s = \left(\frac{\partial \psi}{\partial v}\right)_T$$
$$\left(\frac{\partial h}{\partial p}\right)_s = \left(\frac{\partial g}{\partial p}\right)_T, \qquad \left(\frac{\partial \psi}{\partial T}\right)_v = \left(\frac{\partial g}{\partial T}\right)_p \tag{11.36}$$

Equations 11.24 through 11.36 are 16 property relations obtained from Eqs. 11.18, 11.19, 11.22, and 11.23, using the concept of an exact differential. Since Eqs. 11.19, 11.22, and 11.23 can themselves be derived from Eq. 11.18, the important role of the first $T\,dS$ equation in developing property relations is apparent.

EXAMPLE 11.3

PROBLEM

Evaluate the partial derivative $(\partial s/\partial v)_T$ for water vapor at a state fixed by a temperature of 240°C and a specific volume of 0.4646 m³/kg. Use the Redlich–Kwong equation of state and an appropriate Maxwell relation. Check the value obtained using steam table data.

SOLUTION

Known: The system consists of a fixed amount of water vapor at 240°C and 0.4646 m³/kg.

Find: Determine the partial derivative $(\partial s/\partial v)_T$ employing the Redlich–Kwong equation of state, together with a Maxwell relation. Check the value obtained using steam table data.

Assumptions:

1. The system consists of a fixed amount of water at a known equilibrium state.
2. Accurate values for $(\partial p/\partial T)_v$ in the neighborhood of the given state can be determined from the Redlich–Kwong equation of state.

Analysis: The Maxwell relation given by Eq. 11.34 allows $(\partial s/\partial v)_T$ to be determined from the p–v–T relationship. That is

$$\left(\frac{\partial s}{\partial v}\right)_T = \left(\frac{\partial p}{\partial T}\right)_v$$

The partial derivative $(\partial p/\partial T)_v$ obtained from the Redlich–Kwong equation, Eq. 11.7, is

$$\left(\frac{\partial p}{\partial T}\right)_v = \frac{\bar{R}}{\bar{v} - b} + \frac{a}{2\bar{v}(\bar{v} + b)T^{3/2}}$$

At the specified state, the temperature is 513 K and the specific volume on a molar basis is

$$\bar{v} = 0.4646\ \frac{\text{m}^3}{\text{kg}}\left(\frac{18.02\text{ kg}}{\text{kmol}}\right) = 8.372\ \frac{\text{m}^3}{\text{kmol}}$$

From Table A-23

$$a = 142.59\text{ bar}\left(\frac{\text{m}^3}{\text{kmol}}\right)^2 (\text{K})^{1/2}, \qquad b = 0.0211\ \frac{\text{m}^3}{\text{kmol}}$$

Substituting values into the expression for $(\partial p/\partial T)_v$

$$\left(\frac{\partial p}{\partial T}\right)_v = \frac{\left(8314 \dfrac{\text{N}\cdot\text{m}}{\text{kmol}\cdot\text{K}}\right)}{(8.372 - 0.0211)\dfrac{\text{m}^3}{\text{kmol}}} + \frac{142.59 \text{ bar} \left(\dfrac{\text{m}^3}{\text{kmol}}\right)^2 (\text{K})^{1/2} \left(\dfrac{10^5 \text{ N/m}^2}{\text{bar}}\right)}{2\left(8.372 \dfrac{\text{m}^3}{\text{kmol}}\right)\left(8.3931 \dfrac{\text{m}^3}{\text{kmol}}\right)(513 \text{ K})^{3/2}}$$

$$= 1004.3 \frac{\text{N}\cdot\text{m}}{\text{m}^3\cdot\text{K}} = \left(1004.3 \frac{\text{N}\cdot\text{m}}{\text{m}^3\cdot\text{K}}\right)\left(\frac{\text{kJ}}{10^3 \text{ N}\cdot\text{m}}\right)$$

$$= 1.0043 \frac{\text{kJ}}{\text{m}^3\cdot\text{K}}$$

Accordingly

$$\left(\frac{\partial s}{\partial v}\right)_T = 1.0043 \frac{\text{kJ}}{\text{m}^3\cdot\text{K}}$$

The value obtained for $(\partial s/\partial v)_T$ can be estimated using a graphical approach with steam table data, as follows. At 240°C, Table A-4 provides the values for specific entropy s and specific volume v tabulated below:

T = 240°C		
p (bar)	s (kJ/kg·K)	v (m³/kg)
1.0	7.9949	2.359
1.5	7.8052	1.570
3.0	7.4774	0.781
5.0	7.2307	0.4646
7.0	7.0641	0.3292
10.0	6.8817	0.2275

With the values for s and v listed in the table, the accompanying plot giving s versus v can be prepared. Note that a line representing the tangent to the curve at the given state is shown on the plot. The pressure at this state is 5 bar. The slope of the tangent is closely 1.01 kJ/m³·K. The value for $(\partial s/\partial v)_T$ determined using the Redlich–Kwong equation is within 1% of the value determined graphically with steam table data. An alternative graphical approach would be to evaluate $(\partial p/\partial T)_v$ using steam table data. Then, with Eq. 11.34, $(\partial s/\partial v)_T = (\partial p/\partial T)_v$.

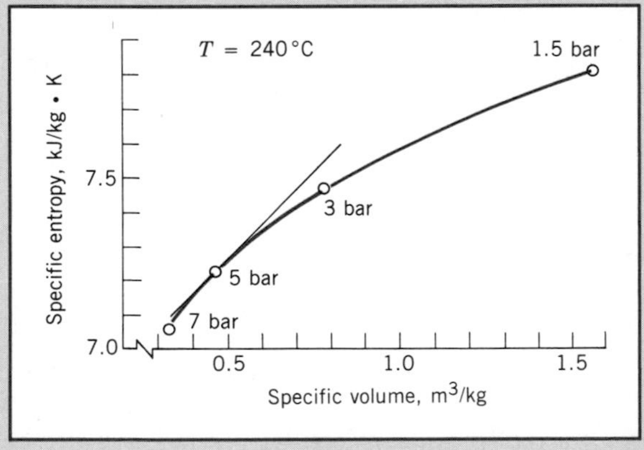

Figure E11.3

11.3.3 Fundamental Thermodynamic Functions

A fundamental thermodynamic function provides a complete description of the thermodynamic state. In the case of a pure substance with two independent properties, the fundamental thermodynamic function can take one of the following four forms:

$$\begin{aligned} u &= u(s, v) \\ h &= h(s, p) \\ \psi &= \psi(T, v) \\ g &= g(T, p) \end{aligned} \tag{11.37}$$

In principle, all properties of interest can be determined from a fundamental thermodynamic function by differentiation and combination. Of the four fundamental functions listed in Eqs. 11.37, the Helmholtz function ψ and the Gibbs function g have the greatest importance for subsequent discussions (see Sec. 11.6.2). Accordingly, let us discuss the fundamental function concept further with reference to ψ and g.

Consider a fundamental function of the form $\psi(T, v)$. The properties v and T, being the independent variables, are specified to fix the state. The pressure p at this state can be determined from Eq. 11.28 by differentiation of $\psi(T, v)$. Similarly, the specific entropy s at the state can be found from Eq. 11.29 by differentiation. By definition, $\psi = u - Ts$, so the specific internal energy is obtained as

$$u = \psi + Ts$$

With u, p, and v known, the specific enthalpy can be found from the definition $h = u + pv$. Similarly, the specific Gibbs function is found from the definition, $g = h - Ts$. The specific heat c_v can be determined by further differentiation, $c_v = (\partial u/\partial T)_v$. Other properties can be calculated with similar operations.

Consider next a fundamental function of the form $g(T, p)$. The properties T and p are specified to fix the state. The specific volume and specific entropy at this state can be determined by differentiation from Eqs. 11.30 and 11.31, respectively. By definition, $g = h - Ts$, so the specific enthalpy is obtained as

$$h = g + Ts$$

With h, p, and v known, the specific internal energy can be found from $u = h - pv$. The specific heat c_p can be determined by further differentiation, $c_p = (\partial h/\partial T)_p$. Other properties can be calculated with similar operations.

Like considerations apply for functions of the form $u(s, v)$ and $h(s, p)$, as can readily be verified. Note that a Mollier diagram provides a graphical representation of the fundamental function $h(s, p)$.

11.4 SOME THERMODYNAMIC RELATIONS INVOLVING ENTROPY, INTERNAL ENERGY, AND ENTHALPY

With the introduction of the Maxwell relations, we are in a position to consider thermodynamic relations that allow changes in entropy, internal energy, and enthalpy to be evaluated from measured property data. The presentation begins with relations applicable to phase changes and then turns to relations for use in single-phase regions.

11.4.1 Clapeyron Equation

The object of this section is to develop relations for evaluating the changes in specific entropy, internal energy, and enthalpy accompanying a change of phase at fixed temperature and pressure. A principal role is played by the *Clapeyron equation,* which allows the change in enthalpy during vaporization, sublimation, or melting at a constant temperature to be evaluated from pressure-specific volume–temperature data pertaining to the phase change. Thus, the present discussion provides important examples of how p–v–T measurements can lead to the determination of other property changes, namely Δs, Δu, and Δh for a change of phase.

Consider a change in phase from saturated liquid to saturated vapor at fixed temperature. For an isothermal phase change, pressure also remains constant, so Eq. 11.19 reduces to

$$dh = T\,ds$$

Integration of this expression gives

$$s_g - s_f = \frac{h_g - h_f}{T} \tag{11.38}$$

Hence, the change in specific entropy accompanying a phase change from saturated liquid to saturated vapor at temperature T can be determined from the temperature and the change in specific enthalpy.

The change in specific internal energy during the phase change can be determined using the definition $h = u + pv$.

$$u_g - u_f = h_g - h_f - p(v_g - v_f) \tag{11.39}$$

To evaluate the change in specific internal energy accompanying a phase change at temperature T, then, requires the temperature and the changes in specific volume and enthalpy.

The change in specific enthalpy required by Eqs. 11.38 and 11.39 can be obtained using the Clapeyron equation. To derive the Clapeyron equation, begin with the Maxwell relation

$$\left(\frac{\partial s}{\partial v}\right)_T = \left(\frac{\partial p}{\partial T}\right)_v \tag{11.34}$$

During a phase change at fixed temperature, the pressure is independent of specific volume and is determined by temperature alone. Thus, the quantity $(\partial p/\partial T)_v$ is determined by the temperature and can be represented as

$$\left(\frac{\partial p}{\partial T}\right)_v = \left(\frac{dp}{dT}\right)_{\text{sat}}$$

where "sat" indicates that the derivative is the slope of the saturation pressure–temperature curve at the point determined by the temperature held constant during the phase change (Sec. 11.2). Combining the last two equations gives

$$\left(\frac{\partial s}{\partial v}\right)_T = \left(\frac{dp}{dT}\right)_{\text{sat}}$$

Since the right side of this equation is fixed when the temperature is specified, the equation can be integrated to give

$$s_g - s_f = \left(\frac{dp}{dT}\right)_{\text{sat}} (v_g - v_f)$$

Introducing Eq. 11.38 into this expression results in the *Clapeyron equation*

$$\left(\frac{dp}{dT}\right)_{\text{sat}} = \frac{h_g - h_f}{T(v_g - v_f)} \tag{11.40}$$

Equation 11.40 allows $(h_g - h_f)$ to be evaluated using only p–v–T data pertaining to the phase change. In instances when the enthalpy change is also measured, the Clapeyron equation can be used to check the consistency of the data. Once the specific enthalpy change is determined, the corresponding changes in specific entropy and specific internal energy can be found from Eqs. 11.38 and 11.39, respectively.

Equations 11.38, 11.39, and 11.40 can also be written for sublimation or melting occurring at constant temperature and pressure. In particular, the Clapeyron equation would take the form

$$\left(\frac{dp}{dT}\right)_{\text{sat}} = \frac{h'' - h'}{T(v'' - v')} \tag{11.41}$$

where $''$ and $'$ denote the respective phases, and $(dp/dT)_{\text{sat}}$ is the slope of the relevant saturation pressure–temperature curve.

The Clapeyron equation shows that the slope of a saturation line on a phase diagram depends on the signs of the specific volume and enthalpy changes accompanying the phase change. In most cases, when a phase change takes place with an increase in specific enthalpy, the specific volume also increases, and $(dp/dT)_{\text{sat}}$ is positive. However, in the case of the melting of ice and a few other substances, the specific volume decreases on melting. The slope of the saturated solid–liquid curve for these few substances is negative, as was pointed out in Sec. 3.2.2 in the discussion of phase diagrams.

An approximate form of Eq. 11.40 can be derived when the following two idealizations are justified: (1) v_f is negligible in comparison to v_g, and (2) the pressure is low enough that v_g can be evaluated from the ideal gas equation of state as $v_g = RT/p$. With these, Eq. 11.40 becomes

$$\left(\frac{dp}{dT}\right)_{\text{sat}} = \frac{h_g - h_f}{RT^2/p}$$

which can be rearranged to read

$$\left(\frac{d \ln p}{dT}\right)_{\text{sat}} = \frac{h_g - h_f}{RT^2} \tag{11.42}$$

Equation 11.42 is called the *Clausius–Clapeyron* equation. A similar expression applies for the case of sublimation.

The use of the Clapeyron equation in any of the foregoing forms requires an accurate representation for the relevant saturation pressure–temperature curve. This must not only depict the pressure–temperature variation accurately but also enable accurate values of the derivative $(dp/dT)_{\text{sat}}$ to be determined. Analytical representations in the form of equations are commonly used. Different equations for different portions of the pressure–temperature curves may be required. These equations can involve several constants. One form that is used for the vapor-pressure curves is the four-constant equation

$$\ln p_{\text{sat}} = A + \frac{B}{T} + C \ln T + DT$$

in which the constants A, B, C, D are determined empirically.

EXAMPLE 11.4

PROBLEM

Using p–v–T data for saturated water from the steam tables, calculate at 100°C (a) $h_g - h_f$, (b) $u_g - u_f$, (c) $s_g - s_f$.

SOLUTION

Known: The system consists of a unit mass of saturated water at 100°C.

Find: Using steam table data, determine at 100°C the change on vaporization of the specific enthalpy, specific internal energy, and specific entropy.

Analysis:

(a) The change on vaporization at fixed temperature of the specific enthalpy can be determined from the Clapeyron equation, Eq. 11.40, expressed as

$$h_g - h_f = T(v_g - v_f)\left(\frac{dp}{dT}\right)_{sat}$$

This equation requires a value for the slope $(dp/dT)_{sat}$ of the saturation pressure–temperature curve at the specified temperature.

The required value for $(dp/dT)_{sat}$ at 100°C can be determined graphically as follows. Using saturation pressure–temperature data from the steam tables, the accompanying plot can be prepared. Note that a line drawn tangent to the curve at 100°C is shown on the plot. The slope of this tangent line is closely 3571 $N/m^2 \cdot K$. Accordingly, at 100°C

$$\left(\frac{dp}{dT}\right)_{sat} = 3571 \frac{N}{m^2 \cdot K}$$

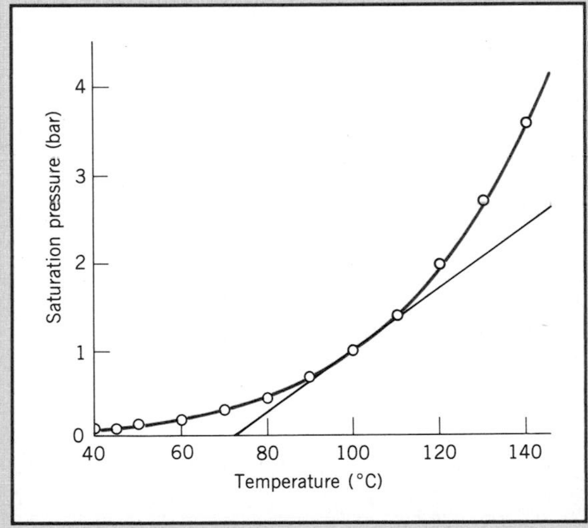

Figure E11.4

Inserting values into the above equation for $h_g - h_f$ gives

$$h_g - h_f = (373 \text{ K})\,(1.673 - 1.0435 \times 10^{-3})\left(\frac{m^3}{kg}\right)\left(3571 \frac{N}{m^2 \cdot K}\right)\left(\frac{kJ}{10^3 \text{ N} \cdot \text{m}}\right)$$

$$= 2227 \text{ kJ/kg}$$

(b) With Eq. 11.39

$$u_g - u_f = h_g - h_f - p_{sat}(v_g - v_f)$$

Inserting the values for $(h_g - h_f)$ from part (a) together with steam table data at 100°C

$$u_g - u_f = 2227 \frac{\text{kJ}}{\text{kg}} - \left(1.014 \times 10^5 \frac{\text{N}}{\text{m}^2}\right)\left(1.672 \frac{\text{m}^3}{\text{kg}}\right)\left(\frac{\text{kJ}}{10^3\ \text{N}\cdot\text{m}}\right)$$

$$= 2057.5 \frac{\text{kJ}}{\text{kg}}$$

(c) With Eq. 11.38 and the value for $h_g - h_f$ from part (a)

$$s_g - s_f = \frac{h_g - h_f}{T} = \frac{2227\ \text{kJ/kg}}{373\ \text{K}} = 5.971 \frac{\text{kJ}}{\text{kg}\cdot\text{K}}$$

Comment: The values of parts (a) to (c) are each within 1.5% of the respective values obtained from the steam tables. The slight differences can be attributed to the graphical method used in the solution and to round-off.

11.4.2 Expressions for Δs, Δu, and Δh for Use in Single-Phase Regions

The objective of the present section is to derive expressions for evaluating Δs, Δu, and Δh between states in single-phase regions. These expressions require both p–v–T data and appropriate specific heat data. Since single-phase regions are under present consideration, any two of the properties pressure, specific volume, and temperature can be regarded as the independent properties that fix the state. Two convenient choices are T, v and T, p.

T and v as Independent Properties. With temperature and specific volume as the independent properties that fix the state, the specific entropy can be regarded as a function of the form $s = s(T, v)$. The differential of this function is

$$ds = \left(\frac{\partial s}{\partial T}\right)_v dT + \left(\frac{\partial s}{\partial v}\right)_T dv$$

The partial derivative $(\partial s/\partial v)_T$ appearing in this expression can be replaced using the Maxwell relation, Eq. 11.34, giving

$$ds = \left(\frac{\partial s}{\partial T}\right)_v dT + \left(\frac{\partial p}{\partial T}\right)_v dv \qquad (11.43)$$

The specific internal energy also can be regarded as a function of T and v: $u = u(T, v)$. The differential of this function is

$$du = \left(\frac{\partial u}{\partial T}\right)_v dT + \left(\frac{\partial u}{\partial v}\right)_T dv$$

With $c_v = (\partial u/\partial T)_v$

$$du = c_v\, dT + \left(\frac{\partial u}{\partial v}\right)_T dv \qquad (11.44)$$

Substituting Eqs. 11.43 and 11.44 into $du = T\,ds - p\,dv$ and collecting terms results in

$$\left[\left(\frac{\partial u}{\partial v}\right)_T + p - T\left(\frac{\partial p}{\partial T}\right)_v\right] dv = \left[T\left(\frac{\partial s}{\partial T}\right)_v - c_v\right] dT \qquad (11.45)$$

Since specific volume and temperature can be varied independently, let us hold specific volume constant and vary temperature. That is, let $dv = 0$ and $dT \neq 0$. It then follows from Eq. 11.45 that

$$\left(\frac{\partial s}{\partial T}\right)_v = \frac{c_v}{T} \tag{11.46}$$

Similarly, suppose that $dT = 0$ and $dv \neq 0$. It then follows that

$$\left(\frac{\partial u}{\partial v}\right)_T = T\left(\frac{\partial p}{\partial T}\right)_v - p \tag{11.47}$$

When Eq. 11.46 is inserted in Eq. 11.43, the following expression results:

$$ds = \frac{c_v}{T}\,dT + \left(\frac{\partial p}{\partial T}\right)_v dv \tag{11.48}$$

Inserting Eq. 11.47 into Eq. 11.44 gives

$$du = c_v\,dT + \left[T\left(\frac{\partial p}{\partial T}\right)_v - p\right] dv \tag{11.49}$$

Observe that the right sides of Eqs. 11.48 and 11.49 are expressed solely in terms of p, v, T, and c_v.

Changes in specific entropy and internal energy between two states are determined by integration of Eqs. 11.48 and 11.49, respectively

$$s_2 - s_1 = \int_1^2 \frac{c_v}{T}\,dT + \int_1^2 \left(\frac{\partial p}{\partial T}\right)_v dv \tag{11.50}$$

$$u_2 - u_1 = \int_1^2 c_v\,dT + \int_1^2 \left[T\left(\frac{\partial p}{\partial T}\right)_v - p\right] dv \tag{11.51}$$

To integrate the first term on the right of each of these expressions, the variation of c_v with temperature at one fixed specific volume (isometric) is required. Integration of the second terms requires knowledge of the p–v–T relation at the states of interest. An equation of state explicit in pressure would be particularly convenient for evaluating the integrals involving $(\partial p/\partial T)_v$. The accuracy of the resulting specific entropy and internal energy changes would depend on the accuracy of this derivative. In cases where the integrands of Eqs. 11.50 and 11.51 are too complicated to be integrated in closed form they may be evaluated numerically.

Let us consider the evaluation of Eq. 11.51 in more detail. Similar considerations apply for Eq. 11.50. Referring to Fig. 11.2, if the specific heat c_v is known as a function of temperature along the isometric passing through the states x and y, one possible path of integration for determining the change in specific internal energy between states 1 and 2 is 1–x–y–2. The integration would be performed in three steps. Since the temperature is constant from state 1 to state x, the first integral of Eq. 11.51 would vanish, so

$$u_x - u_1 = \int_{v_1}^{v_x} \left[T\left(\frac{\partial p}{\partial T}\right)_v - p\right] dv$$

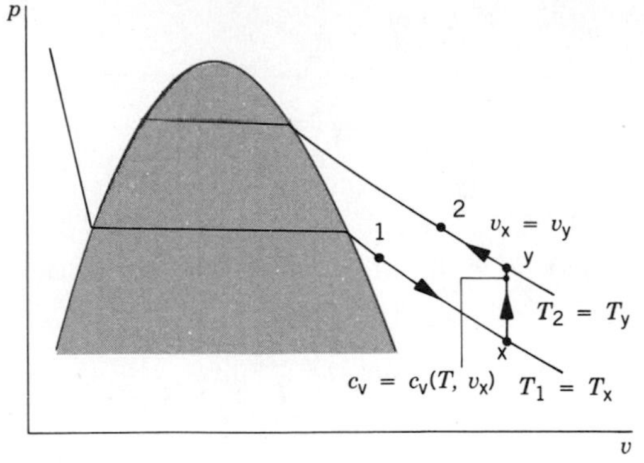

Figure 11.2 Integration path between two vapor states.

From state x to y, the specific volume is constant and c_v is known as a function of temperature only, so

$$u_y - u_x = \int_{T_x}^{T_y} c_v \, dT$$

where $T_x = T_1$ and $T_y = T_2$. From state y to state 2, the temperature is constant once again, and

$$u_2 - u_y = \int_{v_y = v_x}^{v_2} \left[T \left(\frac{\partial p}{\partial T} \right)_v - p \right] dv$$

When these are added together, the result is the change in specific internal energy between states 1 and 2.

Equation 11.47, which expresses the dependence of the specific internal energy on specific volume at fixed temperature, allows us to demonstrate that the internal energy of a gas whose equation of state is $pv = RT$ depends on temperature alone, a result first discussed in Sec. 3.5. Equation 11.47 requires the partial derivative $(\partial p / \partial T)_v$. If $p = RT/v$, the derivative is $(\partial p / \partial T)_v = R/v$. Introducing this, Eq. 11.47 gives

$$\left(\frac{\partial u}{\partial v} \right)_T = T \left(\frac{\partial p}{\partial T} \right)_v - p = T \left(\frac{R}{v} \right) - p = p - p = 0$$

This demonstrates that when $pv = RT$, the specific internal energy is independent of specific volume and depends on temperature alone.

***T* and *p* as Independent Properties.** In this section a presentation parallel to that considered above is provided for the choice of temperature and pressure as the independent properties. With this choice for the independent properties, the specific entropy can be regarded as a function of the form $s = s(T, p)$. The differential of this function is

$$ds = \left(\frac{\partial s}{\partial T} \right)_p dT + \left(\frac{\partial s}{\partial p} \right)_T dp$$

The partial derivative $(\partial s / \partial p)_T$ appearing in this expression can be replaced using the Maxwell relation, Eq. 11.35, giving

$$ds = \left(\frac{\partial s}{\partial T} \right)_p dT - \left(\frac{\partial v}{\partial T} \right)_p dp \tag{11.52}$$

The specific enthalpy can also be regarded as a function of T and p: $h = h(T, p)$. The differential of this function is

$$dh = \left(\frac{\partial h}{\partial T}\right)_p dT + \left(\frac{\partial h}{\partial p}\right)_T dp$$

With $c_p = (\partial h/\partial T)_p$

$$dh = c_p\, dT + \left(\frac{\partial h}{\partial p}\right)_T dp \tag{11.53}$$

Substituting Eqs. 11.52 and 11.53 into $dh = T\, ds + v\, dp$ and collecting terms results in

$$\left[\left(\frac{\partial h}{\partial p}\right)_T + T\left(\frac{\partial v}{\partial T}\right)_p - v\right] dp = \left[T\left(\frac{\partial s}{\partial T}\right)_p - c_p\right] dT \tag{11.54}$$

Since pressure and temperature can be varied independently, let us hold pressure constant and vary temperature. That is, let $dp = 0$ and $dT \neq 0$. It then follows from Eq. 11.54 that

$$\left(\frac{\partial s}{\partial T}\right)_p = \frac{c_p}{T} \tag{11.55}$$

Similarly, when $dT = 0$ and $dp \neq 0$, Eq. 11.54 gives

$$\left(\frac{\partial h}{\partial p}\right)_T = v - T\left(\frac{\partial v}{\partial T}\right)_p \tag{11.56}$$

When Eq. 11.55 is inserted in Eq. 11.52, the following equation results:

$$ds = \frac{c_p}{T}\, dT - \left(\frac{\partial v}{\partial T}\right)_p dp \tag{11.57}$$

Introducing Eq. 11.56 into Eq. 11.53 gives

$$dh = c_p\, dT + \left[v - T\left(\frac{\partial v}{\partial T}\right)_p\right] dp \tag{11.58}$$

Observe that the right sides of Eqs. 11.57 and 11.58 are expressed solely in terms of p, v, T, and c_p.

Changes in specific entropy and enthalpy between two states are found by integrating Eqs. 11.57 and 11.58, respectively

$$s_2 - s_1 = \int_1^2 \frac{c_p}{T}\, dT - \int_1^2 \left(\frac{\partial v}{\partial T}\right)_p dp \tag{11.59}$$

$$h_2 - h_1 = \int_1^2 c_p\, dT + \int_1^2 \left[v - T\left(\frac{\partial v}{\partial T}\right)_p\right] dp \tag{11.60}$$

To integrate the first term on the right of each of these expressions, the variation of c_p with temperature at one fixed pressure (isobar) is required. Integration of the second terms requires knowledge of the p–v–T behavior at the states of interest. An equation of state explicit in v would be particularly convenient for evaluating the integrals involving $(\partial v/\partial T)_p$. The accuracy of the resulting specific entropy and enthalpy changes would depend on the accuracy of this derivative.

Changes in specific enthalpy and internal energy are related through $h = u + pv$ by

$$h_2 - h_1 = (u_2 - u_1) + (p_2 v_2 - p_1 v_1) \tag{11.61}$$

Hence, only one of Δh and Δu need be found by integration. The other can be evaluated from Eq. 11.61. Which of the two property changes is found by integration depends on the information available. Δh would be found using Eq. 11.60 when an equation of state explicit in v and c_p as a function of temperature at some fixed pressure are known. Δu would be found from Eq. 11.51 when an equation of state explicit in p and c_v as a function of temperature at some specific volume are known.

EXAMPLE 11.5

PROBLEM

Using the Redlich–Kwong equation of state, develop expressions for the changes in specific entropy and specific internal energy of a gas between two states where the temperature is the same, $T_1 = T_2$, and the pressures are p_1 and p_2, respectively.

SOLUTION

Known: Two states of a unit mass of a gas as the system are fixed by p_1 and T_1 at state 1 and p_2, $T_2 (= T_1)$ at state 2.

Find: Determine the changes in specific entropy and internal energy between these two states.

Assumption: The Redlich–Kwong equation of state represents the p–v–T behavior at these states and yields accurate values for $(\partial p/\partial T)_v$.

Analysis: The Redlich–Kwong equation of state is explicit in pressure, so Eqs. 11.50 and 11.51 are selected for determining the required property changes. Since $T_1 = T_2$, an isothermal path of integration between the two states is convenient. Thus, these equations reduce to give

$$s_2 - s_1 = \int_1^2 \left(\frac{\partial p}{\partial T}\right)_v dv$$

$$u_2 - u_1 = \int_1^2 \left[T\left(\frac{\partial p}{\partial T}\right)_v - p\right] dv$$

The limits for each of the foregoing integrals are the specific volumes v_1 and v_2 at the two states under consideration. Since the states are fixed by pressure and temperature, it would be necessary to determine these specific volumes from the Redlich–Kwong equation of state. This would require an iterative procedure.

The above integrals involve the partial derivative $(\partial p/\partial T)_v$, which can be determined from the Redlich–Kwong equation of state as

$$\left(\frac{\partial p}{\partial T}\right)_v = \frac{R}{v - b} + \frac{a}{2v(v + b)T^{3/2}}$$

Inserting this into the expression for $(s_2 - s_1)$ gives

$$s_2 - s_1 = \int_{v_1}^{v_2} \left[\frac{R}{v - b} + \frac{a}{2v(v + b)T^{3/2}}\right] dv$$

$$= \int_{v_1}^{v_2} \left[\frac{R}{v - b} + \frac{a}{2bT^{3/2}}\left(\frac{1}{v} - \frac{1}{v + b}\right)\right] dv$$

$$= R \ln\left(\frac{v_2 - b}{v_1 - b}\right) + \frac{a}{2bT^{3/2}}\left[\ln\left(\frac{v_2}{v_1}\right) - \ln\left(\frac{v_2 + b}{v_1 + b}\right)\right]$$

$$= R \ln\left(\frac{v_2 - b}{v_1 - b}\right) + \frac{a}{2bT^{3/2}} \ln\left[\frac{v_2(v_1 + b)}{v_1(v_2 + b)}\right]$$

With the Redlich–Kwong equation, the integrand of the expression for $(u_2 - u_1)$ becomes

$$\left[T\left(\frac{\partial p}{\partial T}\right)_v - p\right] = T\left[\frac{R}{v - b} + \frac{a}{2v(v + b)T^{3/2}}\right] - \left[\frac{RT}{v - b} - \frac{a}{v(v + b)T^{1/2}}\right]$$

$$= \frac{3a}{2v(v + b)T^{1/2}}$$

Accordingly

$$u_2 - u_1 = \int_{v_1}^{v_2} \frac{3a}{2v(v + b)T^{1/2}}\, dv = \frac{3a}{2bT^{1/2}} \int_{v_1}^{v_2} \left(\frac{1}{v} - \frac{1}{v + b}\right) dv$$

$$= \frac{3a}{2bT^{1/2}}\left[\ln \frac{v_2}{v_1} - \ln\left(\frac{v_2 + b}{v_1 + b}\right)\right] = \frac{3a}{2bT^{1/2}} \ln\left[\frac{v_2(v_1 + b)}{v_1(v_2 + b)}\right]$$

11.5 OTHER THERMODYNAMIC RELATIONS

The presentation to this point has been directed mainly at developing thermodynamic relations that allow changes in u, h, and s to be evaluated from measured property data. The object of the present section is to introduce several other thermodynamic relations that are useful for thermodynamic analysis. Each of the properties considered has a common attribute: it is defined in terms of a partial derivative of some other property. The specific heats c_v and c_p are examples of this type of property.

11.5.1 Volume Expansivity, Isothermal, and Isentropic Compressibility

In single-phase regions, pressure and temperature are independent, and we can think of the specific volume as being a function of these two, $v = v(T, p)$. The differential of such a function is

$$dv = \left(\frac{\partial v}{\partial T}\right)_p dT + \left(\frac{\partial v}{\partial p}\right)_T dp$$

Two thermodynamic properties related to the partial derivatives appearing in this differential are the *volume expansivity* β, also called the *coefficient of volume expansion*

$$\beta = \frac{1}{v}\left(\frac{\partial v}{\partial T}\right)_p \qquad (11.62)$$

and the *isothermal compressibility* κ

$$\kappa = -\frac{1}{v}\left(\frac{\partial v}{\partial p}\right)_T \qquad (11.63)$$

TABLE 11.1 Volume Expansivity β and Isothermal Compressibility κ of Liquid Water at 1 atm versus Temperature

T (°C)	Density (kg/m³)	$\beta \times 10^6$ (K)$^{-1}$	$\kappa \times 10^6$ (bar)$^{-1}$
0	999.84	−68.14	50.89
10	999.70	87.90	47.81
20	998.21	206.6	45.90
30	995.65	303.1	44.77
40	992.22	385.4	44.24
50	988.04	457.8	44.18

By inspection, the unit for β is seen to be the reciprocal of that for temperature and the unit for κ is the reciprocal of that for pressure. The volume expansivity is an indication of the change in volume that occurs when temperature changes while pressure remains constant. The isothermal compressibility is an indication of the change in volume that takes place when pressure changes while temperature remains constant. The value of κ is positive for all substances in all phases.

The volume expansivity and isothermal compressibility are thermodynamic properties, and like specific volume are functions of T and p. Values for β and κ are provided in handbooks of engineering data. Table 11.1 gives values of these properties for liquid water at a pressure of 1 atm versus temperature. For a pressure of 1 atm, water has a *state of maximum density* at about 4°C. At this state, the value of β is zero.

The *isentropic compressibility* α is an indication of the change in volume that occurs when pressure changes while entropy remains constant

$$\alpha = -\frac{1}{v}\left(\frac{\partial v}{\partial p}\right)_s \tag{11.64}$$

The unit for α is the reciprocal of that for pressure.

The isentropic compressibility is related to the speed at which sound waves travel in the substance, and such speed measurements can be used to determine α. In Sec. 9.12, the *velocity of sound,* or *sonic velocity,* is introduced as

$$c = \sqrt{-v^2\left(\frac{\partial p}{\partial v}\right)_s} \tag{9.36b}$$

The relationship of the isentropic compressibility and the velocity of sound can be obtained using the relation between partial derivatives expressed by Eq. 11.15. Identifying p with x, v with y, and s with z, we have

$$\left(\frac{\partial p}{\partial v}\right)_s = \frac{1}{(\partial v/\partial p)_s}$$

With this, the previous two equations can be combined to give

$$c = \sqrt{v/\alpha} \tag{11.65}$$

The details are left as an exercise.

11.5.2 Relations Involving the Specific Heats and the Specific Heat Ratio

In this section general relations are derived for the difference between specific heats $(c_p - c_v)$ and the ratio of specific heats c_p/c_v.

An expression for the difference between c_p and c_v can be derived by equating the two differentials for entropy given by Eqs. 11.48 and 11.57 and rearranging to obtain

$$(c_p - c_v)\, dT = T\left(\frac{\partial p}{\partial T}\right)_v dv + T\left(\frac{\partial v}{\partial T}\right)_p dp$$

Considering the equation of state $p = p(T, v)$, the differential dp can be expressed as

$$dp = \left.\frac{\partial p}{\partial T}\right)_v dT + \left.\frac{\partial p}{\partial v}\right)_T dv$$

Eliminating dp between the last two equations and collecting terms gives

$$\left[(c_p - c_v) - T\left(\frac{\partial v}{\partial T}\right)_p \left(\frac{\partial p}{\partial T}\right)_v\right] dT = T\left[\left(\frac{\partial v}{\partial T}\right)_p \left(\frac{\partial p}{\partial v}\right)_T + \left(\frac{\partial p}{\partial T}\right)_v\right] dv$$

Since temperature and specific volume can be varied independently, the coefficients of the differentials in this expression must vanish, so

$$c_p - c_v = T\left(\frac{\partial v}{\partial T}\right)_p \left(\frac{\partial p}{\partial T}\right)_v \tag{11.66}$$

$$\left(\frac{\partial p}{\partial T}\right)_v = -\left(\frac{\partial v}{\partial T}\right)_p \left(\frac{\partial p}{\partial v}\right)_T \tag{11.67}$$

Introducing Eq. 11.67 into Eq. 11.66 gives

$$c_p - c_v = -T\left(\frac{\partial v}{\partial T}\right)_p^2 \left(\frac{\partial p}{\partial v}\right)_T \tag{11.68}$$

This equation allows c_v to be calculated from observed values of c_p or, conversely, knowing only p–v–T data. For the special case of an ideal gas, Eq. 11.68 reduces to Eq. 3.44, as can readily be shown.

The right side of Eq. 11.68 can be expressed in terms of the volume expansivity β and the isothermal compressibility κ. Introducing Eqs. 11.62 and 11.63, we get

$$c_p - c_v = v\frac{T\beta^2}{\kappa} \tag{11.69}$$

In developing this result, the relationship between partial derivatives expressed by Eq. 11.15 has been used.

Several important conclusions about the specific heats c_p and c_v can be drawn from Eq. 11.69. Since the factor β^2 is never negative and κ is known to be positive for all substances in all phases, the value of c_p is always greater than, or equal to, c_v. The specific heats would be equal when $\beta = 0$, as occurs in the case of water at 1 atmosphere and 4°C, where water is at its state of maximum density. The two specific heats also become equal as the temperature approaches absolute zero. For most liquids and solids, the value of β is relatively small, so c_p and c_v of liquids and solids are approximately equal. For this reason, tables often give the specific heat of a liquid or solid without specifying whether it is c_p or c_v. The data reported are normally c_p values, since these are more easily determined for liquids and solids.

Next, let us obtain expressions for the ratio of specific heats k. Employing Eq. 11.16, we can rewrite Eqs. 11.46 and 11.55, respectively, as

$$\frac{c_v}{T} = \left(\frac{\partial s}{\partial T}\right)_v = \frac{-1}{(\partial v/\partial s)_T(\partial T/\partial v)_s}$$

$$\frac{c_p}{T} = \left(\frac{\partial s}{\partial T}\right)_p = \frac{-1}{(\partial p/\partial s)_T(\partial T/\partial p)_s}$$

Forming the ratio of these equations gives

$$\frac{c_p}{c_v} = \frac{(\partial v/\partial s)_T(\partial T/\partial v)_s}{(\partial p/\partial s)_T(\partial T/\partial p)_s} \tag{11.70}$$

Since $(\partial s/\partial p)_T = 1/(\partial p/\partial s)_T$ and $(\partial p/\partial T)_s = 1/(\partial T/\partial p)_s$, Eq. 11.70 can be expressed as

$$\frac{c_p}{c_v} = \left[\left(\frac{\partial v}{\partial s}\right)_T\left(\frac{\partial s}{\partial p}\right)_T\right]\left[\left(\frac{\partial p}{\partial T}\right)_s\left(\frac{\partial T}{\partial v}\right)_s\right] \tag{11.71}$$

Finally, the chain rule from calculus allows us to write $(\partial v/\partial p)_T = (\partial v/\partial s)_T(\partial s/\partial p)_T$ and $(\partial p/\partial v)_s = (\partial p/\partial T)_s(\partial T/\partial v)_s$, so Eq. 11.71 becomes

$$k = \frac{c_p}{c_v} = \left(\frac{\partial v}{\partial p}\right)_T\left(\frac{\partial p}{\partial v}\right)_s \tag{11.72}$$

This can be expressed alternatively in terms of the isothermal and isentropic compressibilities as

$$k = \frac{\kappa}{\alpha} \tag{11.73}$$

Solving Eq. 11.72 for $(\partial p/\partial v)_s$ and substituting the resulting expression into Eq. 9.36*b* gives the following relationship involving the velocity of sound c and the specific heat ratio k:

$$c = \sqrt{-kv^2(\partial p/\partial v)_T} \tag{11.74}$$

Equation 11.74 can be used to determine c knowing the specific heat ratio and p–v–T data, or to evaluate k knowing c and $(\partial p/\partial v)_T$. In the special case of an ideal gas, Eq. 11.74 reduces to Eq. 9.37

$$c = \sqrt{kRT} \quad \text{(ideal gas)} \tag{9.37}$$

as can easily be verified.

EXAMPLE 11.6

PROBLEM

For liquid water at 1 atm and 20°C, determine (a) the percent error in c_v that would result if it were assumed that $c_p = c_v$, (b) the velocity of sound, in m/s.

SOLUTION

Known: The system consists of a fixed amount of liquid water at 1 atm and 20°C.

Find: Determine the percent error in c_v that would result if c_v were approximated by c_p, and the velocity of sound, in m/s.

Analysis:

(a) Equation 11.69 gives the difference between c_p and c_v. Table 11.1 provides the required values for the volume expansivity β, the isothermal compressibility κ, and the specific volume. Thus

$$\begin{aligned} c_p - c_v &= v\frac{T\beta^2}{\kappa} \\ &= \left(\frac{1}{998.21\ \text{kg/m}^3}\right)(293\ \text{K})\left(\frac{206.6 \times 10^{-6}}{\text{K}}\right)^2\left(\frac{\text{bar}}{45.90 \times 10^{-6}}\right) \\ &= \left(272.96 \times 10^{-6}\ \frac{\text{bar}\cdot\text{m}^3}{\text{kg}\cdot\text{K}}\right)\left(\frac{10^5\ \text{N/m}^2}{\text{bar}}\right)\left(\frac{\text{kJ}}{10^3\ \text{N}\cdot\text{m}}\right) \\ &= 0.027\ \frac{\text{kJ}}{\text{kg}\cdot\text{K}} \end{aligned}$$

Interpolating in Table A-13 gives at 20°C, 1 atm the value $c_p = 4.189$ kJ/kg·K. Thus, the value of c_v at this state is

$$c_v = 4.189 - 0.027 = 4.162\ \text{kJ/kg}\cdot\text{K}$$

Using these values, the percent error in approximating c_v by c_p is

$$\left(\frac{c_p - c_v}{c_v}\right)(100) = \left(\frac{0.027}{4.162}\right)(100) = 0.6\%$$

(b) The velocity of sound at this state can be determined using Eq. 11.65. The required value for the isentropic compressibility α is calculable in terms of the specific heat ratio k and the isothermal compressibility κ. With Eq. 11.73, $\alpha = \kappa/k$. Inserting this into Eq. 11.65 results in the following expression for the velocity of sound:

$$c = \sqrt{\frac{kv}{\kappa}}$$

The values of v and κ required by this expression are the same as used in part (a). Also, with the values of c_p and c_v from part (a), the specific heat ratio is $k = 1.006$. Accordingly

$$c = \sqrt{\frac{(1.006)(10^6)\ \text{bar}}{(998.21\ \text{kg/m}^3)(45.90)}\left(\frac{10^5\ \text{N/m}^2}{\text{bar}}\right)\left(\frac{1\ \text{kg}\cdot\text{m/s}^2}{1\ \text{N}}\right)} = 1482\ \text{m/s}$$

Comment: The result of part (a) provides a specific illustration showing that for liquids c_p and c_v are closely equal.

11.5.3 Joule–Thomson Coefficient

The value of the specific heat c_p can be determined from p–v–T data and the Joule–Thomson coefficient. The Joule–Thomson coefficient μ_J is defined as

$$\mu_J = \left(\frac{\partial T}{\partial p}\right)_h \tag{11.75}$$

Like other partial differential coefficients introduced in this section, the Joule–Thomson coefficient is defined in terms of thermodynamic properties only and thus is itself a property. The units of μ_J are those of temperature divided by pressure.

A relationship between the specific heat c_p and the Joule–Thomson coefficient μ_J can be established by using Eq. 11.16 to write

$$\left(\frac{\partial T}{\partial p}\right)_h \left(\frac{\partial p}{\partial h}\right)_T \left(\frac{\partial h}{\partial T}\right)_p = -1$$

The first factor in this expression is the Joule–Thomson coefficient and the third is c_p. Thus

$$c_p = \frac{-1}{\mu_J(\partial p/\partial h)_T}$$

With $(\partial h/\partial p)_T = 1/(\partial p/\partial h)_T$ from Eq. 11.15, this can be written as

$$c_p = -\frac{1}{\mu_J}\left(\frac{\partial h}{\partial p}\right)_T \tag{11.76}$$

The partial derivative $(\partial h/\partial p)_T$, called the *constant-temperature coefficient,* can be eliminated from Eq. 11.76 by use of Eq. 11.56. The following expression results:

$$c_p = \frac{1}{\mu_J}\left[T\left(\frac{\partial v}{\partial T}\right)_p - v\right] \tag{11.77}$$

Equation 11.77 allows the value of c_p at a state to be determined using p–v–T data and the value of the Joule–Thomson coefficient at that state. Let us consider next how the Joule–Thomson coefficient can be found experimentally.

The Joule–Thomson coefficient can be evaluated experimentally using an apparatus like that pictured in Fig. 11.3. Consider first Fig. 11.3*a*, which shows a porous plug through which a gas (or liquid) may pass. During operation at steady state, the gas enters the apparatus at a specified temperature T_1 and pressure p_1 and expands through the plug to a lower pressure p_2, which is controlled by an outlet valve. The temperature T_2 at the exit is measured. The apparatus is designed so that the gas undergoes a *throttling* process as it expands from 1 to 2. Accordingly, the exit state

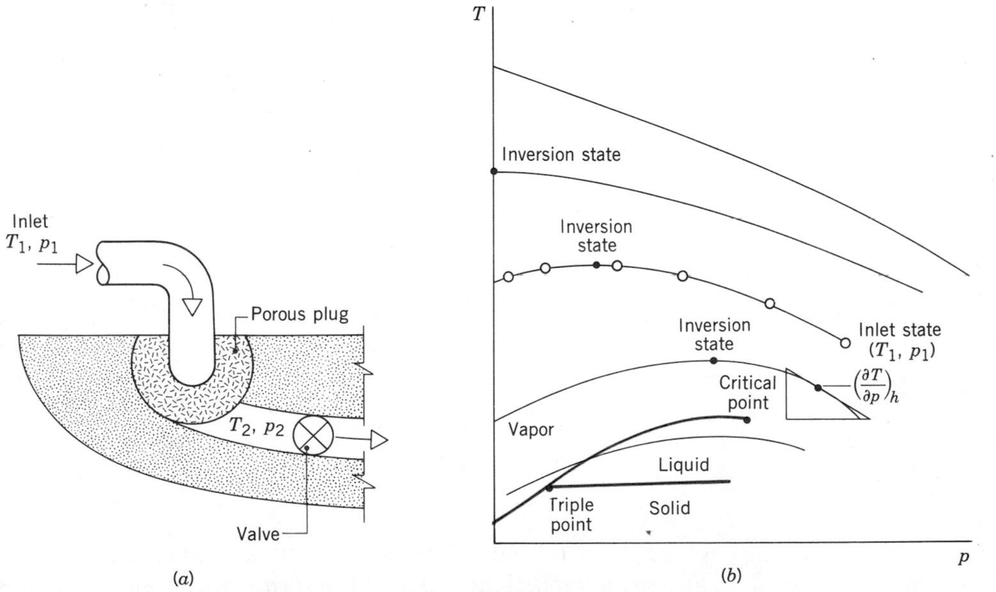

Figure 11.3 Joule–Thomson expansion. (*a*) Apparatus. (*b*) Isenthalpics on a T–p diagram.

fixed by p_2 and T_2 has the same value for the specific enthalpy as at the inlet, $h_2 = h_1$. By progressively lowering the outlet pressure, a finite sequence of such exit states can be visited, as indicated on Fig. 11.3*b*. A curve may be drawn through the set of data points (Fig. 11.3*b*). Such a curve is called an isenthalpic (constant enthalpy) curve. An isenthalpic curve is the locus of all points representing equilibrium states of the same specific enthalpy.

The slope of an isenthalpic curve at any state is the Joule–Thomson coefficient at that state. The slope may be positive, negative, or zero in value. States where the coefficient has a zero value are called *inversion* states. Notice that not all lines of constant h have an inversion state. The uppermost curve of Fig. 11.3*b*, for example, always has a negative slope. Throttling a gas from an initial state on this curve would result in an increase in temperature. However, for isenthalpic curves having an inversion state, the temperature at the exit of the apparatus may be greater than, equal to, or less than the initial temperature, depending on the exit pressure specified. For states to the right of an inversion state, the value of the Joule–Thomson coefficient would be negative. For these states, the temperature would increase as the pressure at the exit of the apparatus is reduced. At states to the left of an inversion state, the value of the Joule–Thomson coefficient would be positive. For these states, the temperature would decrease as the pressure at the exit of the device is reduced. This can be used to advantage in systems designed to *liquefy gases*.

11.6 CONSTRUCTING TABLES OF THERMODYNAMIC PROPERTIES

The objective of this section is to utilize the thermodynamic relations introduced thus far to describe how tables of thermodynamic properties can be constructed. The characteristics of the tables under consideration are embodied in the tables for water and the refrigerants presented in the Appendix.

Two different approaches for constructing property tables are considered. The presentation of Sec. 11.6.1 employs the methods introduced in Sec. 11.4 for assigning specific enthalpy, specific internal energy, and specific entropy to states of pure, simple compressible substances using p–v–T data, together with a limited amount of specific heat data. The principal mathematical operation of this approach is *integration*. The approach of Sec. 11.6.2 utilizes the fundamental thermodynamic function concept introduced in Sec. 11.3.3. Once such a function has been constructed, the principal mathematical operation required to determine all other properties is *differentiation*.

11.6.1 Developing Tables by Integration Using p–v–T and Specific Heat Data

In principle, all properties of present interest can be determined using

$$\begin{gathered} c_p = c_{p0}(T) \\ p = p(v, T), \qquad v = v(p, T) \end{gathered} \tag{11.78}$$

In Eqs. 11.78, $c_{p0}(T)$ is the specific heat c_p for the substance under consideration extrapolated to zero pressure. This function might be determined from data obtained calorimetrically or from spectroscopic data, using equations supplied by statistical mechanics. Specific heat expressions for several gases are given in Tables A-15. The expressions $p(v, T)$ and $v(p, T)$ represent functions that describe the saturation pressure–temperature curves, as well as the p–v–T relations for the single-phase regions. These functions may be tabular, graphical, or analytical in character. Whatever their forms, however, the functions must not only represent the p–v–T data

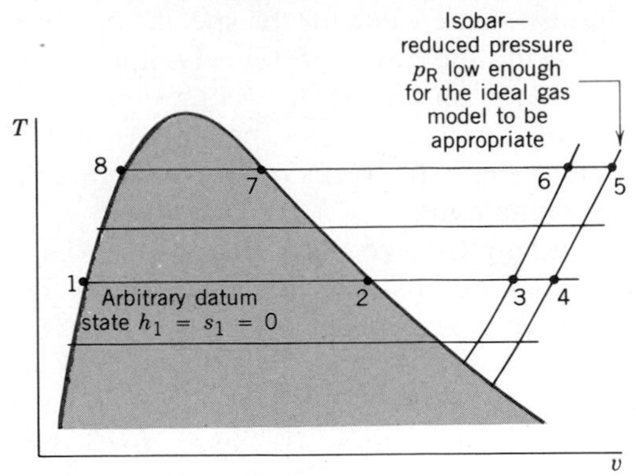

Figure 11.4 T–v diagram used to discuss how h and s can be assigned to liquid and vapor states.

accurately but also yield accurate values for derivatives such as $(\partial v/\partial T)_p$ and $(dp/dT)_{\text{sat}}$.

Figure 11.4 shows eight states of a substance. Let us consider how values can be assigned to specific enthalpy and specific entropy at these states. The same procedures can be used to assign property values at other states of interest. Note that when h has been assigned to a state, the specific internal energy at that state can be found from $u = h - pv$.

Let the state denoted by 1 on Fig. 11.4 be selected as the datum state for enthalpy and entropy. Any value can be assigned to h and s at this state, but a value of zero would be usual. Once a value is assigned to enthalpy at state 1, the enthalpy at the saturated vapor state, state 2, can be determined using the Clapeyron equation, Eq. 11.40

$$h_2 - h_1 = T_1(v_2 - v_1)\left(\frac{dp}{dT}\right)_{\text{sat}}$$

where the derivative $(dp/dT)_{\text{sat}}$ and the specific volumes v_1 and v_2 are obtained from appropriate representations of the p–v–T data for the substance under consideration. The specific entropy at state 2 is found using Eq. 11.38 in the form

$$s_2 - s_1 = \frac{h_2 - h_1}{T_1}$$

Proceeding at constant temperature from state 2 to state 3, the entropy and enthalpy are found by means of Eqs. 11.59 and 11.60, respectively. Since temperature is fixed, these equations reduce to give

$$s_3 - s_2 = -\int_{p_2}^{p_3}\left(\frac{\partial v}{\partial T}\right)_p dp \quad \text{and} \quad h_3 - h_2 = \int_{p_2}^{p_3}\left[v - T\left(\frac{\partial v}{\partial T}\right)_p\right] dp$$

With the same procedure, s_4 and h_4 can be determined.

The isobar passing through state 4 is assumed to be at a pressure sufficiently low for the ideal gas model to be appropriate. Accordingly, to evaluate s and h at states such as 5 on this isobar, the only required information would be $c_{p0}(T)$ and the temperatures at these states. Thus, since pressure is fixed, Eqs. 11.59 and 11.60 give, respectively

$$s_5 - s_4 = \int_{T_4}^{T_5} c_{p0}\frac{dT}{T} \quad \text{and} \quad h_5 - h_4 = \int_{T_4}^{T_5} c_{p0}\, dT$$

Specific entropy and enthalpy values at states 6 and 7 are found from those at state 5 by the same procedure used in assigning values at states 3 and 4 from those at state 2. Finally, s_8 and h_8 are obtained from the values at state 7 using the Clapeyron equation.

It should be noted that the use of an arbitrary datum state and arbitrary reference values for specific enthalpy and specific entropy suffices only for evaluations involving differences in property values between states of the same composition, for then datums cancel.

11.6.2 Developing Tables by Differentiating a Fundamental Thermodynamic Function

Property tables also can be developed using a fundamental thermodynamic function. It is convenient for this purpose to select the independent variables of the fundamental function from among pressure, specific volume (density), and temperature. This indicates the use of the Helmholtz function $\psi(T, v)$ or the Gibbs function $g(T, p)$. The properties of water tabulated in Appendix A have been calculated using a Helmholtz function. Fundamental functions also have been employed successfully to evaluate the properties of other substances.

The development of a fundamental function requires considerable mathematical manipulation and numerical evaluation. Prior to the development of high-speed computers, the evaluation of properties by this means was not feasible, and the approach described in Sec. 11.6.1 was used exclusively. An important step in the development of a fundamental function is the selection of a functional form in terms of the appropriate pair of independent properties and a set of adjustable coefficients, which may number 50 or more. The functional form is specified on the basis of both theoretical and practical considerations. The coefficients in the fundamental function are determined by requiring that a set of carefully selected property values and/or observed conditions be satisfied in a least-squares sense. This generally involves the use of property data requiring the assumed functional form to be differentiated one or more times, such as p–v–T and specific heat data. When all coefficients have been evaluated, the function is carefully tested for accuracy by using it to evaluate properties for which accepted values are known. These may include properties requiring differentiation of the fundamental function two or more times. For example, velocity of sound and Joule–Thomson data might be used. This procedure for developing a fundamental function is not routine and can be accomplished only with the use of a high-speed computer. However, once a suitable fundamental function is established, extreme accuracy in and consistency among the thermodynamic properties is possible.

The form of the Helmholtz function used in constructing the steam tables from which Tables A-1 through A-6 have been extracted is

$$\psi(\rho, T) = \psi_0(T) + RT[\ln \rho + \rho Q(\rho, \tau)] \tag{11.79}$$

where ψ_0 and Q are given as the sums listed in Table 11.2. The independent variables are density and temperature. The variable τ denotes $1000/T$. Values for pressure, specific internal energy, and specific entropy can be determined by differentiation of Eq. 11.79 (see Example 11.7). Values for the specific enthalpy and Gibbs function are found from $h = u + pv$ and $g = \psi + pv$, respectively. The specific heat c_v is evaluated by further differentiation, $c_v = (\partial u/\partial T)_v$. With similar operations, other properties can be evaluated. Property values for water calculated from Eq. 11.79 are in excellent agreement with experimental data over a wide range of conditions.

TABLE 11.2 Fundamental Equation Used to Construct the Steam Tables[a,b]

$$\psi = \psi_0(T) + RT[\ln \rho + \rho Q(\rho, \tau)] \qquad (1)$$

where

$$\psi_0 = \sum_{i=1}^{6} C_i/\tau^{i-1} + C_7 \ln T + C_8 \ln T/\tau \qquad (2)$$

and

$$Q = (\tau - \tau_c) \sum_{j=1}^{7} (\tau - \tau_{aj})^{j-2} \left[\sum_{i=1}^{8} A_{ij}(\rho - \rho_{aj})^{i-1} + e^{-E\rho} \sum_{i=9}^{10} A_{ij}\rho^{i-9} \right] \qquad (3)$$

In (1), (2), and (3), T denotes temperature on the Kelvin scale, τ denotes $1000/T$, ρ denotes density in g/cm^3, $R = 4.6151$ bar·cm^3/g·K or 0.46151 J/g·K, $\tau_c \equiv 1000/T_c = 1.544912$, $E = 4.8$, and

$$\tau_{aj} = \tau_c \; (j = 1) \qquad \rho_{aj} = 0.634 \; (j = 1)$$
$$= 2.5 \; (j > 1) \qquad = 1.0 \; (j > 1)$$

The coefficients for ψ_0 in J/g are given as follows:

$C_1 = 1857.065$	$C_4 = 36.6649$	$C_7 = 46.$
$C_2 = 3229.12$	$C_5 = -20.5516$	$C_8 = -1011.249$
$C_3 = -419.465$	$C_6 = 4.85233$	

Values for the coefficients A_{ij} are listed in the original source.[a]

[a] J. H. Keenan, F. G. Keyes, P. G. Hill, and J. G. Moore, *Steam Tables*, Wiley, New York, 1969.
[b] Also see L. Haar, J. S. Gallagher, and G. S. Kell, *NBS/NRC Steam Tables*, Hemisphere, Washington, D.C., 1984. The properties of water are determined in this reference using a different functional form for the Helmholtz function than given by Eqs. (1)–(3).

EXAMPLE 11.7

PROBLEM

The following expression for the Helmholtz function has been used to determine the properties of water:

$$\psi(\rho, T) = \psi_0(T) + RT[\ln \rho + \rho Q(\rho, \tau)]$$

where ρ denotes density and τ denotes $1000/T$. The functions ψ_0 and Q are sums involving the indicated independent variables and a number of adjustable constants (see Table 11.2). Obtain expressions for (a) pressure, (b) specific entropy, and (c) specific internal energy resulting from this fundamental function.

SOLUTION

Known: An expression for the Helmholtz function ψ is given.

Find: Determine the expressions for pressure, specific entropy, and specific internal energy that result from this fundamental function.

Analysis:

(a) When expressed in terms of density instead of specific volume, Eq. 11.28 becomes

$$p = \rho^2 \left(\frac{\partial \psi}{\partial \rho}\right)_T$$

as can easily be verified. When T is held constant τ is also constant. Accordingly, the following is obtained on differentiation of the given function:

$$\left(\frac{\partial \psi}{\partial \rho}\right)_T = RT\left[\frac{1}{\rho} + Q(\rho, \tau) + \rho\left(\frac{\partial Q}{\partial \rho}\right)_\tau\right]$$

Combining these equations gives an expression for pressure

$$p = \rho RT\left[1 + \rho Q + \rho^2\left(\frac{\partial Q}{\partial \rho}\right)_\tau\right]$$

(b) From Eq. 11.29

$$s = -\left(\frac{\partial \psi}{\partial T}\right)_\rho$$

Differentiation of the given expression for ψ yields

$$\begin{aligned}\left(\frac{\partial \psi}{\partial T}\right)_\rho &= \frac{d\psi_0}{dT} + \left[R(\ln \rho + \rho Q) + RT\rho\left(\frac{\partial Q}{\partial \tau}\right)_\rho \frac{d\tau}{dT}\right] \\ &= \frac{d\psi_0}{dT} + \left[R(\ln \rho + \rho Q) + RT\rho\left(\frac{\partial Q}{\partial \tau}\right)_\rho\left(-\frac{1000}{T^2}\right)\right] \\ &= \frac{d\psi_0}{dT} + R\left[\ln \rho + \rho Q - \rho\tau\left(\frac{\partial Q}{\partial \tau}\right)_\rho\right]\end{aligned}$$

Combining results gives

$$s = -\frac{d\psi_0}{dT} - R\left[\ln \rho + \rho Q - \rho\tau\left(\frac{\partial Q}{\partial \tau}\right)_\rho\right]$$

(c) By definition, $\psi = u - Ts$. Thus, $u = \psi + Ts$. Introducing the given expression for ψ together with the expression for s from part (b) results in

$$\begin{aligned}u &= [\psi_0 + RT(\ln \rho + \rho Q)] + T\left\{-\frac{d\psi_0}{dT} - R\left[\ln \rho + \rho Q - \rho\tau\left(\frac{\partial Q}{\partial \tau}\right)_\rho\right]\right\} \\ &= \psi_0 - T\frac{d\psi_0}{dT} + RT\rho\tau\left(\frac{\partial Q}{\partial \tau}\right)_\rho\end{aligned}$$

This can be written more compactly by noting that

$$T\frac{d\psi_0}{dT} = T\frac{d\psi_0}{d\tau}\frac{d\tau}{dT} = T\frac{d\psi_0}{d\tau}\left(-\frac{1000}{T^2}\right) = -\tau\frac{d\psi_0}{d\tau}$$

Thus

$$\psi_0 - T\frac{d\psi_0}{dT} = \psi_0 + \tau\frac{d\psi_0}{d\tau} = \frac{d(\psi_0\tau)}{d\tau}$$

Finally, the expression for u becomes

$$u = \frac{d(\psi_0\tau)}{d\tau} + RT\rho\tau\left(\frac{\partial Q}{\partial \tau}\right)_\rho$$

Comment: The expressions developed for p, s, and u require only the functions $\psi_0(T)$ and $Q(\rho, \tau)$. Once these functions are determined, p, s, and u can each be determined as a function of density and temperature using elementary mathematical operations.

11.7 GENERALIZED CHARTS FOR ENTHALPY AND ENTROPY

Generalized charts giving the compressibility factor Z in terms of the reduced properties p_R, T_R, and v_R' are introduced in Sec. 3.4. With such charts, estimates of p–v–

T data can be obtained rapidly knowing only the critical pressure and critical temperature for the substance of interest. The objective of the present section is to introduce generalized charts that allow changes in enthalpy and entropy to be closely evaluated.

Generalized Enthalpy Chart. The change in specific enthalpy of a gas (or liquid) between two states fixed by temperature and pressure can be evaluated using the identity

$$h(T_2, p_2) - h(T_1, p_1) = [h^*(T_2) - h^*(T_1)] + \underline{\{[h(T_2, p_2) - h^*(T_2)] - [h(T_1, p_1) - h^*(T_1)]\}} \tag{11.80}$$

The term $[h(T, p) - h^*(T)]$ denotes the specific enthalpy of the substance relative to that of its ideal gas model when both are at the same temperature. The superscript * is used in this section to identify ideal gas property values. Thus, Eq. 11.80 indicates that the change in specific enthalpy between the two states equals the enthalpy change determined using the ideal gas model plus a correction that accounts for the departure from ideal gas behavior. The correction is shown underlined in Eq. 11.80. The ideal gas term can be evaluated using methods introduced in Sec. 3.5. Next, we show how the correction term is evaluated.

The variation of enthalpy with pressure at fixed temperature is given by Eq. 11.56 as

$$\left(\frac{\partial h}{\partial p}\right)_T = v - T\left(\frac{\partial v}{\partial T}\right)_p$$

Integrating from pressure p' to pressure p at fixed temperature T

$$h(T, p) - h(T, p') = \int_{p'}^{p} \left[v - T\left(\frac{\partial v}{\partial T}\right)_p\right] dp$$

This equation is not altered fundamentally by adding and subtracting $h^*(T)$ on the left side. That is

$$[h(T, p) - h^*(T)] - [h(T, p') - h^*(T)] = \int_{p'}^{p} \left[v - T\left(\frac{\partial v}{\partial T}\right)_p\right] dp \tag{11.81}$$

As pressure tends to zero at fixed temperature, the enthalpy of the substance approaches that of its ideal gas model. Accordingly, as p' tends to zero

$$\lim_{p' \to 0} [h(T, p') - h^*(T)] = 0$$

In this limit, the following expression is obtained from Eq. 11.81 for the specific enthalpy of a substance relative to that of its ideal gas model when both are at the same temperature:

$$h(T, p) - h^*(T) = \int_{0}^{p} \left[v - T\left(\frac{\partial v}{\partial T}\right)_p\right] dp \tag{11.82}$$

This can also be thought of as the change in enthalpy as the pressure is increased from zero to the given pressure while temperature is held constant. Using p–v–T data only, Eq. 11.82 can be evaluated at states 1 and 2 and thus the correction term of Eq. 11.80 evaluated. Let us consider next how this procedure can be conducted in terms of compressibility factor data and the reduced properties T_R and p_R.

The integral of Eq. 11.82 can be expressed in terms of the compressibility factor Z and the reduced properties T_R and p_R as follows. Solving $Z = pv/RT$ gives

$$v = \frac{ZRT}{p}$$

On differentiation

$$\left(\frac{\partial v}{\partial T}\right)_p = \frac{RZ}{p} + \frac{RT}{p}\left(\frac{\partial Z}{\partial T}\right)_p$$

With the previous two expressions, the integrand of Eq. 11.81 becomes

$$v - T\left(\frac{\partial v}{\partial T}\right)_p = \frac{ZRT}{p} - T\left[\frac{RZ}{p} + \frac{RT}{p}\left(\frac{\partial Z}{\partial T}\right)_p\right] = -\frac{RT^2}{p}\left(\frac{\partial Z}{\partial T}\right)_p \tag{11.83}$$

Equation 11.83 can be written in terms of reduced properties as

$$v - T\left(\frac{\partial v}{\partial T}\right)_p = -\frac{RT_c}{p_c} \cdot \frac{T_R^2}{p_R}\left(\frac{\partial Z}{\partial T_R}\right)_{p_R}$$

Introducing this into Eq. 11.82 gives on rearrangement

$$\frac{h^*(T) - h(T, p)}{RT_c} = T_R^2 \int_0^{p_R} \left(\frac{\partial Z}{\partial T_R}\right)_{p_R} \frac{dp_R}{p_R}$$

Or, on a per mole basis

$$\frac{\bar{h}^*(T) - \bar{h}(T, p)}{\bar{R}T_c} = T_R^2 \int_0^{p_R} \left(\frac{\partial Z}{\partial T_R}\right)_{p_R} \frac{dp_R}{p_R} \tag{11.84}$$

The right side of Eq. 11.84 depends only on the reduced temperature T_R and reduced pressure p_R. Accordingly, the quantity $(\bar{h}^* - \bar{h})/\bar{R}T_c$, called the *enthalpy departure,* is a function only of these two reduced properties. Using a generalized equation of state giving Z as a function of T_R and p_R, the enthalpy departure can readily be evaluated with a computer. Tabular representations are also found in the literature.[2] Alternatively, the graphical representation provided in Fig. A-4 can be employed.

The change in specific enthalpy between two states can be evaluated by expressing Eq. 11.80 in terms of the enthalpy departure as

$$\bar{h}_2 - \bar{h}_1 = \underline{\bar{h}_2^* - \bar{h}_1^*} - \underline{\bar{R}T_c\left[\left(\frac{\bar{h}^* - \bar{h}}{\bar{R}T_c}\right)_2 - \left(\frac{\bar{h}^* - \bar{h}}{\bar{R}T_c}\right)_1\right]} \tag{11.85}$$

The first underlined term in Eq. 11.85 represents the change in specific enthalpy between the two states assuming ideal gas behavior. The second underlined term is the correction that must be applied to the ideal gas value for the enthalpy change to obtain the actual value for the enthalpy change. The quantity $(\bar{h}^* - \bar{h})/\bar{R}T_c$ at state 1 would be calculated with an equation giving $Z(T_R, p_R)$ or obtained from tables or the generalized enthalpy correction chart, Fig. A-4, using the reduced temperature T_{R1} and reduced pressure p_{R1} corresponding to the temperature T_1 and pressure p_1 at the initial state, respectively. Similarly, $(\bar{h}^* - \bar{h})/\bar{R}T_c$ at state 2 would be evaluated using T_{R2} and p_{R2}.

Generalized Entropy Chart. A generalized chart that allows changes in specific entropy to be evaluated can be developed in a similar manner to the generalized enthalpy chart introduced above. The difference in specific entropy between states 1

[2] Ibid., pp. 596–603.

and 2 of a gas (or liquid) can be expressed as the identity

$$s(T_2, p_2) - s(T_1, p_1) = s^*(T_2, p_2) - s^*(T_1, p_1) + \underline{\{[s(T_2, p_2) - s^*(T_2, p_2)] - [s(T_1, p_1) - s^*(T_1, p_1)]\}} \quad (11.86)$$

where $[s(T, p) - s^*(T, p)]$ denotes the specific entropy of the substance relative to that of its ideal gas model when both are at the same temperature and pressure. Equation 11.86 indicates that the change in specific entropy between the two states equals the entropy change determined using the ideal gas model plus a correction (shown underlined) that accounts for the departure from ideal gas behavior. The ideal gas term can be evaluated using methods introduced in Sec. 6.3.4. Let us consider next how the correction term is evaluated.

The following Maxwell relation gives the variation of entropy with pressure at fixed temperature:

$$\left(\frac{\partial s}{\partial p}\right)_T = -\left(\frac{\partial v}{\partial T}\right)_p \quad (11.35)$$

Integrating from pressure p' to pressure p at fixed temperature T gives

$$s(T, p) - s(T, p') = -\int_{p'}^{p} \left(\frac{\partial v}{\partial T}\right)_p dp \quad (11.87)$$

For an ideal gas, $v = RT/p$, so $(\partial v/\partial T)_p = R/p$. Using this in Eq. 11.87, the change in specific entropy assuming ideal gas behavior is

$$s^*(T, p) - s^*(T, p') = -\int_{p'}^{p} \frac{R}{p}\, dp \quad (11.88)$$

Subtracting Eq. 11.88 from Eq. 11.87 gives

$$[s(T, p) - s^*(T, p)] - [s(T, p') - s^*(T, p')] = \int_{p'}^{p} \left[\frac{R}{p} - \left(\frac{\partial v}{\partial T}\right)_p\right] dp \quad (11.89)$$

Using the idea that the properties of a substance tend to merge into those of its ideal gas model as pressure tends to zero at fixed temperature, we have

$$\lim_{p' \to 0} [s(T, p') - s^*(T, p')] = 0$$

Thus, in the limit as p' tends to zero, Eq. 11.89 becomes

$$s(T, p) - s^*(T, p) = \int_{0}^{p} \left[\frac{R}{p} - \left(\frac{\partial v}{\partial T}\right)_p\right] dp \quad (11.90)$$

Using p–v–T data only, Eq. 11.90 can be evaluated at states 1 and 2 and thus the correction term of Eq. 11.86 evaluated.

Equation 11.90 can be expressed in terms of the compressibility factor Z and the reduced properties T_R and p_R. The result on a per mole basis is

$$\frac{\bar{s}^*(T, p) - \bar{s}(T, p)}{\bar{R}} = \frac{\bar{h}^*(T) - \bar{h}(T, p)}{\bar{R} T_R T_c} + \int_0^{p_R} (Z - 1)\frac{dp_R}{p_R} \quad (11.91)$$

The right side of Eq. 11.91 depends only on the reduced temperature T_R and reduced pressure p_R. Accordingly, the quantity $(\bar{s}^* - \bar{s})/\bar{R}$, called the *entropy departure*, is a function only of these two reduced properties. As for the enthalpy departure, the entropy departure can be evaluated with a computer using a generalized equation of state giving Z as a function of T_R and p_R. Alternatively, the graphical representation provided in Fig. A-5 can be employed.

The change in specific entropy between two states can be evaluated by expressing Eq. 11.86 in terms of the entropy departure as

$$\bar{s}_2 - \bar{s}_1 = \underline{\bar{s}_2^* - \bar{s}_1^*} - \underline{\bar{R}\left[\left(\frac{\bar{s}^* - \bar{s}}{\bar{R}}\right)_2 - \left(\frac{\bar{s}^* - \bar{s}}{\bar{R}}\right)_1\right]} \tag{11.92}$$

The first underlined term in Eq. 11.92 represents the change in specific entropy between the two states assuming ideal gas behavior. The second underlined term is the correction that must be applied to the ideal gas value for entropy change to obtain the actual value for the entropy change. The quantity $(\bar{s}^* - \bar{s})_1/\bar{R}$ appearing in Eq. 11.92 would be calculated with an equation giving $Z(T_R, p_R)$ or obtained from the generalized entropy correction chart using the reduced temperature T_{R1} and reduced pressure p_{R1} corresponding to the temperature T_1 and pressure p_1 at the initial state, respectively. Similarly, $(\bar{s}^* - \bar{s})_2/\bar{R}$ would be evaluated using T_{R2} and p_{R2}.

EXAMPLE 11.8

PROBLEM

Nitrogen enters a turbine operating at steady state at 100 bar and 300 K and exits at 40 bar and 245 K. Using the enthalpy correction chart, determine the work developed, in kJ/kg, if heat transfer with the surroundings can be ignored. Changes in kinetic and potential energy from inlet to exit also can be neglected.

SOLUTION

Known: A turbine operating at steady state has nitrogen entering at 100 bar and 300 K and exiting at 40 bar and 245 K.

Find: Using the enthalpy correction chart, determine the work developed, in kJ/kg.

Schematic and Given Data:

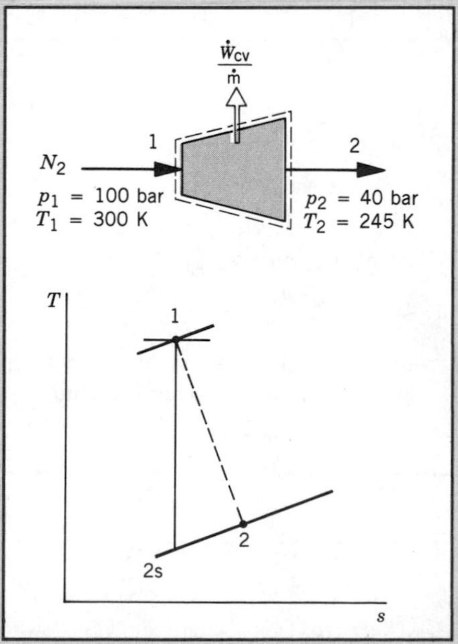

Figure E11.8

Assumptions:

1. The control volume shown on the accompanying figure operates at steady state.
2. There is no significant heat transfer between the control volume and its surroundings.
3. Changes in kinetic and potential energy between inlet and exit can be neglected.
4. Equilibrium property relations apply at the inlet and exit.

Analysis: The mass and energy rate balances reduce at steady state to give

$$0 = \frac{\dot{Q}_{cv}}{\dot{m}} - \frac{\dot{W}_{cv}}{\dot{m}} + \left[h_1 - h_2 + \frac{V_1^2 - V_2^2}{2} + g(z_1 - z_2)\right]$$

where $\dot{m}$ is the mass flow rate. Dropping the heat transfer term by assumption 2 and the kinetic and potential energy terms by assumption 3 gives on rearrangement

$$\frac{\dot{W}_{cv}}{\dot{m}} = h_1 - h_2$$

The term $h_1 - h_2$ can be evaluated as follows:

$$h_1 - h_2 = \frac{1}{M}\left\{\bar{h}_1^* - \bar{h}_2^* - \bar{R}T_c\left[\left(\frac{\bar{h}^* - \bar{h}}{\bar{R}T_c}\right)_1 - \left(\frac{\bar{h}^* - \bar{h}}{\bar{R}T_c}\right)_2\right]\right\}$$

In this expression, M is the molecular weight of nitrogen and the other terms have the same significance as in Eq. 11.85.

With specific enthalpy values from Table A-17 at $T_1 = 300$ K and $T_2 = 245$ K, respectively,

$$\bar{h}_1^* - \bar{h}_2^* = 8723 - 7121 = 1602 \text{ kJ/kmol}$$

The terms $(\bar{h}^* - \bar{h})/\bar{R}T_c$ at states 1 and 2 required by the above expression for $h_1 - h_2$ can be determined from Fig. A-4. First, the reduced temperature and reduced pressure at the inlet and exit must be determined. From Table A-1, $T_c = 126$ K, $p_c = 33.9$ bar. Thus, at the inlet

$$T_{R1} = \frac{300}{126} = 2.38, \qquad p_{R1} = \frac{100}{33.9} = 2.95$$

At the exit

$$T_{R2} = \frac{245}{126} = 1.94, \qquad p_{R2} = \frac{40}{33.9} = 1.18$$

By inspection of Fig. A-4

$$\left(\frac{\bar{h}^* - \bar{h}}{\bar{R}T_c}\right)_1 \approx 0.5, \qquad \left(\frac{\bar{h}^* - \bar{h}}{\bar{R}T_c}\right)_2 \approx 0.31$$

Substituting values

$$\frac{\dot{W}_{cv}}{\dot{m}} = \frac{1}{28\,\dfrac{\text{kg}}{\text{kmol}}}\left[1602\,\frac{\text{kJ}}{\text{kmol}} - \left(8.314\,\frac{\text{kJ}}{\text{kmol}\cdot\text{K}}\right)(126\text{ K})(0.5 - 0.31)\right] = 50.1 \text{ kJ/kg}$$

Comment: If only the ideal gas model was used, the work would be determined as 1602 kJ/kmol or 57.2 kJ/kg. These values are over 14% greater than the respective values determined using the enthalpy correction chart.

EXAMPLE 11.9

PROBLEM

For the case of Example 11.8, determine (a) the rate of entropy production, in kJ/kg·K, and (b) the isentropic turbine efficiency.

SOLUTION

Known: A turbine operating at steady state has nitrogen entering at 100 bar and 300 K and exiting at 40 bar and 245 K.

Find: Determine the rate of entropy production in kJ/kg·K and the isentropic turbine efficiency.

Schematic and Given Data: See Fig. E11.8.

Assumptions: See Example 11.8.

Analysis:

(a) At steady state, the control volume form of the entropy rate equation reduces to give

$$\frac{\dot{\sigma}_{cv}}{\dot{m}} = s_2 - s_1$$

The change in specific entropy required by this expression can be written as

$$s_2 - s_1 = \frac{1}{M}\left\{\bar{s}_2^* - \bar{s}_1^* - \bar{R}\left[\left(\frac{\bar{s}^* - \bar{s}}{\bar{R}}\right)_2 - \left(\frac{\bar{s}^* - \bar{s}}{\bar{R}}\right)_1\right]\right\}$$

where M is the molecular weight of nitrogen and the other terms have the same significance as in Eq. 11.92. The change in specific entropy $s_2^* - s_1^*$ can be evaluated using

$$\bar{s}_2^* - \bar{s}_1^* = \bar{s}^\circ(T_2) - \bar{s}^\circ(T_1) - \bar{R}\ln\frac{p_2}{p_1}$$

With values from Table A-17

$$\bar{s}_2^* - \bar{s}_1^* = 185.775 - 191.682 - 8.314\ln\frac{40}{100} = 1.711\ \frac{\text{kJ}}{\text{kmol}\cdot\text{K}}$$

The terms $(\bar{s}^* - \bar{s})/\bar{R}$ at the inlet and exit can be determined from Fig. A-5. Using the reduced temperature and reduced pressure values calculated in the solution to Example 11.8, inspection of Fig. A-5 gives

$$\left(\frac{\bar{s}^* - \bar{s}}{\bar{R}}\right)_1 \approx 0.19, \qquad \left(\frac{\bar{s}^* - \bar{s}}{\bar{R}}\right)_2 \approx 0.12$$

Substituting values

$$\frac{\dot{\sigma}_{cv}}{\dot{m}} = \frac{1}{(28\ \text{kg/kmol})}\left[1.711\ \frac{\text{kJ}}{\text{kmol}\cdot\text{K}} - 8.314\ \frac{\text{kJ}}{\text{kmol}\cdot\text{K}}(0.12 - 0.19)\right]$$

$$= 0.082\ \frac{\text{kJ}}{\text{kg}\cdot\text{K}}$$

(b) The isentropic turbine efficiency is defined in Sec. 6.8 as

$$\eta_t = \frac{(\dot{W}_{cv}/\dot{m})}{(\dot{W}_{cv}/\dot{m})_s}$$

where the denominator is the work that would be developed by the turbine if the nitrogen expanded isentropically from the specified inlet state to the specified exit pressure. Thus, it is necessary to fix the state, call it 2s, at the turbine exit for an expansion in which there is no change in specific entropy from inlet to exit. With $(\bar{s}_{2s} - \bar{s}_1) = 0$ and procedures similar to those used in part (a)

$$0 = \bar{s}_{2s}^* - \bar{s}_1^* - \bar{R}\left[\left(\frac{\bar{s}^* - \bar{s}}{\bar{R}}\right)_{2s} - \left(\frac{\bar{s}^* - \bar{s}}{\bar{R}}\right)_1\right]$$

$$0 = \left[\bar{s}^\circ(T_{2s}) - \bar{s}^\circ(T_1) - \bar{R}\ln\left(\frac{p_2}{p_1}\right)\right] - \bar{R}\left[\left(\frac{\bar{s}^* - \bar{s}}{\bar{R}}\right)_{2s} - \left(\frac{\bar{s}^* - \bar{s}}{\bar{R}}\right)_1\right]$$

Using values from part (a), the last equation becomes

$$0 = \bar{s}^\circ(T_{2s}) - 191.682 - 8.314\ln\frac{40}{100} - \bar{R}\left(\frac{\bar{s}^* - \bar{s}}{\bar{R}}\right)_{2s} + 1.58$$

or

$$\bar{s}°(T_{2s}) - \bar{R}\left(\frac{\bar{s}^* - \bar{s}}{\bar{R}}\right)_{2s} = 182.484$$

The temperature T_{2s} can be determined in an iterative procedure using $\bar{s}°$ data from Table A-17 and $(\bar{s}^* - \bar{s})/\bar{R}$ from Fig. A-5 as follows: First, a value for the temperature T_{2s} is assumed. The corresponding value of $\bar{s}°$ can then be obtained from Table A-17. The reduced temperature $(T_R)_{2s} = T_{2s}/T_c$, together with $p_{R2} = 1.18$, allows a value for $(\bar{s}^* - \bar{s})/\bar{R}$ to be obtained from Fig. A-5. The procedure continues until agreement with the value on the right side of the above equation is obtained. Using this procedure, T_{2s} is found to be closely 229 K.

With the temperature T_{2s} known, the work that would be developed by the turbine if the nitrogen expanded isentropically from the specified inlet state to the specified exit pressure can be evaluated from

$$\left(\frac{\dot{W}_{cv}}{\dot{m}}\right)_s = h_1 - h_{2s}$$

$$= \frac{1}{M}\left\{(\bar{h}_1^* - \bar{h}_{2s}^*) - \bar{R}T_c\left[\left(\frac{\bar{h}^* - \bar{h}}{\bar{R}T_c}\right)_1 - \left(\frac{\bar{h}^* - \bar{h}}{\bar{R}T_c}\right)_{2s}\right]\right\}$$

From Table A-17, $\bar{h}_{2s}^* = 6654$ kJ/kmol. From Fig. A-4 at $p_{R2} = 1.18$ and $(T_R)_{2s} = 229/126 = 1.82$

$$\left(\frac{\bar{h}^* - \bar{h}}{\bar{R}T_c}\right)_{2s} \approx 0.36$$

Values for the other terms in the expression for $(\dot{W}_{cv}/\dot{m})_s$ are obtained in the solution to Example 11.8. Finally

$$\left(\frac{\dot{W}_{cv}}{\dot{m}}\right)_s = \frac{1}{28}[8723 - 6654 - (8.314)(126)(0.5 - 0.36)] = 68.66 \text{ kJ/kg}$$

With the work value from Example 11.8, the turbine efficiency is

$$\eta_t = \frac{(\dot{W}_{cv}/\dot{m})}{(\dot{W}_{cv}/\dot{m})_s} = \frac{50.1}{68.66} = 0.73\ (73\%)$$

11.8 *p–v–T* RELATIONS FOR GAS MIXTURES

Many systems of interest involve mixtures of two or more components. The principles of thermodynamics introduced thus far are applicable to systems involving mixtures, but to apply them requires that the properties of mixtures be evaluated. Since an unlimited variety of mixtures can be formed from a given set of pure components by varying the relative amounts present, the properties of mixtures are available in tabular, graphical, or equation forms only in special cases such as air. However, means are available for determining the properties of mixtures. In this section, methods for evaluating the p–v–T relations for pure components introduced in previous sections of the book are adapted to obtain plausible estimates for gas mixtures. In Sec. 11.9 some general aspects of property evaluation for multicomponent systems are introduced. The case of ideal gas mixtures is taken up in Chap. 12.

Most techniques for estimating mixture properties are empirical in character and are not derived from fundamental physical principles. The realm of validity of any particular technique can be established only by comparing predicted property values with empirical data. The brief discussion to follow is intended only to show how certain of the procedures for evaluating the p–v–T relations of pure components introduced previously can be extended to gas mixtures.

To evaluate the properties of a mixture requires knowledge of the composition. The composition can be described by giving the *number of moles* (kmol or lbmol) of each component present. The total number of moles, n, is the sum of the number of moles of each of the components

$$n = n_1 + n_2 + \cdots + n_j = \sum_{i=1}^{j} n_i \tag{11.93}$$

The *relative* amounts of the components present can be described in terms of *mole fractions*. The mole fraction y_i of component i is defined as

$$y_i = \frac{n_i}{n} \tag{11.94}$$

Dividing each term of Eq. 11.93 by the total number of moles and using Eq. 11.94

$$1 = \sum_{i=1}^{j} y_i \tag{11.95}$$

That is, the sum of the mole fractions of all components present is equal to unity.

One way in which the p–v–T relation for a gas mixture can be estimated is by applying to the overall mixture an equation of state such as introduced in Sec. 11.1. The constants appearing in the equation selected would be *mixture values* determined with empirical combining rules developed for the equation. For example, mixture values of the constants a and b for use in the van der Waals and Redlich–Kwong equations would be obtained using relations of the form

$$a = \left(\sum_{i=1}^{j} y_i a_i^{1/2}\right)^2, \qquad b = \left(\sum_{i=1}^{j} y_i b_i\right) \tag{11.96}$$

where a_i and b_i are the values of the constants for component i and y_i is the mole fraction. Combination rules for obtaining mixture values for the constants in other equations of state also have been suggested.

The principle of corresponding states method for single components can be extended to mixtures by regarding the mixture as if it were a single pure component having critical properties calculated by one of several mixture rules. *Kay's rule* is perhaps the simplest of these, requiring only the determination of a mole fraction averaged critical temperature T_c and critical pressure p_c

$$T_c = \sum_{i=1}^{j} y_i T_{c,i}, \qquad p_c = \sum_{i=1}^{j} y_i p_{c,i} \tag{11.97}$$

where $T_{c,i}$, $p_{c,i}$, and y_i are the critical temperature, critical pressure, and mole fraction of component i, respectively. Using T_c and p_c, the mixture compressibility factor Z is obtained as for a single pure component. The unknown quantity from among the pressure p, volume V, temperature T, and total number of moles n of the gas mixture can then be obtained by solving

$$Z = \frac{pV}{n\bar{R}T} \tag{11.98}$$

Mixture values for T_c and p_c also can be used to enter the generalized enthalpy departure and entropy departure charts introduced in Sec. 11.7.

Additional means for estimating p–v–T relations for mixtures are provided by empirical mixture rules, of which several are found in the engineering literature. Among these are the *additive pressure* and *additive volume* rules. According to the additive pressure rule, the pressure of a gas mixture occupying volume V at temperature T is expressible as a sum of pressures exerted by the individual components

$$p = p_1 + p_2 + p_3 + \cdot \cdot \cdot]_{T,V} \tag{11.99a}$$

where the pressures p_1, p_2, etc. are evaluated by considering the respective components to be at the volume and temperature of the mixture. These pressures would be determined using tabular or graphical p–v–T data or a suitable equation of state.

An alternative expression of the additive pressure rule in terms of compressibility factors can be obtained. Since component i is considered to be at the volume and temperature of the mixture, the compressibility factor Z_i for this component is $Z_i = p_i V / n_i \bar{R} T$, so the pressure p_i is

$$p_i = \frac{Z_i n_i \bar{R} T}{V}$$

Similarly, for the mixture

$$p = \frac{Z n \bar{R} T}{V}$$

Substituting these expressions into Eq. 11.99a and reducing gives the following relationship between the compressibility factors for the mixture Z and the mixture components Z_i:

$$Z = \sum_{i=1}^{j} y_i Z_i]_{T,V} \tag{11.99b}$$

The compressibility factors Z_i are determined assuming that component i occupies the entire volume of the mixture at the temperature T.

The underlying assumption of the additive volume rule is that the volume V of a gas mixture at temperature T and pressure p is expressible as the sum of volumes occupied by the individual components

$$V = V_1 + V_2 + V_3 + \cdot \cdot \cdot]_{p,T} \tag{11.100a}$$

where the volumes V_1, V_2, etc. are evaluated by considering the respective components to be at the pressure and temperature of the mixture. These volumes would be determined from tabular or graphical p–v–T data or a suitable equation of state.

An alternative expression of the additive volume rule in terms of compressibility factors can be obtained. Since component i is considered to be at the pressure and temperature of the mixture, the compressibility factor Z_i for this component is $Z_i = p V_i / n_i \bar{R} T$, so the volume V_i is

$$V_i = \frac{Z_i n_i \bar{R} T}{p}$$

Similarly, for the mixture

$$V = \frac{Z n \bar{R} T}{p}$$

Substituting these expressions into Eq. 11.100a and reducing gives

$$Z = \sum_{i=1}^{j} y_i Z_i]_{p,T} \tag{11.100b}$$

The compressibility factors Z_i are determined assuming that component i exists at the temperature T and pressure p of the mixture.

EXAMPLE 11.10

PROBLEM

A mixture consisting of 0.18 kmol of methane (CH_4) and 0.274 kmol of butane (C_4H_{10}) occupies a volume of 0.241 m^3 at a temperature of 238°C. The experimental value for the pressure is 68.9 bar. Calculate the pressure, in bar, exerted by the mixture by using (a) the ideal gas equation of state, (b) Kay's rule together with the generalized compressibility chart, (c) the van der Waals equation, (d) the rule of additive pressures employing the generalized compressibility chart.

SOLUTION

Known: A mixture of two specified hydrocarbons with known molar amounts occupies a known volume at a specified temperature.

Find: Determine the pressure, in bar, using four alternative methods.

Schematic and Given Data:

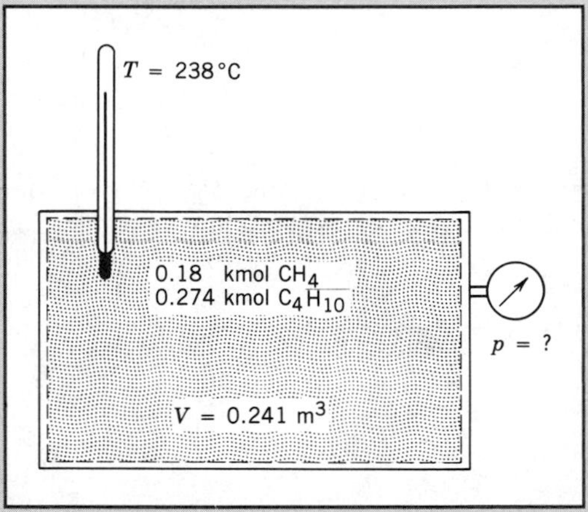

Figure E11.10

Assumption: As shown in the accompanying figure, the system is the mixture.

Analysis: The total number of moles of mixture n is

$$n = 0.18 + 0.274 = 0.454 \text{ kmol}$$

Thus, the mole fractions of the methane and butane are, respectively

$$y_1 = 0.396 \qquad \text{and} \qquad y_2 = 0.604$$

The specific volume of the mixture on a molar basis is

$$\bar{v} = \frac{0.241 \text{ m}^3}{(0.18 + 0.274) \text{ kmol}} = 0.531 \frac{\text{m}^3}{\text{kmol}}$$

(a) Substituting values into the ideal gas equation of state

$$p = \frac{\bar{R}T}{\bar{v}} = \frac{(8314 \text{ N}\cdot\text{m/kmol}\cdot\text{K})(511 \text{ K})}{(0.531 \text{ m}^3/\text{kmol})}\left(\frac{\text{bar}}{10^5 \text{ N/m}^2}\right)$$
$$= 80.01 \text{ bar}$$

(b) To apply Kay's rule, the critical temperature and pressure for each component are required. From Table A-1, for methane

$$T_{c1} = 191 \text{ K}, \qquad p_{c1} = 46.4 \text{ bar}$$

and for butane

$$T_{c2} = 425 \text{ K}, \qquad p_{c2} = 38.0 \text{ bar}$$

Thus, with Eqs. 11.97

$$T_c = y_1 T_{c1} + y_2 T_{c2} = (0.396)(191) + (0.604)(425) = 332.3 \text{ K}$$
$$p_c = y_1 p_{c1} + y_2 p_{c2} = (0.396)(46.4) + (0.604)(38.0) = 41.33 \text{ bar}$$

Treating the mixture as a pure component having the above values for the critical temperature and pressure, the following reduced properties are determined for the mixture:

$$T_R = \frac{T}{T_c} = \frac{511}{332.3} = 1.54$$
$$v_R' = \frac{\bar{v}p_c}{\bar{R}T_c} = \frac{(0.531)(41.33)(10^5)}{(8314)(332.3)}$$
$$= 0.794$$

Turning to Fig. A-2, $Z \approx 0.88$. The mixture pressure is then found from

$$p = \frac{Zn\bar{R}T}{V} = Z\frac{\bar{R}T}{\bar{v}} = 0.88\frac{(8314)(511)}{(0.531)(10^5)}$$
$$= 70.4 \text{ bar}$$

(c) Mixture values for the van der Waals constants can be obtained using Eqs. 11.96. This requires values of the van der Waals constants for each of the two mixture components. Table A-23 gives the following values for methane:

$$a_1 = 2.293 \text{ bar}\left(\frac{\text{m}^3}{\text{kmol}}\right)^2, \qquad b_1 = 0.0428\frac{\text{m}^3}{\text{kmol}}$$

Similarly, from Table A-23 for butane

$$a_2 = 13.86 \text{ bar}\left(\frac{\text{m}^3}{\text{kmol}}\right)^2, \qquad b_2 = 0.1162\frac{\text{m}^3}{\text{kmol}}$$

Then, the first of Eqs. 11.96 gives a mixture value for the constant a as

$$a = (y_1 a_1^{1/2} + y_2 a_2^{1/2})^2 = [0.396(2.293)^{1/2} + 0.604(13.86)^{1/2}]^2$$
$$= 8.113 \text{ bar}\left(\frac{\text{m}^3}{\text{kmol}}\right)^2$$

Substituting into the second of Eqs. 11.96 gives a mixture value for the constant b

$$b = y_1 b_1 + y_2 b_2 = (0.396)(0.0428) + 0.604(0.1162)$$
$$= 0.087\frac{\text{m}^3}{\text{kmol}}$$

Inserting the mixture values for a and b into the van der Waals equation together with known data

$$p = \frac{\bar{R}T}{\bar{v} - b} - \frac{a}{\bar{v}^2}$$

$$= \frac{(8314 \text{ N}\cdot\text{m/kmol}\cdot\text{K})(511 \text{ K})}{(0.531 - 0.087)(\text{m}^3/\text{kmol})}\left(\frac{\text{bar}}{10^5 \text{ N/m}^2}\right) - \frac{8.113 \text{ bar } (\text{m}^3/\text{kmol})^2}{(0.531 \text{ m}^3/\text{kmol})^2}$$

$$= 66.91 \text{ bar}$$

(d) To apply the additive pressure rule with the generalized compressibility chart requires that the compressibility factor for each component be determined assuming that the component occupies the entire volume at the mixture temperature. With this assumption, the following reduced properties are obtained for methane:

$$T_{R1} = \frac{T}{T_{c1}} = \frac{511}{191} = 2.69$$

$$v'_{R1} = \frac{\bar{v}_1 p_{c1}}{\bar{R}T_{c1}} = \frac{(0.241 \text{ m}^3/0.18 \text{ kmol})(46.4 \text{ bar})}{(8314 \text{ N}\cdot\text{m/kmol}\cdot\text{K})(191 \text{ K})}\left(\frac{10^5 \text{ N/m}^2}{\text{bar}}\right) = 3.91$$

With these reduced properties, Fig. A-2 gives $Z_1 \approx 1.0$.

Similarly, for butane

$$T_{R2} = \frac{T}{T_{c2}} = \frac{511}{425} = 1.2$$

$$v'_{R2} = \frac{\bar{v}_2 p_{c2}}{\bar{R}T_{c2}} = \frac{(0.88)(38)(10^5)}{(8314)(425)} = 0.95$$

From Fig. A-2, $Z_2 \approx 0.8$.

The compressibility factor for the mixture determined from Eq. 11.99b is

$$Z = y_1 Z_1 + y_2 Z_2 = (0.396)(1.0) + (0.604)(0.8) = 0.88$$

Accordingly, the same value for pressure as determined in part (b) using Kay's rule results: $p = 70.4$ bar.

Comment: In this particular case, the ideal gas equation of state gives a value for pressure that exceeds the experimental value by nearly 16%. Kay's rule and the rule of additive pressures each give a pressure value about 3% greater than the experimental value. The van der Waals equation with mixture values for the constants gives a pressure value about 3% less than the experimental value.

11.9 MULTICOMPONENT SYSTEMS[3]

In the preceding section we considered means for evaluating the p–v–T relation of gas mixtures by extending methods developed for pure components. The current section is devoted to the development of some general aspects of the properties of systems with two or more components. Primary emphasis is on the case of gas *mixtures,* but the methods developed also apply to *solutions*. When liquids and solids are under consideration, the term *solution* is sometimes used in place of mixture. The present discussion is limited to nonreacting mixtures or solutions in a single phase. The effects of chemical reactions and equilibrium between different phases are taken up in Chaps. 13 and 14.

To describe multicomponent systems, composition must be included in our thermodynamic relations. This leads to the definition and development of several new concepts, including the *partial molal property,* the *chemical potential,* and the *fugacity*.

[3] This section may be deferred until Secs. 12.1–12.5 have been studied.

11.9.1 Partial Molal Properties

In the present discussion we introduce the concept of a *partial molal* property and illustrate its use. This concept plays an important role in subsequent discussions of multicomponent systems.

Any extensive thermodynamic property X of a single-phase, single-component system is a function of two independent intensive properties and the size of the system. Selecting temperature and pressure as the independent properties and the number of moles n as the measure of size, we have $X = X(T, p, n)$. For a single-phase, *multicomponent* system, the extensive property X must then be a function of temperature, pressure, and the number of moles of each component present, $X = X(T, p, n_1, n_2, \ldots, n_j)$.

If each mole number is increased by a factor α, the size of the system increases by the same factor, and so does the value of the extensive property X. That is

$$\alpha X(T, p, n_1, n_2, \ldots, n_j) = X(T, p, \alpha n_1, \alpha n_2, \ldots, \alpha n_j)$$

Differentiating with respect to α while holding temperature, pressure, and the mole numbers fixed and using the chain rule on the right side gives

$$X = \frac{\partial X}{\partial(\alpha n_1)} n_1 + \frac{\partial X}{\partial(\alpha n_2)} n_2 + \cdots + \frac{\partial X}{\partial(\alpha n_j)} n_j$$

This equation holds for all values of α. In particular, it holds for $\alpha = 1$. Setting $\alpha = 1$

$$X = \sum_{i=1}^{j} n_i \left.\frac{\partial X}{\partial n_i}\right)_{T,p,n_l} \tag{11.101}$$

where the subscript n_l denotes that all n's except n_i are held fixed during differentiation.

The *partial molal property* $\bar{X}_i$ is by definition

$$\bar{X}_i = \left.\frac{\partial X}{\partial n_i}\right)_{T,p,n_l} \tag{11.102}$$

The partial molal property $\bar{X}_i$ is a property of the mixture and not simply a property of component i, for $\bar{X}_i$ depends in general on temperature, pressure, *and* mixture composition: $\bar{X}_i(T, p, n_1, n_2, \ldots, n_j)$. Partial molal properties are intensive properties of the mixture.

Introducing Eq. 11.102, Eq. 11.101 becomes

$$X = \sum_{i=1}^{j} n_i \bar{X}_i \tag{11.103}$$

This equation shows that the extensive property X can be expressed as a weighted sum of the partial molal properties $\bar{X}_i$.

Selecting the extensive property X in Eq. 11.103 to be volume, internal energy, enthalpy, and entropy, respectively, gives

$$V = \sum_{i=1}^{j} n_i \bar{V}_i, \qquad U = \sum_{i=1}^{j} n_i \bar{U}_i, \qquad H = \sum_{i=1}^{j} n_i \bar{H}_i, \qquad S = \sum_{i=1}^{j} n_i \bar{S}_i \tag{11.104}$$

where $\bar{V}_i$, $\bar{U}_i$, $\bar{H}_i$, $\bar{S}_i$ denote the partial molal volume, internal energy, enthalpy, and entropy. Similar expressions can be written for the Gibbs function G and the

Helmholtz function Ψ. Moreover, the relations between these extensive properties: $H = U + pV$, $G = H - TS$, $\Psi = U - TS$ can be differentiated with respect to n_i while holding temperature, pressure, and the remaining n's constant to produce corresponding relations among partial molal properties: $\bar{H}_i = \bar{U}_i + p\bar{V}_i$, $\bar{G}_i = \bar{H}_i - T\bar{S}_i$, $\bar{\Psi}_i = \bar{U}_i - T\bar{S}_i$, where $\bar{G}_i$ and $\bar{\Psi}_i$ are the partial molal Gibbs function and Helmholtz function, respectively. Several additional relations involving partial molal properties are developed later in this section.

Partial molal properties can be evaluated by several methods. If the property X can be measured, $\bar{X}_i$ can be found by extrapolating a plot giving $(\Delta X/\Delta n_i)_{T,p,n_l}$ versus Δn_i. That is

$$\bar{X}_i = \left(\frac{\partial X}{\partial n_i}\right)_{T,p,n_l} = \lim_{\Delta n_i \to 0} \left(\frac{\Delta X}{\Delta n_i}\right)_{T,p,n_l}$$

If an expression for X in terms of its independent variables is known, $\bar{X}_i$ can be evaluated by differentiation. The derivative can be determined analytically if the function is expressed analytically or found numerically if the function is in tabular form.

When suitable data are available, a simple graphical procedure known as the *method of intercepts* can be used to evaluate partial molal properties. In principle, the method can be applied for any extensive property. To illustrate this method, let us consider the volume of a system consisting of two components, A and B. For this system, Eq. 11.103 takes the form

$$V = n_A\bar{V}_A + n_B\bar{V}_B$$

where $\bar{V}_A$ and $\bar{V}_B$ are the partial molal volumes of A and B, respectively. Dividing by the number of moles of mixture n

$$\frac{V}{n} = y_A\bar{V}_A + y_B\bar{V}_B$$

where y_A and y_B denote the mole fractions of A and B, respectively. Since $y_A + y_B = 1$, this becomes

$$\frac{V}{n} = (1 - y_B)\bar{V}_A + y_B\bar{V}_B = \bar{V}_A + y_B(\bar{V}_B - \bar{V}_A)$$

This equation provides the basis for the method of intercepts. To illustrate, refer to Fig. 11.5, in which V/n is plotted as a function of y_B at constant T and p. At a

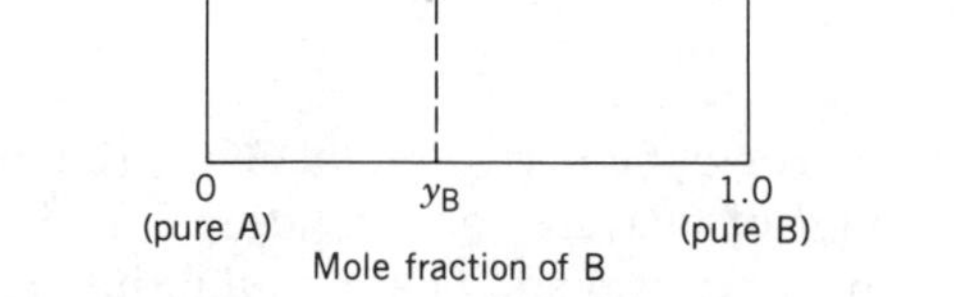

Figure 11.5 Illustration of the evaluation of partial molal volumes by the method of intercepts.

specified value for y_B, a tangent to the curve is shown on the figure. When extrapolated, the tangent line intersects the axis on the left at $\bar{V}_A$ and the axis on the right at $\bar{V}_B$. These values for the partial molal volumes correspond to the particular specifications for T, p, and y_B. At fixed temperature and pressure, $\bar{V}_A$ and $\bar{V}_B$ vary with y_B and are not equal to the molar specific volumes of *pure* A and *pure* B, denoted on the figure as $\bar{v}_A$ and $\bar{v}_B$, respectively. The values of $\bar{v}_A$ and $\bar{v}_B$ are fixed by temperature and pressure only.

Let us conclude the present discussion by evaluating the change in volume on mixing of pure components that are at the same temperature and pressure, a result for which an application is given later in the section. The total volume of the pure components before mixing is

$$V_{\text{components}} = \sum_{i=1}^{j} n_i \bar{v}_i$$

where $\bar{v}_i$ is the molar specific volume of pure component i. The volume of the mixture is

$$V_{\text{mixture}} = \sum_{i=1}^{j} n_i \bar{V}_i$$

where $\bar{V}_i$ is the partial molal volume of component i in the mixture. The volume change on mixing is

$$\Delta V_{\text{mixing}} = V_{\text{mixture}} - V_{\text{components}} = \sum_{i=1}^{j} n_i \bar{V}_i - \sum_{i=1}^{j} n_i \bar{v}_i$$

or

$$\Delta V_{\text{mixing}} = \sum_{i=1}^{j} n_i (\bar{V}_i - \bar{v}_i) \tag{11.105}$$

Similar results can be obtained for other extensive properties, for example,

$$\Delta U_{\text{mixing}} = \sum_{i=1}^{j} n_i (\bar{U}_i - \bar{u}_i)$$

$$\Delta H_{\text{mixing}} = \sum_{i=1}^{j} n_i (\bar{H}_i - \bar{h}_i) \tag{11.106}$$

$$\Delta S_{\text{mixing}} = \sum_{i=1}^{j} n_i (\bar{S}_i - \bar{s}_i)$$

In Eqs. 11.106, $\bar{u}_i$, $\bar{h}_i$, and $\bar{s}_i$ denote the molar internal energy, enthalpy, and entropy of pure component i. The symbols $\bar{U}_i$, $\bar{H}_i$, and $\bar{S}_i$ denote the respective partial molal properties.

11.9.2 Chemical Potential

Of the partial molal properties, the partial molal Gibbs function is particularly useful in describing the behavior of mixtures and solutions. This quantity plays a central role in the criteria for both chemical and phase equilibrium (Chap. 14). Because of its importance in the study of multicomponent systems, the partial molal Gibbs function of component i is given a special name and symbol. It is called the *chemical potential* of component i and symbolized by μ_i

$$\mu_i = \bar{G}_i = \left.\frac{\partial G}{\partial n_i}\right)_{T,p,n_l} \tag{11.107}$$

Like temperature and pressure, the chemical potential μ_i is an *intensive* property.

Applying Eq. 11.103 together with Eq. 11.107, the following expression can be written:

$$G = \sum_{i=1}^{j} n_i \mu_i \tag{11.108}$$

Expressions for the internal energy, enthalpy, and Helmholtz function can be obtained from Eq. 11.108, using the definitions $H = U + pV$, $G = H - TS$, and $\Psi = U - TS$. They are

$$U = TS - pV + \sum_{i=1}^{j} n_i \mu_i$$

$$H = TS + \sum_{i=1}^{j} n_i \mu_i \tag{11.109}$$

$$\Psi = -pV + \sum_{i=1}^{j} n_i \mu_i$$

Other useful relations can be obtained as well. Forming the differential of $G(T, p, n_1, n_2, \ldots, n_j)$

$$dG = \left.\frac{\partial G}{\partial p}\right)_{T,n} dp + \left.\frac{\partial G}{\partial T}\right)_{p,n} dT + \sum_{i=1}^{j} \left(\frac{\partial G}{\partial n_i}\right)_{T,p,n_l} dn_i \tag{11.110}$$

The subscripts n in the first two terms indicate that all n's are held fixed during differentiation. Since this implies fixed composition, it follows from Eqs. 11.30 and 11.31 that

$$V = \left(\frac{\partial G}{\partial p}\right)_{T,n} \quad \text{and} \quad -S = \left(\frac{\partial G}{\partial T}\right)_{p,n} \tag{11.111}$$

With Eqs. 11.107 and 11.111, Eq. 11.110 becomes

$$dG = V\,dp - S\,dT + \sum_{i=1}^{j} \mu_i\,dn_i \tag{11.112}$$

which for a multicomponent system is the counterpart of Eq. 11.23.

Another expression for dG is obtained by forming the differential of Eq. 11.108. That is

$$dG = \sum_{i=1}^{j} n_i\,d\mu_i + \sum_{i=1}^{j} \mu_i\,dn_i$$

Combining this equation with Eq. 11.112 gives

$$\sum_{i=1}^{j} n_i\,d\mu_i = V\,dp - S\,dT \tag{11.113}$$

known as the *Gibbs–Duhem equation*.

11.9.3 Fundamental Thermodynamic Functions for Multicomponent Systems

A *fundamental thermodynamic function* provides a complete description of the thermodynamic state of a system. In principle, all properties of interest can be determined from such a function by differentiation and/or combination. Reviewing the developments of Sec. 11.9.2, we see that a function $G(T, p, n_1, n_2, \ldots, n_j)$ is a fundamental thermodynamic function for a multicomponent system.

Functions of the form $U(S, V, n_1, n_2, \ldots, n_j)$, $H(S, p, n_1, n_2, \ldots, n_j)$, and $\Psi(T, V, n_1, n_2, \ldots, n_j)$ also can serve as fundamental thermodynamic functions for multicomponent systems. To demonstrate this, first form the differential of each of Eqs. 11.109 and use the Gibbs–Duhem equation, Eq. 11.113, to reduce the resultant expressions to obtain

$$d\Psi = T\,dS - p\,dV + \sum_{i=1}^{j} \mu_i\,dn_i \tag{11.114a}$$

$$dH = T\,dS + V\,dp + \sum_{i=1}^{j} \mu_i\,dn_i \tag{11.114b}$$

$$d\Psi = -p\,dV - S\,dT + \sum_{i=1}^{j} \mu_i\,dn_i \tag{11.114c}$$

For multicomponent systems, these are the counterparts of Eqs. 11.18, 11.19, and 11.22, respectively.

The differential of $U(S, V, n_1, n_2, \ldots, n_j)$ is

$$dU = \left.\frac{\partial U}{\partial S}\right)_{V,n} dS + \left.\frac{\partial U}{\partial V}\right)_{S,n} dV + \sum_{i=1}^{j} \left(\frac{\partial U}{\partial n_i}\right)_{S,V,n_l} dn_i$$

Comparing this expression term by term with Eq. 11.114a, we have

$$T = \left.\frac{\partial U}{\partial S}\right)_{V,n}, \qquad -p = \left.\frac{\partial U}{\partial V}\right)_{S,n}, \qquad \mu_i = \left.\frac{\partial U}{\partial n_i}\right)_{S,V,n_l} \tag{11.115a}$$

That is, the temperature, pressure, and chemical potentials can be obtained by differentiation of $U(S, V, n_1, n_2, \ldots, n_j)$. The first two of Eqs. 11.115a are the counterparts of Eqs. 11.24 and 11.25.

A similar procedure using a function of the form $H(S, p, n_1, n_2, \ldots, n_j)$ together with Eq. 11.114b gives

$$T = \left.\frac{\partial H}{\partial S}\right)_{p,n}, \qquad V = \left.\frac{\partial H}{\partial p}\right)_{S,n}, \qquad \mu_i = \left.\frac{\partial H}{\partial n_i}\right)_{S,p,n_l} \tag{11.115b}$$

where the first two of these are the counterparts of Eqs. 11.26 and 11.27. Finally, with $\Psi(S, V, n_1, n_2, \ldots, n_j)$ and Eq. 11.114c

$$-p = \left.\frac{\partial \Psi}{\partial V}\right)_{T,n}, \qquad -S = \left.\frac{\partial \Psi}{\partial T}\right)_{V,n}, \qquad \mu_i = \left.\frac{\partial \Psi}{\partial n_i}\right)_{T,V,n_l} \tag{11.115c}$$

The first two of these are the counterparts of Eqs. 11.28 and 11.29. With each choice of fundamental function, the remaining extensive properties can be found by combination using the definitions $H = U + pV$, $G = H - TS$, $\Psi = U - TS$.

The foregoing discussion of fundamental thermodynamic functions has led to several property relations for multicomponent systems that correspond to relations obtained previously. In addition, counterparts of the Maxwell relations can be obtained by equating mixed second partial derivatives. For example, the first two terms on the right of Eq. 11.112 give

$$\left.\frac{\partial V}{\partial T}\right)_{p,n} = -\left.\frac{\partial S}{\partial p}\right)_{T,n} \tag{11.116}$$

which corresponds to Eq. 11.35. Numerous relationships involving chemical potentials can be derived similarly by equating mixed second partial derivatives. An important example from Eq. 11.112 is

$$\left.\frac{\partial \mu_i}{\partial p}\right)_{T,n} = \left.\frac{\partial V}{\partial n_i}\right)_{T,p,n_l}$$

Recognizing the right side of this equation as the partial molal volume, we have

$$\left.\frac{\partial \mu_i}{\partial p}\right)_{T,n} = \bar{V}_i \tag{11.117}$$

This relationship is applied later in the section.

The present discussion concludes by listing four different expressions derived above for the chemical potential in terms of other properties. In the order obtained, they are

$$\mu_i = \left.\frac{\partial G}{\partial n_i}\right)_{T,p,n_l} = \left.\frac{\partial U}{\partial n_i}\right)_{S,V,n_l} = \left.\frac{\partial H}{\partial n_i}\right)_{S,p,n_l} = \left.\frac{\partial \Psi}{\partial n_i}\right)_{T,V,n_l} \tag{11.118}$$

Only the first of these partial derivatives is a partial molal property, however, for the term *partial molal* applies only to partial derivatives where the independent variables are temperature, pressure, and the number of moles of each component present.

11.9.4 Fugacity

The chemical potential plays an important role in describing multicomponent systems. In some instances, however, it is more convenient to work in terms of a related property, the fugacity. The fugacity is introduced in the present discussion.

Single-Component Systems. Let us begin by taking up the case of a system consisting of a single component. For this case, Eq. 11.108 reduces to give

$$G = n\mu \quad \text{or} \quad \mu = \frac{G}{n} = \bar{g}$$

That is, for a pure component the chemical potential equals the Gibbs function per mole. With this equation, Eq. 11.30 written on a per mole basis becomes

$$\left.\frac{\partial \mu}{\partial p}\right)_T = \bar{v} \tag{11.119}$$

For the special case of an ideal gas, $\bar{v} = \bar{R}T/p$, and Eq. 11.119 assumes the form

$$\left.\frac{\partial \mu^*}{\partial p}\right)_T = \frac{\bar{R}T}{p}$$

where the asterisk denotes ideal gas. Integrating at constant temperature

$$\mu^* = \bar{R}T \ln p + C(T) \tag{11.120}$$

where $C(T)$ is a function of integration. Since the pressure p can take on values from zero to plus infinity, the ln p term of this expression, and thus the chemical potential, has an inconvenient range of values from minus infinity to plus infinity. Equation 11.120 also shows that the chemical potential can be determined only to within an arbitrary constant.

Because of the above considerations, it is advantageous for many types of thermodynamic analyses to use fugacity in place of the chemical potential, for it is a well-behaved function that can be more conveniently evaluated. We introduce the fugacity f by the expression

$$\mu = \bar{R}T \ln f + C(T) \tag{11.121}$$

Comparing Eq. 11.121 with Eq. 11.120, the fugacity is seen to play the same role in the general case as pressure plays in the ideal gas case. Fugacity has the same units as pressure.

Substituting Eq. 11.121 into Eq. 11.119 gives

$$\bar{R}T \left(\frac{\partial \ln f}{\partial p}\right)_T = \bar{v} \tag{11.122}$$

Integration of Eq. 11.122 while holding temperature constant can determine the fugacity only to within a constant term. However, since ideal gas behavior is approached as pressure tends to zero, the constant term can be fixed by requiring that the fugacity of a pure component equals the pressure in the limit of zero pressure. That is

$$\lim_{p \to 0} \frac{f}{p} = 1 \tag{11.123}$$

Equations 11.122 and 11.123 then *completely determine* the fugacity function.

Let us consider next how the fugacity can be evaluated. With $Z = p\bar{v}/\bar{R}T$, Eq. 11.122 becomes

$$\bar{R}T \left(\frac{\partial \ln f}{\partial p}\right)_T = \frac{\bar{R}TZ}{p}$$

or

$$\left(\frac{\partial \ln f}{\partial p}\right)_T = \frac{Z}{p}$$

Subtracting $1/p$ from both sides and integrating from pressure p' to pressure p at fixed temperature T

$$[\ln f - \ln p]_{p'}^{p} = \int_{p'}^{p} (Z - 1)\, d \ln p$$

or

$$\left[\ln \frac{f}{p}\right]_{p'}^{p} = \int_{p'}^{p} (Z - 1)\, d \ln p$$

Taking the limit as p' tends to zero and applying Eq. 11.123 results in

$$\ln \frac{f}{p} = \int_{0}^{p} (Z - 1)\, d \ln p$$

When expressed in terms of the *reduced pressure*, $p_R = p/p_c$, the above equation is

$$\ln \frac{f}{p} = \int_{0}^{p_R} (Z - 1)\, d \ln p_R \tag{11.124}$$

Since the compressibility factor Z depends on the reduced temperature T_R and reduced pressure p_R, it follows that the right side of Eq. 11.124 depends on these properties only. Accordingly, the quantity $\ln f/p$ is a function only of these two reduced properties. Using a generalized equation of state giving Z as a function of T_R and p_R, $\ln f/p$ can readily be evaluated with a computer. Tabular representations are also found in the literature.[4] Alternatively, the generalized graphical representation presented in Fig. A-6 can be employed.

To illustrate the use of Fig. A-6, consider two states of water vapor at the same temperature, 400°C. At state 1 the pressure is 200 bar, and at state 2 the pressure is 240 bar. The change in the chemical potential between these states can be determined using Eq. 11.121 as

$$\mu_2 - \mu_1 = \bar{R}T \ln \frac{f_2}{f_1} = \bar{R}T \ln \left(\frac{f_2}{p_2} \frac{p_2}{p_1} \frac{p_1}{f_1} \right)$$

Using the critical temperature and pressure of water from Table A-1, at state 1 $p_{R1} = 0.91$, $T_{R1} = 1.04$, and at state 2 $p_{R2} = 1.09$, $T_{R2} = 1.04$. By inspection of Fig. A-6, $f_1/p_1 = 0.755$ and $f_2/p_2 = 0.7$. Inserting values in the above equation

$$\mu_2 - \mu_1 = (8.314)(673.15) \ln \left[\frac{0.7}{0.755} \left(\frac{240}{200} \right) \right] = 597 \text{ kJ/kmol}$$

For a pure component, the chemical potential equals the Gibbs function per mole, $\bar{g} = \bar{h} - T\bar{s}$. Since the temperature is the same at states 1 and 2, the change in the chemical potential can be expressed as $\mu_2 - \mu_1 = \bar{h}_2 - \bar{h}_1 - T(\bar{s}_2 - \bar{s}_1)$. Using steam table data, the value obtained with this expression is 597 kJ/kmol, which is the same as that determined from the generalized fugacity coefficient chart.

Multicomponent Systems. The fugacity of a component i in a mixture can be defined by a procedure which parallels that for a pure component. For a pure component, the development begins with Eq. 11.119, and the fugacity is introduced by Eq. 11.121. These are then used to write the pair of equations, Eqs. 11.122 and 11.123, from which the fugacity can be evaluated. For a mixture, the development begins with Eq. 11.117, the counterpart of Eq. 11.119, and the fugacity $\bar{f}_i$ of component i is introduced by

$$\mu_i = \bar{R}T \ln \bar{f}_i + C_i(T) \tag{11.125}$$

which parallels Eq. 11.121. The pair of equations that allow $\bar{f}_i$ to be evaluated are

$$\bar{R}T \left(\frac{\partial \ln \bar{f}_i}{\partial p} \right)_{T,n} = \bar{V}_i \tag{11.126a}$$

$$\lim_{p \to 0} \left(\frac{\bar{f}_i}{y_i p} \right) = 1 \tag{11.126b}$$

The symbol $\bar{f}_i$ denotes the fugacity of component i in the *mixture* and should be carefully distinguished in the presentation to follow from f_i, which denotes the fugacity of *pure i*.

Referring to Eq. 11.126b, note that in the ideal gas limit the fugacity $\bar{f}_i$ is not required to equal the pressure p as for the case of a pure component, but to equal the quantity $y_i p$. To see that this is the appropriate limiting quantity, consider a system consisting of a mixture of gases occupying a volume V at pressure p and temperature

[4] Ibid., pp. 605–611.

T. If the overall mixture behaves as an ideal gas, we can write

$$p = \frac{n\bar{R}T}{V} \tag{11.127}$$

where n is the total number of moles of mixture. Recalling from Sec. 3.5 that an ideal gas can be regarded as composed of molecules that exert negligible forces on one another and whose volume is negligible relative to the total volume, we can think of each component i as behaving as if it were an ideal gas alone at the temperature T and volume V. Thus, the pressure exerted by component i would not be the mixture pressure p but the pressure p_i given by

$$p_i = \frac{n_i\bar{R}T}{V} \tag{11.128}$$

where n_i is the number of moles of component i. Dividing Eq. 11.128 by Eq. 11.127

$$\frac{p_i}{p} = \frac{n_i\bar{R}T/V}{n\bar{R}T/V} = \frac{n_i}{n} = y_i$$

On rearrangement

$$p_i = y_i p \tag{11.129}$$

Summing both sides of Eq. 11.129, we obtain

$$\sum_{i=1}^{j} p_i = \sum_{i=1}^{j} y_i p = p \sum_{i=1}^{j} y_i$$

Or, since the sum of the mole fractions equals unity

$$p = \sum_{i=1}^{j} p_i \tag{11.130}$$

In words, Eq. 11.130 states that the sum of the pressures p_i equals the mixture pressure. This gives rise to the designation *partial pressure* for p_i. With this background, we now see that Eq. 11.126b requires the fugacity of component i to approach the partial pressure of component i as pressure p tends to zero. Comparing Eqs. 11.130 and 11.99a, we also see that the *additive pressure rule* is exact for ideal gas mixtures. This special case is considered further in Sec. 12.2 under the heading *Dalton model*.

Let us consider next how the fugacity of component i in a mixture can be expressed in terms of quantities that can be evaluated. For a pure component i, Eq. 11.122 gives

$$\bar{R}T\left(\frac{\partial \ln f_i}{\partial p}\right)_T = \bar{v}_i \tag{11.131}$$

where $\bar{v}_i$ is the molar specific volume of pure i. Subtracting Eq. 11.131 from Eq. 11.126a

$$\bar{R}T\left[\frac{\partial \ln(\bar{f}_i/f_i)}{\partial p}\right]_{T,n} = \bar{V}_i - \bar{v}_i \tag{11.132}$$

Integrating from pressure p' to pressure p at fixed temperature and mixture composition

$$\bar{R}T\left[\ln\left(\frac{\bar{f}_i}{f_i}\right)\right]_{p'}^{p} = \int_{p'}^{p} (\bar{V}_i - \bar{v}_i)\, dp$$

In the limit as p' tends to zero, this becomes

$$\bar{R}T\left[\ln\left(\frac{\bar{f}_i}{f_i}\right) - \lim_{p'\to 0}\ln\left(\frac{\bar{f}_i}{f_i}\right)\right] = \int_0^p (\bar{V}_i - \bar{v}_i)\,dp$$

Since $f_i \to p'$ and $\bar{f}_i \to y_i p'$ as pressure p' tends to zero

$$\lim_{p'\to 0}\ln\left(\frac{\bar{f}_i}{f_i}\right) \to \ln\left(\frac{y_i p'}{p'}\right) = \ln y_i$$

Therefore, we can write

$$\bar{R}T\left[\ln\left(\frac{\bar{f}_i}{f_i}\right) - \ln y_i\right] = \int_0^p (\bar{V}_i - \bar{v}_i)\,dp$$

or

$$\bar{R}T\ln\left(\frac{\bar{f}_i}{y_i f_i}\right) = \int_0^p (\bar{V}_i - \bar{v}_i)\,dp \qquad (11.133)$$

in which $\bar{f}_i$ is the fugacity of component i at pressure p in a mixture of given composition at a given temperature, and f_i is the fugacity of pure i at the same temperature and pressure. Equation 11.133 expresses the relation between $\bar{f}_i$ and f_i in terms of the difference between $\bar{V}_i$ and $\bar{v}_i$, a measurable quantity.

11.9.5 Ideal Solution

The task of evaluating the fugacities of the components in a mixture is considerably simplified when the mixture can be modeled as an *ideal solution*. An *ideal solution* is a mixture for which

$$\bar{f}_i = y_i f_i \qquad \text{(ideal solution)} \qquad (11.134)$$

Equation 11.134, known as the *Lewis–Randall rule,* states that the fugacity of each component in an ideal solution is equal to the product of its mole fraction and the fugacity of the pure component at the same temperature, pressure, and state of aggregation (gas, liquid, or solid) as the mixture. As consequences of the definition of an ideal solution, the following characteristics are exhibited by such a mixture.

Introducing Eq. 11.134 into Eq. 11.132, the left side vanishes, giving $\bar{V}_i - \bar{v}_i = 0$, or

$$\bar{V}_i = \bar{v}_i \qquad (11.135)$$

Thus, the partial molal volume of each component in an ideal solution is equal to the molar specific volume of the corresponding pure component at the same temperature and pressure. When Eq. 11.135 is introduced in Eq. 11.105, it can be concluded that there is no volume change on mixing pure components to form an ideal solution. It can also be shown that the partial molal internal energy of each component in an ideal solution is equal to the molar internal energy of the corresponding pure component at the same temperature and pressure. A similar result applies for enthalpy. In symbols

$$\bar{U}_i = \bar{u}_i, \qquad \bar{H}_i = \bar{h}_i \qquad (11.136)$$

With these expressions, it can be concluded from Eqs. 11.106 that there is no change in internal energy or enthalpy on mixing pure components to form an ideal solution.

However, we expect an entropy increase to result from the *adiabatic* mixing of different pure components because such a process is irreversible: The separation of the mixture into the pure components would never occur spontaneously. The entropy change on adiabatic mixing is considered further for the special case of ideal gas mixtures in Sec. 12.5.

With Eq. 11.135, the volume of an ideal solution is

$$V = \sum_{i=1}^{j} n_i \bar{V}_i = \sum_{i=1}^{j} n_i \bar{v}_i = \sum_{i=1}^{j} V_i \qquad \text{(ideal solution)} \tag{11.137}$$

where V_i is the volume that pure component i would occupy when at the temperature and pressure of the mixture. Comparing Eqs. 11.137 and 11.100a, the *additive volume rule* is seen to be exact for ideal solutions. With Eqs. 11.136, the internal energy and enthalpy of an ideal solution are

$$U = \sum_{i=1}^{j} n_i \bar{u}_i \quad \text{and} \quad H = \sum_{i=1}^{j} n_i \bar{h}_i \qquad \text{(ideal solution)} \tag{11.138}$$

where $\bar{u}_i$ and $\bar{h}_i$ denote, respectively, the molar internal energy and enthalpy of pure component i at the temperature and pressure of the mixture. Many gaseous mixtures at low to moderate pressures are adequately modeled by the Lewis–Randall rule. The ideal gas mixtures considered in Chap. 12 are an important special class of such mixtures. Some liquid solutions can also be modeled with the Lewis–Randall rule.

The Lewis–Randall rule requires that the fugacity of mixture component i be evaluated in terms of the fugacity of pure component i at the same temperature and pressure as the mixture and in the *same state of aggregation*. For example, if the mixture were a gas at T, p, then f_i would be determined for pure i at T, p and as a gas. However, at certain temperatures and pressures of interest a component of a gaseous mixture may, as a pure substance, be a liquid or solid. An example is an air–water vapor mixture at 20°C (68°F) and 1 atm. At this temperature and pressure, water exists not as a vapor but as a liquid. Though not considered here, means have been developed that allow the ideal solution model to be useful in such cases.

11.9.6 Chemical Potential for Ideal Solutions

The discussion of multicomponent systems concludes with the introduction of expressions for evaluating the chemical potential for ideal solutions of interest in subsequent sections of the book.

Consider a reference state where component i of a multicomponent system is pure at the temperature T of the system and a reference-state pressure p_{ref}. The difference in the chemical potential of i between a specified state of the multicomponent system and the reference state is obtained with Eq. 11.125 as

$$\mu_i - \mu_i^\circ = \bar{R}T \ln \frac{\bar{f}_i}{f_i^\circ} \tag{11.139}$$

where the superscript ° denotes property values at the reference state. The fugacity ratio appearing in the logarithmic term is known as the *activity*, a_i, of component i in the mixture.

$$a_i = \frac{\bar{f}_i}{f_i^\circ} \tag{11.140}$$

For subsequent applications, it suffices to consider the case of gaseous mixtures. For gaseous mixtures, p_{ref} is specified as 1 atm, so μ_i° and f_i° in Eq. 11.140 are, respectively, the chemical potential and fugacity of pure i at temperature T and 1 atm.

Since the chemical potential of a pure component equals the Gibbs function per mole, Eq. 11.139 can be written as

$$\mu_i = \bar{g}_i^\circ + \bar{R}T \ln a_i \tag{11.141}$$

where $\bar{g}_i^\circ$ is the Gibbs function per mole of pure component i evaluated at temperature T and 1 atm: $\bar{g}_i^\circ = \bar{g}_i(T, 1 \text{ atm})$.

For an ideal solution, the Lewis–Randall rule applies and the activity is

$$a_i = \frac{y_i f_i}{f_i^\circ} \tag{11.142}$$

where f_i is the fugacity of pure component i at temperature T and pressure p. Introducing Eq. 11.142 into Eq. 11.141

$$\mu_i = \bar{g}_i^\circ + \bar{R}T \ln \frac{y_i f_i}{f_i^\circ}$$

or

$$\mu_i = \bar{g}_i^\circ + \bar{R}T \ln \left[\underline{\left(\frac{f_i}{p}\right)\left(\frac{p_{\text{ref}}}{f_i^\circ}\right)} \frac{y_i p}{p_{\text{ref}}} \right] \tag{11.143}$$

In principle, the ratios of fugacity to pressure shown underlined in this equation can be evaluated from Eq. 11.124 or the generalized fugacity chart, Fig. A-6, developed from it. If component i behaves as an ideal gas at both T, p and T, p_{ref}, $f_i/p = f_i^\circ/p_{\text{ref}} = 1$, and Eq. 11.143 reduces to

$$\mu_i = \bar{g}_i^\circ + \bar{R}T \ln \frac{y_i p}{p_{\text{ref}}} \tag{11.144}$$

PROBLEMS

11.1 Using the van der Waals equation, determine the pressure, in atm, exerted by 1 lb of CO_2 at a temperature of 212°F if the specific volume is 0.193 ft^3/lb. Compare with the results obtained using the ideal gas equation of state and the generalized compressibility chart.

11.2 One pound of methane has a volume of 0.15 ft^3 and a pressure of 100 atm. Determine the temperature, in °R, using

(a) the ideal gas equation of state.
(b) the van der Waals equation of state.
(c) the compressibility chart.

11.3 Consider the water vapor at 360°C and a specific volume of 23.31×10^{-3} m^3/kg. Estimate the pressure at this state, in bar, using the

(a) ideal gas equation.
(b) van der Waals equation.
(c) Redlich–Kwong equation.
(d) compressibility chart.

Compare with the value from the steam tables.

11.4 Consider Refrigerant 12 vapor at 160°F, 0.5 ft^3/lb. Estimate the pressure at this state, in atm, using the

(a) ideal gas equation.
(b) van der Waals equation.
(c) Redlich–Kwong equation.
(d) compressibility chart.

Compare with the value from the Refrigerant 12 tables.

11.5 Methane gas flows through a pipeline with a mass flow rate of 110 lb/s at a pressure of 183 atm and a temperature of 56°F. Determine the volumetric flow rate, in ft³/s, using the

(a) the ideal gas equation.
(b) van der Waals equation.
(c) compressibility chart.

11.6 Determine the specific volume of water vapor at 10 MPa and 360°C, in m³/kg, using the

(a) steam tables.
(b) compressibility chart.
(c) Redlich–Kwong equation.
(d) ideal gas equation.

11.7 Using the Benedict–Webb–Rubin equation of state, determine the pressure of nitrogen (N_2) gas, in MPa, at 193 K and a specific volume of 4.5 cm³/g. Compare this with the values obtained using the ideal gas equation of state and the generalized compressibility chart.

11.8 Using the Benedict–Webb–Rubin equation of state, determine the pressure of carbon monoxide (CO), in atm, at a state where the specific volume is 0.122 ft³/lb and the temperature is 382°R. Compare with the value obtained from the generalized compressibility chart.

11.9 Using the Benedict–Webb–Rubin equation of state, determine the specific volume for methane, in m³/kmol, at a state where the pressure is 200 atm and the temperature is 400 K. Compare with the results obtained using the ideal gas equation of state and the generalized compressibility chart.

11.10 A rigid tank contains 0.5 kg of oxygen at $p_1 = 40$ bar, $T_1 = 180$ K. The gas is cooled until the pressure drops to 30 bar. Determine the volume of the tank and the final temperature using the

(a) ideal gas equation.
(b) Redlich–Kwong equation.
(c) compressibility chart.

11.11 One pound of air initially occupying a volume of 0.4 ft³ at a temperature of 603°F expands isothermally and without irreversibilities until the pressure is 200 lbf/in.² Using the van der Waals equation of state, determine the

(a) initial pressure, in lbf/in.²
(b) final volume, in ft³.
(c) work developed in the process, in Btu.

11.12 Water vapor initially at 240°C, 1 MPa expands in a piston–cylinder assembly isothermally and without internal irreversibilities to a final pressure of 0.1 MPa. Evaluate the work done, in kJ/kg. Use a truncated virial equation of state with the form

$$Z = 1 + \frac{B}{v} + \frac{C}{v^2}$$

where B and C are evaluated from *steam table* data at 240°C and pressures ranging from 0 to 1 MPa.

11.13 Referring to the virial series, Eqs. 3.27 and 3.28, show that $\hat{B} = B/\bar{R}T$, $\hat{C} = (C - B^2)/\bar{R}^2T^2$.

11.14 Express Eq. 11.5, the van der Waals equation in terms of the compressibility factor Z

(a) as a virial series in v_R'. [*Hint:* Expand the $(v_R' - 1/8)^{-1}$ term of Eq. 11.5 in a series.]
(b) as a virial series in p_R.
(c) Dropping terms involving $(p_R)^2$ and higher in the virial series of part b, obtain the following approximate form:

$$Z = 1 + \left(\frac{1}{8} - \frac{27/64}{T_R}\right)\frac{p_R}{T_R}$$

Compare the compressibility factors determined from this equation with tabulated compressibility factors for $0 < p_R < 0.6$ and each of $T_R = 1.0, 1.2, 1.4, 1.6, 1.8, 2.0$.[5] Comment on the realm of validity of the approximate form.

11.15 The Berthelot equation of state has the form

$$p = \frac{\bar{R}T}{\bar{v} - b} - \frac{a}{T\bar{v}^2}$$

(a) Using Eqs. 11.3, show that

$$a = \frac{27}{64}\frac{\bar{R}^2T_c^3}{p_c}, \qquad b = \frac{1}{8}\frac{\bar{R}T_c}{p_c}$$

(b) Express the equation in terms of the compressibility factor Z, the reduced temperature T_R, and the pseudoreduced specific volume, v_R'.

11.16 The Beattie–Bridgeman equation of state can be expressed as

$$p = \frac{RT(1 - \varepsilon)(v + B)}{v^2} - \frac{A}{v^2}$$

where

$$A = A_0\left(1 - \frac{a}{v}\right), \qquad B = B_0\left(1 - \frac{b}{v}\right)$$

$$\varepsilon = \frac{c}{vT^3}$$

and A_0, B_0, a, b, and c are constants. Express this equation of state in terms of the reduced pressure, p_R, reduced temperature, T_R, pseudoreduced specific volume, v_R', and appropriate dimensionless constants.

11.17 The Dieterici equation of state is

$$p = \left(\frac{RT}{v - b}\right)\exp\left(\frac{-a}{RTv}\right)$$

[5] Ibid., pp. 587–595.

(a) Using Eqs. 11.3, show that

$$a = \frac{4R^2T_c^2}{p_c e^2}, \qquad b = \frac{RT_c}{p_c e^2}$$

(b) Show that the equation of state can be expressed in terms of compressibility chart variables as

$$Z = \left(\frac{v_R'}{v_R' - 1/e^2}\right) \exp\left(\frac{-4}{T_R v_R' e^2}\right)$$

(c) Convert the result of part (b) to a virial series in v_R'. (Use the same approach as in Problem 11.14 and also expand the exponential term in a series.)

11.18 The Peng–Robinson equation of state has the form

$$p = \frac{RT}{v - b} - \frac{a}{v^2 - c^2}$$

Using Eqs. 11.3, evaluate the constants a, b, c in terms of the critical pressure p_c, critical temperature T_c, and critical compressibility factor Z_c.

11.19 The p–v–T relation for chlorofluorinated hydrocarbons can be described by the Carnahan–Starling–DeSantis equation of state

$$\frac{p\bar{v}}{\bar{R}T} = \frac{1 + \beta + \beta^2 - \beta^3}{(1 + \beta)^3} - \frac{a}{\bar{R}T(\bar{v} + b)}$$

where $\beta = b/4\bar{v}$, $a = a_0 \exp(a_1T + a_2T^2)$, and $b = b_0 + b_1T + b_2T^2$. For Refrigerants 12 and 13, the required coefficients for T in K, a in J · L/(mol)^2, and b in L/mol are

	$a_0 \times 10^{-3}$	$a_1 \times 10^3$	$a_2 \times 10^6$	b_0	$b_1 \times 10^4$	$b_2 \times 10^8$
R-12	3.52412	−2.77230	−0.67318	0.15376	−1.84195	−5.03644
R-13	2.29813	−3.41828	−1.52430	0.12814	−1.84474	−10.7951

For Refrigerant 12, determine v in m^3/kg at p = 0.2 MPa, T = 80°C. Compare with the value from Table A-9. Calculate v for Refrigerant 13 at the same state.

11.20 The differential of pressure obtained from a certain equation of state is given by *one* of the following expressions. Determine the equation of state.

$$dp = \frac{2(v - b)}{RT}\,dv + \frac{(v - b)^2}{RT^2}\,dT$$

$$dp = -\frac{RT}{(v - b)^2}\,dv + \frac{R}{v - b}\,dT$$

11.21 Introducing $\delta Q_{\substack{\text{int}\\\text{rev}}} = T\,dS$ into Eq. 6.15 gives

$$\delta Q_{\substack{\text{int}\\\text{rev}}} = dU + p\,dV$$

Using this expression together with the test for exactness, demonstrate that $Q_{\substack{\text{int}\\\text{rev}}}$ is not a property.

11.22 Show that Eq. 11.16 is satisfied by an equation of state with the form $p = RT/(v - b)$.

11.23 For the functions $x = x(y, w)$, $y = y(z, w)$, $z = z(x, w)$, demonstrate that

$$\left(\frac{\partial x}{\partial y}\right)_w \left(\frac{\partial y}{\partial z}\right)_w \left(\frac{\partial z}{\partial x}\right)_w = 1$$

11.24 Using Eq. 11.35, check the consistency of

(a) the steam tables at 2 MPa, 400°C.
(b) the Refrigerant 12 tables at 2 bar, 50°C.

11.25 Using Eq. 11.35, check the consistency of

(a) the steam tables at 100 lbf/in.^{2}, 600°F.
(b) the Refrigerant 12 tables at 40 lbf/in.^{2}, 100°F.

11.26 At a pressure of 1 atm, liquid water has a state of *maximum* density at about 4°C. What can be concluded about $(\partial s/\partial p)_T$ at

(a) 3°C?
(b) 4°C?
(c) 5°C?

11.27 A gas enters a compressor operating at steady state and is compressed isentropically. Does the specific enthalpy increase or decrease as the gas passes from the inlet to the exit?

11.28 Show that T, p, h, ψ, and g can each be determined from a fundamental thermodynamic function of the form $u = u(s, v)$.

11.29 Evaluate p, s, u, h, c_v, and c_p for a substance for which the Helmholtz function has the form

$$\psi = -RT \ln \frac{v}{v'} - cT'\left[1 - \frac{T}{T'} + \frac{T}{T'} \ln \frac{T}{T'}\right]$$

where v' and T' denote specific volume and temperature, respectively, at a reference state, and c is a constant.

11.30 The Mollier diagram provides a graphical representation of the fundamental thermodynamic function $h = h(s, p)$. Show that at any state fixed by s and p the properties T, v, u, ψ, and g can be evaluated using data obtained from the diagram.

11.31 Derive the relation $c_p = -T(\partial^2 g/\partial T^2)_p$.

11.32 Using p–v–T data for saturated Refrigerant 12 from Table A-7E, calculate at 20°F

(a) $h_g - h_f$.
(b) $u_g - u_f$.
(c) $s_g - s_f$.

Compare with the values obtained using table data.

11.33 Using p–v–T data for saturated water from the steam tables, calculate at 30°C

(a) $h_g - h_f$.
(b) $u_g - u_f$.
(c) $s_g - s_f$.

Compare with the values obtained using steam table data.

11.34 Using h_{fg}, v_{fg}, and p_{sat} at 10°F from the Refrigerant 12 tables, estimate the saturation pressure at 20°F. Comment on the accuracy of your estimate.

11.35 Using h_{fg}, v_{fg}, and p_{sat} at 20°C from the steam tables, estimate the saturation pressure at 25°C. Comment on the accuracy of your estimate.

11.36 Using triple-point data for water from Table A-6E, estimate the saturation pressure at −30°F. Compare with the value listed in Table A-6E.

11.37 At 0°C, the specific volumes of saturated solid water (ice) and saturated liquid water are, respectively, $v_i = 1.0911 \times 10^{-3}$ m³/kg and $v_f = 1.0002 \times 10^{-3}$ m³/kg, and the change in specific enthalpy on melting is $h_{if} = 333.4$ kJ/kg. Calculate the melting temperature of ice at (a) 250 bar, (b) 500 bar. Locate your answers on a p–T diagram for water.

11.38 The line representing the two-phase solid–liquid region on the phase diagram slopes to the left for substances that expand on freezing and to the right for substances that contract on freezing (Sec. 3.2.2). Verify this for the cases of lead that contracts on freezing and bismuth that expands on freezing.

11.39 Consider a four-legged chair at rest on an ice rink. The total mass of the chair and a person sitting on it is 80 kg. If the ice temperature is −2°C, determine the minimum total area, in cm², the tips of the chair legs can have before the ice in contact with the legs would melt. Use data from Problem 11.38 and let the local acceleration of gravity be 9.8 m/s².

11.40 Over a certain temperature interval, the saturation pressure–temperature curve of a substance is represented by an equation of the form $\ln p_{sat} = A - B/T$, where A and B are empirically determined constants. Obtain expressions for $h_g - h_f$ and $s_g - s_f$ in terms of p–v–T data and the constant B.

11.41 Using data for water from Table A-2, determine the constants A and B to give the best fit in a least-squares sense to the saturation pressure in the interval from 20 to 30°C by the equation $\ln p_{sat} = A - B/T$. Using this equation, determine dp_{sat}/dT at 25°C. Calculate $h_g - h_f$ at 25°C and compare with the steam table value.

11.42 Over limited intervals of temperature, the saturation pressure–temperature curve for two-phase liquid–vapor states can be represented by an equation of the form $\ln p_{sat} = A - B/T$, where A and B are constants. Derive the following expression relating any three states on such a portion of the curve:

$$\frac{p_{sat,3}}{p_{sat,1}} = \left(\frac{p_{sat,2}}{p_{sat,1}}\right)^{\tau}$$

where $\tau = T_2(T_3 - T_1)/T_3(T_2 - T_1)$.

11.43 Use the result of Problem 11.42 to determine

(a) the saturation pressure at 30°C using saturation pressure-temperature data at 20 and 40°C from Table A-2. Compare with the table value for saturation pressure at 30°C.
(b) the saturation temperature at 0.006 MPa using saturation pressure–temperature data at 20 to 40°C from Table A-2. Compare with the saturation temperature at 0.006 MPa given in Table A-3.

11.44 Complete the following exercises dealing with slopes:

(a) An enthalpy–entropy diagram (Mollier diagram) is often used in analyzing steam turbines. Obtain an expression for the slope of a constant-pressure line on such a diagram in terms of p–v–T data only.
(b) A pressure–enthalpy diagram is often used in the refrigeration industry. Obtain an expression for the slope of an isentropic line on such a diagram in terms of p–v–T data only.
(c) Consider the superheated vapor region of a temperature–entropy diagram. Show that the slope of a constant specific volume line is greater than the slope of a constant-pressure line through the same state.
(d) At the triple point of water, evaluate the ratio of the slope of the vaporization line to the slope of the sublimation line. Use steam table data to obtain a numerical value for the ratio.

11.45 Using only p–v–T data from the steam tables, evaluate the changes in specific enthalpy and entropy for a process from 52 lbf/in.², 400°F to 86 lbf/in.², 400°F. Compare with the steam table values.

11.46 One kmol of argon at 300 K is initially confined to one side of a rigid, insulated container divided into equal volumes of 0.2 m³ by a partition. The other side is initially evacuated. The partition is removed and the argon expands to fill the entire container. Using the van der Waals equation of state, determine the final temperature of the argon, in K. Repeat using the ideal gas equation of state.

11.47 Obtain the relationship between c_p and c_v for a gas that obeys the equation of state $p(v - b) = RT$.

11.48 The p–v–T relation for a certain gas is represented closely by $v = RT/p + B - A/RT$, where R is the gas constant and A and B are constants. Determine expressions for the changes in specific enthalpy and entropy, $[h(p_2, T) - h(p_1, T)]$ and $[s(p_2, T) - s(p_1, T)]$, respectively.

11.49 Develop expressions for the specific enthalpy and entropy changes $[h(v_2, T) - h(v_1, T)]$, $[s(v_2, T) - s(v_1, T)]$, using

(a) the van der Waals equation of state.
(b) the Redlich–Kwong equation of state.

11.50 At certain states, the p–v–T data of a gas can be expressed as $Z = 1 - Ap/T^4$, where Z is the compressibility factor and A is a constant.

(a) Obtain an expression for $(\partial p/\partial T)_v$ in terms of p, T, A, and the gas constant R.
(b) Obtain an expression for the change in specific entropy, $[s(p_2, T) - s(p_1, T)]$.

11.51 For a gas whose p–v–T behavior is described by $Z = 1 + Bp/RT$, where B is a function of temperature, derive expressions for the specific enthalpy and entropy changes, $[h(p_2, T) - h(p_1, T)]$ and $[s(p_2, T) - s(p_1, T)]$.

11.52 For a gas whose p–v–T behavior is described by $Z = 1 + B/v + C/v^2$, where B and C are functions of temperature, derive an expression for the specific entropy change, $[s(v_2, T) - s(v_1, T)]$.

11.53 The pressure exerted on a 1-kg cube of copper is increased from 10 to 1400 bar while the temperature remains constant at 300 K. Determine the volume change in m^3. Average values of ρ, β, and κ are 8888 kg/m^3, 49.2×10^{-6} (K)$^{-1}$, and 0.776×10^{-11} m^2/N, respectively.

11.54 The pressure exerted on a 100-lb cube of copper is reduced from 1200 to 1 atm while keeping the temperature constant at 80°F. Determine the volume change in ft^3. Average values of ρ, β, and κ are 555 lb/ft^3, 2.75×10^{-5} (°R)$^{-1}$, and 3.72×10^{-10} ft^2/lbf, respectively.

11.55 Develop expressions for the volume expansivity β and the isothermal compressibility κ for

(a) an ideal gas.
(b) a gas whose equation of state is $p(v - b) = RT$.
(c) a gas obeying the van der Waals equation.

11.56 Derive expressions for the volume expansivity β and the isothermal compressibility κ in terms of T, p, Z, and the first partial derivatives of Z.

11.57 Show that the isothermal compressibility κ is always greater than or equal to the isentropic compressibility α.

11.58 Prove that $(\partial\beta/\partial p)_T = -(\partial\kappa/\partial T)_p$.

11.59 For copper at 227°C, $\rho = 8930$ kg/m^3, $\beta = 54.2 \times 10^{-6}$ (K)$^{-1}$, $\kappa = 0.837 \times 10^{-11}$ m^2/N. Determine the percent error in c_v that would result if it were assumed that $c_p = c_v$.

11.60 For copper at 80°F, $\rho = 555$ lb/ft^3, $\beta = 2.75 \times 10^{-5}$ (°R)$^{-1}$, $\kappa = 3.72 \times 10^{-10}$ ft^2/lbf. Determine the percent error in c_v that would result if it were assumed that $c_p = c_v$.

11.61 At certain states, the p–v–T data for a particular gas can be represented as $Z = 1 - Ap/T^4$, where Z is the compressibility factor and A is a constant. Obtain an expression for the difference in specific heats, $c_p - c_v$.

11.62 For a gas obeying van der Waals equation of state,

(a) show that $(\partial c_v/\partial v)_T = 0$.
(b) develop expressions for $[u(T_2, v_2) - u(T_1, v_1)]$ and $[s(T_2, v_2) - s(T_1, v_1)]$. What additional information would be required to complete the Δu and Δs evaluations?
(c) develop an expression for $c_p - c_v$.

11.63 If the value of the specific heat c_v of air is 0.1965 Btu/lb·°R at $T_1 = 1000$°F, $v_1 = 36.8$ ft^3/lb, determine the value of c_v at $T_2 = 1000$°F, $v_2 = 0.0555$ ft^3/lb. Assume that air obeys the Berthelot equation of state

$$p = \frac{RT}{v - b} - \frac{a}{Tv^2}$$

where

$$a = \frac{27}{64}\frac{R^2T_c^3}{p_c}, \qquad b = \frac{1}{8}\frac{RT_c}{p_c}$$

11.64 Show that the specific heat ratio k can be expressed as $k = c_p\kappa/(c_p\kappa - Tv\beta^2)$. Using this expression together with data from the steam tables, evaluate k for water vapor at 15 lbf/in.^{2}, 400°F.

11.65 For liquid water at 75°F, 1 atm calculate

(a) c_v, in Btu/lb·°R.
(b) the velocity of sound, in ft/s.

Use data from Table 11.1, as required.

11.66 Using steam table data, estimate the velocity of sound in liquid water at (a) 40°C, 50 bar, (b) 100°F, 1000 lbf/in.^{2}

11.67 At a certain location in a stream of air, the temperature is 500°F, the pressure is 1 atm, and the Mach number is 1.4. Determine the velocity, in ft/s.

11.68 For a gas obeying the equation of state $p(v - b) = RT$, where b is a positive constant, can the temperature be reduced in a Joule–Thomson expansion?

11.69 For a gas described by $v = RT/p - A/T + B$, where A and B are constants, obtain an expression for the temperatures at the Joule–Thomson inversion states.

11.70 Determine the *maximum* Joule–Thomson inversion temperature in terms of the critical temperature T_c predicted by the

(a) van der Waals equation.
(b) Redlich–Kwong equation.
(c) Dieterici equation given in Problem 11.17.

11.71 Derive an equation for the Joule–Thomson coefficient as a function of T and v for a gas that obeys the van der Waals equation of state and whose specific heat c_v is given by $c_v = A + BT + CT^2$, where A, B, C are constants. Evaluate the temperatures at the *inversion* states in terms of R, v, and the van der Waals constants a and b.

11.72 The following expressions for the equation of state and the specific heat c_p are obeyed by a certain gas:

$$v = \frac{RT}{p} + \alpha T^2 \quad \text{and} \quad c_p = A + BT + Cp$$

where α, A, B, C are constants. Obtain an expression for

(a) the Joule–Thomson coefficient.
(b) the specific heat c_v.

11.73 For air, write a computer program that evaluates the change in specific enthalpy from a state where the temperature is 25°C and the pressure is 1 atm to a state where the temperature is T and the pressure is p. Use the van der Waals equation of state and account for the variation of the ideal gas specific heat as in Table A-15.

11.74 Using the Redlich–Kwong equation of state, determine the changes in specific enthalpy, in kJ/kmol, and entropy, in kJ/kmol·K, for argon between 300 K, 1 atm and 300 K, 500 atm.

11.75 Using the Benedict–Webb–Rubin equation of state together with an appropriate specific heat relation, determine the change in specific enthalpy, in kJ/kmol, for methane between 300 K, 1 atm and 400 K, 200 atm.

11.76 Using the Redlich–Kwong equation of state together with an appropriate specific heat relation, determine the final temperature for an isentropic expansion of nitrogen from 400 K, 250 atm to 5 atm.

11.77 A certain pure simple compressible substance has the following property relations. The p–v–T relationship in the vapor phase is

$$v = \frac{RT}{p} - \frac{Bp}{T^2}$$

where v is in ft³/lb, T is in °R, p is in lbf/ft², $R = 50$ ft·lbf/lb·°R, and $B = 100$ ft⁵·(°R)²/lb·lbf. The saturation pressure, in lbf/ft², is described by

$$\ln p_{\text{sat}} = 12 - \frac{2400}{T}$$

The Joule–Thomson coefficient at 10 lbf/in.², 200°F is 0.004°R·ft²/lbf. The ideal gas specific heat c_{p0} is constant over the temperature range 0 to 300°F.

(a) Complete the accompanying table of property values

T	p	v_f	v_g	h_f	h_g	s_f	s_g
0°F		0.03		0		0.000	
100°F		0.03					

for p in lbf/in.², v in ft³/lb, h in Btu/lb, and s in Btu/lb·°R.

(b) Evaluate v, h, s at the state fixed by 15 lbf/in.², 300°F.

11.78 Beginning with Eq. 11.90, derive Eq. 11.91.

11.79 Derive an expression giving

(a) the internal energy of a substance relative to that of its ideal gas model at the same temperature: $[u(T, v) - u^*(T)]$.
(b) the entropy of a substance relative to that of its ideal gas model at the same temperature and specific volume: $[s(T, v) - s^*(T, v)]$.

11.80 Derive expressions for the enthalpy and entropy departures using an equation of state with the form $Z = 1 + Bp_R$, where B is a function of the reduced temperature, T_R.

11.81 Using the equation of state of Problem 11.14 (c), evaluate v and c_p for water vapor at 550°C, 20 MPa and compare with data from Table A-4 and Fig. 3.8, respectively. Discuss.

11.82 Carbon dioxide gas undergoes a process from a state where $T_1 = 80$°C, $p_1 = 7.5$ MPa to a state where $T_2 = 135$°C, $p_2 = 15$ MPa. Evaluate the changes in specific enthalpy and entropy for the process.

11.83 At steady state, methane gas enters a pipe at 65 bar, 70°C and exits at −6°C. The change in pressure is negligible, as are changes in kinetic and potential energy. Determine the rate of heat transfer, in kJ per kg of methane flowing through the pipe. Let $\bar{c}_{p0} = 18.9 + 0.0555T$, where $\bar{c}_{p0}$ is in kJ/kmol·K and T is in K.

11.84 Carbon dioxide gas flowing through a heat exchanger operating at steady state is heated at an essentially constant pressure of 7 MPa from 27 to 427°C. Neglecting kinetic and potential energy changes, determine the heat transfer, in kJ per kg of CO_2 flowing, and compare with the result obtained using the ideal gas model.

11.85 Methane at 109°C, 9.3 MPa enters a turbine operating at steady state, expands adiabatically through a 5 : 1 pressure ratio, and exits at −6°C. Kinetic and potential energy effects are negligible. If $\bar{c}_{po} = 36.5$ kJ/kmol·K, determine, per kmol of methane flowing through the turbine, **(a)** the power developed, and **(b)** the entropy produced.

11.86 Argon gas enters a turbine operating at steady state at 250 bar, 0°C and expands adiabatically to 45 bar, −80°C with no change in kinetic or potential energy. If the isentropic turbine efficiency is 75%, determine the work developed, in kJ per kg of argon flowing.

11.87 Carbon dioxide gas enters a turbine operating at steady state at 50 atm, 200°F and expands adiabatically to 10 atm, 30°F with no significant changes in kinetic or potential energy. Determine **(a)** the isentropic turbine efficiency, **(b)** the turbine effectiveness. Let $T_0 = 77°F$, $p_0 =$ 1 atm.

11.88 Carbon dioxide enters a nozzle operating at steady state at 1000 lbf/in.², 800°F, 10 ft/s and expands isentropically to 500 lbf/in.² Determine at the nozzle exit the

(a) temperature, in °F.
(b) velocity, in ft/s.

11.89 Ethylene at 80°F, 155 lbf/in.² enters a compressor operating at steady state and is compressed isothermally without internal irreversibilities to 15,000 lbf/in.² Kinetic and potential energy changes are negligible. Evaluate in Btu per lb of ethylene flowing through the compressor

(a) the power required.
(b) the heat transfer.

11.90 Ethylene undergoes an isentropic process from 600°R, 200 lbf/in.² to 1500 lbf/in.² Determine the temperature at the final state, in °R.

11.91 Nitrogen is contained in a closed, rigid tank. Initially the temperature is 360°R and the pressure is 98.7 atm. If the nitrogen is heated until the temperature is 630°R, determine the heat transfer, in Btu/lbmol.

11.92 A quantity of nitrogen gas undergoes a process at a constant pressure of 200 atm from 220 to 300 K. Determine the work and heat transfer for the process, each in kJ per kmol of nitrogen.

11.93 A closed, rigid, insulated vessel having a volume of 0.142 m³ contains oxygen initially at 100 atm, 5°C. The oxygen is stirred by a paddle wheel until the pressure becomes 300 atm. Stirring ceases and the gas attains a final equilibrium state. Determine the

(a) final temperature, in °C.
(b) work, in kJ.
(c) amount of availability destroyed in the process, in kJ.

Let $T_0 = 5°C$.

11.94 An initially evacuated, insulated vessel having a volume of 1 m³ is connected by a valve to a large tank filled with carbon dioxide gas maintained at 70 bar, 90°C. The valve is opened with negligible energy transfer to the gas, and carbon dioxide flows into the vessel until the pressure becomes 40 bar. The valve is then closed and the gas within the vessel attains an equilibrium state. Determine the amount of entropy produced within the vessel, in kJ/K.

11.95 A 1 kmol mixture of CO_2 and C_2H_6 (ethane) occupies a volume of 0.15 m³ at a temperature of 400 K. The mole fraction of CO_2 is 0.3. Determine the mixture pressure using

(a) the ideal gas equation of state.
(b) Kay's rule together with the generalized compressibility chart.
(c) the additive pressure rule together with the generalized compressibility chart.

Compare and discuss these results.

11.96 A gaseous mixture at 460°R and 1500 lbf/in.² consists of 70% methane and 30% nitrogen (molar basis). The experimental value for the mixture compressibility factor is 0.82. Estimate the compressibility factor using the generalized compressibility chart and

(a) Kay's rule.
(b) the additive volume rule.

11.97 A gaseous mixture consists of 59.4% carbon dioxide and 40.6% methane (molar basis). The mixture is at 38°C and has a specific volume of 0.221 m³/kmol. Estimate the pressure using

(a) the ideal gas model.
(b) the Redlich–Kwong equation of state together with mixture values for the constants a and b.
(c) Kay's rule together with the generalized compressibility chart.

11.98 A gaseous mixture consisting of 1.5 lbmol of hydrogen and 0.5 lbmol of nitrogen occupies 2.71 ft³ at 77°F. Estimate the pressure using

(a) the ideal gas model.
(b) Kay's rule together with the generalized compressibility chart.
(c) the van der Waals equation together with mixture values for the constants a and b.
(d) the rule of additive pressure together with the generalized compressibility chart.

11.99 A gaseous mixture of 0.5 lbmol of methane and 0.5 lbmol of propane occupies a volume of 7.65 ft³ at a temperature of 194°F. Estimate the pressure using the following procedures and compare each estimate with the measured value of pressure, 50 atm:

(a) the ideal gas equation of state.
(b) Kay's rule together with the generalized compressibility chart.
(c) the van der Waals equation together with mixture values for the constants a and b.
(d) the rule of additive pressures together with the van der Waals equation.
(e) the rule of additive pressures together with the generalized compressibility chart.
(f) the rule of additive volumes together with the van der Waals equation.
(g) the rule of additive volumes together with the generalized compressibility chart.

11.100 One kmol of a gaseous mixture occupies a volume of 0.111 m³ at 100°C. The mixture consists of 69.5% carbon dioxide and 30.5% ethylene (C_2H_4) (molar basis). Estimate the mixture pressure using

(a) the ideal gas equation of state.
(b) Kay's rule together with the generalized compressibility chart.
(c) the additive pressure rule together with the generalized compressibility chart.
(d) the van der Waals equation together with mixture values for the constants a and b.

11.101 A gaseous mixture consisting of 50% argon and 50% nitrogen (molar basis) is contained in a closed tank at 20 atm, $-140°F$. If the total mass of mixture is 3 lb, estimate the volume in ft^3 using

(a) the ideal gas model.
(b) Kay's rule together with the generalized compressibility chart.
(c) the Redlich–Kwong equation with mixture values for a and b.

11.102 Using the Carnahan–Starling–DeSantis equation of state introduced in Problem 11.19, together with the following expressions for the mixture values of a and b:

$$a = y_1^2 a_1 + 2y_1 y_2(1 - f_{12})(a_1 a_2)^{1/2} + y_2^2 a_2$$
$$b = y_1 b_1 + y_2 b_2$$

where f_{12} is an empirical *interaction* parameter, determine the pressure, in kPa, at $v = 0.005\ m^3/kg$, $T = 180°C$ for a mixture of Refrigerants 12 and 13, in which Refrigerant 12 is 40% by mass. For a mixture of Refrigerants 12 and 13, $f_{12} = 0.035$.

11.103 A rigid vessel having a volume of 0.05 m^3 initially contains carbon dioxide gas at temperature 25°C and pressure p. Ethylene gas is allowed to flow into the tank until a mixture consisting of 20% carbon dioxide and 80% ethylene (molar basis) exists within the tank at a temperature of 25°C and a pressure of 100 bar. Determine the pressure p, in bar, using Kay's rule together with the generalized compressibility chart.

11.104 Two tanks each having a volume of 1 ft^3 are connected by a valve. One tank contains carbon dioxide gas at 75°F and pressure p. The other tank contains ethylene gas at 75°F and 1480 $lbf/in.^2$ The valve is opened and the gases mix, eventually attaining equilibrium at 75°F and pressure p' with a composition of 20% carbon dioxide and 80% ethylene (molar basis). Using Kay's rule and the generalized compressibility chart, determine in $lbf/in.^2$

(a) the initial pressure of the carbon dioxide, p.
(b) the final pressure of the mixture, p'.

11.105 A binary solution consists of 20 lb of carbon dioxide (CO_2) and 80 lb of decane ($C_{10}H_{22}$). The partial molal volume of CO_2 is 0.792 ft^3/lbmol and the partial molal volume of $C_{10}H_{22}$ is 3.128 ft^3/lbmol. Determine the total volume, in ft^3, and the specific volume of the solution, in ft^3 per lbmol of solution.

11.106 The following data are for a binary solution of ethane (C_2H_6) and pentane (C_5H_{12}) at a certain temperature and pressure:

mole fraction of ethane	0.2	0.3	0.4	0.5	0.6	0.7	0.8
volume (in m^3) per kmol of solution	0.119	0.116	0.112	0.109	0.107	0.107	0.11

Estimate

(a) the specific volumes of pure ethane and pure pentane, each in m^3/kmol.
(b) the partial molal volumes of ethane and pentane for an equimolar solution, each in m^3/kmol.

11.107 The following data are for a binary mixture of carbon dioxide and methane at a certain temperature and pressure:

mole fraction of methane	0.000	0.204	0.406	0.606	0.847	1.000
volume (in ft^3) per lbmol of mixture	1.506	3.011	3.540	3.892	4.149	4.277

Estimate

(a) the specific volumes of pure carbon dioxide and pure methane, each in ft^3/lbmol.
(b) the partial molal volumes of carbon dioxide and methane for an equimolar mixture, each in ft^3/lbmol.

11.108 Using p–v–T data from the steam tables, determine the fugacity of water as a saturated vapor at 280°C. Compare with the value obtained from the generalized fugacity chart.

11.109 Using p–v–T data from the steam tables, determine the fugacity of water as a saturated vapor at 500°F. Compare with the value obtained from the generalized fugacity chart.

11.110 Using the equation of state of Problem 11.14 (c), evaluate the fugacity of ammonia at 750 K, 100 atm and compare with the value obtained from Fig. A-6.

11.111 Using tabulated compressibility data,[6] evaluate f/p at $T_R = 1.40$ and $p_R = 2.0$. Compare with the value obtained from Fig. A-6.

11.112 Consider the truncated virial expansion

$$Z = 1 + \hat{B}(T_R)p_R + \hat{C}(T_R)p_R^2 + \hat{D}(T_R)p_R^3$$

(a) Using tabulated compressibility data,[6] evaluate the coefficients $\hat{B}$, $\hat{C}$, and $\hat{D}$ for $0 < p_R < 1.0$ and each of $T_R = 1.0, 1.2, 1.4, 1.6, 1.8, 2.0$.

[6] Ibid., pp. 587–595

(b) Obtain an expression for $\ln(f/p)$ in terms of T_R and p_R. Using the coefficients of part (a), evaluate f/p at selected states and compare with tabulated values.[7]

11.113 Derive the following approximation for the fugacity of a liquid at temperature T and pressure p:

$$f(T,p) \approx f^L_{\text{sat}}(T) \exp\left\{\frac{v_f(T)}{RT}[p - p_{\text{sat}}(T)]\right\}$$

where $f^L_{\text{sat}}(T)$ is the fugacity of the saturated liquid at temperature T. For what range of pressures might the approximation $f(T,p) \approx f^L_{\text{sat}}(T)$ apply?

11.114 Consider a one-inlet, one-exit control volume at steady state through which the flow is internally reversible and isothermal. Show that the work per unit of mass flowing can be expressed in terms of the fugacity f as

$$\left(\frac{\dot{W}_{\text{cv}}}{\dot{m}}\right)_{\substack{\text{int}\\ \text{rev}}} = -RT\ln\left(\frac{f_2}{f_1}\right) + \frac{V_1^2 - V_2^2}{2} + g(z_1 - z_2)$$

11.115 Methane is compressed isothermally without irreversibilities from 1.01 bar, 4°C to 23.21 bar. The compressor operates at steady state and kinetic and potential energy changes are negligible. Using the generalized fugacity chart, determine the work required by the compressor, in kJ per kmol of methane flowing.

11.116 Ethane (C_2H_6) is compressed isothermally without irreversibilities at a temperature of 570°R from 1 to 68 atm. Using the generalized fugacity and enthalpy departure charts, determine the work of compression and the heat transfer, each in Btu per lb of ethane flowing. Assume steady-state operation and neglect kinetic and potential energy effects.

11.117 Propane (C_3H_8) enters a turbine operating at steady state at 128 bar, 60°C and expands isothermally without irreversibilities to 12.8 bar. There are no significant changes in kinetic or potential energy. Using the generalized fugacity chart, determine the power developed, in kW, for a mass flow rate of 50 kg/min.

11.118 Methane enters a turbine operating at steady state at 98 atm, 53°F and expands isothermally without irreversibilities to 14 atm. There are no significant changes in kinetic or potential energy. Using the generalized fugacity and enthalpy departure charts, determine the power developed and heat transfer, each in Btu/s, for a mass flow rate of 110 lb/s.

[7] Ibid., pp. 605–611.

11.119 For each case determine the fugacity, in atm, at the state indicated.

(a) butane at 555 K, 150 atm.
(b) methane at 120°F, 800 lbf/in.^{2}.
(c) benzene at 890°R, 135 atm
(d) pure ethane at 310 K, 20.4 atm and as a component with a mole fraction of 0.35 in an ideal solution at the same temperature and pressure.

11.120 A tank having a volume of 1 m^3 contains a gaseous mixture of 70% ethane and 30% nitrogen (molar basis) at 311 K and 170 atm. Determine the mass of mixture in the tank, using the generalized compressibility chart together with **(a)** Kay's rule, **(b)** the ideal solution model. Compare with the experimental value for mixture mass, 312 kg.

11.121 A tank having a volume of 17.5 ft^3 contains a gaseous mixture of 75% argon and 25% ethylene (molar basis) at 77°F, 81.42 atm. Determine the mass of mixture in the tank, using the generalized compressibility chart together with

(a) Kay's rule.
(b) the ideal solution model.

11.122 A tank having a total volume of 1 m^3 contains a gaseous mixture of 70% ethane and 30% nitrogen (molar basis) at 400 K and 200 atm. Determine the mass of mixture in the tank, using

(a) the ideal gas model.
(b) the ideal solution model together with the generalized compressibility chart.
(c) Kay's rule together with the generalized compressibility chart.

11.123 A gaseous mixture consisting of 70% methane and 30% nitrogen (molar basis) enters a compressor operating at steady state at 10 atm and −53°C with a mass flow rate of 10 kg/min. The mixture exits the compressor at 100 atm, −53°C with no significant change in kinetic or potential energy. Heat transfer from the compressor is estimated to occur at the rate of 4000 kJ/min. Using the generalized enthalpy departure chart together with the ideal solution model, determine the power required by the compressor, in kW.

DESIGN AND OPEN ENDED PROBLEMS

11.1D Compressed natural gas (CNG) is under consideration as a fuel to replace gasoline for automobile engines. Aluminum cylinders wrapped in a fibrous composite have been proposed to provide lightweight, economical, and safe on-board storage. The storage vessels should hold enough CNG for 100 to 125 miles of urban travel, at storage pressures up to 3000 lbf/in.^{2}, and with a maximum total weight of 150 lb. Adhering to applicable U.S. Department of Transportation standards, specify both the size and number of cylinders that would meet the above design constraints.

11.2D Write computer programs that allow specific enthalpy and entropy to be evaluated at vapor states for one or more of the cases listed below. Compare calculated results with tabulated data. Write a brief critique comparing and contrasting the underlying *property evaluation approaches* (Sec. 11.6) of the references provided.

(a) water (J. H. Keenan, F. G. Keyes, P. G. Hill, and J. G. Moore, *Steam Tables*, Wiley, New York, 1969).
(b) Refrigerant 134a (D. P. Wilson and R. S. Basu, "Thermodynamic Properties of a New Stratospherically Safe Working Fluid–Refrigerant 134a," *ASHRAE Transactions*, Vol. 94, 1988, Part 2, pp. 2095–2118).
(c) helium (S. Angus and K. M. de Reuck, *Helium International Thermodynamic Tables of the Fluid State-4*, Pergamon Press, Oxford, 1977).

11.3D *Slush hydrogen,* a mixture of liquid and solid hydrogen, will fuel a new aerospace engine currently under development. To produce the quantities of slush required, the power cost is expected to be significant. For *normal* hydrogen at 300 K, 1 atm, estimate the work required, in kJ per kg of fluid, to produce slush continuously with a 0.5-solid mass fraction using the *auger method* described in 1970 U.S. Patent No. 3 521 457. Suggest modifications of the auger method that might lead to a reduced power requirement.

11.4D Natural gas transmission companies employ a network of pipelines to bring gas from distant production sites to users. The pipeline gas is typically at a relatively high pressure, which requires a significant pressure reduction before the gas is distributed to users. The pressure reduction is normally accomplished by a throttling process. Figure P11.4D illustrates an arrangement allowing a combination of throttling and power production that is less wasteful than simple throttling. Specify the details, including one or two turbine stages, heat exchangers, electric generator, and other components as required for the following nominal conditions: gas flow rate range 3 to 4.6 kg/s, inlet pressure range 4.2 to 4.3 MPa, outlet pressure range 0.5 to 0.6 MPa, outlet temperature ≈ 5°C. Discuss the feasibility of this arrangement.

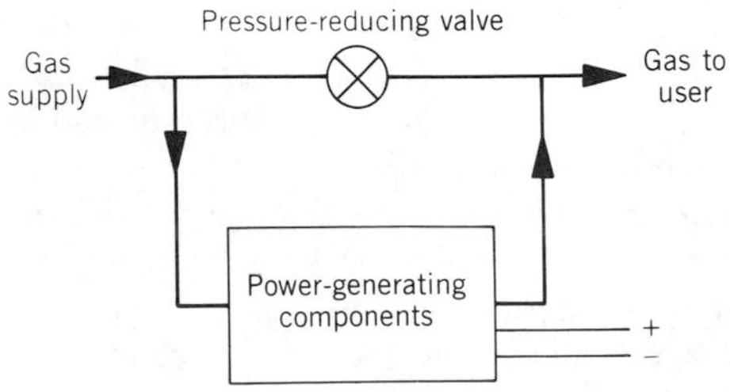

Figure P11.4D

11.5D Liquid hydrogen has been proposed as an environmentally acceptable fuel to replace petroleum-based fuels for long-range jet aircraft. Systems for hydrogen liquefaction typically involve Joule–Thomson expansions and have significant power requirements. According to one study, the power required by the liquid hydrogen facility of a typical major airport would equal the output of two 1000-MW power plants. Write a memorandum in which the pros and cons of using liquid hydrogen for this purpose are critically discussed. Include an assessment of the potential for cost-effective improvement of the processing methods and equipment currently available for liquid hydrogen production.

11.6D In the experiment for the *regelation* of ice, a small-diameter wire weighted at each end is draped over a block of ice. The loaded wire is observed to cut slowly through the ice without leaving a trace. In one such set of experiments, a weighted 1.00-mm diameter wire is reported to have passed through 0°C ice at a rate of 54 mm/h. Perform the regelation experiment and propose a plausible explanation for this phenomenon.

11.7D An inventor has proposed a new type of marine engine *fueled* by fresh water stored on board and seawater drawn in from the surrounding ocean. At steady state, the engine would develop power from freshwater and seawater streams, each entering the engine at the ambient temperature and pressure. A single mixed stream would be discharged at the ambient temperature and pressure. Critically evaluate this proposal.

11.8D The following data are known for a 100-ton ammonia–water absorption system like the one shown in Fig. 10.9. The pump is to handle 570 lb of strong solution per minute. The generator conditions are 175 lbf/in.^{2}, 220°F. The absorber is at 29 lbf/in.^{2} with strong solution exiting at 80°F. For the evaporator, the pressure is 30 lbf/in.^{2} and the exit temperature is 10°F. Specify the type and size, in horsepower, of the pump required. Justify your choices.

CHAPTER 12

NONREACTING IDEAL GAS MIXTURES AND PSYCHROMETRICS

Many systems of interest involve gaseous mixtures of two or more components. For example, the products of combustion of fossil fuels often consist of several gaseous components, including carbon dioxide and water vapor. We will need to know how to handle such mixtures when the subjects of combustion and chemical equilibrium are studied in Chaps. 13 and 14. An understanding of the behavior of mixtures of air and water vapor is also a prerequisite for considering air-conditioning processes later in this chapter.

To apply the principles of thermodynamics introduced thus far to systems involving mixtures requires that the properties of the mixtures be evaluated. Since an unlimited variety of mixtures can be formed from a given set of components simply by varying the relative amounts present, the properties of mixtures are available in tabular, graphical, or equation forms only in certain special cases such as air. Means are available for determining the properties of mixtures from the mixture composition and the properties of the individual pure components from which the mixtures are formed, however. Methods for this purpose are discussed both in Chap. 11 and in the present chapter. In the present chapter, we consider cases where the overall mixture and each mixture component, including water vapor, can be modeled as ideal gases. The methods introduced in Chap. 11 allow mixtures to be treated more generally than is possible using the ideal gas model.

■ IDEAL GAS MIXTURES: GENERAL CONSIDERATIONS

12.1 DESCRIBING MIXTURE COMPOSITION

To specify the state of a mixture requires the composition and the values of two independent intensive properties such as temperature and pressure. The object of the present section is to consider ways for describing mixture composition. In subsequent sections, we show how mixture properties other than composition can be evaluated.

Consider a closed system consisting of a gaseous mixture of two or more components. The composition of the mixture can be described by giving the *mass* or the *number of moles* of each component present. The mass, the number of moles, and the molecular weight of a component i are related by

$$n_i = \frac{m_i}{M_i} \tag{12.1}$$

where m_i is the mass, n_i is the number of moles, and M_i is the molecular weight of component i, respectively. When m_i is expressed in terms of the kilogram, n_i is in kmols. When m_i is in terms of the pound mass, n_i is in lbmols. However, any unit of mass can be used in this relationship.

The total mass of the mixture, m, is the sum of the masses of its components

$$m = m_1 + m_2 + \cdot\cdot\cdot + m_j = \sum_{i=1}^{j} m_i \tag{12.2}$$

The *relative* amounts of the components present in the mixture can be specified in terms of *mass fractions*. The mass fraction mf_i of component i is defined as

$$mf_i = \frac{m_i}{m} \tag{12.3}$$

A listing of the mass fractions of the components of a mixture is sometimes referred to as a *gravimetric* analysis.

Dividing each term of Eq. 12.2 by the total mass of mixture m and using Eq. 12.3

$$1 = \sum_{i=1}^{j} mf_i \tag{12.4}$$

That is, the sum of the mass fractions of all the components in a mixture is equal to unity.

The total number of moles in a mixture, n, is the sum of the number of moles of each of its components

$$n = n_1 + n_2 + \cdot\cdot\cdot + n_j = \sum_{i=1}^{j} n_i \tag{12.5}$$

A listing of the mole fractions of the components of a mixture may be called a *molar* analysis.

The *relative* amounts of the components present in the mixture also can be described in terms of *mole fractions*. The mole fraction y_i of component i is defined as

$$y_i = \frac{n_i}{n} \tag{12.6}$$

Dividing each term of Eq. 12.5 by the total number of moles of mixture n and using Eq. 12.6

$$1 = \sum_{i=1}^{j} y_i \tag{12.7}$$

That is, the sum of the mole fractions of all the components in a mixture is equal to unity.

The apparent (or average) molecular weight of the mixture, M, is defined as the ratio of the total mass of the mixture, m, to the total number of moles of mixture, n

$$M = \frac{m}{n} \tag{12.8}$$

Equation 12.8 can be expressed in a convenient alternative form. With Eq. 12.2, it becomes

$$M = \frac{m_1 + m_2 + \cdots + m_j}{n}$$

Introducing $m_i = n_i M_i$ from Eq. 12.1

$$M = \frac{n_1 M_1 + n_2 M_2 + \cdots + n_j M_j}{n}$$

Finally, with Eq. 12.6, the apparent molecular weight of the mixture can be calculated as a mole-fraction average of the component molecular weights

$$M = \sum_{i=1}^{j} y_i M_i \tag{12.9}$$

As an illustration of the use of Eq. 12.9, consider the case of air. A sample of *atmospheric air* contains several gaseous components including water vapor and contaminants such as dust, pollen, and pollutants. The term *dry air* refers only to the gaseous components when all water vapor and contaminants have been removed. The molar analysis of a typical sample of dry air is given in Table 12.1. Selecting molecular weights for nitrogen, oxygen, argon, and carbon dioxide from Table A-1 and neglecting the trace substances neon, helium, etc., the apparent molecular weight of dry air obtained from Eq. 12.9 is

$$\begin{aligned} M &\approx 0.7808(28.02) + 0.2095(32.00) + 0.0093(39.94) + 0.0003(44.01) \\ &= 28.97 \text{ kg/kmol} = 28.97 \text{ lb/lbmol} \end{aligned}$$

This value, which is the entry for air in Table A-1, is not altered significantly when the trace substances are also included in the calculation.

TABLE 12.1 Approximate Composition of Dry Air

Component	Mole Fraction (%)
Nitrogen	78.08
Oxygen	20.95
Argon	0.93
Carbon dioxide	0.03
Neon, helium, methane, and others	0.01

EXAMPLE 12.1

PROBLEM

The molar analysis of the gaseous products of combustion of a certain hydrocarbon fuel is CO_2, 0.08; H_2O, 0.11; O_2, 0.07; N_2, 0.74. (a) Determine the apparent molecular weight of the mixture. (b) Determine the composition in terms of mass fractions.

SOLUTION

Known: A molar analysis of the gaseous products of combustion of a hydrocarbon fuel is given.

Find: Determine (a) the apparent molecular weight of the mixture, (b) the composition in terms of mass fractions (gravimetric analysis).

Analysis:

(a) Using Eq. 12.9 and approximate molecular weights from Table A-1

$$M = 0.08(44) + 0.11(18) + 0.07(32) + 0.74(28)$$
$$= 28.46 \text{ kg/kmol} = 28.46 \text{ lb/lbmol}$$

(b) Equations 12.1, 12.3, and 12.6 are the key relations required to determine the composition in terms of mass fractions.

Although the actual amount of mixture is not known, the calculations can be based on any convenient amount. Let us base the solution on 1 kmol of mixture. Then, with Eq. 12.6 the amount n_i of each component present in kmol is numerically equal to the mole fraction, as listed in column (ii) of the accompanying table. Column (iii) of the table gives the respective molecular weights of the components.

Column (iv) of the table gives the mass m_i of each component in kg per kmol of mixture obtained with Eq. 12.1 in the form $m_i = M_i n_i$. The values of column (iv) are obtained by multiplying each value of column (ii) by the corresponding value of column (iii). The sum of the values in column (iv) is the mass of the mixture: kg of mixture per kmol of mixture. Note that this sum is just the apparent mixture molecular weight determined in part (a). Finally, using Eq. 12.3, column (v) gives the mass fractions as a percentage. The values of column (v) are obtained by dividing the values of column (iv) by the column (iv) total and multiplying by 100.

(i) Component	(ii)[a] n_i	×	(iii) M_i	=	(iv)[b] m_i	(v) mf_i (%)
CO_2	0.08	×	44	=	3.52	12.37
H_2O	0.11	×	18	=	1.98	6.96
O_2	0.07	×	32	=	2.24	7.87
N_2	0.74	×	28	=	20.72	72.80
	1.00				28.46	100.00

[a] Entries in this column have units of kmol per kmol of mixture. For example, the first entry is 0.08 kmol of CO_2 per kmol of mixture.

[b] Entries in this column have units of kg per kmol of mixture. For example, the first entry is 3.52 kg of CO_2 per kmol of mixture. The column sum, 28.46, has units of kg of mixture per kmol of mixture.

Comment: If the solution to part (b) was conducted on the basis of some other assumed amount of mixture—for example, 100 kmol or 100 lbmol—the same result for the mass fractions would be obtained, as can easily be verified.

EXAMPLE 12.2

PROBLEM

A gas mixture has the following composition in terms of mass fractions: H_2, 0.10; N_2, 0.60; CO_2, 0.30. Determine (a) the composition in terms of mole fractions and (b) the apparent molecular weight of the mixture.

SOLUTION

Known: A gravimetric analysis of a certain gas mixture is known.

Find: Determine the analysis of the mixture in terms of mole fractions (molar analysis) and the apparent molecular weight of the mixture.

Solution:

(a) Equations 12.1, 12.3, and 12.6 are the key relations required to determine the composition in terms of mole fractions.

Although the actual amount of mixture is not known, the calculation can be based on any convenient amount. Let us base the solution on 100 kg. Then, with Eq. 12.3, the amount m_i of each component present in kg is equal to the mass fraction multiplied by 100 kg. The values are listed in column (ii) of the accompanying table. Column (iii) of the table gives the respective molecular weights of the components.

Column (iv) of the table gives the amount n_i of each component in kmol per 100 kg of mixture obtained using Eq. 12.1. The values of column (iv) are obtained by dividing each value of column (ii) by the corresponding value of column (iii). The sum of the values of column (iv) is the total amount of mixture in kmols per 100 kg of mixture. Finally, using Eq. 12.6, column (v) gives the mole fractions as a percentage. The values of column (v) are obtained by dividing the values of column (iv) by the column (iv) total and multiplying by 100.

(i) Component	(ii)[a] m_i	$\div$	(iii) M_i	=	(iv)[b] n_i	(v) y_i (%)
H_2	10	$\div$	2	=	5.00	63.9
N_2	60	$\div$	28	=	2.14	27.4
CO_2	30	$\div$	44	=	0.68	8.7
	100				7.82	100.0

[a] Entries in this column have units of kg per 100 kg of mixture. For example, the first entry is 10 kg of H_2 per 100 kg of mixture.

[b] Entries in this column have units of kmol per 100 kg of mixture. For example, the first entry is 5.00 kmol of H_2 per 100 kg of mixture. The column sum, 7.82, has units of kmol of mixture per 100 kg of mixture.

(b) The apparent molecular weight of the mixture can be found by using Eq. 12.9 and the calculated mole fractions. The value can be determined alternatively by using the column (iv) total giving the total amount of mixture in kmols per 100 kg of mixture. Thus, with Eq. 12.8

$$M = \frac{m}{n} = \frac{100\text{ kg}}{7.82\text{ kmol}} = 12.79\,\frac{\text{kg}}{\text{kmol}} = 12.79\,\frac{\text{lb}}{\text{lbmol}}$$

Comment:

1. If the solution to part (a) was conducted on the basis of some other assumed amount of mixture, the same result for the mass fractions would be obtained, as can be easily verified.
2. Although H_2 has the smallest mass fraction, its mole fraction is the largest.

12.2 p–V–T RELATIONSHIPS FOR IDEAL GAS MIXTURES

The definitions given in Sec. 12.1 apply generally to mixtures. In the present section we are concerned only with *ideal gas* mixtures and introduce two models used in conjunction with this idealization: the *Dalton model* and the *Amagat model.*

Consider a system consisting of a number of gases contained within a closed vessel of volume V as shown in Fig. 12.1. The temperature of the gas mixture is T and the pressure is p. The overall mixture is considered an ideal gas, so p, V, T, and the total number of moles of mixture n are related by the ideal gas equation of state

$$p = n\frac{\bar{R}T}{V} \tag{12.10}$$

With reference to this system, let us consider in turn the Dalton and Amagat models.

Dalton Model. The Dalton model is consistent with the concept of an ideal gas as being made up of molecules that exert negligible forces on one another and whose volume is negligible relative to the volume occupied by the gas (Sec. 3.5). In the absence of significant intermolecular forces, the behavior of each component would be unaffected by the presence of the other components. Moreover, if the volume occupied by the molecules is a very small fraction of the total volume, the molecules of each gas present may be regarded as free to roam throughout the full volume. In keeping with this simple picture, the *Dalton* model premises that each mixture component behaves as an ideal gas as if it were *alone at the temperature T and volume V of the mixture*.

It follows from the Dalton model that the individual components would not exert the mixture pressure p but rather a *partial pressure*. As shown below, the sum of the partial pressures equals the mixture pressure. The partial pressure of component i, p_i, is the pressure that n_i moles of component i would exert if the component were alone in the volume V at the mixture temperature T. The partial pressure can be evaluated using the ideal gas equation of state

$$p_i = \frac{n_i\bar{R}T}{V} \tag{12.11}$$

Dividing Eq. 12.11 by Eq. 12.10

$$\frac{p_i}{p} = \frac{n_i\bar{R}T/V}{n\bar{R}T/V} = \frac{n_i}{n} = y_i$$

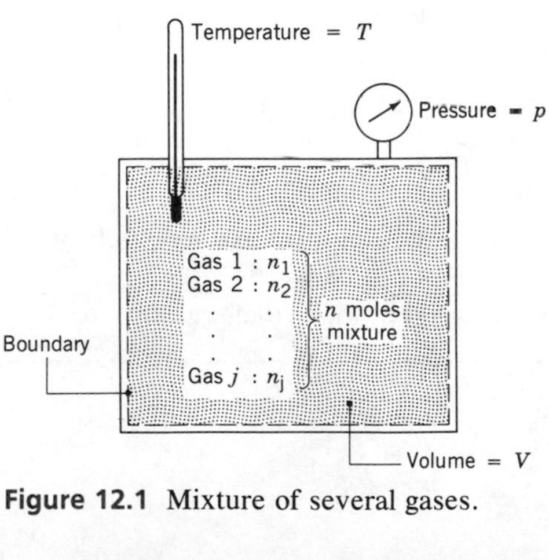

Figure 12.1 Mixture of several gases.

Thus, the partial pressure of component i also can be evaluated in terms of its mole fraction y_i and the mixture pressure p

$$p_i = y_i p \tag{12.12}$$

To show that the sum of partial pressures equals the mixture pressure, sum both sides of Eq. 12.12 to obtain

$$\sum_{i=1}^{j} p_i = \sum_{i=1}^{j} y_i p = p \sum_{i=1}^{j} y_i$$

Since the sum of the mole fractions is unity, this becomes

$$p = \sum_{i=1}^{j} p_i \tag{12.13}$$

Amagat Model. The underlying assumption of the *Amagat* model is that each mixture component behaves as an ideal gas as if it existed separately at the pressure p and temperature T of the mixture. The volume that n_i moles of component i would occupy if the component existed at p and T is called the *partial volume,* V_i, of component i. As shown below, the sum of the partial volumes equals the total volume. The partial volume can be evaluated using the ideal gas equation of state

$$V_i = \frac{n_i \bar{R} T}{p} \tag{12.14}$$

Dividing Eq. 12.14 by the total volume V

$$\frac{V_i}{V} = \frac{n_i \bar{R} T/p}{n \bar{R} T/p} = \frac{n_i}{n} = y_i$$

Thus, the partial volume of component i also can be evaluated in terms of its mole fraction y_i and the total volume

$$V_i = y_i V \tag{12.15}$$

This relationship between volume fraction and mole fraction underlies the use of the term *volumetric analysis* as signifying an analysis of a mixture in terms of mole fractions.

To show that the sum of partial volumes equals the total volume, sum both sides of Eq. 12.15 to obtain

$$\sum_{i=1}^{j} V_i = \sum_{i=1}^{j} y_i V = V \sum_{i=1}^{j} y_i$$

Since the sum of the mole fractions equals unity, this becomes

$$V = \sum_{i=1}^{j} V_i \tag{12.16}$$

The Dalton and Amagat models are special cases, respectively, of the additive pressure and additive volume rules introduced in Sec. 11.8, which do not require the assumption of the ideal gas model for implementation. The concept of an ideal gas mixture is a special case of the *ideal solution* concept introduced in Sec. 11.9.

12.3 *U*, *H*, AND *S* FOR IDEAL GAS MIXTURES

To apply the conservation of energy principle to a system involving an ideal gas mixture requires evaluation of the internal energy or the enthalpy of the mixture at various states. Similarly, to conduct an analysis using the second law normally requires the entropy of the mixture. The objective of the present section is to develop means to evaluate internal energy, enthalpy, and entropy for ideal gas mixtures.

Consider a closed system consisting of an ideal gas mixture. Extensive properties of the mixture, such as U, H, or S, can be found by adding the contribution of each component *at the condition at which the component exists in the mixture*. Let us apply this idea, first to internal energy and enthalpy and then to entropy.

Since the internal energy and enthalpy of ideal gases are functions of temperature only, the values of these properties for each component present in the mixture are determined by the mixture temperature alone. Accordingly,

$$U = U_1 + U_2 + \cdots + U_j = \sum_{i=1}^{j} U_i \tag{12.17}$$

$$H = H_1 + H_2 + \cdots + H_j = \sum_{i=1}^{j} H_i \tag{12.18}$$

where U_i and H_i are the internal energy and enthalpy, respectively, of component i evaluated at the mixture temperature.

Equations 12.17 and 12.18 can be rewritten on a molar basis as

$$n\bar{u} = n_1\bar{u}_1 + n_2\bar{u}_2 + \cdots + n_j\bar{u}_j = \sum_{i=1}^{j} n_i\bar{u}_i \tag{12.19}$$

and

$$n\bar{h} = n_1\bar{h}_1 + n_2\bar{h}_2 + \cdots + n_j\bar{h}_j = \sum_{i=1}^{j} n_i\bar{h}_i \tag{12.20}$$

where $\bar{u}$ and $\bar{h}$ are the specific internal energy and enthalpy of the *mixture* per mole of mixture, and $\bar{u}_i$ and $\bar{h}_i$ are the specific internal energy and enthalpy of *component i* per mole of i. Dividing by the total number of moles of mixture n gives expressions for the specific internal energy and enthalpy of the mixture per mole of mixture, respectively

$$\bar{u} = \sum_{i=1}^{j} y_i\bar{u}_i \tag{12.21}$$

$$\bar{h} = \sum_{i=1}^{j} y_i\bar{h}_i \tag{12.22}$$

Each of the molar internal energy and enthalpy terms appearing in Eqs. 12.19 through 12.22 is evaluated at the mixture temperature only.

Differentiation of Eqs. 12.21 and 12.22 with respect to temperature results, respectively, in the following expressions for the specific heats $\bar{c}_v$ and $\bar{c}_p$ of the mixture on a molar basis:

$$\bar{c}_v = \sum_{i=1}^{j} y_i\bar{c}_{v,i} \tag{12.23}$$

$$\bar{c}_p = \sum_{i=1}^{j} y_i \bar{c}_{p,i} \tag{12.24}$$

That is, the mixture specific heats $\bar{c}_p$ and $\bar{c}_v$ are mole-fraction averages of the respective component specific heats. The specific heat ratio for the mixture is $k = \bar{c}_p/\bar{c}_v$.

The entropy of a mixture can be found, as for U and H, by adding the contribution of each component at the condition at which the component exists in the mixture. The entropy of an ideal gas depends on two properties, not on temperature alone as for internal energy and enthalpy. Accordingly, for the mixture

$$S = S_1 + S_2 + \cdots + S_j = \sum_{i=1}^{j} S_i \tag{12.25}$$

where S_i is the entropy of component i evaluated at the mixture temperature T and partial pressure p_i (or at temperature T and total volume V).

Equation 12.25 can be written on a molar basis as

$$n\bar{s} = n_1\bar{s}_1 + n_2\bar{s}_2 + \cdots + n_j\bar{s}_j = \sum_{i=1}^{j} n_i\bar{s}_i \tag{12.26}$$

where $\bar{s}$ is the entropy of the *mixture* per mole of mixture and $\bar{s}_i$ is the entropy of *component i* per mole of i. Dividing by the total number of moles of mixture gives an expression for the entropy of the mixture per mole of mixture

$$\bar{s} = \sum_{i=1}^{j} y_i\bar{s}_i \tag{12.27}$$

The specific entropies $\bar{s}_i$ of Eqs. 12.26 and 12.27 are evaluated at the mixture temperature T and the partial pressure p_i.

In cases where it is convenient to work on a mass basis, the foregoing expressions would be written with the mass of the mixture, m, and the mass of component i in the mixture, m_i, replacing, respectively, the number of moles of mixture, n, and the number of moles of component i, n_i. Similarly, the mass fraction of component i, mf_i, would replace the mole fraction, y_i. All specific internal energies, enthalpies, and entropies would be evaluated on a unit mass basis rather than on a per mole basis as above. The development of the appropriate expressions is left as an exercise. By using the molecular weight of the mixture or of component i, as appropriate, data can be converted from a mass basis to a molar basis, or conversely, with relations of the form

$$\bar{u} = Mu, \quad \bar{h} = Mh, \quad \bar{c}_p = Mc_p, \quad \bar{c}_v = Mc_v, \quad \bar{s} = Ms \tag{12.28}$$

for the mixture, and

$$\bar{u}_i = M_iu_i, \quad \bar{h}_i = M_ih_i, \quad \bar{c}_{p,i} = M_ic_{p,i}, \quad \bar{c}_{v,i} = M_ic_{v,i}, \quad \bar{s}_i = M_is_i \tag{12.29}$$

for component i.

To perform thermodynamic analyses of systems involving nonreacting ideal gas mixtures requires no new fundamental principles. The conservation of mass and energy principles and the second law of thermodynamics are applicable in the forms previously introduced. The only new aspect is the proper evaluation of the required property data for the mixtures involved. This is illustrated in the next two sections, which deal with two classes of problems involving mixtures. In Sec. 12.4 the mixture is already formed, and we study processes in which there is no change in composition. Section 12.5 considers the formation of mixtures from individual components that are initially separate.

12.4 MIXTURE PROCESSES AT CONSTANT COMPOSITION

In the present section we are concerned with the case of ideal gas mixtures that undergo processes during which the composition remains constant. The number of moles of each component present and thus the total number of moles of mixture remain the same throughout the process. This case is shown schematically in Fig. 12.2, which is labeled with expressions for U, H, and S of a mixture at the initial and final states of a process undergone by the mixture. In accordance with the discussion of Sec. 12.3, the specific internal energies and enthalpies of the components are evaluated at the temperature of the mixture. The specific entropies of the components are evaluated at the mixture temperature and the partial pressure of the component in the mixture.

The changes in the internal energy and enthalpy of the mixture during the process are given, respectively, by

$$U_2 - U_1 = \sum_{i=1}^{j} n_i[\bar{u}_i(T_2) - \bar{u}_i(T_1)] \tag{12.30}$$

$$H_2 - H_1 = \sum_{i=1}^{j} n_i[\bar{h}_i(T_2) - \bar{h}_i(T_1)] \tag{12.31}$$

where T_1 and T_2 denote the temperature at the initial and final states. Dividing by the number of moles of mixture, n, expressions for the change in internal energy and enthalpy of the mixture per mole of mixture result

$$\Delta\bar{u} = \sum_{i=1}^{j} y_i[\bar{u}_i(T_2) - \bar{u}_i(T_1)] \tag{12.32}$$

$$\Delta\bar{h} = \sum_{i=1}^{j} y_i[\bar{h}_i(T_2) - \bar{h}_i(T_1)] \tag{12.33}$$

Similarly, the change in entropy for the mixture is

$$S_2 - S_1 = \sum_{i=1}^{j} n_i[\bar{s}_i(T_2, p_{i2}) - \bar{s}_i(T_1, p_{i1})] \tag{12.34}$$

$$U_1 = \sum_{i=1}^{j} n_i\bar{u}_i(T_1) \qquad U_2 = \sum_{i=1}^{j} n_i\bar{u}_i(T_2)$$

$$H_1 = \sum_{i=1}^{j} n_i\bar{h}_i(T_1) \qquad H_2 = \sum_{i=1}^{j} n_i\bar{h}_i(T_2)$$

$$S_1 = \sum_{i=1}^{j} n_i\bar{s}_i(T_1, p_{i1}) \qquad S_2 = \sum_{i=1}^{j} n_i\bar{s}_i(T_2, p_{i2})$$

Figure 12.2 Process of an ideal gas mixture.

where p_{i1} and p_{i2} denote, respectively, the initial and final partial pressures of component i. Dividing by the total moles of mixture, Eq. 12.34 becomes

$$\Delta \bar{s} = \sum_{i=1}^{j} y_i[\bar{s}_i(T_2, p_{i2}) - \bar{s}_i(T_1, p_{i1})] \tag{12.35}$$

Companion expressions for Eqs. 12.30 through 12.35 on a mass basis also can be written. This is left as an exercise.

The foregoing expressions giving the changes in internal energy, enthalpy, and entropy of the mixture are written in terms of the respective property changes of the components. Accordingly, different datums might be used to assign specific enthalpy values to the various components because the datums would cancel when the component enthalpy changes are calculated. Similar remarks apply to the cases of internal energy and entropy.

Ideal Gas Tables. For several common gases modeled as ideal gases, the quantities $\bar{u}_i$ and $\bar{h}_i$ appearing in the foregoing expressions can be evaluated as functions of temperature only from Tables A-16 through A-22. Tables A-17 through A-22 give these quantities on a *molar* basis. Tables A-16 for air give them on a *mass* basis. The ideal gas tables also can be used to evaluate the entropy change. Thus, the change in specific entropy of component i required by Eqs. 12.34 and 12.35 can be determined with Eq. 6.25b as

$$\Delta \bar{s}_i = \bar{s}_i^\circ(T_2) - \bar{s}_i^\circ(T_1) - \bar{R} \ln \frac{p_{i2}}{p_{i1}}$$

Since the mixture composition remains constant, the ratio of the partial pressures in this expression is the same as the ratio of the mixture pressures, as can be shown by using Eq. 12.12 to write

$$\frac{p_{i2}}{p_{i1}} = \frac{y_i p_2}{y_i p_1} = \frac{p_2}{p_1}$$

Accordingly, when the composition is constant, the change in the specific entropy of component i is simply

$$\Delta \bar{s}_i = \bar{s}_i^\circ(T_2) - \bar{s}_i^\circ(T_1) - \bar{R} \ln \frac{p_2}{p_1} \tag{12.36}$$

where p_1 and p_2 denote, respectively, the initial and final *mixture* pressures. The terms $\bar{s}_i^\circ$ of Eq. 12.36 can be obtained as functions of temperature for several common gases from Tables A-17 through A-22. Tables A-16 for air give s° versus temperature.

Constant Specific Heat Evaluations. When the component specific heats $\bar{c}_{v,i}$ and $\bar{c}_{p,i}$ are taken as constants, the specific internal energy, enthalpy, and entropy changes of the mixture and the components of the mixture are given by

$$\Delta \bar{u} = \bar{c}_v(T_2 - T_1), \qquad \Delta \bar{u}_i = \bar{c}_{v,i}(T_2 - T_1) \tag{12.37}$$

$$\Delta \bar{h} = \bar{c}_p(T_2 - T_1), \qquad \Delta \bar{h}_i = \bar{c}_{p,i}(T_2 - T_1) \tag{12.38}$$

$$\Delta \bar{s} = \bar{c}_p \ln \frac{T_2}{T_1} - \bar{R} \ln \frac{p_2}{p_1}, \qquad \Delta \bar{s}_i = \bar{c}_{p,i} \ln \frac{T_2}{T_1} - \bar{R} \ln \frac{p_2}{p_1} \tag{12.39}$$

where the mixture specific heats $\bar{c}_v$ and $\bar{c}_p$ are evaluated from Eqs. 12.23 and 12.24, respectively. The expression for $\Delta \bar{u}$ can be obtained formally by substituting the above expression for $\Delta \bar{u}_i$ into Eq. 12.32 and using Eq. 12.23 to simplify the result.

Similarly, the expressions for $\Delta\bar{h}$ and $\Delta\bar{s}$ can be obtained by inserting $\Delta\bar{h}_i$ and $\Delta\bar{s}_i$ into Eqs. 12.33 and 12.35, respectively, and using Eq. 12.24 to simplify. In the equations for entropy change, the ratio of mixture pressures replaces the ratio of partial pressures as discussed above. Similar expressions can be written for the mixture specific internal energy, enthalpy, and entropy changes on a mass basis. This is left as an exercise.

The examples to follow illustrate the use of the relations developed in this section.

EXAMPLE 12.3

PROBLEM

A mixture of 0.3 lb of carbon dioxide and 0.2 lb of nitrogen is compressed from $p_1 = 1$ atm, $T_1 = 540°R$ to $p_2 = 3$ atm in a polytropic process for which $n = 1.25$. Determine (a) the final temperature, in °R, (b) the work, in Btu, (c) the heat transfer, in Btu, (d) the change in entropy of the mixture, in Btu/°R.

SOLUTION

Known: A mixture of 0.3 lb of CO_2 and 0.2 lb of N_2 is compressed in a polytropic process for which $n = 1.25$. At the initial state, $p_1 = 1$ atm, $T_1 = 540°R$. At the final state, $p_2 = 3$ atm.

Find: Determine the final temperature, in °R, the work, in Btu, the heat transfer, in Btu, and the entropy change of the mixture in, Btu/°R.

Schematic and Given Data:

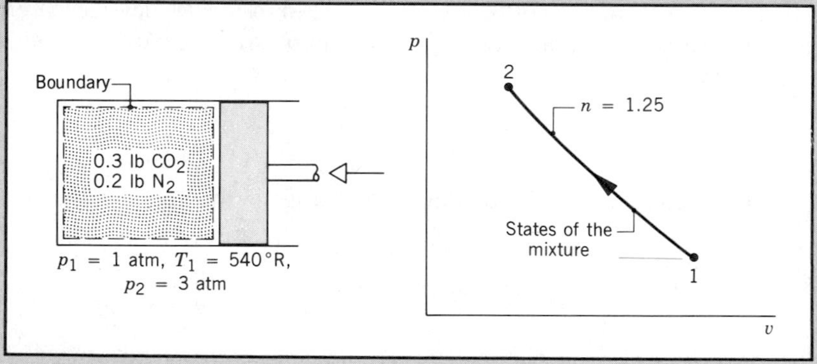

Figure E12.3

Assumptions:

1. As shown in the accompanying figure, the system is the mixture of CO_2 and N_2. The mixture composition remains constant during the compression.
2. Each mixture component behaves as if it were an ideal gas occupying the entire system volume at the mixture temperature. The overall mixture acts as an ideal gas.
3. The compression process is a polytropic process for which $n = 1.25$.
4. The changes in kinetic and potential energy between the initial and final states can be ignored.

Analysis:

(a) For an ideal gas, the temperatures and pressures at the end states of a polytropic process are related by

$$T_2 = T_1 \left(\frac{p_2}{p_1}\right)^{(n-1)/n}$$

Inserting values

$$T_2 = 540 \left(\frac{3}{1}\right)^{0.2} = 673°\text{R}$$

(b) The work for the compression process is given by

$$W = \int_1^2 p\, dV$$

Introducing pV^n = constant and performing the integration

$$W = \frac{p_2V_2 - p_1V_1}{1 - n}$$

With the ideal gas equation of state, this reduces to

$$W = \frac{m(\bar{R}/M)(T_2 - T_1)}{1 - n}$$

The mass of the mixture is $m = 0.3 + 0.2 = 0.5$ lb. The apparent molecular weight of the mixture can be calculated using $M = m/n$, where n is the total number of moles of mixture. With Eq. 12.1, the number of moles of CO_2 and N_2 are, respectively,

$$n_{CO_2} = \frac{0.3}{44} = 0.0068 \text{ lbmol}, \qquad n_{N_2} = \frac{0.2}{28} = 0.0071 \text{ lbmol}$$

The total number of moles of mixture is then $n = 0.0139$ lbmol. The apparent molecular weight of the mixture is $M = 0.5/0.0139 = 35.97$.

Calculating the work

$$W = \frac{(0.5 \text{ lb})\left(\frac{1545}{35.97}\frac{\text{ft}\cdot\text{lbf}}{\text{lb}\cdot°\text{R}}\right)(673°\text{R} - 540°\text{R})}{1 - 1.25}\left(\frac{1 \text{ Btu}}{778 \text{ ft}\cdot\text{lbf}}\right)$$
$$= -14.69 \text{ Btu}$$

where the minus sign indicates that work is done on the mixture in the process.

(c) With assumption 4, the closed system energy balance can be placed in the form

$$Q = \Delta U + W$$

where ΔU is the change in internal energy of the mixture.

The change in internal energy of the mixture equals the sum of the internal energy changes of the components. With Eq. 12.30

$$\Delta U = n_{CO_2}[\bar{u}_{CO_2}(T_2) - \bar{u}_{CO_2}(T_1)] + n_{N_2}[\bar{u}_{N_2}(T_2) - \bar{u}_{N_2}(T_1)]$$

This form is convenient because Tables A-17E and 21E give internal energy values for N_2 and CO_2, respectively, on a molar basis. With values from these tables

$$\Delta U = (0.0068)(3954 - 2984) + (0.0071)(3340 - 2678)$$
$$= 11.3 \text{ Btu}$$

Inserting values for ΔU and W into the expression for Q

$$Q = +11.3 - 14.69 = -3.39 \text{ Btu}$$

where the minus sign signifies a heat transfer from the system.

(d) The change in entropy of the mixture equals the sum of the entropy changes of the components. With Eq. 12.34

$$\Delta S = n_{CO_2}\, \Delta\bar{s}_{CO_2} + n_{N_2}\, \Delta\bar{s}_{N_2}$$

where $\Delta\bar{s}_{N_2}$ and $\Delta\bar{s}_{CO_2}$ are evaluated with Eq. 12.36 and values of $\bar{s}^\circ$ from Tables A-17E and 21E, respectively. Hence

$$\Delta S = 0.0068\left(53.123 - 51.082 - 1.986 \ln \frac{3}{1}\right)$$
$$+ 0.0071\left(47.313 - 45.781 - 1.986 \ln \frac{3}{1}\right)$$
$$= -0.0056 \text{ Btu/°R}$$

Entropy decreases in the process because entropy is transferred from the system accompanying heat transfer.

Comments:

1. Since the composition remains constant, the ratio of mixture pressures equals the ratio of partial pressures, so Eq. 12.36 can be used to evaluate the component specific entropy changes required in part (d).
2. In view of the relatively small temperature change, the changes in the internal energy and entropy of the mixture can be evaluated using the constant specific heat relations, Eqs. 12.37 and 12.39, respectively. In these equations, $\bar{c}_v$ and $\bar{c}_p$ are specific heats for the mixture determined by using Eqs. 12.23 and 12.24 together with appropriate specific heat values for the components chosen from Table A-14E.

EXAMPLE 12.4

PROBLEM

A gas mixture consisting of CO_2 and O_2 with mole fractions 0.8 and 0.2, respectively, expands isentropically and at steady state through a nozzle from 700 K, 5 atm, 3 m/s to an exit pressure of 1 atm. Determine (a) the temperature at the nozzle exit, in K, (b) the entropy changes of the CO_2 and O_2 from inlet to exit, in kJ/kmol·K, (c) the exit velocity, in m/s.

SOLUTION

Known: A gas mixture consisting of CO_2 and O_2 in specified proportions expands isentropically through a nozzle from specified inlet conditions to a given exit pressure.

Find: Determine the temperature at the nozzle exit, in K, the entropy changes of the CO_2 and O_2 from inlet to exit, in kJ/kmol·K, and the exit velocity, in m/s.

Schematic and Given Data:

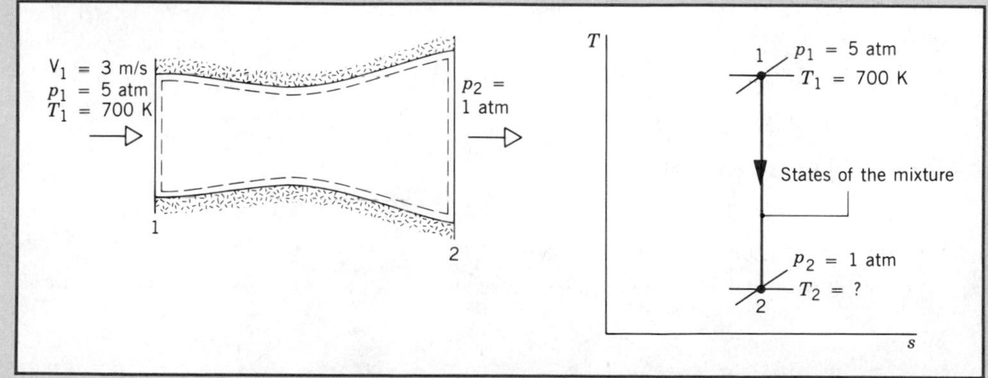

Figure E12.4

Assumptions:

1. The control volume shown by the dashed line on the accompanying figure operates at steady state.
2. The mixture composition remains constant as the mixture expands isentropically through the nozzle. The overall mixture and each mixture component act as ideal gases. The state of each component is defined by the temperature and the partial pressure of the component.
3. The change in potential energy between inlet and exit can be ignored.

Analysis:

(a) The temperature at the exit can be determined with an iterative procedure using the fact that the expansion occurs isentropically: $\bar{s}_2 - \bar{s}_1 = 0$. As there is no change in the specific entropy *of the mixture* between inlet and exit, Eq. 12.35 can be used to write

$$\bar{s}_2 - \bar{s}_1 = y_{O_2}\,\Delta\bar{s}_{O_2} + y_{CO_2}\,\Delta\bar{s}_{CO_2} = 0$$

The change in specific entropy of each component can be determined using Eq. 12.36 and values of $\bar{s}^\circ$ from Tables A-18 and A-21, as appropriate.

Evaluating $\Delta\bar{s}_{O_2}$ and $\Delta\bar{s}_{CO_2}$ with Eq. 12.36, the foregoing equation becomes

$$y_{O_2}\left[\bar{s}^\circ_{O_2}(T_2) - \bar{s}^\circ_{O_2}(T_1) - \bar{R}\ln\frac{p_2}{p_1}\right] + y_{CO_2}\left[\bar{s}^\circ_{CO_2}(T_2) - \bar{s}^\circ_{CO_2}(T_1) - \bar{R}\ln\frac{p_2}{p_1}\right] = 0$$

On rearrangement

$$y_{O_2}\bar{s}^\circ_{O_2}(T_2) + y_{CO_2}\bar{s}^\circ_{CO_2}(T_2) = y_{O_2}\bar{s}^\circ_{O_2}(T_1) + y_{CO_2}\bar{s}^\circ_{CO_2}(T_1) + (y_{O_2} + y_{CO_2})\bar{R}\ln\frac{p_2}{p_1}$$

The sum of mole fractions equals unity, so the coefficient of the last term on the right side is $(y_{O_2} + y_{CO_2}) = 1$.

Introducing given data and values for $\bar{s}^\circ$ at $T_1 = 700$ K from Tables A-18 and A-21, respectively

$$0.2\bar{s}^\circ_{O_2}(T_2) + 0.8\bar{s}^\circ_{CO_2}(T_2) = 0.2(231.358) + 0.8(250.663) + 8.314\ln\frac{1}{5}$$

or

$$0.2\bar{s}^\circ_{O_2}(T_2) + 0.8\bar{s}^\circ_{CO_2}(T_2) = 233.42\ \text{kJ/kmol}\cdot\text{K}$$

To determine the temperature T_2 requires an iterative approach with the above equation: A final temperature T_2 is assumed, and the $\bar{s}^\circ$ values for O_2 and CO_2 are found from the respective ideal gas tables. If these two values do not satisfy the equation, another temperature is assumed. The procedure continues until the desired agreement is attained. In the present case

$$\text{at } T = 510\ \text{K:}\qquad 0.2(221.206) + 0.8(235.700) = 232.80$$

$$\text{at } T = 520\ \text{K:}\qquad 0.2(221.812) + 0.8(236.575) = 233.62$$

Linear interpolation between these values gives $T_2 = 517.6$ K.

(b) The change in the specific entropy for each of the components can be determined using Eq. 12.36. For O_2

$$\Delta\bar{s}_{O_2} = \bar{s}^\circ_{O_2}(T_2) - \bar{s}^\circ_{O_2}(T_1) - \bar{R}\ln\frac{p_2}{p_1}$$

Inserting $\bar{s}^\circ$ values from Table A-18

$$\Delta\bar{s}_{O_2} = 221.667 - 231.358 - 8.314\ln(0.2) = 3.69\ \text{kJ/kmol}\cdot\text{K}$$

Similarly, for CO_2

$$\Delta\bar{s}_{CO_2} = \bar{s}^\circ_{CO_2}(T_2) - \bar{s}^\circ_{CO_2}(T_1) - \bar{R}\ln\frac{p_2}{p_1}$$

$$= 236.365 - 250.663 - 8.314\ln(0.2)$$

$$= -0.92\ \text{kJ/kmol}\cdot\text{K}$$

(c) Reducing the energy rate balance for the one-inlet, one-exit control volume at steady state

$$0 = h_1 - h_2 + \frac{V_1^2 - V_2^2}{2}$$

where h_1 and h_2 are the enthalpy *of the mixture,* per unit mass of mixture, at the inlet and exit, respectively. Solving for V_2

$$V_2 = \sqrt{V_1^2 + 2(h_1 - h_2)}$$

The term $(h_1 - h_2)$ in the expression for V_2 can be evaluated as

$$h_1 - h_2 = \frac{\bar{h}_1 - \bar{h}_2}{M} = \frac{1}{M}[y_{O_2}(\bar{h}_1 - \bar{h}_2)_{O_2} + y_{CO_2}(\bar{h}_1 - \bar{h}_2)_{CO_2}]$$

where M is the apparent molecular weight of mixture, and the molar specific enthalpies are from Tables A-18 and A-21 as appropriate.

With Eq. 12.9, the apparent molecular weight of the mixture is

$$M = 0.8(44) + 0.2(32) = 41.6 \text{ kg/kmol}$$

Then, with enthalpy values at $T_1 = 700$ K and $T_2 = 517.6$ K from Tables A-18 and A-21

$$h_1 - h_2 = \frac{1}{41.6}[0.2(21{,}184 - 15{,}320) + 0.8(27{,}125 - 18{,}468)]$$
$$= 194.7 \text{ kJ/kg}$$

Finally,

$$V_2 = \sqrt{\left(3\frac{\text{m}}{\text{s}}\right)^2 + 2\left(194.7\frac{\text{kJ}}{\text{kg}}\right)\left(\frac{1 \text{ kg}\cdot\text{m/s}^2}{1 \text{ N}}\right)\left(\frac{10^3 \text{ N}\cdot\text{m}}{1 \text{ kJ}}\right)} = 624 \text{ m/s}$$

Comments:

1. Note the use of unit conversion factors in the calculation of V_2.
2. Inspection of Table A-14 shows that the specific heats of CO_2 and O_2 increase slightly with temperature over the interval from 518 to 700 K. Accordingly, suitable constant values for c_p for the components and the overall mixture can be readily determined. The changes in specific entropy of part (b) can then be found using Eq. 12.39. The change in specific enthalpy of part (c) can be determined with Eq. 12.38. The solution of parts (b) and (c) using constant c_p is left as an exercise.
3. Each component experiences an entropy change as it passes from inlet to exit. The increase in entropy of the oxygen and the decrease in entropy of the carbon dioxide are due to an entropy transfer accompanying heat transfer from the CO_2 to the O_2 as they expand through the nozzle. This occurs reversibly, so the total entropy change is zero.

12.5 MIXING OF IDEAL GASES

Thus far, we have considered only mixtures that have already been formed. Now let us take up cases where ideal gas mixtures are formed by mixing gases that initially exist separately. Like all spontaneous processes, the mixing of different gases is an irreversible process. Mixing is a source of irreversibility because the mixture forms spontaneously. An energy input from the surroundings would be required to separate the gases and return them to their respective initial states.

In this section the irreversibility of mixing is demonstrated through calculations of the entropy production. We also consider how to determine the properties of a mixture that results from mixing components whose properties are known. Since no new principles are required, these objectives are met through consideration of examples.

EXAMPLE 12.5

PROBLEM

Consider a system consisting initially of n_A moles of gas A at pressure p and temperature T and n_B moles of gas B separate from gas A but at the same pressure and temperature. The gases are allowed to mix with no heat or work interactions with the surroundings. The final equilibrium pressure and temperature are p and T, respectively, and the mixing occurs with no change in total volume. Evaluate the amount of entropy produced in the process, assuming ideal gases.

SOLUTION

Known: Two gases, A and B, that are initially separate at temperature T and pressure p are allowed to mix with no change in total volume to form a mixture at T, p. The amounts of the gases are specified, and no interactions with the surroundings occur during the process.

Find: Evaluate the amount of entropy produced in the process.

Schematic and Given Data:

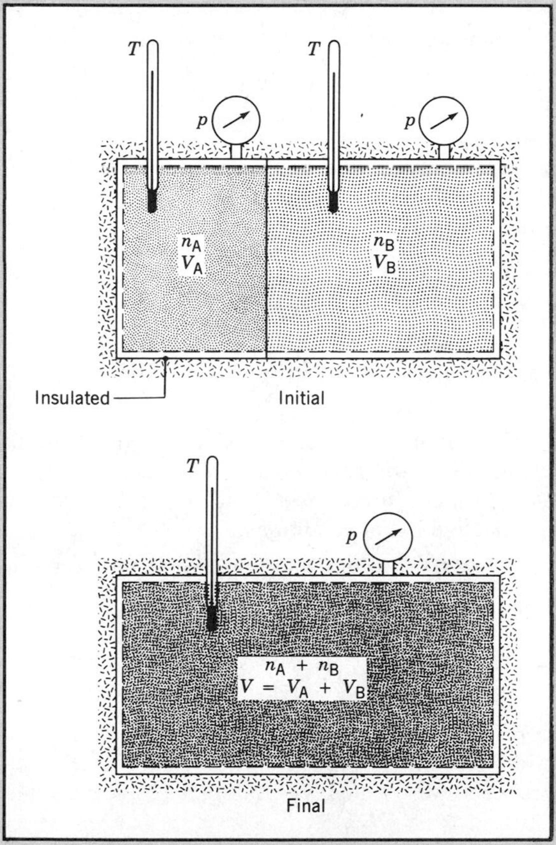

Figure E12.5

Assumptions:

1. The system consists of gases A and B together.
2. Each gas behaves as an ideal gas. The final mixture also acts as an ideal gas with each mixture component occupying the total volume and exhibiting the mixture temperature.
3. No heat or work interactions with the surroundings occur.

Analysis: An entropy balance for the closed system reduces to

$$S_2 - S_1 = \int_1^2 \left(\frac{\delta Q}{T}\right)_b^{\cancel{0}} + \sigma$$

where the entropy transfer term drops out in the adiabatic process. S_1 and S_2 denote the initial and final entropies of the system, respectively.

The initial entropy of the system is the sum of the entropies of the gases *when separate*. That is

$$S_1 = n_A \bar{s}_A(T, p) + n_B \bar{s}_B(T, p)$$

Since each gas initially exists separately at T and p, each of the specific entropies appearing in this expression is evaluated at T, p.

The entropy of the system at the final state equals the sum of the entropies of the gases A and B evaluated at the conditions at which they exist in the mixture. That is

$$S_2 = n_A \bar{s}_A(T, y_A p) + n_B \bar{s}_B(T, y_B p)$$

In this expression, the specific entropy of each gas is evaluated at its partial pressure in the mixture and at the mixture temperature.

Combining the last three equations and solving for the entropy production

$$\sigma = n_A[\bar{s}_A(T, y_A p) - \bar{s}_A(T, p)] + n_B[\bar{s}_B(T, y_B p) - \bar{s}_B(T, p)]$$

The specific entropy changes can be evaluated using Eq. 6.25b. Since there is no change in temperature

$$\bar{s}_A(T, y_A p) - \bar{s}_A(T, p) = -\bar{R} \ln \frac{y_A p}{p} = -\bar{R} \ln y_A$$

$$\bar{s}_B(T, y_B p) - \bar{s}_B(T, p) = -\bar{R} \ln \frac{y_B p}{p} = -\bar{R} \ln y_B$$

Introducing the foregoing into the entropy production expression gives

$$\sigma = -\bar{R}(n_A \ln y_A + n_B \ln y_B)$$

Comments:

1. Since the mole fractions y_A and y_B are each less than unity, the value of the entropy production would be positive, in accordance with the second law.
2. The irreversibility in this case is due to the *free expansion* each individual gas would undergo from its initial volume, V_A or V_B, and initial pressure p to the total volume V and its partial pressure in the mixture, p_A or p_B. The mixture would be formed spontaneously; however, once formed, energy would be required from the surroundings to separate the gases and return them to their respective initial conditions.
3. For the case of mixing several components, each initially at the same temperature and pressure, the expression for the entropy production takes the form

$$\sigma = -\bar{R} \sum_{i=1}^{j} n_i \ln y_i$$

where n_i and y_i are the number of moles and the mole fraction, respectively, of component i. The result obtained in the solution is a special case of this equation. These expressions for entropy production apply only if the gases are distinguishable from one another. No entropy would be produced were samples of the *same* gas at the *same* pressure and temperature allowed to mix adiabatically.

EXAMPLE 12.6

PROBLEM

Two rigid, insulated tanks are interconnected by a valve. Initially 0.79 lbmol of nitrogen at 2 atm and 460°R fills one tank. The other tank contains 0.21 lbmol of oxygen at 1 atm and 540°R. The valve is opened and the gases are allowed to mix until a final equilibrium state is attained. During this process, there are no heat or work interactions between the tank contents and the surroundings. Determine (a) the final temperature of the mixture, in °R, (b) the final pressure of the mixture, in atm, (c) the amount of entropy produced in the mixing process, in Btu/°R.

SOLUTION

Known: Nitrogen and oxygen, initially separate at different temperatures and pressures, are allowed to mix without heat or work interactions with the surroundings until a final equilibrium state is attained.

Find: Determine the final temperature of the mixture, in °R, the final pressure of the mixture, in atm, and the amount of entropy produced in the mixing process, in Btu/°R.

Schematic and Given Data:

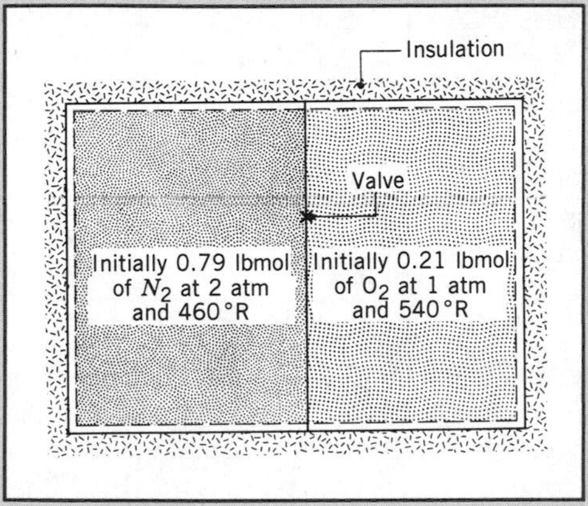

Figure E12.6

Assumptions:

1. The system is taken to be the nitrogen and the oxygen together.
2. When separate, each of the gases behaves as an ideal gas. The final mixture also acts as an ideal gas. Each mixture component occupies the total volume and exhibits the mixture temperature.
3. No heat or work interactions occur with the surroundings, and there are no changes in kinetic and potential energy.

Analysis:

(a) The final temperature of the mixture can be determined from an energy balance. With assumption 3, the closed system energy balance reduces to

$$\Delta U = \cancelto{0}{Q} - \cancelto{0}{W} \quad \text{or} \quad U_2 - U_1 = 0$$

The initial internal energy of the system, U_1, equals the sum of the internal energies of the two gases when separate

$$U_1 = n_{N_2}\bar{u}_{N_2}(T_{N_2}) + n_{O_2}\bar{u}_{O_2}(T_{O_2})$$

where $T_{N_2} = 460°R$ is the initial temperature of the nitrogen and $T_{O_2} = 540°R$ is the initial temperature of the oxygen. The final internal energy of the system, U_2, equals the sum of the internal energies of the two gases evaluated at the final mixture temperature T_2

$$U_2 = n_{N_2}\bar{u}_{N_2}(T_2) + n_{O_2}\bar{u}_{O_2}(T_2)$$

Collecting the last three equations

$$n_{N_2}[\bar{u}_{N_2}(T_2) - \bar{u}_{N_2}(T_{N_2})] + n_{O_2}[\bar{u}_{O_2}(T_2) - \bar{u}_{O_2}(T_{O_2})] = 0$$

The temperature T_2 can be determined using specific internal energy data from Tables A-17 and A-18E and an iterative procedure like that employed in part (a) of Example 12.4. However, since the specific heats of N_2 and O_2 vary little over the temperature interval from 460 to 540°R, the solution can be conducted accurately on the basis of constant specific heats. Hence, the foregoing equation becomes

$$n_{N_2}\bar{c}_{v,N_2}(T_2 - T_{N_2}) + n_{O_2}\bar{c}_{v,O_2}(T_2 - T_{O_2}) = 0$$

Solving for T_2

$$T_2 = \frac{n_{N_2}\bar{c}_{v,N_2}T_{N_2} + n_{O_2}\bar{c}_{v,O_2}T_{O_2}}{n_{N_2}\bar{c}_{v,N_2} + n_{O_2}\bar{c}_{v,O_2}}$$

Selecting c_v values for N_2 and O_2 from Table A-14E at the average of the initial temperatures of the gases, 500°R, and using the respective molecular weights to convert to a molar basis

$$\bar{c}_{v,N_2} = \left(28.02\,\frac{\text{lb}}{\text{lbmol}}\right)\left(0.177\,\frac{\text{Btu}}{\text{lb}\cdot{}^\circ\text{R}}\right) = 4.96\,\frac{\text{Btu}}{\text{lbmol}\cdot{}^\circ\text{R}}$$

$$\bar{c}_{v,O_2} = \left(32.0\,\frac{\text{lb}}{\text{lbmol}}\right)\left(0.156\,\frac{\text{Btu}}{\text{lb}\cdot{}^\circ\text{R}}\right) = 4.99\,\frac{\text{Btu}}{\text{lbmol}\cdot{}^\circ\text{R}}$$

Substituting values into the expression for T_2

$$T_2 = \frac{(0.79\text{ lbmol})\left(4.96\,\frac{\text{Btu}}{\text{lbmol}\cdot{}^\circ\text{R}}\right)(460^\circ\text{R}) + (0.21\text{ lbmol})\left(4.99\,\frac{\text{Btu}}{\text{lbmol}\cdot{}^\circ\text{R}}\right)(540^\circ\text{R})}{(0.79\text{ lbmol})\left(4.96\,\frac{\text{Btu}}{\text{lbmol}\cdot{}^\circ\text{R}}\right) + (0.21\text{ lbmol})\left(4.99\,\frac{\text{Btu}}{\text{lbmol}\cdot{}^\circ\text{R}}\right)}$$

$$= 477^\circ\text{R}$$

(b) The final mixture pressure p_2 can be determined using the ideal gas equation of state, $p_2 = n\bar{R}T_2/V$, where n is the total number of moles of mixture and V is the total volume occupied by the mixture. The volume V is the sum of the volumes of the two tanks obtained with the ideal gas equation of state

$$V = \frac{n_{N_2}\bar{R}T_{N_2}}{p_{N_2}} + \frac{n_{O_2}\bar{R}T_{O_2}}{p_{O_2}}$$

where $p_{N_2} = 2$ atm is the initial pressure of the nitrogen and $p_{O_2} = 1$ atm is the initial pressure of the oxygen. Combining results and reducing

$$p_2 = \frac{(n_{N_2} + n_{O_2})T_2}{\left(\dfrac{n_{N_2}T_{N_2}}{p_{N_2}} + \dfrac{n_{O_2}T_{O_2}}{p_{O_2}}\right)}$$

Substituting values

$$p_2 = \frac{(1.0\text{ lbmol})(477^\circ\text{R})}{\left[\dfrac{(0.79\text{ lbmol})(460^\circ\text{R})}{2\text{ atm}} + \dfrac{(0.21\text{ lbmol})(540^\circ\text{R})}{1\text{ atm}}\right]}$$

$$= 1.62\text{ atm}$$

(c) Reducing the closed system form of the entropy balance

$$S_2 - S_1 = \cancelto{0}{\int_1^2 \left(\frac{\delta Q}{T}\right)_b} + \sigma$$

where the entropy transfer term drops out for the adiabatic mixing process. The initial entropy of the system, S_1, is the sum of the entropies of the gases at the respective initial states

$$S_1 = n_{N_2}\bar{s}_{N_2}(T_{N_2}, p_{N_2}) + n_{O_2}\bar{s}_{O_2}(T_{O_2}, p_{O_2})$$

The final entropy of the system, S_2, is the sum of the entropies of the individual components evaluated at the final mixture temperature and the partial pressure of the component in the mixture

$$S_2 = n_{N_2}\bar{s}_{N_2}(T_2, y_{N_2}p_2) + n_{O_2}\bar{s}_{O_2}(T_2, y_{O_2}p_2)$$

Collecting the last three equations

$$\sigma = n_{N_2}[\bar{s}_{N_2}(T_2, y_{N_2}p_2) - \bar{s}_{N_2}(T_{N_2}, p_{N_2})] + n_{O_2}[\bar{s}_{O_2}(T_2, y_{O_2}p_2) - \bar{s}_{O_2}(T_{O_2}, p_{O_2})]$$

Evaluating the change in specific entropy of each gas in terms of a constant specific heat $\bar{c}_p$, this becomes

$$\sigma = n_{N_2}\left(\bar{c}_{p,N_2} \ln \frac{T_2}{T_{N_2}} - \bar{R} \ln \frac{y_{N_2}p_2}{p_{N_2}}\right) + n_{O_2}\left(\bar{c}_{p,O_2} \ln \frac{T_2}{T_{O_2}} - \bar{R} \ln \frac{y_{O_2}p_2}{p_{O_2}}\right)$$

The required values for $\bar{c}_p$ can be found by adding $\bar{R}$ to the $\bar{c}_v$ values found previously (Eq. 3.45)

$$\bar{c}_{p,N_2} = 6.95 \frac{\text{Btu}}{\text{lbmol} \cdot {}^\circ\text{R}}, \qquad \bar{c}_{p,O_2} = 6.98 \frac{\text{Btu}}{\text{lbmol} \cdot {}^\circ\text{R}}$$

Since the total number of moles of mixture $n = 0.79 + 0.21 = 1.0$, the mole fractions of the two gases are $y_{N_2} = 0.79$ and $y_{O_2} = 0.21$.

Substituting values into the expression for σ gives

$$\sigma = 0.79 \text{ lbmol}\left[6.95 \frac{\text{Btu}}{\text{lbmol} \cdot {}^\circ\text{R}} \ln\left(\frac{477^\circ\text{R}}{460^\circ\text{R}}\right) - 1.986 \frac{\text{Btu}}{\text{lbmol} \cdot {}^\circ\text{R}} \ln\left(\frac{(0.79)(1.62 \text{ atm})}{2 \text{ atm}}\right)\right]$$
$$+ 0.21 \text{ lbmol}\left[6.98 \frac{\text{Btu}}{\text{lbmol} \cdot {}^\circ\text{R}} \ln\left(\frac{477^\circ\text{R}}{540^\circ\text{R}}\right) - 1.986 \frac{\text{Btu}}{\text{lbmol} \cdot {}^\circ\text{R}} \ln\left(\frac{(0.21)(1.62 \text{ atm})}{1 \text{ atm}}\right)\right]$$
$$= 1.168 \text{ Btu}/{}^\circ\text{R}$$

Comment: Three factors contribute to the production of entropy in this mixing process: (1) The gases are initially at different temperatures. (2) The gases are initially at different pressures. (3) The gases are distinguishable from one another. Entropy is produced when any of these factors is present during a mixing process.

EXAMPLE 12.7

PROBLEM

100 m³/min of dry air at 32°C and 1 bar is mixed with a stream of hydrogen at 127°C and 1 bar to form a mixed stream at 47°C and 1 bar. The mixing occurs adiabatically and at steady state. Kinetic and potential energy effects can be ignored. Determine (a) the mass flow rates of the dry air and hydrogen, in kg/min, (b) the mole fractions of the dry air and hydrogen in the exiting mixture, and (c) the time rate of entropy production, in kJ/K per min.

SOLUTION

Known: 100 m³/min of dry air at 32°C and 1 bar is mixed adiabatically and at steady state with a hydrogen stream at 127°C and 1 bar to form a mixed stream at 47°C and 1 bar.

Find: Determine the mass flow rates of the dry air and hydrogen, in kg/min, the mole fractions of the dry air and hydrogen in the exiting mixture, and the time rate of entropy production, in kJ/K per min.

Schematic and Given Data:

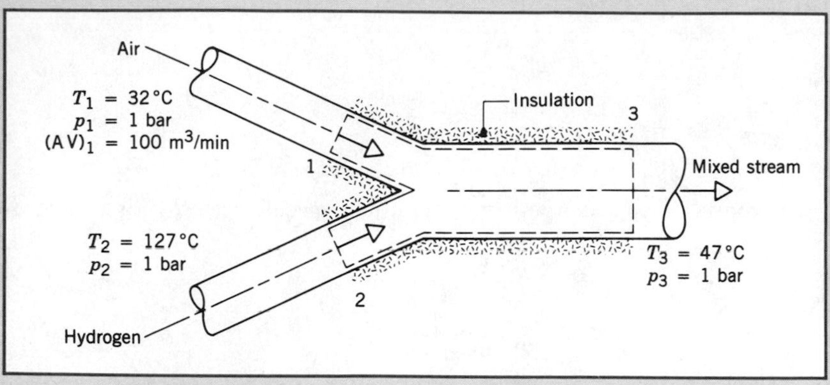

Figure E12.7

Assumptions:

1. The control volume identified by the dashed line on the accompanying figure operates at steady state.
2. No heat transfer occurs with the surroundings.
3. Kinetic and potential energy effects can be ignored, and $\dot{W}_{cv} = 0$.
4. The entering gases can be regarded as ideal gases. The exiting mixture can be regarded as an ideal gas mixture.
5. The dry air is treated as a pure component.

Analysis:

(a) The mass flow rate of the dry air entering the control volume can be determined from the given volumetric flow rate $(AV)_1$

$$\dot{m}_{a1} = \frac{(AV)_1}{v_{a1}}$$

where v_{a1} is the specific volume of the air at 1. Using the ideal gas equation of state

$$v_{a1} = \frac{(\bar{R}/M_a)T_1}{p_1} = \frac{\left(\frac{8314}{28.97}\frac{\text{N}\cdot\text{m}}{\text{kg}\cdot\text{K}}\right)(305\ \text{K})}{10^5\ \text{N/m}^2} = 0.875\ \frac{\text{m}^3}{\text{kg}}$$

The mass flow rate of the dry air is then

$$\dot{m}_{a1} = \frac{100\ \text{m}^3/\text{min}}{0.875\ \text{m}^3/\text{kg}} = 114.29\ \frac{\text{kg}}{\text{min}}$$

The mass flow rate of the hydrogen can be determined using mass and energy rate balances. At steady state, the amounts of dry air and hydrogen contained within the control volume do not vary. Thus, for each component individually it is necessary for the incoming and outgoing mass flow rates to be equal. That is

$$\dot{m}_{a1} = \dot{m}_{a3} \quad \text{(dry air)}$$

$$\dot{m}_{h2} = \dot{m}_{h3} \quad \text{(hydrogen)}$$

Using assumptions 1–3 together with the foregoing mass flow rate relations, the energy rate balance reduces to

$$0 = \dot{m}_a h_a(T_1) + \dot{m}_h h_h(T_2) - [\dot{m}_a h_a(T_3) + \dot{m}_h h_h(T_3)]$$

where $\dot{m}_a$ and $\dot{m}_h$ denote the mass flow rates of the dry air and hydrogen, respectively. The enthalpy of the mixture at the exit is evaluated by summing the contributions of the air and hydrogen at their respective conditions within the mixture. Solving for $\dot{m}_h$

$$\dot{m}_h = \dot{m}_a \left[\frac{h_a(T_3) - h_a(T_1)}{h_h(T_2) - h_h(T_3)}\right]$$

The specific enthalpies can be obtained from Tables A-16 and A-22. Since Table A-22 for hydrogen gives values on a molar basis, the molecular weight of hydrogen is introduced into the denominator to convert the molar enthalpy values to a mass basis

$$\dot{m}_h = \frac{(114.29\ \text{kg/min})(320.29\ \text{kJ/kg} - 305.22\ \text{kJ/kg})}{\left(\dfrac{1}{2\ \text{kg/kmol}}\right)(11{,}426\ \text{kJ/kmol} - 9{,}100\ \text{kJ/kmol})}$$

$$= 1.48\ \frac{\text{kg}}{\text{min}}$$

(b) To obtain the mole fractions of the dry air and hydrogen in the exiting mixture, first convert the mass flow rates to molar flow rates using the respective molecular weights

$$\dot{n}_a = \frac{\dot{m}_a}{M_a} = \frac{114.29\ \text{kg/min}}{28.97\ \text{kg/kmol}} = 3.95\ \text{kmol/min}$$

$$\dot{n}_h = \frac{\dot{m}_h}{M_h} = \frac{1.48\ \text{kg/min}}{2\ \text{kg/kmol}} = 0.74\ \text{kmol/min}$$

where $\dot{n}$ denotes molar flow rate. The molar flow rate of the mixture $\dot{n}$ is the sum

$$\dot{n} = \dot{n}_a + \dot{n}_h = 3.95 + 0.74 = 4.69\ \text{kmol/min}$$

The mole fractions of the air and hydrogen are, respectively

$$y_a = \frac{\dot{n}_a}{\dot{n}} = \frac{3.95}{4.69} = 0.842 \qquad \text{and} \qquad y_h = \frac{\dot{n}_h}{\dot{n}} = \frac{0.74}{4.69} = 0.158$$

(c) For the control volume at steady state, the entropy rate balance reduces to

$$0 = \dot{m}_a s_a(T_1, p_1) + \dot{m}_h s_h(T_2, p_2) - [\dot{m}_a s_a(T_3, y_a p_3) + \dot{m}_h s_h(T_3, y_h p_3)] + \dot{\sigma}$$

The specific entropy of each component in the exiting ideal gas mixture is evaluated at its partial pressure in the mixture and at the mixture temperature. Solving for $\dot{\sigma}$

$$\dot{\sigma} = \dot{m}_a[s_a(T_3, y_a p_3) - s_a(T_1, p_1)] + \dot{m}_h[s_h(T_3, y_h p_3) - s_h(T_2, p_2)]$$

Since $p_1 = p_3$, the specific entropy change of the dry air is

$$s_a(T_3, y_a p_3) - s_a(T_1, p_1) = s_a^\circ(T_3) - s_a^\circ(T_1) - \frac{\bar{R}}{M_a} \ln \frac{y_a p_3}{p_1}$$

$$= s_a^\circ(T_3) - s_a^\circ(T_1) - \frac{\bar{R}}{M_a} \ln y_a$$

The s_a° terms are evaluated from Table A-16. Similarly, since $p_2 = p_3$, the specific entropy change of the hydrogen is

$$s_h(T_3, y_h p_3) - s_h(T_2, p_2) = \frac{1}{M_h}[\bar{s}_h^\circ(T_3) - \bar{s}_h^\circ(T_2) - \bar{R} \ln y_h]$$

The $\bar{s}_h^\circ$ terms are evaluated from Table A-22. Note the use of the molecular weights M_a and M_h in the last two equations to obtain the respective entropy changes on a mass basis.

The expression for the rate of entropy production becomes

$$\dot{\sigma} = \dot{m}_a\left[s_a^\circ(T_3) - s_a^\circ(T_1) - \frac{\bar{R}}{M_a} \ln y_a\right] + \frac{\dot{m}_h}{M_h}[\bar{s}_h^\circ(T_3) - \bar{s}_h^\circ(T_2) - \bar{R} \ln y_a]$$

Substituting values

$$\dot{\sigma} = \left(114.29\ \frac{\text{kg}}{\text{min}}\right)\left[1.7669\ \frac{\text{kJ}}{\text{kg}\cdot\text{K}} - 1.71865\ \frac{\text{kJ}}{\text{kg}\cdot\text{K}} - \left(\frac{8.314}{28.97}\ \frac{\text{kJ}}{\text{kg}\cdot\text{K}}\right) \ln 0.842\right]$$

$$+ \left(\frac{1.48\ \text{kg/min}}{2\ \text{kg/kmol}}\right)\left[132.621\ \frac{\text{kJ}}{\text{kmol}\cdot\text{K}} - 139.106\ \frac{\text{kJ}}{\text{kmol}\cdot\text{K}} - \left(8.314\ \frac{\text{kJ}}{\text{kmol}\cdot\text{K}}\right) \ln 0.158\right]$$

$$= 17.71\ \frac{\text{kJ}}{\text{K}\cdot\text{min}}$$

Comment: The entropy production in this case is associated with the same factors as were listed in the comment following the solution of Example 12.6.

PSYCHROMETRIC APPLICATIONS

The remainder of this chapter is concerned with the study of systems involving mixtures of dry air and water vapor. A condensed water phase may also be present. Knowledge of the behavior of such systems is essential for the analysis and design of air-conditioning devices, cooling towers, and industrial processes requiring close control of the vapor content in air. The study of systems consisting of dry air and water is known as *psychrometrics*.

12.6 INTRODUCTORY PSYCHROMETRIC PRINCIPLES

The object of the present section is to introduce some important terms, definitions, and principles used in the study of systems consisting of dry air and water.

12.6.1 Moist Air

The term *moist air* refers to a mixture of dry air and water vapor in which the dry air is treated as if it were a pure component. As can be verified by reference to appropriate property data, the overall mixture and each mixture component behave as ideal gases at the states under present consideration. Accordingly, for the applications to be considered, the ideal gas mixture concepts introduced previously apply directly.

Shown in Fig. 12.3 is a closed system consisting of moist air occupying a volume V at mixture pressure p and mixture temperature T. The overall mixture is assumed to obey the ideal gas equation of state. Thus

$$p = \frac{n\bar{R}T}{V} = \frac{m(\bar{R}/M)T}{V} \tag{12.40}$$

where n, m, and M denote the moles, mass, and molecular weight of the mixture, respectively. Each mixture component is considered to act as if it existed alone in the volume V at the mixture temperature T while exerting a part of the pressure. The mixture pressure is the sum of the partial pressures of the dry air and the water vapor. Using the ideal gas equation of state, the partial pressures p_a and p_v of the dry air and water vapor are, respectively

$$p_a = \frac{n_a\bar{R}T}{V} = \frac{m_a(\bar{R}/M_a)T}{V}, \qquad p_v = \frac{n_v\bar{R}T}{V} = \frac{m_v(\bar{R}/M_v)T}{V} \tag{12.41}$$

where n_a and n_v denote the moles of dry air and water vapor, respectively; m_a, m_v, M_a, and M_v are the respective masses and molecular weights. The amount of water

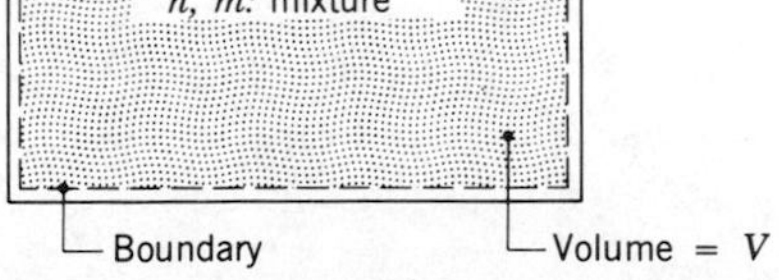

Figure 12.3 Mixture of dry air and water vapor.

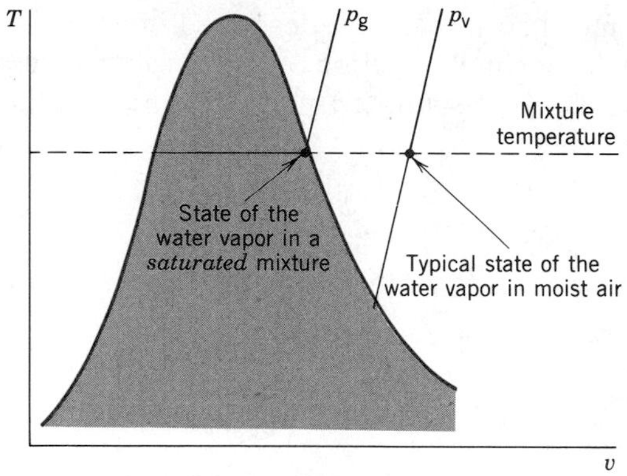

Figure 12.4 T–v diagram for water vapor in an air–water mixture.

vapor present is normally much less than the amount of dry air. Accordingly, the values of n_v, m_v, and p_v are small relative to the corresponding values of n_a, m_a, and p_a.

A typical state of water vapor in moist air is shown in Fig. 12.4. At this state, fixed by the partial pressure p_v and the mixture temperature T, the vapor is superheated. When the partial pressure of the water vapor corresponds to the saturation pressure of water at the mixture temperature, p_g of Fig. 12.4, the mixture is said to be *saturated*. *Saturated air* is a mixture of dry air and saturated water vapor. The amount of water vapor in moist air varies from zero in dry air to a maximum depending on the pressure and temperature when the mixture is saturated.

12.6.2 Humidity Ratio and Relative Humidity

The composition of a given moist air sample can be described in a number of ways. The mixture can be described in terms of the moles of dry air and water vapor present or in terms of the respective mole fractions. Alternatively, the mass of dry air and water vapor, or the respective mass fractions, can be specified. The composition also can be indicated by means of the *humidity ratio* ω, defined as the ratio of the mass of the water vapor to the mass of dry air

$$\omega = \frac{m_v}{m_a} \tag{12.42}$$

The humidity ratio is sometimes referred to as the *specific humidity*.

The humidity ratio can be expressed in terms of partial pressures and molecular weights by solving Eqs. 12.41 for m_a and m_v, respectively, and substituting the resulting expressions into Eq. 12.42 to obtain

$$\omega = \frac{m_v}{m_a} = \frac{M_v p_v V/\bar{R}T}{M_a p_a V/\bar{R}T} = \frac{M_v p_v}{M_a p_a}$$

Introducing $p_a = p - p_v$ and noting that the ratio of the molecular weight of water to that of dry air is approximately 0.622, this expression can be written as

$$\omega = 0.622 \frac{p_v}{p - p_v} \tag{12.43}$$

The makeup of moist air also can be described in terms of the *relative humidity* ϕ, defined as the ratio of the mole fraction of water vapor y_v in a given moist air sample to the mole fraction $y_{v,sat}$ in a saturated moist air sample at the same mixture temperature and pressure

$$\phi = \left.\frac{y_v}{y_{v,sat}}\right)_{T,p}$$

Since $p_v = y_v p$ and $p_g = y_{v,sat} p$, this can be expressed alternatively as

$$\phi = \left.\frac{p_v}{p_g}\right)_{T,p} \tag{12.44}$$

The pressures in this expression for the relative humidity are labeled on Fig. 12.4.

The humidity ratio and relative humidity can be measured experimentally. The humidity ratio can be determined using a *hygrometer* in which a moist air sample is exposed to suitable chemicals until the moisture present is absorbed. The amount of water vapor absorbed is determined by weighing the chemicals. Continuous recording of the relative humidity can be accomplished by means of electric transducers consisting of resistance elements that change resistance with relative humidity.

The values of U, H, and S for moist air can be found by adding the contribution of each component at the condition at which the component exists in the mixture. For example, the enthalpy H of a given moist air sample is

$$H = H_a + H_v = m_a h_a + m_v h_v \tag{12.45}$$

Dividing Eq. 12.45 by m_a and introducing the humidity ratio gives the enthalpy of the mixture *per unit mass of dry air*

$$\frac{H}{m_a} = h_a + \frac{m_v}{m_a} h_v = h_a + \omega h_v \tag{12.46}$$

The enthalpies of the dry air and water vapor appearing in Eq. 12.46 are evaluated at the mixture temperature.

Reference to steam table data or a Mollier diagram for water shows that the enthalpy of superheated water vapor at low vapor pressures is very closely given by the saturated vapor value corresponding to the given temperature. Hence, the enthalpy of the water vapor h_v in Eq. 12.46 can be taken as h_g at the mixture temperature. That is

$$h_v \approx h_g(T) \tag{12.47}$$

This approximation is used in the remainder of the chapter. Tables A-19 for water vapor as an ideal gas should be used cautiously to obtain h_v because the enthalpy datum differs from that of the steam tables. These different datums can lead to error when studying systems that contain both water vapor and a liquid or solid phase of water. The enthalpy of dry air, h_a, can be obtained from the appropriate ideal gas table, Table A-16 or Table A-16E, however, because air is a gas at all states under present consideration and is closely modeled as an ideal gas at these states.

An approach similar to that for enthalpy also applies to the evaluation of the internal energy of moist air. When evaluating the entropy of moist air, the contribution of each component is determined at the mixture temperature and the partial pressure of the component in the mixture. Using Eq. 6.23, it can readily be shown that the specific entropy of the water vapor is given by $s_v(T, p_v) = s_g(T) - R \ln \phi$, where s_g is the specific entropy of saturated vapor at temperature T from the steam tables and ϕ is the relative humidity.

12.6.3 Moist Air in Contact with Liquid Water

Thus far, our study of psychrometrics has been conducted as an application of the ideal gas mixture principles introduced in the first part of this chapter. However, many systems of interest are composed of a mixture of dry air and water vapor in contact with a liquid (or solid) water phase. To study these systems requires additional considerations.

Shown in Fig. 12.5 is a vessel containing liquid water, above which is a mixture of water vapor and dry air. If no interactions with the surroundings are allowed, liquid will evaporate until eventually the gas phase becomes saturated and the system attains an equilibrium state. For many engineering applications, systems consisting of moist air in *equilibrium* with a liquid water phase can be described simply and accurately with the following idealizations: (1) The dry air and water vapor behave as independent ideal gases. (2) The equilibrium between the liquid phase and the water vapor is not significantly disturbed by the presence of the air. Accordingly, the partial pressure of the water vapor equals the saturation pressure of water corresponding to the temperature of the mixture. Similar considerations apply for systems consisting of moist air in equilibrium with a solid water phase. The presence of the air actually alters the partial pressure of the vapor from the saturation pressure by a small amount whose magnitude is calculated in Sec. 14.6.

12.6.4 Dew Point

A significant aspect of the behavior of moist air is that partial condensation of the water vapor can occur when the temperature is reduced. This type of phenomenon is commonly encountered in the condensation of vapor on windowpanes in winter and on pipes carrying cold water. The formation of dew on grass is another familiar example. To study this, consider a closed system consisting of a sample of moist air that is cooled at *constant* pressure, as shown in Fig. 12.6. The property diagram given on this figure locates states of the water vapor. Initially, the water vapor is superheated at state 1. In the first part of the cooling process, both the system pressure *and* the composition of the moist air would remain constant. Accordingly, since $p_v = y_v p$, the *partial pressure* of the water vapor would remain constant, and the water vapor would cool at constant p_v from state 1 to state d, called the *dew*

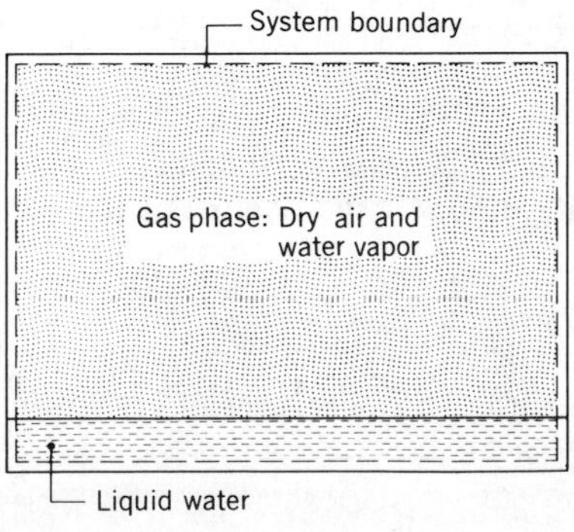

Figure 12.5 System consisting of moist air in contact with liquid water.

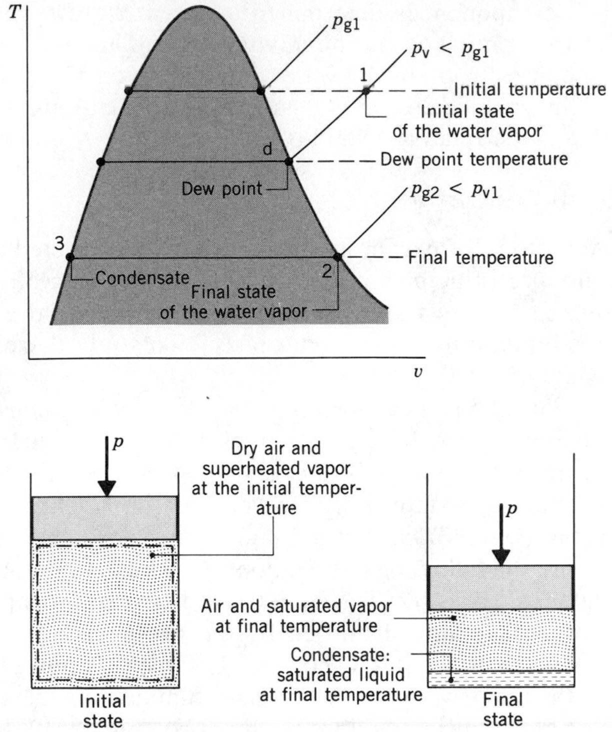

Figure 12.6 States of water for moist air cooled at constant mixture pressure.

point. The saturation temperature corresponding to p_v is called the *dew point temperature*. This temperature is labeled on Fig. 12.6.

In the next part of the cooling process, the system would be cooled *below* the dew point temperature and some of the water vapor initially present would condense. At the final state, the system would consist of a gas phase of dry air and water vapor in equilibrium with a liquid water phase. The vapor that remains can be regarded as saturated at the final temperature, state 2 of Fig. 12.6, with a partial pressure equal to the saturation pressure p_{g2} corresponding to this temperature. The condensate would be a saturated liquid at the final temperature, state 3 of Fig. 12.6. Note that the partial pressure of the water vapor at the final state, p_{g2}, is less than the initial value, p_v. The partial pressure decreases because the mole fraction of the water vapor present at the final state is less than the initial value since condensation occurs.

EXAMPLE 12.8

PROBLEM

A 1-lb sample of moist air initially at 70°F, 14.7 lbf/in.^{2}, and 70% relative humidity is cooled to 40°F while keeping the pressure constant. Determine (a) the initial humidity ratio, (b) the dew point temperature, in °F, and (c) the amount of water vapor that condenses, in lb.

SOLUTION

Known: A 1-lb sample of moist air is cooled at a constant mixture pressure of 14.7 lbf/in.^{2} from 70 to 40°F. The initial relative humidity is 70%.

Find: Determine the initial humidity ratio, the dew point temperature, in °F, and the amount of water vapor that condenses, in lb.

Schematic and Given Data:

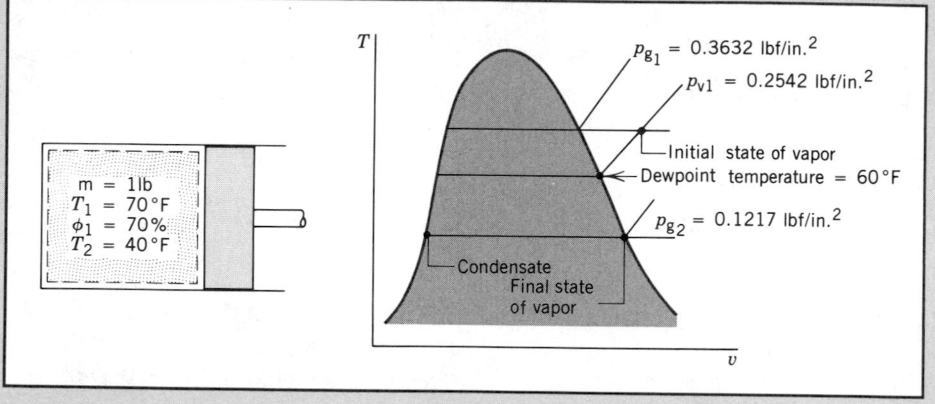

Figure E12.8

Assumptions:

1. The 1-lb sample of moist air is taken as the closed system. The system pressure remains constant at 14.7 lbf/in.^{2}
2. The gas phase present can be treated as an ideal gas mixture. Each mixture component acts as an ideal gas existing alone in the volume occupied by the gas phase at the mixture temperature.
3. When a liquid water phase is present, the water vapor exists as a saturated vapor at the system temperature. The liquid present is a saturated liquid at the system temperature.

Analysis:

(a) The initial humidity ratio can be evaluated from Eq. 12.43. This requires the partial pressure of the water vapor p_{v1} which can be found from the given relative humidity and p_g from Table A-2E at 70°F

$$p_{v1} = \phi p_g = (0.7)\left(0.3632\ \frac{\text{lbf}}{\text{in.}^2}\right) = 0.2542\ \frac{\text{lbf}}{\text{in.}^2}$$

Inserting values in Eq. 12.43

$$\omega_1 = 0.622\left(\frac{0.2542}{14.7 - 0.2542}\right) = 0.011\ \frac{\text{lb (vapor)}}{\text{lb (dry air)}}$$

(b) The dew point temperature is the saturation temperature corresponding to the partial pressure, p_{v1}. Interpolation in Table A-2E gives $T = 60°F$. The dew point temperature is labeled on the accompanying property figure.

(c) The amount of condensate, m_w, equals the difference between the initial amount of water vapor in the sample, m_{v1}, and the final amount of water vapor, m_{v2}. That is

$$m_w = m_{v1} - m_{v2}$$

To evaluate m_{v1}, note that the system initially consists of 1 lb of dry air and water vapor, so 1 lb $= m_a + m_{v1}$, where m_a is the mass of dry air present in the sample. With $\omega_1 = m_{v1}/m_a$, this becomes

$$1\text{ lb} = \frac{m_{v1}}{\omega_1} + m_{v1} = m_{v1}\left(\frac{1}{\omega_1} + 1\right)$$

Solving for m_{v1}

$$m_{v1} = \frac{1\text{ lb}}{(1/\omega_1) + 1}$$

Inserting the value of ω_1 determined in part (a)

$$m_{v1} = \frac{1\text{ lb}}{(1/0.011) + 1} = 0.0109\text{ lb (vapor)}$$

The mass of dry air present is then $m_a = 1 - 0.0109 = 0.9891$ lb (dry air).

Next, let us evaluate m_{v2}. With assumption 3, the partial pressure of the water vapor remaining in the system at the final state is the saturation pressure corresponding to 40°F: $p_g = 0.1217$ lbf/in.² Accordingly, the humidity ratio after cooling is found from Eq. 12.43 as

$$\omega_2 = 0.622\left(\frac{0.1217}{14.7 - 0.1217}\right) = 0.0052\ \frac{\text{lb (vapor)}}{\text{lb (dry air)}}$$

The mass of the water vapor present at the final state is then

$$m_{v2} = \omega_2 m_a = (0.0052)(0.9891) = 0.0051 \text{ lb (vapor)}$$

Finally, the amount of water vapor that condenses is

$$m_w = m_{v1} - m_{v2} = 0.0109 - 0.0051 = 0.0058 \text{ lb (condensate)}$$

Comments:

1. The amount of water vapor present in a typical moist air mixture is considerably less than the amount of dry air present.
2. At the final state, the *quality* of the two-phase liquid–vapor mixture of water is $x = 0.0051/0.0109 = 0.47$ (47%). The relative humidity of the gas phase is 100%.

EXAMPLE 12.9

PROBLEM

An air–water vapor mixture is contained in a rigid, closed vessel with a volume of 35 m³ at 1.5 bar, 120°C, and ϕ = 10%. The mixture is cooled at constant volume until its temperature is reduced to 22°C. Determine (a) the dew point temperature corresponding to the initial state, in °C, (b) the temperature at which condensation begins, in °C, and (c) the amount of water condensed, in kg.

SOLUTION

Known: A rigid, closed tank with a volume of 35 m³ containing moist air initially at 1.5 bar, 120°C, and ϕ = 10% is cooled to 22°C.

Find: Determine the dew point temperature at the initial state, in °C, the temperature at which condensation begins, in °C, and the amount of water condensed, in kg.

Schematic and Given Data:

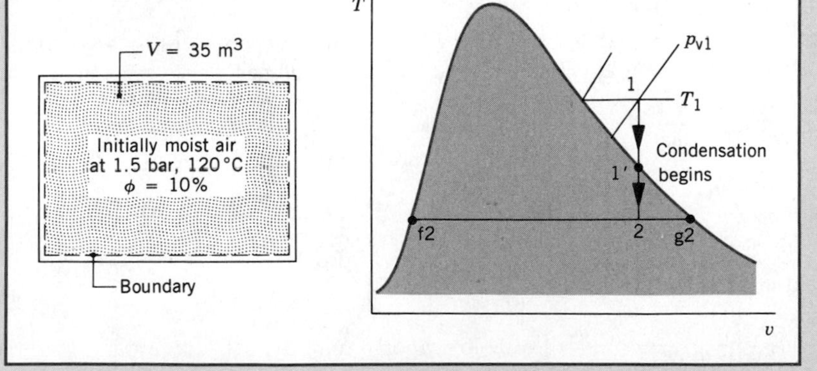

Figure E12.9

Assumptions:

1. The contents of the tank are taken as a closed system. The system volume remains constant.
2. The gas phase present can be treated as an ideal gas mixture. Each mixture component acts as an ideal gas existing alone in the volume occupied by the gas phase at the mixture temperature.
3. When a liquid water phase is present, the water vapor exists as a saturated vapor at the system temperature. The liquid is a saturated liquid at the system temperature.

Analysis:

(a) The dew point temperature at the initial state is the saturation temperature corresponding to the partial pressure p_{v1}. With the given relative humidity and the saturation pressure at 120°C from Table A-2

$$p_{v1} = \phi_1 p_{g1} = (0.10)(1.985) = 0.1985 \text{ bar}$$

Interpolating in Table A-2 gives the dew point temperature as 60°C.

(b) Whether the water exists as a vapor only, or as liquid *and* vapor, it occupies the full volume, which remains constant. Accordingly, since the total mass of the water present is also constant, the water undergoes the constant specific volume process illustrated on the accompanying T–v diagram. In the process from state 1 to state 1′, the water exists as a vapor only. For the process from state 1′ to state 2, the water exists as a two-phase liquid–vapor mixture. Note that pressure does not remain constant during the cooling process from state 1 to state 2.

State 1′ on the T–v diagram denotes the state where the water vapor first becomes saturated. The saturation temperature at this state is denoted as T'. Cooling to a temperature less than T' would result in condensation of some of the water vapor present. Since state 1′ is a saturated vapor state, the temperature T' can be found by interpolating in Table A-2 with the specific volume of the water at this state. The specific volume of the vapor at state 1′ equals the specific volume of the vapor at state 1, which can be evaluated from the ideal gas equation

$$v_{v1} = \frac{(\bar{R}/M_v)T_1}{p_{v1}} = \left(\frac{8314}{18}\,\frac{\text{N}\cdot\text{m}}{\text{kg}\cdot\text{K}}\right)\left(\frac{393 \text{ K}}{0.1985 \times 10^5 \text{ N/m}^2}\right)$$

$$= 9.145 \,\frac{\text{m}^3}{\text{kg}}$$

Interpolation in Table A-2 with $v_{v1} = v_g$ gives $T' = 56$°C. Note that this temperature is *not* the dew point temperature determined in part (a).

(c) The amount of condensate equals the difference between the initial and final amounts of water vapor present. The mass of the water vapor present initially is

$$m_{v1} = \frac{V}{v_{v1}} = \frac{35 \text{ m}^3}{9.145 \text{ m}^3/\text{kg}} = 3.827 \text{ kg}$$

The mass of water vapor present finally can be determined from the *quality*. At the final state, the water forms a two-phase liquid–vapor mixture having a specific volume of 9.145 m³/kg. Using this specific volume value, the quality x_2 of the liquid–vapor mixture can be found as

$$x_2 = \frac{v_{v2} - v_{f2}}{v_{g2} - v_{f2}} = \frac{9.145 - 1.0022 \times 10^{-3}}{51.447 - 1.0022 \times 10^{-3}} = 0.178$$

where v_{f2} and v_{g2} are the saturated liquid and saturated vapor specific volumes at $T_2 = 22$°C, respectively.

Using the quality together with the known total amount of water present, 3.827 kg, the mass of the water vapor contained in the system at the final state is

$$m_{v2} = (0.178)(3.827) = 0.681 \text{ kg}$$

The mass of the condensate, m_{w2}, is then

$$m_{w2} = m_{v1} - m_{v2} = 3.827 - 0.681 = 3.146 \text{ kg}$$

Comment: This example shows that when a moist air mixture is cooled at constant mixture volume, the temperature at which condensation begins is not the dew point temperature corresponding to the initial state.

12.7 CONSERVATION OF MASS AND CONSERVATION OF ENERGY APPLIED TO PSYCHROMETRIC SYSTEMS

The object of this section is to illustrate the use of the conservation of mass and conservation of energy principles in analyzing systems involving mixtures of dry air and water vapor in which a condensed water phase may be present. Both control volumes and closed systems are considered. The same basic solution approach that has been used in thermodynamic analyses considered thus far is applicable. The only new aspect is the use of the special vocabulary and parameters of psychrometrics.

Devices that accomplish air-conditioning processes such as heating, cooling, humidification, or dehumidification are normally analyzed on a control volume basis. To consider a typical analysis, refer to Fig. 12.7, which shows a two-inlet, single-exit device at steady state. A moist air stream enters at 1, a moist air stream exits at 2, and a water-only stream enters at 3. The water-only stream may be a liquid or a vapor. Heat transfer at the rate $\dot{Q}_{cv}$ can occur between the control volume and its surroundings. Depending on the application, the value of $\dot{Q}_{cv}$ might be positive, negative, or zero.

At steady state, the amounts of dry air and water vapor contained within the control volume cannot vary. Thus, for each component individually it is necessary for the total incoming and outgoing mass flow rates to be equal. That is

$$\dot{m}_{a1} = \dot{m}_{a2} \qquad \text{(dry air)}$$

$$\dot{m}_{v1} + \dot{m}_{w} = \dot{m}_{v2} \qquad \text{(water)}$$

For simplicity, the constant mass flow rate of the dry air is denoted by $\dot{m}_a$. The mass flow rates of the water vapor can be expressed conveniently in terms of humidity ratios as $\dot{m}_{v1} = \omega_1 \dot{m}_a$ and $\dot{m}_{v2} = \omega_2 \dot{m}_a$. With these expressions, the mass balance for water becomes

$$\dot{m}_w = \dot{m}_a(\omega_2 - \omega_1) \tag{12.48}$$

When water is added at 3, ω_2 would be greater than ω_1.

Assuming $\dot{W}_{cv} = 0$ and ignoring all kinetic and potential energy effects, the energy rate balance reduces at steady state to

$$0 = \dot{Q}_{cv} + (\dot{m}_a h_{a1} + \dot{m}_{v1} h_{v1}) + \dot{m}_w h_w - (\dot{m}_a h_{a2} + \dot{m}_{v2} h_{v2}) \tag{12.49}$$

In this equation, the entering and exiting moist air streams are regarded as ideal gas mixtures of dry air and water vapor.

Equation 12.49 can be cast into a form that is particularly convenient for the analysis of air-conditioning devices. First, with Eq. 12.47 the enthalpies of the entering and exiting water vapor can be evaluated as the saturated vapor enthalpies corresponding to the temperatures T_1 and T_2, respectively, giving

$$0 = \dot{Q}_{cv} + (\dot{m}_a h_{a1} + \dot{m}_{v1} h_{g1}) + \dot{m}_w h_w - (\dot{m}_a h_{a2} + \dot{m}_{v2} h_{g2})$$

Then, with $\dot{m}_{v1} = \omega_1 \dot{m}_a$ and $\dot{m}_{v2} = \omega_2 \dot{m}_a$, the equation can be expressed as

$$0 = \dot{Q}_{cv} + \dot{m}_a(h_{a1} + \omega_1 h_{g1}) + \dot{m}_w h_w - \dot{m}_a(h_{a2} + \omega_2 h_{g2}) \tag{12.50}$$

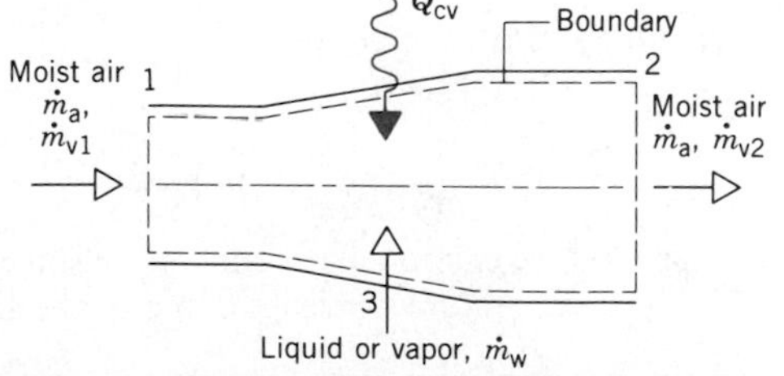

Figure 12.7 System for conditioning moist air.

Finally, introducing Eq. 12.48, the energy rate balance becomes

$$0 = \dot{Q}_{cv} + \dot{m}_a[\underline{(h_{a1} - h_{a2})} + \underline{\omega_1 h_{g1} + (\omega_2 - \omega_1)h_w - \omega_2 h_{g2}}] \qquad (12.51)$$

The first underlined term of Eq. 12.51 can be evaluated from Tables A-16 giving the ideal gas properties of air. Alternatively, since relatively small temperature differences are normally encountered in the class of devices under present consideration, this term can be evaluated as $h_{a1} - h_{a2} = c_{pa}(T_1 - T_2)$, where c_{pa} is a constant value for the specific heat of dry air. The second underlined term of Eq. 12.51 can be evaluated using steam table data together with known values for ω_1 and ω_2.

As suggested by the foregoing development, several simplifying assumptions are commonly used when analyzing the class of devices under present consideration. In addition to the assumption of steady-state operation, equilibrium property relations are assumed to apply at locations where matter crosses the boundary of the control volume, and the effects of kinetic and potential energy at these locations are neglected. In most cases there is no work, except for flow work where matter crosses the boundary. Further simplifications also may be required in particular cases. Example 12.10 gives an elementary illustration of the use of the foregoing methodology. Several additional examples are provided in Sec. 12.10.

No additional fundamental concepts are required for the study of closed systems involving mixtures of dry air and water vapor. Example 12.11 brings out some special features of the use of conservation of mass and conservation of energy in analyzing this kind of system. Similar considerations can be used to study other closed systems involving moist air.

EXAMPLE 12.10

PROBLEM

Moist air enters a duct at 10°C, 80% relative humidity, and a volumetric flow rate of 150 m³/min. The mixture is heated as it flows through the duct and exits at 30°C. No moisture is added or removed, and the mixture pressure remains approximately constant at 1 bar. For steady-state operation, determine (a) the rate of heat transfer, in kJ/min, and (b) the relative humidity at the exit. Changes in kinetic and potential energy can be ignored.

SOLUTION

Known: Moist air that enters a duct at 10°C and ϕ = 80% with a volumetric flow rate of 150 m³/min is heated at constant pressure and exits at 30°C. No moisture is added or removed.

Find: Determine the rate of heat transfer, in kJ/min, and the relative humidity at the exit.

Schematic and Given Data:

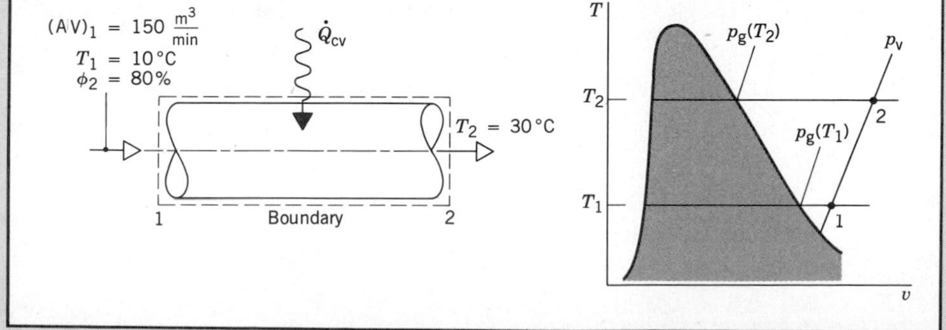

Figure E12.10

Assumptions:

1. The control volume shown in the accompanying figure operates at steady state.
2. The changes in kinetic and potential energy between inlet and exit can be ignored and $\dot{W}_{cv} = 0$.
3. The entering and exiting moist air streams can be regarded as ideal gas mixtures.

Analysis:

(a) The heat transfer rate $\dot{Q}_{cv}$ can be determined from the mass and energy rate balances. At steady state, the amounts of dry air and water vapor contained within the control volume cannot vary. Thus, for each component individually it is necessary for the incoming and outgoing mass flow rates to be equal. That is

$$\dot{m}_{a1} = \dot{m}_{a2} \qquad \text{(dry air)}$$

$$\dot{m}_{v1} = \dot{m}_{v2} \qquad \text{(water)}$$

For simplicity, the constant mass flow rates of the dry air and water vapor are denoted, respectively, by $\dot{m}_a$ and $\dot{m}_v$. From these considerations, it can be concluded that the humidity ratio is the same at the inlet and exit: $\omega_1 = \omega_2$.

The steady-state form of the energy rate balance reduces with assumption 2 to

$$0 = \dot{Q}_{cv} - \cancelto{0}{\dot{W}_{cv}} + (\dot{m}_a h_{a1} + \dot{m}_v h_{v1}) - (\dot{m}_a h_{a2} + \dot{m}_v h_{v2})$$

In writing this equation, the incoming and outgoing moist air streams are regarded as ideal gas mixtures of dry air and water vapor.

Solving for $\dot{Q}_{cv}$

$$\dot{Q}_{cv} = \dot{m}_a(h_{a2} - h_{a1}) + \dot{m}_v(h_{v2} - h_{v1})$$

Noting that $\dot{m}_v = \omega \dot{m}_a$, where ω is the humidity ratio, the expression for $\dot{Q}_{cv}$ can be written in the form

$$\dot{Q}_{cv} = \dot{m}_a[(h_{a2} - h_{a1}) + \omega(h_{v2} - h_{v1})]$$

To evaluate $\dot{Q}_{cv}$ from this expression requires the specific enthalpies of the dry air and water vapor at the inlet and exit, the mass flow rate of the dry air, and the humidity ratio.

The specific enthalpies of the dry air are read from Table A-16 at the inlet and exit temperatures T_1 and T_2: $h_{a1} = 283.1$ kJ/kg, $h_{a2} = 303.2$ kJ/kg. The specific enthalpies of the water vapor are found using $h_v \approx h_g$ and data from Table A-2 at T_1 and T_2: $h_{g1} = 2519.8$ kJ/kg, $h_{g2} = 2556.3$ kJ/kg.

The mass flow rate of the dry air can be determined from the volumetric flow rate at the inlet $(AV)_1$

$$\dot{m}_a = \frac{(AV)_1}{v_{a1}}$$

In this equation, v_{a1} is the specific volume of the dry air evaluated at T_1 and the partial pressure of the dry air p_{a1}. Using the ideal gas equation of state

$$v_{a1} = \frac{(\bar{R}/M)T_1}{p_{a1}}$$

The partial pressure p_{a1} can be determined from the mixture pressure p and the partial pressure of the water vapor p_{v1}: $p_{a1} = p - p_{v1}$. To find p_{v1}, use the given inlet relative humidity and the saturation pressure at 10°C from Table A-2

$$p_{v1} = \phi_1 p_{g1} = (0.8)(0.01228 \text{ bar}) = 0.0098 \text{ bar}$$

Since the mixture pressure is 1 bar, it follows that $p_{a1} = 0.9902$ bar. The specific volume of the dry air is then

$$v_{a1} = \frac{\left(\dfrac{8314}{28.97}\dfrac{\text{N}\cdot\text{m}}{\text{kg}\cdot\text{K}}\right)(283 \text{ K})}{(0.9902 \times 10^5 \text{ N/m}^2)} = 0.82 \text{ m}^3/\text{kg}$$

Using this value, the mass flow rate of the dry air is

$$\dot{m}_a = \frac{150 \text{ m}^3/\text{min}}{0.82 \text{ m}^3/\text{kg}} = 182.9 \text{ kg/min}$$

The humidity ratio ω can be found from

$$\omega = 0.622 \left(\frac{p_{v1}}{p - p_{v1}}\right) = 0.622 \left(\frac{0.0098}{1 - 0.0098}\right)$$

$$= 0.00616 \frac{\text{kg (vapor)}}{\text{kg (dry air)}}$$

Finally, substituting values into the expression for $\dot{Q}_{cv}$ gives

$$\dot{Q}_{cv} = 182.9[(303.2 - 283.1) + (0.00616)(2556.3 - 2519.8)]$$

$$= 3717 \text{ kJ/min}$$

(b) The states of the water vapor at the duct inlet and exit are located on the accompanying T–v diagram. Both the composition of the moist air and the mixture pressure remain constant, so the partial pressure of the water vapor at the exit equals the partial pressure of the water vapor at the inlet: $p_{v2} = p_{v1} = 0.0098$ bar. The relative humidity at the exit is then

$$\phi_2 = \frac{p_{v2}}{p_{g2}} = \frac{0.0098}{0.04246} = 0.231 \text{ (23.1\%)}$$

where p_{g2} is from Table A-2 at 30°C.

Comments:

1. The first underlined term in the equation for $\dot{Q}_{cv}$ in part (a) is evaluated with specific enthalpies from the ideal gas table for air, Table A-16. Steam table data are used to evaluate the second underlined term. Note that the different datums for enthalpy underlying these tables cancel because each of the two terms involves enthalpy *differences* only.
2. The specific heat c_{pa} for dry air varies only slightly over the interval from 10 to 30°C (Table A-14). Hence, the specific enthalpy change of the dry air can be evaluated accurately with $c_{pa} = 1.005$ kJ/kg · K.
3. No water is added or removed as the moist air passes through the duct at constant pressure; accordingly, both the humidity ratio ω and the partial pressures p_v and p_a remain constant. However, because the saturation pressure increases as the temperature increases from inlet to exit, the relative humidity decreases: $\phi_2 < \phi_1$.

EXAMPLE 12.11

PROBLEM

An air–water vapor mixture is contained in a rigid, closed vessel with a volume of 35 m³ at 1.5 bar, 120°C, and ϕ = 10%. The mixture is cooled until its temperature is reduced to 22°C. Determine the heat transfer during the process, in kJ.

SOLUTION

Known: A rigid, closed tank with a volume of 35 m³ containing moist air initially at 1.5 bar, 120°C, and ϕ = 10% is cooled to 22°C.

Find: Determine the heat transfer for the process, in kJ.

Schematic and Given Data: See the figure for Example 12.9.

Assumptions:

1. The contents of the tank are taken as a closed system. The system volume remains constant.

2. The gas phase present can be treated as an ideal gas mixture. Each component acts as an ideal gas existing alone in the volume occupied by the gas phase at the mixture temperature.
3. When a liquid water phase is present, the water vapor exists as a saturated vapor and the liquid is a saturated liquid, each at the system temperature.
4. There is no work during the cooling process and no change in kinetic or potential energy.

Analysis: Reduction of the closed system energy balance using assumption 4 results in

$$\Delta U = Q - \overset{0}{\cancel{W}}$$

or

$$Q = U_2 - U_1$$

where

$$U_1 = m_a u_{a1} + m_{v1} u_{v1} = m_a u_{a1} + m_{v1} u_{g1}$$

and

$$U_2 = m_a u_{a2} + m_{v2} u_{v2} + m_{w2} u_{w2} = m_a u_{a2} + m_{v2} u_{g2} + m_{w2} u_{f2}$$

In these equations, the subscripts a, v, and w denote, respectively, dry air, water vapor, and liquid water. The specific internal energy of the water vapor at the initial state can be approximated as the saturated vapor value at T_1. At the final state, the water vapor is assumed to exist as a saturated vapor, so its specific internal energy is u_g at T_2. The liquid water at the final state is saturated, so its specific internal energy is u_f at T_2.

Collecting the last three equations

$$Q = \underline{m_a(u_{a2} - u_{a1})} + \underline{m_{v2} u_{g2} + m_{w2} u_{f2} - m_{v1} u_{g1}}$$

The mass of dry air, m_a, can be found using the ideal gas equation of state together with the partial pressure of the dry air at the initial state obtained from the solution to Example 12.9

$$m_a = \frac{p_{a1} V}{(\bar{R}/M_a)(T_1)} = \frac{[(1.5 - 0.1985) \times 10^5 \text{ N/m}^2](35 \text{ m}^3)}{(8314/28.97 \text{ N}\cdot\text{m/kg}\cdot\text{K})(393 \text{ K})}$$
$$= 40.389 \text{ kg}$$

Then, evaluating internal energies of dry air and water from Tables A-16 and A-2, respectively

$$Q = 40.389(210.49 - 281.1) + 0.681(2405.7) + 3.146(92.32) - 3.827(2529.3)$$
$$= -2851.87 + 1638.28 + 290.44 - 9679.63 = -10{,}603 \text{ kJ}$$

The values for m_{v1}, m_{v2}, and m_{w2} are from the solution to Example 12.9.

Comments:

1. The first underlined term in the equation for Q is evaluated with specific internal energies from the ideal gas table for air, Table A-16. Steam table data are used to evaluate the second underlined term. The different datums for internal energy underlying these tables cancel because each of these two terms involves internal energy *differences.*
2. The specific heat c_{va} for dry air varies only slightly over the interval from 120 to 22°C (Table A-14); hence, the specific internal energy change of the dry air can be evaluated alternatively using a constant c_{va} value. This is left as an exercise.

12.8 ADIABATIC-SATURATION AND WET-BULB TEMPERATURES

The humidity ratio ω of an air–water vapor mixture can be determined knowing the values of three mixture properties: the pressure p, the temperature T, and the *adiabatic-saturation* temperature T_{as} introduced in this section. The relationship among these quantities is given by

$$\omega = \frac{h_a(T_{as}) - h_a(T) + \omega'[h_g(T_{as}) - h_f(T_{as})]}{h_g(T) - h_f(T_{as})} \tag{12.52}$$

In this equation, h_f and h_g denote the enthalpies of saturated liquid water and saturated water vapor, respectively, obtained from the steam tables at the indicated temperatures. The enthalpies of the dry air h_a can be obtained from the ideal gas table for air. Alternatively, $h_a(T_{as}) - h_a(T) = c_{pa}(T_{as} - T)$, where c_{pa} is an appropriate constant value for the specific heat of dry air. The humidity ratio ω' appearing in Eq. 12.52 is

$$\omega' = 0.622 \frac{p_g(T_{as})}{p - p_g(T_{as})} \tag{12.53}$$

where $p_g(T_{as})$ is the saturation pressure at the adiabatic-saturation temperature and p is the mixture pressure.

The object of this section is to develop Eq. 12.52 and discuss its use. We begin by considering the operation of a device known as an *adiabatic saturator.*

12.8.1 Adiabatic Saturation Process

Figure 12.8 shows the schematic and process representations of an adiabatic saturator, which is a two-inlet, single-exit device through which moist air passes. The device is assumed to operate at steady state and without significant heat transfer with its surroundings. An air–water vapor mixture of *unknown* humidity ratio ω enters the adiabatic saturator at a known pressure p and temperature T. As the mixture passes through the device, it comes into contact with a pool of water. If the entering mixture is not saturated ($\phi < 100\%$), some of the water would evaporate. The energy required to evaporate the water would come from the moist air, so the mixture temperature would decrease as the air passes through the duct. For a sufficiently long duct, the mixture would be saturated as it exits ($\phi = 100\%$). Since a saturated mixture would be achieved without heat transfer with the surroundings, the temperature of the exiting mixture is called the *adiabatic-saturation* temperature. As indicated on Fig. 12.8, a steady flow of makeup water at temperature T_{as} is added at the same rate

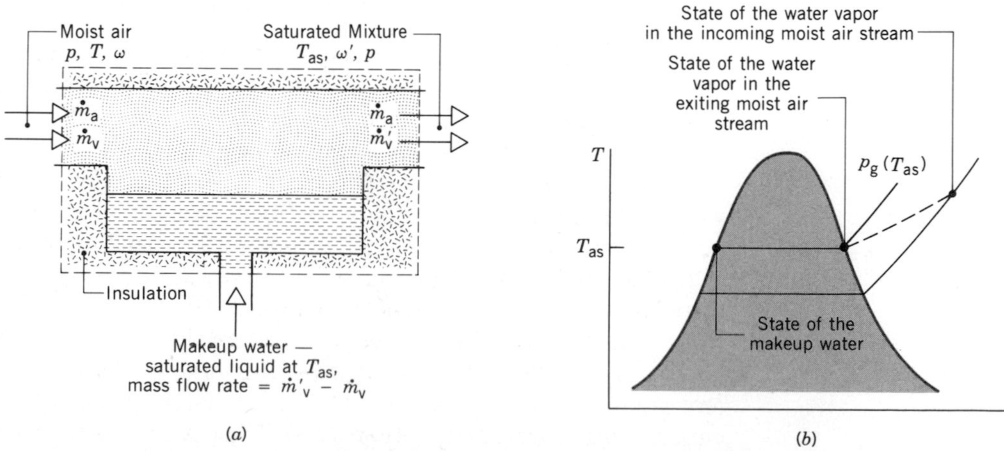

Figure 12.8 Adiabatic saturator. (*a*) Schematic. (*b*) Process representation.

at which water is evaporated. The pressure of the mixture is assumed to remain constant as it passes through the device.

A relationship for the humidity ratio ω of the entering moist air in terms of p, T, and T_{as} can be obtained by applying the control volume approach introduced in Sec. 12.7. At steady state, the mass flow rate of the dry air entering the device, $\dot{m}_a$, must equal the mass flow rate of the dry air exiting. The mass flow rate of the makeup water is the difference between the exiting and entering vapor flow rates denoted by $\dot{m}_v$ and $\dot{m}'_v$, respectively. These flow rates are labeled on Fig. 12.8*a*. At steady state, the energy rate balance reduces to

$$\underset{\text{entering}}{(\dot{m}_a h_a + \dot{m}_v h_v)_{\text{moist air}}} + \underset{\text{water}}{[(\dot{m}'_v - \dot{m}_v)h_w]_{\text{makeup}}} = \underset{\text{exiting}}{(\dot{m}_a h_a + \dot{m}'_v h_v)_{\text{moist air}}}$$

Several assumptions underlie this expression: Each of the two moist air streams is modeled as an ideal gas mixture of dry air and water vapor. Heat transfer with the surroundings is assumed to be negligible. There is no work $\dot{W}_{cv}$, and changes in kinetic and potential energy are ignored.

Dividing by the mass flow rate of the dry air, $\dot{m}_a$, the energy rate balance can be written on the basis of a unit mass of dry air passing through the device as

$$\underset{\text{entering}}{(h_a + \omega h_g)_{\text{moist air}}} + \underset{\text{water}}{[(\omega' - \omega)h_f]_{\text{makeup}}} = \underset{\text{exiting}}{(h_a + \omega' h_g)_{\text{moist air}}} \tag{12.54}$$

where $\omega = \dot{m}_v/\dot{m}_a$ and $\omega' = \dot{m}'_v/\dot{m}_a$. For the exiting saturated mixture, the partial pressure of the water vapor is the saturation pressure corresponding to the adiabatic-saturation temperature, $p_g(T_{as})$. Accordingly, the humidity ratio ω' can be evaluated knowing T_{as} and the mixture pressure p, as indicated by Eq. 12.53. In writing Eq. 12.54, the specific enthalpy of the entering water vapor has been evaluated as that of saturated water vapor at the temperature of the incoming mixture, in accordance with Eq. 12.47. Since the exiting mixture is saturated, the enthalpy of the water vapor at the exit is given by the saturated vapor value at T_{as}. The enthalpy of the makeup water is evaluated as that of saturated liquid at T_{as}.

When Eq. 12.54 is solved for ω, Eq. 12.52 results. The details of the solution are left as an exercise. Although derived with reference to an adiabatic saturator, the relationship provided by Eq. 12.52 applies generally to moist air mixtures and is not restricted to this type of system or even to control volumes. The relationship allows the humidity ratio ω to be determined for *any* moist air mixture for which the pressure p, temperature T, and adiabatic-saturation temperature T_{as} are known.

12.8.2 Wet-Bulb and Dry-Bulb Temperatures

For air–water vapor mixtures in the normal pressure and temperature range of atmospheric air, the adiabatic-saturation temperature is closely approximated by the *wet-bulb* temperature T_{wb}. Accordingly, *to determine the humidity ratio ω for such mixtures, the wet-bulb temperature can be used in Eq. 12.52 in place of the adiabatic-saturation temperature*. Close agreement between the adiabatic-saturation and wet-bulb temperatures is not generally found for moist air departing from these normal conditions or for gas–vapor mixtures other than moist air.

The wet-bulb temperature is read from a wet-bulb thermometer, which is an ordinary liquid-in-glass thermometer whose bulb is enclosed by a wick moistened with water. The term *dry-bulb temperature* refers simply to the temperature that would be measured by a thermometer placed in the mixture. Often a wet-bulb thermometer is mounted together with a dry-bulb thermometer on a holder to form an instrument called a *sling psychrometer* (Fig. 12.9). The psychrometer is whirled around in the air whose humidity ratio ω is to be determined. This induces air to flow over the two

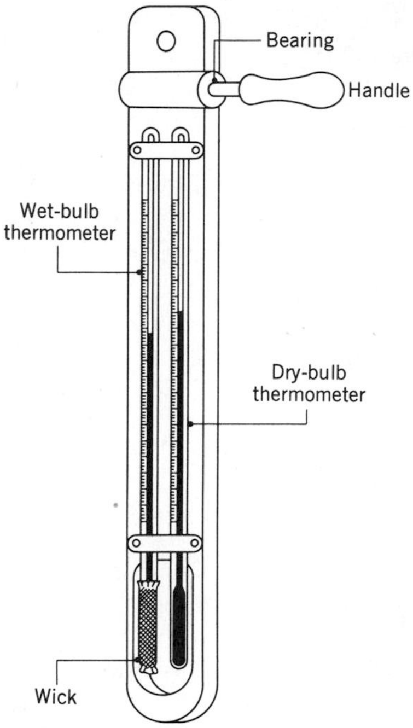

Figure 12.9 Sling psychrometer.

thermometers. If the surrounding air is not saturated, the water in the wick of the wet-bulb thermometer evaporates and the water temperature falls below the dry-bulb temperature. Eventually a steady-state condition is attained by the wet-bulb thermometer. The wet- and dry-bulb temperatures are then read from the respective thermometers. The wet-bulb temperature depends on the rates of heat and mass transfer between the moistened wick and the air. Since these depend in turn on the geometry of the thermometer, air velocity, supply water temperature, and other factors, the wet-bulb temperature is not a mixture property.

EXAMPLE 12.12

PROBLEM

A sling psychrometer indicates that in a classroom the dry-bulb temperature is 68°F and the wet-bulb temperature is 60°F. Determine the humidity ratio and relative humidity if the pressure is 14.7 lbf/in.^{2}

SOLUTION

Known: The dry-bulb and wet-bulb temperatures are known in a classroom where the pressure is 14.7 lbf/in.^{2}

Find: Determine the humidity ratio and relative humidity.

Assumptions: The measured wet-bulb temperature can be used in place of the adiabatic-saturation temperature to evaluate ω.

Analysis: The humidity ratio ω can be determined from Eq. 12.52. The value of ω' in this equation is found using p_g at the wet-bulb temperature from Table A-2E

$$\omega' = 0.622\left(\frac{0.2563}{14.7 - 0.2563}\right) = 0.011 \frac{\text{lb (vapor)}}{\text{lb (dry air)}}$$

Then, substituting into Eq. 12.52, the humidity ratio ω is

$$\omega = \frac{(124.3 - 126.2) + 0.011(1059.6)}{1091.2 - 28.08} = 0.0092 \frac{\text{lb (vapor)}}{\text{lb (dry air)}}$$

where the enthalpies are obtained from Tables A-2E and A-16E, as appropriate, and the wet-bulb temperature is used in place of the adiabatic-saturation temperature.

The relative humidity can be found from Eq. 12.44. First, an expression for the partial pressure of the water vapor is obtained by solving Eq. 12.43

$$p_v = \frac{\omega p}{\omega + 0.622}$$

Substituting known values

$$p_v = \frac{(0.0092)(14.7)}{(0.0092 + 0.622)} = 0.2143 \text{ lbf/in.}^2$$

Then, with p_g at 68°F from Table A-2E

$$\phi = \frac{p_v}{p_g} = \frac{0.2143}{0.3391} = 0.63\ (63\%)$$

Comment: In the present case, the relationship provided by Eq. 12.52 is used to obtain ω for a fixed quantity of moist air, a closed system. The same expression can be applied, however, to determine the humidity ratio for moist air entering or exiting a control volume. All that is required for any of these applications is the temperature T, the pressure p, and the wet-bulb temperature T_{wb}, which is used in place of T_{as} in Eq. 12.52.

12.9 PSYCHROMETRIC CHARTS

Graphical representations of several important properties of moist air are provided by *psychrometric charts*. The main features of one form of chart are shown in Fig. 12.10. Complete charts in SI and English units are given in Figs. A-9 and A-9E. These charts are limited to a mixture pressure of 1 atm, but charts for other mixture pressures also can be constructed.

Referring to Fig. 12.10, note that the abscissa gives the dry-bulb temperature and the ordinate provides the humidity ratio. For charts in SI, the temperature is in °C

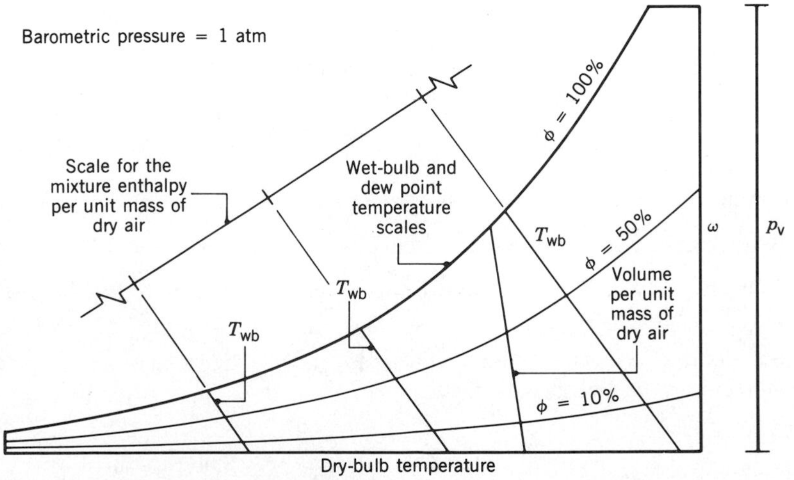

Figure 12.10 Psychrometric chart.

and ω is expressed in kg, or g, of water vapor per kg of dry air. Charts in English units give temperature in °F and ω in lb, or *grains*, of water vapor per lb of dry air, where 1 lb = 7000 grains.

Equation 12.43 shows that for fixed mixture pressure there is a direct correspondence between the partial pressure of the water vapor and the humidity ratio. Accordingly, the vapor pressure also can be shown on the ordinate, as illustrated on Fig. 12.10.

Curves of constant relative humidity are also drawn on psychrometric charts. On Fig. 12.10, curves labeled ϕ = 100, 50, and 10% are indicated. Since the dew point is the state where the mixture becomes saturated when cooled at constant vapor pressure, the dew point temperature corresponding to a given moist air state can be determined by following a line of constant ω (constant p_v) to the saturation line, ϕ = 100%. The dew point temperature and dry-bulb temperature are identical for states on the saturation curve.

Psychrometric charts also give values of the mixture enthalpy per unit mass of dry air in the mixture: $h_a + \omega h_v$. In Figs. A-9 and A-9E, the mixture enthalpy has units of kJ per kg of dry air and Btu per lb of dry air, respectively. The numerical values provided on these charts are determined relative to the following *special* reference states and reference values. In Fig. A-9, the enthalpy of the dry air h_a is determined relative to a zero value at 0°C, and not 0 K as in Tables A-16. Accordingly, in place of Eq. 3.49 used to develop the enthalpy data of Tables A-16, the following expression is employed to evaluate the enthalpy of the dry air for use on the psychrometric chart:

$$h_a = \int_{273.15\ \mathrm{K}}^{T} c_{pa}\, dT = c_{pa}T(^\circ\mathrm{C}) \tag{12.55}$$

where c_{pa} is a constant value for the specific heat c_p of dry air and T(°C) denotes the temperature in °C. For the chart in English units, Fig. A-9E, h_a is determined relative to a datum of 0°F, using $h_a = c_{pa}T(^\circ\mathrm{F})$, where T(°F) denotes the temperature in °F. In the temperature ranges of Figs. A-9 and A-9E, c_{pa} can be taken as 1.005 kJ/kg · K and 0.24 Btu/lb · °R, respectively. On Figs. A-9 the enthalpy of the water vapor h_v is evaluated as h_g at the dry-bulb temperature of the mixture from Table A-2 or A-2E, as appropriate.

Another important parameter on psychrometric charts is the wet-bulb temperature. As illustrated by Fig. 12.10, constant T_{wb} lines run from the upper left to the lower right of the chart. The relationship between the wet-bulb temperature and other chart quantities is provided by Eq. 12.52. The wet-bulb temperature can be used in this equation in place of the adiabatic-saturation temperature for the states of moist air located on Figs. A-9.

Lines of constant wet-bulb temperature are approximately lines of constant mixture enthalpy per unit mass of dry air. This feature can be brought out by study of the energy balance for the adiabatic saturator, Eq. 12.54. Since the contribution of the energy entering the adiabatic saturator with the makeup water is normally much smaller than that of the moist air, the enthalpy of the entering moist air is very nearly equal to the enthalpy of the saturated mixture exiting. Accordingly, all states with the same value of the wet-bulb temperature (adiabatic-saturation temperature) have nearly the same value for the mixture enthalpy per unit mass of dry air. Although Figs. A-9 ignore this slight effect, some psychrometric charts are drawn to show the departure of lines of constant wet-bulb temperature from lines of constant mixture enthalpy.

As indicated on Fig. 12.10, psychrometric charts also provide lines representing volume per unit mass of dry air, V/m_a, in units of m³/kg or ft³/lb. These specific volume lines can be interpreted as giving the volume of dry air or of water vapor, per unit mass of dry air, since each mixture component is considered to fill the entire volume.

EXAMPLE 12.13

PROBLEM

Solve Example 12.10 using the psychrometric chart.

SOLUTION

Known: Moist air that enters a duct at 10°C and $\phi = 80\%$ with a volumetric flow rate of 150 m^3/min is heated at a constant pressure of 1 bar and exits at 30°C. No moisture is added or removed.

Find: Determine the rate of heat transfer $\dot{Q}_{cv}$, in kJ/min.

Schematic and Given Data:

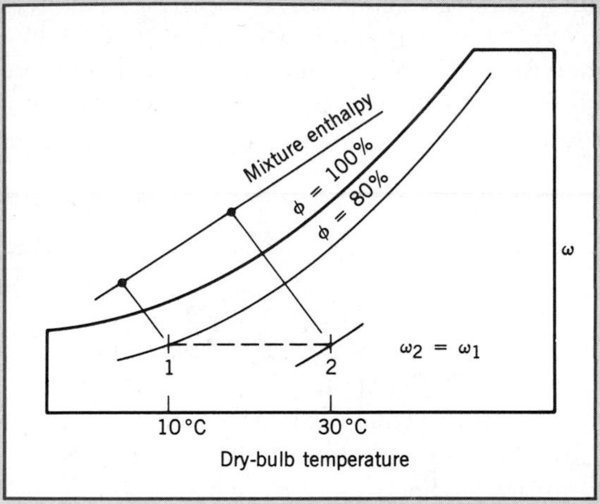

Figure E12.13

Assumptions: See the solution to Example 12.10.

Analysis: The heat transfer rate $\dot{Q}_{cv}$ can be determined from mass and energy rate balances, as in Example 12.10. The details of the present solution using the psychrometric chart differ, however.

As shown on the accompanying sketch of the psychrometric chart, the state of the moist air at the inlet is defined by $\phi_1 = 80\%$ and a dry-bulb temperature of 10°C. From the solution to Example 12.10, we know that the humidity ratio has the same value at the exit as at the inlet. Accordingly, the state of the moist air at the exit is fixed by $\omega_2 = \omega_1$ and a dry-bulb temperature of 30°C.

Reducing the energy rate balance as in Example 12.10 results in

$$\dot{Q}_{cv} = (\dot{m}_a h_{a2} + \dot{m}_v h_{v2}) - (\dot{m}_a h_{a1} + \dot{m}_v h_{v1})$$

With $\dot{m}_v = \omega \dot{m}_a$

$$\dot{Q}_{cv} = \dot{m}_a[(h_a + \omega h_v)_2 - (h_a + \omega h_v)_1]$$

To evaluate $\dot{Q}_{cv}$ from this expression requires values for the mixture enthalpy per unit mass of dry air $(h_a + \omega h_v)$ at the inlet and exit. These can be determined by inspection of the psychrometric chart, Fig. A-9, as $(h_a + \omega h_v)_1 =$ 25.7 kJ/kg (dry air), $(h_a + \omega h_v)_2 = 45.9$ kJ/kg (dry air).

Using the specific volume value from the chart at the inlet state together with the given volumetric flow rate at the inlet, the mass flow rate of the dry air is found as

$$\dot{m}_a = \frac{150 \text{ m}^3/\text{min}}{0.81 \text{ m}^3/\text{kg(dry air)}} = 185 \frac{\text{kg (dry air)}}{\text{min}}$$

Substituting values into the energy rate balance

$$\dot{Q}_{cv} = 185(45.9 - 25.7) = 3737 \text{ kJ/min}$$

Comment: The value for $\dot{Q}_{cv}$ found with the psychrometric chart agrees with the result of Example 12.10 despite the fact that the chart is based on a mixture pressure of 1 atm.

12.10 PSYCHROMETRIC APPLICATIONS

The purpose of the present section is to study typical air-conditioning processes using the psychrometric principles developed in this chapter. Specific illustrations are provided in the form of solved examples involving control volumes at steady state. In each case, the methodology introduced in Sec. 12.7 is employed to arrive at the solution. This material should be reviewed now, if necessary. To reinforce the psychrometric principles developed in the present chapter, the required psychrometric parameters are determined in most cases using tabular data provided in the appendix. It is left as an exercise to check these values by means of a psychrometric chart.

12.10.1 Dehumidification

When a moist air stream is cooled at constant mixture pressure to a temperature below its dew point temperature, some condensation of the water vapor initially present would occur. Figure 12.11 shows the schematic of a dehumidifier using this principle. Moist air enters at state 1 and flows across a cooling coil through which a refrigerant or chilled water circulates. Some of the water vapor initially present in

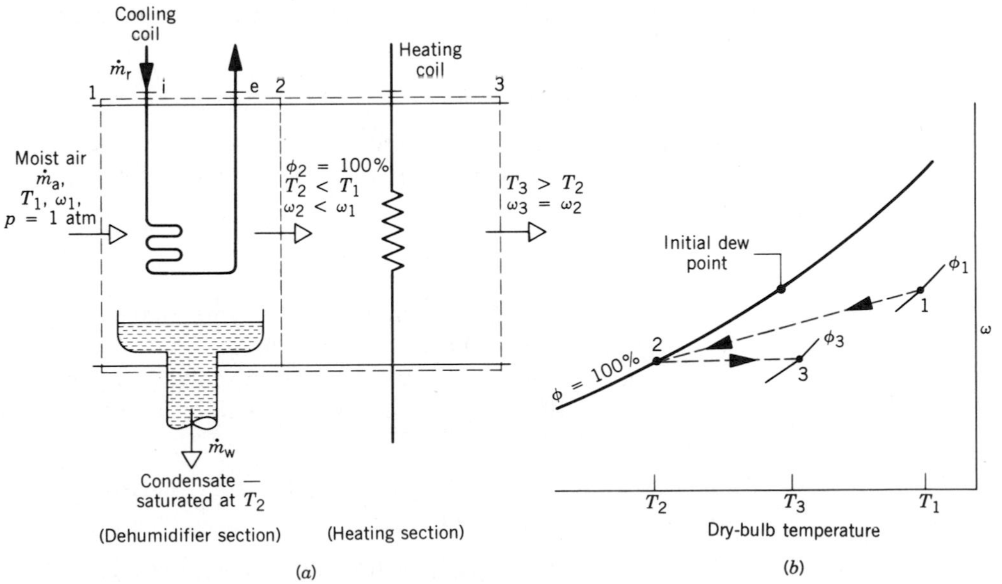

Figure 12.11 Dehumidification. (*a*) Equipment schematic. (*b*) Psychrometric chart representation.

the moist air condenses, and a saturated moist air mixture exits the dehumidifier section at state 2. Although water would condense at various temperatures, the condensed water is assumed to be cooled to T_2 before it exits the dehumidifier. Since the moist air leaving the humidifier is saturated at a temperature lower than the temperature of the moist air entering, the moist air stream might be unsuitable for direct use in occupied spaces. However, by passing the stream through a following heating section, it can be brought to a condition most occupants would regard as comfortable. Let us sketch the procedure for evaluating the rates at which condensate exits and refrigerant circulates.

The mass flow rate of the condensate $\dot{m}_w$ can be related to the mass flow rate of the dry air $\dot{m}_a$ by applying conservation of mass separately for the dry air and water passing through the dehumidifier section. At steady state

$$\dot{m}_{a1} = \dot{m}_{a2} \qquad \text{(dry air)}$$

$$\dot{m}_{v1} = \dot{m}_w + \dot{m}_{v2} \qquad \text{(water)}$$

The common mass flow rate of the dry air is denoted as $\dot{m}_a$. Solving for the mass flow rate of the condensate

$$\dot{m}_w = \dot{m}_{v1} - \dot{m}_{v2}$$

Introducing $\dot{m}_{v1} = \omega_1 \dot{m}_a$ and $\dot{m}_{v2} = \omega_2 \dot{m}_a$, the amount of water condensed per unit mass of dry air passing through the device is

$$\frac{\dot{m}_w}{\dot{m}_a} = \omega_1 - \omega_2$$

This expression requires the humidity ratios ω_1 and ω_2. Because no moisture is added or removed in the heating section, it can be concluded from conservation of mass that $\omega_2 = \omega_3$, so ω_3 can be used in the above equation in place of ω_2.

The mass flow rate of the refrigerant through the cooling coil $\dot{m}_r$ can be related to the mass flow rate of the dry air $\dot{m}_a$ by means of an energy rate balance applied to the dehumidifier section. With $\dot{W}_{cv} = 0$, negligible heat transfer with the surroundings, and no significant kinetic and potential energy changes, the energy rate balance reduces at steady state to

$$0 = \dot{m}_r(h_i - h_e) + (\dot{m}_a h_{a1} + \dot{m}_{v1} h_{v1}) - \dot{m}_w h_w - (\dot{m}_a h_{a2} + \dot{m}_{v2} h_{v2})$$

where h_i and h_e denote the specific enthalpy values of the refrigerant entering and exiting the dehumidifier section, respectively. Introducing $\dot{m}_{v1} = \omega_1 \dot{m}_a$, $\dot{m}_{v2} = \omega_2 \dot{m}_a$, and $\dot{m}_w = (\omega_1 - \omega_2)\dot{m}_a$

$$0 = \dot{m}_r(h_i - h_e) + \dot{m}_a[(h_{a1} - h_{a2}) + \omega_1 h_{g1} - \omega_2 h_{g2} - (\omega_1 - \omega_2)h_{f2}]$$

where the specific enthalpies of the water vapor at 1 and 2 are evaluated at the saturated vapor values corresponding to T_1 and T_2, respectively. Since the condensate is assumed to exit as a saturated liquid at T_2, $h_w = h_{f2}$. This equation can be solved for the refrigerant mass flow rate per unit mass of dry air flowing through the device

$$\frac{\dot{m}_r}{\dot{m}_a} = \frac{(h_{a1} - h_{a2}) + \omega_1 h_{g1} - \omega_2 h_{g2} - (\omega_1 - \omega_2)h_{f2}}{h_e - h_i}$$

The accompanying psychrometric chart, Fig. 12.11*b*, illustrates important features of the processes involved. As indicated by the chart, the moist air first cools from state 1, where the temperature is T_1 and the humidity ratio is ω_1, to state 2, where the mixture is saturated ($\phi_2 = 100\%$), the temperature is $T_2 < T_1$, and the humidity ratio is $\omega_2 < \omega_1$. During the subsequent heating process, the humidity ratio would remain constant, $\omega_2 = \omega_3$, and the temperature would increase to T_3. Since the states visited would not be equilibrium states, these processes are indicated on the psychrometric chart by dashed lines. The example to follow provides a specific illustration.

EXAMPLE 12.14

PROBLEM

Moist air at 30°C and 50% relative humidity enters a dehumidifier operating at steady state with a volumetric flow rate of 280 m³/min. The moist air passes over a cooling coil and water vapor condenses. Condensate exits the dehumidifier saturated at 10°C. Saturated moist air exits in a separate stream at the same temperature. There is no significant loss of energy by heat transfer to the surroundings and pressure remains constant at 1.013 bar. Determine (a) the mass flow rate of the dry air, in kg/min, (b) the rate at which water is condensed, in kg per kg of dry air flowing through the control volume, and (c) the required refrigerating capacity, in tons.

SOLUTION

Known: Moist air enters a dehumidifier at 30°C and 50% relative humidity with a volumetric flow rate of 280 m³/min. Condensate and moist air exit in separate streams at 10°C.

Determine: Find the mass flow rate of the dry air, in kg/min, the rate at which water is condensed, in kg per kg of dry air, and the required refrigerating capacity, in tons.

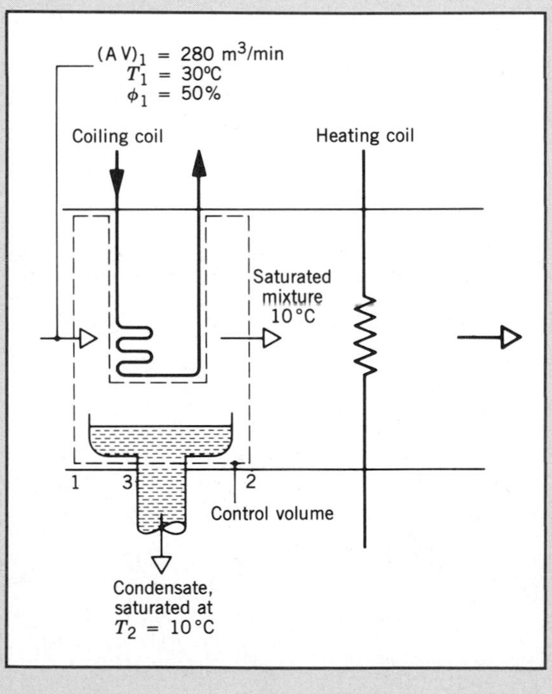

Figure E12.14

Assumptions:

1. The control volume shown in the accompanying figure operates at steady state. Changes in kinetic and potential energy can be neglected, and $\dot{W}_{cv} = 0$.
2. There is no significant heat transfer to the surroundings.
3. The pressure remains constant throughout at 1.013 bar.

4. At location 2, the moist air is saturated. The condensate exits at location 3 as a saturated liquid at temperature T_2.

Analysis:

(a) At steady state, the mass flow rates of the dry air entering and exiting are equal. The common mass flow rate of the dry air can be determined from the volumetric flow rate at the inlet

$$\dot{m}_a = \frac{(AV)_1}{v_{a1}}$$

The specific volume of the dry air at inlet 1, v_{a1}, can be evaluated using the ideal gas equation of state, so

$$\dot{m}_a = \frac{(AV)_1}{(\bar{R}/M_a)(T_1/p_{a1})}$$

The partial pressure of the dry air p_{a1} can be determined from $p_{a1} = p_1 - p_{v1}$. Using the relative humidity at the inlet ϕ_1 and the saturation pressure at 30°C from Table A-2

$$p_{v1} = \phi_1 p_{g1} = (0.5)(0.04246) = 0.02123 \text{ bar}$$

Thus, $p_{a1} = 1.013 - 0.02123 = 0.99177$ bar. Inserting values into the expression for $\dot{m}_a$ gives

$$\dot{m}_a = \frac{(280 \text{ m}^3/\text{min})(0.99177 \times 10^5 \text{ N/m}^2)}{(8314/28.97 \text{ N}\cdot\text{m/kg}\cdot\text{K})(303 \text{ K})} = 319.35 \text{ kg/min}$$

(b) Conservation of mass for the water requires $\dot{m}_{v1} = \dot{m}_{v2} + \dot{m}_w$. With $\dot{m}_{v1} = \omega_1 \dot{m}_a$ and $\dot{m}_{v2} = \omega_2 \dot{m}_a$, the rate at which water is condensed per unit mass of dry air is

$$\frac{\dot{m}_w}{\dot{m}_a} = \omega_1 - \omega_2$$

The humidity ratios ω_1 and ω_2 can be evaluated using Eq. 12.43. Thus, ω_1 is

$$\omega_1 = 0.622\left(\frac{p_{v1}}{p_1 - p_{v1}}\right) = 0.622\left(\frac{0.02123}{0.99177}\right) = 0.0133 \frac{\text{kg(vapor)}}{\text{kg(dry air)}}$$

Since the moist air is saturated at 10°C, p_{v2} equals the saturation pressure at 10°C: $p_g = 0.01228$ bar from Table A-2. Equation 12.43 then gives $\omega_2 = 0.0076$ kg(vapor)/kg(dry air). With these values for ω_1 and ω_2

$$\frac{\dot{m}_w}{\dot{m}_a} = 0.0133 - 0.0076 = 0.0057 \frac{\text{kg(condensate)}}{\text{kg(dry air)}}$$

(c) The rate of heat transfer $\dot{Q}_{cv}$ between the moist air stream and the refrigerant coil can be determined using an energy rate balance. With assumptions 1 and 2, the steady-state form of the energy rate balance reduces to

$$0 = \dot{Q}_{cv} + (\dot{m}_a h_{a1} + \dot{m}_{v1} h_{v1}) - \dot{m}_w h_w - (\dot{m}_a h_{a2} + \dot{m}_{v2} h_{v2})$$

With $\dot{m}_{v1} = \omega_1 \dot{m}_a$, $\dot{m}_{v2} = \omega_2 \dot{m}_a$, and $\dot{m}_w = (\omega_1 - \omega_2)\dot{m}_a$, this becomes

$$\dot{Q}_{cv} = \dot{m}_a[(h_{a2} - h_{a1}) - \omega_1 h_{g1} + \omega_2 h_{g2} + (\omega_1 - \omega_2) h_{f2}]$$

where the specific enthalpies of the water vapor at 1 and 2 are evaluated at the saturated vapor values corresponding to T_1 and T_2, respectively, and the specific enthalpy of the exiting condensate is evaluated as h_f at T_2. Selecting enthalpies from Tables A-2 and A-16, as appropriate

$$\begin{aligned}\dot{Q}_{cv} &= (319.35)[(283.1 - 303.2) - 0.0133(2556.3) \\ &\quad + 0.0076(2519.8) + 0.0057(42.01)] \\ &= -11{,}084 \text{ kJ/min}\end{aligned}$$

Since 1 ton of refrigeration equals a heat transfer rate of 211 kJ/min (Sec. 10.2), the required refrigerating capacity is 52.5 tons.

Comment: If a psychrometric chart were used to obtain a solution, the expression for $\dot{Q}_{cv}$ in part (c) would be rearranged to read

$$\dot{Q}_{cv} = \dot{m}_a[\underline{(h_a + \omega h_g)_2} - \underline{(h_a + \omega h_g)_1} + (\omega_1 - \omega_2) h_{f2}]$$

The underlined terms and humidity ratios ω_1 and ω_2 would be read directly from the chart; the specific enthalpy h_{f2} would be obtained from Table A-2.

12.10.2 Humidification

It is often necessary to increase the moisture content of the air circulated through occupied spaces. One way to accomplish this is to inject steam. Alternatively, liquid water can be sprayed into the air. Both cases are shown schematically in Fig. 12.12*a*. The temperature of the moist air as it exits the humidifier depends on the condition of the water introduced. When relatively high-temperature steam is injected, both the humidity ratio and the dry-bulb temperature would be increased. This is illustrated by the accompanying psychrometric chart of Fig. 12.12*b*. If liquid water was injected instead of steam, the moist air may exit the humidifier with a *lower* temperature than at the inlet. This is illustrated in Fig. 12.12*c*. The example to follow illustrates the case of steam injection. The case of liquid water injection is considered further in the next section.

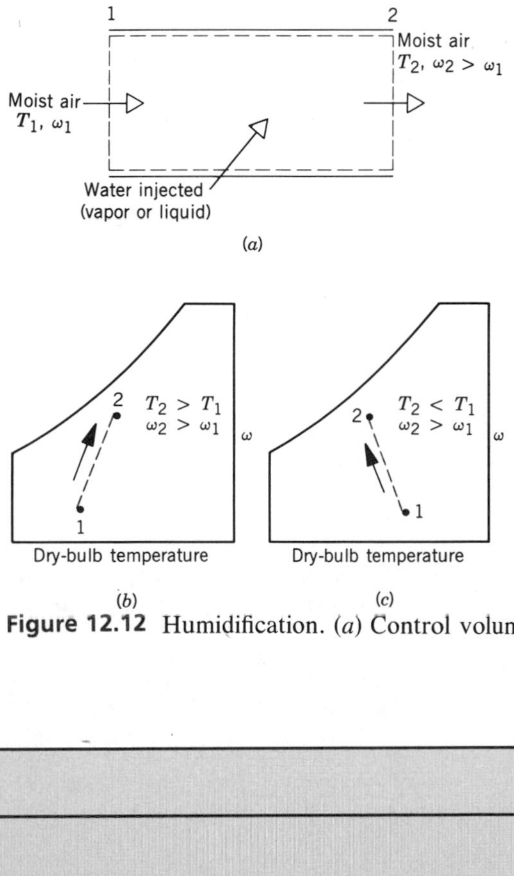

Figure 12.12 Humidification. (*a*) Control volume. (*b*) Steam injected. (*c*) Liquid injected.

EXAMPLE 12.15

PROBLEM

Moist air with a temperature of 22°C and a wet-bulb temperature of 9°C enters a steam-spray humidifier. The mass flow rate of the dry air is 90 kg/min. Saturated water vapor at 110°C is injected into the mixture at a rate of 52 kg/h. There is no heat transfer with the surroundings, and the pressure is constant throughout at 1 atm. Determine at the exit (a) the humidity ratio and (b) the temperature, in °C.

SOLUTION

Known: Moist air enters a humidifier at a temperature of 22°C and a wet-bulb temperature of 9°C. The mass flow rate of the dry air is 90 kg/min. Saturated water vapor at 110°C is injected into the mixture at a rate of 52 kg/h.

Find: Determine at the exit the humidity ratio and the temperature, in °C.

Schematic and Given Data:

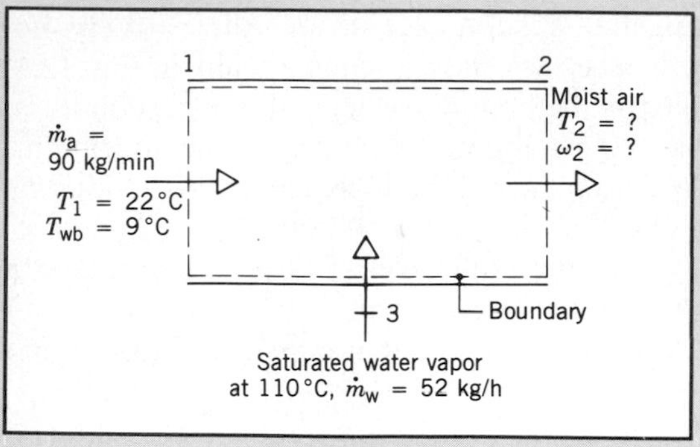

Figure E12.15

Assumptions:

1. The control volume shown in the accompanying figure operates at steady state. Changes in kinetic and potential energy can be neglected and $\dot{W}_{cv} = 0$.
2. There is no heat transfer with the surroundings.
3. The pressure remains constant throughout at 1 atm.

Solution:

(a) The humidity ratio at the exit ω_2 can be found from mass rate balances on the dry air and water individually. Thus

$$\dot{m}_{a1} = \dot{m}_{a2} \quad \text{(dry air)}$$

$$\dot{m}_{v1} + \dot{m}_w = \dot{m}_{v2} \quad \text{(water)}$$

With $\dot{m}_{v1} = \omega_1 \dot{m}_a$ and $\dot{m}_{v2} = \omega_2 \dot{m}_a$, where $\dot{m}_a$ is the mass flow rate of the air, the second of these becomes

$$\omega_2 = \omega_1 + \frac{\dot{m}_w}{\dot{m}_a}$$

The value of the humidity ratio ω_1 can be determined from Eq. 12.52 using the given wet-bulb temperature in place of the adiabatic-saturation temperature. Alternatively, the value can be found by inspection of the psychrometric chart, Fig. A-9. The result is $\omega_1 = 0.0019$ kg(vapor)/kg(dry air). This value should be verified as an exercise. Inserting values into the expression for ω_2

$$\omega_2 = 0.0019 + \frac{(52 \text{ kg/h})(\text{h}/60 \text{ min})}{90 \text{ kg/min}} = 0.0115 \frac{\text{kg(vapor)}}{\text{kg(dry air)}}$$

(b) The temperature at the exit can be determined by an iterative procedure using an energy rate balance. With assumptions 1 and 2, the steady-state form of the energy rate balance reduces to

$$0 = h_{a1} - h_{a2} + \omega_1 h_{g1} + (\omega_2 - \omega_1) h_{g3} - \omega_2 h_{g2}$$

In writing this, the specific enthalpies of the water vapor at 1 and 2 are evaluated as the respective saturated vapor values, and h_{g3} denotes the enthalpy of the saturated vapor injected into the moist air.

Evaluating enthalpies from Tables A-2 and A-16, as appropriate

$$0 = 295.17 - h_{a2} + 0.0019(2541.7) + (0.0115 - 0.0019)(2691.5) - 0.0115 h_{g2}$$

On rearrangement

$$h_{a2} + 0.0115 h_{g2} = 325.8 \text{ kJ/kg(dry air)}$$

The temperature at the exit can be found by the following iterative procedure. For each assumed value of T_2, the enthalpy of the dry air h_{a2} is obtained from Table A-16 and the enthalpy of the water vapor h_{g2} is obtained from Table A-2. The left side of the equation is evaluated with these values and compared with the value on the right. The procedure continues until satisfactory agreement is reached. The result is $T_2 = 24°C$, as can be verified.

Comment: An iterative solution to part (b) can be avoided by using the psychrometric chart, Fig. A-9. This is facilitated by writing the energy balance as

$$(h_a + \omega h_g)_2 = (h_a + \omega h_g)_1 + (\omega_2 - \omega_1)h_{g3}$$

The first term on the right can be obtained from Fig. A-9 at the state defined by the intersection of the inlet dry-bulb temperature, 22°C, and the inlet wet-bulb temperature, 9°C. The second term on the right can be evaluated with the known humidity ratios ω_1 and ω_2 and the value of h_{g3} from Table A-2. The state at the exit is fixed by ω_2 and the value of $(h_{a2} + \omega_2 h_{g2})$ as determined above. The temperature at the exit can then be read directly from the chart. The psychrometric chart solution is left as an exercise.

12.10.3 Evaporative Cooling

Cooling in hot, relatively dry climates can be accomplished by *evaporative cooling*. This involves either spraying liquid water into air or forcing air through a soaked pad that is kept replenished with water, as shown in Fig. 12.13. Owing to the low humidity of the moist air entering at state 1, part of the injected water evaporates. The energy for evaporation is provided by the air stream, which is reduced in temperature and exits at state 2 with a lower temperature than the entering stream. Because the incoming air is relatively dry, the additional moisture carried by the exiting moist air stream is normally beneficial.

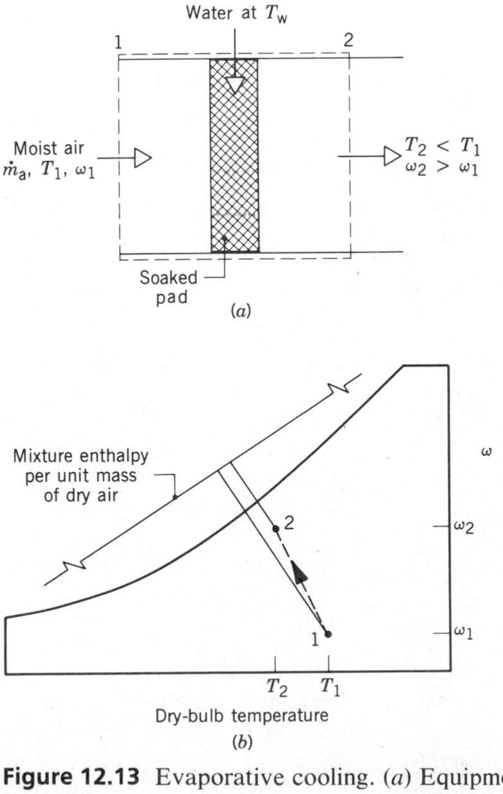

Figure 12.13 Evaporative cooling. (*a*) Equipment schematic. (*b*) Psychrometric chart representation.

For negligible heat transfer with the surroundings, no work $\dot{W}_{cv}$, and no significant changes in kinetic and potential energy, the steady-state forms of the mass and energy rate balances reduce for the control volume of Fig. 12.13*a* to

$$(h_{a2} + \omega_2 h_{g2}) = \underline{(\omega_2 - \omega_1)h_f} + (h_{a1} + \omega_1 h_{g1})$$

where h_f denotes the specific enthalpy of the liquid stream entering the control volume. All the injected water is assumed to evaporate into the moist air stream. The underlined term accounts for the energy carried in with the injected liquid water. This term is normally much smaller in magnitude than either of the two moist air enthalpy terms. Accordingly, the enthalpy of the moist air remains nearly constant, as illustrated on the psychrometric chart of Fig. 12.13*b*. Recalling that lines of constant mixture enthalpy are closely lines of constant wet-bulb temperature (Sec. 12.9), it follows that evaporative cooling takes place at a nearly constant wet-bulb temperature.

EXAMPLE 12.16

PROBLEM

Air at 100°F and 10% relative humidity enters an evaporative cooler with a volumetric flow rate of 5000 ft³/min. Moist air exits the cooler at 70°F. Water is added to the soaked pad of the cooler as a liquid at 70°F and evaporates fully into the moist air. There is no heat transfer with the surroundings and the pressure is constant throughout at 1 atm. Determine (a) the mass flow rate of the water to the soaked pad, in lb/h, and (b) the relative humidity of the moist air at the exit to the evaporative cooler.

SOLUTION

Known: Air at 100°F and ϕ = 10% enters an evaporative cooler with a volumetric flow rate of 5000 ft³/min. Moist air exits the cooler at 70°F. Water is added to the soaked pad of the cooler at 70°F.

Find: Determine the mass flow rate of the water to the soaked pad, in lb/h, and the relative humidity of the moist air at the exit of the cooler.

Schematic and Given Data:

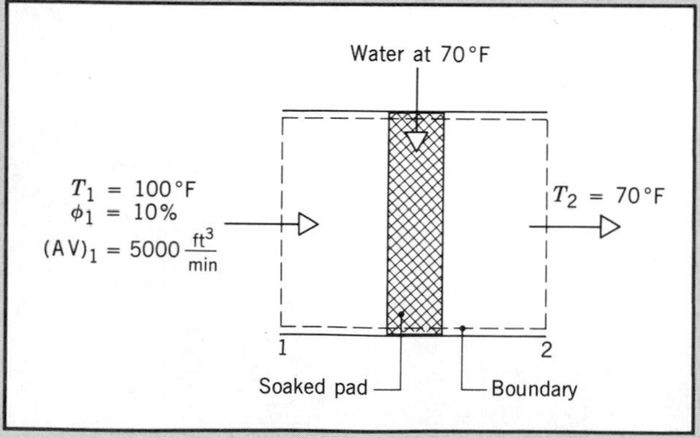

Figure E12.16

Assumptions:

1. The control volume shown in the accompanying figure operates at steady state. Changes in kinetic and potential energy can be neglected and $\dot{W}_{cv} = 0$.
2. There is no heat transfer with the surroundings.

3. The water added to the soaked pad enters as a liquid and evaporates fully into the moist air.
4. The pressure remains constant throughout at 1 atm.

Analysis:

(a) Applying conservation of mass to the dry air and water individually as in previous examples gives

$$\dot{m}_w = \dot{m}_a(\omega_2 - \omega_1)$$

where $\dot{m}_w$ is the mass flow rate of the water to the soaked pad. To find $\dot{m}_w$ requires ω_1, $\dot{m}_a$, and ω_2. These will now be evaluated in turn.

The humidity ratio ω_1 can be found from Eq. 12.43, which requires p_{v1}, the partial pressure of the moist air entering the control volume. Using the given relative humidity ϕ_1 and p_g at T_1 from Table A-2E, we have $p_{v1} = \phi_1 p_{g1} = 0.095$ lbf/in.^{2} With this, $\omega_1 = 0.00405$ lb(vapor)/lb(dry air).

The mass flow rate of the dry air $\dot{m}_a$ can be found as in previous examples, using the volumetric flow rate and specific volume of the dry air. Thus

$$\dot{m}_a = \frac{(\mathrm{AV})_1}{v_{a1}}$$

The specific volume of the dry air can be evaluated from the ideal gas equation of state. The result is $v_{a1} = 14.2$ ft^3/lb(dry air). Inserting values, the mass flow rate of the dry air is

$$\dot{m}_a = \frac{5000 \text{ ft}^3/\text{min}}{14.2 \text{ ft}^3/\text{lb(dry air)}} = 352.1 \frac{\text{lb(dry air)}}{\text{min}}$$

To find the humidity ratio ω_2, reduce the steady-state forms of the mass and energy rate balances using assumption 1 to obtain

$$0 = (\dot{m}_a h_{a1} + \dot{m}_{v1} h_{v1}) + \dot{m}_w h_w - (\dot{m}_a h_{a2} + \dot{m}_{v2} h_{v2})$$

With the same reasoning as in previous examples, this can be expressed as

$$0 = (h_a + \omega h_g)_1 + (\omega_2 - \omega_1) h_f - (h_a + \omega h_g)_2$$

where h_f denotes the specific enthalpy of the water entering the control volume at 70°F. Solving for ω_2

$$\omega_2 = \frac{h_{a1} - h_{a2} + \omega_1(h_{g1} - h_f)}{h_{g2} - h_f} = \frac{c_{pa}(T_1 - T_2) + \omega_1(h_{g1} - h_f)}{h_{g2} - h_f}$$

where $c_{pa} = 0.24$ Btu/lb·°R. With h_f, h_{g1}, and h_{g2} from Table A-2E

$$\omega_2 = \frac{0.24(100 - 70) + 0.00405(1105 - 38.1)}{(1092 - 38.1)}$$

$$= 0.0109 \frac{\text{lb(vapor)}}{\text{lb(dry air)}}$$

Substituting values for $\dot{m}_a$, ω_1, and ω_2 into the expression for $\dot{m}_w$

$$\dot{m}_w = \left[352.1 \frac{\text{lb(dry air)}}{\text{min}} \frac{60 \text{ min}}{\text{h}}\right] (0.0109 - 0.00405) \frac{\text{lb(water)}}{\text{lb(dry air)}}$$

$$= 144.7 \frac{\text{lb(water)}}{\text{h}}$$

(b) The relative humidity of the moist air at the exit can be determined using Eq. 12.44. The partial pressure of the water vapor required by this expression can be found by solving Eq. 12.43 to obtain

$$p_{v2} = \frac{\omega_2 p}{\omega_2 + 0.622}$$

Inserting values

$$p_{v2} = \frac{(0.0109)(14.696 \text{ lbf/in.}^2)}{(0.0109 + 0.622)} = 0.253 \text{ lbf/in.}^2$$

At 70°F, the saturation pressure is 0.3632 lbf/in.[2] Thus, the relative humidity at the exit is

$$\phi_2 = \frac{0.253}{0.3632} = 0.697(69.7\%)$$

Comments:

1. For illustrative purposes, a constant value of the specific heat c_{pa} has been used to evaluate the term $(h_{a1} - h_{a2})$. As shown in previous examples, this term can be evaluated alternatively using the ideal gas table for air.
2. It is left as an exercise to locate the incoming and outgoing moist air states on a psychrometric chart and verify that the evaporative cooling process takes place at a nearly constant wet-bulb temperature.

12.10.4 Adiabatic Mixing of Two Moist Air Streams

A common process in air-conditioning systems is the mixing of moist air streams, as shown in Fig. 12.14. The objective of the thermodynamic analysis of such a process is normally to fix the flow rate and state of the exiting stream for specified flow rates and states of each of the two inlet streams. The case of adiabatic mixing is governed by Eqs. 12.56 to follow.

The mass rate balances for the dry air and water vapor at steady state are, respectively,

$$\dot{m}_{a1} + \dot{m}_{a2} = \dot{m}_{a3} \qquad \text{(dry air)} \tag{12.56a}$$

and

$$\dot{m}_{v1} + \dot{m}_{v2} = \dot{m}_{v3} \qquad \text{(water vapor)}$$

With $\dot{m}_v = \omega \dot{m}_a$, the last equation becomes

$$\omega_1 \dot{m}_{a1} + \omega_2 \dot{m}_{a2} = \omega_3 \dot{m}_{a3} \tag{12.56b}$$

Assuming $\dot{Q}_{cv} = \dot{W}_{cv} = 0$ and ignoring the effects of kinetic and potential energy, the energy rate balance reduces at steady state to

$$\dot{m}_{a1}(h_{a1} + \omega_1 h_{g1}) + \dot{m}_{a2}(h_{a2} + \omega_2 h_{g2}) = \dot{m}_{a3}(h_{a3} + \omega_3 h_{g3}) \tag{12.56c}$$

where the enthalpies of the entering and exiting water vapor are evaluated as the saturated vapor values at the respective dry-bulb temperatures.

If the inlet flow rates and states are known, Eqs. 12.56 are three equations in three unknowns: $\dot{m}_{a3}$, ω_3, and $(h_{a3} + \omega_3 h_{g3})$. The solution of these equations is straightforward, as illustrated by the next example.

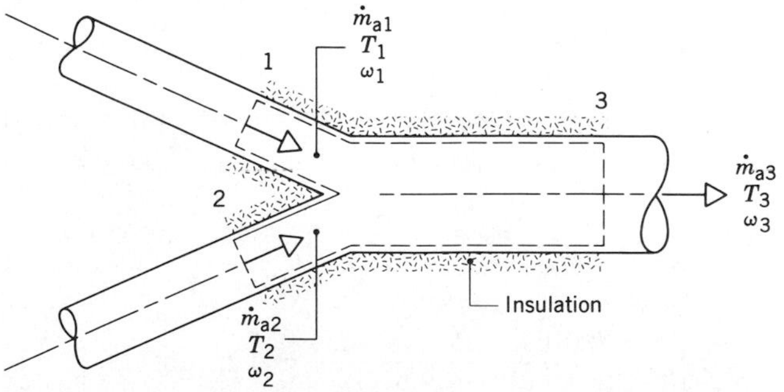

Figure 12.14 Adiabatic mixing of two moist air streams.

EXAMPLE 12.17

PROBLEM

A stream consisting of 142 m³/min of moist air at a temperature of 5°C and a humidity ratio of 0.002 kg(vapor)/kg(dry air) is mixed adiabatically with a second stream consisting of 425 m³/min of moist air at 24°C and 50% relative humidity. The pressure is constant throughout at 1 atm. Determine (a) the humidity ratio and (b) the temperature of the exiting mixed stream, in °C.

SOLUTION

Known: A moist air stream at 5°C, $\omega = 0.002$ kg(vapor)/kg(dry air), and a volumetric flow rate of 142 m³/min is mixed adiabatically with a stream consisting of 425 m³/min of moist air at 24°C and $\phi = 50\%$.

Find: Determine the humidity ratio and the temperature, in °C, of the mixed stream exiting the control volume.

Schematic and Given Data:

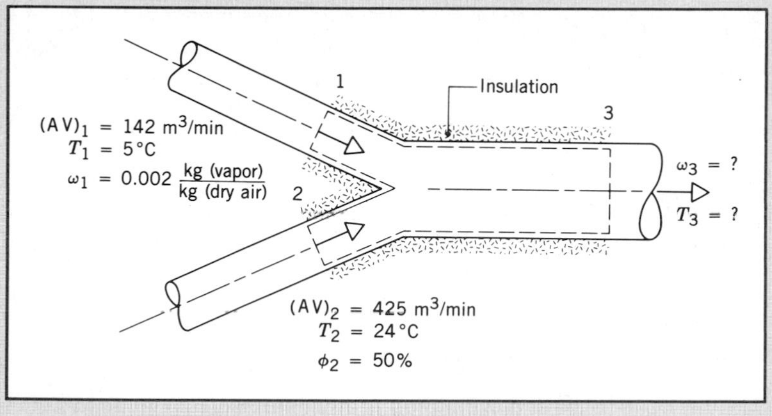

Figure E12.17

Assumptions:

1. The control volume shown in the accompanying figure operates at steady state. Changes in kinetic and potential energy can be neglected and $\dot{W}_{cv} = 0$.
2. There is no heat transfer with the surroundings.
3. The pressure remains constant throughout at 1 atm.

Analysis:

(a) The humidity ratio ω_3 can be found by means of mass rate balances for the dry air and water vapor, respectively

$$\dot{m}_{a1} + \dot{m}_{a2} = \dot{m}_{a3} \qquad \text{(dry air)}$$

$$\dot{m}_{v1} + \dot{m}_{v2} = \dot{m}_{v3} \qquad \text{(water vapor)}$$

With $\dot{m}_{v1} = \omega_1 \dot{m}_{a1}$, $\dot{m}_{v2} = \omega_2 \dot{m}_{a2}$, and $\dot{m}_{v3} = \omega_3 \dot{m}_{a3}$, the second of these balances becomes

$$\omega_1 \dot{m}_{a1} + \omega_2 \dot{m}_{a2} = \omega_3 \dot{m}_{a3}$$

Solving

$$\omega_3 = \frac{\omega_1 \dot{m}_{a1} + \omega_2 \dot{m}_{a2}}{\dot{m}_{a3}}$$

Since $\dot{m}_{a3} = \dot{m}_{a1} + \dot{m}_{a2}$, this can be expressed as

$$\omega_3 = \frac{\omega_1 \dot{m}_{a1} + \omega_2 \dot{m}_{a2}}{\dot{m}_{a1} + \dot{m}_{a2}}$$

To determine ω_3 requires values for ω_2, $\dot{m}_{a1}$, and $\dot{m}_{a2}$. The mass flow rates of the dry air, $\dot{m}_{a1}$ and $\dot{m}_{a2}$, can be found as in previous examples using the given volumetric flow rates

$$\dot{m}_{a1} = \frac{(AV)_1}{v_{a1}}, \qquad \dot{m}_{a2} = \frac{(AV)_2}{v_{a2}}$$

The values of v_{a1}, and v_{a2}, and ω_2 are readily found from the psychrometric chart, Fig. A-9. Thus, at $\omega_1 = 0.002$ and $T_1 = 5°C$, $v_{a1} = 0.79$ m³/kg(dry air). At $\phi_2 = 50\%$ and $T_2 = 24°C$, $v_{a2} = 0.854$ m³/kg(dry air) and $\omega_2 = 0.0093$. The mass flow rates of the dry air are then $\dot{m}_{a1} = 179.7$ kg/(dry air)/min and $\dot{m}_{a2} = 497.7$ kg(dry air)/min. Inserting values into the expression for ω_3

$$\omega_3 = \frac{(0.002)(179.7) + (0.0093)(497.7)}{179.7 + 497.7} = 0.0074 \frac{\text{kg(vapor)}}{\text{kg(dry air)}}$$

(b) The temperature T_3 of the exiting mixed stream can be found from an energy rate balance. Reduction of the energy rate balance using assumptions 1 and 2 gives

$$\dot{m}_{a1}(h_a + \omega h_v)_1 + \dot{m}_{a2}(h_a + \omega h_v)_2 = \dot{m}_{a3}(h_a + \omega h_v)_3$$

Solving

$$(h_{a3} + \omega_3 h_v)_3 = \frac{\dot{m}_{a1}(h_a + \omega h_v)_1 + \dot{m}_{a2}(h_a + \omega h_v)_2}{\dot{m}_{a1} + \dot{m}_{a2}}$$

With $(h_a + \omega h_v)_1 = 10$ kJ/kg(dry air) and $(h_a + \omega h_v)_2 = 47.8$ kJ/kg(dry air) from Fig. A-9 and other known values

$$(h_a + \omega h_v)_3 = \frac{179.7(10) + 497.7(47.8)}{179.7 + 497.7} = 37.8 \frac{\text{kJ}}{\text{kg(dry air)}}$$

This value for the enthalpy of the moist air at the exit, together with the previously determined value for ω_3, fixes the state of the exiting moist air. From inspection of Fig. A-9, $T_3 = 19°C$.

Comment: Use of the psychrometric chart facilitates the solution of this example.

12.10.5 Cooling Towers

Power plants invariably discharge considerable energy to their surroundings by heat transfer (Chap. 8). Although water drawn from a nearby river or lake is often employed to carry away this energy, cooling towers provide an alternative in locations where sufficient cooling water cannot be obtained from natural sources or where concerns for the environment place a limit on the temperature at which cooling water can be returned to the surroundings. Cooling towers also are frequently employed to provide chilled water for applications other than those involving power plants.

Cooling towers can operate by *natural* or *forced* convection. Also they may be *counterflow, cross-flow,* or a combination of these. A schematic diagram of a forced-convection, counterflow cooling tower is shown in Fig. 12.15. The warm water to be cooled enters at 1 and is sprayed from the top of the tower. The falling water usually passes through a series of baffles intended to keep it broken up into fine drops to promote evaporation. Atmospheric air drawn in at 3 by the fan flows upward, counter to the direction of the falling water droplets. As the two streams interact, a small fraction of the water stream evaporates into the moist air, which exits at 4 with a greater humidity ratio than the incoming moist air at 3. The energy required for evaporation is provided mainly by the portion of the incoming water stream that does not evaporate, with the result that the water exiting at 2 is at a lower temperature than the water entering at 1. Since some of the incoming water is evaporated into the moist air stream, an equivalent amount of makeup water is added at 5 so that the return mass flow rate of the cool water equals the mass flow rate of the warm water entering at 1.

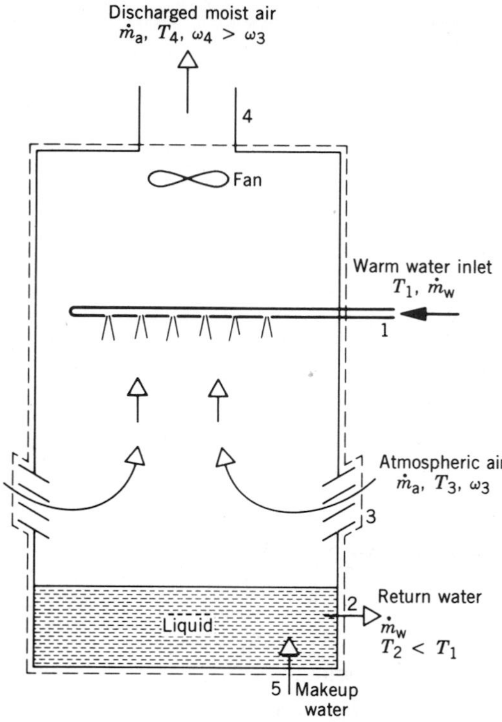

Figure 12.15 Schematic of a cooling tower.

For operation at steady state, mass balances for the dry air and water and an energy balance on the overall cooling tower provide information about cooling tower performance. In applying the energy balance, heat transfer with the surroundings is usually neglected. The power input to the fan of forced-convection towers also may be negligible relative to other energy rates involved. The example to follow illustrates the analysis of a cooling tower using conservation of mass and energy together with property data for the dry air and water.

EXAMPLE 12.18

PROBLEM

Water exiting the condenser of a power plant at 38°C enters a cooling tower with a mass flow rate of 4.5×10^7 kg/h. A stream of cooled water is returned to the condenser from a cooling tower with a temperature of 30°C and the same flow rate. Makeup water is added in a separate stream at 20°C. Atmospheric air enters the cooling tower at 25°C and 35% relative humidity. Moist air exits the tower at 35°C and 90% relative humidity. Determine the mass flow rates of the dry air and the makeup water, in kg/h. The cooling tower operates at steady state. Heat transfer with the surroundings can be neglected, as can changes in kinetic and potential energy. The pressure remains constant throughout at 1 atm.

SOLUTION

Known: A liquid water stream enters a cooling tower from a condenser at 38°C with a known mass flow rate. A stream of cooled water is returned to the condenser at 30°C and the same flow rate. Makeup water is added at 20°C. Atmospheric air enters the tower at 25°C and $\phi = 35\%$. Moist air exits the tower at 35°C and $\phi = 90\%$.

Find: Determine the mass flow rates of the dry air and the makeup water, in kg/h.

Schematic and Given Data:

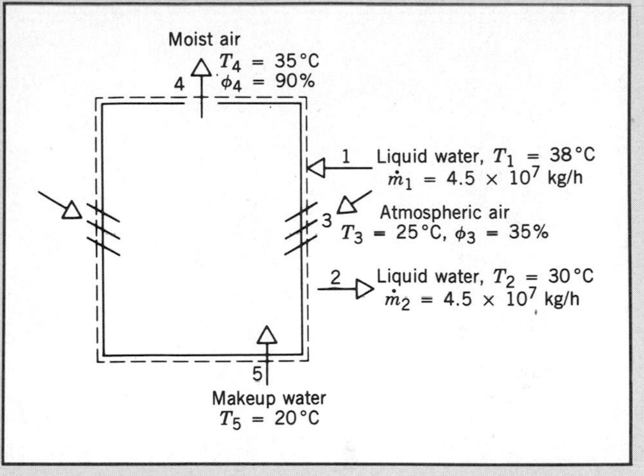

Figure E12.18

Assumptions:

1. The control volume shown in the accompanying figure operates at steady state. Heat transfer with the surroundings can be neglected, as can changes in kinetic and potential energy; also $\dot{W}_{cv} = 0$.
2. To evaluate specific enthalpies, each liquid stream is regarded as a saturated liquid at the corresponding specified temperature.
3. The pressure is constant throughout at 1 atm.

Analysis: The required mass flow rates can be found from mass and energy rate balances. Mass balances for the dry air and water individually reduce at steady state to

$$\dot{m}_{a3} = \dot{m}_{a4} \qquad \text{(dry air)}$$

$$\dot{m}_1 + \dot{m}_5 + \dot{m}_{v3} = \dot{m}_2 + \dot{m}_{v4} \qquad \text{(water)}$$

The common mass flow rate of the dry air is denoted as $\dot{m}_a$. Since $\dot{m}_1 = \dot{m}_2$, the second of these equations becomes

$$\dot{m}_5 = \dot{m}_{v4} - \dot{m}_{v3}$$

With $\dot{m}_{v3} = \omega_3 \dot{m}_a$ and $\dot{m}_{v4} = \omega_4 \dot{m}_a$

$$\dot{m}_5 = \dot{m}_a(\omega_4 - \omega_3)$$

Accordingly, the two required mass flow rates, $\dot{m}_a$ and $\dot{m}_5$, are related by this equation. Another equation relating the flow rates is provided by the energy rate balance.

Reducing the energy rate balance with assumption 1 results in

$$0 = \dot{m}_1 h_{w1} + (\dot{m}_a h_{a3} + \dot{m}_{v3} h_{v3}) + \dot{m}_5 h_{w5} - \dot{m}_2 h_{w2} - (\dot{m}_a h_{a4} + \dot{m}_{v4} h_{v4})$$

Evaluating the enthalpies of the water vapor as the saturated vapor values at the respective temperatures and the enthalpy of each liquid stream as the saturated liquid enthalpy at the respective temperature, the energy rate equation becomes

$$0 = \dot{m}_1 h_{f1} + (\dot{m}_a h_{a3} + \dot{m}_{v3} h_{g3}) + \dot{m}_5 h_{f5} - \dot{m}_2 h_{f2} - (\dot{m}_a h_{a4} + \dot{m}_{v4} h_{g4})$$

Introducing $\dot{m}_1 = \dot{m}_2$, $\dot{m}_5 = \dot{m}_a(\omega_4 - \omega_3)$, $\dot{m}_{v3} = \omega_3 \dot{m}_a$, and $\dot{m}_{v4} = \omega_4 \dot{m}_a$ and solving for $\dot{m}_a$

$$\dot{m}_a = \frac{\dot{m}_1(h_{f1} - h_{f2})}{h_{a4} - h_{a3} + \omega_4 h_{g4} - \omega_3 h_{g3} - (\omega_4 - \omega_3) h_{f5}}$$

The humidity ratios ω_3 and ω_4 required by this expression can be determined from Eq. 12.43, using the partial pressure of the water vapor obtained with the respective relative humidity. Thus, $\omega_3 = 0.00688$ kg(vapor)/kg(dry air) and $\omega_4 = 0.0327$ kg(vapor)/kg(dry air).

With enthalpies from Tables A-2 and A-16, as appropriate, and the known values for ω_3, ω_4, and $\dot{m}_1$, the expression for $\dot{m}_a$ becomes

$$\dot{m}_a = \frac{(4.5 \times 10^7)(159.21 - 125.79)}{(308.2 - 298.2) + (0.0327)(2565.3) - (0.00688)(2547.2) - (0.0258)(83.96)}$$

$$= 2.03 \times 10^7 \text{ kg/h}$$

Finally, inserting known values into the expression for $\dot{m}_5$ results in

$$\dot{m}_5 = (2.03 \times 10^7)(0.0327 - 0.00688) = 5.24 \times 10^5 \text{ kg/h}$$

Comment: The expression obtained above for $\dot{m}_a$ can be rearranged to read

$$\dot{m}_a = \frac{\dot{m}_1(h_{f1} - h_{f2})}{\underline{(h_{a4} + \omega_4 h_{g4})} - \underline{(h_{a3} + \omega_3 h_{g3})} - (\omega_4 - \omega_3)h_{f5}}$$

The underlined terms can be obtained by inspection of a suitable psychrometric chart.

PROBLEMS

12.1 Answer the following questions involving a mixture of two gases:

(a) When would the analysis of the mixture in terms of mass fractions be *identical* to the analysis in terms of mole fractions?
(b) When would the apparent molecular weight of the mixture equal the average of the molecular weights of the two gases?

12.2 The molar analysis of a gas mixture at 60°C, 1 bar is 70% N_2, 20% CO_2, 10% O_2. Determine

(a) the analysis in terms of mass fractions.
(b) the partial pressure of each component, in bar.
(c) the volume occupied by 20 kg of mixture, in m³.

12.3 The analysis on a molar basis of a gas mixture at 80°F, 30 lbf/in.² is 60% N_2, 33% CO_2, 7% O_2. Determine

(a) the analysis in terms of mass fractions.
(b) the partial pressure of each component, in lbf/in.²
(c) the volume occupied by 175 lb of mixture, in ft³.

12.4 The molar analysis of a gas mixture at 27°C, 0.1 MPa is 60% N_2, 30% CO_2, 10% O_2. Determine

(a) the analysis in terms of mass fractions.
(b) the partial pressure of each component, in MPa.
(c) the volume occupied by 40 kg of the mixture, in m³.

12.5 The analysis on a mass basis of a gas mixture at 40°F, 14.7 lbf/in.² is 60% CO_2, 25% CO, 15% O_2. Determine

(a) the analysis in terms of mole fractions.
(b) the partial pressure of each component, in lbf/in.²
(c) the volume occupied by 10 lb of the mixture, in ft³.

12.6 Natural gas at 27°C, 1 bar enters a furnace with the following molar analysis: 40% propane (C_3H_8), 40% ethane (C_2H_6), 20% methane (CH_4). Determine

(a) the analysis in terms of mass fractions.
(b) the partial pressure of each component, in bar.
(c) the mass flow rate, in kg/s, for a volumetric flow rate of 10 m³/s.

12.7 A rigid vessel having a volume of 0.028 m³ initially contains a mixture at 21°C, 1 bar consisting of 79% N_2 and 21% O_2 on a molar basis. Helium at 21°C is allowed to flow into the vessel until the pressure is 1.36 bar. If the final temperature of the mixture within the vessel is 21°C, determine the mass, in kg, of each component present.

12.8 A vessel having a volume of 0.28 m³ contains a mixture at 40°C, 6.9 bar with a molar analysis of 70% O_2, 30% CH_4. Determine the mass of methane that would have to be added and the mass of oxygen that would have to be removed, each in kg, to obtain a mixture having a molar analysis of 30% O_2, 70% CH_4 at the same temperature and pressure.

12.9 Four pounds of oxygen (O_2) is mixed with 8 lb of another gas to form a mixture that occupies 45 ft³ at 150°F, 40 lbf/in.² Applying ideal gas mixture principles, determine

(a) the molecular weight of the unspecified gas.
(b) the analysis of the mixture in terms of mole fractions.

12.10 A closed, rigid tank having a volume of 0.1 m³ contains 0.7 kg of N_2 and 1.1 kg of CO_2 at 27°C. Determine

(a) the composition in terms of mass fractions.
(b) the composition in terms of mole fractions.

(c) the partial pressure of each component, in bar.
(d) the mixture pressure, in bar.
(e) the heat transfer, in kJ, required to bring the mixture to 127°C.
(f) the entropy change of the mixture for the process of part (e), in kJ/K.

12.11 A gas mixture consisting of 3 lb of N_2 and 5 lb of CO_2 is maintained at a pressure of 40 lbf/in.² Determine

(a) the composition in terms of mass fractions.
(b) the composition in terms of mole fractions.
(c) the heat transfer, in Btu, required to increase the mixture temperature from 70 to 100°F.
(d) the change in entropy of the mixture for the process of part (c), in Btu/°R.

Solve parts (c) and (d) on the basis that the specific heats of the gases are constant.

12.12 A mixture having an analysis on a mass basis of 50% N_2, 30% CO_2, 20% O_2 is compressed adiabatically from 1 bar, 7°C to 5 bar, 177°C. Determine

(a) the work, in kJ per kg of mixture.
(b) the amount of entropy produced, in kJ/K per kg of mixture.

12.13 A mixture consisting of 0.6 lbmol of N_2 and 0.4 lbmol of O_2 is compressed isothermally at 1000°R from 1 to 3 atm. For the mixture, determine

(a) the work, in Btu.
(b) the heat transfer, in Btu.
(c) the amount of entropy produced, in Btu/°R.

If the mixture interacts thermally only with the surrounding atmosphere at 40°F, is entropy produced during the process? Where? Evaluate the amount of entropy produced, in Btu/°R.

12.14 A mixture of 5 lb of nitrogen and 2 lb of helium is compressed from 20 lbf/in.², 80°F to 80 lbf/in.², 240°F. Can this process be accomplished adiabatically? Explain.

12.15 A mixture having a molar analysis of 50% CO_2, 33.3% CO, and 16.7% O_2 enters a compressor at 37°C, 1 bar, 50 m/s with a mass flow rate of 5 kg/s and exits at 237°C, 80 m/s. Heat transfer from the compressor to its surroundings occurs at the rate of 3 kJ per kg of mixture flowing.

(a) Neglecting potential energy effects, determine the power input required by the compressor at steady state, in kW.
(b) If the compression is polytropic, evaluate the polytropic exponent n and the exit pressure, in bar.

12.16 A mixture of 2 kg of H_2 and 4 kg of N_2 is compressed in a piston–cylinder assembly in a polytropic process for which $n = 1.2$. The temperature increases from 22 to 150°C. Using *mean* values for the specific heats, determine

(a) the heat transfer, in kJ.
(b) the entropy change, in kJ/K.

12.17 A gas turbine receives a mixture having the following molar analysis: 19% CO_2, 1.95% O_2, 79.05% N_2 at 500 K, 0.3 MPa and a volumetric flow rate of 1800 m³/h. Products exit the turbine at 350 K, 0.11 MPa. For adiabatic operation, determine the power developed at steady state, in kW.

12.18 A mixture having a molar analysis of 66% N_2, 17% CO_2, and 17% water vapor enters a turbine at 1000 K, 8 bar with a mass flow rate of 2 kg/s and expands isentropically to a pressure of 1 bar. Ignoring kinetic and potential energy effects, determine for steady-state operation

(a) the temperature at the exit, in K.
(b) the power developed by the turbine, in kW.

12.19 An equimolar mixture of helium and carbon dioxide enters an insulated nozzle at 260°F, 5 atm, 100 ft/s and expands isentropically to a velocity of 1110 ft/s. Determine the temperature, in °F, and the pressure, in atm, at the nozzle exit. Neglect potential energy effects.

12.20 An equimolar mixture of helium and argon gases enters a turbine at 2340°F and expands adiabatically through a 7.5 : 1 pressure ratio. If the isentropic turbine efficiency is 80%, determine at steady state the power developed, in Btu per lb of mixture flowing.

12.21 A gas mixture having a molar analysis of 60% O_2 and 40% N_2 enters an insulated compressor operating at steady state at 1 bar, 20°C with a mass flow rate of 1 kg/s and is compressed to 5 bar. Kinetic and potential energy effects are negligible. For an isentropic compressor efficiency of 75%, determine

(a) the temperature at the exit, in °C.
(b) the power required, in kW.
(c) the rate of entropy production, in kW/K.

12.22 A mixture having an analysis on a mass basis of 80% N_2, 20% CO_2 enters a nozzle at 1000 K with a velocity of 5 m/s and expands adiabatically through a 7.5 : 1 pressure ratio, exiting with a velocity of 900 m/s. Determine the nozzle efficiency.

12.23 A mixture having a molar analysis of 60% N_2 and 40% CO_2 enters an insulated compressor operating at steady state at 1.5 bar, 30°C with a mass flow rate of 2 kg/s and is compressed to 4.5 bar, 142°C. Neglecting kinetic and potential energy effects, determine

(a) the power required, in kW.
(b) the isentropic compressor efficiency.
(c) the irreversibility rate, in kW, for $T_0 = 300$ K.

12.24 An equimolar mixture of N_2 and CO_2 enters a heat exchanger at −40°F, 500 lbf/in.² and exits at 500°F, 500 lbf/in.² The heat exchanger operates at steady state, and kinetic and potential energy effects are negligible.

(a) Using the ideal gas mixture concepts of the present chapter, determine the rate of heat transfer to the mixture, in Btu per lbmol of mixture flowing.
(b) Compare with the value of the heat transfer determined using the generalized enthalpy chart (Fig. A-4), together with Kay's rule (see Sec. 11.8).

12.25 Natural gas having a molar analysis of 60% methane (CH_4) and 40% ethane (C_2H_6) enters a compressor at 67°C, 6.8 atm and is compressed isothermally without internal irreversibilities to 23.8 atm. The compressor operates at steady state, and kinetic and potential energy effects are negligible.

(a) Using the ideal gas mixture concepts of the present chapter, determine the work input required by the compressor, in kJ per kmol of mixture flowing.
(b) Compare with the value for the work input determined using the generalized enthalpy and entropy charts (Figs. A-4 and A-5), together with Kay's rule (see Sec. 11.8).

12.26 One kilogram of argon at 27°C, 1 bar is contained in a rigid tank connected by a valve to another rigid tank containing 0.8 kg of O_2 at 127°C, 5 bar. The valve is opened, and the gases are allowed to mix, achieving an equilibrium state at 77°C. Determine

(a) the volume of each tank, in m^3.
(b) the final pressure, in bar.
(c) the heat transfer to or from the gases during the process, in kJ.
(d) the entropy change of each gas, in kJ/K.

12.27 Two kilograms of N_2 at 450 K, 7 bar are contained in a rigid tank connected by a valve to another rigid tank holding 1 kg of O_2 at 300 K, 3 bar. The valve is opened and the gases are allowed to mix, achieving an equilibrium state at 370 K. Determine

(a) the volume of each tank, in m^3.
(b) the final pressure, in bar.
(c) the heat transfer to or from the gases during the process, in kJ.
(d) the entropy change of each gas, in kJ/K.

12.28 Determine the amount of entropy produced, in Btu/°R, when 2 lb of O_2 at 150°F, 1 atm are allowed to mix adiabatically to a final equilibrium state with 6 lb of **(a)** N_2 and **(b)** O_2, initially at the same temperature and pressure.

12.29 An insulated tank having a total volume of 0.6 m^3 is divided into two compartments. Initially one compartment contains 0.4 m^3 of hydrogen (H_2) at 127°C, 2 bar and the other contains nitrogen (N_2) at 27°C, 4 bar. The gases are allowed to mix until an equilibrium state is attained. On the basis of constant specific heats, determine

(a) the final temperature, in °C.
(b) the final pressure, in bar.
(c) the amount of entropy produced, in kJ/K.

12.30 An insulated tank having a total volume of 3 ft^3 is divided into two compartments. Initially one compartment having a volume of 1 ft^3 contains 0.2 lb of carbon monoxide (CO) at 500°F and the other contains 0.04 lb of helium (He) at 60°F. The gases are allowed to mix until an equilibrium state is attained. Determine

(a) the final temperature, in °F.
(b) the final pressure, in lbf/in.^{2}
(c) the irreversibility, in Btu, for T_0 = 60°F.

12.31 A rigid insulated tank has two compartments. Initially one compartment is filled with 5 lbmol of argon at 100°F, 20 lbf/in.^{2} and the other is filled with 10 lbmol of helium at 200°F, 30 lbf/in.^{2} The gases are allowed to mix until an equilibrium state is attained. Determine

(a) the final temperature, in °F.
(b) the final pressure, in lbf/in.^{2}
(c) the amount of entropy produced, in Btu/°R.

12.32 A rigid insulated tank has two compartments. Initially one contains 1 kmol of carbon dioxide (CO_2) at 27°C, 2 bar and the other contains 2 kmol of oxygen (O_2) at 152°C, 5 bar. The gases are allowed to mix while 1000 kJ of energy are added by electrical work. Determine

(a) the final temperature, in °C.
(b) the final pressure, in bar.
(c) the change in availability, in kJ.
(d) the irreversibility, in kJ, for T_0 = 27°C.

12.33 Hydrogen (H_2) at 77°C, 4 bar enters an insulated chamber at steady state where it mixes with nitrogen (N_2) entering at 277°C, 4 bar. The mixture exits at 3.8 bar with the molar analysis 75% H_2, 25% N_2. Kinetic and potential energy effects can be ignored. Determine

(a) the temperature of the exiting mixture, in °C.
(b) the rate at which entropy is produced, in kJ/K per kmol of mixture exiting.

12.34 A stream of oxygen (O_2) at 100°F, 2 atm enters an insulated chamber at steady state with a mass flow rate of 1 lb/min and mixes with a stream of air entering at 200°F, 1.5 atm with a mass flow rate of 2 lb/min. The mixture exits at a pressure of 1 atm. Kinetic and potential energy effects can be ignored. On the basis of constant specific heats, determine

(a) the temperature of the exiting mixture, in °F.
(b) the irreversibility rate, in Btu/min, for T_0 = 40°F.

12.35 Air at 77°C, 1 bar enters an insulated chamber at steady state and mixes with water vapor entering at 277°C, 1 bar. The mixture exits at 1 bar with the molar analysis 75% water vapor, 25% dry air. Kinetic and potential energy effects and be ignored. Determine

(a) the temperature of the exiting mixture, in °C.
(b) the rate at which entropy is produced, in kJ/K per kmol of exiting mixture.

12.36 A gas mixture required in an industrial process is prepared by first allowing carbon monoxide (CO) at 80°F, 18 lbf/in.^{2} to enter an insulated mixing chamber at steady-state and mix with argon (Ar) entering at 380°F, 18 lbf/in.^{2}. The mass flow rate of the carbon monoxide is twice that of the argon. The mixture exits the chamber at 16 lbf/in.^{2} and is then allowed to expand in a throttling process through a valve to 10 lbf/in.^{2} Determine

(a) the temperature of the mixture at the exit of the mixing chamber, in °F.
(b) the temperature of the mixture at the exit of the valve, in °F.
(c) the irreversibility rates for the mixing chamber and the valve, each in Btu per lb of mixture, for T_0 = 40°F.

Kinetic and potential energy effects can be ignored.

12.37 Helium at 400 K, 1 bar enters an insulated mixing chamber, where it mixes with argon entering at 300 K, 1 bar. The mixture exits at a pressure of 1 bar. If the argon mass flow rate is x times that of helium, determine for selected values of x

(a) the exit temperature, in K.
(b) the irreversibility rate, in kJ per kg of helium entering.

Kinetic and potential energy effects can be ignored. Let T_0 = 300 K.

12.38 A stream of dry air (21% O_2 and 79% N_2 on a molar basis) at 300 K and 0.1 MPa is to be separated into pure oxygen and nitrogen streams, each at 300 K and 0.1 MPa. A device to achieve the separation is claimed to require a work input at steady state of 1200 kJ per kmol of air while heat transfer occurs with the surroundings. Ignoring kinetic and potential energy effects, evaluate whether the work value can be as claimed.

12.39 A device is being designed to separate a natural gas having a molar analysis of 94% CH_4 and 6% C_2H_6 into components. The device will receive natural gas at 20°C, 1 atm with a volumetric flow rate of 100 m^3/s. Separate streams of CH_4 and C_2H_6 will exit, each at 20°C, 1 atm. The device will operate essentially isothermally at 20°C. Ignoring kinetic and potential energy effects, determine the minimum theoretical work input required at steady state, in kW.

12.40 A 1000-ft^3 tank initially filled with N_2 at 70°F, 5 lbf/in.^{2} is connected by a valve to a large vessel containing O_2 at 70°F, 20 lbf/in.^{2} Oxygen is allowed to flow into the tank until the pressure in the tank becomes 15 lbf/in.^{2} If heat transfer with the surroundings maintains the tank contents at a constant temperature, determine

(a) the mass of oxygen that enters the tank, in lb.
(b) the heat transfer, in Btu.

12.41 An insulated, rigid tank initially contains 1 lb of H_2 at 60°F, 200 lbf/in.^{2} The tank is connected by a valve to a large vessel containing N_2 at 160°F, 400 lbf/in.^{2} If 5.6 lb of nitrogen is allowed to flow slowly into the tank, determine

(a) the temperature and pressure of the final mixture of hydrogen and nitrogen, in °F and lbf/in.^{2}, respectively.
(b) the amount of entropy produced, in Btu/°R.

12.42 A water pipe at 55°F passes through a basement in which the air is at 70°F. What is the maximum relative humidity the air can have before condensation occurs on the pipe?

12.43 A can of soft drink at a temperature of 40°F is taken from a refrigerator into a room where the temperature is 70°F and the relative humidity is 70%. Explain why beads of moisture form on the can's outer surface. Provide supporting calculations.

12.44 On entering a dwelling maintained at 20°C from the outdoors where the temperature is 10°C, a person's eyeglasses are observed *not* to become fogged. A humidity gauge indicates that the relative humidity in the dwelling is 55%. Can this reading be correct? Provide supporting calculations.

12.45 A fixed amount of air initially at 1 bar, 25°C, and a relative humidity of 60% is compressed isothermally until condensation of water begins. Determine the pressure of the mixture at the onset of condensation, in bar.

12.46 A fixed amount of air initially at 14.5 lbf/in.^{2}, 80°F, and a relative humidity of 50% is compressed isothermally until condensation of water begins. Determine the pressure of the mixture at the onset of condensation, in lbf/in.^{2}

12.47 One pound of moist air initially at 68°F, 1 atm, 75% relative humidity is compressed isothermally to 2 atm. If condensation occurs, determine the amount of water condensed, in lb. If there is no condensation, determine the final relative humidity.

12.48 A vessel whose volume is 0.3 m^3 initially contains dry air at 0.1 MPa and 31°C. Water is added to the vessel until the air is saturated at 31°C. Determine the

(a) mass of water added, in kg.
(b) final pressure in the vessel, in MPa.

12.49 A mixture of helium and water vapor at 60°C, 1 bar has the molar analysis 90.4% helium, 9.6% water vapor. If the mixture is cooled at constant pressure, determine the temperature at which water vapor begins to condense.

12.50 A mixture of nitrogen and water vapor at 200°F, 14.8 lbf/in.^{2} has the molar analysis 85% N_2, 15% water vapor. If the mixture is cooled at constant pressure, determine the temperature at which water vapor begins to condense.

12.51 A system consisting initially of 1 m^3 of air at 35°C, 1 bar, and 70% relative humidity is cooled at constant pressure to 30°C. Determine the work and heat transfer for the process, each in kJ.

12.52 A closed, rigid tank having a volume of 3 m^3 initially contains air at 100°C, 4.4 bar, and 40% relative humidity. If the tank contents are cooled to 20°C, determine

(a) the temperature at which condensation begins, in °C.
(b) the heat transfer, in kJ.

If the tank contents are cooled to 80°C, determine the heat transfer, in kJ.

12.53 A closed, rigid tank initially contains 0.24 m^3 of moist air in equilibrium with 0.06 m^3 of liquid water at 29°C and 0.1 MPa. If the tank contents are heated to 140°C, determine

(a) the final pressure, in MPa.
(b) the heat transfer, in kJ.

12.54 A closed, rigid tank having a volume of 100 ft^3 initially contains air at 300°F, 80 $lbf/in.^2$, and 10% relative humidity. If the tank contents are cooled to 170°F, determine

(a) the final pressure, in $lbf/in.^2$
(b) the heat transfer, in Btu.
(c) the change in entropy, in Btu/°R.

12.55 Combustion products with the molar analysis of 10% CO_2, 20% H_2O, 70% N_2 enter an engine's exhaust pipe at 1000°F, 1 atm and are cooled as they pass through the pipe, exiting at 100°F, 1 atm. Determine the heat transfer rate at steady state, in Btu per lb of entering mixture.

12.56 Air at 60°F, 14.7 $lbf/in.^2$, and 75% relative humidity enters an insulated compressor operating at steady state and is compressed to 100 $lbf/in.^2$ For an isentropic compressor efficiency of 80%, determine

(a) the temperature, in °F, of the exiting air.
(b) the work input required, in Btu per lb of dry air.

Neglect kinetic and potential energy effects.

12.57 Air at 35°C, 3 bar, 30% relative humidity, and a velocity of 50 m/s expands isentropically through a nozzle. Determine the lowest exit pressure, in bar, that can be attained without condensation. For this exit pressure, determine the exit velocity, in m/s. The nozzle operates at steady state and without significant potential energy effects.

12.58 A closed, rigid tank having a volume of 1 m^3 contains a mixture of nitrogen (N_2) and water vapor at 50°C, 20 bar. The respective masses are 20.8 kg of nitrogen and 0.034 kg of water vapor. If the tank contents are cooled to 10°C, determine the heat transfer, in kJ.

12.59 Air at 12°C, 1 atm, and 40% relative humidity enters a heat exchanger with a volumetric flow rate of 1 m^3/s. A separate stream of dry air enters at 280°C, 1 atm with a mass flow rate of 0.875 kg/s and exits at 220°C. Neglecting heat transfer between the heat exchanger and its surroundings, pressure drops of each stream, and kinetic and potential energy effects, determine

(a) the temperature of the exiting moist air, in °C.
(b) the irreversibility rate, in kW, for T_0 = 12°C.

12.60 Moist air at 20°C, 1.05 bar, 85% relative humidity and a volumetric flow rate of 0.3 m^3/s enters a well-insulated compressor operating at steady state. If moist air exits at 100°C, 2.0 bar, determine

(a) the relative humidity at the exit.
(b) the power input, in kW.
(c) the rate of entropy production, in kW/K.

12.61 Moist air at 90°F, 1 atm, 60% relative humidity and a volumetric flow rate of 2000 ft^3/min enters a control volume at steady state and flows along a surface maintained at 40°F. Saturated moist air and condensate, each at 54°F, exit the control volume. Determine

(a) the rate of heat transfer, in Btu/min.
(b) the rate of entropy production, in Btu/min°R.

12.62 Moist air at 20°C, 1 atm, 43% relative humidity and a volumetric flow rate of 900 m^3/h enters a control volume at steady state and flows along a surface maintained at 65°C. Liquid water at 20°C is injected at a rate of 5 kg/h and evaporates into the flowing stream. Moist air exits at 32°C, 1 atm. Determine

(a) the rate of heat transfer, in kW.
(b) the rate of entropy production, in kW/K.

12.63 The dry-bulb and wet-bulb temperatures in a conference room are 24 and 16°C, respectively. Determine the humidity ratio and relative humidity if the pressure is 1 atm.

12.64 The dry-bulb and wet-bulb temperatures within a factory space are 75 and 60°F, respectively. Determine the humidity ratio and relative humidity if the pressure is 1 atm.

12.65 Using the psychrometric chart, determine

(a) the relative humidity, the humidity ratio, and the specific enthalpy of the mixture, in kJ per kg of dry air, corresponding to dry-bulb and wet-bulb temperatures of 30 and 20°C, respectively.
(b) the humidity ratio, mixture specific enthalpy, and wet-bulb temperature corresponding to a dry-bulb temperature of 20°C and 60% relative humidity.
(c) the dew point temperature corresponding to dry-bulb and wet-bulb temperatures of 35 and 26°C, respectively.

12.66 Using the psychrometric chart, determine

(a) the dew point temperature corresponding to dry-

bulb and wet-bulb temperatures of 80 and 65°F, respectively.
(b) the humidity ratio, the specific enthalpy of the mixture, in Btu per lb of dry air, and the wet-bulb temperature corresponding to a dry-bulb temperature of 70°F and 70% relative humidity.
(c) the relative humidity, humidity ratio, and mixture specific enthalpy corresponding to dry-bulb and wet-bulb temperatures of 75 and 60°F, respectively.

12.67 Solve Problem 12.63 using the psychrometric chart.

12.68 Solve Problem 12.64 using the psychrometric chart.

12.69 A fixed amount of air initially at 25°C, 1 atm, and 50% relative humidity is cooled at constant pressure to 10°C. Using the psychrometric chart, determine whether condensation occurs. Evaluate the amount of water condensed, in kg per kg of dry air, or if there is no condensation determine the relative humidity at the final state.

12.70 A fan within an insulated duct delivers moist air at the duct exit at 22°C, 60% relative humidity, and a volumetric flow rate of 0.5 m³/s. At steady state, the power input to the fan is 1.3 kW. Using the psychrometric chart, determine the temperature and relative humidity at the duct inlet.

12.71 The mixture enthalpy per unit mass of dry air, in kJ/kg(a), represented on Fig. A-9 can be approximated closely from the expression

$$\frac{H}{m_a} = 1.005T(°C) + \omega[2501.7 + 1.82T(°C)]$$

When using Fig. A-9E, the corresponding expression, in Btu/lb(a), is

$$\frac{H}{m_a} = 0.24T(°F) + \omega[1061 + 0.444T(°F)]$$

Noting all significant assumptions, develop the above expressions.

12.72 Air at 1 atm with dry-bulb and wet-bulb temperatures of 82 and 68°F, respectively, enters a duct with a mass flow rate of 7 lb/min and is cooled at essentially constant pressure to 61°F. For steady-state operation and negligible kinetic and potential energy effects, determine

(a) the relative humidity at the duct inlet.
(b) the rate of heat transfer, in Btu/min.

Check your answers using the psychrometric chart.

12.73 Air at 25°C, 1 atm, and 65% relative humidity enters a dehumidifier operating at steady state. Saturated moist air and condensate exit in separate streams, each at 12°C. Neglecting kinetic and potential energy effects, determine

(a) the heat transfer from the moist air, in kJ per kg of dry air.
(b) the amount of water condensed, in kg per kg of dry air.

Check your answers using the psychrometric chart.

12.74 Air at 80°F, 1 atm, and 70% relative humidity enters a dehumidifier operating at steady state. Saturated moist air and condensate exit in separate streams, each at 50°F. Neglecting kinetic and potential energy effects, determine

(a) the heat transfer from the moist air, in Btu per lb of dry air.
(b) the amount of water condensed, in lb per lb of dry air.

Check your answers using the psychrometric chart.

12.75 An air conditioner operating at steady state takes in moist air at 28°C, 1 bar, and 70% relative humidity with a volumetric flow rate of 50 m³/min. The moist air first passes over a cooling coil and some water vapor is condensed. Saturated moist air and condensate streams exit the dehumidifier unit at the same temperature. The moist air then passes through a heating unit, exiting at 24°C, 1 bar, and 40% relative humidity. Neglecting kinetic and potential energy effects, determine

(a) the temperature of the moist air exiting the dehumidifier unit, in °C.
(b) the rate at which water is condensed, in kg/min.
(c) the rate of heat transfer between the moist air and the cooling coil, in kJ/min.
(d) the rate of heat transfer to the air passing through the heating unit, in kJ/min.

12.76 Figure P12.76 shows a compressor followed by an aftercooler. Atmospheric air at 14.7 lbf/in.², 90°F, and

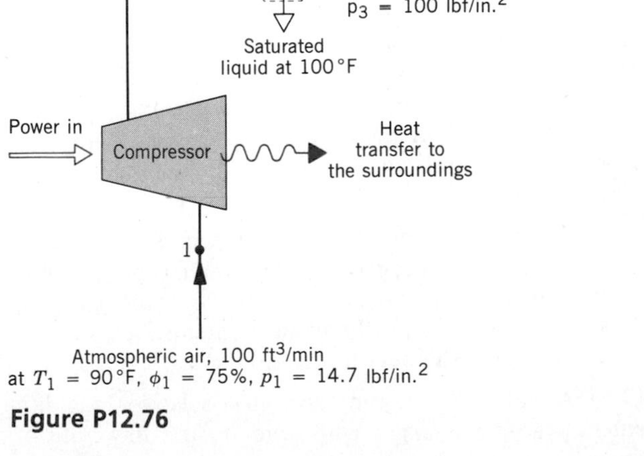

Figure P12.76

75% relative humidity enters the compressor with a volumetric flow rate of 100 ft^3/min. The moist air exiting the compressor at 100 lbf/in.^{2}, 400°F flows through the aftercooler, where it is cooled at constant pressure, exiting saturated at 100°F. Condensate also exits the aftercooler at 100°F. Heat transfer from the compressor to its surroundings occurs at a rate of 800 Btu/min. For steady-state operation and negligible kinetic and potential energy effects, determine

(a) the power input required by the compressor, in Btu/min.
(b) the mass flow rate of the condensate, in lb/min.
(c) the rate of heat transfer from the moist air to the refrigerant circulating in the cooling coil, in tons of refrigeration.

12.77 Outside air at 50°F, 1 atm, and 40% relative humidity enters a conditioning device operating at steady state. Liquid water is injected at 45°F and a moist air stream exits at 70°F, 1 atm, and 50% relative humidity. Neglecting kinetic and potential energy effects, determine

(a) the amount of water injected, in lb per lb of dry air.
(b) the rate of heat transfer to the moist air, in Btu per lb of dry air.

12.78 Moist air enters a device operating at steady state at 1 atm with a dry-bulb temperature of 55°C and a wet-bulb temperature of 25°C. Liquid water at 20°C is sprayed into the air stream, bringing it to 40°C, 1 atm at the exit. Determine

(a) the relative humidities at the inlet and exit.
(b) the amount of liquid water sprayed in, in kg per kg of dry air.

Ignore both heat transfer between the device and its surroundings and kinetic and potential energy effects.

12.79 Figure P12.79 shows a steam-spray humidification device at steady state. Heat transfer between the device and its surroundings can be ignored, as can kinetic and potential energy effects. Determine the irreversibility rate, in Btu/min, for T_0 = 95°F.

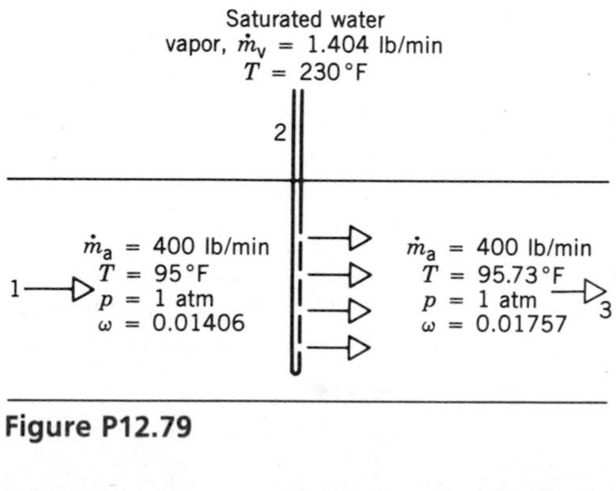

Figure P12.79

12.80 Outside air at 50°F, 1 atm, and 40% relative humidity enters an air conditioner operating at steady state. The air is first heated at essentially constant pressure to 90°F. Liquid water at 60°F is then injected bringing the air to 70°F, 1 atm. Determine

(a) the rate of heat transfer to the air passing through the heating section, in Btu per lb of dry air.
(b) the relative humidity at the exit of the humidification section.

Kinetic and potential energy effects can be ignored.

12.81 In an industrial drier operating at steady state, atmospheric air at 80°F, 1 atm, and 60% relative humidity is first heated to 250°F at constant pressure. The heated air is then allowed to pass over the materials being dried, exiting the drier at 150°F, 1 atm, and 30% relative humidity. If moisture is to be removed from the materials at a rate of 3000 lb/h, determine

(a) the mass flow rate of dry air required, in lb/h.
(b) the rate of heat transfer to the air as it passes through the heating section, in Btu/h.

Neglect kinetic and potential energy effects.

12.82 At steady state, moist air enters a classroom at a temperature of 16°C and is removed in a separate stream at a temperature of 27°C and 50% relative humidity. Moisture is added to the air in the room from the occupants at a rate of 4.5 kg/h. The moisture can be regarded as saturated water vapor at 33°C. Heat transfer into the occupied space from all sources is estimated to occur at a rate of 34,000 kJ/h. The pressure remains uniform at 1 atm. For the incoming moist air, determine the humidity ratio, the relative humidity, and the volumetric flow rate, in m^3/min. Neglect kinetic and potential energy effects.

12.83 At steady state, a device for heating and humidifying air has 300 ft^3/min of air at 40°F, 1 atm, and 80% relative humidity entering at one location, 1200 ft^3/min of air at 60°F, 1 atm, and 80% relative humidity entering at another location, and liquid water injected at 60°F. A single moist air stream exits at 85°F, 1 atm, and 35% relative humidity. Determine

(a) the rate of heat transfer to the device, in Btu/min.
(b) the rate at which liquid water is introduced, in lb/min.

Neglect kinetic and potential energy effects.

12.84 Air at 35°C, 1 bar, and 10% relative humidity enters an evaporative cooler operating at steady state. Liquid water at 20°C enters the cooler and fully evaporates. Moist air exits the cooler at 25°C, 1 bar. If there is no significant heat transfer between the device and its surroundings, determine

(a) the rate at which liquid enters, in kg per kg of dry air.
(b) the relative humidity at the exit.

(c) the irreversibility rate, in kJ per kg of dry air, for $T_0 = 20°C$.

Neglect kinetic and potential energy effects.

12.85 Figure P12.85 shows two options for conditioning atmospheric air at steady state. In each case, air enters at 15°C, 1 atm, and 20% relative humidity with a volumetric flow rate of 150 m³/min and exits at 30°C, 1 atm, and 40% relative humidity. One method conditions the air by injecting saturated water vapor at 1 atm. The other method allows the entering air to pass through a soaked pad kept replenished by water entering at 20°C. The moist air stream is then heated by an electric resistor. Which of the two options is preferable? Discuss.

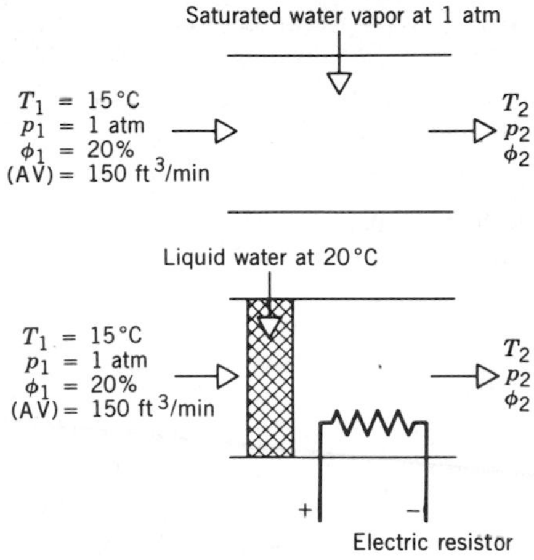

Figure P12.85

12.86 Using Eqs. 12.56b and 12.56c, show that

$$\frac{\dot{m}_{a1}}{\dot{m}_{a2}} = \frac{\omega_3 - \omega_2}{\omega_1 - \omega_3} = \frac{(h_{a3} + \omega_3 h_{g3}) - (h_{a2} + \omega_2 h_{g2})}{(h_{a1} + \omega_1 h_{g1}) - (h_{a3} + \omega_3 h_{g3})}$$

Employ this relation to demonstrate on a psychrometric chart that state 3 of the mixture lies on a straight line connecting the initial states of the two streams before mixing.

12.87 A stream consisting of 50 m³/min of saturated air at 14°C, 1 atm mixes adiabatically with a stream consisting of 20 m³/min of moist air at 32°C, 1 atm, 60% relative humidity, giving a single mixed stream at 1 atm. Using the psychrometric chart together with the procedure of Problem 12.86, determine the relative humidity and temperature, in °C, of the exiting stream.

12.88 At steady state, a stream of air at 60°F, 1 atm, 30% relative humidity is mixed adiabatically with a stream of air at 90°F, 1 atm, 70% relative humidity. The mass flow rate of the higher-temperature stream is twice that of the other stream. A single mixed stream exits at 1 atm. Determine for the exiting stream

(a) the relative humidity.
(b) the temperature, in °F.

Neglect kinetic and potential energy effects.

12.89 Atmospheric air having dry-bulb and wet-bulb temperatures of 33 and 29°C, respectively, enters a well-insulated chamber operating at steady state and mixes with air entering with dry-bulb and wet-bulb temperatures of 16 and 12°C, respectively. The volumetric flow rate of the lower temperature stream is twice that of the other stream. A single mixed stream exits. Determine for the exiting stream

(a) the relative humidity.
(b) the temperature, in °C.

Pressure is uniform throughout at 1 atm. Neglect kinetic and potential energy effects.

12.90 At steady state, a stream consisting of 650 ft³/min of air at 55°F, 1 atm, 20% relative humidity is mixed adiabatically with a stream consisting of 900 ft³/min of air at 75°F, 1 atm, 80% relative humidity. A single mixed stream exits at 1 atm. Determine for the exiting stream

(a) the relative humidity.
(b) the temperature, in °F.

Neglect kinetic and potential energy effects.

12.91 At steady state, air at 42°C, 1 atm, 30% relative humidity is mixed adiabatically with a second air stream entering at 1 atm. The mass flow rates of the two streams are the same. A single mixed stream exits at 29°C, 1 atm, 40% relative humidity with a mass flow rate of 2 kg/s. For the second entering stream, determine

(a) the relative humidity.
(b) the temperature, in °C.

Neglect kinetic and potential energy effects.

12.92 Air at 36°C, 1 bar, 40% relative humidity enters an insulated chamber operating at steady state with a mass flow rate of 2.5 kg/min and mixes with a saturated moist air stream entering at 5°C, 1 bar with a mass flow rate of 5 kg/min. A single mixed stream exits at 1 bar. Determine

(a) the relative humidity and temperature, in °C, of the exiting stream.
(b) the irreversibility rate, in kW, for $T_0 = 20°C$.

Neglect kinetic and potential energy effects.

12.93 A stream of air at 60°F, 1 atm, 30% relative humidity is mixed adiabatically with a stream of air at 90°F, 1 atm, 80% relative humidity. A single mixed stream exits at 1 atm. Using the result of Problem 12.71, plot the temperature of the exiting stream, in °F, as a function of the ratio of the mass flow rates of the dry air present in the two incoming streams. Assume steady state and ignore kinetic and potential energy effects.

12.94 Figure P12.94 shows the adiabatic mixing of two moist air streams at steady state. Kinetic and potential energy effects are negligible. Determine the irreversibility rate, in Btu/min, for $T_0 = 95°F$.

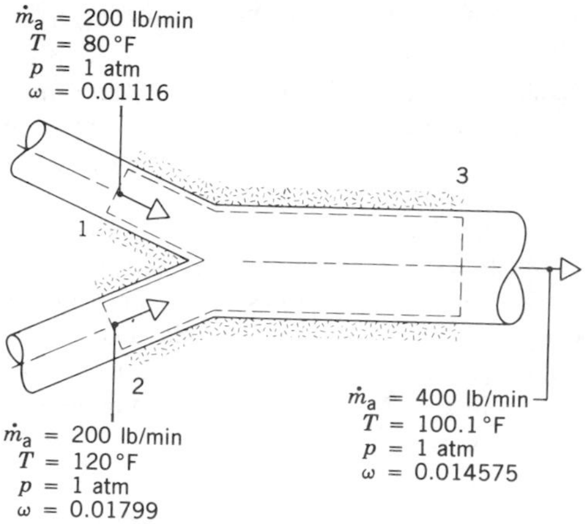

Figure P12.94

12.95 Water from a power plant condenser enters a cooling tower operating at steady state at 40°C with a mass flow rate of 10,000 kg/s. Cooled water at 20°C is returned to the condenser at the same flow rate. Atmospheric air enters the tower at 25°C, 1 atm, 35% relative humidity. Moist air exits at 35°C, 1 atm, 90% relative humidity. Makeup water is supplied at 20°C. Determine the mass flow rate, in kg/s, of

(a) the entering atmospheric air.
(b) the makeup water.

Ignore kinetic and potential energy effects.

12.96 Liquid water at 120°F enters a cooling tower operating at steady state with a mass flow rate of 140 lb/s. Atmospheric air enters at 80°F, 1 atm, 30% relative humidity. Saturated air exits at 100°F, 1 atm. Makeup water is not provided. Determine the mass flow rate of dry air required, in lb/h, if cooled water exits the tower at **(a)** 80°F and **(b)** 60°F. Ignore kinetic and potential energy effects.

12.97 Liquid water at 120°F and a mass flow rate of 3×10^5 lb/h enters a cooling tower operating at steady state. Liquid water exits the tower at 80°F. No makeup water is provided. Atmospheric air enters at 1 atm with a dry-bulb temperature of 70°F and a wet-bulb temperature of 60°F. Saturated air exits at 110°F, 1 atm. Ignoring kinetic and potential energy effects, determine the mass flow rate of the cooled water stream exiting the tower, in lb/h.

12.98 Liquid water at 50°C enters a forced draft cooling tower operating at steady state. Cooled water exits the tower with a mass flow rate of 80 kg/min. No makeup water is provided. A fan located within the tower draws in atmospheric air at 17°C, 0.098 MPa, 60% relative humidity with a volumetric flow rate of 110 m^3/min. Saturated air exits the tower at 30°C, 0.098 MPa. The power input to the fan is 8 kW. Ignoring kinetic and potential energy effects, determine

(a) the mass flow rate of the liquid stream entering, in kg/min.
(b) the temperature of the cooled liquid stream exiting, in °C.

12.99 Water enters a cooling tower operating at steady state at 40°C with a mass flow rate of 10^5 kg/h. Cooled water at 25°C exits the cooling tower at the same mass flow rate. Makeup water is supplied at 23°C. Atmospheric air enters the tower at 30°C, 1 bar, 35% relative humidity. A saturated moist air stream exits at 34°C, 1 bar. Determine

(a) the mass flow rates of the dry air and makeup water, each in kg/h.
(b) the irreversibility rate, in kW, for T_0 = 23°C.

Ignore kinetic and potential energy effects.

DESIGN AND OPEN ENDED PROBLEMS

12.1D For comfort in indoor natatoriums, air and water temperatures should be between 75 and 85°F, and the relative humidity in the range 50 to 60%. A major contributor to discomfort *and* inefficient energy use is water evaporating from the pool surface. At 80°F, an olympic-size pool can evaporate up to 200 lb/h. The airborne water vapor has energy that can be extracted during dehumidification and used to heat the pool water. Devise a practical means for accomplishing this energy-saving measure. Also propose means for preventing condensation on the natatorium windows for outdoor temperatures as low as −10°F. For each case, provide an equipment schematic and appropriate supporting calculations.

12.2D Computer rooms require year-round cooling to protect sensitive electronic components. Often, air is discharged at 55°F, or less, to various parts of the computer room through perforations in a raised floor. Humidification also must be provided to protect components from *electrostatic discharge*. The acceptable relative humidity range is 45 to 55%. Using appropriate instrumentation, determine the relative humidity in the computer room of your campus computer center. If the relative humidity is out of the acceptable range, check for improper humidifier operation and suggest means for correcting it. Consider the option of a separate recirculating air system that would remove relatively warm air, add moisture, and return the conditioned air to the computer room. Summarize your findings in the form of a written report.

12.3D Savings can be realized on domestic water heating bills by using a heat pump that exhausts to the hot water storage tank of a conventional water heater. When the heat pump evaporator is located within the dwelling living space, cooling *and* dehumidification of the room air also result. For a hot water (140°F) requirement of 64 gal per day, compare the annual cost of electricity for a heat-pump water heater with the annual cost for a conventional electrical water heater. Provide a schematic showing how the heat-pump water heater might be configured. Where might the heat pump evaporator be located during periods of the year when cooling and dehumidifying are not required?

12.4D Control of microbial contaminants is crucial in hospital operating rooms. Figure P12.4D illustrates the proposed design of a hospital air-handling system aimed at effective infection control. Write a critique of this design, taking into consideration issues including, but not necessarily limited to the following: Why is 100% outside air specified rather than a recirculated air-filtration arrangement requiring minimum outside air? Why is steam humidification used rather than the more common water-spray method? Why are three filters with increasing efficiency specified, and why are they located as shown? What additional features might be incorporated to promote air quality?

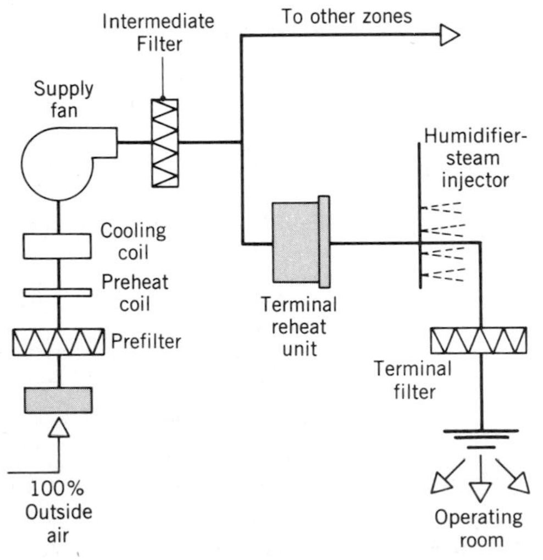

Figure P12.4D

12.5D A 147,000-ft^2 zone within an industrial plant must have a dry-bulb temperature of 76°F ± 2°F and a relative humidity of 45% ± 5% when the outside summer condition is a dry-bulb temperature of 94°F and a wet-bulb temperature of 77°F. Under these conditions, the zone receives 2025 kW by heat transfer from machinery, lighting, people, and through the windows, roof, and walls. A conventional air-conditioning system would provide air leaving the cooling coil at 53.5°F and at a flow rate of 284,000 ft^3/min. A proposed system that would achieve a substantial reduction in the air flow rate involves spraying water into the room from several ceiling locations at an estimated total rate of 3 gal/min. In this design, the air would leave the cooling coil nearly saturated at 51.7°F. Specify any special design features that such a water-spray application should have. Where would cost savings be realized with the proposed system: first-cost savings, annual operating cost savings, both? Discuss.

12.6D *Weatherization* of existing homes and the tighter construction of new homes can lead to winter heating season indoor moisture levels that exceed health and comfort standards. In locales such as the northwestern part of the U.S. where winters tend to be damp, infiltration cannot be relied on to reduce indoor moisture levels, and portable electric dehumidifiers are often employed for this purpose. These dehumidifiers not only remove moisture but also act like heat pumps, since they provide the dwelling with the energy supplied to drive the humidifier plus the energy released when water vapor condenses. Owing to this heat-pumping feature, it has been claimed that the cost to operate the humidifier is offset by a reduction in the space heating cost. Evaluate this claim.

12.7D Buildings in the humid southeastern portion of the U.S. frequently require both cooling and dehumidification. Figure 12.11 shows one means to accomplish this objective. A more effective use of energy is possible, however, with a system in which ambient air is first passed through a *desiccant dehumidification system* and then cooled by an internal combustion-engine-driven vapor-compression refrigeration system arranged so that the desiccant is regenerated using *waste heat* from the internal combustion engine. Develop a schematic diagram of the integrated system. Specify the desiccant dehumidification system, including its geometry and the desiccant media.

CHAPTER 13

REACTING MIXTURES AND COMBUSTION

The objective of this chapter is to study systems involving chemical reactions. Since the *combustion* of hydrocarbon fuels occurs in most power-producing devices (Chaps. 8 and 9), this type of reaction is emphasized.

The thermodynamic analysis of reactive systems is primarily an extension of principles introduced thus far. The fundamental concepts remain the same: conservation of mass, conservation of energy, and the second law. It is necessary, though, to modify the methods used to evaluate specific enthalpy, internal energy, and entropy. Only the manner in which these properties are evaluated represents a departure from previous practice, for once appropriate values are determined they are used as in earlier chapters in the energy and entropy balances for the system under consideration.

The principles developed in this chapter allow the equilibrium composition of a mixture of chemical substances to be determined. This topic is studied in the next chapter. The subject of *dissociation* is also deferred until then. Prediction of *reaction rates* is not within the scope of classical thermodynamics, so the topic of chemical kinetics, which deals with reaction rates, is not discussed in this text.

13.1 COMBUSTION PROCESS

When a chemical reaction occurs, the bonds within molecules of the *reactants* are broken, and atoms and electrons rearrange to form *products*. In combustion reactions, rapid oxidation of combustible elements of the fuel results in energy release as combustion products are formed. The three major combustible chemical elements in most common fuels are carbon, hydrogen, and sulfur. Sulfur is usually a relatively unimportant contributor to the energy released, but it can be a significant cause of pollution and corrosion problems. A fuel is said to have burned *completely* if all the carbon present in the fuel is burned to carbon dioxide, all the hydrogen is burned to water, and all the sulfur is burned to sulfur dioxide. If these conditions are not fulfilled, combustion is *incomplete*.

In this chapter we deal with combustion reactions expressed by chemical equations of the form

$$\text{reactants} \rightarrow \text{products}$$

or

$$\text{fuel} + \text{oxidizer} \rightarrow \text{products}$$

When dealing with chemical reactions, it is necessary to remember that mass is conserved, so the mass of the products equals the mass of the reactants. The total mass of each chemical *element* must be the same on both sides of the equation, even though the elements exist in different chemical compounds in the reactants and products. However, the number of moles of products may differ from the number of moles of reactants.

Consider as an elementary example the complete combustion of hydrogen with oxygen

$$1H_2 + \tfrac{1}{2}O_2 \rightarrow 1H_2O \tag{13.1}$$

In this case, the reactants are hydrogen and oxygen. Hydrogen is the fuel and oxygen is the oxidizer. Water is the only product of the reaction. The numerical coefficients in the equation, which precede the chemical symbols to give equal amounts of each chemical element on both sides of the equation, are called *stoichiometric* coefficients. In words, Eq. 13.1 states

$$1 \text{ kmol } H_2 + \tfrac{1}{2} \text{ kmol } O_2 \rightarrow 1 \text{ kmol } H_2O$$

or in English units

$$1 \text{ lbmol } H_2 + \tfrac{1}{2} \text{ lbmol } O_2 \rightarrow 1 \text{ lbmol } H_2O$$

Note that the total numbers of moles on the left and right sides of Eq. 13.1 are not equal. However, because mass is conserved, the total mass of reactants must equal the total mass of products. Since 1 kmol of H_2 equals 2 kg, $\frac{1}{2}$ kmol of O_2 equals 16 kg, and 1 kmol of H_2O equals 18 kg, Eq. 13.1 can be interpreted as stating

$$2 \text{ kg } H_2 + 16 \text{ kg } O_2 \rightarrow 18 \text{ kg } H_2O$$

or in English units

$$2 \text{ lb } H_2 + 16 \text{ lb } O_2 \rightarrow 18 \text{ lb } H_2O$$

In the remainder of this section, consideration is given to the makeup of the fuel, oxidizer, and combustion products typically involved in engineering combustion applications.

13.1.1 Fuels

A *fuel* is simply a combustible substance. In this chapter emphasis is on hydrocarbon fuels, which contain hydrogen and carbon. Sulfur and other chemical substances may also be present. Hydrocarbon fuels can exist as liquids, gases, and solids.

Liquid hydrocarbon fuels are commonly derived from crude oil through distillation and cracking processes. Examples are gasoline, diesel fuel, kerosene, and other types of fuel oils. Most liquid fuels are mixtures of hydrocarbons for which compositions are usually given in terms of mass fractions. For convenience in combustion calculations, gasoline is often considered to be octane, C_8H_{18}, and diesel fuel is considered to be dodecane, $C_{12}H_{26}$.

Gaseous hydrocarbon fuels are obtained from natural gas wells or are produced in certain chemical processes. Natural gas normally consists of several different hydrocarbons, with the major constituent being methane, CH_4. The compositions of gaseous fuels are usually given in terms of mole fractions. Both gaseous and liquid hydrocarbon fuels can be synthesized from coal, oil shale, and tar sands.

Coal is a familiar solid fuel. Its composition varies considerably with the location from which it is mined. For combustion calculations, the composition of coal is usually expressed as an *ultimate* analysis. The ultimate analysis gives the composition on a mass basis in terms of the relative amounts of chemical elements (carbon, sulfur, hydrogen, nitrogen, oxygen) and ash.

13.1.2 Combustion Air

Oxygen is required in every combustion reaction. Pure oxygen is used only in special applications such as cutting and welding. In most combustion applications, air provides the needed oxygen. The composition of a typical sample of dry air is given in Table 12.1. For the combustion calculations of this book, however, all components of air other than oxygen are lumped together with nitrogen for simplicity. Accordingly, air is considered to be 21% oxygen and 79% nitrogen on a molar basis. With this idealization the molar ratio of the nitrogen to the oxygen is 0.79/0.21 = 3.76. When air supplies the oxygen in a combustion reaction, therefore, every mole of oxygen is accompanied by 3.76 moles of nitrogen. The air considered here contains no water vapor. When *moist* air is involved in combustion, the water vapor present would be considered in writing the combustion equation.

We will also assume for simplicity that the nitrogen present in the combustion air does *not* undergo chemical reaction. That is, nitrogen is regarded as inert. The nitrogen in the products is at the same temperature as the other products, however, so the nitrogen undergoes a change of state if the products are at a temperature other than the air temperature before combustion. If high enough temperatures are attained, nitrogen can form compounds such as nitric oxide and nitrogen dioxide. Even trace amounts of oxides of nitrogen appearing in the exhaust of internal combustion engines can be a source of air pollution.

Two parameters that are frequently used to quantify the amounts of fuel and air in a particular combustion process are the air–fuel ratio and its reciprocal, the fuel–air ratio. The air–fuel ratio is simply the ratio of the amount of air in a reaction to the amount of fuel. The ratio can be written on a molar basis (moles of air divided by moles of fuel) or on a mass basis (mass of air divided by mass of fuel). Conversion between these values is accomplished using the molecular weights of the air, M_{air}, and fuel, M_{fuel},

$$\frac{\text{mass of air}}{\text{mass of fuel}} = \frac{\text{moles of air} \times M_{air}}{\text{moles of fuel} \times M_{fuel}} = \frac{\text{moles of air}}{\text{moles of fuel}}\left(\frac{M_{air}}{M_{fuel}}\right)$$

or

$$AF = \overline{AF}\left(\frac{M_{air}}{M_{fuel}}\right) \tag{13.2}$$

where $\overline{AF}$ is the air–fuel ratio on a molar basis and AF is the ratio on a mass basis. For the combustion calculations of this book the molecular weight of air is taken as 28.97. Tables A-1 provide the molecular weights of several important hydrocarbons.

Since AF is a ratio, it has the same value whether the quantities of air and fuel are expressed in SI units or English units.

The minimum amount of air that supplies sufficient oxygen for the complete combustion of all the carbon, hydrogen, and sulfur present in the fuel is called the *theoretical,* or *stoichiometric,* amount of air. The products of combustion would then consist of carbon dioxide, water, sulfur dioxide, the nitrogen accompanying the oxygen in the air, and any nitrogen contained in the fuel. No free oxygen would appear in the products.

As an illustration, let us determine the theoretical amount of air for the complete combustion of methane. For this reaction, the products contain only carbon dioxide, water, and nitrogen. The reaction is

$$CH_4 + a(O_2 + 3.76N_2) \rightarrow bCO_2 + cH_2O + dN_2 \quad (13.3)$$

where a, b, c, and d represent the numbers of moles of oxygen, carbon dioxide, water, and nitrogen. In writing the left side of Eq. 13.3, 3.76 moles of nitrogen are considered to accompany each mole of oxygen. Applying the conservation of mass principle to the carbon, hydrogen, oxygen, and nitrogen, respectively, results in four equations among the four unknowns

$$\begin{aligned} &\text{C:} & b &= 1 \\ &\text{H:} & 2c &= 4 \\ &\text{O:} & 2b + c &= 2a \\ &\text{N:} & d &= 3.76a \end{aligned}$$

Solving these equations, the *balanced* chemical equation is

$$CH_4 + 2(O_2 + 3.76N_2) \rightarrow CO_2 + 2H_2O + 7.52N_2 \quad (13.4)$$

The coefficient 2 before the term $(O_2 + 3.76N_2)$ in Eq. 13.4 is the number of moles of *oxygen* in the combustion air, per mole of fuel, and *not* the amount of air. The amount of combustion air is 2 moles of oxygen *plus* 2 × 3.76 moles of nitrogen, giving a total of 9.52 moles of air per mole of fuel. Thus, for the reaction given by Eq. 13.4 the air–fuel ratio on a molar basis is 9.52. To calculate the air–fuel ratio on a mass basis, use Eq. 13.2 to write

$$AF = \overline{AF}\left(\frac{M_{air}}{M_{fuel}}\right) = 9.52\left(\frac{28.97}{16.04}\right) = 17.19$$

Normally the amount of air supplied is either greater or less than the theoretical amount. The amount of air actually supplied is commonly expressed in terms of the *percent of theoretical air*. For example, 150% of theoretical air means that the air actually supplied is 1.5 times the theoretical amount of air. The amount of air supplied can be expressed alternatively as a *percent excess* or a *percent deficiency* of air. For example, 150% of theoretical air is equivalent to 50% excess air, and 80% of theoretical air is the same as a 20% deficiency of air.

As an illustration, consider the *complete* combustion of methane with 150% theoretical air (50% excess air). The balanced chemical reaction equation is

$$CH_4 + (1.5)(2)(O_2 + 3.76N_2) \rightarrow CO_2 + 2H_2O + O_2 + 11.28N_2 \quad (13.5)$$

In this equation, the amount of air per mole of fuel is 1.5 times the theoretical amount determined by Eq. 13.4. Accordingly, the air–fuel ratio is 1.5 times the air–fuel ratio determined for Eq. 13.4. Since complete combustion is assumed, the products contain only carbon dioxide, water, nitrogen, and oxygen. The excess air supplied appears in the products as uncombined oxygen and a greater amount of nitrogen than in Eq. 13.4, based on the theoretical amount of air.

The *equivalence ratio* is the ratio of the actual fuel–air ratio to the fuel–air ratio for complete combustion with the theoretical amount of air. The reactants are said to form a *lean* mixture when the equivalence ratio is less than unity. When the ratio is greater than unity, the reactants are said to form a *rich* mixture.

EXAMPLE 13.1

PROBLEM

Determine the air–fuel ratio on both a molar and mass basis for the complete combustion of octane, C_8H_{18}, with (a) the theoretical amount of air, (b) 150% theoretical air (50% excess air).

SOLUTION

Known: Octane, C_8H_{18}, is burned completely with (a) the theoretical amount of air, (b) 150% theoretical air.

Find: Determine the air–fuel ratio on a molar and a mass basis.

Assumptions:

1. Each mole of oxygen in the combustion air is accompanied by 3.76 moles of nitrogen only.

2. The nitrogen is inert.

3. Combustion is complete.

Analysis:

(a) For complete combustion of C_8H_{18} with the theoretical amount of air, the products contain carbon dioxide, water, and nitrogen only. That is

$$C_8H_{18} + a(O_2 + 3.76N_2) \rightarrow bCO_2 + cH_2O + dN_2$$

Applying the conservation of mass principle to the carbon, hydrogen, oxygen, and nitrogen, respectively, gives

$$\begin{aligned} &\text{C:} \quad & b &= 8 \\ &\text{H:} \quad & 2c &= 18 \\ &\text{O:} \quad & 2b + c &= 2a \\ &\text{N:} \quad & d &= 3.76a \end{aligned}$$

Solving these equations, $a = 12.5$, $b = 8$, $c = 9$, $d = 47$. The balanced chemical equation is

$$C_8H_{18} + 12.5(O_2 + 3.76N_2) \rightarrow 8CO_2 + 9H_2O + 47N_2$$

The air–fuel ratio on a molar basis is

$$\overline{AF} = \frac{12.5 + 12.5(3.76)}{1} = \frac{12.5(4.76)}{1} = 59.5\,\frac{\text{kmol (air)}}{\text{kmol (fuel)}}$$

The air–fuel ratio expressed on a mass basis is

$$AF = \left[59.5\,\frac{\text{kmol (air)}}{\text{kmol (fuel)}}\right]\left[\frac{28.97\,\dfrac{\text{kg (air)}}{\text{kmol (air)}}}{114.22\,\dfrac{\text{kg (fuel)}}{\text{kmol (fuel)}}}\right] = 15.1\,\frac{\text{kg (air)}}{\text{kg (fuel)}}$$

(b) For 150% theoretical air, the chemical equation for complete combustion takes the form

$$C_8H_{18} + 1.5(12.5)(O_2 + 3.76N_2) \rightarrow bCO_2 + cH_2O + dN_2 + eO_2$$

Applying conservation of mass

$$\begin{aligned} &C: && b = 8 \\ &H: && 2c = 18 \\ &O: && 2b + c + 2e = (1.5)(12.5)(2) \\ &N: && d = (1.5)(12.5)(3.76) \end{aligned}$$

Solving this set of equations, $b = 8$, $c = 9$, $d = 70.5$, $e = 6.25$, giving a balanced chemical equation

$$C_8H_{18} + 18.75(O_2 + 3.76N_2) \rightarrow 8CO_2 + 9H_2O + 70.5N_2 + 6.25O_2$$

The air–fuel ratio on a molar basis is

$$\overline{AF} = \frac{18.75(4.76)}{1} = 89.25 \frac{\text{kmol (air)}}{\text{kmol (fuel)}}$$

On a mass basis, the air–fuel ratio is 22.6 kg (air)/kg (fuel).

Comment: In the excess air case (part b), oxygen appears in the products, in addition to carbon dioxide, water, and nitrogen.

13.1.3 Products of Combustion

In each of the illustrations given above, complete combustion is assumed. For a hydrocarbon fuel, this means that the only allowed products are CO_2, H_2O, and N_2, with O_2 also present when excess air is supplied. If the fuel is specified and combustion is complete, the respective amounts of the products can be determined by applying the conservation of mass principle to the chemical equation. The procedure for obtaining the balanced reaction equation of an *actual* reaction where combustion is incomplete is not always so straightforward.

Combustion is the result of a series of very complicated and rapid chemical reactions, and the products that are formed depend on many factors. When fuel is burned in the cylinder of an internal combustion engine, the products of the reaction vary with the temperature and pressure in the cylinder. In combustion equipment of all kinds, the degree of mixing of the fuel and air is a controlling factor in the reactions that occur once the fuel and air mixture is ignited. Although the amount of air supplied in an actual combustion process may exceed the theoretical amount, it is not uncommon for some carbon monoxide and unburned oxygen to appear in the products. This can be due to incomplete mixing, insufficient time for complete combustion, and other factors. When the amount of air supplied is less than the theoretical amount of air, the products may include both CO_2 and CO, and there also may be unburned fuel in the products. Unlike the complete combustion cases considered above, the products of combustion of an actual combustion process and their relative amounts can be determined only *by experiment*.

Among several devices for the experimental determination of the composition of products of combustion are the *Orsat analyzer, gas chromatograph, infrared analyzer,* and *flame ionization detector*. Data from these devices can be used to determine the mole fractions of the gaseous products of combustion. The analyses are usually reported on a "dry" basis. This means that the mole fractions are given for all gaseous products *except* the water vapor. Some experimental procedures give an analysis including the water vapor. In the examples to follow, we show how analyses of the products of combustion can be used to determine the balanced chemical reaction equations.

Since water is formed when hydrocarbon fuels are burned, the mole fraction of water vapor in the gaseous products of combustion can be significant. If the gaseous products of combustion are cooled at constant mixture pressure, the *dew point temperature* is reached when water vapor begins to condense. Since corrosion of duct work, mufflers, and other metal parts can occur when water is deposited on them, knowledge of the dew point temperature is important.

EXAMPLE 13.2

PROBLEM

Methane, CH_4, is burned with dry air. The molar analysis of the products on a dry basis is CO_2, 9.7%; CO, 0.5%; O_2, 2.95%; and N_2, 86.85%. Determine (a) the air–fuel ratio on both a molar and a mass basis, (b) the percent theoretical air, (c) the dew point temperature of the products, in °F, if the pressure is 1 atm.

SOLUTION

Known: Methane is burned with dry air. The molar analysis of the products on a dry basis is provided.

Find: Determine (a) the air–fuel ratio on both a molar and a mass basis, (b) the percent theoretical air, and (c) the dew point temperature of the products, in °F, if the pressure is 1 atm.

Assumptions:

1. Each mole of oxygen in the combustion air is accompanied by 3.76 moles of nitrogen, which is inert.

2. The products form an ideal gas mixture.

Analysis:

(a) The solution is conveniently conducted on the basis of 100 lbmol of dry products. The chemical equation then reads

$$aCH_4 + b(O_2 + 3.76N_2) \rightarrow 9.7CO_2 + 0.5CO + 2.95O_2 + 86.85N_2 + cH_2O$$

In addition to the assumed 100 lbmol of dry products, water must be included as a product.

Applying conservation of mass to carbon, hydrogen, and oxygen, respectively

$$\text{C:} \quad 9.7 + 0.5 = a$$

$$\text{H:} \quad 2c = 4a$$

$$\text{O:} \quad (9.7)(2) + 0.5 + 2(2.95) + c = 2b$$

Solving this set of equations gives $a = 10.2$, $b = 23.1$, $c = 20.4$. The balanced chemical equation is

$$10.2CH_4 + 23.1(O_2 + 3.76N_2) \rightarrow 9.7CO_2 + 0.5CO + 2.95O_2 + 86.85N_2 + 20.4H_2O$$

On a molar basis, the air–fuel ratio is

$$\overline{AF} = \frac{23.1(4.76)}{10.2} = 10.78 \frac{\text{lbmol (air)}}{\text{lbmol (fuel)}}$$

On a mass basis

$$AF = (10.78)\left(\frac{28.97}{16.04}\right) = 19.47 \frac{\text{lb (air)}}{\text{lb (fuel)}}$$

(b) The balanced chemical equation for the complete combustion of methane with the *theoretical* amount of air is

$$CH_4 + 2(O_2 + 3.76N_2) \rightarrow CO_2 + 2H_2O + 7.52N_2$$

The theoretical air–fuel ratio on a molar basis is

$$(\overline{AF})_{\text{theo}} = \frac{2(4.76)}{1} = 9.52\ \frac{\text{lbmol (air)}}{\text{lbmol (fuel)}}$$

The percent theoretical air is then found from

$$\%\ \text{theoretical air} = \frac{(\overline{AF})}{(\overline{AF})_{\text{theo}}}$$

$$= \frac{10.78\ \text{lbmol (air)/lbmol (fuel)}}{9.52\ \text{lbmol (air)/lbmol (fuel)}} = 1.13\ (113\%)$$

(c) To determine the dew point temperature requires the partial pressure of the water vapor p_v. The partial pressure p_v is found from $p_v = y_v p$, where y_v is the mole fraction of the water vapor in the combustion products and p is 1 atm.

Referring to the balanced chemical equation of part (a), the mole fraction of the water vapor is

$$y_v = \frac{20.4}{100 + 20.4} = 0.169$$

Thus, $p_v = 0.169$ atm $= 2.484$ lbf/in.^{2} Interpolating in Table A-2E, $T = 134°F$.

Comments:

1. The solution to part (a) could be obtained on the basis of any assumed amount of dry products—for example, 1 lbmol. With some other assumed amount, the values of the coefficients of the balanced chemical equation would differ from those obtained in the solution, but the air–fuel ratio, the value for the percent of theoretical air, and the dew point temperature would be unchanged.
2. The three unknown coefficients, a, b, and c, are evaluated in part (a) by application of conservation of mass to carbon, hydrogen, and oxygen. As a check, note that the nitrogen also balances

$$\text{N:} \quad b(3.76) = 86.85$$

   This confirms the accuracy of both the given product analysis and the calculations conducted to determine the unknown coefficients.
3. If the products of combustion were cooled at constant pressure below the dew point temperature of 134°F, some condensation of the water vapor would occur.

EXAMPLE 13.3

PROBLEM

A natural gas has the following molar analysis: CH_4, 80.62%; C_2H_6, 5.41%; C_3H_8, 1.87%; C_4H_{10}, 1.60%; N_2, 10.50%. The gas is burned with dry air, giving products having a molar analysis on a dry basis: CO_2, 7.8%; CO, 0.2%; O_2, 7%; N_2, 85%. (a) Determine the air–fuel ratio on a molar basis. (b) Assuming ideal gas behavior for the fuel mixture, determine the amount of products in kmol that would be formed from 100 m^3 of fuel mixture at 300 K and 1 bar. (c) Determine the percent of theoretical air.

SOLUTION

Known: A natural gas with a specified molar analysis burns with dry air to produce products having a known molar analysis on a dry basis.

Find: Determine the air–fuel ratio on a molar basis, the amount of products in kmol that would be formed from 100 m^3 of natural gas at 300 K and 1 bar, and the percent of theoretical air.

Assumptions:

1. Each mole of oxygen in the combustion air is accompanied by 3.76 moles of nitrogen, which is inert.
2. The fuel mixture can be modeled as an ideal gas.

Analysis:

(a) The solution can be conducted on the basis of an assumed amount of fuel mixture or on the basis of an assumed amount of dry products. Let us illustrate the first procedure, basing the solution on 1 kmol of fuel mixture. The chemical equation then takes the form

$$(0.8062CH_4 + 0.0541C_2H_6 + 0.0187C_3H_8 + 0.0160C_4H_{10} + 0.1050N_2)$$
$$+ a(O_2 + 3.76N_2) \rightarrow b(0.078CO_2 + 0.002CO + 0.07O_2 + 0.85N_2) + cH_2O$$

The products consist of b kmol of dry products and c kmol of water vapor, each per kmol of fuel mixture.
Applying conservation of mass to carbon

$$b(0.078 + 0.002) = 0.8062 + 2(0.0541) + 3(0.0187) + 4(0.0160)$$

Solving gives $b = 12.931$. Conservation of mass for hydrogen results in

$$2c = 4(0.8062) + 6(0.0541) + 8(0.0187) + 10(0.0160)$$

which gives $c = 1.93$. The unknown coefficient a can be found from either an oxygen balance or a nitrogen balance. Applying conservation of mass to oxygen

$$12.931[2(0.078) + 0.002 + 2(0.07)] + 1.93 = 2a$$

giving $a = 2.892$.
The balanced chemical equation is then

$$(0.8062CH_4 + 0.0541C_2H_6 + 0.0187C_3H_8 + 0.0160C_4H_{10} + 0.1050N_2)$$
$$+ 2.892(O_2 + 3.76N_2) \rightarrow 12.931(0.078CO_2 + 0.002CO + 0.07O_2 + 0.85N_2)$$
$$+ 1.93H_2O$$

The air–fuel ratio on a molar basis is

$$\overline{AF} = \frac{(2.892)(4.76)}{1} = 13.77 \frac{\text{kmol (air)}}{\text{kmol (fuel)}}$$

(b) By inspection of the chemical reaction equation, the total amount of products is $b + c = 12.931 + 1.93 = 14.861$ kmol of products per kmol of fuel. The amount of fuel in moles, n_F, present in 100 m^3 of fuel mixture at 300 K and 1 bar can be determined from the ideal gas equation of state as

$$n_F = \frac{pV}{\bar{R}T}$$
$$= \frac{(10^5 \text{ N/m}^2)(100 \text{ m}^3)}{(8314 \text{ N}\cdot\text{m/kmol}\cdot\text{K})(300 \text{ K})} = 4.01 \text{ kmol (fuel)}$$

Accordingly, the amount of product mixture that would be formed from 100 m^3 of fuel mixture is $(14.861)(4.01) = 59.59$ kmol of product gas.

(c) The balanced chemical equation for the *complete combustion* of the fuel mixture with the *theoretical amount* of air is

$$(0.8062CH_4 + 0.0541C_2H_6 + 0.0187C_3H_8 + 0.0160C_4H_{10} + 0.1050N_2)$$
$$+ 2(O_2 + 3.76N_2) \rightarrow 1.0345CO_2 + 1.93H_2O + 7.625N_2$$

The theoretical air–fuel ratio on a molar basis is

$$(\overline{AF})_{\text{theo}} = \frac{2(4.76)}{1} = 9.52 \frac{\text{kmol (air)}}{\text{kmol (fuel)}}$$

The percent theoretical air is then

$$\% \text{ theoretical air} = \frac{13.77 \text{ kmol (air)/kmol (fuel)}}{9.52 \text{ kmol (air)/kmol (fuel)}} = 1.45\ (145\%)$$

Comment: A check on both the accuracy of the given molar analyses and the calculations conducted to determine the unknown coefficients in the solution to part (a) is obtained by applying conservation of mass to nitrogen. The amount of nitrogen in the reactants is

$$0.105 + (3.76)(2.892) = 10.98 \text{ kmol/kmol of fuel}$$

The amount of nitrogen in the products is (0.85)(12.931) = 10.99 kmol/kmol of fuel. The difference can be attributed to round-off.

13.2 CONSERVATION OF ENERGY FOR REACTING SYSTEMS

The objective of the present section is to illustrate the application of the conservation of energy principle to reacting systems. The forms of the conservation of energy principle introduced previously remain valid whether a chemical reaction occurs within the system or not. However, the methods used for evaluating the properties of reacting systems differ somewhat from the practices used to this point.

13.2.1 Enthalpy of Formation

In each of the tables of thermodynamic properties used thus far, values for the specific internal energy, enthalpy, and entropy are given relative to some arbitrary datum state where the enthalpy (or alternatively the internal energy) and entropy are set to zero. This approach is satisfactory for evaluations involving *differences* in property values between states of the same composition, for then arbitrary datums cancel. However, when a chemical reaction occurs, reactants disappear and products are formed, so differences cannot be calculated for all substances involved. For reacting systems, it is necessary to evaluate u, h, and s in such a way that there are no subsequent ambiguities or inconsistencies in evaluating properties. In this section, we will consider how this is accomplished for u and h. The case of entropy is handled differently and is taken up in Sec. 13.4.

An enthalpy datum for the study of reacting systems can be established by assigning arbitrarily a value of zero to the enthalpy of the *stable elements* at a state called the *standard reference state* and defined by $T_{ref} = 298.15$ K (25°C) and $p_{ref} = 1$ atm. In English units the temperature at the standard reference state is closely 537°R (77°F). Note that only *stable* elements are assigned a value of zero enthalpy at the standard state. The term stable simply means that the particular element is chemically stable. For example, at the standard state the stable forms of hydrogen, oxygen, and nitrogen are H_2, O_2, and N_2 and not the monatomic H, O, and N. No ambiguities or conflicts result with this choice of datum.

Using the datum introduced above, enthalpy values can be assigned to *compounds* for use in the study of reacting systems. The enthalpy of a compound at the standard state equals its *enthalpy of formation,* symbolized by $\bar{h}_f^\circ$. The enthalpy of formation is the energy released or absorbed when the compound is formed from its elements, the compound and elements all being at T_{ref} and p_{ref}. The enthalpy of formation is usually determined by application of procedures from statistical thermodynamics using observed spectroscopic data. The enthalpy of formation also can be found in principle by measuring the heat transfer in a reaction in which the compound is formed from the elements. This is illustrated in the next paragraph. Tables A-24 and A-24E give

Figure 13.1 Illustration used to discuss the enthalpy of formation concept.

values of the enthalpy of formation for several compounds in units of kJ/kmol and Btu/lbmol, respectively. In this text, the superscript ° is used to denote properties at 1 atm. For the case of the enthalpy of formation, the reference temperature T_{ref} is also intended by this symbol.

To consider the enthalpy of formation concept further, refer to the simple reactor shown in Fig. 13.1, in which carbon and oxygen each enter at T_{ref} and p_{ref} and react completely at steady state to form carbon dioxide at the same temperature and pressure. Carbon dioxide is *formed* from carbon and oxygen according to

$$C + O_2 \rightarrow CO_2 \tag{13.6}$$

This reaction would be *exothermic,* so for the carbon dioxide to exit at the same temperature as the entering elements, there would have to be a heat transfer from the reactor to its surroundings. The rate of heat transfer and the enthalpies of the incoming and exiting streams are related by the energy rate balance

$$0 = \dot{Q}_{cv} + \dot{m}_C h_C + \dot{m}_{O_2} h_{O_2} - \dot{m}_{CO_2} h_{CO_2}$$

where $\dot{m}$ and h denote, respectively, mass flow rate and specific enthalpy. In writing this equation, we have assumed no work $\dot{W}_{cv}$ and negligible effects of kinetic and potential energy. For enthalpies on a molar basis, the energy rate balance appears as

$$0 = \dot{Q}_{cv} + \dot{n}_C \bar{h}_C + \dot{n}_{O_2} \bar{h}_{O_2} - \dot{n}_{CO_2} \bar{h}_{CO_2}$$

where $\dot{n}$ and $\bar{h}$ denote, respectively, the molar flow rate and enthalpy per mole. Solving for the specific enthalpy of carbon dioxide and noting from Eq. 13.6 that all molar flow rates are equal

$$\bar{h}_{CO_2} = \frac{\dot{Q}_{cv}}{\dot{n}_{CO_2}} + \frac{\dot{n}_C}{\dot{n}_{CO_2}} \bar{h}_C + \frac{\dot{n}_{O_2}}{\dot{n}_{CO_2}} \bar{h}_{O_2} = \frac{\dot{Q}_{cv}}{\dot{n}_{CO_2}} + \bar{h}_C + \bar{h}_{O_2} \tag{13.7}$$

Carbon and oxygen are stable elements at the standard state, so $\bar{h}_C = \bar{h}_{O_2} = 0$. Equation 13.7 becomes

$$\bar{h}_{CO_2} = \frac{\dot{Q}_{cv}}{\dot{n}_{CO_2}} \tag{13.8}$$

Accordingly, the value assigned to the specific enthalpy of carbon dioxide at the standard state equals the heat transfer, per mole of CO_2, between the reactor and its surroundings. If the heat transfer could be measured accurately, it would be found to equal −393,520 kJ per kmol of carbon dioxide formed (−169,300 Btu per lbmol of CO_2 formed). This is the value listed for CO_2 in Table A-24 (Table A-24E).

The sign associated with the enthalpy of formation values appearing in Tables A-24 corresponds to the sign convention for heat transfer. If there is heat transfer *from* a reactor in which a compound is formed from its elements (an *exothermic* reaction), the enthalpy of formation has a negative sign. If a heat transfer *to* the reactor is required (an *endothermic* reaction), the enthalpy of formation is positive.

The specific enthalpy of a compound at a state other than the standard state is found by adding the specific enthalpy change $\Delta\bar{h}$ between the standard state and the state of interest to the enthalpy of formation

$$\bar{h}(T, p) = \bar{h}_f^\circ + [\bar{h}(T, p) - \bar{h}(T_{ref}, p_{ref})] = \bar{h}_f^\circ + \Delta\bar{h} \tag{13.9}$$

That is, the enthalpy of a compound is composed of $\bar{h}_f^\circ$, associated with the formation of the compound from its elements, and $\Delta\bar{h}$, associated with a change of state at constant composition. An arbitrary choice of datum can be used to determine $\Delta\bar{h}$, since it is a *difference* at constant composition. Accordingly, $\Delta\bar{h}$ can be evaluated from tabular sources such as the steam tables, the ideal gas tables when appropriate, and so on. Note that as a consequence of the enthalpy datum adopted for the stable elements, the specific enthalpy determined from Eq. 13.9 is often negative.

Tables A-24 provide two values of the enthalpy of formation of water. One is for liquid water and the other is for water vapor. Under equilibrium conditions, water exists only as a liquid at 25°C (77°F) and 1 atm. The vapor value listed is for a *hypothetical* state in which water is a vapor at 25°C (77°F) and 1 atm obeying the ideal gas model. The difference between the two enthalpy of formation values is given closely by the enthalpy of vaporization $\bar{h}_{fg}$ at T_{ref}

$$\bar{h}_f^\circ(g) - \bar{h}_f^\circ(l) \approx \bar{h}_{fg} \tag{13.10}$$

The ideal gas value for the enthalpy of formation of water is convenient for use in Eq. 13.9 when the $\Delta\bar{h}$ term is found from the ideal gas table, Table A-19 or A-19E. Similar considerations apply to other substances for which liquid and vapor values for $\bar{h}_f^\circ$ are listed in Tables A-24.

13.2.2 Energy Balances for Reacting Systems

Several considerations enter when writing energy balances for systems involving combustion. Some of these apply generally, without regard for whether combustion takes place. For example, it is necessary to consider if significant work and heat transfers take place and if the respective values are known or unknown. Also, the effects of kinetic and potential energy must be assessed. Other considerations are related directly to the occurrence of combustion, however. For example, it is important to know the state of the fuel before combustion occurs. Whether the fuel is a liquid, a gas, or a solid is important. It is necessary to consider whether the fuel is premixed with the combustion air or the fuel and air enter a reactor separately. The state of the combustion products also must be assessed. It is important to know whether the products of combustion are a gaseous mixture or whether some of the water formed on combustion has been condensed.

Because of these many considerations, it is best to illustrate the writing of energy balances for reacting systems by referring to special cases of broad interest, highlighting the underlying assumptions. Let us begin by considering the steady-state reactor shown in Fig. 13.2, in which a hydrocarbon fuel C_aH_b burns completely with

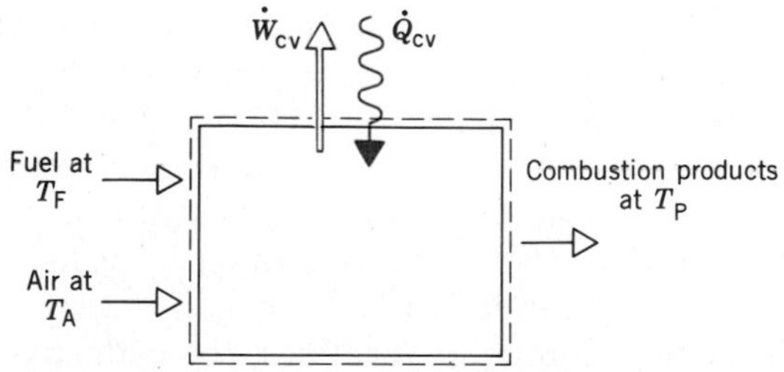

Figure 13.2 Reactor at steady state.

the theoretical amount of air according to

$$C_aH_b + \left(a + \frac{b}{4}\right)(O_2 + 3.76N_2) \rightarrow aCO_2 + \frac{b}{2}H_2O + \left(a + \frac{b}{4}\right)3.76N_2 \tag{13.11}$$

The fuel enters the reactor in a stream separate from the combustion air, which is regarded as an ideal gas mixture. The products of combustion also are assumed to form an ideal gas mixture. Kinetic and potential energy effects are ignored.

With the foregoing idealizations, the mass and energy rate balances for the two-inlet, single-exit reactor can be used to obtain the following equation on a *per mole of fuel basis:*

$$\frac{\dot{Q}_{cv}}{\dot{n}_F} - \frac{\dot{W}_{cv}}{\dot{n}_F} = \underline{\left[a\bar{h}_{CO_2} + \frac{b}{2}\bar{h}_{H_2O} + \left(a + \frac{b}{4}\right)3.76\bar{h}_{N_2}\right]}$$

$$- \bar{h}_F - \underline{\left[\left(a + \frac{b}{4}\right)\bar{h}_{O_2} + \left(a + \frac{b}{4}\right)3.76\bar{h}_{N_2}\right]} \tag{13.12a}$$

where $\dot{n}_F$ denotes the molar flow rate of the fuel. Note that each coefficient on the right side of this equation is the same as the coefficient of the corresponding substance in the reaction equation.

The first underlined term on the right side of Eq. 13.12a is the enthalpy of the exiting gaseous products of combustion *per mole of fuel.* The second underlined term on the right side is the enthalpy of the combustion air *per mole of fuel.* Note that the enthalpies of the combustion products and the air have been evaluated by adding the contribution of each component present in the respective ideal gas mixtures. The symbol $\bar{h}_F$ denotes the molar enthalpy of the fuel. Equation 13.12a can be expressed more concisely as

$$\frac{\dot{Q}_{cv}}{\dot{n}_F} - \frac{\dot{W}_{cv}}{\dot{n}_F} = \bar{h}_P - \bar{h}_R \tag{13.12b}$$

where $\bar{h}_P$ and $\bar{h}_R$ denote, respectively, the enthalpies of the products and reactants per mole of fuel.

Once the energy balance has been written, the next step is to evaluate the individual enthalpy terms. Since each component of the combustion products is assumed to behave as an ideal gas, its contribution to the enthalpy of the products depends solely on the temperature of the products, T_P. Accordingly, for each component of the products, Eq. 13.9 takes the form

$$\bar{h} = \bar{h}_f^\circ + [\bar{h}(T_P) - \bar{h}(T_{ref})] \tag{13.13}$$

In Eq. 13.13, $\bar{h}_f^\circ$ is the enthalpy of formation from Table A-24 or A-24E, as appropriate. The second term accounts for the change in enthalpy from the temperature T_{ref} to the temperature T_P. For several common gases, this term can be evaluated from tabulated values of enthalpy versus temperature in Tables A-17 through A-22 (Tables A-17E through A-22E). Alternatively, the term can be obtained by integration of the ideal gas specific heat $\bar{c}_p$ obtained from Tables A-15 or some other source of data. A similar approach would be employed to evaluate the enthalpies of the oxygen and nitrogen in the combustion air. For these

$$\bar{h} = \cancelto{0}{\bar{h}_f^\circ} + [\bar{h}(T_A) - \bar{h}(T_{ref})] \tag{13.14}$$

where T_A is the temperature of the air entering the reactor. Note that the enthalpy of formation for oxygen and nitrogen is zero by definition and thus drops out of Eq.

13.14 as indicated. The evaluation of the enthalpy of the fuel is also based on Eq. 13.9. If the fuel can be modeled as an ideal gas, the fuel enthalpy is obtained using an expression of the same form as Eq. 13.13 with the temperature of the incoming fuel replacing T_P.

With the foregoing considerations, Eq. 13.12a takes the form

$$\frac{\dot{Q}_{cv}}{\dot{n}_F} - \frac{\dot{W}_{cv}}{\dot{n}_F} = a(\bar{h}_f^\circ + \Delta\bar{h})_{CO_2} + \frac{b}{2}(\bar{h}_f^\circ + \Delta\bar{h})_{H_2O} + \left(a + \frac{b}{4}\right) 3.76(\cancelto{0}{\bar{h}_f^\circ} + \Delta\bar{h})_{N_2}$$

$$- (\bar{h}_f^\circ + \Delta\bar{h})_F - \left(a + \frac{b}{4}\right)(\cancelto{0}{\bar{h}_f^\circ} + \Delta\bar{h})_{O_2} - \left(a + \frac{b}{4}\right) 3.76(\cancelto{0}{\bar{h}_f^\circ} + \Delta\bar{h})_{N_2} \quad (13.15a)$$

The terms set to zero in this expression are the enthalpies of formation of oxygen and nitrogen.

Equation 13.15a can be written more concisely as

$$\frac{\dot{Q}_{cv}}{\dot{n}_F} - \frac{\dot{W}_{cv}}{\dot{n}_F} = \sum_P n_e(\bar{h}_f^\circ + \Delta\bar{h})_e - \sum_R n_i(\bar{h}_f^\circ + \Delta\bar{h})_i \quad (13.15b)$$

where i denotes the incoming fuel and air streams and e the exiting combustion products. The coefficients n_i and n_e correspond to the respective coefficients of the reaction equation giving the moles of reactants and products *per mole of fuel,* respectively. Although Eqs. 13.15 have been developed with reference to the reaction of Eq. 13.11, equations having the same general forms would be obtained for other combustion reactions.

Let us consider next a closed system involving a combustion process. In the absence of kinetic and potential energy effects, the appropriate form of the energy balance is

$$U_P - U_R = Q - W$$

where U_R denotes the internal energy of the reactants and U_P denotes the internal energy of the products. If the reactants and products form ideal gas mixtures, the energy balance can be expressed as

$$\sum_P n\bar{u} - \sum_R n\bar{u} = Q - W \quad (13.16)$$

where the coefficients n on the left side are the coefficients of the reaction equation giving the moles of each reactant or product.

Since each component of the reactants and products behaves as an ideal gas, the respective specific internal energies of Eq. 13.16 can be evaluated as $\bar{u} = \bar{h} - \bar{R}T$, so the equation becomes

$$Q - W = \sum_P n(\bar{h} - \bar{R}T_P) - \sum_R n(\bar{h} - \bar{R}T_R) \quad (13.17a)$$

where T_P and T_R denote the temperature of the products and reactants, respectively. With expressions of the form of Eq. 13.13 for each of the reactants and products, Eq. 13.17a can be written alternatively as

$$Q - W = \sum_P n(\bar{h}_f^\circ + \Delta\bar{h} - \bar{R}T_P) - \sum_R n(\bar{h}_f^\circ + \Delta\bar{h} - \bar{R}T_R)$$

$$= \sum_P n(\bar{h}_f^\circ + \Delta\bar{h}) - \sum_R n(\bar{h}_f^\circ + \Delta\bar{h}) - \bar{R}T_P \sum_P n + \bar{R}T_R \sum_R n \quad (13.17b)$$

The enthalpy of formation terms are obtained from Table A-24 or Table A-24E. The $\Delta\bar{h}$ terms are evaluated as discussed above.

EXAMPLE 13.4

PROBLEM

Liquid octane enters an internal combustion engine operating at steady state with a mass flow rate of 0.004 lb/s and is mixed with the theoretical amount of air. The fuel and air enter the engine at 77°F and 1 atm. The mixture burns completely and combustion products leave the engine at 1140°F. The engine develops a power output of 50 horsepower. Determine the rate of heat transfer from the engine, in Btu/s, neglecting kinetic and potential energy effects.

SOLUTION

Known: Liquid octane and the theoretical amount of air enter an internal combustion engine operating at steady state in separate streams at 77°F, 1 atm. Combustion is complete and the products exit at 1140°F. The power developed by the engine and fuel mass flow rate are specified.

Find: Determine the rate of heat transfer from the engine, in Btu/s.

Schematic and Given Data:

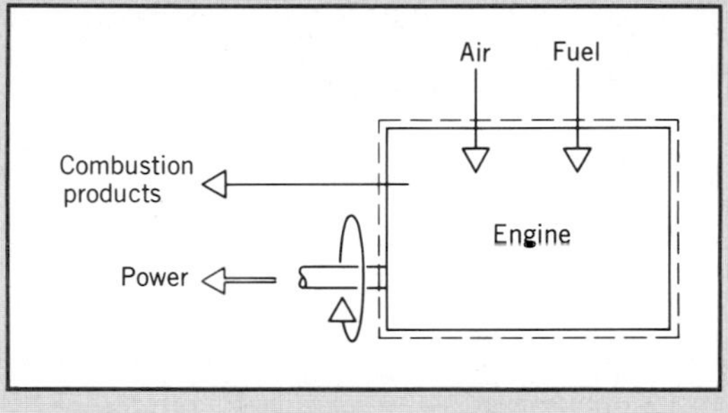

Figure E13.4

Assumptions:

1. The control volume identified by a dashed line on the accompanying figure operates at steady state.
2. Kinetic and potential energy effects can be ignored.
3. The combustion air and the products of combustion each form ideal gas mixtures.
4. Each mole of oxygen in the combustion air is accompanied by 3.76 moles of nitrogen only. The nitrogen is inert and combustion is complete.

Analysis: The balanced chemical equation for complete combustion with the theoretical amount of air is obtained from the solution to Example 13.1 as

$$C_8H_{18} + 12.5O_2 + 47N_2 \rightarrow 8CO_2 + 9H_2O + 47N_2$$

The energy rate balance reduces, with assumptions 1–3, to give

$$\frac{\dot{Q}_{cv}}{\dot{n}_F} = \frac{\dot{W}_{cv}}{\dot{n}_F} + \bar{h}_P - \bar{h}_R$$

$$= \frac{\dot{W}_{cv}}{\dot{n}_F} + \{8[\bar{h}_f^\circ + \Delta\bar{h}]_{CO_2} + 9[\bar{h}_f^\circ + \Delta\bar{h}]_{H_2O(g)} + 47[\overset{0}{\cancel{\bar{h}_f^\circ}} + \Delta\bar{h}]_{N_2}\}$$

$$- \{[\bar{h}_f^\circ + \overset{0}{\cancel{\Delta\bar{h}}}]_{C_8H_{18}(l)} + 12.5[\overset{0}{\cancel{\bar{h}_f^\circ}} + \overset{0}{\cancel{\Delta\bar{h}}}]_{O_2} + 47[\overset{0}{\cancel{\bar{h}_f^\circ}} + \overset{0}{\cancel{\Delta\bar{h}}}]_{N_2}\}$$

where each coefficient is the same as the corresponding term of the balanced chemical equation. The enthalpy of formation terms for oxygen and nitrogen are zero, and $\Delta\bar{h} = 0$ for each of the reactants, because the fuel and combustion air enter at 77°F.

With the enthalpy of formation for $C_8H_{18}(l)$ from Table A-24E

$$\bar{h}_R = (\bar{h}_f^\circ)_{C_8H_{18}(l)} = -107{,}530 \text{ Btu/lbmol(fuel)}$$

With enthalpy of formation values for CO_2 and $H_2O(g)$ from Table A-24E, and enthalpy values for N_2, H_2O, and CO_2 from Tables A-17E, 19E, and 21E, respectively

$$\begin{aligned}\bar{h}_P &= 8[-169{,}300 + (15{,}829 - 4027.5)] + 9[-104{,}040 + (13{,}494.4 - 4258)] \\ &\quad + 47[11{,}409.7 - 3729.5] \\ &= -1{,}752{,}251 \text{ Btu/lbmol(fuel)}\end{aligned}$$

Using the molecular weight of the fuel from Table A-1, the molar flow rate of the fuel is

$$\dot{n}_F = \frac{0.004 \text{ lb(fuel)/s}}{114.22 \text{ lb(fuel)/lbmol(fuel)}} = 3.5 \times 10^{-5} \text{ lbmol(fuel)/s}$$

Inserting values into the expression for the rate of heat transfer

$$\begin{aligned}\dot{Q}_{cv} &= \dot{W}_{cv} + \dot{n}_F(\bar{h}_P - \bar{h}_R) \\ &= (50 \text{ horsepower})\left(\frac{2545 \text{ Btu/h}}{\text{horsepower}}\right)\left(\frac{\text{h}}{3600 \text{ s}}\right) \\ &\quad + \left[3.5 \times 10^{-5} \frac{\text{lbmol(fuel)}}{\text{s}}\right][-1{,}752{,}251 - (-107{,}530)] \frac{\text{Btu}}{\text{lbmol(fuel)}} \\ &= -22.22 \text{ Btu/s}\end{aligned}$$

EXAMPLE 13.5

PROBLEM

Methane gas at 400 K and 1 atm enters a combustion chamber, where it is mixed with air entering at 500 K and 1 atm. The products of combustion exit at 1800 K and 1 atm with the product analysis given in Example 13.2. For operation at steady state, determine the rate of heat transfer from the combustion chamber in kJ per kmol of fuel. Neglect kinetic and potential energy effects. The average value for the specific heat $\bar{c}_p$ of methane between 298 and 400 K is 38 kJ/kmol·K.

SOLUTION

Known: Methane gas, air, and combustion products enter and exit a combustion chamber at steady state in separate streams with specified temperatures and a pressure of 1 atm. An analysis of the dry products of combustion is provided.

Find: Determine the rate of heat transfer, in kJ per kmol of fuel.

Schematic and Given Data:

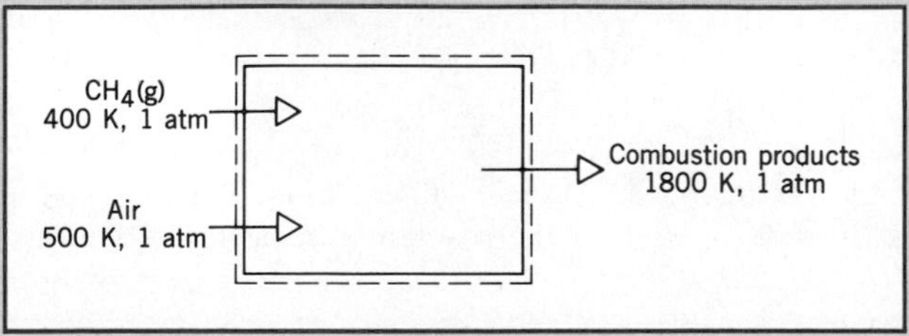

Figure E13.5

Assumptions:

1. The control volume identified by a dashed line on the accompanying figure operates at steady state with $\dot{W}_{cv} = 0$.
2. Kinetic and potential energy effects can be ignored.
3. The fuel is modeled as an ideal gas with $\bar{c}_p = 38$ kJ/kmol·K. The combustion air and the products of combustion each form ideal gas mixtures.
4. Each mole of oxygen in the combustion air is accompanied by 3.76 moles of nitrogen, which is inert.

Analysis: When expressed on a per mole of fuel basis, the balanced chemical equation obtained in the solution to Example 13.2 takes the form

$$CH_4 + 2.265O_2 + 8.515N_2 \rightarrow 0.951CO_2 + 0.049CO + 0.289O_2 + 8.515N_2 + 2H_2O$$

The energy rate balance reduces with assumptions 1–3 to give

$$\frac{\dot{Q}_{cv}}{\dot{n}_F} = \bar{h}_P - \bar{h}_R$$

$$= [0.951(\bar{h}_f^\circ + \Delta\bar{h})_{CO_2} + 0.049(\bar{h}_f^\circ + \Delta\bar{h})_{CO} + 0.289(\cancelto{0}{\bar{h}_f^\circ} + \Delta\bar{h})_{O_2}$$

$$+ 8.515(\cancelto{0}{\bar{h}_f^\circ} + \Delta\bar{h})_{N_2} + 2(\bar{h}_f^\circ + \Delta\bar{h})_{H_2O(g)}] - [(\bar{h}_f^\circ + \Delta\bar{h})_{CH_4}$$

$$+ 2.265(\cancelto{0}{\bar{h}_f^\circ} + \Delta\bar{h})_{O_2} + 8.515(\cancelto{0}{\bar{h}_f^\circ} + \Delta\bar{h})_{N_2}]$$

Each coefficient in this equation is the same as the coefficient of the corresponding term of the balanced chemical equation. The enthalpies of formation for oxygen and nitrogen are set to zero.

Consider first the reactants. With the enthalpy of formation for methane from Table A-24, the given $\bar{c}_p$ value for methane, and enthalpy values for nitrogen and oxygen from Tables A-17 and 18, respectively

$$\bar{h}_R = (\bar{h}_f^\circ + \bar{c}_p\,\Delta T)_{CH_4} + 2.265(\Delta\bar{h})_{O_2} + 8.515(\Delta\bar{h})_{N_2}$$
$$= [-74{,}850 + 38(400 - 298)] + 2.265[14{,}770 - 8682] + 8.515[14{,}581 - 8669]$$
$$= -6844 \text{ kJ/kmol } (CH_4)$$

Next consider the products. With enthalpy of formation values for CO_2, CO, and $H_2O(g)$ from Table A-24 and enthalpy values from Tables A-17 through A-21

$$\bar{h}_P = 0.951[-393{,}520 + (88{,}806 - 9364)]$$
$$+ 0.049[-110{,}530 + (58{,}191 - 8669)]$$
$$+ 0.289(60{,}371 - 8682) + 8.515(57{,}651 - 8669)$$
$$+ 2[-241{,}820 + (72{,}513 - 9904)]$$
$$= -228{,}079 \text{ kJ/kmol } (CH_4)$$

Inserting values into the expression for the rate of heat transfer

$$\frac{\dot{Q}_{cv}}{\dot{n}_F} = \bar{h}_P - \bar{h}_R = -228{,}079 - (-6844) = -221{,}235 \text{ kJ/kmol } (CH_4)$$

EXAMPLE 13.6

PROBLEM

A mixture of 1 kmol of gaseous methane and 2 kmol of oxygen initially at 25°C and 1 atm burns completely in a closed, rigid container. Heat transfer occurs until the products are cooled to 900 K. If the reactants and products each form ideal gas mixtures, determine (a) the amount of heat transfer, in kJ, and (b) the final pressure, in atm.

SOLUTION

Known: A mixture of gaseous methane and oxygen, initially at 25°C and 1 atm, burns completely within a closed rigid container. The products are cooled to 900 K.

Find: Determine the amount of heat transfer, in kJ, and the final pressure of the combustion products, in atm.

Schematic and Given Data:

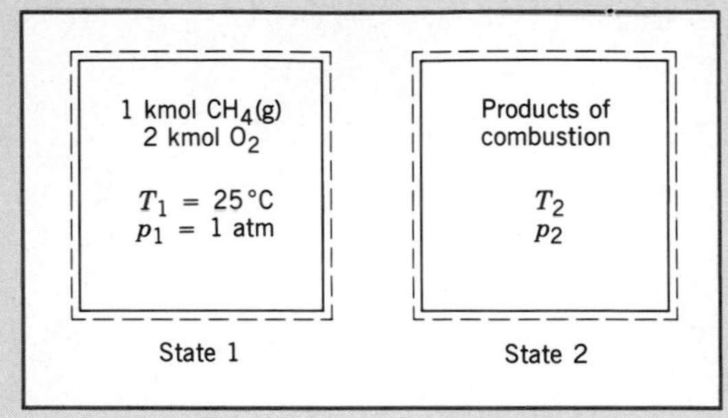

Figure E13.6

Assumptions:

1. The contents of the container are taken as the system.
2. Kinetic and potential energy effects are absent, and $W = 0$.
3. The initial mixture and the products of combustion each form ideal gas mixtures.
4. The initial and final states are equilibrium states.

Analysis: The chemical reaction equation for the complete combustion of methane with oxygen is

$$CH_4 + 2O_2 \rightarrow CO_2 + 2H_2O$$

(a) With assumptions 2 and 3, the closed system energy balance takes the form

$$U_P - U_R = Q - \overset{0}{\cancel{W}}$$

or

$$Q = U_P - U_R = (1\bar{u}_{CO_2} + 2\bar{u}_{H_2O(g)}) - (1\bar{u}_{CH_4(g)} + 2\bar{u}_{O_2})$$

Each coefficient in this equation is the same as the corresponding term of the balanced chemical equation.

Since each reactant and product behaves as an ideal gas, the respective specific internal energies can be evaluated as $\bar{u} = \bar{h} - \bar{R}T$. The energy balance then becomes

$$Q = [1(\bar{h}_{CO_2} - \bar{R}T_2) + 2(\bar{h}_{H_2O(g)} - \bar{R}T_2)] - [1(\bar{h}_{CH_4(g)} - \bar{R}T_1) + 2(\bar{h}_{O_2} - \bar{R}T_1)]$$

where T_1 and T_2 denote, respectively, the initial and final temperatures. Collecting like terms

$$Q = (\bar{h}_{CO_2} + 2\bar{h}_{H_2O(g)} - \bar{h}_{CH_4(g)} - 2\bar{h}_{O_2}) + 3\bar{R}(T_1 - T_2)$$

The specific enthalpies are evaluated in terms of the respective enthalpies of formation to give

$$Q = [(\bar{h}_f^\circ + \Delta\bar{h})_{CO_2} + 2(\bar{h}_f^\circ + \Delta\bar{h})_{H_2O(g)}$$

$$- (\bar{h}_f^\circ + \overset{0}{\cancel{\Delta\bar{h}}})_{CH_4(g)} - 2(\overset{0}{\cancel{\bar{h}_f^\circ}} + \overset{0}{\cancel{\Delta\bar{h}}})_{O_2}] + 3\bar{R}(T_1 - T_2)$$

Since the methane and oxygen are initially at 25°C, $\Delta\bar{h} = 0$ for each of these reactants. Also, $\bar{h}_f^\circ = 0$ for oxygen.

With enthalpy of formation values for CO_2, $H_2O(g)$, and $CH_4(g)$ from Table A-24 and enthalpy values for H_2O and CO_2 from Tables A-19 and A-21, respectively

$$\begin{aligned} Q &= [-393{,}520 + (37{,}405 - 9364)] + 2[-241{,}820 + (31{,}828 - 9904)] \\ &\quad - (-74{,}850) + 3(8.314)(298 - 900) \\ &= -745{,}436 \text{ kJ} \end{aligned}$$

(b) By assumption 3, the initial mixture and the products of combustion each form ideal gas mixtures. Thus, for the reactants

$$p_1 V = n_R \bar{R} T_1$$

where n_R is the total number of moles of reactants and p_1 is the initial pressure. Similarly, for the products

$$p_2 V = n_P \bar{R} T_2$$

where n_P is the total number of moles of products and p_2 is the final pressure.

Since $n_R = n_P = 3$ and volume is constant, these equations combine to give

$$p_2 = \frac{T_2}{T_1} p_1 = \left(\frac{900 \text{ K}}{298 \text{ K}}\right)(1 \text{ atm}) = 3.02 \text{ atm}$$

13.2.3 Enthalpy of Combustion and Heating Values

Although the enthalpy of formation concept underlies the formulations of the energy balances for reactive systems presented thus far, the enthalpy of formation of the fuel is not always known. For example, fuel oil and coal are normally composed of a variety of individual chemical substances, the relative amounts of which may vary considerably, depending on the source. Owing to the wide variation in composition that such fuels can exhibit, we do not find their enthalpies of formation listed in Tables A-24 or similar compilations of thermophysical data. In many cases of practical interest, however, the *enthalpy of combustion,* which is accessible experimentally, can be used to conduct an energy analysis when enthalpy of formation data are not directly available.

The enthalpy of combustion $\bar{h}_{RP}$ is defined as the difference between the enthalpy of the products and the enthalpy of the reactants when *complete* combustion occurs at a given temperature and pressure. That is,

$$\bar{h}_{RP} = \sum_P n_e \bar{h}_e - \sum_R n_i \bar{h}_i \tag{13.18}$$

where the n's correspond to the respective coefficients of the reaction equation giving the moles of reactants and products per mole of fuel. When the enthalpy of combustion is expressed on a unit mass of fuel basis, it is designated h_{RP}. Tabulated values are usually given at the standard temperature T_{ref} and pressure p_{ref} introduced in Sec. 13.2.1. The symbol $\bar{h}^\circ_{RP}$ or h°_{RP} is used for data at this temperature and pressure.

When enthalpy of formation data are available for *all* the reactants and products, the enthalpy of combustion can be calculated directly from Eq. 13.18. Otherwise, it must be obtained experimentally using devices known as *calorimeters*. Both constant-volume (bomb calorimeters) and flow-through devices are employed for this purpose. Consider as an illustration a reactor operating at steady state in which the fuel is burned completely with air. For the products to be returned to the same temperature as the reactants, a heat transfer from the reactor would be required.

From an energy rate balance, the required heat transfer is

$$\frac{\dot{Q}_{cv}}{\dot{n}_F} = \sum_P n_e \bar{h}_e - \sum_R n_i \bar{h}_i \tag{13.19}$$

where the symbols have the same significance as in previous discussions. The heat transfer per mole of fuel, $\dot{Q}_{cv}/\dot{n}_F$, would be determined from measured data. Comparing Eq. 13.19 with the defining equation, Eq. 13.18, we have $\bar{h}_{RP} = \dot{Q}_{cv}/\dot{n}_F$. In accord with the usual sign convention for heat transfer, the enthalpy of combustion would be negative.

To illustrate how the enthalpy of combustion might be used for the energy analysis of reacting systems, consider a control volume at steady state for which the energy rate balance takes the form

$$\frac{\dot{Q}_{cv}}{\dot{n}_F} - \frac{\dot{W}_{cv}}{\dot{n}_F} = \sum_P n_e(\bar{h}_f^\circ + \Delta\bar{h})_e - \sum_R n_i(\bar{h}_f^\circ + \Delta\bar{h})_i$$

All symbols have the same significance as in previous discussions. This equation can be rearranged to read

$$\frac{\dot{Q}_{cv}}{\dot{n}_F} - \frac{\dot{W}_{cv}}{\dot{n}_F} = \underline{\sum_P n_e(\bar{h}_f^\circ)_e - \sum_R n_i(\bar{h}_f^\circ)_i} + \sum_P n_e(\Delta\bar{h})_e - \sum_R n_i(\Delta\bar{h})_i$$

For a complete reaction, the underlined term is just the enthalpy of combustion at T_{ref} and p_{ref}, $\bar{h}_{RP}^\circ$. Thus, the equation becomes

$$\frac{\dot{Q}_{cv}}{\dot{n}_F} - \frac{\dot{W}_{cv}}{\dot{n}_F} = \bar{h}_{RP}^\circ + \sum_P n_e(\Delta\bar{h})_e - \sum_R n_i(\Delta\bar{h})_i \tag{13.20}$$

The right side of Eq. 13.20 can be evaluated with an experimentally determined value for $\bar{h}_{RP}^\circ$ and $\Delta\bar{h}$ values for the reactants and products determined as discussed previously.

The term *heating value* is commonly used. The heating value of a fuel is a positive number equal to the magnitude of the enthalpy of combustion. Two heating values are recognized by name: the *higher* heating value (HHV) and the *lower* heating value (LHV). The higher heating value is obtained when all the water formed by combustion is a liquid; the lower heating value is obtained when all the water formed by combustion is a vapor. The higher heating value exceeds the lower heating value by the energy that would be required to vaporize the liquid formed. Values for the HHV and LHV also depend on whether the fuel is a liquid or a gas.

EXAMPLE 13.7

PROBLEM

Calculate the enthalpy of combustion of gaseous methane, in kJ per kg of fuel, (a) at 25°C, 1 atm with liquid water in the products, (b) at 25°C, 1 atm with water vapor in the products. (c) Repeat part (b) at 1000 K, 1 atm.

SOLUTION

Known: The fuel is gaseous methane.

Find: Determine the enthalpy of combustion, in kJ per kg of fuel, (a) at 25°C, 1 atm with liquid water in the products, (b) at 25°C, 1 atm with water vapor in the products, (c) at 1000 K, 1 atm with water vapor in the products.

Assumptions:

1. Each mole of oxygen in the combustion air is accompanied by 3.76 moles of nitrogen only, which is inert.
2. Combustion is complete, and both reactants and products are at the same temperature and pressure.
3. The ideal gas model applies for methane, the combustion air, and the gaseous products of combustion.

Analysis: The combustion equation is

$$CH_4 + 2O_2 + 7.52N_2 \rightarrow CO_2 + 2H_2O + 7.52N_2$$

The enthalpy of combustion is, from Eq. 13.18

$$\bar{h}_{RP} = \sum_P n_e(\bar{h}_f^\circ + \Delta\bar{h})_e - \sum_R n_i(\bar{h}_f^\circ + \Delta\bar{h})_i$$

Introducing the coefficients of the combustion equation and evaluating the specific enthalpies in terms of the respective enthalpies of formation

$$\bar{h}_{RP} = \bar{h}_{CO_2} + 2\bar{h}_{H_2O} - \bar{h}_{CH_4(g)} - 2\bar{h}_{O_2}$$

$$= (\bar{h}_f^\circ + \Delta\bar{h})_{CO_2} + 2(\bar{h}_f^\circ + \Delta\bar{h})_{H_2O} - (\bar{h}_f^\circ + \Delta\bar{h})_{CH_4(g)} - 2(\overset{0}{\cancel{\bar{h}_f^\circ}} + \Delta\bar{h})_{O_2}$$

For nitrogen, the enthalpy terms of the reactants and products cancel. Also, the enthalpy of formation of oxygen is zero by definition. On rearrangement, the enthalpy of combustion expression becomes

$$\bar{h}_{RP} = (\bar{h}_f^\circ)_{CO_2} + 2(\bar{h}_f^\circ)_{H_2O} - (\bar{h}_f^\circ)_{CH_4(g)} + [(\Delta\bar{h})_{CO_2} + 2(\Delta\bar{h})_{H_2O} - (\Delta\bar{h})_{CH_4(g)} - 2(\Delta\bar{h})_{O_2}]$$

$$= \bar{h}_{RP}^\circ + [(\Delta\bar{h})_{CO_2} + 2(\Delta\bar{h})_{H_2O} - (\Delta\bar{h})_{CH_4(g)} - 2(\Delta\bar{h})_{O_2}]$$

The values for $\bar{h}_{RP}^\circ$ and $(\Delta\bar{h})_{H_2O}$ depend on whether the water in the products is a liquid or a vapor.

(a) Since the reactants and products are at 25°C, 1 atm in this case, the $\Delta\bar{h}$ terms drop out of the above expression for $\bar{h}_{RP}$. Thus, for liquid water in the products the enthalpy of combustion is

$$\bar{h}_{RP}^\circ = (\bar{h}_f^\circ)_{CO_2} + 2(\bar{h}_f^\circ)_{H_2O(l)} - (\bar{h}_f^\circ)_{CH_4(g)}$$

With enthalpy of formation values from Table A-24

$$\bar{h}_{RP}^\circ = -393{,}520 + 2(-285{,}830) - (-74{,}850) = -890{,}330 \text{ kJ/kmol (fuel)}$$

Dividing by the molecular weight of methane places this result on a unit mass of fuel basis

$$h_{RP}^\circ = \frac{-890{,}330 \text{ kJ/kmol (fuel)}}{16.04 \text{ kg (fuel)/kmol (fuel)}} = -55{,}507 \text{ kJ/kg (fuel)}$$

(b) As in part (a), the $\Delta\bar{h}$ terms drop out of the above expression for $\bar{h}_{RP}$, which for water vapor in the products reduces to $\bar{h}_{RP}^\circ$, where

$$\bar{h}_{RP}^\circ = (\bar{h}_f^\circ)_{CO_2} + 2(\bar{h}_f^\circ)_{H_2O(g)} - (\bar{h}_f^\circ)_{CH_4(g)}$$

With enthalpy of formation values from Table A-24

$$\bar{h}_{RP}^\circ = -393{,}520 + 2(-241{,}820) - (-74{,}850) = -802{,}310 \text{ kJ/kmol (fuel)}$$

On a unit of mass of fuel basis, the enthalpy of combustion for this case is

$$h_{RP}^\circ = \frac{-802{,}310}{16.04} = -50{,}019 \text{ kJ/kg fuel}$$

(c) For the case where the reactants and products are at 1000 K, 1 atm, the term $\bar{h}_{RP}^\circ$ in the above expression for $\bar{h}_{RP}$ has the value determined in part (b): $\bar{h}_{RP}^\circ = -802{,}310$ kJ/kmol (fuel), and the $\Delta\bar{h}$ terms for O_2, $H_2O(g)$, and CO_2 can be evaluated using specific enthalpies at 298 and 1000 K from Tables A-18, A-19, A-21, respectively. The results are

$$(\Delta\bar{h})_{O_2} = 31{,}389 - 8682 = 22{,}707 \text{ kJ/kmol}$$

$$(\Delta\bar{h})_{H_2O(g)} = 35{,}882 - 9904 = 25{,}978 \text{ kJ/kmol}$$

$$(\Delta\bar{h})_{CO_2} = 42{,}769 - 9364 = 33{,}405 \text{ kJ/kmol}$$

For methane, the $\bar{c}_p$ expression of Table A-15 can be used to obtain

$$(\Delta\bar{h})_{CH_4(g)} = \int_{298}^{1000} \bar{c}_p \, dT$$

$$= \bar{R}\left(3.826T - \frac{3.979}{10^3}\frac{T^2}{2} + \frac{24.558}{10^6}\frac{T^3}{3} - \frac{22.733}{10^9}\frac{T^4}{4} + \frac{6.963}{10^{12}}\frac{T^5}{5}\right)_{298}^{1000}$$

$$= 38{,}189 \text{ kJ/kmol (fuel)}$$

Substituting values into the expression for the enthalpy of combustion

$$\bar{h}_{RP} = -802{,}310 + [33{,}405 + 2(25{,}978) - 38{,}189 - 2(22{,}707)]$$

$$= -800{,}552 \text{ kJ/kmol (fuel)}$$

On a unit mass basis

$$h_{RP} = \frac{-800{,}552}{16.04} = -49{,}910 \text{ kJ/kg (fuel)}$$

Comment: Comparing the values of parts (b) and (c), the enthalpy of combustion of methane is seen to vary little with temperature. The same is true for many hydrocarbon fuels. This fact is often used to simplify combustion calculations.

13.3 ADIABATIC FLAME TEMPERATURE

Let us reconsider the reactor at steady state pictured in Fig. 13.2. In the absence of work $\dot{W}_{cv}$ and appreciable kinetic and potential energy effects, the energy liberated on combustion is transferred from the reactor in two ways only: by energy accompanying the exiting combustion products and by heat transfer to the surroundings. The smaller the heat transfer, the greater the energy carried out with the combustion products and thus the greater the temperature of the products. The temperature that would be achieved by the products in the limit of adiabatic operation of the reactor is called the *adiabatic flame* or *adiabatic combustion* temperature.

The adiabatic flame temperature can be determined by use of the conservation of mass and conservation of energy principles. To illustrate the procedure, let us suppose that the combustion air and the combustion products each form ideal gas mixtures. Then, with the other assumptions stated above, the energy rate balance reduces to the form

$$\bar{h}_P = \bar{h}_R$$

$$\sum_P n_e(\bar{h}_f^\circ + \Delta\bar{h})_e = \sum_R n_i(\bar{h}_f^\circ + \Delta\bar{h})_i$$

or

$$\sum_P n_e(\Delta\bar{h})_e = \sum_R n_i(\Delta\bar{h})_i + \sum_R n_i\bar{h}_{fi}^\circ - \sum_P n_e\bar{h}_{fe}^\circ \tag{13.21}$$

The n's are obtained from the balanced chemical reaction equation. The enthalpies of formation of the reactants and products are obtained from Table A-24 or A-24E. Enthalpy of combustion data might be employed in situations where the enthalpy of formation for the fuel is not available. Knowing the states of the reactants as they enter the reactor, the $\Delta\bar{h}$ terms for the reactants can be evaluated as discussed previously. Thus, all terms on the right side of Eq. 13.21 can be evaluated. The terms $(\Delta\bar{h})_e$ on the left side account for the changes in enthalpy of the products from T_{ref} to the unknown adiabatic flame temperature. Since the unknown temperature appears in each term of the sum on the left side of the equation, determination of the adiabatic flame temperature requires *iteration:* A temperature for the products is assumed and used to evaluate the left side of Eq. 13.21. The value obtained is compared with the previously determined value for the right side of the equation. The procedure continues until satisfactory agreement is attained. Example 13.8 gives an illustration.

For a specified fuel and specified temperature and pressure of the reactants, the *maximum* adiabatic flame temperature is for complete combustion with the theoretical amount of air. The measured value of the temperature of the combustion products may be several hundred degrees below the calculated maximum adiabatic flame temperature, however, for several reasons. For example, once adequate oxygen has been provided to permit complete combustion, bringing in more air dilutes the combustion products, lowering their temperature. Incomplete combustion tends to reduce the temperature of the products, and combustion is seldom complete. Also, heat losses can be reduced but not altogether eliminated. As a result of the high temperatures achieved, some of the combustion products may dissociate. Endothermic dissociation reactions lower the product temperature. The effect of dissociation on the adiabatic flame temperature is considered in Sec. 14.4.

EXAMPLE 13.8

PROBLEM

Liquid octane at 25°C, 1 atm enters a well-insulated reactor and reacts with air entering at the same temperature and pressure. For steady-state operation and negligible effects of kinetic and potential energy, determine the temperature of the combustion products for complete combustion with (a) the theoretical amount of air, (b) 400% theoretical air.

SOLUTION

Known: Liquid octane and air, each at 25°C and 1 atm, burn completely within a well-insulated reactor operating at steady state.

Find: Determine the temperature of the combustion products for (a) the theoretical amount of air and (b) 400% theoretical air.

Schematic and Given Data:

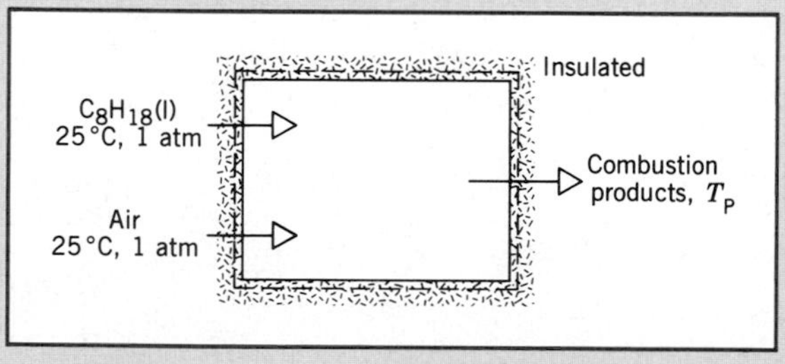

Figure E13.8

Assumptions:

1. The control volume indicated on the accompanying figure by a dashed line operates at steady state.
2. For the control volume, $\dot{Q}_{cv} = 0$, $\dot{W}_{cv} = 0$, and kinetic and potential effects are negligible.
3. The combustion air and the products of combustion each form ideal gas mixtures.
4. Combustion is complete.

5. Each mole of oxygen in the combustion air is accompanied by 3.76 moles of nitrogen only and the nitrogen is inert.

Analysis: At steady state, the control volume energy rate balance reduces with assumptions 2 and 3 to give

$$0 = \frac{\dot{Q}_{cv}\,^{0}}{\dot{n}_F} - \frac{\dot{W}_{cv}\,^{0}}{\dot{n}_F} + \sum_R n_i(\bar{h}_f^\circ + \Delta\bar{h})_i - \sum_P n_e(\bar{h}_f^\circ + \Delta\bar{h})_e$$

On rearrangement,

$$\sum_P n_e(\Delta\bar{h})_e = \sum_R n_i(\Delta\bar{h})_i + \sum_R n_i\bar{h}_{fi}^\circ - \sum_P n_e\bar{h}_{fe}^\circ$$

where all symbols have the same significance as in Eq. 13.21. Since the reactants enter at 25°C, the $(\Delta\bar{h})_i$ terms on the right side vanish, and the energy rate equation becomes

$$\sum_P n_e(\Delta\bar{h})_e = \sum_R n_i\bar{h}_{fi}^\circ - \sum_P n_e\bar{h}_{fe}^\circ$$

(a) For combustion of liquid octane with the theoretical amount of air, the chemical equation is

$$C_8H_{18}(l) + 12.5O_2 + 47N_2 \rightarrow 8CO_2 + 9H_2O(g) + 47N_2$$

Introducing the coefficients of this equation, the energy rate balance takes the form

$$8(\Delta\bar{h})_{CO_2} + 9(\Delta\bar{h})_{H_2O(g)} + 47(\Delta\bar{h})_{N_2}$$
$$= [(\bar{h}_f^\circ)_{C_8H_{18}(l)} + 12.5(\bar{h}_f^\circ)_{O_2}^{\,0} + 47(\bar{h}_f^\circ)_{N_2}^{\,0}]$$
$$- [8(\bar{h}_f^\circ)_{CO_2} + 9(\bar{h}_f^\circ)_{H_2O(g)} + 47(\bar{h}_f^\circ)_{N_2}^{\,0}]$$

The right side of the above equation can be evaluated with enthalpy of formation data from Table A-24, giving

$$8(\Delta\bar{h})_{CO_2} + 9(\Delta\bar{h})_{H_2O(g)} + 47(\Delta\bar{h})_{N_2} = 5{,}074{,}630 \text{ kJ/kmol (fuel)}$$

Each $\Delta\bar{h}$ term on the left side of this equation depends on the temperature of the products, T_P. This temperature can be determined by an iterative procedure.

The following table gives a summary of the iterative procedure for three trial values of T_P. Since the summation of the enthalpies of the products equals 5,074,630 kJ/kmol, the actual value of T_P is in the interval from 2350 to 2400 K. Interpolation between these temperatures gives $T_P = 2395$ K.

	2500 K	**2400 K**	**2350 K**
$8(\Delta\bar{h})_{CO_2}$	975,408	926,304	901,816
$9(\Delta\bar{h})_{H_2O(g)}$	890,676	842,436	818,478
$47(\Delta\bar{h})_{N_2}$	3,492,664	3,320,597	3,234,869
$\sum_P n_e(\Delta\bar{h})_e$	5,358,748	5,089,337	4,955,163

(b) For complete combustion of liquid octane with 400% theoretical air, the chemical equation is

$$C_8H_{18}(l) + 50O_2 + 188N_2 \rightarrow 8CO_2 + 9H_2O + 37.5O_2 + 188N_2$$

The energy rate balance for this case reduces to give

$$8(\Delta\bar{h})_{CO_2} + 9(\Delta\bar{h})_{H_2O(g)} + 37.5(\Delta\bar{h})_{O_2} + 188(\Delta\bar{h})_{N_2} = 5{,}074{,}630 \text{ kJ/kmol (fuel)}$$

Observe that the right side has the same value as in part (a). Proceeding iteratively as above, the temperature of the products is $T_P = 962$ K.

Comment: The temperature determined in part (b) is considerably lower than the value found in part (a). This illustrates that once enough oxygen has been provided for complete combustion, bringing in more air dilutes the combustion products, lowering their temperature.

13.4 ABSOLUTE ENTROPY AND THE THIRD LAW OF THERMODYNAMICS

Thus far our analyses of reacting systems have been conducted using the conservation of mass and conservation of energy principles. In the present section some of the implications of the second law of thermodynamics for reacting systems are considered. The discussion continues in Sec. 13.6, dealing with the availability concept, and in the next chapter, where the subject of chemical equilibrium is taken up.

13.4.1 Evaluation of Entropy for Reacting Systems

The property entropy plays an important part in quantitative evaluations using the second law of thermodynamics. When reacting systems are under consideration, the same basic problem arises for entropy as for enthalpy and internal energy: A common datum must be used to assign entropy values for each substance involved in the reaction. This is accomplished using the *third law* of thermodynamics and the *absolute entropy* concept.

The *third law* deals with the entropy of substances at the absolute zero of temperature. Based on experimental observations, this law states that the entropy of a pure crystalline substance is zero at the absolute zero of temperature, 0 K or 0°R. Substances not having a pure crystalline structure at absolute zero have a nonzero value of entropy at absolute zero. The experimental evidence on which the third law is based is obtained primarily from studies of chemical reactions at low temperatures and specific heat measurements at temperatures approaching absolute zero.

For present considerations, the importance of the third law is that it provides a datum relative to which the entropy of each substance participating in a reaction can be evaluated so that no ambiguities or conflicts arise. The entropy relative to this datum is called the *absolute* entropy. The change in entropy of a substance between absolute zero and any given state can be determined from precise measurements of energy transfers and specific heat data or from procedures based on statistical thermodynamics and observed molecular data.

Tables A-24 and A-24E give the value of the absolute entropy for selected substances at the standard reference state, $T_{ref} = 298.15$ K, $p_{ref} = 1$ atm. Two values of absolute entropy for water are provided. One is for liquid water and the other is for water vapor. As for the case of the enthalpy of formation of water considered previously, the vapor value listed is for a hypothetical state in which water is a vapor at 25°C (77°F) and 1 atm obeying the ideal gas model. Tables A-16 through A-22 and Tables A-16E through A-22E give tabulations of absolute entropy versus temperature at a pressure of 1 atm for selected gases. The absolute entropy at 1 atm and temperature T is designated as $s°(T)$ or $\bar{s}°(T)$, depending on whether the value is on a unit mass or per mole basis. In all these tables, ideal gas behavior is assumed for the gases.

When the absolute entropy is known at the standard state, the specific entropy at any other state can be found by adding the specific entropy change between the two states to the absolute entropy at the standard state. Similarly, when the absolute entropy is known at the pressure p_{ref} and temperature T, the absolute entropy at the same temperature and *any* pressure p can be found from

$$\bar{s}(T, p) = \bar{s}(T, p_{ref}) + [\bar{s}(T, p) - \bar{s}(T, p_{ref})]$$

The second term on the right side of this equation can be evaluated for an *ideal gas* by using Eq. 6.25b, giving

$$\bar{s}(T, p) = \bar{s}°(T) - \bar{R} \ln \frac{p}{p_{ref}} \quad \text{(ideal gas)} \tag{13.22}$$

where $\bar{s}°(T)$ is the absolute entropy at temperature T and pressure p_{ref}.

The entropy of the ith component of an ideal gas mixture is evaluated at the mixture temperature T and the *partial* pressure p_i: $\bar{s}_i(T, p_i)$. The partial pressure is given by $p_i = y_i p$, where y_i is the mole fraction of component i and p is the mixture pressure. Thus, Eq. 13.22 takes the form

$$\bar{s}_i(T, p_i) = \bar{s}_i^\circ(T) - \bar{R} \ln \frac{p_i}{p_{ref}}$$

or

$$\bar{s}_i(T, p_i) = \bar{s}_i^\circ(T) - \bar{R} \ln \frac{y_i p}{p_{ref}} \quad \left(\begin{matrix}\text{component } i \text{ of an} \\ \text{ideal gas mixture}\end{matrix}\right) \tag{13.23}$$

where $\bar{s}_i^\circ(T)$ is the absolute entropy of component i at temperature T and p_{ref}.

In subsequent discussions, the thermodynamic property known as the Gibbs function is found to be useful. The specific Gibbs function $\bar{g}$, introduced in Sec. 11.3, is

$$\bar{g} = \bar{h} - T\bar{s} \tag{13.24}$$

The procedure followed in setting a datum for the Gibbs function closely parallels that used in defining the enthalpy of formation: To each stable element at the standard state is assigned a zero value of the Gibbs function. The *Gibbs function of formation* of a compound equals the change in the Gibbs function for the reaction in which the compound is formed from its elements. Tables A-24 and A-24E give the Gibbs function of formation, $\bar{g}_f^\circ$, at 25°C (77°F) and 1 atm for selected substances.

The Gibbs function at a state other than the standard state is found by adding to the Gibbs function of formation the change in the specific Gibbs function $\Delta\bar{g}$ between the standard state and the state of interest

$$\bar{g}(T, p) = \bar{g}_f^\circ + [\bar{g}(T, p) - \bar{g}(T_{ref}, p_{ref})] = \bar{g}_f^\circ + \Delta\bar{g} \tag{13.25}$$

With Eq. 13.24, $\Delta\bar{g}$ can be written as

$$\Delta\bar{g} = [\bar{h}(T, p) - \bar{h}(T_{ref}, p_{ref})] - [T\bar{s}(T, p) - T_{ref}\bar{s}(T_{ref}, p_{ref})]$$

The Gibbs function of component i in an ideal gas mixture is evaluated at the partial pressure of component i and the mixture temperature.

EXAMPLE 13.9

PROBLEM

Determine the Gibbs function of formation of methane at the standard state, in kJ/kmol.

SOLUTION

Known: The compound is methane.

Find: Determine the Gibbs function of formation at the standard state, in kJ/kmol.

Assumptions: In the formation of methane from carbon and hydrogen (H_2), the carbon and hydrogen are each initially at 25°C and 1 atm pressure. The methane formed is also at 25°C and 1 atm pressure.

Analysis: Methane is formed from carbon and hydrogen according to $C + 2H_2 \rightarrow CH_4$. The change in the Gibbs function for this reaction is

$$\begin{aligned}\bar{g}_P - \bar{g}_R &= (\bar{h} - T\bar{s})_{CH_4} - (\bar{h} - T\bar{s})_C - 2(\bar{h} - T\bar{s})_{H_2} \\ &= (\bar{h}_{CH_4} - \bar{h}_C - 2\bar{h}_{H_2}) - T(\bar{s}_{CH_4} - \bar{s}_C - 2\bar{s}_{H_2})\end{aligned}$$

where $\bar{g}_P$ and $\bar{g}_R$ denote, respectively, the Gibbs functions of the reactants and products, each per kmol of methane.

In the present case, all substances are at the same temperature and pressure, 25°C and 1 atm, which correspond to the standard reference state values. At the standard reference state, the enthalpy and Gibbs function for carbon and hydrogen are zero by definition. Thus, the above equation reduces to

$$(\bar{g}_f^\circ)_{CH_4} = (\bar{h}_f^\circ)_{CH_4} - T_{ref}(\bar{s}^\circ_{CH_4} - \bar{s}^\circ_C - 2\bar{s}^\circ_{H_2})$$

With enthalpy of formation and absolute entropy data from Table A-24

$$(\bar{g}_f^\circ)_{CH_4} = -74{,}850 - 298.15[186.16 - 5.74 - 2(130.57)] = -50{,}783 \text{ kJ/kmol}$$

Comment: The slight difference between the calculated value for the Gibbs function of formation of methane and the value of Table A-24 can be attributed to round-off.

13.4.2 Entropy Balances for Reacting Systems

Many of the considerations that enter when energy balances are written for reacting systems also apply to entropy balances. The writing of entropy balances for reacting systems will be illustrated by referring to special cases of broad interest.

Let us begin by reconsidering the steady-state reactor of Fig. 13.2, for which the combustion reaction is given by Eq. 13.11. The combustion air and the products of combustion are each assumed to form ideal gas mixtures. The entropy rate balance for the two-inlet, single-exit reactor can be expressed on a *per mole of fuel* basis as

$$\begin{aligned}0 = \sum_j \frac{\dot{Q}_j/\dot{n}_F}{T_j} + \bar{s}_F + \left[\underline{\left(a + \frac{b}{4}\right)\bar{s}_{O_2} + \left(a + \frac{b}{4}\right) 3.76\bar{s}_{N_2}}\right] \\ - \left[\underline{a\bar{s}_{CO_2} + \frac{b}{2}\,\bar{s}_{H_2O} + \left(a + \frac{b}{4}\right) 3.76\bar{s}_{N_2}}\right] + \frac{\dot{\sigma}_{cv}}{\dot{n}_F}\end{aligned} \tag{13.26}$$

where $\dot{n}_F$ is the molar flow rate of the fuel and the coefficients appearing in the underlined terms are the same as those for the corresponding substances in the reaction equation.

The specific entropies of Eq. 13.26 are absolute entropies. Let us consider how the entropies are evaluated for the combustion products and the combustion air. The entropies of the combustion products would be evaluated from Eq. 13.23, using the temperature, pressure, and composition of the products. The entropies of the entering oxygen and nitrogen would be evaluated similarly, using the temperature, pressure, and composition of the combustion air. If the fuel and air entered the reactor as an ideal gas mixture, the entropies of the mixture components would be evaluated from Eq. 13.23 using the appropriate partial pressures.

Next consider an entropy balance for a process of a closed system during which a chemical reaction occurs

$$S_P - S_R = \int \left(\frac{\delta Q}{T}\right)_b + \sigma \tag{13.27}$$

S_R and S_P denote, respectively, the entropy of the reactants and the entropy of the products. When the reactants and products form ideal gas mixtures, the entropy balance can be expressed on a *per mole of fuel* basis as

$$\sum_{P} n\bar{s} - \sum_{R} n\bar{s} = \frac{1}{n_F}\int\left(\frac{\delta Q}{T}\right)_b + \frac{\sigma}{n_F} \tag{13.28}$$

where the coefficients n on the left are the coefficients of the reaction equation giving the moles of each reactant or product *per mole of fuel*. The entropy terms would be evaluated from Eq. 13.23 using the temperature, pressure, and composition of the reactants or products, as appropriate. The fuel would be mixed with the oxidizer, so this must be taken into account when determining the partial pressures of the reactants.

EXAMPLE 13.10

PROBLEM

Determine the change in entropy of the system of Example 13.6 in kJ/K.

SOLUTION

Known: A mixture of gaseous methane and oxygen, initially at 25°C and 1 atm, burns completely within a closed rigid container. The products are cooled to 900 K, 3.02 atm.

Find: Determine the change in entropy for the process in kJ/K.

Schematic and Given Data: See Fig. E13.6.

Assumptions:

1. The contents of the container are taken as the system.
2. The initial mixture can be modeled as an ideal gas mixture, as can the products of combustion.
3. Combustion is complete.

Analysis: The chemical equation for the complete combustion of methane with oxygen is

$$CH_4 + 2O_2 \rightarrow CO_2 + 2H_2O$$

The change in entropy for the process of the closed system is $\Delta S = S_P - S_R$, where S_R and S_P denote, respectively, the initial and final entropies of the system. Since the initial mixture forms an ideal gas mixture (assumption 2), the entropy of the reactants can be expressed as the sum of the contributions of the components, each evaluated at the mixture temperature and the partial pressure of the component. That is

$$S_R = 1\bar{s}_{CH_4}(T_1, y_{CH_4}p_1) + 2\bar{s}_{O_2}(T_1, y_{O_2}p_1)$$

where $y_{CH_4} = 1/3$ and $y_{O_2} = 2/3$ denote, respectively, the mole fractions of the methane and oxygen in the initial mixture. Similarly, since the products of combustion form an ideal gas mixture (assumption 2)

$$S_P = 1\bar{s}_{CO_2}(T_2, y_{CO_2}p_2) + 2\bar{s}_{H_2O}(T_2, y_{H_2O}p_2)$$

where $y_{CO_2} = 1/3$ and $y_{H_2O} = 2/3$ denote, respectively, the mole fractions of the carbon dioxide and water vapor in the products of combustion. In these equations, p_1 and p_2 denote the pressure at the initial and final states, respectively.

The specific entropies required to determine S_R can be calculated from Eq. 13.23 using absolute entropy data from Table A-24. Since $T_1 = T_{ref}$ and $p_1 = p_{ref}$

$$\bar{s}_{CH_4}(T_1, y_{CH_4}p_1) = \bar{s}^{\circ}_{CH_4}(T_{ref}) - \bar{R}\ln\frac{y_{CH_4}p_{ref}}{p_{ref}}$$

$$= 186.16 - 8.314\ln\frac{1}{3} = 195.294 \text{ kJ/kmol}\cdot\text{K}$$

Similarly

$$\bar{s}_{O_2}(T_1, y_{O_2}p_1) = \bar{s}^{\circ}_{O_2}(T_{ref}) - \bar{R} \ln \frac{y_{O_2}p_{ref}}{p_{ref}}$$

$$= 205.03 - 8.314 \ln \frac{2}{3} = 208.401 \text{ kJ/kmol} \cdot \text{K}$$

At the final state, the products are at $T_2 = 900$ K and $p_2 = 3.02$ atm. With Eq. 13.23 and absolute entropy data from Tables A-19 and A-21, as appropriate

$$\bar{s}_{CO_2}(T_2, y_{CO_2}p_2) = \bar{s}^{\circ}_{CO_2}(T_2) - \bar{R} \ln \frac{y_{CO_2}p_2}{p_{ref}}$$

$$= 263.559 - 8.314 \ln \frac{(1/3)(3.02)}{1} = 263.504 \text{ kJ/kmol} \cdot \text{K}$$

$$\bar{s}_{H_2O}(T_2, y_{H_2O}p_2) = \bar{s}^{\circ}_{H_2O}(T_2) - \bar{R} \ln \frac{y_{H_2O}p_2}{p_{ref}}$$

$$= 228.321 - 8.314 \ln \frac{(2/3)(3.02)}{1} = 222.503 \text{ kJ/kmol} \cdot \text{K}$$

Finally, the entropy change for the process is

$$\Delta S = S_P - S_R$$

$$= [263.504 + 2(222.503)] - [195.294 + 2(208.401)]$$

$$= 96.414 \text{ kJ/K}$$

EXAMPLE 13.11

PROBLEM

Liquid octane at 25°C, 1 atm enters a well-insulated reactor and reacts with air entering at the same temperature and pressure. The products of combustion exit at 1 atm pressure. For steady-state operation and negligible effects of kinetic and potential energy, determine the rate of entropy production, in kJ/K per kmol of fuel, for complete combustion with (a) the theoretical amount of air, (b) 400% theoretical air.

SOLUTION

Known: Liquid octane and air, each at 25°C and 1 atm, burn completely in a well-insulated reactor operating at steady state. The products of combustion exit at 1 atm pressure.

Find: Determine the rate of entropy production, in kJ/K per kmol of fuel, for combustion with (a) the theoretical amount of air, (b) 400% theoretical air.

Schematic and Given Data:

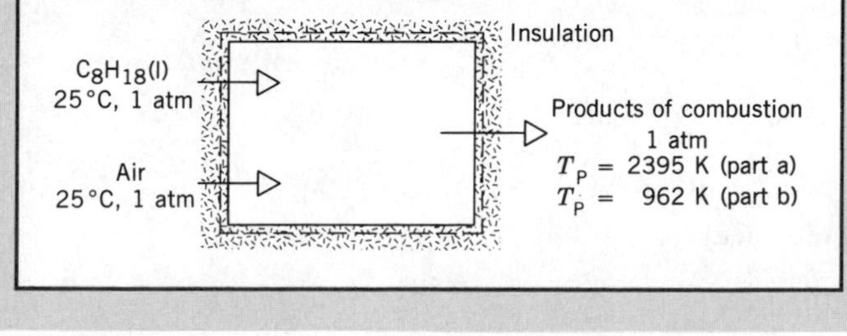

Figure E13.11

Assumptions:

1. The control volume shown on the accompanying figure by a dashed line operates at steady state and without heat transfer with its surroundings.
2. Combustion is complete. Each mole of oxygen in the combustion air is accompanied by 3.76 moles of nitrogen, which is inert.
3. The combustion air can be modeled as an ideal gas mixture, as can the products of combustion.
4. The reactants enter at 25°C, 1 atm. The products exit at a pressure of 1 atm.

Analysis: The temperature of the exiting products of combustion T_P was evaluated in Example 13.8 for each of the two cases. For combustion with the theoretical amount of air, $T_P = 2395$ K. For complete combustion with 400% theoretical air, $T_P = 962$ K.

(a) For combustion of liquid octane with the theoretical amount of air, the chemical equation is

$$C_8H_{18}(l) + 12.5O_2 + 47N_2 \rightarrow 8CO_2 + 9H_2O(g) + 47N_2$$

With assumptions 1 and 3, the entropy rate balance on a per mole of fuel basis takes the form

$$0 = \cancelto{0}{\sum_j \frac{\dot{Q}_j/T_j}{\dot{n}_F}} + \bar{s}_F + (12.5\bar{s}_{O_2} + 47\bar{s}_{N_2}) - (8\bar{s}_{CO_2} + 9\bar{s}_{H_2O(g)} + 47\bar{s}_{N_2}) + \frac{\dot{\sigma}_{cv}}{\dot{n}_F}$$

or on rearrangement

$$\frac{\dot{\sigma}_{cv}}{\dot{n}_F} = (8\bar{s}_{CO_2} + 9\bar{s}_{H_2O(g)} + 47\bar{s}_{N_2}) - \bar{s}_F - (12.5\bar{s}_{O_2} + 47\bar{s}_{N_2})$$

Each coefficient of this equation is the same as for the corresponding term of the balanced chemical equation.

The fuel enters the reactor separately at T_{ref}, p_{ref}. The absolute entropy of liquid octane required by the entropy balance is obtained from Table A-24 as 360.79 kJ/kmol·K.

The oxygen and nitrogen in the combustion air enter the reactor as components of an ideal gas mixture at T_{ref}, p_{ref}. With Eq. 13.23 and absolute entropy data from Table A-24

$$\bar{s}_{O_2} = \bar{s}^{\circ}_{O_2}(T_{ref}) - \bar{R} \ln \frac{y_{O_2} p_{ref}}{p_{ref}}$$

$$= 205.03 - 8.314 \ln 0.21 = 218.01 \text{ kJ/kmol}\cdot\text{K}$$

$$\bar{s}_{N_2} = \bar{s}^{\circ}_{N_2}(T_{ref}) - \bar{R} \ln \frac{y_{N_2} p_{ref}}{p_{ref}}$$

$$= 191.5 - 8.314 \ln 0.79 = 193.46 \text{ kJ/kmol}\cdot\text{K}$$

The product gas exits as an ideal gas mixture at 1 atm, 2395 K with the following composition: $y_{CO_2} = 8/64 = 0.125$, $y_{H_2O(g)} = 9/64 = 0.1406$, $y_{N_2} = 47/64 = 0.7344$. With Eq. 13.23 and absolute entropy data at 2395 K from Tables A-17, A-19, A-21, as appropriate

$$\bar{s}_{CO_2} = \bar{s}^{\circ}_{CO_2} - \bar{R} \ln y_{CO_2}$$

$$= 320.173 - 8.314 \ln 0.125 = 337.46 \text{ kJ/kmol}\cdot\text{K}$$

$$\bar{s}_{H_2O} = 273.986 - 8.314 \ln 0.1406 = 290.30 \text{ kJ/kmol}\cdot\text{K}$$

$$\bar{s}_{N_2} = 258.503 - 8.314 \ln 0.7344 = 261.07 \text{ kJ/kmol}\cdot\text{K}$$

Inserting values into the expression for the rate of entropy production

$$\frac{\dot{\sigma}_{cv}}{\dot{n}_F} = 8(337.46) + 9(290.30) + 47(261.07)$$

$$- 360.79 - 12.5(218.01) - 47(193.46)$$

$$= 5404 \text{ kJ/kmol (octane)}\cdot\text{K}$$

(b) The complete combustion of liquid octane with 400% theoretical air is described by the following chemical equation:

$$C_8H_{18}(l) + 50O_2 + 188N_2 \rightarrow 8CO_2 + 9H_2O(g) + 37.5O_2 + 188N_2$$

The entropy rate balance on a per mole of fuel basis takes the form

$$\frac{\dot{\sigma}_{cv}}{\dot{n}_F} = (8\bar{s}_{CO_2} + 9\bar{s}_{H_2O(g)} + 37.5\bar{s}_{O_2} + 188\bar{s}_{N_2}) - \bar{s}_F - (50\bar{s}_{O_2} + 188\bar{s}_{N_2})$$

The specific entropies of the reactants have the same values as in part (a). The product gas exits as an ideal gas mixture at 1 atm, 962 K with the following composition: $y_{CO_2} = 8/242.5 = 0.033$, $y_{H_2O(g)} = 9/242.5 = 0.0371$, $y_{O_2} = 37.5/242.5 = 0.1546$, $y_{N_2} = 0.7753$. With the same approach as in part (a)

$$\bar{s}_{CO_2} = 267.12 - 8.314 \ln 0.033 = 295.481 \text{ kJ/kmol}\cdot\text{K}$$

$$\bar{s}_{H_2O} = 231.01 - 8.314 \ln 0.0371 = 258.397 \text{ kJ/kmol}\cdot\text{K}$$

$$\bar{s}_{O_2} = 242.12 - 8.314 \ln 0.1546 = 257.642 \text{ kJ/kmol}\cdot\text{K}$$

$$\bar{s}_{N_2} = 226.795 - 8.314 \ln 0.7753 = 228.911 \text{ kJ/kmol}\cdot\text{K}$$

Inserting values into the expression for the rate of entropy production

$$\frac{\dot{\sigma}_{cv}}{\dot{n}_F} = 8(295.481) + 9(258.397) + 37.5(257.642) + 188(228.911)$$

$$- 360.79 - 50(218.01) - 188(193.46)$$

$$= 9754 \text{ kJ/kmol (octane)}\cdot\text{K}$$

Comments:

1. Once enough oxygen has been provided for complete combustion, mixing a greater amount of air with the fuel prior to combustion results in a lower product gas temperature and a greater rate of entropy production.
2. The rates of entropy production calculated in this example are positive, as required by the second law. This does not necessarily mean that the proposed reactions would occur, however, for the results are based on the assumption of *complete* combustion. The possibility of achieving complete combustion with specified reactants at a given temperature and pressure can be investigated with the methods of Chap. 14, dealing with chemical equilibrium. For further discussion, see Sec. 14.4.

13.5 FUEL CELLS

A *fuel cell* is a device in which fuel and an oxidizer undergo a controlled chemical reaction, producing products and providing electrical current directly to an external circuit. The fuel and oxidizer do not react in a rapid combustion process, but react in stages on separate electrodes: a positive electrode (*cathode*) and a negative electrode (*anode*). An electrolyte separates the two electrodes. The rates of reaction are limited by the time it takes for diffusion of chemical species through the electrodes and the electrolyte and by the reaction kinetics.

There are many different kinds of fuel cells. A schematic of one type, the hydrogen–oxygen fuel cell, is shown in Fig. 13.3. Hydrogen supplied to the cell diffuses through the porous anode and reacts on the anode surface with OH^- ions, forming water and yielding free electrons according to

$$H_2 + 2OH^- \rightarrow 2H_2O + 2e^-$$

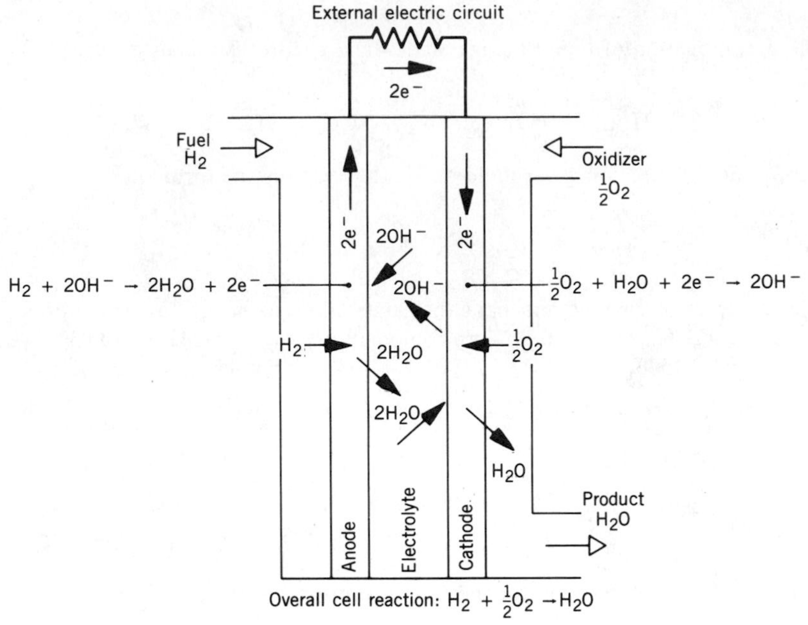

Figure 13.3 Hydrogen–oxygen fuel cell.

The electrons enter the external circuit, and the water goes into the electrolyte. On the cathode surface, oxygen combines with water from the electrolyte and electrons from the external circuit to produce OH^- ions and water according to

$$\tfrac{1}{2}O_2 + 2H_2O + 2e^- \rightarrow 2OH^- + H_2O$$

The electrolyte separating the electrodes transports the OH^- ions, completing the circuit, and the water is removed from the cell. The overall cell reaction is

$$H_2 + \tfrac{1}{2}O_2 \rightarrow H_2O$$

which is the same as the equation for the highly exothermic combustion reaction that opened the present chapter, Eq. 13.1. In the fuel cell, however, only a relatively small amount of heat transfer between the cell and its surroundings takes place. Moreover, the temperature within a fuel cell is low in comparison to the temperature of the products of rapid combustion reactions.

In a fuel cell, the chemical reaction is harnessed to produce electric power without moving parts or the utilization of intermediate heat transfers, as in the power plants studied in Chaps. 8 and 9. Since a fuel cell does not operate as a thermodynamic power cycle, the notion of a limiting thermal efficiency imposed by the second law is not applicable. Given a supply of fuel and oxidizer, more power can be obtained with a fuel cell than if the gases were allowed to react spontaneously (and thus highly irreversibly) and the hot products of combustion supplied to a power cycle for which the thermal efficiency is strictly second law-limited.

In recent years, hydrogen–oxygen fuel cells have been used to produce power for spaceflight vehicles. Fuel cells operating on natural gas are under development by electric utilities and in the future may be used to power electric cars and trucks. Fuel cells do not produce large quantities of undesirable emissions, nor do they reject great quantities of waste heat to the surroundings. However, many technical problems remain to be solved before fuel cells can be made economically attractive for more general use.

13.6 CHEMICAL AVAILABILITY[1]

The object of the present section is to extend the availability concept introduced in Chap. 7 for use in studying systems in which combustion occurs. To distinguish the current considerations from those introduced previously, let the system be a specified amount of a hydrocarbon fuel C_aH_b at temperature T_0 and pressure p_0 and the environment consist of a gas phase at T_0, p_0 involving nitrogen, oxygen, water vapor, and carbon dioxide. Since the system is in thermal and mechanical equilibrium with the environment, the value of availability as defined in Sec. 7.2 would be zero. More precisely, we should say that the *thermomechanical* contribution to the availability magnitude has a value of zero, for a *chemical* contribution not previously identified can be defined that has a nonzero value. This aspect of the availability concept is the subject of the present section.

13.6.1 Evaluation of Chemical Availability

Availability is introduced in Chap. 7 through study of a combined system consisting of a closed system of interest and an environment. The object of the development is an expression for the maximum theoretical work obtainable from the combined system as the closed system comes into thermal and mechanical equilibrium with the environment. The thermomechanical availability is this value for work. In the present section, we study a combined system formed by an environment and a system consisting of a specified amount of fuel at T_0, p_0. The object is to evaluate the work obtainable by allowing the fuel to react with oxygen from the environment to produce the environmental components carbon dioxide and water. The *chemical availability* is, by definition, the maximum theoretical work that could be developed by the combined system. The sum of the thermomechanical and chemical availabilities is the total availability associated with a given system at a specified state, relative to a specified environment.

Derivation. To help us visualize how work might be obtained through the reaction of a fuel with environmental components, a fuel cell operating at steady state is included in the combined system, as shown in Fig. 13.4. Referring to this figure, fuel enters the cell at temperature T_0 and pressure p_0. At another location, oxygen enters from the environment. Assuming the environment consists of an ideal gas mixture, the oxygen would enter at its condition within the environment: temperature T_0 and partial pressure $y^e_{O_2}p_0$, where $y^e_{O_2}$ is the mole fraction of the oxygen in the environment. The fuel and oxygen react completely within the cell to produce carbon dioxide and water vapor, which exit in separate streams at their conditions within the environment: temperature T_0 and the partial pressures $y^e_{CO_2}p_0$ and $y^e_{H_2O}p_0$, respectively. The reaction is given by

$$C_aH_b + \left(a + \frac{b}{4}\right) O_2 \rightarrow aCO_2 + \frac{b}{2} H_2O \tag{13.29}$$

For steady-state operation, the energy rate balance for a control volume enclosing the fuel cell reduces to give

$$\frac{\dot{W}_{cv}}{\dot{n}_F} = \frac{\dot{Q}_{cv}}{\dot{n}_F} + \bar{h}_F + \left(a + \frac{b}{4}\right) \bar{h}_{O_2} - a\bar{h}_{CO_2} - \frac{b}{2} \bar{h}_{H_2O} \tag{13.30}$$

[1] Study of Chap. 7 is a prerequisite for this section.

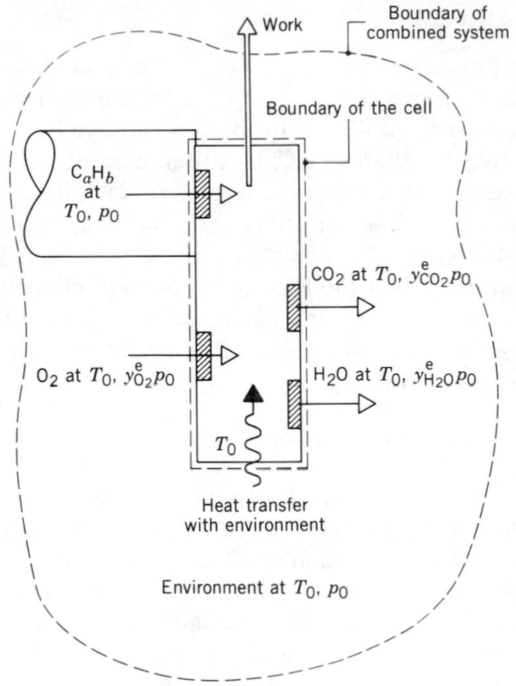

Figure 13.4 Illustration used to introduce the fuel chemical availability concept.

Kinetic and potential energy effects are regarded as negligible. Since the fuel cell is at steady state, its volume does not change with time, so no portion of $\dot{W}_{cv}$ is required to displace the environment. Thus, Eq. 13.30 gives the work developed by the combined system of system plus environment. Heat transfer is assumed to occur with the environment only at temperature T_0. An entropy rate balance for the control volume therefore takes the form

$$0 = \frac{\dot{Q}_{cv}/\dot{n}_F}{T_0} + \bar{s}_F + \left(a + \frac{b}{4}\right)\bar{s}_{O_2} - a\bar{s}_{CO_2} - \frac{b}{2}\bar{s}_{H_2O} + \frac{\dot{\sigma}_{cv}}{\dot{n}_F} \tag{13.31}$$

Eliminating the heat transfer rate between Eqs. 13.30 and 13.31 results in

$$\frac{\dot{W}_{cv}}{\dot{n}_F} = \underline{\left[\bar{h}_F + \left(a + \frac{b}{4}\right)\bar{h}_{O_2} - a\bar{h}_{CO_2} - \frac{b}{2}\bar{h}_{H_2O}\right]}$$

$$-T_0\underline{\left[\bar{s}_F + \left(a + \frac{b}{4}\right)\bar{s}_{O_2} - a\bar{s}_{CO_2} - \frac{b}{2}\bar{s}_{H_2O}\right]} - T_0\frac{\dot{\sigma}_{cv}}{\dot{n}_F} \tag{13.32}$$

In Eq. 13.32, the specific enthalpy and entropy of the fuel are evaluated at T_0 and p_0. The values of the specific enthalpies in the first underlined term can be determined knowing only the temperature T_0. The specific entropies in the second underlined term can be determined knowing the temperature, pressure, and composition of the environment. Accordingly, once the environment is specified, all enthalpy and entropy terms of Eq. 13.32 can be regarded as known and independent of the nature of the processes occurring within the control volume. The term $T_0\dot{\sigma}_{cv}$ depends explicitly on the nature of such processes, however. In accordance with the second law, $T_0\dot{\sigma}_{cv}$ is positive whenever internal irreversibilities are present; vanishes in the limiting case of no irreversibilities, and is never negative. The maximum theoretical value for the work developed is obtained when no irreversibilities are present. Setting $T_0\dot{\sigma}_{cv}$ to zero in Eq. 13.32, the following expression for the chemical availability results:

$$\bar{a}^{\text{ch}} = \left[\bar{h}_{\text{F}} + \left(a + \frac{b}{4}\right)\bar{h}_{\text{O}_2} - a\bar{h}_{\text{CO}_2} - \frac{b}{2}\bar{h}_{\text{H}_2\text{O}}\right] - T_0\left[\bar{s}_{\text{F}} + \left(a + \frac{b}{4}\right)\bar{s}_{\text{O}_2} - a\bar{s}_{\text{CO}_2} - \frac{b}{2}\bar{s}_{\text{H}_2\text{O}}\right] \tag{13.33}$$

The superscript ch is used to distinguish this contribution to the availability magnitude from the thermomechanical availability introduced in Chap. 7.

13.6.2 Alternative Expressions for Chemical Availability

For computational convenience, the chemical availability given by Eq. 13.33 can be written as Eq. 13.35 by replacing the specific entropy terms for O_2, CO_2, and H_2O using the following expression obtained by reduction of Eq. 6.25b:

$$\bar{s}_i(T_0, y_i^{\text{e}} p_0) = \bar{s}_i(T_0, p_0) - \bar{R} \ln y_i^{\text{e}} \tag{13.34}$$

The first term on the right is the absolute entropy at T_0 and p_0, and y_i^{e} is the mole fraction of component i in the environment.

Applying Eq. 13.34, Eq. 13.33 becomes

$$\bar{a}^{\text{ch}} = \left[\bar{h}_{\text{F}} + \left(a + \frac{b}{4}\right)\bar{h}_{\text{O}_2} - a\bar{h}_{\text{CO}_2} - \frac{b}{2}\bar{h}_{\text{H}_2\text{O}}\right](T_0, p_0) - T_0\left[\bar{s}_{\text{F}} + \left(a + \frac{b}{4}\right)\bar{s}_{\text{O}_2} - a\bar{s}_{\text{CO}_2} - \frac{b}{2}\bar{s}_{\text{H}_2\text{O}}\right](T_0, p_0) + \bar{R}T_0 \ln\left[\frac{(y_{\text{O}_2}^{\text{e}})^{a+b/4}}{(y_{\text{CO}_2}^{\text{e}})^{a}(y_{\text{H}_2\text{O}}^{\text{e}})^{b/2}}\right] \tag{13.35}$$

The specific enthalpy terms are determined using the enthalpies of formation for the respective substances. The specific entropies appearing in the equation are absolute entropies determined as described in Sec. 13.4. The logarithmic term normally contributes only a few percent to the chemical availability magnitude.

Equation 13.35 can be expressed alternatively in terms of the Gibbs functions of the respective substances as

$$\bar{a}^{\text{ch}} = \left[\bar{g}_{\text{F}} + \left(a + \frac{b}{4}\right)\bar{g}_{\text{O}_2} - a\bar{g}_{\text{CO}_2} - \frac{b}{2}\bar{g}_{\text{H}_2\text{O}}\right](T_0, p_0) + \bar{R}T_0 \ln\left[\frac{(y_{\text{O}_2}^{\text{e}})^{a+b/4}}{(y_{\text{CO}_2}^{\text{e}})^{a}(y_{\text{H}_2\text{O}}^{\text{e}})^{b/2}}\right] \tag{13.36}$$

The specific Gibbs functions are evaluated at the temperature T_0 and pressure p_0 of the environment. These terms can be determined with Eq. 13.25 as

$$\bar{g}(T_0, p_0) = \bar{g}_{\text{f}}^{\circ} + [\bar{g}(T_0, p_0) - \bar{g}(T_{\text{ref}}, p_{\text{ref}})] \tag{13.37}$$

where $\bar{g}_{\text{f}}^{\circ}$ is the Gibbs function of formation. For the *special case* where T_0 and p_0 are the same as T_{ref} and p_{ref}, respectively, the second term on the right of Eq. 13.37 vanishes and the specific Gibbs function is just the Gibbs function of formation.

The method introduced in this section for evaluating the chemical availability of pure hydrocarbons can also be used in principle for substances other than hydrocarbons: The chemical availability is the maximum theoretical work that could be developed by a fuel cell into which a substance of interest enters at T_0, p_0 and reacts completely with environmental components to produce environmental components. All environmental components involved enter and exit the cell at their conditions within the environment. By describing the environment appropriately, this method can be applied to all compounds of practical interest.[2] In the following discussion, we limit consideration to the compounds CO, H_2O, N_2, O_2, and CO_2, for these participate in the elementary combustion reactions serving as the focus of the present chapter. Of these five compounds, only carbon monoxide is not among the substances present in the environment we have been considering. Let us take up the compounds in the order listed.

Paralleling the development of Eq. 13.36 for the case of pure carbon monoxide, at T_0, p_0, the reaction within the cell is $CO + \frac{1}{2}O_2 \rightarrow CO_2$, and the chemical availability is given by

$$\bar{a}^{\text{ch}}_{\text{CO}} = [\bar{g}_{\text{CO}} + \tfrac{1}{2}\bar{g}_{\text{O}_2} - \bar{g}_{\text{CO}_2}](T_0, p_0) + \bar{R}T_0 \ln \left[\frac{(y^{\text{e}}_{\text{O}_2})^{1/2}}{y^{\text{e}}_{\text{CO}_2}}\right] \tag{13.38}$$

If carbon monoxide is not pure but a component of an ideal gas mixture at T_0, p_0, each component i of the mixture would enter the cell at temperature T_0 and the respective partial pressure $y_i p_0$. The contribution of carbon monoxide to the chemical availability of the mixture, per mole of CO, is then given by Eq. 13.38 with the mole fraction of carbon monoxide in the mixture, y_{CO}, appearing in the numerator of the logarithmic term that then reads $\ln[y_{\text{CO}}(y^{\text{e}}_{\text{O}_2})^{1/2}/y^{\text{e}}_{\text{CO}_2}]$. This consideration is important when evaluating the availability of combustion products involving carbon monoxide.

Next, consider the case of pure water at T_0, p_0. Water is present as a vapor within the environment under present consideration but normally is a liquid when at T_0, p_0. Thus, water would enter the cell as a liquid at T_0, p_0 and exit as a vapor at T_0, $y^{\text{e}}_{\text{H}_2\text{O}}p_0$, with no cell reaction occurring. The chemical availability is

$$\bar{a}^{\text{ch}}_{\text{H}_2\text{O}} = [\bar{g}_{\text{H}_2\text{O(l)}} - \bar{g}_{\text{H}_2\text{O(g)}}](T_0, p_0) + \bar{R}T_0 \ln \left(\frac{1}{y^{\text{e}}_{\text{H}_2\text{O}}}\right) \tag{13.39}$$

Finally, consider N_2, O_2, CO_2, each pure at T_0, p_0. Nitrogen, oxygen, and carbon dioxide are present within the environment, and normally are gases when at T_0, p_0. For each of these, the chemical availability reduces simply to a logarithmic term of the form

$$\bar{a}^{\text{ch}} = \bar{R}T_0 \ln \left(\frac{1}{y^{\text{e}}}\right) \tag{13.40}$$

where y^{e} is the mole fraction of N_2, O_2, or CO_2 in the environment, as appropriate. For an ideal gas mixture at T_0, p_0 consisting *only* of substances present as gases in the environment, the chemical availability is obtained by summing the contributions of each of the components. The result, per mole of mixture, is

$$\bar{a}^{\text{ch}} = \bar{R}T_0 \sum_i y_i \ln \left(\frac{y_i}{y^{\text{e}}_i}\right) \tag{13.41}$$

where y_i and y_i^{e} denote, respectively, the mole fraction of component i in the mixture at T_0, p_0 and in the environment.

[2] For further discussion see M. J. Moran, *Availability Analysis: A Guide to Efficient Energy Use*, ASME Press, New York, 1989, pp. 169–170.

The development of Eqs. 13.38 through 13.41 is left as an exercise. Subsequent examples and end-of-chapter problems illustrate their use.

13.6.3 Availability Summary

The availability associated with a specified state is the sum of two contributions: the thermomechanical contribution introduced in Chap. 7 and the chemical contribution introduced in this section. When evaluating availability, we can think of a two-step process: The system is brought without change in composition from the specified state to the state where it is in thermal and mechanical equilibrium with the environment. Depending on the nature of the system, this may be a hypothetical state. Next, the components of the system are allowed to react with environmental components to produce environmental components. On a unit mass basis, the specific availability is the sum

$$a = \underline{(u - u_0) + p_0(v - v_0) - T_0(s - s_0) + \frac{V^2}{2} + gz} + a^{ch} \tag{13.42}$$

where the underlined term is the thermomechanical contribution (Eq. 7.9) associated with the first step and a^{ch} is the chemical contribution associated with the second step.

Similarly, the specific flow availability associated with a specified state is the sum

$$a_f = \underline{h - h_0 - T_0(s - s_0) + \frac{V^2}{2} + gz} + a^{ch} \tag{13.43}$$

where the underlined term is the thermomechanical contribution (Eq. 7.31) and a^{ch} is the chemical contribution.

When a difference in availability or flow availability between states of the same composition is evaluated, the chemical contribution cancels, leaving just the thermomechanical contribution. This has been the case in all availability analyses considered in previous sections of the book. However, for many evaluations, usually involving chemical reaction, it is necessary to account explicitly for the chemical availability contribution.

13.6.4 Illustrations

Several examples are given in this section to illustrate the principles introduced above.

EXAMPLE 13.12

PROBLEM

Determine the chemical availability of liquid octane, in kJ/kg. Let the environment consist of a gas phase at 25°C and 1 atm obeying the ideal gas model with the following composition on a molar basis: N_2, 75.67%; O_2, 20.35%; H_2O, 3.12%; CO_2, 0.03%; other, 0.83%.

SOLUTION

Known: The fuel is liquid octane.

Find: Determine the chemical availability relative to an environment consisting of a gas phase at 25°C, 1 atm with a specified composition.

Schematic and Given Data:

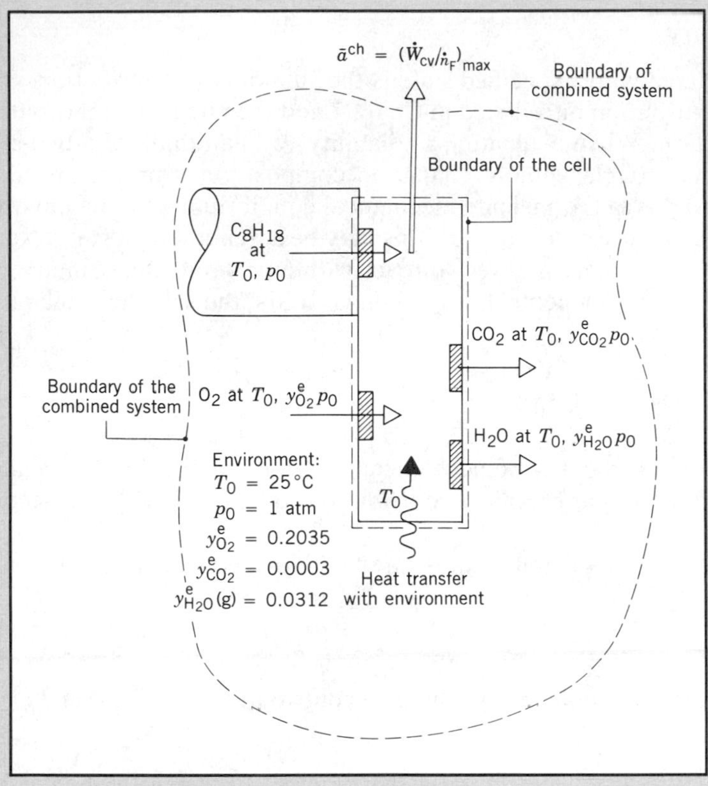

Figure E13.12

Assumptions: The environment consists of an ideal gas mixture with the molar analysis: N_2, 75.67%; O_2, 20.35%; H_2O, 3.12%; CO_2, 0.03%; other, 0.83%.

Analysis: Complete combustion of liquid octane with O_2 is described by

$$C_8H_{18}(l) + 12.5O_2 \rightarrow 8CO_2 + 9H_2O(g)$$

For this case, Eq. 13.36 takes the form

$$\bar{a}^{ch} = [\bar{g}_{C_8H_{18}(l)} + 12.5\bar{g}_{O_2} - 8\bar{g}_{CO_2} - 9\bar{g}_{H_2O(g)}](T_0, p_0) + \bar{R}T_0 \ln\left[\frac{(y^e_{O_2})^{12.5}}{(y^e_{CO_2})^8 (y^e_{H_2O(g)})^9}\right]$$

Since $T_0 = T_{ref}$ and $p_0 = p_{ref}$, the required specific Gibbs functions are just the Gibbs functions of formation from Table A-24. With the given composition of the environment and data from Table A-24, the above equation gives

$$\bar{a}^{ch} = [6610 + 12.5(0) - 8(-394{,}380) - 9(-228{,}590)] + 8.314(298.15)\ln\left[\frac{(0.2035)^{12.5}}{(0.0003)^8(0.0312)^9}\right]$$

$$= 5{,}218{,}960 + 188{,}883 = 5{,}407{,}843 \text{ kJ/kmol}$$

Dividing by the molecular weight, the chemical availability is obtained on a unit mass basis

$$a^{ch} = \frac{5{,}407{,}843}{114.22} = 47{,}346 \text{ kJ/kg}$$

Comments:

1. The value of the logarithmic term of Eq. 13.36 is dependent on the composition of the environment. In the present case, this term contributes about 3% to the magnitude of the chemical availability. Since the contribution of the logarithmic term is usually small, a satisfactory approximation to the chemical availability can often be obtained by omitting the term.
2. A molar analysis of the assumed environment on a *dry* basis reads O_2, 21%; N_2 and CO_2 and the other dry components, 79%. This is consistent with the dry air analysis used throughout the chapter. The water vapor present in the assumed environment corresponds to the amount of vapor that would be present were the gas phase saturated with water at the specified temperature and pressure.

EXAMPLE 13.13

PROBLEM

Steam at 5 bar, 240°C leaks from a line in a vapor power plant. Evaluate the flow availability of the steam, in kJ/kg, relative to an environment at 25°C, 1 atm in which the mole fraction of water vapor is $y^e_{H_2O} = 0.0303$.

SOLUTION

Known: Water vapor at a known state is specified. The environment is also described.

Find: Determine the flow availability of the water vapor, in kJ/kg.

Assumptions:

1. The environment consists of a gas phase that obeys the ideal gas model. The mole fraction of water vapor in the environment is 0.0303.
2. Neglect the effects of motion and gravity.

Analysis: With assumption 2, the specific flow availability is given by Eq. 13.43 as

$$a_f = \underline{(h - h_0) - T_0(s - s_0)} + a^{ch}$$

The underlined term is the thermomechanical contribution to the flow availability, evaluated as in Chap. 7. With data from the steam tables

$$\begin{aligned} h - h_0 - T_0(s - s_0) &= (2939.9 - 104.9) - 298(7.2307 - 0.3674) \\ &= 789.7 \text{ kJ/kg} \end{aligned}$$

where h_0 and s_0 are approximated as the saturated liquid values at T_0.

The chemical availability contribution to the flow availability is evaluated using Eq. 13.39. With data from Table A-24

$$\begin{aligned} a^{ch} &= \frac{1}{M}\left\{[\bar{g}_{H_2O(l)} - \bar{g}_{H_2O(g)}](T_0, p_0) + \bar{R}T_0 \ln\left(\frac{1}{y^e_{H_2O}}\right)\right\} \\ &= \frac{1}{18}\left\{[-237{,}180 - (-228{,}590)] + (8.314)(298) \ln\left(\frac{1}{0.0303}\right)\right\} \\ &= \frac{73.1 \text{ kJ/kmol}}{18 \text{ kg/kmol}} = 4.1 \text{ kJ/kg} \end{aligned}$$

Adding the thermomechanical and chemical contributions, the flow availability of steam at the specified state is

$$a_f = 789.7 + 4.1 = 793.8 \text{ kJ/kg}$$

Comment: In this case, the chemical availability contribution is less than 1% of the flow availability magnitude.

EXAMPLE 13.14

PROBLEM

Methane gas enters a reactor and burns completely with 140% theoretical air. Combustion products exit as a mixture at temperature T and a pressure of 1 atm. For T = 865 and 2820°R, evaluate the flow availability of the combustion products, in Btu per lbmol of fuel. Perform calculations relative to an environment consisting of an ideal gas mixture at 77°F, 1 atm with the molar analysis, $y^e_{N_2} = 0.7567$, $y^e_{O_2} = 0.2035$, $y^e_{H_2O} = 0.0303$, $y^e_{CO_2} = 0.0003$.

SOLUTION

Known: Methane gas reacts completely with 140% of the theoretical amount of air. Combustion products exit the reactor at 1 atm and a specified temperature. The environment is also specified.

Find: Determine the flow availability of the combustion products, in Btu per lbmol of fuel, for each of two given temperatures.

Assumptions:

1. The combustion products are modeled as an ideal gas mixture at all states considered.
2. Neglect the effects of motion and gravity.

Analysis: For 140% theoretical air, the reaction equation for complete combustion of methane is

$$CH_4 + 2.8(O_2 + 3.76N_2) \rightarrow CO_2 + 2H_2O + 10.53N_2 + 0.8O_2$$

The flow availability is given by Eq. 13.43, which involves chemical and thermomechanical contributions. Since the combustion products form an ideal gas mixture when at T_0, p_0 and each component is present within the environment, the chemical availability contribution, per mole of fuel, is obtained from the following expression patterned after Eq. 13.41:

$$\bar{a}^{ch} = \bar{R}T_0 \left[1 \ln\left(\frac{y_{CO_2}}{y^e_{CO_2}}\right) + 2 \ln\left(\frac{y_{H_2O}}{y^e_{H_2O}}\right) + 10.53 \ln\left(\frac{y_{N_2}}{y^e_{N_2}}\right) + 0.8 \ln\left(\frac{y_{O_2}}{y^e_{O_2}}\right)\right]$$

From the reaction equation, the mole fractions of the components of the products are $y_{CO_2} = 0.0698$, $y_{H_2O} = 0.1396$, $y_{N_2} = 0.7348$, $y_{O_2} = 0.0558$. Substituting these values together with the respective environmental mole fractions, we obtain $\bar{a}^{ch}$ = 7637 Btu per lbmol of fuel.

Applying ideal gas mixture principles, the thermomechanical contribution to the flow availability, per mole of fuel, is

$$\begin{aligned}\bar{h} - \bar{h}_0 - T_0(\bar{s} - \bar{s}_0) = {} & [\bar{h}(T) - \bar{h}(T_0) - T_0(\bar{s}^\circ(T) - \bar{s}^\circ(T_0) - \bar{R} \ln(y_{CO_2}p/y_{CO_2}p_0))]_{CO_2} \\ & + 2[\bar{h}(T) - \bar{h}(T_0) - T_0(\bar{s}^\circ(T) - \bar{s}^\circ(T_0) - \bar{R} \ln(y_{H_2O}p/y_{H_2O}p_0))]_{H_2O} \\ & + 10.53[\bar{h}(T) - \bar{h}(T_0) - T_0(\bar{s}^\circ(T) - \bar{s}^\circ(T_0) - \bar{R} \ln(y_{N_2}p/y_{N_2}p_0))]_{N_2} \\ & + 0.8[\bar{h}(T) - \bar{h}(T_0) - T_0(\bar{s}^\circ(T) - \bar{s}^\circ(T_0) - \bar{R} \ln(y_{O_2}p/y_{O_2}p_0))]_{O_2}\end{aligned}$$

With $p = p_0$ and values from Tables A-17E, 18E, 19E, and 21E

$$\begin{aligned}\bar{h} - \bar{h}_0 - T_0(\bar{s} - \bar{s}_0) = {} & [\bar{h}(T) - 4027.5 - 537(\bar{s}^\circ(T) - 51.032)]_{CO_2} \\ & + 2[\bar{h}(T) - 4258 - 537(\bar{s}^\circ(T) - 45.079)]_{H_2O} \\ & + 10.53[\bar{h}(T) - 3729.5 - 537(\bar{s}^\circ(T) - 45.743)]_{N_2} \\ & + 0.8[\bar{h}(T) - 3725.1 - 537(\bar{s}^\circ(T) - 48.982)]_{O_2}\end{aligned}$$

Then, with $\bar{h}$ and $\bar{s}^\circ$ from Tables A-17E, 18E, 19E, and 21E at T = 865 and 2820°R, the following results are obtained:

$$T = 865°\text{R}: \bar{h} - \bar{h}_0 - T_0(\bar{s} - \bar{s}_0) = 7622 \text{ Btu per lbmol of fuel}$$

$$T = 2820°\text{R}: \bar{h} - \bar{h}_0 - T_0(\bar{s} - \bar{s}_0) = 169{,}319 \text{ Btu per lbmol of fuel}$$

Adding the two contributions, the flow availability of the combustion products at each of the specified states is

$$T = 865°\text{R}: \bar{a}_f = 15{,}259 \text{ Btu per lbmol of fuel}$$

$$T = 2820°\text{R}: \bar{a}_f = 176{,}956 \text{ Btu per lbmol of fuel}$$

Comments:

1. The chemical contribution to the flow availability is relatively unimportant in the higher-temperature case, amounting only to about 4% of the flow availability. Chemical availability accounts for about half of the availability in the lower-temperature case, however.
2. When evaluating the thermomechanical contribution, we think of bringing a sample of the exiting combustion products to the dead state without a change in composition.

13.7 SECOND LAW EFFICIENCIES OF REACTING SYSTEMS

For a fuel at T_0, p_0, the chemical availability is the maximum theoretical work that could be obtained through reaction with environmental substances. This amount of work would be obtained only in the idealized circumstances considered in the derivation given above. Actual devices designed to do work by utilization of a combustion process, such as vapor and gas power plants, and internal combustion engines do not fulfill the conditions of the derivation and invariably have irreversibilities and losses associated with their operation. Accordingly, actual devices produce work equal to only a fraction of the theoretical value. The vapor power plant availability analysis of Sec. 8.6 provides an illustration.

The performance of devices intended to do work can be evaluated as the ratio of the actual work developed to the maximum theoretical work. This ratio is a type of *second law efficiency*. The relatively low second law efficiency exhibited by many common power-producing devices suggests that thermodynamically more thrifty ways of utilizing the fuel to develop power might be possible. However, efforts in this direction must be tempered by the economic imperatives that govern the practical application of all devices. The trade-off between fuel savings and the additional costs required to achieve those savings must be carefully weighed. The fuel cell provides an illustration. We noted previously (Sec. 13.5) that the chemical reactions in fuel cells are more controlled than the rapidly occurring, relatively unconstrained combustion reactions taking place in conventional power-producing devices. Being less dissipative, fuel cells can achieve greater second law efficiencies than such devices. Future advances in fuel cell technology may allow them to be more widely used for power production. But for now their high cost limits them to applications where cost is secondary.

The examples to follow provide illustrations of the evaluation of second law efficiencies for reacting systems.

EXAMPLE 13.15

PROBLEM

Devise and evaluate a second law efficiency for the internal combustion engine of Example 13.4. Use the chemical availability value determined in Example 13.12.

SOLUTION

Known: Liquid octane and the theoretical amount of air enter an internal combustion engine operating at steady state in separate streams at 77°F, 1 atm, and burn completely. The combustion products exit at 1140°F. The power developed by the engine is 50 horsepower, and the fuel mass flow rate is 0.004 lb/s.

Find: Devise and evaluate a second law efficiency for the engine using the chemical availability value determined in Example 13.12.

Schematic and Given Data: See Fig. E13.4.

Assumptions:

1. See the assumptions listed in the solution to Example 13.4.
2. The environment is the same as used in Example 13.12.
3. The combustion air enters at the condition of the environment.
4. Kinetic and potential energy effects are negligible.

Analysis: An availability balance can be used in formulating a second law efficiency for the engine. At steady state, the rate at which availability enters the engine equals the rate at which availability exits plus the rate at which availability is destroyed within the engine. As the combustion air enters at the condition of the environment, and thus with zero availability, availability enters the engine only with the fuel. Availability exits the engine accompanying heat and work, and with the products of combustion.

If the power developed is taken to be the *product* of the engine, and the heat transfer and exiting product gas are regarded as *losses,* a second law efficiency expression that gauges the extent to which the availability entering the engine with the fuel is converted to the product is

$$\varepsilon = \frac{\dot{W}_{cv}}{\dot{A}_F}$$

where $\dot{A}_F$ denotes the rate at which availability enters with the fuel.

Since the fuel enters the engine at 77°F and 1 atm, which correspond to the values of T_0 and p_0 of the environment, and kinetic and potential energy effects are negligible, the availability of the fuel is just the chemical availability evaluated in Example 13.12. There is no thermomechanical contribution. Thus

$$\dot{A}_F = \dot{m}_F a^{ch} = \left(0.004 \frac{\text{lb}}{\text{s}}\right)\left(47{,}346 \frac{\text{kJ}}{\text{kg}}\right)\left(\frac{\text{Btu/lb}}{2.326\ \text{kJ/kg}}\right) = 81.42 \frac{\text{Btu}}{\text{s}}$$

The second law efficiency is then

$$\varepsilon = \frac{(50)(2545/3600)}{81.42} = 0.434\ (43.4\%)$$

Comment: The "waste heat" from large engines may be utilizable by some other device, for example, an absorption heat pump. In such cases, the availability accompanying heat transfer might be included in the numerator of the second law efficiency expression. Since a greater portion of the entering fuel availability would be utilized in such arrangements, the value of ε would be greater than that evaluated in the solution.

EXAMPLE 13.16

PROBLEM

For the reactor of Example 13.11, determine the irreversibility in kJ per kmol of fuel, and devise and evaluate a second law efficiency. Consider (a) complete combustion with the theoretical amount of air and (b) complete combustion with 400% theoretical air. Use the chemical availability value determined in Example 13.12.

SOLUTION

Known: Liquid octane and air, each at 25°C and 1 atm, burn completely in a well-insulated reactor operating at steady state. The products of combustion exit at 1 atm pressure.

Find: Determine the irreversibility, in kJ per kmol of fuel, and evaluate a second law efficiency for complete combustion with (a) the theoretical amount of air, (b) 400% theoretical air.

Schematic and Given Data: See Fig. E13.11.

Assumptions:

1. See assumptions listed in Example 13.11.
2. The environment is the same as that in Example 13.12.
3. The combustion air enters at the condition of the environment.
4. Kinetic and potential energy effects are negligible.

Analysis: The irreversibility can be found from the relation

$$\frac{\dot{I}_{cv}}{\dot{n}_F} = T_0 \frac{\dot{\sigma}_{cv}}{\dot{n}_F}$$

where T_0 is the temperature of the environment and $\dot{\sigma}_{cv}$ is the rate of entropy production. The rate of entropy production is evaluated in the solution to Example 13.11 for each of the two cases. For the case of complete combustion with the theoretical amount of air

$$\frac{\dot{I}_{cv}}{\dot{n}_F} = (298 \text{ K})\left(5404 \frac{\text{kJ}}{\text{kmol} \cdot \text{K}}\right) = 1{,}610{,}392 \frac{\text{kJ}}{\text{kmol} \cdot \text{K}}$$

Similarly, for the case of complete combustion with 400% of the theoretical amount of air

$$\frac{\dot{I}_{cv}}{\dot{n}_F} = (298)(9754) = 2{,}906{,}692 \frac{\text{kJ}}{\text{kmol} \cdot \text{K}}$$

At steady state, the rate at which availability enters the reactor equals the rate at which availability exits plus the rate at which availability is destroyed within the reactor. As the combustion air enters at the condition of the environment, and thus with zero availability, availability enters the reactor only with the fuel. The reactor is well insulated, so there is no availability transfer accompanying heat transfer. There is no work $\dot{W}_{cv}$. Accordingly, availability exits only with the combustion products. A second law efficiency can be written as

$$\varepsilon = \frac{\dot{A}_{\text{products}}}{\dot{A}_F}$$

where $\dot{A}_F$ is the rate at which availability enters with the fuel and $\dot{A}_{\text{products}}$ is the rate at which availability exits with the combustion products.

The second law efficiency expression can be written alternatively as

$$\varepsilon = \frac{\dot{A}_F - \dot{I}_{cv}}{\dot{A}_F} = 1 - \frac{\dot{I}_{cv}}{\dot{A}_F}$$

Since the fuel enters the reactor at 25°C, 1 atm, which correspond to the values of T_0 and p_0 of the environment, and kinetic and potential effects are negligible, the availability of the fuel is just the chemical availability evaluated in Example 13.12. There is no thermomechanical contribution. Thus, for the case of complete combustion with the theoretical amount of air

$$\varepsilon = 1 - \frac{1{,}610{,}392}{5{,}407{,}843} = 0.702 \ (70.2\%)$$

Similarly, for the case of complete combustion with 400% of the theoretical amount of air

$$\varepsilon = 1 - \frac{2{,}906{,}692}{5{,}407{,}843} = 0.463 \ (46.3\%)$$

Comment: The calculated efficiency values show that a substantial portion of the fuel availability is destroyed in the combustion process. In the case of combustion with the theoretical amount of air, about 30% of the fuel availability is destroyed. In the excess air case, over 50% of the fuel availability is destroyed. Further availability destructions would take place as the hot gases are utilized. It might be evident, therefore, that the overall conversion from fuel input to end use would have a relatively low second law efficiency. The vapor power plant availability analysis of Sec. 8.6 illustrates this point.

PROBLEMS

13.1 A vessel contains a mixture of 34% O_2 and 66% CO on **(a)** a mass basis and **(b)** a molar basis. For each case, determine whether there is enough oxygen to support complete combustion of the carbon monoxide.

13.2 Estimate the amount of CO_2 produced, in lb, for every gallon of gasoline burned by an automobile. In a year, how much CO_2 would be produced, in lb, by a typical automobile in the U.S.?

13.3 Determine the balanced reaction equation for complete combustion of octane (C_8H_{18}) with

(a) the theoretical amount of oxygen.
(b) the theoretical amount of air.
(c) 50% excess air.

13.4 Butane (C_4H_{10}) burns with air. If the air–fuel ratio on a mass basis is 12, determine

(a) the air–fuel ratio on a molar basis.
(b) the percent excess or percent deficiency of air, whichever is appropriate.

13.5 Methane (CH_4) burns completely with the *stoichiometric* amount of hydrogen peroxide (H_2O_2). Determine the balanced reaction equation.

13.6 A gaseous fuel mixture has a molar analysis of 14% H_2, 3% CH_4, 27% CO, 0.6% O_2, 4.5% CO_2, 50.9% N_2. Determine the theoretical air–fuel ratio for complete combustion on a

(a) molar basis.
(b) mass basis.

13.7 A fuel mixture having a molar analysis of 60% CH_4, 30% C_2H_6, 10% N_2 is supplied to a furnace where it burns completely with 120% of theoretical air. Determine the air–fuel ratio on a

(a) molar basis.
(b) mass basis.

13.8 A fuel mixture having a molar analysis of 20% CH_4, 40% H_2, 40% NH_3 burns completely with 150% of theoretical oxygen. Determine the balanced reaction equation.

13.9 Coal with a mass analysis of 80% carbon, 3% sulfur, 17% noncombustible ash burns completely with 120% of theoretical air. Determine the amount of SO_2 produced, in kg per kg of coal.

13.10 A coal sample has a mass analysis of 80.1% carbon, 4.3% hydrogen (H_2), 4.2% oxygen (O_2), 1.7% nitrogen (N_2), 1.0% sulfur, and the rest is noncombustible ash. For complete combustion with the theoretical amount of air, determine the air–fuel ratio on a mass basis.

13.11 A sample of shredded municipal waste with a mass analysis of 80.7% C, 4.9% H_2, 1.8% S, 5.3% O_2, 1.1% N_2, 6.2% noncombustible ash burns completely with 120% of theoretical air. Determine

(a) the balanced reaction equation.
(b) the air–fuel ratio on a mass basis.

13.12 Coal with a mass analysis of 88% C, 6% H, 4% O, 1% N, 1% S burns completely with the theoretical amount of air. Determine

(a) the amount of SO_2 produced, in kg per kg of coal.
(b) the air–fuel ratio on a mass basis.

13.13 Ethane (C_2H_6) burns completely with air. If the air–fuel ratio on a mass basis is 18, determine

(a) the percent excess, or percent deficiency of air, whichever is appropriate.
(b) the dew point temperature of the combustion products when cooled at 1 atm.

13.14 Butane (C_4H_{10}) burns completely with 150% of theoretical air and the products are cooled to 20°C at 1 atm. Determine the amount of water vapor condensed, in kmol per kmol of fuel.

13.15 Ethylene (C_2H_4) burns completely with air and the combustion products are cooled to 70°F at 1 atm. If the air–fuel ratio on a mass basis is 17, determine

(a) the percent excess air.
(b) the amount of water vapor condensed, in lbmol per lbmol of fuel.

13.16 A gaseous fuel mixture with a molar analysis of 72% CH_4, 9% H_2, 14% N_2, 2% O_2, and 3% CO_2 burns completely with moist air to form *gaseous* products at 1 atm consisting of CO_2, H_2O, and N_2 only. If the dew point temperature of the products is 60°C, determine the amount of water vapor present in the combustion air, in kmol per kmol of fuel mixture.

13.17 A fuel with the mass analysis of 85% C, 13% H_2, 2% O_2, burns completely with the theoretical amount of air and the products of combustion are cooled to 70°F at 1 atm.

(a) Determine the molar analysis of the dry products of combustion prior to cooling and the air–fuel ratio on a mass basis.
(b) Determine the amount of water vapor that condenses, in kmol per kmol of fuel.

13.18 Propane (C_3H_8) enters a combustion chamber and burns completely with 180% of theoretical air entering at 40°C, 1 atm, 60% relative humidity. Obtain the balanced reaction equation.

13.19 Butane (C_4H_{10}) enters a combustion chamber and burns completely with 150% of theoretical air entering at 68°F, 1 atm, 75% relative humidity. Obtain the balanced reaction equation.

13.20 Methane (CH_4) enters a furnace and burns completely with 150% of theoretical air entering at 25°C, 94.5 kPa, 75% relative humidity. Obtain the balanced reaction equation.

13.21 Pentane (C_5H_{12}) burns completely with the theoretical amount of air at 75°F, 1 atm, 75% relative humidity. Determine

(a) the balanced reaction equation.
(b) the dew point temperature of the combustion products at 1 atm.

13.22 Decane ($C_{10}H_{22}$) is burned completely with 10% excess air at 25°C, 1 atm, 90% relative humidity. If the combustion products are cooled to 25°C at a pressure of 1 atm, determine the amount of water vapor that condenses, in kg per kg of fuel.

13.23 Hydrogen (H_2) enters a combustion chamber with a mass flow rate of 2 kg/h and burns with air entering at 30°C, 1 atm with a volumetric flow rate of 120 m^3/h. Determine the percent of theoretical air used.

13.24 Carbon burns with 85% theoretical air yielding CO_2, CO, and N_2 only. Determine

(a) the balanced reaction equation.
(b) the air–fuel ratio on a mass basis.

13.25 Propane (C_3H_8) reacts with 80% of theoretical air to form products including CO_2, CO, H_2O, and N_2 only. Determine

(a) the balanced reaction equation.
(b) the analysis of the products on a dry molar basis.

13.26 Dodecane ($C_{12}H_{26}$) enters an engine and burns with air to give products with the dry molar analysis of CO_2, 12.1%; CO, 3.4%; O_2, 0.5%; H_2, 1.5%; N_2, 82.5%. Determine the air–fuel ratio on a molar basis.

13.27 The components of the exhaust gas of a spark-ignition engine using a fuel mixture represented as C_8H_{17} have a dry molar analysis of 8.7% CO_2, 8.9% CO, 0.3% O_2, 3.7% H_2, 0.3% CH_4, and 78.1% N_2. Determine the air–fuel ratio on a mass basis and compare with the measured value of 11 : 1.

13.28 Octane (C_8H_{18}) burns with 95% of theoretical air, producing a gaseous mixture of CO_2, CO, H_2O, and N_2. Determine

(a) the air–fuel ratio on a molar basis.
(b) the analysis of the products on a dry molar basis.

13.29 Natural gas having a molar analysis of 20% CH_4, 40% C_2H_6, and 40% C_3H_8 burns with air, giving products having a dry molar analysis of 11.4% CO_2, 1.7% O_2, 1.2% CO, and 85.7% N_2. Determine the air–fuel ratio on a

(a) molar basis.
(b) mass basis.

13.30 A gas mixture having a molar analysis of 30% CH_4, 20% C_2H_6, 40% H_2, and 10% N_2 burns with air, giving products having a dry molar analysis of 8.2% CO_2, 0.6% CO, 4.1% O_2, and 87.1% N_2. Determine

(a) the air–fuel ratio on a molar basis.
(b) the percent theoretical air.

13.31 A fuel oil having an analysis on a mass basis of 85.7% C, 14.2% H, 0.1% inert matter burns with air to give products with a dry molar analysis of 12.29% CO_2; 3.76% O_2; 83.95% N_2. Determine the air–fuel ratio on a mass basis.

13.32 Pentane (C_5H_{12}) burns with air so that a fraction x of the carbon is converted to CO_2. The remaining carbon appears as CO. There is no free O_2 in the products.

(a) Obtain an expression for the percent of theoretical air in terms of x.
(b) Plot for x ranging from zero to unity.

13.33 Determine the equivalence ratio for each of the following cases and indicate whether the reactants form a lean or rich mixture:

(a) Problem 13.24.
(b) Problem 13.25.

13.34 For each of the following mixtures, determine the equivalence ratio and indicate if the mixture is lean or rich:

(a) 1 lbmol of methane (CH_4) and 8 lbmol of air.
(b) 1 kg of ethane (C_2H_6) and 17.2 kg of air.

13.35 Octane (C_8H_{18}) enters an engine and burns with air to give products with the dry molar analysis of CO_2,

10.5%; CO, 5.8%; CH_4, 0.9%; H_2, 2.6%; O_2, 0.3%; N_2, 79.9%. Determine the equivalence ratio.

13.36 Methane (CH_4) burns with air to form products consisting of CO_2, CO, H_2O, and N_2 only. If the equivalence ratio is 1.25, determine the balanced reaction equation.

13.37 Acetylene gas (C_2H_2) at 25°C, 1.6 bar enters a reactor operating at steady state and burns completely with 150% of theoretical air entering at the same conditions. If the products exit at 25°C, 1.5 bar, determine the rate of heat transfer from the reactor, in kJ per kmol of fuel. Kinetic and potential energy effects are negligible.

13.38 Ethane (C_2H_6) at 77°F, 1 atm enters a combustion chamber operating at steady state and burns completely with the theoretical amount of air entering at the same conditions. If the products exit at 77°F, 1 atm, determine the rate of heat transfer from the combustion chamber, in Btu per lbmol of fuel. Kinetic and potential energy effects are negligible.

13.39 Methane gas (CH_4) at 25°C, 1 atm and a volumetric flow rate of 27 m^3/h enters a furnace operating at steady state. The methane burns completely with 140% of theoretical air, entering at 127°C, 1 atm. Products of combustion exit at 427°C, 1 atm. Determine

(a) the volumetric flow rate of the air, in m^3/h.
(b) the rate of heat transfer from the furnace, in kJ/h.

Kinetic and potential energy effects are negligible.

13.40 Liquid ethanol (C_2H_5OH) at 25°C, 1 atm enters a combustion chamber operating at steady state and burns with air entering at 227°C, 1 atm. The fuel flow rate is 25 kg/s and the air–fuel ratio on a mass basis is 7.5. Products of combustion, consisting of CO_2, CO, $H_2O(g)$, and N_2, exit at 1000 K. Ignoring kinetic and potential energy effects, determine

(a) the rate of heat transfer from the combustion chamber, in kJ/s.
(b) the equivalence ratio.

13.41 Benzene gas (C_6H_6) at 25°C, 1 atm enters a combustion chamber operating at steady state and burns with 95% theoretical air entering at 25°C, 1 atm. The combustion products exit at 1000 K and include only CO_2, CO, H_2O, and N_2. Determine the mass flow rate of the fuel, in kg/s, to provide heat transfer at a rate of 1000 kW.

13.42 A rigid tank initially contains 16.04 lb of CH_4 and 96 lb of O_2 at 77°F, 1 atm. After complete combustion, the pressure in the tank is 3.352 atm. Determine the heat transfer, in Btu.

13.43 A closed rigid vessel initially contains a gaseous mixture at 25°C, 1 atm with the molar analysis of 25% ethylene (C_2H_4), 75% oxygen (O_2). The mixture burns completely and the products are cooled to 500 K. Determine the heat transfer between the vessel and its surroundings, in kJ per kmol of fuel present initially.

13.44 A closed, rigid vessel initially contains a gaseous mixture of 1 kmol of octane (C_8H_{18}) and 300% of theoretical air at 25°C, 1 atm. If the mixture burns completely, determine the heat transfer from the vessel, in kJ, and the final pressure, in atm, for a final temperature of 1000 K.

13.45 The energy required to vaporize the working fluid passing through the boiler of a simple vapor power plant is provided by the complete combustion of methane with 110% of theoretical air. The fuel and air enter in separate streams at 25°C, 1 atm. Products of combustion exit the stack at 150°C, 1 atm. If the thermal efficiency of the power plant is 33%, determine the mass flow rate of fuel required, in kg/h per MW of power developed by the plant. Kinetic and potential energy effects are negligible.

13.46 Liquid octane (C_8H_{18}) at 25°C, 1 atm enters the combustor of a simple open gas turbine power plant and burns completely with 400% of theoretical air entering the compressor at 25°C, 1 atm. Products of combustion exit the turbine at 627°C, 1 atm. If the rate of heat transfer from the gas turbine is estimated as 15% of the net power developed, determine the net power developed, in kJ per kmol of fuel. Kinetic and potential energy effects are negligible.

13.47 Octane gas (C_8H_{18}) at 25°C, 1 atm enters a jet engine and burns completely with 300% of theoretical air entering at 25°C, 1 atm. Products of combustion exit at 990 K, 1 atm. If the fuel and air enter with relatively low velocities, determine the velocity of the exiting combustion products, in m/s. Neglect potential energy effects and heat transfer between the engine and its surroundings.

13.48 Determine the enthalpy of combustion for gaseous butane (C_4H_{10}), in kJ per kmol of fuel, at 25°C, 1 atm, assuming

(a) water vapor in the products.
(b) liquid water in the products.

13.49 Determine the enthalpy of combustion for gaseous propane (C_3H_8), in Btu per lbmol of fuel, at

(a) 77°F, 1 atm.
(b) 500°F, 1 atm.

Assume water vapor in the products. For propane, let $c_p = 0.41$ Btu/lb·°R.

13.50 Determine the enthalpy of combustion for gaseous methane (CH_4), in Btu per lbmol of fuel, at

(a) 77°F, 1 atm.
(b) 900°R, 1 atm.
(c) 1800°R, 1 atm.

Assume water vapor in the products. For methane, let $\bar{c}_p = 4.52 + 7.37(T/1000)$ Btu/lbmol·°R, where T is in °R.

13.51 For a natural gas with a molar analysis of 40% pentane (C_5H_{12}), 35% methane (CH_4), and 25% propane (C_3H_8), determine the enthalpy of combustion, in kJ per kmol of mixture, at 25°C, 1 atm, assuming water vapor in the products.

13.52 For gaseous propane (C_3H_8) at 25°C, 1 atm, determine the higher and lower heating values, in kJ/kg.

13.53 Determine the higher heating value, in kJ per kmol of fuel, at 25°C, 1 atm for

(a) liquid octane (C_8H_{18}).
(b) gaseous hydrogen (H_2).
(c) liquid methanol (CH_3OH).
(d) gaseous butane (C_4H_{10}).

13.54 Liquid octane (C_8H_{18}) at 25°C, 1 atm enters an insulated reactor operating at steady state and burns with 90% of theoretical air at 25°C, 1 atm to form products consisting of CO_2, CO, H_2O, and N_2 only. Determine the temperature of the exiting products, in K. Compare with the results of Example 13.8 and comment.

13.55 For each of the following fuels, determine the adiabatic flame temperature, in K, for complete combustion with 200% of theoretical air in a combustor operating at steady state. The reactants enter at 25°C, 1 atm.

(a) carbon.
(b) hydrogen (H_2).
(c) liquid octane (C_8H_{18}).

13.56 Propane gas (C_3H_8) at 25°C, 1 atm enters an insulated reactor operating at steady state and burns completely with air entering at 25°C, 1 atm. Determine the adiabatic flame temperature if combustion occurs with

(a) 120% of theoretical air.
(b) 140% of theoretical air.
(c) 160% of theoretical air.

How does the adiabatic flame temperature vary with increasing combustion air?

13.57 Hydrogen (H_2) at 77°F, 1 atm enters an insulated reactor operating at steady state and burns completely with x% of theoretical air entering at 77°F, 1 atm. Plot the adiabatic flame temperature for x ranging from 100 to 400%.

13.58 Methane gas (CH_4) at 25°C, 1 atm enters an insulated reactor operating at steady state and burns completely with x% of theoretical air entering at 25°C, 1 atm. Plot the adiabatic flame temperature for x ranging from 100 to 400%.

13.59 Methane (CH_4) at 25°C, 1 atm enters an insulated reactor operating at steady state and burns with the theoretical amount of air entering at 25°C, 1 atm. Determine the temperature of the exiting combustion products if

(a) combustion is complete.
(b) 90% of the carbon in the fuel burns to CO_2 and the rest burns to CO.

Neglect kinetic and potential energy effects.

13.60 Propane gas (C_3H_8) at 77°F, 1 atm enters an insulated reactor operating at steady state and burns completely with air entering at 77°F, 1 atm. If the combustion products exit at 2240°F, determine the percent of theoretical air used. Neglect kinetic and potential energy effects.

13.61 Liquid octane (C_8H_{18}) at 25°C, 1 atm enters an insulated reactor operating at steady state and burns completely with air entering at 227°C, 1 atm. If the combustion products exit at 1127°C, determine the percent excess air used. Neglect kinetic and potential energy effects.

13.62 Methane (CH_4) at 77°F, 1 atm enters the combustor of a gas turbine power plant operating at steady state and burns completely with air entering at 400°F. Owing to metallurgical limitations, the temperature of the products of combustion flowing from the combustor to the turbine can be no higher than 1600°F. Determine the percent excess air that allows this constraint to be met. There is no significant heat transfer between the power plant and its surroundings, and kinetic and potential energy effects can be ignored.

13.63 Air enters the compressor of a simple gas turbine power plant at 70°F, 1 atm, is compressed adiabatically to 40 lbf/in.^{2}, and then enters the combustion chamber where it burns completely with propane gas (C_3H_8) entering at 77°F, 40 lbf/in.^{2} The combustion products at 1340°F, 40 lbf/in.^{2} enter the turbine and expand adiabatically to a pressure of 1 atm. The isentropic compressor efficiency is 83.3% and the isentropic turbine efficiency is 90%. If the power plant develops 100 horsepower, determine at steady state

(a) the percent of theoretical air required.
(b) the mass flow rate of the propane, in lb/h.

13.64 A mixture of gaseous octane (C_8H_{18}) and the theoretical amount of air, initially at 25°C, 1 atm, reacts completely in a rigid, insulated vessel. Determine the temperature, in °C, and the pressure, in atm, of the combustion products.

13.65 A mixture of methane (CH_4) and 200% of theoretical air, initially at 77°F, 1 atm, reacts completely in a rigid, insulated vessel. Determine the temperature, in °F, and the pressure, in atm, of the combustion products.

13.66 Repeat Problem 13.65, but assume the reaction occurs at constant pressure in a piston–cylinder assembly.

13.67 Determine the change in the Gibbs function, in kJ per kmol of carbon, at 25°C, 1 atm for $C(s) + O_2(g) \rightarrow CO_2(g)$, using

(a) Gibbs function of formation data.
(b) enthalpy of formation data, together with absolute entropy data.

13.68 Determine the change in the Gibbs function, in Btu per lbmol of hydrogen, at 77°F, 1 atm for $H_2(g) + \frac{1}{2}O_2(g) \rightarrow H_2O(g)$, using

(a) Gibbs function of formation data.
(b) enthalpy of formation data, together with absolute entropy data.

13.69 Carbon monoxide (CO) at 25°C, 1 atm enters an insulated reactor operating at steady state and reacts completely with the theoretical amount of air entering in a separate stream at 25°C, 1 atm. The products of combustion exit as a mixture at 1 atm. For the reactor, determine the rate of entropy production, in kJ/K per kmol of CO entering. Neglect kinetic and potential energy effects.

13.70 Methane (CH_4) at 77°F, 1 atm enters an insulated reactor operating at steady state and burns completely with air entering in a separate stream at 77°F, 1 atm. The products of combustion exit as a mixture at 1 atm. For the reactor, determine the rate of entropy production, in Btu/°R per lbmol of methane entering, for combustion with

(a) the theoretical amount of air.
(b) 200% of theoretical air.

Neglect kinetic and potential energy effects.

13.71 Carbon monoxide (CO) reacts with water vapor in an insulated reactor operating at steady state to form hydrogen (H_2) and carbon dioxide (CO_2). The products exit as a mixture at 1 atm. For the reactor, determine the rate of entropy production, in kJ/K per kmol of carbon monoxide entering. Neglect kinetic and potential energy effects. Consider two cases:

(a) the carbon monoxide and water vapor enter the reactor in separate streams, each at 400 K, 1 atm.
(b) the carbon monoxide and water vapor enter the reactor as a mixture at 400 K, 1 atm.

Explain why the answers in parts (a) and (b) differ.

13.72 A gaseous mixture of butane (C_4H_{10}) and 80% excess air at 25°C, 3 atm enters a reactor fitted with a cooling water jacket. Complete combustion occurs, and the products exit as a mixture at 1200 K, 3 atm. Water enters the jacket as a saturated liquid at 1 atm and saturated vapor exits at essentially the same pressure. No significant heat transfer occurs from the outer surface of the jacket, and kinetic and potential energy effects are negligible. Determine for the water-jacketed reactor

(a) the mass flow rate of the cooling water, in kg per kmol of fuel.
(b) the rate of entropy production, in kJ/K per kmol of fuel.
(c) the irreversibility rate, in kJ per kmol of fuel, for T_0 = 25°C.

13.73 Liquid enthanol (C_2H_5OH) at 25°C, 1 atm enters a reactor operating at steady state and burns completely with 130% of theoretical air entering in a separate stream at 25°C, 1 atm. Combustion products exit at 227°C, 1 atm. For a control volume enclosing the reactor and enough of the nearby surroundings that heat transfer occurs at the ambient temperature, 25°C, determine the irreversibility rate, in kJ per kmol of fuel, for T_0 = 25°C. Kinetic and potential energy effects are negligible.

13.74 A gaseous mixture of ethane (C_2H_6) and the theoretical amount of air at 25°C, 1 atm enters a reactor operating at steady state and burns completely. Combustion products exit at 627°C, 1 atm. For a control volume enclosing the reactor and enough of the nearby surroundings that heat transfer occurs at the ambient temperature, 25°C, determine the irreversibility rate, in kJ per kmol of fuel, for T_0 = 25°C. Kinetic and potential energy effects are negligible.

13.75 Separate streams of hydrogen (H_2) and oxygen (O_2) at 25°C, 1 atm enter a fuel cell operating at steady state, and liquid water exits at 25°C, 1 atm. If the fuel cell operates isothermally at 25°C, determine the maximum theoretical work it can develop, in kJ per kmol of hydrogen. Kinetic and potential energy effects are negligible.

13.76 Streams of methane (CH_4) and oxygen (O_2), each at 25°C, 1 atm, enter a fuel cell operating at steady state. Streams of carbon dioxide and water exit separately at 25°C, 1 atm. If the fuel cell operates isothermally at 25°C, 1 atm, determine the maximum theoretical work that it can develop, in kJ per kmol of methane. Ignore kinetic and potential energy effects.

13.77 Streams of hydrogen (H_2) and oxygen (O_2), each at 1 atm, enter a fuel cell operating at steady state and water vapor exits at 1 atm. If the cell operates isothermally at **(a)** 500 K, **(b)** 1000 K, and **(c)** 2000 K, determine the maximum theoretical work that can be developed by the cell in each case, in kJ per kmol of hydrogen flowing. Present the results graphically and comment. Heat transfer with the surroundings takes place at the cell temperature, and kinetic and potential energy effects can be ignored.

13.78 An inventor has developed a device that at steady state takes in liquid water at 25°C, 1 atm with a mass flow rate of 5 kg/h and produces separate streams of hydrogen (H_2) and oxygen (O_2), each at 25°C, 1 atm. The inventor claims that the device requires an electrical power input of 16 kW when operating isothermally at 25°C. Kinetic and potential energy effects can be ignored. Evaluate the inventor's claim.

13.79 For environmental reasons, it is desired to separate the SO_2 from the combustion products of Problem 13.12 by supplying the products at 340K, 1 atm to a device operating isothermally at steady state. A stream of SO_2 and a stream of the remaining gases exit, each at 340K, 1 atm. If the coal is burned at a rate of 10 kg/s, determine the minimum theoretical power input required by the device, in kW. Ignore kinetic and potential energy effects.

13.80 Figure P13.80 shows a system for heating a stream of liquid water from 77 to 140°F at an essentially constant pressure of 1 atm. An *electrolysis generator* requires an electric power input to convert liquid water entering at 77°F, 1 atm into hydrogen (H_2) and oxygen (O_2) gas streams, each at 77°F, 1 atm. The hydrogen and oxygen are provided in stoichiometric proportions to a combus-

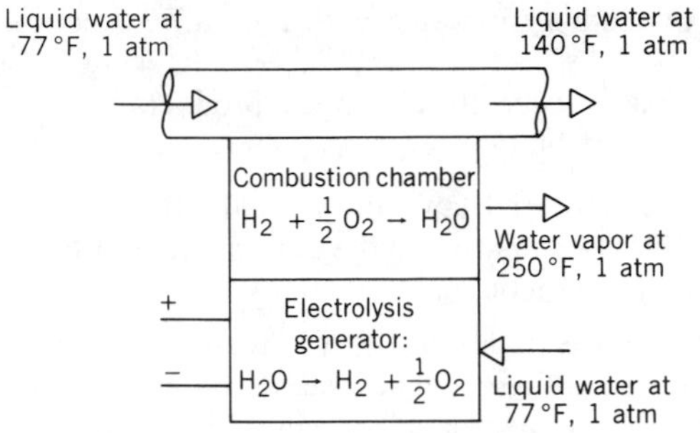

Figure P13.80

tion chamber and burned to yield the energy required for heating the water stream. Water vapor at 250°F, 1 atm exits the combustion chamber. The system operates at steady state, all kinetic and potential energy effects can be ignored, and no significant heat transfer occurs with the surroundings. Compare this system thermodynamically with the use of an electrical resistance heating element for achieving the same purpose.

13.81 For **(a)** carbon, **(b)** hydrogen (H_2), **(c)** methane (CH_4), **(d)** carbon monoxide, **(e)** liquid methanol (CH_3OH), **(f)** nitrogen (N_2), **(g)** oxygen (O_2), **(h)** carbon dioxide, and **(i)** water, determine the chemical availability, in kJ/kg, relative to the following environment in which the gas phase obeys the ideal gas model:

Environment
T_0 = 298.15 K (25°C), p_0 = 1 atm

Gas Phase:	Component	y^e (%)
	N_2	75.67
	O_2	20.35
	$H_2O(g)$	3.12
	CO_2	0.03
	Other	0.83

13.82 The accompanying table shows an environment consisting of a gas phase and a condensed water phase. The gas phase forms an ideal gas mixture.

Environment
T_0 = 298.15 K (25°C), p_0 = 1 atm

Condensed Phase: $H_2O(l)$ at T_0, p_0

Gas Phase:	Component	y^e (%)
	N_2	75.67
	O_2	20.35
	H_2O	3.12
	CO_2	0.03
	Other	0.83

Show that the chemical availability of the hydrocarbon C_aH_b can be determined as

$$\bar{a}^{ch} = \left[\bar{g}_F + \left(a + \frac{b}{4}\right)\bar{g}_{O_2} - a\bar{g}_{CO_2} - \frac{b}{2}\bar{g}_{H_2O(l)}\right] + \bar{R}T_0 \ln\left[\frac{(y^e_{O_2})^{a+b/4}}{(y^e_{CO_2})^a}\right]$$

13.83 Using the result of Problem 13.82, repeat parts **(a)** through **(c)** of Problem 13.81.

13.84 Showing all important steps, derive **(a)** Eq. 13.38, **(b)** Eq. 13.39, **(c)** Eq. 13.40, **(d)** Eq. 13.41.

13.85 The chemical availabilities of common hydrocarbons C_aH_b can be represented in terms of their respective lower heating value, $\overline{\text{LHV}}$, by an expression of the form

$$\frac{\bar{a}^{ch}}{(\overline{\text{LHV}})} = c_1 + c_2(b/a) + c_3/a$$

where c_1, c_2, and c_3 are constants. Evaluate the constants relative to the environment of Problem 13.81 to obtain an expression valid for several selected **(a)** gaseous hydrocarbons, **(b)** liquid hydrocarbons.

13.86 Evaluate the specific flow availability of nitrogen (N_2), in Btu/lb, at 200°F, 4 atm, relative to the environment of Problem 13.81. Neglect the effects of motion and gravity.

13.87 Evaluate the specific flow availability of water vapor, in kJ/kg, at 200°C, 1 bar, relative to the environment of Problem 13.81. Neglect the effects of motion and gravity.

13.88 Evaluate the specific flow availability of an equimolar mixture of oxygen (O_2) and nitrogen (N_2), in kJ/kg, at 20°C, 1 atm, relative to the environment of Problem 13.81. Neglect the effects of motion and gravity.

13.89 A mixture of methane gas (CH_4) and 150% of theoretical air enters a combustion chamber at 77°F, 1 atm. Determine the specific flow availability of the entering mixture, in Btu per lbmol of methane. Perform calculations relative to the environment of Problem 13.81. Ignore the effects of motion and gravity.

13.90 A mixture having an analysis on a molar basis of 85% dry air, 15% CO enters a device at 125°C, 2.1 atm, and a velocity of 250 m/s. If the mass flow rate is 1.0 kg/s, determine the rate availability enters, in kW. Perform calculations relative to the environment of Problem 13.81. Neglect the effect of gravity.

13.91 The following flow rates in lb/h are reported for the exiting SNG (substitute natural gas) stream in a certain process for producing SNG from bituminous coal:

CH_4	429,684 lb/h
CO_2	9,093 lb/h
N_2	3,741 lb/h
H_2	576 lb/h
CO	204 lb/h
H_2O	60 lb/h

If the SNG stream is at 77°F, 1 atm, determine the rate at which availability exits. Perform calculations relative to the environment of Problem 13.81. Neglect the effects of motion and gravity.

13.92 Carbon at 25°C, 1 atm enters an insulated reactor operating at steady state and reacts completely with the theoretical amount of air entering separately at 25°C, 1 atm. For the reactor, **(a)** determine the irreversibility rate, in kJ per kmol of carbon, and **(b)** evaluate a second law efficiency. Perform calculations relative to the environment of Problem 13.81. Neglect the effects of motion and gravity.

13.93 Propane gas (C_3H_8) at 25°C, 1 atm and a volumetric flow rate of 0.03 m^3/min enters a furnace operating at steady state and burns completely with 200% of theoretical air entering at 25°C, 1 atm. Combustion products exit at 227°C, 1 atm. The furnace provides energy by heat transfer at 227°C for some industrial process. For the furnace, compare the rate of availability transfer accompanying heat transfer with the rate of availability destruction, each in kJ/min. Let $T_0 = 25$°C and ignore kinetic and potential energy effects.

13.94 Figure P13.94 shows a coal gasification reactor making use of the carbon–steam process. The energy required for the endothermic reaction is supplied by an electrical resistor. The reactor operates at steady state, with no stray heat transfers and negligible kinetic and potential energy effects. Evaluate in Btu per lbmol of carbon entering

(a) the required electrical input.
(b) the availability entering with the carbon.
(c) the availability entering with the steam.
(d) the availability exiting with the product gas.
(e) the irreversibility for the reactor.

Perform calculations relative to the environment of Problem 13.81. Ignore kinetic and potential energy effects.

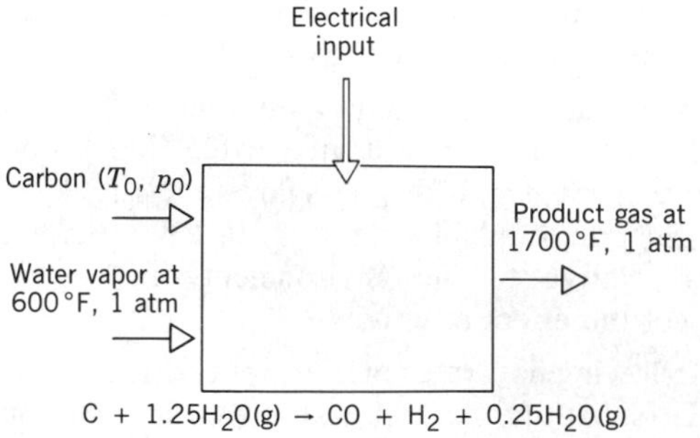

Figure P13.94

13.95 Carbon monoxide at 25°C, 1 atm enters an insulated reactor operating at steady state and reacts completely with the theoretical amount of air entering in a separate stream at 25°C, 1 atm. The products exit as a mixture at 1 atm. Determine in kJ per kmol of CO

(a) the availability entering with the carbon monoxide.
(b) the availability exiting with the products.
(c) the irreversibility rate.

Also, evaluate a second law efficiency for the reactor. Perform calculations relative to the environment of Problem 13.81. Ignore kinetic and potential energy effects.

13.96 Acetylene gas (C_2H_2) at 77°F, 1 atm enters an insulated reactor operating at steady state and burns completely with 180% of theoretical air, entering in a separate stream at 77°F, 1 atm. The products exit as a mixture at 1 atm. Determine in Btu per lbmol of fuel

(a) the availability of the fuel entering the reactor.
(b) the availability exiting with the products.
(c) the irreversibility rate.

Also, evaluate a second law efficiency for the reactor. Perform calculations relative to the environment of Problem 13.81. Ignore kinetic and potential energy effects.

13.97 Liquid octane (C_8H_{18}) at 25°C, 1 atm and a mass flow rate of 0.57 kg/h enters a small internal combustion engine operating at steady state. The fuel burns with air entering the engine in a separate stream at 25°C, 1 atm. Combustion products exit at 670 K, 1 atm with a dry molar analysis of 11.4% CO_2, 2.9% CO, 1.6% O_2, and 84.1% N_2. If the engine develops power at the rate of 1 kW, determine

(a) the rate of heat transfer from the engine, in kW.
(b) a second law efficiency for the engine.

Use the environment of Problem 13.81 and neglect kinetic and potential energy effects.

13.98 Figure P13.98 shows a simple vapor power plant. The fuel is methane that enters at 77°F, 1 atm and burns completely with 200% theoretical air entering at 77°F, 1 atm. Steam exits the steam generator at 900°F, 500 lbf/in.^{2} The vapor expands through the turbine and exits at 1 lbf/in.^{2}, and a quality of 97%. At the condenser exit, the pressure is 1 lbf/in.^{2} and the water is a saturated liquid. The plant operates at steady state with no stray heat transfers from any plant component. Pump work and all kinetic and potential energy effects are negligible. Determine

(a) the balanced reaction equation.
(b) the vapor mass flow rate, in lb per lbmol of fuel.
(c) the cooling water mass flow rate, in lb per lbmol of fuel.
(d) each of the following, expressed as a percent of the availability entering the steam generator with the fuel, (i) the availability exiting with the stack gases, (ii) the availability destroyed in the steam generator, (iii) the power developed by the turbine, (iv) the availability destroyed in the turbine, (v) the availability exiting with the cooling water, (vi) the availability destroyed in the condenser.

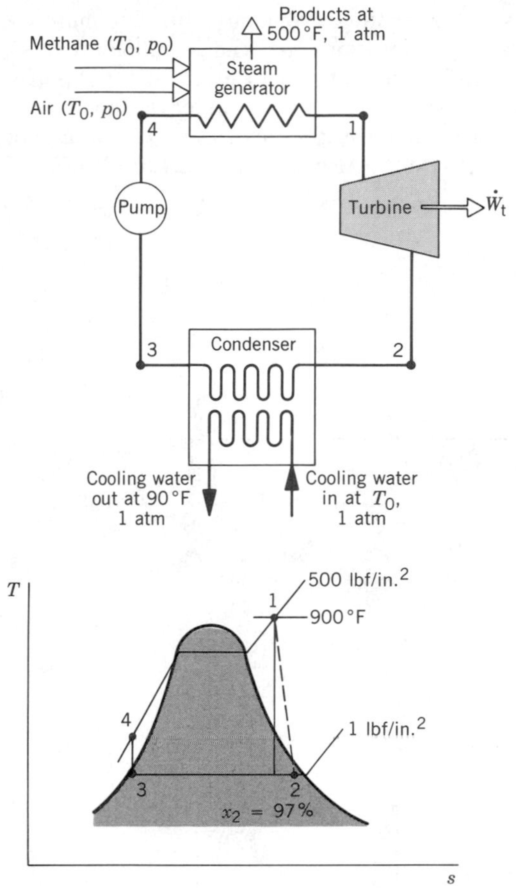

Figure P13.98

Base availability values on the environment of Problem 13.81.

13.99 Consider a furnace operating at steady state idealized as shown in Fig. P13.99. The fuel is methane, which enters at 25°C, 1 atm and burns completely with 200% theoretical air entering at the same temperature and pressure. The furnace delivers energy by heat transfer at an average temperature of 60°C. Combustion products exit at 600 K, 1 atm. There are no stray heat transfers, and kinetic and potential energy effects can be ignored. Determine in kJ per kmol of fuel

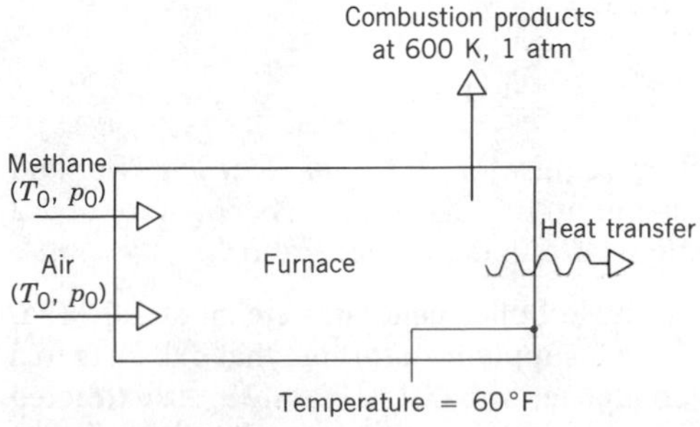

Figure P13.99

(a) the availability entering the furnace with the fuel.
(b) the availability exiting with the products.
(c) the irreversibility rate.

Also, evaluate a second law efficiency for the furnace. Perform calculations relative to the environment of Problem 13.81.

13.100 Coal enters the combustor of a power plant with a mass analysis of 49.8% C, 3.5% H, 6.8% O, 6.4% S, 14.1% H_2O, and 19.4% noncombustible ash. The higher heating value of the coal, as received, is measured as 21,220 kJ/kg, and the lower heating value on a dry basis, $(\text{LHV})_d$, is 20,450 kJ/kg. The following expression can be used to estimate the chemical availability of the coal, in kJ/kg:

$$a^{ch} = (\text{LHV})_d \left(1.0438 + 0.0013\,\frac{h}{c} + 0.1083\,\frac{o}{c} + 0.0549\,\frac{n}{c}\right) + 6740s$$

where h/c, o/c, and n/c denote, respectively, the mass ratio of hydrogen to carbon, oxygen to carbon, and nitrogen to carbon, and s is the mass fraction of sulfur in kg per kg of fuel.[3] The environment is closely the same as in Problem 13.82, but extended appropriately to account for the presence of sulfur in the coal.

(a) Using the above expression, calculate the chemical availability of the coal, in kJ/kg.
(b) Compare the answer of part (a) with the values that would result by approximating the chemical availability with each of the measured heating values.
(c) What data would be required to determine the chemical availability in this case using the methodology of Sec. 13.6.1? Discuss.

13.101 For psychrometric applications such as those considered in Chap. 12, the environment often can be modeled simply as an ideal gas mixture of water vapor and dry air at temperature T_0 and pressure p_0. The composition of the environment is defined by the dry air and water vapor mole fractions y_a^e, y_v^e, respectively. Show that relative to such an environment the flow availability of a moist air stream at temperature T and pressure p with dry air and water vapor mole fractions y_a and y_v, respectively, can be expressed on a molar basis as

$$\bar{a}_f = T_0 \left\{ (y_a \bar{c}_{pa} + y_v \bar{c}_{pv}) \left[\left(\frac{T}{T_0}\right) - 1 - \ln\left(\frac{T}{T_0}\right) \right] + \bar{R} \ln\left(\frac{p}{p_0}\right) \right\} + \bar{R}T_0 \left[y_a \ln\left(\frac{y_a}{y_a^e}\right) + y_v \ln\left(\frac{y_v}{y_v^e}\right) \right]$$

[3] Ibid., pp. 192–193.

where $\bar{c}_{pa}$ and $\bar{c}_{pv}$ denote the molar specific heats of dry air and water vapor, respectively. Neglect the effects of motion and gravity.

13.102 For each of the following, use the result of Problem 13.101 to determine the specific flow availability, in kJ/kg, relative to an environment consisting of moist air at 20°C, 1 atm, $\phi = 100\%$.

(a) moist air at 20°C, 1 atm, $\phi = 100\%$.
(b) moist air at 20°C, 1 atm, $\phi = 50\%$.
(c) dry air at 20°C, 1 atm.

13.103 Using the result of Problem 13.101 determine the flow availability at locations 1, 2, and 3 and the irreversibility, each in Btu/min, for the device of **(a)** Problem 12.79 and **(b)** Problem 12.94. Let the environment be a mixture of dry air and water vapor at 95°F, 1 atm with $y_v^e = 0.022$, $y_a^e = 0.978$. Also, let $c_{pa} = 0.24$ Btu/lb·°R and $c_{pv} = 0.44$ Btu/lb·°R.

DESIGN AND OPEN ENDED PROBLEMS

13.1D The term *acid rain* is frequently used today. Define what is meant by the term. Discuss the origin and consequences of acid rain. Also discuss options for its control.

13.2D For a city in your locale having a population of at least 100,000, estimate the production of CO_2 annually, in lb, from each of the following sources: transportation, power plants, and heating systems. Provide sample calculations to support your estimates. Why is the amount of CO_2 in the atmosphere of interest? Discuss.

13.3D In steam power plants, the energy rejected to the condenser cooling water is normally several times as great as the energy lost up the stack. Still, greater efforts are made to reduce the stack loss than to reduce the condenser loss. Discuss this, giving sample calculations using actual data from a power plant in your locale to supplement the discussion.

13.4D With too much excess air for combustion, considerable energy is lost as the combustion products exit the stack, and with too much fuel, excess carbon monoxide and combustibles may exit. To facilitate control of excess air, microprocessor-based flue gas analyzers have been developed. Specify a commercially available stack-mounted sensor unit that would be suitable for monitoring the oxygen and carbon monoxide in the products of a natural gas-fired boiler at temperatures up to 1200°F. Consider ease of operation, calibration, and cost.

13.5D Figure P13.5D shows a natural gas-fired boiler for steam generation integrated with a *direct-contact condensing* heat exchanger that discharges warm water for tasks such as space heating, water heating, and combustion air preheating. For 7200 h of operation annually, estimate the annual savings for fuel, in dollars, by integrating these functions. What other cost considerations would enter into the decision to install such a condensing heat exchanger? Discuss.

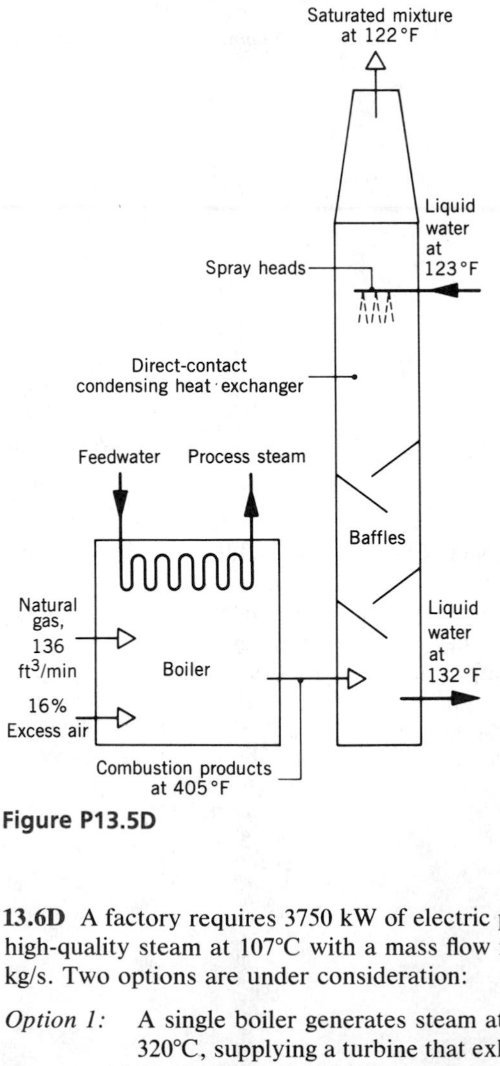

Figure P13.5D

13.6D A factory requires 3750 kW of electric power and high-quality steam at 107°C with a mass flow rate of 2.2 kg/s. Two options are under consideration:

Option 1: A single boiler generates steam at 2.0 MPa, 320°C, supplying a turbine that exhausts to a condenser at 0.007 MPa. Steam is extracted from the turbine at 107°C, returning as a liquid to the boiler after use.

Option 2: A boiler generates steam at 2.0 MPa, 320°C, supplying a turbine that exhausts to a condenser at 0.007 MPa. A separate process steam boiler generates the required steam at 107°C, which is returned as a liquid to the boiler after use.

The boilers are fired with natural gas and 20% excess air. For 7200 h of operation annually, evaluate the two options by estimating and comparing the fuel costs.

13.7D Using data obtained from your local electric utility, estimate the unit cost of the availability entering its power plant with fuel, in \$ per kW·h. Compare this with the amount charged by the utility for electricity. What factors account for the difference between these values? Present your results as memorandum.

13.8D Devise means for estimating the chemical availability of coals containing sulfur in terms of the *heating value, standard chemical availabilities,*[4] and the coal *ultimate analysis.*[5] Prepare a table giving data for several common coal types.

[4] See, for example, J. Szargut, D. R. Morris, and F. R. Steward, *Exergy Analysis of Thermal, Chemical, and Metallurgical Processes,* Hemisphere Publishing Corporation, New York, 1988, pp. 101–110.

[5] See for example, M. L. Gorbaty, J. W. Larsen, and I. Wender, *Coal Science,* Vol. 1, Academic Press, New York, 1982, p. 52.

CHAPTER 14

CHEMICAL AND PHASE EQUILIBRIUM

The object of the present chapter is to consider the concept of equilibrium in greater depth than has been done thus far. In the first part of the chapter, the study of reactive systems initiated in Chap. 13 is continued with a discussion of *chemical* equilibrium in a single phase. Particular emphasis is placed on the case of reactive ideal gas mixtures. The second part of the chapter concerns *phase* equilibrium. The equilibrium of multicomponent, multiphase, nonreacting systems is considered and the *phase rule* is introduced.

14.1 PRELIMINARY CONSIDERATIONS

In this section, concepts are developed that are useful in the study of chemical and phase equilibrium. Among these are equilibrium criteria and the chemical potential concept.

14.1.1 Equilibrium Criteria

A system is said to be in thermodynamic equilibrium if, when it is isolated from its surroundings, there would be no macroscopically observable changes. An important requirement for equilibrium is that the temperature be uniform throughout the system or each part of the system in thermal contact. If this condition were not met, spontaneous heat transfer from one location to another could occur while the system is isolated. There must also be no unbalanced forces between parts of the system. These conditions ensure that the system is in thermal and mechanical equilibrium, but there is still the possibility that complete equilibrium does not exist. A process might occur involving a chemical reaction, a transfer of mass between phases, or both. The object of this section is to devise criteria that can be applied to decide whether a system in a particular state is in equilibrium. These criteria are developed using the conservation of energy principle and the second law of thermodynamics.

Consider states of a simple compressible system of fixed mass at which temperature and pressure are uniform with position throughout. In the absence of overall system motion and ignoring the influence of gravity, the energy balance in differential form is

$$dU = \delta Q - \delta W$$

If volume change is the only possible work mode and pressure is uniform with position throughout the system, $\delta W = p\, dV$. Introducing this and solving for δQ gives

$$\delta Q = dU + p\, dV$$

Since temperature is uniform with position throughout the system, the entropy balance in differential form is

$$dS = \frac{\delta Q}{T} + \delta\sigma$$

Eliminating δQ between the last two equations

$$T\, dS - dU - p\, dV = T\, \delta\sigma \tag{14.1}$$

Entropy is produced in all actual processes and conserved only in the absence of irreversibilities. Hence, Eq. 14.1 provides a constraint on the direction of processes. The only processes allowed are those for which

$$T\, dS - dU - p\, dV \geq 0 \tag{14.2}$$

For example, a process taking place in an insulated, constant-volume vessel, where $dU = 0$ and $dV = 0$, must be such that

$$dS]_{U,V} \geq 0 \tag{14.3}$$

Equation 14.3 suggests that changes of state of a closed system at constant internal energy and volume can occur only in the direction of *increasing entropy*. The expression also implies that entropy approaches a *maximum* as a state of equilibrium is approached. This is a special case of the increase of entropy principle introduced in Sec. 6.5.

Equation 14.2 can be used to study equilibrium under conditions other than fixed U and fixed V. A useful case for the study of chemical and phase equilibria is one in which temperature and pressure are fixed. For this, it is convenient to employ the *Gibbs function* in extensive form

$$G = H - TS = U + pV - TS$$

Forming the differential

$$dG = dU + p\, dV + V\, dp - T\, dS - S\, dT$$

or on rearrangement

$$dG - V\, dp + S\, dT = -(T\, dS - dU - p\, dV)$$

Except for the minus sign, the right side of this equation is the same as the expression appearing in Eq. 14.2. Accordingly, Eq. 14.2 can be written as

$$dG - V\, dp + S\, dT \leq 0 \tag{14.4}$$

where the inequality reverses direction because of the minus sign noted above.

It can be concluded from Eq. 14.4 that any process taking place at a specified temperature and pressure ($dT = 0$ and $dp = 0$) must be such that

$$dG]_{T,p} \leq 0 \tag{14.5}$$

This inequality indicates that the Gibbs function of a system at fixed T and p *decreases* during an irreversible process. Each step of such a process results in a decrease in the Gibbs function of the system and brings the system closer to equilibrium. The equilibrium state is the one having the *minimum* value of the Gibbs function. Therefore, when

$$dG]_{T,p} = 0 \tag{14.6}$$

we have equilibrium.

Equation 14.6 provides a relationship among the properties of a system when it is *at* an equilibrium state. The manner in which the equilibrium state is reached is unimportant, however, for once an equilibrium state is obtained, a system exists at a particular T and p and no further changes can take place. When applying Eq. 14.6, therefore, we may specify the temperature T and pressure p, but it is unnecessary to require additionally that the system actually achieves equilibrium at fixed T and fixed p.

14.1.2 Gibbs Function and the Chemical Potential

In the present discussion, the Gibbs function is considered further as a prerequisite for application of the criterion $dG]_{T,p} = 0$ introduced above. Let us begin by noting that any *extensive* property of a single-phase, single-component system is a function of two independent intensive properties and the size of the system. Selecting temperature and pressure as the independent properties and the number of moles n as the measure of size, the Gibbs function can be expressed in the form $G = G(T, p, n)$. For a single-phase, *multicomponent* system, G may then be considered a function of temperature, pressure, and the number of moles of each component present, written $G = G(T, p, n_1, n_2, \ldots, n_j)$.

If each mole number is multiplied by α, the size of the system changes by the same factor and so does the value of every extensive property. Thus, for the Gibbs function we may write

$$\alpha G(T, p, n_1, n_2, \ldots, n_j) = G(T, p, \alpha n_1, \alpha n_2, \ldots, \alpha n_j)$$

Differentiating with respect to α while holding temperature, pressure, and the mole numbers fixed and using the chain rule on the right side gives

$$G = \frac{\partial G}{\partial(\alpha n_1)} n_1 + \frac{\partial G}{\partial(\alpha n_2)} n_2 + \cdots + \frac{\partial G}{\partial(\alpha n_j)} n_j$$

This equation holds for all values of α. In particular, it holds for $\alpha = 1$. Setting $\alpha = 1$, the following expression results:

$$G = \sum_{i=1}^{j} n_i \left(\frac{\partial G}{\partial n_i}\right)_{T,p,n_l} \tag{14.7}$$

where the subscript n_l denotes that all n's except n_i are held fixed during differentiation.

The partial derivatives appearing in Eq. 14.7 have such importance for our study of chemical and phase equilibrium that they are given a special name and symbol. The *chemical potential* of component i, symbolized by μ_i, is defined as

$$\mu_i = \left(\frac{\partial G}{\partial n_i}\right)_{T,p,n_l} \tag{14.8}$$

The chemical potential is an *intensive property*. With Eq. 14.8, Eq. 14.7 becomes

$$G = \sum_{i=1}^{j} n_i \mu_i \tag{14.9}$$

The equilibrium criterion given by Eq. 14.6 can be written in terms of chemical potentials, providing an expression of fundamental importance for subsequent discussions of equilibrium. Forming the differential of $G(T, p, n_1, \ldots, n_j)$ while holding temperature and pressure fixed results in

$$dG]_{T,p} = \sum_{i=1}^{j} \left(\frac{\partial G}{\partial n_i}\right)_{T,p,n_l} dn_i$$

The partial derivatives are recognized from Eq. 14.8 as the chemical potentials, so

$$dG]_{T,p} = \sum_{i=1}^{j} \mu_i \, dn_i \tag{14.10}$$

With Eq. 14.10, the equilibrium criterion $dG]_{T,p} = 0$ can be placed in the form

$$\sum_{i=1}^{j} \mu_i \, dn_i = 0 \tag{14.11}$$

Like Eq. 14.6, from which it is obtained, this equation provides a relationship among properties of a system when the system is *at* an equilibrium state where the temperature is T and the pressure is p. Like Eq. 14.6, this equation applies to a particular state, and the manner in which that state is attained is not important.

14.1.3 Evaluation of Chemical Potentials

Means for evaluating the chemical potentials for certain cases of present interest are introduced in this section. An elementary case considered later in this chapter is that of equilibrium between two phases involving a pure substance. For a single phase of a pure substance, Eq. 14.9 becomes simply

$$G = n\mu$$

or

$$\mu = \frac{G}{n} = \bar{g} \tag{14.12}$$

That is, the chemical potential is just the Gibbs function per mole.

Another case of importance in the study of chemical equilibrium is that of an ideal gas mixture. The enthalpy and entropy of an ideal gas mixture are given by

$$H = \sum_{i=1}^{j} n_i \bar{h}_i(T) \quad \text{and} \quad S = \sum_{i=1}^{j} n_i \bar{s}_i(T, p_i)$$

where $p_i = y_i p$ is the partial pressure of component i. Accordingly, the Gibbs function takes the form

$$G = H - TS = \sum_{i=1}^{j} n_i \bar{h}_i(T) - T \sum_{i=1}^{j} n_i \bar{s}_i(T, p_i)$$

$$= \sum_{i=1}^{j} n_i [\bar{h}_i(T) - T\bar{s}_i(T, p_i)] \quad \text{(ideal gas)} \tag{14.13}$$

Introducing the molar Gibbs function of component i

$$\bar{g}_i(T, p_i) = \bar{h}_i(T) - T\bar{s}_i(T, p_i) \quad \text{(ideal gas)} \tag{14.14}$$

Equation 14.13 can be expressed as

$$G = \sum_{i=1}^{j} n_i \bar{g}_i(T, p_i) \quad \text{(ideal gas)} \tag{14.15}$$

Comparing Eq. 14.15 to Eq. 14.9 suggests that

$$\mu_i = \bar{g}_i(T, p_i) \quad \text{(ideal gas)} \tag{14.16}$$

That is, the chemical potential of component i in an ideal gas mixture is equal to its Gibbs function per mole of i, evaluated at the mixture temperature and the partial pressure of i in the mixture. Equation 14.16 can be obtained formally by taking the partial derivative of Eq. 14.15 with respect to n_i, holding temperature, pressure, and the remaining n's constant, and then applying the definition of chemical potential, Eq. 14.8.

The chemical potential of component i in an ideal gas mixture can be expressed in an alternative form that is somewhat more convenient for subsequent applications. Using Eq. 13.23, Eq. 14.14 becomes

$$\mu_i = \bar{h}_i(T) - T\bar{s}_i(T, p_i)$$

$$= \bar{h}_i(T) - T\left(\bar{s}_i^\circ(T) - \bar{R} \ln \frac{y_i p}{p_{\text{ref}}}\right)$$

$$= \bar{h}_i(T) - T\bar{s}_i^\circ(T) + \bar{R}T \ln \frac{y_i p}{p_{\text{ref}}}$$

where p_{ref} is 1 atm and y_i is the mole fraction of component i in a mixture at temperature T and pressure p. The last equation can be written compactly as

---

$$\mu_i = \bar{g}_i^\circ + \bar{R}T \ln \frac{y_i p}{p_{\text{ref}}} \quad \text{(ideal gas)} \tag{14.17}$$

---

where $\bar{g}_i^\circ$ is the Gibbs function of component i evaluated at temperature T and a pressure of 1 atm. Additional details concerning the chemical potential concept are provided in Sec. 11.9. Equation 14.17 is the same as Eq. 11.144 developed there.

CHEMICAL EQUILIBRIUM

In this part of the chapter, the equilibrium criterion $dG]_{T,p} = 0$ introduced in Sec. 14.1 is used to study the equilibrium of reacting mixtures. The objective is to establish the composition present at equilibrium for a specified temperature and pressure. An important parameter for determining the equilibrium composition is the *equilibrium constant*. The equilibrium constant is introduced and its use illustrated by several solved examples. The discussion is concerned only with equilibrium states of reacting systems, and no information can be deduced about the *rates of reaction*. Whether an equilibrium mixture would form quickly or slowly can be determined only by considering the *chemical kinetics* of the reaction, a topic that is not treated in this text.

14.2 EQUATION OF REACTION EQUILIBRIUM

In Chap. 13 the conservation of mass and conservation of energy principles are applied to reacting systems by assuming that the reactions can occur as written. However, the extent to which a chemical reaction proceeds is limited by many factors. In general, the composition of the products actually formed from a given set of reactants, and the relative amounts of the products, can be determined only from experiment. Knowledge of the composition that would be present were a reaction to proceed to equilibrium is frequently useful, however. The *equation of reaction equilibrium* introduced in the present section provides the basis for determining the equilibrium composition of a reacting mixture.

Consider a closed system consisting initially of a gaseous mixture of hydrogen and oxygen. A number of reactions might take place, including

$$1H_2 + \tfrac{1}{2}O_2 \rightleftarrows 1H_2O \tag{14.18}$$

$$1H_2 \rightleftarrows 2H \tag{14.19}$$

$$1O_2 \rightleftarrows 2O \tag{14.20}$$

Let us consider for illustration purposes only the first of the reactions given above, in which hydrogen and oxygen combine to form water. At equilibrium, the system will consist in general of three components: H_2, O_2, and H_2O, for not all of the hydrogen and oxygen initially present need be reacted. *Changes* in the amounts of these components during each differential step of the reaction leading to the formation of an equilibrium mixture are governed by Eq. 14.18. That is

$$dn_{H_2} = -dn_{H_2O}, \qquad dn_{O_2} = -\tfrac{1}{2}dn_{H_2O} \tag{14.21a}$$

where dn denotes a differential change in the respective component. The minus signs signal that the amounts of hydrogen and oxygen present decrease when the reaction proceeds toward the right. Equations 14.21a can be expressed alternatively as

$$\frac{-dn_{H_2}}{1} = \frac{-dn_{O_2}}{\frac{1}{2}} = \frac{dn_{H_2O}}{1} \tag{14.21b}$$

which emphasizes that increases and decreases in the components are proportional to the stoichiometric coefficients of Eq. 14.18.

Equilibrium is a condition of *balance*. Accordingly, as suggested by the direction of the arrows in Eq. 14.18, when the system is at equilibrium, the tendency of the hydrogen and oxygen to form water is just balanced by the tendency of water to dissociate into oxygen and hydrogen. The equilibrium criterion $dG]_{T,p} = 0$ can be used to determine the composition at an equilibrium state where the temperature is T

and the pressure is p. This requires evaluation of the differential $dG]_{T,p}$ in terms of system properties. For the present case, Eq. 14.10 giving the difference in the Gibbs function of the mixture between two states having the same temperature and pressure, but compositions that differ infinitesimally, takes the following form:

$$dG]_{T,p} = \mu_{H_2}\, dn_{H_2} + \mu_{O_2}\, dn_{O_2} + \mu_{H_2O}\, dn_{H_2O} \tag{14.22}$$

The changes in the mole numbers are related by Eqs. 14.21. Hence

$$dG]_{T,p} = (-1\mu_{H_2} - \tfrac{1}{2}\mu_{O_2} + 1\mu_{H_2O})\, dn_{H_2O}$$

At equilibrium, $dG]_{T,p} = 0$, so the term in parentheses must be zero. That is,

$$-1\mu_{H_2} - \tfrac{1}{2}\mu_{O_2} + 1\mu_{H_2O} = 0$$

When expressed in a form that resembles Eq. 14.18, this becomes

$$1\mu_{H_2} + \tfrac{1}{2}\mu_{O_2} = 1\mu_{H_2O} \tag{14.23}$$

Equation 14.23 is the *equation of reaction equilibrium* for the case under consideration. The chemical potentials are functions of temperature, pressure, and composition. Thus, the composition that would be present at equilibrium for a given temperature and pressure can be determined, in principle, by solving this equation. The solution procedure is described in the next section.

The foregoing development can be repeated for reactions involving any number of components. Consider a closed system containing *five* components, A, B, C, D, and E, at a given temperature and pressure, subject to a chemical reaction of the form

$$\nu_A A + \nu_B B \rightleftarrows \nu_C C + \nu_D D \tag{14.24}$$

where the ν's are stoichiometric coefficients. Component E is assumed to be inert and thus does not appear in the reaction equation. As we will see, component E does influence the equilibrium composition even though it is not involved in the chemical reaction. The form of Eq. 14.24 suggests that at equilibrium the tendency of A and B to form C and D is just balanced by the tendency of C and D to form A and B.

The stoichiometric coefficients ν_A, ν_B, ν_C, and ν_D do not correspond to the respective number of moles of the components present. The amounts of the components present are designated n_A, n_B, n_C, n_D, and n_E. However, *changes* in the amounts of the components present do bear a definite relationship to the values of the stoichiometric coefficients. That is,

$$\frac{-dn_A}{\nu_A} = \frac{-dn_B}{\nu_B} = \frac{dn_C}{\nu_C} = \frac{dn_D}{\nu_D} \tag{14.25a}$$

where the minus signs indicate that A and B would be consumed when C and D are produced. Since E is inert, the amount of this component remains constant, so $dn_E = 0$.

By introducing a proportionality factor $d\varepsilon$, Eqs. 14.25a may be replaced by the following expressions:

$$\begin{aligned} dn_A &= -\nu_A\, d\varepsilon, & dn_B &= -\nu_B\, d\varepsilon \\ dn_C &= \nu_C\, d\varepsilon, & dn_D &= \nu_D\, d\varepsilon \end{aligned} \tag{14.25b}$$

The parameter ε is sometimes referred to as the *degree* (or *extent*) *of reaction*.

For the system under present consideration, Eq. 14.10 takes the form

$$dG]_{T,p} = \mu_A\, dn_A + \mu_B\, dn_B + \mu_C\, dn_C + \mu_D\, dn_D + \mu_E\, dn_E$$

Introducing Eqs. 14.25b and noting that $dn_E = 0$, this becomes

$$dG]_{T,p} = (-\nu_A\mu_A - \nu_B\mu_B + \nu_C\mu_C + \nu_D\mu_D)\, d\varepsilon$$

At equilibrium, $dG]_{T,p} = 0$, so the term in parentheses must be zero. That is

$$-\nu_A\mu_A - \nu_B\mu_B + \nu_C\mu_C + \nu_D\mu_D = 0$$

or when written in a form resembling Eq. 14.24,

$$\nu_A\mu_A + \nu_B\mu_B = \nu_C\mu_C + \nu_D\mu_D \tag{14.26}$$

For the present case, Eq. 14.26 is the *equation of reaction equilibrium*. In principle, the composition that would be present at equilibrium for a given temperature and pressure can be determined by solving this equation. The solution procedure is simplified through the *equilibrium constant* concept introduced in the next section.

14.3 CALCULATION OF EQUILIBRIUM COMPOSITIONS

The object of the present section is to show how the equilibrium composition of a system at a specified temperature and pressure can be determined by solving the equation of reaction equilibrium. An important part is played in this by the *equilibrium constant*.

14.3.1 Equilibrium Constant for Ideal Gas Mixtures

The first step in solving the equation of reaction equilibrium, Eq. 14.26, for the equilibrium composition is to introduce expressions for the chemical potentials in terms of temperature, pressure, and composition. For an ideal gas mixture, Eq. 14.17 can be used for this purpose. When this expression is introduced for each of the components A, B, C, and D, Eq. 14.26 becomes

$$\nu_A\left(\bar{g}_A^\circ + \bar{R}T\ln\frac{y_A p}{p_{ref}}\right) + \nu_B\left(\bar{g}_B^\circ + \bar{R}T\ln\frac{y_B p}{p_{ref}}\right)$$
$$= \nu_C\left(\bar{g}_C^\circ + \bar{R}T\ln\frac{y_C p}{p_{ref}}\right) + \nu_D\left(\bar{g}_D^\circ + \bar{R}T\ln\frac{y_D p}{p_{ref}}\right) \tag{14.27}$$

where $\bar{g}_i^\circ$ is the Gibbs function of component i evaluated at temperature T and the pressure $p_{ref} = 1$ atm. Equation 14.27 is the basic working relation for chemical equilibrium in a mixture of ideal gases. For a specified temperature and pressure, it can be solved for the composition. However, subsequent calculations are facilitated if it is written in an alternative form as follows.

Collect like terms and rearrange the equation as

$$(\nu_C\bar{g}_C^\circ + \nu_D\bar{g}_D^\circ - \nu_A\bar{g}_A^\circ - \nu_B\bar{g}_B^\circ) = -\bar{R}T\left(\nu_C\ln\frac{y_C p}{p_{ref}} + \nu_D\ln\frac{y_D p}{p_{ref}}\right.$$
$$\left. - \nu_A\ln\frac{y_A p}{p_{ref}} - \nu_B\ln\frac{y_B p}{p_{ref}}\right) \tag{14.28}$$

The term on the left side of Eq. 14.28 can be expressed concisely as ΔG°. That is

$$\Delta G^\circ = \nu_C\bar{g}_C^\circ + \nu_D\bar{g}_D^\circ - \nu_A\bar{g}_A^\circ - \nu_B\bar{g}_B^\circ \tag{14.29a}$$

which is the change in the Gibbs function for the reaction given by Eq. 14.24 if each reactant and product is separate at temperature T and a pressure of 1 atm. This expression can be written alternatively in terms of specific enthalpies and entropies as

$$\begin{aligned}\Delta G^\circ &= \nu_C(\bar{h}_C - T\bar{s}_C^\circ) + \nu_D(\bar{h}_D - T\bar{s}_D^\circ) - \nu_A(\bar{h}_A - T\bar{s}_A^\circ) - \nu_B(\bar{h}_B - T\bar{s}_B^\circ) \\ &= (\nu_C\bar{h}_C + \nu_D\bar{h}_D - \nu_A\bar{h}_A - \nu_B\bar{h}_B) - T(\nu_C\bar{s}_C^\circ + \nu_D\bar{s}_D^\circ - \nu_A\bar{s}_A^\circ - \nu_B\bar{s}_B^\circ) \quad (14.29b)\end{aligned}$$

Since the enthalpy of an ideal gas depends on temperature only, the $\bar{h}$'s of Eq. 14.29b are evaluated at temperature T. As indicated by the superscript °, each of the entropies is evaluated at temperature T and a pressure of 1 atm.

Introducing Eq. 14.29a into Eq. 14.28 and combining the terms involving logarithms into a single expression gives

$$-\frac{\Delta G^\circ}{\bar{R}T} = \ln\left[\frac{y_C^{\nu_C} y_D^{\nu_D}}{y_A^{\nu_A} y_B^{\nu_B}}\left(\frac{p}{p_{ref}}\right)^{\nu_C+\nu_D-\nu_A-\nu_B}\right] \quad (14.30)$$

Equation 14.30 is simply the form taken by the equation of reaction equilibrium, Eq. 14.26, for an ideal gas mixture subject to the reaction Eq. 14.24. As illustrated by subsequent examples, similar expressions can be written for other reactions.

Equation 14.30 can be expressed in an especially concise way as

$$-\frac{\Delta G^\circ}{\bar{R}T} = \ln K(T) \quad (14.31)$$

where K is the *equilibrium constant* defined by

$$K(T) = \frac{y_C^{\nu_C} y_D^{\nu_D}}{y_A^{\nu_A} y_B^{\nu_B}}\left(\frac{p}{p_{ref}}\right)^{\nu_C+\nu_D-\nu_A-\nu_B} \quad (14.32)$$

Given the values of the stoichiometric coefficients ν_A, ν_B, ν_C, and ν_D and the temperature T, the left side of Eq. 14.31 can be evaluated using either of Eqs. 14.29 together with the appropriate property data. The equation can then be solved for the value of the equilibrium constant K. Accordingly, for selected reactions K can be evaluated and tabulated against temperature. It is common to tabulate $\log_{10}K$ versus temperature, however. A tabulation of $\log_{10}K$ values over a range of temperatures for several reactions is provided in Table A-25, which is extracted from a more extensive compilation. Example 14.1 illustrates how the $\log_{10}K$ values are determined. Subsequent examples show how the table values can be used to evaluate equilibrium compositions.

The terms in the numerator and denominator of Eq. 14.32 correspond, respectively, to the products and reactants of the reaction given by Eq. 14.24 as it proceeds from left to right as written. For the *inverse* reaction $\nu_C C + \nu_D D \rightleftarrows \nu_A A + \nu_B B$, the equilibrium constant takes the form

$$K^* = \frac{y_A^{\nu_A} y_B^{\nu_B}}{y_C^{\nu_C} y_D^{\nu_D}}\left(\frac{p}{p_{ref}}\right)^{\nu_A+\nu_B-\nu_C-\nu_D} \quad (14.33)$$

Comparing Eqs. 14.32 and 14.33, it follows that the value of K^* is just the reciprocal of K: $K^* = 1/K$. Accordingly,

$$\log_{10}K^* = -\log_{10}K \quad (14.34)$$

Hence, Table A-25 can be used to evaluate K for the reactions listed proceeding in the direction left to right or to evaluate K^* for the inverse reactions proceeding in the direction right to left.

EXAMPLE 14.1

PROBLEM

Evaluate the equilibrium constant, expressed as $\log_{10}K$, for the reaction $CO + \frac{1}{2}O_2 \leftrightarrows CO_2$ at (a) 298 K and (b) 2000 K. Compare with the value obtained from Table A-25.

SOLUTION

Known: The reaction is $CO + \frac{1}{2}O_2 \rightleftarrows CO_2$.

Find: Determine the equilibrium constant for $T = 298$ K (25°C) and $T = 2000$ K.

Assumption: The ideal gas model applies.

Analysis: The equilibrium constant requires the evaluation of $\Delta G°$ for the reaction. Invoking Eq. 14.29b for this purpose, we have

$$\Delta G° = (\bar{h}_{CO_2} - \bar{h}_{CO} - \tfrac{1}{2}\bar{h}_{O_2}) - T(\bar{s}^\circ_{CO_2} - \bar{s}^\circ_{CO} - \tfrac{1}{2}\bar{s}^\circ_{O_2})$$

where the enthalpies are evaluated at temperature T and the absolute entropies are evaluated at temperature T and a pressure of 1 atm. Evaluating the enthalpies in terms of the respective enthalpies of formation gives

$$\begin{aligned}\Delta G° = {} & [(\bar{h}^\circ_f)_{CO_2} - (\bar{h}^\circ_f)_{CO} - \tfrac{1}{2}\overset{0}{\cancel{(\bar{h}^\circ_f)_{O_2}}}] + [(\Delta\bar{h})_{CO_2} - (\Delta\bar{h})_{CO} - \tfrac{1}{2}(\Delta\bar{h})_{O_2}] \\ & - T(\bar{s}^\circ_{CO_2} - \bar{s}^\circ_{CO} - \tfrac{1}{2}\bar{s}^\circ_{O_2})\end{aligned}$$

where the $\Delta\bar{h}$ terms account for the change in specific enthalpy from $T_{ref} = 298$ K to the specified temperature T. The enthalpy of formation of oxygen is zero by definition.

(a) When $T = 298$ K, the $\Delta\bar{h}$ terms of the above expression for $\Delta G°$ vanish. The required enthalpy of formation and absolute entropy values can be read from Table A-24, giving

$$\begin{aligned}\Delta G° &= [(-393{,}520) - (-110{,}530) - \tfrac{1}{2}(0)] - 298[213.69 - 197.54 - \tfrac{1}{2}(205.03)] \\ &= -257{,}253 \text{ kJ/kmol}\end{aligned}$$

With this value for $\Delta G°$, Eq. 14.31 gives

$$\ln K = -\frac{(-257{,}253 \text{ kJ/kmol})}{(8.314 \text{ kJ/kmol}\cdot\text{K})(298 \text{ K})} = 103.83$$

which corresponds to $\log_{10}K = 45.093$.

Table A-25 gives the logarithm to the base 10 of the equilibrium constant for the inverse reaction: $CO_2 \rightleftarrows CO + \frac{1}{2}O_2$. That is, $\log_{10}K^* = -45.066$. Thus, with Eq. 14.34, $\log_{10}K = 45.066$, which agrees well with the calculated value.

(b) When $T = 2000$ K, the $\Delta\bar{h}$ and $\bar{s}°$ terms for O_2, CO, and CO_2 required by the above expression for $\Delta G°$ are evaluated from Tables A-18, A-20, and A-21, respectively. The enthalpy of formation values are the same as in part (a). Thus

$$\begin{aligned}\Delta G° &= [(-393{,}520) - (-110{,}530) - \tfrac{1}{2}(0)] + [(100{,}804 - 9364) - (65408 - 8669) \\ &\quad - \tfrac{1}{2}(67{,}881 - 8682)] - 2000[309.210 - 258.600 - \tfrac{1}{2}(268.655)] \\ &= -282{,}990 + 5102 + 167{,}435 = -110{,}453 \text{ kJ/kmol}\end{aligned}$$

With this value, Eq. 14.31 gives

$$\ln K = -\frac{(-110{,}453)}{(8.314)(2000)} = 6.643$$

which corresponds to $\log_{10} K = 2.885$.

At 2000 K, Table A-25 gives $\log_{10} K^* = -2.884$. With Eq. 14.34, $\log_{10} K = 2.884$, which is in agreement with the calculated value.

Comment: Using the procedures described above, it is straightforward to determine $\log_{10} K$ versus temperature for each of several specified reactions and tabulate the results as in Table A-25.

14.3.2 Illustrations of the Calculation of Equilibrium Compositions for Reacting Ideal Gas Mixtures

It is often convenient to express Eq. 14.32 explicitly in terms of the number of moles that would be present at equilibrium. Each mole fraction appearing in the equation has the form $y_i = n_i/n$, where n_i is the amount of component i in the equilibrium mixture and n is the total number of moles of mixture. Hence, Eq. 14.32 can be rewritten as

$$K = \frac{n_C^{\nu_C} n_D^{\nu_D}}{n_A^{\nu_A} n_B^{\nu_B}} \left(\frac{p/p_{\text{ref}}}{n}\right)^{\nu_C + \nu_D - \nu_A - \nu_B} \tag{14.35}$$

The value of n must include not only the reacting components A, B, C, and D but also all inert components present. Since inert component E has been assumed present, we would write $n = n_A + n_B + n_C + n_D + n_E$.

Equation 14.35 provides a relationship among the temperature, pressure, and composition of a reacting ideal gas mixture at equilibrium. Accordingly, if any two of temperature, pressure, and composition are known, the third can be found by solving this equation. As an example, suppose that the temperature T and pressure p are known and the object is the equilibrium composition. With temperature known, the value of K can be obtained from Table A-25. The n's of the reacting components A, B, C, and D can be expressed in terms of a single unknown variable through application of the conservation of mass principle to the various chemical species present. Then, since the pressure is known, Eq. 14.35 constitutes a single equation in a single unknown. The solution normally requires iteration. This procedure involving Eq. 14.35 is illustrated below by examples.

EXAMPLE 14.2

PROBLEM

One kilomole of carbon monoxide, CO, reacts with $\frac{1}{2}$ kmol of oxygen, O_2, to form an equilibrium mixture of CO_2, CO, and O_2 at 2500 K and (a) 1 atm, (b) 10 atm. Determine the equilibrium composition in terms of mole fractions.

SOLUTION

Known: A system initially consisting of 1 kmol of CO and $\frac{1}{2}$ kmol of O_2 reacts to form an equilibrium mixture of CO_2, CO, and O_2. The temperature of the mixture is 2500 K and the pressure is (a) 1 atm, (b) 10 atm.

Find: Determine the equilibrium composition in terms of mole fractions.

Assumption: The equilibrium mixture is modeled as an ideal gas mixture.

Analysis: Applying the conservation of mass principle to the system, the overall balanced chemical reaction equation is

$$1CO + \frac{1}{2}O_2 \rightarrow zCO + \frac{z}{2}O_2 + (1 - z)CO_2$$

where z is the amount of CO, in kmol, present in the equilibrium mixture.

The total number of moles n in the equilibrium mixture is

$$n = z + \frac{z}{2} + (1 - z) = \frac{2 + z}{2}$$

Accordingly, the molar analysis of the equilibrium mixture is

$$y_{CO} = \frac{2z}{2 + z}, \qquad y_{O_2} = \frac{z}{2 + z}, \qquad y_{CO_2} = \frac{2(1 - z)}{2 + z}$$

At equilibrium, the tendency of CO and O_2 to form CO_2 is just balanced by the tendency of CO_2 to form CO and O_2, so we must have $CO_2 \rightleftarrows CO + \frac{1}{2}O_2$. Accordingly, Eq. 14.35 takes the form

$$K = \frac{z(z/2)^{1/2}}{(1 - z)}\left[\frac{p/p_{ref}}{(2 + z)/2}\right]^{1+1/2-1} = \frac{z}{1 - z}\left(\frac{z}{2 + z}\right)^{1/2}\left(\frac{p}{p_{ref}}\right)^{1/2}$$

At 2500 K, Table A-25 gives $\log_{10}K = -1.44$. Thus, $K = 0.0363$. Inserting this value into the last expression

$$0.0363 = \frac{z}{1 - z}\left(\frac{z}{2 + z}\right)^{1/2}\left(\frac{p}{p_{ref}}\right)^{1/2}$$

(a) When $p = 1$ atm, the equilibrium constant expression is

$$0.0363 = \frac{z}{1 - z}\left(\frac{z}{2 + z}\right)^{1/2}$$

Solving iteratively, $z = 0.129$. The equilibrium composition in terms of mole fractions is then

$$y_{CO} = \frac{2(0.129)}{2.129} = 0.121, \qquad y_{O_2} = \frac{0.129}{2.129} = 0.061, \qquad y_{CO_2} = \frac{2(1 - 0.129)}{2.129} = 0.818$$

(b) When $p = 10$ atm, the equilibrium constant expression is

$$0.0363 = \frac{z}{1 - z}\left(\frac{z}{2 + z}\right)^{1/2}(10)^{1/2}$$

Solving iteratively, $z = 0.062$. The corresponding equilibrium composition in terms of mole fractions is $y_{CO} = 0.06$, $y_{O_2} = 0.03$, $y_{CO_2} = 0.91$.

Comments:

1. Comparing the results of parts (a) and (b), we conclude that the extent to which the reaction proceeds toward completion (the extent to which CO_2 is formed) is increased by increasing the pressure.
2. If any two of temperature, pressure, and composition are known for a mixture at equilibrium, the third can be found by solving the equilibrium constant expression. To illustrate this, show that if the mole fraction of CO_2 in the equilibrium mixture at 2500 K was 0.93, the corresponding pressure would be closely 22.4 atm.

EXAMPLE 14.3

PROBLEM

Measurements show that at a temperature T and a pressure of 1 atm, the equilibrium mixture for the system of Example 14.2 has the composition $y_{CO} = 0.298$, $y_{O_2} = 0.149$, $y_{CO_2} = 0.553$. Determine the temperature T of the mixture, in K.

SOLUTION

Known: The pressure and composition of an equilibrium mixture of CO, O_2, and CO_2 are specified.

Find: Determine the temperature of the mixture, in K.

Assumption: The mixture can be modeled as an ideal gas mixture.

Analysis: When $p = 1$ atm, the equilibrium constant of the solution to Example 14.2 can be expressed as

$$K(T) = \frac{z}{1 - z}\left(\frac{z}{2 + z}\right)^{1/2}$$

where z is the amount of CO, in kmol, present in the equilibrium mixture and T is the temperature of the mixture.

The solution to Example 14.2 gives the following expression for the mole fraction of the CO in the mixture: $y_{CO} = 2z/(2 + z)$. Since $y_{CO} = 0.298$, $z = 0.35$.

Inserting this value for z into the expression for the equilibrium constant gives $K = 0.2078$. Thus, $\log_{10}K = -0.6824$. Interpolation in Table A-25 then gives $T = 2881$ K.

Comments:

1. Comparing this example with part (a) of Example 14.2, we conclude that the extent to which the reaction proceeds to completion (the extent to which CO_2 is formed) is decreased by increasing the temperature.
2. Equation 14.35 provides a relationship among the temperature, pressure, and composition of a reacting ideal gas mixture at equilibrium. If any two of temperature, pressure, and composition are known, the third can be found using this equation. In the present case, composition and pressure are known, and the temperature is the unknown. Example 14.2 illustrates cases where temperature and pressure are known and the composition is the unknown.

EXAMPLE 14.4

PROBLEM

One kilomole of carbon monoxide reacts with the theoretical amount of air to form an equilibrium mixture of CO_2, CO, O_2, and N_2 at 2500 K and 1 atm. Determine the equilibrium composition in terms of mole fractions.

SOLUTION

Known: A system initially consisting of 1 kmol of CO and the theoretical amount of air reacts to form an equilibrium mixture of CO_2, CO, O_2, and N_2. The temperature and pressure of the mixture are 2500 K and 1 atm.

Find: Determine the equilibrium composition in terms of mole fractions.

Assumption: The equilibrium mixture can be modeled as an ideal gas mixture.

Analysis: For a complete reaction of CO with the theoretical amount of air

$$CO + \tfrac{1}{2}O_2 + 1.88N_2 \rightarrow CO_2 + 1.88N_2$$

Accordingly, the reaction of CO with the theoretical amount of air to form CO_2, CO, O_2, and N_2 is

$$CO + \tfrac{1}{2}O_2 + 1.88N_2 \rightarrow zCO + \frac{z}{2}O_2 + (1 - z)CO_2 + 1.88N_2$$

where z is the amount of CO, in kmol, present in the equilibrium mixture.

The total number of moles n in the equilibrium mixture is

$$n = z + \frac{z}{2} + (1 - z) + 1.88 = \frac{5.76 + z}{2}$$

The composition of the equilibrium mixture in terms of mole fractions is

$$y_{CO} = \frac{2z}{5.76 + z}, \quad y_{O_2} = \frac{z}{5.76 + z}, \quad y_{CO_2} = \frac{2(1 - z)}{5.76 + z}, \quad y_{N_2} = \frac{3.76}{5.76 + z}$$

At equilibrium we must have $CO_2 \rightleftarrows CO + \tfrac{1}{2}O_2$. So Eq. 14.35 takes the form

$$K = \frac{z(z/2)^{1/2}}{(1 - z)}\left[\frac{p/p_{ref}}{(5.76 + z)/2}\right]^{1/2}$$

The value of K is the same as in the solution to Example 14.2, $K = 0.0363$. Thus, since $p = 1$ atm, we have

$$0.0363 = \frac{z}{1 - z}\left(\frac{z}{5.76 + z}\right)^{1/2}$$

Solving iteratively, $z = 0.175$. The corresponding equilibrium composition is $y_{CO} = 0.059$, $y_{CO_2} = 0.278$, $y_{O_2} = 0.029$, $y_{N_2} = 0.634$.

Comment: Comparing this example with Example 14.2, we conclude that the presence of the inert component nitrogen reduces the extent to which the reaction proceeds toward completion at the specified temperature and pressure (reduces the extent to which CO_2 is formed).

EXAMPLE 14.5

PROBLEM

A control volume operating at steady state has carbon dioxide entering at one location at 25°C, 1 atm and an equilibrium mixture of CO_2, CO, and O_2 exiting at another location at 3200 K, 1 atm. Determine the rate of heat transfer to the control volume, in kJ per kmol of CO_2 entering. The effects of kinetic and potential energy can be ignored and $\dot{W}_{cv} = 0$.

SOLUTION

Known: Carbon dioxide at 25°C, 1 atm enters a control volume at steady state. An equilibrium mixture of CO_2, CO, and O_2 exits at 3200 K, 1 atm.

Find: Determine the rate of heat transfer to the control volume, in kJ per kmol of CO_2 entering.

Schematic and Given Data:

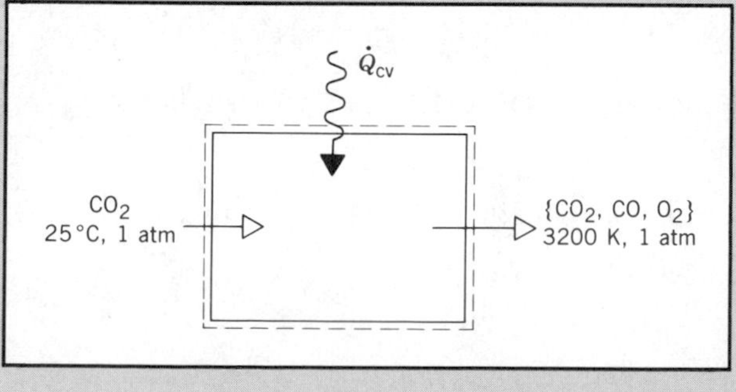

Figure E14.5

Assumptions:

1. The control volume shown on the accompanying sketch by a dashed line operates at steady state with $\dot{W}_{cv} = 0$. Kinetic energy and potential energy effects can be ignored.
2. The entering CO_2 is modeled as an ideal gas.
3. The exiting mixture of CO_2, CO, and O_2 is an equilibrium ideal gas mixture.

Analysis: The required heat transfer can be determined from an energy rate balance for the control volume, but first the composition of the exiting equilibrium mixture must be determined.

Applying the conservation of mass principle, the overall reaction is described by

$$CO_2 \rightarrow zCO_2 + (1 - z)CO + \left(\frac{1 - z}{2}\right) O_2$$

where z is the amount of CO_2, in kmol, present in the mixture exiting the control volume, per kmol of CO_2 entering. The total number of moles n in the mixture is then

$$n = z + (1 - z) + \left(\frac{1 - z}{2}\right) = \frac{3 - z}{2}$$

The exiting mixture is assumed to be an equilibrium mixture (assumption 3). Thus, for the mixture we must have $CO_2 \rightleftarrows CO + \frac{1}{2}O_2$. Equation 14.35 takes the form

$$K = \frac{(1 - z)[(1 - z)/2]^{1/2}}{z}\left[\frac{p/p_{\text{ref}}}{(3 - z)/2}\right]^{1+1/2-1}$$

Rearranging and noting that $p = 1$ atm

$$K = \left(\frac{1 - z}{z}\right)\left(\frac{1 - z}{3 - z}\right)^{1/2}$$

At 3200 K, Table A-25 gives $\log_{10}K = -0.189$. Thus, $K = 0.647$, and the equilibrium constant expression becomes

$$0.647 = \left(\frac{1 - z}{z}\right)\left(\frac{1 - z}{3 - z}\right)^{1/2}$$

Solving iteratively, $z = 0.422$. The composition of the exiting equilibrium mixture in kmol per kmol of CO_2 entering is then ($0.422CO_2$, $0.578CO$, $0.289O_2$).

When expressed per kmol of CO_2 entering the control volume, the energy rate balance at steady state is

$$0 = \frac{\dot{Q}_{cv}}{\dot{n}_{CO_2}} - \frac{\cancelto{0}{\dot{W}_{cv}}}{\dot{n}_{CO_2}} + \bar{h}_{CO_2} - (0.422\bar{h}_{CO_2} + 0.578\bar{h}_{CO} + 0.289\bar{h}_{O_2})$$

By assumption 1, $\dot{W}_{cv} = 0$, and kinetic and potential energy effects are ignored.

Solving for the heat transfer rate and evaluating each enthalpy in terms of the respective enthalpy of formation

$$\frac{\dot{Q}_{cv}}{\dot{n}_{CO_2}} = 0.422(\bar{h}_f^\circ + \Delta\bar{h})_{CO_2} + 0.578(\bar{h}_f^\circ + \Delta\bar{h})_{CO} + 0.289(\cancelto{0}{\bar{h}_f^\circ} + \Delta\bar{h})_{O_2} - (\bar{h}_f^\circ + \cancelto{0}{\Delta\bar{h}})_{CO_2}$$

The enthalpy of formation of O_2 is zero by definition; $\Delta\bar{h}$ for the CO_2 at the inlet vanishes because CO_2 enters at 25°C.

With enthalpy of formation values from Tables A-24 and $\Delta\bar{h}$ values for O_2, CO, and CO_2 from Tables A-18, 20, 21, respectively,

$$\begin{aligned}\frac{\dot{Q}_{cv}}{\dot{n}_{CO_2}} &= 0.422[-393{,}520 + (174{,}695 - 9364)] + 0.578[-110{,}530 + (109{,}667 - 8669)] \\ &\quad + 0.289(114{,}809 - 8682) - (-393{,}520) \\ &= 322{,}385 \text{ kJ/kmol } (CO_2)\end{aligned}$$

Comment: For comparison, let us determine the heat transfer in the absence of dissociation. In this case, only CO_2 exits the control volume. With data from Table A-21, the heat transfer is

$$\frac{\dot{Q}_{cv}}{\dot{n}_{CO_2}} = \bar{h}_{CO_2}(3200 \text{ K}) - \bar{h}_{CO_2}(298 \text{ K})$$

$$= 174{,}695 - 9364 = 165{,}331 \text{ kJ/kmol } (CO_2)$$

The heat transfer value found in the solution is greater than this value because the dissociation of CO_2 requires energy.

14.3.3 Equilibrium Constant for Ideal Solutions[1]

The procedures that led to the equilibrium constant for reacting ideal gas mixtures can be followed for the general case of reacting mixtures by using the fugacity and activity concepts introduced in Sec. 11.9. In principle, equilibrium compositions of such mixtures can be determined with an approach paralleling the one for ideal gas mixtures.

Equation 11.141 can be used to evaluate the chemical potentials appearing in the equation of reaction equilibrium (Eq. 14.26). The result is

$$\nu_A(\bar{g}_A^\circ + \bar{R}T \ln a_A) + \nu_B(\bar{g}_B^\circ + \bar{R}T \ln a_B) = \nu_C(\bar{g}_C^\circ + \bar{R}T \ln a_C) + \nu_D(\bar{g}_D^\circ + \bar{R}T \ln a_D) \tag{14.36}$$

where $\bar{g}_i^\circ$ is the Gibbs function of pure component i at temperature T and the pressure $p_{ref} = 1$ atm, and a_i is the *activity* of that component.

Collecting terms and employing Eq. 14.29a, Eq. 14.36 becomes

$$-\frac{\Delta G^\circ}{\bar{R}T} = \ln\left(\frac{a_C^{\nu_C} a_D^{\nu_D}}{a_A^{\nu_A} a_B^{\nu_B}}\right) \tag{14.37}$$

This equation can be expressed in the same form as Eq. 14.31 by defining the equilibrium constant as

$$K = \frac{a_C^{\nu_C} a_D^{\nu_D}}{a_A^{\nu_A} a_B^{\nu_B}} \tag{14.38}$$

Since Table A-25 and similar compilations are constructed simply by evaluating $-\Delta G^\circ/\bar{R}T$ for specified reactions at several temperatures, such tables can be employed to evaluate the more general equilibrium constant given above. However, before Eq. 14.38 can be used to determine the equilibrium composition for a known value of K, it is necessary to evaluate the activity of the various mixture components. Let us illustrate this for the case of mixtures that can be modeled as *ideal solutions*.

For an ideal solution, the activity of component i is given by

$$a_i = \frac{y_i f_i}{f_i^\circ} \tag{11.142}$$

where f_i is the fugacity of pure i at the temperature T and pressure p of the mixture, and f_i° is the fugacity of pure i at temperature T and the pressure p_{ref}. Using this

[1] This section requires study of Sec. 11.9.

expression to evaluate a_A, a_B, a_C, and a_D, Eq. 14.38 becomes

$$K = \frac{(y_C f_C/f_C^\circ)^{\nu_C}(y_D f_D/f_D^\circ)^{\nu_D}}{(y_A f_A/f_A^\circ)^{\nu_A}(y_B f_B/f_B^\circ)^{\nu_B}} \tag{14.39a}$$

which can be expressed alternatively as

$$K = \left[\frac{(f_C/p)^{\nu_C}(f_D/p)^{\nu_D}}{(f_A/p)^{\nu_A}(f_B/p)^{\nu_B}}\right]\left[\frac{(f_A^\circ/p_{ref})^{\nu_A}(f_B^\circ/p_{ref})^{\nu_B}}{(f_C^\circ/p_{ref})^{\nu_C}(f_D^\circ/p_{ref})^{\nu_D}}\right]\underline{\left[\frac{y_C^{\nu_C}y_D^{\nu_D}}{y_A^{\nu_A}y_B^{\nu_B}}\left(\frac{p}{p_{ref}}\right)^{\nu_C+\nu_D-\nu_A-\nu_B}\right]} \tag{14.39b}$$

The ratios of fugacity to pressure in this equation can be evaluated, in principle, from Eq. 11.124 or the generalized fugacity chart, Fig. A-6, developed from it. In the special case when each component behaves as an ideal gas at both T, p and T, p_{ref}, these ratios equal unity and Eq. 14.39b reduces to the underlined term, which is just Eq. 14.32.

14.4 FURTHER EXAMPLES OF THE USE OF THE EQUILIBRIUM CONSTANT

Some additional aspects of the use of the equilibrium constant are introduced in this section: the equilibrium flame temperature, the van't Hoff equation, and chemical equilibrium for ionization reactions and simultaneous reactions. To keep the presentation at an introductory level, only the case of ideal gas mixtures is considered.

14.4.1 Equilibrium Flame Temperature

In this section, the effect of incomplete combustion on the adiabatic flame temperature is considered using concepts developed in the present chapter. We begin with a review of some ideas related to the adiabatic flame temperature by considering a reactor operating at steady state for which no significant heat transfer with the surroundings takes place.

Let carbon monoxide gas entering at one location react *completely* with air entering at another location as follows:

$$CO + \tfrac{1}{2}O_2 + 1.88N_2 \rightarrow CO_2 + 1.88N_2$$

As discussed in Sec. 13.3, the products would exit the reactor at a temperature we have designated the *maximum* adiabatic flame temperature. This temperature can be determined by solving a *single* equation, the energy equation. At such an elevated temperature, however, there would be a tendency for CO_2 to dissociate

$$CO_2 \rightarrow CO + \tfrac{1}{2}O_2$$

Since dissociation requires energy (an endothermic reaction), the temperature of the products would be *less than* the maximum adiabatic temperature found under the assumption of complete combustion.

When dissociation takes place, the gaseous products exiting the reactor would be a mixture of CO_2, CO, O_2, and N_2, and the balanced chemical reaction equation would read

$$CO + \tfrac{1}{2}O_2 + 1.88N_2 \rightarrow zCO + (1 - z)CO_2 + \frac{z}{2}O_2 + 1.88N_2 \tag{14.40}$$

where z is the amount of CO, in kmol, present in the exiting mixture for each kmol of CO entering the reactor. Accordingly, there are *two* unknowns: z and the temperature of the exiting stream. To solve a problem with two unknowns requires two equations. One is provided by an energy equation. If the exiting gas mixture is in equilibrium, the other equation is provided by the equilibrium constant, Eq. 14.35. The temperature of the products may then be called the *equilibrium* flame temperature. The equilibrium constant used to evaluate the equilibrium flame temperature would be determined with respect to $CO_2 \rightleftarrows CO + \frac{1}{2}O_2$.

Although only the dissociation of CO_2 has been discussed, other products of combustion may dissociate, for example

$$H_2O \rightleftarrows H_2 + \tfrac{1}{2}O_2$$
$$H_2O \rightleftarrows OH + \tfrac{1}{2}H_2$$
$$O_2 \rightleftarrows 2O$$
$$H_2 \rightleftarrows 2H$$
$$N_2 \rightleftarrows 2N$$

Generally, the inclusion of many possible dissociation reactions is facilitated by the use of computers to solve the *simultaneous* equations that result. Simultaneous reactions are considered in Sec. 14.4.4. The following example illustrates how the equilibrium flame temperature is determined when one dissociation reaction occurs.

EXAMPLE 14.6

PROBLEM

Carbon monoxide at 25°C, 1 atm enters a well-insulated reactor and reacts with the theoretical amount of air entering at the same temperature and pressure. An equilibrium mixture of CO_2, CO, O_2, and N_2 exits the reactor at a pressure of 1 atm. For steady-state operation and negligible effects of kinetic and potential energy, determine the composition and temperature of the exiting mixture in K.

SOLUTION

Known: Carbon monoxide at 25°C, 1 atm reacts with the theoretical amount of air at 25°C, 1 atm to form an equilibrium mixture of CO_2, CO, O_2, and N_2 at temperature T and a pressure of 1 atm.

Find: Determine the composition and temperature of the exiting mixture.

Schematic and Given Data:

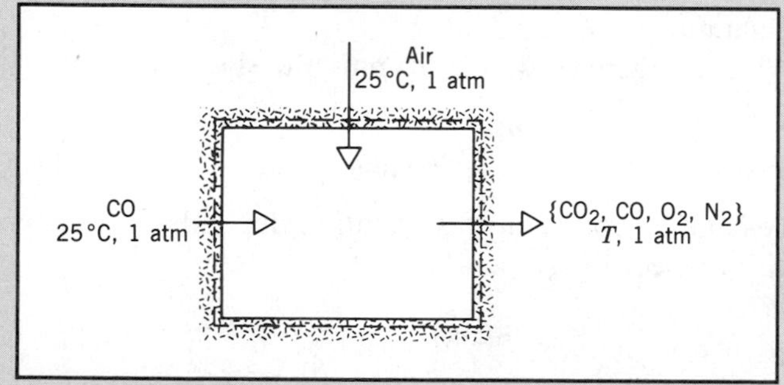

Figure E14.6

Assumptions:

1. The control volume shown on the accompanying sketch by a dashed line operates at steady state with $\dot{Q}_{cv} = 0$, $\dot{W}_{cv} = 0$, and negligible effects of kinetic and potential energy.
2. The entering gases are modeled as ideal gases.
3. The exiting mixture is an ideal gas mixture at equilibrium.

Analysis: The overall reaction is the same as in the solution to Example 14.4

$$CO + \tfrac{1}{2}O_2 + 1.88N_2 \rightarrow zCO + (1-z)CO_2 + \frac{z}{2}O_2 + 1.88N_2$$

The total number of moles n in the equilibrium mixture is

$$n = \frac{5.76 + z}{2}$$

By assumption 3, the exiting mixture is an equilibrium ideal gas mixture. The equilibrium constant expression is also developed in the solution to Example 14.4. Since $p = 1$ atm, the expression reduces to

$$K(T) = \frac{z}{1-z}\left(\frac{z}{5.76 + z}\right)^{1/2}$$

This equation involves two unknowns: z and the temperature T of the exiting equilibrium mixture.

Another equation involving the two unknowns is obtained from the energy rate balance, which reduces to

$$0 = \cancelto{0}{\frac{\dot{Q}_{cv}}{\dot{n}_{CO}}} - \cancelto{0}{\frac{\dot{W}_{cv}}{\dot{n}_{CO}}} + (\bar{h}_f^\circ + \cancelto{0}{\Delta\bar{h}})_{CO} + \frac{1}{2}(\cancelto{0}{\bar{h}_f^\circ} + \cancelto{0}{\Delta\bar{h}})_{O_2} + 1.88(\cancelto{0}{\bar{h}_f^\circ} + \cancelto{0}{\Delta\bar{h}})_{N_2}$$

$$- \left[z(\bar{h}_f^\circ + \Delta\bar{h})_{CO} + \frac{z}{2}(\cancelto{0}{\bar{h}_f^\circ} + \Delta\bar{h})_{O_2} + (1-z)(\bar{h}_f^\circ + \Delta\bar{h})_{CO_2} + 1.88(\cancelto{0}{\bar{h}_f^\circ} + \Delta\bar{h})_{N_2}\right]$$

The enthalpy of formation terms set to zero are those for oxygen and nitrogen. Since the reactants enter at 25°C, the corresponding $\Delta\bar{h}$ terms also vanish in this equation. On rearrangement

$$z(\Delta\bar{h})_{CO} + \frac{z}{2}(\Delta\bar{h})_{O_2} + (1-z)(\Delta\bar{h})_{CO_2} + 1.88(\Delta\bar{h})_{N_2} + (1-z)[(\bar{h}_f^\circ)_{CO_2} - (\bar{h}_f^\circ)_{CO}] = 0$$

When the above two equations for z and T are solved simultaneously, the result is $z = 0.12$ and $T = 2388$ K, as can readily be verified. The composition of the equilibrium mixture, in kmol per kmol of CO entering the reactor, is then $0.12CO$, $0.06O_2$, $0.88CO_2$, $1.88N_2$.

14.4.2 van't Hoff Equation

The dependence of the equilibrium constant on temperature exhibited by the values of Table A-25 follows from Eq. 14.31. An alternative way to express this dependence is given by the van't Hoff equation.

The development begins by introducing Eq. 14.29b into Eq. 14.31 to obtain on rearrangement

$$\bar{R}T \ln K = -[(\nu_C\bar{h}_C + \nu_D\bar{h}_D - \nu_A\bar{h}_A - \nu_B\bar{h}_B) - T(\nu_C\bar{s}_C^\circ + \nu_D\bar{s}_D^\circ - \nu_A\bar{s}_A^\circ - \nu_B\bar{s}_B^\circ)] \tag{14.41}$$

Each of the specific enthalpies and entropies in this equation depends on temperature alone. Differentiating with respect to temperature

$$\bar{R}T\frac{d\ln K}{dT} + \bar{R}\ln K = -\left[\nu_C\underline{\left(\frac{d\bar{h}_C}{dT} - T\frac{d\bar{s}_C^\circ}{dT}\right)} + \nu_D\underline{\left(\frac{d\bar{h}_D}{dT} - T\frac{d\bar{s}_D^\circ}{dT}\right)}\right.$$

$$\left. - \nu_A\underline{\left(\frac{d\bar{h}_A}{dT} - T\frac{d\bar{s}_A^\circ}{dT}\right)} - \nu_B\underline{\left(\frac{d\bar{h}_B}{dT} - T\frac{d\bar{s}_B^\circ}{dT}\right)}\right]$$

$$+ (\nu_C\bar{s}_C^\circ + \nu_D\bar{s}_D^\circ - \nu_A\bar{s}_A^\circ - \nu_B\bar{s}_B^\circ)$$

From the definition of $\bar{s}^\circ(T)$ (Eq. 6.24), we have $d\bar{s}^\circ/dT = \bar{c}_p/T$. Moreover, $d\bar{h}/dT = \bar{c}_p$. Accordingly, each of the underlined terms in the above equation vanishes identically, leaving

$$\bar{R}T\frac{d\ln K}{dT} + \bar{R}\ln K = (\nu_C\bar{s}_C^\circ + \nu_D\bar{s}_D^\circ - \nu_A\bar{s}_A^\circ - \nu_B\bar{s}_B^\circ) \tag{14.42}$$

Using Eq. 14.41 to evaluate the second term on the left and simplifying the resulting expression, Eq. 14.42 becomes

$$\frac{d\ln K}{dT} = \frac{(\nu_C\bar{h}_C + \nu_D\bar{h}_D - \nu_A\bar{h}_A - \nu_B\bar{h}_B)}{\bar{R}T^2} \tag{14.43a}$$

or, expressed more concisely

$$\frac{d\ln K}{dT} = \frac{\Delta H}{\bar{R}T^2} \tag{14.43b}$$

which is the *van't Hoff equation*. In Eq. 14.43b, ΔH is the *enthalpy of reaction* at temperature T. The van't Hoff equation shows that when ΔH is negative (exothermic reaction), K decreases with temperature, whereas for ΔH positive (endothermic reaction), K increases with temperature.

The enthalpy of reaction ΔH is often very nearly constant over a rather wide interval of temperature. In such cases, Eq. 14.43b can be integrated to yield

$$\ln\frac{K_2}{K_1} = -\frac{\Delta H}{\bar{R}}\left(\frac{1}{T_2} - \frac{1}{T_1}\right) \tag{14.44}$$

where K_1 and K_2 denote the equilibrium constants at temperatures T_1 and T_2, respectively. This equation shows that $\ln K$ is linear in $1/T$. Accordingly, plots of $\ln K$ versus $1/T$ can be used to determine ΔH from experimental equilibrium composition data. Alternatively, the equilibrium constant can be determined using enthalpy data.

14.4.3 Ionization

The methods developed for determining the equilibrium composition of a reactive ideal gas mixture can be applied to systems involving ionized gases, also known as *plasmas*. In previous sections we considered the chemical equilibrium of systems where dissociation is a factor. For example, the dissociation reaction of diatomic nitrogen

$$N_2 \rightleftarrows 2N$$

can occur at elevated temperatures. At still higher temperatures, ionization may take place according to

$$N \rightleftarrows N^+ + e^- \tag{14.45}$$

That is, a nitrogen atom loses an electron, yielding a singly ionized nitrogen atom N^+ and a free electron e^-. Further heating can result in the loss of additional electrons until all electrons have been removed from the atom.

For some cases of practical interest, it is reasonable to think of the neutral atoms, positive ions, and electrons as forming an ideal gas mixture. With this idealization, ionization equilibrium can be treated in the same manner as the chemical equilibrium of reacting ideal gas mixtures. The change in the Gibbs function for the equilibrium ionization reaction required to evaluate the ionization-equilibrium constant can be calculated as a function of temperature by using the procedures of statistical thermodynamics. In general, the extent of ionization increases as the temperature is raised and the pressure is lowered. An example serves best to illustrate the approach.

EXAMPLE 14.7

PROBLEM

Consider an equilibrium mixture at 3600°R consisting of Cs, Cs^+, and e^-, where Cs denotes neutral cesium atoms, Cs^+ singly ionized cesium ions, and e^- free electrons. The ionization-equilibrium constant at this temperature for

$$Cs \rightleftarrows Cs^+ + e^-$$

is $K = 15.63$. Determine the pressure in atmospheres if the ionization of Cs is (a) 95%, (b) 97%, (c) 99% complete.

SOLUTION

Known: An equilibrium mixture of Cs, Cs^+, e^- is at 3600°R. The value of the equilibrium constant at this temperature is known.

Find: Determine the pressure of the mixture in atmospheres if the ionization of Cs is (a) 95%, (b) 97%, (c) 99% complete.

Assumption: Equilibrium can be treated in this case using ideal gas mixture equilibrium considerations.

Analysis: The ionization of cesium to form a mixture of Cs, Cs^+, and e^- is described by

$$Cs \rightarrow (1 - z)Cs + zCs^+ + ze^-$$

where z denotes the extent of the ionization. The total number of moles of mixture n is

$$n = (1 - z) + z + z = 1 + z$$

At equilibrium, we have $Cs \rightleftarrows Cs^+ + e^-$, so Eq. 14.35 takes the form

$$K = \frac{(z)(z)}{(1 - z)}\left[\frac{p/p_{ref}}{(1 + z)}\right]^{1+1-1} = \left(\frac{z^2}{1 - z^2}\right)\left(\frac{p}{p_{ref}}\right)$$

Solving for the ratio p/p_{ref} and introducing the known value of K

$$\frac{p}{p_{ref}} = (15.63)\left(\frac{1 - z^2}{z^2}\right)$$

(a) When $z = 0.95$, $p/p_{ref} = 1.69$, or $p = 1.69$ atm.

(b) When $z = 0.97$, $p/p_{ref} = 0.98$, or $p = 0.98$ atm.

(c) When $z = 0.99$, $p/p_{ref} = 0.32$, or $p = 0.32$ atm.

Comment: This example illustrates that ionization tends to occur to a greater extent as pressure is lowered. Ionization also tends to occur to a greater extent as temperature is raised at fixed pressure.

14.4.4 Simultaneous Reactions

Let us return to the discussion of Sec. 14.2 and consider the possibility of more than one reaction among the substances present within a system. For the present application, the closed system is assumed to contain a mixture of *eight* components A, B, C, D, E, L, M, and N, subject to *two* independent reactions

$$(1) \qquad \nu_A A + \nu_B B \rightleftarrows \nu_C C + \nu_D D \tag{14.24}$$

$$(2) \qquad \nu_{A'} A + \nu_L L \rightleftarrows \nu_M M + \nu_N N \tag{14.46}$$

As in Sec. 14.2, component E is inert. Also, note that component A has been taken as common to both reactions but with a possibly different stoichiometric coefficient ($\nu_{A'}$ is not necessarily equal to ν_A).

The stoichiometric coefficients of the above equations do not correspond to the numbers of moles of the respective components present within the system, but *changes* in the amounts of the components are related to the stoichiometric coefficients by

$$\frac{-dn_A}{\nu_A} = \frac{-dn_B}{\nu_B} = \frac{dn_C}{\nu_C} = \frac{dn_D}{\nu_D} \tag{14.25a}$$

following from Eq. 14.24, and

$$\frac{-dn_A}{\nu_{A'}} = \frac{-dn_L}{\nu_L} = \frac{dn_M}{\nu_M} = \frac{dn_N}{\nu_N} \tag{14.47a}$$

following from Eq. 14.46. Introducing a proportionality factor $d\varepsilon_1$, Eqs. 14.25a may be represented by

$$\begin{aligned} dn_A &= -\nu_A\, d\varepsilon_1, & dn_B &= -\nu_B\, d\varepsilon_1 \\ dn_C &= \nu_C\, d\varepsilon_1, & dn_D &= \nu_D\, d\varepsilon_1 \end{aligned} \tag{14.25b}$$

Similarly, with the proportionality factor $d\varepsilon_2$, Eqs. 14.47a may be represented by

$$\begin{aligned} dn_A &= -\nu_{A'}\, d\varepsilon_2, & dn_L &= -\nu_L\, d\varepsilon_2 \\ dn_M &= \nu_M\, d\varepsilon_2, & dn_N &= \nu_N\, d\varepsilon_2 \end{aligned} \tag{14.47b}$$

Component A is involved in both reactions, so the total change in A is given by

$$dn_A = -\nu_A\, d\varepsilon_1 - \nu_{A'}\, d\varepsilon_2 \tag{14.48}$$

Also, we have $dn_E = 0$ because component E is inert.

For the system under present consideration, Eq. 14.10 is

$$\begin{aligned} dG]_{T,p} = {} & \mu_A\, dn_A + \mu_B\, dn_B + \mu_C\, dn_C + \mu_D\, dn_D \\ & + \mu_E\, dn_E + \mu_L\, dn_L + \mu_M\, dn_M + \mu_N\, dn_N \end{aligned} \tag{14.49}$$

Introducing the above expressions giving the changes in the n's, this becomes

$$\begin{aligned} dG]_{T,p} = {} & (-\nu_A\mu_A - \nu_B\mu_B + \nu_C\mu_C + \nu_D\mu_D)\, d\varepsilon_1 \\ & + (-\nu_{A'}\mu_A - \nu_L\mu_L + \nu_M\mu_M + \nu_N\mu_N)\, d\varepsilon_2 \end{aligned} \tag{14.50}$$

Since the two reactions are independent, $d\varepsilon_1$ and $d\varepsilon_2$ can be independently varied. Accordingly, when $dG]_{T,p} = 0$, the terms in parentheses must be zero and *two* equations of reaction equilibrium result, one corresponding to each of the foregoing reactions

$$\nu_A\mu_A + \nu_B\mu_B = \nu_C\mu_C + \nu_D\mu_D \tag{14.26b}$$

$$\nu_{A'}\mu_A + \nu_L\mu_L = \nu_M\mu_M + \nu_N\mu_N \tag{14.51}$$

The first of these equations is exactly the same as that obtained in Sec. 14.2. For the case of reacting ideal gas mixtures, this equation can be expressed as

$$-\left(\frac{\Delta G^\circ}{\bar{R}T}\right)_1 = \ln\left[\frac{y_C^{\nu_C} y_D^{\nu_D}}{y_A^{\nu_A} y_B^{\nu_B}}\left(\frac{p}{p_{\text{ref}}}\right)^{\nu_C+\nu_D-\nu_A-\nu_B}\right] \tag{14.52}$$

Similarly, Eq. 14.51 can be expressed as

$$-\left(\frac{\Delta G^\circ}{\bar{R}T}\right)_2 = \ln\left[\frac{y_M^{\nu_M} y_N^{\nu_N}}{y_A^{\nu_{A'}} y_L^{\nu_L}}\left(\frac{p}{p_{\text{ref}}}\right)^{\nu_M+\nu_N-\nu_{A'}-\nu_L}\right] \tag{14.53}$$

In each of these equations, the ΔG° term is evaluated as the change in Gibbs function for the respective reaction, regarding each reactant and product as separate at temperature T and a pressure of 1 atm.

From Eq. 14.52 follows the equilibrium constant

$$K_1 = \frac{y_C^{\nu_C} y_D^{\nu_D}}{y_A^{\nu_A} y_B^{\nu_B}} \left(\frac{p}{p_{ref}}\right)^{\nu_C+\nu_D-\nu_A-\nu_B} \tag{14.54}$$

and from Eq. 14.53 follows

$$K_2 = \frac{y_M^{\nu_M} y_N^{\nu_N}}{y_A^{\nu_{A'}} y_L^{\nu_L}} \left(\frac{p}{p_{ref}}\right)^{\nu_M+\nu_N-\nu_{A'}-\nu_L} \tag{14.55}$$

The equilibrium constants K_1 and K_2 can be determined from Table A-25 or a similar compilation. The mole fractions appearing in these expressions must be evaluated by considering *all* the substances present within the system, including the inert substance E. Each mole fraction has the form $y_i = n_i/n$, where n_i is the amount of component i in the equilibrium mixture and

$$n = n_A + n_B + n_C + n_D + n_E + n_L + n_M + n_N \tag{14.56}$$

The n's appearing in Eq. 14.56 can be expressed in terms of *two* unknown variables through application of the conservation of mass principle to the various chemical species present. Accordingly, for a specified temperature and pressure, Eqs. 14.54 and 14.55 give *two* equations in *two* unknowns. The composition of the system at equilibrium can be determined by solving these equations simultaneously. This procedure is illustrated by Example 14.8.

The procedure discussed in this section can be extended to systems involving several simultaneous independent reactions. The number of simultaneous equilibrium constant expressions that results equals the number of independent reactions. As these equations are nonlinear and require simultaneous solution, the use of a computer is usually required.

EXAMPLE 14.8

PROBLEM

As a result of heating, a system consisting initially of 1 kmol of CO_2, $\frac{1}{2}$ kmol of O_2, and $\frac{1}{2}$ kmol of N_2 forms an equilibrium mixture of CO_2, CO, O_2, N_2, and NO at 3000 K, 1 atm. Determine the composition of the equilibrium mixture.

SOLUTION

Known: A system consisting of specified amounts of CO_2, O_2, and N_2 is heated to 3000 K, 1 atm, forming an equilibrium mixture of CO_2, CO, O_2, N_2, and NO.

Find: Determine the equilibrium composition.

Assumption: The final mixture is an equilibrium mixture of ideal gases.

Analysis: The overall reaction has the form

$$1CO_2 + \tfrac{1}{2}O_2 + \tfrac{1}{2}N_2 \rightarrow aCO + bNO + cCO_2 + dO_2 + eN_2$$

Applying conservation of mass to carbon, oxygen, and nitrogen, the five unknown coefficients can be expressed in terms of any two of the coefficients. Selecting a and b as the unknowns, the following balanced equation results:

$$1CO_2 + \tfrac{1}{2}O_2 + \tfrac{1}{2}N_2 \rightarrow aCO + bNO + (1-a)CO_2 + \tfrac{1}{2}(1+a-b)O_2 + \tfrac{1}{2}(1-b)N_2$$

The total number of moles n in the mixture formed by the products is

$$n = a + b + (1-a) + \tfrac{1}{2}(1+a-b) + \tfrac{1}{2}(1-b) = \frac{4+a}{2}$$

At equilibrium, two independent reactions relate the components of the product mixture.

1. $CO_2 \rightleftarrows CO + \frac{1}{2}O_2$
2. $\frac{1}{2}O_2 + \frac{1}{2}N_2 \rightleftarrows NO$

For the first of these reactions, the form taken by the equilibrium constant when $p = 1$ atm is

$$K_1 = \frac{a[\frac{1}{2}(1 + a - b)]^{1/2}}{(1 - a)}\left[\frac{1}{(4 + a)/2}\right]^{1+1/2-1} = \frac{a}{1 - a}\left(\frac{1 + a - b}{4 + a}\right)^{1/2}$$

Similarly, the equilibrium constant for the second of the reactions is

$$K_2 = \frac{b}{[\frac{1}{2}(1 + a - b)]^{1/2}[\frac{1}{2}(1 - b)]^{1/2}}\left[\frac{1}{(4 + a)/2}\right]^{1-1/2-1/2} = \frac{2b}{[(1 + a - b)(1 - b)]^{1/2}}$$

At 3000 K, Table A-25 provides $\log_{10}K_1 = -0.485$ and $\log_{10}K_2 = -0.913$, giving $K_1 = 0.3273$ and $K_2 = 0.1222$. Accordingly, the two equations that must be solved simultaneously for the two unknowns a and b are

$$0.3273 = \frac{a}{1 - a}\left(\frac{1 + a - b}{4 + a}\right)^{1/2}, \qquad 0.1222 = \frac{2b}{[(1 + a - b)(1 - b)]^{1/2}}$$

The solution is $a = 0.3745$, $b = 0.0675$, as can readily be verified. The composition of the equilibrium mixture, in kmol per kmol of CO_2 present initially, is then 0.3745CO, 0.0675NO, 0.6255CO_2, 0.6535O_2, 0.4663N_2.

Comment: If high enough temperatures are attained, nitrogen can combine with oxygen to form components such as nitric oxide. Even trace amounts of oxides of nitrogen in products of combustion can be a source of air pollution.

PHASE EQUILIBRIUM

In this part of the chapter the equilibrium condition $dG]_{T,p} = 0$ introduced in Sec. 14.1 is used to study the equilibrium of multicomponent, multiphase, nonreacting systems. The discussion begins with the elementary case of equilibrium between two phases of a pure substance and then turns to the general case of several components present in several phases.

14.5 EQUILIBRIUM BETWEEN TWO PHASES OF A PURE SUBSTANCE

Consider the case of a system consisting of two phases of a pure substance at equilibrium. Since the system is at equilibrium, each phase is at the same temperature and pressure. The Gibbs function for the system is

$$G = n'\bar{g}'(T, p) + n''\bar{g}''(T, p) \tag{14.57}$$

where the primes ′ and ″ denote phases 1 and 2, respectively.

Forming the differential of G at fixed T and p

$$dG]_{T,p} = \bar{g}'\, dn' + \bar{g}''\, dn'' \tag{14.58}$$

Since the total amount of the pure substance remains constant, an increase in the amount present in one of the phases must be compensated by an equivalent decrease in the amount present in the other phase. Thus, we have $dn'' = -dn'$, and Eq. 14.58 becomes

$$dG]_{T,p} = (\bar{g}' - \bar{g}'')\, dn'$$

At equilibrium, $dG]_{T,p} = 0$, so

$$\bar{g}' = \bar{g}'' \tag{14.59}$$

At equilibrium, the molar Gibbs functions of the phases are equal.

Equation 14.59 can be used to derive the *Clapeyron* equation, obtained by other means in Sec. 11.4. For two phases at equilibrium, variations in pressure are uniquely related to variations in temperature: $p = p_{\text{sat}}(T)$; thus, differentiation of Eq. 14.59 with respect to temperature gives

$$\left.\frac{\partial \bar{g}'}{\partial T}\right)_p + \left.\frac{\partial \bar{g}'}{\partial p}\right)_T \frac{dp_{\text{sat}}}{dT} = \left.\frac{\partial \bar{g}''}{\partial T}\right)_p + \left.\frac{\partial \bar{g}''}{\partial p}\right)_T \frac{dp_{\text{sat}}}{dT}$$

With Eqs. 11.30 and 11.31, this becomes

$$-\bar{s}' + \bar{v}' \frac{dp_{\text{sat}}}{dT} = -\bar{s}'' + \bar{v}'' \frac{dp_{\text{sat}}}{dT}$$

Or on rearrangement

$$\frac{dp_{\text{sat}}}{dT} = \frac{\bar{s}'' - \bar{s}'}{\bar{v}'' - \bar{v}'}$$

This can be expressed alternatively by noting that, with $\bar{g} = \bar{h} - T\bar{s}$, Eq. 14.59 becomes

$$\bar{h}' - T\bar{s}' = \bar{h}'' - T\bar{s}''$$

or

$$\bar{s}'' - \bar{s}' = \frac{\bar{h}'' - \bar{h}'}{T} \tag{14.60}$$

Combining results, the Clapeyron equation is obtained

$$\frac{dp_{\text{sat}}}{dT} = \frac{1}{T}\left(\frac{\bar{h}'' - \bar{h}'}{\bar{v}'' - \bar{v}'}\right) \tag{14.61}$$

Applications of the Clapeyron equation are provided in Chap. 11.

A special form of Eq. 14.61 for a system at equilibrium consisting of a liquid or solid phase and a vapor phase can be obtained simply. If the specific volume of the liquid or solid, $\bar{v}'$, is negligible compared with the specific volume of the vapor, $\bar{v}''$, and the vapor can be treated as an ideal gas, $\bar{v}'' = \bar{R}T/p_{\text{sat}}$, Eq. 14.61 becomes

$$\frac{dp_{\text{sat}}}{dT} = \frac{\bar{h}'' - \bar{h}'}{\bar{R}T^2/p_{\text{sat}}}$$

or

$$\frac{d \ln p_{\text{sat}}}{dT} = \frac{\bar{h}'' - \bar{h}'}{\bar{R}T^2} \tag{14.62}$$

This is the *Clausius–Clapeyron* equation. The similarity in form of Eq. 14.62 and the van't Hoff equation, Eq. 14.43b, may be noted. The van't Hoff equation for chemical equilibrium is the counterpart of the Clausius–Clapeyron equation for phase equilibrium.

14.6 EQUILIBRIUM OF MULTICOMPONENT, MULTIPHASE SYSTEMS

The equilibrium of systems that may involve several phases, each having a number of components present, is considered in this section. The principal result is the Gibbs phase rule, which summarizes important limitations on multicomponent, multiphase systems at equilibrium.

14.6.1 Preliminaries

Figure 14.1 shows a system consisting of *two* components A and B in *two* phases 1 and 2 that are at the same temperature and pressure. Applying Eq. 14.10 to each of the phases

$$\begin{aligned} dG']_{T,p} &= \mu'_A\, dn'_A + \mu'_B\, dn'_B \\ dG'']_{T,p} &= \mu''_A\, dn''_A + \mu''_B\, dn''_B \end{aligned} \tag{14.63}$$

where as before the primes identify the two phases.

Although matter might be transferred between the two phases, in the absence of chemical reactions the total amounts of A and B must remain constant. Thus, the increase in the amount present in one of the phases must be compensated by an equivalent decrease in the amount present in the other phase. That is,

$$dn''_A = -dn'_A, \qquad dn''_B = -dn'_B \tag{14.64}$$

With Eqs. 14.63 and 14.64, the change in the Gibbs function for the system is

$$\begin{aligned} dG]_{T,p} &= dG']_{T,p} + dG'']_{T,p} \\ &= (\mu'_A - \mu''_A)\, dn'_A + (\mu'_B - \mu''_B)\, dn'_B \end{aligned} \tag{14.65}$$

Since n'_A and n'_B can be varied independently, it follows that when $dG]_{T,p} = 0$, the terms in parentheses are zero, resulting in

$$\mu'_A = \mu''_A \quad \text{and} \quad \mu'_B = \mu''_B \tag{14.66}$$

At equilibrium, the chemical potential of each component is the same in each phase.

The significance of the chemical potential for phase equilibrium can be brought out simply by reconsidering the system of Fig. 14.1 in the special case when the chemical potential of component B is the same in both phases: $\mu'_B = \mu''_B$. With this constraint,

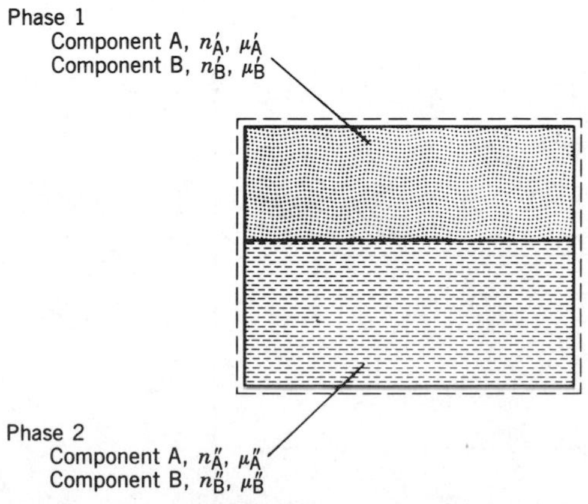

Figure 14.1 System consisting of two components in two phases.

Eq. 14.65 reduces to

$$dG]_{T,p} = (\mu'_{\rm A} - \mu''_{\rm A})\, dn'_{\rm A}$$

Any spontaneous process of the system taking place at a fixed temperature and pressure must be such that the Gibbs function decreases: $dG]_{T,p} < 0$. Thus, with the above expression we have

$$(\mu'_{\rm A} - \mu''_{\rm A})\, dn'_{\rm A} < 0$$

Accordingly, when the chemical potential of A is greater in phase 1 than in phase 2 ($\mu'_{\rm A} > \mu''_{\rm A}$), it follows that $dn'_{\rm A} < 0$. That is, substance A passes from phase 1 to phase 2. However, when the chemical potential of A is greater in phase 2 than in phase 1 ($\mu''_{\rm A} > \mu'_{\rm A}$), it follows that $dn'_{\rm A} > 0$. That is, substance A passes from phase 2 to phase 1. At equilibrium, the chemical potentials are equal, and there is no net transfer of A between the two phases. With this reasoning, we see that the chemical potential can be regarded as a measure of the *escaping tendency* of a component. If the chemical potential of a component is not the same in each phase, there will be a tendency for that component to pass from the phase having the higher chemical potential for that component to the phase having the lower chemical potential. When the chemical potential is the same in both phases, there is no tendency for a net transfer to occur from one phase to the other.

EXAMPLE 14.9

PROBLEM

A closed system at a temperature of 70°F and a pressure of 1 atm consists of a pure liquid water phase in equilibrium with a vapor phase composed of water vapor and dry air. Determine the departure, in percent, of the partial pressure of the water vapor from the saturation pressure of water at 70°F.

SOLUTION

Known: A phase of liquid water only is in equilibrium with *moist air* at 70°F and 1 atm.

Find: Determine the percentage departure of the partial pressure of the water vapor in the moist air from the saturation pressure of water at 70°F.

Schematic and Given Data:

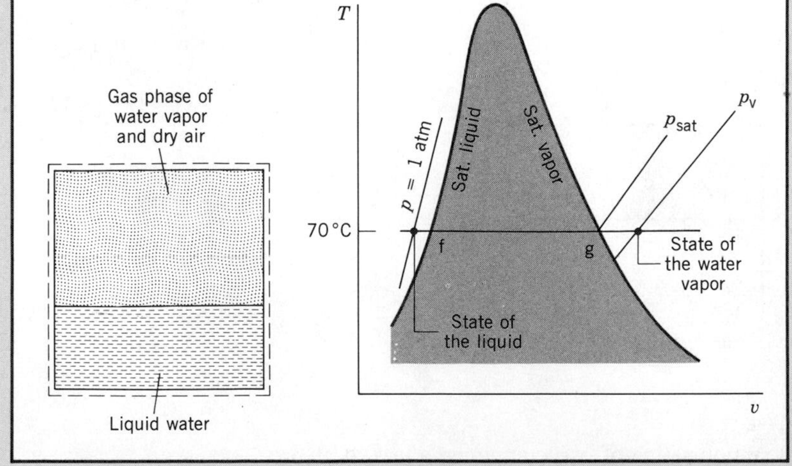

Figure E14.9

Assumptions:

1. The gas phase can be modeled as an ideal gas mixture.
2. The liquid phase is pure water only.

Analysis: For phase equilibrium, the chemical potential of the water must have the same value in both phases: $\mu_l = \mu_v$, where μ_l and μ_v denote, respectively, the chemical potentials of the pure liquid water in the liquid phase and the water vapor in the vapor phase.

The chemical potential μ_l is the Gibbs function per mole of pure liquid water (Eq. 14.12)

$$\mu_l = \bar{g}(T, p)$$

Since the vapor phase is assumed to form an ideal gas mixture, the chemical potential μ_v equals the Gibbs function per mole evaluated at temperature T and the partial pressure p_v of the water vapor (Eq. 14.16)

$$\mu_v = \bar{g}(T, p_v)$$

For phase equilibrium, $\mu_l = \mu_v$, or

$$\bar{g}(T, p_v) = \bar{g}(T, p)$$

With $\bar{g} = \bar{h} - T\bar{s}$, this can be expressed alternatively as

$$\bar{h}(T, p_v) - T\bar{s}(T, p_v) = \bar{h}(T, p) - T\bar{s}(T, p)$$

The water vapor is modeled as an ideal gas. Thus, the enthalpy is given closely by the saturated vapor value at temperature T

$$\bar{h}(T, p_v) \approx \bar{h}_g$$

Furthermore, with Eq. 6.25b, the difference between the specific entropy of the water vapor and the specific entropy at the corresponding saturated vapor state is

$$\bar{s}(T, p_v) - \bar{s}_g(T) = -\bar{R} \ln \frac{p_v}{p_{sat}}$$

or

$$\bar{s}(T, p_v) = \bar{s}_g(T) - \bar{R} \ln \frac{p_v}{p_{sat}}$$

where p_{sat} is the saturation pressure at temperature T.

With Eq. 3.14, the enthalpy of the liquid is closely

$$\bar{h}(T, p) \approx \bar{h}_f + \bar{v}_f(p - p_{sat})$$

where $\bar{v}_f$ and $\bar{h}_f$ are the saturated liquid specific volume and enthalpy at temperature T. Furthermore, with Eq. 6.10

$$\bar{s}(T, p) \approx \bar{s}_f(T)$$

where $\bar{s}_f$ is the saturated liquid specific entropy at temperature T.

Collecting the foregoing expressions, we have

$$\bar{h}_g - T\left(\bar{s}_g - \bar{R} \ln \frac{p_v}{p_{sat}}\right) = \bar{h}_f + \bar{v}_f(p - p_{sat}) - T\bar{s}_f$$

or

$$\bar{R}T \ln \frac{p_v}{p_{sat}} = \bar{v}_f(p - p_{sat}) - \underline{[(\bar{h}_g - \bar{h}_f) - T(\bar{s}_g - \bar{s}_f)]}$$

The underlined term vanishes by Eq. 14.60, leaving

$$\ln \frac{p_v}{p_{sat}} = \frac{\bar{v}_f(p - p_{sat})}{\bar{R}T} \qquad \text{or} \qquad \frac{p_v}{p_{sat}} = \exp \frac{\bar{v}_f(p - p_{sat})}{\bar{R}T}$$

With data from Table A-2E at 70°F, $v_f = 0.01605 \text{ ft}^3/\text{lb}$ and $p_{sat} = 0.3632 \text{ lbf/in.}^2$, we have

$$\frac{v_f(p - p_{sat})}{RT} = \frac{0.01605 \text{ ft}^3/\text{lb}(14.696 - 0.3632)(\text{lbf/in.}^2)(144 \text{ in.}^2/\text{ft}^2)}{\left(\frac{1545}{18.02}\frac{\text{ft}\cdot\text{lbf}}{\text{lb}\cdot{}^\circ\text{R}}\right)(530^\circ\text{R})}$$

$$= 7.29 \times 10^{-4}$$

Finally

$$\frac{p_v}{p_{sat}} = \exp(7.29 \times 10^{-4}) = 1.00073$$

When expressed as a percentage, the departure of p_v from p_{sat} is

$$\left(\frac{p_v - p_{sat}}{p_{sat}}\right)(100) = (1.00073 - 1)(100) = 0.073\%$$

Comments:

1. The departure of p_v from p_g is negligible at the specified conditions. This suggests that at normal temperatures and pressures the equilibrium between the liquid water phase and the water vapor is not significantly disturbed by the presence of the dry air. Accordingly, the partial pressure of the water vapor can be taken as equal to the saturation pressure of the water at the system temperature. This model, introduced without proof in Sec. 12.6, is used extensively in Chap. 12.
2. For the liquid phase to consist of *pure* water only as assumed, it is necessary for the chemical potential of air in the liquid phase to be equal to (or greater than) the equilibrium value; otherwise, dry air would diffuse into the liquid phase.

14.6.2 Gibbs Phase Rule

The requirement for equilibrium of a system consisting of two components and two phases, given by Eqs. 14.66, can be extended with similar reasoning to nonreacting multicomponent, multiphase systems. At equilibrium, the chemical potential of each component must be the same in all phases. For the case of N components that are present in P phases we have, therefore, the following set of $N(P - 1)$ equations:

$$N\left\{\begin{array}{l} \overbrace{\mu_1^1 = \mu_1^2 = \mu_1^3 = \cdots = \mu_1^P}^{P-1} \\ \mu_2^1 = \mu_2^2 = \mu_2^3 = \cdots = \mu_2^P \\ \vdots \\ \mu_N^1 = \mu_N^2 = \mu_N^3 = \cdots = \mu_N^P \end{array}\right. \tag{14.67}$$

where μ_i^j denotes the chemical potential of the ith component in the jth phase. This set of equations provides the basis for the *Gibbs phase rule,* which allows the determination of the number of *independent intensive* properties that may be arbitrarily specified in order to fix the *intensive* state of the system. The number of independent intensive properties is called the *degrees of freedom* (or the *variance*).

Since the chemical potential is an intensive property, its value depends on the relative proportions of the components present and not on the amounts of the components. In other words, in a given phase involving N components at temperature T and pressure p, the chemical potential is determined by the *mole fractions* of the components present and not the respective n's. However, as the mole fractions add up to unity, at most $N - 1$ of the mole fractions can be independent. Thus, for a system involving N components, there are at most $N - 1$ independently variable mole fractions for each phase. For P phases, therefore, there are at most $P(N - 1)$ independently variable mole fractions. In addition, the temperature and pressure, which are the same in each phase, are two further intensive properties, giving a maximum of $P(N - 1) + 2$ independently variable intensive properties for the system. But because of the $N(P - 1)$ equilibrium conditions represented by Eqs. 14.67 among these properties, the number of intensive properties that are freely variable, the degrees of freedom F, is

$$F = [P(N - 1) + 2] - N(P - 1) = 2 + N - P \tag{14.68}$$

The relationship given by Eq. 14.68 is the Gibbs phase rule. F is the number of intensive properties that may be arbitrarily specified and that must be specified to fix the intensive state of a nonreacting system at equilibrium. The phase rule given here must be modified for application to systems in which chemical reactions occur.

The phase rule summarizes important limitations on various types of systems. For example, for a system involving a single component such as water, $N = 1$ and Eq. 14.67 becomes

$$F = 3 - P \tag{14.69}$$

The minimum number of phases is one, corresponding to $P = 1$. For this case, Eq. 14.69 gives $F = 2$. That is, *two* intensive properties must be specified to fix the intensive state of the system. This requirement is familiar from our use of the steam tables and similar property tables. To obtain properties of superheated vapor, say, from such tables requires that we give values for *any two* of the tabulated properties, for example, T and p.

When two phases are present in a system involving a single component, $N = 1$ and $P = 2$. Equation 14.69 then gives $F = 1$. That is, the intensive state is determined by a single intensive property value. For example, the intensive states of the separate phases of an equilibrium mixture of liquid water and water vapor are completely determined by specifying the temperature.

The minimum allowable value for the degrees of freedom is zero: $F = 0$. For a single-component system, Eq. 14.69 shows that this corresponds to $P = 3$, a three-phase system. Thus, *three* is the maximum number of different phases of pure component that can coexist in equilibrium. Since there are no degrees of freedom, both temperature and pressure are fixed at equilibrium. For example, there is only a single temperature 0.01°C (32.02°F) and a single pressure 0.6113 kPa (0.006 atm) for which ice, liquid water, and water vapor are in equilibrium.

As another illustration, let us apply the Gibbs phase rule to a liquid solution consisting of water and ammonia such as considered in the discussion of absorption refrigeration (Sec. 10.5). This solution involves two components and a single phase: $N = 2$ and $P = 1$. Equation 14.68 then gives $F = 3$, so the intensive state is fixed by giving the values of *three* intensive properties, such as temperature, pressure, and the ammonia (or water) mole fraction.

The system of equations, Eqs. 14.67, giving the requirements for phase equilibrium at a specified temperature and pressure can be expressed alternatively in terms of partial molal Gibbs functions, fugacities, and activities, all of which are introduced in Sec. 11.9. To use any such expression to determine the equilibrium composition of the different phases present within a system at equilibrium requires a model for each phase that allows the relevant quantities—the chemical potentials, fugacities, etc.—to be evaluated for the components present in terms of system properties that can be determined. For example, a gas phase might be modeled as an ideal gas mixture or, at higher pressures, as an ideal solution.

PROBLEMS

14.1 Determine the change in the Gibbs function $\Delta G°$ at 25°C, in kJ/kmol, for the reaction

$$CH_4(g) + 2O_2 \rightleftarrows CO_2 + 2H_2O(g)$$

using

(a) Gibbs function of formation data.
(b) enthalpy of formation and absolute entropy data.

14.2 Calculate the equilibrium constant, expressed as $\log_{10}K$, for $CO_2 \rightleftarrows CO + \frac{1}{2}O_2$ at **(a)** 1000 K, **(b)** 3600°R. Compare with values from Table A-25.

14.3 Calculate the equilibrium constant, expressed as $\log_{10}K$, for the *water-gas* reaction $CO + H_2O(g) \rightleftarrows CO_2 + H_2$ at 25°C. Compare with the value from Table A-25.

14.4 Calculate the equilibrium constant, expressed as $\log_{10}K$, for $H_2O \rightleftarrows H_2 + \frac{1}{2}O_2$ at **(a)** 298 K, **(b)** 2000 K. Compare with values from Table A-25.

14.5 Using data from Table A-25, determine $\log_{10}K$ at 2500 K for

(a) $H_2O \rightleftarrows H_2 + \frac{1}{2}O_2$.
(b) $H_2 + \frac{1}{2}O_2 \rightleftarrows H_2O$.
(c) $2H_2O \rightleftarrows 2H_2 + O_2$.

14.6 An equilibrium mixture at 1 atm has a molar analysis of 86.53% CO_2, 8.98% CO, and 4.49% O_2. Determine $\log_{10}K$ for $CO_2 \rightleftarrows CO + \frac{1}{2}O_2$.

14.7 Determine the relationship between the ideal gas equilibrium constants K_1 and K_2 for the following two alternative ways of expressing the ammonia synthesis reaction:

1. $\frac{1}{2}N_2 + \frac{3}{2}H_2 \rightleftarrows NH_3$
2. $N_2 + 3H_2 \rightleftarrows 2NH_3$

14.8 Consider the reactions

1. $CO + H_2O \rightleftarrows H_2 + CO_2$
2. $2CO_2 \rightleftarrows 2CO + O_2$
3. $2H_2O \rightleftarrows 2H_2 + O_2$

Show that $K_1 = (K_3/K_2)^{1/2}$.

14.9 Consider the reactions

1. $CO_2 + H_2 \rightleftarrows CO + H_2O$
2. $CO_2 \rightleftarrows CO + \frac{1}{2}O_2$
3. $H_2O \rightleftarrows H_2 + \frac{1}{2}O_2$

(a) Show that $K_1 = K_2/K_3$.
(b) Evaluate $\log_{10}K_1$ at 298 K, 1 atm using the expression from part (a), together with $\log_{10}K$ data from Table A-25.
(c) Check the value for $\log_{10}K_1$ obtained in part (b) by applying Eq. 14.31 to reaction 1.

14.10 One kmol of N_2O_4 dissociates to form an equilibrium ideal gas mixture of N_2O_4 and NO_2 at 25°C, 2 atm. Determine the equilibrium composition. For $N_2O_4 \rightleftarrows 2NO_2$, $\Delta G° = 5400$ kJ/kmol at 25°C.

14.11 One kmol of CO_2 dissociates to form an equilibrium mixture of CO_2, CO, and O_2. Determine the equilibrium composition for a pressure of 1 atm if the temperature is 3000 K. Repeat for a pressure of 10 atm.

14.12 One lbmol of CO_2 dissociates to form an equilibrium mixture of CO_2, CO, and O_2 at 4710°F. Determine the equilibrium composition if the pressure is

(a) 1 atm.
(b) 10 atm.

14.13 One kmol of H_2O dissociates to form an equilibrium mixture of H_2O, H_2, and O_2. Determine the equilibrium composition for a pressure of 1 atm if the temperature is 3000 K. Repeat for a pressure of 10 atm.

14.14 One kmol of water dissociates to form an equilibrium mixture of H_2O, H_2, and O_2. Determine the equilibrium composition at 2727°C if the pressure is 1 atm. Repeat if 1 kmol of N_2 (inert) is also present.

14.15 Determine the extent to which dissociation occurs in the following cases. One lbmol of H_2O dissociates to form an equilibrium mixture of H_2O, H_2, and O_2 at 4740°F, 1.25 atm. One lbmol of CO_2 dissociates to form an equilibrium mixture of CO_2, CO, and O_2 at the same temperature and pressure.

14.16 One lbmol of carbon reacts with 2 lbmol of oxygen (O_2) to form an equilibrium mixture of CO_2, CO, and O_2 at 4940°F, 1 atm. Determine the equilibrium composition.

14.17 An equimolar mixture of CO and O_2 reacts to form an equilibrium mixture of CO_2, CO, and O_2 at 3000 K, 5 atm. Determine the composition of the equilibrium mixture. Will lowering the pressure while keeping the temperature fixed increase or decrease the amount of CO_2 present? Explain.

14.18 An equimolar mixture of CO and $H_2O(g)$ reacts to form an equilibrium mixture of CO_2, CO, H_2O, and H_2 at 1727°C, 1 atm.

(a) Will lowering the temperature increase or decrease the amount of H_2 present? Explain.
(b) Will decreasing the pressure while keeping the temperature constant increase or decrease the amount of H_2 present? Explain.

14.19 One kmol of N_2O_4 dissociates at 25°C, 1 atm to form an equilibrium ideal gas mixture of N_2O_4 and NO_2 in which the amount of N_2O_4 present is 0.8154 kmol. Determine the amount of N_2O_4 that would be present in an equilibrium mixture at 25°C, 0.5 atm.

14.20 A gaseous mixture with a molar analysis of 20% CO_2, 40% CO, and 40% O_2 enters a heat exchanger and is heated at constant pressure. An equilibrium mixture of CO_2, CO, and O_2 exits at 3000 K, 1.5 bar. Determine the molar analysis of the exiting mixture.

14.21 Two kmol of CO_2 dissociate to form an equilibrium mixture of CO_2, CO, and O_2 in which 1.8 kmol of CO_2 is present. Determine the temperature of the mixture, in K, if the pressure is 1 atm.

14.22 One kmol of $H_2O(g)$ dissociates to form an equilibrium mixture of $H_2O(g)$, H_2, and O_2 in which the amount of water vapor present is 0.95 kmol. Determine the temperature of the mixture, in K, if the pressure is 10 atm.

14.23 Determine the temperature, in K, at which 1 kmol of CO_2 dissociates to form an equilibrium mixture of CO_2, CO, and O_2 containing 0.9 kmol of CO_2 at a pressure of 2 atm.

14.24 One kmol of CO and 1 kmol of O_2 react to form an equilibrium mixture at 3000 K consisting of CO_2, CO, and O_2 in which 0.807 kmol of CO_2 is present. Determine the pressure, in atm.

14.25 Determine the temperature, in K, at which 9% of diatomic hydrogen (H_2) dissociates into monatomic hydrogen (H) at a pressure of 10 atm. For a greater percentage of H_2 at the same pressure, would the temperature be *higher* or *lower*? Explain.

14.26 A vessel initially containing 1 kmol of $H_2O(g)$ and x kmol of N_2 forms an equilibrium mixture at 3143°C and 1 atm consisting of $H_2O(g)$, H_2, O_2, and N_2 in which 0.5 kmol of $H_2O(g)$ is present. Determine x.

14.27 A vessel initially containing 1 kmol of CO and 4.76 kmol of dry air forms an equilibrium mixture of CO_2, CO, O_2, and N_2 at 3000 K, 1 atm. Determine the equilibrium composition.

14.28 A vessel initially containing 1 kmol of O_2, 2 kmol of N_2, and 1 kmol of Ar forms an equilibrium mixture of O_2, N_2, NO, and Ar at 2727°C, 1 atm. Determine the equilibrium composition.

14.29 One kmol of CO and 0.5 kmol of O_2 react to form a mixture at temperature T and pressure p consisting of CO_2, CO, and O_2. If 0.35 kmol of CO is present in an equilibrium mixture when the pressure is 1 atm, determine the amount of CO present in an equilibrium mixture at the same temperature if the pressure were 10 atm.

14.30 A vessel initially contains 1 kmol of H_2 and 4 kmol of N_2. An equilibrium mixture of H_2, H, and N_2 forms at 3000 K, 1 atm. Determine the equilibrium composition. If the pressure were increased while keeping the temperature fixed, would the amount of monatomic hydrogen in the equilibrium mixture increase or decrease? Explain.

14.31 Dry air enters a heat exchanger. An equilibrium mixture of N_2, O_2, and NO exits at 3882°F, 1 atm. Determine the mole fraction of NO in the exiting mixture. Will the amount of NO increase or decrease as temperature decreases at fixed pressure? Explain.

14.32 A mixture of 1 kmol CO and 0.5 kmol O_2 in a closed vessel, initially at 1 atm and 300 K, reacts to form an equilibrium mixture of CO_2, CO, and O_2 at 2500 K. Determine the final pressure, in atm.

14.33 Methane burns with 90% of theoretical air to form an equilibrium mixture of CO_2, CO, $H_2O(g)$, H_2, and N_2 at 1000 K, 1 atm. Determine the composition of the equilibrium mixture, per kmol of mixture.

14.34 Octane (C_8H_{18}) burns with air to form an equilibrium mixture of CO_2, H_2, CO, $H_2O(g)$, and N_2 at 1700 K, 1 atm. Determine the composition of the products, in kmol per kmol of fuel, for an equivalence ratio of 1.2.

14.35 Acetylene gas (C_2H_2) at 25°C, 1 atm enters a reactor operating at steady state and burns with 40% excess air entering at 25°C, 1 atm, 80% relative humidity. An equilibrium mixture of CO_2, H_2O, O_2, NO, and N_2 exits at 2200 K, 0.9 atm. Determine, per kmol of C_2H_2 entering, the composition of exiting mixture.

14.36 A vessel initially containing 2 lbmol of N_2 and 1 lbmol of O_2 forms an equilibrium mixture at 1 atm consisting of N_2, O_2, and NO. Plot the amount of NO formed versus temperature T for $3600 \le T \le 6300$°R.

14.37 One 1 kmol of CO_2 dissociates to form an equilibrium mixture of CO_2, CO, and O_2 at temperature T and pressure p.

(a) For $T = 3000$ K, plot the amount of CO present, in kmol, versus pressure for $1 \le p \le 10$ atm.
(b) For $p = 1$ atm, plot the amount of CO present, in kmol, versus temperature for $2000 \le T \le 3500$ K.

14.38 Carbon dioxide gas at 25°C, 5.1 atm enters a heat exchanger operating at steady state. An equilibrium mixture of CO_2, CO, and O_2 exits at 2527°C, 5 atm. Determine, per kmol of CO_2 entering,

(a) the composition of the exiting mixture.
(b) the heat transfer to the gas stream, in kJ.

Neglect kinetic and potential energy effects.

14.39 Steam at 220°F, 1.05 atm enters a heat exchanger operating at steady state. An equilibrium mixture of $H_2O(g)$, H_2, and O_2 exits at 4040°F, 1 atm. Determine, per kmol of steam entering,

(a) the composition of the exiting mixture.
(b) the heat transfer to the flowing stream, in Btu.

Neglect kinetic and potential energy effects.

14.40 Carbon at 25°C, 1 atm enters a reactor operating at steady state and burns with oxygen entering at 25°C, 1 atm. The entering streams have equal molar flow rates. An equilibrium mixture of CO_2, CO, and O_2 exits at 2727°C, 1 atm. Determine, per kmol of carbon,

(a) the composition of the exiting mixture.
(b) the heat transfer between the reactor and its surroundings, in kJ.

Neglect kinetic and potential energy effects.

14.41 An equimolar mixture of carbon monoxide and water vapor at 700°R, 1 atm enters a reactor operating at steady state. An equilibrium mixture of CO_2, CO, $H_2O(g)$, and H_2 exits at 2700°R, 1 atm. Determine the heat transfer between the reactor and its surroundings, in Btu per lbmol of CO entering. Neglect kinetic and potential energy effects.

14.42 Carbon dioxide (CO_2) and oxygen (O_2) in a 1 : 2 molar ratio enter a reactor operating at steady state in separate streams each at 25°C, 1 atm. An equilibrium mixture of CO_2, CO, and O_2 exits at 1 atm. If the mole fraction of CO in the exiting mixture is 0.1, determine the rate of heat transfer from the reactor, in kJ per kmol of CO_2 entering. Ignore kinetic and potential energy effects.

14.43 Methane gas at 25°C, 1 atm enters a reactor operating at steady state and burns with 80% of theoretical air entering at 25°C, 1 atm. An equilibrium mixture of CO_2, CO, $H_2O(g)$, H_2, and N_2 exits at 1427°C, 1 atm. Determine, per kmol of methane entering,

(a) the composition of the exiting mixture.
(b) the heat transfer between the reactor and its surroundings, in kJ.

Neglect kinetic and potential energy effects.

14.44 Gaseous propane (C_3H_8) at 25°C, 1 atm enters a reactor operating at steady state and burns with 80% of theoretical air entering separately at 25°C, 1 atm. An equilibrium mixture of CO_2, CO, $H_2O(g)$, H_2, and N_2 exits at 1227°C, 1 atm. Determine the heat transfer between the reactor and its surroundings, in kJ per kmol of propane flowing. Neglect kinetic and potential energy effects.

14.45 Gaseous propane (C_3H_8) at 77°F, 1 atm enters a reactor operating at steady state and burns with the theoretical amount of air entering separately at 77°F, 1 atm. An equilibrium mixture of CO_2, CO, $H_2O(g)$, O_2, and N_2 exits at 3140°F, 1 atm. Determine the heat transfer between the reactor and its surroundings, in Btu per lbmol of propane flowing. Neglect kinetic and potential energy effects.

14.46 One kmol of CO_2 initially at 25°C, 1 atm is heated at constant pressure until a final state is attained consisting of an equilibrium mixture of CO_2, CO, and O_2 in which the amount of CO_2 present is 0.422 kmol. Determine the heat transfer and the work, each in kJ.

14.47 Methane at 77°F, 1 atm enters an insulated reactor operating at steady state and burns with 90% of theoretical air entering separately at 77°F, 1 atm. The products exit at 1 atm as an equilibrium mixture of CO_2, CO, $H_2O(g)$, H_2, and N_2. Determine the temperature of the exiting products, in °R. Kinetic and potential energy effects are negligible.

14.48 Hydrogen gas (H_2) at 25°C, 1 atm enters an insulated reactor operating at steady state and reacts with 250% excess oxygen entering at 227°C, 1 atm. The products of combustion exit at 1 atm. Determine the temperature of the products, in K, if

(a) combustion is complete.
(b) an equilibrium mixture of H_2O, H_2, and O_2 exits.

Kinetic and potential energy effects are negligible.

14.49 Carbon monoxide at 77°F, 1 atm enters an insulated reactor and burns with 80% of theoretical air entering at 77°F, 1 atm. The products exit at 1 atm as an equilibrium mixture of CO_2, CO, O_2, and N_2. Determine the temperature of the exiting products, in °R. Kinetic and potential energy effects are negligible.

14.50 Carbon monoxide at 25°C, 1 atm enters an insulated reactor and burns with excess oxygen (O_2) entering at 25°C, 1 atm. The products exit at 2950 K, 1 atm as an equilibrium mixture of CO_2, CO, and O_2. Determine the percent excess oxygen. Kinetic and potential energy effects are negligible.

14.51 Carbon monoxide at 25°C, 1 atm enters an insulated reactor and burns with the theoretical amount of air entering at 25°C, 1 atm. The products exit at 1 atm as an equilibrium mixture of CO_2, CO, O_2, and N_2. Determine the temperature of the exiting mixture, in K. Kinetic and potential energy effects are negligible.

14.52 A gaseous mixture of carbon monoxide and the theoretical amount of air at 260°F, 1.5 atm enters an insulated reactor operating at steady state. An equilibrium mixture of CO_2, CO, O_2, and N_2 exits at 1.5 atm. Determine the temperature of the exiting mixture, in °R. Kinetic and potential energy effects are negligible.

14.53 Methane at 25°C, 1 atm enters an insulated reactor operating at steady state and burns with oxygen entering at 127°C, 1 atm. An equilibrium mixture of CO_2, CO, O_2, and $H_2O(g)$ exits at 3250 K, 1 atm. Determine the rate at which oxygen enters the reactor, in kmol per kmol of methane. Kinetic and potential energy effects are negligible.

14.54 Methane at 77°F, 1 atm enters an insulated reactor operating at steady state and burns with the theoretical amount of air entering at 77°F, 1 atm. An equilibrium mixture of CO_2, CO, O_2, $H_2O(g)$, and N_2 exits at 1 atm. Determine the temperature of the exiting products, in °R. Kinetic and potential energy effects are negligible.

14.55 A mixture consisting of 1 kmol of carbon monoxide (CO), 0.5 kmol of oxygen (O_2), and 1.88 kmol of nitrogen (N_2), initially at 227°C, 1 atm, reacts in a closed, rigid, insulated vessel, forming an equilibrium mixture of CO_2, CO, O_2, and N_2. Determine the final equilibrium pressure, in atm.

14.56 A mixture consisting of 1 kmol of CO and the theoretical amount of air, initially at 60°C, 1 atm, reacts in a closed, rigid, insulated vessel to form an equilibrium mixture. An analysis of the products shows that there are 0.808 kmol of CO_2, 0.192 kmol of CO, and 0.096 kmol of O_2 present. The temperature of the final mixture is measured as 2465°C. Check the feasibility of these data.

14.57 For each case of Problem 14.48, determine the rate of entropy production, in kJ/K per kmol of H_2 entering. What can be concluded about the possibility of achieving complete combustion?

14.58 For Problem 14.49, determine the rate of entropy production, in Btu/°R per lbmol of CO entering.

14.59 For Problem 14.51, determine the irreversibility rate, in kJ per kmol of CO entering, for $T_0 = 298$ K.

14.60 For Problem 14.54, determine the irreversibility rate, in Btu per lbmol of methane entering, for $T_0 = 537$°R.

14.61 Methane gas at 25°C, 1 atm enters an insulated reactor operating at steady state, where it burns with x times the theoretical amount of air ($x \geq 1$) entering at 25°C, 1 atm. An equilibrium mixture of CO_2, CO, O_2, and N_2 exits at 1 atm. Write a computer program to determine the temperature of the exiting equilibrium mixture as a function of x. Kinetic and potential energy effects are negligible.

14.62 Estimate the enthalpy of reaction at 2000 K, in kJ/kmol, for $CO_2 \rightleftarrows CO + \frac{1}{2}O_2$ using the van't Hoff equation and equilibrium constant data. Compare with the value obtained for the enthalpy of reaction with enthalpy data.

14.63 Estimate the enthalpy of reaction at 4500°R, in Btu/lbmol, for $CO_2 \rightleftarrows CO + \frac{1}{2}O_2$ using the van't Hoff equation and equilibrium constant data. Compare with the value obtained for the enthalpy of reaction with enthalpy data.

14.64 Estimate the equilibrium constant at 3000 K for $CO_2 \rightleftarrows CO + \frac{1}{2}O_2$ using the equilibrium constant at 2000 K from Table A-25, together with the van't Hoff equation and enthalpy data. Compare with the value for the equilibrium constant obtained from Table A-25.

14.65 Estimate the equilibrium constant at 3000 K for the reaction $H_2O \rightleftarrows H_2 + \frac{1}{2}O_2$ using the equilibrium constant at 2500 K from Table A-25, together with the van't Hoff equation and enthalpy data. Compare with the value for the equilibrium constant obtained from Table A-25.

14.66 At 25°C, $\log_{10}K = 8.9$ for $C + 2H_2 \rightleftarrows CH_4$. Assuming that the enthalpy of reaction does not vary much with temperature, estimate the value of $\log_{10}K$ at 500°C.

14.67 If the ionization-equilibrium constants for $Cs \rightleftarrows Cs^+ + e^-$ at 1600 and 2000 K are $K = 0.78$ and $K = 15.63$, respectively, estimate the enthalpy of ionization, in kJ/kmol, at 1800 K using the van't Hoff equation.

14.68 An equilibrium mixture at 2000 K, 1 atm consists of Cs, Cs^+, and e^-. Based on 1 kmol of Cs present initially, determine the percent ionization of cesium. At 2000 K, the ionization-equilibrium constant for $Cs \rightleftarrows Cs^+ + e^-$ is $K = 15.63$.

14.69 An equilibrium mixture at 18,000°R consists of Ar, Ar^+, and e^-. Based on 1 lbmol of neutral argon present initially, determine the percent ionization of argon for a pressure of

(a) 0.01 atm.
(b) 0.03 atm.
(c) 0.05 atm.

At 18,000°R, the ionization-equilibrium constant for $Ar \rightleftarrows Ar^+ + e^-$ is $K = 4.2 \times 10^{-4}$.

14.70 At 2000 K and pressure p, 1 kmol of Na ionizes to form an equilibrium mixture of Na, Na^+, and e^- in which the amount of Na present is 0.24 kmol. Determine p, in atm. At 2000 K, the ionization-equilibrium constant for $Na \rightleftarrows Na^+ + e^-$ is $K = 0.668$.

14.71 At 12,000 K and 6 atm, 1 kmol of N ionizes to form an equilibrium mixture of N, N^+, and e^- in which the amount of N present is 0.95 kmol. Determine the ionization-equilibrium constant at this temperature for $N \rightleftarrows N^+ + e^-$.

14.72 An equimolar mixture of carbon monoxide and water vapor enters a heat exchanger operating at steady state. An equilibrium mixture of CO, CO_2, O_2, H_2O(g), and H_2 exits at 2227°C, 1 atm. Determine the molar analysis of the exiting equilibrium mixture.

14.73 A closed vessel initially contains a gaseous mixture consisting of 3 lbmol of CO_2, 6 lbmol of CO, and 1 lbmol of H_2. An equilibrium mixture at 4220°F, 1 atm is formed containing CO_2, CO, H_2O, H_2, and O_2. Determine the composition of the equilibrium mixture.

14.74 Butane (C_4H_{10}) burns with 100% excess air to form an equilibrium mixture at 1400 K, 20 atm consisting of CO_2, O_2, H_2O(g), N_2, NO, and NO_2. Determine the balanced reaction equation. For $N_2 + 2O_2 \rightleftarrows 2NO_2$ at 1400 K, $K = 8.4 \times 10^{-10}$.

14.75 One lbmol of H_2O(g) dissociates to form an equilibrium mixture at 5000°R, 1 atm consisting of H_2O(g), H_2, O_2, and OH. Determine the equilibrium composition.

14.76 Steam enters a heat exchanger operating at steady state. An equilibrium mixture of H_2O, H_2, O_2, H, and OH exits at temperature T, 1 atm. Determine the molar analysis of the exiting equilibrium mixture for

(a) $T = 2800$ K.
(b) $T = 3000$ K.

14.77 For a two-phase liquid–vapor mixture of water at 100°C, use tabulated property data to show that the specific Gibbs functions of the saturated liquid and saturated vapor are equal. Repeat for a two-phase liquid–vapor mixture of Refrigerant 12 at 20°C.

14.78 Using the Clapeyron equation, solve the following problems from Chap. 11: **(a)** 11.32, **(b)** 11.33, **(c)** 11.34, **(d)** 11.35, **(e)** 11.40.

14.79 A closed system at 20°C, 1 atm consists of a pure liquid water phase in equilibrium with a vapor phase composed of water vapor and dry air. Determine the departure, in percent, of the partial pressure of the water vapor from the saturation pressure of pure water at 20°C.

14.80 An isolated system has two phases, denoted by A and B, each of which consists of the same two substances, denoted by 1 and 2. Show that necessary conditions for equilibrium are

1. the temperature of each phase is the same, $T_A = T_B$.
2. the pressure of each phase is the same, $p_A = p_B$.
3. the chemical potential of each component has the same value in each phase, $\mu_1^A = \mu_1^B$, $\mu_2^A = \mu_2^B$.

14.81 An isolated system has two phases, denoted by A and B, each of which consists of the same two substances, denoted by 1 and 2. The phases are separated by a freely moving, *thin* wall permeable only by substance 2. Determine the necessary conditions for equilibrium.

14.82 Referring to Problem 14.81, let each phase be a binary mixture of argon and helium and the wall be permeable only to argon. If the phases initially are at the conditions tabulated below, determine the final equilibrium temperature, pressure, and composition in the two phases.

	T(K)	p(MPa)	n(kmol)	y_{Ar}	y_{He}
Phase A	300	0.2	6	0.5	0.5
Phase B	400	0.1	5	0.8	0.2

14.83 Figure P14.83 shows an ideal gas mixture at temperature T and pressure p containing substance k, separated from a gas phase of pure k at temperature T and pressure p' by a semipermeable membrane that allows only k to pass through. Assuming the ideal gas model also applies to the pure gas phase, determine the relationship between p and p' for there to be no net transfer of k through the membrane.

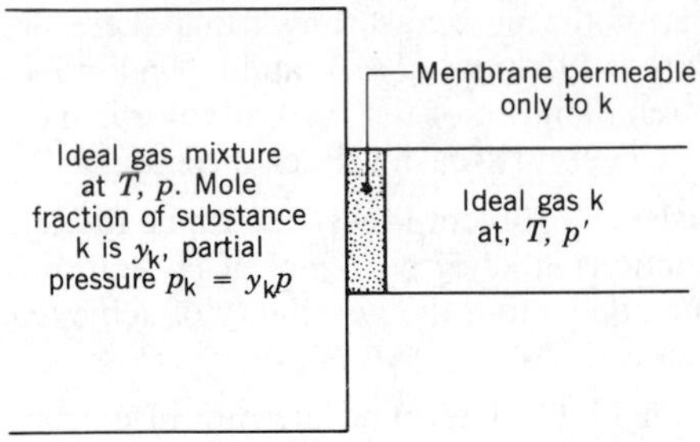

Figure P14.83

14.84 What is the maximum number of homogeneous phases that can exist at equilibrium for a system involving

(a) one component?
(b) two components?
(c) three components?

14.85 Determine the number of degrees of freedom for systems composed of

(a) ice, liquid water, and water vapor.
(b) liquid water and water vapor.
(c) water vapor only.
(d) water vapor and dry air.
(e) liquid water, water vapor, and dry air.
(f) ice, water vapor, and dry air.
(g) N_2 and O_2 at 20°C, 1 atm.
(h) a liquid phase and a vapor phase, each of which contains ammonia and water.
(i) liquid mercury, liquid water, and a vapor phase of mercury and water.

14.86 Develop the *phase rule* for chemically reacting systems.

DESIGN AND OPEN ENDED PROBLEMS

14.1D A reactor is being designed in which a mixture of carbon monoxide (CO) and steam enters and a gaseous mixture of H_2, CO_2, CO, and H_2O exits at a pressure of 1 atm and an elevated temperature. At the preliminary design stage, a plausible estimate is required of the molar analysis of the entering mixture that would correspond to an exiting mixture in which the hydrogen mole fraction is the greatest. Obtain the required estimate. Is this process involving the *water–gas reaction* a promising means for producing hydrogen as a fuel? Discuss.

14.2D The amount of sulfur dioxide (SO_2) present in *off gases* from industrial processes can be reduced by oxidizing the SO_2 to SO_3 at an elevated temperature in a catalytic reactor. The SO_3 can be reacted in turn with water to form sulfuric acid that has economic value. For an off gas at 1 atm having the molar analysis of 12% SO_2, 8% O_2, 80% N_2, estimate the range of temperatures at which a *substantial* conversion of SO_2 to SO_3 might be realized.

14.3D A gaseous mixture of hydrogen (H_2) and carbon monoxide (CO) enters a catalytic reactor and a gaseous mixture of methanol (CH_3OH), hydrogen, and carbon monoxide exits. At the preliminary process design stage, a plausible estimate is required of the inlet hydrogen mole fraction, y_{H_2}, the temperature of the exiting mixture, T_e, and the pressure of the exiting mixture, p_e, subject to the following four constraints: (1) $0.5 \leq y_{H_2} \leq 0.75$, (2) $300 \leq$

$T_e \leq 400$ K, (3) $1 \leq p_e \leq 10$ atm, and (4) the exiting mixture contains at least 75% methanol on a molar basis. Obtain this estimate.

14.4D A computer program developed by NASA is an example of computer programs available for the calculation of chemical equilibrium compositions for conditions typical in internal combustion engines.[2] Using the *NASA program* or a comparable code, develop plots giving the variation with equivalence ratio of the equilibrium products of octane–air mixtures at 30 atm and selected temperatures ranging from 1700 to 2800 K. Consider equivalence ratios in the interval from 0.2 to 1.4 and equilibrium products including, but not necessarily limited to, CO_2, CO, H_2O, O_2, O, H_2, N_2, NO, OH. Under what conditions is the formation of nitric oxide (NO) and carbon monoxide (CO) most significant? Discuss.

14.5D Spark-ignition engine exhaust gases contain several air pollutants including the oxides of nitrogen, NO and NO_2, collectively known as NOX. Additionally, the exhaust gases may contain carbon monoxide (CO) and unburned or partially burned hydrocarbons (HC). The amounts actually present depend on engine design and operating conditions, and typically differ significantly from values calculated on the basis of chemical equilibrium. Discuss both the reasons for these discrepancies and possible mechanisms by which such pollutants are formed in an actual engine.

14.6D For spark-ignition engines, the average production of pollutants upstream of the catalyst, in g per mile of vehicle travel, are nitric oxides (NOX), 1.5; hydrocarbons (HC), 2; carbon monoxide (CO), 20. For a city in your locale having a population of 100,000 or more, estimate the annual amount, in kg, of each pollutant that would be discharged if automobiles had no emission control devices. Repeat if the vehicles adhere to current U.S. government emissions standards.

14.7D A newspaper article states that a small gasoline engine powering a weed trimmer, lawn mower, chain saw, or snow blower typically emits as much as 50 times more pollution than the average truck engine. The article also states that a chain saw engine operated for 2 h emits as much unburned hydrocarbon as a new car driven 3,000 miles. Confirm or disprove these statements, providing supporting references and/or calculations as appropriate.

[2] R. A. Svehla and B. J. McBride, "Fortran IV Computer Program for Calculation of Thermodynamic and Transport Properties of Complex Chemical Systems," *NASA Technical Note TN D-7506*, NASA Lewis Research Center, Cleveland, OH, 1973.

INDEX TO TABLES IN SI UNITS

TABLE A-1 Atomic or Molecular Weights and Critical Properties of Some Common Elements and Compounds

Substance	Chemical formula	M	T_c, K	p_c, bars	$Z_c = \frac{p_c v_c}{RT_c}$
Acetylene	C_2H_2	26.04	309	62.8	0.274
Air (equivalent)	—	28.97	133	37.7	0.284
Ammonia	NH_3	17.04	406	112.8	0.242
Argon	Ar	39.94	151	48.6	0.290
Benzene	C_6H_6	78.11	563	49.3	0.274
Butane	C_4H_{10}	58.12	425	38.0	0.274
Carbon	C	12.01	—	—	—
Carbon dioxide	CO_2	44.01	304	73.9	0.276
Carbon monoxide	CO	28.01	133	35.0	0.294
Copper	Cu	63.54	—	—	—
Ethane	C_2H_6	30.07	305	48.8	0.285
Ethyl alcohol	C_2H_5OH	46.07	516	63.8	0.249
Ethylene	C_2H_4	28.05	283	51.2	0.270
Helium	He	4.003	5.2	2.3	0.300
Hydrogen	H_2	2.018	33.2	13.0	0.304
Methane	CH_4	16.04	191	46.4	0.290
Methyl alcohol	CH_3OH	32.05	513	79.5	0.220
Nitrogen	N_2	28.01	126	33.9	0.291
Octane	C_8H_{18}	114.22	569	24.9	0.258
Oxygen	O_2	32.00	154	50.5	0.290
Propane	C_3H_8	44.09	370	42.7	0.276
Propylene	C_3H_6	42.08	365	46.2	0.276
Refrigerant 12	CCl_2F_2	120.92	385	41.2	0.278
Refrigerant 134a	CF_3CH_2F	102.03	374	40.7	0.260
Sulfur dioxide	SO_2	64.06	431	78.7	0.268
Water	H_2O	18.02	647.3	220.9	0.233

Sources: Adapted from *International Critical Tables* and L. C. Nelson and E. F. Obert, Generalized Compressibility Charts, *Chem. Eng.*, *61:* 203 (1954).

TABLE A-2 Properties of Saturated Water (Liquid–Vapor): Temperature Table

		Specific Volume m^3/kg		Internal Energy kJ/kg		Enthalpy kJ/kg			Entropy kJ/kg · K		
Temp. °C	Press. bars	Sat. Liquid $v_f \times 10^3$	Sat. Vapor v_g	Sat. Liquid u_f	Sat. Vapor u_g	Sat. Liquid h_f	Evap. h_{fg}	Sat. Vapor h_g	Sat. Liquid s_f	Sat. Vapor s_g	Temp. °C
.01	0.00611	1.0002	206.136	0.00	2375.3	0.01	2501.3	2501.4	0.0000	9.1562	.01
4	0.00813	1.0001	157.232	16.77	2380.9	16.78	2491.9	2508.7	0.0610	9.0514	4
5	0.00872	1.0001	147.120	20.97	2382.3	20.98	2489.6	2510.6	0.0761	9.0257	5
6	0.00935	1.0001	137.734	25.19	2383.6	25.20	2487.2	2512.4	0.0912	9.0003	6
8	0.01072	1.0002	120.917	33.59	2386.4	33.60	2482.5	2516.1	0.1212	8.9501	8
10	0.01228	1.0004	106.379	42.00	2389.2	42.01	2477.7	2519.8	0.1510	8.9008	10
11	0.01312	1.0004	99.857	46.20	2390.5	46.20	2475.4	2521.6	0.1658	8.8765	11
12	0.01402	1.0005	93.784	50.41	2391.9	50.41	2473.0	2523.4	0.1806	8.8524	12
13	0.01497	1.0007	88.124	54.60	2393.3	54.60	2470.7	2525.3	0.1953	8.8285	13
14	0.01598	1.0008	82.848	58.79	2394.7	58.80	2468.3	2527.1	0.2099	8.8048	14
15	0.01705	1.0009	77.926	62.99	2396.1	62.99	2465.9	2528.9	0.2245	8.7814	15
16	0.01818	1.0011	73.333	67.18	2397.4	67.19	2463.6	2530.8	0.2390	8.7582	16
17	0.01938	1.0012	69.044	71.38	2398.8	71.38	2461.2	2532.6	0.2535	8.7351	17
18	0.02064	1.0014	65.038	75.57	2400.2	75.58	2458.8	2534.4	0.2679	8.7123	18
19	0.02198	1.0016	61.293	79.76	2401.6	79.77	2456.5	2536.2	0.2823	8.6897	19
20	0.02339	1.0018	57.791	83.95	2402.9	83.96	2454.1	2538.1	0.2966	8.6672	20
21	0.02487	1.0020	54.514	88.14	2404.3	88.14	2451.8	2539.9	0.3109	8.6450	21
22	0.02645	1.0022	51.447	92.32	2405.7	92.33	2449.4	2541.7	0.3251	8.6229	22
23	0.02810	1.0024	48.574	96.51	2407.0	96.52	2447.0	2543.5	0.3393	8.6011	23
24	0.02985	1.0027	45.883	100.70	2408.4	100.70	2444.7	2545.4	0.3534	8.5794	24
25	0.03169	1.0029	43.360	104.88	2409.8	104.89	2442.3	2547.2	0.3674	8.5580	25
26	0.03363	1.0032	40.994	109.06	2411.1	109.07	2439.9	2549.0	0.3814	8.5367	26
27	0.03567	1.0035	38.774	113.25	2412.5	113.25	2437.6	2550.8	0.3954	8.5156	27
28	0.03782	1.0037	36.690	117.42	2413.9	117.43	2435.2	2552.6	0.4093	8.4946	28
29	0.04008	1.0040	34.733	121.60	2415.2	121.61	2432.8	2554.5	0.4231	8.4739	29
30	0.04246	1.0043	32.894	125.78	2416.6	125.79	2430.5	2556.3	0.4369	8.4533	30
31	0.04496	1.0046	31.165	129.96	2418.0	129.97	2428.1	2558.1	0.4507	8.4329	31
32	0.04759	1.0050	29.540	134.14	2419.3	134.15	2425.7	2559.9	0.4644	8.4127	32
33	0.05034	1.0053	28.011	138.32	2420.7	138.33	2423.4	2561.7	0.4781	8.3927	33
34	0.05324	1.0056	26.571	142.50	2422.0	142.50	2421.0	2563.5	0.4917	8.3728	34
35	0.05628	1.0060	25.216	146.67	2423.4	146.68	2418.6	2565.3	0.5053	8.3531	35
36	0.05947	1.0063	23.940	150.85	2424.7	150.86	2416.2	2567.1	0.5188	8.3336	36
38	0.06632	1.0071	21.602	159.20	2427.4	159.21	2411.5	2570.7	0.5458	8.2950	38
40	0.07384	1.0078	19.523	167.56	2430.1	167.57	2406.7	2574.3	0.5725	8.2570	40
45	0.09593	1.0099	15.258	188.44	2436.8	188.45	2394.8	2583.2	0.6387	8.1648	45

TABLE A-2 *(Continued)*

		Specific Volume m³/kg		Internal Energy kJ/kg		Enthalpy kJ/kg			Entropy kJ/kg · K		
Temp. °C	Press. bars	Sat. Liquid $v_f \times 10^3$	Sat. Vapor v_g	Sat. Liquid u_f	Sat. Vapor u_g	Sat. Liquid h_f	Evap. h_{fg}	Sat. Vapor h_g	Sat. Liquid s_f	Sat. Vapor s_g	Temp. °C
50	.1235	1.0121	12.032	209.32	2443.5	209.33	2382.7	2592.1	.7038	8.0763	50
55	.1576	1.0146	9.568	230.21	2450.1	230.23	2370.7	2600.9	.7679	7.9913	55
60	.1994	1.0172	7.671	251.11	2456.6	251.13	2358.5	2609.6	.8312	7.9096	60
65	.2503	1.0199	6.197	272.02	2463.1	272.06	2346.2	2618.3	.8935	7.8310	65
70	.3119	1.0228	5.042	292.95	2469.6	292.98	2333.8	2626.8	.9549	7.7553	70
75	.3858	1.0259	4.131	313.90	2475.9	313.93	2321.4	2635.3	1.0155	7.6824	75
80	.4739	1.0291	3.407	334.86	2482.2	334.91	2308.8	2643.7	1.0753	7.6122	80
85	.5783	1.0325	2.828	355.84	2488.4	355.90	2296.0	2651.9	1.1343	7.5445	85
90	.7014	1.0360	2.361	376.85	2494.5	376.92	2283.2	2660.1	1.1925	7.4791	90
95	.8455	1.0397	1.982	397.88	2500.6	397.96	2270.2	2668.1	1.2500	7.4159	95
100	1.014	1.0435	1.673	418.94	2506.5	419.04	2257.0	2676.1	1.3069	7.3549	100
110	1.433	1.0516	1.210	461.14	2518.1	461.30	2230.2	2691.5	1.4185	7.2387	110
120	1.985	1.0603	0.8919	503.50	2529.3	503.71	2202.6	2706.3	1.5276	7.1296	120
130	2.701	1.0697	0.6685	546.02	2539.9	546.31	2174.2	2720.5	1.6344	7.0269	130
140	3.613	1.0797	0.5089	588.74	2550.0	589.13	2144.7	2733.9	1.7391	6.9299	140
150	4.758	1.0905	0.3928	631.68	2559.5	632.20	2114.3	2746.5	1.8418	6.8379	150
160	6.178	1.1020	0.3071	674.86	2568.4	675.55	2082.6	2758.1	1.9427	6.7502	160
170	7.917	1.1143	0.2428	718.33	2576.5	719.21	2049.5	2768.7	2.0419	6.6663	170
180	10.02	1.1274	0.1941	762.09	2583.7	763.22	2015.0	2778.2	2.1396	6.5857	180
190	12.54	1.1414	0.1565	806.19	2590.0	807.62	1978.8	2786.4	2.2359	6.5079	190
200	15.54	1.1565	0.1274	850.65	2595.3	852.45	1940.7	2793.2	2.3309	6.4323	200
210	19.06	1.1726	0.1044	895.53	2599.5	897.76	1900.7	2798.5	2.4248	6.3585	210
220	23.18	1.1900	0.08619	940.87	2602.4	943.62	1858.5	2802.1	2.5178	6.2861	220
230	27.95	1.2088	0.07158	986.74	2603.9	990.12	1813.8	2804.0	2.6099	6.2146	230
240	33.44	1.2291	0.05976	1033.2	2604.0	1037.3	1766.5	2803.8	2.7015	6.1437	240
250	39.73	1.2512	0.05013	1080.4	2602.4	1085.4	1716.2	2801.5	2.7927	6.0730	250
260	46.88	1.2755	0.04221	1128.4	2599.0	1134.4	1662.5	2796.6	2.8838	6.0019	260
270	54.99	1.3023	0.03564	1177.4	2593.7	1184.5	1605.2	2789.7	2.9751	5.9301	270
280	64.12	1.3321	0.03017	1227.5	2586.1	1236.0	1543.6	2779.6	3.0668	5.8571	280
290	74.36	1.3656	0.02557	1278.9	2576.0	1289.1	1477.1	2766.2	3.1594	5.7821	290
300	85.81	1.4036	0.02167	1332.0	2563.0	1344.0	1404.9	2749.0	3.2534	5.7045	300
320	112.7	1.4988	0.01549	1444.6	2525.5	1461.5	1238.6	2700.1	3.4480	5.5362	320
340	145.9	1.6379	0.01080	1570.3	2464.6	1594.2	1027.9	2622.0	3.6594	5.3357	340
360	186.5	1.8925	0.006945	1725.2	2351.5	1760.5	720.5	2481.0	3.9147	5.0526	360
374.14	220.9	3.155	0.003155	2029.6	2029.6	2099.3	0	2099.3	4.4298	4.4298	374.14

Source: Tables A-2 through A-5 are adapted from K. Wark, *Thermodynamics, 4th ed.*, McGraw-Hill, New York, 1983, as extracted from J. H. Keenan, F. G. Keyes, P. G. Hill, and J. G. Moore, *Steam Tables*, Wiley, New York, 1969.

TABLE A-3 Properties of Saturated Water (Liquid–Vapor): Pressure Table

		Specific Volume m^3/kg		Internal Energy kJ/kg		Enthalpy kJ/kg			Entropy kJ/kg · K		
Press. bars	**Temp. °C**	**Sat. Liquid** $v_f \times 10^3$	**Sat. Vapor** v_g	**Sat. Liquid** u_f	**Sat. Vapor** u_g	**Sat. Liquid** h_f	**Evap.** h_{fg}	**Sat. Vapor** h_g	**Sat. Liquid** s_f	**Sat. Vapor** s_g	**Press. bars**
0.04	28.96	1.0040	34.800	121.45	2415.2	121.46	2432.9	2554.4	0.4226	8.4746	0.04
0.06	36.16	1.0064	23.739	151.53	2425.0	151.53	2415.9	2567.4	0.5210	8.3304	0.06
0.08	41.51	1.0084	18.103	173.87	2432.2	173.88	2403.1	2577.0	0.5926	8.2287	0.08
0.10	45.81	1.0102	14.674	191.82	2437.9	191.83	2392.8	2584.7	0.6493	8.1502	0.10
0.20	60.06	1.0172	7.649	251.38	2456.7	251.40	2358.3	2609.7	0.8320	7.9085	0.20
0.30	69.10	1.0223	5.229	289.20	2468.4	289.23	2336.1	2625.3	0.9439	7.7686	0.30
0.40	75.87	1.0265	3.993	317.53	2477.0	317.58	2319.2	2636.8	1.0259	7.6700	0.40
0.50	81.33	1.0300	3.240	340.44	2483.9	340.49	2305.4	2645.9	1.0910	7.5939	0.50
0.60	85.94	1.0331	2.732	359.79	2489.6	359.86	2293.6	2653.5	1.1453	7.5320	0.60
0.70	89.95	1.0360	2.365	376.63	2494.5	376.70	2283.3	2660.0	1.1919	7.4797	0.70
0.80	93.50	1.0380	2.087	391.58	2498.8	391.66	2274.1	2665.8	1.2329	7.4346	0.80
0.90	96.71	1.0410	1.869	405.06	2502.6	405.15	2265.7	2670.9	1.2695	7.3949	0.90
1.00	99.63	1.0432	1.694	417.36	2506.1	417.46	2258.0	2675.5	1.3026	7.3594	1.00
1.50	111.4	1.0528	1.159	466.94	2519.7	467.11	2226.5	2693.6	1.4336	7.2233	1.50
2.00	120.2	1.0605	0.8857	504.49	2529.5	504.70	2201.9	2706.7	1.5301	7.1271	2.00
2.50	127.4	1.0672	0.7187	535.10	2537.2	535.37	2181.5	2716.9	1.6072	7.0527	2.50
3.00	133.6	1.0732	0.6058	561.15	2543.6	561.47	2163.8	2725.3	1.6718	6.9919	3.00
3.50	138.9	1.0786	0.5243	583.95	2546.9	584.33	2148.1	2732.4	1.7275	6.9405	3.50
4.00	143.6	1.0836	0.4625	604.31	2553.6	604.74	2133.8	2738.6	1.7766	6.8959	4.00
4.50	147.9	1.0882	0.4140	622.25	2557.6	623.25	2120.7	2743.9	1.8207	6.8565	4.50
5.00	151.9	1.0926	0.3749	639.68	2561.2	640.23	2108.5	2748.7	1.8607	6.8212	5.00
6.00	158.9	1.1006	0.3157	669.90	2567.4	670.56	2086.3	2756.8	1.9312	6.7600	6.00
7.00	165.0	1.1080	0.2729	696.44	2572.5	697.22	2066.3	2763.5	1.9922	6.7080	7.00
8.00	170.4	1.1148	0.2404	720.22	2576.8	721.11	2048.0	2769.1	2.0462	6.6628	8.00
9.00	175.4	1.1212	0.2150	741.83	2580.5	742.83	2031.1	2773.9	2.0946	6.6226	9.00
10.0	179.9	1.1273	0.1944	761.68	2583.6	762.81	2015.3	2778.1	2.1387	6.5863	10.0
15.0	198.3	1.1539	0.1318	843.16	2594.5	844.84	1947.3	2792.2	2.3150	6.4448	15.0
20.0	212.4	1.1767	0.09963	906.44	2600.3	908.79	1890.7	2799.5	2.4474	6.3409	20.0
25.0	224.0	1.1973	0.07998	959.11	2603.1	962.11	1841.0	2803.1	2.5547	6.2575	25.0
30.0	233.9	1.2165	0.06668	1004.8	2604.1	1008.4	1795.7	2804.2	2.6457	6.1869	30.0
35.0	242.6	1.2347	0.05707	1045.4	2603.7	1049.8	1753.7	2803.4	2.7253	6.1253	35.0
40.0	250.4	1.2522	0.04978	1082.3	2602.3	1087.3	1714.1	2801.4	2.7964	6.0701	40.0
45.0	257.5	1.2692	0.04406	1116.2	2600.1	1121.9	1676.4	2798.3	2.8610	6.0199	45.0
50.0	264.0	1.2859	0.03944	1147.8	2597.1	1154.2	1640.1	2794.3	2.9202	5.9734	50.0
60.0	275.6	1.3187	0.03244	1205.4	2589.7	1213.4	1571.0	2784.3	3.0267	5.8892	60.0
70.0	285.9	1.3513	0.02737	1257.6	2580.5	1267.0	1505.1	2772.1	3.1211	5.8133	70.0
80.0	295.1	1.3842	0.02352	1305.6	2569.8	1316.6	1441.3	2758.0	3.2068	5.7432	80.0
90.0	303.4	1.4178	0.02048	1350.5	2557.8	1363.3	1378.9	2742.1	3.2858	5.6772	90.0
100.	311.1	1.4524	0.01803	1393.0	2544.4	1407.6	1317.1	2724.7	3.3596	5.6141	100.
110.	318.2	1.4886	0.01599	1433.7	2529.8	1450.1	1255.5	2705.6	3.4295	5.5527	110.

H_2O

TABLE A-3 (*Continued*)

		Specific Volume m³/kg		Internal Energy kJ/kg		Enthalpy kJ/kg			Entropy kJ/kg · K		
Press. bars	Temp. °C	Sat. Liquid $v_f \times 10^3$	Sat. Vapor v_g	Sat. Liquid u_f	Sat. Vapor u_g	Sat. Liquid h_f	Evap. h_{fg}	Sat. Vapor h_g	Sat. Liquid s_f	Sat. Vapor s_g	Press. bars
120.	324.8	1.5267	0.01426	1473.0	2513.7	1491.3	1193.6	2684.9	3.4962	5.4924	120.
130.	330.9	1.5671	0.01278	1511.1	2496.1	1531.5	1130.7	2662.2	3.5606	5.4323	130.
140.	336.8	1.6107	0.01149	1548.6	2476.8	1571.1	1066.5	2637.6	3.6232	5.3717	140.
150.	342.2	1.6581	0.01034	1585.6	2455.5	1610.5	1000.0	2610.5	3.6848	5.3098	150.
160.	347.4	1.7107	0.009306	1622.7	2431.7	1650.1	930.6	2580.6	3.7461	5.2455	160.
170.	352.4	1.7702	0.008364	1660.2	2405.0	1690.3	856.9	2547.2	3.8079	5.1777	170.
180.	357.1	1.8397	0.007489	1698.9	2374.3	1732.0	777.1	2509.1	3.8715	5.1044	180.
190.	361.5	1.9243	0.006657	1739.9	2338.1	1776.5	688.0	2464.5	3.9388	5.0228	190.
200.	365.8	2.036	0.005834	1785.6	2293.0	1826.3	583.4	2409.7	4.0139	4.9269	200.
220.9	374.1	3.155	0.003155	2029.6	2029.6	2099.3	0	2099.3	4.4298	4.4298	220.9

TABLE A-4 Properties of Superheated Water Vapor

T °C	*v* m³/kg	*u* kJ/kg	*h* kJ/kg	*s* kJ/kg · K	*v* m³/kg	*u* kJ/kg	*h* kJ/kg	*s* kJ/kg · K
	p = 0.06 bar = 0.006 MPa (T_{sat} = 36.16°C)				**p = 0.35 bar = 0.035 MPa (T_{sat} = 72.69°C)**			
Sat.	23.739	2425.0	2567.4	8.3304	4.526	2473.0	2631.4	7.7158
80	27.132	2487.3	2650.1	8.5804	4.625	2483.7	2645.6	7.7564
120	30.219	2544.7	2726.0	8.7840	5.163	2542.4	2723.1	7.9644
160	33.302	2602.7	2802.5	8.9693	5.696	2601.2	2800.6	8.1519
200	36.383	2661.4	2879.7	9.1398	6.228	2660.4	2878.4	8.3237
240	39.462	2721.0	2957.8	9.2982	6.758	2720.3	2956.8	8.4828
280	42.540	2781.5	3036.8	9.4464	7.287	2780.9	3036.0	8.6314
320	45.618	2843.0	3116.7	9.5859	7.815	2842.5	3116.1	8.7712
360	48.696	2905.5	3197.7	9.7180	8.344	2905.1	3197.1	8.9034
400	51.774	2969.0	3279.6	9.8435	8.872	2968.6	3279.2	9.0291
440	54.851	3033.5	3362.6	9.9633	9.400	3033.2	3362.2	9.1490
500	59.467	3132.3	3489.1	10.1336	10.192	3132.1	3488.8	9.3194
	p = 0.70 bar = 0.07 MPa (T_{sat} = 89.95°C)				**p = 1.0 bar = 0.10 MPa (T_{sat} = 99.63°C)**			
Sat.	2.365	2494.5	2660.0	7.4797	1.694	2506.1	2675.5	7.3594
100	2.434	2509.7	2680.0	7.5341	1.696	2506.7	2676.2	7.3614
120	2.571	2539.7	2719.6	7.6375	1.793	2537.3	2716.6	7.4668
160	2.841	2599.4	2798.2	7.8279	1.984	2597.8	2796.2	7.6597
200	3.108	2659.1	2876.7	8.0012	2.172	2658.1	2875.3	7.8343
240	3.374	2719.3	2955.5	8.1611	2.359	2718.5	2954.5	7.9949
280	3.640	2780.2	3035.0	8.3162	2.546	2779.6	3034.2	8.1445
320	3.905	2842.0	3115.3	8.4504	2.732	2841.5	3114.6	8.2849
360	4.170	2904.6	3196.5	8.5828	2.917	2904.2	3195.9	8.4175
400	4.434	2968.2	3278.6	8.7086	3.103	2967.9	3278.2	8.5435
440	4.698	3032.9	3361.8	8.8286	3.288	3032.6	3361.4	8.6636
500	5.095	3131.8	3488.5	8.9991	3.565	3131.6	3488.1	8.8342
	p = 1.5 bars = 0.15 MPa (T_{sat} = 111.37°C)				**p = 3.0 bars = 0.30 MPa (T_{sat} = 133.55°C)**			
Sat.	1.159	2519.7	2693.6	7.2233	0.606	2543.6	2725.3	6.9919
120	1.188	2533.3	2711.4	7.2693				
160	1.317	2595.2	2792.8	7.4665	0.651	2587.1	2782.3	7.1276
200	1.444	2656.2	2872.9	7.6433	0.716	2650.7	2865.5	7.3115
240	1.570	2717.2	2952.7	7.8052	0.781	2713.1	2947.3	7.4774
280	1.695	2778.6	3032.8	7.9555	0.844	2775.4	3028.6	7.6299
320	1.819	2840.6	3113.5	8.0964	0.907	2838.1	3110.1	7.7722
360	1.943	2903.5	3195.0	8.2293	0.969	2901.4	3192.2	7.9061
400	2.067	2967.3	3277.4	8.3555	1.032	2965.6	3275.0	8.0330
440	2.191	3032.1	3360.7	8.4757	1.094	3030.6	3358.7	8.1538
500	2.376	3131.2	3487.6	8.6466	1.187	3130.0	3486.0	8.3251
600	2.685	3301.7	3704.3	8.9101	1.341	3300.8	3703.2	8.5892

H_2O

TABLE A-4 *(Continued)*

T °C	v m³/kg	u kJ/kg	h kJ/kg	s kJ/kg · K	v m³/kg	u kJ/kg	h kJ/kg	s kJ/kg · K
	p = 5.0 bars = 0.50 MPa (T_{sat} = 151.86°C)				**p = 7.0 bars = 0.70 MPa (T_{sat} = 164.97°C)**			
Sat.	0.3749	2561.2	2748.7	6.8213	0.2729	2572.5	2763.5	6.7080
180	0.4045	2609.7	2812.0	6.9656	0.2847	2599.8	2799.1	6.7880
200	0.4249	2642.9	2855.4	7.0592	0.2999	2634.8	2844.8	6.8865
240	0.4646	2707.6	2939.9	7.2307	0.3292	2701.8	2932.2	7.0641
280	0.5034	2771.2	3022.9	7.3865	0.3574	2766.9	3017.1	7.2233
320	0.5416	2834.7	3105.6	7.5308	0.3852	2831.3	3100.9	7.3697
360	0.5796	2898.7	3188.4	7.6660	0.4126	2895.8	3184.7	7.5063
400	0.6173	2963.2	3271.9	7.7938	0.4397	2960.9	3268.7	7.6350
440	0.6548	3028.6	3356.0	7.9152	0.4667	3026.6	3353.3	7.7571
500	0.7109	3128.4	3483.9	8.0873	0.5070	3126.8	3481.7	7.9299
600	0.8041	3299.6	3701.7	8.3522	0.5738	3298.5	3700.2	8.1956
700	0.8969	3477.5	3925.9	8.5952	0.6403	3476.6	3924.8	8.4391
	p = 10.0 bars = 1.0 MPa (T_{sat} = 179.91°C)				**p = 15.0 bars = 1.5 MPa (T_{sat} = 198.32°C)**			
Sat.	0.1944	2583.6	2778.1	6.5865	0.1318	2594.5	2792.2	6.4448
200	0.2060	2621.9	2827.9	6.6940	0.1325	2598.1	2796.8	6.4546
240	0.2275	2692.9	2920.4	6.8817	0.1483	2676.9	2899.3	6.6628
280	0.2480	2760.2	3008.2	7.0465	0.1627	2748.6	2992.7	6.8381
320	0.2678	2826.1	3093.9	7.1962	0.1765	2817.1	3081.9	6.9938
360	0.2873	2891.6	3178.9	7.3349	0.1899	2884.4	3169.2	7.1363
400	0.3066	2957.3	3263.9	7.4651	0.2030	2951.3	3255.8	7.2690
440	0.3257	3023.6	3349.3	7.5883	0.2160	3018.5	3342.5	7.3940
500	0.3541	3124.4	3478.5	7.7622	0.2352	3120.3	3473.1	7.5698
540	0.3729	3192.6	3565.6	7.8720	0.2478	3189.1	3560.9	7.6805
600	0.4011	3296.8	3697.9	8.0290	0.2668	3293.9	3694.0	7.8385
640	0.4198	3367.4	3787.2	8.1290	0.2793	3364.8	3783.8	7.9391
	p = 20.0 bars = 2.0 MPa (T_{sat} = 212.42°C)				**p = 30.0 bars = 3.0 MPa (T_{sat} = 233.90°C)**			
Sat.	0.0996	2600.3	2799.5	6.3409	0.0667	2604.1	2804.2	6.1869
240	0.1085	2659.6	2876.5	6.4952	0.0682	2619.7	2824.3	6.2265
280	0.1200	2736.4	2976.4	6.6828	0.0771	2709.9	2941.3	6.4462
320	0.1308	2807.9	3069.5	6.8452	0.0850	2788.4	3043.4	6.6245
360	0.1411	2877.0	3159.3	6.9917	0.0923	2861.7	3138.7	6.7801
400	0.1512	2945.2	3247.6	7.1271	0.0994	2932.8	3230.9	6.9212
440	0.1611	3013.4	3335.5	7.2540	0.1062	3002.9	3321.5	7.0520
500	0.1757	3116.2	3467.6	7.4317	0.1162	3108.0	3456.5	7.2338
540	0.1853	3185.6	3556.1	7.5434	0.1227	3178.4	3546.6	7.3474
600	0.1996	3290.9	3690.1	7.7024	0.1324	3285.0	3682.3	7.5085
640	0.2091	3362.2	3780.4	7.8035	0.1388	3357.0	3773.5	7.6106
700	0.2232	3470.9	3917.4	7.9487	0.1484	3466.5	3911.7	7.7571

TABLE A-4 (*Continued*)

T °C	*v* m³/kg	*u* kJ/kg	*h* kJ/kg	*s* kJ/kg · K	*v* m³/kg	*u* kJ/kg	*h* kJ/kg	*s* kJ/kg · K
	p = 40 bars = 4.0 MPa (T_{sat} = 250.4°C)				**p = 60 bars = 6.0 MPa (T_{sat} = 275.64°C)**			
Sat.	0.04978	2602.3	2801.4	6.0701	0.03244	2589.7	2784.3	5.8892
280	0.05546	2680.0	2901.8	6.2568	0.03317	2605.2	2804.2	5.9252
320	0.06199	2767.4	3015.4	6.4553	0.03876	2720.0	2952.6	6.1846
360	0.06788	2845.7	3117.2	6.6215	0.04331	2811.2	3071.1	6.3782
400	0.07341	2919.9	3213.6	6.7690	0.04739	2892.9	3177.2	6.5408
440	0.07872	2992.2	3307.1	6.9041	0.05122	2970.0	3277.3	6.6853
500	0.08643	3099.5	3445.3	7.0901	0.05665	3082.2	3422.2	6.8803
540	0.09145	3171.1	3536.9	7.2056	0.06015	3156.1	3517.0	6.9999
600	0.09885	3279.1	3674.4	7.3688	0.06525	3266.9	3658.4	7.1677
640	0.1037	3351.8	3766.6	7.4720	0.06859	3341.0	3752.6	7.2731
700	0.1110	3462.1	3905.9	7.6198	0.07352	3453.1	3894.1	7.4234
740	0.1157	3536.6	3999.6	7.7141	0.07677	3528.3	3989.2	7.5190

T °C	*v* m³/kg	*u* kJ/kg	*h* kJ/kg	*s* kJ/kg · K	*v* m³/kg	*u* kJ/kg	*h* kJ/kg	*s* kJ/kg · K
	p = 80 bars = 8.0 MPa (T_{sat} = 295.06°C)				**p = 100 bars = 10.0 MPa (T_{sat} = 311.06°C)**			
Sat.	0.02352	2569.8	2758.0	5.7432	0.01803	2544.4	2724.7	5.6141
320	0.02682	2662.7	2877.2	5.9489	0.01925	2588.8	2781.3	5.7103
360	0.03089	2772.7	3019.8	6.1819	0.02331	2729.1	2962.1	6.0060
400	0.03432	2863.8	3138.3	6.3634	0.02641	2832.4	3096.5	6.2120
440	0.03742	2946.7	3246.1	6.5190	0.02911	2922.1	3213.2	6.3805
480	0.04034	3025.7	3348.4	6.6586	0.03160	3005.4	3321.4	6.5282
520	0.04313	3102.7	3447.7	6.7871	0.03394	3085.6	3425.1	6.6622
560	0.04582	3178.7	3545.3	6.9072	0.03619	3164.1	3526.0	6.7864
600	0.04845	3254.4	3642.0	7.0206	0.03837	3241.7	3625.3	6.9029
640	0.05102	3330.1	3738.3	7.1283	0.04048	3318.9	3723.7	7.0131
700	0.05481	3443.9	3882.4	7.2812	0.04358	3434.7	3870.5	7.1687
740	0.05729	3520.4	3978.7	7.3782	0.04560	3512.1	3968.1	7.2670

T °C	*v* m³/kg	*u* kJ/kg	*h* kJ/kg	*s* kJ/kg · K	*v* m³/kg	*u* kJ/kg	*h* kJ/kg	*s* kJ/kg · K
	p = 120 bars = 12.0 MPa (T_{sat} = 324.75°C)				**p = 140 bars = 14.0 MPa (T_{sat} = 336.75°C)**			
Sat.	0.01426	2513.7	2684.9	5.4924	0.01149	2476.8	2637.6	5.3717
360	0.01811	2678.4	2895.7	5.8361	0.01422	2617.4	2816.5	5.6602
400	0.02108	2798.3	3051.3	6.0747	0.01722	2760.9	3001.9	5.9448
440	0.02355	2896.1	3178.7	6.2586	0.01954	2868.6	3142.2	6.1474
480	0.02576	2984.4	3293.5	6.4154	0.02157	2962.5	3264.5	6.3143
520	0.02781	3068.0	3401.8	6.5555	0.02343	3049.8	3377.8	6.4610
560	0.02977	3149.0	3506.2	6.6840	0.02517	3133.6	3486.0	6.5941
600	0.03164	3228.7	3608.3	6.8037	0.02683	3215.4	3591.1	6.7172
640	0.03345	3307.5	3709.0	6.9164	0.02843	3296.0	3694.1	6.8326
700	0.03610	3425.2	3858.4	7.0749	0.03075	3415.7	3846.2	6.9939
740	0.03781	3503.7	3957.4	7.1746	0.03225	3495.2	3946.7	7.0952

H_2O

TABLE A-4 (*Continued*)

T °C	v m³/kg	u kJ/kg	h kJ/kg	s kJ/kg · K	v m³/kg	u kJ/kg	h kJ/kg	s kJ/kg · K
	p = 160 bars = 16.0 MPa (T_{sat} = 347.44°C)				**p = 180 bars = 18.0 MPa (T_{sat} = 357.06°C)**			
Sat.	0.00931	2431.7	2580.6	5.2455	0.00749	2374.3	2509.1	5.1044
360	0.01105	2539.0	2715.8	5.4614	0.00809	2418.9	2564.5	5.1922
400	0.01426	2719.4	2947.6	5.8175	0.01190	2672.8	2887.0	5.6887
440	0.01652	2839.4	3103.7	6.0429	0.01414	2808.2	3062.8	5.9428
480	0.01842	2939.7	3234.4	6.2215	0.01596	2915.9	3203.2	6.1345
520	0.02013	3031.1	3353.3	6.3752	0.01757	3011.8	3378.0	6.2960
560	0.02172	3117.8	3465.4	6.5132	0.01904	3101.7	3444.4	6.4392
600	0.02323	3201.8	3573.5	6.6399	0.02042	3188.0	3555.6	6.5696
640	0.02467	3284.2	3678.9	6.7580	0.02174	3272.3	3663.6	6.6905
700	0.02674	3406.0	3833.9	6.9224	0.02362	3396.3	3821.5	6.8580
740	0.02808	3486.7	3935.9	7.0251	0.02483	3478.0	3925.0	6.9623

T °C	v m³/kg	u kJ/kg	h kJ/kg	s kJ/kg · K	v m³/kg	u kJ/kg	h kJ/kg	s kJ/kg · K
	p = 200 bars = 20.0 MPa (T_{sat} = 365.81°C)				**p = 240 bars = 24.0 MPa**			
Sat.	0.00583	2293.0	2409.7	4.9269				
400	0.00994	2619.3	2818.1	5.5540	0.00673	2477.8	2639.4	5.2393
440	0.01222	2774.9	3019.4	5.8450	0.00929	2700.6	2923.4	5.6506
480	0.01399	2891.2	3170.8	6.0518	0.01100	2838.3	3102.3	5.8950
520	0.01551	2992.0	3302.2	6.2218	0.01241	2950.5	3248.5	6.0842
560	0.01689	3085.2	3423.0	6.3705	0.01366	3051.1	3379.0	6.2448
600	0.01818	3174.0	3537.6	6.5048	0.01481	3145.2	3500.7	6.3875
640	0.01940	3260.2	3648.1	6.6286	0.01588	3235.5	3616.7	6.5174
700	0.02113	3386.4	3809.0	6.7993	0.01739	3366.4	3783.8	6.6947
740	0.02224	3469.3	3914.1	6.9052	0.01835	3451.7	3892.1	6.8038
800	0.02385	3592.7	4069.7	7.0544	0.01974	3578.0	4051.6	6.9567

T °C	v m³/kg	u kJ/kg	h kJ/kg	s kJ/kg · K	v m³/kg	u kJ/kg	h kJ/kg	s kJ/kg · K
	p = 280 bars = 28.0 MPa				**p = 320 bars = 32.0 MPa**			
400	0.00383	2223.5	2330.7	4.7494	0.00236	1980.4	2055.9	4.3239
440	0.00712	2613.2	2812.6	5.4494	0.00544	2509.0	2683.0	5.2327
480	0.00885	2780.8	3028.5	5.7446	0.00722	2718.1	2949.2	5.5968
520	0.01020	2906.8	3192.3	5.9566	0.00853	2860.7	3133.7	5.8357
560	0.01136	3015.7	3333.7	6.1307	0.00963	2979.0	3287.2	6.0246
600	0.01241	3115.6	3463.0	6.2823	0.01061	3085.3	3424.6	6.1858
640	0.01338	3210.3	3584.8	6.4187	0.01150	3184.5	3552.5	6.3290
700	0.01473	3346.1	3758.4	6.6029	0.01273	3325.4	3732.8	6.5203
740	0.01558	3433.9	3870.0	6.7153	0.01350	3415.9	3847.8	6.6361
800	0.01680	3563.1	4033.4	6.8720	0.01460	3548.0	4015.1	6.7966
900	0.01873	3774.3	4298.8	7.1084	0.01633	3762.7	4285.1	7.0372

TABLE A-5 Properties of Compressed Liquid Water

T °C	$v \times 10^3$ m³/kg	u kJ/kg	h kJ/kg	s kJ/kg · K	$v \times 10^3$ m³/kg	u kJ/kg	h kJ/kg	s kJ/kg · K
	p = 25 bars = 2.5 MPa (T_{sat} = 223.99°C)				p = 50 bars = 5.0 MPa (T_{sat} = 263.99°C)			
20	1.0006	83.80	86.30	.2961	.9995	83.65	88.65	.2956
40	1.0067	167.25	169.77	.5715	1.0056	166.95	171.97	.5705
80	1.0280	334.29	336.86	1.0737	1.0268	333.72	338.85	1.0720
100	1.0423	418.24	420.85	1.3050	1.0410	417.52	422.72	1.3030
140	1.0784	587.82	590.52	1.7369	1.0768	586.76	592.15	1.7343
180	1.1261	761.16	763.97	2.1375	1.1240	759.63	765.25	2.1341
200	1.1555	849.9	852.8	2.3294	1.1530	848.1	853.9	2.3255
220	1.1898	940.7	943.7	2.5174	1.1866	938.4	944.4	2.5128
Sat.	1.1973	959.1	962.1	2.5546	1.2859	1147.8	1154.2	2.9202
	p = 75 bars = 7.5 MPa (T_{sat} = 290.59°C)				p = 100 bars = 10.0 MPa (T_{sat} = 311.06°C)			
20	.9984	83.50	90.99	.2950	.9972	83.36	93.33	.2945
40	1.0045	166.64	174.18	.5696	1.0034	166.35	176.38	.5686
80	1.0256	333.15	340.84	1.0704	1.0245	332.59	342.83	1.0688
100	1.0397	416.81	424.62	1.3011	1.0385	416.12	426.50	1.2992
140	1.0752	585.72	593.78	1.7317	1.0737	584.68	595.42	1.7292
180	1.1219	758.13	766.55	2.1308	1.1199	756.65	767.84	2.1275
220	1.1835	936.2	945.1	2.5083	1.1805	934.1	945.9	2.5039
260	1.2696	1124.4	1134.0	2.8763	1.2645	1121.1	1133.7	2.8699
Sat.	1.3677	1282.0	1292.2	3.1649	1.4524	1393.0	1407.6	3.3596
	p = 150 bars = 15.0 MPa (T_{sat} = 342.24°C)				p = 200 bars = 20.0 MPa (T_{sat} = 365.81°C)			
20	.9950	83.06	97.99	.2934	.9928	82.77	102.62	.2923
40	1.0013	165.76	180.78	.5666	.9992	165.17	185.16	.5646
80	1.0222	331.48	346.81	1.0656	1.0199	330.40	350.80	1.0624
100	1.0361	414.74	430.28	1.2955	1.0337	413.39	434.06	1.2917
140	1.0707	582.66	598.72	1.7242	1.0678	580.69	602.04	1.7193
180	1.1159	753.76	770.50	2.1210	1.1120	750.95	773.20	2.1147
220	1.1748	929.9	947.5	2.4953	1.1693	925.9	949.3	2.4870
260	1.2550	1114.6	1133.4	2.8576	1.2462	1108.6	1133.5	2.8459
300	1.3770	1316.6	1337.3	3.2260	1.3596	1306.1	1333.3	3.2071
Sat.	1.6581	1585.6	1610.5	3.6848	2.036	1785.6	1826.3	4.0139
	p = 250 bars = 25 MPa				p = 300 bars = 30 MPa			
20	.9907	82.47	107.24	.2911	.9886	82.17	111.84	.2899
40	.9971	164.60	189.52	.5626	.9951	164.04	193.89	.5607
100	1.0313	412.08	437.85	1.2881	1.0290	410.78	441.66	1.2844
200	1.1344	834.5	862.8	2.2961	1.1302	831.4	865.3	2.2893
300	1.3442	1296.6	1330.2	3.1900	1.3304	1287.9	1327.8	3.1741

TABLE A-6 Properties of Saturated Water (Solid–Vapor): Temperature Table

Temp. °C	Pressure kPa	Specific Volume m^3/kg		Internal Energy kJ/kg			Enthalpy kJ/kg			Entropy kJ/kg · K		
		Sat. Solid $v_i \times 10^3$	Sat. Vapor v_g	Sat. Solid u_i	Subl. u_{ig}	Sat. Vapor u_g	Sat. Solid h_i	Subl. h_{ig}	Sat. Vapor h_g	Sat. Solid s_i	Subl. s_{ig}	Sat. Vapor s_g
.01	.6113	1.0908	206.1	−333.40	2708.7	2375.3	−333.40	2834.8	2501.4	−1.221	10.378	9.156
0	.6108	1.0908	206.3	−333.43	2708.8	2375.3	−333.43	2834.8	2501.3	−1.221	10.378	9.157
−2	.5176	1.0904	241.7	−337.62	2710.2	2372.6	−337.62	2835.3	2497.7	−1.237	10.456	9.219
−4	.4375	1.0901	283.8	−341.78	2711.6	2369.8	−341.78	2835.7	2494.0	−1.253	10.536	9.283
−6	.3689	1.0898	334.2	−345.91	2712.9	2367.0	−345.91	2836.2	2490.3	−1.268	10.616	9.348
−8	.3102	1.0894	394.4	−350.02	2714.2	2364.2	−350.02	2836.6	2486.6	−1.284	10.698	9.414
−10	.2602	1.0891	466.7	−354.09	2715.5	2361.4	−354.09	2837.0	2482.9	−1.299	10.781	9.481
−12	.2176	1.0888	553.7	−358.14	2716.8	2358.7	−358.14	2837.3	2479.2	−1.315	10.865	9.550
−14	.1815	1.0884	658.8	−362.15	2718.0	2355.9	−362.15	2837.6	2475.5	−1.331	10.950	9.619
−16	.1510	1.0881	786.0	−366.14	2719.2	2353.1	−366.14	2837.9	2471.8	−1.346	11.036	9.690
−18	.1252	1.0878	940.5	−370.10	2720.4	2350.3	−370.10	2838.2	2468.1	−1.362	11.123	9.762
−20	.1035	1.0874	1128.6	−374.03	2721.6	2347.5	−374.03	2838.4	2464.3	−1.377	11.212	9.835
−22	.0853	1.0871	1358.4	−377.93	2722.7	2344.7	−377.93	2838.6	2460.6	−1.393	11.302	9.909
−24	.0701	1.0868	1640.1	−381.80	2723.7	2342.0	−381.80	2838.7	2456.9	−1.408	11.394	9.985
−26	.0574	1.0864	1986.4	−385.64	2724.8	2339.2	−385.64	2838.9	2453.2	−1.424	11.486	10.062
−28	.0469	1.0861	2413.7	−389.45	2725.8	2336.4	−389.45	2839.0	2449.5	−1.439	11.580	10.141
−30	.0381	1.0858	2943.	−393.23	2726.8	2333.6	−393.23	2839.0	2445.8	−1.455	11.676	10.221
−32	.0309	1.0854	3600.	−396.98	2727.8	2330.8	−396.98	2839.1	2442.1	−1.471	11.773	10.303
−34	.0250	1.0851	4419.	−400.71	2728.7	2328.0	−400.71	2839.1	2438.4	−1.486	11.872	10.386
−36	.0201	1.0848	5444.	−404.40	2729.6	2325.2	−404.40	2839.1	2434.7	−1.501	11.972	10.470
−38	.0161	1.0844	6731.	−408.06	2730.5	2322.4	−408.06	2839.0	2430.9	−1.517	12.073	10.556
−40	.0129	1.0841	8354.	−411.70	2731.3	2319.6	−411.70	2838.9	2427.2	−1.532	12.176	10.644

Source: J. H. Keenan, F. G. Keyes, P. G. Hill, and J. G. Moore, *Steam Tables,* Wiley, New York, 1978.

TABLE A-7 Properties of Saturated Refrigerant 12 (Liquid–Vapor): Temperature Table

		Specific Volume m^3/kg		Internal Energy kJ/kg		Enthalpy kJ/kg			Entropy kJ/kg · K		
Temp. °C	Press. bars	Sat. Liquid $v_f \times 10^3$	Sat. Vapor v_g	Sat. Liquid u_f	Sat. Vapor u_g	Sat. Liquid h_f	Evap. h_{fg}	Sat. Vapor h_g	Sat. Liquid s_f	Sat. Vapor s_g	Temp. °C
−40	0.6417	0.6595	0.24191	−0.04	154.07	0.00	169.59	169.59	0.0000	0.7274	−40
−35	0.8071	0.6656	0.19540	4.37	156.13	4.42	167.48	171.90	0.0187	0.7219	−35
−30	1.0041	0.6720	0.15938	8.79	158.20	8.86	165.33	174.20	0.0371	0.7170	−30
−28	1.0927	0.6746	0.14728	10.58	159.02	10.65	164.46	175.11	0.0444	0.7153	−28
−26	1.1872	0.6773	0.13628	12.35	159.84	12.43	163.59	176.02	0.0517	0.7135	−26
−25	1.2368	0.6786	0.13117	13.25	160.26	13.33	163.15	176.48	0.0552	0.7126	−25
−24	1.2880	0.6800	0.12628	14.13	160.67	14.22	162.71	176.93	0.0589	0.7119	−24
−22	1.3953	0.6827	0.11717	15.92	161.48	16.02	161.82	177.83	0.0660	0.7103	−22
−20	1.5093	0.6855	0.10885	17.72	162.31	17.82	160.92	178.74	0.0731	0.7087	−20
−18	1.6304	0.6883	0.10124	19.51	163.12	19.62	160.01	179.63	0.0802	0.7073	−18
−15	1.8260	0.6926	0.09102	22.20	164.35	22.33	158.64	180.97	0.0906	0.7051	−15
−10	2.1912	0.7000	0.07665	26.72	166.39	26.87	156.31	183.19	0.1080	0.7019	−10
−5	2.6096	0.7078	0.06496	31.27	168.42	31.45	153.93	185.37	0.1251	0.6991	−5
0	3.0861	0.7159	0.05539	35.83	170.44	36.05	151.48	187.53	0.1420	0.6965	0
4	3.5124	0.7227	0.04895	39.51	172.04	39.76	149.47	189.23	0.1553	0.6946	4
8	3.9815	0.7297	0.04340	43.21	173.63	43.50	147.41	190.91	0.1686	0.6929	8
12	4.4962	0.7370	0.03860	46.93	175.20	47.26	145.30	192.56	0.1817	0.6913	12
16	5.0591	0.7446	0.03442	50.67	176.78	51.05	143.14	194.19	0.1948	0.6898	16
20	5.6729	0.7525	0.03078	54.44	178.32	54.87	140.91	195.78	0.2078	0.6884	20
24	6.3405	0.7607	0.02759	58.25	179.85	58.73	138.61	197.34	0.2207	0.6871	24
26	6.6954	0.7650	0.02614	60.17	180.61	60.68	137.44	198.11	0.2271	0.6865	26
28	7.0648	0.7694	0.02478	62.09	181.36	62.63	136.24	198.87	0.2335	0.6859	28
30	7.4490	0.7739	0.02351	64.01	182.11	64.59	135.03	199.62	0.2400	0.6853	30
32	7.8485	0.7785	0.02231	65.96	182.85	66.57	133.79	200.36	0.2463	0.6847	32
34	8.2636	0.7832	0.02118	67.90	183.59	68.55	132.53	201.09	0.2527	0.6842	34
36	8.6948	0.7880	0.02012	69.86	184.31	70.55	131.25	201.80	0.2591	0.6836	36
38	9.1423	0.7929	0.01912	71.84	185.03	72.56	129.94	202.51	0.2655	0.6831	38
40	9.6065	0.7980	0.01817	73.82	185.74	74.59	128.61	203.20	0.2718	0.6825	40
42	10.088	0.8033	0.01728	75.82	186.45	76.63	127.25	203.88	0.2782	0.6820	42
44	10.587	0.8086	0.01644	77.82	187.13	78.68	125.87	204.54	0.2845	0.6814	44
48	11.639	0.8199	0.01488	81.88	188.51	82.83	123.00	205.83	0.2973	0.6802	48
52	12.766	0.8318	0.01349	86.00	189.83	87.06	119.99	207.05	0.3101	0.6791	52
56	13.972	0.8445	0.01224	90.18	191.10	91.36	116.84	208.20	0.3229	0.6779	56
60	15.259	0.8581	0.01111	94.43	192.31	95.74	113.52	209.26	0.3358	0.6765	60
112	41.155	1.792	0.00179	175.98	175.98	183.35	0	183.35	0.5687	0.5687	112

Source: Tables A-7 through A-9 are adapted from K. Wark, *Thermodynamics, 4th ed.*, McGraw-Hill, New York, 1983, as based on data from Freon Products Division, E. I. du Pont de Nemours & Company, 1969.

R-12

TABLE A-8 Properties of Saturated Refrigerant 12 (Liquid–Vapor): Pressure Table

		Specific Volume m^3/kg		Internal Energy kJ/kg		Enthalpy kJ/kg			Entropy kJ/kg · K		
Press. bars	Temp. °C	Sat. Liquid $v_f \times 10^3$	Sat. Vapor v_g	Sat. Liquid u_f	Sat. Vapor u_g	Sat. Liquid h_f	Evap. h_{fg}	Sat. Vapor h_g	Sat. Liquid s_f	Sat. Vapor s_g	Press. bars
0.6	−41.42	0.6578	0.2575	−1.29	153.49	−1.25	170.19	168.94	−0.0054	0.7290	0.6
1.0	−30.10	0.6719	0.1600	8.71	158.15	8.78	165.37	174.15	0.0368	0.7171	1.0
1.2	−25.74	0.6776	0.1349	12.58	159.95	12.66	163.48	176.14	0.0526	0.7133	1.2
1.4	−21.91	0.6828	0.1168	15.99	161.52	16.09	161.78	177.87	0.0663	0.7102	1.4
1.6	−18.49	0.6876	0.1031	19.07	162.91	19.18	160.23	179.41	0.0784	0.7076	1.6
1.8	−15.38	0.6921	0.09225	21.86	164.19	21.98	158.82	180.80	0.0893	0.7054	1.8
2.0	−12.53	0.6962	0.08354	24.43	165.36	24.57	157.50	182.07	0.0992	0.7035	2.0
2.4	−7.42	0.7040	0.07033	29.06	167.44	29.23	155.09	184.32	0.1168	0.7004	2.4
2.8	−2.93	0.7111	0.06076	33.15	169.26	33.35	152.92	186.27	0.1321	0.6980	2.8
3.2	1.11	0.7177	0.05351	36.85	170.88	37.08	150.92	188.00	0.1457	0.6960	3.2
4.0	8.15	0.7299	0.04321	43.35	173.69	43.64	147.33	190.97	0.1691	0.6928	4.0
5.0	15.60	0.7438	0.03482	50.30	176.61	50.67	143.35	194.02	0.1935	0.6899	5.0
6.0	22.00	0.7566	0.02913	56.35	179.09	56.80	139.77	196.57	0.2142	0.6878	6.0
7.0	27.65	0.7686	0.02501	61.75	181.23	62.29	136.45	198.74	0.2324	0.6860	7.0
8.0	32.74	0.7802	0.02188	66.68	183.13	67.30	133.33	200.63	0.2487	0.6845	8.0
9.0	37.37	0.7914	0.01942	71.22	184.81	71.93	130.36	202.29	0.2634	0.6832	9.0
10.0	41.64	0.8023	0.01744	75.46	186.32	76.26	127.50	203.76	0.2770	0.6820	10.0
12.0	49.31	0.8237	0.01441	83.22	188.95	84.21	122.03	206.24	0.3015	0.6799	12.0
14.0	56.09	0.8448	0.01222	90.28	191.11	91.46	116.76	208.22	0.3232	0.6778	14.0
16.0	62.19	0.8660	0.01054	96.80	192.95	98.19	111.62	209.81	0.3329	0.6758	16.0

TABLE A-9 Properties of Superheated Refrigerant 12

T °C	v m³/kg	u kJ/kg	h kJ/kg	s kJ/kg · K	v m³/kg	u kJ/kg	h kJ/kg	s kJ/kg · K
	p = 0.6 bar = 0.06 MPa (T_{sat} = −41.42°C)				**p = 1.0 bar = 0.10 MPa (T_{sat} = −30.10°C)**			
Sat.	0.2575	153.49	168.94	0.7290	0.1600	158.15	174.15	0.7171
−40	0.2593	154.16	169.72	0.7324				
−20	0.2838	163.91	180.94	0.7785	0.1677	163.22	179.99	0.7406
0	0.3079	174.05	192.52	0.8225	0.1827	173.50	191.77	0.7854
10	0.3198	179.26	198.45	0.8439	0.1900	178.77	197.77	0.8070
20	0.3317	184.57	204.47	0.8647	0.1973	184.12	203.85	0.8281
30	0.3435	189.96	210.57	0.8852	0.2045	189.57	210.02	0.8488
40	0.3552	195.46	216.77	0.9053	0.2117	195.09	216.26	0.8691
50	0.3670	201.02	223.04	0.9251	0.2188	200.70	222.58	0.8889
60	0.3787	206.69	229.41	0.9444	0.2260	206.38	228.98	0.9084
80	0.4020	218.25	242.37	0.9822	0.2401	218.00	242.01	0.9464

T °C	v m³/kg	u kJ/kg	h kJ/kg	s kJ/kg · K	v m³/kg	u kJ/kg	h kJ/kg	s kJ/kg · K
	p = 1.4 bars = 0.14 MPa (T_{sat} = −21.91°C)				**p = 1.8 bars = 0.18 MPa (T_{sat} = −15.38°C)**			
Sat.	0.1168	161.52	177.87	0.7102	0.0922	164.20	180.80	0.7054
−20	0.1179	162.50	179.01	0.7147				
−10	0.1235	167.69	184.97	0.7378	0.0925	164.39	181.03	0.7181
0	0.1289	172.94	190.99	0.7602	0.0991	172.37	190.21	0.7408
10	0.1343	178.28	197.08	0.7821	0.1034	177.77	196.38	0.7630
20	0.3197	183.67	203.23	0.8035	0.1076	183.23	202.60	0.7846
30	0.1449	189.17	209.46	0.8243	0.1118	188.77	208.89	0.8057
40	0.1502	194.72	215.75	0.8447	0.1160	194.35	215.23	0.8263
50	0.1553	200.38	222.12	0.8648	0.1201	200.02	221.64	0.8464
60	0.1605	206.08	228.55	0.8844	0.1241	205.78	228.12	0.8662
80	0.1707	217.74	241.64	0.9225	0.1322	217.47	241.27	0.9045
100	0.1809	229.67	255.00	0.9593	0.1402	229.45	254.69	0.9414

T °C	v m³/kg	u kJ/kg	h kJ/kg	s kJ/kg · K	v m³/kg	u kJ/kg	h kJ/kg	s kJ/kg · K
	p = 2.0 bars = 0.20 MPa (T_{sat} = −12.53°C)				**p = 2.4 bars = 0.24 MPa (T_{sat} = −7.42°C)**			
Sat.	0.0835	165.37	182.07	0.7035	0.0703	167.45	184.32	0.7004
0	0.0886	172.08	189.08	0.7325	0.0729	171.49	188.99	0.7177
10	0.0926	177.50	196.02	0.7548	0.0763	176.98	195.29	0.7404
20	0.0964	183.00	202.28	0.7766	0.0796	182.53	201.63	0.7624
30	0.1002	188.56	208.60	0.7978	0.0828	188.14	208.01	0.7838
40	0.1040	194.17	214.97	0.8184	0.0860	193.80	214.44	0.8047
50	0.1077	199.86	221.40	0.8387	0.0892	199.51	220.92	0.8251
60	0.1114	205.62	227.90	0.8585	0.0923	205.31	227.46	0.8450
80	0.1187	217.35	241.09	0.8969	0.0985	217.07	240.71	0.8836
100	0.1259	229.35	254.53	0.9339	0.1045	229.12	254.20	0.9208
120	0.1331	241.59	268.21	0.9696	0.1105	241.41	267.93	0.9566

R-12

TABLE A-9 (*Continued*)

T °C	v m³/kg	u kJ/kg	h kJ/kg	s kJ/kg · K	v m³/kg	u kJ/kg	h kJ/kg	s kJ/kg · K
	p = 2.8 bars = 0.28 MPa (T_{sat} = −2.93°C)				**p = 3.2 bars = 0.32 MPa (T_{sat} = 1.11°C)**			
Sat.	0.06076	169.26	186.27	0.6980	0.05351	170.88	188.00	0.6960
0	0.06166	170.89	188.15	0.7049				
10	0.06464	176.45	194.55	0.7279	0.05590	175.90	193.79	0.7167
20	0.06755	182.06	200.97	0.7502	0.05852	181.57	200.30	0.7393
30	0.07040	187.71	207.42	0.7718	0.06106	187.28	206.82	0.7612
40	0.07319	193.42	213.91	0.7928	0.06355	193.02	213.36	0.7824
50	0.07594	199.18	220.44	0.8134	0.06600	198.82	219.94	0.8031
60	0.07865	205.00	227.02	0.8334	0.06841	204.68	226.57	0.8233
80	0.08399	216.82	240.34	0.8722	0.07314	216.55	239.96	0.8623
100	0.08924	228.29	253.88	0.9095	0.07778	228.66	253.55	0.8997
120	0.09443	241.21	267.65	0.9455	0.08236	241.00	267.36	0.9358
	p = 4.0 bars = 0.40 MPa (T_{sat} = 8.15°C)				**p = 5.0 bars = 0.50 MPa (T_{sat} = 15.60°C)**			
Sat.	0.04321	173.69	190.97	0.6928	0.03482	176.61	194.02	0.6899
10	0.04363	174.76	192.21	0.6972				
20	0.04584	180.57	198.91	0.7204	0.03565	179.26	197.08	0.7004
30	0.04797	186.39	205.58	0.7428	0.03746	185.23	203.96	0.7235
40	0.05005	192.23	212.25	0.7645	0.03922	191.20	210.81	0.7457
50	0.05207	198.11	218.94	0.7855	0.04091	197.19	217.64	0.7672
60	0.05406	204.03	225.65	0.8060	0.04257	203.20	224.48	0.7881
80	0.05791	216.03	239.19	0.8454	0.04578	215.32	238.21	0.8281
100	0.06173	228.20	252.89	0.8831	0.04889	227.61	252.05	0.8662
120	0.06546	240.61	266.79	0.9194	0.05193	240.10	266.06	0.9028
140	0.06913	253.23	280.88	0.9544	0.05492	252.77	280.23	0.9379
	p = 6.0 bars = 0.60 MPa (T_{sat} = 22.00°C)				**p = 7.0 bars = 0.70 MPa (T_{sat} = 27.65°C)**			
Sat.	0.02913	179.09	196.57	0.6878	0.02501	181.23	198.74	0.6860
30	0.03042	184.01	202.26	0.7068	0.02535	182.72	200.46	0.6917
40	0.03197	190.13	209.31	0.7297	0.02676	189.00	207.73	0.7153
50	0.03345	196.23	216.30	0.7516	0.02810	195.23	214.90	0.7378
60	0.03489	202.34	223.27	0.7729	0.02939	201.45	222.02	0.7595
80	0.03765	214.61	237.20	0.8135	0.03184	213.88	236.17	0.8008
100	0.04032	227.01	251.20	0.8520	0.03419	226.40	250.33	0.8398
120	0.04291	239.57	265.32	0.8889	0.03646	239.05	264.57	0.8769
140	0.04545	252.31	279.58	0.9243	0.03867	251.85	278.92	0.9125
160	0.04794	265.25	294.01	0.9584	0.04085	264.83	293.42	0.9468

TABLE A-9 (*Continued*)

p = 8.0 bars = 0.80 MPa (T_{sat} = 32.74°C) | p = 9.0 bars = 0.90 MPa (T_{sat} = 37.37°C)

T °C	v m³/kg	u kJ/kg	h kJ/kg	s kJ/kg · K	v m³/kg	u kJ/kg	h kJ/kg	s kJ/kg · K
Sat.	0.02188	183.13	200.63	0.6845	0.01942	184.81	202.29	0.6832
40	0.02283	187.81	206.07	0.7021	0.01974	186.55	204.32	0.6897
50	0.02407	194.19	213.45	0.7253	0.02091	193.10	211.92	0.7136
60	0.02525	200.52	220.72	0.7474	0.02201	199.56	219.37	0.7363
80	0.02748	213.13	235.11	0.7894	0.02407	212.37	234.03	0.7790
100	0.02959	225.77	249.44	0.8289	0.02601	225.13	248.54	0.8190
120	0.03162	238.51	263.81	0.8664	0.02785	237.97	263.03	0.8569
140	0.03359	251.39	278.26	0.9022	0.02964	250.90	277.58	0.8930
160	0.03552	264.41	292.83	0.9367	0.03138	263.99	292.23	0.9276
180	0.03742	277.60	307.54	0.9699	0.03309	277.23	307.01	0.9609

p = 10.0 bars = 1.00 MPa (T_{sat} = 41.64°C) | p = 12.0 bars = 1.20 MPa (T_{sat} = 49.31°C)

T °C	v m³/kg	u kJ/kg	h kJ/kg	s kJ/kg · K	v m³/kg	u kJ/kg	h kJ/kg	s kJ/kg · K
Sat.	0.01744	186.32	203.76	0.6820	0.01441	188.95	206.24	0.6799
50	0.01837	191.95	210.32	0.7026	0.01448	189.43	206.81	0.6816
60	0.01941	198.56	217.97	0.7259	0.01546	196.41	214.96	0.7065
80	0.02134	211.57	232.91	0.7695	0.01722	209.91	230.57	0.7520
100	0.02313	224.48	247.61	0.8100	0.01881	223.13	245.70	0.7937
120	0.02484	237.41	262.25	0.8482	0.02030	236.27	260.63	0.8326
140	0.02647	250.43	276.90	0.8845	0.02172	249.45	275.51	0.8696
160	0.02807	263.56	291.63	0.9193	0.02309	263.70	290.41	0.9048
180	0.02963	276.84	306.47	0.9528	0.02443	276.05	305.37	0.9385
200	0.03116	290.26	321.42	0.9851	0.02574	289.55	320.44	0.9711

p = 14.0 bars = 1.40 MPa (T_{sat} = 56.09°C) | p = 16.0 bars = 1.60 MPa (T_{sat} = 62.19°C)

T °C	v m³/kg	u kJ/kg	h kJ/kg	s kJ/kg · K	v m³/kg	u kJ/kg	h kJ/kg	s kJ/kg · K
Sat.	0.01222	191.11	208.22	0.6778	0.01054	192.95	209.81	0.6758
60	0.01258	194.00	211.61	0.6881				
80	0.01425	208.11	228.06	0.7360	0.01198	206.17	225.34	0.7209
100	0.01571	221.70	243.69	0.7791	0.01337	220.19	241.58	0.7656
120	0.01705	235.09	258.96	0.8189	0.01461	233.84	257.22	0.8065
140	0.01832	248.43	274.08	0.8564	0.01577	247.38	272.61	0.8447
160	0.01954	261.80	289.16	0.8921	0.01686	260.90	287.88	0.8808
180	0.02071	275.27	304.26	0.9262	0.01792	274.47	303.14	0.9152
200	0.02186	288.84	319.44	0.9589	0.01895	288.11	318.43	0.9482
220	0.02299	302.51	334.70	0.9905	0.01996	301.84	333.78	0.9800

R-134a

TABLE A-10 Properties of Saturated Refrigerant 134a (Liquid–Vapor): Temperature Table

Temp. °C	Press. bars	Specific Volume m³/kg Sat. Liquid $v_f \times 10^3$	Specific Volume m³/kg Sat. Vapor v_g	Internal Energy kJ/kg Sat. Liquid u_f	Internal Energy kJ/kg Sat. Vapor u_g	Enthalpy kJ/kg Sat. Liquid h_f	Enthalpy kJ/kg Evap. h_{fg}	Enthalpy kJ/kg Sat. Vapor h_g	Entropy kJ/kg · K Sat. Liquid s_f	Entropy kJ/kg · K Sat. Vapor s_g	Temp. °C
−40	0.5164	0.7055	0.3569	−0.04	204.45	0.00	222.88	222.88	0.0000	0.9560	−40
−36	0.6332	0.7113	0.2947	4.68	206.73	4.73	220.67	225.40	0.0201	0.9506	−36
−32	0.7704	0.7172	0.2451	9.47	209.01	9.52	218.37	227.90	0.0401	0.9456	−32
−28	0.9305	0.7233	0.2052	14.31	211.29	14.37	216.01	230.38	0.0600	0.9411	−28
−26	1.0199	0.7265	0.1882	16.75	212.43	16.82	214.80	231.62	0.0699	0.9390	−26
−24	1.1160	0.7296	0.1728	19.21	213.57	19.29	213.57	232.85	0.0798	0.9370	−24
−22	1.2192	0.7328	0.1590	21.68	214.70	21.77	212.32	234.08	0.0897	0.9351	−22
−20	1.3299	0.7361	0.1464	24.17	215.84	24.26	211.05	235.31	0.0996	0.9332	−20
−18	1.4483	0.7395	0.1350	26.67	216.97	26.77	209.76	236.53	0.1094	0.9315	−18
−16	1.5748	0.7428	0.1247	29.18	218.10	29.30	208.45	237.74	0.1192	0.9298	−16
−12	1.8540	0.7498	0.1068	34.25	220.36	34.39	205.77	240.15	0.1388	0.9267	−12
−8	2.1704	0.7569	0.0919	39.38	222.60	39.54	203.00	242.54	0.1583	0.9239	−8
−4	2.5274	0.7644	0.0794	44.56	224.84	44.75	200.15	244.90	0.1777	0.9213	−4
0	2.9282	0.7721	0.0689	49.79	227.06	50.02	197.21	247.23	0.1970	0.9190	0
4	3.3765	0.7801	0.0600	55.08	229.27	55.35	194.19	249.53	0.2162	0.9169	4
8	3.8756	0.7884	0.0525	60.43	231.46	60.73	191.07	251.80	0.2354	0.9150	8
12	4.4294	0.7971	0.0460	65.83	233.63	66.18	187.85	254.03	0.2545	0.9132	12
16	5.0416	0.8062	0.0405	71.29	235.78	71.69	184.52	256.22	0.2735	0.9116	16
20	5.7160	0.8157	0.0358	76.80	237.91	77.26	181.09	258.36	0.2924	0.9102	20
24	6.4566	0.8257	0.0317	82.37	240.01	82.90	177.55	260.45	0.3113	0.9089	24
26	6.8530	0.8309	0.0298	85.18	241.05	85.75	175.73	261.48	0.3208	0.9082	26
28	7.2675	0.8362	0.0281	88.00	242.08	88.61	173.89	262.50	0.3302	0.9076	28
30	7.7006	0.8417	0.0265	90.84	243.10	91.49	172.00	263.50	0.3396	0.9070	30
32	8.1528	0.8473	0.0250	93.70	244.12	94.39	170.09	264.48	0.3490	0.9064	32
34	8.6247	0.8530	0.0236	96.58	245.12	97.31	168.14	265.45	0.3584	0.9058	34
36	9.1168	0.8590	0.0223	99.47	246.11	100.25	166.15	266.40	0.3678	0.9053	36
38	9.6298	0.8651	0.0210	102.38	247.09	103.21	164.12	267.33	0.3772	0.9047	38
40	10.164	0.8714	0.0199	105.30	248.06	106.19	162.05	268.24	0.3866	0.9041	40
42	10.720	0.8780	0.0188	108.25	249.02	109.19	159.94	269.14	0.3960	0.9035	42
44	11.299	0.8847	0.0177	111.22	249.96	112.22	157.79	270.01	0.4054	0.9030	44
48	12.526	0.8989	0.0159	117.22	251.79	118.35	153.33	271.68	0.4243	0.9017	48
52	13.851	0.9142	0.0142	123.31	253.55	124.58	148.66	273.24	0.4432	0.9004	52
56	15.278	0.9308	0.0127	129.51	255.23	130.93	143.75	274.68	0.4622	0.8990	56
60	16.813	0.9488	0.0114	135.82	256.81	137.42	138.57	275.99	0.4814	0.8973	60
70	21.162	1.0027	0.0086	152.22	260.15	154.34	124.08	278.43	0.5302	0.8918	70
80	26.324	1.0766	0.0064	169.88	262.14	172.71	106.41	279.12	0.5814	0.8827	80
90	32.435	1.1949	0.0046	189.82	261.34	193.69	82.63	276.32	0.6380	0.8655	90
100	39.742	1.5443	0.0027	218.60	248.49	224.74	34.40	259.13	0.7196	0.8117	100

Source: Tables A-10 through A-12 are calculated based on equations from D. P. Wilson and R. S. Basu, "Thermodynamic Properties of a New Stratospherically Safe Working Fluid—Refrigerant 134a," *ASHRAE Trans.*, Vol. 94, Pt. 2, 1988, pp. 2095–2118.

TABLE A-11 Properties of Saturated Refrigerant 134a (Liquid–Vapor): Pressure Table

		Specific Volume m^3/kg		Internal Energy kJ/kg		Enthalpy kJ/kg			Entropy kJ/kg · K		
Press. bars	Temp. °C	Sat. Liquid $v_f \times 10^3$	Sat. Vapor v_g	Sat. Liquid u_f	Sat. Vapor u_g	Sat. Liquid h_f	Evap. h_{fg}	Sat. Vapor h_g	Sat. Liquid s_f	Sat. Vapor s_g	Press. bars
0.6	−37.07	0.7097	0.3100	3.41	206.12	3.46	221.27	224.72	0.0147	0.9520	0.6
0.8	−31.21	0.7184	0.2366	10.41	209.46	10.47	217.92	228.39	0.0440	0.9447	0.8
1.0	−26.43	0.7258	0.1917	16.22	212.18	16.29	215.06	231.35	0.0678	0.9395	1.0
1.2	−22.36	0.7323	0.1614	21.23	214.50	21.32	212.54	233.86	0.0879	0.9354	1.2
1.4	−18.80	0.7381	0.1395	25.66	216.52	25.77	210.27	236.04	0.1055	0.9322	1.4
1.6	−15.62	0.7435	0.1229	29.66	218.32	29.78	208.19	237.97	0.1211	0.9295	1.6
1.8	−12.73	0.7485	0.1098	33.31	219.94	33.45	206.26	239.71	0.1352	0.9273	1.8
2.0	−10.09	0.7532	0.0993	36.69	221.43	36.84	204.46	241.30	0.1481	0.9253	2.0
2.4	−5.37	0.7618	0.0834	42.77	224.07	42.95	201.14	244.09	0.1710	0.9222	2.4
2.8	−1.23	0.7697	0.0719	48.18	226.38	48.39	198.13	246.52	0.1911	0.9197	2.8
3.2	2.48	0.7770	0.0632	53.06	228.43	53.31	195.35	248.66	0.2089	0.9177	3.2
3.6	5.84	0.7839	0.0564	57.54	230.28	57.82	192.76	250.58	0.2251	0.9160	3.6
4.0	8.93	0.7904	0.0509	61.69	231.97	62.00	190.32	252.32	0.2399	0.9145	4.0
5.0	15.74	0.8056	0.0409	70.93	235.64	71.33	184.74	256.07	0.2723	0.9117	5.0
6.0	21.58	0.8196	0.0341	78.99	238.74	79.48	179.71	259.19	0.2999	0.9097	6.0
7.0	26.72	0.8328	0.0292	86.19	241.42	86.78	175.07	261.85	0.3242	0.9080	7.0
8.0	31.33	0.8454	0.0255	92.75	243.78	93.42	170.73	264.15	0.3459	0.9066	8.0
9.0	35.53	0.8576	0.0226	98.79	245.88	99.56	166.62	266.18	0.3656	0.9054	9.0
10.0	39.39	0.8695	0.0202	104.42	247.77	105.29	162.68	267.97	0.3838	0.9043	10.0
12.0	46.32	0.8928	0.0166	114.69	251.03	115.76	155.23	270.99	0.4164	0.9023	12.0
14.0	52.43	0.9159	0.0140	123.98	253.74	125.26	148.14	273.40	0.4453	0.9003	14.0
16.0	57.92	0.9392	0.0121	132.52	256.00	134.02	141.31	275.33	0.4714	0.8982	16.0
18.0	62.91	0.9631	0.0105	140.49	257.88	142.22	134.60	276.83	0.4954	0.8959	18.0
20.0	67.49	0.9878	0.0093	148.02	259.41	149.99	127.95	277.94	0.5178	0.8934	20.0
25.0	77.59	1.0562	0.0069	165.48	261.84	168.12	111.06	279.17	0.5687	0.8854	25.0
30.0	86.22	1.1416	0.0053	181.88	262.16	185.30	92.71	278.01	0.6156	0.8735	30.0

TABLE A-12 Properties of Superheated Refrigerant 134a

T °C	v m³/kg	u kJ/kg	h kJ/kg	s kJ/kg · K	v m³/kg	u kJ/kg	h kJ/kg	s kJ/kg · K
	p = 0.6 bars = 0.06 MPa (T_{sat} = −37.07°C)				**p = 1.0 bars = 0.10 MPa (T_{sat} = −26.43°C)**			
Sat.	0.31003	206.12	224.72	0.9520	0.19170	212.18	231.35	0.9395
−20	0.33536	217.86	237.98	1.0062	0.19770	216.77	236.54	0.9602
−10	0.34992	224.97	245.96	1.0371	0.20686	224.01	244.70	0.9918
0	0.36433	232.24	254.10	1.0675	0.21587	231.41	252.99	1.0227
10	0.37861	239.69	262.41	1.0973	0.22473	238.96	261.43	1.0531
20	0.39279	247.32	270.89	1.1267	0.23349	246.67	270.02	1.0829
30	0.40688	255.12	279.53	1.1557	0.24216	254.54	278.76	1.1122
40	0.42091	263.10	288.35	1.1844	0.25076	262.58	287.66	1.1411
50	0.43487	271.25	297.34	1.2126	0.25930	270.79	296.72	1.1696
60	0.44879	279.58	306.51	1.2405	0.26779	279.16	305.94	1.1977
70	0.46266	288.08	315.84	1.2681	0.27623	287.70	315.32	1.2254
80	0.47650	296.75	325.34	1.2954	0.28464	296.40	324.87	1.2528
90	0.49031	305.58	335.00	1.3224	0.29302	305.27	334.57	1.2799
	p = 1.4 bars = 0.14 MPa (T_{sat} = −18.80°C)				**p = 1.8 bars = 0.18 MPa (T_{sat} = −12.73°C)**			
Sat.	0.13945	216.52	236.04	0.9322	0.10983	219.94	239.71	0.9273
−10	0.14549	223.03	243.40	0.9606	0.11135	222.02	242.06	0.9362
0	0.15219	230.55	251.86	0.9922	0.11678	229.67	250.69	0.9684
10	0.15875	238.21	260.43	1.0230	0.12207	237.44	259.41	0.9998
20	0.16520	246.01	269.13	1.0532	0.12723	245.33	268.23	1.0304
30	0.17155	253.96	277.97	1.0828	0.13230	253.36	277.17	1.0604
40	0.17783	262.06	286.96	1.1120	0.13730	261.53	286.24	1.0898
50	0.18404	270.32	296.09	1.1407	0.14222	269.85	295.45	1.1187
60	0.19020	278.74	305.37	1.1690	0.14710	278.31	304.79	1.1472
70	0.19633	287.32	314.80	1.1969	0.15193	286.93	314.28	1.1753
80	0.20241	296.06	324.39	1.2244	0.15672	295.71	323.92	1.2030
90	0.20846	304.95	334.14	1.2516	0.16148	304.63	333.70	1.2303
100	0.21449	314.01	344.04	1.2785	0.16622	313.72	343.63	1.2573
	p = 2.0 bars = 0.20 MPa (T_{sat} = −10.09°C)				**p = 2.4 bars = 0.24 MPa (T_{sat} = −5.37°C)**			
Sat.	0.09933	221.43	241.30	0.9253	0.08343	224.07	244.09	0.9222
−10	0.09938	221.50	241.38	0.9256				
0	0.10438	229.23	250.10	0.9582	0.08574	228.31	248.89	0.9399
10	0.10922	237.05	258.89	0.9898	0.08993	236.26	257.84	0.9721
20	0.11394	244.99	267.78	1.0206	0.09399	244.30	266.85	1.0034
30	0.11856	253.06	276.77	1.0508	0.09794	252.45	275.95	1.0339
40	0.12311	261.26	285.88	1.0804	0.10181	260.72	285.16	1.0637
50	0.12758	269.61	295.12	1.1094	0.10562	269.12	294.47	1.0930
60	0.13201	278.10	304.50	1.1380	0.10937	277.67	303.91	1.1218
70	0.13639	286.74	314.02	1.1661	0.11307	286.35	313.49	1.1501
80	0.14073	295.53	323.68	1.1939	0.11674	295.18	323.19	1.1780
90	0.14504	304.47	333.48	1.2212	0.12037	304.15	333.04	1.2055
100	0.14932	313.57	343.43	1.2483	0.12398	313.27	343.03	1.2326

TABLE A-12 *(Continued)*

T °C	v m^3/kg	u kJ/kg	h kJ/kg	s kJ/kg · K	v m^3/kg	u kJ/kg	h kJ/kg	s kJ/kg · K
	p = 2.8 bars = 0.28 MPa (T_{sat} = −1.23°C)				**p = 3.2 bars = 0.32 MPa (T_{sat} = 2.48°C)**			
Sat.	0.07193	226.38	246.52	0.9197	0.06322	228.43	248.66	0.9177
0	0.07240	227.37	247.64	0.9238				
10	0.07613	235.44	256.76	0.9566	0.06576	234.61	255.65	0.9427
20	0.07972	243.59	265.91	0.9883	0.06901	242.87	264.95	0.9749
30	0.08320	251.83	275.12	1.0192	0.07214	251.19	274.28	1.0062
40	0.08660	260.17	284.42	1.0494	0.07518	259.61	283.67	1.0367
50	0.08992	268.64	293.81	1.0789	0.07815	268.14	293.15	1.0665
60	0.09319	277.23	303.32	1.1079	0.08106	276.79	302.72	1.0957
70	0.09641	285.96	312.95	1.1364	0.08392	285.56	312.41	1.1243
80	0.09960	294.82	322.71	1.1644	0.08674	294.46	322.22	1.1525
90	0.10275	303.83	332.60	1.1920	0.08953	303.50	332.15	1.1802
100	0.10587	312.98	342.62	1.2193	0.09229	312.68	342.21	1.2076
110	0.10897	322.27	352.78	1.2461	0.09503	322.00	352.40	1.2345
120	0.11205	331.71	363.08	1.2727	0.09774	331.45	362.73	1.2611

T °C	v m^3/kg	u kJ/kg	h kJ/kg	s kJ/kg · K	v m^3/kg	u kJ/kg	h kJ/kg	s kJ/kg · K
	p = 4.0 bars = 0.40 MPa (T_{sat} = 8.93°C)				**p = 5.0 bars = 0.50 MPa (T_{sat} = 15.74°C)**			
Sat.	0.05089	231.97	252.32	0.9145	0.04086	235.64	256.07	0.9117
10	0.05119	232.87	253.35	0.9182				
20	0.05397	241.37	262.96	0.9515	0.04188	239.40	260.34	0.9264
30	0.05662	249.89	272.54	0.9837	0.04416	248.20	270.28	0.9597
40	0.05917	258.47	282.14	1.0148	0.04633	256.99	280.16	0.9918
50	0.06164	267.13	291.79	1.0452	0.04842	265.83	290.04	1.0229
60	0.06405	275.89	301.51	1.0748	0.05043	274.73	299.95	1.0531
70	0.06641	284.75	311.32	1.1038	0.05240	283.72	309.92	1.0825
80	0.06873	293.73	321.23	1.1322	0.05432	292.80	319.96	1.1114
90	0.07102	302.84	331.25	1.1602	0.05620	302.00	330.10	1.1397
100	0.07327	312.07	341.38	1.1878	0.05805	311.31	340.33	1.1675
110	0.07550	321.44	351.64	1.2149	0.05988	320.74	350.68	1.1949
120	0.07771	330.94	362.03	1.2417	0.06168	330.30	361.14	1.2218
130	0.07991	340.58	372.54	1.2681	0.06347	339.98	371.72	1.2484
140	0.08208	350.35	383.18	1.2941	0.06524	349.79	382.42	1.2746

R-134a

TABLE A-12 (*Continued*)

T °C	v m³/kg	u kJ/kg	h kJ/kg	s kJ/kg · K	v m³/kg	u kJ/kg	h kJ/kg	s kJ/kg · K
	p = 6.0 bars = 0.60 MPa (T_{sat} = 21.58°C)				p = 7.0 bars = 0.70 MPa (T_{sat} = 26.72°C)			
Sat.	0.03408	238.74	259.19	0.9097	0.02918	241.42	261.85	0.9080
30	0.03581	246.41	267.89	0.9388	0.02979	244.51	265.37	0.9197
40	0.03774	255.45	278.09	0.9719	0.03157	253.83	275.93	0.9539
50	0.03958	264.48	288.23	1.0037	0.03324	263.08	286.35	0.9867
60	0.04134	273.54	298.35	1.0346	0.03482	272.31	296.69	1.0182
70	0.04304	282.66	308.48	1.0645	0.03634	281.57	307.01	1.0487
80	0.04469	291.86	318.67	1.0938	0.03781	290.88	317.35	1.0784
90	0.04631	301.14	328.93	1.1225	0.03924	300.27	327.74	1.1074
100	0.04790	310.53	339.27	1.1505	0.04064	309.74	338.19	1.1358
110	0.04946	320.03	349.70	1.1781	0.04201	319.31	348.71	1.1637
120	0.05099	329.64	360.24	1.2053	0.04335	328.98	359.33	1.1910
130	0.05251	339.38	370.88	1.2320	0.04468	338.76	370.04	1.2179
140	0.05402	349.23	381.64	1.2584	0.04599	348.66	380.86	1.2444
150	0.05550	359.21	392.52	1.2844	0.04729	358.68	391.79	1.2706
160	0.05698	369.32	403.51	1.3100	0.04857	368.82	402.82	1.2963
	p = 8.0 bars = 0.80 MPa (T_{sat} = 31.33°C)				p = 9.0 bars = 0.90 MPa (T_{sat} = 35.53°C)			
Sat.	0.02547	243.78	264.15	0.9066	0.02255	245.88	266.18	0.9054
40	0.02691	252.13	273.66	0.9374	0.02325	250.32	271.25	0.9217
50	0.02846	261.62	284.39	0.9711	0.02472	260.09	282.34	0.9566
60	0.02992	271.04	294.98	1.0034	0.02609	269.72	293.21	0.9897
70	0.03131	280.45	305.50	1.0345	0.02738	279.30	303.94	1.0214
80	0.03264	289.89	316.00	1.0647	0.02861	288.87	314.62	1.0521
90	0.03393	299.37	326.52	1.0940	0.02980	298.46	325.28	1.0819
100	0.03519	308.93	337.08	1.1227	0.03095	308.11	335.96	1.1109
110	0.03642	318.57	347.71	1.1508	0.03207	317.82	346.68	1.1392
120	0.03762	328.31	358.40	1.1784	0.03316	327.62	357.47	1.1670
130	0.03881	338.14	369.19	1.2055	0.03423	337.52	368.33	1.1943
140	0.03997	348.09	380.07	1.2321	0.03529	347.51	379.27	1.2211
150	0.04113	358.15	391.05	1.2584	0.03633	357.61	390.31	1.2475
160	0.04227	368.32	402.14	1.2843	0.03736	367.82	401.44	1.2735
170	0.04340	378.61	413.33	1.3098	0.03838	378.14	412.68	1.2992
180	0.04452	389.02	424.63	1.3351	0.03939	388.57	424.02	1.3245

TABLE A-12 (*Continued*)

T °C	v m³/kg	u kJ/kg	h kJ/kg	s kJ/kg · K	v m³/kg	u kJ/kg	h kJ/kg	s kJ/kg · K
	p = 10.0 bars = 1.00 MPa (T_{sat} = 39.39°C)				**p = 12.0 bars = 1.20 MPa (T_{sat} = 46.32°C)**			
Sat.	0.02020	247.77	267.97	0.9043	0.01663	251.03	270.99	0.9023
40	0.02029	248.39	268.68	0.9066				
50	0.02171	258.48	280.19	0.9428	0.01712	254.98	275.52	0.9164
60	0.02301	268.35	291.36	0.9768	0.01835	265.42	287.44	0.9527
70	0.02423	278.11	302.34	1.0093	0.01947	275.59	298.96	0.9868
80	0.02538	287.82	313.20	1.0405	0.02051	285.62	310.24	1.0192
90	0.02649	297.53	324.01	1.0707	0.02150	295.59	321.39	1.0503
100	0.02755	307.27	334.82	1.1000	0.02244	305.54	332.47	1.0804
110	0.02858	317.06	345.65	1.1286	0.02335	315.50	343.52	1.1096
120	0.02959	326.93	356.52	1.1567	0.02423	325.51	354.58	1.1381
130	0.03058	336.88	367.46	1.1841	0.02508	335.58	365.68	1.1660
140	0.03154	346.92	378.46	1.2111	0.02592	345.73	376.83	1.1933
150	0.03250	357.06	389.56	1.2376	0.02674	355.95	388.04	1.2201
160	0.03344	367.31	400.74	1.2638	0.02754	366.27	399.33	1.2465
170	0.03436	377.66	412.02	1.2895	0.02834	376.69	410.70	1.2724
180	0.03528	388.12	423.40	1.3149	0.02912	387.21	422.16	1.2980
	p = 14.0 bars = 1.40 MPa (T_{sat} = 52.43°C)				**p = 16.0 bars = 1.60 MPa (T_{sat} = 57.92°C)**			
Sat.	0.01405	253.74	273.40	0.9003	0.01208	256.00	275.33	0.8982
60	0.01495	262.17	283.10	0.9297	0.01233	258.48	278.20	0.9069
70	0.01603	272.87	295.31	0.9658	0.01340	269.89	291.33	0.9457
80	0.01701	283.29	307.10	0.9997	0.01435	280.78	303.74	0.9813
90	0.01792	293.55	318.63	1.0319	0.01521	291.39	315.72	1.0148
100	0.01878	303.73	330.02	1.0628	0.01601	301.84	327.46	1.0467
110	0.01960	313.88	341.32	1.0927	0.01677	312.20	339.04	1.0773
120	0.02039	324.05	352.59	1.1218	0.01750	322.53	350.53	1.1069
130	0.02115	334.25	363.86	1.1501	0.01820	332.87	361.99	1.1357
140	0.02189	344.50	375.15	1.1777	0.01887	343.24	373.44	1.1638
150	0.02262	354.82	386.49	1.2048	0.01953	353.66	384.91	1.1912
160	0.02333	365.22	397.89	1.2315	0.02017	364.15	396.43	1.2181
170	0.02403	375.71	409.36	1.2576	0.02080	374.71	407.99	1.2445
180	0.02472	386.29	420.90	1.2834	0.02142	385.35	419.62	1.2704
190	0.02541	396.96	432.53	1.3088	0.02203	396.08	431.33	1.2960
200	0.02608	407.73	444.24	1.3338	0.02263	406.90	443.11	1.3212

TABLE A-13 Specific Heats of Selected Liquids and Solids

A. Liquid	State	c (kJ/kg · K)
Ammonia	sat. liq., −20°C	4.52
	sat. liq., 10°C	4.67
	sat. liq., 50°C	5.10
Ethyl alcohol	1 atm, 25°C	2.43
Glycerin	1 atm, 10°C	2.32
	1 atm, 50°C	2.58
Mercury	1 atm, 10°C	0.138
	1 atm, 315°C	0.134
Refrigerant 12	sat. liq., −20°C	0.90
	sat. liq., 20°C	0.96
Water	1 atm, 0°C	4.217
	1 atm, 27°C	4.179
	1 atm, 100°C	4.218

B. Solid	Temperature (K)	c (kJ/kg · K)
Aluminum	300	0.903
Copper	300	0.385
	400	0.393
Ice	200	1.56
	240	1.86
	273	2.11
Iron	300	0.447
Lead	300	0.129
Silver	300	0.235

TABLE A-14 Ideal Gas Specific Heats of Some Common Gases (kJ/kg · K)

Temp. K	Air c_p	Air c_v	Air k	Nitrogen, N_2 c_p	Nitrogen, N_2 c_v	Nitrogen, N_2 k	Oxygen, O_2 c_p	Oxygen, O_2 c_v	Oxygen, O_2 k	Temp. K
250	1.003	0.716	1.401	1.039	0.742	1.400	0.913	0.653	1.398	250
300	1.005	0.718	1.400	1.039	0.743	1.400	0.918	0.658	1.395	300
350	1.008	0.721	1.398	1.041	0.744	1.399	0.928	0.668	1.389	350
400	1.013	0.726	1.395	1.044	0.747	1.397	0.941	0.681	1.382	400
450	1.020	0.733	1.391	1.049	0.752	1.395	0.956	0.696	1.373	450
500	1.029	0.742	1.387	1.056	0.759	1.391	0.972	0.712	1.365	500
550	1.040	0.753	1.381	1.065	0.768	1.387	0.988	0.728	1.358	550
600	1.051	0.764	1.376	1.075	0.778	1.382	1.003	0.743	1.350	600
650	1.063	0.776	1.370	1.086	0.789	1.376	1.017	0.758	1.343	650
700	1.075	0.788	1.364	1.098	0.801	1.371	1.031	0.771	1.337	700
750	1.087	0.800	1.359	1.110	0.813	1.365	1.043	0.783	1.332	750
800	1.099	0.812	1.354	1.121	0.825	1.360	1.054	0.794	1.327	800
900	1.121	0.834	1.344	1.145	0.849	1.349	1.074	0.814	1.319	900
1000	1.142	0.855	1.336	1.167	0.870	1.341	1.090	0.830	1.313	1000

Temp. K	Carbon dioxide, CO_2 c_p	Carbon dioxide, CO_2 c_v	Carbon dioxide, CO_2 k	Carbon monoxide, CO c_p	Carbon monoxide, CO c_v	Carbon monoxide, CO k	Hydrogen, H_2 c_p	Hydrogen, H_2 c_v	Hydrogen, H_2 k	Temp. K
250	0.791	0.602	1.314	1.039	0.743	1.400	14.051	9.927	1.416	250
300	0.846	0.657	1.288	1.040	0.744	1.399	14.307	10.183	1.405	300
350	0.895	0.706	1.268	1.043	0.746	1.398	14.427	10.302	1.400	350
400	0.939	0.750	1.252	1.047	0.751	1.395	14.476	10.352	1.398	400
450	0.978	0.790	1.239	1.054	0.757	1.392	14.501	10.377	1.398	450
500	1.014	0.825	1.229	1.063	0.767	1.387	14.513	10.389	1.397	500
550	1.046	0.857	1.220	1.075	0.778	1.382	14.530	10.405	1.396	550
600	1.075	0.886	1.213	1.087	0.790	1.376	14.546	10.422	1.396	600
650	1.102	0.913	1.207	1.100	0.803	1.370	14.571	10.447	1.395	650
700	1.126	0.937	1.202	1.113	0.816	1.364	14.604	10.480	1.394	700
750	1.148	0.959	1.197	1.126	0.829	1.358	14.645	10.521	1.392	750
800	1.169	0.980	1.193	1.139	0.842	1.353	14.695	10.570	1.390	800
900	1.204	1.015	1.186	1.163	0.866	1.343	14.822	10.698	1.385	900
1000	1.234	1.045	1.181	1.185	0.888	1.335	14.983	10.859	1.380	1000

Source: Adapted from K. Wark, *Thermodynamics, 4th ed.*, McGraw-Hill, New York, 1983, as based on "Tables of Thermal Properties of Gases," NBS Circular 564, 1955.

TABLE A-15 Variation of $\bar{c}_p$ with Temperature for Various Ideal Gases

$$\frac{\bar{c}_p}{\bar{R}} = \alpha + \beta T + \gamma T^2 + \delta T^3 + \varepsilon T^4$$

T is in K, equations valid from 300 to 1000 K

Gas	α	$\beta \times 10^3$	$\gamma \times 10^6$	$\delta \times 10^9$	$\epsilon \times 10^{12}$
CO	3.710	−1.619	3.692	−2.032	0.240
CO_2	2.401	8.735	−6.607	2.002	0
H_2	3.057	2.677	−5.810	5.521	−1.812
H_2O	4.070	−1.108	4.152	−2.964	0.807
O_2	3.626	−1.878	7.055	−6.764	2.156
N_2	3.675	−1.208	2.324	−0.632	−0.226
Air	3.653	−1.337	3.294	−1.913	0.2763
SO_2	3.267	5.324	0.684	−5.281	2.559
CH_4	3.826	−3.979	24.558	−22.733	6.963
C_2H_2	1.410	19.057	−24.501	16.391	−4.135
C_2H_4	1.426	11.383	7.989	−16.254	6.749
Monatomic gases[a]	2.5	0	0	0	0

[a] For monatomic gases, such as He, Ne, and Ar, $\bar{c}_p$ is constant over a wide temperature range and is very nearly equal to $5/2\ \bar{R}$.

Source: Adapted from K. Wark, *Thermodynamics, 4th ed.*, McGraw-Hill, New York, 1983, as based on NASA SP-273, U.S. Government Printing Office, Washington, DC, 1971.

Table A-15

TABLE A-16 Ideal Gas Properties of Air

T (K), *h* and *u* (kJ/kg), s^o (kJ/kg · K)

T	h	p_r	u	v_r	s^o	T	h	p_r	u	v_r	s^o
200	199.97	0.3363	142.56	1707.	1.29559	450	451.80	5.775	322.62	223.6	2.11161
210	209.97	0.3987	149.69	1512.	1.34444	460	462.02	6.245	329.97	211.4	2.13407
220	219.97	0.4690	156.82	1346.	1.39105	470	472.24	6.742	337.32	200.1	2.15604
230	230.02	0.5477	164.00	1205.	1.43557	480	482.49	7.268	344.70	189.5	2.17760
240	240.02	0.6355	171.13	1084.	1.47824	490	492.74	7.824	352.08	179.7	2.19876
250	250.05	0.7329	178.28	979.	1.51917	500	503.02	8.411	359.49	170.6	2.21952
260	260.09	0.8405	185.45	887.8	1.55848	510	513.32	9.031	366.92	162.1	2.23993
270	270.11	0.9590	192.60	808.0	1.59634	520	523.63	9.684	374.36	154.1	2.25997
280	280.13	1.0889	199.75	738.0	1.63279	530	533.98	10.37	381.84	146.7	2.27967
285	285.14	1.1584	203.33	706.1	1.65055	540	544.35	11.10	389.34	139.7	2.29906
290	290.16	1.2311	206.91	676.1	1.66802	550	554.74	11.86	396.86	133.1	2.31809
295	295.17	1.3068	210.49	647.9	1.68515	560	565.17	12.66	404.42	127.0	2.33685
300	300.19	1.3860	214.07	621.2	1.70203	570	575.59	13.50	411.97	121.2	2.35531
305	305.22	1.4686	217.67	596.0	1.71865	580	586.04	14.38	419.55	115.7	2.37348
310	310.24	1.5546	221.25	572.3	1.73498	590	596.52	15.31	427.15	110.6	2.39140
315	315.27	1.6442	224.85	549.8	1.75106	600	607.02	16.28	434.78	105.8	2.40902
320	320.29	1.7375	228.42	528.6	1.76690	610	617.53	17.30	442.42	101.2	2.42644
325	325.31	1.8345	232.02	508.4	1.78249	620	628.07	18.36	450.09	96.92	2.44356
330	330.34	1.9352	235.61	489.4	1.79783	630	638.63	19.84	457.78	92.84	2.46048
340	340.42	2.149	242.82	454.1	1.82790	640	649.22	20.64	465.50	88.99	2.47716
350	350.49	2.379	250.02	422.2	1.85708	650	659.84	21.86	473.25	85.34	2.49364
360	360.58	2.626	257.24	393.4	1.88543	660	670.47	23.13	481.01	81.89	2.50985
370	370.67	2.892	264.46	367.2	1.91313	670	681.14	24.46	488.81	78.61	2.52589
380	380.77	3.176	271.69	343.4	1.94001	680	691.82	25.85	496.62	75.50	2.54175
390	390.88	3.481	278.93	321.5	1.96633	690	702.52	27.29	504.45	72.56	2.55731
400	400.98	3.806	286.16	301.6	1.99194	700	713.27	28.80	512.33	69.76	2.57277
410	411.12	4.153	293.43	283.3	2.01699	710	724.04	30.38	520.23	67.07	2.58810
420	421.26	4.522	300.69	266.6	2.04142	720	734.82	32.02	528.14	64.53	2.60319
430	431.43	4.915	307.99	251.1	2.06533	730	745.62	33.72	536.07	62.13	2.61803
440	441.61	5.332	315.30	236.8	2.08870	740	756.44	35.50	544.02	59.82	2.63280

TABLE A-16 *(Continued)*

T (K), h and u (kJ/kg), s^o (kJ/kg · K)

T	h	p_r	u	v_r	s^o	T	h	p_r	u	v_r	s^o
750	767.29	37.35	551.99	57.63	2.64737	1300	1395.97	330.9	1022.82	11.275	3.27345
760	778.18	39.27	560.01	55.54	2.66176	1320	1419.76	352.5	1040.88	10.747	3.29160
770	789.11	41.31	568.07	53.39	2.67595	1340	1443.60	375.3	1058.94	10.247	3.30959
780	800.03	43.35	576.12	51.64	2.69013	1360	1467.49	399.1	1077.10	9.780	3.32724
790	810.99	45.55	584.21	49.86	2.70400	1380	1491.44	424.2	1095.26	9.337	3.34474
800	821.95	47.75	592.30	48.08	2.71787	1400	1515.42	450.5	1113.52	8.919	3.36200
820	843.98	52.59	608.59	44.84	2.74504	1420	1539.44	478.0	1131.77	8.526	3.37901
840	866.08	57.60	624.95	41.85	2.77170	1440	1563.51	506.9	1150.13	8.153	3.39586
860	888.27	63.09	641.40	39.12	2.79783	1460	1587.63	537.1	1168.49	7.801	3.41247
880	910.56	68.98	657.95	36.61	2.82344	1480	1611.79	568.8	1186.95	7.468	3.42892
900	932.93	75.29	674.58	34.31	2.84856	1500	1635.97	601.9	1205.41	7.152	3.44516
920	955.38	82.05	691.28	32.18	2.87324	1520	1660.23	636.5	1223.87	6.854	3.46120
940	977.92	89.28	708.08	30.22	2.89748	1540	1684.51	672.8	1242.43	6.569	3.47712
960	1000.55	97.00	725.02	28.40	2.92128	1560	1708.82	710.5	1260.99	6.301	3.49276
980	1023.25	105.2	741.98	26.73	2.94468	1580	1733.17	750.0	1279.65	6.046	3.50829
1000	1046.04	114.0	758.94	25.17	2.96770	1600	1757.57	791.2	1298.30	5.804	3.52364
1020	1068.89	123.4	776.10	23.72	2.99034	1620	1782.00	834.1	1316.96	5.574	3.53879
1040	1091.85	133.3	793.36	22.39	3.01260	1640	1806.46	878.9	1335.72	5.355	3.55381
1060	1114.86	143.9	810.62	21.14	3.03449	1660	1830.96	925.6	1354.48	5.147	3.56867
1080	1137.89	155.2	827.88	19.98	3.05608	1680	1855.50	974.2	1373.24	4.949	3.58335
1100	1161.07	167.1	845.33	18.896	3.07732	1700	1880.1	1025	1392.7	4.761	3.5979
1120	1184.28	179.7	862.79	17.886	3.09825	1750	1941.6	1161	1439.8	4.328	3.6336
1140	1207.57	193.1	880.35	16.946	3.11883	1800	2003.3	1310	1487.2	3.944	3.6684
1160	1230.92	207.2	897.91	16.064	3.13916	1850	2065.3	1475	1534.9	3.601	3.7023
1180	1254.34	222.2	915.57	15.241	3.15916	1900	2127.4	1655	1582.6	3.295	3.7354
1200	1277.79	238.0	933.33	14.470	3.17888	1950	2189.7	1852	1630.6	3.022	3.7677
1220	1301.31	254.7	951.09	13.747	3.19834	2000	2252.1	2068	1678.7	2.776	3.7994
1240	1324.93	272.3	968.95	13.069	3.21751	2050	2314.6	2303	1726.8	2.555	3.8303
1260	1348.55	290.8	986.90	12.435	3.23638	2100	2377.4	2559	1775.3	2.356	3.8605
1280	1372.24	310.4	1004.76	11.835	3.25510	2150	2440.3	2837	1823.8	2.175	3.8901
						2200	2503.2	3138	1872.4	2.012	3.9191
						2250	2566.4	3464	1921.3	1.864	3.9474

Source: Adapted from K. Wark, *Thermodynamics, 4th ed.*, McGraw-Hill, New York, 1983, as based on J. H. Keenan and J. Kaye, "Gas Tables," Wiley, New York, 1945.

TABLE A-17 Ideal Gas Properties of Nitrogen, N_2

T(K), $\bar{h}$ and $\bar{u}$(kJ/kmol), $\bar{s}^\circ$(kJ/kmol · K)

[$\bar{h}_f^\circ$ = 0 kJ/kmol]

T	$\bar{h}$	$\bar{u}$	$\bar{s}^\circ$	T	$\bar{h}$	$\bar{u}$	$\bar{s}^\circ$
0	0	0	0	600	17,563	12,574	212.066
220	6,391	4,562	182.639	610	17,864	12,792	212.564
230	6,683	4,770	183.938	620	18,166	13,011	213.055
240	6,975	4,979	185.180	630	18,468	13,230	213.541
250	7,266	5,188	186.370	640	18,772	13,450	214.018
260	7,558	5,396	187.514	650	19,075	13,671	214.489
270	7,849	5,604	188.614	660	19,380	13,892	214.954
280	8,141	5,813	189.673	670	19,685	14,114	215.413
290	8,432	6,021	190.695	680	19,991	14,337	215.866
298	8,669	6,190	191.502	690	20,297	14,560	216.314
300	8,723	6,229	191.682	700	20,604	14,784	216.756
310	9,014	6,437	192.638	710	20,912	15,008	217.192
320	9,306	6,645	193.562	720	21,220	15,234	217.624
330	9,597	6,853	194.459	730	21,529	15,460	218.059
340	9,888	7,061	195.328	740	21,839	15,686	218.472
350	10,180	7,270	196.173	750	22,149	15,913	218.889
360	10,471	7,478	196.995	760	22,460	16,141	219.301
370	10,763	7,687	197.794	770	22,772	16,370	219.709
380	11,055	7,895	198.572	780	23,085	16,599	220.113
390	11,347	8,104	199.331	790	23,398	16,830	220.512
400	11,640	8,314	200.071	800	23,714	17,061	220.907
410	11,932	8,523	200.794	810	24,027	17,292	221.298
420	12,225	8,733	201.499	820	24,342	17,524	221.684
430	12,518	8,943	202.189	830	24,658	17,757	222.067
440	12,811	9,153	202.863	840	24,974	17,990	222.447
450	13,105	9,363	203.523	850	25,292	18,224	222.822
460	13,399	9,574	204.170	860	25,610	18,459	223.194
470	13,693	9,786	204.803	870	25,928	18,695	223.562
480	13,988	9,997	205.424	880	26,248	18,931	223.927
490	14,285	10,210	206.033	890	26,568	19,168	224.288
500	14,581	10,423	206.630	900	26,890	19,407	224.647
510	14,876	10,635	207.216	910	27,210	19,644	225.002
520	15,172	10,848	207.792	920	27,532	19,883	225.353
530	15,469	11,062	208.358	930	27,854	20,122	225.701
540	15,766	11,277	208.914	940	28,178	20,362	226.047
550	16,064	11,492	209.461	950	28,501	20,603	226.389
560	16,363	11,707	209.999	960	28,826	20,844	226.728
570	16,662	11,923	210.528	970	29,151	21,086	227.064
580	16,962	12,139	211.049	980	29,476	21,328	227.398
590	17,262	12,356	211.562	990	29,803	21,571	227.728

TABLE A-17 *(Continued)*

T	$\bar{h}$	$\bar{u}$	$\bar{s}°$	T	$\bar{h}$	$\bar{u}$	$\bar{s}°$
1000	30,129	21,815	228.057	1760	56,227	41.594	247.396
1020	30,784	22,304	228.706	1780	56,938	42,139	247.798
1040	31,442	22,795	229.344	1800	57,651	42,685	248.195
1060	32,101	23,288	229.973	1820	58,363	43,231	248.589
1080	32,762	23,782	230.591	1840	59,075	43,777	248.979
1100	33,426	24,280	231.199	1860	59,790	44,324	249.365
1120	34,092	24,780	231.799	1880	60,504	44,873	249.748
1140	34,760	25,282	232.391	1900	61,220	45,423	250.128
1160	35,430	25,786	232.973	1920	61,936	45,973	250.502
1180	36,104	26,291	233.549	1940	62,654	46,524	250.874
1200	36,777	26,799	234.115	1960	63,381	47,075	251.242
1220	37,452	27,308	234.673	1980	64,090	47,627	251.607
1240	38,129	27,819	235.223	2000	64,810	48,181	251.969
1260	38,807	28,331	235.766	2050	66,612	49,567	252.858
1280	39,488	28,845	236.302	2100	68,417	50,957	253.726
1300	40,170	29,361	236.831	2150	70,226	52,351	254.578
1320	40,853	29,878	237.353	2200	72,040	53,749	255.412
1340	41,539	30,398	237.867	2250	73,856	55,149	256.227
1360	42,227	30,919	238.376	2300	75,676	56,553	257.027
1380	42,915	31,441	238.878	2350	77,496	57,958	257.810
1400	43,605	31,964	239.375	2400	79,320	59,366	258.580
1420	44,295	32,489	239.865	2450	81,149	60,779	259.332
1440	44,988	33,014	240.350	2500	82,981	62,195	260.073
1460	45,682	33,543	240.827	2550	84,814	63,613	260.799
1480	46,377	34,071	241.301	2600	86,650	65,033	261.512
1500	47,073	34,601	241.768	2650	88,488	66,455	262.213
1520	47,771	35,133	242.228	2700	90,328	67,880	262.902
1540	48,470	35,665	242.685	2750	92,171	69,306	263.577
1560	49,168	36,197	243.137	2800	94,014	70,734	264.241
1580	49,869	36,732	243.585	2850	95,859	72,163	264.895
1600	50,571	37,268	244.028	2900	97,705	73,593	265.538
1620	51,275	37,806	244.464	2950	99,556	75,028	266.170
1640	51,980	38,344	244.896	3000	101,407	76,464	266.793
1660	52,686	38,884	245.324	3050	103,260	77,902	267.404
1680	53,393	39,424	245.747	3100	105,115	79,341	268.007
1700	54,099	39,965	246.166	3150	106,972	80,782	268.601
1720	54,807	40,507	246.580	3200	108,830	82,224	269.186
1740	55,516	41,049	246.990	3250	110,690	83,668	269.763

Source: Tables A-14 through A-19 are adapted from K. Wark, *Thermodynamics, 4th ed.*, McGraw-Hill, New York, 1983, as based on the JANAF Thermochemical Tables, NSRDS-NBS-37, 1971.

TABLE A-18 Ideal Gas Properties of Oxygen, O_2

T(K), $\bar{h}$ and $\bar{u}$(kJ/kmol), $\bar{s}°$(kJ/kmol · K)

[$\bar{h}_f^\circ$ = 0 kJ/kmol]

T	$\bar{h}$	$\bar{u}$	$\bar{s}°$	T	$\bar{h}$	$\bar{u}$	$\bar{s}°$
0	0	0	0	600	17,929	12,940	226.346
220	6,404	4,575	196.171	610	18,250	13,178	226.877
230	6,694	4,782	197.461	620	18,572	13,417	227.400
240	6,984	4,989	198.696	630	18,895	13,657	227.918
250	7,275	5,197	199.885	640	19,219	13,898	228.429
260	7,566	5,405	201.027	650	19,544	14,140	228.932
270	7,858	5,613	202.128	660	19,870	14,383	229.430
280	8,150	5,822	203.191	670	20,197	14,626	229.920
290	8,443	6,032	204.218	680	20,524	14,871	230.405
298	8,682	6,203	205.033	690	20,854	15,116	230.885
300	8,736	6,242	205.213	700	21,184	15,364	231.358
310	9,030	6,453	206.177	710	21,514	15,611	231.827
320	9,325	6,664	207.112	720	21,845	15,859	232.291
330	9,620	6,877	208.020	730	22,177	16,107	232.748
340	9,916	7,090	208.904	740	22,510	16,357	233.201
350	10,213	7,303	209.765	750	22,844	16,607	233.649
360	10,511	7,518	210.604	760	23,178	16,859	234.091
370	10,809	7,733	211.423	770	23,513	17,111	234.528
380	11,109	7,949	212.222	780	23,850	17,364	234.960
390	11,409	8,166	213.002	790	24,186	17,618	235.387
400	11,711	8,384	213.765	800	24,523	17,872	235.810
410	12,012	8,603	214.510	810	24,861	18,126	236.230
420	12,314	8,822	215.241	820	25,199	18,382	236.644
430	12,618	9,043	215.955	830	25,537	18,637	237.055
440	12,923	9,264	216.656	840	25,877	18,893	237.462
450	13,228	9,487	217.342	850	26,218	19,150	237.864
460	13,535	9,710	218.016	860	26,559	19,408	238.264
470	13,842	9,935	218.676	870	26,899	19,666	238.660
480	14,151	10,160	219.326	880	27,242	19,925	239.051
490	14,460	10,386	219.963	890	27,584	20,185	239.439
500	14,770	10,614	220.589	900	27,928	20,445	239.823
510	15,082	10,842	221.206	910	28,272	20,706	240.203
520	15,395	11,071	221.812	920	28,616	20,967	240.580
530	15,708	11,301	222.409	930	28,960	21,228	240.953
540	16,022	11,533	222.997	940	29,306	21,491	241.323
550	16,338	11,765	223.576	950	29,652	21,754	241.689
560	16,654	11,998	224.146	960	29,999	22,017	242.052
570	16,971	12,232	224.708	970	30,345	22,280	242.411
580	17,290	12,467	225.262	980	30,692	22,544	242.768
590	17,609	12,703	225.808	990	31,041	22,809	243.120

TABLE A-18 (*Continued*)

T	$\bar{h}$	$\bar{u}$	$\bar{s}°$	T	$\bar{h}$	$\bar{u}$	$\bar{s}°$
1000	31,389	23,075	243.471	1760	58,880	44,247	263.861
1020	32,088	23,607	244.164	1780	59,624	44,825	264.283
1040	32,789	24,142	244.844	1800	60,371	45,405	264.701
1060	33,490	24,677	245.513	1820	61,118	45,986	265.113
1080	34,194	25,214	246.171	1840	61,866	46,568	265.521
1100	34,899	25,753	246.818	1860	62,616	47,151	265.925
1120	35,606	26,294	247.454	1880	63,365	47,734	266.326
1140	36,314	26,836	248.081	1900	64,116	48,319	266.722
1160	37,023	27,379	248.698	1920	64,868	48,904	267.115
1180	37,734	27,923	249.307	1940	65,620	49,490	267.505
1200	38,447	28,469	249.906	1960	66,374	50,078	267.891
1220	39,162	29,018	250.497	1980	67,127	50,665	268.275
1240	39,877	29,568	251.079	2000	67,881	51,253	268.655
1260	40,594	30,118	251.653	2050	69,772	52,727	269.588
1280	41,312	30,670	252.219	2100	71,668	54,208	270.504
1300	42,033	31,224	252.776	2150	73,573	55,697	271.399
1320	42,753	31,778	253.325	2200	75,484	57,192	272.278
1340	43,475	32,334	253.868	2250	77,397	58,690	273.136
1360	44,198	32,891	254.404	2300	79,316	60,193	273.981
1380	44,923	33,449	254.932	2350	81,243	61,704	274.809
1400	45,648	34,008	255.454	2400	83,174	63,219	275.625
1420	46,374	34,567	255.968	2450	85,112	64,742	276.424
1440	47,102	35,129	256.475	2500	87,057	66,271	277.207
1460	47,831	35,692	256.978	2550	89,004	67,802	277.979
1480	48,561	36,256	257.474	2600	90,956	69,339	278.738
1500	49,292	36,821	257.965	2650	92,916	70,883	279.485
1520	50,024	37,387	258.450	2700	94,881	72,433	280.219
1540	50,756	37,952	258.928	2750	96,852	73,987	280.942
1560	51,490	38,520	259.402	2800	98,826	75,546	281.654
1580	52,224	39,088	259.870	2850	100,808	77,112	282.357
1600	52,961	39,658	260.333	2900	102,793	78,682	283.048
1620	53,696	40,227	260.791	2950	104,785	80,258	283.728
1640	54,434	40,799	261.242	3000	106,780	81,837	284.399
1660	55,172	41,370	261.690	3050	108,778	83,419	285.060
1680	55,912	41,944	262.132	3100	110,784	85,009	285.713
1700	56,652	42,517	262.571	3150	112,795	86,601	286.355
1720	57,394	43,093	263.005	3200	114,809	88,203	286.989
1740	58,136	43,669	263.435	3250	116,827	89,804	287.614

TABLE A-19 Ideal Gas Properties of Water Vapor, H_2O

T(K), $\bar{h}$ and $\bar{u}$(kJ/kmol), $\bar{s}°$(kJ/kmol · K)

[$\bar{h}_f° = -241,820$ kJ/kmol]

T	$\bar{h}$	$\bar{u}$	$\bar{s}°$	T	$\bar{h}$	$\bar{u}$	$\bar{s}°$
0	0	0	0	600	20,402	15,413	212.920
220	7,295	5,466	178.576	610	20,765	15,693	213.529
230	7,628	5,715	180.054	620	21,130	15,975	214.122
240	7,961	5,965	181.471	630	21,495	16,257	214.707
250	8,294	6,215	182.831	640	21,862	16,541	215.285
260	8,627	6,466	184.139	650	22,230	16,826	215.856
270	8,961	6,716	185.399	660	22,600	17,112	216.419
280	9,296	6,968	186.616	670	22,970	17,399	216.976
290	9,631	7,219	187.791	680	23,342	17,688	217.527
298	9,904	7,425	188.720	690	23,714	17,978	218.071
300	9,966	7,472	188.928	700	24,088	18,268	218.610
310	10,302	7,725	190.030	710	24,464	18,561	219.142
320	10,639	7,978	191.098	720	24,840	18,854	219.668
330	10,976	8,232	192.136	730	25,218	19,148	220.189
340	11,314	8,487	193.144	740	25,597	19,444	220.707
350	11,652	8,742	194.125	750	25,977	19,741	221.215
360	11,992	8,998	195.081	760	26,358	20,039	221.720
370	12,331	9,255	196.012	770	26,741	20,339	222.221
380	12,672	9,513	196.920	780	27,125	20,639	222.717
390	13,014	9,771	197.807	790	27,510	20,941	223.207
400	13,356	10,030	198.673	800	27,896	21,245	223.693
410	13,699	10,290	199.521	810	28,284	21,549	224.174
420	14,043	10,551	200.350	820	28,672	21,855	224.651
430	14,388	10,813	201.160	830	29,062	22,162	225.123
440	14,734	11,075	201.955	840	29,454	22,470	225.592
450	15,080	11,339	202.734	850	29,846	22,779	226.057
460	15,428	11,603	203.497	860	30,240	23,090	226.517
470	15,777	11,869	204.247	870	30,635	23,402	226.973
480	16,126	12,135	204.982	880	31,032	23,715	227.426
490	16,477	12,403	205.705	890	31,429	24,029	227.875
500	16,828	12,671	206.413	900	31,828	24,345	228.321
510	17,181	12,940	207.112	910	32,228	24,662	228.763
520	17,534	13,211	207.799	920	32,629	24,980	229.202
530	17,889	13,482	208.475	930	33,032	25,300	229.637
540	18,245	13,755	209.139	940	33,436	25,621	230.070
550	18,601	14,028	209.795	950	33,841	25,943	230.499
560	18,959	14,303	210.440	960	34,247	26,265	230.924
570	19,318	14,579	211.075	970	34,653	26,588	231.347
580	19,678	14,856	211.702	980	35,061	26,913	231.767
590	20,039	15,134	212.320	990	35,472	27,240	232.184

H_2O

TABLE A-19 (*Continued*)

T	$\bar{h}$	$\bar{u}$	$\bar{s}°$	T	$\bar{h}$	$\bar{u}$	$\bar{s}°$
1000	35,882	27,568	232.597	1760	70,535	55,902	258.151
1020	36,709	28,228	233.415	1780	71,523	56,723	258.708
1040	37,542	28,895	234.223	1800	72,513	57,547	259.262
1060	38,380	29,567	235.020	1820	73,507	58,375	259.811
1080	39,223	30,243	235.806	1840	74,506	59,207	260.357
1100	40,071	30,925	236.584	1860	75,506	60,042	260.898
1120	40,923	31,611	237.352	1880	76,511	60,880	261.436
1140	41,780	32,301	238.110	1900	77,517	61,720	261.969
1160	42,642	32,997	238.859	1920	78,527	62,564	262.497
1180	43,509	33,698	239.600	1940	79,540	63,411	263.022
1200	44,380	34,403	240.333	1960	80,555	64,259	263.542
1220	45,256	35,112	241.057	1980	81,573	65,111	264.059
1240	46,137	35,827	241.773	2000	82,593	65,965	264.571
1260	47,022	36,546	242.482	2050	85,156	68,111	265.838
1280	47,912	37,270	243.183	2100	87,735	70,275	267.081
1300	48,807	38,000	243.877	2150	90,330	72,454	268.301
1320	49,707	38,732	244.564	2200	92,940	74,649	269.500
1340	50,612	39,470	245.243	2250	95,562	76,855	270.679
1360	51,521	40,213	245.915	2300	98,199	79,076	271.839
1380	52,434	40,960	246.582	2350	100,846	81,308	272.978
1400	53,351	41,711	247.241	2400	103,508	83,553	274.098
1420	54,273	42,466	247.895	2450	106,183	85,811	275.201
1440	55,198	43,226	248.543	2500	108,868	88,082	276.286
1460	56,128	43,989	249.185	2550	111,565	90,364	277.354
1480	57,062	44,756	249.820	2600	114,273	92,656	278.407
1500	57,999	45,528	250.450	2650	116,991	94,958	279.441
1520	58,942	46,304	251.074	2700	119,717	97,269	280.462
1540	59,888	47,084	251.693	2750	122,453	99,588	281.464
1560	60,838	47,868	252.305	2800	125,198	101,917	282.453
1580	61,792	48,655	252.912	2850	127,952	104,256	283.429
1600	62,748	49,445	253.513	2900	130,717	106,605	284.390
1620	63,709	50,240	254.111	2950	133,486	108,959	285.338
1640	64,675	51,039	254.703	3000	136,264	111,321	286.273
1660	65,643	51,841	255.290	3050	139,051	113,692	287.194
1680	66,614	52,646	255.873	3100	141,846	116,072	288.102
1700	67,589	53,455	256.450	3150	144,648	118,458	288.999
1720	68,567	54,267	257.022	3200	147,457	120,851	289.884
1740	69,550	55,083	257.589	3250	150,272	123,250	290.756

TABLE A-20 Ideal Gas Properties of Carbon Monoxide, CO

T(K), $\bar{h}$ and $\bar{u}$(kJ/kmol), $\bar{s}^\circ$(kJ/kmol · K)

$[\bar{h}_f^\circ = -110{,}530$ kJ/kmol$]$

T	$\bar{h}$	$\bar{u}$	$\bar{s}^\circ$	T	$\bar{h}$	$\bar{u}$	$\bar{s}^\circ$
0	0	0	0	600	17,611	12,622	218.204
220	6,391	4,562	188.683	610	17,915	12,843	218.708
230	6,683	4,771	189.980	620	18,221	13,066	219.205
240	6,975	4,979	191.221	630	18,527	13,289	219.695
250	7,266	5,188	192.411	640	18,833	13,512	220.179
260	7,558	5,396	193.554	650	19,141	13,736	220.656
270	7,849	5,604	194.654	660	19,449	13,962	221.127
280	8,140	5,812	195.173	670	19,758	14,187	221.592
290	8,432	6,020	196.735	680	20,068	14,414	222.052
298	8,669	6,190	197.543	690	20,378	14,641	222.505
300	8,723	6,229	197.723	700	20,690	14,870	222.953
310	9,014	6,437	198.678	710	21,002	15,099	223.396
320	9,306	6,645	199.603	720	21,315	15,328	223.833
330	9,597	6,854	200.500	730	21,628	15,558	224.265
340	9,889	7,062	201.371	740	21,943	15,789	224.692
350	10,181	7,271	202.217	750	22,258	16,022	225.115
360	10,473	7,480	203.040	760	22,573	16,255	225.533
370	10,765	7,689	203.842	770	22,890	16,488	225.947
380	11,058	7,899	204.622	780	23,208	16,723	226.357
390	11,351	8,108	205.383	790	23,526	16,957	226.762
400	11,644	8,319	206.125	800	23,844	17,193	227.162
410	11,938	8,529	206.850	810	24,164	17,429	227.559
420	12,232	8,740	207.549	820	24,483	17,665	227.952
430	12,526	8,951	208.252	830	24,803	17,902	228.339
440	12,821	9,163	208.929	840	25,124	18,140	228.724
450	13,116	9,375	209.593	850	25,446	18,379	229.106
460	13,412	9,587	210.243	860	25,768	18,617	229.482
470	13,708	9,800	210.880	870	26,091	18,858	229.856
480	14,005	10,014	211.504	880	26,415	19,099	230.227
490	14,302	10,228	212.117	890	26,740	19,341	230.593
500	14,600	10,443	212.719	900	27,066	19,583	230.957
510	14,898	10,658	213.310	910	27,392	19,826	231.317
520	15,197	10,874	213.890	920	27,719	20,070	231.674
530	15,497	11,090	214.460	930	28,046	20,314	232.028
540	15,797	11,307	215.020	940	28,375	20,559	232.379
550	16,097	11,524	215.572	950	28,703	20,805	232.727
560	16,399	11,743	216.115	960	29,033	21,051	233.072
570	16,701	11,961	216.649	970	29,362	21,298	233.413
580	17,003	12,181	217.175	980	29,693	21,545	233.752
590	17,307	12,401	217.693	990	30,024	21,793	234.088

CO

TABLE A-20 (*Continued*)

T	$\bar{h}$	$\bar{u}$	$\bar{s}°$	T	$\bar{h}$	$\bar{u}$	$\bar{s}°$
1000	30,355	22,041	234.421	1760	56,756	42,123	253.991
1020	31,020	22,540	235.079	1780	57,473	42,673	254.398
1040	31,688	23,041	235.728	1800	58,191	43,225	254.797
1060	32,357	23,544	236.364	1820	58,910	43,778	255.194
1080	33,029	24,049	236.992	1840	59,629	44,331	255.587
1100	33,702	24,557	237.609	1860	60,351	44,886	255.976
1120	34,377	25,065	238.217	1880	61,072	45,441	256.361
1140	35,054	25,575	238.817	1900	61,794	45,997	256.743
1160	35,733	26,088	239.407	1920	62,516	46,552	257.122
1180	36,406	26,602	239.989	1940	63,238	47,108	257.497
1200	37,095	27,118	240.663	1960	63,961	47,665	257.868
1220	37,780	27,637	241.128	1980	64,684	48,221	258.236
1240	38,466	28,426	241.686	2000	65,408	48,780	258.600
1260	39,154	28,678	242.236	2050	67,224	50,179	259.494
1280	39,884	29,201	242.780	2100	69,044	51,584	260.370
1300	40,534	29,725	243.316	2150	70,864	52,988	261.226
1320	41,266	30,251	243.844	2200	72,688	54,396	262.065
1340	41,919	30,778	244.366	2250	74,516	55,809	262.887
1360	42,613	31,306	244.880	2300	76,345	57,222	263.692
1380	43,309	31,836	245.388	2350	78,178	58,640	264.480
1400	44,007	32,367	245.889	2400	80,015	60,060	265.253
1420	44,707	32,900	246.385	2450	81,852	61,482	266.012
1440	45,408	33,434	246.876	2500	83,692	62,906	266.755
1460	46,110	33,971	247.360	2550	85,537	64,335	267.485
1480	46,813	34,508	247.839	2600	87,383	65,766	268.202
1500	47,517	35,046	248.312	2650	89,230	67,197	268.905
1520	48,222	35,584	248.778	2700	91,077	68,628	269.596
1540	48,928	36,124	249.240	2750	92,930	70,066	270.285
1560	49,635	36,665	249.695	2800	94,784	71,504	270.943
1580	50,344	37,207	250.147	2850	96,639	72,945	271.602
1600	51,053	37,750	250.592	2900	98,495	74,383	272.249
1620	51,763	38,293	251.033	2950	100,352	75,825	272.884
1640	52,472	38,837	251.470	3000	102,210	77,267	273.508
1660	53,184	39,382	251.901	3050	104,073	78,715	274.123
1680	53,895	39,927	252.329	3100	105,939	80,164	274.730
1700	54,609	40,474	252.751	3150	107,802	81,612	275.326
1720	55,323	41,023	253.169	3200	109,667	83,061	275.914
1740	56,039	41,572	253.582	3250	111,534	84,513	276.494

TABLE A-21 Ideal Gas Properties of Carbon Dioxide, CO_2

T(K), $\bar{h}$ and $\bar{u}$(kJ/kmol), $\bar{s}°$(kJ/kmol · K)

[$\bar{h}_f^\circ$ = −393,520 kJ/kmol]

T	$\bar{h}$	$\bar{u}$	$\bar{s}°$	T	$\bar{h}$	$\bar{u}$	$\bar{s}°$
0	0	0	0	600	22,280	17,291	243.199
220	6,601	4,772	202.966	610	22,754	17,683	243.983
230	6,938	5,026	204.464	620	23,231	18,076	244.758
240	7,280	5,285	205.920	630	23,709	18,471	245.524
250	7,627	5,548	207.337	640	24,190	18,869	246.282
260	7,979	5,817	208.717	650	24,674	19,270	247.032
270	8,335	6,091	210.062	660	25,160	19,672	247.773
280	8,697	6,369	211.376	670	25,648	20,078	248.507
290	9,063	6,651	212.660	680	26,138	20,484	249.233
298	9,364	6,885	213.685	690	26,631	20,894	249.952
300	9,431	6,939	213.915	700	27,125	21,305	250.663
310	9,807	7,230	215.146	710	27,622	21,719	251.368
320	10,186	7,526	216.351	720	28,121	22,134	252.065
330	10,570	7,826	217.534	730	28,622	22,552	252.755
340	10,959	8,131	218.694	740	29,124	22,972	253.439
350	11,351	8,439	219.831	750	29,629	23,393	254.117
360	11,748	8,752	220.948	760	30,135	23,817	254.787
370	12,148	9,068	222.044	770	30,644	24,242	255.452
380	12,552	9,392	223.122	780	31,154	24,669	256.110
390	12,960	9,718	224.182	790	31,665	25,097	256.762
400	13,372	10,046	225.225	800	32,179	25,527	257.408
410	13,787	10,378	226.250	810	32,694	25,959	258.048
420	14,206	10,714	227.258	820	33,212	26,394	258.682
430	14,628	11,053	228.252	830	33,730	26,829	259.311
440	15,054	11,393	229.230	840	34,251	27,267	259.934
450	15,483	11,742	230.194	850	34,773	27,706	260.551
460	15,916	12,091	231.144	860	35,296	28,125	261.164
470	16,351	12,444	232.080	870	35,821	28,588	261.770
480	16,791	12,800	233.004	880	36,347	29,031	262.371
490	17,232	13,158	233.916	890	36,876	29,476	262.968
500	17,678	13,521	234.814	900	37,405	29,922	263.559
510	18,126	13,885	235.700	910	37,935	30,369	264.146
520	18,576	14,253	236.575	920	38,467	30,818	264.728
530	19,029	14,622	237.439	930	39,000	31,268	265.304
540	19,485	14,996	238.292	940	39,535	31,719	265.877
550	19,945	15,372	239.135	950	40,070	32,171	266.444
560	20,407	15,751	239.962	960	40,607	32,625	267.007
570	20,870	16,131	240.789	970	41,145	33,081	267.566
580	21,337	16,515	241.602	980	41,685	33,537	268.119
590	21,807	16,902	242.405	990	42,226	33,995	268.670

TABLE A-21 *(Continued)*

T	$\bar{h}$	$\bar{u}$	$\bar{s}°$	T	$\bar{h}$	$\bar{u}$	$\bar{s}°$
1000	42,769	34,455	269.215	1760	86,420	71,787	301.543
1020	43,859	35,378	270.293	1780	87,612	72,812	302.271
1040	44,953	36,306	271.354	1800	88,806	73,840	302.884
1060	46,051	37,238	272.400	1820	90,000	74,868	303.544
1080	47,153	38,174	273.430	1840	91,196	75,897	304.198
1100	48,258	39,112	274.445	1860	92,394	76,929	304.845
1120	49,369	40,057	275.444	1880	93,593	77,962	305.487
1140	50,484	41,006	276.430	1900	94,793	78,996	306.122
1160	51,602	41,957	277.403	1920	95,995	80,031	306.751
1180	52,724	42,913	278.362	1940	97,197	81,067	307.374
1200	53,848	43,871	279.307	1960	98,401	82,105	307.992
1220	54,977	44,834	280.238	1980	99,606	83,144	308.604
1240	56,108	45,799	281.158	2000	100,804	84,185	309.210
1260	57,244	46,768	282.066	2050	103,835	86,791	310.701
1280	58,381	47,739	282.962	2100	106,864	89,404	312.160
1300	59,522	48,713	283.847	2150	109,898	92,023	313.589
1320	60,666	49,691	284.722	2200	112,939	94,648	314.988
1340	61,813	50,672	285.586	2250	115,984	97,277	316.356
1360	62,963	51,656	286.439	2300	119,035	99,912	317.695
1380	64,116	52,643	287.283	2350	122,091	102,552	319.011
1400	65,271	53,631	288.106	2400	125,152	105,197	320.302
1420	66,427	54,621	288.934	2450	128,219	107,849	321.566
1440	67,586	55,614	289.743	2500	131,290	110,504	322.808
1460	68,748	56,609	290.542	2550	134,368	113,166	324.026
1480	69,911	57,606	291.333	2600	137,449	115,832	325.222
1500	71,078	58,606	292.114	2650	140,533	118,500	326.396
1520	72,246	59,609	292.888	2700	143,620	121,172	327.549
1540	73,417	60,613	292.654	2750	146,713	123,849	328.684
1560	74,590	61,620	294.411	2800	149,808	126,528	329.800
1580	76,767	62,630	295.161	2850	152,908	129,212	330.896
1600	76,944	63,741	295.901	2900	156,009	131,898	331.975
1620	78,123	64,653	296.632	2950	159,117	134,589	333.037
1640	79,303	65,668	297.356	3000	162,226	137,283	334.084
1660	80,486	66,592	298.072	3050	165,341	139,982	335.114
1680	81,670	67,702	298.781	3100	168,456	142,681	336.126
1700	82,856	68,721	299.482	3150	171,576	145,385	337.124
1720	84,043	69,742	300.177	3200	174,695	148,089	338.109
1740	85,231	70,764	300.863	3250	177,822	150,801	339.069

TABLE A-22 Ideal Gas Properties of Hydrogen, H_2

T(K), $\bar{h}$ and $\bar{u}$(kJ/kmol), $\bar{s}°$(kJ/kmol · K)

[$\bar{h}_f^\circ$ = 0 kJ/kmol]

T	$\bar{h}$	$\bar{u}$	$\bar{s}°$	T	$\bar{h}$	$\bar{u}$	$\bar{s}°$
0	0	0	0	1440	42,808	30,835	177.410
260	7,370	5,209	126.636	1480	44,091	31,786	178.291
270	7,657	5,412	127.719	1520	45,384	32,746	179.153
280	7,945	5,617	128.765	1560	46,683	33,713	179.995
290	8,233	5,822	129.775	1600	47,990	34,687	180.820
298	8,468	5,989	130.574	1640	49,303	35,668	181.632
300	8,522	6,027	130.754	1680	50,662	36,654	182.428
320	9,100	6,440	132.621	1720	51,947	37,646	183.208
340	9,680	6,853	134.378	1760	53,279	38,645	183.973
360	10,262	7,268	136.039	1800	54,618	39,652	184.724
380	10,843	7,684	137.612	1840	55,962	40,663	185.463
400	11,426	8,100	139.106	1880	57,311	41,680	186.190
420	12,010	8,518	140.529	1920	58,668	42,705	186.904
440	12,594	8,936	141.888	1960	60,031	43,735	187.607
460	13,179	9,355	143.187	2000	61,400	44,771	188.297
480	13,764	9,773	144.432	2050	63,119	46,074	189.148
500	14,350	10,193	145.628	2100	64,847	47,386	189.979
520	14,935	10,611	146.775	2150	66,584	48,708	190.796
560	16,107	11,451	148.945	2200	68,328	50,037	191.598
600	17,280	12,291	150.968	2250	70,080	51,373	192.385
640	18,453	13,133	152.863	2300	71,839	52,716	193.159
680	19,630	13,976	154.645	2350	73,608	54,069	193.921
720	20,807	14,821	156.328	2400	75,383	55,429	194.669
760	21,988	15,669	157.923	2450	77,168	56,798	195.403
800	23,171	16,520	159.440	2500	78,960	58,175	196.125
840	24,359	17,375	160.891	2550	80,755	59,554	196.837
880	25,551	18,235	162.277	2600	82,558	60,941	197.539
920	26,747	19,098	163.607	2650	84,386	62,335	198.229
960	27,948	19,966	164.884	2700	86,186	63,737	198.907
1000	29,154	20,839	166.114	2750	88,008	65,144	199.575
1040	30,364	21,717	167.300	2800	89,838	66,558	200.234
1080	31,580	22,601	168.449	2850	91,671	67,976	200.885
1120	32,802	23,490	169.560	2900	93,512	69,401	201.527
1160	34,028	24,384	170.636	2950	95,358	70,831	202.157
1200	35,262	25,284	171.682	3000	97,211	72,268	202.778
1240	36,502	26,192	172.698	3050	99,065	73,707	203.391
1280	37,749	27,106	173.687	3100	100,926	75,152	203.995
1320	39,002	28,027	174.652	3150	102,793	76,604	204.592
1360	40,263	28,955	175.593	3200	104,667	78,061	205.181
1400	41,530	29,889	176.510	3250	106,545	79,523	205.765

TABLE A-23 Constants for the van der Waals, Redlich–Kwong, and Benedict–Webb–Rubin Equations of State

1. van der Waals and Redlich-Kwong: Constants for pressure in bars, specific volume in m^3/kmol, and temperature in K

Substance	van der Waals		Redlich-Kwong	
	a bar $\left(\frac{m^3}{kmol}\right)^2$	b $\frac{m^3}{kmol}$	a bar $\left(\frac{m^3}{kmol}\right)^2 K^{1/2}$	b $\frac{m^3}{kmol}$
Air	1.368	0.0367	15.989	0.02541
Butane (C_4H_{10})	13.86	0.1162	289.55	0.08060
Carbon dioxide (CO_2)	3.647	0.0428	64.43	0.02963
Carbon monoxide (CO)	1.474	0.0395	17.22	0.02737
Methane (CH_4)	2.293	0.0428	32.11	0.02965
Nitrogen (N_2)	1.366	0.0386	15.53	0.02677
Oxygen (O_2)	1.369	0.0317	17.22	0.02197
Propane (C_3H_8)	9.349	0.0901	182.23	0.06242
Refrigerant 12	10.49	0.0971	208.59	0.06731
Sulfur dioxide (SO_2)	6.883	0.0569	144.80	0.03945
Water (H_2O)	5.531	0.0305	142.59	0.02111

Source: Calculated from critical data.

2. Benedict-Webb-Rubin: Constants for pressure in bars, specific volume in m^3/kmol, and temperature in K

Substance	a	A	b	B	c	C	α	γ
C_4H_{10}	1.9073	10.218	0.039998	0.12436	3.206×10^5	1.006×10^6	1.101×10^{-3}	0.0340
CO_2	0.1386	2.7737	0.007210	0.04991	1.512×10^4	1.404×10^5	8.47×10^{-5}	0.00539
CO	0.0371	1.3590	0.002632	0.05454	1.054×10^3	8.676×10^3	1.350×10^{-4}	0.0060
CH_4	0.0501	1.8796	0.003380	0.04260	2.579×10^3	2.287×10^4	1.244×10^{-4}	0.0060
N_2	0.0254	1.0676	0.002328	0.04074	7.381×10^2	8.166×10^3	1.272×10^{-4}	0.0053

Source: H. W. Cooper, and J. C. Goldfrank, *Hydrocarbon Processing, 46* (12): 141 (1967).

TABLE A-24 Enthalpy of Formation, Gibbs Function of Formation, and Absolute Entropy of Various Substances at 298 K and 1 atm

$\bar{h}_f^\circ$ and $\bar{g}_f^\circ$ (kJ/kmol), $\bar{s}^\circ$ (kJ/kmol · K)

Substance	Formula	$\bar{h}_f^\circ$	$\bar{g}_f^\circ$	$\bar{s}^\circ$
Carbon	$C(s)$	0	0	5.74
Hydrogen	$H_2(g)$	0	0	130.57
Nitrogen	$N_2(g)$	0	0	191.50
Oxygen	$O_2(g)$	0	0	205.03
Carbon monoxide	$CO(g)$	−110,530	−137,150	197.54
Carbon dioxide	$CO_2(g)$	−393,520	−394,380	213.69
Water	$H_2O(g)$	−241,820	−228,590	188.72
Water	$H_2O(l)$	−285,830	−237,180	69.95
Hydrogen peroxide	$H_2O_2(g)$	−136,310	−105,600	232.63
Ammonia	$NH_3(g)$	−46,190	−16,590	192.33
Oxygen	$O(g)$	249,170	231,770	160.95
Hydrogen	$H(g)$	218,000	203,290	114.61
Nitrogen	$N(g)$	472,680	455,510	153.19
Hydroxyl	$OH(g)$	39,460	34,280	183.75
Methane	$CH_4(g)$	−74,850	−50,790	186.16
Acetylene	$C_2H_2(g)$	226,730	209,170	200.85
Ethylene	$C_2H_4(g)$	52,280	68,120	219.83
Ethane	$C_2H_6(g)$	−84,680	−32,890	229.49
Propylene	$C_3H_6(g)$	20,410	62,720	266.94
Propane	$C_3H_8(g)$	−103,850	−23,490	269.91
Butane	$C_4H_{10}(g)$	−126,150	−15,710	310.03
Pentane	$C_5H_{12}(g)$	−146,440	−8,200	348.40
Octane	$C_8H_{18}(g)$	−208,450	17,320	463.67
Octane	$C_8H_{18}(l)$	−249,910	6,610	360.79
Benzene	$C_6H_6(g)$	82,930	129,660	269.20
Methyl alcohol	$CH_3OH(g)$	−200,890	−162,140	239.70
Methyl alcohol	$CH_3OH(l)$	−238,810	−166,290	126.80
Ethyl alcohol	$C_2H_5OH(g)$	−235,310	−168,570	282.59
Ethyl alcohol	$C_2H_5OH(l)$	−277,690	174,890	160.70

Sources: Adapted from K. Wark, *Thermodynamics, 4th ed.*, McGraw-Hill, New York, 1983, as based on JANAF Thermochemical Tables, NSRDS-NBS-37, 1971; *Selected Values of Chemical Thermodynamic Properties*, NBS Tech. Note 270-3, 1968; and *API Research Project 44*, Carnegie Press, 1953.

TABLE A-25 Logarithms to the Base 10 of the Equilibrium Constant *K*

Temp. K	$\log_{10} K$								Temp. °R
	$H_2 \leftrightharpoons 2H$	$O_2 \leftrightharpoons 2O$	$N_2 \leftrightharpoons 2N$	$\frac{1}{2}O_2 + \frac{1}{2}N_2 \leftrightharpoons NO$	$H_2O \leftrightharpoons H_2 + \frac{1}{2}O_2$	$H_2O \leftrightharpoons OH + \frac{1}{2}H_2$	$CO_2 \leftrightharpoons CO + \frac{1}{2}O_2$	$CO_2 + H_2 \leftrightharpoons CO + H_2O$	
298	−71.224	−81.208	−159.600	−15.171	−40.048	−46.054	−45.066	−5.018	537
500	−40.316	−45.880	−92.672	−8.783	−22.886	−26.130	−25.025	−2.139	900
1000	−17.292	−19.614	−43.056	−4.062	−10.062	−11.280	−10.221	−0.159	1800
1200	−13.414	−15.208	−34.754	−3.275	−7.899	−8.811	−7.764	+0.135	2160
1400	−10.630	−12.054	−28.812	−2.712	−6.347	−7.021	−6.014	+0.333	2520
1600	−8.532	−9.684	−24.350	−2.290	−5.180	−5.677	−4.706	+0.474	2880
1700	−7.666	−8.706	−22.512	−2.116	−4.699	−5.124	−4.169	+0.530	3060
1800	−6.896	−7.836	−20.874	−1.962	−4.270	−4.613	−3.693	+0.577	3240
1900	−6.204	−7.058	−19.410	−1.823	−3.886	−4.190	−3.267	+0.619	3420
2000	−5.580	−6.356	−18.092	−1.699	−3.540	−3.776	−2.884	+0.656	3600
2100	−5.016	−5.720	−16.898	−1.586	−3.227	−3.434	−2.539	+0.688	3780
2200	−4.502	−5.142	−15.810	−1.484	−2.942	−3.091	−2.226	+0.716	3960
2300	−4.032	−4.614	−14.818	−1.391	−2.682	−2.809	−1.940	+0.742	4140
2400	−3.600	−4.130	−13.908	−1.305	−2.443	−2.520	−1.679	+0.764	4320
2500	−3.202	−3.684	−13.070	−1.227	−2.224	−2.270	−1.440	+0.784	4500
2600	−2.836	−3.272	−12.298	−1.154	−2.021	−2.038	−1.219	+0.802	4680
2700	−2.494	−2.892	−11.580	−1.087	−1.833	−1.823	−1.015	+0.818	4860
2800	−2.178	−2.536	−10.914	−1.025	−1.658	−1.624	−0.825	+0.833	5040
2900	−1.882	−2.206	−10.294	−0.967	−1.495	−1.438	−0.649	+0.846	5220
3000	−1.606	−1.898	−9.716	−0.913	−1.343	−1.265	−0.485	+0.858	5400
3100	−1.348	−1.610	−9.174	−0.863	−1.201	−1.103	−0.332	+0.869	5580
3200	−1.106	−1.340	−8.664	−0.815	−1.067	−0.951	−0.189	+0.878	5760
3300	−0.878	−1.086	−8.186	−0.771	−0.942	−0.809	−0.054	+0.888	5940
3400	−0.664	−0.846	−7.736	−0.729	−0.824	−0.674	+0.071	+0.895	6120
3500	−0.462	−0.620	−7.312	−0.690	−0.712	−0.547	+0.190	+0.902	6300

Source: Based on data from the JANAF Thermochemical Tables, NSRDS-NBS-37, 1971.

Table A-25

INDEX TO TABLES IN ENGLISH UNITS

TABLE A-1E Atomic or Molecular Weights and Critical Properties of Some Common Elements and Compounts

Substance	Chemical formula	M	T_c, °R	p_c, atm	$Z_c = \frac{p_c v_c}{RT_c}$
Acetylene	C_2H_2	26.04	556	62	0.274
Air (equivalent)	—	28.97	239	37.2	0.284
Ammonia	NH_3	17.04	730	111.3	0.242
Argon	Ar	39.94	272	47.97	0.290
Benzene	C_6H_6	78.11	1013	48.7	0.274
Butane	C_4H_{10}	58.12	765	37.5	0.274
Carbon	C	12.01	—	—	—
Carbon dioxide	CO_2	44.01	548	72.9	0.276
Carbon monoxide	CO	28.01	239	34.5	0.294
Copper	Cu	63.54	—	—	—
Ethane	C_2H_6	30.07	549	48.2	0.285
Ethyl alcohol	C_2H_5OH	46.07	929	63.0	0.249
Ethylene	C_2H_4	28.05	510	50.5	0.270
Helium	He	4.003	9.33	2.26	0.300
Hydrogen	H_2	2.018	59.8	12.8	0.304
Methane	CH_4	16.04	344	45.8	0.290
Methyl alcohol	CH_3OH	32.05	924	78.5	0.220
Nitrogen	N_2	28.01	227	33.5	0.291
Octane	C_8H_{18}	114.22	1025	24.6	0.258
Oxygen	O_2	32.00	278	49.8	0.290
Propane	C_3H_8	44.09	666	42.1	0.276
Propylene	C_3H_6	42.08	657	45.6	0.276
Refrigerant 12	CCl_2F_2	120.92	693	40.6	0.278
Refrigerant 134a	CF_3CH_2F	102.03	673	40.2	0.260
Sulfur dioxide	SO_2	64.06	775	77.7	0.268
Water	H_2O	18.02	1165	218.0	0.233

Sources: Adapted from *International Critical Tables* and L. C. Nelson and E. F. Obert, Generalized Compressibility Charts, *Chem. Eng.*, *61:* 203 (1954).

$M = \frac{\text{lbmol}}{\text{sub lbm}}$

TABLE A-2E Properties of Saturated Water (Liquid–Vapor): Temperature Table

Temp., °F	Press., lbf/in.^{2}	Specific volume ft^3/lb Sat. liquid v_f	Sat. vapor v_g	Internal energy Btu/lb Sat. liquid u_f	Sat. vapor u_g	Enthalpy Btu/lb Sat. liquid h_f	Evap. h_{fg}	Sat. vapor h_g	Entropy Btu/lb · °R Sat. liquid s_f	Sat. vapor s_g	Temp. °F
32	0.0886	0.01602	3305	−.01	1021.2	−.01	1075.4	1075.4	−.00003	2.1870	32
35	0.0999	0.01602	2948	2.99	1022.2	3.00	1073.7	1076.7	0.00607	2.1764	35
40	0.1217	0.01602	2445	8.02	1023.9	8.02	1070.9	1078.9	0.01617	2.1592	40
45	0.1475	0.01602	2037	13.04	1025.5	13.04	1068.1	1081.1	0.02618	2.1423	45
50	0.1780	0.01602	1704	18.06	1027.2	18.06	1065.2	1083.3	0.03607	2.1259	50
52	0.1917	0.01603	1589	20.06	1027.8	20.07	1064.1	1084.2	0.04000	2.1195	52
54	0.2064	0.01603	1482	22.07	1028.5	22.07	1063.0	1085.1	0.04391	2.1131	54
56	0.2219	0.01603	1383	24.08	1029.1	24.08	1061.9	1085.9	0.04781	2.1068	56
58	0.2386	0.01603	1292	26.08	1029.8	26.08	1060.7	1086.8	0.05159	2.1005	58
60	0.2563	0.01604	1207	28.08	1030.4	28.08	1059.6	1087.7	0.05555	2.0943	60
62	0.2751	0.01604	1129	30.09	1031.1	30.09	1058.5	1088.6	0.05940	2.0882	62
64	0.2952	0.01604	1056	32.09	1031.8	32.09	1057.3	1089.4	0.06323	2.0821	64
66	0.3165	0.01604	988.4	34.09	1032.4	34.09	1056.2	1090.3	0.06704	2.0761	66
68	0.3391	0.01605	925.8	36.09	1033.1	36.09	1055.1	1091.2	0.07084	2.0701	68
70	0.3632	0.01605	867.7	38.09	1033.7	38.09	1054.0	1092.0	0.07463	2.0642	70
72	0.3887	0.01606	813.7	40.09	1034.4	40.09	1052.8	1092.9	0.07839	2.0584	72
74	0.4158	0.01606	763.5	42.09	1035.0	42.09	1051.7	1093.8	0.08215	2.0526	74
76	0.4446	0.01606	716.8	44.09	1035.7	44.09	1050.6	1094.7	0.08589	2.0469	76
78	0.4750	0.01607	673.3	46.09	1036.3	46.09	1049.4	1095.5	0.08961	2.0412	78
80	0.5073	0.01607	632.8	48.08	1037.0	48.09	1048.3	1096.4	0.09332	2.0356	80
82	0.5414	0.01608	595.0	50.08	1037.6	50.08	1047.2	1097.3	0.09701	2.0300	82
84	0.5776	0.01608	559.8	52.08	1038.3	52.08	1046.0	1098.1	0.1007	2.0245	84
86	0.6158	0.01609	527.0	54.08	1038.9	54.08	1044.9	1099.0	0.1044	2.0190	86
88	0.6562	0.01609	496.3	56.07	1039.6	56.07	1043.8	1099.9	0.1080	2.0136	88
90	0.6988	0.01610	467.7	58.07	1040.2	58.07	1042.7	1100.7	0.1117	2.0083	90
92	0.7439	0.01611	440.9	60.06	1040.9	60.06	1041.5	1101.6	0.1153	2.0030	92
94	0.7914	0.01611	415.9	62.06	1041.5	62.06	1040.4	1102.4	0.1189	1.9977	94
96	0.8416	0.01612	392.4	64.05	1041.2	64.06	1039.2	1103.3	0.1225	1.9925	96
98	0.8945	0.01612	370.5	66.05	1042.8	66.05	1038.1	1104.2	0.1261	1.9874	98
100	0.9503	0.01613	350.0	68.04	1043.5	68.05	1037.0	1105.0	0.1296	1.9822	100
110	1.276	0.01617	265.1	78.02	1046.7	78.02	1031.3	1109.3	0.1473	1.9574	110
120	1.695	0.01621	203.0	87.99	1049.9	88.00	1025.5	1113.5	0.1647	1.9336	120
130	2.225	0.01625	157.2	97.97	1053.0	97.98	1019.8	1117.8	0.1817	1.9109	130
140	2.892	0.01629	122.9	107.95	1056.2	107.96	1014.0	1121.9	0.1985	1.8892	140
150	3.722	0.01634	97.0	117.95	1059.3	117.96	1008.1	1126.1	0.2150	1.8684	150
160	4.745	0.01640	77.2	127.94	1062.3	127.96	1002.2	1130.1	0.2313	1.8484	160
170	5.996	0.01645	62.0	137.95	1065.4	137.97	996.2	1134.2	0.2473	1.8293	170
180	7.515	0.01651	50.2	147.97	1068.3	147.99	990.2	1138.2	0.2631	1.8109	180
190	9.343	0.01657	41.0	158.00	1071.3	158.03	984.1	1142.1	0.2787	1.7932	190
200	11.529	0.01663	33.6	168.04	1074.2	168.07	977.9	1145.9	0.2940	1.7762	200

TABLE A-2E *(Continued)*

		Specific volume ft^3/lb_m		Internal energy Btu/lb_m		Enthalpy Btu/lb_m			Entropy $Btu/lb \cdot °R$		
Temp. °F	Press. lbf/in.^{2}	Sat. liquid v_f	Sat. vapor v_g	Sat. liquid u_f	Sat. vapor u_g	Sat. liquid h_f	Evap. h_{fg}	Sat. vapor h_g	Sat. liquid s_f	Sat. vapor s_g	Temp. °F
210	14.13	0.01670	27.82	178.1	1077.0	178.1	971.6	1149.7	0.3091	1.7599	210
212	14.70	0.01672	26.80	180.1	1077.6	180.2	970.3	1150.5	0.3121	1.7567	212
220	17.19	0.01677	23.15	188.2	1079.8	188.2	965.3	1153.5	0.3241	1.7441	220
230	20.78	0.01685	19.39	198.3	1082.6	198.3	958.8	1157.1	0.3388	1.7289	230
240	24.97	0.01692	16.33	208.4	1085.3	208.4	952.3	1160.7	0.3534	1.7143	240
250	29.82	0.01700	13.83	218.5	1087.9	218.6	945.6	1164.2	0.3677	1.7001	250
260	35.42	0.01708	11.77	228.6	1090.5	228.8	938.8	1167.6	0.3819	1.6864	260
270	41.85	0.01717	10.07	238.8	1093.0	239.0	932.0	1170.9	0.3960	1.6731	270
280	49.18	0.01726	8.65	249.0	1095.4	249.2	924.9	1174.1	0.4099	1.6602	280
290	57.53	0.01735	7.47	259.3	1097.7	259.4	917.8	1177.2	0.4236	1.6477	290
300	66.98	0.01745	6.472	269.5	1100.0	269.7	910.4	1180.2	0.4372	1.6356	300
310	77.64	0.01755	5.632	279.8	1102.1	280.1	903.0	1183.0	0.4507	1.6238	310
320	89.60	0.01765	4.919	290.1	1104.2	290.4	895.3	1185.8	0.4640	1.6123	320
330	103.00	0.01776	4.312	300.5	1106.2	300.8	887.5	1188.4	0.4772	1.6010	330
340	117.93	0.01787	3.792	310.9	1108.0	311.3	879.5	1190.8	0.4903	1.5901	340
350	134.53	0.01799	3.346	321.4	1109.8	321.8	871.3	1193.1	0.5033	1.5793	350
360	152.92	0.01811	2.961	331.8	1111.4	332.4	862.9	1195.2	0.5162	1.5688	360
370	173.23	0.01823	2.628	342.4	1112.9	343.0	854.2	1197.2	0.5289	1.5585	370
380	195.60	0.01836	2.339	353.0	1114.3	353.6	845.4	1199.0	0.5416	1.5483	380
390	220.2	0.01850	2.087	363.6	1115.6	364.3	836.2	1200.6	0.5542	1.5383	390
400	247.1	0.01864	1.866	374.3	1116.6	375.1	826.8	1202.0	0.5667	1.5284	400
410	276.5	0.01878	1.673	385.0	1117.6	386.0	817.2	1203.1	0.5792	1.5187	410
420	308.5	0.01894	1.502	395.8	1118.3	396.9	807.2	1204.1	0.5915	1.5091	420
430	343.3	0.01909	1.352	406.7	1118.9	407.9	796.9	1204.8	0.6038	1.4995	430
440	381.2	0.01926	1.219	417.6	1119.3	419.0	786.3	1205.3	0.6161	1.4900	440
450	422.1	0.01943	1.1011	428.6	1119.5	430.2	775.4	1205.6	0.6282	1.4806	450
460	466.3	0.01961	0.9961	439.7	1119.6	441.4	764.1	1205.5	0.6404	1.4712	460
470	514.1	0.01980	0.9025	450.9	1119.4	452.8	752.4	1205.2	0.6525	1.4618	470
480	565.5	0.02000	0.8187	462.2	1118.9	464.3	740.3	1204.6	0.6646	1.4524	480
490	620.7	0.02021	0.7436	473.6	1118.3	475.9	727.8	1203.7	0.6767	1.4430	490
500	680.0	0.02043	0.6761	485.1	1117.4	487.7	714.8	1202.5	0.6888	1.4335	500
520	811.4	0.02091	0.5605	508.5	1114.8	511.7	687.3	1198.9	0.7130	1.4145	520
540	961.5	0.02145	0.4658	532.6	1111.0	536.4	657.5	1193.8	0.7374	1.3950	540
560	1131.8	0.02207	0.3877	548.4	1105.8	562.0	625.0	1187.0	0.7620	1.3749	560
580	1324.3	0.02278	0.3225	583.1	1098.9	588.6	589.3	1178.0	0.7872	1.3540	580
600	1541.0	0.02363	0.2677	609.9	1090.0	616.7	549.7	1166.4	0.8130	1.3317	600
620	1784.4	0.02465	0.2209	638.3	1078.5	646.4	505.0	1151.4	0.8398	1.3075	620
640	2057.1	0.02593	0.1805	668.7	1063.2	678.6	453.4	1131.9	0.8681	1.2803	640
660	2362	0.02767	0.1446	702.3	1042.3	714.4	391.1	1105.5	0.8990	1.2483	660
680	2705	0.03032	0.1113	741.7	1011.0	756.9	309.8	1066.7	0.9350	1.2068	680
700	3090	0.03666	0.0744	801.7	947.7	822.7	167.5	990.2	0.9902	1.1346	700
705.4	3204	0.05053	0.05053	872.6	872.6	902.5	0	902.5	1.0580	1.0580	705.4

Source: Tables A-2E through A-5E are adapted from K. Wark, *Thermodynamics, 4th ed.*, McGraw-Hill, New York, 1983, as extracted from J. H. Keenan, F. G. Keyes, P. G. Hill, and J. G. Moore, "Steam Tables," Wiley, New York, 1969.

TABLE A-3E Properties of Saturated Water (Liquid–Vapor): Pressure Table

Press. lbf/in.^{2}	Temp., °F	Specific volume ft^3/lb Sat. liquid v_f	Sat. vapor v_g	Internal energy Btu/lb Sat. liquid u_f	Sat. vapor u_g	Enthalpy Btu/lb Sat. liquid h_f	Evap. h_{fg}	Sat. vapor h_g	Entropy Btu/lb · °R Sat. liquid s_f	Evap. s_{fg}	Sat. vapor s_g	Press. lbf/in.^{2}
0.4	72.84	0.01606	792.0	40.94	1034.7	40.94	1052.3	1093.3	0.0800	1.9760	2.0559	0.4
0.6	85.19	0.01609	540.0	53.26	1038.7	53.27	1045.4	1098.6	0.1029	1.9184	2.0213	0.6
0.8	94.35	0.01611	411.7	62.41	1041.7	62.41	1040.2	1102.6	0.1195	1.8773	1.9968	0.8
1.0	101.70	0.01614	333.6	69.74	1044.0	69.74	1036.0	1105.8	0.1327	1.8453	1.9779	1.0
1.2	107.88	0.01616	280.9	75.90	1046.0	75.90	1032.5	1108.4	0.1436	1.8190	1.9626	1.2
1.5	115.65	0.01619	227.7	83.65	1048.5	83.65	1028.0	1111.7	0.1571	1.7867	1.9438	1.5
2.0	126.04	0.01623	173.75	94.02	1051.8	94.02	1022.1	1116.1	0.1750	1.7448	1.9198	2.0
3.0	141.43	0.01630	118.72	109.38	1056.6	109.39	1013.1	1122.5	0.2009	1.6852	1.8861	3.0
4.0	152.93	0.01636	90.64	120.88	1060.2	120.89	1006.4	1127.3	0.2198	1.6426	1.8624	4.0
5.0	162.21	0.01641	73.53	130.15	1063.0	130.17	1000.9	1131.0	0.2349	1.6093	1.8441	5.0
6.0	170.03	0.01645	61.98	137.98	1065.4	138.00	996.2	1134.2	0.2474	1.5819	1.8292	6.0
7.0	176.82	0.01649	53.65	144.78	1067.4	144.80	992.1	1136.9	0.2581	1.5585	1.8167	7.0
8.0	182.84	0.01653	47.35	150.81	1069.2	150.84	988.4	1139.3	0.2675	1.5383	1.8058	8.0
9.0	188.26	0.01656	42.41	156.25	1070.8	156.27	985.1	1141.4	0.2760	1.5203	1.7963	9.0
10	193.19	0.01659	38.42	161.20	1072.2	161.23	982.1	1143.3	0.2836	1.5041	1.7877	10
14.696	211.99	0.01672	26.80	180.10	1077.6	180.15	970.4	1150.5	0.3121	1.4446	1.7567	14.696
15	213.03	0.01672	26.29	181.14	1077.9	181.19	969.7	1150.9	0.3137	1.4414	1.7551	15
20	227.96	0.01683	20.09	196.19	1082.0	196.26	960.1	1156.4	0.3358	1.3962	1.7320	20
25	240.08	0.01692	16.31	208.44	1085.3	208.52	952.2	1160.7	0.3535	1.3607	1.7142	25
30	250.34	0.01700	13.75	218.84	1088.0	218.93	945.4	1164.3	0.3682	1.3314	1.6996	30
35	259.30	0.01708	11.90	227.93	1090.3	228.04	939.3	1167.4	0.3809	1.3064	1.6873	35
40	267.26	0.01715	10.50	236.03	1092.3	236.16	933.8	1170.0	0.3921	1.2845	1.6767	40
45	274.46	0.01721	9.40	243.37	1094.0	243.51	928.8	1172.3	0.4022	1.2651	1.6673	45
50	281.03	0.01727	8.52	250.08	1095.6	250.24	924.2	1174.4	0.4113	1.2476	1.6589	50
55	287.10	0.01733	7.79	256.28	1097.0	256.46	919.9	1176.3	0.4196	1.2317	1.6513	55
60	292.73	0.01738	7.177	262.1	1098.3	262.2	915.8	1178.0	0.4273	1.2170	1.6443	60
65	298.00	0.01743	6.647	267.5	1099.5	267.7	911.9	1179.6	0.4345	1.2035	1.6380	65
70	302.96	0.01748	6.209	272.6	1100.6	272.8	908.3	1181.0	0.4412	1.1909	1.6321	70
75	307.63	0.01752	5.818	277.4	1101.6	277.6	904.8	1182.4	0.4475	1.1790	1.6265	75
80	312.07	0.01757	5.474	282.0	1102.6	282.2	901.4	1183.6	0.4534	1.1679	1.6213	80
85	316.29	0.01761	5.170	286.3	1103.5	286.6	898.2	1184.8	0.4591	1.1574	1.6165	85
90	320.31	0.01766	4.898	290.5	1104.3	290.8	895.1	1185.9	0.4644	1.1475	1.6119	90
95	324.16	0.01770	4.654	294.5	1105.0	294.8	892.1	1186.9	0.4695	1.1380	1.6075	95
100	327.86	0.01774	4.434	298.3	1105.8	298.6	889.2	1187.8	0.4744	1.1290	1.6034	100
110	334.82	0.01781	4.051	305.5	1107.1	305.9	883.7	1189.6	0.4836	1.1122	1.5958	110
120	341.30	0.01789	3.730	312.3	1108.3	312.7	878.5	1191.1	0.4920	1.0966	1.5886	120
130	347.37	0.01796	3.457	318.6	1109.4	319.0	873.5	1192.5	0.4999	1.0822	1.5821	130
140	353.08	0.01802	3.221	324.6	1110.3	325.1	868.7	1193.8	0.5073	1.0688	1.5761	140
150	358.48	0.01809	3.016	330.2	1111.2	330.8	864.2	1194.9	0.5142	1.0562	1.5704	150
160	363.60	0.01815	2.836	335.6	1112.0	336.2	859.8	1196.0	0.5208	1.0443	1.5651	160

TABLE A-3E *(Continued)*

		Specific volume ft³/lb		Internal energy Btu/lb		Enthalpy Btu/lb			Entropy Btu/lb · °R			
Press. lbf/in.²	Temp., °F	Sat. liquid v_f	Sat. vapor v_g	Sat. liquid u_f	Sat. vapor u_g	Sat. liquid h_f	Evap. h_{fg}	Sat. vapor h_g	Sat. liquid s_f	Evap. s_{fg}	Sat. vapor s_g	Press. lbf/in.²
170	368.47	0.01821	2.676	340.8	1112.7	341.3	855.6	1196.9	0.5270	1.0330	1.5600	170
180	373.13	0.01827	2.553	345.7	1113.4	346.3	851.5	1197.8	0.5329	1.0223	1.5552	180
190	377.59	0.01833	2.405	350.4	1114.0	351.0	847.5	1198.6	0.5386	1.0122	1.5508	190
200	381.86	0.01839	2.289	354.9	1114.6	355.6	843.7	1199.3	0.5440	1.0025	1.5465	200
250	401.04	0.01865	1.845	375.4	1116.7	376.2	825.8	1202.1	0.5680	0.9594	1.5274	250
300	417.43	0.01890	1.544	393.0	1118.2	394.1	809.8	1203.9	0.5883	0.9232	1.5115	300
350	431.82	0.01912	1.327	408.7	1119.0	409.9	795.0	1204.9	0.6060	0.8917	1.4977	350
400	444.70	0.01934	1.162	422.8	1119.5	424.2	781.2	1205.5	0.6218	0.8638	1.4856	400
450	456.39	0.01955	1.033	435.7	1119.6	437.4	768.2	1205.6	0.6360	0.8385	1.4745	450
500	467.13	0.01975	0.928	447.7	1119.4	449.5	755.8	1205.3	0.6490	0.8154	1.4644	500
550	477.07	0.01994	0.842	458.9	1119.1	460.9	743.9	1204.8	0.6611	0.7941	1.4551	550
600	486.33	0.02013	0.770	469.4	1118.6	471.7	732.4	1204.1	0.6723	0.7742	1.4464	600
700	503.23	0.02051	0.656	488.9	1117.0	491.5	710.5	1202.0	0.6927	0.7378	1.4305	700
800	518.36	0.02087	0.569	506.6	1115.0	509.7	689.6	1199.3	0.7110	0.7050	1.4160	800
900	532.12	0.02123	0.501	523.0	1112.6	526.6	669.5	1196.0	0.7277	0.6750	1.4027	900
1000	544.75	0.02159	0.446	538.4	1109.9	542.4	650.0	1192.4	0.7432	0.6471	1.3903	1000
1100	556.45	0.02195	0.401	552.9	1106.8	557.4	631.0	1188.3	0.7576	0.6209	1.3786	1100
1200	567.37	0.02232	0.362	566.7	1103.5	571.7	612.3	1183.9	0.7712	0.5961	1.3673	1200
1300	577.60	0.02269	0.330	579.9	1099.8	585.4	593.8	1179.2	0.7841	0.5724	1.3565	1300
1400	587.25	0.02307	0.302	592.7	1096.0	598.6	575.5	1174.1	0.7964	0.5497	1.3461	1400
1500	596.39	0.02346	0.277	605.0	1091.8	611.5	557.2	1168.7	0.8082	0.5276	1.3359	1500
1600	605.06	0.02386	0.255	616.9	1087.4	624.0	538.9	1162.9	0.8196	0.5062	1.3258	1600
1700	613.32	0.02428	0.236	628.6	1082.7	636.2	520.6	1156.9	0.8307	0.4852	1.3159	1700
1800	621.21	0.02472	0.218	640.0	1077.7	648.3	502.1	1150.4	0.8414	0.4645	1.3060	1800
1900	628.76	0.02517	0.203	651.3	1072.3	660.1	483.4	1143.5	0.8519	0.4441	1.2961	1900
2000	636.00	0.02565	0.188	662.4	1066.6	671.9	464.4	1136.3	0.8623	0.4238	1.2861	2000
2250	652.90	0.02698	0.157	689.9	1050.6	701.1	414.8	1115.9	0.8876	0.3728	1.2604	2250
2500	668.31	0.02860	0.131	717.7	1031.0	730.9	360.5	1091.4	0.9131	0.3196	1.2327	2500
2750	682.46	0.03077	0.107	747.3	1005.9	763.0	297.4	1060.4	0.9401	0.2604	1.2005	2750
3000	695.52	0.03431	0.084	783.4	968.8	802.5	213.0	1015.5	0.9732	0.1843	1.1575	3000
3203.6	705.44	0.05053	0.0505	872.6	872.6	902.5	0	902.5	1.0580	0	1.0580	3203.6

TABLE A-4E Properties of Superheated Water Vapor

T °F	v ft³/lb	u Btu/lb	h Btu/lb	s Btu/lb · °R	v ft³/lb	u Btu/lb	h Btu/lb	s Btu/lb · °R
	p = 1 lbf/in.² (T_{sat} = 101.7°F)				p = 5 lbf/in.² (T_{sat} = 162.2°F)			
Sat.	333.6	1044.0	1105.8	1.9779	73.53	1063.0	1131.0	1.8441
150	362.6	1060.4	1127.5	2.0151				
200	392.5	1077.5	1150.1	2.0508	78.15	1076.0	1148.6	1.8715
250	422.4	1094.7	1172.8	2.0839	84.21	1093.8	1171.7	1.9052
300	452.3	1112.0	1195.7	2.1150	90.24	1111.3	1194.8	1.9367
400	511.9	1147.0	1241.8	2.1720	102.24	1146.6	1241.2	1.9941
500	571.5	1182.8	1288.5	2.2235	114.20	1182.5	1288.2	2.0458
600	631.1	1219.3	1336.1	2.2706	126.15	1219.1	1335.8	2.0930
700	690.7	1256.7	1384.5	2.3142	138.08	1256.5	1384.3	2.1367
800	750.3	1294.4	1433.7	2.3550	150.01	1294.7	1433.5	2.1775
900	809.9	1333.9	1483.8	2.3932	161.94	1333.8	1483.7	2.2158
1000	869.5	1373.9	1534.8	2.4294	173.86	1373.9	1534.7	2.2520

T °F	v ft³/lb	u Btu/lb	h Btu/lb	s Btu/lb · °R	v ft³/lb	u Btu/lb	h Btu/lb	s Btu/lb · °R
	p = 10 lbf/in.² (T_{sat} = 193.2°F)				p = 14.7 lbf/in.² (T_{sat} = 212.0°F)			
Sat.	38.42	1072.2	1143.3	1.7877	26.80	1077.6	1150.5	1.7567
200	38.85	1074.7	1146.6	1.7927				
250	41.95	1092.6	1170.2	1.8272	28.42	1091.5	1168.8	1.7832
300	44.99	1110.4	1193.7	1.8592	30.52	1109.6	1192.6	1.8157
400	51.03	1146.1	1240.5	1.9171	34.67	1145.6	1239.9	1.8741
500	57.04	1182.2	1287.7	1.9690	38.77	1181.8	1287.3	1.9263
600	63.03	1218.9	1335.5	2.0164	42.86	1218.6	1335.2	1.9737
700	69.01	1256.3	1384.0	2.0601	46.93	1256.1	1383.8	2.0175
800	74.98	1294.6	1433.3	2.1009	51.00	1294.4	1433.1	2.0584
900	80.95	1333.7	1483.5	2.1393	55.07	1333.6	1483.4	2.0967
1000	86.91	1373.8	1534.6	2.1755	59.13	1373.7	1534.5	2.1330
1100	92.88	1414.7	1586.6	2.2099	63.19	1414.6	1586.4	2.1674

T °F	v ft³/lb	u Btu/lb	h Btu/lb	s Btu/lb · °R	v ft³/lb	u Btu/lb	h Btu/lb	s Btu/lb · °R
	p = 20 lbf/in.² (T_{sat} = 228.0°F)				p = 40 lbf/in.² (T_{sat} = 267.3°F)			
Sat.	20.09	1082.0	1156.4	1.7320	10.50	1093.3	1170.0	1.6767
250	20.79	1090.3	1167.2	1.7475				
300	22.36	1108.7	1191.5	1.7805	11.04	1105.1	1186.8	1.6993
350	23.90	1126.9	1215.4	1.8110	11.84	1124.2	1211.8	1.7312
400	25.43	1145.1	1239.2	1.8395	12.62	1143.0	1236.4	1.7606
500	28.46	1181.5	1286.8	1.8919	14.16	1180.1	1284.9	1.8140
600	31.47	1218.4	1334.8	1.9395	15.69	1217.3	1333.4	1.8621
700	34.47	1255.9	1383.5	1.9834	17.20	1255.1	1382.4	1.9063
800	37.46	1294.3	1432.9	2.0243	18.70	1293.7	1432.1	1.9474
900	40.45	1333.5	1483.2	2.0627	20.20	1333.0	1482.5	1.9859
1000	43.44	1373.5	1534.3	2.0989	21.70	1373.1	1533.8	2.0223
1100	46.42	1414.5	1586.3	2.1334	23.20	1414.2	1585.9	2.0568

H_2O

TABLE A-4E (*Continued*)

T °F	v ft³/lb	u Btu/lb	h Btu/lb	s Btu/lb · °R	v ft³/lb	u Btu/lb	h Btu/lb	s Btu/lb · °R
	p = 60 lbf/in.² (T_{sat} = 292.7°F)				p = 80 lbf/in.² (T_{sat} = 312.1°F)			
Sat.	7.17	1098.3	1178.0	1.6444	5.47	1102.6	1183.6	1.6214
300	7.26	1101.3	1181.9	1.6496				
350	7.82	1121.4	1208.2	1.6830	5.80	1118.5	1204.3	1.6476
400	8.35	1140.8	1233.5	1.7134	6.22	1138.5	1230.6	1.6790
500	9.40	1178.6	1283.0	1.7678	7.02	1177.2	1281.1	1.7346
600	10.43	1216.3	1332.1	1.8165	7.79	1215.3	1330.7	1.7838
700	11.44	1254.4	1381.4	1.8609	8.56	1253.6	1380.3	1.8285
800	12.45	1293.0	1431.2	1.9022	9.32	1292.4	1430.4	1.8700
900	13.45	1332.5	1481.8	1.9408	10.08	1332.0	1481.2	1.9087
1000	14.45	1372.7	1533.2	1.9773	10.83	1372.3	1532.6	1.9453
1100	15.45	1413.8	1585.4	2.0119	11.58	1413.5	1584.9	1.9799
1200	16.45	1455.8	1638.5	2.0448	12.33	1455.5	1638.1	2.0130

T °F	v ft³/lb	u Btu/lb	h Btu/lb	s Btu/lb · °R	v ft³/lb	u Btu/lb	h Btu/lb	s Btu/lb · °R
	p = 100 lbf/in.² (T_{sat} = 327.8°F)				p = 120 lbf/in.² (T_{sat} = 341.3°F)			
Sat.	4.434	1105.8	1187.8	1.6034	3.730	1108.3	1191.1	1.5886
350	4.592	1115.4	1200.4	1.6191	3.783	1112.2	1196.2	1.5950
400	4.934	1136.2	1227.5	1.6517	4.079	1133.8	1224.4	1.6288
450	5.265	1156.2	1253.6	1.6812	4.360	1154.3	1251.2	1.6590
500	5.587	1175.7	1279.1	1.7085	4.633	1174.2	1277.1	1.6868
600	6.216	1214.2	1329.3	1.7582	5.164	1213.2	1327.8	1.7371
700	6.834	1252.8	1379.2	1.8033	5.682	1252.0	1378.2	1.7825
800	7.445	1291.8	1429.6	1.8449	6.195	1291.2	1428.7	1.8243
900	8.053	1331.5	1480.5	1.8838	6.703	1330.9	1479.8	1.8633
1000	8.657	1371.9	1532.1	1.9204	7.208	1371.5	1531.5	1.9000
1100	9.260	1413.1	1584.5	1.9551	7.711	1412.8	1584.0	1.9348
1200	9.861	1455.2	1637.7	1.9882	8.213	1454.9	1637.3	1.9679

T °F	v ft³/lb	u Btu/lb	h Btu/lb	s Btu/lb · °R	v ft³/lb	u Btu/lb	h Btu/lb	s Btu/lb · °R
	p = 140 lbf/in.² (T_{sat} = 353.1°F)				p = 160 lbf/in.² (T_{sat} = 363.6°F)			
Sat.	3.221	1110.3	1193.8	1.5761	2.836	1112.0	1196.0	1.5651
400	3.466	1131.4	1221.2	1.6088	3.007	1128.8	1217.8	1.5911
450	3.713	1152.4	1248.6	1.6399	3.228	1150.5	1246.1	1.6230
500	3.952	1172.7	1275.1	1.6682	3.440	1171.2	1273.0	1.6518
550	4.184	1192.5	1300.9	1.6945	3.646	1191.3	1299.2	1.6785
600	4.412	1212.1	1326.4	1.7191	3.848	1211.1	1325.0	1.7034
700	4.860	1251.2	1377.1	1.7648	4.243	1250.4	1376.0	1.7494
800	5.301	1290.5	1427.9	1.8068	4.631	1289.9	1427.0	1.7916
900	5.739	1330.4	1479.1	1.8459	5.015	1329.9	1478.4	1.8308
1000	6.173	1371.0	1531.0	1.8827	5.397	1370.6	1530.4	1.8677
1100	6.605	1412.4	1583.6	1.9176	5.776	1412.1	1583.1	1.9026
1200	7.036	1454.6	1636.9	1.9507	6.154	1454.3	1636.5	1.9358

TABLE A-4E *(Continued)*

T °F	v ft³/lb	u Btu/lb	h Btu/lb	s Btu/lb · °R	v ft³/lb	u Btu/lb	h Btu/lb	s Btu/lb · °R
	p = 180 lbf/in.² (T_{sat} = 373.1°F)				p = 200 lbf/in.² (T_{sat} = 381.8°F)			
Sat.	2.533	1113.4	1197.8	1.5553	2.289	1114.6	1199.3	1.5464
400	2.648	1126.2	1214.4	1.5749	2.361	1123.5	1210.8	1.5600
450	2.850	1148.5	1243.4	1.6078	2.548	1146.4	1240.7	1.5938
500	3.042	1169.6	1270.9	1.6372	2.724	1168.0	1268.8	1.6239
550	3.228	1190.0	1297.5	1.6642	2.893	1188.7	1295.7	1.6512
600	3.409	1210.0	1323.5	1.6893	3.058	1208.9	1322.1	1.6767
700	3.763	1249.6	1374.9	1.7357	3.379	1248.8	1373.8	1.7234
800	4.110	1289.3	1426.2	1.7781	3.693	1288.6	1425.3	1.7660
900	4.453	1329.4	1477.7	1.8174	4.003	1328.9	1477.1	1.8055
1000	4.793	1370.2	1529.8	1.8545	4.310	1369.8	1529.3	1.8425
1100	5.131	1411.7	1582.6	1.8894	4.615	1411.4	1582.2	1.8776
1200	5.467	1454.0	1636.1	1.9227	4.918	1453.7	1635.7	1.9109
	p = 250 lbf/in.² (T_{sat} = 401.0°F)				p = 300 lbf/in.² (T_{sat} = 417.4°F)			
Sat.	1.845	1116.7	1202.1	1.5274	1.544	1118.2	1203.9	1.5115
450	2.002	1141.1	1233.7	1.5632	1.636	1135.4	1226.2	1.5365
500	2.150	1163.8	1263.3	1.5948	1.766	1159.5	1257.5	1.5701
550	2.290	1185.3	1291.3	1.6233	1.888	1181.9	1286.7	1.5997
600	2.426	1206.1	1318.3	1.6494	2.004	1203.2	1314.5	1.6266
700	2.688	1246.7	1371.1	1.6970	2.227	1244.0	1368.3	1.6751
800	2.943	1287.0	1423.2	1.7301	2.442	1285.4	1421.0	1.7187
900	3.193	1327.6	1475.3	1.7799	2.653	1326.3	1473.6	1.7589
1000	3.440	1368.7	1527.9	1.8172	2.860	1367.7	1526.5	1.7964
1100	3.685	1410.5	1581.0	1.8524	3.066	1409.6	1579.8	1.8317
1200	3.929	1453.0	1634.8	1.8858	3.270	1452.2	1633.8	1.8653
1300	4.172	1496.3	1689.3	1.9177	3.473	1495.6	1688.4	1.8973
	p = 350 lbf/in.² (T_{sat} = 431.8°F)				p = 400 lbf/in.² (T_{sat} = 444.7°F)			
Sat.	1.327	1119.0	1204.9	1.4978	1.162	1119.5	1205.5	1.4856
450	1.373	1129.2	1218.2	1.5125	1.175	1122.6	1209.6	1.4901
500	1.491	1154.9	1251.5	1.5482	1.284	1150.1	1245.2	1.5282
550	1.600	1178.3	1281.9	1.5790	1.383	1174.6	1277.0	1.5605
600	1.703	1200.3	1310.6	1.6068	1.476	1197.3	1306.6	1.5892
700	1.898	1242.5	1365.4	1.6562	1.650	1240.4	1362.5	1.6397
800	2.085	1283.8	1418.8	1.7004	1.816	1282.1	1416.6	1.6844
900	2.267	1325.0	1471.8	1.7409	1.978	1323.7	1470.1	1.7252
1000	2.446	1366.6	1525.0	1.7787	2.136	1365.5	1523.6	1.7632
1100	2.624	1408.7	1578.6	1.8142	2.292	1407.8	1577.4	1.7989
1200	2.799	1451.5	1632.8	1.8478	2.446	1450.7	1631.8	1.8327
1300	2.974	1495.0	1687.6	1.8799	2.599	1494.3	1686.8	1.8648

H_2O

TABLE A-4E (*Continued*)

T °F	v ft³/lb	u Btu/lb	h Btu/lb	s Btu/lb · °R	v ft³/lb	u Btu/lb	h Btu/lb	s Btu/lb · °R
	p = 450 lbf/in.² (T_{sat} = 456.4°F)				p = 500 lbf/in.² (T_{sat} = 467.1°F)			
Sat.	1.033	1119.6	1205.6	1.4746	0.928	1119.4	1205.3	1.4645
500	1.123	1145.1	1238.5	1.5097	0.992	1139.7	1231.5	1.4923
550	1.215	1170.7	1271.9	1.5436	1.079	1166.7	1266.6	1.5279
600	1.300	1194.3	1302.5	1.5732	1.158	1191.1	1298.3	1.5585
700	1.458	1238.2	1359.6	1.6248	1.304	1236.0	1356.7	1.6112
800	1.608	1280.5	1414.4	1.6701	1.441	1278.8	1412.1	1.6571
900	1.752	1322.4	1468.3	1.7113	1.572	1321.0	1466.5	1.6987
1000	1.894	1364.4	1522.2	1.7495	1.701	1363.3	1520.7	1.7471
1100	2.034	1406.9	1576.3	1.7853	1.827	1406.0	1575.1	1.7731
1200	2.172	1450.0	1630.8	1.8192	1.952	1449.2	1629.8	1.8072
1300	2.308	1493.7	1685.9	1.8515	2.075	1493.1	1685.1	1.8395
1400	2.444	1538.1	1741.7	1.8823	2.198	1537.6	1741.0	1.8704

T °F	v ft³/lb	u Btu/lb	h Btu/lb	s Btu/lb · °R	v ft³/lb	u Btu/lb	h Btu/lb	s Btu/lb · °R
	p = 600 lbf/in.² (T_{sat} = 486.3°F)				p = 700 lbf/in.² (T_{sat} = 503.2°F)			
Sat.	0.770	1118.6	1204.1	1.4464	0.656	1117.0	1202.0	1.4305
500	0.795	1128.0	1216.2	1.4592				
550	0.875	1158.2	1255.4	1.4990	0.728	1149.0	1243.2	1.4723
600	0.946	1184.5	1289.5	1.5320	0.793	1177.5	1280.2	1.5081
700	1.073	1231.5	1350.6	1.5872	0.907	1226.9	1344.4	1.5661
800	1.190	1275.4	1407.6	1.6343	1.011	1272.0	1402.9	1.6145
900	1.302	1318.4	1462.9	1.6766	1.109	1315.6	1459.3	1.6576
1000	1.411	1361.2	1517.8	1.7155	1.204	1358.9	1514.9	1.6970
1100	1.517	1404.2	1572.7	1.7519	1.296	1402.4	1570.2	1.7337
1200	1.622	1447.7	1627.8	1.7861	1.387	1446.2	1625.8	1.7682
1300	1.726	1491.7	1683.4	1.8186	1.476	1490.4	1681.7	1.8009
1400	1.829	1536.5	1739.5	1.8497	1.565	1535.3	1738.1	1.8321

T °F	v ft³/lb	u Btu/lb	h Btu/lb	s Btu/lb · °R	v ft³/lb	u Btu/lb	h Btu/lb	s Btu/lb · °R
	p = 800 lbf/in.² (T_{sat} = 518.3°F)				p = 900 lbf/in.² (T_{sat} = 532.1°F)			
Sat.	0.569	1115.0	1199.3	1.4160	0.501	1112.6	1196.0	1.4027
550	0.615	1138.8	1229.9	1.4469	0.527	1127.5	1215.2	1.4219
600	0.677	1170.1	1270.4	1.4861	0.587	1162.2	1260.0	1.4652
650	0.732	1197.2	1305.6	1.5186	0.639	1191.1	1297.5	1.4999
700	0.783	1222.1	1338.0	1.5471	0.686	1217.1	1331.4	1.5297
800	0.876	1268.5	1398.2	1.5969	0.772	1264.9	1393.4	1.5810
900	0.964	1312.9	1455.6	1.6408	0.851	1310.1	1451.9	1.6257
1000	1.048	1356.7	1511.9	1.6807	0.927	1354.5	1508.9	1.6662
1100	1.130	1400.5	1567.8	1.7178	1.001	1398.7	1565.4	1.7036
1200	1.210	1444.6	1623.8	1.7526	1.073	1443.0	1621.7	1.7386
1300	1.289	1489.1	1680.0	1.7854	1.144	1487.8	1687.3	1.7717
1400	1.367	1534.2	1736.6	1.8167	1.214	1533.0	1735.1	1.8031

H_2O

TABLE A-4E *(Continued)*

T °F	v ft³/lb	u Btu/lb	h Btu/lb	s Btu/lb · °R	v ft³/lb	u Btu/lb	h Btu/lb	s Btu/lb · °R
	p = 1000 lbf/in.² (T_{sat} = 544.7°F)				p = 1200 lbf/in.² (T_{sat} = 567.4°F)			
Sat.	0.446	1109.0	1192.4	1.3903	0.362	1103.5	1183.9	1.3673
600	0.514	1153.7	1248.8	1.4450	0.402	1134.4	1223.6	1.4054
650	0.564	1184.7	1289.1	1.4822	0.450	1170.9	1270.8	1.4490
700	0.608	1212.0	1324.6	1.5135	0.491	1201.3	1310.2	1.4837
800	0.688	1261.2	1388.5	1.5665	0.562	1253.7	1378.4	1.5402
900	0.761	1307.3	1448.1	1.6120	0.626	1301.5	1440.4	1.5876
1000	0.831	1352.2	1505.9	1.6530	0.685	1347.5	1499.7	1.6297
1100	0.898	1396.8	1562.9	1.6908	0.743	1393.0	1557.9	1.6682
1200	0.963	1441.5	1619.7	1.7261	0.798	1438.3	1615.5	1.7040
1300	1.027	1486.5	1676.5	1.7593	0.853	1483.8	1673.1	1.7377
1400	1.091	1531.9	1733.7	1.7909	0.906	1529.6	1730.7	1.7696
1600	1.215	1624.4	1849.3	1.8499	1.011	1622.6	1847.1	1.8290
	p = 1400 lbf/in.² (T_{sat} = 587.2°F)				p = 1600 lbf/in.² (T_{sat} = 605.1°F)			
Sat.	0.302	1096.0	1174.1	1.3461	0.255	1087.4	1162.9	1.3258
600	0.318	1110.9	1193.1	1.3641				
650	0.367	1155.5	1250.5	1.4171	0.303	1137.8	1227.4	1.3852
700	0.406	1189.6	1294.8	1.4562	0.342	1177.0	1278.1	1.4299
800	0.471	1245.8	1367.9	1.5168	0.403	1237.7	1357.0	1.4953
900	0.529	1295.6	1432.5	1.5661	0.466	1289.5	1424.4	1.5468
1000	0.582	1342.8	1493.5	1.6094	0.504	1338.0	1487.1	1.5913
1100	0.632	1389.1	1552.8	1.6487	0.549	1385.2	1547.7	1.6315
1200	0.681	1435.1	1611.4	1.6851	0.592	1431.8	1607.1	1.6684
1300	0.728	1481.1	1669.6	1.7192	0.634	1478.3	1666.1	1.7029
1400	0.774	1527.2	1727.8	1.7513	0.675	1524.9	1724.8	1.7354
1600	0.865	1620.8	1844.8	1.8111	0.755	1619.0	1842.6	1.7955
	p = 1800 lbf/in.² (T_{sat} = 621.2°F)				p = 2000 lbf/in.² (T_{sat} = 636.0°F)			
Sat.	0.218	1077.7	1150.4	1.3060	0.188	1066.6	1136.3	1.2861
650	0.251	1117.0	1200.4	1.3517	0.206	1091.1	1167.2	1.3141
700	0.291	1163.1	1259.9	1.4042	0.249	1147.7	1239.8	1.3782
750	0.322	1198.6	1305.9	1.4430	0.280	1187.3	1291.1	1.4216
800	0.350	1229.1	1345.7	1.4753	0.307	1220.1	1333.8	1.4562
900	0.399	1283.2	1416.1	1.5291	0.353	1276.8	1407.6	1.5126
1000	0.443	1333.1	1480.7	1.5749	0.395	1328.1	1474.1	1.5598
1100	0.484	1381.2	1542.5	1.6159	0.433	1377.2	1537.2	1.6017
1200	0.524	1428.5	1602.9	1.6534	0.469	1425.2	1598.6	1.6398
1300	0.561	1475.5	1662.5	1.6883	0.503	1472.7	1659.0	1.6751
1400	0.598	1522.5	1721.8	1.7211	0.537	1520.2	1718.8	1.7082
1600	0.670	1617.2	1840.4	1.7817	0.602	1615.4	1838.2	1.7692

H_2O

TABLE A-4E *(Continued)*

T °F	v ft³/lb	u Btu/lb	h Btu/lb	s Btu/lb · °R	v ft³/lb	u Btu/lb	h Btu/lb	s Btu/lb · °R
	p = 2500 lbf/in.² (T_{sat} = 668.3°F)				p = 3000 lbf/in.² (T_{sat} = 695.5°F)			
Sat.	0.1306	1031.0	1091.4	1.2327	0.0840	968.8	1015.5	1.1575
700	0.1684	1098.7	1176.6	1.3073	0.0977	1003.9	1058.1	1.1944
750	0.2030	1155.2	1249.1	1.3686	0.1483	1114.7	1197.1	1.3122
800	0.2291	1195.7	1301.7	1.4112	0.1757	1167.6	1265.2	1.3675
900	0.2712	1259.9	1385.4	1.4752	0.2160	1241.8	1361.7	1.4414
1000	0.3069	1315.2	1457.2	1.5262	0.2485	1301.7	1439.6	1.4967
1100	0.3393	1366.8	1523.8	1.5704	0.2772	1356.2	1510.1	1.5434
1200	0.3696	1416.7	1587.7	1.6101	0.3086	1408.0	1576.6	1.5848
1300	0.3984	1465.7	1650.0	1.6465	0.3285	1458.5	1640.9	1.6224
1400	0.4261	1514.2	1711.3	1.6804	0.3524	1508.1	1703.7	1.6571
1500	0.4531	1562.5	1772.1	1.7123	0.3754	1557.3	1765.7	1.6896
1600	0.4795	1610.8	1832.6	1.7424	0.3978	1606.3	1827.1	1.7201

T °F	v ft³/lb	u Btu/lb	h Btu/lb	s Btu/lb · °R	v ft³/lb	u Btu/lb	h Btu/lb	s Btu/lb · °R
	p = 3500 lbf/in.²				p = 4000 lbf/in.²			
650	0.0249	663.5	679.7	0.8630	0.0245	657.7	675.8	0.8574
700	0.0306	759.5	779.3	0.9506	0.0287	742.1	763.4	0.9345
750	0.1046	1058.4	1126.1	1.2440	0.0633	960.7	1007.5	1.1395
800	0.1363	1134.7	1223.0	1.3226	0.1052	1095.0	1172.9	1.2740
900	0.1763	1222.4	1336.5	1.4096	0.1462	1201.5	1309.7	1.3789
1000	0.2066	1287.6	1421.4	1.4699	0.1752	1272.9	1402.6	1.4449
1100	0.2328	1345.2	1496.0	1.5193	0.1995	1333.9	1481.6	1.4973
1200	0.2566	1399.2	1565.3	1.5624	0.2213	1390.1	1553.9	1.5423
1300	0.2787	1451.1	1631.7	1.6012	0.2414	1443.7	1622.4	1.5823
1400	0.2997	1501.9	1696.1	1.6368	0.2603	1495.7	1688.4	1.6188
1500	0.3199	1552.0	1759.2	1.6699	0.2784	1546.7	1752.8	1.6526
1600	0.3395	1601.7	1831.6	1.7010	0.2959	1597.1	1816.1	1.6841

T °F	v ft³/lb	u Btu/lb	h Btu/lb	s Btu/lb · °R	v ft³/lb	u Btu/lb	h Btu/lb	s Btu/lb · °R
	p = 4400 lbf/in.²				p = 4800 lbf/in.²			
650	0.0242	653.6	673.3	0.8535	0.0237	649.8	671.0	0.8499
700	0.0278	732.7	755.3	0.9257	0.0271	725.1	749.1	0.9187
750	0.0415	870.8	904.6	1.0513	0.0352	832.6	863.9	1.0154
800	0.0844	1056.5	1125.3	1.2306	0.0668	1011.2	1070.5	1.1827
900	0.1270	1183.7	1287.1	1.3548	0.1109	1164.8	1263.4	1.3310
1000	0.1552	1260.8	1387.2	1.4260	0.1385	1248.3	1317.4	1.4078
1100	0.1784	1324.7	1469.9	1.4809	0.1608	1315.3	1458.1	1.4653
1200	0.1989	1382.8	1544.7	1.5274	0.1802	1375.4	1535.4	1.5133
1300	0.2176	1437.7	1614.9	1.5685	0.1979	1431.7	1607.4	1.5555
1400	0.2352	1490.7	1682.3	1.6057	0.2143	1485.7	1676.1	1.5934
1500	0.2520	1542.7	1747.6	1.6399	0.2300	1538.2	1742.5	1.6282
1600	0.2681	1593.4	1811.7	1.6718	0.2450	1589.8	1807.4	1.6605

H_2O

TABLE A-5E Properties of Compressed Liquid Water

T °F	v ft^3/lb	u Btu/lb	h Btu/lb	s Btu/lb · °R	v ft^3/lb	u Btu/lb	h Btu/lb	s Btu/lb · °R
	p = 500 lbf/in.^{2} (T_{sat} = 467.1°F)				p = 1000 lbf/in.^{2} (T_{sat} = 544.7°F)			
32	0.015994	0.00	1.49	0.00000	0.015967	0.03	2.99	0.00005
50	0.015998	18.02	19.50	0.03599	0.015972	17.99	20.94	0.03592
100	0.016106	67.87	69.36	0.12932	0.016082	67.70	70.68	0.12901
150	0.016318	117.66	119.17	0.21457	0.016293	117.38	120.40	0.21410
200	0.016608	167.65	169.19	0.29341	0.016580	167.26	170.32	0.29281
300	0.017416	268.92	270.53	0.43641	0.017379	268.24	271.46	0.43552
400	0.018608	373.68	375.40	0.56604	0.018550	372.55	375.98	0.56472
Sat.	0.019748	447.70	449.53	0.64904	0.021591	538.39	542.38	0.74320

T °F	v ft^3/lb	u Btu/lb	h Btu/lb	s Btu/lb · °R	v ft^3/lb	u Btu/lb	h Btu/lb	s Btu/lb · °R
	p = 1500 lbf/in.^{2} (T_{sat} = 596.4°F)				p = 2000 lbf/in.^{2} (T_{sat} = 636.0°F)			
32	0.015939	0.05	4.47	0.00007	0.015912	0.06	5.95	0.00008
50	0.015946	17.95	22.38	0.03584	0.015920	17.91	23.81	0.03575
100	0.016058	67.53	71.99	0.12870	0.016034	67.37	73.30	0.12839
150	0.016268	117.10	121.62	0.21364	0.016244	116.83	122.84	0.21318
200	0.016554	166.87	171.46	0.29221	0.016527	166.49	172.60	0.29162
300	0.017343	267.58	272.39	0.43463	0.017308	266.93	273.33	0.43376
400	0.018493	371.45	376.59	0.56343	0.018439	370.38	377.21	0.56216
500	0.02024	481.8	487.4	0.6853	0.02014	479.8	487.3	0.6832
Sat.	0.02346	605.0	611.5	0.8082	0.02565	662.4	671.9	0.8623

T °F	v ft^3/lb	u Btu/lb	h Btu/lb	s Btu/lb · °R	v ft^3/lb	u Btu/lb	h Btu/lb	s Btu/lb · °R
	p = 3000 lbf/in.^{2} (T_{sat} = 695.5°F)				p = 4000 lbf/in.^{2}			
32	0.015859	0.09	8.90	0.00009	0.015807	0.10	11.80	0.00005
50	0.015870	17.84	26.65	0.03555	0.015821	17.76	29.47	0.03534
100	0.015987	67.04	75.91	0.12777	0.015942	66.72	78.52	0.12714
150	0.016196	116.30	125.29	0.21226	0.016150	115.77	127.73	0.21136
200	0.016476	165.74	174.89	0.29046	0.016425	165.02	177.18	0.28931
300	0.017240	265.66	275.23	0.43205	0.017174	264.43	277.15	0.43038
400	0.018334	368.32	378.50	0.55970	0.018235	366.35	379.85	0.55734
500	0.019944	476.2	487.3	0.6794	0.019766	472.9	487.5	0.6758
Sat.	0.034310	783.5	802.5	0.9732				

TABLE A-6E Properties of Saturated Water (Solid–Vapor): Temperature Table

Temp. °F	Press. lbf/in²	Specific volume ft³/lb		Internal energy Btu/lb			Enthalpy Btu/lb			Entropy Btu/lb · °R		
		Sat. Solid v_i	Sat. Vapor $v_g \times 10^{-3}$	Sat. Solid u_i	Subl. u_{ig}	Sat. Vapor u_g	Sat. Solid h_i	Subl. h_{ig}	Sat. Vapor h_g	Sat. Solid s_i	Subl. s_{ig}	Sat. Vapor s_g
32.018	.0887	.01747	3.302	−143.34	1164.6	1021.2	−143.34	1218.7	1075.4	−.292	2.479	2.187
32	.0886	.01747	3.305	−143.35	1164.6	1021.2	−143.35	1218.7	1075.4	−.292	2.479	2.187
30	.0808	.01747	3.607	−144.35	1164.9	1020.5	−144.35	1218.9	1074.5	−.294	2.489	2.195
25	.0641	.01746	4.506	−146.84	1165.7	1018.9	−146.84	1219.1	1072.3	−.299	2.515	2.216
20	.0505	.01745	5.655	−149.31	1166.5	1017.2	−149.31	1219.4	1070.1	−.304	2.542	2.238
15	.0396	.01745	7.13	−151.75	1167.3	1015.5	−151.75	1219.7	1067.9	−.309	2.569	2.260
10	.0309	.01744	9.04	−154.17	1168.1	1013.9	−154.17	1219.9	1065.7	−.314	2.597	2.283
5	.0240	.01743	11.52	−156.56	1168.8	1012.2	−156.56	1220.1	1063.5	−.320	2.626	2.306
0	.0185	.01743	14.77	−158.93	1169.5	1010.6	−158.93	1220.2	1061.2	−.325	2.655	2.330
−5	.0142	.01742	19.03	−161.27	1170.2	1008.9	−161.27	1220.3	1059.0	−.330	2.684	2.354
−10	.0109	.01741	24.66	−163.59	1170.9	1007.3	−163.59	1220.4	1056.8	−.335	2.714	2.379
−15	.0082	.01740	32.2	−165.89	1171.5	1005.6	−165.89	1220.5	1054.6	−.340	2.745	2.405
−20	.0062	.01740	42.2	−168.16	1172.1	1003.9	−168.16	1220.6	1052.4	−.345	2.776	2.431
−25	.0046	.01739	55.7	−170.40	1172.7	1002.3	−170.40	1220.6	1050.2	−.351	2.808	2.457
−30	.0035	.01738	74.1	−172.63	1173.2	1000.6	−172.63	1220.6	1048.0	−.356	2.841	2.485
−35	.0026	.01737	99.2	−174.82	1173.8	998.9	−174.82	1220.6	1045.8	−.361	2.874	2.513
−40	.0019	.01737	133.8	−177.00	1174.3	997.3	−177.00	1220.6	1043.6	−.366	2.908	2.542

Source: J. H. Keenan, F. G. Keyes, P. G. Hill, and J. G. Moore, *Steam Tables,* Wiley, New York, 1969.

TABLE A-7E Properties of Saturated Refrigerant 12 (Liquid–Vapor): Temperature Table

		Specific volume ft³/lb		Internal energy Btu/lb		Enthalpy Btu/lb			Entropy Btu/lb · °R		
Temp. °F	Press. lbf/in.²	Sat. liquid v_f	Sat. vapor v_g	Sat. liquid u_f	Sat. vapor u_g	Sat. liquid h_f	Evap. h_{fg}	Sat. vapor h_g	Sat. liquid s_f	Sat. vapor s_g	Temp. °F
−40	9.308	0.01056	3.8750	−0.02	66.24	0	72.91	72.91	0	0.1737	−40
−30	11.999	0.01067	3.0585	1.93	67.22	2.11	71.90	74.01	0.0050	0.1723	−30
−20	15.267	0.01079	2.4429	4.21	68.21	4.24	70.87	75.11	0.0098	0.1710	−20
−15	17.141	0.01085	2.1924	5.27	68.69	5.30	70.35	75.65	0.0122	0.1704	−15
−10	19.189	0.01091	1.9727	6.33	69.19	6.37	69.82	76.19	0.0146	0.1699	−10
−5	21.422	0.01097	1.7794	7.40	69.67	7.44	69.29	76.73	0.0170	0.1694	−5
0	23.849	0.01103	1.6089	8.47	70.17	8.52	68.75	77.27	0.0193	0.1689	0
5	26.483	0.01109	1.4580	9.55	70.65	9.60	68.20	77.80	0.0216	0.1684	5
10	29.335	0.01116	1.3241	10.62	71.15	10.68	67.65	78.33	0.0240	0.1680	10
20	35.736	0.01130	1.0988	12.79	72.12	12.86	66.52	79.38	0.0285	0.1672	20
25	39.310	0.01137	1.0039	13.88	72.61	13.96	65.95	79.91	0.0308	0.1668	25
30	43.148	0.01144	0.9188	14.97	73.08	15.06	65.36	80.42	0.0330	0.1665	30
40	51.667	0.01159	0.7736	17.16	74.04	17.27	64.16	81.43	0.0375	0.1659	40
50	61.394	0.01175	0.6554	19.38	74.99	19.51	62.93	82.44	0.0418	0.1653	50
60	72.433	0.01191	0.5584	21.61	75.92	21.77	61.64	83.41	0.0462	0.1648	60
70	84.89	0.01209	0.4782	23.86	76.85	24.05	60.31	84.36	0.0505	0.1643	70
80	98.87	0.01228	0.4114	26.14	77.76	26.37	58.92	85.29	0.0548	0.1639	80
85	106.47	0.01238	0.3821	27.29	78.20	27.53	58.20	85.73	0.0569	0.1637	85
90	114.49	0.01248	0.3553	28.45	78.65	28.71	57.46	86.17	0.0590	0.1635	90
95	122.95	0.01258	0.3306	29.61	79.09	29.90	56.71	86.61	0.0611	0.1633	95
100	131.86	0.01269	0.3079	30.79	79.51	31.10	55.93	87.03	0.0632	0.1632	100
105	141.25	0.01281	0.2870	31.98	79.94	32.31	55.13	87.44	0.0653	0.1630	105
110	151.11	0.01292	0.2677	33.16	80.36	33.53	54.31	87.84	0.0675	0.1628	110
115	161.47	0.01305	0.2498	34.37	80.76	34.76	53.47	88.23	0.0696	0.1626	115
120	172.35	0.01317	0.2333	35.59	81.17	36.01	52.60	88.61	0.0717	0.1624	120
140	221.32	0.01375	0.1780	40.60	82.68	41.16	48.81	89.97	0.0802	0.1616	140
160	279.82	0.01445	0.1360	45.88	83.96	46.63	44.37	91.00	0.0889	0.1605	160
180	349.00	0.01536	0.1033	51.57	84.89	52.56	39.00	91.56	0.0980	0.1590	180
200	430.09	0.01666	0.0767	57.87	85.17	59.20	32.08	91.28	0.1079	0.1565	200
233.6	596.9	0.0287	0.0287	75.69	75.69	78.86	0	78.86	0.1359	0.1359	233.6

Source: Tables A-7E through A-9E are adapted from K. Wark, *Thermodynamics, 4th ed.*, McGraw-Hill, New York, 1983, as based on data from Freon Products Division, E. I. du Pont de Nemours & Company, 1956.

TABLE A-8E Properties of Saturated Refrigerant 12 (Liquid–Vapor): Pressure Table

		Specific volume ft³/lb		Internal energy Btu/lb		Enthalpy Btu/lb			Entropy Btu/lb · °R		
Press. lbf/in.²	Temp. °F	Sat. liquid v_f	Sat. vapor v_g	Sat. liquid u_f	Sat. vapor u_g	Sat. liquid h_f	Evap. h_{fg}	Sat. vapor h_g	Sat. liquid s_f	Sat. vapor s_g	Press. lbf/in.²
5	−62.35	0.0103	6.9069	−4.69	64.04	−4.68	75.11	70.43	−0.0114	0.1776	5
10	−37.23	0.0106	3.6246	0.56	66.51	0.58	72.64	73.22	0.0014	0.1733	10
15	−20.75	0.0108	2.4835	4.05	68.13	4.08	70.95	75.03	0.0095	0.1711	15
20	−8.13	0.0109	1.8977	6.73	69.37	6.77	69.63	76.40	0.0155	0.1697	20
30	11.11	0.0112	1.2964	10.86	71.25	10.93	67.53	78.45	0.0245	0.1679	30
40	25.93	0.0114	0.9874	14.08	72.69	14.16	65.84	80.00	0.0312	0.1668	40
50	38.15	0.0116	0.7982	16.75	73.86	16.86	64.39	81.25	0.0366	0.1660	50
60	48.64	0.0117	0.6701	19.07	74.86	19.20	63.10	82.30	0.0413	0.1654	60
70	57.90	0.0119	0.5772	21.13	75.73	21.29	61.92	83.21	0.0453	0.1649	70
80	66.21	0.0120	0.5068	23.00	76.50	23.18	60.82	84.00	0.0489	0.1645	80
90	73.79	0.0122	0.4514	24.72	77.20	24.92	59.79	84.71	0.0521	0.1642	90
100	80.76	0.0123	0.4067	26.31	77.82	26.54	58.81	85.35	0.0551	0.1639	100
120	93.29	0.0126	0.3389	29.21	78.93	29.49	56.97	86.46	0.0604	0.1634	120
140	104.35	0.0128	0.2896	31.82	79.89	32.15	55.24	87.39	0.0651	0.1630	140
160	114.30	0.0130	0.2522	34.21	80.71	34.59	53.59	88.18	0.0693	0.1626	160
180	123.38	0.0133	0.2228	36.42	81.44	36.86	52.00	88.86	0.0731	0.1623	180
200	131.74	0.0135	0.1989	38.50	82.08	39.00	50.44	89.44	0.0767	0.1620	200
220	139.51	0.0137	0.1792	40.48	82.08	41.03	48.90	89.94	0.0816	0.1616	220
240	146.77	0.0140	0.1625	42.35	83.14	42.97	47.39	90.36	0.0831	0.1613	240
260	153.60	0.0142	0.1483	44.16	83.58	44.84	45.88	90.72	0.0861	0.1609	260
280	160.06	0.0145	0.1359	45.90	83.97	46.65	44.36	91.01	0.0890	0.1605	280
300	166.18	0.0147	0.1251	47.59	84.30	48.41	42.83	91.24	0.0917	0.1601	300

TABLE A-9E Properties of Superheated Refrigerant 12

T °F	v ft³/lb	u Btu/lb	h Btu/lb	s Btu/lb · °R	v ft³/lb	u Btu/lb	h Btu/lb	s Btu/lb · °R
	p = 10 lbf/in.² (T_{sat} = −37.23°F)				p = 15 lbf/in.² (T_{sat} = −20.75°F)			
Sat.	3.6246	66.512	73.219	0.1733	2.4835	68.134	75.028	0.1711
0	3.9809	70.879	78.246	0.1847	2.6201	70.629	77.902	0.1775
20	4.1691	73.299	81.014	0.1906	2.7494	73.080	80.712	0.1835
40	4.3556	75.768	83.828	0.1964	2.8770	75.575	83.561	0.1893
60	4.5408	78.286	86.689	0.2020	3.0031	78.115	86.451	0.1950
80	4.7248	80.853	89.596	0.2075	3.1281	80.700	89.383	0.2005
100	4.9079	83.466	92.548	0.2128	3.2521	83.330	92.357	0.2059
120	5.0903	86.126	95.546	0.2181	3.3754	86.004	95.373	0.2112
140	5.2720	88.830	98.586	0.2233	3.4981	88.719	98.429	0.2164
160	5.4533	91.578	101.669	0.2283	3.6202	91.476	101.525	0.2215
180	5.6341	94.367	104.793	0.2333	3.7419	94.274	104.661	0.2265
200	5.8145	97.197	107.957	0.2381	3.8632	97.112	107.835	0.2314
	p = 20 lbf/in.² (T_{sat} = −8.13°F)				p = 30 lbf/in.² (T_{sat} = 11.11°F)			
Sat.	1.8977	69.374	76.397	0.1697	1.2964	71.255	78.452	0.1679
20	2.0391	72.856	80.403	0.1783	1.3278	72.394	79.765	0.1707
40	2.1373	75.379	83.289	0.1842	1.3969	74.975	82.730	0.1767
60	2.2340	77.942	86.210	0.1899	1.4644	77.586	85.716	0.1826
80	2.3295	80.546	89.168	0.1955	1.5306	80.232	88.729	0.1883
100	2.4241	83.192	92.164	0.2010	1.5957	82.911	91.770	0.1938
120	2.5179	85.879	95.198	0.2063	1.6600	85.627	94.843	0.1992
140	2.6110	88.607	98.270	0.2115	1.7237	88.379	97.948	0.2045
160	2.7036	91.374	101.380	0.2166	1.7868	91.166	101.086	0.2096
180	2.7957	94.181	104.528	0.2216	1.8494	93.991	104.258	0.2146
200	2.8874	97.026	107.712	0.2265	1.9116	96.852	107.464	0.2196
220	2.9789	99.907	110.932	0.2313	1.9735	99.746	110.702	0.2244
	p = 40 lbf/in.² (T_{sat} = 25.93°F)				p = 50 lbf/in.² (T_{sat} = 38.15°F)			
Sat.	0.9874	72.691	80.000	0.1668	0.7982	73.863	81.249	0.1660
40	1.0258	74.555	82.148	0.1711	0.8025	74.115	81.540	0.1666
60	1.0789	77.220	85.206	0.1771	0.8471	76.838	84.676	0.1727
80	1.1306	79.908	88.277	0.1829	0.8903	79.574	87.811	0.1786
100	1.1812	82.624	91.367	0.1885	0.9322	82.328	90.953	0.1843
120	1.2309	85.369	94.480	0.1940	0.9731	85.106	94.110	0.1899
140	1.2798	88.147	97.620	0.1993	1.0133	87.910	97.286	0.1953
160	1.3282	90.957	100.788	0.2045	1.0529	90.743	100.485	0.2005
180	1.3761	93.800	103.985	0.2096	1.0920	93.604	103.708	0.2056
200	1.4236	96.674	107.212	0.2146	1.1307	96.496	106.958	0.2106
220	1.4707	99.583	110.469	0.2194	1.1690	99.419	110.235	0.2155
240	1.5176	102.524	113.757	0.2242	1.2070	102.371	113.539	0.2203

R-12

TABLE A-9E *(Continued)*

T °F	v ft³/lb	u Btu/lb	h Btu/lb	s Btu/lb·°R	v ft³/lb	u Btu/lb	h Btu/lb	s Btu/lb·°R
	p = 60 lbf/in.² (T_{sat} = 48.64°F)				p = 70 lbf/in.² (T_{sat} = 57.90°F)			
Sat.	0.6701	74.859	82.299	0.1654	0.5772	75.729	83.206	0.1649
60	0.6921	76.442	84.126	0.1689	0.5809	76.027	83.552	0.1656
80	0.7296	79.229	87.330	0.1750	0.6146	78.871	86.832	0.1718
100	0.7659	82.024	90.528	0.1808	0.6469	81.712	90.091	0.1777
120	0.8011	84.836	93.731	0.1864	0.6780	84.560	93.343	0.1834
140	0.8355	87.668	96.945	0.1919	0.7084	87.421	96.597	0.1889
160	0.8693	90.524	100.776	0.1972	0.7380	90.302	99.862	0.1943
180	0.9025	93.406	103.427	0.2023	0.7671	93.205	103.141	0.1995
200	0.9353	96.315	106.700	0.2074	0.7957	96.132	106.439	0.2046
220	0.9678	99.252	109.997	0.2123	0.8240	99.083	109.756	0.2095
240	0.9998	102.217	113.319	0.2171	0.8519	102.061	113.096	0.2144
260	1.0318	105.210	116.666	0.2218	0.8796	105.065	116.459	0.2191
	p = 80 lbf/in.² (T_{sat} = 66.21°F)				p = 90 lbf/in.² (T_{sat} = 73.79°F)			
Sat.	0.5068	76.500	84.003	0.1645	0.4514	77.194	84.713	0.1642
80	0.5280	78.500	86.316	0.1689	0.4602	78.115	85.779	0.1662
100	0.5573	81.389	89.640	0.1749	0.4875	81.056	89.175	0.1723
120	0.5856	84.276	92.945	0.1807	0.5135	83.984	92.536	0.1782
140	0.6129	87.169	96.242	0.1863	0.5385	86.911	95.879	0.1839
160	0.6394	90.076	99.542	0.1917	0.5627	89.845	99.216	0.1894
180	0.6654	93.000	102.851	0.1970	0.5863	92.793	102.557	0.1947
200	0.6910	95.945	106.174	0.2021	0.6094	95.755	105.905	0.1998
220	0.7161	98.912	109.513	0.2071	0.6321	98.739	109.267	0.2049
240	0.7409	101.904	112.872	0.2119	0.6545	101.743	112.644	0.2098
260	0.7654	104.919	116.251	0.2167	0.6766	104.771	116.040	0.2146
280	0.7898	107.960	119.652	0.2214	0.6985	107.823	119.456	0.2192
	p = 100 lbf/in.² (T_{sat} = 80.76°F)				p = 120 lbf/in.² (T_{sat} = 93.29°F)			
Sat.	0.4067	77.824	85.351	0.1639	0.3389	78.933	86.459	0.1634
100	0.4314	80.711	88.694	0.1700	0.3466	79.978	87.675	0.1656
120	0.4556	83.685	92.116	0.1760	0.3684	83.056	91.237	0.1718
140	0.4788	86.647	95.507	0.1817	0.3890	86.098	94.736	0.1778
160	0.5012	89.610	98.884	0.1873	0.4087	89.123	98.199	0.1835
180	0.5229	92.580	102.257	0.1926	0.4277	92.144	101.642	0.1889
200	0.5441	95.564	105.633	0.1978	0.4461	95.170	105.076	0.1942
220	0.5649	98.564	109.018	0.2029	0.4640	98.205	108.509	0.1993
240	0.5854	101.582	112.415	0.2078	0.4816	101.253	111.948	0.2043
260	0.6055	104.622	115.828	0.2126	0.4989	104.317	115.396	0.2092
280	0.6255	107.684	119.258	0.2173	0.5159	107.401	118.857	0.2139
300	0.6452	110.768	122.707	0.2219	0.5327	110.504	122.333	0.2186

R-12

TABLE A-9E *(Continued)*

T °F	v ft³/lb	u Btu/lb	h Btu/lb	s Btu/lb · °R	v ft³/lb	u Btu/lb	h Btu/lb	s Btu/lb · °R
	p = 140 lbf/in.² (T_{sat} = 104.35°F)				p = 160 lbf/in.² (T_{sat} = 114.30°F)			
Sat.	0.2896	79.886	87.389	0.1630	0.2522	80.713	88.180	0.1626
120	0.3055	82.382	90.297	0.1681	0.2576	81.656	89.283	0.1645
140	0.3245	85.516	93.923	0.1742	0.2756	84.899	93.059	0.1709
160	0.3423	88.615	97.483	0.1801	0.2922	88.080	96.732	0.1770
180	0.3594	91.692	101.003	0.1857	0.3080	91.221	100.340	0.1827
200	0.3758	94.765	104.501	0.1910	0.3230	94.344	103.907	0.1882
220	0.3918	97.837	107.987	0.1963	0.3375	97.457	107.450	0.1935
240	0.4073	100.918	111.470	0.2013	0.3516	100.570	110.980	0.1986
260	0.4226	104.008	114.956	0.2062	0.3653	103.690	114.506	0.2036
280	0.4375	107.115	118.449	0.2110	0.3787	106.820	118.033	0.2084
300	0.4523	110.235	121.953	0.2157	0.3919	109.964	121.567	0.2131
320	0.4668	113.376	125.470	0.2202	0.4049	113.121	125.109	0.2177

T °F	v ft³/lb	u Btu/lb	h Btu/lb	s Btu/lb · °R	v ft³/lb	u Btu/lb	h Btu/lb	s Btu/lb · °R
	p = 180 lbf/in.² (T_{sat} = 123.38°F)				p = 200 lbf/in.² (T_{sat} = 131.74°F)			
Sat.	0.2228	81.436	88.857	0.1623	0.1989	82.077	89.439	0.1620
140	0.2371	84.238	92.136	0.1678	0.2058	83.521	91.137	0.1648
160	0.2530	87.513	95.940	0.1741	0.2212	86.913	95.100	0.1713
180	0.2678	90.727	99.647	0.1800	0.2354	90.211	98.921	0.1774
200	0.2818	93.904	103.291	0.1856	0.2486	93.451	102.652	0.1831
220	0.2952	97.063	106.896	0.1910	0.2612	96.659	106.325	0.1886
240	0.3081	100.215	110.478	0.1961	0.2732	99.850	109.962	0.1939
260	0.3207	103.364	114.046	0.2012	0.2849	103.032	113.576	0.1990
280	0.3329	106.521	117.610	0.2061	0.2962	106.214	117.178	0.2039
300	0.3449	109.686	121.174	0.2108	0.3073	109.402	120.775	0.2087
320	0.3567	112.863	124.744	0.2155	0.3182	112.598	124.373	0.2134
340	0.3683	116.053	128.321	0.2200	0.3288	115.805	127.974	0.2179

T °F	v ft³/lb	u Btu/lb	h Btu/lb	s Btu/lb · °R	v ft³/lb	u Btu/lb	h Btu/lb	s Btu/lb · °R
	p = 300 lbf/in.² (T_{sat} = 166.18°F)				p = 400 lbf/in.² (T_{sat} = 192.93°F)			
Sat.	0.1251	84.295	91.240	0.1601	0.0856	85.178	91.513	0.1576
180	0.1348	87.071	94.556	0.1654				
200	0.1470	90.816	98.975	0.1722	0.0910	86.982	93.718	0.1609
220	0.1577	94.379	103.136	0.1784	0.1032	91.410	99.046	0.1689
240	0.1676	97.835	107.140	0.1842	0.1130	95.371	103.735	0.1757
260	0.1769	101.225	111.043	0.1897	0.1216	99.102	108.105	0.1818
280	0.1856	104.574	114.879	0.1950	0.1295	102.701	112.286	0.1876
300	0.1940	107.899	118.670	0.2000	0.1368	106.217	116.343	0.1930
320	0.2021	111.208	122.430	0.2049	0.1437	109.680	120.318	0.1981
340	0.2100	114.512	126.171	0.2096	0.1503	113.108	124.235	0.2031
360	0.2177	117.814	129.900	0.2142	0.1567	116.514	128.112	0.2079

TABLE A-10E Properties of Saturated Refrigerant 134a (Liquid–Vapor): Temperature Table

		Specific Volume ft³/lb		Internal Energy Btu/lb		Enthalpy Btu/lb			Entropy Btu/lb · °R		
Temp. °F	Press. lbf/in.²	Sat. Liquid v_f	Sat. Vapor v_g	Sat. Liquid u_f	Sat. Vapor u_g	Sat. Liquid h_f	Evap. h_{fg}	Sat. Vapor h_g	Sat. Liquid s_f	Sat. Vapor s_g	Temp. °F
−40	7.490	0.01130	5.7173	−0.02	87.90	0.00	95.82	95.82	0.0000	0.2283	−40
−30	9.920	0.01143	4.3911	2.81	89.26	2.83	94.49	97.32	0.0067	0.2266	−30
−20	12.949	0.01156	3.4173	5.69	90.62	5.71	93.10	98.81	0.0133	0.2250	−20
−15	14.718	0.01163	3.0286	7.14	91.30	7.17	92.38	99.55	0.0166	0.2243	−15
−10	16.674	0.01170	2.6918	8.61	91.98	8.65	91.64	100.29	0.0199	0.2236	−10
−5	18.831	0.01178	2.3992	10.09	92.66	10.13	90.89	101.02	0.0231	0.2230	−5
0	21.203	0.01185	2.1440	11.58	93.33	11.63	90.12	101.75	0.0264	0.2224	0
5	23.805	0.01193	1.9208	13.09	94.01	13.14	89.33	102.47	0.0296	0.2219	5
10	26.651	0.01200	1.7251	14.60	94.68	14.66	88.53	103.19	0.0329	0.2214	10
15	29.756	0.01208	1.5529	16.13	95.35	16.20	87.71	103.90	0.0361	0.2209	15
20	33.137	0.01216	1.4009	17.67	96.02	17.74	86.87	104.61	0.0393	0.2205	20
25	36.809	0.01225	1.2666	19.22	96.69	19.30	86.02	105.32	0.0426	0.2200	25
30	40.788	0.01233	1.1474	20.78	97.35	20.87	85.14	106.01	0.0458	0.2196	30
40	49.738	0.01251	0.9470	23.94	98.67	24.05	83.34	107.39	0.0522	0.2189	40
50	60.125	0.01270	0.7871	27.14	99.98	27.28	81.46	108.74	0.0585	0.2183	50
60	72.092	0.01290	0.6584	30.39	101.27	30.56	79.49	110.05	0.0648	0.2178	60
70	85.788	0.01311	0.5538	33.68	102.54	33.89	77.44	111.33	0.0711	0.2173	70
80	101.37	0.01334	0.4682	37.02	103.78	37.27	75.29	112.56	0.0774	0.2169	80
85	109.92	0.01346	0.4312	38.72	104.39	38.99	74.17	113.16	0.0805	0.2167	85
90	118.99	0.01358	0.3975	40.42	105.00	40.72	73.03	113.75	0.0836	0.2165	90
95	128.62	0.01371	0.3668	42.14	105.60	42.47	71.86	114.33	0.0867	0.2163	95
100	138.83	0.01385	0.3388	43.87	106.18	44.23	70.66	114.89	0.0898	0.2161	100
105	149.63	0.01399	0.3131	45.62	106.76	46.01	69.42	115.43	0.0930	0.2159	105
110	161.04	0.01414	0.2896	47.39	107.33	47.81	68.15	115.96	0.0961	0.2157	110
115	173.10	0.01429	0.2680	49.17	107.88	49.63	66.84	116.47	0.0992	0.2155	115
120	185.82	0.01445	0.2481	50.97	108.42	51.47	65.48	116.95	0.1023	0.2153	120
140	243.86	0.01520	0.1827	58.39	110.41	59.08	59.57	118.65	0.1150	0.2143	140
160	314.63	0.01617	0.1341	66.26	111.97	67.20	52.58	119.78	0.1280	0.2128	160
180	400.22	0.01758	0.0964	74.83	112.77	76.13	43.78	119.91	0.1417	0.2101	180
200	503.52	0.02014	0.0647	84.90	111.66	86.77	30.92	117.69	0.1575	0.2044	200
210	563.51	0.02329	0.0476	91.84	108.48	94.27	19.18	113.45	0.1684	0.1971	210

Source: Tables A-10E through A-12E are calculated based on equations from D. P. Wilson and R. S. Basu, "Thermodynamic Properties of a New Stratospherically Safe Working Fluid—Refrigerant 134a," *ASHRAE Trans.*, Vol. 94, Pt. 2, 1988, pp. 2095–2118.

TABLE A-11E Properties of Saturated Refrigerant 134a (Liquid–Vapor): Pressure Table

		Specific Volume ft³/lb		Internal Energy Btu/lb		Enthalpy Btu/lb			Entropy Btu/lb · °R		
Press. lbf/in.²	Temp. °F	Sat. Liquid v_f	Sat. Vapor v_g	Sat. Liquid u_f	Sat. Vapor u_g	Sat. Liquid h_f	Evap. h_{fg}	Sat. Vapor h_g	Sat. Liquid s_f	Sat. Vapor s_g	Press. lbf/in.²
5	−53.48	0.01113	8.3508	−3.74	86.07	−3.73	97.53	93.79	−.0090	0.2311	5
10	−29.71	0.01143	4.3581	2.89	89.30	2.91	94.45	97.37	0.0068	0.2265	10
15	−14.25	0.01164	2.9747	7.36	91.40	7.40	92.27	99.66	0.0171	0.2242	15
20	−2.48	0.01181	2.2661	10.84	93.00	10.89	90.50	101.39	0.0248	0.2227	20
30	15.38	0.01209	1.5408	16.24	95.40	16.31	87.65	103.96	0.0364	0.2209	30
40	29.04	0.01232	1.1692	20.48	97.23	20.57	85.31	105.88	0.0452	0.2197	40
50	40.27	0.01252	0.9422	24.02	98.71	24.14	83.29	107.43	0.0523	0.2189	50
60	49.89	0.01270	0.7887	27.10	99.96	27.24	81.48	108.72	0.0584	0.2183	60
70	58.35	0.01286	0.6778	29.85	101.05	30.01	79.82	109.83	0.0638	0.2179	70
80	65.93	0.01302	0.5938	32.33	102.02	32.53	78.28	110.81	0.0686	0.2175	80
90	72.83	0.01317	0.5278	34.62	102.89	34.84	76.84	111.68	0.0729	0.2172	90
100	79.17	0.01332	0.4747	36.75	103.68	36.99	75.47	112.46	0.0768	0.2169	100
120	90.54	0.01360	0.3941	40.61	105.06	40.91	72.91	113.82	0.0839	0.2165	120
140	100.56	0.01386	0.3358	44.07	106.25	44.43	70.52	114.95	0.0902	0.2161	140
160	109.56	0.01412	0.2916	47.23	107.28	47.65	68.26	115.91	0.0958	0.2157	160
180	117.74	0.01438	0.2569	50.16	108.18	50.64	66.10	116.74	0.1009	0.2154	180
200	125.28	0.01463	0.2288	52.90	108.98	53.44	64.01	117.44	0.1057	0.2151	200
220	132.27	0.01489	0.2056	55.48	109.68	56.09	61.96	118.05	0.1101	0.2147	220
240	138.79	0.01515	0.1861	57.93	110.30	58.61	59.96	118.56	0.1142	0.2144	240
260	144.92	0.01541	0.1695	60.28	110.84	61.02	57.97	118.99	0.1181	0.2140	260
280	150.70	0.01568	0.1550	62.53	111.31	63.34	56.00	119.35	0.1219	0.2136	280
300	156.17	0.01596	0.1424	64.71	111.72	65.59	54.03	119.62	0.1254	0.2132	300
350	168.72	0.01671	0.1166	69.88	112.45	70.97	49.03	120.00	0.1338	0.2118	350
400	179.95	0.01758	0.0965	74.81	112.77	76.11	43.80	119.91	0.1417	0.2102	400
450	190.12	0.01863	0.0800	79.63	112.60	81.18	38.08	119.26	0.1493	0.2079	450
500	199.38	0.02002	0.0657	84.54	111.76	86.39	31.44	117.83	0.1570	0.2047	500

R-134a

TABLE A-12E Properties of Superheated Refrigerant 134a Vapor

T °F	v ft³/lb	u Btu/lb	h Btu/lb	s Btu/lb · °R	v ft³/lb	u Btu/lb	h Btu/lb	s Btu/lb · °R
	p = 10 lbf/in.² (T_{sat} = −29.71°F)				**p = 15 lbf/in.² (T_{sat} = −14.25°F)**			
Sat.	4.3581	89.30	97.37	0.2265	2.9747	91.40	99.66	0.2242
−20	4.4718	90.89	99.17	0.2307				
0	4.7026	94.24	102.94	0.2391	3.0893	93.84	102.42	0.2303
20	4.9297	97.67	106.79	0.2472	3.2468	97.33	106.34	0.2386
40	5.1539	101.19	110.72	0.2553	3.4012	100.89	110.33	0.2468
60	5.3758	104.80	114.74	0.2632	3.5533	104.54	114.40	0.2548
80	5.5959	108.50	118.85	0.2709	3.7034	108.28	118.56	0.2626
100	5.8145	112.29	123.05	0.2786	3.8520	112.10	122.79	0.2703
120	6.0318	116.18	127.34	0.2861	3.9993	116.01	127.11	0.2779
140	6.2482	120.16	131.72	0.2935	4.1456	120.00	131.51	0.2854
160	6.4638	124.23	136.19	0.3009	4.2911	124.09	136.00	0.2927
180	6.6786	128.38	140.74	0.3081	4.4359	128.26	140.57	0.3000
200	6.8929	132.63	145.39	0.3152	4.5801	132.52	145.23	0.3072
	p = 20 lbf/in.² (T_{sat} = −2.48°F)				**p = 30 lbf/in.² (T_{sat} = 15.38°F)**			
Sat.	2.2661	93.00	101.39	0.2227	1.5408	95.40	103.96	0.2209
0	2.2816	93.43	101.88	0.2238				
20	2.4046	96.98	105.88	0.2323	1.5611	96.26	104.92	0.2229
40	2.5244	100.59	109.94	0.2406	1.6465	99.98	109.12	0.2315
60	2.6416	104.28	114.06	0.2487	1.7293	103.75	113.35	0.2398
80	2.7569	108.05	118.25	0.2566	1.8098	107.59	117.63	0.2478
100	2.8705	111.90	122.52	0.2644	1.8887	111.49	121.98	0.2558
120	2.9829	115.83	126.87	0.2720	1.9662	115.47	126.39	0.2635
140	3.0942	119.85	131.30	0.2795	2.0426	119.53	130.87	0.2711
160	3.2047	123.95	135.81	0.2869	2.1181	123.66	135.42	0.2786
180	3.3144	128.13	140.40	0.2922	2.1929	127.88	140.05	0.2859
200	3.4236	132.40	145.07	0.3014	2.2671	132.17	144.76	0.2932
220	3.5323	136.76	149.83	0.3085	2.3407	136.55	149.54	0.3003
	p = 40 lbf/in.² (T_{sat} = 29.04°F)				**p = 50 lbf/in.² (T_{sat} = 40.27)**			
Sat.	1.1692	97.23	105.88	0.2197	0.9422	98.71	107.43	0.2189
40	1.2065	99.33	108.26	0.2245				
60	1.2723	103.20	112.62	0.2331	0.9974	102.62	111.85	0.2276
80	1.3357	107.11	117.00	0.2414	1.0508	106.62	116.34	0.2361
100	1.3973	111.08	121.42	0.2494	1.1022	110.65	120.85	0.2443
120	1.4575	115.11	125.90	0.2573	1.1520	114.74	125.39	0.2523
140	1.5165	119.21	130.43	0.2650	1.2007	118.88	129.99	0.2601
160	1.5746	123.38	135.03	0.2725	1.2484	123.08	134.64	0.2677
180	1.6319	127.62	139.70	0.2799	1.2953	127.36	139.34	0.2752
200	1.6887	131.94	144.44	0.2872	1.3415	131.71	144.12	0.2825
220	1.7449	136.34	149.25	0.2944	1.3873	136.12	148.96	0.2897
240	1.8006	140.81	154.14	0.3015	1.4326	140.61	153.87	0.2969
260	1.8561	145.36	159.10	0.3085	1.4775	145.18	158.85	0.3039
280	1.9112	149.98	164.13	0.3154	1.5221	149.82	163.90	0.3108

TABLE A-12E (*Continued*)

T °F	v ft³/lb	u Btu/lb	h Btu/lb	s Btu/lb · °R	v ft³/lb	u Btu/lb	h Btu/lb	s Btu/lb · °R
	p = 60 lbf/in.² (T_{sat} = 49.89°F)				**p = 70 lbf/in.² (T_{sat} = 58.35°F)**			
Sat.	0.7887	99.96	108.72	0.2183	0.6778	101.05	109.83	0.2179
60	0.8135	102.03	111.06	0.2229	0.6814	101.40	110.23	0.2186
80	0.8604	106.11	115.66	0.2316	0.7239	105.58	114.96	0.2276
100	0.9051	110.21	120.26	0.2399	0.7640	109.76	119.66	0.2361
120	0.9482	114.35	124.88	0.2480	0.8023	113.96	124.36	0.2444
140	0.9900	118.54	129.53	0.2559	0.8393	118.20	129.07	0.2524
160	1.0308	122.79	134.23	0.2636	0.8752	122.49	133.82	0.2601
180	1.0707	127.10	138.98	0.2712	0.9103	126.83	138.62	0.2678
200	1.1100	131.47	143.79	0.2786	0.9446	131.23	143.46	0.2752
220	1.1488	135.91	148.66	0.2859	0.9784	135.69	148.36	0.2825
240	1.1871	140.42	153.60	0.2930	1.0118	140.22	153.33	0.2897
260	1.2251	145.00	158.60	0.3001	1.0448	144.82	158.35	0.2968
280	1.2627	149.65	163.67	0.3070	1.0774	149.48	163.44	0.3038
300	1.3001	154.38	168.81	0.3139	1.1098	154.22	168.60	0.3107
	p = 80 lbf/in.² (T_{sat} = 65.93°F)				**p = 90 lbf/in.² (T_{sat} = 72.83°F)**			
Sat.	0.5938	102.02	110.81	0.2175	0.5278	102.89	111.68	0.2172
80	0.6211	105.03	114.23	0.2239	0.5408	104.46	113.47	0.2205
100	0.6579	109.30	119.04	0.2327	0.5751	108.82	118.39	0.2295
120	0.6927	113.56	123.82	0.2411	0.6073	113.15	123.27	0.2380
140	0.7261	117.85	128.60	0.2492	0.6380	117.50	128.12	0.2463
160	0.7584	122.18	133.41	0.2570	0.6675	121.87	132.98	0.2542
180	0.7898	126.55	138.25	0.2647	0.6961	126.28	137.87	0.2620
200	0.8205	130.98	143.13	0.2722	0.7239	130.73	142.79	0.2696
220	0.8506	135.47	148.06	0.2796	0.7512	135.25	147.76	0.2770
240	0.8803	140.02	153.05	0.2868	0.7779	139.82	152.77	0.2843
260	0.9095	144.63	158.10	0.2940	0.8043	144.45	157.84	0.2914
280	0.9384	149.32	163.21	0.3010	0.8303	149.15	162.97	0.2984
300	0.9671	154.06	168.38	0.3079	0.8561	153.91	168.16	0.3054
320	0.9955	158.88	173.62	0.3147	0.8816	158.73	173.42	0.3122
	p = 100 lbf/in.² (T_{sat} = 79.17°F)				**p = 120 lbf/in.² (T_{sat} = 90.54°F)**			
Sat.	0.4747	103.68	112.46	0.2169	0.3941	105.06	113.82	0.2165
80	0.4761	103.87	112.68	0.2173				
100	0.5086	108.32	117.73	0.2265	0.4080	107.26	116.32	0.2210
120	0.5388	112.73	122.70	0.2352	0.4355	111.84	121.52	0.2301
140	0.5674	117.13	127.63	0.2436	0.4610	116.37	126.61	0.2387
160	0.5947	121.55	132.55	0.2517	0.4852	120.89	131.66	0.2470
180	0.6210	125.99	137.49	0.2595	0.5082	125.42	136.70	0.2550
200	0.6466	130.48	142.45	0.2671	0.5305	129.97	141.75	0.2628
220	0.6716	135.02	147.45	0.2746	0.5520	134.56	146.82	0.2704
240	0.6960	139.61	152.49	0.2819	0.5731	139.20	151.92	0.2778
260	0.7201	144.26	157.59	0.2891	0.5937	143.89	157.07	0.2850
280	0.7438	148.98	162.74	0.2962	0.6140	148.63	162.26	0.2921
300	0.7672	153.75	167.95	0.3031	0.6339	153.43	167.51	0.2991
320	0.7904	158.59	173.21	0.3099	0.6537	158.29	172.81	0.3060

R-134a

TABLE A-12E (*Continued*)

T °F	v ft³/lb	u Btu/lb	h Btu/lb	s Btu/lb · °R	v ft³/lb	u Btu/lb	h Btu/lb	s Btu/lb · °R
	p = 140 lbf/in.² (T_{sat} = 100.56°F)				p = 160 lbf/in.² (T_{sat} = 109.55°F)			
Sat.	0.3358	106.25	114.95	0.2161	0.2916	107.28	115.91	0.2157
120	0.3610	110.90	120.25	0.2254	0.3044	109.88	118.89	0.2209
140	0.3846	115.58	125.54	0.2344	0.3269	114.73	124.41	0.2303
160	0.4066	120.21	130.74	0.2429	0.3474	119.49	129.78	0.2391
180	0.4274	124.82	135.89	0.2511	0.3666	124.20	135.06	0.2475
200	0.4474	129.44	141.03	0.2590	0.3849	128.90	140.29	0.2555
220	0.4666	134.09	146.18	0.2667	0.4023	133.61	145.52	0.2633
240	0.4852	138.77	151.34	0.2742	0.4192	138.34	150.75	0.2709
260	0.5034	143.50	156.54	0.2815	0.4356	143.11	156.00	0.2783
280	0.5212	148.28	161.78	0.2887	0.4516	147.92	161.29	0.2856
300	0.5387	153.11	167.06	0.2957	0.4672	152.78	166.61	0.2927
320	0.5559	157.99	172.39	0.3026	0.4826	157.69	171.98	0.2996
340	0.5730	162.93	177.78	0.3094	0.4978	162.65	177.39	0.3065
360	0.5898	167.93	183.21	0.3162	0.5128	167.67	182.85	0.3132
	p = 180 lbf/in.² (T_{sat} = 117.74°F)				p = 200 lbf/in.² (T_{sat} = 125.28°F)			
Sat.	0.2569	108.18	116.74	0.2154	0.2288	108.98	117.44	0.2151
120	0.2595	108.77	117.41	0.2166				
140	0.2814	113.83	123.21	0.2264	0.2446	112.87	121.92	0.2226
160	0.3011	118.74	128.77	0.2355	0.2636	117.94	127.70	0.2321
180	0.3191	123.56	134.19	0.2441	0.2809	122.88	133.28	0.2410
200	0.3361	128.34	139.53	0.2524	0.2970	127.76	138.75	0.2494
220	0.3523	133.11	144.84	0.2603	0.3121	132.60	144.15	0.2575
240	0.3678	137.90	150.15	0.2680	0.3266	137.44	149.53	0.2653
260	0.3828	142.71	155.46	0.2755	0.3405	142.30	154.90	0.2728
280	0.3974	147.55	160.79	0.2828	0.3540	147.18	160.28	0.2802
300	0.4116	152.44	166.15	0.2899	0.3671	152.10	165.69	0.2874
320	0.4256	157.38	171.55	0.2969	0.3799	157.07	171.13	0.2945
340	0.4393	162.36	177.00	0.3038	0.3926	162.07	176.60	0.3014
360	0.4529	167.40	182.49	0.3106	0.4050	167.13	182.12	0.3082
	p = 300 lbf/in.² (T_{sat} = 156.17°F)				p = 400 lbf/in.² (T_{sat} = 179.95°F)			
Sat.	0.1424	111.72	119.62	0.2132	0.0965	112.77	119.91	0.2102
160	0.1462	112.95	121.07	0.2155				
180	0.1633	118.93	128.00	0.2265	0.0965	112.79	119.93	0.2102
200	0.1777	124.47	134.34	0.2363	0.1143	120.14	128.60	0.2235
220	0.1905	129.79	140.36	0.2453	0.1275	126.35	135.79	0.2343
240	0.2021	134.99	146.21	0.2537	0.1386	132.12	142.38	0.2438
260	0.2130	140.12	151.95	0.2618	0.1484	137.65	148.64	0.2527
280	0.2234	145.23	157.63	0.2696	0.1575	143.06	154.72	0.2610
300	0.2333	150.33	163.28	0.2772	0.1660	148.39	160.67	0.2689
320	0.2428	155.44	168.92	0.2845	0.1740	153.69	166.57	0.2766
340	0.2521	160.57	174.56	0.2916	0.1816	158.97	172.42	0.2840
360	0.2611	165.74	180.23	0.2986	0.1890	164.26	178.26	0.2912
380	0.2699	170.94	185.92	0.3055	0.1962	169.57	184.09	0.2983
400	0.2786	176.18	191.64	0.3122	0.2032	174.90	189.94	0.3051

TABLE A-13E Specific Heats of Selected Liquids and Solids

A. Liquid	State	c (Btu/lb · °R)
Ammonia	sat. liq., 0°F	1.08
	sat. liq., 50°F	1.12
	sat. liq., 120°F	1.22
Ethyl alcohol	1 atm, 76°F	0.580
Glycerin	1 atm, 50°F	0.554
	1 atm, 120°F	0.617
Mercury	1 atm, 50°F	0.033
	1 atm, 600°F	0.032
Refrigerant 12	sat. liq., 0°F	0.216
	sat. liq., 70°F	0.232
Water	1 atm, 32°F	1.007
	1 atm, 80°F	0.998
	1 atm, 212°F	1.007

B. Solid	Temperature (°R)	c (Btu/lb · °R)
Aluminum	540	0.216
Copper	540	0.0920
	720	0.0939
Ice	360	0.373
	460	0.471
	492	0.504
Iron	540	0.107
Lead	540	0.0308
Silver	540	0.0561

Table A-14E

Ideal Gas Specific Heats of Some Common Gases (Btu/lb · °R)

Temp. °F	Air c_p	Air c_v	Air k	Nitrogen, N_2 c_p	Nitrogen, N_2 c_v	Nitrogen, N_2 k	Oxygen, O_2 c_p	Oxygen, O_2 c_v	Oxygen, O_2 k	Temp. °F
40	0.240	0.171	1.401	0.248	0.177	1.400	0.219	0.156	1.397	40
100	0.240	0.172	1.400	0.248	0.178	1.399	0.220	0.158	1.394	100
200	0.241	0.173	1.397	0.249	0.178	1.398	0.223	0.161	1.387	200
300	0.243	0.174	1.394	0.250	0.179	1.396	0.226	0.164	1.378	300
400	0.245	0.176	1.389	0.251	0.180	1.393	0.230	0.168	1.368	400
500	0.248	0.179	1.383	0.254	0.183	1.388	0.235	0.173	1.360	500
600	0.250	0.182	1.377	0.256	0.185	1.383	0.239	0.177	1.352	600
700	0.254	0.185	1.371	0.260	0.189	1.377	0.242	0.181	1.344	700
800	0.257	0.188	1.365	0.262	0.191	1.371	0.246	0.184	1.337	800
900	0.259	0.191	1.358	0.265	0.194	1.364	0.249	0.187	1.331	900
1000	0.263	0.195	1.353	0.269	0.198	1.359	0.252	0.190	1.326	1000
1500	0.276	0.208	1.330	0.283	0.212	1.334	0.263	0.201	1.309	1500
2000	0.286	0.217	1.312	0.293	0.222	1.319	0.270	0.208	1.298	2000

Temp. °F	Carbon dioxide, CO_2 c_p	Carbon dioxide, CO_2 c_v	Carbon dioxide, CO_2 k	Carbon monoxide, CO c_p	Carbon monoxide, CO c_v	Carbon monoxide, CO k	Hydrogen, H_2 c_p	Hydrogen, H_2 c_v	Hydrogen, H_2 k	Temp. °F
40	0.195	0.150	1.300	0.248	0.177	1.400	3.397	2.412	1.409	40
100	0.205	0.160	1.283	0.249	0.178	1.399	3.426	2.441	1.404	100
200	0.217	0.172	1.262	0.249	0.179	1.397	3.451	2.466	1.399	200
300	0.229	0.184	1.246	0.251	0.180	1.394	3.461	2.476	1.398	300
400	0.239	0.193	1.233	0.253	0.182	1.389	3.466	2.480	1.397	400
500	0.247	0.202	1.223	0.256	0.185	1.384	3.469	2.484	1.397	500
600	0.255	0.210	1.215	0.259	0.188	1.377	3.473	2.488	1.396	600
700	0.262	0.217	1.208	0.262	0.191	1.371	3.477	2.492	1.395	700
800	0.269	0.224	1.202	0.266	0.195	1.364	3.494	2.509	1.393	800
900	0.275	0.230	1.197	0.269	0.198	1.357	3.502	2.519	1.392	900
1000	0.280	0.235	1.192	0.273	0.202	1.351	3.513	2.528	1.390	1000
1500	0.298	0.253	1.178	0.287	0.216	1.328	3.618	2.633	1.374	1500
2000	0.312	0.267	1.169	0.297	0.226	1.314	3.758	2.773	1.355	2000

Source: Adapted from K. Wark, *Thermodynamics, 4th ed.*, McGraw-Hill, New York, 1983, as based on "Tables of Thermal Properties of Gases," NBS Circular 564, 1955.

TABLE A-15E Variation of $\bar{c}_p$ with Temperature for Various Ideal Gases

$$\frac{\bar{c}_p}{\bar{R}} = \alpha + \beta T + \gamma T^2 + \delta T^3 + \varepsilon T^4$$

T is in °R, equations valid from 540 to 1800 °R

Gas	α	$\beta \times 10^3$	$\gamma \times 10^6$	$\delta \times 10^9$	$\epsilon \times 10^{12}$
CO	3.710	−0.899	1.140	−0.348	0.0228
CO_2	2.401	4.853	−2.039	0.343	0
H_2	3.057	1.487	−1.793	0.947	−0.1726
H_2O	4.070	−0.616	1.281	−0.508	0.0769
O_2	3.626	−1.043	2.178	−1.160	0.2053
N_2	3.675	−0.671	0.717	−0.108	−0.0215
Air	3.653	−0.7428	1.017	−0.328	0.02632
NH_3	3.591	0.274	2.576	−1.437	0.2601
NO	4.046	−1.899	2.464	−1.048	0.1517
NO_2	3.459	1.147	2.064	−1.639	0.3448
SO_2	3.267	2.958	0.211	−0.906	0.2438
SO_3	2.578	8.087	−2.832	−0.136	0.1878
CH_4	3.826	−2.211	7.580	−3.898	0.6633
C_2H_2	1.410	10.587	−7.562	2.811	−0.3939
C_2H_4	1.426	6.324	2.466	−2.787	0.6429
Monatomic gases[a]	2.5	0	0	0	0

[a] For monatomic gases, such as He, Ne, and Ar, $\bar{c}_p$ is constant over a wide temperature range and is very nearly equal to 5/2 $\bar{R}$.

Source: Adapted from K. Wark, *Thermodynamics, 4th ed.*, McGraw-Hill, New York, 1983, as based on NASA SP-273, U.S. Government Printing Office, Washington, DC, 1971.

Air

TABLE A-16E Ideal Gas Properties of Air

T(°R), h and u (Btu/lb), s^o (Btu/lb · °R)

T	h	p_r	u	v_r	s^o	T	h	p_r	u	v_r	s^o
360	85.97	0.3363	61.29	396.6	0.50369	940	226.11	9.834	161.68	35.41	0.73509
380	90.75	0.4061	64.70	346.6	0.51663	960	231.06	10.61	165.26	33.52	0.74030
400	95.53	0.4858	68.11	305.0	0.52890	980	236.02	11.43	168.83	31.76	0.74540
420	100.32	0.5760	71.52	270.1	0.54058	1000	240.98	12.30	172.43	30.12	0.75042
440	105.11	0.6776	74.93	240.6	0.55172	1040	250.95	14.18	179.66	27.17	0.76019
460	109.90	0.7913	78.36	215.33	0.56235	1080	260.97	16.28	186.93	24.58	0.76964
480	114.69	0.9182	81.77	193.65	0.57255	1120	271.03	18.60	194.25	22.30	0.77880
500	119.48	1.0590	85.20	174.90	0.58233	1160	281.14	21.18	201.63	20.29	0.78767
520	124.27	1.2147	88.62	158.58	0.59172	1200	291.30	24.01	209.05	18.51	0.79628
537	128.34	1.3593	91.53	146.34	0.59945	1240	301.52	27.13	216.53	16.93	0.80466
540	129.06	1.3860	92.04	144.32	0.60078	1280	311.79	30.55	224.05	15.52	0.81280
560	133.86	1.5742	95.47	131.78	0.60950	1320	322.11	34.31	231.63	14.25	0.82075
580	138.66	1.7800	98.90	120.70	0.61793	1360	332.48	38.41	239.25	13.12	0.82848
600	143.47	2.005	102.34	110.88	0.62607	1400	342.90	42.88	246.93	12.10	0.83604
620	148.28	2.249	105.78	102.12	0.63395	1440	353.37	47.75	254.66	11.17	0.84341
640	153.09	2.514	109.21	94.30	0.64159	1480	363.89	53.04	262.44	10.34	0.85062
660	157.92	2.801	112.67	87.27	0.64902	1520	374.47	58.78	270.26	9.578	0.85767
680	162.73	3.111	116.12	80.96	0.65621	1560	385.08	65.00	278.13	8.890	0.86456
700	167.56	3.446	119.58	75.25	0.66321	1600	395.74	71.73	286.06	8.263	0.87130
720	172.39	3.806	123.04	70.07	0.67002	1650	409.13	80.89	296.03	7.556	0.87954
740	177.23	4.193	126.51	65.38	0.67665	1700	422.59	90.95	306.06	6.924	0.88758
760	182.08	4.607	129.99	61.10	0.68312	1750	436.12	101.98	316.16	6.357	0.89542
780	186.94	5.051	133.47	57.20	0.68942	1800	449.71	114.0	326.32	5.847	0.90308
800	191.81	5.526	136.97	53.63	0.69558	1850	463.37	127.2	336.55	5.388	0.91056
820	196.69	6.033	140.47	50.35	0.70160	1900	477.09	141.5	346.85	4.974	0.91788
840	201.56	6.573	143.98	47.34	0.70747	1950	490.88	157.1	357.20	4.598	0.92504
860	206.46	7.149	147.50	44.57	0.71323	2000	504.71	174.0	367.61	4.258	0.93205
880	211.35	7.761	151.02	42.01	0.71886	2050	518.61	192.3	378.08	3.949	0.93891
900	216.26	8.411	154.57	39.64	0.72438	2100	532.55	212.1	388.60	3.667	0.94564
920	221.18	9.102	158.12	37.44	0.72979	2150	546.54	233.5	399.17	3.410	0.95222

TABLE A-16E *(Continued)*

T(°R), h and u(Btu/lb), s^o(Btu/lb · °R)

T	h	p_r	u	v_r	s^o	T	h	p_r	u	v_r	s^o
2200	560.59	256.6	409.78	3.176	0.95868	3700	998.11	2330	744.48	.5882	1.10991
2250	574.69	281.4	420.46	2.961	0.96501	3750	1013.1	2471	756.04	.5621	1.11393
2300	588.82	308.1	431.16	2.765	0.97123	3800	1028.1	2618	767.60	.5376	1.11791
2350	603.00	336.8	441.91	2.585	0.97732	3850	1043.1	2773	779.19	.5143	1.12183
2400	617.22	367.6	452.70	2.419	0.98331	3900	1058.1	2934	790.80	.4923	1.12571
2450	631.48	400.5	463.54	2.266	0.98919	3950	1073.2	3103	802.43	.4715	1.12955
2500	645.78	435.7	474.40	2.125	0.99497	4000	1088.3	3280	814.06	.4518	1.13334
2550	660.12	473.3	485.31	1.996	1.00064	4050	1103.4	3464	825.72	.4331	1.13709
2600	674.49	513.5	496.26	1.876	1.00623	4100	1118.5	3656	837.40	.4154	1.14079
2650	688.90	556.3	507.25	1.765	1.01172	4150	1133.6	3858	849.09	.3985	1.14446
2700	703.35	601.9	518.26	1.662	1.01712	4200	1148.7	4067	860.81	.3826	1.14809
2750	717.83	650.4	529.31	1.566	1.02244	4300	1179.0	4513	884.28	.3529	1.15522
2800	732.33	702.0	540.40	1.478	1.02767	4400	1209.4	4997	907.81	.3262	1.16221
2850	746.88	756.7	551.52	1.395	1.03282	4500	1239.9	5521	931.39	.3019	1.16905
2900	761.45	814.8	562.66	1.318	1.03788	4600	1270.4	6089	955.04	.2799	1.17575
2950	776.05	876.4	573.84	1.247	1.04288	4700	1300.9	6701	978.73	.2598	1.18232
3000	790.68	941.4	585.04	1.180	1.04779	4800	1331.5	7362	1002.5	.2415	1.18876
3050	805.34	1011	596.28	1.118	1.05264	4900	1362.2	8073	1026.3	.2248	1.19508
3100	820.03	1083	607.53	1.060	1.05741	5000	1392.9	8837	1050.1	.2096	1.20129
3150	834.75	1161	618.82	1.006	1.06212	5100	1423.6	9658	1074.0	.1956	1.20738
3200	849.48	1242	630.12	.9546	1.06676	5200	1454.4	10539	1098.0	.1828	1.21336
3250	864.24	1328	641.46	.9069	1.07134	5300	1485.3	11481	1122.0	.1710	1.21923
3300	879.02	1418	652.81	.8621	1.07585						
3350	893.83	1513	664.20	.8202	1.08031						
3400	908.66	1613	675.60	.7807	1.08470						
3450	923.52	1719	687.04	.7436	1.08904						
3500	938.40	1829	698.48	.7087	1.09332						
3550	953.30	1946	709.95	.6759	1.09755						
3600	968.21	2068	721.44	.6449	1.10172						
3650	983.15	2196	732.95	.6157	1.10584						

Source: Tables A-13E through A-19E are adapted from K. Wark, *Thermodynamics, 4th ed.*, McGraw-Hill, New York, 1983, as extracted from J. H. Keenan and J. Kaye, "Gas Tables," Wiley, New York, 1945.

TABLE A-17E Ideal Gas Properties of Nitrogen, N_2

T(°R), $\bar{h}$ and $\bar{u}$(Btu/lbmol), $\bar{s}°$(Btu/lbmol · °R)

[$\bar{h}_f^\circ$ = 0 Btu/lbmol]

T	$\bar{h}$	$\bar{u}$	$\bar{s}°$	T	$\bar{h}$	$\bar{u}$	$\bar{s}°$
300	2082.0	1486.2	41.695	1080	7551.0	5406.2	50.651
320	2221.0	1585.5	42.143	1100	7695.0	5510.5	50.783
340	2360.0	1684.4	42.564	1120	7839.3	5615.2	50.912
360	2498.9	1784.0	42.962	1140	7984.0	5720.1	51.040
380	2638.0	1883.4	43.337	1160	8129.0	5825.4	51.167
400	2777.0	1982.6	43.694	1180	8274.4	5931.0	51.291
420	2916.1	2082.0	44.034	1200	8420.0	6037.0	51.413
440	3055.1	2181.3	44.357	1220	8566.1	6143.4	51.534
460	3194.1	2280.6	44.665	1240	8712.6	6250.1	51.653
480	3333.1	2379.9	44.962	1260	8859.3	6357.2	51.771
500	3472.2	2479.3	45.246	1280	9006.4	6464.5	51.887
520	3611.3	2578.6	45.519	1300	9153.9	6572.3	51.001
537	3729.5	2663.1	45.743	1320	9301.8	6680.4	52.114
540	3750.3	2678.0	45.781	1340	9450.0	6788.9	52.225
560	3889.5	2777.4	46.034	1360	9598.6	6897.8	52.335
580	4028.7	2876.9	46.278	1380	9747.5	7007.0	52.444
600	4167.9	2976.4	46.514	1400	9896.9	7116.7	52.551
620	4307.1	3075.9	46.742	1420	10046.6	7226.7	52.658
640	4446.4	3175.5	46.964	1440	10196.6	7337.0	52.763
660	4585.8	3275.2	47.178	1460	10347.0	7447.6	52.867
680	4725.3	3374.9	47.386	1480	10497.8	7558.7	52.969
700	4864.9	3474.8	47.588	1500	10648.0	7670.1	53.071
720	5004.5	3574.7	47.785	1520	10800.4	7781.9	53.171
740	5144.3	3674.7	47.977	1540	10952.2	7893.9	53.271
760	5284.1	3774.9	48.164	1560	11104.3	8006.4	53.369
780	5424.2	3875.2	48.345	1580	11256.9	8119.2	53.465
800	5564.4	3975.7	48.522	1600	11409.7	8232.3	53.561
820	5704.7	4076.3	48.696	1620	11562.8	8345.7	53.656
840	5845.3	4177.1	48.865	1640	11716.4	8459.6	53.751
860	5985.9	4278.1	49.031	1660	11870.2	8573.6	53.844
880	6126.9	4379.4	49.193	1680	12024.3	8688.1	53.936
900	6268.1	4480.8	49.352	1700	12178.9	8802.9	54.028
920	6409.6	4582.6	49.507	1720	12333.7	8918.0	54.118
940	6551.2	4684.5	49.659	1740	12488.8	9033.4	54.208
960	6693.1	4786.7	49.808	1760	12644.3	9149.2	54.297
980	6835.4	4889.3	49.955	1780	12800.2	9265.3	54.385
1000	6977.9	4992.0	50.099	1800	12956.3	9381.7	54.472
1020	7120.7	5095.1	50.241	1820	13112.7	9498.4	54.559
1040	7263.8	5198.5	50.380	1840	13269.5	9615.5	54.645
1060	7407.2	5302.2	50.516	1860	13426.5	9732.8	54.729

TABLE A-17E (*Continued*)

T	$\bar{h}$	$\bar{u}$	$\bar{s}°$	T	$\bar{h}$	$\bar{u}$	$\bar{s}°$
1900	13,742	9,968	54.896	3500	27,016	20,065	59.944
1940	14,058	10,205	55.061	3540	27,359	20,329	60.041
1980	14,375	10,443	55.223	3580	27,703	20,593	60.138
2020	14,694	10,682	55.383	3620	28,046	20,858	60.234
2060	15,013	10,923	55.540	3660	28,391	21,122	60.328
2100	15,334	11,164	55.694	3700	28,735	21,387	60.422
2140	15,656	11,406	55.846	3740	29,080	21,653	60.515
2180	15,978	11,649	55.995	3780	29,425	21,919	60.607
2220	16,302	11,893	56.141	3820	29,771	22,185	60.698
2260	16,626	12,138	56.286	3860	30,117	22,451	60.788
2300	16,951	12,384	56.429	3900	30,463	22,718	60.877
2340	17,277	12,630	56.570	3940	30,809	22,985	60.966
2380	17,604	12,878	56.708	3980	31,156	23,252	61.053
2420	17,932	13,126	56.845	4020	31,503	23,520	61.139
2460	18,260	13,375	56.980	4060	31,850	23,788	61.225
2500	18,590	13,625	57.112	4100	32,198	24,056	61.310
2540	18,919	13,875	57.243	4140	32,546	24,324	61.395
2580	19,250	14,127	57.372	4180	32,894	24,593	61.479
2620	19,582	14,379	57.499	4220	33,242	24,862	61.562
2660	19,914	14,631	57.625	4260	33,591	25,131	61.644
2700	20,246	14,885	57.750	4300	33,940	25,401	61.726
2740	20,580	15,139	57.872	4340	34,289	25,670	61.806
2780	20,914	15,393	57.993	4380	34,638	25,940	61.887
2820	21,248	15,648	58.113	4420	34,988	26,210	61.966
2860	21,584	15,905	58.231	4460	35,338	26,481	62.045
2900	21,920	16,161	58.348	4500	35,688	26,751	62.123
2940	22,256	16,417	58.463	4540	36,038	27,022	62.201
2980	22,593	16,675	58.576	4580	36,389	27,293	62.278
3020	22,930	16,933	58.688	4620	36,739	27,565	62.354
3060	23,268	17,192	58.800	4660	37,090	27,836	62.429
3100	23,607	17,451	58.910	4700	37,441	28,108	62.504
3140	23,946	17,710	59.019	4740	37,792	28,379	62.578
3180	24,285	17,970	59.126	4780	38,144	28,651	62.652
3220	24,625	18,231	59.232	4820	38,495	28,924	62.725
3260	24,965	18,491	59.338	4860	38,847	29,196	62.798
3300	25,306	18,753	59.442	4900	39,199	29,468	62.870
3340	25,647	19,014	59.544	5000	40,080	30,151	63.049
3380	25,989	19,277	59.646	5100	40,962	30,834	63.223
3420	26,331	19,539	59.747	5200	41,844	31,518	63.395
3460	26,673	19,802	59.846	5300	42,728	32,203	63.563

TABLE A-18E Ideal Gas Properties of Oxygen, O_2

T(°R), $\bar{h}$ and $\bar{u}$(Btu/lbmol), $\bar{s}^\circ$(Btu/lbmol · °R)

[$\bar{h}_f^\circ$ = 0 Btu/lbmol]

T	$\bar{h}$	$\bar{u}$	$\bar{s}^\circ$	T	$\bar{h}$	$\bar{u}$	$\bar{s}^\circ$
300	2073.5	1477.8	44.927	1080	7696.8	5552.1	54.064
320	2212.6	1577.1	45.375	1100	7850.4	5665.9	54.204
340	2351.7	1676.5	45.797	1120	8004.5	5780.3	54.343
360	2490.8	1775.9	46.195	1140	8159.1	5895.2	54.480
380	2630.0	1875.3	46.571	1160	8314.2	6010.6	54.614
400	2769.1	1974.8	46.927	1180	8469.8	6126.5	54.748
420	2908.3	2074.3	47.267	1200	8625.8	6242.8	54.879
440	3047.5	2173.8	47.591	1220	8782.4	6359.6	55.008
460	3186.9	2273.4	47.900	1240	8939.4	6476.9	55.136
480	3326.5	2373.3	48.198	1260	9096.7	6594.5	55.262
500	3466.2	2473.2	48.483	1280	9254.6	6712.7	55.386
520	3606.1	2573.4	48.757	1300	9412.9	6831.3	55.508
537	3725.1	2658.7	48.982	1320	9571.6	6950.2	55.630
540	3746.2	2673.8	49.021	1340	9730.7	7069.6	55.750
560	3886.6	2774.5	49.276	1360	9890.2	7189.4	55.867
580	4027.3	2875.5	49.522	1380	10050.1	7309.6	55.984
600	4168.3	2976.8	49.762	1400	10210.4	7430.1	56.099
620	4309.7	3078.4	49.993	1420	10371.0	7551.1	56.213
640	4451.4	3180.4	50.218	1440	10532.0	7672.4	56.326
660	4593.5	3282.9	50.437	1460	10693.3	7793.9	56.437
680	4736.2	3385.8	50.650	1480	10855.1	7916.0	56.547
700	4879.3	3489.2	50.858	1500	11017.1	8038.3	56.656
720	5022.9	3593.1	51.059	1520	11179.6	8161.1	56.763
740	5167.0	3697.4	51.257	1540	11342.4	8284.2	56.869
760	5311.4	3802.2	51.450	1560	11505.4	8407.4	56.975
780	5456.4	3907.5	51.638	1580	11668.8	8531.1	57.079
800	5602.0	4013.3	51.821	1600	11832.5	8655.1	57.182
820	5748.1	4119.7	52.002	1620	11996.6	8779.5	57.284
840	5894.8	4226.6	52.179	1640	12160.9	8904.1	57.385
860	6041.9	4334.1	52.352	1660	12325.5	9029.0	57.484
880	6189.6	4442.0	52.522	1680	12490.4	9154.1	57.582
900	6337.9	4550.6	52.688	1700	12655.6	9279.6	57.680
920	6486.7	4659.7	52.852	1720	12821.1	9405.4	57.777
940	6636.1	4769.4	53.012	1740	12986.9	9531.5	57.873
960	6786.0	4879.5	53.170	1760	13153.0	9657.9	57.968
980	6936.4	4990.3	53.326	1780	13319.2	9784.4	58.062
1000	7087.5	5101.6	53.477	1800	13485.8	9911.2	58.155
1020	7238.9	5213.3	53.628	1820	13652.5	10038.2	58.247
1040	7391.0	5325.7	53.775	1840	13819.6	10165.6	58.339
1060	7543.6	5438.6	53.921	1860	13986.8	10293.1	58.428

TABLE A-18E (*Continued*)

T	$\bar{h}$	$\bar{u}$	$\bar{s}°$	T	$\bar{h}$	$\bar{u}$	$\bar{s}°$
1900	14,322	10,549	58.607	3500	28,273	21,323	63.914
1940	14,658	10,806	58.782	3540	28,633	21,603	64.016
1980	14,995	11,063	58.954	3580	28,994	21,884	64.114
2020	15,333	11,321	59.123	3620	29,354	22,165	64.217
2060	15,672	11,581	59.289	3660	29,716	22,447	64.316
2100	16,011	11,841	59.451	3700	30,078	22,730	64.415
2140	16,351	12,101	59.612	3740	30,440	23,013	64.512
2180	16,692	12,363	59.770	3780	30,803	23,296	64.609
2220	17,036	12,625	59.926	3820	31,166	23,580	64.704
2260	17,376	12,888	60.077	3860	31,529	23,864	64.800
2300	17,719	13,151	60.228	3900	31,894	24,149	64.893
2340	18,062	13,416	60.376	3940	32,258	24,434	64.986
2380	18,407	13,680	60.522	3980	32,623	24,720	65.078
2420	18,572	13,946	60.666	4020	32,989	25,006	65.169
2460	19,097	14,212	60.808	4060	33,355	25,292	65.260
2500	19,443	14,479	60.946	4100	33,722	25,580	65.350
2540	19,790	14,746	61.084	4140	34.089	25,867	64.439
2580	20,138	15,014	61.220	4180	34,456	26,155	65.527
2620	20,485	15,282	61.354	4220	34,824	26,444	65.615
2660	20,834	15,551	61.486	4260	35,192	26,733	65.702
2700	21,183	15,821	61.616	4300	35,561	27,022	65.788
2740	21,533	16,091	61.744	4340	35,930	27,312	65.873
2780	21,883	16,362	61.871	4380	36,300	27,602	65.958
2820	22,232	16,633	61.996	4420	36,670	27,823	66.042
2860	22,584	16,905	62.120	4460	37,041	28,184	66.125
2900	22,936	17,177	62.242	4500	37,412	28,475	66.208
2940	23,288	17,450	62.363	4540	37,783	28,768	66.290
2980	23,641	17,723	62.483	4580	38,155	29,060	66.372
3020	23,994	17,997	62.599	4620	38,528	29,353	66.453
3060	24,348	18,271	62.716	4660	38,900	29,646	66.533
3100	24,703	18,546	62.831	4700	39,274	29,940	66.613
3140	25,057	18,822	62.945	4740	39,647	30,234	66.691
3180	25,413	19,098	63.057	4780	40,021	30,529	66.770
3220	25,769	19,374	63.169	4820	40,396	30,824	66.848
3260	26,175	19,651	63.279	4860	40,771	31,120	66.925
3300	26,412	19,928	63.386	4900	41,146	31,415	67.003
3340	26,839	20,206	63.494	5000	42,086	32,157	67.193
3380	27,197	20,485	63.601	5100	43,021	32,901	67.380
3420	27,555	20,763	63.706	5200	43,974	33,648	67.562
3460	27,914	21,043	63.811	5300	44,922	34,397	67.743

TABLE A-19E Ideal Gas Properties of Water Vapor, H_2O

T(°R), $\bar{h}$ and $\bar{u}$(Btu/lbmol), $\bar{s}°$(Btu/lbmol · °R)

[$\bar{h}_f^\circ = -104{,}040$ Btu/lbmol]

T	$\bar{h}$	$\bar{u}$	$\bar{s}°$	T	$\bar{h}$	$\bar{u}$	$\bar{s}°$
300	2367.6	1771.8	40.439	1080	8768.2	6623.5	50.854
320	2526.8	1891.3	40.952	1100	8942.0	6757.5	51.013
340	2686.0	2010.8	41.435	1120	9116.4	6892.2	51.171
360	2845.1	2130.2	41.889	1140	9291.4	7027.5	51.325
380	3004.4	2249.8	42.320	1160	9467.1	7163.5	51.478
400	3163.8	2369.4	42.728	1180	9643.4	7300.1	51.630
420	3323.2	2489.1	43.117	1200	9820.4	7437.4	51.777
440	3482.7	2608.9	43.487	1220	9998.0	7575.2	51.925
460	3642.3	2728.8	43.841	1240	10176.1	7713.6	52.070
480	3802.0	2848.8	44.182	1260	10354.9	7852.7	52.212
500	3962.0	2969.1	44.508	1280	10534.4	7992.5	52.354
520	4122.0	3089.4	44.821	1300	10714.5	8132.9	52.494
537	4258.0	3191.9	45.079	1320	10895.3	8274.0	52.631
540	4282.4	3210.0	45.124	1340	11076.6	8415.5	52.768
560	4442.8	3330.7	45.415	1360	11258.7	8557.9	52.903
580	4603.7	3451.9	45.696	1380	11441.4	8700.9	53.037
600	4764.7	3573.2	45.970	1400	11624.8	8844.6	53.168
620	4926.1	3694.9	46.235	1420	11808.8	8988.9	53.299
640	5087.8	3816.8	46.492	1440	11993.4	9133.8	53.428
660	5250.0	3939.3	46.741	1460	12178.8	9279.4	53.556
680	5412.5	4062.1	46.984	1480	12364.8	9425.7	53.682
700	5575.4	4185.3	47.219	1500	12551.4	9572.7	53.808
720	5738.8	4309.0	47.450	1520	12738.8	9720.3	53.932
740	5902.6	4433.1	47.673	1540	12926.8	9868.6	54.055
760	6066.9	4557.6	47.893	1560	13115.6	10017.6	54.117
780	6231.7	4682.7	48.106	1580	13305.0	10167.3	54.298
800	6396.9	4808.2	48.316	1600	13494.4	10317.6	54.418
820	6562.6	4934.2	48.520	1620	13685.7	10468.6	54.535
840	6728.9	5060.8	48.721	1640	13877.0	10620.2	54.653
860	6895.6	5187.8	48.916	1660	14069.2	10772.7	54.770
880	7062.9	5315.3	49.109	1680	14261.9	10925.6	54.886
900	7230.9	5443.6	49.298	1700	14455.4	11079.4	54.999
920	7399.4	5572.4	49.483	1720	14649.5	11233.8	55.113
940	7568.4	5701.7	49.665	1740	14844.3	11388.9	55.226
960	7738.0	5831.6	49.843	1760	15039.8	11544.7	55.339
980	7908.2	5962.0	50.019	1780	15236.1	11701.2	55.449
1000	8078.9	6093.0	50.191	1800	15433.0	11858.4	55.559
1020	8250.4	6224.8	50.360	1820	15630.6	12016.3	55.668
1040	8422.4	6357.1	50.528	1840	15828.7	12174.7	55.777
1060	8595.0	6490.0	50.693	1860	16027.6	12333.9	55.884

TABLE A-19E (*Continued*)

T	$\bar{h}$	$\bar{u}$	$\bar{s}°$	T	$\bar{h}$	$\bar{u}$	$\bar{s}°$
1900	16,428	12,654	56.097	3500	34,324	27,373	62.876
1940	16,830	12,977	56.307	3540	34,809	27,779	63.015
1980	17,235	13,303	56.514	3580	35,296	28,187	63.153
2020	17,643	13,632	56.719	3620	35,785	28,596	63.288
2060	18,054	13,963	56.920	3660	36,274	29,006	63.423
2100	18,467	14,297	57.119	3700	36,765	29,418	63.557
2140	18,883	14,633	57.315	3740	37,258	29,831	63.690
2180	19,301	14,972	57.509	3780	37,752	30,245	63.821
2220	19,722	15,313	57.701	3820	38,247	30,661	63.952
2260	20,145	15,657	57.889	3860	38,743	31,077	64.082
2300	20,571	16,003	58.077	3900	39,240	31,495	64.210
2340	20,999	16,352	58.261	3940	39,739	31,915	64.338
2380	21,429	16,703	58.445	3980	40,239	32,335	64.465
2420	21,862	17,057	58.625	4020	40,740	32,757	64.591
2460	22,298	17,413	58.803	4060	41,242	33,179	64.715
2500	22,735	17,771	58.980	4100	41,745	33,603	64.839
2540	23,175	18,131	59.155	4140	42,250	34,028	64.962
2580	23,618	18,494	59.328	4180	42,755	34,454	65.084
2620	24,062	18,859	59.500	4220	43,267	34,881	65.204
2660	24,508	19,226	59.669	4260	43,769	35,310	65.325
2700	24,957	19,595	59.837	4300	44,278	35,739	65.444
2740	25,408	19,967	60.003	4340	44,788	36,169	65.563
2780	25,861	20,340	60.167	4380	45,298	36,600	65.680
2820	26,316	20,715	60.330	4420	45,810	37,032	65.797
2860	26,773	21,093	60.490	4460	46,322	37,465	65.913
2900	27,231	21,472	60.650	4500	46,836	37,900	66.028
2940	27,692	21,853	60.809	4540	47,350	38,334	66.142
2980	28,154	22,237	60.965	4580	47,866	38,770	66.255
3020	28,619	22,621	61.120	4620	48,382	39,207	66.368
3060	29,085	23,085	61.274	4660	48,899	39,645	66.480
3100	29,553	23,397	61.426	4700	49,417	40,083	66.591
3140	30,023	23,787	61.577	4740	49,936	40,523	66.701
3180	30,494	24,179	61.727	4780	50,455	40,963	66.811
3220	30,967	24,572	61.874	4820	50,976	41,404	66.920
3260	31,442	24,968	62.022	4860	51,497	41,856	67.028
3300	31,918	25,365	62.167	4900	52,019	42,288	67.135
3340	32,396	25,763	62.312	5000	53,327	43,398	67.401
3380	32,876	26,164	62.454	5100	54,640	44,512	67.662
3420	33,357	26,565	62.597	5200	55,957	45,631	67.918
3460	33,839	26,968	62.738	5300	57,279	46,754	68.172

TABLE A-20E Ideal Gas Properties of Carbon Monoxide, CO

T(°R), $\bar{h}$ and $\bar{u}$(Btu/lbmol), $\bar{s}°$(Btu/lbmol · °R)

[$\bar{h}_f^\circ$ = −47,540 Btu/lbmol]

T	$\bar{h}$	$\bar{u}$	$\bar{s}°$	T	$\bar{h}$	$\bar{u}$	$\bar{s}°$
300	2081.9	1486.1	43.223	1080	7571.1	5426.4	52.203
320	2220.9	1585.4	43.672	1100	7716.8	5532.3	52.337
340	2359.9	1684.7	44.093	1120	7862.9	5638.7	52.468
360	2498.8	1783.9	44.490	1140	8009.2	5745.4	52.598
380	2637.9	1883.3	44.866	1160	8156.1	5851.5	52.726
400	2776.9	1982.6	45.223	1180	8303.3	5960.0	52.852
420	2916.0	2081.9	45.563	1200	8450.8	6067.8	52.976
440	3055.0	2181.2	45.886	1220	8598.8	6176.0	53.098
460	3194.0	2280.5	46.194	1240	8747.2	6284.7	53.218
480	3333.0	2379.8	46.491	1260	8896.0	6393.8	53.337
500	3472.1	2479.2	46.775	1280	9045.0	6503.1	53.455
520	3611.2	2578.6	47.048	1300	9194.6	6613.0	53.571
537	3725.1	2663.1	47.272	1320	9344.6	6723.2	53.685
540	3750.3	2677.9	47.310	1340	9494.8	6833.7	53.799
560	3889.5	2777.4	47.563	1360	9645.5	6944.7	53.910
580	4028.7	2876.9	47.807	1380	9796.6	7056.1	54.021
600	4168.0	2976.5	48.044	1400	9948.1	7167.9	54.129
620	4307.4	3076.2	48.272	1420	10100.0	7280.1	54.237
640	4446.9	3175.9	48.494	1440	10252.2	7392.6	54.344
660	4586.6	3275.8	48.709	1460	10404.8	7505.4	54.448
680	4726.2	3375.8	48.917	1480	10557.8	7618.7	54.522
700	4866.0	3475.9	49.120	1500	10711.1	7732.3	54.665
720	5006.1	3576.3	49.317	1520	10864.9	7846.4	54.757
740	5146.4	3676.9	49.509	1540	11019.0	7960.8	54.858
760	5286.8	3777.5	49.697	1560	11173.4	8075.4	54.958
780	5427.4	3878.4	49.880	1580	11328.2	8190.5	55.056
800	5568.2	3979.5	50.058	1600	11483.4	8306.0	55.154
820	5709.4	4081.0	50.232	1620	11638.9	8421.8	55.251
840	5850.7	4182.6	50.402	1640	11794.7	8537.9	55.347
860	5992.3	4284.5	50.569	1660	11950.9	8654.4	55.411
880	6134.2	4386.6	50.732	1680	12107.5	8771.2	55.535
900	6276.4	4489.1	50.892	1700	12264.3	8888.3	55.628
920	6419.0	4592.0	51.048	1720	12421.4	9005.7	55.720
940	6561.7	4695.0	51.202	1740	12579.0	9123.6	55.811
960	6704.9	4798.5	51.353	1760	12736.7	9241.6	55.900
980	6848.4	4902.3	51.501	1780	12894.9	9360.0	55.990
1000	6992.2	5006.3	51.646	1800	13053.2	9478.6	56.078
1020	7136.4	5110.8	51.788	1820	13212.0	9597.7	56.166
1040	7281.0	5215.7	51.929	1840	13371.0	9717.0	56.253
1060	7425.9	5320.9	52.067	1860	13530.2	9836.5	56.339

TABLE A-20E (*Continued*)

T	$\bar{h}$	$\bar{u}$	$\bar{s}°$	T	$\bar{h}$	$\bar{u}$	$\bar{s}°$
1900	13,850	10,077	56.509	3500	27,262	20,311	61.612
1940	14,170	10,318	56.677	3540	27,608	20,576	61.710
1980	14,492	10,560	56.841	3580	27,954	20,844	61.807
2020	14,815	10,803	57.007	3620	28,300	21,111	61.903
2060	15,139	11,048	57.161	3660	28,647	21,378	61.998
2100	15,463	11,293	57.317	3700	28,994	21,646	62.093
2140	15,789	11,539	57.470	3740	29,341	21,914	62.186
2180	16,116	11,787	57.621	3780	29,688	22,182	62.279
2220	16,443	12,035	57.770	3820	30,036	22,450	62.370
2260	16,722	12,284	57.917	3860	30,384	22,719	62.461
2300	17,101	12,534	58.062	3900	30,733	22,988	62.511
2340	17,431	12,784	58.204	3940	31,082	23,257	62.640
2380	17,762	13,035	58.344	3980	31,431	23,527	62.728
2420	18,093	13,287	58.482	4020	31,780	23,797	62.816
2460	18,426	13,541	58.619	4060	32,129	24,067	62.902
2500	18,759	13,794	58.754	4100	32,479	24,337	62.988
2540	19,093	14,048	58.885	4140	32,829	24,608	63.072
2580	19,427	14,303	59.016	4180	33,179	24,878	63.156
2620	19,762	14,559	59.145	4220	33,530	25,149	63.240
2660	20,098	14,815	59.272	4260	33,880	25,421	63.323
2700	20,434	15,072	59.398	4300	34,231	25,692	63.405
2740	20,771	15,330	59.521	4340	34,582	25,934	63.486
2780	21,108	15,588	59.644	4380	34,934	26,235	63.567
2820	21,446	15,846	59.765	4420	35,285	26,508	63.647
2860	21,785	16,105	59.884	4460	35,637	26,780	63.726
2900	22,124	16,365	60.002	4500	35,989	27,052	63.805
2940	22,463	16,625	60.118	4540	36,341	27,325	63.883
2980	22,803	16,885	60.232	4580	36,693	27,598	63.960
3020	23,144	17,146	60.346	4620	37,046	27,871	64.036
3060	23,485	17,408	60.458	4660	37,398	28,144	64.113
3100	23,826	17,670	60.569	4700	37,751	28,417	64.188
3140	24,168	17,932	60.679	4740	38,104	28,691	64.263
3180	24,510	18,195	60.787	4780	38,457	28,965	64.337
3220	24,853	18,458	60.894	4820	38,811	29,239	64.411
3260	25,196	18,722	61.000	4860	39,164	29,513	64.484
3300	25,539	18,986	61.105	4900	39,518	29,787	64.556
3340	25,883	19,250	61.209	5000	40,403	30,473	64.735
3380	26,227	19,515	61.311	5100	41,289	31,161	64.910
3420	26,572	19,780	61.412	5200	42,176	31,849	65.082
3460	26,917	20,045	61.513	5300	43,063	32,538	65.252

TABLE A-21E Ideal Gas Properties of Carbon Dioxide, CO_2

T(°R), $\bar{h}$ and $\bar{u}$(Btu/lbmol), $\bar{s}^\circ$(Btu/lbmol · °R)

[$\bar{h}_f^\circ$ = −169,300 Btu/lbmol]

T	$\bar{h}$	$\bar{u}$	$\bar{s}^\circ$	T	$\bar{h}$	$\bar{u}$	$\bar{s}^\circ$
300	2108.2	1512.4	46.353	1080	9575.8	7431.1	58.072
320	2256.6	1621.1	46.832	1100	9802.6	7618.1	58.281
340	2407.3	1732.1	47.289	1120	10030.6	7806.4	58.485
360	2560.5	1845.6	47.728	1140	10260.1	7996.2	58.689
380	2716.4	1961.8	48.148	1160	10490.6	8187.0	58.889
400	2874.7	2080.4	48.555	1180	10722.3	8379.0	59.088
420	3035.7	2201.7	48.947	1200	10955.3	8572.3	59.283
440	3199.4	2325.6	49.329	1220	11189.4	8766.6	59.477
460	3365.7	2452.2	49.698	1240	11424.6	8962.1	59.668
480	3534.7	2581.5	50.058	1260	11661.0	9158.8	59.858
500	3706.2	2713.3	50.408	1280	11898.4	9356.5	60.044
520	3880.3	2847.7	50.750	1300	12136.9	9555.3	60.229
537	4027.5	2963.8	51.032	1320	12376.4	9755.0	60.412
540	4056.8	2984.4	51.082	1340	12617.0	9955.9	60.593
560	4235.8	3123.7	51.408	1360	12858.5	10157.7	60.772
580	4417.2	3265.4	51.726	1380	13101.0	10360.5	60.949
600	4600.9	3409.4	52.038	1400	13344.7	10564.5	61.124
620	4786.6	3555.6	52.343	1420	13589.1	10769.2	61.298
640	4974.9	3704.0	52.641	1440	13834.5	10974.8	61.469
660	5165.2	3854.6	52.934	1460	14080.8	11181.4	61.639
680	5357.6	4007.2	53.225	1480	14328.0	11388.9	61.800
700	5552.0	4161.9	53.503	1500	14576.0	11597.2	61.974
720	5748.4	4318.6	53.780	1520	14824.9	11806.4	62.138
740	5946.8	4477.3	54.051	1540	15074.7	12016.5	62.302
760	6147.0	4637.9	54.319	1560	15325.3	12227.3	62.464
780	6349.1	4800.1	54.582	1580	15576.7	12439.0	62.624
800	6552.9	4964.2	54.839	1600	15829.0	12651.6	62.783
820	6758.3	5129.9	55.093	1620	16081.9	12864.8	62.939
840	6965.7	5297.6	55.343	1640	16335.7	13078.9	63.095
860	7174.7	5466.9	55.589	1660	16590.2	13293.7	63.250
880	7385.3	5637.7	55.831	1680	16845.5	13509.2	63.403
900	7597.6	5810.3	56.070	1700	17101.4	13725.4	63.555
920	7811.4	5984.4	56.305	1720	17358.1	13942.4	63.704
940	8026.8	6160.1	56.536	1740	17615.5	14160.1	63.853
960	8243.8	6337.4	56.765	1760	17873.5	14378.4	64.001
980	8462.2	6516.1	56.990	1780	18132.2	14597.4	64.147
1000	8682.1	6696.2	57.212	1800	18391.5	14816.9	64.292
1020	8903.4	6877.8	57.432	1820	18651.5	15037.2	64.435
1040	9126.2	7060.9	57.647	1840	18912.2	15258.2	64.578
1060	9350.3	7245.3	57.861	1860	19173.4	15479.7	64.719

TABLE A-21E (*Continued*)

T	$\bar{h}$	$\bar{u}$	$\bar{s}°$	T	$\bar{h}$	$\bar{u}$	$\bar{s}°$
1900	19,698	15,925	64.999	3500	41,965	35,015	73.462
1940	20,224	16,372	65.272	3540	42,543	35,513	73.627
1980	20,753	16,821	65.543	3580	43,121	36,012	73.789
2020	21,284	17,273	65.809	3620	43,701	36,512	73.951
2060	21,818	17,727	66.069	3660	44,280	37,012	74.110
2100	22,353	18,182	66.327	3700	44,861	37,513	74.267
2140	22,890	18,640	66.581	3740	45,442	38,014	74.423
2180	23,429	19,101	66.830	3780	46,023	38,517	74.578
2220	23,970	19,561	67.076	3820	46,605	39,019	74.732
2260	24,512	20,024	67.319	3860	47,188	39,522	74.884
2300	25,056	20,489	67.557	3900	47,771	40,026	75.033
2340	25,602	20,955	67.792	3940	48,355	40,531	75.182
2380	26,150	21,423	68.025	3980	48,939	41,035	75.330
2420	26,699	21,893	68.253	4020	49,524	41,541	75.477
2460	27,249	22,364	68.479	4060	50,109	42,047	75.622
2500	27,801	22,837	68.702	4100	50,695	42,553	75.765
2540	28,355	23,310	68.921	4140	51,282	43,060	75.907
2580	28,910	23,786	69.138	4180	51,868	43,568	76.048
2620	29,465	24,262	69.352	4220	52,456	44,075	76.188
2660	30,023	24,740	69.563	4260	53,044	44,584	76.327
2700	30,581	25,220	69.771	4300	53,632	45,093	76.464
2740	31,141	25,701	69.977	4340	54,221	45,602	76.601
2780	31,702	26,181	70.181	4380	54,810	46,112	76.736
2820	32,264	26,664	70.382	4420	55,400	46,622	76.870
2860	32,827	27,148	70.580	4460	55,990	47,133	77.003
2900	33,392	27,633	70.776	4500	56,581	47,645	77.135
2940	33,957	28,118	70.970	4540	57,172	48,156	77.266
2980	34,523	28,605	71.160	4580	57,764	48,668	77.395
3020	35,090	29,093	71.350	4620	58,356	49,181	77.581
3060	35,659	29,582	71.537	4660	58,948	49,694	77.652
3100	36,228	30,072	71.722	4700	59,541	50,208	77.779
3140	36,798	30,562	71.904	4740	60,134	50,721	77.905
3180	37,369	31,054	72.085	4780	60,728	51,236	78.029
3220	37,941	31,546	72.264	4820	61,322	51,750	78.153
3260	38,513	32,039	72.441	4860	61,916	52,265	78.276
3300	39,087	32,533	72.616	4900	62,511	52,781	78.398
3340	39,661	33,028	72.788	5000	64,000	54,071	78.698
3380	40,236	33,524	72.960	5100	65,491	55,363	78.994
3420	40,812	34,020	73.129	5200	66,984	56,658	79.284
3460	41,388	34,517	73.297	5300	68,471	57,954	79.569

TABLE A-22E Ideal Gas Properties of Hydrogen, H_2

T(°R), $\bar{h}$ and $\bar{u}$(Btu/lbmol), $\bar{s}°$(Btu/lbmol · °R)

[$\bar{h}_f^\circ$ = 0 Btu/lbmol]

T	$\bar{h}$	$\bar{u}$	$\bar{s}°$	T	$\bar{h}$	$\bar{u}$	$\bar{s}°$
300	2063.5	1467.7	27.337	1400	9673.8	6893.6	37.883
320	2189.4	1553.9	27.742	1500	10381.5	7402.7	38.372
340	2317.2	1642.0	28.130	1600	11092.5	7915.1	38.830
360	2446.8	1731.9	28.501	1700	11807.4	8431.4	39.264
380	2577.8	1823.2	28.856	1800	12526.8	8952.2	39.675
400	2710.2	1915.8	29.195	1900	13250.9	9477.8	40.067
420	2843.7	2009.6	29.520	2000	13980.1	10008.4	40.441
440	2978.1	2104.3	29.833	2100	14714.5	10544.2	40.799
460	3113.5	2200.0	30.133	2200	15454.4	11085.5	41.143
480	3249.4	2296.2	30.424	2300	16199.8	11632.3	41.475
500	3386.1	2393.2	30.703	2400	16950.6	12184.5	41.794
520	3523.2	2490.6	30.972	2500	17707.3	12742.6	42.104
537	3640.3	2573.9	31.194	2600	18469.7	13306.4	42.403
540	3660.9	2588.5	31.232	2700	19237.8	13876.0	42.692
560	3798.8	2686.7	31.482	2800	20011.8	14451.4	42.973
580	3937.1	2785.3	31.724	2900	20791.5	15032.5	43.247
600	4075.6	2884.1	31.959	3000	21576.9	15619.3	43.514
620	4214.3	2983.1	32.187	3100	22367.7	16211.5	43.773
640	4353.1	3082.1	32.407	3200	23164.1	16809.3	44.026
660	4492.1	3181.4	32.621	3300	23965.5	17412.1	44.273
680	4631.1	3280.7	32.829	3400	24771.9	18019.9	44.513
700	4770.2	3380.1	33.031	3500	25582.9	18632.4	44.748
720	4909.5	3479.6	33.226	3600	26398.5	19249.4	44.978
740	5048.8	3579.2	33.417	3700	27218.5	19870.8	45.203
760	5188.1	3678.8	33.603	3800	28042.8	20496.5	45.423
780	5327.6	3778.6	33.784	3900	28871.1	21126.2	45.638
800	5467.1	3878.4	33.961	4000	29703.5	21760.0	45.849
820	5606.7	3978.3	34.134	4100	30539.8	22397.7	46.056
840	5746.3	4078.2	34.302	4200	31379.8	23039.2	46.257
860	5885.9	4178.0	34.466	4300	32223.5	23684.3	46.456
880	6025.6	4278.0	34.627	4400	33070.9	24333.1	46.651
900	6165.3	4378.0	34.784	4500	33921.6	24985.2	46.842
920	6305.1	4478.1	34.938	4600	34775.7	25640.7	47.030
940	6444.9	4578.1	35.087	4700	35633.0	26299.4	47.215
960	6584.7	4678.3	35.235	4800	36493.4	26961.2	47.396
980	6724.6	4778.4	35.379	4900	35356.9	27626.1	47.574
1000	6864.5	4878.6	35.520	5000	38223.3	28294.0	47.749
1100	7564.6	5380.1	36.188	5100	39092.8	28964.9	47.921
1200	8265.8	5882.8	36.798	5200	39965.1	29638.6	48.090
1300	8968.7	6387.1	37.360	5300	40840.2	30315.1	48.257

TABLE A-23E Constants for the van der Waals, Redlich–Kwong, and Benedict–Webb–Rubin Equations of State

1. van der Waals and Redlich-Kwong: Constants for pressure in atm, specific volume in ft^3/lbmol, and temperature in °R

	van der Waals		Redlich-Kwong	
Substance	a atm $\left(\frac{ft^3}{lbmol}\right)^2$	b $\frac{ft^3}{lbmol}$	a atm $\left(\frac{ft^3}{lbmol}\right)^2 (°R)^{1/2}$	b $\frac{ft^3}{lbmol}$
Air	345	0.586	5,409	0.4064
Butane (C_4H_{10})	3,509	1.862	98,349	1.2903
Carbon dioxide (CO_2)	926	0.686	21,972	0.4755
Carbon monoxide (CO)	372	0.632	5,832	0.4382
Methane (CH_4)	581	0.685	10,919	0.4751
Nitrogen (N_2)	346	0.618	5,280	0.4286
Oxygen (O_2)	349	0.509	5,896	0.3531
Propane (C_3H_8)	2,369	1.444	61,952	1.0006
Refrigerant 12	2,660	1.558	70,951	1.0796
Sulfur dioxide (SO_2)	1,738	0.910	49,032	0.6309
Water (H_2O)	1,400	0.488	48,418	0.3380

Source: Calculated from critical data.

2. Benedict-Webb-Rubin: Constants for pressure in atm, specific volume in ft^3/lbmol, and temperature in °R

Substance	a	A	b	B	c	C	α	γ
C_4H_{10}	7736.7	2587.6	10.26	1.9921	4.214×10^9	8.254×10^8	4.527	8.724
CO_2	562.3	702.4	1.850	0.7995	1.987×10^8	1.152×10^8	0.348	1.384
CO	150.6	344.1	0.675	0.8737	1.385×10^7	7.118×10^6	0.555	1.540
CH_4	203.0	476.0	0.867	0.6824	3.389×10^7	1.876×10^7	0.511	1.540
N_2	103.2	270.4	0.597	0.6526	9.700×10^6	6.700×10^6	0.523	1.360

Source: H. W. Cooper, and J. C. Goldfrank, *Hydrocarbon Processing, 46* (12): 141 (1967).

TABLE A-24E Enthalpy of Formation, Gibbs Function of Formation, and Absolute Entropy of Various Substances at 537°R and 1 atm

$\bar{h}_f^\circ$ and $\bar{g}_f^\circ$ (Btu/lbmol), $\bar{s}^\circ$ (Btu/lbmol · °R)

Substance	Formula	$\bar{h}_f^\circ$	$\bar{g}_f^\circ$	$\bar{s}^\circ$
Carbon	$C(s)$	0	0	1.36
Hydrogen	$H_2(g)$	0	0	31.19
Nitrogen	$N_2(g)$	0	0	45.74
Oxygen	$O_2(g)$	0	0	48.98
Carbon monoxide	$CO(g)$	−47,540	−59,010	47.27
Carbon dioxide	$CO_2(g)$	−169,300	−169,680	51.03
Water	$H_2O(g)$	−104,040	−98,350	45.08
Water	$H_2O(l)$	−122,970	−102,040	16.71
Hydrogen peroxide	$H_2O_2(g)$	−58,640	−45,430	55.60
Ammonia	$NH_3(g)$	−19,750	−7,140	45.97
Oxygen	$O(g)$	107,210	99,710	38.47
Hydrogen	$H(g)$	93,780	87,460	27.39
Nitrogen	$N(g)$	203,340	195,970	36.61
Hydroxyl	$OH(g)$	16,790	14,750	43.92
Methane	$CH_4(g)$	−32,210	−21,860	44.49
Acetylene	$C_2H_2(g)$	97,540	87,990	48.00
Ethylene	$C_2H_4(g)$	22,490	29,306	52.54
Ethane	$C_2H_6(g)$	−36,420	−14,150	54.85
Propylene	$C_3H_6(g)$	8,790	26,980	63.80
Propane	$C_3H_8(g)$	−44,680	−10,105	64.51
Butane	$C_4H_{10}(g)$	−54,270	−6,760	74.11
Pentane	$C_5H_{12}(g)$	−62,960	−3,530	83.21
Octane	$C_8H_{18}(g)$	−89,680	7,110	111.55
Octane	$C_8H_{18}(l)$	−107,530	2,840	86.23
Benzene	$C_6H_6(g)$	35,680	55,780	64.34
Methyl alcohol	$CH_3OH(g)$	−86,540	−69,700	57.29
Methyl alcohol	$CH_3OH(l)$	−102,670	−71,570	30.30
Ethyl alcohol	$C_2H_5OH(g)$	−101,230	−72,520	67.54
Ethyl alcohol	$C_2H_5OH(l)$	−119,470	−75,240	38.40

Sources: Adapted from K. Wark, *Thermodynamics, 4th ed.*, McGraw-Hill, New York, 1983, as based on JANAF Thermochemical Tables, NSRDS-NBS-37, 1971; *Selected Values of Chemical Thermodynamic Properties*, NBS Tech. Note 270-3, 1968; and *API Research Project 44*, Carnegie Press, 1953.

INDEX TO FIGURES AND CHARTS

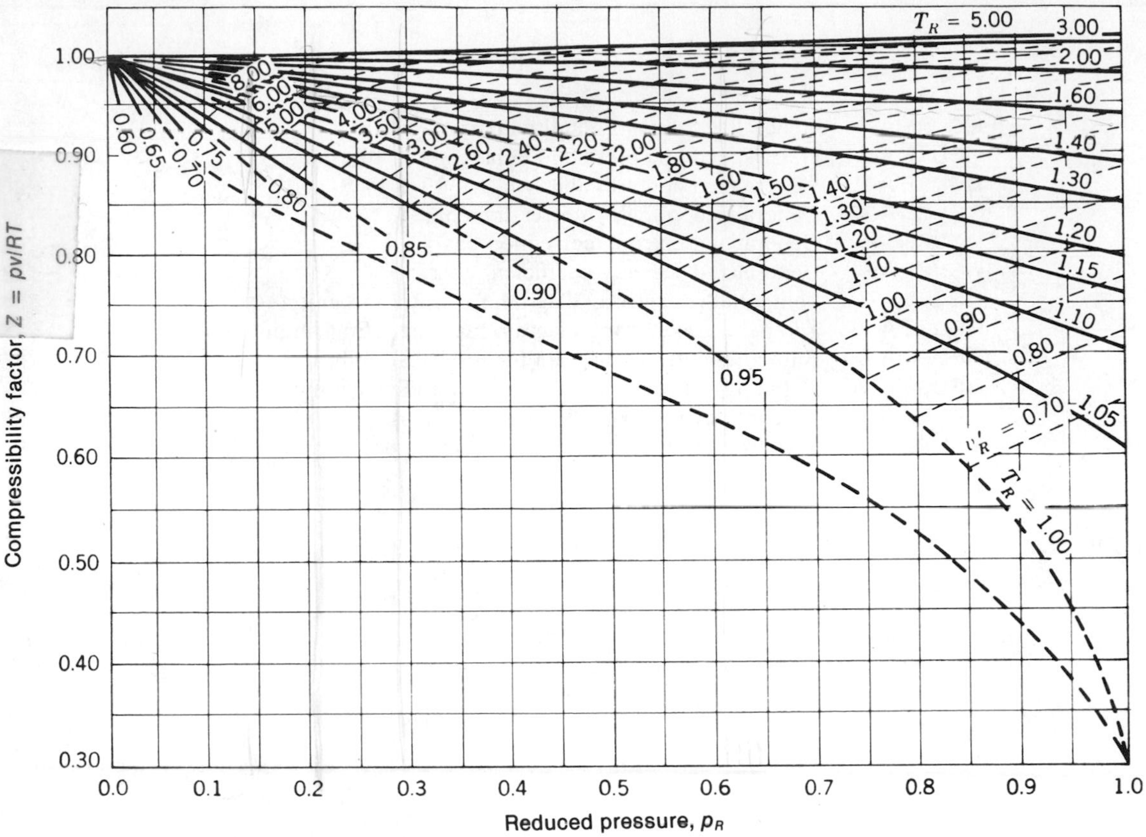

Figure A-1 Generalized compressibility chart, $p_R \leq 1.0$. *Source:* E. F. Obert, *Concepts of Thermodynamics,* McGraw-Hill, New York, 1960.

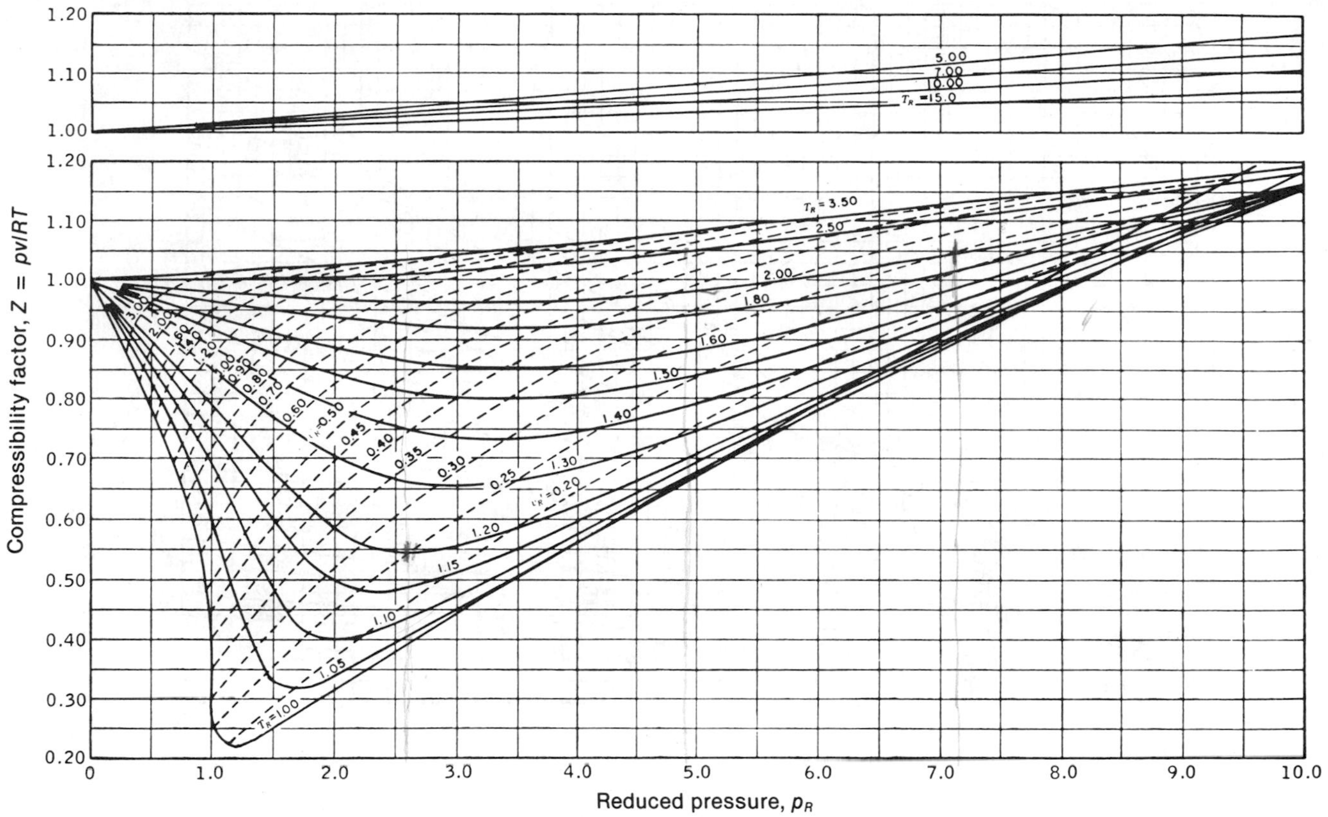

Figure A-2 Generalized compressibility chart, $p_R \leq 10.0$. *Source:* E. F. Obert, *Concepts of Thermodynamics,* McGraw-Hill, New York, 1960.

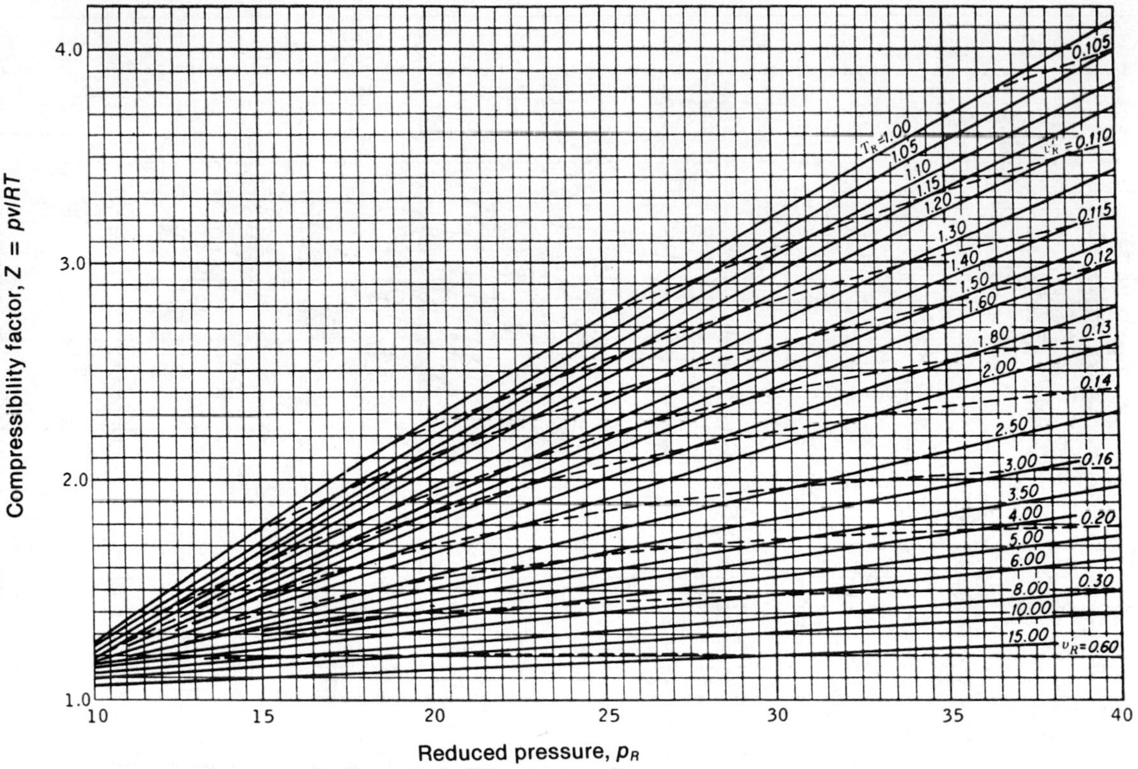

Figure A-3 Generalized compressibility chart, $10 \leq p_R \leq 40$. *Source:* E. F. Obert, *Concepts of Thermodynamics,* McGraw-Hill, New York, 1960.

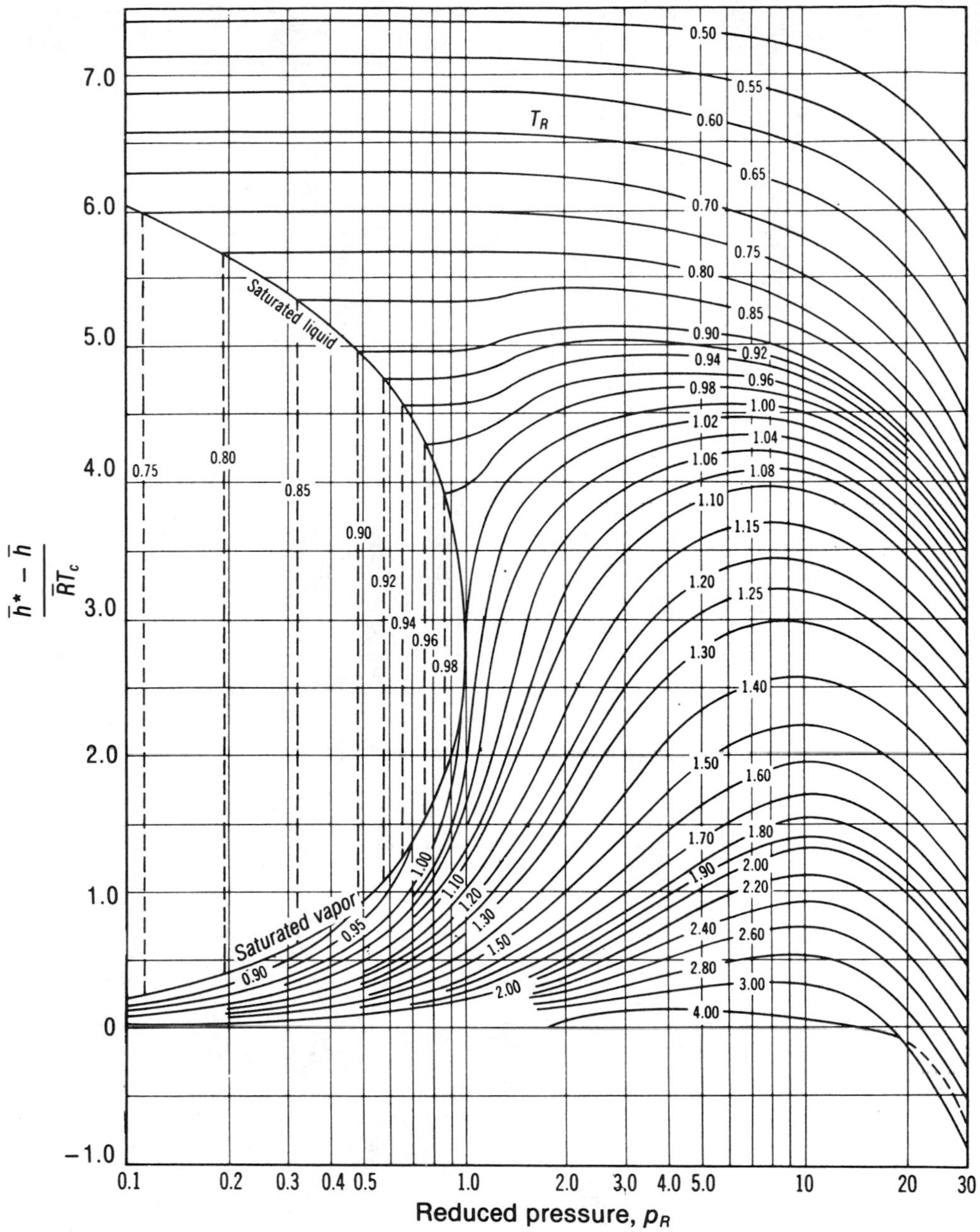

Figure A-4 Generalized enthalpy correction chart. *Source:* Adapted from G. J. Van Wylen and R. E. Sonntag, *Fundamentals of Classical Thermodynamics,* 3rd. ed., English/SI, Wiley, New York, 1986.

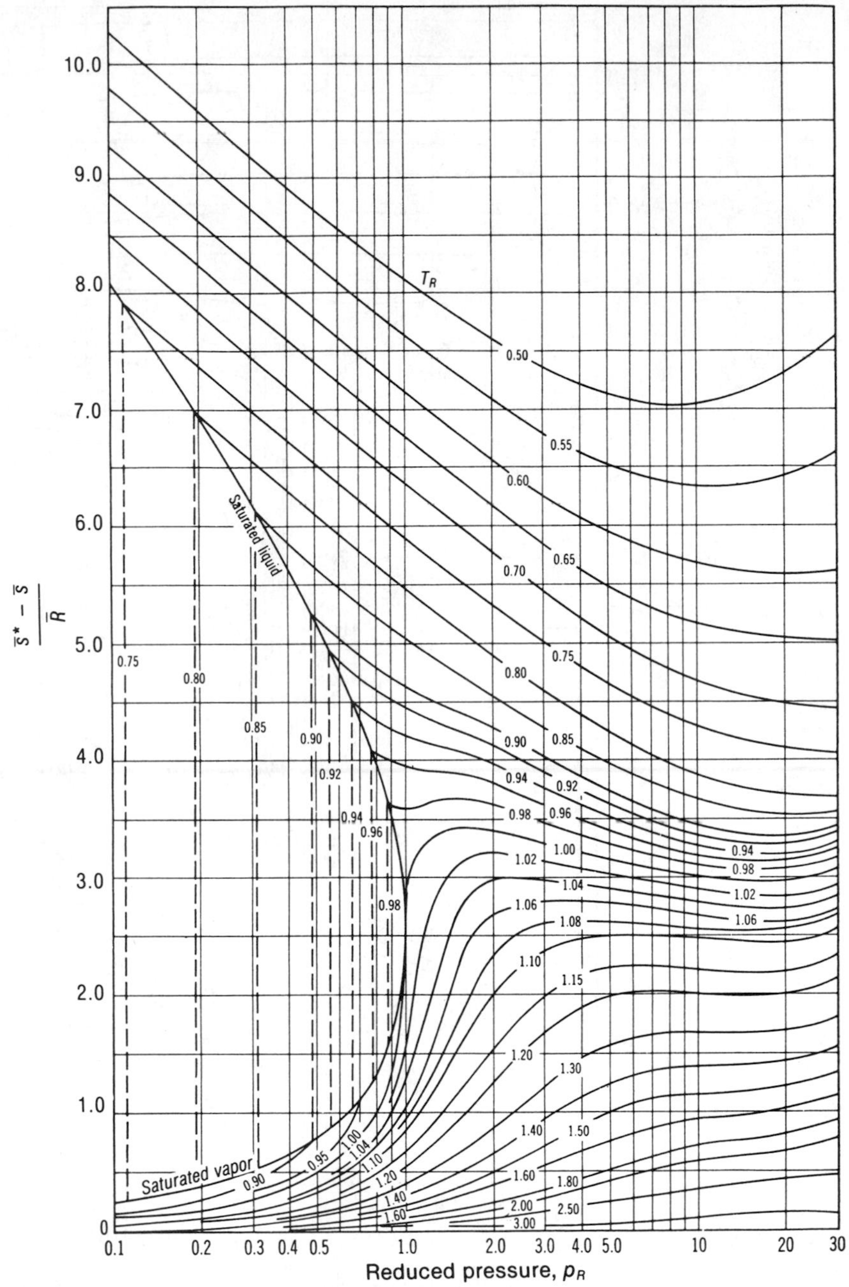

Figure A-5 Generalized entropy correction chart. *Source:* Adapted from G. J. Van Wylen and R. E. Sonntag, *Fundamentals of Classical Thermodynamics,* 3rd ed., English/SI, Wiley, New York, 1986.

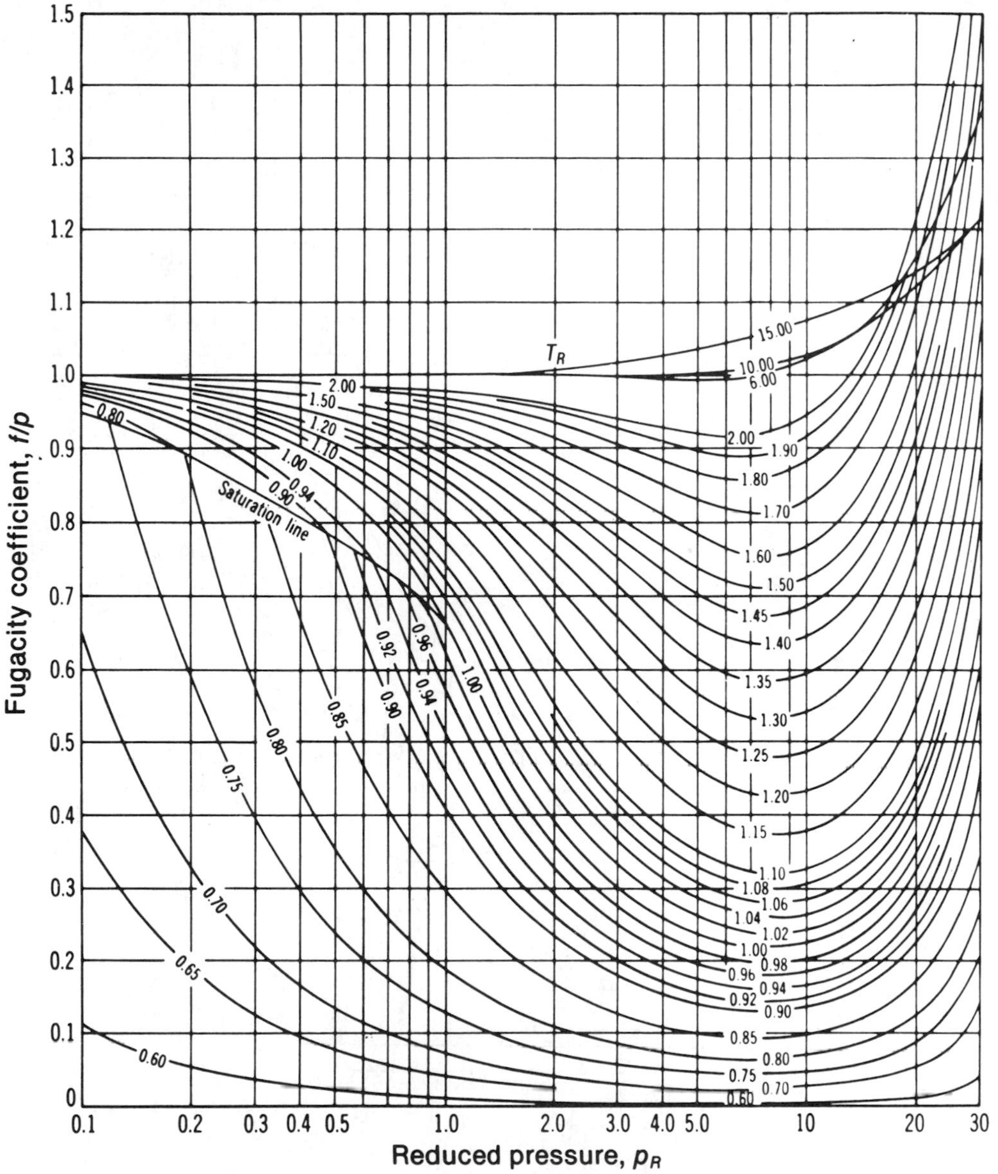

Figure A-6 Generalized fugacity coefficient chart. *Source:* G. J. Van Wylen and R. E. Sonntag, *Fundamentals of Classical Thermodynamics,* 3rd. ed., English/SI, Wiley, New York, 1986.

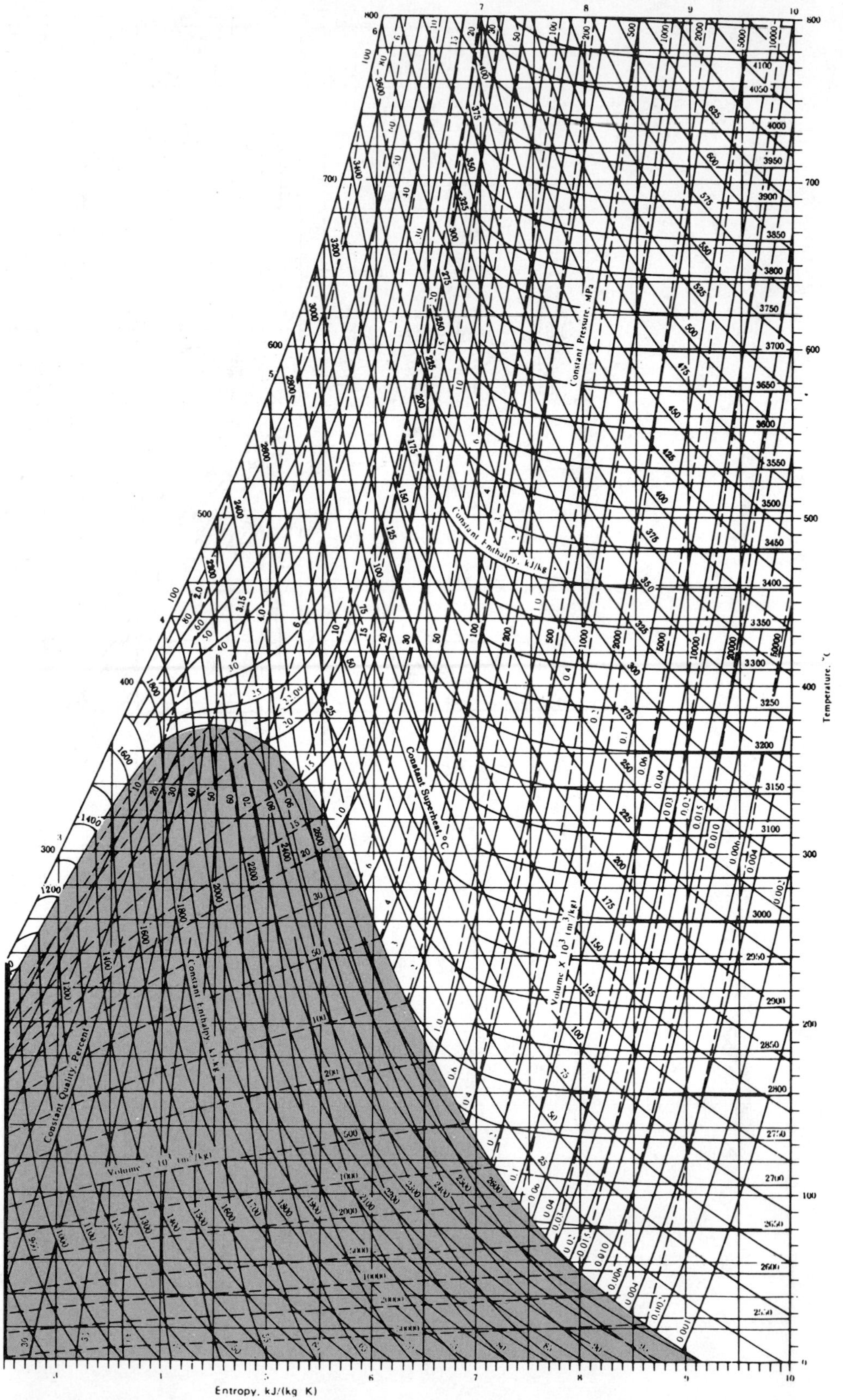

Figure A-7 Temperature–entropy diagram for water (SI units). *Source:* J. H. Keenan, F. G. Keyes, P. G. Hill, and J. G. Moore, *Steam Tables,* Wiley, New York, 1978.

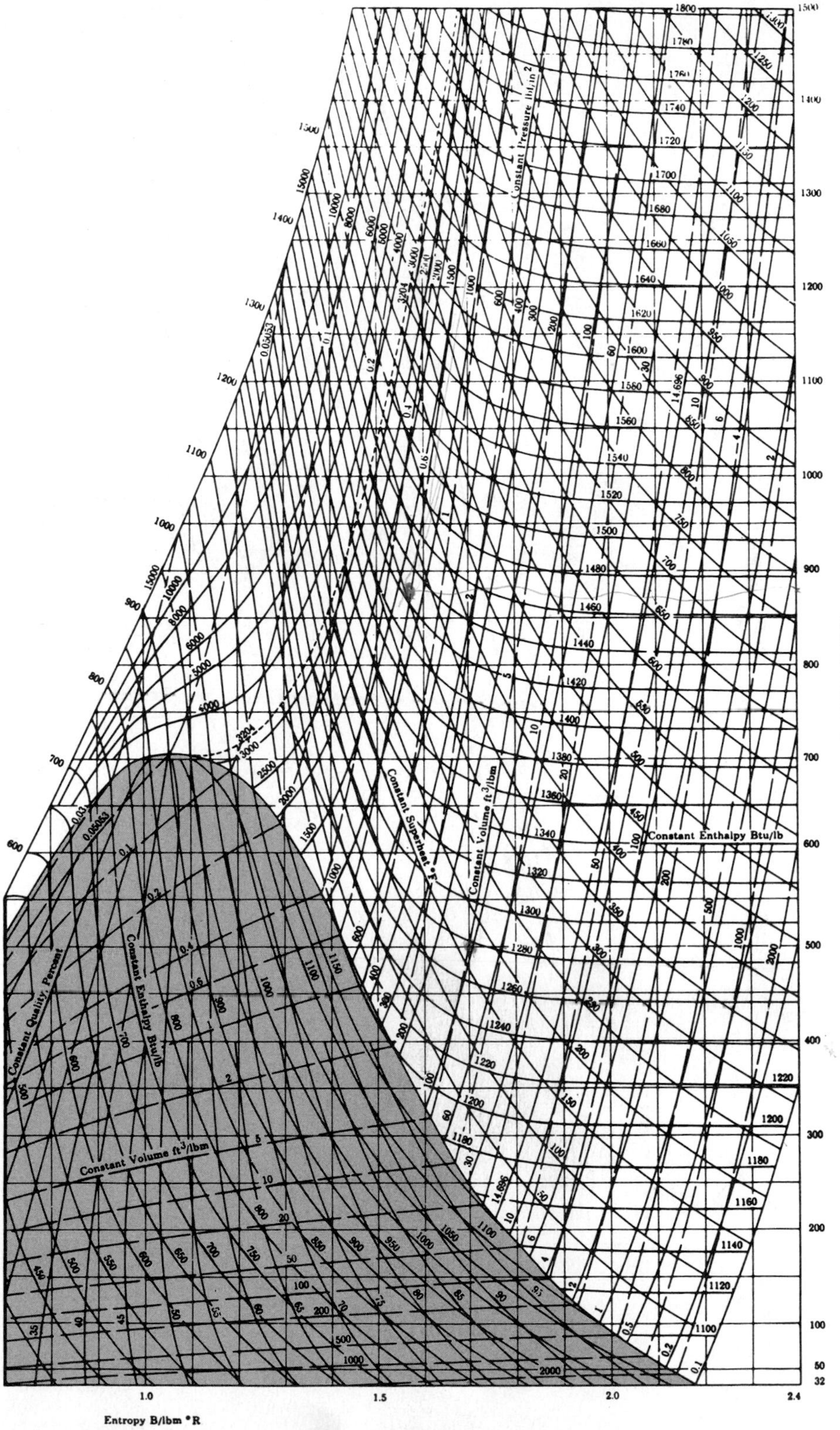

Figure A-7E Temperature–entropy diagram for water (English units). *Source:* J. H. Keenan, F. G. Keyes, P. G. Hill, and J. G. Moore, *Steam Tables,* Wiley, New York, 1969.

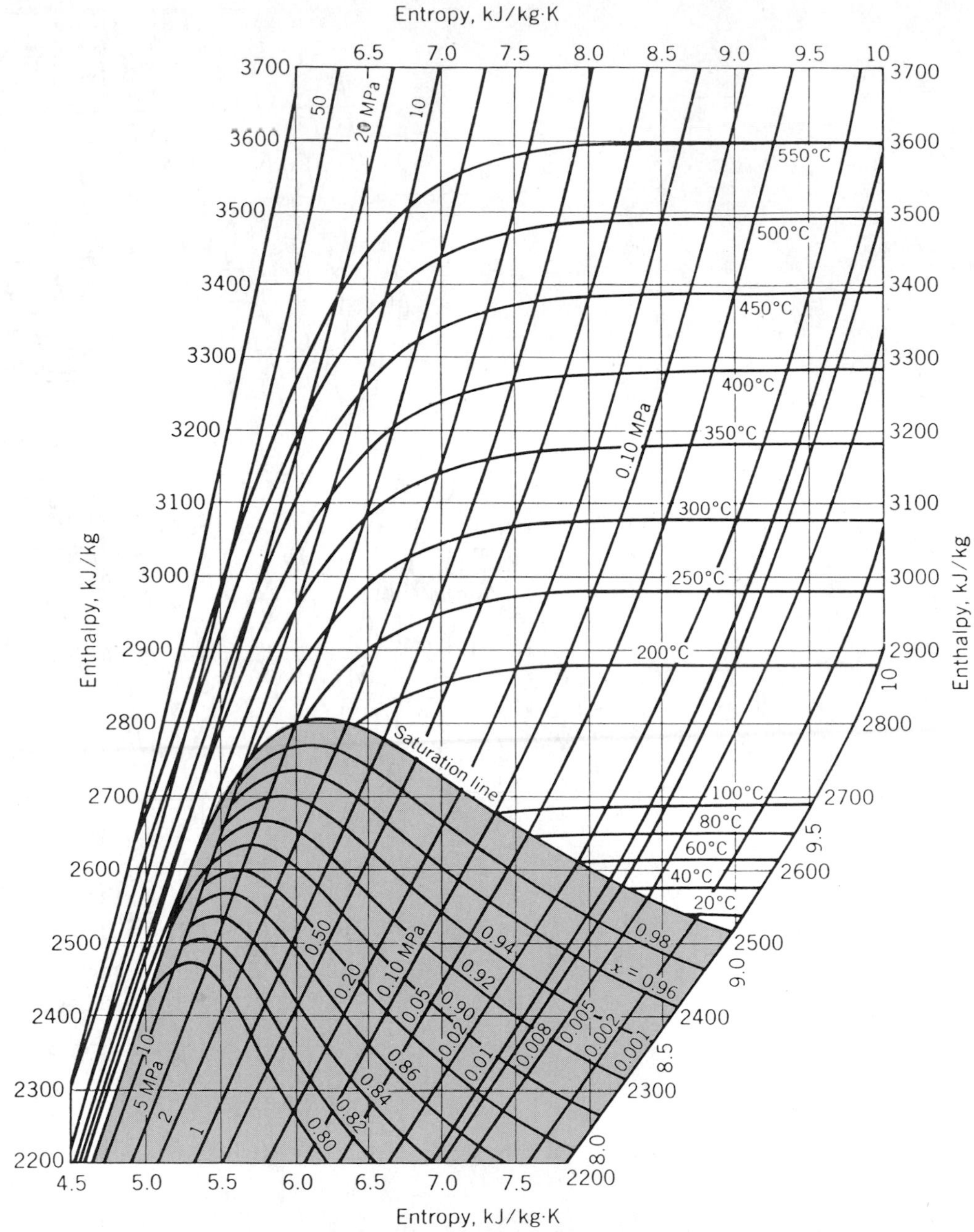

Figure A-8 Enthalpy–entropy diagram for water (SI units). *Source:* J. B. Jones and G. A. Hawkins, *Engineering Thermodynamics*, 2nd ed., Wiley, New York, 1986.

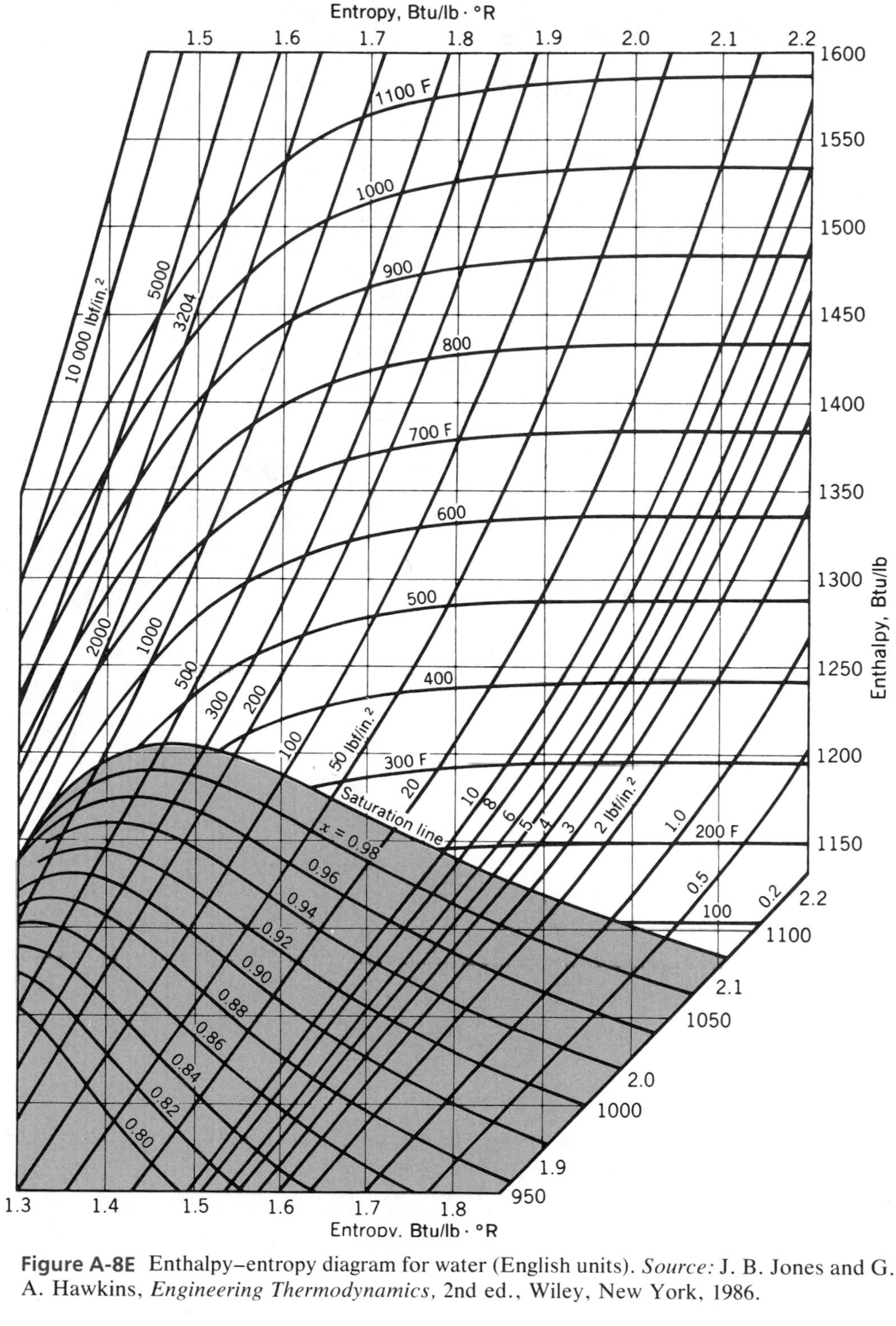

Figure A-8E Enthalpy–entropy diagram for water (English units). *Source:* J. B. Jones and G. A. Hawkins, *Engineering Thermodynamics,* 2nd ed., Wiley, New York, 1986.

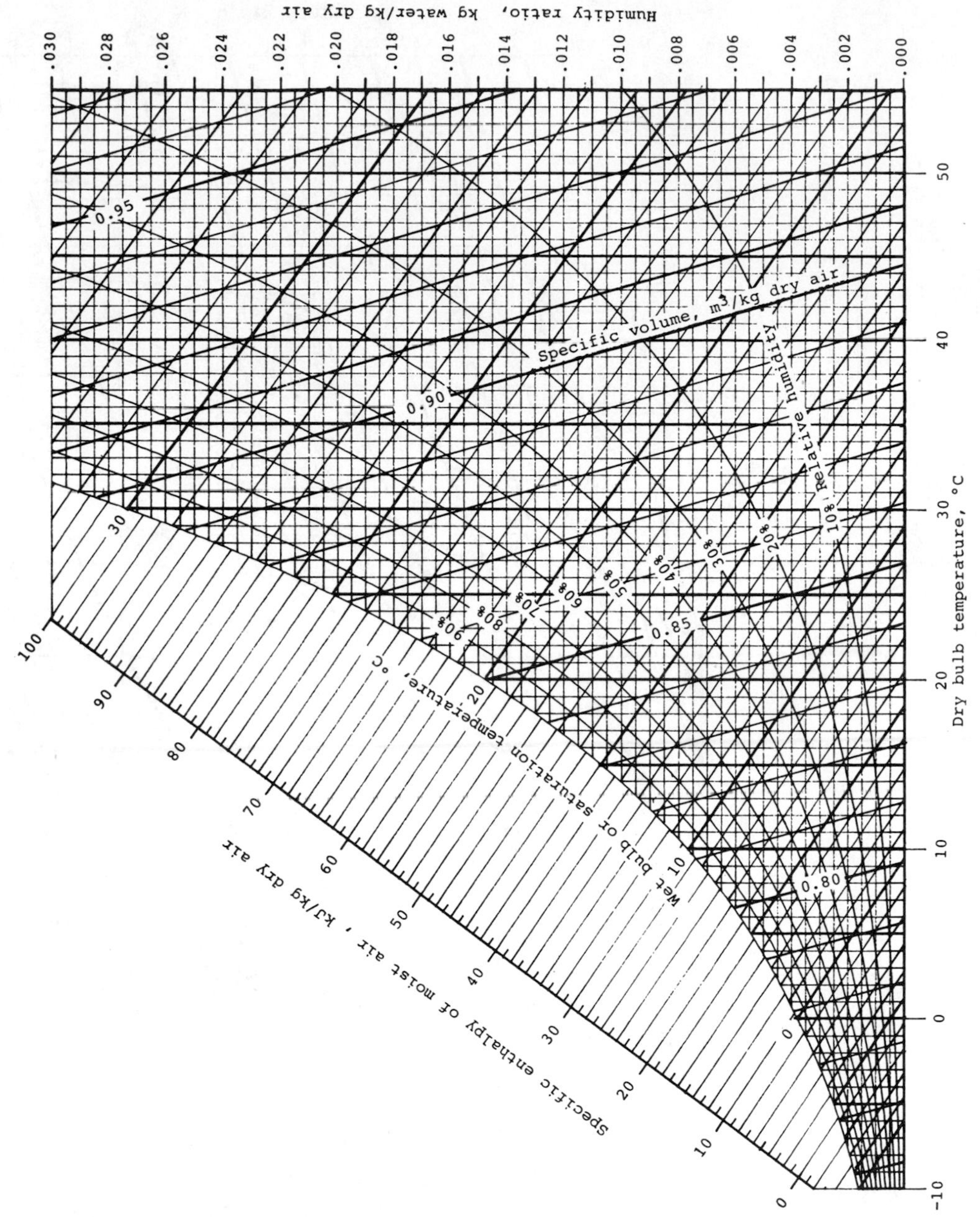

Figure A-9 Psychrometric chart for 1 atm (SI units). *Source:* Z. Zhang and M. B. Pate, "A Methodology for Implementing a Psychrometric Chart in a Computer Graphics System," *ASHRAE Transactions,* Vol. 94, Pt. 1, 1988.

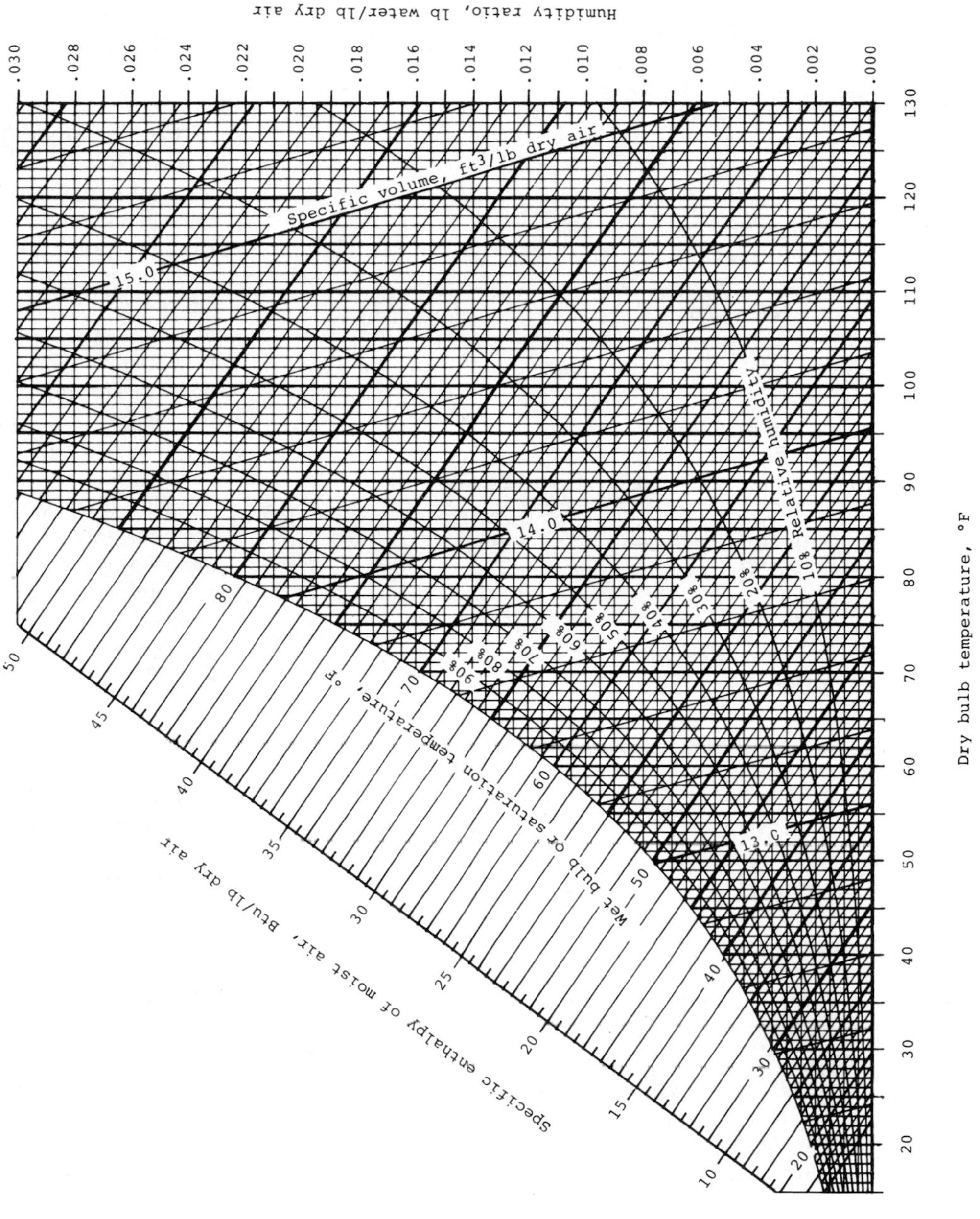

Figure A-9E Psychrometric chart for 1 atm (English units). *Source:* Z. Zhang and M. B. Pate, "A Methodology for Implementing a Psychrometric Chart in a Computer Graphics System," *ASHRAE Transactions,* Vol. 94, Pt. 1, 1988.

SYMBOLS

a	acceleration, activity
a, A	specific availability, availability
a_f	flow availability
A	area
AF	air–fuel ratio
bwr	back work ratio
c	specific heat of an incompressible substance, velocity of sound
c_p	specific heat at constant pressure, $\partial h/\partial T)_p$
c_v	specific heat at constant volume, $\partial u/\partial T)_v$
c_{p0}	specific heat c_p at zero pressure
e, E	energy per unit of mass, energy
E	electric field strength
$\mathcal{E}$	electrical potential, electromotive force (emf)
f	fugacity
$\bar{f}_i$	fugacity of component i in a mixture
F	degrees of freedom in the phase rule
F, F	force
FA	fuel–air ratio
g	acceleration of gravity
g, G	Gibbs function per unit of mass, Gibbs function
g_c	constant that relates force, mass, length, and time
$\bar{g}_f^\circ$	Gibbs function of formation per mole at standard state
h, H	enthalpy per unit of mass, enthalpy
h	heat transfer coefficient
H	magnetic field strength
$\bar{h}_f^\circ$	enthalpy of formation per mole at standard state
$\bar{h}_{RP}$	enthalpy of combustion per mole
HHV	higher heating value
i	electric current
I, $\dot{I}$	irreversibility, irreversibility rate
k	specific heat ratio: c_p/c_v
k	Boltzmann constant
K	equilibrium constant
ke, KE	kinetic energy per unit of mass, kinetic energy
l, L	length
LHV	lower heating value
m	mass
$\dot{m}$	mass flow rate
M	molecular weight
M	Mach number
M	magnetic dipole moment per unit volume
mep	mean effective pressure
mf	mass fraction
n	number of moles
n	polytropic exponent
N	number of components in the phase rule
p	pressure
p_{atm}	atmospheric pressure
p_i	pressure associated with mixture component i, partial pressure of i
p_r	relative pressure as used in gas tables
p_R	reduced pressure: p/p_c
P	number of phases in the phase rule
P	electric dipole moment per unit volume
pe, PE	potential energy per unit of mass, potential energy
$\dot{q}$	heat flux
Q	heat transfer
$\dot{Q}$	heat transfer rate
$\dot{Q}_e$	thermal radiation emission rate
r	compression ratio
r_c	cutoff ratio
R	resultant force, electric resistance
R	gas constant: $\bar{R}/M$
$\bar{R}$	universal gas constant
s, S	entropy per unit of mass, entropy
s°	entropy function as used in gas tables, absolute entropy at the standard reference state
t	time
T	temperature
T_R	reduced temperature: T/T_c
$\mathcal{T}$	torque
u, U	internal energy per unit of mass, internal energy
v, V	specific volume, volume
V, V	Velocity
v_r	relative volume as used in gas tables
v'_R	pseudoreduced specific volume: $\bar{v}/(\bar{R}T_c/p_c)$
V_i	volume associated with mixture component i, partial volume of i
W	work
$\dot{W}$	rate of work, or power
x	quality
x	position
X	extensive property
y	mass flow rate ratio
y	mole fraction
z	elevation
Z	compressibility factor
Z	electric charge

Greek Letters

α	isentropic compressibility
β	coefficient of performance for a refrigerator, volume expansivity
γ	coefficient of performance for a heat pump
Δ	change = final minus initial
ε	second law efficiency
ε	emissivity, extent of reaction
η	efficiency, effectiveness
θ	temperature

κ	thermal conductivity, isothermal compressibility
μ	chemical potential
μ_J	Joule–Thomson coefficient
ν	stoichiometric coefficient
ρ	density
σ, $\dot{\sigma}$	entropy production, rate of entropy production
σ	normal stress, Stefan–Boltzmann constant
Σ	summation
τ	surface tension
ϕ	relative humidity
ψ, Ψ	Helmholtz function per unit of mass, Helmholtz function
ω	angular velocity
ω	humidity ratio (specific humidity)

Subscripts

a	dry air
ad	adiabatic
as	adiabatic saturation
b	boundary surface
c	property at the critical point, compressor, combined system
cv	control volume
cw	cooling water
C	cold reservoir, low temperature
db	dry bulb
e	state of a substance exiting a control volume
e	availability reference environment
f	property of saturated liquid, final value
F	fuel
fg	difference in property for saturated vapor and saturated liquid
g	property of saturated vapor
H	hot reservoir, high temperature
i	state of a substance entering a control volume, mixture component
i	initial value, property of saturated solid
I	irreversible
ig	difference in property for saturated vapor and saturated solid
isol	isolated
int rev	internally reversible
j	portion of the boundary, number of components present in a mixture
n	normal component
p	pump
ref	reference value or state
reg	regenerator
res	reservoir
P	products
R	reversible, reactants
s	isentropic
sat	saturated
surr	surroundings
t	turbine
tp	triple point
0	property at the dead state, property of the surroundings
o	stagnation property
v	vapor
w	water
wb	wet bulb
x	upstream of a normal shock
y	downstream of a normal shock
1,2,3	different states of a system, different locations in space

Superscripts

ch	chemical contribution to availability
e	component of the availability reference environment
‾	bar over symbol denotes property on a molar basis (over X, V, H, S, U, Ψ, G, the bar denotes partial molal property)
.	dot over symbol denotes time rate
°	property at standard state
*	ideal gas
*	quantity corresponding to sonic velocity

ANSWERS TO SELECTED PROBLEMS

1.10 98.1

1.13 3.59, 0.831

1.15 1.53

1.20 20.04

1.23 341.6

1.26 (a) 1.935, (b) 210.7, (c) 522

1.29 3.75

1.33 1.07

1.36 (a) 9.81×10^{-4}, (b) 0.01333

1.39 0.2 (vacuum)

1.42 99.92

1.47 (a) 21.11, (b) −17.78, (c) −34.44, (d) 260, (e) 100, (f) −273.15

1.50 51.57

1.53 (a) 2.78, (b) 4, (c) 693, (d) 50

2.4 −51.9

2.8 (a) 2.82, (b) 1

2.13 109.4

2.15 (a) 350, (b) 294, (c) 56

2.18 32

2.22 −9.6

2.25 1.277, −1.01

2.28 −9.57

2.31 0.0455, 191

2.35 13.2

2.38 (a) $= 2.5 \times 10^6(x - 10)$, (b) −12.5

2.41 45

2.44 0.729

2.48 (a) +90, +70, (b) +50, +20, (c) +105, +55, (d) −90, +50, (e) +50, −80

2.51 359, 0.461

2.54 130

2.57 (a) 0.833, (b) 20

2.60 −5.45

2.64 (b) −8.28, (c) 36.9

2.67 150

2.71 hot body: outside air, cold body: inside air

2.75 12,500, 17,500

3.3 (a) two-phase, liquid-vapor mixture, (b) superheated vapor, (c) superheated vapor, (d) solid, (e) subcooled (compressed) liquid

3.10 0.713

3.13 0.1521

3.17 0.123

3.19 0.753, 2.259

3.22 35.42, 1.824

3.26 110.2

3.28 89.35, 319.8

3.33 0.0105

3.37 −0.423

3.39 (a) 28.46, 1181.5, (b) 227.96, 813.4, (c) 4.3705, 1426.6, (d) 1025, 1151.3 (e) 0.959, 1373.7, (f) 247.1, 1.497, (g) 0.0396, −151.75, (h) 0.01765, 289.9

3.43 430.28

3.47 −991.5

3.50 −1.752, −12.81

3.53 −61.93, 0.0683, 0.0746

3.55 6.994

3.58 (a) State 1: 4.525, 1166.3, State 2: 4.525, 849.5, State 3: 10, 1104.1, (b) −633.6, 90.8

3.61 −41.58, 69.12

3.64 8.39, 1-2: 0, 2-3: 316.2, 3-1: −307.8

3.68 $mc(dT/dt) = \mathrm{hA}(T_0 - T) - \dot{W}$, $T(t) = T_0 - (\dot{W}/\mathrm{hA})\{1 - \exp[-(\mathrm{hA}/mc)t]\}$

3.72 5.99

3.75 450.7, 120

3.80 (a) $p < 3940$, (b) $p < 2730$, (c) $p < 2190$

3.83 10,509, 12,730

3.85 41.48

3.88 300.7, 1.16

3.92 801.5

3.95 425.6, 2.462

3.98 1.041

3.100 1.5

3.104 (b) 11.95%

4.3 (a) −0.01, (b) 0.3095

4.7 $\dfrac{dL}{dt} = \dfrac{D_1^2\mathrm{V}_1 - D_2^2\mathrm{V}_2}{D^2}$

4.9 $= (60/49)\ \mathrm{V}_1$

4.13 (a) 2.014, (b) −0.91

4.18 2.94, 0.601

4.22 (a) 248, (b) 6.71

4.25 174.9

4.28 (a) 342, (b) 120.9

4.31 (a) 631.8 (b) 1.389
4.35 0.82
4.38 2307
4.41 −735
4.45 −2.14
4.48 880, 0.1805
4.51 −48.9
4.55 (a) 1.16×10^7, (b) 2.088×10^8
4.59 (a) 11.21, (b) 984.2
4.62 6264
4.64 22.6
4.68 0.961
4.72 394, 88
4.75 (a) −632.2, (b) −20.34, (c) 526.5
4.78 (a) 3426, (b) 1.744
4.81 (a) 0.0505, (b) 15.53
4.86 (a) 44.83, (b) 154
5.3 58.285, no
5.21 47.95%, 479.5
5.23 2000
5.26 (a) $= (T_H + T_C)/2$, (b) $= (T_H T_C)^{1/2}$
5.29 (a) no, (b) no, (c) yes
5.31 feasible
5.35 4.8%
5.38 13,833 m^2
5.41 (a) irreversible, (b) impossible, (c) reversible, (d) impossible
5.44 (a) 3.48, (b) 0.36
5.47 6.37
5.50 $= \dfrac{T_C[T_H - T_0]}{T_H[T_0 - T_C]}$
5.53 2.08, 12.87
5.55 $3.63
5.60 (a) $= \dfrac{T'_H[T_H - T_C]}{T_H[T'_H - T'_C]}$, (b) $\dfrac{T_C}{T'_C} < \dfrac{T_H}{T'_H}$
5.62 (b) 1-2: 1460.53, 146.98, 2-3: 0, 403.8, 3-4: −1156.6, −86.68, 4-1: 0, −159.73, (c) 20.8%
5.65 (a) 49.99, (b) −13.4, (c) 1-2: 49.99, 2-3: 276.08, 3-4: −13.4, 4-1: −276.08, (d) 73.2%
5.68 (a) 3.81, (b) 0.83, (c) 1.00, (d) 0.37
6.2 (a) $= (W_R - W_I)/T_C$
6.7 (a) 7.024, (b) 2.3294, (c) 3.7729
6.9 (a) −4.9128, (b) 0.3262, (c) 0.97515, (d) −22.1377
6.12 (a) 0.961, (b) 0.96
6.15 (a) 1017, (b) 6.897, (c) 4.119, (d) 39, (e) 0.2167
6.19 (a) −8.88, (b) −0.02137
6.23 (a) −0.4969, (b) −0.5028, (c) −0.5039
6.25 (a) 6.178, (b) 242.2
6.28 (a) −35.25, (b) −139.92
6.31 (b) 383 K, 242 K, (c) −33.1
6.34 (b) 1359.2, 345.3, (c) 25.4%
6.37 (b) power cycle, 33.3%
6.39 (a) positive, (b) zero, (c) negative, (d) positive, (e) positive, (f) indeterminate
6.45 47.22
6.48 no
6.50 no, energy removed
6.53 no
6.56 (a) −0.005, (b) 1.54×10^{-5}
6.59 −0.4, 1.24×10^{-3}
6.62 (a) 48, (b) 0.0192
6.63 (b) 0.2482
6.67 (a) 495.6, (b) 9.9%, (c) 1.0597
6.70 (a) 1.5, (b) 600, (c) 7.7×10^{-3}
6.75 (b) $T_H \geq T'_H$, $T'_C \geq T_C$, (c) $= Q_H\left[1 - \dfrac{T'_C}{T'_H}\right]$, $= Q_H\left[1 - \dfrac{T_C}{T_H}\right]$
6.80 0.59
6.82 no
6.85 impossible
6.87 no
6.90 possible
6.92 0.0245, 5.97
6.94 (a) 22, (b) 0.818
6.99 1.89
6.102 (a) 0.0224, (b) 0.675, 0.0238
6.104 532.6, 0.4845
6.107 0.0608
6.110 (a) 380 K, (b) 0.0729
6.112 0.0318
6.115 (a) 21.88, (b) 0.459, (c) 0.6121
6.120 (a) 1.766, −5.146, (b) 1.764, −5.15, (c) 1.351, −4.166
6.123 $= \left(V_1^2 + \dfrac{2kRT_1}{k-1}\left[1 - \left(\dfrac{p_2}{p_1}\right)^{(k-1)/k}\right]\right)^{1/2}$
6.125 (a) 377.4, (b) 0.874, 0.296
6.130 (a) 4383, 183.1, (b) 81.3%
6.133 Superheated vapor, 124.9°C
6.137 94.1%
6.140 (a) 645.06, (b) 92%, (c) 0.0209
6.143 (a) 76.8, (b) 40.03

6.147 4.07×10^7

6.149 6.4, 1.5, 19.2

6.152 −300.9

6.156 −6.75

6.159 33.92 lbf/in.^{2}

6.164 (a) 1952.13, (b) −0.62

7.3 (a) 2056.9, (b) 99.66, (c) 340.08

7.6 27601.5

7.9 1.97

7.12 (a) 37.32, (b) 11.16

7.15 $= c_p T_0 \left[\frac{T}{T_0} - 1 - \ln \frac{T}{T_0}\right]$

7.18 (a) 32.16, (b) 4.48, (c) −27.68

7.21 (a) 123, (b) 710

7.24 25.84, −85, 59.16

7.27 (a) 353 K, 300, (b) 345.2

7.31 (a) yes, (b) no

7.33 0.363, −0.037

7.36 0.481, 0.044

7.39 (a) 2361.5, 11963.2, 1703.6, 3728.6, (b) 4271.5

7.43 469.15, 592.4

7.48 $= (h - h_0) - T_0(s - s_0) + V^2/2 + gz$

7.51 (a) 89.5, (b) 57.37

7.54 157.44

7.57 135.4

7.60 (a) 537.4, (b) 715.12

7.63 −32.13, 3.43

7.68 16.17, 92.51

7.72 (a) 645.3, (b) 92.2%, (c) 5.82

7.74 (a) −0.146, −3.005, 0, (b) 0, −3.005, 0.146

7.81 compressor: 25.06 (17.3%), turbine: 27.73 (19.1%)

7.83 (a) 52.5%, (b) 40%, (c) 30%

7.88 87.2%, 84%

7.91 (a) 33, (b) 74.4%, (c) 80.6%

7.94 (a) 812 K, (b) 18.87, (c) 86.4%

7.97 (a) −236.3, (b) 85.3%

7.100 (a) 3.829, (b) 894.2, (c) 45.9%

8.3 (a) 44.63%, (b) 0.332, (c) 1441.4, (d) 798.34

8.6 (a) 0.00715, (b) 413.6, (c) 1119.7, (d) 36.9%, (e) 35.28

8.9 (a) 8.21×10^8, (b) 41.05%, (c) 5.89×10^7

8.15 (a) 24%, (b) 0.597, (c) 1304.2, (d) 990.9

8.18 (a) 0.0082, (b) 461.9, (c) 1374.5, (d) 33.6%, (e) 45.61

8.22 (a) 29.8%, (b) 76.9%, (c) 78.4%

8.25 $\leq$100 bars

8.28 (a) 101×10^3, (b) 36.5%, (c) 175.9×10^3

8.33 (a) 235×10^3, (b) 42.5%, (c) 5.825×10^6

8.36 (a) 42.4%, (b) 381

8.39 (a) 6.606×10^8, (b) 39.4%, (c) 5.08×10^7

8.42 (a) 243×10^3, (b) 41.2%, (c) 6.144×10^6

8.45 (a) 36.8%, (b) 273.3×10^3

8.48 (a) 6.437×10^8, (b) 38.3%, (c) 5.183×10^7

8.51 (a) 43.18%, (b) 3.375×10^5

8.54 (a) 39.2%, (b) 290.7×10^3

8.57 47.5%

8.60 45.5%

8.63 (b) 44.0%, (c) 5.286×10^6

8.67 (a) 9.529×10^8, (b) 236.5×10^3, (c) 3.032×10^9

8.71 (a) 191×10^3, (b) 136.6×10^3, (c) 54.4×10^3, (d) 71.5%

8.74 (a) 2.63×10^8, (b) 9.185×10^7, 4.74×10^5, (c) 7.102×10^7. Cooling water: 2.36×10^7

8.77 Steam generator: 3.681×10^9, net work output: 3×10^9, condenser loss: 2.838×10^8, total irreversibility rate (turbine stages, feedwater heaters, traps): 3.978×10^8

9.3 (a) 0.6715, (b) 0.3397, (c) 50.6%, (d) 9.71, (e) −0.1823

9.6 (a) 3861, (b) 634.5, (c) 45.7%

9.9 $r = 6$: 46.7%, 90.66; $r = 8$: 51.9%, 90.75; $r = 10$: 55.6%, 89.76

9.12 (a) 83.08, (b) 54.3%, (c) 138.5

9.14 0.778, 88

9.21 (a) 54.87, (b) 53.1%, (c) 157.5

9.23 (a) 1879.5, (b) 11,959, (c) 1.66, (d) 543.5, (e) 60.4%, (f) −174.1

9.26 $r_c = 1.5$: (a) 1301.5, (b) 189.2, (c) 297.5, (d) 59.4%; $r_c = 2$: (a) 1735.4, (b) 293.1, (c) 566.3, (d) 55.2%; $r_c = 2.5$: (a) 2169.2, (b) 404.0, (c) 808.6, (d) 51.6%

9.32 (a) 409.6, 226.4, (b) 390.1, (c) 245.9, (d) 61.3%

9.37 pressure ratio = 6: (a) 37.7%, (b) 0.345, (c) 2218.6

9.41 (a) 372.6, 833.3, (b) 55.3%

9.47 (a) 246.3 (ideal: 446.7), (b) 612.7 (ideal: 478.2), (c) 28.7% (ideal: 48.3%)

9.50 (a) 31.9%, (b) 1641.5

9.56 (a) 246.3, $\eta_{reg} = 100\%$: (b) 366.23, (c) 40.2%; $\eta_{reg} = 80\%$: (b) 415.52, (c) 37.2%; $\eta_{reg} = 60\%$: (b) 464.81, (c) 34.6%

9.59 (a) 0.562, (b) 43.2%, (c) 523.2

9.62 (a) 219.6, (b) 219.6, (c) 8.7%

9.65 44.2%

9.68 53.46, 63.71

9.71 (a) 42.5%, (b) 4085

9.74 (a) 45.65% (b) 4913

9.77 961

9.81 (a) 23.93, 3.041, (b) 20.24, (c) 49.4%

9.84 Combustor: 5.231×10^7, net work output: 3.766×10^7, destructions (turbines, compressor, pump): 4.625×10^6, destruction (condenser): 1.226×10^6, destruction (heat exchanger): 5.254×10^6, losses (exiting air, cooling water): 3.543×10^6

9.87 (a) 13.134, (b) 70.96%, (c) 5.254

9.90 10,034 (opposite to the direction of flow)

9.94 (a) 619.2, (b) 340.7, (c) 1019.2

9.97 (a) 0.541, (b) 359.8, (c) 122

9.100 (a) 1.285, (b) 0.578, (c) converging-diverging

9.103 (a) 800.5, 0.577, 0.01868, (b) 1186.5, 0.889, 0.0155, (c) 1364.9, 1.048, 0.01535, (d) 1544.8, 1.224, 0.01591

9.105 497.2, 549.3

9.109 (a) 25.56, (b) 576, (c) 0.672

9.112 (a) 3.107, (b) 2.386, 2055.2, 2.37

10.2 (a) 5.61, (b) 0.821, (c) 6.83

10.6 (a) 1.714, (b) 2.86, (c) 5.86

10.9 (a) 5.526, (b) 5.75, (c) 3.66

10.12 (a) 4.449, 1.582, (b) 2.633, 1.454, (c) 2.084, 1.388

10.15 (a) 6.712, (b) 1284.9, (c) 3.51, (d) 39.25, 40.46

10.18 (a) 23.92, (b) 486.3, (c) 2.056, (d) 78.45%

10.23 (a) R-13: 586.3, R-12: 1177.2, (b) 5763.7, (c) 2.268, (d) 2.214

10.26 (a) Evaporator 1: 6.124, Evaporator 2: 3.777, (b) 5.85, (c) 23.44

10.29 (a) 6.209, (b) 3.01, (c) 79.2%, (d) 3.14

10.33 (a) 4.231×10^4, (b) 171.7, (c) 4.576

10.37 (a) 16.0, (b) 43.86, (c) 2.74, (d) 6.75

10.39 (a) 481.97, (b) 2.075, (c) 3.95, 1.132

10.43 (a) 204.3, (b) 24.1, (c) 35.76, (d) 1.484

10.46 (a) 138.8, (b) 0.569

11.3 (a) 125.29, (b) 103.76, (c) 101.34, (d) 101.48

11.5 (a) 14.08, (b) 11.04, (c) 11.41

11.8 72.12

11.11 (a) 998.6, (b) 1.974, (c) 117.1

11.15 (b) $= \dfrac{v'_R}{v'_R - 1/8} - \dfrac{27/64}{v'_R T_R^2}$

11.20 $p = \dfrac{RT}{v - b} + \text{constant}$

11.26 (a) positive, (b) zero, (c) negative

11.32 Table data: (a) 66.62, (b) 59.33, (c) 0.1387

11.35 0.03169 bar (table data)

11.39 0.292

11.45 Table data: −5 Btu/lb, −0.06 Btu/lb · °R

11.47 $c_p = c_v + R$

11.50 (a) $= \dfrac{p^2}{RT}\left[\dfrac{R}{p} + \dfrac{3AR}{T^4}\right]$,

(b) $= -R \ln \dfrac{p_2}{p_1} + \dfrac{3AR}{T^4}[p_2 - p_1]$

11.53 -1.21×10^{-7}

11.56 $\beta = \dfrac{1}{T} + \dfrac{1}{Z}\left(\dfrac{\partial Z}{\partial T}\right)_p, \ \kappa = \dfrac{1}{p} - \dfrac{1}{Z}\left(\dfrac{\partial Z}{\partial p}\right)_T$

11.59 5.2%

11.62 (b) $u_2 - u_1 = \int_1^2 c_v(T)dT - a\left[\dfrac{1}{v_2} - \dfrac{1}{v_1}\right]$,

(c) $= \dfrac{R}{1 - 2a(v - b)^2/RTv^3}$

11.65 (a) 0.991, (b) 4896

11.68 no

11.71 $T_{inv} = \dfrac{2a}{bR}\left[1 - \left(\dfrac{b}{v}\right)\right]^2$

11.74 −57.07, −1797.9

11.77 (b) 17.2 ft³/lb, 245.5 Btu/lb, 0.43 Btu/lb · °R

11.80 $= T_R^2\left(\dfrac{dB}{dT_R}\right) p_R, = p_R \dfrac{d(T_R B)}{dT_R}$

11.84 610.5

11.86 57.66

11.88 (a) 194, (b) 1338

11.93 (a) 432, (b) −6469, (c) 4099

11.96 (a) 0.815, (b) 0.792

11.99 (a) 62.4, (b) 51.48, (c) 49.9, (d) 54.5, (e) 54.6, (f) 38.8, (g) range of values

11.103 46.6

11.105 2.118, 2.083

11.108 52.1 bar, 52.9 bar

11.115 −7113

11.118 14,756, 10,071

11.122 (a) 179.46, (b) 220.45, (c) 206.28

12.4 (a) N_2: 50.61%, CO_2: 39.76%, O_2: 9.64%, (b) N_2: 0.06, CO_2: 0.03, O_2: 0.01, (c) 30.04

12.7 O_2: 0.00768, N_2: 0.02549, He: 0.00164

12.10 (a) N_2: 38.9%, (b) 0.5, (c) 6.236, (d) 12.472, (e) 129.8, (f) 0.3729

12.13 (a) −2181.7, (b) −2181.7, (c) 0, 2.182
12.15 (a) −996.6, (b) 1.2944, 8.93
12.18 (a) 618, (b) 950.3
12.21 (a) 271, (b) −217.2, (c) 0.1103
12.25 (a) −3541.3, (b) −3743.3
12.27 (a) 0.3816, 0.2598, (b) 4.93, (c) −73, (d) N_2: 0.01484, O_2: 0.37432
12.30 (a) 301, (b) 46.622, (c) 13.77
12.33 (a) 400.5 K, (b) 5.7131
12.36 (a) 600°R, (b) 600°R, (c) 23.8, 14.98
12.39 2299.2
12.42 59%
12.45 1.67
12.50 129.4°F
12.54 (a) 66.32, (b) −651.49, (c) −2.5587
12.57 228.1
12.59 (a) 56, (b) 21.35
12.63 0.0081, 43.6
12.65 (a) 40%, 0.0108, 57.5, (b) 0.0088, 42.2, 15, (c) 23
12.67 0.0081, 44%
12.72 (a) 48.9%, (b) −35.57
12.76 (a) −618.22, (b) 0.1185, (c) 3.22
12.80 (a) 9.65, (b) 48.7%
12.83 (a) 952.1, (b) 0.129
12.88 (a) 69.5, (b) 80
12.90 (a) 63.2%, (b) 68
12.95 (a) 11,261, (b) 290.8

13.4 (a) 24.07, (b) 22.2% deficiency
13.9 0.0602
13.12 (a) 0.0192, (b) 12.05
13.15 (a) 15.27%, (b) 1.609
13.22 1.383
13.25 (b) CO_2: 5.54%, CO: 11.09%, N_2: 83.37%
13.28 (a) 56.525, (b) CO_2: 12.82%, CO: 2.37%, N_2: 84.81%
13.30 (a) 8.578, (b) 120.1%
13.33 (a) 1.176, (b) 1.25
13.38 −661,822
13.40 (a) −375,894, (b) 1.2
13.43 −1,299,202
13.46 439,750
13.49 (a) −879,380, (b) −875,763
13.52 50,352, 46,359
13.55 (a) 1506, (b) 1647, (c) 1507
13.60 202%
13.62 271%
13.65 3260°R, 6.07
13.68 −98,350
13.70 (a) 188.87, (b) 263.97
13.73 1.29×10^6
13.78 impossible
13.83 (a) 34,212, (b) 116,554, (c) 51,756
13.86 66.46
13.91 $95{,}885 \times 10^5$ Btu/h
13.95 (a) 275,364, (b) 226,105, (c) 49,259, $\varepsilon = 82.1\%$
13.98 (b) 203.1, (c) 15,724, (d) 6.25%, 62.23%, 22.28%, 6.84%, 0.52%, 1.96%
13.100 (a) 22,081

14.3 5.018
14.6 −1.658
14.9 (b) −5.018
14.12 (a) CO_2: 0.6568, CO: 0.3432, O_2: 0.1716, (b) CO_2: 0.82, CO: 0.18, O_2: 0.09
14.15 0.108, 0.336
14.18 (a) increase, (b) no change
14.20 CO_2: 45.54%, CO: 21.43%, O_2: 33.04%
14.23 2490
14.27 CO_2: 0.528, CO: 0.472, O_2: 0.736, N_2: 3.76
14.31 0.02, decrease
14.39 (a) H_2O: 0.9594, H_2: 0.0406, O_2: 0.0203, (b) 45,832
14.42 441,423
14.46 322,386, 31,816
14.50 108%
14.53 4.0373
14.57 (a) 139.18, (b) 139.66
14.60 101,643
14.63 118,192, 118,314
14.66 0.84
14.69 (a) 20.1%, (b) 11.8%, (c) 9.1%
14.72 CO_2: 0.28808, H_2O: 0.71105, CO: 0.71192, O_2: 0.000435, H_2: 0.28895
14.79 0.073%
14.85 (a) 0, (b) 1, (c) 2, (d) 3, (e) 2, (f) 2, (g) 3, (h) 2, (i) 1

INDEX